ELASTIC AND PLASTIC FRACTURE

Metals, Polymers, Ceramics, Composites, Biological Materials

ELLIS HORWOOD SERIES IN MECHANICAL ENGINEERING

Alexander, J. M.	**Strength of Materials: Vol. 1: Fundamentals; Vol.2: Applications**
Alexander, J. M. & Brewer, R. C.	**Technology of Engineering Manufacture**
Atkins, A. G. & Mai, Y. W.	**Elastic and Plastic Fracture**
Beards, C.	**Vibration Analysis and Control System Dynamics**
Beards, C.	**Structural Vibration Analysis**
Besant, C. B.	**Computer Aided Design and Manufacture, 3rd Edition**
Cameron, A.	**Basic Lubrication Theory, 3rd Edition**
Collar, A. R. & Simpson, A.	**Matrices and Engineering Dynamics***
Dowling, A. P. & Ffowcs-Williams	**Sound and Sources of Sound**
Edmunds, H. G.	**Mechanical Foundations of Engineering Science**
Fenner, R. T.	**Engineering Elasticity***
Ford, Sir Hugh, F.R.S. & Alexander, J. M.	**Advanced Mechanics of Materials, 2nd Edition**
Ford, Sir Hugh, F.R.S. & Fenner, R. T.	**Elasticity and Plasticity in Engineering**
Haddad, S. D. & Watson, N.	**Principles and Performance in Diesel Engineering**
Haddad, S. D. & Watson, N.	**Design and Applications in Diesel Engineering**
Irons, B. M. & Ahmad, S.	**Techniques of Finite Elements**
Irons, B. M. & Shrive, N.	**Finite Element Primer**
Launder, Gosman & Reece	**Computer-aided Engineering: Heat Transfer and Fluid Flow***
Leech, D. J. & Turner, B. T.	**Engineering Design for Profit**
McCloy, D. & Martin, H. R.	**Control of Fluid Power: Analysis and Design 2nd (Revised) Edition**
Parker, R.	**Unsteady Fluid Flow**
Prentis, J. M.	**Dynamics of Mechanical Systems, 2nd Edition**
Richards, T. H.	**Energy Methods in Vibration Analysis**
Richards, T. H.	**Energy Methods in Stress Analysis: With Introduction to Finite Element Techniques**
Ross, C. T. F.	**Computational Methods in Structural and Continuum Mechanics**
Ross, C. T. F.	**Finite Element Programs for Axisymmetric Problems in Engineering**
Ross, C. T. F.	**Finite Element Methods in Structural Mechanics***
Sherwin, K.	**Engineering Design for Performance**
Thring, M. W.	**Robots and Telechirs**
Walshaw, A. C.	**Mechanical Vibrations with Applications**

**In preparation*

ELASTIC AND PLASTIC FRACTURE

Metals, Polymers, Ceramics, Composites, Biological Materials

A. G. ATKINS, B.Sc., M.A., Ph.D., C.Eng., M.I.Mech.E., F.I.M., A.C.Inst.Arb.
Professor of Mechanical Engineering, University of Reading
and
Y. W. MAI, B.Sc., Ph.D., M.A.S.M.E., F.I.E. Aust.
Associate Professor
Department of Mechanical Engineering, University of Sydney

ELLIS HORWOOD LIMITED
Publishers · Chichester
Halsted Press: a division of
JOHN WILEY & SONS
New York · Chichester · Brisbane · Toronto

First published in 1985 by

ELLIS HORWOOD LIMITED
Market Cross House, Cooper Street, Chichester, West Sussex, PO19 1EB, England

The publisher's colophon is reproduced from James Gillison's drawing of the ancient Market Cross, Chichester.

Distributors:

Australia, New Zealand, South-east Asia:
Jacaranda-Wiley Ltd., Jacaranda Press,
JOHN WILEY & SONS INC.,
G.P.O. Box 859, Brisbane, Queensland 40001, Australia

Canada:
JOHN WILEY & SONS CANADA LIMITED
22 Worcester Road, Rexdale, Ontario, Canada.

Europe, Africa:
JOHN WILEY & SONS LIMITED
Baffins Lane, Chichester, West Sussex, England.

North and South America and the rest of the world:
Halsted Press: a division of
JOHN WILEY & SONS
605 Third Avenue, New York, N.Y. 10016, U.S.A.

British Library Cataloguing in Publication Data
Atkins, A.G.
Elastic and plastic fracture. –
(Ellis Horwood series in mechanical engineering)
1. Fracture mechanics
I. Title II. Mai, Y.W.
620.1'126 TA409

Library of Congress Card No. 85–5568

ISBN 0–85312–562–7 (Ellis Horwood Limited)
ISBN 0–470–20195–9 (Halsted Press)

Typeset by Ellis Horwood Limited
Printed in Great Britain by The Camelot Press, Southampton

Table of Contents

Preface 13

List of Symbols 21

Chapter 1 Introduction

1.1 Conventional Wisdom on Fracture 29
1.2 Micromechanisms of Fracture 33
1.2.1 Crystalline Solids 33
1.2.2 Amorphous Solids 42
1.2.3 Composite Materials 46
1.2.4 Biological Materials 51
1.2.5 Pre-existing Cracks 56
1.3 Crack Tip Process Zone 59
1.4 Mechanics 60
1.5 Mechanics-to-Microstructure Link 67
1.6 Application to Design: Scaling and Size Effects 69

Chapter 2 The Mechanics of Elastic Fracture

2.1 Strain Energy 75
2.2 The Mechanics of Elastic Cracking: the work rate viewpoint 78
2.2.1 Graphical approach 80
2.2.2 Analytical approach 89

2.3 R-curves 98
2.4 Displacement Reversibilities and Apparent Irreversibilities in Elastic Fracture 100
2.4.1 Elastic Tearing and Peeling of Thin Sheets 101
2.4.2 Elastic Guillotining and Microtoming 105
2.4.3 Elastic Recovery and Residual Displacements: The General Case. 108
2.5 The Mechanics of Elastic Cracking: the stress intensity factor viewpoint 113
2.6 K_C as a Parameter for Crack Propagation 123
2.7 Relation between R and K_C 128
2.8 Crack Tip Zones of Irreversibility 130
2.9 Crack Sharpness and COD. 143
2.10 Fracture of Highly Deformable Materials 149
2.10.1 Graphical Methods for R. 150
2.10.2 Algebraic Formulae for Cracking Loads. 151
2.10.3 Cutting of Extensible Materials 157
Problems 158

Chapter 3 Special Topics in Quasi-Static Elastic Fracture
3.1 Crack Velocity 168
3.1.1 Changes in Crack Speed at Interfaces. 171
3.2 Elastic Fracture in Compression 174
3.2.1 Indentation Hardness of Brittle Solids 179
3.3 Fracture in the Presence of Many Cracks: Porous Bodies. 184
3.4 Cracking under Multiaxial Loading 187
3.5 Crack Paths 198
3.5.1 The Kinked Angled Crack Problem 211
3.5.2 Crack Paths in Anisotropic Solids 214
3.6 Cracking of Layered Structures 219
3.7 Cracking in the Presence of Residual Stresses 228
3.7.1 Tensile Residual Stresses. 230
3.7.2 Compressive Residual Stresses 239
3.7.3 More General Residual Stress Problems after Considerable Crack Propagation 240
3.7.4 Analysis in Terms of Stress Intensity Factors 241
3.7.5 Determination of Residual Stress using K-Formulae 242
3.8 Crack Propagation from Notches 244
3.9 Avoidance of Remote Yielding. 247
3.9.1 Grooving 250
3.9.2 Gurney Rig 250
3.9.3 Weitzmann and Finnie, and Seth and Page Rigs 256
3.9.4 Suppression of Yielding in Layered Bodies. 258
Problems 260

Chapter 4 Combined Flow and Fracture

4.1 Introduction. 269
4.2 Elastoplastic Topics for Application to Elastic Fracture 274
4.2.1 Use of 'Small' Specimens for K_{IC} 274
4.2.2 Empirical correlations between K_C and other measurements. 275
4.2.3 Crack Opening Displacement (COD) 281
4.2.4 R-curves and Elastic-Plastic Fracture 286
4.3 Interfacial Adhesive Fractures in Presence of Plastic Flow. 289
4.3.1 Peeling of Glued Soft Metal Strips 289
4.3.2 The Soldered Can Lid Problem. 291
4.3.3 Elastoplastic Deformation of the Adherand 292
4.4 Cracking in Monolithic Bodies in the Presence of Extensive Remote Flow 297
4.4.1 Fracture in Ductile DEN Sheets 298
4.4.2 Tearing of Ductile Metal Sheets 312
4.5 Eshelby–Rice J-integral 316
4.5.1 Derivation of J 316
4.5.2 J_C as a Criterion for Crack Initiation. 321
4.5.3 J during Propagation: J_R curves. 334
4.5.4 Finite Element/Finite Difference Elastoplastic Fracture Solutions. 337
4.6 Partition of Work during Combined Flow and Fracture. 341
4.6.1 Work Components for the Glued Cantilever Problem: Residual Strain Energy 343
4.6.2 Why does Cracking plus Flow Supervene Flow Alone? 348
4.6.3 Correspondence between R and J_C for Glued Cantilever Problem 351
4.6.4 General (X, u, R, A) relations in Presence of Remote Flow. 353
4.6.5 Interpretation of Unloaded Xu Diagrams after Elastoplastic Flow. 357
Problems 361

Chapter 5 Crack and Craze Nucleation

5.1 Introduction. 369
5.2 Experimental Observations of Crack Nucleation in Metalforming 370
5.3 Criteria for Ductile Fracture 381
5.3.1 Empirical Criteria 381
5.3.2 Models of Void Coalescence and Growth 385
5.4 Applications of Void-growth Models 389

5.5 Mechanics-to-microstructure Link 397
5.6 Craze Initiation in Amorphous Polymers 401
5.6.1 Theories and Mechanisms for Craze Initiation 403
5.6.2 Craze Growth 407
5.7 Propagation of Initiated Cracks in Forming Operations 408
5.8 Possible Link between Crack Initiation and Propagation: Fracture in Solids with and without Starter Cracks 419
5.8.1 J-void growth model connexion 422
Problems 426

Chapter 6 Rate, Environment and Temperature Effects on Toughness
6.1 Introduction and Representative Behaviour 432
6.1.1 'Slow Crack Growth' 449
6.1.2 Expected Effects of Rate, Temperature and Environment 460
6.2 Crack Velocity 462
6.2.1 Crack Tip Strain Rate 463
6.2.2 Local Heating at Moving Crack Tip 465
6.3 Relaxation and Creep Effects on Toughness 469
6.4 Slow Crack Growth: Environment-assisted Delayed Fracture and Stress Corrosion Cracking 478
6.5 Mechanisms of Environment-assisted Fracture 488
6.5.1 Changes in R 488
6.5.2 Electrochemical Effects 489
6.5.3 Plasticisation Theories 493
6.5.4 Liquid and Solid Metal Embrittlement 499
6.6 Structural Life Predictions using K_C–$\dot{a}$ Diagrams 499
6.6.1 Proof Testing and Statistical Variation of Structural Lives 503
6.7 Thermal Stresses and Shock 509
Problems 520

Chapter 7 Fracture under Repeated Loading
7.1 Crack Growth under Repeated Loading in Mineral Glass 529
7.2 Elastic (High Cycle) Fatigue in Metals and Polymers 536
7.2.1 Long Cracks 541
7.2.2 Threshold ΔK_0 554
7.2.3 Short Cracks 562
7.3 Effect of Mean Stress: ΔG rather than ΔK 568
7.4 Fatigue in the Plastic Range (Low Cycle Fatigue) 574
7.4.1 Cyclic Workhardening 574
7.4.2 da/dN in Low Cycle Fatigue 575
7.5 Cumulative Damage and Fatigue Crack Propagation under Random Loading 580
7.5.1 Palmgren–Miner Rule 581

7.5.2 Simple and Complex Variable Amplitude Cyclic Loading . . . 585
7.5.3 Overload Effects on Fatigue Crack Growth . . . 585
7.5.4 Prediction of Fatigue Life due to Variable Amplitude Loading . . . 592
7.6 Elastic and Plastic Fatigue under Multiaxial Loading . . . 594
7.7 Fatigue from Notches . . . 605
7.8 Effect of Residual Stresses on Fatigue . . . 611
7.9 Fatigue at High Homologous Temperatures: Creep Fatigue . . . 615
7.10 Stress Corrosion Fatigue: Creep Stress Corrosion Fatigue . . . 617
7.11 Selection of Materials for Design against Fatigue . . . 626
Problems . . . 630

Chapter 8 Energetic Stability
8.1 Introduction . . . 638
8.2 Crack Stability in Elastic Fracture . . . 640
8.2.1 Graphical Interpretation of Crack Stability in Elastic Fracture . . . 640
8.2.2 Geometrical Stability Factors in Elastic Fracture . . . 644
8.3 Testing Machine Stiffness . . . 648
8.4 Stability in Terms of Crack Velocity . . . 650
8.5 R-curves and Stability . . . 652
8.6 Practical Examples of Energetically Stable and Unstable Elastic Cracking . . . 660
8.6.1 Choice of Specimen . . . 661
8.6.2 Instabilities caused by $\mathrm{d}R/\mathrm{d}A < 0$ or $\mathrm{d}R/\mathrm{d}\dot{A} < 0$ . . . 663
8.6.3 Instabilities caused by Local Crack Tip Heating: Limiting Quasi-static Crack Velocities in $\mathrm{d}R/\mathrm{d}T$-negative materials . . . 664
8.6.4 Instabilities caused by Blunt Crack Tips . . . 669
8.7 Methods of Promoting Stable Cracking . . . 670
8.7.1 Maximising $\mathrm{d}R/\mathrm{d}A$ . . . 670
8.7.2 Minimising $(\partial J/\partial A)_u$ or X . . . 672
8.8 Turner's η Factor and Upper Bound Toughnesses . . . 678
8.9 Crack Arrest . . . 682
8.10 Crack Stability in the Presence of Remote Plastic Flow . . . 694
8.10.1 Gurney-type Analysis . . . 695
8.10.2 J_R formulation and the Tearing Modulus . . . 696
8.10.3 Comparison between J_R-based stability criteria and 'separable-R'-based criteria . . . 701
8.10.4 Turner's Instability Criterion . . . 702
8.11 Stability in Fatigue . . . 702
Problems . . . 703

Chapter 9 Scaling in Static Fracture and Fatigue Fracture
9.1 Introduction . . . 707
9.2 Geometrically Similar Linear Elastic Structures under Monotonic Loading . . . 708
9.2.1 Application to Ice-tank Towing Studies . . . 712
9.2.2 Non-proportionately Scaled Linear Bodies . . . 715
9.3 Non-linear Elastic Structures . . . 718
9.4 Scale Effects on Elastic Crack Velocity . . . 719
9.5 Fracture Predictions Using Equivalent Energies from Scaled Testpieces . . . 721
9.6 Scaling in Elastoplastic Fracture . . . 727
9.7 Scale Effects in Fatigue . . . 731
Problems . . . 735

Chapter 10 Deformation Transitions
10.1 Flow or Fracture? . . . 737
10.2 Experimental Evidence . . . 749
10.3 Abrasive Machining . . . 758
10.4 Wear . . . 764
10.5 Cutting . . . 771
10.6 Deformation Transitions and Design . . . 776
Problems . . . 786

Appendices
1. Tougness Equations and Stability Conditions for Various Specimen Geometries . . . 794
2. Representative Toughness Values for Various Materials . . . 798
3. Relations between Stresses and Strains in Plasticity Incremental and Deformation Theories . . . 801

Author Index . . . 804

Subject Index . . . 811

DEDICATION

to
Charles Gurney,
engineer, poet, philosopher
and our teacher

Alan Arnold Griffith (1893–1963)

Photograph taken at the Royal Aircraft Establishment, Farnborough, April, 1937. (Crown Copyright, with permission.)

Preface

'When *I* use a word,' Humpty Dumpty said, in rather a scornful tone, 'it means just what I choose it to mean – neither more nor less.'

This is an introductory text on the mechanics of fracture. The definitions of *elastic, elastoplastic,* and *plastic* fracture used by us concern the size of the crack tip plastic zone relative to the crack length and the magnitude of the far-field general strains in the cracked body remote from the crack faces, together with the relative magnitudes of the components of incremental work done, i.e. reversible elastic strain energy, irreversible work of fracture, and irreversible work of remote flow (and the likes of irreversible work against friction if fracture is produced by tooling, as in cutting operations).

In *elastic* fracture the general strains are below the yield level and the plastic zone at the crack tip is much smaller than the size of crack or body. Under these conditions, the body behaves elastically and reversibly in a global sense with the irreversible fracture work being confined to thin boundary layers contiguous with the crack faces; this means that severed parts can be fitted back together to regain the original component. External work input relates to elastic strain energy and fracture work. Elastic fracture is also called brittle fracture, but the micromechanisms of such fracture are not always 'brittle' especially in large structures made from materials which are ductile in the laboratory – such as brittle fractures in large ships, bridges, pipelines, storage tanks and so on made from low carbon steel.

FOOTNOTE TO ABOVE QUOTATIONS

(All quotations for all chapters are from *Alice's Adventures in Wonderland,* and *Through the Looking-Glass,* by Lewis Carroll.)

In *elastoplastic* fracture, the size of the overall plastic zone is comparable with the dimensions of the crack. External work input relates to elastic strain energy, fracture work and work of remote flow, and incremental work of remote flow is comparable in magnitude to the other work components. The strains remote from the crack may be of the same order of magnitude as the yield strain, and yielding may be contained and not be spread to the boundaries of the cracked body. Owing to the fact that the increment of plastic work involves the product of work/volume and volume, however, the same magnitude of incremental plastic work done could arise either from small strains diffused over a large plastic volume or from large strains concentrated in a small volume undergoing irreversible flow, see Fig. 1.19. The lack of conformity with the original shape on refitting broken pieces depends clearly on the extent of the plastic flow.

When remote plastic flow predominates and the cracked body experiences general yielding and hence extensive permanent deformation prior to (and perhaps during) subsequent crack spreading, so that it is impossible to refit the broken parts, we have *plastic* fracture. The incremental component of plastic work swamps the incremental component of elastic strain energy. Examples of plastic fracture arise in metal forming processes (e.g. cracking in forging, extrusion and rolling) and homely examples are the opening of beer, sardine or corned-beef cans, where the curled-up portion and the rest of the opened tin cannot be put back together to give the original container.

Fig. 1.19 illustrates schematically increasing degrees of plastic deformation in cracked bodies. Clearly the boundaries between these classifications are blurred in many problems and furthermore may change during crack spreading. A simple, but sensible, viewpoint for distinguishing between elastic and plastic fracture, is to see if the original body would be regained were the broken parts to be glued back together. Of course, from the point of view of atoms on fracture surfaces, few fractures are truly 'reversible' but for engineering purposes the fact that broken ships have been rewelded back together is an adequate demonstration of reversibility and elastic fracture.

Of the four time-independent modes of deformation, i.e. simple elasticity, simple plastic flow, elastic cracking and plastic cracking, the mechanics of the first three are well-established.

The mechanics of elastic fracture is what is often popularly called 'fracture mechanics'. An important realisation from the mechanics of elastic fracture is that (unlike simple elastic or plastic deformations) there is a size-effect, whereby the stresses required to cause fracture in large structures can often be lower than those to cause fracture in small structures. Thus brittle fractures (i.e. elastic fractures) can occur in large bodies made of 'ductile' solids even though in small bodies, made of the same material, plastic flow would precede fracture. Furthermore crack velocities in the larger bodies are usually higher than in the small bodies, which has led to the description 'fast fracture' for brittle fracture. Chapters 9 and 10 are devoted to problems in scaling and deformation transitions, which are topics not fully appreciated in the literature.

Now the mechanics of crack propagation accompanied by (or preceded by) extensive yielding throughout the body (i.e. plastic fracture) are ill-understood and the prediction of loads and deflexions during plastic fracture is something of a lacuna. A major difficulty which arises in such problems is the partition of total work input into elastic strain energy, irreversible fracture and irreversible flow components (and irreversible friction components too if fracture is produced by tooling; also, if we are dealing with work areas *after unloading*, into a residual elastic strain energy component after non-uniform plastic flow). The degree of interest in elastoplastic and plastic fracture centres on the end-use to which the information is to be put. Often plasticity is seen as a nuisance and complication as, for example, it is *elastic* fracture parameters that are really required for the design of large structures in solids possessing high toughness-to-strength ratios, but where the scale-effect causes bodies to behave elastoplastically or fully plastically in standard laboratory-sized testpieces. Most treatments of 'post-yield fracture' or 'yielding fracture mechanics' are extensions of elastic fracture methods in the elastoplastic range, and plastic strains are comparatively small; until recently relatively little analysis had been done starting at the fully plastic end of the spectrum where the strains are large, as found in metalforming operations. Of course both approaches should coincide in elastoplastic problems if a unified theory of fracture exists. We show that there is a host of technologically important plastic fracture problems (where plastic strains swamp elastic strains) to which adequate solutions may sometimes be obtained by neglecting elastic effects. Even so, we also show that residual elastic strain energy, which occurs on unloading from the plastic range, can significantly affect subsequent cracking loads.

In plastic fracture also, important questions regarding crack-initiation have to be investigated, which are similar to questions that are relevant to fatigue in elastic fracture and to the so-called 'short crack' problem (where the size of cracks is comparable with microstructural features and where the application of continuum mechanics is open to question). The fracture mechanics treatment of fatigue (and indeed of monotonic loading) concerns the propagation of *existing* cracks – but we must ask where those cracks come from in the first place in supposedly crack-free bodies. Thus we shall be concerned with void formation and coalescence in commercial metals which contain inclusions or hard second phases, with craze formation in polymers and so on.

Unlike many texts on 'fracture mechanics', the emphasis throughout this book is on energy methods. There is a number of reasons for this, the foremost of which is simply pedagogical. In our experience students grasp readily the concept of a requirement to do work to initiate and propagate cracks (whether in elastic, elastoplastic or plastic fracture), and thereby they get an easy physical 'feel' for the subject. In contrast, the critical stress intensity factor K_C of linear elastic fracture mechanics is mysterious to a beginner; indeed with K_C's peculiar units of Pa$\sqrt{m}$, the student finds it difficult to understand what he is calculating or determining experimentally – but the learner *can* understand K_C as a combination of two fundamental mechanical properties of a solid, i.e. $K_C^2 \approx E^*R$, where E^* is the Young's modulus (for plane stress or plane strain as appropriate) and R is the specific work of fracture. It seems to us also that certain advanced topics in elastic fracture are more readily attacked via energy methods than K-methods; among these are crack stability, crack arrest, cracking in layered structures and cracking in the presence of residual stresses. Furthermore – and this is important for development of the subject – work methods (and methods which may be thought of in terms of energy, such as the J-integral and indeed the Hutchinson–Rice–Rosengren crack tip characterisation) seem to be the common thread running through solutions of elastoplastic and fully plastic fracture problems. For these problems, the K parameter of linear elastic fracture is no longer appropriate (indeed it is not appropriate even for *non-linear* elastic fracture) but the specific work of fracture R retains its meaning throughout. Of course, R may depend on rate, temperature, environment, degree of crack tip constraint and size of the crack tip process zone.

We follow Gurney's formulation of crack propagation in terms of work, using Newtonian mechanics. Of course, the seminal Griffith treatment of cracking is energy-rate-based and the Irwin–Kies critical strain energy release rate (G_C) and 'compliance calibration' methods were developed alongside K-methods in the evolution of elastic fracture mechanics. Gurney built on these energy approaches to give a generalised Castigliano-type theory for linear and non-linear elastic fracture of cracked bodies and introduced a new method for the experimental determination of the specific work of fracture or fracture toughness during stable crack propagation which avoids all calibration procedures. The connexion between the energy solution and the stress intensity factor solution for the same problem is of course the same as the connexion in crack-free bodies between a 'theory of elasticity' solution (solving for stresses from the equilibrium equations and boundary conditions, strains from the compatibility equations, and integrating up the incremental strains to give displacements) and a Castigliano solution for loads or displacements obtained directly from differentiation of strain energy functions. The experimental observation of a critical K_C at fracture in a linear elastic cracked body may be interpreted as the existence of an integrated strain energy field in the body which is available to provide the necessary work for fracture, given also a mechanism at the crack tip which permits fracture to take place. We have extended the Gurney analysis to elastoplastic and fully plastic fracture, in which events prior to initiation, and during propagation, are separated.

Some comment should be made regarding nomenclature and symbols. Throughout this text R, the specific work of fracture (whether applicable to elastic, elastoplastic or plastic fracture) is called the *fracture toughness*. Some writers call K_C or K_{IC} the fracture toughness in linear elastic fracture, but we employ the alternative name of *critical stress intensity factor*. R refers to a *fracture* event and is a material property (in general dependent on rate, temperature and environment, etc.). Before fracture occurs we can only define an *energy rate* term, for which we use the general symbol J, and $J < R$ unless fracture occurs, at which time $J_C = R$ where J_C is the critical energy rate change. Stability is controlled by rates of change of J and R. The symbol J has been chosen for energy release rate owing to the existence of the (equivalent) Eshelby–Rice J-integral. In linear elastic fracture J_C is synonymous with the well-known G_C.

Similarly K_C refers to fracture and is a criterion for linear elastic fracture, whereas K (the stress intensity factor at a stationary crack produced by external loading) is *not* a fracture criterion. In elastoplastic and plastic fracture, a critical J_C is used for crack initiation (i.e. first propagation). Within the limitations of Hencky deformation theory of plasticity, a J_C-criterion for elastoplastic fracture coincides with a Gurney formulation for non-linear elastic fracture. Subsequent propagation with the accompaniment of plastic flow is often described by so-called J_R curves where the total work during propagation is *not* partitioned into flow and fracture components. This approach leads to difficulties as J_R is (not unexpectedly) *geometry-dependent* but, as we show, appropriate separation of flow and fracture components in the Gurney formulation provides a sound basis for analysis which is supported by experiment. Most often J_C and J_R are determined from work areas and not from the contour integral definition. While such experimental results are *called* J_C and J_R they are *not* formally the same as J, particularly if unloading has occurred since then the contour integral definition is lost. J_C and J_R derived from work areas really relate to R and (partitioned) plastic work rate components.

Griffith, in addition to his reversible thermodynamic energy arguments, discussed the historical notion that propagation ensued when the stress at the crack tip reached a critical value, and we must ask whether an energy approach is both necessary and sufficient, in comparison with merely a stress condition. Whether an energy criterion is equivalent to a stress criterion in elastic fracture depends on *crack tip sharpness* and crack orientation. When the crack is adequately sharp and aligned perpendicular to the axis of loading, the (theoretical elastic) crack tip stress is elevated to cohesive strength levels, the work of fracture takes the lowest possible value (the Gibbs surface free energy) and crack healing may even occur as required of thermodynamicallly reversible processes (see Obreimoff's 1930 experiments on the cleavage of mica†). In practical materials where cracking is irreversible (i.e. entropy-generating in the thermodynamics sense), it is still possible to measure a specific work of fracture during globally reversible cracking which incorporates not only the surface free energy required for the creation of new surfaces, but also non-linear elastic deformation at the crack tip, plastic and viscous deformation at the crack surfaces, and delayed reversible and irreversible effects associated with possible atomic configuaration rearrangements such as phase changes at the crack surface. These much larger works of fracture may be compared with the requirements of models of the crack tip process zone (e.g. Dugdale–Cottrell–Bilby–Swinden models), exactly like Barenblatt's solution for brittle solids, and it is found that energy and stress approaches are indeed equivalent for cracks aligned perpendicular to the axis of loading. For inclined cracks in elastic fracture, energy and stress criteria do *not* give the same predictions as explained in Chapters 2 and 3. In the case of crack initiation in ductile solids by void coalescence and growth, crack sharpness is of less concern and the criterion for the accumulation of crack tip 'damage', and hence fracture, concerns both stress *and* strain; for many loading paths this is the same as a

---

†J. W. Obreimoff, 1930, *Proc. Roy. Soc.* **A127**, 290.

criterion of a critical work done per unit volume for initiation as affected by hydrostatic stress. The connexion between initiation and propagation parameters for fracture, and whether they are the same under elastic, elastoplastic and plastic fracture conditions, and whether a one-parameter characterization of fracture is always possible, are all explored in the book.

As an introductory text, the book is suitable for all beginners in the subject, including those practising engineers and scientists who may not have received formal instruction in the mechanics of fracture. Our coverage is broader and deeper than could be covered in the usual sort of lecture courses, but much of the book is suitable for the later years of first-degree courses, and Master's courses, in both British-derived and US-derived university systems. In addition to end-of-chapter problems, a feature of the book is the great number of worked problems sprinkled through the chapters. Our experience tells us that these are of great help – not only to 'self-taught' readers. It will be discovered that our presentation of some of the, by now, 'standard' fracture mechanics topics is different from conventional treatments (but standard topics are thoroughly treated even so). In the case of elastoplastic, and particularly plastic fracture, we bring together material scattered throughout the literature and introduce a number of new approaches to problems. We also highlight the vital link between mechanics and microstructure in fracture. It is hoped that the alternative viewpoints and the alternative lines of attack given in this book will appeal to experts in fracture as well as to learners; for example, crack stability may be of interest to the more mathematically inclined; design of testpieces to determine the dependence of toughness on rate, temperature and environment may be of interest to metallurgists and materials scientists; those concerned with metal-forming should be interested in the prediction of limiting fracture strains in working processes; structural engineers should be interested in problems of scaling between model and full-size bodies.

Details of certain traditional topics found in books on elastic fracture are omitted in this book – for example, on the one hand, Westergaard and other stress functions and methods of deriving elastic stress intensity factors from first principles, and on the other hand, detailed treatment of all aspects of the microstructural mechanisms involved in fracture, such as the traditional extensive coverage of micromode deformation transitions in structural steels which spurred on the development of 'fracture mechanics'. There are admirable books dealing in detail with the fracture of specific materials, but a feature of this book is that examples of cracking are provided from a wide variety of solids (metals, polymers, ceramics, glasses and rocks, composites, glued structures, paper, wood and biological materials) and their behaviours are interrelated. Space is also taken to describe analyses and experiments on the propagation of cracks in the burgeoning field of plastic fracture. Furthermore, in contrast to many books on fracture, we pay as much attention to situations where fracture is *required* as to those where it must be prevented. Professor Bilby remarked at the fourth Tewksbury Symposium on Fracture† that things breaking and tearing is a much older

† *Fracture at Work*, 1979, University of Melbourne.

concern of mankind than keeping aircraft flying, or ensuring the safety of nuclear reactors or offshore structures. Flint knapping and the development of sharp tools and weapons are elements in the very history of man, and the products of the mill and grindstone were his earliest manufactures.

We are indebted to many people for help, guidance and encouragement in the preparation of this book. Professor Charles Gurney taught us both and we wish to pay tribute to him. His papers and ideas are less well-known than they should be. The idea for the book arose in the early 1970s when we were both at the University of Michigan at Ann Arbor. Discussions at the time with R. M. Caddell, D. K. Felbeck, T. U. Marston, G. Wilkowsky and others were valuable in formulating our ideas. Since then, colleagues at Delta Research, Imperial College, the University of Sydney, the University of Reading and the University of Birmingham have commented on draft manuscripts. As the field of plastic fracture in particular is comparatively new, and as our aim was a systematised 'structured' text, we had to discover things as we went along and sound out the new ideas on colleagues. Again, as we discuss fracture in all types of materials and bring together experience from a variety of fields including that of biological materials, a deal of effort had to be expended on providing, as far as possible, a *unified approach* to the mechanics of fracture. In this light, discussions with C. R. Chaplin, B. Cotterell, B. Dodd, J. E. Gordon, G. Jeronimidis, N. Jones, K. J. Miller, K. Pascoe, A. J. Pretlove, P. Purslow, K. E. Puttick, G. W. Rowe, C. Ruiz, R. A. Smith, C. E. N. Sturgress, B. Tomkins, J. F. V. Vincent and J. G. Williams have been particularly fruitful. Likewise we owe a debt to those many friends and others authors who have permitted us to reproduce diagrams and photographs from their work. Mrs A. M. Watson and Miss T. Santos have typed the draft manuscript with greater fortitude, and Jan Stevens, I. Maclean and their colleagues prepared many drawings and photographs. Finally we must thank J. M. Alexander who introduced us to our publishers who, in turn, have been very patient with us for some time, and our reviewer who made a number of valuable suggestions for improvement of the text.

The literature on fracture and the mechanics of fracture is extensive and rapidly growing. We do not pretent that this book is everywhere in detail at the frontiers of the subject. We do, however, point the reader in the direction of current thinking; many aspects of the field of fracture are still contentious and will not be resolved for some time yet. We hope we have given due acknowledgement to pioneers and thinkers in the field, but it is impossible to give references to all work on a given topic. The responsibility for mistakes and omissions is ours: our excuse is that of Dr Samuel Johnson when he was asked by a lady 'why in your dictionary do you call a horse's pastern its knee?'. He replied, 'Pure ignorance, dear lady, pure ignorance'.

Reading and Sydney AGA and YWM
1985

List of Symbols

(Where the same symbol has had to be used for different things, the particular meaning should be clear from the context.)

A	crack area
$\bar{A}$	constant in toughness-biased Ree–Eyring rate relation for $R(\dot{a}, T)$
a	crack length
	depth of surface (thumbnail) crack
	major semi-axis of ellipse representing notch or crack
$\dot{a}$	crack velocity
a_0	bond or interchain spacing
B	thickness
B^*	crack path thickness in presence of grooves
b	minor semi-axis of ellipse representing notch or craze thickness of Dugdale crack tip zone
C	compliance (u/X)
	constant in Paris Stage II fatigue crack growth equation using ΔK
	Cauchy number
	Total length of crack in ductile shearing and guillotining $(= a + (u - u_{cr}))$
C^*	time-dependent J contour integral
$\bar{C}$	constant in high cycle fatigue crack growth equation using ΔG or ΔJ
$\bar{\bar{C}}$	constant in low cycle fatigue crack growth equation using $\Delta\epsilon_p$.
c	half length of surface (thumbnail) crack
	length of Dugdale crack tip zone
	half focal length of ellipse representing crack

D	constant in $\dot{a} = D\,K_C^{n^*}$
d	height of region of cracked sheet forming process zone
d_y	extent of crack tip plastic zone
E	Young's modulus
	elliptic integral of first kind
E', $\bar{E}$	apparent Young's modulus of cracked body
E^*	is $E/(1-\nu^2)$ for plane strain; is E for plane stress
e	engineering strain
	proportion of notch size to be added to sharp crack length to give equivalent sharp crack size.
F	Faraday constant
	elliptic integral of the second kind
F_n	Froude number
f	factor of safety in traditional strength-of-materials design
G	Irwin–Kies potential energy release rate
G_C, G_{IC} etc.	critical potential energy release rate (see comments under K_C, K_{IC} et. (for use of subscripts)
ΔG	range of G in cyclic loading ($= G_{max} - G_{min}$)
G^{Δ}	Kfouri–Miller parameter
H	kinetic energy
	hardness
h	height
	depth of cut in machining
	convective heat transfer coefficient
I	second moment of area
	anodic current areal density
I_n	non-dimensional number for ice towing tank scaling
I_1	first stress invariant ($= \sigma_1 + \sigma_2 + \sigma_3 = 3\,\sigma_H$)
J	potential energy release rate in non-linear elastic cracking
	Eshelby–Rice contour integral
J_C, J_{IC} etc.	critical value of J-integral at crack initiation (see comments under K_{IC} etc. for use of subscripts)
J_i	specific essential work of fracture initiation
J_p	specific essential work of fracture propagation
J_M	Ernst's modified J-integral
J_R	J during crack propagation in presence of remote flow
K	applied stress intensity factor
K_C, K_{IC}, etc.	critical value of stress intensity factor at fracture (see sections 2.6 and 6.1.1 for further comments on the use of K_C for plane stress K_c for sub-critical growth and so on)
K_{Ia}	crack arrest toughness determined by quasi-static analysis
K_{Id}	critical stress intensity factor during dynamic (fast) crack tip loading
K_{Im}	crack arrest toughness determined by dynamic analysis
K_{ISCC}	threshold static plane strain stress intensity factor in stress corrosion cracking

K_{Cl}	crack closure stress intensity factor caused by residual stresses
K_F	applied stress intensity factor in fatigue pre-cracking
K_f	stress concentration factor (reduction factor) in fatigue
K_0	lower threshold stress intensity factor for stress corrosion cracking
K_Q	candidate K_{IC} value obtained from secant line cracking load
K_R	stress intensity factor converted from R-curve in early crack propagation
K_r	ordinate of R-6 diagram ($= K_{applied}/K_{IC}$)
K_t, K_T	stress concentration factor
ΔK	range of stress intensity factor during cyclic loading ($= K_{max} - K_{min}$)
ΔK_{eff}	range of effective stress intensity factor during cyclic loading
ΔK_0	threshold range of stress intensity factor during cyclic loading
k	shear yield stress
	proportionality constant
	thermal conductivity
	stiffness of elastic foundation
	$\sigma_\infty \sqrt{a}$ grouping for stress intensity factor
$\bar{k}$	Boltzmann constant
L	gauge length
	equivalent crack length for notch–sharp crack combination
	ligament length between cracks
	strain ratio $d\epsilon_1/d\epsilon_2$
l	length
	spacing of voids
M	moment
	molecular mass
	stress ratio σ_1/σ_2
	exponent in power law creep relation
M_e	elastic magnification factor on stress intensity (surface crack K-formula)
M_n	Mach number
m	coefficient in R or $J = m\,\sigma_y\,\delta_C$
	exponent in $da/dN - \Delta G$ high cycle fatigue crack growth relation
	exponent in σ–t relaxation relation
m^*	Weibull modulus
N	number of cycles in fatigue
	number of voids in craze
	exponent in Ramberg–Osgood σ–ϵ relation
N_D	period of fatigue growth delay
n	exponent in non-linear elastic X–u relation
	exponent in Ludwik σ–ϵ relation for plastic flow
	geometric stability factor
	exponent in Paris high cycle $da/dN - \Delta K$ fatigue crack growth relation

	exponent in E–t relaxation relation
n'	exponent in cyclic σ–ϵ relation
n^*	exponent in K_c–$\dot{a}$ relations
p	pressure
p^*	weakening factor of hole in shell structure
Q	elastic shape factor for semi-elliptical crack (surface crack K-formula)
	activation energy
	factor relating true toughness to initiation and arrest toughness values.
q	exponent in K_c–σ_y empirical relation
R	specific work of fracture (fracture toughness)
R_n	Reynolds number
R^*	specific work of fracture of layered structure
r	radius, radius of void
	distance ahead of crack tip
	mean stress factor ($= \sigma_{min}/\sigma_{max}$) in fatigue
r_y	additional crack length in Irwin's 'embedded crack tip' model
S	nominal stress (S–N curves in fatigue)
	strain energy density
S_r	abscissa of R-6 diagram ($= X/X_{collapse}$)
s	half-width of process zone in tearing of ductile metal sheets
	specific heat
T	temperature
	period of cycling in fatigue
	traction force
	cyclic 'true stress at ultimate'
	tearing modulus ($= (E/\sigma_y^2)\, dJ/da$)
	optical transmittance
T_g	glass transition temperature
T_m	homologous temperature
t	time
U_T	total work area under Xu plot under load
U^*	total work area rates under Xu plots
u	displacement associated with load X
u_{xx}	local displacement parallel to crack
V	volume
V_S	molar volume
V^P	volume of body undergoing plastic flow
v	velocity
v_f	volume fraction
v_{yy}	local displacement perpendicular to crack
W	elastic strain energy density
W_T	total work done
W^P	plastic work/volume ($= \int \sigma d\epsilon$)
w	width

w_e	specific essential work of ductile fracture
w_i	specific work of ductile crack initiation
w_p	specific non-essential work of ductile fracture
w_T	total specific work of ductile fracture
X	load
x	coordinate distance
Y	shape correction factor in $K = \sigma\sqrt{\pi a} \cdot Y(a/W)$
y	distance from neutral axis in beam bending
Z	shape correction factor in $K = \sigma\sqrt{a} \cdot Z(a/W)$
z	valency of solvated ions
α	constant of proportionality
	coefficient of thermal expansion
	exponent in low cycle $da/dN - \Delta\epsilon_p$ relation
	major semi-axis of elliptical notch
	craze stress reduction factor in liquid environment
	ratio of E_1B_1/E_2B_2 in cracking of layered structures
$\bar{\alpha}$	tool rake angle in cutting
β	minor semi-axis of elliptical notch
	relief angle of blade in microtoming
	'attack angle' of abrasive particle ($= 90° + \bar{\alpha}$)
	fraction of R used to break bonds
	ratio of R_1B_1/R_2B_2 in cracking of layered structures
Γ	irreversible work
γ	shear strain
	surface free energy
γ_p	Orowan–Irwin surface work term incorporating crack tip irreversible flow
δ	solubility parameter
	crack tip opening or stretch
δ_C, δ_c	critical crack opening displacement (COD, CTOD)
ϵ	normal true strain
ϵ_f	fracture strain
ϵ_y	yield strain
$\bar{\epsilon}$	effective strain
ϵ^0_{1f}	plane strain workability index
$\Delta\epsilon$	strain range in fatigue
$\dot{\epsilon}_m$	strain rate at tip of moving crack
$\dot{\epsilon}_s$	strain rate at tip of stationary crack
η	Turner's factor relating Λ (elastic) or U_T (elastoplastic) and ligament area to G or J
	anodic overpotential
	elliptic curvilinear coordinate
θ	angle, rotation
κ	factor in elasticity ($= (3 - \nu)/(1 + \nu)$ in plane stress or $= (3 - 4\nu)$ in plane strain)

Λ	elastic strain energy of cracked body
Λ_0	elastic strain energy of body without crack
λ	'extension ratio' $(= 1 + e)$
	constant in toughness-biased Ree–Eyring $R(\dot{a}, T)$ relation.
	stress ratio in biaxial fatigue $(= \sigma_{x\infty}/\sigma_{y\infty})$
	scaling factor
λ_B	scaling factor on thickness
λ_h	scaling factor on 'height'
λ_w	scaling factor on 'width'
λ_c	stress ratio in biaxial fatigue where both remote stresses fluctuate
λ_m	stress ratio in biaxial fatigue where remote stress parallel to crack remains constant
μ	shear modulus $(= E/2(1 + \nu)$ for isotropic solid)
	viscosity
ν	Poisson's ratio
	kinematic viscosity $(= \mu/\bar{\rho})$
ξ	dummy variable
	elliptical curvilinear coordinate
	proportion of beam depth remaining elastic when bending into the plastic range
Π	potential energy $(= \Lambda - \int X\mathrm{d}u)$
$\rho, \bar{\rho}$	density
ρ	radius of curvature at crack end
	parameter connected with J_R-controlled elastoplastic fracture
ρ_c	crack 'natural' sharpness
	length of necked zone in ductile fracture of sheets
σ	normal true stress
$\bar{\sigma}$	effective stress
σ_c	crazing stress, fracture stress
σ_d	reference stress in creep fracture
σ_H	hydrostatic (mean) stress $(= (\sigma_1 + \sigma_2 + \sigma_3)/3)$
σ_0	fatigue limit of plain (unnotched) testpeices
	constraint in Ludwik σ–ϵ relation for plastic flow
	constant in σ–t relaxation relation
σ_0'	constant in cyclic σ–ϵ relation
σ_y	yield stress
$\Delta\sigma$	stress range in fatigue $(= \sigma_{max} - \sigma_{min})$
τ	repassivation time in stress corrosion cracking
	shear stress
τ_y	shear yield stress
Φ	loss function
ϕ	shear plane angle in cutting
	volume fraction
Ψ	chemical, electrochemical energy
Ω	complementary elastic strain energy $(= Xu - \Lambda)$
ω	parameter connected with J_R – controlled elastoplastic fracture.

$\mathscr{G}$	symbol formerly used for Irwin–Kies G
$\mathscr{K}$	symbol formerly used for $\sigma\sqrt{a}$
$\mathscr{T}$	generalised toughness in Andrews's theory

Subscripts and Superscripts

a, app	applied
c, cr	critical, cracking event
cw	cold work
e	elastic
elpl	elastoplastic
env	environment
f	fracture
	final
H	hydrostatic
i	generalised variable
	initial value
m	model
m,max	maximum, greatest
min	minimum, least
P, p	plastic
SL	shear lip
s	solvent
th	thermal
u	ultimate
x, xx	on normal stresses parallel to two-dimensional crack, aligned along x-axis.
y, yy	on normal stresses perpendicular to two-dimensional crack aligned along x-axis.
xy	on shear stresses
z, zz	axial direction
θ, $\theta\theta$	hoop direction
$r\theta$	on shear stresses
∞	on remotely applied stresses
0	minimum, least, starting value
I, II, III	modes of cracking
//	parallel
$\perp$	perpendicular

1

Introduction

'Ahem!', said the Mouse with an important air. 'Are you all ready? This is the driest thing I know. Silence all round, if you please!'

1.1 CONVENTIONAL WISDOM ON FRACTURE

Common experience tells us that under similar conditions of loading, different materials deform and break in different ways. At the one extreme ice, glass, blackboard chalk, bone, cello-strings, rubber bands and the like break suddenly without much warning; and some materials will shatter into many pieces. At the other extreme soft metal wires, lead sheet, pitch, chewing gum, polyethylene film and so on first experience plastic yielding or viscoelastic flow, which produces extensive deformation before eventual fracture – and sometimes fracture is difficult to achieve in materials like modelling clay. In between these extremes are materials such as some cold-worked metals and several rubber-modified polymers which, under similar conditions of loading, suffer a limited amount of permanent deformation before fracture.

On the basis of this common experience, it is customary to classify the behaviour of the materials at room temperature broadly into two types of fracture, namely *brittle* and *ductile,* the distinguishing feature between which is whether the broken parts may be refitted together to regain essentially the original dimensions. Archaeologists can reconstruct ancient vases from pottery shards because the fracture was brittle, but it is impossible to refit together the portions of an open beer or sardine can unless the wound-up bit is plastically uncurled.

When conditions change, the same size pieces of material may behave differently, depending on how the relative resistance to fracture and flow change. For example, increase in temperature usually softens materials, so that flow takes over from fracture (consider the well-known brittle–ductile transitions in Charpy tests on bcc and some hcp metals; the manufacture of glass bottles and the drawing of glass fibres at high temperatures compared with the brittle behaviour of glass at room temperature – or the extensive hot-stamping industry based on the hot ductility of β-brasses which at room temperature are comparatively brittle). Increase of temperature which removes moisture on the other hand promotes brittleness (consider wet and dry mud, wet and dry wood). Increases in rate of deformation usually promote brittle behaviour in a given material (consider pitch which flows at low stresses over long periods of time, but which may be snapped brittlely when stressed in short times, and also 'silly putty'). Changes in environment may promote either brittleness (consider hydrogen embrittlement in steels and liquid-metal embrittlement in steels and aluminium) or flow (consider the Joffé effect) and a given material may be affected in different ways by different environments (high impact polystyrene is made brittle by sunlight, but its threshold resistance to fatigue cracking is increased by soaking in alcohols).

Changes in microstructure of materials with similar compositions can also be important. In metals, for example, coarse-grained zinc is brittle at room temperature whereas fine-grained zinc can be deformed to 90 per cent reduction of area in a tensile test. This comes about because small grains produce small glide bands and hence slip-induced microcracks of short length in comparison with those produced in large grains; as explained later shorter cracks require higher stresses to cause fracture, and in the case of zinc of adequately small grain size, since the fracture stress in a laboratory tensile specimen exceeds the yield stress, flow will precede fracture. High temperature creep-resisting alloys, which typically possess large grain diameters so as to avoid a lot of grain boundary sliding, are not particularly tough at room temperature unless special alloying additions are used. The effects of ageing and irradiation on metallic microstructure cause embrittlement. Some alloying elements tend to promote ductile behaviour, either because of their effect on gross microstructural features, or their effect on the metallic lattice. For those microstructural variables that are well understood in crystalline solids, the one common factor in increasing ductility appears to be the suppression of void formation and growth of microcracks. The 'cleanliness' of the microstructure of commercial alloys (inclusion content and its geometric distribution and shape) is vital in this context. For example, less than 0.01% of bismuth has a potent effect on the ductility of copper, and the effect of the geometric shape and distribution of second phases on ductility is strikingly displayed between flake and spheroidal cast irons. Manganese is a desirable alloying element in steels containing sulphur since MnS inclusions are relatively harmless, unlike FeS which encourages grain boundary brittleness.

Fracture in 'damage-free' glass (i.e. glass in which best attempts have been made to remove the surface microcracks from which normal fractures start) is also affected by micro-inclusions. For example, NiS changes its crystal structure at 300°C giving an increase in volume, so that some glass failures have been attributed to residual stresses around NiS inclusions. (Heavy metal sulphides are among the few minerals insoluble in soda-lime glass and not wetted by it).

A major factor controlling the occurrence of fracture or flow is the state of stress, an indicator of which is the hydrostatic or mean stresss $\sigma_H = (\sigma_1 + \sigma_2 + \sigma_3)/3$ where σ_1 etc. are principal stresses. Notches encourage brittle fracture owing to increased tensile σ_H near the root of the notch; thus Charpy and Izod notched impact tests arguably measure the resistance to fracture under high tensile σ_H, as well as some high strain rate 'impact strength'. In thick structural sections, the constraint of the surrounding material promotes high tensile mean stresses in the internal centre region of the body which thereby has a lower resistance to fracture than thin pieces of the same material. In contrast, experiments on stone under large compressive σ_H at room temperature produce plastic flow rather than brittle fracture, which may explain some of the tortuous flow of otherwise brittle rocks in the earth's crust. 'Brittle' metals may be plastically formed successfully by means of hydrostatic extrusion where the billet is surrounded by fluid under high pressure in the container, and extrude into fluids which themselves are under pressure. The effect of residual or manufacturing stresses can also be thought of in terms of the sign and magnitude of σ_H. Residual tensile stresses significantly affect stress-corrosion cracking, and residual compressive stresses deliberately introduced into the surface of machine parts (by, for example, shot-peening) inhibit fatigue crack initiation and propagation.

Ludwik in 1909 described the role of changing stress states in terms of raising or lowering the stress–strain curve relative to a 'fracture stress' curve characteristic of the material, as shown schematically in Fig. 1.1. Increase of

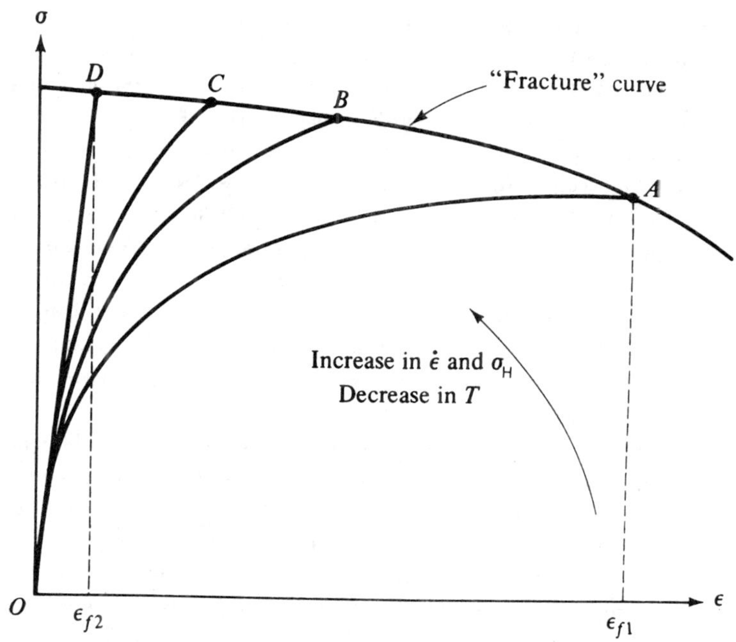

Fig. 1.1 – Ludwik's early idea of a 'fracture stress' curve which was supposed to be a material property. (Reproduced from Caddell (1980), with permission.)

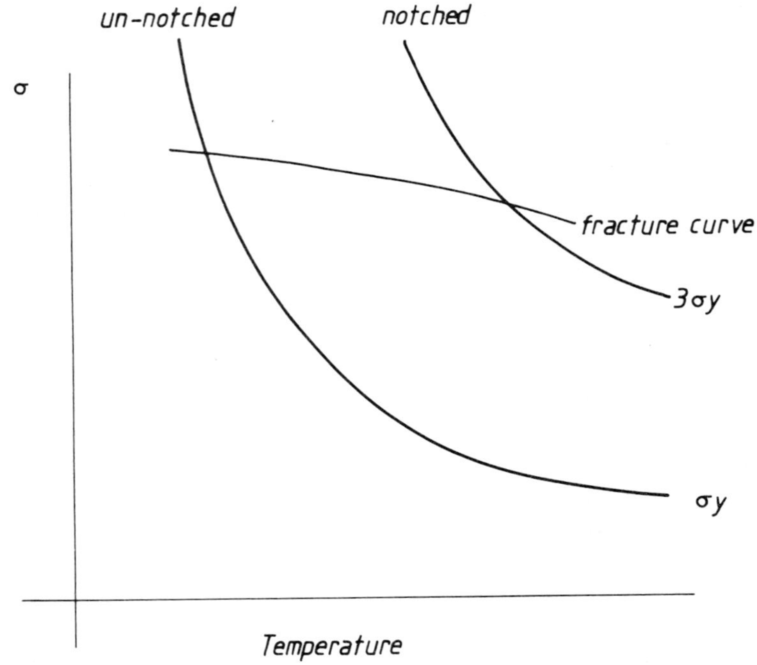

Fig. 1.2 – Orowan's classification (based upon a material-property fracture stress curve) of solids into 'simply brittle', 'notch-brittle' and 'simply ductile'.

strain rate, increase of σ_H (notches and/or thick sections), or decrease of temperature tend to raise the σ–ϵ curve as shown, thus giving smaller strains to fracture (points A . . . D where the σ–ϵ curve intersects the fracture curve). *Notch-brittleness* may be explained if the characteristic fracture stress σ_f is approximately constant with temperature (Fig. 1.2) and the variations of uniaxial yield stress (σ_y) and fully-constrained uniaxial yield stress ($3\sigma_y$ in the presence of deep notches) are plotted on the same diagram. Materials are simply ductile at high temperatures where $\sigma_f > 3\sigma_y$; at low temperatures where $\sigma_f < \sigma_y$ the material is simply brittle; when $\sigma_f < 3\sigma_y$, Orowan called the material notch-brittle, i.e. ductile in plain tensile tests but brittle in notched tests having large plastic constraint. As shown later, the idea of a single 'fracture stress curve', which is representative of the material is *not* correct and although Figs. 1.1 and 1.2 are useful conceptually, they are wrong in general.

Notch-sensitivity has a different meaning from *notch-brittleness.* The presence of a notch or crack reduces the load-bearing area of a component and two possibilities arise, (i) the 'strength' of the solid changes directly as the change in cross-sectional area (failure load = failure stress of unnotched component × smaller area in presence of crack) or (ii) the strength changes (reduces) by a proportionately greater amount than merely the reduction in load-bearing area. Well-lubricated wire ropes, in which a few strands are severed, more or less obey (i), but rusty wire ropes follow (ii) and are less strong. Grass, leaves and leather dog-leashes bitten by puppies are notch-sensitive, but fresh apples, potatoes and carrots lose a large proportion of their strength with only small notches. The clue, as to the reasons why, is given by the behaviour of wire rope in the well-lubricated and rusty conditions. When there is little mechanical connexion (particularly shear coupling) between elements in a body, individual elements may break without influencing the other elements. On the other hand, when there is some shear connexion, fracture of one element influences the other elements and released energy from the fracture of one element can be transmitted to other elements which break in a 'domino' fashion. The overall failure stress is then lower than that based on the reduction of area alone. Indeed, much of 'fracture mehanics' is concerned with by how much the size of cracks, notches, flaws, defects and so on affect failure stresses. Continua which are anisotropic (as regards elastic properties and perhaps also as regards properties determining fracture resistance) may have reduced notch sensitivity in certain directions, e.g. timber and many biological materials discussed later.

1.2 MICROMECHANISMS OF FRACTURE

1.2.1 Crystalline solids

Fig. 1.3 shows a number of ways in which a cylindrical tensile bar of a crystalline solid may fail. At low temperatures the material may fail by cleavage (i.e.

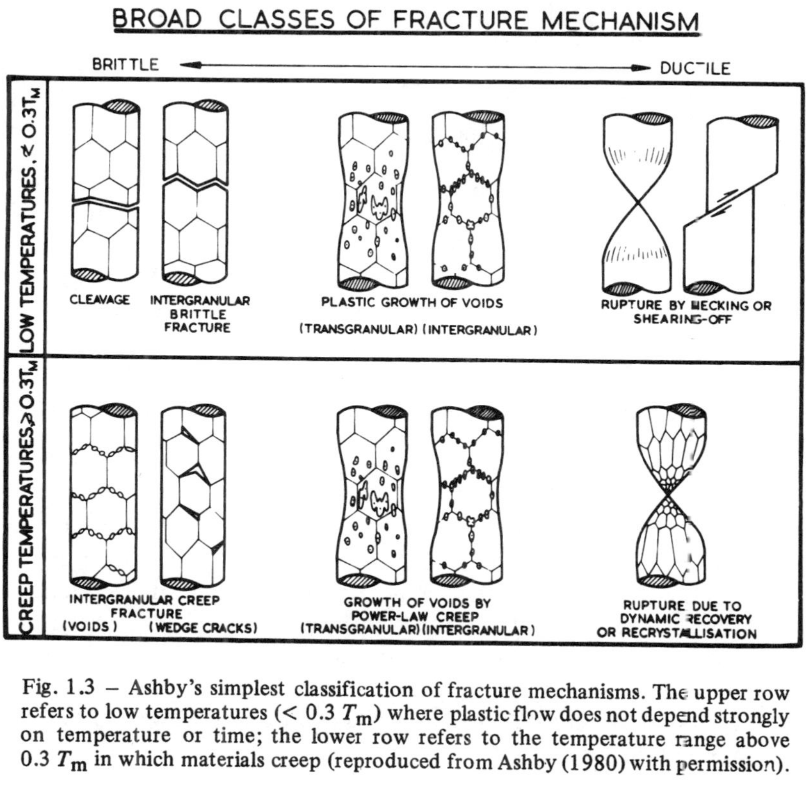

Fig. 1.3 – Ashby's simplest classification of fracture mechanisms. The upper row refers to low temperatures ($< 0.3\ T_m$) where plastic flow does not depend strongly on temperature or time; the lower row refers to the temperature range above $0.3\ T_m$ in which materials creep (reproduced from Ashby (1980) with permission).

splitting along natural crystallographic planes), or by brittle fracture along grain boundaries (intergranular fracture); or it may fail in a ductile fashion by hole coalescence or by shearing off. At high temperatures it may fail by a variety of creep fracture mechanisms, some of which are transgranular, some intergranular.

As explained by Ashby in 1979 (Ashby, 1980) three types of fracture may be identified at lower temperatures in which cleavage is predominant. They are cleavages which relate to: (1) fracture from pre-existing small cracks (arising from abrasion or corrosion or as growth defects) in the absence of any plasticity (Fig. 1.4(a)); (2) a similar kind of fracture, but at higher stresses, when pre-existing flaws are small or absent and have to be nucleated by localised twinning or slip (Fig. 1.4(b)); (3) final fracture by cleavage, but at higher temperatures than cleavages described under (1) or (2), which is preceded by extensive yielding with plastic strains up to about 0.1. In many metals and ceramics there is a

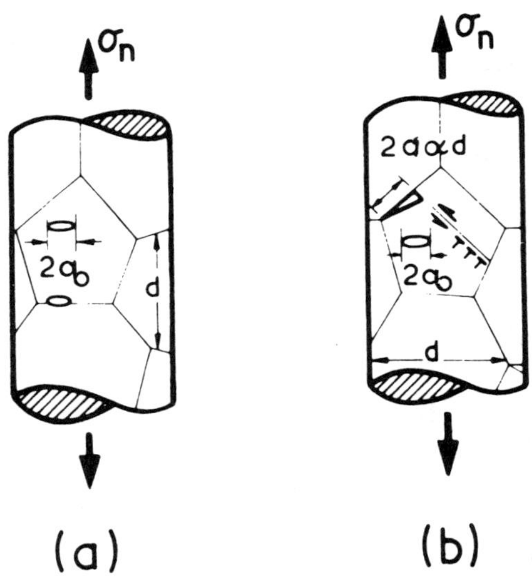

Fig. 1.4 – (a) Brittle solids contain incipient cracks of length $2a_0$. (b) Cracks can be generated by slip. Their length often scales as the grain size. Either sort of crack may propagate to give a cleavage fracture (reproduced from Ashby (1980) with permission).

delicate balance between cleavage across grains and brittle fracture along grain boundaries (and mixed fractures with the crack path switching from one to the other are often observed). Ashby refers to fractures which are similar to cleavage (1), (2), (3) but which run along grain boundaries as brittle intergranular fracture (BIF) (1), (2), (3). Cleavage fracture surfaces are faceted and shiny under the microscope; as fracture sometimes occurs on a number of closely aligned planes, which join up by steps between cleavage planes, cleavage surfaces have so-called 'river-markings'.

When polycrystalline solids do not cleave, they may fail in a transgranular, ductile manner. Holes nucleate at inclusions; further plastic flow makes them grow; and when large enough they coalesce by expanding into one another or by flowing unstably into one another and the specimen fractures (Fig. 1.5). If the density of inclusions, or of pre-existing holes, is higher on grain boundaries than it is within the grains, then the fracture path may follow the boundaries, giving

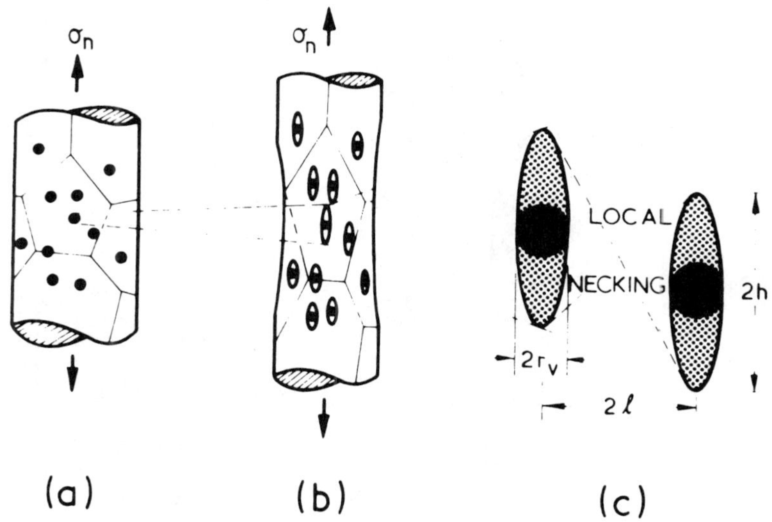

Fig. 1.5 – (a) Ductile fracture, and transgranular creep fracture, requires either that holes pre-exist or that they nucleate at inclusions which concentrate stress and strain. (b) The holes elongate as the specimen is extended. (c) They link, when their length $2h$ is about equal to their separation $(2l - 2r_V)$. (Reproduced from Ashby (1980) with permission).

alternatively a ductile intergranular fracture. In both cases the stretched-out ligaments, standing proud of coalesced voids, give ductile fractures a characteristic fibrous, dimpled appearance (Fig. 1.6(a)). In shear, the dimples are elongated in the direction of shear (Fig. 1.6(b)) and failure occurs by 'sheets' of voids linking together (Rogers, 1960). In the absence of foreign particles, but when plastic flow is easy, ductile fractures can initiate from 'cavity dislocations' produced from pile-ups or intersecting slip planes (cf. the different behaviour of coarse- and fine-grained zinc already mentioned in section 1.1).

In his Bakerian and Tewksbury lectures, Cottrell (1963, 1965) described in some detail the factors controlling stable and unstable crack advance in nominally ductile metals. Fig. 1.7 summarises the mechanisms of cracking which he examined for simple elastic-plastic materials. Fig. 1.7(a) shows a crack opened by applied tensile load under plane strain conditions. Crack advance can be thought of as being produced by the injection of dislocations into slip lines at ±45° to the

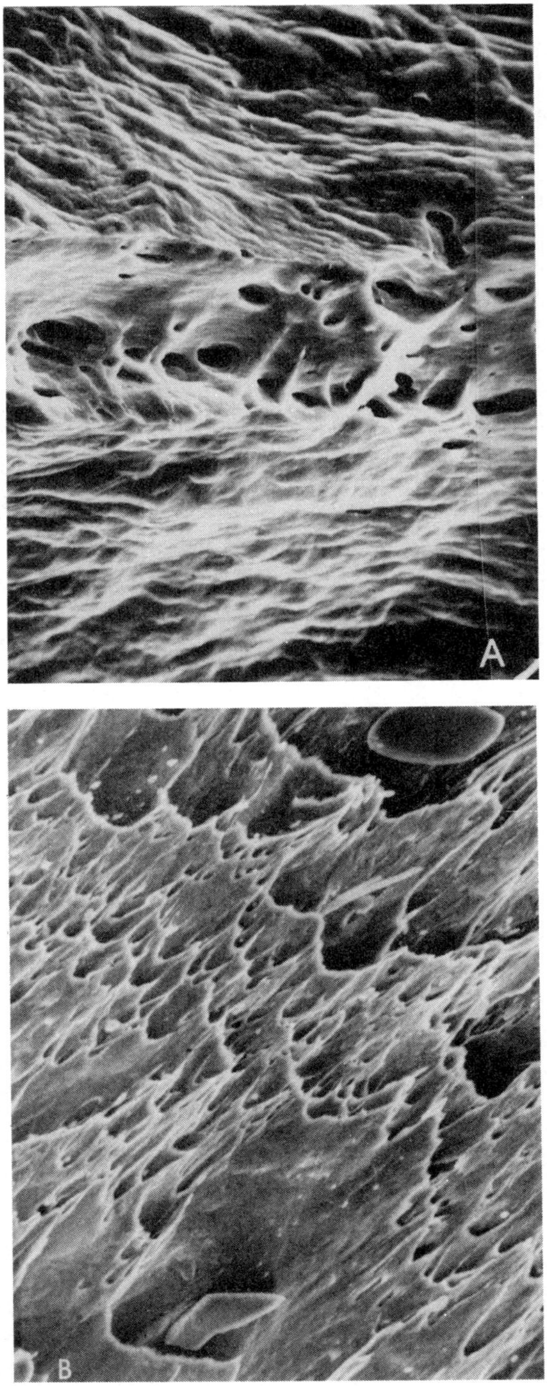

Fig. 1.6 – (a) Tensile fracture in necked region of aluminium sheet, showing void coalescence and growth. (b) Shear fracture in low carbon steel, showing 'void sheets'.

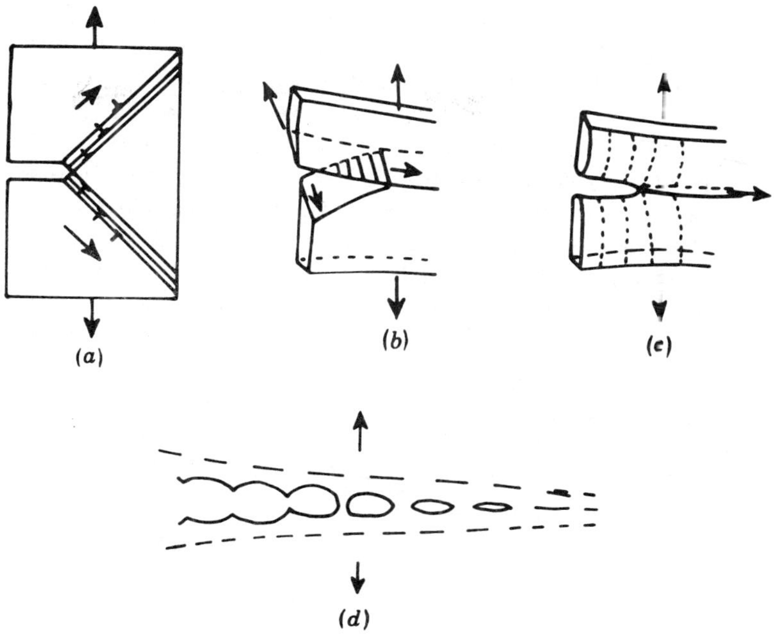

Fig. 1.7 – Continuous, (a), (b), (c), and discontinuous (d), modes of crack advance under tensile loading. Mode (a) gives stable crack advance below general yield, whereas (b), (c) and (d) can give unstable cracking. (After Cottrell, 1963.)

crack plane at the tip. New crack surface is produced by 'sliding off' or 'shear decohesion' at the tip and the crack advance step, Δa, is equal to $\frac{1}{2}\delta$, where δ is the crack tip opening. Crack growth by this process is stable and leads to a high local ductility failure with the fracture stress, σ_f, not less than the yield stress, σ_Y. If the same cracked material is loaded under plane stress conditions, however, crack advance can be achieved at applied stresses below σ_Y. Crack tip deformation is confined to a local region in the plane of the crack, and crack advance occurs by the continuous operation of a simple array of screw (Fig. 1.7(b)) or edge (Fig. 1.7(c)) 'dislocations'. Cottrell termed these fractures 'cumulative' because the contribution of a single dislocation to fracture increases continuously with distance travelled by the dislocation. In contrast, the stable crack advance of Fig. 1.7(a) he termed 'non-cumulative', as each new amount of growth requires the operation of new dislocations. The plane stress fractures of Fig. 1.7(b), (c) are unstable because the applied stress needed for crack growth decreases with crack advance. Finally, Cottrell noted crack advance situations where some material cracking occurs ahead of the main crack tip (Fig. 1.7(d)). This is observed in plane strain and can result in discontinuous unstable fracture at stresses below σ_Y. As discussed by Tomkins (1981), the modes of crack growth shown in Fig. 1.7(a), (d) are important in several time-dependent stable crack growth processes (Chapter 6).

Changes from cleavage and twinning to plastic flow and fracture by void formation, as the temperature is raised, are reflected in 'brittle' stress–strain $\sigma-\epsilon$ curves at low temperatures (or high strain rates) gradually changing to 'ductile' $\sigma-\epsilon$ curves (Fig. 1.8(a)). There is a marked increase in tensile strain to fracture, and also in the work of fracture, at about 0.20–0.25 of the melting point (T_m); similar changes occur in other measures of ductility such as Charpy values (CVN) (Fig. 1.8(b)). In those metals and alloys that do not normally cleave, viz. fcc metals (Ag, Al, Au, Cu, Ni, Pb, Pt, although Ir and Rh do cleave), there is no transition in values, which merely gradually rise with temperature from 0 K. The increase in ductility over the 'transition temperature range' is followed by a gradual drop beyond about $0.35T_m$. This comes about from the continuing fall in the Peierls–Nabarro stress which opposes dislocation movement, coupled with the emergence of cross-slip (as opposed to Frank–Read sources) as a dislocation generator as the temperature is raised. However, at temperatures around $0.5-0.6T_m$, recovery and recrystallisation processes set in.

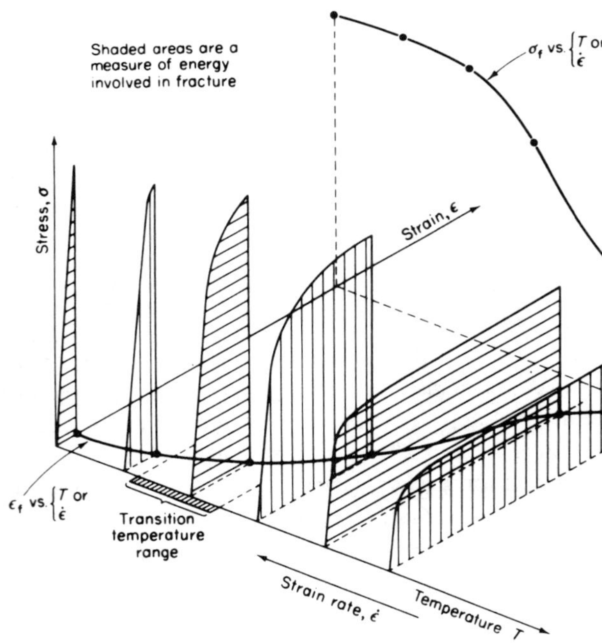

Fig. 1.8(a) – Relations between stress, strain, strain rate, and temperature in fracture. Brittle stress–strain curves transform to ductile curves at increasing temperature or decreasing strain rate (and vice versa) for most materials. Shaded areas under $\sigma-\epsilon$ curves are some measure of the energy involved in fracture. Change from brittle to ductile behaviour takes place over a reasonably narrow span of temperature, called the transition temperature range; the position of this temperature range can vary with type of test peice and conditions of measurement. (Reproduced from Felbeck and Atkins (1984), with permission.)

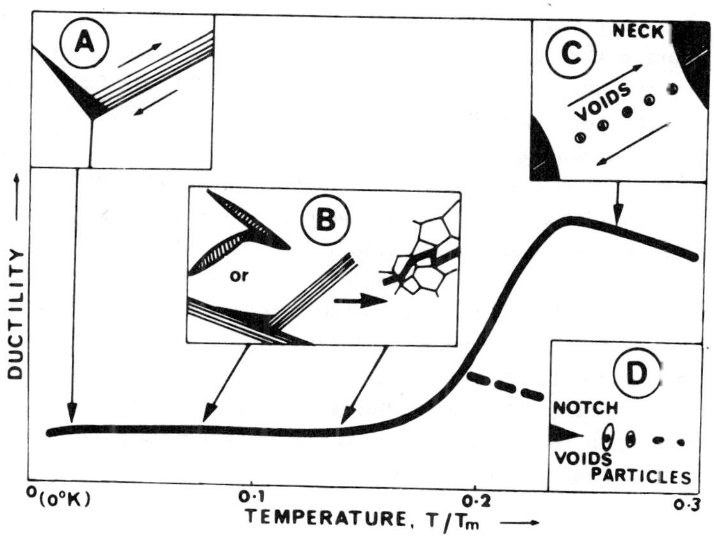

Fig. 1.8(b) – Changes in ductility and typical associated mechanisms of fracture in for a bcc material, in the temperature range < 0.3 T_m. (A) low-temperature intergrannular cracks, (B) twinning or slip leading to cleavage, (C) shear fracture (after necking) at particles, (D) low energy shear at particles. (Reproduced from Gifkins (1969), with permission.)

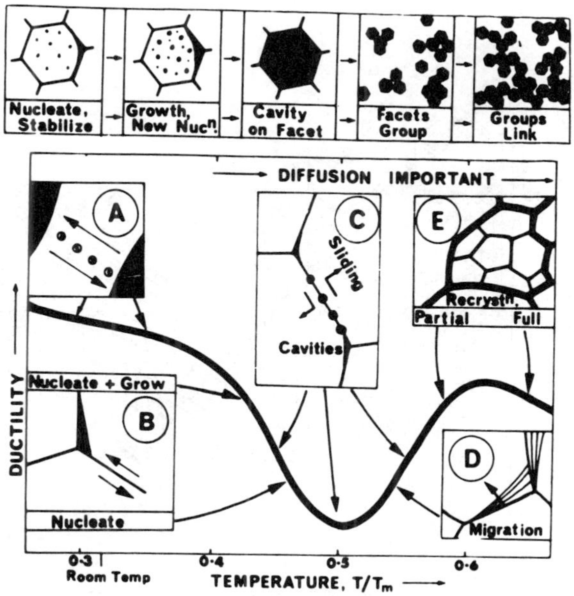

Fig. 1.8(c) – Sequence of events leading to fracture by growth and linkage of grain-boundary cavities. Changes of ductility with temperature for fcc, bcc and hcp alloys at temperatures > 0.3 T_m, for creep rates of strain. (A) Shear at particles (after necking), (B) Zener or w-type cavities (grain boundary sliding), (C) r-type cavities along grain faces (grain boundary sliding), (D) grain-boundary migration reduces cavitation, (E) recrystallisation suppresses cavitation. (reproduced from Gifkins (1969) with permission.)

(Recovery relates to a redistribution of dislocation sources so that dislocation movement is easier, and in recrystallisation the energy of dislocations generated during prior deformation is used to nucleate and grow new grains, thus effecting an annealed structure over long times.) The net effect is to increase 'ductility', causing a bump in the Charpy value vs. temperature curve (Fig. 1.8(c)).

At high temperatures, grain boundaries become significant. Below about $0.45T_m$, grain boundaries act principally as barriers, inhibiting cleavage and causing dislocation pile-ups. At higher temperatures the regions of intense deformation, which are contained *within* the grains at lower temperatures, now shift to the grain boundaries themselves. Voids are nucleated and cracks then develop on the grain boundaries. Shear stresses on the boundaries cause relative sliding of the grains, and voids are produced in regions of stress concentration (e.g. grain-boundary triple junctions, grain-boundary steps, or precipitate particles in the grain boundary) unless the sliding displacements are accommodated by plastic deformation elsehwere. Mobility of vacancies (i.e. missing atoms in the lattice) is

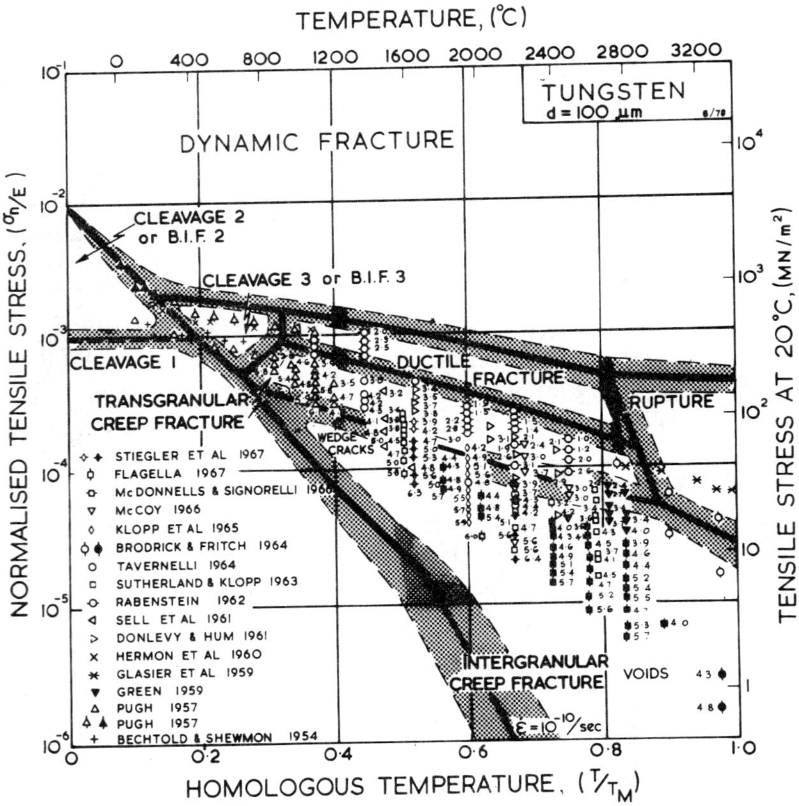

Fig. 1.9 – Fracture-mechanism map for tungsten tested in tension. There are three modes of cleavage at low temperatures; other fields resemble those of fcc metals (reproduced from Ashby, (1980) with permission.)

important in void formation under these conditions, since they diffuse through the crystal at these high temperatures and are captured by the voids. They may serve initially to 'blunt' the wedge-shaped voids formed by grain-boundary sliding and thus temporarily inhibit the loss of ductility with increase of temperature, but the final effect is to aid the growth and linking of grain-boundary cavities, in a conventional creep or stress-rupture experiment, and fracture occurs. Indeed, beyond $0.8T_m$, deformation occurs essentially by diffusion alone, grain-boundary voids eventually covering whole grain-boundary facets given enough time and an adequate stress level.

Recognising that a solid may fail by any one of a number of competing micromechanisms, the overall behaviour of a given material in a given type of test may be summarised in an Ashby fracture-mechanism map. Such a diagram has, usually, stress as one axis and temperature as the other and shows the field of dominance of each mechanism; it is also possible to plot stress and time, or strain and temperature. The maps are constructed by assembling and analysing fracture observations and data for the material, usually from uniaxial tensile tests. By way of illustration, Fig. 1.9 shows the map for tungsten, which has been chosen as it displays all the modes of fracture shown in Fig. 1.3, and is typical of bcc refractory metals. The highest stress shown on the ordinate of the diagrams ($\sigma_n = 0.1E$) corresponds with the theoretical tensile strength of all perfect crystals, i.e. the stress required to overcome interatomic forces, on a plane normal to the stress axis, which a variety of calculations shows to be about $E/10$ (see Example 3.2).

The effect of alloying is to move the fields of dominance of the mechanisms for a particular metal to higher stress levels and to change their size, but distinct breaks in mechanisms are still found. Boundaries between fields have a finite width, in some instances they are broad, in others they are narrow. Within a field there can be a number of subdivisions of mechanism, e.g. in that of intergranular creep fracture there can be an area of wedge crack formation and an area of discrete cavity formation. Fig. 1.10(a) shows the elevated temperature data for Nimonic 80A on a fracture map, whilst Fig. 1.10(b) shows the same information cross plotted on to the more familiar 'stress-rupture' axes of log stress and log time. This is a creep-resistant alloy and the transition from transgranular to intergranular creep fracture occurs only at long times. Fig. 1.10(b) shows how difficult extrapolation of stress-rupture data can be within fields where subdivisions of fracture mechanism are present. At lower temperatures where data have not been generated for fully-intergranular fracture, Ashby (1980) points out that such a transition would take place given long enough time and that this must be taken into account when extrapolating. It is also worth noting the broad boundary of overlap between the wedge cracking and cavitation regions over certain temperature ranges. In such a boundary area simple extrapolation is deceptive.

Strictly speaking, these diagrams apply to one purity of a crystal, or one composition of an alloy, with one grain size and in one state of heat treatment. Also specimen type is important: rupture is favoured in very thin sheet, for example (but otherwise thin sheet data and round bar tensile data are similar).

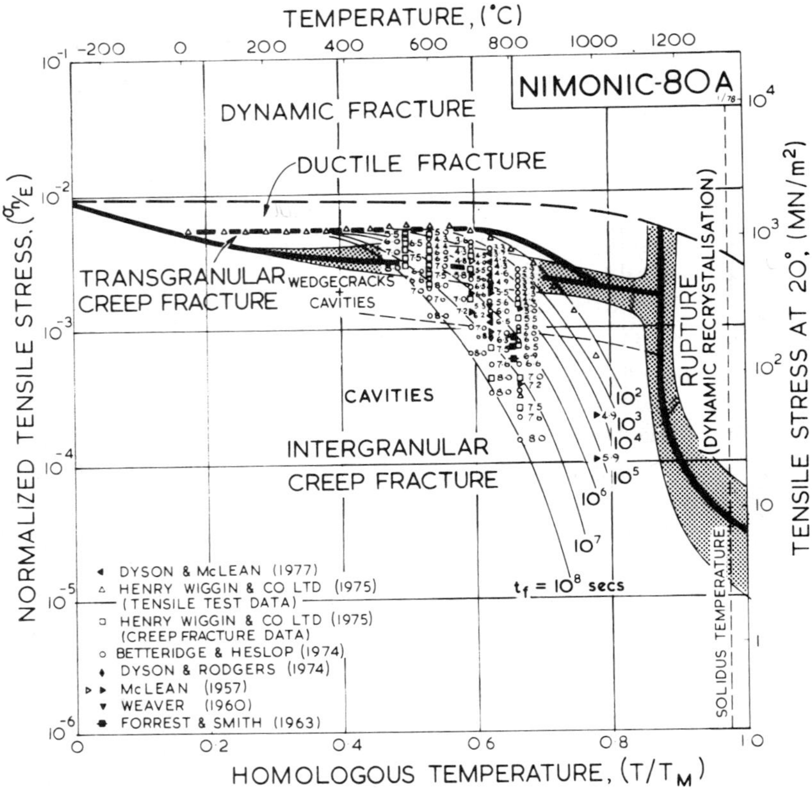

Fig. 1.10(a) – Fracture mechanism map for Nimonic 80A (0.1% C, 4 Fe, 18.5 Cr, 18.5 Co, 2.5 Ti, balance nickel) creep-resistant alloy. (Reproduced from Ashby (1980), with permission.)

In thick plate, however, and particularly in the presence of notches, plane strain prevails giving enhanced σ_H. This has the effect of expanding the cleavage fields at the expense of the ductile fields thus producing a 'more brittle' material, as already discussed in section 1.1. The relation between Ashby maps and 'deformation transitions' is discussed in Chapter 10. Table 1.1 gives the specific work of fracture (work per unit area of crack advance) associated with different mechanisms of fracture.

1.2.2 Amorphous solids

Below the glass transition temperature T_g, crosslinked polymers (such as polyesters and epoxies) and glasses exhibit brittle behaviour, and uncrosslinked polymers such as Perspex (PMMA) and Polystyrene (PS) exhibit crazing, which is a rate-sensitive phenomenon giving rise to macroscopic behaviour reminiscent of yielding but the net result of which is still often 'brittleness' in the sense that broken parts can be refitted together. Time dependency in amorphous solids is

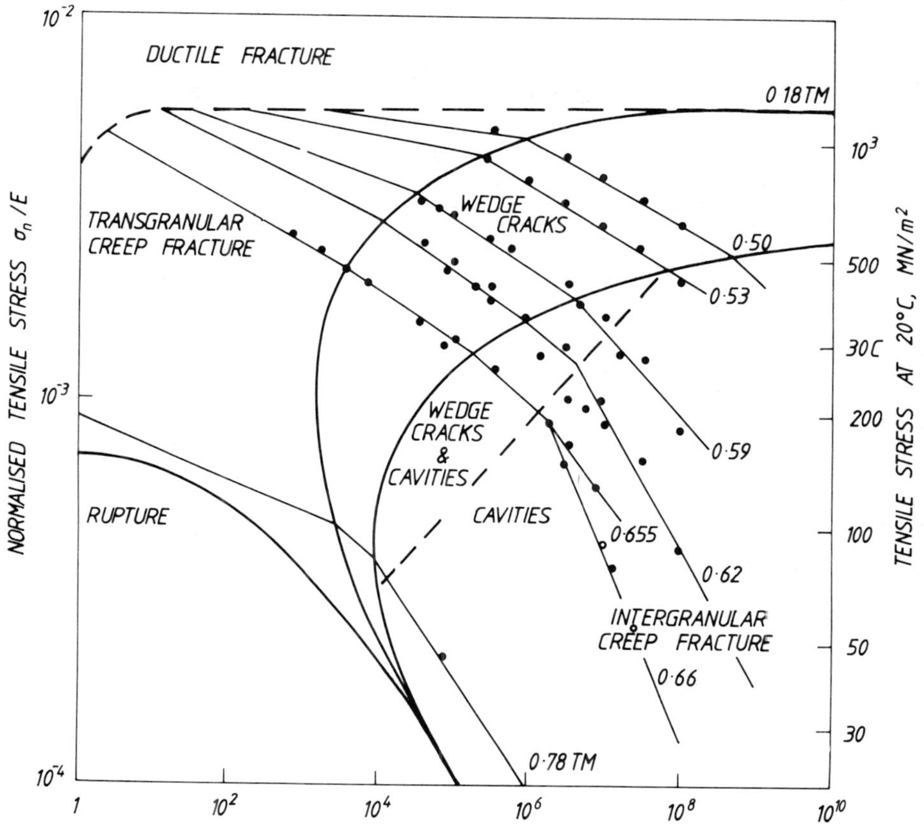

Fig. 1.10(b) – Stress-rupture cross-plot of data in Fig. 1.10(a) on to axes of stress and time. The difficulty of extrapolation of short-time stress-rupture data through fields having sub-divisions of fracture mechanisms is highlighted (reproduced from Ashby, (1980) with permission.)

marked above T_g, where crosslinked polymers (such as rubbers) can suffer large time-dependent deformations, and where uncrosslinked polymers may behave as very viscous liquids at high temperatures and sufficiently low strain rates, with fracture occurring by viscous fluid flow and cavity growth, rather like the behaviour of creeping metals.

Fracture in crosslinked organic glasses, and inorganic glasses, is controlled by microscopic inclusions, voids or defects. Thus, the practical strength of inorganic glass is controlled by surface microcracks. Such cracks are impossible to see by conventional optical or electron microscopy because their surfaces are essentially in contact with one another. Ion-exchange decoration techniques (first performed by Andrade and Tsien in 1937) provide evidence, however, of 'Griffith flaws' where patterns (similar to cracks in the glaze of old pottery) are observed, each crack in the pattern orginating from a surface microcrack. Depending upon the grade of glass, the density of microcracks is some tens per square millimetre. Griffith flaws are not an inherent property of the material, but are the result of mechanical surface abrasion. Freshly drawn glass fibres have high strength approaching the theoretical strength, but are easily damaged by contact with other fibres. Again, plate glass is known to have flaws introduced by grinding operation which precedes polishing, and the latter process does not remove all fissures remaining after the 'microchipping' grind. However, glass etched in hydrofluoric acid to remove the surface flaws gives no crack pattern in ion-exchange experiments. Fracture in 'damage-free' glass is then initiated by various mechanisms such as micro-inclusions (bits of refractory, micro-bubbles etc.) or, after moderate heat treatment, devitrification nuclei (regions of recrystallised glass produced by contamination with airborne dust or moisture, or finger marks).

Fracture is dominated by crazing in uncrosslinked amorphous polymers. Crazes are thin crack-like domains which form under tension. Crazes differ from cracks in that they are filled with mechanically transformed polymer of density lower than the bulk material. Crazed material has an index of refraction different from that of the bulk material and thus crazes appear as shiny cracks in an otherwise clear solid. They are induced by stress concentrations owing to inclusions or suface defects and their formation is sensitive to the presence of solvents (see Chapters 5 and 6)). Crazes are the precursors of cracks, the latter being generated by progressive deterioration of the viscoelastic craze material until macroscopic voids appear in the craze itself, the voids then grow into a crack within a craze; the tips of such cracks advance through the further conversion of bulk solid into craze material (Fig. 1.11).

In highly extensible polymeric solids, the molecules become orientated and strain crystallisation sometimes occurs. Thus not only are cracks blunted in highly extensible polymers, but the morphology of the crack tip material also changes – often over quite large regions. Both these events affect the resistance to fracture. Furthermore, for fracture to occur, strain energy has to get to the crack tip region. The effectiveness of transmission depends on shear modulus, which itself is most likely to be rate dependent, and on 'coupling' between different regions of the body. Extensible biological materials have a characteristic J-shaped σ–ϵ curve (section 1.2.4, Fig. 1.17 later) which inhibits energy transfer in the early stages of deformation. This is quite different from rubber-elasticity, where the σ–ϵ curve is sigmoid-shaped (Fig. 1.17). Even so, many polymeric sheets are difficult to tear (often they are used for packaging food!) owing no doubt to changes in local morphology as stressing occurs and to the shape of the σ–ϵ curve which determines energy transfer to potential crack sites. It is of interest to observe that a pin-prick in an inflated rubber ballon will cause it to burst, but if the ballon is pricked *before* inflation it will sustain a ***higher*** pressure before bursting. Rates of orientation *vis-à-vis* speed of crack development must feature in the explanation of the phenomenon which, it may be noted, occurs with water-filled ballons as well as air-filled ballons. Hydraulic or pneumatic means of applying pressure will aslo dictate how rapidly the crack 'driving force' falls as the ballon fractures.

Since amorphous solids contain examples of very brittle and very tough solids, the associated specific work of fracture covers a range as wide as that

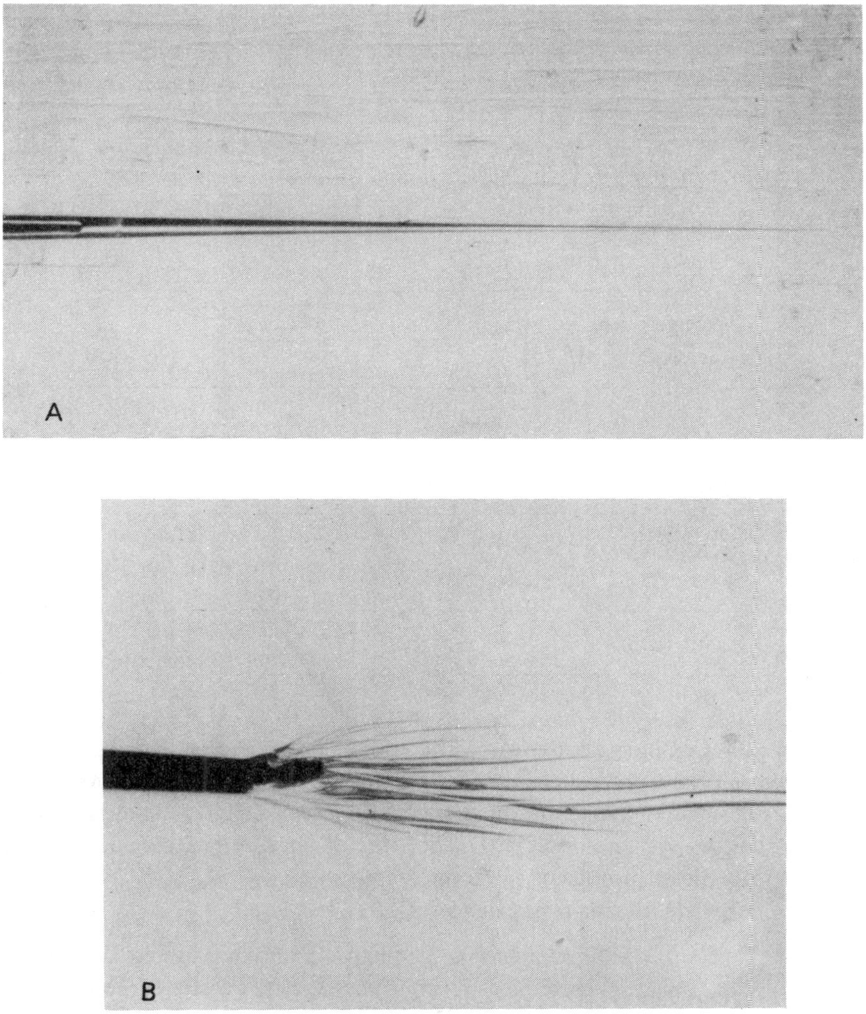

Fig. 1.11 – (a) At low speeds a crack propagating in polystyrene is preceded by a wedge-shaped layer of craze; the tip of the craze is sharp while that of the crack within the craze is blunt. (b) At lower temperatures, or if the molecular weight is greater, the stress to rupture the craze at the crack tip is higher and subsidiary or satellite crazes may be initiated above and below the craze layer containing the fracture plane. (Courtesy M. J. Doyle.)

shown in Table 1.1 for crystalline solids, although, of course, the micromechanisms of crack initiation and propagation may be quite different. Much research is being carried out to improve the toughness of ceramic solids at the brittle end of the range by, for example, clever microstructural effects such as found in partially stabilised zirconia.

Table 1.1

Specific works of fracture associated with mechanisms of crack advance

	Specific work of fracture (J/m^2)
Cleavage without slip ('Cleavage or BIF (1) and (2)')	1–100
Cleavage preceded by slip ('Cleavage or BIF (3)')	30–300
Mixed fibrous and cleavage	10^3–2×10^4
Unstable shear: void sheet linkage	10^3–10^5
Fibrous, by void growth and coalescence ('Ductile fracture')	10^4–10^6
Plastic rupture	10^6–10^7

1.2.3 Composite materials

Structural composites include any mixtures of separate material components combined to give a useful material. Such a composite often possesses the best qualities of the individual constituents and sometimes displays composite properties that none of the components possess.

Composite materials may be classified as follows:

(a) *fibrous composites*, having continuous or discontinuous filaments or whiskers in a matrix. Examples are wood, bone, glass-fibre or carbon-fibre reinforced plastics or boron-fibre reinforced metals;

(b) *laminated composites*, having layers of various solids (such as clad metals, laminated glass, plastic-impregnated cloths and papers, laminated fibrous composites);

(c) *particulate composites*, having particles in a matrix. Examples are concrete (stones in mortar), ABS polymers and some rocket propellants (inorganic particles in a rubber matrix). Particulate composites having metal matrices may have either non-metallic particles (such as ceramic particles giving so-called *cermets*) or particles of other metals (which encompass inclusions in metals).

The components of many practical filament-reinforced composites are often brittle. Consequently, although they have high fracture strength, and are stiff, such composites are not always very tough. One peculiarity of fracture in filamentary composites is that the fracture surfaces are rarely simply planar, and have bits of fibre sticking out of the matrix (Fig. 1.12). This feature helps to produce some extra toughness in composites. The cause of the 'stubble-like' feature is as follows: if the fibres are not continuous, and a matrix crack (orginating perhaps at a bubble, dust particles or notch in the matrix) approaches a discontinuous fibre such that the distance between the end of the fibre and the fracture plane is less than the critical transfer length (i.e. the length over which matrix loads are transferred by shear into the filaments), we expect that filament end to be *pulled out* of the matrix rather than broken. This comes about simply because within the transfer length at the ends of the filaments, the breaking stress is not reached. On the other hand, continuous filaments of uniform strength should fail in the plane of the matrix crack, with no pull-out; this applies to discontinuous filaments whose ends are embedded at distances *greater* than the transfer length below the main fracture plane. Now real fibres have weak points from place to place along their length, and vary statistically in strength from fibre to fibre, so that a fibre in the path of the crack will be broken either where it crosses the plane of the crack (i.e. where the stress is greatest) or at a flaw in the fibre which is near, but not necessarily on, the plane. Thus fracture surfaces

Fig. 1.12 – Characteristic brushy fracture surface of filament-reinforced composites showing fibre 'pull-out' (Courtesy B. Harris).

appear 'whiskery', and real composites (having continuous filaments, but weak points) possess properties in between the two extremes shown by those with full pull-out and no pull-out.

When a filament fractures off the main fracture plane, debonding has to take place 'up/down and around' the filament fracture to allow the main crack to propagate (Fig. 1.13). After the interfacial bond has been broken, there is an interfacial frictional stress opposing the fibres coming out of the holes. The friction stress may be as large as the interfacial stress was before debonding (if the interfacial break occurs by slip in the matrix adjacent to the fibres), but usually it is rather less. Even so, work has to be done pulling fibres out as the original main crack front propagates. Cottrell and Kelly in 1964 showed that pull-out work contributes *significantly* to resistance against cracking in composites.

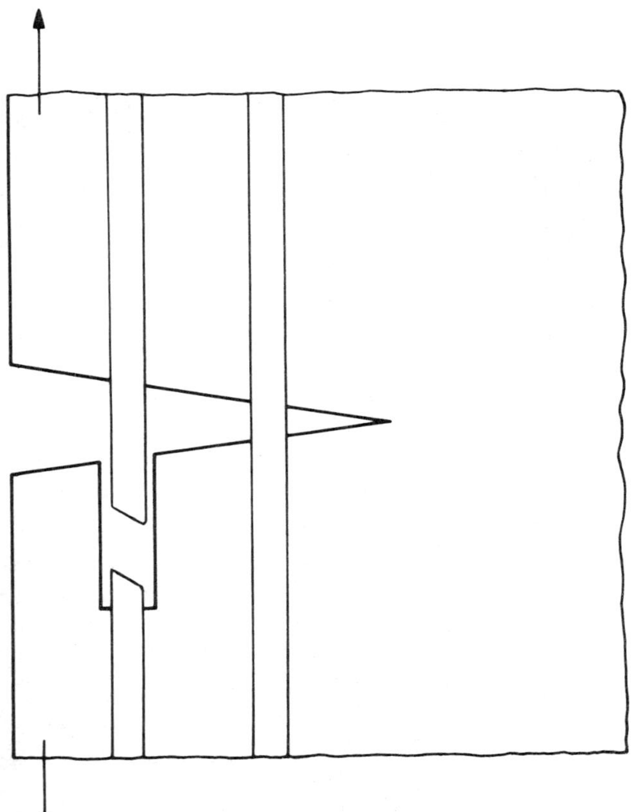

Fig. 1.13 – Passage of crack in fibre-reinforced composites involves interfacial debonding and filament fractures off the main fracture plane. Fibres bridge the crack faces in the wake of the crack front.

Some other contributions to composite toughness come from the work of debonding and the creation of new surfaces, both in the main fracture plane and also in the cylindrical areas around the pulled-out filaments. Composite fracture toughness is the sum of all the various dissipative work components per unit cross-sectional area of fracture *referenced to the area of the main crack.* As cylindrical debond areas and pull-out lengths are related to events off the main fracture plane, a consequence of dividing the total work of crack propagation by merely the projected area of the main matrix crack – and not the actual total area of new surfaces which would include the cylindrical debond areas – is that a *synergism* in toughness occurs. That is, the toughness of a composite is greater than the sum of the toughness of the components. Thus, even though many filamentary composites are made from components which are individually brittle, the composites can have respectable resistance to crack propagation. The total toughness of typically laid-up glass or carbon-fibre-type composites with 50 or so per cent volume fraction of fibres may be about 50 kJ/m^2.

A feature of the cracking behaviour of filamentary composites which is different from most other situations is that partially pulled-out fibres bridge the crack-opening in the wake of the advancing matrix crack (Fig. 1.13). The effect is enhanced the longer the pull-out lengths. Cracked composites thus have an ability to hang together in situations where other solids of similar toughness would breack apart. Note that the full contribution to toughness given by pull-out is achieved only when the filaments have pulled all the way out, at which time they no longer bridge the crack faces.

Strong interfacial bonding between the components of composites is necessary in order rapidly to transfer the load into the reinforcing components and thereby achieve high values of composite modulus and strength. However, it may be shown that high fracture toughness results when the interfacial bonding is weak (weak bonding gives long pull-out lengths and corresponding greater work). When a filament fractures in a composite having a strong interfacial bonding, a crack is formed in the matrix around the fracture, and usually this crack is energetic enough to run through the composite, breaking fibres as they are met in a 'zipper' action. The work of fracture can be very low.

One means of alleviating the problem of low toughness is to introduce crack stoppers which arrest the running crack. Cook and Gordon in 1964 made use of the fact that tensile stresses exist *parallel* to a running crack in order to produce arrest. The magnitude of these stresses is about one-fifth of the normal stress concentrated at the crack front in an isotropic continuum (see Fig. 2.23), so if there are locally weak interfaces in the path of the crack which are about one-fifth as strong as the main body, debonding ahead of the crack should occur and the crack be blunted into a T-shape (Fig. 1.14). Cook–Gordon debonding occurs in many biological systems (e.g. teeth) and may be applied with success to anisotropic man-made composites (section 3.5.2).

The occurrence of Cook-Gordon debonding gives a bonus to overall toughness work in the method of achieving simultaneous high strength and high toughness introduced by Atkins and Marston in 1974. High strength and high toughness appear to be mutally exclusive following the arguments above about

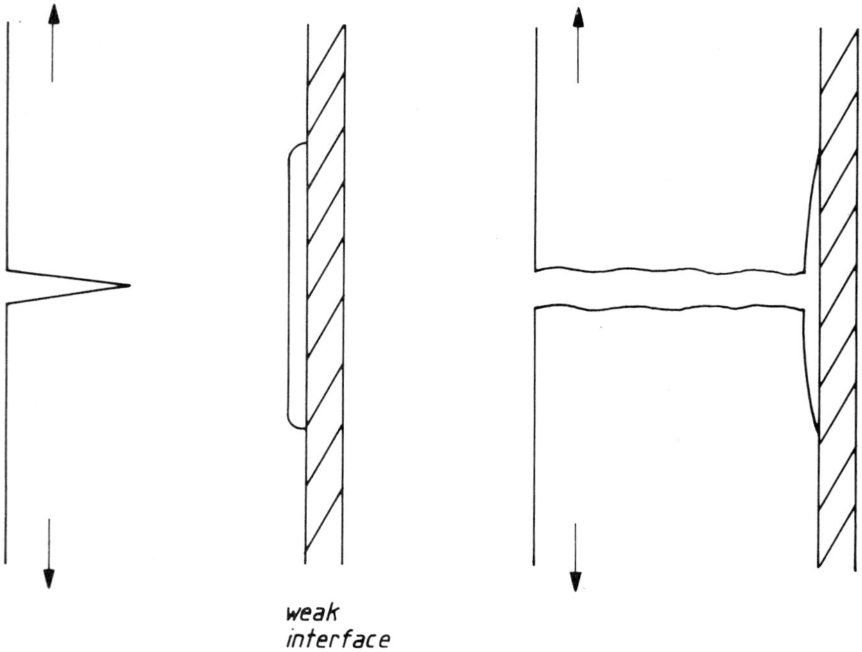

Fig. 1.14 – Cook–Gordon debonding. Weak interfaces in the path of a crack debond ahead of the propagating crack tip, owing to a stress concentration of the stress *parallel* to the crack. Crack runs into debonded region, is blunted and at least slowed down, if not arrested.

strong or weak interfacial bonding, but it was argued that, if the filaments were arranged to have random alternate sections of high and low shear stress (and locally low and high interfacial toughness), the desired result should be achieved. Weak and strong areas can be produced by intermittent coating of the fibres with appropriate substances. The strong regions ensure that the filament strength is picked up; weak areas randomly in the path of running cracks provide the source of toughness and pull-out work. Toughnesses of 300 kJ/m^2 are achieved in boron–epoxy composites using the method of intermittent bonding. Toughnesses of 400 kJ/m^2 are realised in model solids consisting of helically-wound fibre-reinforced tubes in a different method of improving composite toughness reported by Jeronimidis and Gordon in 1974. This achievement appeals to Nature's way of achieving high toughness in wood which is described in section 1.2.4 following.

Another approach to produce combined high strength and toughness by Morley and coworkers (e.g. Morley 1974) employs a new type of reinforcing member, consisting of a core and a surrounding sheath. In the laboratory, hypodermic tubes have been used which are filled with an inner core of helical wire, the unstressed helix diameter of which is greater than the internal bore of the tube so that there is a frictional bond between the core and sheath. Under normal loading conditions, the reinforcing members stiffen and strengthen the matrix in the usual way as with any other fibre-reinforced solid. The shear strength of the interface is reduced, however, as tensile loads on the core member increases, which gives rise to a simple stress-controlled interfacial debonding mechanism. The core helix wire is prevented from fracturing and is merely drawn through the composite against the friction at the walls of the sheath. As the decoupling mechanism is a result of small elastic deformations, it is possible to use high strength brittle metal core elements, or even bundles of polymer-bonded glass or carbon fibres, in the construction of the core-reinforcing elements. Thus the conflict between high yield strengths and large elongations to failure encountered particularly in metals can be avoided.

1.2.4 Biological materials

Most animal and plant materials are cellular composite structures formed of fibrous or plate-like components aggregated together. Biological materials are based on comparatively few chemicals. Indeed proteins (polymers of amino acids) and polysacchrides (polymers of sugars) make up nearly all the skeletal tissues, both compliant and stiff, of animals and plants. There is often a mineral phase present (consisting for example of calcium salts) and also water. Although small in number, the variety of combinations of these ingredients is immense as we move through the hiearchical structure from the basic macromolecules, through microfibrils, fibrils and structured fibres to the final 'material'. Depending on the biological requirement, almost any arrangement of components may be found. It should be remembered that biological materials are grown under stress by living organisms which are capable of detecting changes in load patterns and hence may alter the properties of the growing materials. For example, conditions of growth have a profound effect on the mechanical behaviour of wood; tree branches grow under stress. Jeronimidis in 1980 remarks that although 'reaction wood' (i.e. wood grown under non-symmetrical load patterns, as in a tree leaning over in a prevailing wind, which is brittle if grown under compression and has little fibre coupling if grown under tension) is generally a nuisance for the technological use of timber, in trees it is the logical answer to a particular set of circumstances. Again local changes in properties may be deliberately effected by the living organism to improve properties (water is used by organisms to 'pre-stress' biological materials, and the transfer of fluid from one site to another seems to be a mechanism for toughening teeth and probably other materials). Biological materials have evolved over aeons of time and are truly 'designed' materials. Man-made composites, in contrast, are much less complex and rarely contain more than two or three identifiable phases.

Soft biological tissues are composites of a pliant elastometer filled with stiffer fibrous or particulate material. Families of structural proteins are: keratins, collages, silks and elastin; fibrous polysaccharides include chitin and cellulose. Thus meat is comprised of myocin fibres helically-wrapped by collagen rather like an armoured cable. Insect cuticle is a composite of chitin and protein plus a few other components (mostly lipids and some salts) which contribute only a few per cent to the total dry matter; the chitin is present as microfibril bundles typically 2.8 nm in diameter containing some 20 chitin chains. Silk is the protein fibre produced by the larvae of the silkworm. The major chemical constituent of plants is cellulose, a high molecular weight polysaccharide which is directly responsible for stiffness and strength. The cellulose molecules are arranged in microfibrils (units about 10–20 nm in diameter) in which crystalline and amorphous regions coexist. There are also varying quantities of low molecular weight

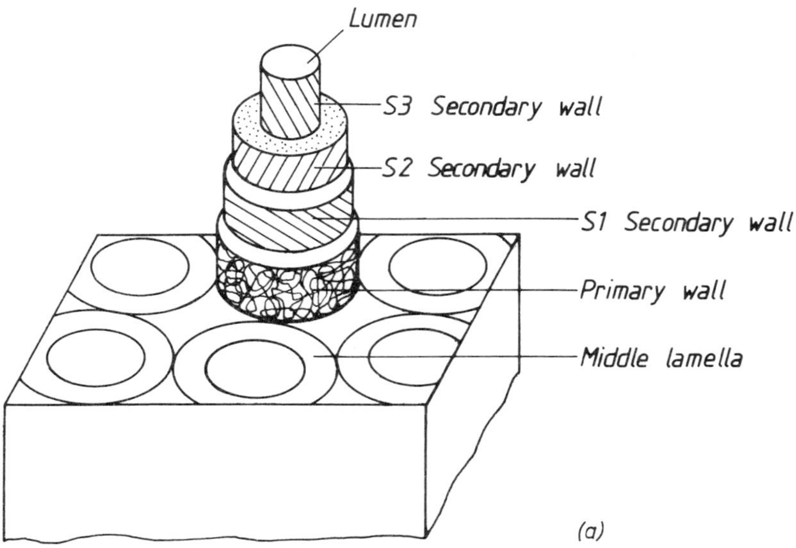

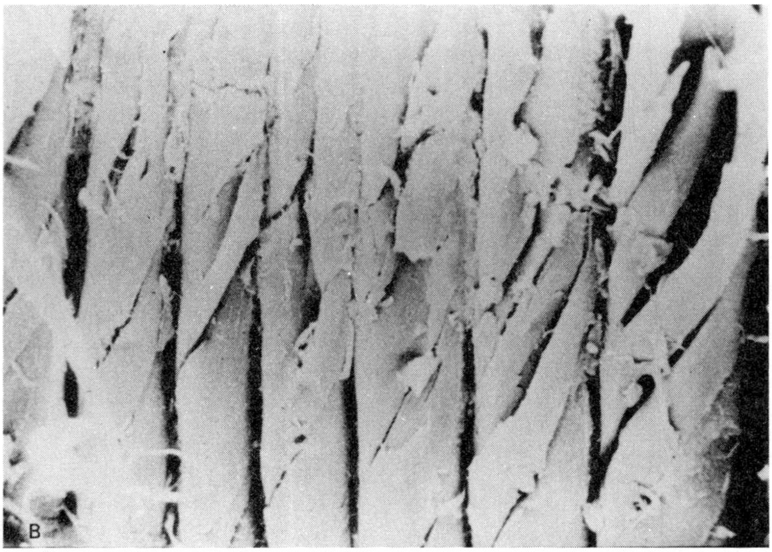

Fig. 1.15 – (a) Simplified model of a wood tracheid. (b) Inside of sitka spruce cell exposed by tensile fracture. Development of helical cracks in the S_2 wall leads to Euler buckling in *tension*. The decohesion and buckling gives high specific work of fracture in timber. (Courtesy G. Jeronimidis.)

sugars, hemicelluloses and lignin which can be considered as a binding matrix. During growth, the cellulose produced by the cells is deposited forming different walls at different stages, each being characterised by a distinct microfibrillar pattern. Fig. 1.15 shows a simplified and generalized model of a wood tracheid. The structure and behaviour of grass and similar materials may be interpreted in this way.

When biological materials have to withstand compression and bending loads, the interfacial coupling between components in the biological composite (at possibly all levels of hierarchical complexity from macromolecules upwards) must be increased in order to inhibit the fibres from sliding past one another. This is achieved either by stiffening the matrix (as in insect cuticle and wood, for example) or by the introduction of a new ceramic phase (as in the host of shell-like biological structures, teeth, tusks and so on). Thus mollusc shell has an outer layer of protein and an inner calcified layer. Mother-of-pearl (nacre) has an arrangement of aragonite ($CaCO_3$) crystals looking rather like a brick wall, with growth steps some 1/3 μm high which gives the characteristic iridescence from interference fringes (Fig. 1.16).

As may be expected, interfacial behaviour plays a predominant role in the mechanisms of fracture of biological materials. Unlike artificial composites, however, it is not a case of just the 'one' fibre–matrix interface, but rather the interactions at the multiplicity of interfaces between the different structural levels of molecules, cells and macroscopic fibres or platelets. In addition to the local orientation and shape of the composite components, the content and distribution of the mineral phase has to be considered and – of great importance – the water content. The problems are complex and complicated.

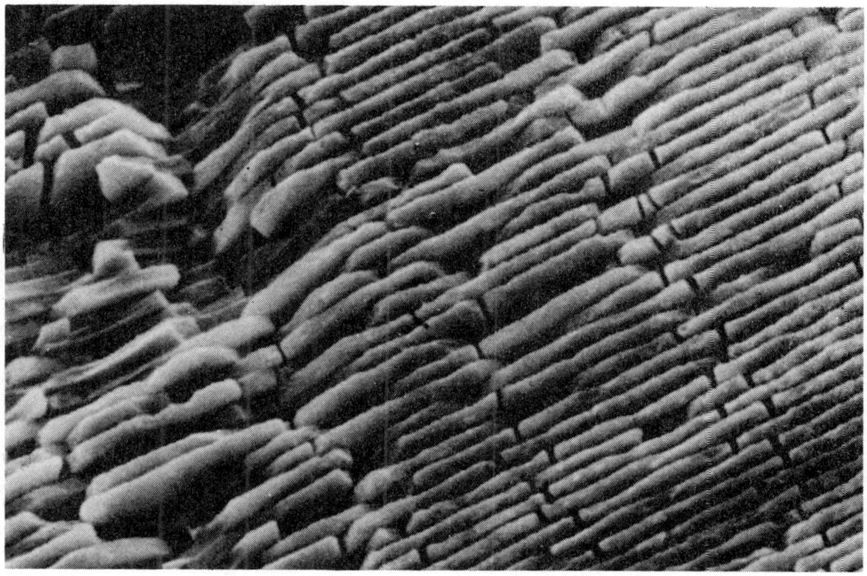

Fig. 1.16 – Fracture surface of nacre (mother-of-pearl) showing platelet structure. Platelets are rather less than 1 μm thick. (Courtesy J. D. Currey.)

In general the strength of the components making up biological materials decreases as the components increase in size from molecules to macroscopic components. Toughness is controlled by interfacial adhesion, and low shear moduli for the interfaces is commonly found. This means that fibres act independently and if individual fractures occur, the load is redistributed evenly among the remainder, i.e. many biological solids are notch-insensitive. Furthermore, whereas man-made fibre-reinforcing elements are usually isotropic *in themselves,* the cells in biological materials are usually 'structured'. Only the artificial du Pont 'Kevlar' fibre, being made up of other microfibres, vaguely resembles natural composites.

The 'structural foam' construction of biological materials is well-illustrated by wood, in which the tubular cell walls themselves have a structure of spirally-wound cellulose fibrillae embedded in a matrix of lignin (Fig. 1.15). Eighty per cent of the wall layer in a wood cell consists of S_2 structure which therefore controls the load-bearing capacity. The fracture toughness of most softwoods and hardwoods (~ 10 kJ/m^2) is exceptionally high in relation to other mechanical properties. This cannot be explained on the basis of the 'pull-out' contribution to work of fracture (section 1.2.3 which predicts at most 1 kJ/m^2 for wood) – indeed wood fracture surfaces do not have an excessively stubble-like appearance. The 'sub-structure' of biological materials is vital in this case for, as discovered by Jeronimidis and Gordon in 1974, it is the helical arrangement of the cellulose fibres in the load-bearing S_2 walls that provides the clue. When pulled in tension the fibres decohere, and the cell walls buckle inwards into the lumen in such a way that longitudinal extensions of the cell up to 20 per cent are possible. Thus, although the elastic fracture strain in wood as a whole seldom much exceeds 1 per cent, cells close to the fracture surface extend far more and absorb much energy as they do so. Arrays of model cellulose fibres, made by winding glass and carbon filaments into hollow helices with resin, exhibit experimental toughnesses up to 400 kJ/m^2, which is much improved over the usual level of toughness in man-made composites. The orientation of fibres in the S_2 wall is a compromise between stiffness and toughness. In grass the fibres are more parallel to the cell axis than in wood, so the cell is stiffer but less tough; even so grass when trodden on buckles, deflects and eventually recovers.

The Cook–Gordon effect (section 1.2.3) plays an important role in toughening biological materials. Currey showed in 1979 (Currey, 1980) that in mother-of-pearl (the material of mussel shells) the work of fracture is exceptionally high (~ 1.5 kJ/m^2), yet 95% of it is calcium carbonate (chalk) with a toughness of a few joules per square metre (Fig. 1.16). The missing 5 per cent is a mixture of protein and polysaccharides within the interfaces at which the spreading of cracks is inhibited by controlled debonding. Kendall (1981) has interpreted the behaviour as that of interfacial dislocations (section 2.4.1). In teeth, Cook-Gordon debonded areas will heal eventually in healthy bodies. In diving, however, it has recently been suggested that gases become trapped in the microcrevices of damaged teeth, prevent healing and induce necrosis in the bone. Many diseases of course are associated with biological materials no longer correctly performing their mechanical functions.

In addition to interfacial effects, the shape of load–deflexion curves plays an important role in determining the resistance to fracture of biological materials. Materials such as wood and bone are essentially Hookean in their response (as are artificial membranes like cloth and paper). They obey Griffith-style fracture mechanics (Chapter 2 *et seq.*) with, in the case of bone for example, blood vessels or the holes and tubes containing the osteocytes acting as inherent starter cracks (~ 25 to 50 μm in size). In contrast most biological materials are highly deformable and at much lower stresses experience extensions perhaps a thousand times greater than commonly found in traditional engineering solids. Even though the stresses in the soft tissues of living organisms may be low, yet, because the strains are so high, the area under a load–deflexion diagram giving the stored strain energy may be even higher than that found in engineering materials. (J. E. Gordon has calculated that the strain energy in the tissues of one's cheek when one yawns is approximately equal, per unit volume of material, to that in mild steel at ordinary engineering working stresses – say 10^6 J/m^3). Even in the widest yawn, however, one's cheek does not split across like a box girder bridge; most animal tissues are in fact exceptionally resistant to cracking – animals do not go pop, like a balloon, when pins are stuck into them. It is often difficult to paunch a rabbit.

The extensibility of skin, gut wall, artery, bladder and most other collagenous tissue is *not*, as might be first thought, like rubber elasticity. Rather these biological materials have a J-shaped elastic curve (Fig. 1.17) and this characteristic shape plays a crucial role in fracture behaviour. Rubber has a sigmoid-shaped

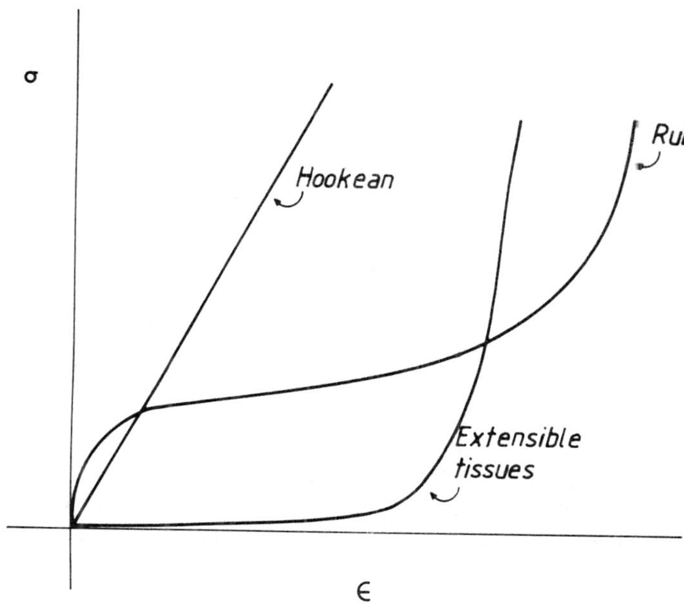

Fig. 1.17 – Schematic stress (σ)-strain (ϵ) curves for Hookean solids, rubber-type materials and extensible biological materials. (not to scale).

curve. Difficult-to-tear biological membranes do not in fact possess exceptionally high works of fracture (about 1–10 kJ/m^2, Appendix 2) which is comparable with many other solids which can be torn comparatively easily under comparable conditions. So, are the mechanics of fracture of biological materials solely determined by 'work of fracture'? As Gordon has pointed out on a number of occasions, where rat skin, worm cuticle or human arteries differ from metal sheet, or indeed from rubber sheet, is not in work of fracture but in the shape of the stress–strain curve. (Amniotic membranes which are intended to fracture in childbirth, and egg shells, do not have a J-shaped curve and more or less follow Hooke's law). The initial part of the J-shaped σ–ϵ curve is almost horizontal, rather like surface tension in a liquid, and hence the material has a very low shear modulus. There is thus virtually *no* communication of strain energy between parts of the stressed body and it is difficult to feed energy into potential crack sites. Such materials therefore do not tear. A J-shaped curve is produced when a knitted or woven fabric (at 45° to the warp or weft) is pulled in tension, the progressively increasing stiffness at larger extensions arising from progressive orientation of the fibres towards the pulling axis. These materials are difficult to fracture and Gordon remarks that tears in trousers are L-shaped along the warp and weft of the cloth, and are not along the 45° axis into which direction it is difficult to transfer strain energy. Progressive orientation of fibres is similarly the cause of J-curves in biological materials (e.g. X-ray diffraction studies show that randomly-orientated collagen fibres in the wall of pig aorta progressively line up with increasing deformation). Traditional pegged, bolted or riveted joints in engineering also display J-shaped curves, unlike welded structures. As welded steel structures are often full of cracks, welded ships do break in two; traditionally-built ships and similar structures rarely broke in this way owing, no doubt, to the wobbly characteristics of their joints.

The different characteristic J-shape of the σ–ϵ curves of many biological materials compared with those of man-made materials presents difficulties sometimes with artificial implants and prostheses. The J-shape is there for a purpose, and must be matched when replacing biological tissue with man-made substitute materials – one example concerns the stability of artery walls, and aneurisms (local 'ballooning' and fracture).

The study of fracture in biological materials is in its infancy, yet already it is clear that much can be learnt from nature. The mechanics of fracture of biological materials is considered elsewhere in this text, and the interested reader should also consult the book by Vincent (1982) and other references listed at the end of this chapter.

1.2.5 Pre-existing cracks

Few materials are 'perfect', even carefully prepared laboratory samples. Commercial materials usually have flaws of one sort or the other, and many fabricated components in engineering are probably full of defects which may take the form of short cracks or be connected with microstructural features which under stress will become cracks. The deformation and fracture of bodies may be primarily controlled by such inherent flaws (depending on their size, shape and distribution) rather than by fine details of microstructure. Much effort is spent on the inspection and elimination of flaws in critical engineering components and structures.

The size of flaws which are present is determined by the way they arise, i.e. whether from processing, fabrication, service stressing or environmental attack, or from some combination of these acting simultaneously (as in fretting). Tomkins (1983) has reviewed the formation of flaws in engineering components as follows.

Most metallic products are made by one of three main processing routes: wrought (or cast and wrought), cast, powder-formed. For wrought products, the degree of deoxidation, and composition of the initial cast form, determines the amount of porosity, segregation and distribution of inclusions prior to working. Working modifies the shape of these defects as well as introducing grain texture and sometimes anisotropy. This modification of local mechanical properties can dominate the development of short cracks on a similar or smaller scale. However, the degree to which defects behave as cracks depends on their morphology. For example, sulphide inclusions in structural steels which are plastic at hot working temperatures can be rolled out to produce thin crack-like platelets up to 1 mm in length. Depending on orientation, these inclusions can behave as cracks or crack-arresters. On cooling, residual stress fields are set up around inclusions by differential thermal contraction, MnS inclusions being left in tension whilst oxides and silicates are left in compression. Final defect sizes in wrought products depend on material quality and working process and can range from several micrometres to a millimetre or so.

Cast products can contain a variety of defects such as shrinkage cavities, gas porosity, slag entrapment and solid non-metallic deoxidation products. Defect sizes are often of order 0.1–1 mm. The recognition that defects will exist in casting results in higher safety factors being used than for wrought products.

For some applications, particularly of complex alloys which might have segregation problems when conventionally cast and wrought, a powder compaction route is taken. This can involve hot isostatic pressing or forging. The process may result in foreign particles, e.g. of refractory, being present in the final product of a size comparable with the initial powder size. The resulting defects could be of order 0.1 mm.

Tomkins concludes that small defects with a variety of sizes in the sub-millimetre range may be present in all products. Their subsequent behaviour as cracks during service will depend on their detailed morphology, location (e.g. surface-breaking or buried) and orientation to the local applied stress field.

As regards fabrication, while soldering, brazing and various forms of pressure or friction welding may introduce defects through the formation of brittle intermetallic compounds or incompletely bonded interfaces, it is probable that fusion welding is the process most likely to produce the greatest variety of defects in metal–metal joints. Solidification cracking, slag entrapment and lack of fusion may all produce significant weld metal defects whilst residual stresses are readily induced as the cast weld metal cools. Hydrogen uptake may lead to heat affected zone cracking whilst inclusions normal to transverse stresses induced by the weld can give rise to lamellar tearing. Finally stress relief (reheat) cracking may be produced in heat-affected zones of quenched and tempered steels, particularly in the presence of trace impurities. The scale of defects introduced by welding is large but with care, they can be kept in the sub-millimetre range. Ahmad and Kanninen (1984) present a valuable overview of analysis methods for cracks in welds. They remark that it is likely that the great majority of structural failures originate from weld defects. Sub-millimetre deep cracks can also be induced by other fabrication processes such as grinding and machining, particularly in hard alloys, whilst heat treatment can give rise to quench cracks by differential transformation strains in high-C steels. Hard surfaces produced by chemical treatment alone are prone to cracking. Machining itself gives rise to a surface finish with grooves on a scale which can cover a wide range of effective notch depths, typically 1 μm for a fine turned finish and 10–20 μm for a rough finish.

Assembly of components may induce additional surface damage on a sub-millimetre scale by scratches, arc strikes, etc.

Cracks may appear during loading and in service. For example, in otherwise perfect material, short fatigue cracks can initiate naturally by dislocation processes. These are normally confined to the material surface where lack of restraint aids crack formation. The slip characteristics, e.g. planar or wavy, determines the ease of initiation but microscopically discernible crack sizes are of the order 1–10 μm (Chapter 7). Note that this is considerably below inherent material, processing or fabrication defect sizes which can be present in large components, so that these latter inherent flaws often dominate life rather than the fine-scale flaws. Surface-breaking defects are particularly dangerous fatigue crack initiators in that they either behave as a crack *per se* or act as a stress raiser which readily initiates a crack preferentially. In this regard, surface scratches are effective crack starters, including rough machining marks.

As naturally initiated cracks are of order a few micrometres, surface 'notches' can be on a fine scale and still be effective in giving a local 'high strain fatigue' condition at the root. Constrained thermal cycling as experienced under thermal shock conditions can be a similar source of surface cracking; 'thermal checking' in high temperature plant is a result of this process (Chapter 6 and 7).

If components operate in a chemically aggressive environment, there may be some form of local surface attack, e.g. intergranular corrosion or pitting. Pits are not cracks, but act as local stress raisers which give rise to cracks of a depth somewhat in excess of the pit depth. This may range from a fraction of a millimetre to several millimetres. Intergranular attack by corrosion or selective oxidation may readily produce effective irregular cracks a few grains deep.

It follows from this discussion that all components are likely to contain defects with a range of sizes on entering service. From the fatigue point of view, in particular, it is surface-breaking defects which usually give most cause for concern. Fabrication and environmental damage is the most likely source of surface defects, with many components likely to have such defects in the 0.1–1 mm range.

1.3 CRACK TIP PROCESS ZONE

To produce fracture, irreversible work has to be performed to nucleate, initiate and propagate a crack. The *nucleation* of an incipient crack concerns the formation of a void or cavity from an inclusion, crazed zone, pile-up of dislocations etc. in an originally 'flawless' solid; the 'sharpening' of voids is the process of *initiation* which is a necessary, but not sufficient, prerequisite for crack *propagation.* While these events are distinct, the differences between them are often blurred, and the separate terms are used loosely in the literature. As we have already seen, sometimes there are small pre-existing sharp cracks present in solids, in which case work of fracture is essentially work of propagation of the 'already initiated' crack. Even when cracks or crazes have to be nucleated by some localised flow, the work associated with nucleation and initiation may still be a minor component of the total work involved in complete fracture. On the other hand, crack nucleation in ductile solids may involve quite large plastic strains, albeit contained in a small volume in which the crack is nucleated. Propagation under such conditions may be thought of as 'continuous nucleation' of the crack ahead of itself as it propagates. Table 1.1 gives orders of magnitude of the work of fracture per unit area of crack associated with the various micromechanisms of fracture. Here we do not distinguish between the nucleation, initiation and propagation phases, but later in the book we do do so.

Since the work of fracture is irreversible, when in the Preface we talk of 'elastic' fracture, where broken pieces may be refitted to regain the original size piece body, we mean fractures which occur while the body deforms elastically in a *global* sense. The non-elastic irreversibilities making up the fracture work in the broken body are confined to a small *process zone* at the crack tip which, as propagation ensues, becomes a thin boundary layer contiguous with both faces of the crack, Fig. 1.18. It is in the process zone that the various mechanisms controlling toughness, described in sections 1.2.1–1.2.4, act.

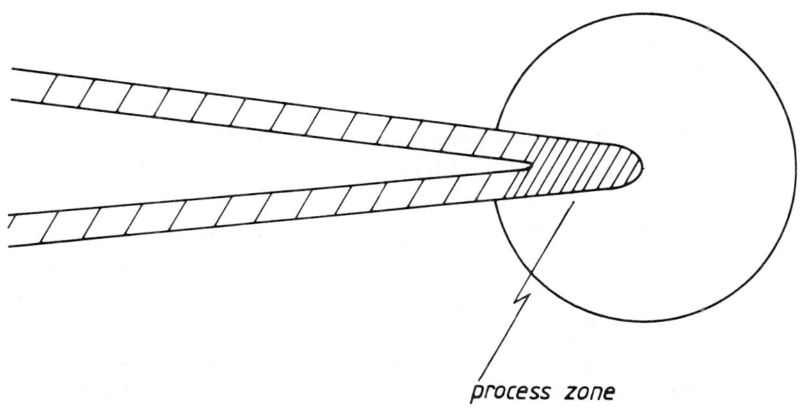

Fig. 1.18 – Process zone at crack tip in which fracture mechanisms take place.

The lowest values for the specific work of fracture in Table 1.1 are not much different from the Gibbs' surface free energy of solids, which is the thermodynamic minimum work of crack propagation. This means that the process zone and hence boundary layer of irreversibility is extremely thin, – 'vanishingly thin' in an ideal continuum. The increasing values of work of fracture in Table 1.1 for crystalline solids experiencing different micromechanisms of fracture reflect additional contributions from localised plasticity which now becomes an integral part of the process of cracking. The process zone size increases, but even in ductile fibrous fracture of a low carbon steel displaying $R = 100\ \text{kJ/m}^2$, the boundary layer is not more than about 0.5 mm thick, which is quite thin in the context of brittle fracture of a large ship or long pipeline.

Fig. 1.11 shows related behaviour in polystyrene, where a propagating crack is preceded by a very long wedge-shaped layer of craze; the tip of the craze is very sharp while that of the crack which propagates within the craze is blunt. As the craze is parted at or near its median plane, the surfaces of fracture are then covered by a smooth continuous layer of crazed material, akin to the plastically deformed 'wake' left behind the crack tip plastic zone in metals. At lower temperatures, or with higher molecular weight, the stress to rupture the craze at the crack tip is increased and subsidiary satellite crazes may be initiated in the high-stress regions ahead of the crack tip (e.g. Bevis and Hull, 1970). These secondary cracks need not necessarily occur in the plane of crack propagation, so their union with the main crack occurs through steps in the fracture surface (cf. the origins of river markings in cleavage), which gives characteristic striations in the plane of the fracture. Fracture surfaces then also appear iridescent owing to increased reflexion of light from the many parallel crazes below the crack surface.

1.4 MECHANICS

When all irreversibilities associated with fracture are contained within regions contiguous with the crack faces, and deformations elsewhere are below the yield strain (so that all severed pieces may be refitted to regain essentially the original size of the body), we have *elastic fracture,* the mechanics of which may be analysed in terms of bulk elastic parameters: the various work irreversibilities may be lumped together in a specific macroscopic work of fracture term within the process zone called the fracture toughness (R) which is the work required to propagate a crack by unit area. R includes the Gibbs' thermodynamic minimum work required for the creation of new surfaces, together with non-linear elastic strains at the crack front, plastic and viscous deformation at the crack surfaces, and delayed reversible and irreversible effects associated with possible atomic configurational rearrangements such as phase changes at the crack surface. When elastic cracking occurs, the work consumed in the process zone is provided from the potential energy of the elastic material which surrounds the crack. We use the general symbol J for the rate of change of potential energy with respect to crack area; in *linear* elastic fracture J is the more-familiarly known 'strain energy release rate' G. When quasi-static fracture occurs, J or G have the critical values J_C or G_C which exactly match the specific work of fracture required in the process zone given by R†. Experiment shows that R is a reproducible quantity over a useful range of variables (stress, strain, rate, temperature, environment, etc.) and we may write for a quasi-static increment of crack propagation in a body, containing an existing crack of area A, being loaded by a pair of self-equilibrating external forces X with associated displacements u,

$$\left.\begin{aligned} &X\mathrm{d}u = \mathrm{d}\Lambda + J\mathrm{d}A \\ &J = J_C = R \quad \text{at equilibrium fracture generally} \\ &J \equiv G \text{ in linear elastic bodies} \end{aligned}\right\} \quad (1.1)$$

where Λ is the elastic strain energy. The continuum equation (1.1) is simply the work theorem of statics, $X\mathrm{d}u$ being the external work done and $(\mathrm{d}\Lambda + R\mathrm{d}A)$ the internal work done in the event. Equation (1.1) is the basis of the mechanics of elastic fracture, and it may be used to characterise the crack-resistance of materials and to calculate the strength of cracked structures, as discussed in Chapters 2–6 of this book. When $\partial J/\partial A)_{X \text{ or } u} > \mathrm{d}R/\mathrm{d}A$, fracture is unstable, as discussed in Chapter 8.

Note that use of equation (1.1) does not require detailed knowledge of the micromechanisms of fracture. The often complicated crack tip events during fracture are conveniently glossed over in the macroscopic R-term. The acid test of applicability of equation (1.1) is whether the body behaves in a globally elastic manner, and whether there is confidence and reliability in the particular values of R employed in the equation. These R values will in turn be established, over a range of variables, from experiments which are known to satisfy elastic reversibility.

In so far as R is a propagation term, the values of specific work of fracture in Table 1.1 are identical with R under those conditions where work of initiation is a small proportion of the total work required for fracture. It should be apprecia-ted that equation (1.1) is not limited to the fracture of materials like glass or pottery for which R is small. It applies to all *reversible* cases, whatever the magnitude of R, e.g. fracture of PMMA for which $R \approx 500\ \mathrm{J/m^2}$ at room tempera-ture, and even the brittle fracture of large structures made of low carbon steel for which R is over $100\ \mathrm{kJ/m^2}$.

When the broken pieces cannot be refitted together, to regain the original size of the body, there must be zones of plastic flow near the crack tip region and also distributed throughout the body. The strains in these yielded regions may be of the same order of magnitude as the yield strain (in *elastoplastic* fracture) or they may be so large that they swamp the elastic strains (in *plastic* fracture), (Fig. 1.19). The zones of remote flow need not be uniformly distri-buted through the body, and there may be extensive regions remaining, in which

† Alternative measures for the resistance to fracture (critical stress intensity factors, critical crack opening displacements, etc.) follow later in the text.

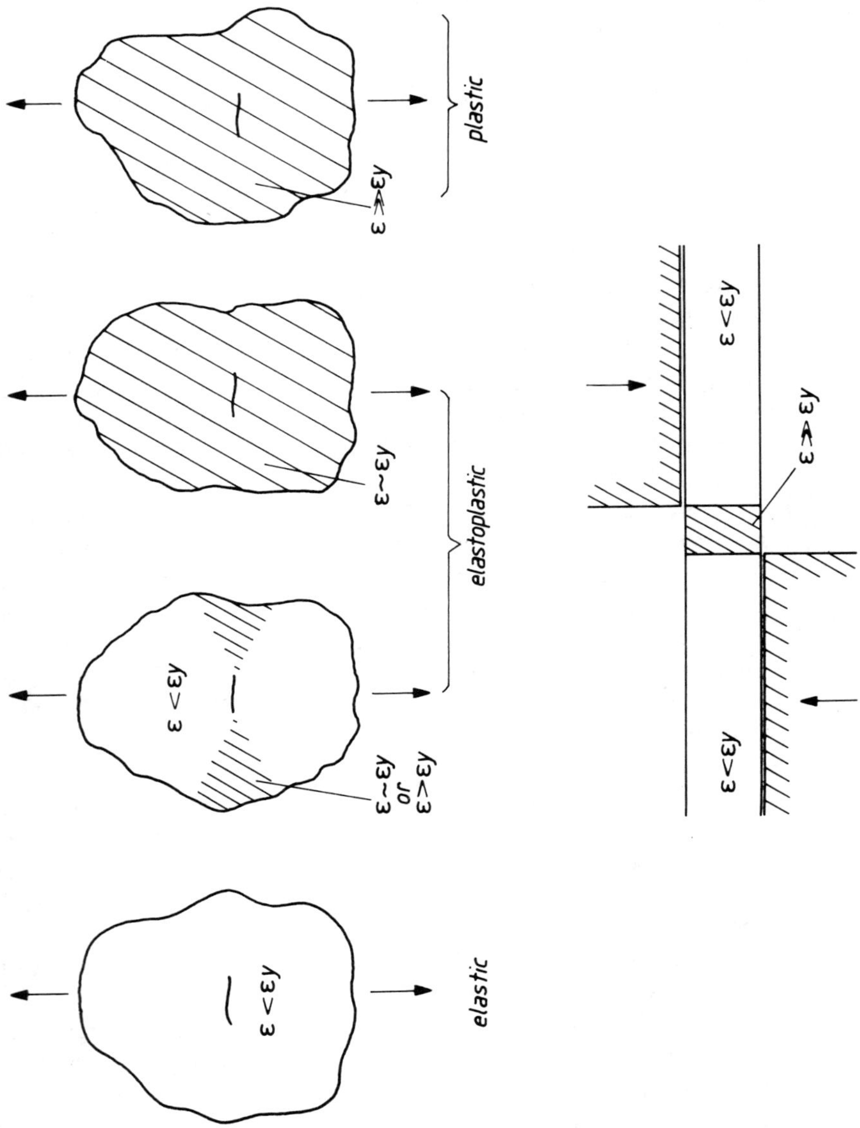
elastic
elastoplastic
plastic

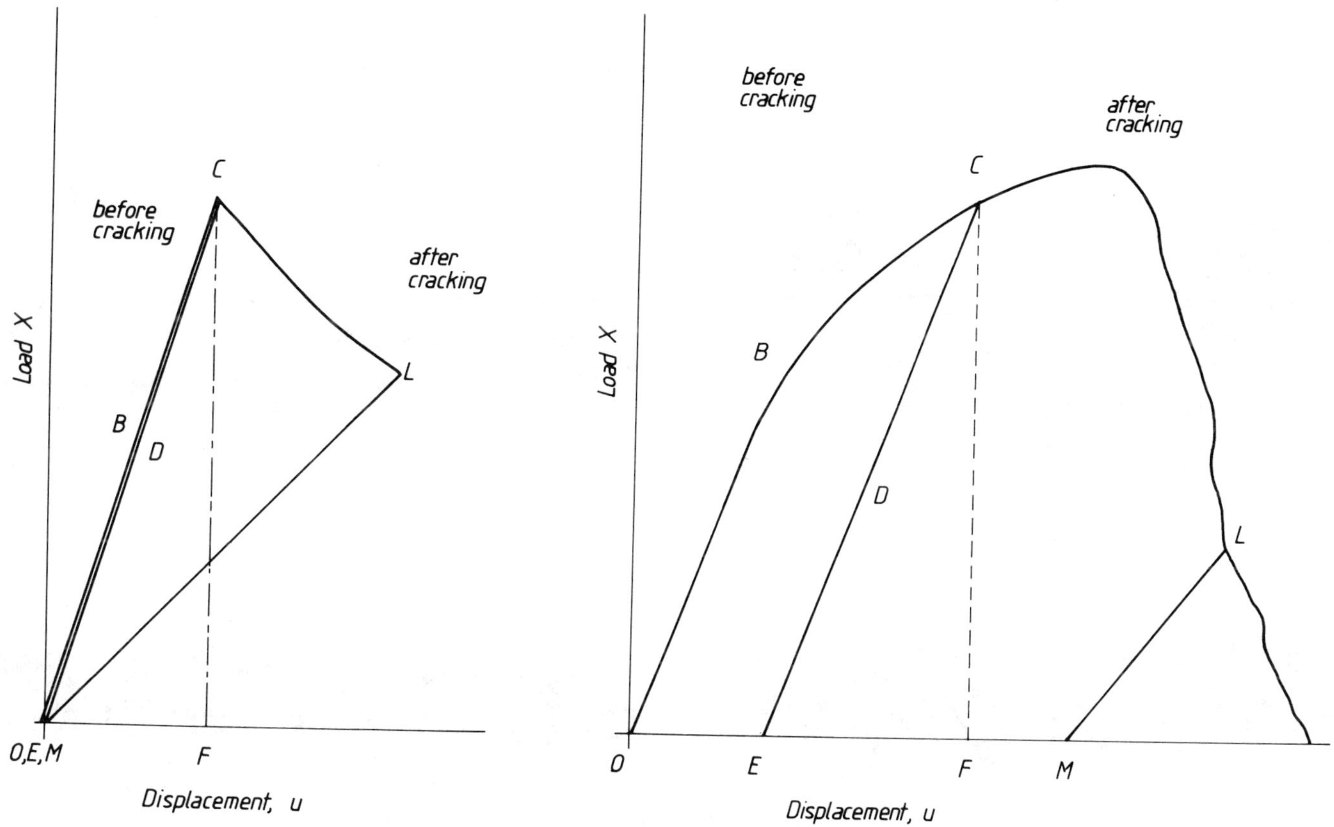

Fig. 1.19 – Different types of elastic, elastoplastic and plastic fracture stituations. Volume of material V^P undergoing irreversible plastic deformation shown shaded. Since incremental plastic work $d\Gamma = W^P dV^P$ where $W^P = \int \sigma d\epsilon$ (the plastic work per volume), $d\Gamma$ can have the same value from strains only just greater than the initial yield strain but spread over a large volume, or from large strains contained in a small plastic vlume. Load (X)-displacement (u) plots for displacement-reversible (globally elastic) fracture and displacement-irreversible (elastoplastic or plastic) fracture.

Fig. 1.20 – Different degrees of combined flow and fracture shown by Charpy tests on low carbon steel samples quenched and tempered by different amounts.

the strains are elastic, but the important point at issue is that the zones of irreversibility are large and no longer condensed into thin boundary layers alongside the faces of the propagated crack (as in elastic fracture) so that refitting is impossible. Fig. 1.20 shows a progression of these different types of fracture in Charpy V-notch testpieces made from a low carbon steel in different quenched and tempered states.

Clearly, in addition to work of fracture required by the micromechanisms operating in the crack tip process zone, there will be irreversible work in the yielded regions aways from the crack. In the energy balance for crack propagation there will be an additional term Γ that represents the energy dissipation due to remote plastic flow. Hence,

$$X\mathrm{d}u = \mathrm{d}\Lambda + J\mathrm{d}A + \mathrm{d}\Gamma \qquad (1.2)$$

Within the process zone R is balanced by J. It is crucial to realise that the remote plastic flow has nothing to do with the process of fracture as such, and that the magnitude of Γ depends on the size and shape of the deforming body, and the level of the strains in the body as a whole during fracture. The work during loading, given by the area under a quasi-static load-displacement X–u diagram, represents the combined contributions to RA and Γ, and what makes the mechanics of elastoplastic fracture difficult is that it is often impossible to separate out the two work terms. Indeed, there is a further complication for unloaded work areas after non-uniform plastic flow, since they also contain residual elastic strain energy (Fig. 1.21(a); Chapter 4) and friction work if cracking is

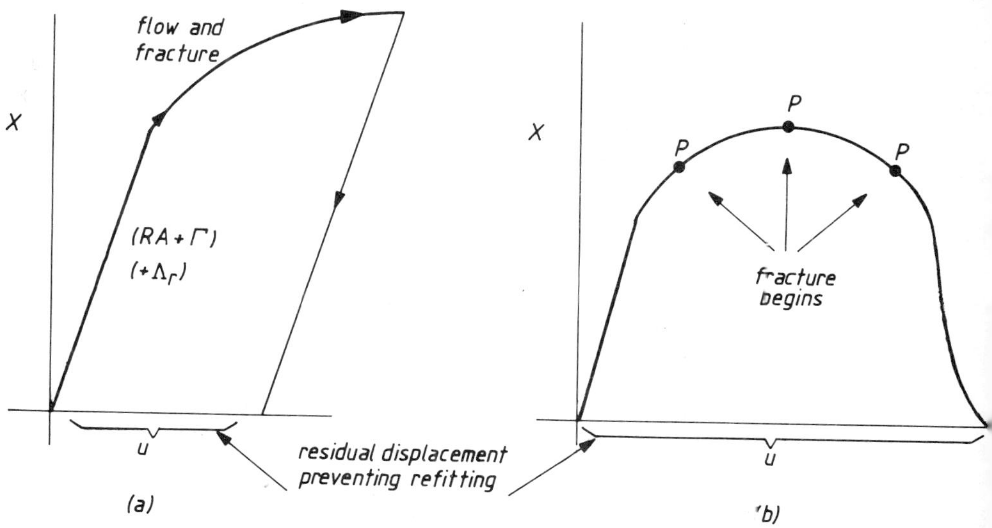

Fig. 1.21 – Load (X) – displacement (u) diagrams for combined flow and fracture. Unloaded work areas represent fracture work, remote plastic work and residual elastic strain energy. Residual displacements prevent refitting broken pieces to regain original-sized articles.

produced by tooling (as in bar shearing, guillotining, etc.) as discussed in Chapter 5. Under some circumstances $\mathrm{d}\Gamma$ and $R\mathrm{d}A$ are coupled (such as in the steady-state tearing of soft metal sheets (Chapter 4)) but in general they are not necessarily linked. Depending on conditions at the crack tip and micromechanisms of fracture, it could be that the separate terms in equation (1.2) (and similar equations following) do affect one another (Rice, 1965). Furthermore, it is possible in elastoplastic fracture for regions that at first plastically deform to later elastically unload as the crack propagates. The mathematics of combined flow and fracture, represented by equation (1.2) can be formidable, as discussed in Chapters 4 and 5, but if experiments show that the work components can be partitioned (i.e. mathematically, that the variables are 'separable'), useful solutions to problems are attainable, as shown later. We also discuss later the validity of the so-called J_R-approach to the mechanics of elastoplastic fracture, in which the $J\mathrm{d}A$ and $\mathrm{d}\Gamma$ terms of equation (1.2) are lumped together.

A complicating issue – particularly in plastic fracture – is that extensive remote plastic flow often precedes crack propagation. That can happen in the presence of pre-existing cracks (see the Charpy specimens in Fig. 1.20), and it most certainly occurs in supposedly crack-free ductile bodies before a crack is nucleated. Consequently the work of nucleation becomes significant, and is part of the initial plastic flow, in a localised part of which deformation becomes highly concentrated. The events leading to crack nucleation are hence concerned with an equation of the form

$$X\mathrm{d}u = \mathrm{d}\Lambda + \mathrm{d}\Gamma \qquad (1.3)$$

which has nothing whatsoever to do with crack propagation. The appearance of first observable cracking in supposedly crack-free bodies is discussed in Chapter 5, particularly as it applies to metal-working processes. The aim of such calculations is to predict where, on a yielding X–u diagram, cracking first starts, i.e. point P in Fig. 1.21(b); (N.B. P may occur before, at, or after a load maximum in the X–u diagram, it depends on the circumstances – P should not automatically be identified with a maximum).

Similarly, if crazes in polymers first appear in strain fields which are more or less elastic, the events leading to craze nucleation are concerned with

$$X\mathrm{d}u = \mathrm{d}\Lambda \qquad (1.4)$$

Crazes are the precursors of fracture and eventually equation (1.1) takes over.

As explained in Chapter 5 (and also in Chapter 7 on fatigue) there is a gap in our knowledge to date concerning the nucleation and early growth of cracks, and the meaning of a 'nucleated crack' or a short crack is not always clear. (For example, in deep drawing steel sheets, it is likely that most of the carbide particles are already cracked from previous rolling, when presswork begins.)

In creep fractures, we are concerned with rates of deformation and the relevant continuum equation takes the form

$$X\dot{u} = \mathrm{d}\dot{\Lambda} + \dot{\overline{(J\mathrm{d}A)}} + \mathrm{d}\dot{\Gamma} \qquad (1.5)$$

The mathematics involved are more complex, and as there is no general approach available yet, we shall not consider the mechanics of high temperature creep fracture at any length in this book. An area of interest however where rate effects are important in the low-to-medium homologous temperature range is *dynamic fracture,* where kinetic energy H is important in the energy balance. Then, for elastic fracture, we have

$$X\mathrm{d}u = \mathrm{d}\Lambda + J\mathrm{d}A + \mathrm{d}H \tag{1.6}$$

This relation is applied to the topic of crack arrest in Chapter 8.

In environmental elastic cracking, we write

$$X\mathrm{d}u = \mathrm{d}\Lambda + J\mathrm{d}A + \mathrm{d}\psi \tag{1.7}$$

where $\mathrm{d}\psi$ relates to chemical energy, electrochemical energy and so on. Chapter 6 discusses the application of this relation.

The influence of residual strain energy on equations (1.1), (1.2), (1.4) and (1.7) is discussed later in the book.

An alternative formulation for crack mechanics has been presented by Andrews and co-workers (1974, 1976, 1977). Instead of summing the work inputs, as in this section, a multiplicative approach is used for reversible, irreversible and time-dependent energy densities. Then $\mathscr{T} = \mathscr{T}_0\ \Phi\ (\dot{a}, T, \epsilon_0)$ where $\mathscr{T}_0$ is identified with surface free energy γ, $\mathscr{T}$ is related to G_c R or J, and with a crack velocity, T temperature and ϵ_0 the remote representative strain Φ is the 'loss function' by which the specific work actually required for fracture is greater than the thermodynamically least work. Clearly Φ incorporates $\Gamma, \dot{\Gamma}, H, \psi$ (and environmental effects) as appropriate.

1.5 MECHANICS-TO-MICROSTRUCTURE LINK

The solutions to equations (1.1), (1.2), (1.5), (1.6) and (1.7) give relations between load X, displacement u and crack area A (or stress σ, strain ϵ and A) during crack propagation; the solution to equation (1.3) gives the relation between X and u (σ and ϵ) at crack nucleation. X and u are *bulk* continuum parameters of interest to engineers, and as already mentioned the results of computations do not require detailed knowledge of the micromechanisms of fracture propagation since the reproducible R term conveniently lumps together all events at the crack tip. In the case of equations (1.3) and (1.4) however, a criterion for crack or craze nucleation has to be invoked. The criterion is expressed in terms of bulk continuum parameters (such as, for example, the attainment possibly of a critical stress or critical strain or some path-dependent combination of the two – details are given in Chapter 5) but unless it is entirely empirical, the criterion will reflect what is thought to be going on in the localised region of intensified deformation in which a crack or craze nucleates.

Material behaviour in such a region may be modelled in a variety of ways. It is sometimes convenient to think of a solid on the one hand as a collection of atoms or molecules with individual properties, or on the other hand as a continuum with properties which represent the macroscopic average of the finer-scale properties. Mechanisms of crack nucleation, initiation and propagation may be viewed at various mangifications, since sometimes the behaviour of the atomic lattice is important (dimensions of ≈ 1 nm being relevant), at other times microscopic regions (≈ 10 μm) over which inclusions can play a part, and at other times again macroscopic regions (~ 1 mm) such as the plastic zones found at the tips of cracks in many solids. These different viewpoints broadly reflect the interests of, on the one hand, metal physicists and materials scientists and, on the other hand, engineers and designers. The one group studies the deformation behaviour of materials leading up to the nucleation and propagation of cracks by examining, for example, changes in structure and dislocation morphology; the other group will define a nucleated crack as one that can be ***measured,*** which may mean in practice a length of a few micrometres to a millimetre depending on the performance of the detection instruments being used.

Models of cleavage relate directly and quantitatively to equation (1.1) since in ideal cleavage on a single crystallographic plane, the boundary layer of irreversibility is vanishingly thin and R is then the surface free energy. The Griffith formula for fracture of brittle solids (Chapter 2) is the solution of equation (1.1) for the special case of a small crack in a very large sheet loaded at its external boundaries. Models have been developed for other micromechanisms of fracture – in particular for nucleation, growth and coalescence of internal voids, to produce a crack where growth takes place by plastic flow below about $T_m/3$, and by diffusion at higher homologous temperatures T_m. These models take into account rates of nucleation, rates of growth and so on, and relate local stress and strain history to microstructure. They are valuable in that they indicate the important variables controlling fracture by the particular micromechanism being modelled (such as inclusion spacing-to-size ratios). Microstructural control may therefore be applied to improve a material's resistance to fracture, or particular stress–strain histories that should be avoided in a deformation operation in order to prevent fracture may be highlighted. Examples are given in Chapter 5.

Models of localised fracture however do not always relate quantitatively to the bulk parameters X, σ, u, or ϵ. Difficulties are to be expected in that, unlike a continuum, a practical microstructure has properties varying locally and sometimes a statistical approach is required. Even so, experiments show that results of value in engineering can often be obtained if some of the finer details of the micromechanisms of fracture are by-passed. The R parameter does just that (as does the related stress intensity facture K of linear-elastic fracture mechanics which is introduced in Chapter 2). While most of the mechanics in this book are concerned with bulk continuum parameters, they are related back to local micromechanisms where appropriate, such as in the effects of environment on resistance to cracking (Chapter 6). Similarly the mechanics-to-microstructure link is vitally important in processes of crack nucleation prior to ductile fracture: in this case it must be better to relate bulk behaviour to the likely local microstructural behaviour rather than rely on empirical relations of dubious pedigree between X, σ, u or ϵ (see Chapter 5).

1.6 APPLICATION TO DESIGN: SCALING AND SIZE EFFECTS

The basis of traditional 'strength-of-materials' engineering design is that the working stresses should be kept below some proportion (σ_Y/f) of the yield stress, where f is the factor of safety. Due allowance is made in stress calculations for the effects of combined stresses, changes in section, notches and other stress raisers, fatigue loading, etc. Materials used in manufacture are chosen to have acceptable mechanical properties and minimum levels of, for example, yield strength are specified in codes used for certain types of construction (bridges, pressure vessels, ships and so on).

J. E. Gordon, in the first of his splendid books on materials and structures, (1968) said that the worst sin in an engineering material is not lack of *strength* or lack of *stiffness,* desirable as these properties are, but lack of *toughness,* i.e. lack of resistance to the propagation of cracks, and that the history of the attempts to prevent cracks spreading, or to evade the consequences, is almost the history of engineering. In modern engineering, the accepted way of preventing things from breaking is to use a material with a high work of fracture – hence the use of ductile metals like mild steel. But, of course, this approach almost rules out the use of really strong materials because they are nearly always brittle. The situation is not helped by the use of monocoque or plate-welded construction, which provides easy shear connexion throughout the structure and hence easy transmission of strain energy to crack sites leading to easy fracture.

We have seen an alternative way of ensuring safety provided by Nature: it is to prevent released energy from reaching fracture zones by interrupting or controlling the elastic communication between parts of the system. In this way suspension bridges, using cables made up of decoupled strands of 'dangerous' high strength/low toughness wires, can have spans about ten times longer than other designs of bridge built from 'safe' low strength/high toughness materials, but Nature's design methods for structures and materials are not yet in general use by man. Instead there is a quest for materials which are simultaneously strong, tough and stiff. This requirement is difficult to achieve, and usually one of the three properties has to be traded-off against the others. For example, the success of certain thermoplastics, such as polyethylene and nylon, in replacing metals is partly due to their inherent toughness. But they are often insufficiently stiff, particularly in large components. Conventional composites are stiff, but are rarely tough enough.

In many applications design is against excessive deflexions, i.e. is on the basis of stiffness rather than strength. In practical structures, the compressive members usually constitute a far higher proportion of the mass of the structure than the tensile members owing to design against buckling. Geometry of cross-section becomes important. In those circumstances optimum design for minimum mass no longer depends simply on E/ρ, but may be shown to depend on $\sqrt{E}/\rho$ (columns), or $\sqrt[3]{E}/\rho$ (panels); an E/ρ^3 parameter also arises in the case of maximum bending stiffness for fixed mass. The form of the $\sqrt[3]{E}/\rho$ parameter explains why, in the case of conventional composites structural weight are likely to be higher than for an equivalent stiff *wooden* component: the specific gravity of most fibre-reinforced composites is about three times that of timber. It does not seem to be widely realised that the theoretical upper limit for composite toughness based on fibre pull-out mechanisms is too low if composites are to compete successfully against metals in many engineering applications. In order to achieve a fracture toughness of some hundreds of kilojoules per square metre in composites, energy must be absorbed in regions which, on a molecular scale, are extremely remote from the fracture plane. The Gordon–Jeronimidis method of producing simultaneous high stiffness, high strength and high toughness (based on the behaviour of wood (section 1.2.3)) does give works of fracture comparable with those of ductile metals. In fact this helical-tube structural solution gives the best of both worlds since high $\sqrt[3]{E}/\rho$ can just as well be achieved by lowering ρ as increasing E. Holes are cheaper than carbon fibres!

Implicit in the traditional strength-of-materials approach to design is that yielding will always precede fracture and that the material used is 'simply ductile'. Yet some designs have failed by elastic fracture when all the traditional design safeguards had been met. Books (by authors such as W. D. Biggs (1960) and C. F. Tipper (1962)) have been written on the story of the brittle fracture of structures made from steel which satisfied the metallurgical specification and which gave satisfactory mechanical properties in laboratory tests. There were, for example, fractures of Vierendeel truss bridges over canals in Belgium just before the Second World War; the well-known fractures of Liberty ships and T2 tankers during the war; and a number of examples of storage tanks, pressure vessels, penstocks, pipelines, and bridges since then. An article by Shank (1953) and the book by Boyd (1970) list many cases of brittle fracture, including those of riveted (not welded) constructions, dating back to the turn of the century.

When a failure occurs, either (i) the designer's calculations are at fault or (ii) the material of construction does not have the correct properties or (iii) the manufacture and fabrication procedures are at fault. Furthermore, conventional thinking would say that if 'overstressing' occurs in a structure, owing to unforeseen circumstances, yielding should occur in the ductile steel and the stresses should be relaxed by plastic flow. Indeed traditional engineering materials have sometimes been called 'forgiving' materials as they are often able to accommodate errors in design, manufacture and fabrication, by localised plastic flow. In this way failure is supposedly guarded against. But the brittle fractures experienced in large structures made from 'ductile' materials were *elastic* fractures – the whole ship or storage tank or whatever had not yielded, or locally necked down, prior to fracture as would have been expected perhaps on the basis of the performance of laboratory-size testpieces. Some of the fractured ships were simply rewelded back together again – an excellent example of reversible deformation and elastic fracture. Clearly something else was going on which traditional strength-of-materials was incapable of explaining for large structures. The eventual understanding of the missing link gave rise to the new disciplines of elastic 'fracture mechanics'. A valuable review of the role played by fracture mechanics in design is given in the 1984 report on slow crack growth prepared by a Working Party of the Royal Society. A variety of design applications is contained in the following chapters of this book.

While low temperatures, below the transition temperature range (Fig. 1.8) was sometimes a contributory factor, a key to the explanation for brittle fractures in bodies made from ductile materials is found to be a *size-effect*. The one crucial aspect in which the elastic mechanics of fracture differs from normal elasticity or plasticity is that the stress to cause fracture, i.e. the solution to equation (1.1), depends on the *size* of the body and the size of cracks in it. For a given material the size effect is, in broad terms, that the stress to cause fracture is not constant (as proposed by Ludwik, Fig. 1.1) and *may* be *smaller* in *larger* pieces of the same material. In contrast, the stress to cause yielding at the same degree of deformation is usually taken to be independent of the size of a body. Consequently, even though a laboratory-size testpiece may flow plastically before fracturing, a large structure made from the same material may fracture first and never yield plastically at the same temperature since the size-effect may bring down the fracture stress in the large body below the general yield stress.

The size-effect is shown schematically in Fig. 1.22; detailed calculations will be found in Chapter 9. A simple explanation for the size-effect comes from appreciating that the work *required* for fracture increases as the square of some characteristic dimension (representing crack area) but the energy *available* through the volume of the body to produce fracture increases as the cube of that length.

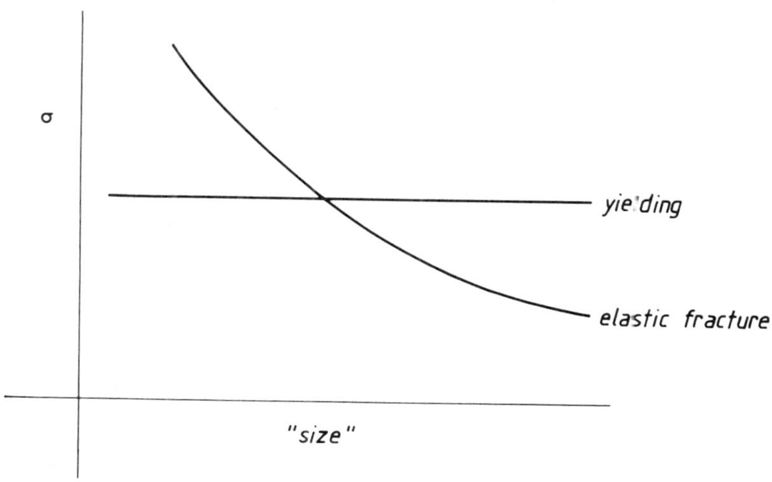

Fig. 1.22 – Size (or scale) effect in the mechanics of fracture. The stress to cause cracking often falls with increase in size of the cracked body (it depends on the geometry, scaling and method of loading and is the result of cube/square scaling). The stress to cause plastic flow, on the other hand, is usually considered to be independent of size. Hence large structures, made of materials which behave in a ductile fashion in the laboratory, can fracture in a brittle fashion if they are large enough. Vice versa it is difficult to comminute powders below certain sizes, owing to flow supervening fracture in small sizes.

The position of the fracture stress curve is governed by a material's inherent resistance to crack nucleation and propagation. The more brittle the material, the closer the curve is to the ordinate and, except for extremely small sample sizes, $\sigma_{cr} < \sigma_y$. The fact that it seems impossible to avoid flow even at very small sizes explains why brittle solids such as diamond and glass can be indented plastically when the size of the hardness test deformation-zone is sufficiently small. Similarly glass may be ground or even turned in the lathe when the depth of cut is small (a few micrometres). Again it is the size effect which causes difficulties in the crushing and comminution of materials into very fine powders: below a certain size it is more favourable for the particles to flow plastically rather than break down further (i.e. the process of comminution from large pieces to small proceeds from right to left along the abscissa of Fig. 1.22 and eventually $\sigma_y < \sigma_{cr}$).

The position of the fracture stress curve is also influenced by the presence of notches, pre-existing cracks and so on. Such features tend to promote 'less ductile' behaviour, so that badly finished details, improperly treated heat-affected zones in welds, and other features in large structures exacerbate the size effect.

It should be remarked that an explanation for the size-effect, which (along with unfavourably low temperatures) caused the brittle fracture of large structures that caused so much consternation in the Second World War, was already available in the Griffith analysis of brittle fracture published in the early 1920s (Chapter 2). It is easy, however, to have hindsight and it is understandable that it was far from clear in those days that the Griffith theory (developed for glass) should also apply to *all* elastic fractures – including those in large structures made from 'ductile' steel. It is now realised that, although in low carbon steels local rupture of the ligaments bridging voids involves local heavily strained plastic regions (Fig. 1.5, Fig. 1.7(d)), discontinuous ductile fracture at the same temperature *can* readily occur when the bulk of the surrounding material is still *elastic*. Another way of looking at the size effect is that, if fracture is characterised by a certain size-independent 'crack tip opening displacement' (COD – see Chapters 2 and 4), it may not be possible in small pieces in the laboratory to accommodate large COD at the crack tip whilst the bulk of the body remains elastic; instead large yield zones span the cracked structure, and cracking occurs with generalised yielding. Large COD though can be accommodated elastically in sufficiently large pieces, especially in the presence of long cracks.

It is important to appreciate that microstructural features remain fixed in size – inclusions are the same whether the structure is small or large. That the inclusions remain the same size permits a large COD to be encapsulated in an elastic strain field when the structure gets very large. Were the size of specimens to be the same as the crack tip plastic zone, the material would behave ductilely, as it would if the crack-nucleating inclusions and stretching ligaments were scaled up to the size of footballs. It is unfortunate that we customarily classify materials as being 'ductile' or 'brittle' solely on the basis of small-size laboratory tests.

REFERENCES

Ahmed, J. and Kanninen, M. F. (1984) *Proc. Conf. Numerical Methods in Fract. Mech.* University College, Swansea.

Andrade, E. N. da C. and Tsien, L. C. (1937) *Proc. Roy. Soc.* **A159**, 346.

Andrews, E. H. (1974) *J. Mater. Sci.* **9**, 887.

Andrews, E. H. and Billington, E. W. (1976) *J. Mater. Sci.* **11**, 1354.

Andrews, E. H. and Fukahori, Y. (1977) *J. Mater. Sci.* **12**, 1307.

Ashby, M. F. and Tomkins, B. (1980) *Proceedings, ICM-3 (1979)*, Smith and Miller (Eds.). Oxford: Pergamon, Vol 1, p. 47.

Atkins, A. G. (1975) *J. Mater. Sci.* **10**, 819.

Averbach, B. L., Felbeck, D. K., Hahn, G. T. and Thomas, D. A. (Eds.) (1959) 'Fracture', *Proc. Swampscott Conference.* Cambridge, Mass: MIT Press.

Bevis, M. and Hull, D. (1970) *J. Mater. Sci.* **5**, 983.

Biggs, W. D. (1960) *The Brittle Fracture of Steel.* London: Macdonald & Evans.

Boyd, G. M. (1970) *Brittle Fracture in Steel Structures.* London: Butterworths.

Caddell, R. M. (1980) *Deformation and Fracture of Solids.* Englewood Cliffs, N.J.: Prentice-Hall.

Cook, J. and Gordon, J. E. (1964) *Proc. Royal Soc.* **A282**, 508.

Cottrell, A. H. (1963) *Proc. Roy. Soc.* **A276**, 1.

Cottrell, A. H. (1965) *Proceedings, First Tewksbury Symposium* University of Melbourne/Butterworths, 1.

Cottrell, A. H. (1964), *Proc. Roy. Soc.* **A282**, 2.

Currey, J. (1980) The mechanical properties of biological materials. *Symp. Soc. Exp. Biol. 34 (1979).* Cambridge: Cambridge University Press.

Felbeck, D. K. and Atkins, A. G. (1984) *Strength and Fracture of Engineering Solids.* Englewood Cliffs, N. J.: Prentice-Hall.

Gifkins, R. C. (1969) *Proc. 2nd. Tewksbury Fracture Symposium, 31.* University of Melbourne/Butterworths.

Gordon, J. E. (1968) *The New Science of Strong Materials – or Why you don't Fall through the Floor.* London: Penguin Books.

Gordon, J. E. (1978) *Structures – or Why Things don't Fall Down.* London: Penguin Books.

Gordon, J. E. (1979) *Catastrophe design – or how to behave like a worm,* Review of Progress in Quantitative NDE, La Jolla, California: Rockwell International.

Griffith, A. A. (1921) *Phil. Trans. Roy. Soc.* **A221**, 163.

Griffith, A. A. (1924) *Proc. 1st Int. Conf. Appl. Mech. (Delft),* p. 55.

Gurney, C. and Hunt, J. (1967) *Proc. Roy. Soc.* **A299**, 508.

Jeronimidis, G. (1980) The mechanical properties of biological materials. *Symp. Soc. Exp. Biol. 34 (1979).* Cambridge: Cambridge University Press.

Jeronimidis, G. and Gordon, J. E. (1974) *Nature* **252**, 116 (see also *Proc. Roy. Soc.* **B208**, 449. (1980)).

Kelly, A. (1964) *Proc. Roy. Soc.* **A282**, 63.

Kendall, K. (1981) *Int. J. Adhesion and Adhesives,* **1**, 301.

Ludwik, P. (1909) *Elemente der tech. Mech.* Berlin.

Morley, J. G. (1974) *Chemistry in Britain,* **10**, 478.

Rice, J. R. (1965) *Proceedings, First International Congress on Fracture, Sendai, Japan,* p. 309.

Rogers, H. C. (1960) *Trans. AIME* **218**, 498 (see also *Ductility*. Metals Park, OH: ASM).

Royal Society (1979) *Fracture Mechanics in Design and Service.* London: Royal Society.

Royal Society (1984) Report of Working Group on Subcritical Crack Growth (Brit. National C'ttee on Theor. and Appl. Mech.) London: Royal Society.

Shank, M. E. (1953) National Academy of Sciences/National Research Council Report. Welding Research Council (USA) Bulletin Series No. 17.

Tipper, C. F. (1962) *The Brittle Fracture Story.* Cambridge: Cambridge University Press.

Tomkins, B. (1981) *Phil. Trans. Roy. Soc.* **A299**, 31.

Tomkins, B. (1983) *Trans. ASME,* Paper 83-PVP-98.

Vincent, J. F. V. (1982) *Structural Biomaterials.* London: Macmillan.

2

The mechanics of elastic fracture

'What *is* the use of repeating all that stuff?' the Mock Turtle interrupted, 'if you don't explain it as you go on? It's by far the most confusing thing *I* ever heard!'

2.1 STRAIN ENERGY

When a body is loaded elastically, the work done in deforming the body is stored as elastic strain energy, Λ. In Fig. 2.1, if OB is the load–deflexion response of the body, Λ at load X is given by the shaded work area under the load–deflexion diagram, and $\Lambda = \frac{1}{2}Xu$ in the case of linear Xu behaviour. On unloading, the linear Xu line is retraced backwards and all the energy is recovered as in loading and unloading a spring (in elasticity theory there is no hysteresis or other losses). Under the action of many loads X_i and moments M_i (which produce rotations θ_i) the total reversible strain energy is given by

$$\Lambda = \Sigma \tfrac{1}{2} X_i u_i + \Sigma \tfrac{1}{2} M_i \theta_i \tag{2.1}$$

It is shown in books on elasticity and strength-of-materials that the strain energy stored per unit volume of the body in uniaxial deformation is given by $\frac{1}{2}\sigma\epsilon = \frac{1}{2}\sigma^2/E = \frac{1}{2}E\epsilon^2$ where σ is stress, ϵ is strain and E is Young's modulus. In the presence of normal and shear stresses in an xyz system, the generalised expression for the elastic strain energy stored in an incremental volume $\mathrm{d}V = \mathrm{d}x\mathrm{d}y\mathrm{d}z$ of the solid is

$$\mathrm{d}\Lambda = \begin{Bmatrix} (\Sigma \tfrac{1}{2}\, \sigma_{ii}\epsilon_{ii} + \Sigma \tfrac{1}{2}\, \tau_{ij}\gamma_{ij})\, \mathrm{d}V \\ (\Sigma \tfrac{1}{2}\, \sigma_{ii}^2/E + \Sigma \tfrac{1}{2}\, \tau_{ij}^2/\mu)\, \mathrm{d}V \\ (\Sigma \tfrac{1}{2}\, E\epsilon_{ii}^2 + \Sigma \tfrac{1}{2}\, \mu\gamma_{ij}^2)\, \mathrm{d}V \end{Bmatrix} = W\mathrm{d}V \tag{2.2}$$

where σ_{ii} ϵ_{ii} are normal stresses and strains, and τ_{ij}, γ_{ij} are shear stresses and strains, μ is the shear modulus and W is the (elastic) strain energy density. In a

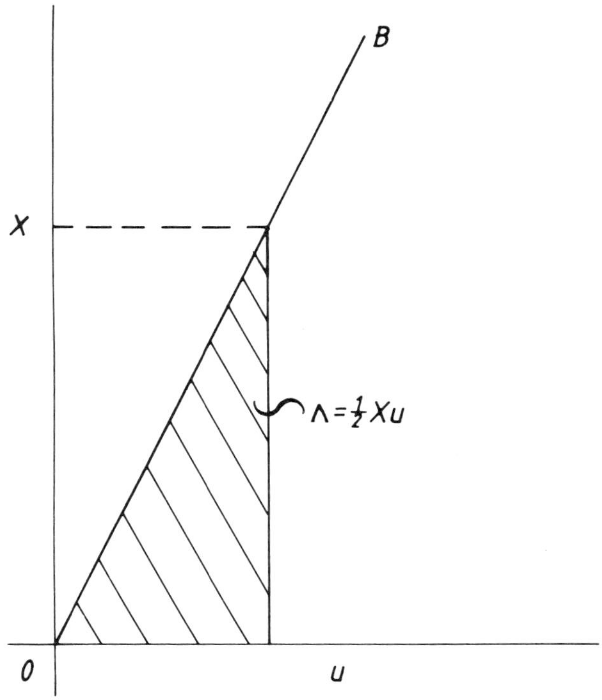

Fig. 2.1 – Load (X)-deflexion (u) curve for a linear elastic solid.

complicated state of stress, produced by a variety of external loads where all stress components vary from point to point, the total strain energy is obtained by integration over the volume. The result is the same as the total external work done, given by equation (2.1) (where external X and M produce internal stresses; and u and θ may be identified with strains).

In a crack-free body, under simple conditions of loading, giving a uniform state of stress, the slope of the Xu diagram may be used to determine E (e.g. in uniaxial tension, $E = (L/wB)\ (X/u)$, where L is the gauge length of the testpiece with cross-sectional area wB. (X/u) is called the *stiffness* of the body. When a body contains holes, cracks, or notches, E may not be determined directly from the slope of an Xu diagram, as the body deflects more for a given load the longer the cracks, or the greater the number of smaller cracks, i.e. the stiffness is decreased (Fig. 2.2). We are concerned here with holes, cracks, notches and other flaws in the body which are at least macroscopic in size (i.e. drilled holes, sawcuts, casting cracks, heat-treatment cracks, splits in wood, even hatches in ships' decks), which affect the continuum behaviour of the body. We are not concerned – at least at this stage – with microstructural variations, but we note in passing that the Young's modulus of grey cast iron is less than that of its ferritic or pearlitic matrix because of the volume fraction and size of the graphite flakes, with typically 85 GPa $< E <$ 140 GPa, the stiffer irons corresponding with smaller flake sizes.

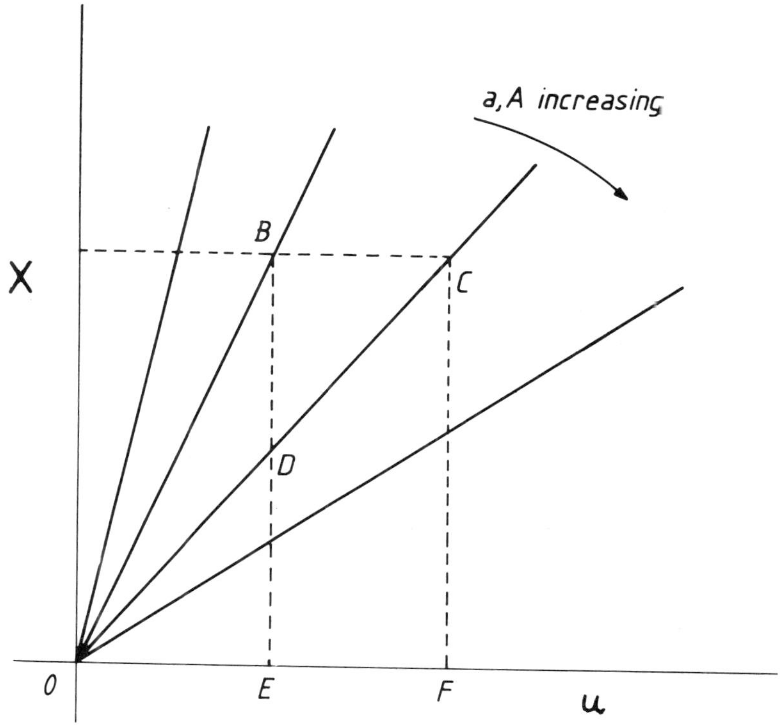

Fig. 2.2 – Lines representing varying stiffnesses (X/u) for a linear elastic solid containing cracks of increasing crack length (a) or crack area (A).

Cracks and flaws in a body disturb uniform stress patterns, and lead to localised concentrations of stress and strain. Although often tremendously complicated, it is possible in principle to compute the strain energy of cracked bodies by integration of equation (2.2), where now σ and τ vary particularly in the vicinity of the notch, crack etc. At first in this chapter, we consider strain energy simply from the area under the external load/deflexion diagram (which is of course the numerical result of integrating equation (2.2)) without consideration of detailed stress–strain fields. Later we shall be concerned with the variation of σ, ϵ and Λ around cracks.

The size of a crack is more properly described by its area A rather than its length a, but as we shall consider many two-dimensional cracking problems where uniform thickness is implied, length will often be used instead of area. The stiffness X/u of the cracked body decreases as A or a increases. Instead of stiffness, it is sometimes convenient to use its reciprocal u/X (called the *compliance*); obviously the longer the crack the bigger the compliance. For every A or a there is a different Xu line emanating from the origin of load displacement coordinates (Fig. 2.2). For a number of bodies of the same material, identical in all respects except for the length of pre-existing cracks, it is clear that Λ is greater at longer a for the same X, (ΔOCF vs. ΔOBE). Similarly, at the same u, the stiffer the body, the greater the Λ (ΔOBE vs. ΔODE). This means that when crack propagation takes place and a increases, the capacity of the cracked body to store elastic strain-energy changes. If, for the sake of argument, crack propagation took place at constant u (loading up to point B at initial crack area A_1 and dropping in load to D, say, for crack growth ($A_2 - A_1$) the strain energy in the body is *decreased* since ΔODE < ΔOBE; on the other hand, if propagation occurred at constant X (loading up to point B and cracking to point C for the same change in crack area) the elastic strain energy is *increased* after the same amount of crack propagation since ΔOCF > ΔOBE. Energy interchanges such as these are crucial to the mechanics of elastic fracture as we shall see in the next section.

Example 2.1 The two arms of a long, split beam-like testpiece may be considered to a first approximation as encastre cantilevers (the specimen is often called a double cantilever beam (DCB) testpiece). Determine an expression for the strain energy of the specimen when loaded by a pair of self-equilibrating forces X.
Ans. In Fig. 2.3, the deflexion of each arm is, from strength-of-materials $u = Xa^3/3EI$, where I is the second moment of area of the arms. The total deflexion of the cracked body is the sum of each upwards and downwards movement, i.e. $u_{\text{total}} = 2a^3/3EI$. The strain energy at some (X,u) is given by $\Lambda = Xu/2 = X^2a^3/3EI = 4X^2a^3/BEh^3$ since $I = Bh^3/12$.

The compliance of the DCB testpiece from Example 2.1 is $u/X = 2a^3/3EI$ and we see that a plot of u/X against a takes the shape shown in Fig. 2.4. Experimental compliances (from load–deflexion diagrams of bodies containing cracks of different length deliberately put into the body) may be checked against the theoretical results on this plot. In other cases, where there is no simple relation for u/X as a function of a, experimental data provide a 'compliance calibration' for that geometry of cracked body.

2.2 THE MECHANICS OF ELASTIC CRACKING: THE WORK RATE VIEWPOINT

So far in this chapter we have simply asked questions about strain energy in loaded bodies containing cracks. At some 'critical' load, the body may crack and a increase. We require to know what happens during cracking. How do X and u change? Is the cracking catastrophic with an uncontrolled 'bang', or is it controlled and well-behaved? What are the crack velocities, and so on? These are questions that will be answered in the remainder of this chapter and in succeeding chapters.

As already mentioned in section 1.3, the crucial thing to realise in studies of cracking is that *work always has to be done to propagate a crack.* As a cracked body is loaded, a critical state can be reached when the crack begins to extend. During the ensuing propagation, (i) the strain energy of the system changes,

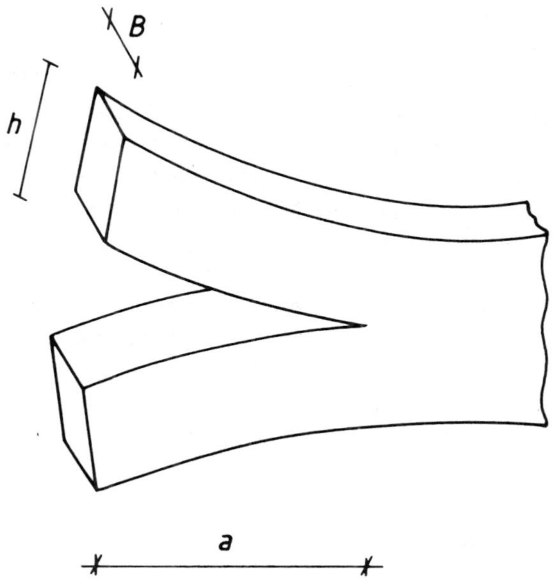

Fig. 2.3 – Geometry of a double-cantilever beam (DCB) test piece.

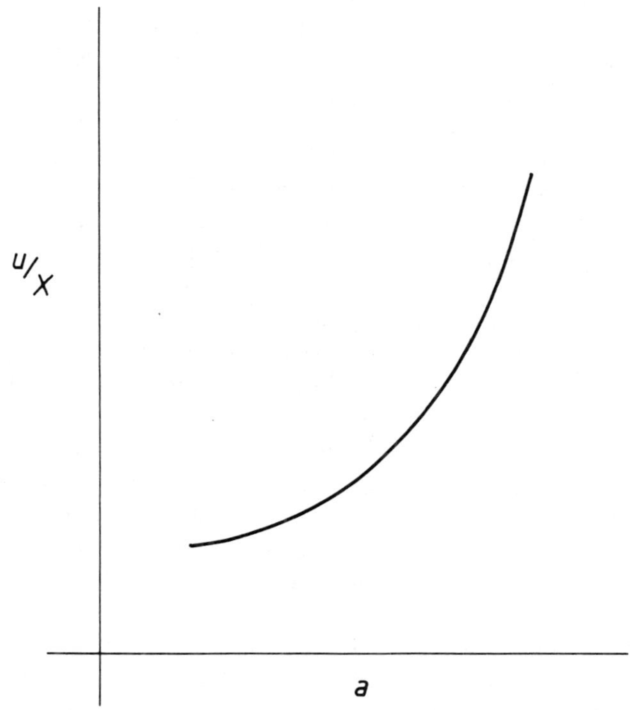

Fig. 2.4 – Compliance (u/X) as a function of crack length (a) for the DCB geometry shown in Fig. 2.3.

(ii) external work may be done by the loading forces moving their points of application, (iii) kinetic energy may be generated by the inertia of the faces of the crack moving apart if the crack velocity is very fast. If all the interchanges of energy are balanced out, an amount will be found to be lost after crack extension. This is the fracture work, which when divided by the increase of the area of one side of the crack is called the fracture toughness, R. Fracture toughness is a characteristic material property, like yield strength or elastic modulus, and like those properties can vary with strain rate (crack velocity), temperature, environment, microstructure, radiation dosage, etc.

The inter-relation between X, u, R and A during crack propagation may be determined from the energy balance during propagation. In turn, this may be done graphically or analytically. We start with the graphical approach, as it helps to understand what is taking place physically, and it is also the basis of Gurney's important experimental method for the determination of R.

2.2.1 Graphical approach

2.2.1.1 *Linear elastic behaviour*

Consider a body of arbitrary shape with a starter crack of length a_1 (Fig. 2.5(a)). When the body is loaded (by, for example, pulling on it or by driving a wedge into the crack), a typical load/deflexion plot would be OB before the crack propagates (Fig. 2.5(b)). Let us assume that at point B, cracking commences, and let us for the moment consider the special possibility of cracking with $u = u_B =$ constant. This may happen when driving a wedge into the crack, the crack moving off at some instant without further penetration of the wedge, rather like driving a nail into wood. During this crack propagation, the load on the wedge would drop. The crack may stop at some longer length a_2 when the load has dropped to D, and were the wedge or nail then removed, the Xu plot would go back elastically to the origin along the less steep stiffness line DO of the extended crack, the faces of the split closing back up again. It is important to appreciate that the unloading line *does* return to the origin under these circumstances. Simple experiments on chopping blocks of knot-free wood confirm these ideas. Glue can be inserted into the wedged-open gap of partially split wood before unloading, which later will give an identical glued-up version of the original-sized piece of material; and certainly after chopping right through, the pieces can be glued back together again. Broken furniture may be repaired owing to this behaviour. We realise that whatever is going on in the body during cracking must be concentrated right at the crack tip and crack faces, for the rest of the body is evidently simply behaving elastically. This behaviour can occur with all materials (metals, polymers, ceramics, wood, biological materials and so on) under the right circumstances.

The strain energy in the body before the crack ran was the area OBE; after propagation, it is area ODE. No extra work has been supplied to the body, since the load at the wedge has not moved its point of application, i.e. u_B is fixed. The work area OBD represents a reduction in the strain energy of the body

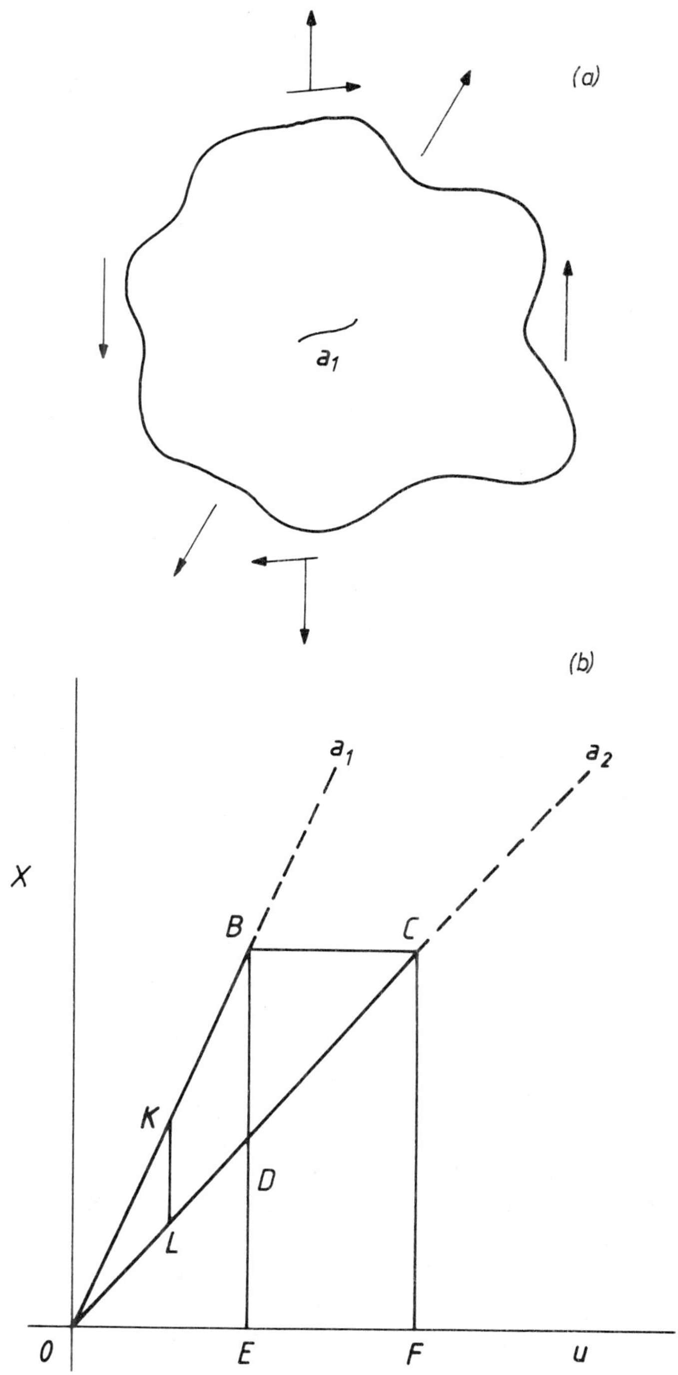

Fig. 2.5 – (a) A cracked solid containing an initial crack size (a_1). (b) Load-deflexion (X-u) diagram for crack propagation from crack length a_1 to a_2 under constant load along BC.

during crack propagation. It is helpful to normalise this energy release by the growth in area of the crack to give

$$\left.\begin{aligned} G &= \text{area OBD}/(A_2 - A_1) \\ &= \text{area OBD}/B(a_2 - a_1) \end{aligned}\right\} \quad (2.3)$$

for a testpiece of uniform thickness B. G is known as the *elastic energy release rate* ('rate' with respect to crack area). In quasi-static deformation, where kinetic energy is negligible, the strain energy release is exactly *balanced* by the work *absorbed* in the fracture process zone, i.e. $R(A_2 - A_1)$. Hence

$$\left.\begin{aligned} G = R &= \text{area OBD}/(A_2 - A_1) \\ &= \text{area OBD}/B(a_2 - a_1) \end{aligned}\right\} \quad (2.4a)$$

In unstable fracture, the work required is less than the work available, so $dR < dG$ and the excess energy accelerates the crack. This topic forms part of the subject matter of Chapter 8 and for the remainder of this chapter we shall be concerned with quasi-static situations where G and R balance one another. Again, in the absence of actual cracking, it is possible to talk of (hypothetical) values for G as if the crack were to propagate over the same distance, e.g. area OKL/$(A_2 - A_1)$ were the crack to propagate at load K. Under these circumstances, it is self-evident that as the crack does not propagate, the work required is greater than that available, and $R > G$. In this way we see that quasi-static crack spreading corresponds with the *critical* strain energy release rate G_C that exactly balances R. Hence we should write

$$\left.\begin{aligned} G_C = R &= \text{area OBD}/(A_2 - A_1) \\ &= \text{area OBD}/B(a_2 - a_1) \end{aligned}\right\} \quad (2.4b)$$

Let us now consider what happens if the crack were to propagate at constant load X over the same increase in crack length (by hanging masses on to the body, for example). The displacement increases from B to C as the crack lengthens from a_1 to a_2. Let the load be removed at the longer crack length. (This operation is quite feasible when the crack velocity is not too fast, and the velocity can often be controlled readily as explained in Chapters 3 and 6). Again, the energy at the start is area OBE; at the finish it is now OCF (the elastic stiffness line when crack length has grown to a_2 is still the line ODC). This time however, since X has moved through a distance $(u_F - u_E)$ external work has been done, represented by area BCFE, and energy has been fed into the system as a whole (i.e. to crack *and* testpiece). The energy actually released for work of fracture in this case is the difference between (Λ before cracking plus the external work done) less (Λ recovered) i.e. (area OBE + area BCFE − area OCF) = area OBC. Hence for quasi-static fracture

$$\left.\begin{aligned} G_C = R &= \text{area OBC}/(A_2 - A_1) \\ &= \text{area OBC}/B(a_2 - a_1) \end{aligned}\right\} \quad (2.4c)$$

Notice that the constant load case (Relation (2.4c) and the constant deflexion case (Relation (2.4a, b) approach the same value of work done to open the crack, as the increments of crack opening or deflexion are made smaller. The segmental work area OBD for the constant load case differs from the segmental work area OBC only by the amount BDC, which disappears as BD and BC go to zero.

In general, cracking takes place with both u and X changing during propagation. Usually X diminishes as the crack extends: the DCB testpiece discussed in Example 2.1 when loaded by end tensile forces in a stiff testing machine usually cracks with X decreasing and u increasing, along some path BM (Fig. 2.6). There are examples of crack geometries where propagation does follow the X–u paths already discussed, e.g. 3-point bend specimens with short starter cracks follow paths like OBL (Fig. 2.6(a)); DCB specimens with appropriately tapering arms crack at constant load (Fig. 2.6(c)). There are even cases where crack propagation occurs under increasing load, along paths OBP, which occurs when a hard ball is indented into a glass block and a cone crack develops (Fig. 2.6(d)). There is a host of other examples of crack propagation under decreasing

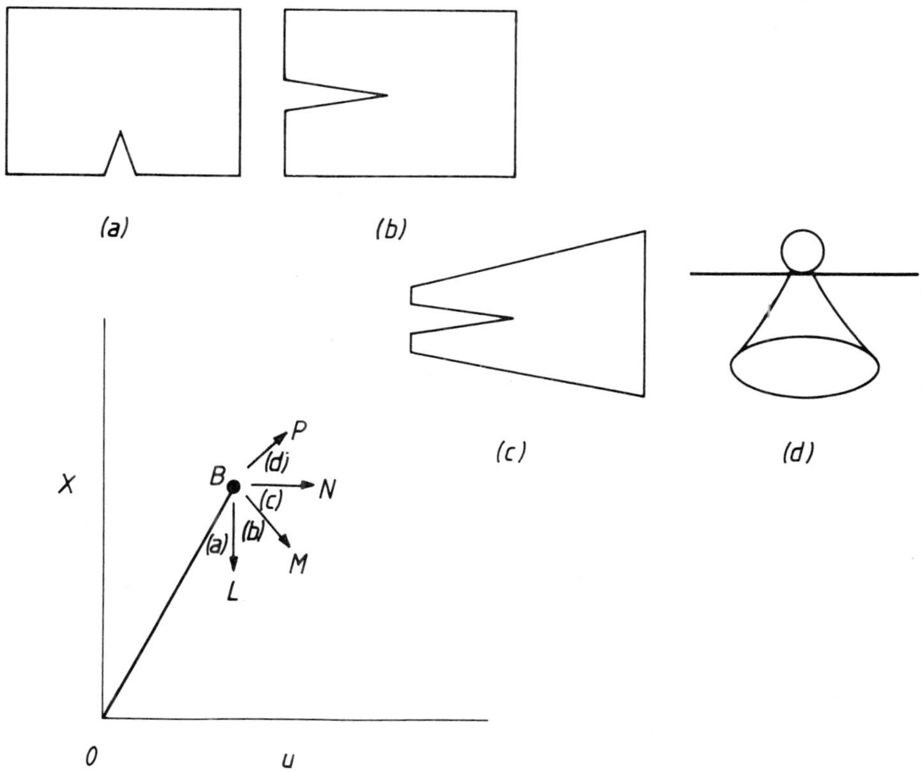

Fig. 2.6 – Load traces for cracking in different test geometries: (a) 3-point bend speciment containing a short crack; (b) DCB; (c) tapered DCB; and (d) cone crack by indentation. Lower diagram shows the different Xu responses.

or increasing load both with increasing displacement – and some even with apparently *decreasing* displacement (but that concerns the special topic of crack stability, which is deferred until Chapter 8).

Whatever the particular variation of X and u during cracking, the energy interchange described in Fig. 2.5 suggests a way of determining R. If experiments show that unloading lines revert to the origin of Xu coordinates, the segmental areas contained between the radial lines of constant crack area represent the works of fracture for the particular changes in crack area. This method of fracture toughness determination was introduced by Gurney in 1967. If the unloading lines do not revert to the origin, then other events may be taking place and a different line of attack may be required – see section 2.4 and Chapters 3 and 4 later. Fig. 2.7 shows quasi-static results reported by Gurney and Hunt (1967) for a 0.7%C, 2.0%Ni high tensile steel sheet of 0.28 mm thickness having 170 MPa proof stress, using a 'double-ended' DCB type of testpiece. Both repeated loading/unloading and a monotonic test are shown. The reloading lines lie along the unloading lines for most of their extent, and all emanate from the origin, and the falling Xu paths during fracture agree well with each other. The numbers on the radial lines are crack lengths and the R values are determined by dividing the segmental work areas by the associated change in crack area. For example, the first segment is found to be 386 mJ so

$$R = 386 \times 10^{-3}/(65–42)10^{-3} \times 0.28 \times 10^{-3} = 60 \text{ kJ/m}^2$$

and so on. We should say from these results that crack propagation occurs under essentially constant R (60 to 65 kJ/m^2) at the average crack velocity of this experiment (about 2 mm/s which was determined by the particular crosshead velocity employed in the test).

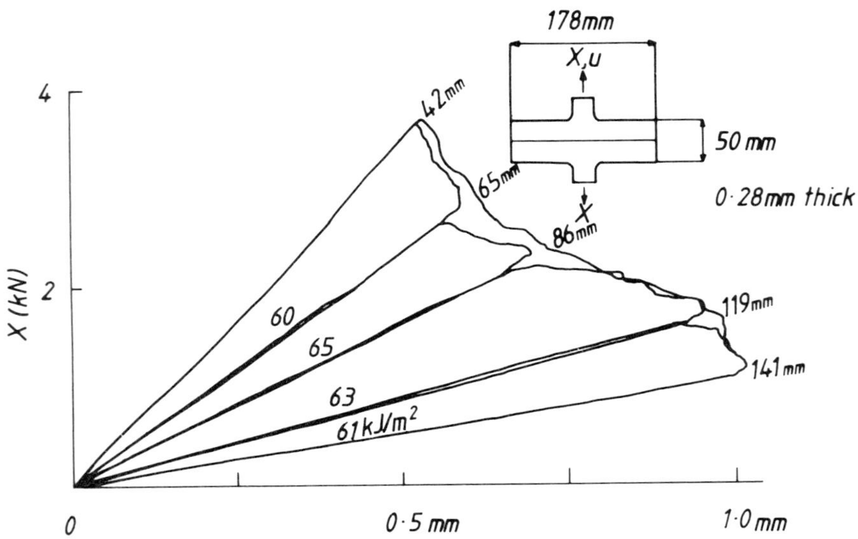

Fig. 2.7 – *X-u* relation for quasi-static cracking in a high tensile steel. (After Gurney and Hunt (1967), with permission.)

It is not necessary to keep unloading and reloading during a test to check displacement-reversibility: as is clear from Fig. 2.7, providing that reversibility is proved towards the end of crack propagation, all intermediate behaviour must also be displacement-reversible. If intermediate crack lengths can be marked on the autographic *Xu* diagram on the testing machine, (by 'pipping' the record as the crack front is seen to pass markers in the path of the crack, for example) the stiffness lines for those crack lengths are simply obtained by joining the pips to the origin.

Some workers measure the total area under the *Xu* diagram after the load has fallen to zero and the specimen has broken, and divide by the completed fracture area to give the so-called 'work of fracture' given the symbol γ (sometimes they divide by twice the fracture area as this follows the definition of surface free energy). While this may seem to be an acceptable averaging procedure for *R*, which avoids Gurney's incremental approach, we shall learn later that if *R* varies with crack velocity, the incremental approach permits many estimates for *R* as a function of velocity to be obtained economically from a single specimen which has been deliberatley chosen to give changing crack velocity. Furthermore – and this is the principal danger of taking the total area – there is no check of displacement reversibility in most 'work of fracture' total area determinations, and it could be that spurious work irreversibilities which have nothing to do with fracture, and which vary with testpiece geometry, are being included in the 'average' *R*. This point will be clearer after further discussion in Chapters 4 and 5.

The *Xu* path during quasi-static crack propagation is part of the '*R* locus' for the particular geometry of testpiece being used. Having established the *R* locus, we have the load at which any crack will propagate, given by the intersection of the stiffness line appropriate to the known crack length and the *R* locus. Thus, in Fig. 2.8 for crack propagation at 423 μm/s in a glued joint, a 180-mm

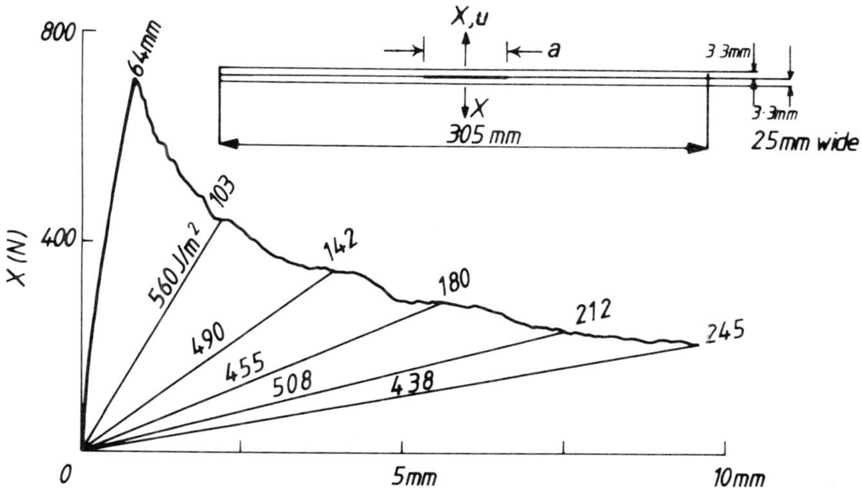

Fig. 2.8 – *X-u* relation for crack propagation in a glued-joint. (After Gurney and Hunt (1967), with permission.)

crack would start to propagate when $X = 311$ N (or equivalently $u = 5.65$ mm). Likewise those are the unique (X, u) coordinates when any shorter crack in that system has grown to 180 mm long.

Thus if we have load/extension/crack-length data we can find the fracture toughness. Vice versa if we know something about a material's fracture toughness, it is possible to predict the likelihood of cracking in a given situation, or, in other words, predict the *strength of cracked parts.*

***Example** 2.2* The load and displacement at which cracking commences are X_1 and u_1 when the crack area is A_1. Later, the load and displacement are X_2 and u_2 when the area is A_2. Assuming displacement reversibility, and that the R locus is a straight line joining $u_1 X_1$ and $u_2 X_2$, show that the fracture toughness R is given by $R = (X_1 u_2 - X_2 u_1)/2\,(A_2 - A_1)$. Apply the result to data in Fig. 2.7 and 2.8 and see how accurate the predictions are in comparison with results from work areas measured graphically.

Ans. In Fig. 2.9, area OLP + area PLMN = $R(A_2 - A_1)$ + area OMN, i.e. $\frac{1}{2} X_1 u_1 + \frac{1}{2}(X_1 + X_2)(u_2 - u_1) = R(A_2 - A_1) + \frac{1}{2} X_2 u_2$. Simplifying, there results $R = [X_1 u_2 - X_2 u_1]/2(A_2 - A_1)$. By way of illustration in Fig. 2.7, consider crack spreading from 42 to 65 mm length, i.e. the first segmental work area. We have $X_1 = 3.65$ kN, $u_1 = 0.53$ mm, $X_2 = 2.88$ kN, $u_2 = 0.60$ mm. So $R = (3.65 \times 0.6 - 2.88 \times 0.53)/2\,(65 - 42)\,0.28 \times 10^{-6} = 55$ kJ/m^2. This compares with 60 kJ/m^2 obtained with a planimeter or by counting squares. Again, in Fig. 2.8, consider propagation between lengths 142 mm and 180 mm, for which $X_1 = 357$ N, $u_1 = 3.9$ mm, $X_2 = 295$ N, $u_2 = 5.7$ mm. $R \rightarrow 465$ J/m^2 compared with 455 J/m^2 in Fig. 2.8.

2.2.1.2 *Non-linear elastic behaviour*

As the name implies, non-linear elastic structures display curved X/u stiffness lines, but they are *displacement-reversible.* Non-linear effects in bodies made from common Hookean engineering materials are often produced by geometric effects when bending resistance (which may predominate at small deflexions, as in the DCB example) is reinforced at higher deflexions by membrane stresses. Other examples occur with the inverse situation (buckling) in which resistance to deformation from membrance stresses at small deflexions is reinforced at larger loads by bending resistance. Further examples occur in elastomeric structures such as pneumatic tyres, and many biological materials are inherently non-Hookean in their behaviour. It is important to emphasise that in these cases curvature in the Xu diagram is *not* caused by irreversible remote flow, such as plasticity. Cracking in the presence of remote flow is dealt with later in the book (Chapters 4, 5 and 10).

Non-linear elastic behaviour in cracked plates is readily produced by loading out of the plane of the shear centre. Fig. 2.10 shows the results of eccentrically loading a 6 mm thick double-ended DCB testpiece of PMMA *in compression.*

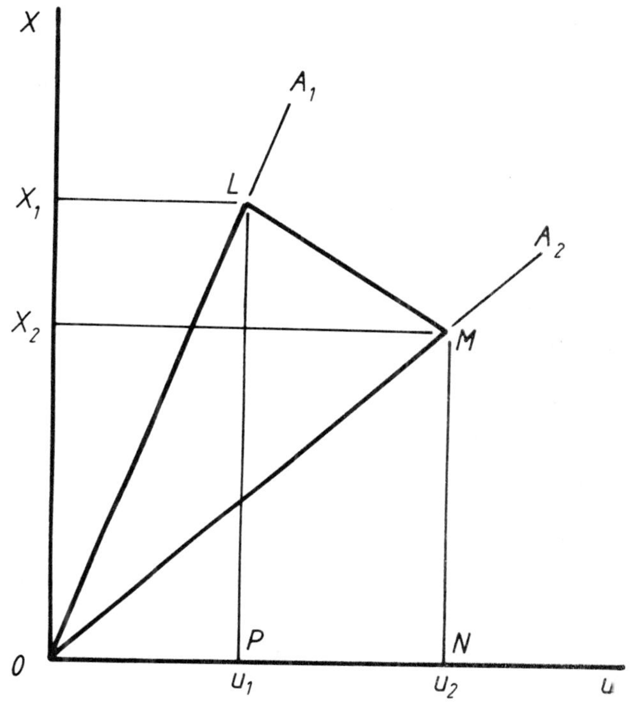

Fig. 2.9 – Load-displacement diagram for a crack propagation experiment.

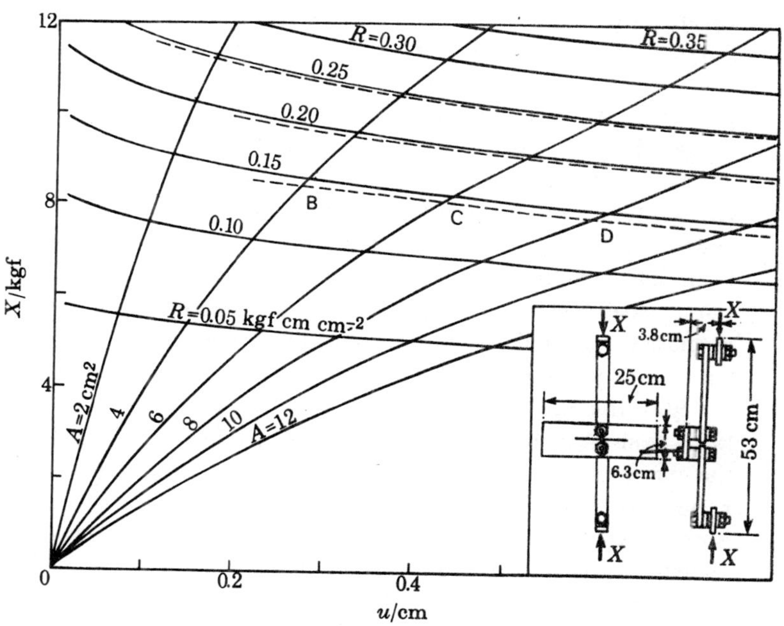

Fig. 2.10 – An off-axis compression experiment in PMMA showing non-linear elastic behaviour. (After Gurney and Ngan (1971), with permission.) (1 Kgf/cm = 1 kJ/m²^{2}.)

The behaviour is non-linear (with stiffness drecreasing at large deflexions) owing to bending and lateral deflexion of the plate out of its plane. Cracking experiments follow paths such as OBCDO, with the *curved* unloading lines returning to the origin of displacement; were the curvature in the Xu plots to have been produced by plastic flow or other remote irreversibilities, there would have been residual displacement on unloading (cf. Fig. 1.21). Since the system is displacement reversible, the segmental work areas (such as OBC) contained between the curved stiffness lines at constant crack area, *still* represent $R(A_2 - A_1)$. In this way R may be determined from non-linear elastic Xu plots, and also cracking loads predicted for non-linear cracked elastic structures. Of course, since R is a material property, the *same* R will apply irrespective of whether the body is linear or non-linear in its behaviour, i.e. R for PMMA under the same conditions of rate, temperature and environment, is invariant.

Even when loaded in its own plane by end tensile forces, a DCB-type specimen becomes non-linear in its load/deflexion response at large displacements, if it can remain elastic at large deflexions (because for example the modulus is relatively low and the yield stress relatively high). The small-deflexion beam theory ($u = \dfrac{2Xa^3}{3EI}$) is not accurate when appreciable curvature occurs (i.e. in the strength-of-materials expression 'curvature $= d^2y/dx^2 \div \sqrt{1 + (dy/dx)^2}$', dy/dx is not negligible). In fact the beams stiffen up at large rotations, and X/u lines at constant crack length curve up in the opposite direction to those in Fig. 2.10.

The cracking of thin cylindrical shells provides another example of non-linear elastic effects. Strength-of-materials thin-cylinder theory tells us that internal pressure is withstood almost entirely by membrane stresses in the cylinder wall ('$\sigma_{hoop} = pD/2B$'). However, if a crack is introduced along a generator, the hoop stress is interrupted and the stiffness of the structure (given conveniently by internal pressure/volume of cylinder rather than X/u) becomes very small at long crack lengths, since the resistance to the applied pressure in the cracked region is supplied by bending actions only. (The internal pressure is visualised as being contained by an internal flexible membrane inside the cylinder, i.e. a rubber patch over the crack.) As the pressure increases, a bulge forms in the cracked region. Axial membrane forces acting on this swollen region then come into play and contain the pressure, so that the stiffness rapidly increases with increasing pressure. Fig. 2.11 shows the pressure/volume stiffness relations for five different crack lengths cut in experimental 50 mm diameter cylinders made from 0.05 mm thick shim brass. All p/v lines were reversible on release of the pressure, indicating that the yield strength of the brass had not been exceeded away from the crack surfaces. Actual cracking tests on other cylinders resulted in the crack propagation (p, v) coordinates shown by the filled circles which lie, within experimental accuracy, on the $R = 36$ kJ/m^2 locus. Other experiments on the same brass sheet (using the type of flat specimen shown in Fig. 2.7) also gave $R = 36$ kJ/m^2 from measurement of the work areas. These data agree with observations that pressurised cracked thin cyclinders fail at hoop stresses lower than would be predicted from extrapolation of tests on flat specimens, owing

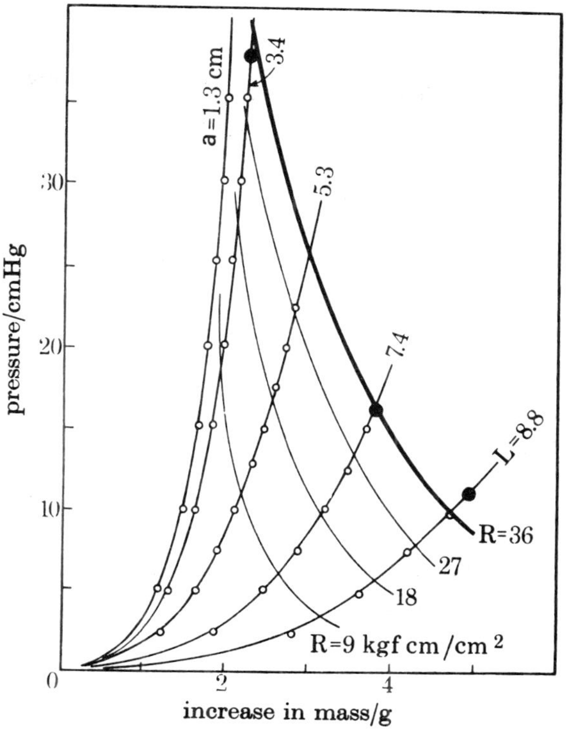

Fig. 2.11 – Non-linear pressure (p) versus increase in mass of water (i.e. effectively increase in volume) for a thin-walled cylinder containing different crack lengths (a). (After Gurney and Ngan (1971), with permission.) (1 kgf/cm = 1kJ/m².)

to the non-linearity in behaviour, which is important in connexion with the integrity of aeroplane pressure cabins. The interested reader is referred to the papers by Peters and Kuhn (1957), and by Parry and Lazzeri (1969). Chow (1982) has shown how similar methods may be employed to determine the toughness of pneumatic tyres *in situ*.

In all non-linear elastic cracking, it is vital to appreciate that the body is displacement reversible, and *not* as shown in Fig. 1.21: curvature in an Xu diagram caused by remote irreversibilities, (a Γ term in equations (1.3) or (1.2) giving displacement irreversibility), should never be confused with curvature in an Xu plot caused by elastic geometric effects or non-Hookean, but reversible, behaviour.

2.2.2 Analytical approach

2.2.2.1 *Linear elastic behaviour*

The quasi-static incremental energy interchanges described graphically in section 2.2.1 are described algebraically by equation (1.1), i.e.

$$\left.\begin{aligned} \Sigma\, X\mathrm{d}u &= \mathrm{d}\Lambda + J\mathrm{d}A \\ &= \mathrm{d}\Lambda + G\mathrm{d}a \end{aligned}\right\} \quad (1.1)\ bis$$

in a linear elastic system where the Irwin–Kies (1952) symbol G is traditionally used for the strain energy release rate.

In a linear system $d\Lambda = d(\frac{1}{2}Xu)$, i.e. the triangular area under a given stiffness line. Thus for the resultant force X and associated displacement u,

$$X\mathrm{d}u = \tfrac{1}{2}(X\mathrm{d}u + u\mathrm{d}X) + G\mathrm{d}a$$

or
$$X\mathrm{d}u - u\mathrm{d}X = 2G\mathrm{d}A \tag{2.5}$$

Dividing both sides by X^2, we have

$$\mathrm{d}/\mathrm{d}A\ (u/X) = 2G/X^2$$

or
$$X^2 = 2G/\mathrm{d}/\mathrm{d}A\ (u/X) \tag{2.6}$$

Equation (2.6) gives the single-valued relationship between X, u and A along a locus of constant release of strain energy G in a linear elastic cracked body. If cracking is quasi-static, the critical value G_C balances the specific work of fracture R in the process zone at the crack tip. (As before, if $G > R$ unstable fracture ensues; if $G < R$ fracture does not occur.) Thus we write for quasi-static cracking

$$X_c^2 = 2R/d/\mathrm{d}A\ (u/X) \tag{2.7}$$

where the subscript c denotes the value of the variable when cracking occurs. Equation (2.7) describes the X, u, A, R relations along lines such as BL, BM, BN or BP in Fig. 2.6. There are five other derivative relations from equation (1.1) for quasi-static linear elastic cracking. They are

$$u_c^2 = -\ 2R/d/\mathrm{d}A\ (X/u). \tag{2.8}$$

$$X_c = 2R/(\partial u/\partial A)_{X=X_c} \tag{2.9}$$

$$u_c = -\ 2R/(\partial X/\partial A)_{u=u_c} \tag{2.10}$$

$$R = \partial/\partial A\ (\tfrac{1}{2}\ Xu)_{X=X_c} = (X_c/2)\ (\mathrm{d}u/\mathrm{d}A)_{X=X_c} \tag{2.11}$$

$$R = -\ \partial/\partial A\ (\tfrac{1}{2}Xu)_{u=u_c} = -\ (u_c/2)\ (\mathrm{d}X/\mathrm{d}A)_{u=u_c} \tag{2.12}$$

Equivalent formulae for non-linear elastic cracking are given later. Relations (2.7–2.12) are special cases of those more general relations.

Of these expressions, (2.7) is most well-known, and is the basis of the so-called 'compliance calibration' method of determining R from experimental Xu plots, and is also a means of predicting the cracking load in a given situation, knowing A and R (Irwin and Kies, 1952). Fig. 2.4 has shown how the compliance (u/X) may vary with crack length in a body. The denominator of equation (2.7) is the rate of change of (u/X) with respect to crack area which, in a plate testpiece of thickness B, is $(1/B)$ times the rate of change of (u/X) with respect to crack length. In other words it is the slope of Fig. 2.4.

As already discussed in section 2.1, the (u/X) vs. a plot may be entirely experimental or it may be based on elasticity theory. Some experimental measurements of compliance for a DCB specimen whose arms taper and become deeper towards the backface of the testpiece (Fig. 2.12(a)) are shown in Fig. 2.12(b).

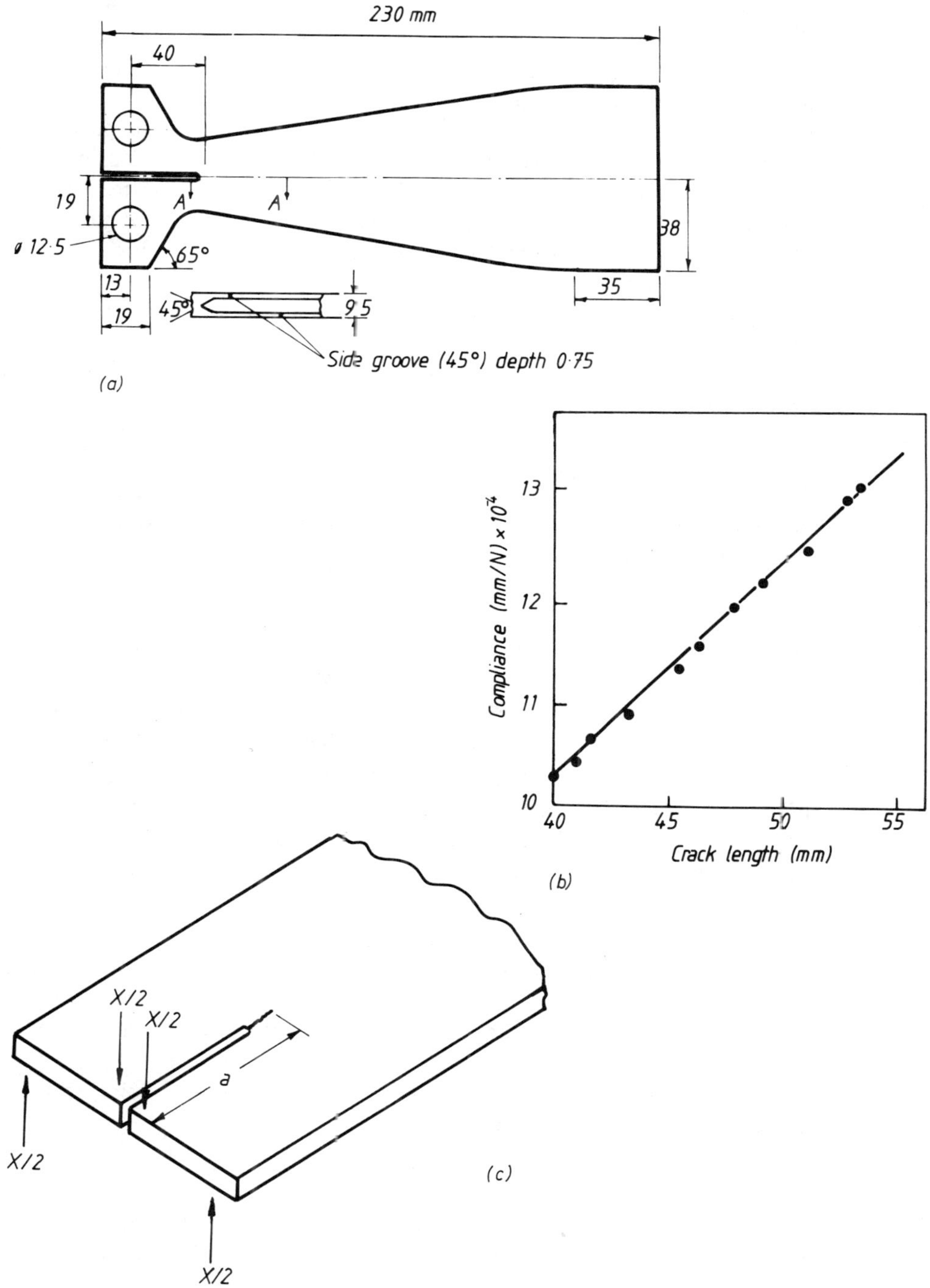

Fig. 2.12 – (a) A contoured double cantilever beam specimen. (b) Typical compliance versus crack length curve for the specimen in (a) with a nominal thickness of 9.5 mm and made from an aluminium alloy. (c) An Outwater double-torsion (D-T) specimen. ((a) and (b) after J. C. Radon, with permission.)

Over quite a long range of crack length, u/X increases linearly with a, so that in this region $d/da\ (u/X)$ is constant ($\sim 2.2 \times 10^{-5}\ N^{-1}$). In equation (2.7), $X^2 = 2R/(2.2 \times 10^{-5}/8 \times 10^{-3}) = 727R\ N^2$ as the net section is 8 mm thick. We therefore expect crack propagation to take place at constant load over the range of a in which $d/dA\ (u/X)$ is constant, and knowing the 'compliance calibration' of the testpiece we may convert directly from X to R. Mostovoy and Ripling introduced the 'contoured' DCB specimen in 1966. A testpiece displaying $d/dA\ (u/X)$ constant is known as a 'constant rate of change of compliance' specimen. The Outwater double torsion testpiece shown in Fig. 2.12(c) is another such specimen.

When $d/dA\ (u/X)$ varies with a, a different conversion factor has to be employed at every crack length to convert X at cracking into R. Even so, many workers perform compliance calibrations and measure $d/dA\ (u/X)$ in order to obtain one-off R values from the X when propagation commences. However, we note that Gurney's segmental work area method requires *no* compliance calibration, gives many R values from one test, and also provides a check on displacement reversibility. (The meaning of compliance calibration R-values obtained from unstable fracture tests, is discussed in Chapter 8). Example 2.3 shows how an analytical expression for u/X is used in equation (2.7).

Example 2.3 Consider the splitting of a thin slender beam of thickness B and depth h, as shown in Fig. 2.3, the load/deflexion response of which can be adequately represented by strength-of-materials beam theory. Determine an expression for the load at crack propagation when the crack length is a.

Ans. $u/X = 2a^3/3EI$ from Example 2.1, $A = Ba$ so $d/dA\ (u/X) = 2a^2/EIB$ and in (2.7) $X_c^2 = REIB/a^2$. If the material cracks at constant R, X diminishes as a increases (path BM in Fig. 2.6), but note that the bending moment at the crack tip (Xa) remains constant during cracking.

Example 2.4 Fig. 2.13(a) shows a 'compliance calibration' of pre-cracked 3-point bending testpieces made of 7075–T6 aluminium. The beams were of constant depth W and various thickness B so the compliance ordinate is normalised by multiplying by the thickness (i.e. $u/X \propto 1/I \propto 1/B$ for constant W). Fig. 2.13(b) shows an Xu plot for a bar with $a = 16.77$ mm. Determine R. Compare it with a calculation based on work areas (ASTM).

Ans. Through the crack length $a = 16.7$ mm in Fig. 2.13(a), $d/dA\ (u/X) = (1/B)\ d/da\ (u/X) \approx 3.3 \times 10^{-7}\ N^{-1}\ mm^{-1}$, by drawing a tangent. $G_C = \frac{1}{2} X^2\ d/dA\ (u/X) = \frac{1}{2}\ (10{,}400)^2 \times 2 \times 10^{-7}$ N/mm ≈ 18 kJ/m^2 since the cracking load is 10.4 kN. Hence $R = 18$ kJ/m^2. The whole work area in Fig. 2.13(b) is some $(10{,}400 \times 1.5)/2 + 10{,}400 \times (3 - 1.5)/2$ (assuming two triangular areas) $\approx 15{,}600$ N mm or 156.6 J. The ligament crack area is $(50 - 16.7)\ 19 = 633$ mm^2, so another estimate for R is $156.6/633 \times 10^{-6} \approx 24$ kJ/m^2. (In section 8.8, equation (8.30), it is shown that R is related to the triangular work area Λ up to

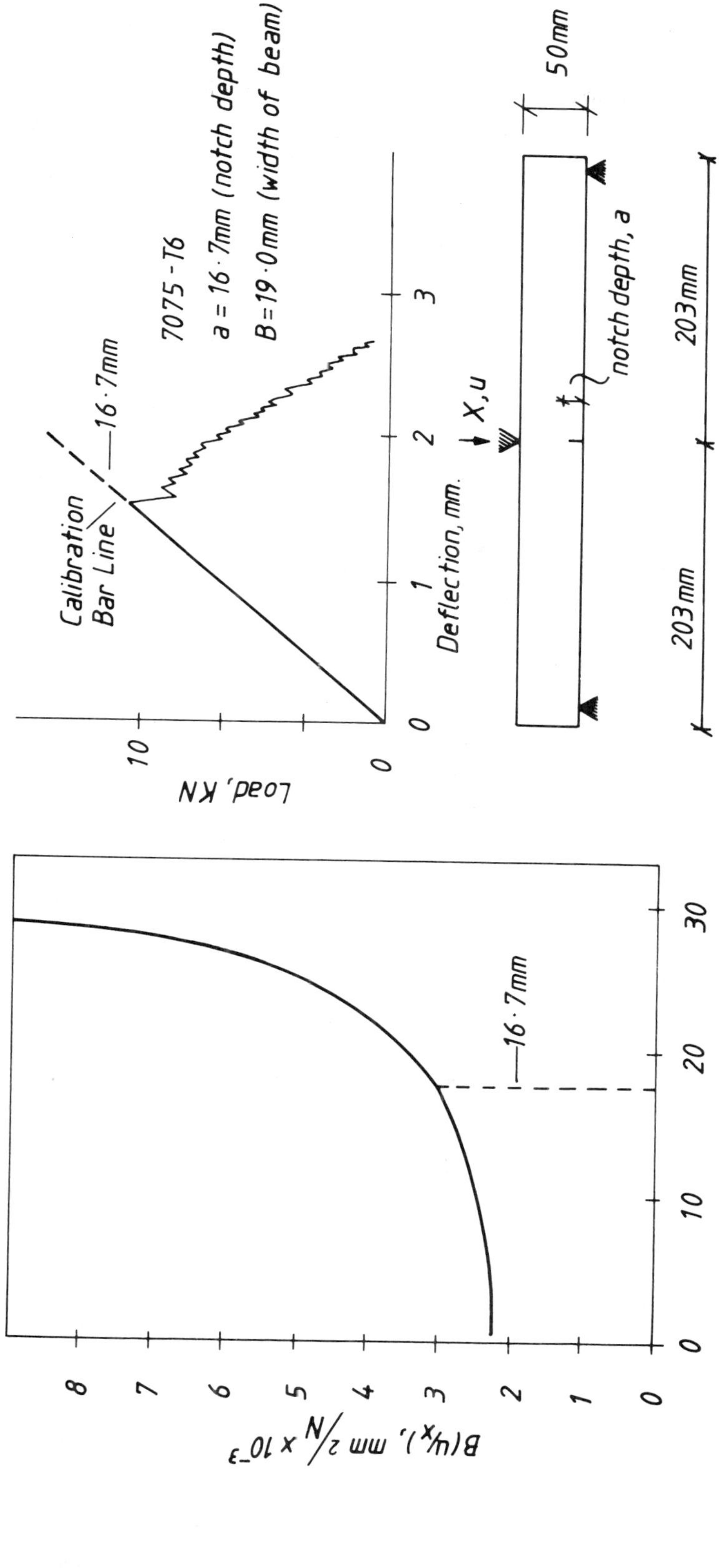

Fig. 2.13 – (a) Compliance calibration curve for three-point bend 7075-T6 aluminium specimen. (b) Load-deflexion curve for crack propagation.

the first fracture by Turner's η factor where $R = \eta\Lambda/B\,(W - a)$. For a deeply notched bend bar $\eta = 2$ and we have already noted that the total area in Fig. 2.13(b) is approximately twice the initiation area (10,400 × 1.5/2) N mm which agrees very well.) The disagreement between the compliance calibration value for R (18 kJ/m^2) and the work area methods (24 kJ/m^2) highlights the imprecision in drawing tangents to compliance/crack length curves.

Example 2.5 Griffith in 1920 (Griffith, 1921) introduced the scientific study of cracks. He recognised that a high stress concentration was necessary but not in itself sufficient for crack growth, and he identified the local work of crack spreading with the free energy of the surfaces formed. For his work on glass, Griffith considered the spreading of a small crack of length $2a$ in a very large sheet loaded by uniform uniaxial stresses σ remote from the crack. Using Inglis's mathematics for the complex stress distribution around the crack in the large sheet (section 2.4), Griffith showed that the increase in the strain energy storage capacity of sheet due to the presence of the crack was equal to $\pi\sigma^2a^2B/E$ where B is the thickness of the sheet and E is the Young's modulus of the material. Using (2.7), show that the 'Griffith formula' is $\sigma_{cr} = \sqrt{ER/\pi a}$.

Ans. Imagine the large plate to be of extent $h \times w \times B$ (= volume V). $X = \sigma wB$, so (2.6) gives $\sigma^2 = 4RB/V\;\mathrm{d}/\mathrm{d}a\,(1/E^1)$ where E^1 is the apparent Young's modulus of the cracked sheet. ($E^1 < E$). Λ for the crack-free sheet would be $\sigma^2 V/2E$, so adding the increase in Λ due to the crack $\sigma^2 V/2E + \pi\sigma^2a^2B/E = \sigma^2 V/2E^1$ since the cracked body stores energy as if it had a modulus E^1. Hence $\mathrm{d}/\mathrm{d}a\,(1/E^1) = 4\,\pi aB/EV$ and in (2.7) we get $\sigma_{cr}^2 = ER/\pi a$ or $\sigma_{cr} = \sqrt{ER/\pi a}$.

The Griffith formula was first applied to fracture in glass, where the process zone of irreversible flow at the crack tip is extremely small and where therefore the work of fracture has, for all practical purposes, the value of the surface free energy γ. Since γ is traditionally defined with respect to *both* sides of a surface and R with respect to *one*, we have $R = 2\gamma$; the Griffith formula is then written $\sigma = \sqrt{2E\gamma/\pi a}$. When the surface energies of many other materials were estimated from Griffith's expression however, they were substantially greater than theoretical values. In the 1940s Orowan and Irwin independently realised that the discrepancies were caused by irreversible work dissipated during plastic flow at the tip. In the historical development of the mechanics of fracture, Orowan heuristically modified the the Griffith formula by the inclusion of a term γ_p to represent the irreversible work over and above γ, viz. $\sigma = \sqrt{2E(\gamma + \gamma_p)/\pi a}$. For $2\,(\gamma + \gamma_p)$ we nowadays write R, and in practice $\gamma_p/\gamma \simeq 1000$, i.e. the surface free energy is a minor part of the total work of fracture†. The process of cracking to which Orowan referred had still to be displacement-reversible, of course.

---

† An alternative treatment for the relationship between R and γ will be found in section 6.5.1.

Poisson's ratio ν may appear in the Griffith formula (and indeed as we shall see all other expressions relating the loads, deflexions and crack lengths to the resistance to fracture), thus

$$\sigma_{cr} = \sqrt{2E\gamma/(1-\nu^2)\,\pi a}$$

This comes about because, as shown in books on elasticity, $E/(1-\nu^2)$ should be used in problems of plane strain in place of E alone which applies strictly to plane stress. In this book we often write E^* for Young's modulus, which is to be read as E or $E/(1-\nu^2)$ as appropriate. For those many solids with $\nu \approx 1/3$, $E/(1-\nu^2) \approx 1.1E$ and the 5 per cent error (on taking the square root) may not be important in many engineering calculations where other factors may not be known as accurately.

Example 2.6 A sheet of glass measuring 2 m × 200 mm × 2 mm contains a central slit in the lengthwise direction, i.e. the slit is parallel to the 200 mm side. A mass of 500 kg induces a direct tensile stress in the sheet and it is assumed here that any effect of the mass of the glass is included in the 500 kg value. Consider the following property values: $E = 60$ GPa, $\gamma = 0.5$ J/m^2, $\nu = 0.25$, $\sigma_f = 170$ MPa (fracture stength of sound glass). What is the minimum length of the slit that would lead to fracture under these conditions?

Ans. If considered as a condition of plane stress, then $a = 2E\gamma/\pi\sigma^2$ where $\sigma = 500 \times 9.8/200 \times 2 \times 10^{-6} = 12.3$ MPa, $a = 2 \times 60 \times 10^9 \times 0.5/1 \times (12.3)^2 \times 10^{12} = 0.126$ mm. So the crack length $= 2a = 0.252$ mm. If instead plane strain conditions prevail $a = 2E\gamma/\pi\sigma^2\,(1-\nu^2) = 0.126/(1-0.25^2) = 0.134$ mm; $2a = 0.268$ mm as the slit length.

2.2.2.2 *Non-linear elastic behaviour*

In non-linear reversible cracking the area contained between a stiffness line and the displacement abscissa is still called the elastic strain energy $\Lambda = \int X du$ (Fig. 2.14), but to do mechanics for non-linear cracking we must introduce the complementary strain energy function $\Omega = \int u.dX$ which is the work area contained between a stiffness line and the load ordinate. We have therefore

$$\Omega = Xu - \Lambda \tag{2.13}$$

In bodies with linear stiffness, Ω and Λ are equal in value ($= \frac{1}{2}Xu$) and for that reason it is not common practice to distinguish between them. In the mechanics of non-linear cracking, however, we must.

A set of general non-linear relations equivalent to relations (2.9)–(2.12) may be obtained from equation (1.1) which is now understood to apply to non-linear (but reversible) behaviour. Before noting these expressions, it must be pointed out that the Irwin–Kies strain energy release rate parameter G applies only to *linear* elastic systems and, by the same token that a distinction has to

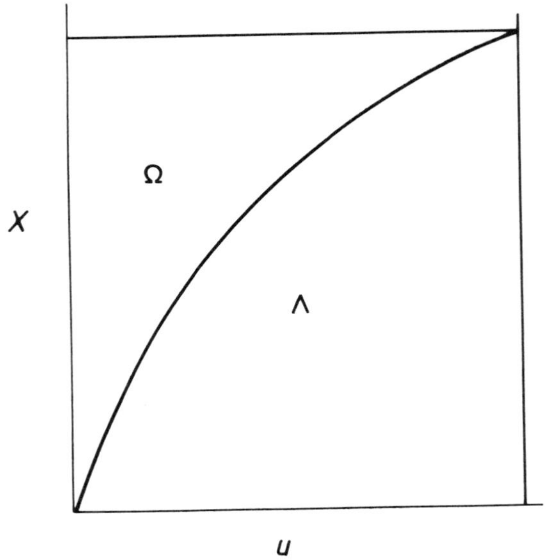

Fig. 2.14 – Load-deflexion curve for a non-linear elastic solid showing the definition of strain-energy (Λ) and complementary-strain energy (Ω).

be made between Λ and Ω, we use the more general symbol J in non-linear systems for the rate of change of potential energy were a crack to run, i.e. the rate of change of the areas bounded by *curved* stiffness lines. Thus manipulation of

$$X\mathrm{d}u = \mathrm{d}\Lambda + J\mathrm{d}A \tag{1.1a}$$

and the equivalent expression written in terms of Ω obtained by substitution, i.e.

$$u\mathrm{d}X = \mathrm{d}\Omega - J\mathrm{d}A \tag{1.1b}$$

gives general relations equivalent to Relations (2.7)–(2.12). They are†

† At constant crack area, $X = (\partial\Lambda/\partial u)_A$ and $u = (\partial\Omega/\partial X)_A$ which are Castigliano's theorems. As $\Omega = \Lambda$ for linear elastic systems, many books write $u = \partial\Lambda/\partial X$ and often do not introduce the concept of complementary strain energy. This leads many authors to state that Castigliano's theorem for displacements *cannot* be applied to non-linear elastic systems for which the principle of superposition does not apply. A popular example used to illustrate this point is that of the 3-pinned horizontal beam of total length $2l$, loaded vertically through the centre pin, for which the correct solution is $u = l\,(X/AE)^{1/3}$ (e.g. Timoshenko, 1955, vol. I, pp. 360–361). When $u = \mathrm{d}\Lambda/\mathrm{d}X$ is used, $u \to (l/3)\,(X/AE)^{1/3}$ which is incorrect; however, using $u = \mathrm{d}\Omega/\mathrm{d}X$ gives the correct answer. Curiously, although even Timoshenko (1955, vol. I, pp. 359–361), and Timoshenko and Goodier (1951, p. 165) suggest that Castigliano *cannot* be used in non-linear cases, Timoshenko and Young (1945, pp. 240–241) *do* explain the proper procedure required to get the correct answer.

$$\left(\frac{\partial u}{\partial A}\right)_X = \left(\frac{\partial J}{\partial X}\right)_A \quad (2.9a)$$

$$\left(\frac{\partial X}{\partial A}\right)_u = -\left(\frac{\partial J}{\partial u}\right)_A \quad (2.10a)$$

$$J = \left(\frac{\partial \Omega}{\partial A}\right)_X = \int \left(\frac{\partial u}{\partial A}\right)_X \mathrm{d}X \quad (2.11a)$$

$$J = -\left(\frac{\partial \Lambda}{\partial A}\right)_u = -\int \left(\frac{\partial X}{\partial A}\right)_u \mathrm{d}u \quad (2.12a)$$

(see also Chapter 4)†. For quasi-static cracking $J = J_C = R$ and the semental work areas contained by curved stiffness lines represent work of fracture as already described.

Relations (2.7) and (2.8) relate only to linear bodies, and there are no 'rate of change of compliance' expressions for non-linear bodies.

Problems (2.19–2.21) show that these relations form part of a pattern which indicates that there is a one-to-one correspondence between the crack mechanics variables X, u, J and A and the classical thermodynamics variables pressure, volume, temperature and entropy respectively, together with correspondence with Maxwell's relations and the well-known named thermodynamic functions (e.g. Ω corresponds with enthalpy).

A simple non-linear algebraic relation between X, u and A, which has wide applicability, is

$$X = k(A)u^n \quad (2.14)$$

(see Chapters 3 and 8). The non-linear behaviour of the double-ended DCB testpiece loaded eccentrically in compression (Fig. 2.10) is represented by the series

$$X = \sum_m C_m u^m / A^{(m+1)/2} \quad (2.15)$$

where the coefficients C_m have dimensions of work, and $m = 1, 2, 3 \ldots$ This particular series has the interesting property that $JA = \frac{1}{2}Xu$ (Problems 2.12), i.e. in quasi-static cracking $R = J_C$ is given directly from the instantaneous values of X and u during cracking for known A – which is a sort of compliance-calibration result for this particular form of non-linear representation. Again, elimination of A from (2.15) using $JA = \frac{1}{2}Xu$ gives a simple Xu relation at

† Sometimes the group $(\Lambda - \int X\mathrm{d}u) = \Pi$ is called the 'potential energy' in which case $J(= G$ in a linear system$) = -\mathrm{d}\Pi/\mathrm{d}A$. Further discussion on this formation will be found in Chapter 4 in connexion with the J-integral. Also, this formulation appears in variational methods for approximate lower and upper bounding solutions to problems.

constant J, thus allowing constant J loci to be plotted easily, which are identical with constant R loci in quasi-static fracture as in Fig. 2.10 (see Problem 2.13).

In passing we observe that the first term of the general series ($m = 1$) corresponds with constant rate of change of compliance linear specimens, since $X = Cu/A$ or $u/X = A/C$ and $\mathrm{d}/\mathrm{d}A(u/X) = 1/C =$ constant. Since $JA = Xu/2$ for the series generally, $J = Xu/2A$ equally for the first term, i.e. in quasi-static cracking the toughness in constant rate of change of compliance specimens must be given by instantaneous value of Λ ($= Xu/2$) divided by the current crack area. Constant $\mathrm{d}/\mathrm{d}A(u/X)$ specimens crack at constant load, and for the Gurney segmental work area we write $R(A_2 - A_1) = X(u_2 - u_1)/2$ by geometry which is, of course, the same as $Xu/2A$.

***Example** 2.7* Non-linear load–deflexion diagrams at constant crack area were curve-fitted by $X = 350\ (u/A) - 1890\ (u^2/A^{3/2}) + 5250\ (u^3/A^2)$ (X in N, u in mm, A in mm^2) in a plate specimen. A crack of area 44 mm^2 started to propagate when $X = 102$ N. What was the fracture toughness?

Ans. Non-linear Xu stiffness lines could be constructed on graph paper for various A using the algebraic expression for stiffness and constant J loci located by trial-and-error by counting squares to give constant '$R\Delta A$' between Xu lines of constant crack area. However, we may integrate the Xu relation with respect to u at constant A to give the strain energy function Λ as follows: $\Lambda = \int X\mathrm{d}u = 175u^2/A - 630\,u^3/A^{3/2} + 1312\,u^4/A^2$. Using equation (2.12a), $J = -\,\partial\Lambda/\partial A)_u = 175u^2/A^2 - 945u^3/A^{5/2} + 2625u^4/A^3$. If $X = 102$ N, $u \cong 4$ mm so with $A = 44$ mm^2, $R = J_C = 777$ J/m^2.

2.3 *R*-CURVES

The (X, u, A, J) and (X, u, A, R) family of data in crack propagation comprises four variables, of which only two can be employed as axes for the usual Cartesian plot. As already shown in this chapter, it is customary to employ X, u coordinates and project on to the X, u plane lines of constant J and constant A. This is sensible, because Xu plots are obtained directly from testing machines.

It is possible, however, to employ J–A coordinates (G–A in the linear case) and superpose lines of constant X and u. In quasi-static cracking, G or J will be the critical values corresponding with the material R during fracture. The shapes of the constant X and u loci depend entirely on the geometry of the specimen, and whether it is linear or non-linear (section 8.5). In contrast, in quasi-static cracking the J–A or G–A function is 'fixed' for all geometries since it represents the material R response. For equilibrium cracking under constant R, the RA 'curve' is a step-function BCD (Fig. 2.15(a)), where OB is the starting crack length and BC the value of the constant R during propagation. But for various reasons (change in crack tip constraint, velocity, temperature, environmental effects) R may change during propagation, leading to the situation shown in Figs. 2.15(b) and 2.15(c). In Fig. 2.15(b), constant J loci fall to the

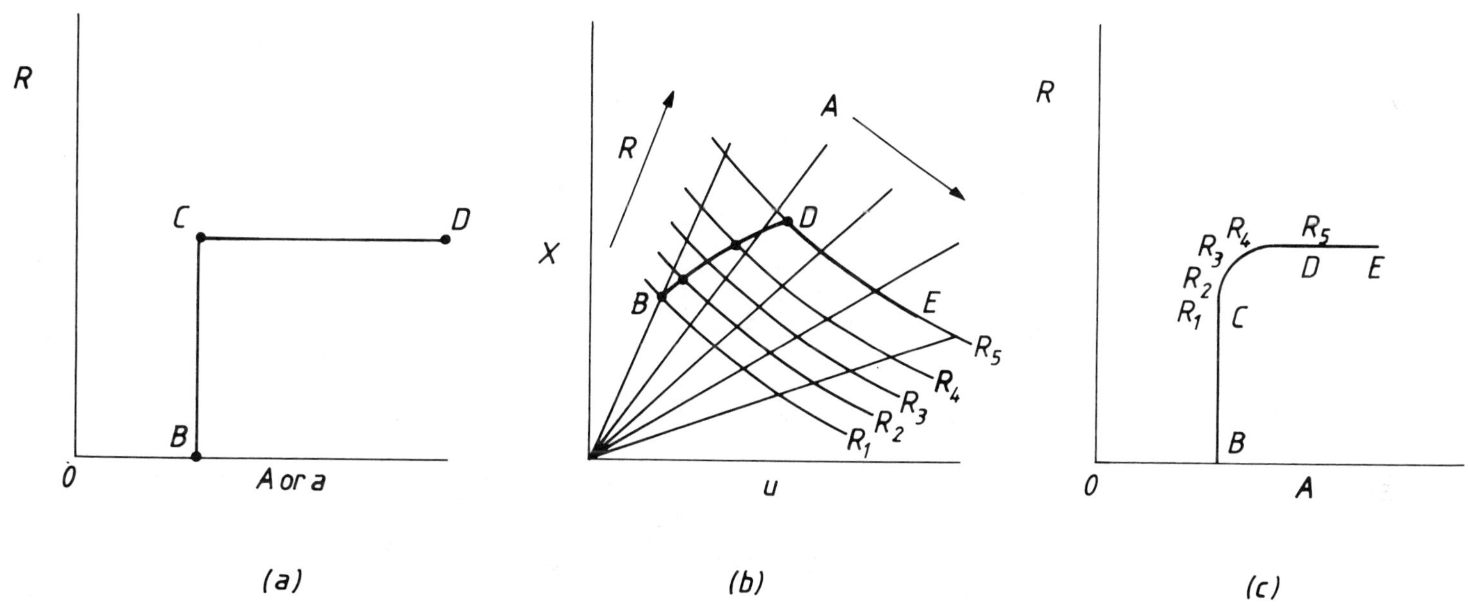

Fig. 2.15 – (a) A step-function R-curve. (b) Load-deflexion curve for a material exhibiting rising crack growth resistance from B to D and a steady value from D to E. (c) Crack growth resistance curve for the material in (b).

right, which corresponds with equilibrium cracking in the DCB testpiece for example. If cracking is seen to start at B and continue quasi-statically as the load rises past C to eventually follow a constant J locus DE, it is clear that J is increasing in the early part of the propagation. Thus the material fracture process zone must be requiring an increasing rate of work input in those early stages and hence the R–A function curves over to a plateau as shown in Fig. 2.15(c). The plateau value would be the 'critical' G_C or J_C in the absence of the earlier crack propagation, which is therefore called 'sub-critical' crack growth.

When R is constant, the R-curve approach is trivial of course, but when R does vary, the R–A visualisation is useful. It has found particular appeal in thin sheet studies, where growth of the comparatively large crack tip zones is often not completed before the crack starts to run, leading to the terminology 'crack growth resistance curves' for R-curves. Results seem to show that the R-curve is characteristic of the material and independent of starting crack length. Details follow in Chapters 4 and 8. Furthermore, as discussed in Chapter 8, when R does change with A the relative values of $\mathrm{d}R/\mathrm{d}A$ or $\mathrm{d}G/\mathrm{d}A$ are crucial in questions of crack stability.

2.4 DISPLACEMENT REVERSIBILITIES AND APPARENT IRREVERSIBILITIES IN ELASTIC FRACTURE

We have emphasised so far in this chapter that elastic cracking is displacement reversible on an Xu plot, i.e. that after some propagation (before final severance) the crack faces are sprung back together by the surrounding elastic stress field, Figs. 2.7–2.11. Gurney's segmental area method for calculating R in quasi-static cracking follows from that and also all the various analytical relations 2.8–2.12.

There are, however, certain examples of elastic fracture which have residual displacements u_r on unloading which are yet globally elastic and in which the severed pieces can be refitted together. Ripping paper or cutting it with scissors are examples: since the offcut is so 'floppy', it stores no strain energy in bending and there is no restoring force to close up the crack. Consequently there is a residual displacement in the direction of the tearing or cutting force (equal to the length of tear) but the cut pieces can be laid flat and fitted together to regain the original piece of paper. Furthermore, and this is of great importance, is that the total work area on the Xu plot is work of fracture (from which valid R values can be obtained) *even though the unloading line does not return through the origin of displacement.*

In this section, the mechanics of tearing, peeling and guillotining are investigated and a general analytical expression (a generalisation of (2.8)–(2.12)) is derived to account for residual displacements in elastic cracking. As shown in Chapter 3, this applies to cracking in the presence of residual stresses which is always accompanied by residual displacements on relief of residual stresses. The relevance of this treatment to situations where residual displacements are due to *other* causes (such as large crack tip zones, remote flow, time-dependent recovery and so on) is discussed and reference made to other sections in the book where those matters are dealt with more fully.

2.4.1 Elastic tearing and peeling of thin sheets

When a cracked body stores negligible strain energy in comparison with the work done by the external forces that produce cracking, the area under the Xu diagram is a proper measure of R even though the system is displacement-irreversible in terms of u. Consider, for example, a *trousers test* (called the Elmendorf tear test in the paper industry) of a tape, paper or thin rubber sheet or other floppy material which has negligible bending stiffness (Fig. 2.16(a)). There is no contribution to $d\Lambda$ in equation (1.1) from bending resistance and often the contribution from simple stretching of the trouser legs is much smaller than the work done by the external forces. Then

$$2\,(X\mathrm{d}u/2) = J\mathrm{d}A = JB\mathrm{d}u/2 \tag{2.16}$$

assuming that the legs move up and down the same amount ($\mathrm{d}u/2$), and neglecting any stretch in the legs so that $\mathrm{d}A = B\mathrm{d}u/2$. Hence

$$J_C = R = 2X/B \tag{2.17}$$

i.e. the tearing load is constant for constant toughness. Although floppy in bending (where the bending stiffness is proportional to EwB^3) many sheet materials can be stiff in tension so that the elastic stretch of the legs of the specimen ($XL/EwB = RL/2Ew$) is small and the combined effects together give Xu plots as in Fig. 2.16(b). There is essentially no elastic recovery to the origin of displacement u, yet the blocks of work area, based on ever-increasing u, *all* go to toughness work from which valid R measurements can be made.

Clearly Λ owing to the stretch legs might be sometimes significant, and indeed the stress in the legs ($\sigma = X/wB = R/2w$) might be greater than the tensile yield stress or viscoelastic flow stress of the tape material (using small w), but such complications can often be avoided and a very simple test results. What needs to be done to the analysis in the presence of elastic stretch, or even flow in bending or tension is discussed later (Example 2.8, Chapter 4, etc.).

The mechanics of peeling tapes from substrates are essentially the same as those in the trousers test, providing that the tape material has negligible bending resistance, and does not stretch appreciably. Care has to be taken in experiments with common sticky tapes, as their vicoelastic flow stress can be exceeded, thus producing irreversible work in the stretched tapes. Nevertheless, it is found that at sufficiently high peeling speeds, the simple analysis given above works well.

Example 2.8 Analyse the trousers test when the legs are linear elastic and can store strain energy in tension.

Ans. Consider Fig. 2.16(c). In contrast to Fig. 2.16(b) in which the legs had infinite stiffness in stretching, the stiffness lines OL, NM depend on the length of material torn. As the load is increased from zero, there will be a finite displacement δ_1 in each leg before cracking commences. During subsequent tearing let $u = 2(a + \delta)$, where a is the (unloaded) current length of each leg and δ the current stretch in each leg. For tearing load X, the stress in the leg is $\sigma = X/wB$

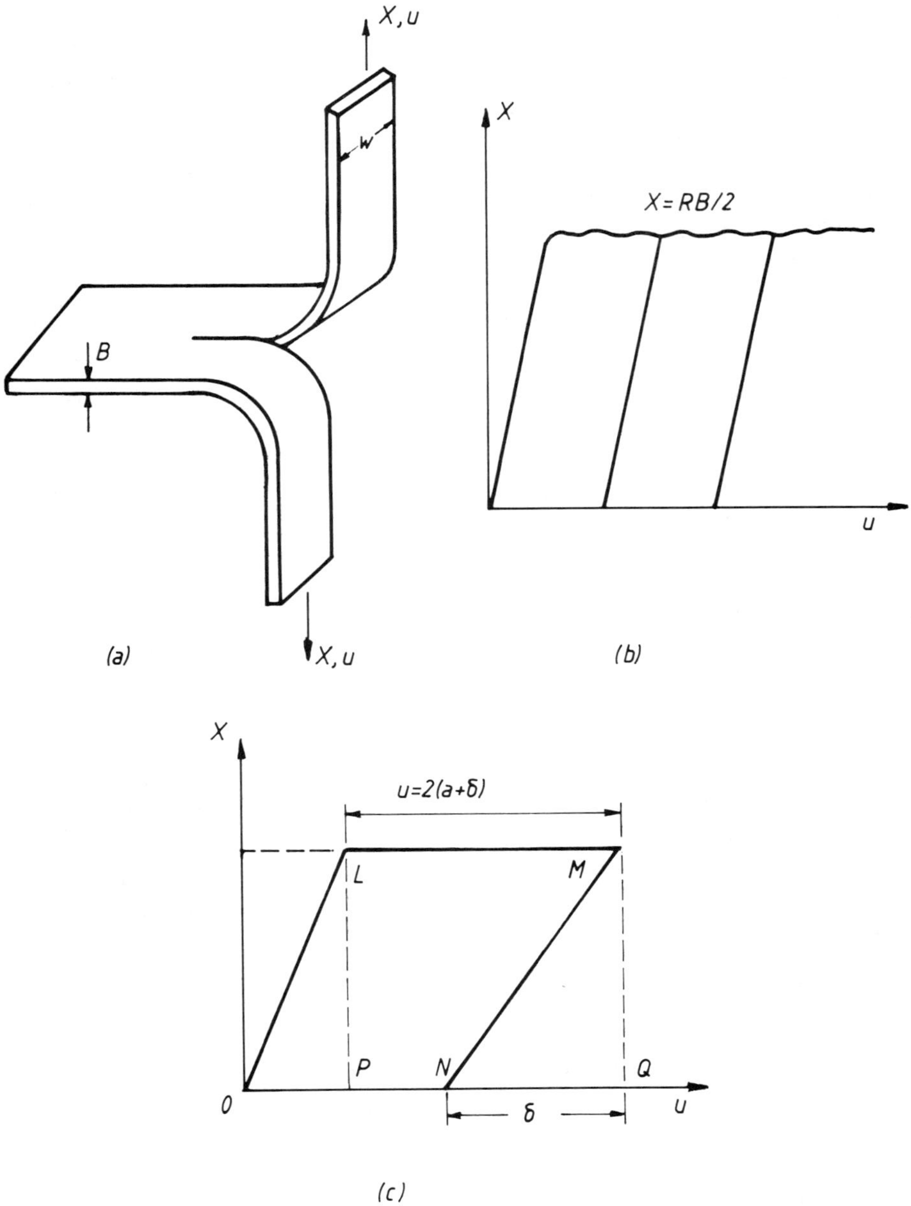

Fig. 2.16 – (a) Test piece geometry for the trousers fracture test. (b) Associated load-deflexion curve for fracturing. (c) Load-deflexion curve for trousers test when the legs can store strain energy in tension.

and the elastic strain is $e = X/wBE$. The stretch δ is given by $\int_0^a e dx$ where $wBdx$ is an elemental volume in a leg. Hence $\delta = Xa/wBE$. In the basic work equation (1.1), we have

$$Xdu = Xd[2(a + \delta)] = d(2 \cdot \tfrac{1}{2} X\delta) + RdA$$

or

$$d/dA\ (\delta/X) = R/X^2 - 2/XB = 1/wEB^2 \text{ as } \delta/X = a/EwB$$

for each leg. Hence $R = \dfrac{X^2}{wEB^2} + 2X/B$. For $E = \infty$, the expression reverts to equation (2.17).

The analysis given in Example 2.8 is due to Caddell (1980) who applied it to the peeling of masking tapes from rolls (including the case of peeling tapered sections which often frustratingly come off rolls of clear sticky tape! – see Problems 2.9 and 2.11; see also Isherwood and Williams (1978)). Note that as X is constant, and therefore the stress in the elastic trouser leg or tape is constant, the strain in the (growing) leg is constant, which means that the extension increases in proportion as peeling or tearing proceeds.

In the case of tearing rubber sheets, the response is not linear elastic, since the $\sigma\epsilon$ curve has the shape shown in Fig. 1.17. The pioneering study by Rivlin and Thomas (1953) discusses how such non-linear stretch in the legs of trousers

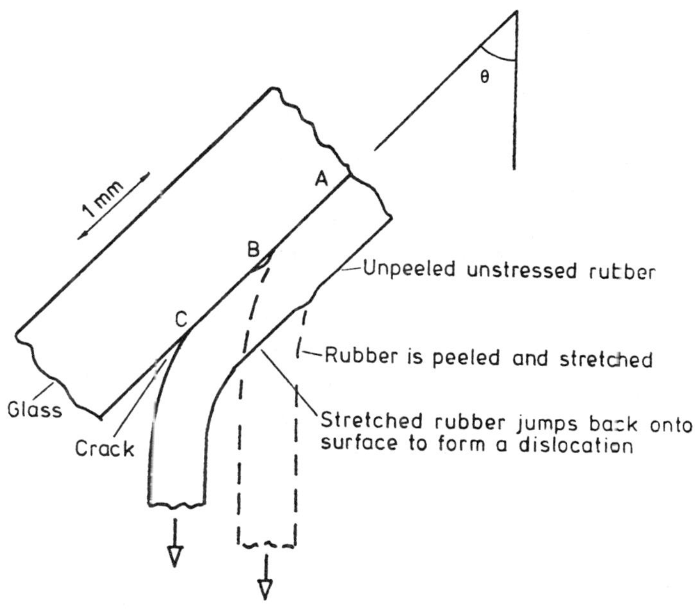

Fig. 2.17 – Peeling of rubber from glass at low angle. (After Kendall (1978), with permission.)

tests on rubber sheet may be analysed (section 2.10). For many materials and specimen proportions, the external work predominates and $R \approx 2X/B$. Thus Gent and Kim (1978) report values for R of 1 ~ 20 kJ/m^2 for a variety of filled and unfilled rubber compounds; these are comparable with Rivlin's and Thomas's results for rubber vulcanisates. Again, sheets of paper have similar values for R (Atkins and Mai, 1979; Mai, 1983).

Consider now a thin strip of width w and thickness B stuck to a substrate which is peeled off at an angle θ (Fig. 2.17). The net displacement in the direction of force if a length da is peeled off is d$a(1 - \cos\theta)$ assuming that there is no stretch in the tape. Since dA = wda, a similar argument to that which produced equation (2.17) gives

$$J_C = R = X(1 - \cos\theta)/w \tag{2.18}$$

(When $\theta = \pi/2$, $R = X/w$ which is the peeling equivalent of the tearing relation (2.17).)

If the strip is stretchy, Kendall in 1975 showed that peeling is governed by the following equation

$$(X/w)^2\ (1/2EB) + (X/w)\,(1 - \cos\theta) = R \tag{2.19}$$

where the first term arises from elastic energy released on stretching the strip. Equation (2.19) is a generalisation of the relations in Example 2.8. If peeling takes place at a low angle, $(1 - \cos\theta) \approx 0$ and $X^2 = 2REw^2B$.

Kendall (1978) discussed a curious phenomenon which may arise during low angle peeling of rubber from glass under appropriate conditions, where some of the peeled rubber jumps back on to the glass, partially healing the crack and arresting further peeling, even though the load may satisfy the above relation. Such 'dislocation formation' results from peeling, stretching and healing processes driven by both the applied load and the attractive surface forces between rubber and glass (Fig. 2.17). These dislocations have little to do with crystallinity and atomic defects; rather they correspond with textbook pictures of Volterra dislocations formed by partially cutting a block of material, moving the cut surfaces relative to one another and gluing back together (e.g. Cottrell, 1964). Owing to the residual stresses arising from the process of cutting and healing, the force required to peel a dislocated interface is greater, i.e. the joint is strengthened (Problem 2.15). Clearly this has implications for composites. It also explains why limestone is so much weaker and more brittle than the prehistorical mollusc shells of which it is composed: Currey's investigations (e.g. 1977) show that nacre (mother-of-pearl) is five times stronger and thirty times tougher than limestone. According to Kendall, interfacial dislocations form at the adhesive joints between nacre platelets and thereby increase the toughness and the strength.

These mechanics are also relevant to paint peeling from walls and cracking in the presence of residual or thermal stresses (Chapter 3 and 7).

Kendall's analysis for peeling along interfaces which possess different bending stiffness and the consequent changes in crack velocity, is discussed in section 3.1.1.

2.4.2 Elastic guillotining and microtoming

In the cutting of thin paper and similar materials by a guillotine or scissors-like tool, the offcut stores negligible strain energy. Under these conditions the total external work performed comprises the two components (i) work of separating the surfaces (i.e. the fracture toughness of the material) and (ii) work of friction of blade against work material and more particularly blade against blade or guillotine base plate (especially when the blades are 'set' and scrape against one another). Measurements of the fracture toughness of thin flexible sheets may be made by loading a guillotine blade and picking up the *Xu* plot with a testing machine working in compression. An estimate of the friction work is given by running back the crosshead of the machine to the original starting position and reloading against the (now cut) edge and baseplate.

Figure 2.18 shows a typical *Xu* trace for cutting a single layer of manila folder 0.26 mm thick. (The load fluctuations were caused by blade/baseplate roughnesses and irregularities, their periodicity coinciding on successive cuts). The toughness *R* may be determined from the differences in work areas ABCD and the cut edge area. Table 2.1 summarises Atkins and Mai's 1979 results for a variety of materials. In the case of the metal shimstock, burring sometimes occurs at the cut edge (when a blunt section of the blade finds it easier at first to bend over the material between blade and baseplate rather than cut it). Estimates for the associated plastic work may be made from the sheared angle of the burr and deducted from the work area (Problem 4.10; section 4.4.2).

Measurement of work in the microtomy of biological materials presents a powerful means of determining the resistance to fracture in highly extensible tissues. Standard toughness testpieces in many biological materials are difficult

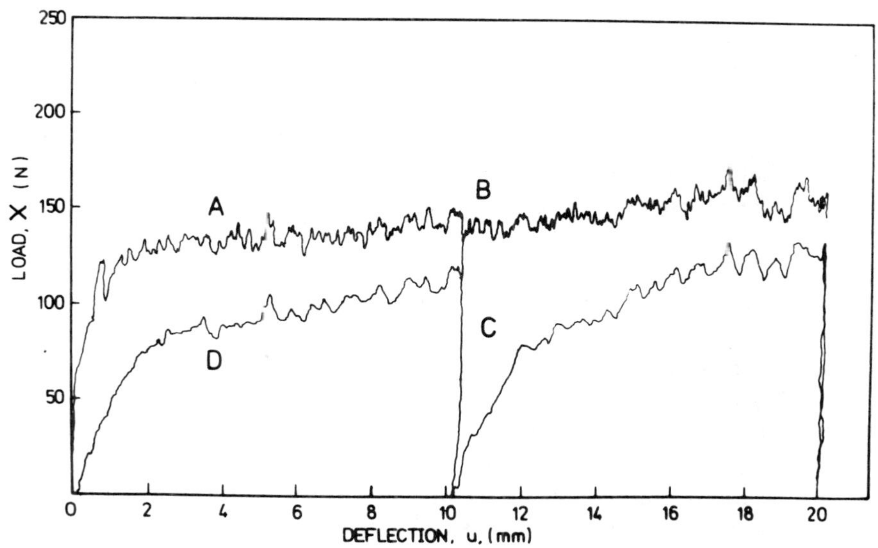

Fig. 2.18 – Load-deflexion traces for guillotining a thin layer of manilla folder, 0.26 mm thick. (After Atkins and Mai (1979), with permission.)

Table 2.1
Fracture toughness values determined from guillotine experiments

Material	t(mm)	R(kJm^{-2})	Remarks
Interleaving paper	0.06	16.93	1 layer
	0.24	10.32	4 layers
Copier paper	0.36	12.70	4 layers
Cardboard paper	0.48	13.60	
Drawing paper	0.27	18.50	0° angle of cut with fibre
		26.00	45° direction
		29.00	90°
Manila folder	0.26	15.16	1 layer
	0.52	13.63	2 layers
Waxed paper	0.18	15.10	2 layers
Rubber reinforced with cotton fabric	1.05	4.35	
Aluminium foil	0.105	10.68	8 layers
Shim brass	0.05	53.18	1 layer, ⊥ rolled direction
	0.10	33.38	2 layers, ⊥ rolled direction
	0.05	63.93	1 layer, ‖ rolled direction
	0.10	42.86	2 layers, ‖ rolled direction
	0.12	41.45	⊥ rolled direction
	0.11	41.39	‖ rolled direction
Copper foil	0.11	18.47	⊥ rolled direction
	0.11	25.95	‖ rolled direction
	0.22	21.41	2 layers, ⊥ rolled direction
	0.22	22.59	2 layers, ‖ rolled direction
Shim steel	0.16	42.06	⊥ rolled direction
	0.16	47.15	‖ rolled direction

(Atkins and Mai, 1979)

to make and to grip. Furthermore, as discussed in Chapter 1 and section 2.10, it is far from clear that all 'fracture mechanics' methods described in this book (which has been developed for comparatively stiff, Hookean, solids) can be directly applied to extensible tissues. In microtoming and guillotining we are not concerned with energy release rates, as the external work alone provides the energy for fracture and the intensification of stress is produced by the sharp tooling.

Owing to the geometry of microtome blades (where the included angle of glass knives can be up to 45°) and the ability to alter the rake angle of the knife, a third work component (over and above fracture and friction work) enters the picture; it is work of slice curling and is similar in kind to plastic flow in the chips produced in metal cutting (Chapter 10) (but in biological materials the deformation is certainly not as extensive as in metalcutting). Since friction has a major influence when the relief angle of the blade β is small, and 'redundant' work of chip curl is greater when β is large, there is an optimum cutting angle when the cutting force is least; $4° < \beta_{opt} < 7°$ for many biological materials.

The cutting force in microtoming is constant for given slice thickness (unlike Fig. 2.18 where the force rises slightly owing to the scissors geometry). At the optimum blade angle, the work areas contain work components of

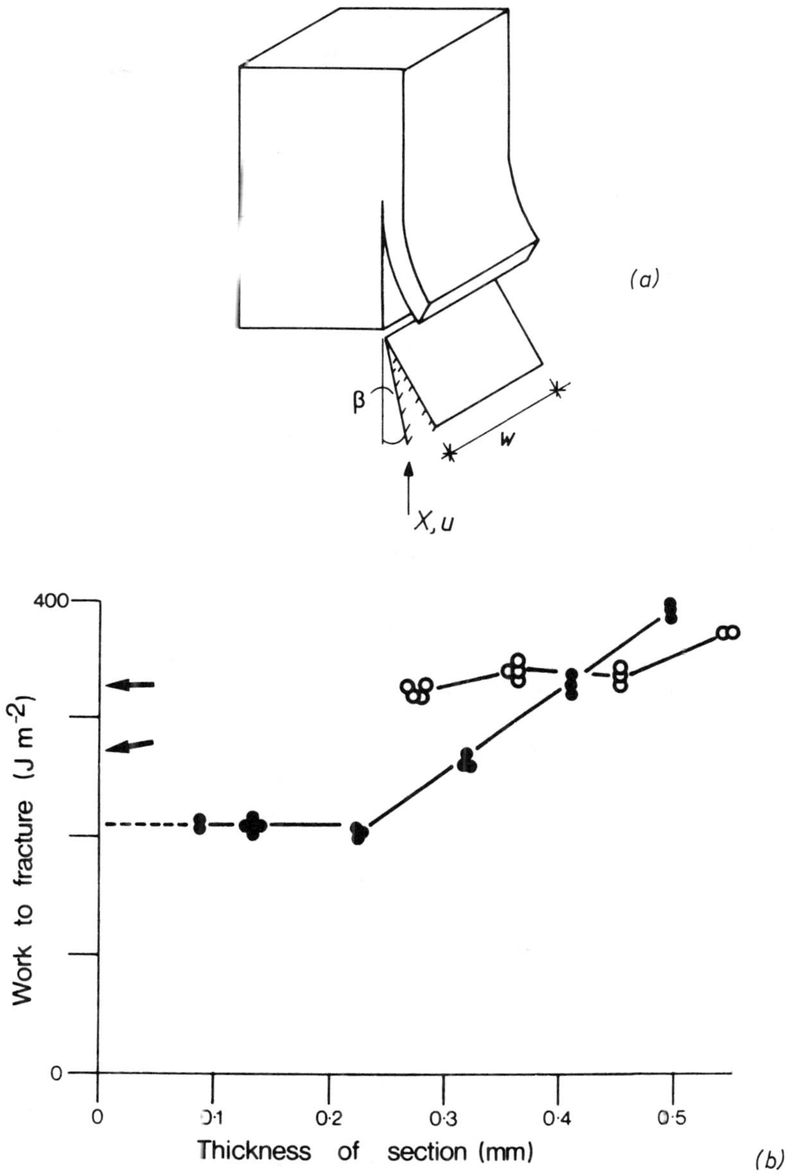

Fig. 2.19 – (Upper) microtome blade cutting specimen of width w. External work = Xu. Area cut = Wu. Therefore toughness (for rectangular specimens) = Xu/wu = X/u. (Lower) variation in apparent work to fracture with thickness of section in turgid (•) and flaccid (○) carrot cortex. At thicknesses greater than 220 μm the section curls aways from the knife as it is cut. This requires additional work which increases in proportion to the thickness of the section. Flaccid carrot does not curl at any of the thicknesses sectioned: its fracture toughness as calculated from this graph depends upon whether the extrapolation assumes constant slope (giving about 270 Jm^{-2}), or that a plateau has been reached (giving about 320 Jm^{-2}). Either way it is higher than turgid carrot. (After Atkins and Vincent (1984), with permission.)

fracture, flow and friction with the combined influence of the latter two being least. In practice many biological materials are watery when cut owing to the opening-up of cells, and friction is negligible. Then, only the work of chip curl has to be deducted from the total work area to give the work of fracture. A suitable treatment of data is to plot the cutting forces (obtained at the optimum angle) versus slice thickness. A typical result for carrot is shown in Fig. 2.19 where we observe that the force levels out at thin slices, where the slices are extremely floppy. The least force divided by the width of cut w (in rectangular specimens) then gives the fracture toughness, $R = X/w$. (Rather like the derivation of equation (2.17), work input is Xu and fracture area is wu; see inset to Fig. 2.17).

The reader interested in elastic guillotining and microtoming is referred to the original papers by Atkins and Mai (1979) and Atkins and Vincent (1984). Guillotining in the presence of plastic flow (cropping, bar shearing) is discussed in Chapters 4 and 5.

2.4.3 Elastic recovery and residual displacements: The General Case

Consider Fig. 2.20 in which a single loading/cracking/unloading cycle OLMN is shown schematically with a residual unloaded displacement ON = u_r. Cracking occurs over LM which is shown downward-pointing, but LM could follow other courses. The total dissipated work within area OLMN is given by

$$\Delta\text{OLMN} = \tfrac{1}{2}\, X_L\, u_P + \frac{(X_L + X_M)}{2}\,(u_Q - u_P) - \tfrac{1}{2}\, X_M\,(u_Q - u_N)$$

assuming that LM is a straight line. Writing $C_1 = (u/X)_{A_1}$ for the compliance of the body containing crack area A_1 and similarly C_2 for the compliance $(u/X)_{A_2}$ = NQ/QM of the body with propagated crack of area A_2, we obtain

$$\Delta\text{OLMN} = \tfrac{1}{2}\, X_1\, X_2\,(C_2 - C_1) + \left(\frac{X_1 + X_2}{2}\right) u_r$$

remembering that u_r + NQ = OP + PQ, i.e. $u_r + X_M\, C_2 = X_L C_1 + (u_Q - u_P)$ (cf. Example 2.2 corresponding with $u_r = 0$). In the limit, therefore, when the area OLMN between stiffness lines is small as X_L and X_M approach one another, and u_r between stiffness lines becomes small, we may write generally

$$\text{d(OLMN)} = \tfrac{1}{2}\, X^2\, \text{d}C + X\text{d}u_r \tag{2.20}$$

This relation presumes that loading/unloading lines are linear which permits X and u to be simply related via C.

If it is argued that the whole work area OLMN goes to fracture work we have, by analogy with the treatment in section 2.2

$$R = G_C = \text{d(OLMN)}/\text{d}A$$
$$= \tfrac{1}{2}\, X^2\, \text{d}C/\text{d}A + X\text{d}u_r/\text{d}A \tag{2.21}$$

Of course if $u_r = 0$ or is constant, equation (2.21) reduces to the 'compliance

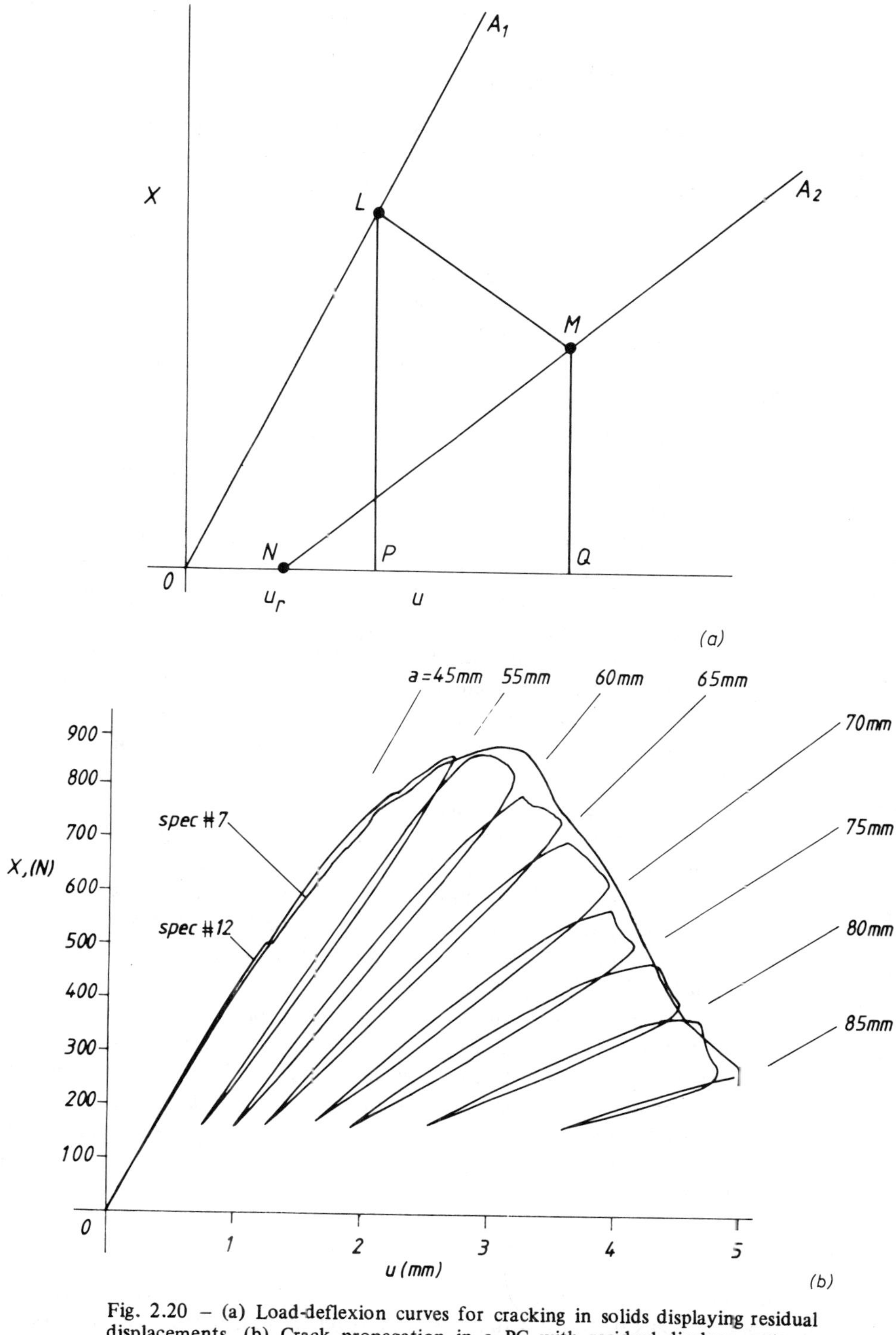

Fig. 2.20 – (a) Load-deflexion curves for cracking in solids displaying residual displacements. (b) Crack propagation in a PC with residual displacements on complete unloading. (After Uete (1981), with permission.)

calibration' equation (2.7) for G_C as $C = u/X$. Also, if there is no elastic strain energy ($C = 0$),

$$R = G_C = X\mathrm{d}u_r/\mathrm{d}a \qquad (2.22)$$

Equation (2.22) applies to the tearing and guillotining examples in sections 2.4.1 and 2.4.2: in the trousers test $u_r = 2A/B$ so $\mathrm{d}u_r/\mathrm{d}A = 2/B$ leading to equation (2.17) and in microtoming $u_r = A/w$ giving $R = X/w$.

Equation (2.21) is the general expression for the trousers test with elasticity in the legs: $u_r = 2a$ where a is the current unloaded length of each leg, so $\mathrm{d}u_r/\mathrm{d}a = 2/B$ and as the total compliance is given by $2a/wBE$, $\mathrm{d}C/\mathrm{d}A = 2/wB^2E$ and $R = X^2/wEB^2 + 2X/B$ as in Example 2.8.

Note that equation (2.21) may be recast in terms of the displacement under load (u_Q at M in Fig. 2.20) instead of the unloaded residual displacement u_r at N. It is easily shown that

$$R = G_C = X\mathrm{d}u_Q/\mathrm{d}A - \tfrac{1}{2}X^2\,\mathrm{d}C/\mathrm{d}A - XC\mathrm{d}X/\mathrm{d}A \qquad (2.23)$$

(When $u_r = 0$, $u_Q = XC$ and equation (2.23) reduces to the familiar compliance calibration equation (2.7).)

Tearing and guillotining are truly reversible cases which display u_r owing to geometric effects. In Chapter 3 we shall see that equations (2.21) and (2.23) apply to problems of cracking in the presence of residual stresses where residual displacements also occur. It is well-known that components can spring open owing to the relief of residual stresses thus producing a u_r. It is interesting in this class of problem that as broken components have to be (elastically) forced back together, 'reversibility' means the restoration not only of geometric size and shape but also an internal stress distribution at least equipollent with the original residual stress distribution.

There are other cases in which residual displacements occur owing to events at the crack tip which prevent the crack faces from closing up on unloading. Crack tip zones (section 2.8) which are large in comparison with the crack length are one example, and fibre-reinforced composites with filaments bridging the crack faces after propagation and preventing closure is another. Residual unloaded displacements in such cases are caused by the body behaving 'more stiffly' than it should for a given crack length owing to crack tip interference – think of the behaviour of a propped cantilever. Experimental proof as to whether the surrounding stress field is elastic, and whether it is merely geometric interference at crack tip that prevents displacement irreversibility, may be obtained by sawcutting along the crack line beyond the crack tip zone into supposedly elastic material. If on so doing the crack faces then do close up, it may be argued that valid fracture toughness information can be derived from the Xu plot using equation (2.21); see Problem 2.18.

Example 2.9 Describe how heat treatment of a PC specimen with residual crack opening helps to confirm that the irreversibility is associated with the crack tip zone.

Ans. If an unloaded polycarbonate specimen, stuck open by a necked-down large crack tip zone is annealed in an oven at about 150°C, it will be found that the zone disappears, the necked section regains its former thickness, and the crack closes up eliminating the residual opening. The driving force for crack closure is the surrounding elastic field. A surrounding plastic field would not produce crack displacement reversibility.

In Uete's 1981 experiments using edge-crack specimens of 6 mm thick polycarbonate, the loading–unloading–reloading Xu plots shown in Fig. 2.20(b) were obtained. Necked-down crack tip zones prevented the surrounding elastic field from closing back up the crack faces, which gave rise to the residual displacements u_r along the abscissa. For a range of crack lengths ($50 < a < 80$ mm), in specimens having a distance of 105 mm in the path of the crack for which $u_r < 1$ mm, it was found that u_r and a were linearly related by $du_r/da = 0.33$ (independent of starting crack length). For those cases where the plastic flow was contained, and apparently did not extend to the specimen boundaries, application of equation (2.21) to Uete's data gives $R \approx 8$ kJ/m^2 during propagation with a fully-developed crack tip zone.

It will be noticed that there is a kink or blip in the Xu trace at the end of the initial linear load build-up in Fig. 2.20(b). This behaviour is called 'pop-in' (section 3.6) and corresponds with a small amount of crack propagation in the middle thickness of the testpiece (when the crack tip zone size is still small) at lower toughness values (ca. 2.8 kJ/m^2) than the 8 kJ/m^2 at which the fully developed crack tip zone propagates (see Caddell and Uete, 1983). The difference is caused by differences in constraint to flow: the middle thickness deforms under approximately plane strain conditions with high triaxial mean stress which encourages early fracture; however, the outside regions deform under relaxed plane stress conditions, stretch more and inhibit cracking in the centre regions; the process zone thereby grows and more work is required to propagate the crack. This type of behaviour is discussed further in section 2.8 (see Fig. 2.31) and section 3.4).

More often than not, residual u_r displacements in Xu plots are caused *not* by elastic effects but rather by *remote flow* extending to the boundaries of the cracked body, where relief of the crack tip interference by sawcutting (say) would not make u_r become zero. Since, however, the Xu loading–unloading plots in these cases often looks Fig. 2.20(b), some workers have employed equations (2.21) and (2.23) to analyse experiments, even though the cause of u_r is different. While events on cursory inspection may seem similar to the residual stress examples in Chapter 3, forcing fibres back up holes to restore the original state of affairs or attempting to compress necked-down large crack tip zones to regain original conditions is *not* the same, as such processes involve irreversible work.

Fig. 2.21 shows X–u plots for double-torsion specimens of Portland cement mortar reported by Wecharantana and Shah (1980). In a globally elastic specimen

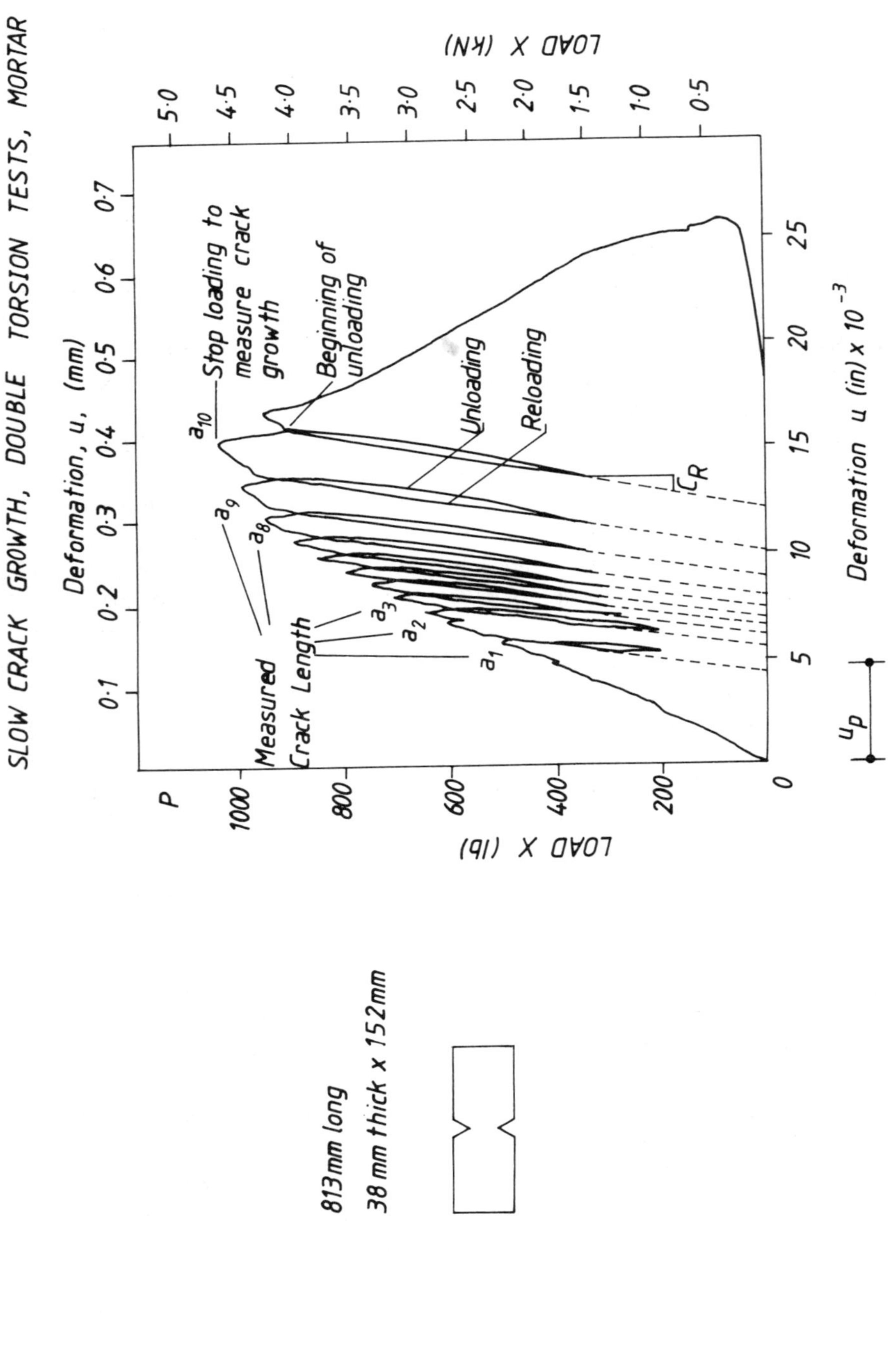

Fig. 2.21 – Typical load-deflexion curve for the fracture of a 'double torsion' cement mortar specimen. (After Wecharantana and Shah (1980), with permission.)

with crack length much longer than the size of the process zone, cracking should occur at constant load, in this type of testpiece, if R is constant. Clearly that behaviour is *not* being followed as cracking takes place under increasing load and there are marked displacement irreversibilities. The data in Fig. 2.21 were analysed by Wecharantana and Shah using equation (2.21) in the form

$$R = G_C = \tfrac{1}{2} X^2 \, dC_m/dA \tag{2.24}$$

with $C_m = [dC/dA + (2/X) \, du_r/dA]$ a 'modified' compliance to reduce the relation to the compliance calibration equation (2.7). At the same time Hodgkinson and Williams (1981) independently employed a version of equation (2.21) to analyse the tearing behaviour of highly ductile low density polyethylene sheet (see section 2.10.2).

One of the difficulties in employing equations (2.21) and (2.23) and versions of them to cases like reinforced solids is that the whole area under the Xu plot has genuinely to be fracture work and nothing else. (In Chapter 4 we shall see that attempts are made to apply these equations to cases where u_r is caused by remote plastic flow and where the area under the Xu plot is certainly *not* wholly concerned with fracture work (see Fig. 1.19 and Example 4.5). For a fixed geometry, of course, with displacement irreversibilities caused by non-elastic events, equations (2.21) and (2.23) may reproducibly describe cracking events and be used for predictive purposes *in that same geometry alone,* but derived R values will probably be *geometry-dependent* through the du_r/dA term in equation (2.21). Although R varying with crack length, for example, is acceptable in terms of R-curves (section 2.3, Chapters 7 and 8) the R-curve will not be unique as it will be different for different starter crack lengths in a given testpiece, and different in different testpieces. Only if the size of the crack tip zone (in composites, PC specimens and so on) is very much smaller than the crack length (which is tantamount to saying that u_r is very small) will a geometry-independent toughness parameter result. Thus it is advisable to employ specimens that are as large as possible when determining toughness. Indeed the data in Fig. 2.21 were obtained from 813 mm long × 152 mm wide × 38 mm thick Outwater testpieces as the process zone in cracked cement mortar bodies is influenced by the size of inclusions (aggregates, fibres), which in concrete could be some 25 mm in size.

2.5 THE MECHANICS OF ELASTIC CRACKING: THE STRESS INTENSITY FACTOR VIEWPOINT

It is well-known that notches, keyways and other changes in body cross-section act as local 'stress-risers'. Photoelastic experiments well demonstrate such stress concentration effects. The traditional stress concentration facture (K_t) is defined as the ratio of the locally largest stress at the discontinuity to the remote average stress; 'remote' in this context, by St Venant's principle, means at a distance greater than say 10 times the size of the discontinuity (i.e. size of hole, radius of fillet joining two section sizes, etc). Values of K_t (and K_f the equivalent factor for fatigue) are found in books on design (for example those by Peterson (1974), Neuber (1961) and Savin (1961) listed at the end of the chapter).

It is important to note that K_t gives the greatest stress magnification at *one point* and no indication is given of the manner in which the locally high stresses fall off and diffuse down to the stresses at regions which are essentially undisturbed by the presence of the hole, crack, notch, flaw and so on. It will be anticipated from the comments on strain energy in section 2.1 and the mechanics of cracking in section 2.2 that we shall need to know the whole *field* of stress in the flawed body in order to write down expressions for the mechanics of cracking approaching it from the stress and strain point of view rather than directly via energy. For this reason K_t is not appropriate and another parameter, called the stress intensity factor, is employed. Stress *intensity* factors should not be confused with stress *concentration* factors; there is a relation between them, however, given in Problem 2.25, which depends on what happens to K_t as the root radius (or other appropriate geometrical factor) diminishes in size.

How may cracks be represented and the stress distribution around them determined? It is customary to consider continuum cracks as flat ellipses which may (a) go right through the body, or (b) go only partly into the body from the surface (thumbnail cracks) or (c) be interior cracks (penny cracks). The 'sharp ends' of the ellipse represent the tips of cracks (Fig. 2.22). Cracks can also be modelled as arrays of dislocations and their behaviour described in terms of dislocation mechanics (see the Bilby–Cottrell–Swinden (1963) model). The classical work on continuum linear elastic stress fields around elliptical through-holes was done by Inglis (1913) in England and, according to Timoshenko and Goodier (1951), Kolossof in Estonia before the First World War. Inglis's paper entitled 'Stresses in a plate due to the presence of cracks and sharp corners' appeared in the 1913 *Transactions of the Institute of Naval Architects* and was juxtaposed with an early photoelastic study by Coker and Scoble of stresses around rivet holes. Fig. 2.23 taken from the paper, shows the distribution of stresses along the axes of a 3:1 ellipse in a very large linear elastic plate loaded at its boundaries by a stress σ_∞ perpendicular to the major axis. The stress σ_{yy} is locally very high across the ends of the major axis, ($\sim 7\sigma_\infty$, so $K_t = 7$) and is the greatest stress in the plate, but it falls off quite rapidly to the mean stress σ_∞. The tensile stress σ_{xx} along the major axis is zero at the end of the ellipse (there is fresh air in the elliptical hole), but it has a local peak value just greater than σ_∞ within a short distance of the end of the major axis.† Owing to bending action around the ellipse, a compressive stress $\sigma_{xx} = \sigma_\infty$ is set up at the minor axis; the complementary σ_{yy} at the minor axis is zero, becomes slightly compressive, and slowly rises to σ_∞ at distances far removed from the crack. Inglis showed that the stress concentration factors for ellipses loaded perpendicular to the major axis are $K_t = -1$ at the end of the minor axis, and $K_t = (1 + 2a/b)$ at the ends of the major axis (hence $K_t = 1 + (2 \times 3) = 7$ in Fig. 2.23). Notice that K_t involves the ratio a/b, not the absolute sizes, so pinholes in a structure are just as bad stress raisers as are very large holes. Noting for an ellipse that the

---

†The Cook–Gordon mechanism of toughening materials by placing weak interfaces in the path of a crack relies on σ_{xx} debonding the weak interface (Cook and Gordon, 1964, Chapter 1).

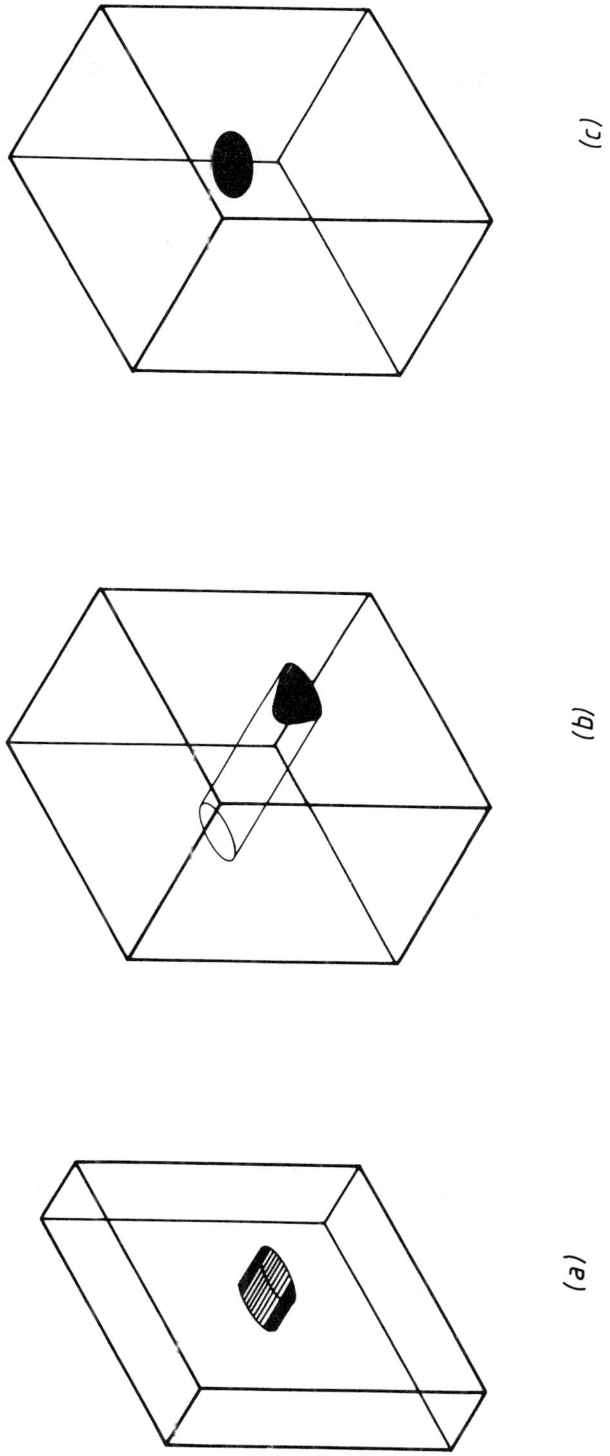

Fig. 2.22 – Solids containing (a) through-the-thickness elliptical cracks; (b) part-through thumbnail cracks and (c) penny-shaped cracks.

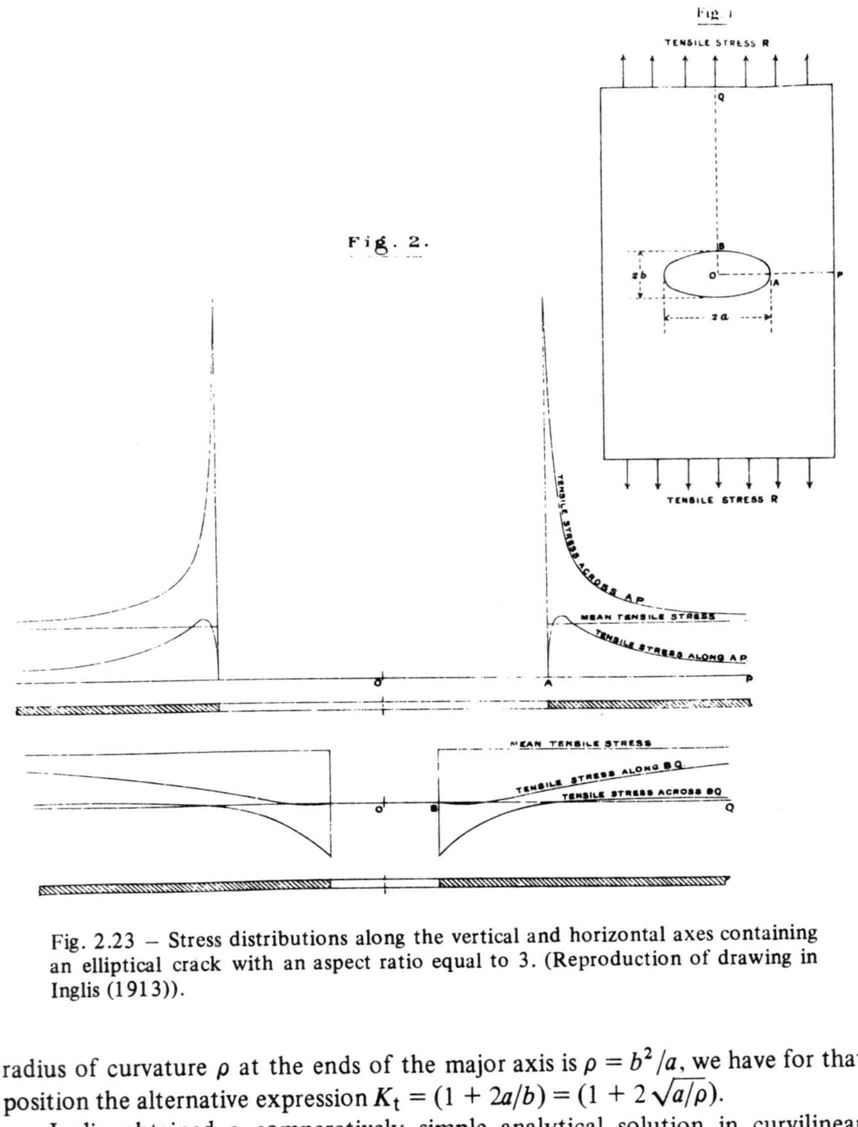

Fig. 2.23 – Stress distributions along the vertical and horizontal axes containing an elliptical crack with an aspect ratio equal to 3. (Reproduction of drawing in Inglis (1913)).

radius of curvature ρ at the ends of the major axis is $\rho = b^2/a$, we have for that position the alternative expression $K_t = (1 + 2a/b) = (1 + 2\sqrt{a/\rho})$.

Inglis obtained a comparatively simple analytical solution in curvilinear coordinates to the problem of remote radial (not uniaxial) stresses σ_∞; the algebra for the uniaxial problem in Fig. 2.23 is more cumbersome to write out, but the general solutions for ellipses will not concern us (unless we are interested in blunt cracks or notches) as sharp dangerous cracks are more like thin slits, for which a $\gg b$. Then the stresses σ_{yy} σ_{xx} and τ_{xy} in the line of the crack (major axis of the thin ellipse) may be written

$$
\begin{aligned}
\sigma_{yy} &= \sigma_\infty \coth \xi \\
\sigma_{xx} &= \sigma_\infty \{1 - [1 + 4\exp(-\xi)] \sin 2\xi + [2\exp(\xi) \sinh 2\xi - 1] \cosh 2\xi\} \\
\tau_{xy} &= 0
\end{aligned}
\tag{2.25}
$$

where $x = a \cosh \xi$ measured from the centre of the ellipse.† There are corresponding expressions for displacements.

It is useful to write out the stresses in terms of the distance r ahead of the crack tip. In the line of the crack $r = (x - a) = a(\cosh \xi - 1)$ or $\cosh \xi = \frac{r}{a} + 1$. It follows that

$$\sinh \xi = (\cosh^2 \xi - 1)^{\frac{1}{2}} = \left(\frac{r^2}{a^2} + 2\frac{r}{a}\right)^{\frac{1}{2}} = \sqrt{\frac{r}{a}} \cdot \sqrt{\frac{r}{a} + 2}$$

Consequently, for σ_{yy},

$$\sigma_{yy} = \sigma_\infty \coth \xi = \sigma_\infty \cosh \xi / \sinh \xi = \frac{\sigma_\infty (r/a + 1)}{\sqrt{\frac{r}{a}} \cdot \sqrt{\frac{r}{a} + 2}} \tag{2.26}$$

The fairly rapid fall-off in σ_{yy} from the edge of the 3:1 ellipse in Fig. 2.23 is even more pronounced in slender ellipses of greater a/b ratio. If the strong intensification of stress in the region of the crack tip is thought to control fracture (and this is discussed in detail in the next section), it is reasonable to write expression (2.26) as an expansion in $(r/2a)$ of which only the first few terms will be important as the remote stresses (and displacements) will not be influencing matters. Then

$$\sigma_{yy} = \sigma_\infty \coth \xi = \frac{\sigma_\infty \sqrt{a}}{\sqrt{2r}} \left(1 + \frac{3}{2}(r/2a) - \frac{5}{8}(r/2a) + \frac{7}{16}(r/2a)^3 \ldots\right) \tag{2.27}$$

Notice that an infinite stress is predicted for the end of the flaw at $r = 0$, i.e. there is a mathematical 'singularity' in stress. This occurs because in a thin ellipse, for which the foci and end-points coincide, the radius of curvature at the tip is zero and in $K_t = [1 + 2\sqrt{a/\rho}]$, $K_t \to \infty$. In practical materials there will of course be a finite ρ – at least at big as the interatomic distance in brittle materials which cleave, and bigger in materials which may plastically flow locally (Chapter 1).

†Strictly $x = c \cosh \xi$ where $2c$ is the focal length of the ellipse ($< 2a$), but in a thin slender ellipse, the ends of the major axis may be assumed to coincide with the foci and $a = c$.

The ratio of approximate stress to exact stress is given by

$$\frac{\sigma_\infty\sqrt{a}}{\sqrt{2r}}\left[1+\frac{3}{2}\left(\frac{r}{2a}\right)-\frac{5}{8}\left(\frac{r}{2a}\right)^2+\frac{7}{16}\left(\frac{r}{2a}\right)^3\ldots\right]$$

$$\div\ \frac{\sigma_\infty(r/a+1)}{\sqrt{\dfrac{r}{a}}\cdot\sqrt{\dfrac{r}{a}+2}}$$

$$\text{i.e.}\quad \frac{\sqrt{(r/a)+2}}{\sqrt{2}\,(r/a+1)}\left[1+\frac{3}{2}\left(\frac{r}{2a}\right)\ldots\text{etc.}\right] \tag{2.28}$$

Fig. 2.24 shows this ratio plotted against (r/a) taking successively more terms in the series. (Of course the *shape* of the curves is steeply falling as in Fig. 2.23 (in fact from infinity at the crack tip); Fig. 2.24 describes how the steeply falling curves *differ in magnitude* at various r/a). The first term of the series approximation agrees with the exact solution certainly out to $r/a = 0.01$, and perhaps adequately enough out to $r/a = 0.1$, but thereafter there is considerable discrepancy, with the exact solution being greater than the approximate solution. In fact, at $r/a = 100$ the inexact value is only 0.1 of the exact solution. Taking further terms in the series improves agreement up to $r/a = 0.3 \sim 1.0$ but thereafter the predictions diverge (the first and second terms together giving an overestimate of σ_{yy}; the first, second and third terms giving an underestimate again).

Although σ_{yy} is infinite at the leading edge of the crack, the associated displacement is bounded. It is given by

$$v_{yy} = \frac{2\sigma_\infty\sqrt{a}}{E^*}\cdot\sqrt{\frac{r}{2}} \tag{2.29}$$

where $E^* = E/(1-\nu^2)$ for plane strain conditions at the crack tip or $E^* = E$ for plane stress. Strain energy is also bounded.

The product of applied stress and crack half-length in equation (2.27) etc. (i.e. $\sigma_\infty\sqrt{a}$) is grouped into one parameter which is the fundamental parameter of linear elastic fracture mechanics. In the early days the symbol $\mathcal{K}$ was used for $\sigma_\infty\sqrt{a}$, (and today k is used), but usually $\sqrt{\pi}$ is customarily introduced into the numerator and denominator to obtain similarlity with the Griffith formula (Example 2.5) and the symbol K used. Thus we have for the stress perpendicular to the direction of a thin slit in a large sheet

$$\sigma_{yy} \approx \frac{\sigma_\infty\sqrt{\pi a}}{\sqrt{2\pi a}} = \frac{K}{\sqrt{2\pi a}} \tag{2.30}$$

where K is now called the *stress intensity factor.* The region over which the first term of the series is an adequate approximation for σ_{yy} is called the 'K-dominant' region of the field. Equation (2.29) for displacement may also be written in terms of K. Note the utility of K: it is a single parameter, incorporating *both*

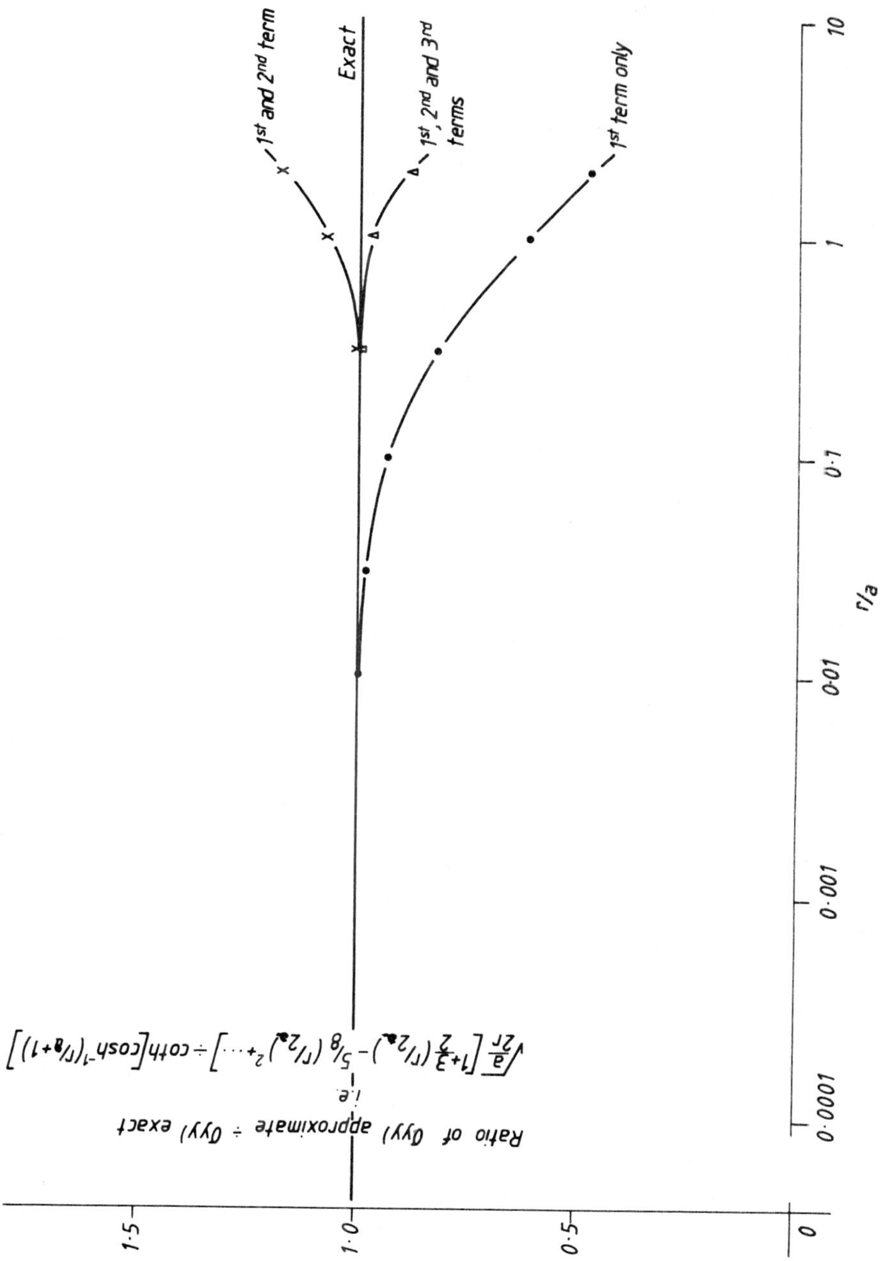

Fig. 2.24 – Ratio of σ_{yy} (approximate) to σ_{yy} (exact) from Inglis's solution.

the applied stress (σ_∞) *and* the half-length of the slit-like crack a, that describes the stress *field* adjacent to the crack tip. As such it should not be confused with K_t which gives the magnification of only at *one* point. K does, however, have peculiar units of (stress) (length)$^{1/2}$ or force/(length)$^{3/2}$, such as Pa$\sqrt{\text{m}}$ or N/m$^{3/2}$. The relationship between K and more easily appreciated physical parameters is given in section 2.7.

Irwin's general equations for normal and shear stresses at locations (r,θ) near the crack tip in a uniaxially stressed body are given in end-of-chapter problem 4.3. There are related expressions for displacements which may be found in standard texts. Stresses at (r,θ) may also be given in the form σ_{rr} $\sigma_{\theta\theta}$ $\tau_{r\theta}$ and so on (cf. Example 2.18 and Williams (1957)).

Thus far we have used the Inglis solution for a small crack in large flat plate. What of other geometries and other boundary conditions? Irwin realised in 1957 that with the same 'first term of a series' approximation, the local crack tip stresses in the line of the crack for other geometries may be expressed in the same way but with a 'shape correction factor' Y which takes account of different methods of loading, shapes of external boundaries of the cracked body, nearness of external boundaries and so on. Thus instead of simply $K = \sigma_\infty \sqrt{\pi a}$ in the stress field expressions,

$$K = Y \,.\, \sigma\sqrt{\pi a} \tag{2.31}$$

where σ is the applied stress (the ∞ subscript is dropped as the applied loading could now be quite close to the crack), a is some size of the crack (often half the total length, but sometimes the total length especially for edge cracks; also sometimes the ***radius,*** or perhaps *depth* of a 'three-dimensional' penny crack or part-through crack) (Fig. 2.22). Y generally varies with crack size and with the circumstances of loading. Some authors do not incorporate $\sqrt{\pi}$ in the K formulae and write $K = Z \,.\, \sigma\sqrt{a}$. Clearly $Y\sqrt{\pi} = Z$ (see Example 2.11).

Although we limit ourselves here to expressions using the crack tip stress σ_{yy} normal to the line of the crack it is shown in books on fracture mechanics how other components of stress at other points in the deformation field may be written in terms of K; likewise for the displacements.

Appendix 1 contains a variety of expressions involving K, and a number of handbooks (listed at the end of the chapter) are available which have collections of K formulae. K in biaxial and multiaxial loading situations is discussed in Chapter 3. The various mathematical methods of determining K formulae are described in books on fracture mechanics. Some solutions are available in closed-form; others are expressed as simple empirical formulae or as algebraic series; others are given as curves (sometimes over limited ranges of size of crack as in Fig. 2.25); others again may be calculated analytically by finite element techniques or experimentally by the methods of photoelasticity and caustics (e.g. Theocaris, 1971).

Example 2.10 If a stucture is said to contain a crack of size 20 mm, say, students often are unclear as to whether this is a or $2a$ for use in equation (2.31). Discuss

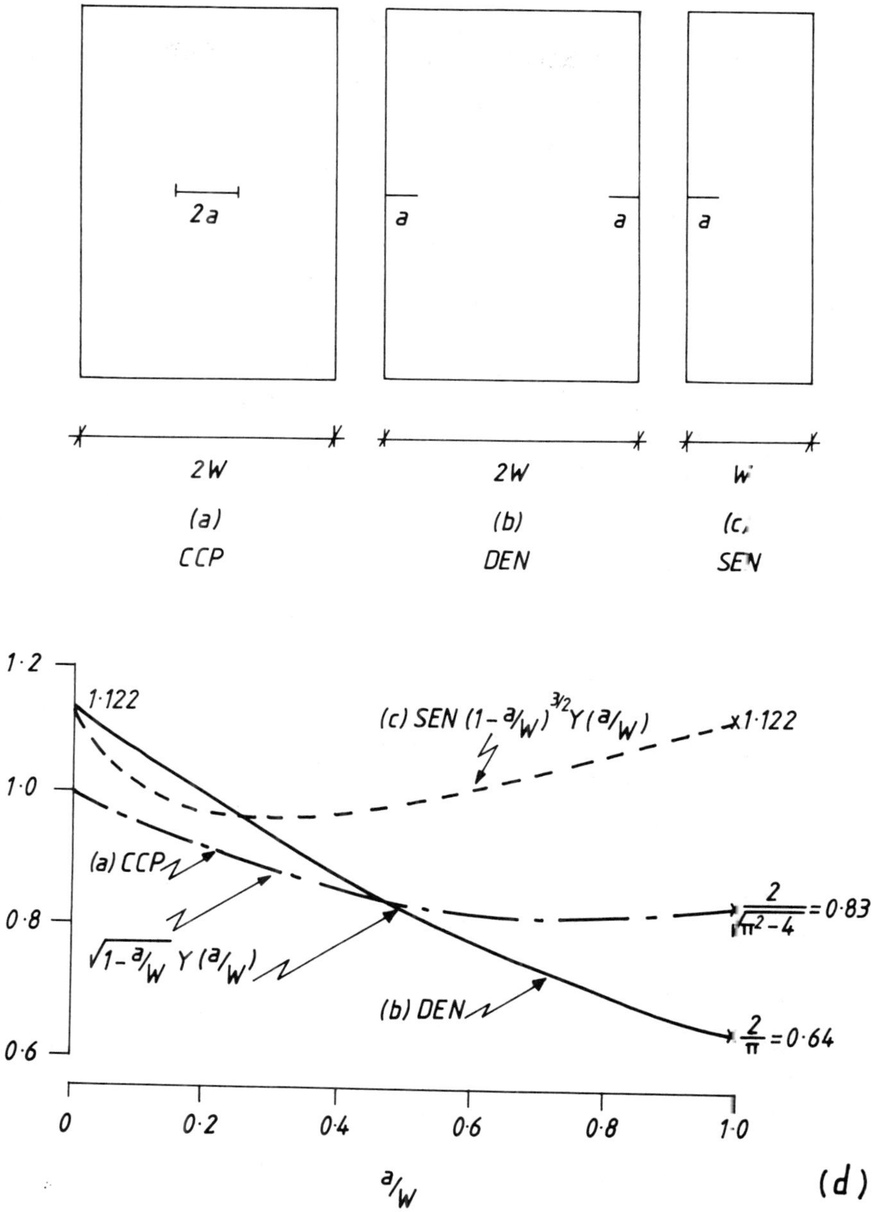

Fig. 2.25 – Three common test pieces: (a) centre-cracked panel (CCP); (b) double-edge notched (DEN) specimen; and (c) single-edge notched (SEN) specimen. Variations of $Y(a/W)$ with a/W are given in (d) for these test geometries. (After Tada, Paris and Irwin (1973), with permission.)

the difficulty in relation to edge and central cracks in a plate and show that in the case of cracks of total length 20 mm, the stress intensity of a small edge crack is elevated above that for a central crack of the same total length by a factor of nearly 1.6.

Ans. For all three crack geometries shown in Fig. 2.25 we have $K = \sigma Y(a/W)\sqrt{\pi a}$ where a is the half-length of crack in Figs. 2.25(a) and (b), where also W is the half-width of the body, but in Fig. 2.25(c), a is the total length and W the total width. $Y(a/W)$ are given in each case as calibration curves. For small (a/W) we see that $Y = 1.12$ in cases (b) and (c) but $Y = 1$ in case (a); the elevated Y comes about owing to the lack of constraint to crack opening on the free edges in comparison with a central crack. Thus if crack length were said to be 20 mm, and a/W were small, (i) a would be set at 20 mm for an edge crack and the formula would have the 1.12 correction factor for edge effects, i.e. $K = 1.12\sigma\sqrt{\pi}\sqrt{20 \times 10^{-3}}$, but (ii) a would be set at 10 mm for a central crack and $K = \sigma\sqrt{\pi}\sqrt{10 \times 10^{-3}}$. The ratio of the two stress intensities is therefore $1.12\sqrt{2} \approx 1.6$. (cf. the Griffith formula in Example 2.5 where $2a$ is the total crack length, and where a *second* factor of 2 enters the picture if $\gamma = R/2$ is used, i.e. $\sigma_{cr} = \sqrt{2E\gamma/\pi a}$).

Since K-formulae are linear elastic expressions, the principle of superposition applies so that under a variety of external loads the final stress intensifications may be obtained by adding the separate effects (Problem 2.32).

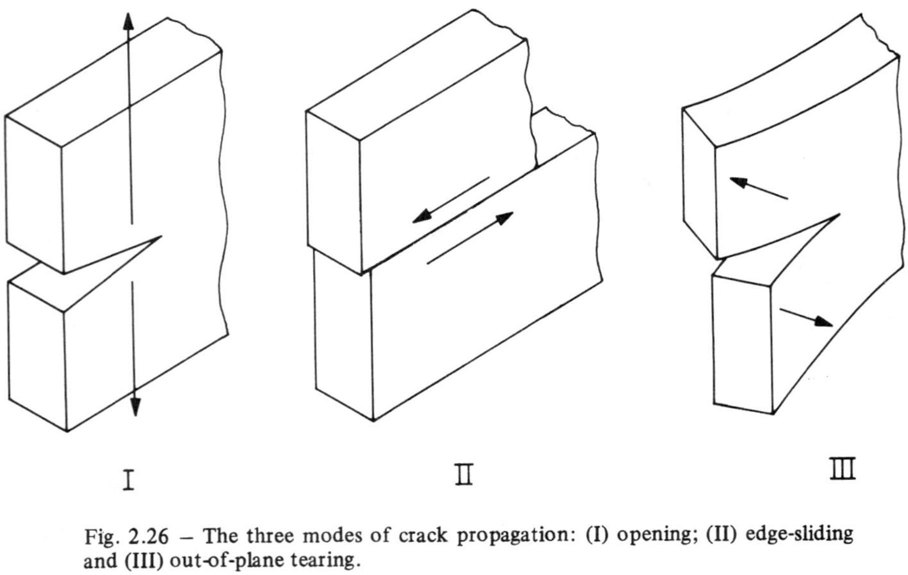

Fig. 2.26 – The three modes of crack propagation: (I) opening; (II) edge-sliding and (III) out-of-plane tearing.

K-formulae (stresses or displacements) are usually associated with a 'mode' of crack opening, indicated by Roman numerical subscripts I, II or III, for which separate stress intensification expressions are appropriate. These correspond with whether the body is being deformed so as to encourage the crack to (I) split open along its length, (II) shear along its length, or (III) propagate by twisting across the crack front (as in tearing paper); see Fig. 2.26. Experimentally, crack propagation sometimes comprises a mixture of these modes (hence one use of the term 'mixed mode' fracture). All cracks that turn or twist have more than one mode of opening, but we are most often concerned with mode I (tensile) crack opening; pure mode III deformation is rearely achieved in practice as it is difficult to eliminate a mode I tensile component at the crack tip when twisting across the crack front. Crack propagation in mode II and in mode III are similar to the motions of edge and screw dislocations respectively.

2.6 K_C AS A PARAMETER FOR CRACK PROPAGATION

For a given starting crack size in a given body K increases as the level of applied loading increases according to the appropriate form of $K = Y\sigma\sqrt{\pi a}$; if the applied load is lowered, K likewise decreases. K, as such, has *nothing* to do with fracture, and is merely a convenient means of describing the stress and displacement fields around discontinuities, notches, flaws and so on.

As already explained, the total strain energy Λ given by the area under the elastic stiffness line is the result of integration of equation (2.2) throughout the cracked body. Now, in principle, σ may be substituted in terms of $K(\sigma = K/Y\sqrt{\pi a})$ to give energy terms such as $\frac{1}{2}K^2/EY^2\pi a$ per unit volume and integrated for the whole body. It follows that if cracking actually occurs, the energy rate balance equation (1.1) may also be written in terms of K, where K is obtained from the current values of load and crack length during propagation. Since therefore $(X\mathrm{d}u - \mathrm{d}\Lambda) = G\mathrm{d}A$ may be written in terms of K of the running crack, the potential energy release rate G, and hence specific work of fracture R in quasi-static propagation, must also be expressible in terms of K via equation (1.1). In other words K of the propagating crack could serve to characterise a material's resistance to crack propagation, and would be an alternative parameter to R. Fracture events themselves take place in the process zone, so we anticipate that a K-characterisation of fracture events will be successful provided that the process zone is small and lies wholly within the 'K-dominated' region of the actual crack tip stress and strain field, i.e. the region in which the first term of the series is a sufficiently good approximation.

Experiments on given testpieces of the same material with varying crack lengths do indeed show that $Y\sigma\sqrt{\pi a}$ during propagation is constant within experimental error. Moreover experiments on different testpiece geometries of the same material with completely different Y factors give $K = (\sigma\sqrt{\pi a})_1\ Y_1 = (\sigma\sqrt{\pi a})_2\ Y_2 = \ldots =$ constant, (under comparable environmental conditions, crack velocities and so on). While these relations can apply to running cracks, most attention has been paid in the literature to (σ, a) combinations at the start of cracking, i.e. the a is the starter crack length and σ is converted from the load at which this crack starts to move. The reasonably constant K obtained experimentally under such conditions is called the *critical stress intensity factor* that causes fracture and, like R, is considered to be a material property (see, for example, ASTM (1981); British Standards Institution (1977)). Exactly like G_C and J_C, the critical stress intensity factor is written K_C. K_C and R are related via the Young's modulus of the material, as explained in section 2.7. Where appropriate, the mode of crack opening is specified thus: K_{IC}, K_{IIC}, K_{IIIC}, although we do not always use subscripts on the corresponding R†. It should be noted that K_C, K_{IC}, etc. are sometimes called the 'fracture toughness' in the literature. In this book that name is given to the specific work of fracture and we call K_C, etc. 'critical stress intensity factors'.

Confusion sometimes arises between the stress intensity factor $K = \sigma Y\sqrt{\pi a}$ *produced in the cracked body* by the applied stress σ (which is a strength-of-materials expression) and the critical stress intensity factor K_C *of the material.* Only when cracking takes place has the applied stress intensity factor reached the critical value. A body with a given crack length loaded at a stress lower than that to cause fracture still has a K given by $\sigma Y\sqrt{\pi a}$, but it is lower than K_C since σ has not reached the critical stress σ_{cr} for that crack length given by $K_C/Y\sqrt{\pi a}$. In relation to traditional strength-of-materials design based on yield stress σ_y, K is to K_C as the working stress σ is to σ_y.

When K_C is known for a material, the relation $K = K_C = \sigma_{cr}\ Y\sqrt{\pi a}$ gives σ_{cr} the fracture stress (or X_{cr} the cracking load).

Example 2.11 The 'compact tension' (CT) testpiece for measurement of K_{IC} is shown in Fig. 2.27 with ASTM recommended proportions. $K_I = \sigma\sqrt{\pi a}\ Y(a/W)$ where, for this geometry, $Y(a/W) = 16.70 - 104.66\,(a/W) + 369\,(a/W)^2 - 573.78\,(a/W)^3 + 360.46\,(a/W)^4$; alternatively it is more convenient to write the expression directly in terms of the applied load using $\sigma = X/WB$, so $K_I = X/B\sqrt{W}\,[29.6\,(a/W)^{1/2} - 185.5\,(a/W)^{3/2} + 655.7\,(a/W)^{5/2} - 1017.0\,(a/W)^{7/2} + 638.9\,(a/W)^{9/2}]$ (Wessel, 1968). Experiments were performed on a 1.25 Cr/0.5 Mo steel and fracture occurred at 12 kN in testpieces with $(a/W) = 0.6$, $B = 6.4$ mm and $W = 79$ mm (N.B. the thickness is less than that recommended; see Example 2.15 later). Determine the stress intensity factor (a) when $X = 5$ kN and (b) at fracture.

Ans. Using the second expression for K_I, the series expansion for (a/W) in brackets when $(a/W) = 0.6$ is equal to 17.48. Hence at 5 kN load $K_I = 5 \times 10^3 \times 17.48/6.4 \times 10^{-3}\sqrt{79 \times 10^{-3}} = 48.6$ MPa$\sqrt{\text{m}}$. At fracture ($K_I = K_{IC}$) $K_{IC} = 48.6 \times 12/5 = 116.6$ MPa$\sqrt{\text{m}}$.

---

† There is inconsistency in the literature over the use of K_C and K_{IC}. Both are applied to opening modes (mode I), but K_C is used for plane stress conditions at the crack tip (as in thin sheets) and K_{IC} for plane strain (in thick sections). Since $K_{IC} < K_C$ owing to triaxiality, the plane strain critical stress intensity factor K_{IC} is the more important parameter in conservative design (see section 2.8). Sometimes in this book we write K_C to indicate the critical stress intensity factor which may be thickness-dependent (see Fig. 2.30 and associated text). Again, in the context of rate, temperature and environmental effects on toughness, K_c is used in the literature for sub-critical crack growth at levels below K_{IC} (see Chapter 6).

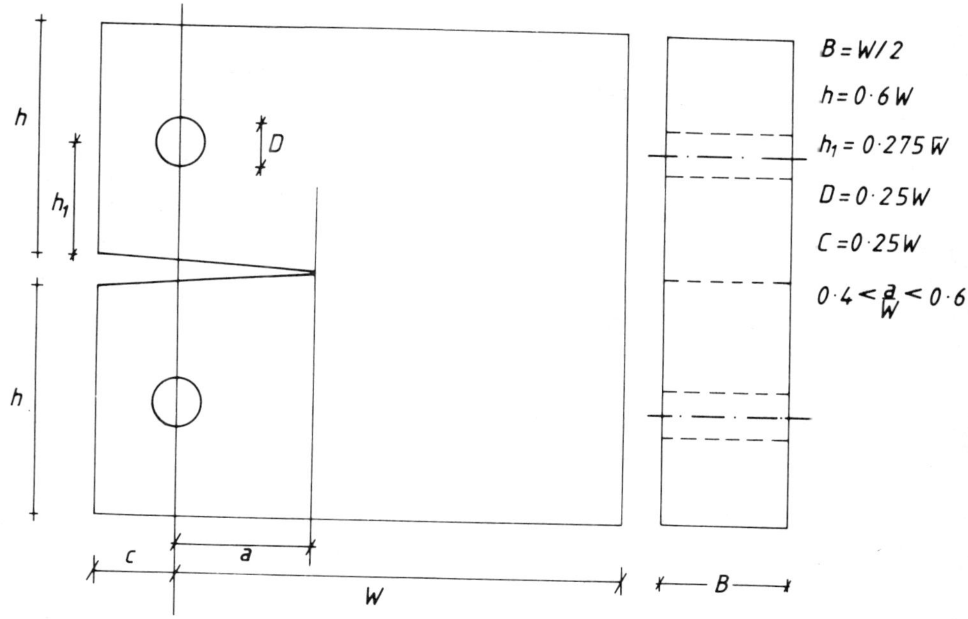

Fig. 2.27 – Standard compact tension (CT) specimen for fracture toughness testing.

K-formulae apply to cracked bodies which are linear elastic in behaviour, but what of non-linear elastic bodies that fracture? Section 2.2 shows how to analyse such problems in terms of specific work of fracture R, where it is realised that the *same* R (a property of the material) relates to fracture irrespective of whether, owing to geometric peculiarities of cracked bodies, the behaviour is linear or non-linear. Are there, consequently, K-formulae for non-linear bodies? In fact there are such expressions as explained in section 4.5, but they are not used in the way anticipated in this chapter and we make no further reference to them at this point.

A three-dimensional geometry of great importance is that of a surface crack, which may be modelled as a semi-elliptical part-through crack, where the crack depth and width may grow at different rates. The fracture mechanics formula for crack propagation produced by remote boundary stresses σ_∞ (Fig. 2.28(a)) is given by Newman (1971) as

$$K = \sigma_\infty \sqrt{\pi a}\,(M_e/\sqrt{Q}) \tag{2.32}$$

where M_e is called an elastic magnification factor on stress intensity, and Q is an elastic shape factor for an elliptical crack; a is the *depth* of the crack in this case and $2c$ the surface length. M_e and Q are functions of a/B and a/c; plots of the factor $M_e/\sqrt{Q}$ are given in Fig. 2.28(b). In general terms we see that long, shallow cracks have high $M_e/\sqrt{Q}$ increasing with a/B whereas short, deep cracks have essentially constant low $M_e/\sqrt{Q}$.

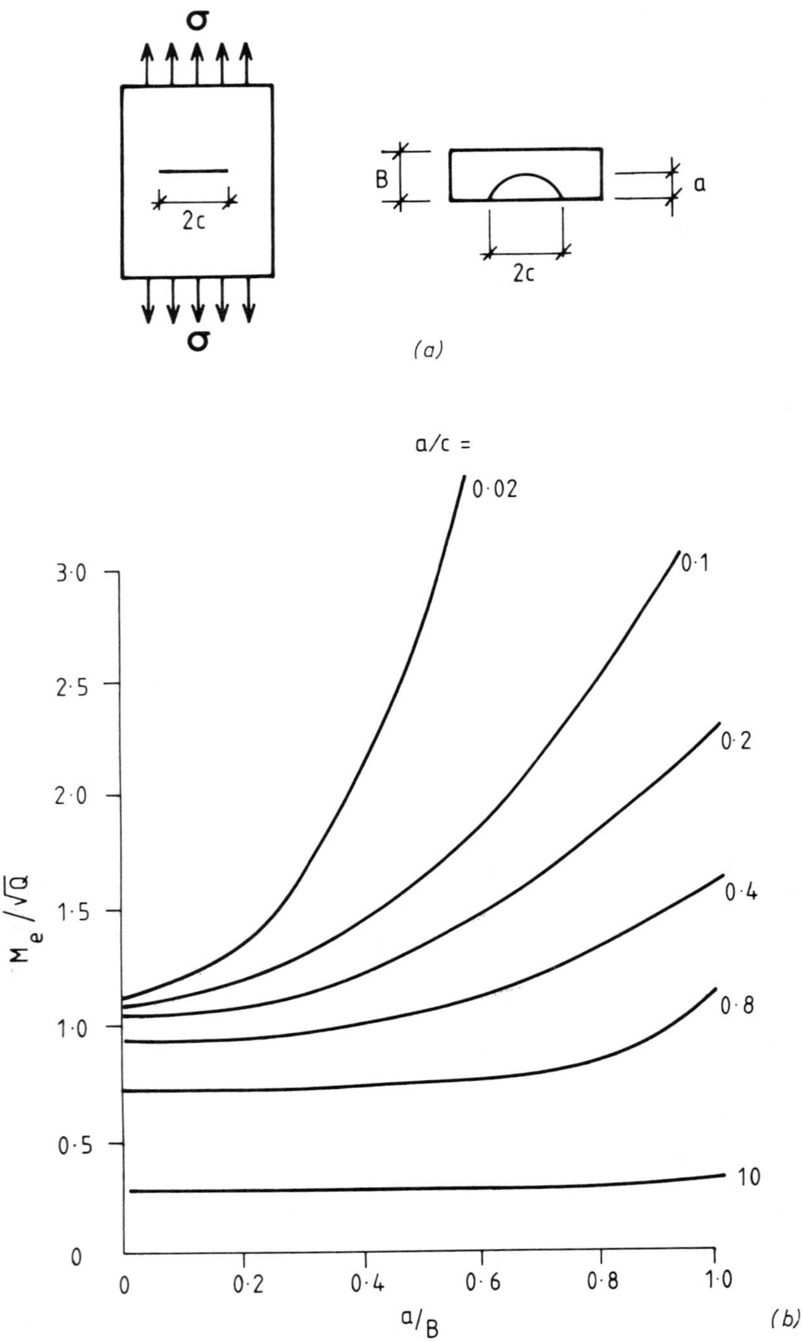

Fig. 2.28 – (a) A surface crack subject to uniaxial tension and (b) the associated elastic magnification factors on stress intensity. (After Newman.)

Equation (2.32) is of great use in the design of pressure vessels where part-through cracks are involved. An example follows.

Example 2.12 A small beer barrel of diameter 380 mm and wall thickness 3.2 mm made of an aluminium alloy exploded when a pressure-reduction valve malfunctioned and the barrel experienced the 4.2 MPa full pressure of the CO_2 cylinder supplying it with gas. Afterwards, cracks 75–125 mm long by (probably) 1–1.5 mm deep were discovered on the inside surfaces of salvaged pieces of the barrel (it was impossible to measure their depth). Independent tests gave some 56 MPa$\sqrt{\text{m}}$ for the critical stress intensity of the aluminium alloy. Are cracks of the size quoted 'critical' at a pressure of 4.2 MPa?

Ans. The hoop stress is $\sigma = pD/2B = 4.2 \times 10^6 \times 380 \times 10^{-3}/2 \times 3.2 \times 10^{-3}$ $= 250$ MPa. This may be used for σ_∞ in equation (2.32) as the circumference/thickness ratio of the barrel is over 350. (Corrections for curvature of the vessel can be made if desired; see section 9.2.1.) Thus for $K = K_C$, $56 \times 10^6 = 250 \times 10^6$ $\sqrt{\pi a}\,(M_e/\sqrt{Q})$ or $0.126 = \sqrt{a}\,(M_e/\sqrt{Q})$. Choosing values for a (and hence specifying a/B), the associated a/c (and hence $2c$) may be found by interpolation on Fig. 2.28(b) which satisfy the $K = K_C$ condition. We obtain

a (mm)	a/B	$0.126/\sqrt{a}$	a/c	$2c$ (mm)
1.5	0.47	3.25	<0.02	>150
2	0.63	2.81	0.03	133
2.5	0.78	2.52	0.1	50

These results suggest that surface cracks 75–125 mm long are some 2.5 ~ 1.5 mm deep when critical for a pressure of 4.2 MPa. The depths of crack in the salvaged pieces of barrel are about that size. These cracks probably came from rough handling of empty casks, bouncing them on their corners.

It is possible to design pressure vessels according to a 'leak-before-burst' criterion. Such a concept applies when the critical depth of flaw is greater than the wall thickness, for by the time the flaw has become critical (growing perhaps by fatigue from a smaller starting size, see Chapter 7) the pressure driving the crack should have been relieved by leakage. None of the $(a, 2c)$ pairs in Example 2.12 satisfies this criterion, as they clearly satisfy 'burst-before-leak'. We must establish what flaw aspect ratio would permit leakage at 4.2 MPa in this particular example. We have $0.126/\sqrt{a} = M_e/\sqrt{Q}$ and when $a = B = 3.2$ mm, $M_e/\sqrt{Q} = 2.2$ and the aspect ratio $a/c \approx 0.2$. Then $2c \approx 32$ mm. This aspect ratio is greater than any in Example 2.12.

The problem can be worked the other way to establish what toughness is required for the barrel material to leak-before-burst at the overload pressure in the presence of existing flaws with given aspect ratio. For $a/c \approx 0.03$, and

$a/B = 1, M_e/\sqrt{Q} \approx 6$ (?) (extrapolating from Fig. 2.28 (b)). Hence $K = K_C = 250 \times 10^6 \sqrt{\pi \times 3.2 \times 10^{-3}}$ (6) = 150 MPa$\sqrt{\text{m}}$. This is about three times the toughness of the alloy given in Example 2.12.

These calculations employing equation (2.32) presume elastic fracture: often pressure vessels made from solids with medium-to-high toughness–strength ratios flow plastically before fracture or even may be considered to fail by plastic collapse with no fracture. These considerations are discussed in Chapters 4 and 10 (see particularly Example 10.8).

An unusual adaptation of Newman's part-through crack formula is to the penetration of wood by nails (Atkins and Mai, 1975). While the work of actual fracture in nailing many woods is negligible ($R < 100$ J/m^2 along the grain or $K_C \sim 200$ kPa$\sqrt{\text{m}}$), toughness determines the amount of splitting around the nail (Problem 2.33). This in turn determines the lateral contact pressure between nail and wood, and hence the frictional force which is the major component of the resistance to nailing (nails stay in place because of friction). It is well-known that deliberately blunted nails (and 'cut nails') are less likely to split timber than sharp nails. This is because blunt nails sever the wood fibres across the grain in a cutting action whereas sharp nails cause splits along the grain; the mode of fracture is different.

2.7 RELATION BETWEEN R AND K_C

$K_C = \sigma_{cr} Y\sqrt{\pi a}$ gives the stress or load to cause fracture, but so also do relations (2.6)–(2.11) which are written out in terms of R and not K_C. There must therefore be a relation between R and K_C in linear elastic fracture. If the crack extends normal to the direction of applied stress (i.e. along the major axis of the slender ellipse), equations (2.25) say that the sides of the newly formed crack separate under pure tension. Then only stress σ_{yy} contributes to the *release* of energy which supplies the work required in the process zone for propagation. Assuming that reversible thermodynamics apply, we conclude that if a propagated crack were to be closed back up by stresses σ_{yy} acting 'in reverse', the associated *increase* in strain energy would also match the toughness work.

Consider crack opening or closure over a distance Δa ahead or behind the current crack tip in a cracked plate of thickness B. The work done within Δa is

$$B \int_0^{\Delta a} \sigma_{yy} \,.\, v_{yy} \,.\, \mathrm{d}r$$

assuming fixed grip conditions where the external loads do no work. Thus equation (2.12) in incremental form must be

$$R = -\operatorname*{Lt}_{\Delta a \to 0} \frac{\Delta \Lambda}{\Delta A} = \operatorname*{Lt}_{\Delta a \to 0} \frac{1}{\Delta a} \int_0^{\Delta a} \sigma_{yy} \,.\, v_{yy} \,.\, \mathrm{d}r \qquad (2.33)$$

σ_{yy} and v_{yy} may be substituted from equations (2.30) and (2.29) using K_C and there results (Problem 2.34)

$$R = K_C^2/E^* \qquad (2.34)$$

which for plane strain (using K_{IC} in place of K_C) is $K_{IC}{}^2 = ER/(1 - \nu^2)$. The relation is seen quickly by comparing the basic Griffith formula, Example 2.5, with the critical K formula for a small flaw in large sheet loaded at the remote boundaries.

Immediately we see the reason for K's and K_C's peculiar units of stress$\sqrt{}$length: the stress intensity factor is the square root of the product of two simpler parameters, viz. Young's modulus (force/area) and the strain energy release rate or specific work of fracture (work/area = force/length).

The ability to relate energy arguments to considerations of stress by means of simple transformations (which makes the mechanics of *linear* elastic fracture comparatively straightforward) depends on two factors, (i) the defining of crack tip stresses by means of the single paramter K, and (ii) the ability to relate K_C to R in a simple form. Regarding the latter, we remember that we assumed that the crack propagated in its own path (called 'self-similar' propagation): it did not curve out of that path which would introduce shear terms into relation (2.2) and complicate the mathematics. We take note that the usual K-formulae presume self-similar crack advance (i.e. a crack gets longer in its original direction) and do not deal with kinked crack paths to permit the calculations of the conversion; as discussed in Chapter 3, the question of crack paths can be difficult. In contrast R, as a directly measured material property, can be determined irrespective of whether the crack curves or not (it may of course happen that R varies with direction of propagation). Notice also that owing to the way in which the K, R conversion was derived, predictions about cracking under conditions of combined loading employing K-superposition (as in Problem 2.32) are properly valid only for non-turning cracks.

The formulae for cracking loads given in Appendix 1 which are written in terms of R may be recast readily in terms of K_C by writing $K_C{}^2 = E^*R$. For example, the cracking formula $M^2 = REIB$ for a DCB specimen derived in Example 2.3 using engineer's bending theory may be written $K_{IC} = M/\sqrt{IB}$. In the case of double-ended encastre beam models, as would apply to the testpiece geometry in Fig. 2.8, it is easily shown that $X^2a^2 = 64\ REIB$, using $(u/X) = 2a^3/192EI$ in equation (2.7), so $K_{IC} = Xa/8\sqrt{IB}$.

Example 2.13 Calculate K_{IC} for the glued beam testpiece, the Xu diagram for which is shown in Fig. 2.8, using $K_{IC} = Xa/8\sqrt{IB}$. Derive the corresponding R, using $K_{IC}{}^2 = E^*R$, and comment on this value compared with the value determined directly by Gurney's segmental area method, viz. ~ 560 J/m^2.

Ans. When $a = 64$ mm, $X = 714$ N and with $B = 25$ mm and $I = 25 \times 10^{-3} \times (3.2 \times 10^{-3})^3/12 = 68.3 \times 10^{-12}\,\text{m}^4$ we obtain

$$K_{IC} = 714 \times 64 \times 10^{-3}/8\sqrt{68.3 \times 10^{-12} \times 25 \times 10^{-3}} = 4.37\ \text{MPa}\sqrt{\text{m}}.$$

As the adherands were made of L73 aluminium alloy for which $E = 72$ GPa, $R = K_{IC}^2/E = 265$ J/m^2. This should be contrasted with the measured value of 560 J/m^2. The discrepancy arises owing to two causes, (i) the assumption of simple beam theory for relatively short crack lengths i.e. an inexact K formula and (ii) the lack of consideration of different elastic properties in the adhesive and adherand. For the first effect, at the longest of crack (245 mm), $X = 215$ N from which $K_{IC} = 5.04$ MPa$\sqrt{m}$ and $R = 352$ J/m^2. Although closer to the experimental value, an underestimate still exists between the converted R and the directly measured R. Note that whereas the final Gurney segmental work area is lower than the value obtained for the first area in Fig. 2.8, these converted values show an increasing trend. For the second effect, if the glued testpiece is analysed in terms of 'beam-on-elastic foundation' models of crack propagation (section 8.7.2), we find that $M^2 = \alpha\, REIB$ or $K_{IC} = Xa/8\sqrt{\alpha IB}$ where $\alpha < 1$, so the actual K_{IC} are greater than those calculated using simple beam theory with $\alpha = 1$; similarly $R = K_{IC}{}^2/E^* = X^2a^2/64\, \alpha IBE^*$ are greater than before. For $a = 64$ mm, $\alpha \to 0.72$ in order for the converted R to agree with the directly measured R; for $a = 245$ mm, $\alpha \to 0.9$. Both of these values are sensible for the geometry involved (see Mai, 1976).

Note that in Example 2.13, we use E^* for the aluminium alloy adherands in the conversion from K_{IC} to R via $K_{IC}{}^2 = E^*R$, *not* the E for the adhesive since it is presumed that it is the adherand (the beams) that possesses all the strain energy which feeds the crack. In truth the glue line also stores energy which contributes to the discrepancy between converted and experimental R, hence the beam-on-elastic foundation analysis. Example 2.13 has been chosen to illustrate the complications in K-formulae which can arise in glued joints and in layered structures. For cracking in monolithic bodies (which is by far the more common occurrence) we use of course the E of the solid in question in the conversion between K_{IC} and R, and these difficulties do not arise. The agreement between measured and converted values will be as good as the K-formula (i.e. simplified beam analyses will only give good agreement for cases where they are appropriate, but exact K-formulae give agreement with R determined by work area methods – see, for example, Chow and Lau (1977)). All such comparisons, of course, must be under comparable conditions of crack velocity, temperature and environment as explained in Chapter 6.

2.8 CRACK TIP ZONES OF IRREVERSIBILITY

According to equation (2.30) the stress right at the crack tip is infinite owing to the continuum assumption of an infinitely sharp-ended crack with zero crack-tip radius. If this were really true, bodies containing cracks would not be able to sustain any loads and would have no strength at all. But there is a lower limit on the effective crack tip radius given by the inter-atomic spacing. Thus even solids which cleave in an ideal fashion require finite loads to cause crack propagation. In other solids, the high crack tip stresses are relieved by highly localised plastic flow, craze formation and the like, with associated crack tip stretching which lowers the crack tip radius further before propagation ensues (section 2.9).

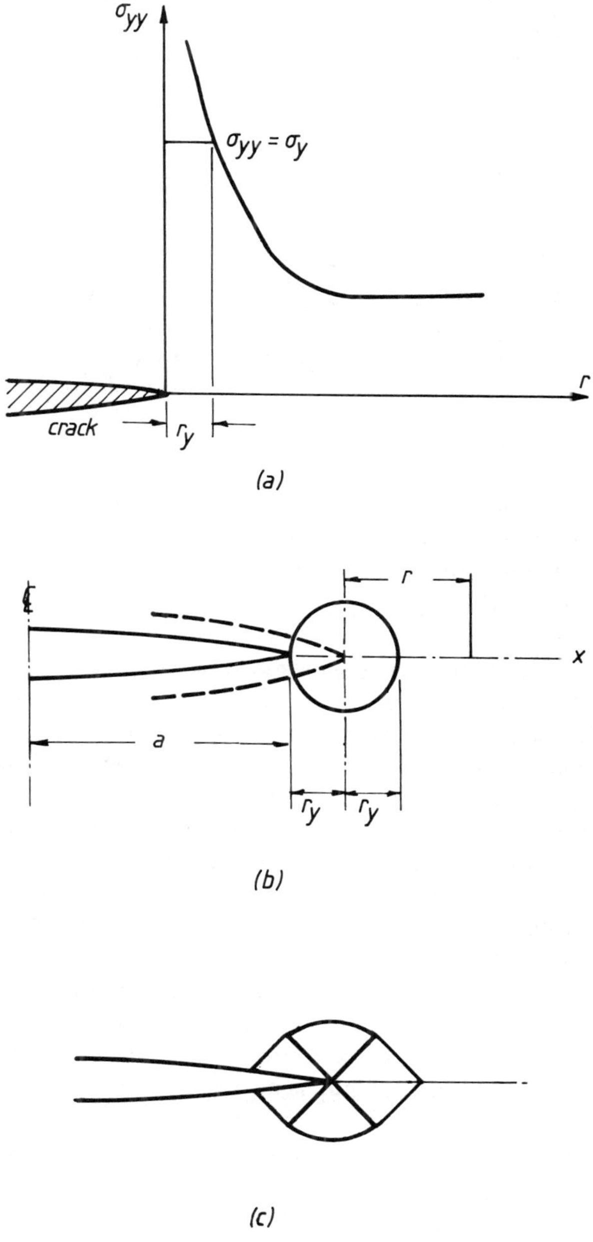

Fig. 2.29 – (a) Load shedding in a region close to the crack tip. (b) The embedded crack tip model. (c) Prandtl-Hill slip line field for constrained plastic flow at the crack tip.

The size of these zones of local irreversibility may be estimated as follows. The simplest approach is to say that they extend out to the point r_y where σ_{yy} equals the yield stress σ_y in metals (or the stress (and strain) combination necessary for yielding and cavitation crazing in polymers). Then $\sigma_{yy} = K/\sqrt{2\pi r_y} = \sigma_y$

or $$r_y = K^2/2\pi\sigma_y{}^2 \qquad (2.35)$$

Clearly r_y increases as the load applied to the cracked body is increased via the particular appropriate K-formula.

However, as illustrated in Fig. 2.29(a) this approach has shed the load sustained (supposedly elastically) by that part of the σ_{yy} curve cut off by $\sigma_{yy} = \sigma_y$, and there really is an imbalance in the load-carrying capacity of the body. This may be accounted for in a plate structure of constant thickness by noting that the total area under the σ_{yy}, r stress distribution curve out to r_y is $\int_0^{r_y} (K/\sqrt{2\pi r})\,\mathrm{d}r = K\sqrt{2r_y/\pi} = 2\sigma_y r_y$. But the truncated distribution with $\sigma_{yy} = \sigma_y$ = constant out to r_y has area $\sigma_y\, r_y$. Therefore it is customary to argue that the plastic zone must extend out not just to r_y but to $2r_y$ (called d_y). Hence the size of the zone is given by

$$d_y = K^2/\pi\sigma_y{}^2. \qquad (2.36)$$

Zones are often shown schematically as being circular with diameter d_y ahead of the crack tip (or 'radius' d_y centred on the crack tip since the zone actually spreads back to the flank of the crack rather like a Prandtl–Hill slip line field for constrained plastic flow at the tip of the crack (Fig. 2.29(b) and (c)). Zones are not really circular as, in metals, slip occurs on preferred slip bands. Example 2.18 later discusses inclined slip band models, and extends the analysis to predict the pair of intensely strained 'wings' which are a feature of many crack tip zones. An alternative treatment (the Dugdale 'strip yield' model) has a single band of plastically-flowed material in the path of the crack, see section 4.1.2.

The sizes of the zones at crack initiation (and during propagation) in monotonic fracture are given by equation (2.36) using the value of the critical stress intensity factor. As explained in Chapter 6 this will vary in general with rate, temperature and environment, but even when these parameters are fixed, it is found that resistance to cracking changes with thickness of the body. It was hinted in section 2.6 that variations in toughness with thickness came about owing to the difference between the low constraint to flow at the outsides of a body (which deform under approximately plane stress conditions, owing to the stress-free surfaces) and the high constraint to flow in the interior of a body (where plane strain conditions are more likely to prevail). Fig. 2.30(a) shows the thickness variation of fracture toughness R (and hence critical K by implication) for 7075-T6 aluminium alloy reported by Krafft, Sullivan and Boyle in 1961. For similar microstructures, there is a factor of 10 difference between the toughness for 2-mm thick sheet and, say, 15-mm thick material.

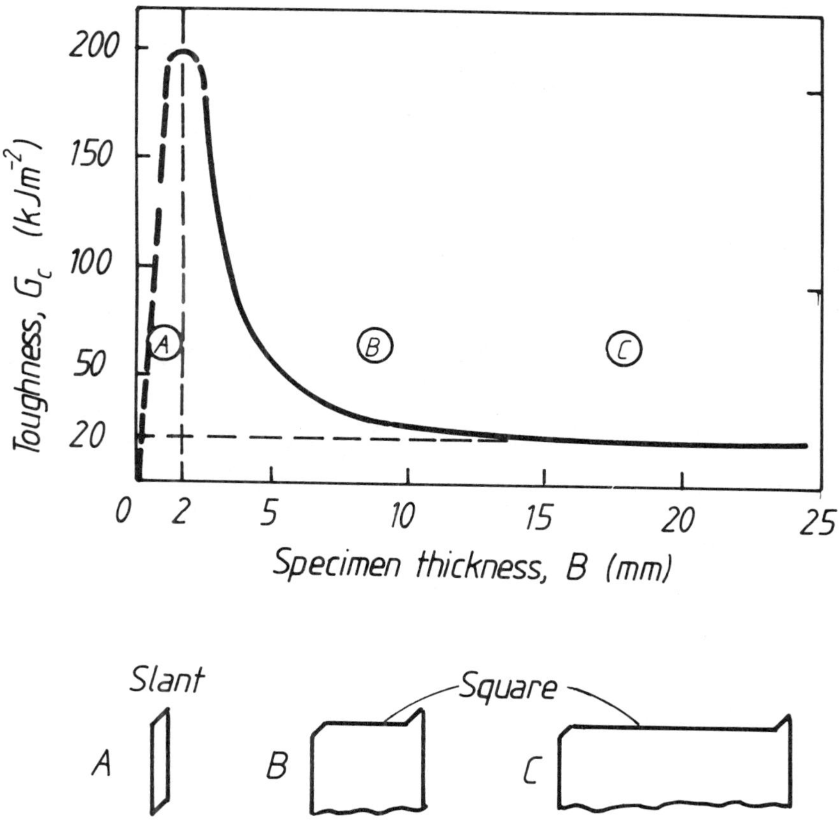

Fig. 2.30(a) – Effects of specimen thickness on fracture toughness and types of fracture surface in aluminium. (Reproduced for Knott (1973), with permission.)

As $K^2 = E^*R$, there is a corresponding difference of $\sqrt{10/(1-\nu^2)} \approx 3$ in the critical Ks. In this case, using the nomenclature of the footnote in section 2.6, we write $K_C \approx \sqrt{71 \times 10^9 \times 200 \times 10^3} = 120$ MPa$\sqrt{m}$ and $K_{IC} \approx \sqrt{71 \times 10^9 \times 20 \times 10^3/(1-0.3^2)} \approx 38$ MPa$\sqrt{m}$.

Fig. 2.30(a) shows that the change of restraint with thickness affects the type of fracture surface obtained in the aluminium alloy. In very thick sections, the fracture is nearly completely flat having small 'shear lips' at the edges. In thinner sections, the shear lips (which often may be considered to be fixed in size) take a larger proportion of the cross-section and the peak in the curve of toughness corresponds more-or-less with where the shear lips meet to give a completely slant fracture (cf. Example 4.5 but see section 3.9.3). For even smaller thicknesses, the resistance to fracture decreases again owing to a change in mode of fracture: in place of mode I, mode III slip-off occurs with the original set of dislocations going (literally) through the thickness of the sheet from one side to the other, producing fracture propagation by gliding forward (Fig. 2.30(b)).

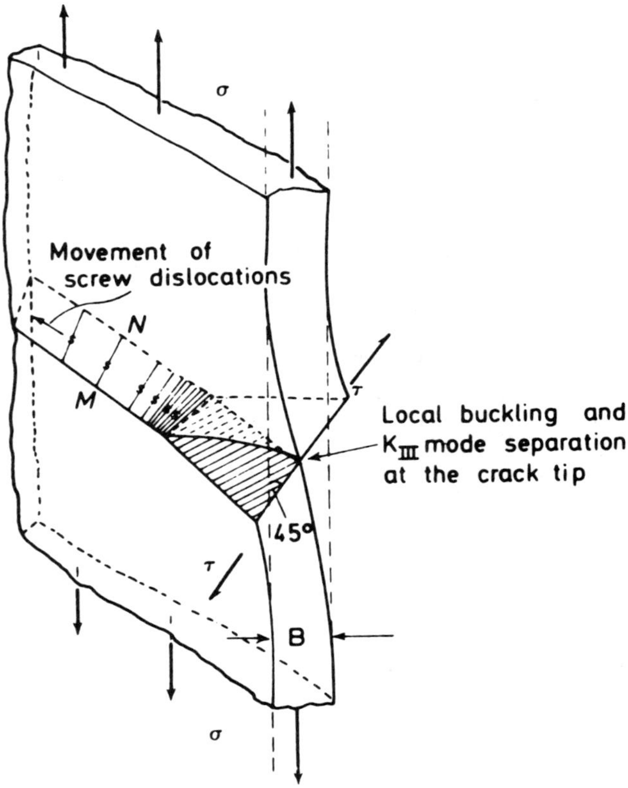

Fig. 2.30(b) – Mode III separation in thin sheet under uniaxial tension. (Reproduced from Knott (1973), with permission.)

Such out-of-plane tearing fractures are familiar in aluminium foil and chocolate wrappers; cf. Fig. 1.7.

In the range of thicknesses between plane stress and plane strain, the fracture surface after some propagation does not necessarily maintain the same starting proportion of shear lip and flat fracture as implied by the middle sketch below Fig. 2.30(a). As shown in Fig. 2.30(c), the flat portion can get consumed by the shear lips and a completely slant fracture surface result. This is likely to occur when the initial flat fracture surface and shear lips have comparable dimensions. The behaviour is complicated as it involves early propagation of the low toughness central regions (which 'tunnel' with a flat fracture surface into the specimen) while the surface regions merely plastically stretch; later they are dragged along behind as they too crack. Such propagation occurs under *rising* load, whereas if crack resistance were uniform along the crack front there would be one unique starting fracture load for the material and testpiece geometry. As the load increases, the size of the plastic zone increases (in equation (2.36), the applied K increases) and this helps to relax plastically the through-thicknesses stresses in the interior regions. Consequently the whole behaviour tends towards

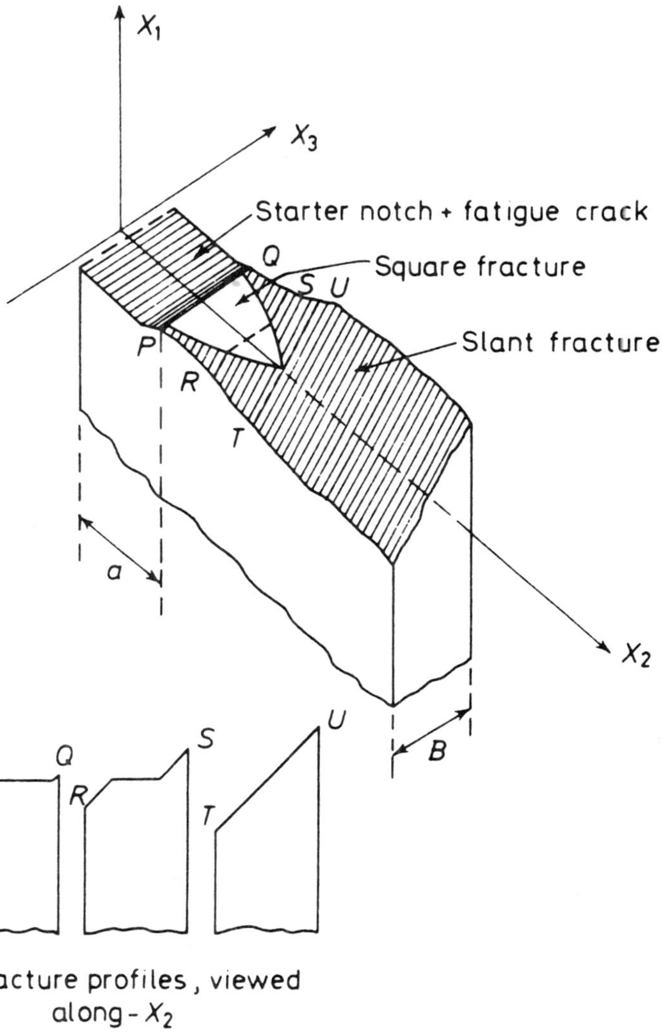

Fig. 2.30(c) – Details of a fracture surface displaying varying fracture characteristics along the propagation direction. (Reproduced from Knot (1973), with permission.)

a plane stress response and an originally-flat starter crack may twist into a slant fracture. Of course, if the stress relaxation is inadequate, flat fracture surfaces will be retained in the central regions. Cracking under rising load in this way is one cause of R changing with propagation as discussed in section 2.3 on R-curves. Tunnelling and so-called sub-critical crack growth, and the phenomenon of 'pop-in' which sometimes occurs with the initial propagation of central flat regions in this type of problem, are discussed in sections 2.4.3 and 3.6.

Variations of toughness with thickness are found in most materials, although the pattern of shear lip influence in Fig. 2.30(a) is not always shown by non-metals. For example, Fig. 2.30(d) shows data for high impact polystyrene

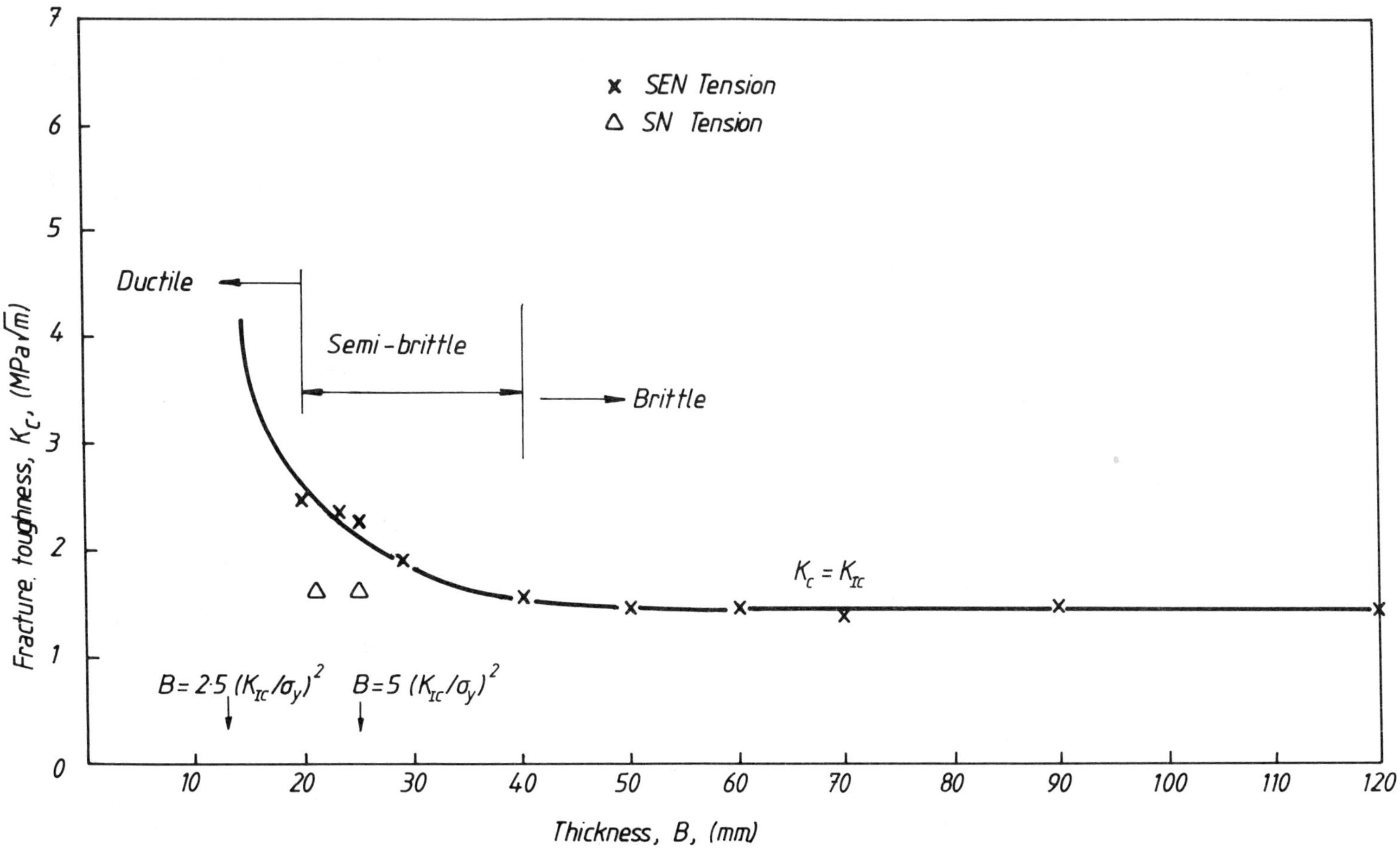

Fig. 2.30(d) – Fracture toughness as a function of specimen thickness for high impact polystyrene (HIPS). (After Yap *et al.* (1983), with permission.)

(HIPS) at room temperature which, like PMMA, displays no shear lips at all (Yap *et al.*, 1983).

It is clear that the plastic zone size given by equation (2.36) will vary with thickness depending on the variation of fracture resistance with thickness (and the value of σ_y). Some typical sizes for d_y at fracture are: PMMA 0.1 mm ($K_{IC}/\sigma_y \sim 0.02 \sqrt{m}$) PC, 0.8 mm ($K_{IC}/\sigma_y \sim 0.05$); high strength aluminium alloys, 1.5 mm ($K_{IC}/\sigma_y \sim 0.07$); high strength steels and titanium alloys, 4.5 mm ($K_{IC}/\sigma_y \sim 0.12$ for both); brass and maraging steels 12 mm ($K_{IC}/\sigma_y \sim 0.2$); low carbon steels 115 mm ($K_{IC}/\sigma_y \sim 0.6$).

In fatigue, as explained in Chapter 7, the crack tip zones are smaller owing to the applied stress intensity being smaller than in monotonic loading. Also in fatigue, σ_y is the *cyclic* yield stess.

Example 2.14 Irwin argued that the stress relaxation caused by the local crack tip yielding is equivalent to a small additional crack extension of magnitude r_y $(= d_y/2)$, see Fig. 2.29(b). The stress intensity distribution given by the K solution is then supposed to be located at an 'embedded crack tip' where the crack of real length a has notional length $(a + d_y/2)$. Determine what difference this makes to derived K_C values.

Ans. Instead of $K_C = \sigma Y \sqrt{\pi a}$, we have $K_C = \sigma Y \sqrt{\pi(a + d_y/2)}$. Assuming for simplicity a small crack in a large sheet so that $Y = 1$, the ratio of the two K_Cs determined from the same test (same applied σ or X) is $\sqrt{(a - d_y/2)/a} = \sqrt{(a/W + d_y/2W)/(a/W)}$. So in centrally cracked plates with $a = 10$ mm and $W = 1000$ mm, say, $(a/W) = 0.01$ and the ratios for a variety of materials are 1.00 (PMMA): 1.02 (PC); 1.04 (high strength Al alloys); 1.11 (high strength steels and titanium alloys); 1.26 (brass and maraging steels); 2.6 (low carbon steels) – using the values for d_y quoted above. [Values for r_y in a workhardening solid are given in section 4.5.4 on elastoplastic fracture.]

The embedded crack tip length corrections for low carbon steel (57 mm), for brass and maraging steels (6 mm) have little meaning in Example 2.14 as the size of the plastic zone correction is comparable with the crack length of 20 mm. Plasticity at the crack tip extends backwards so that overall displacement reversibility is doubtful and therefore the application of linear elastic fracture mechanics is suspect. In thick sheets undergoing plane strain deformations the crack tip zones are smaller owing to constraint on the flow stress which elevates the flow stress. It is found experimentally that the constrained yield stress at crack tips in thick plates is some $(1.5 \sim 2)\sigma_y$. Irwin used a factor of $\sqrt{2\sqrt{2}} \approx 1.68$. Thus at the *same* value of applied K, (before propagation)

$$(d_y)_{\text{plane strain}} = \frac{1}{\pi} \frac{1}{2.83} (K/\sigma_y)^2 \approx \frac{1}{3} (d_y)_{\text{plane stress}} \qquad (2.37)$$

On the other hand, at the critical propagation K values (which are different in plane stress and plane strain)

$$(d_y)_{crit} \text{ plane strain} = (1/3)\,(1/\pi)\,(K_{IC}/\sigma_y)^2$$

and

$$(d_y)_{crit} \text{ plane stress} = (1/\pi)\,(K_C/\sigma_y)^2$$

from equations (2.36). Thus

$$\frac{(d_y)_{crit} \text{ plane strain}}{(d_y)_{crit} \text{ plane stress}} = \frac{1}{3}\left(\frac{K_{IC}}{K_C}\right)^2$$

Thus, *at fracture,* the plane strain crack tip zone size is *less* than one-third the plane stress value, since $(K_{IC}/K_C) < 1$.

Reduced plastic zone sizes permit the crack more readily to be encapsulated in an essentially elastic stress and strain K field, but clearly there must be restrictions on the size of testpieces to which K-solutions can be applied, both for the determination of critical stress intensity factors and in the use of K-formulae for design against fracture. This is a most important point, which is related to the 'short crack' problem discussed in Chapter 7 in relation to fatigue, which concerns what happens when crack lengths are comparable in size to microstructural features.

Example 2.15 The ASTM E-399 Standard for K_{IC} testing recommends a minimum thickness for metal plate testpieces to achieve plane strain (essentially flat) fracture given by $B > 2.5\,(K_{IC}/\sigma_y)^2$. Discuss this empirical condition in relation to the relative sizes of crack tip plastic zone and crack length.

Ans. If fracture events are to be characterised successfully in tems of critical K values, the conditions at fracture must agree with the conditions under which K-formulae are valid. Fig. 2.24 shows that the first term of the series approximation (which we are employing when we use K-formulae) agrees with the exact solution for $r/a \rightarrow 0.01$ and possibly accurately enough for $r/a \rightarrow 0.1$. It is clear then that the plastic zone size must be within this range of applicability, i.e. $r_y < 0.01a \sim 0.1a$, say $< a/50$. Thus, in plane strain, $(1/2 \times \pi \times 2.83)(K_{IC}/\sigma_y)^2 < a/50$ or $a \geqslant 2.8\,(K_{IC}/\sigma_y)^2$. In for example, the compact tension specimen $a = B$, so the least thickness for plane strain corresponds with the least r_y/a, i.e. $B \approx 47\,r_y$. The recommended factor of 2.5 is empirical and corresponds with choice of values for r_y/a and constraint slightly different from 1/50 and $\sqrt{2\sqrt{2}}$, but the detail is not important. In Fig. 2.30(a) the K_{IC} low plateau is supposed to set in for $B > 2.5\,(K_{IC}/\sigma_y)^2$. It is true for many metals, but there are exceptions even with metals (see Fig. 3.45), and certainly with tough polymers. Fig. 2.30(d) shows that the least fracture toughness for high impact polystyrene (HIPS) does not occur until $B > 5\,(K_{IC}/\sigma_y)^2$. In the design of toughness testpieces, other factors which concern the avoidance of remote yielding have also to be considered. As explained in section 3.9, such limiting minimum sizes can also be expressed in terms of $(K_C/\sigma_y)^2$.

Example 2.16 Fig. 2.31 shows the Xu plot obtained from a standard ASTM 3-point bend K_{IC} testpiece made of 0.35 C/1.25 Cr/0.5 Mo rotor steel of 0.2% proof stress 950 MPa. The beam span was 191 mm, the depth 50 mm, thickness 25 mm and fatigued starter crack length 25 mm. Determine whether K_Q (the value for the critical stress intensity factor derived from the experiment) is a 'valid' measure of K_{IC}.

Ans. Superimposed on Fig. 2.31 is a stiffness line which is 5 per cent less steep than the elastic stiffness line of the testpiece and which intersects the Xu trace at Q. The purpose of this so-called 'secant line', and checks made from it, is twofold, viz. to aid the detection of crack initiation and to test whether the size of the plastic zone size is indeed small and less than about $a/50$ as in Example 2.15. If cracking commenced uniformly across the crack front and R were

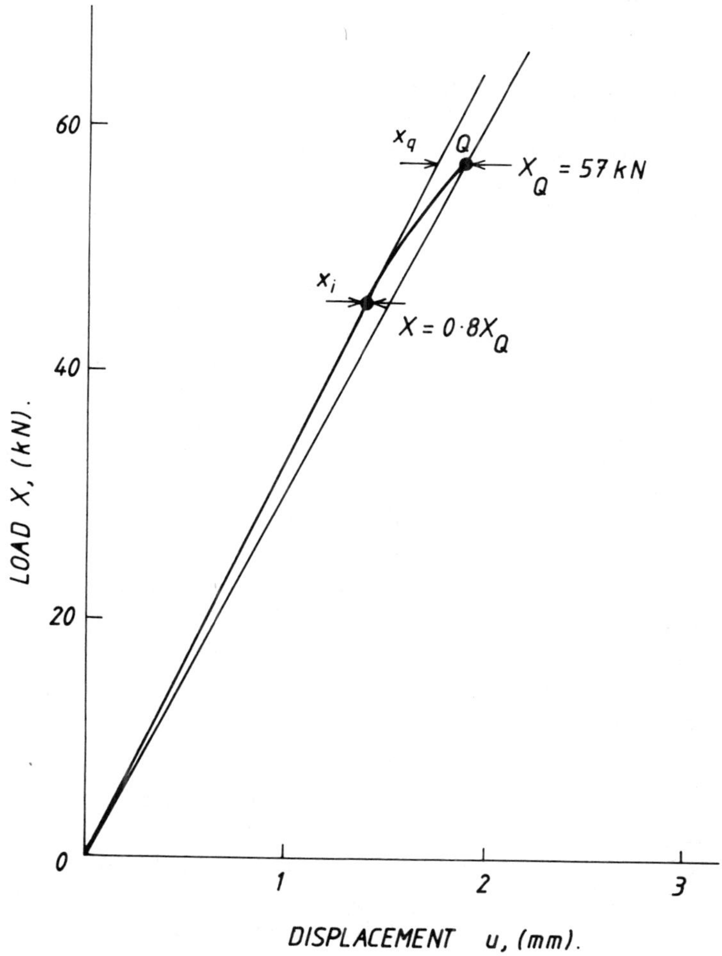

Fig. 2.31 – Load-deflexion record for the fracture of a three-point bend rotor steel.

constant, the load would drop straight away with propagation in 3-point bend testpieces. Owing to the elevated hydrostatic stress under plane strain conditions at the centre of the testpiece, cracking commences there before it commences at the edges, and the crack front takes up a bowed or thumbnail shape. The crack is said to have 'tunnelled' and propagation occurs under increasing load while this happens. The average length of the crack has increased, so the stiffness decreases. It may be shown (see Knott (1973), p. 140 *et seq.*) and Problem 2.35 here, that an average crack extension equal in magnitude to the radius of the plane strain plastic zone size is equivalent to a reduction in stiffness of 5 per cent. Hence the superimposed stiffness line. A check that must be carried out is a comparison of the deviation from linearity at $X = 0.8\,X_Q$ (x_i) with that at $X = X_Q$ (x_q). If $x_i < x_q/4$ it is argued that the increase in displacement may be associated with cracking; if $x_i > x_q/4$ it is argued that remote plastic flow is taking place and that the test is therefore invalid for estimating elastic fracture parameters. Inspection of Fig. 2.31 shows that $x_i < x_q/4$ in this case. We have for 3-point bend testpieces $K_Q = (X_Q S/B\ W^{3/2})Y\ (a/W)$ where $Y(a/W) = Y\,(0.5) = 2.66$. Since $X_Q = 57$ kN, we have $K_Q = [57 \times 10^3 \times 191 \times 10^{-3}/25 \times 10^{-3}\ (50 \times 10^{-3})^{3/2}]\ 2.66 = 92$ MPa$\sqrt{\text{m}}$. We note that $2.5\,(K_Q/\sigma_y)^2 = 2.5\,(92 \times 10^6/950 \times 10^6)^2 = 23$ mm and $B = 25$ mm which is just larger. We also note that $X_{max}/X_Q \approx 62/57 = 1.08$ which is below the limit of 1.1 set in testing codes of practice. It is concluded that K_Q is a valid estimate for K_{IC}.

The 3-point bend testpiece described in Example 2.16 and the compact tension testpiece of Example 2.11 are popular standard testpieces for the determination of K_{IC}. The CT specimen, in particular, is efficient in the (comparatively) limited amount of metal required for 'valid' tests. Standard K_{IC} testing is more difficult and more expensive than traditional stress–strain mechanical property acquisition. Barker in 1977 introduced a 'short rod' method which is said to be simpler and less expensive than the standard procedures in ASTM E399-81 for metallic materials. The testpiece is shown in Fig. 2.32(a) where the growing crack area is seen to be triangular in plan view, starting from a sharp point (crack fronts with such chevron or swallow-tailed shapes are sometimes used in toughness testpieces to ensure a gradual start to propagation). Barker found that crack propagation occurred under increasing load and eventually passed through a maximum, and he correlated the peak load X_{max} with K_{IC} by the following equation

$$K_{IC} = C\,.\,X_{max}/B^{3/2} \qquad (2.38)$$

where C (≈ 22) is a calibration constant for a specimen with the proportions given in Fig. 2.32(a). The value of Barker's method is that experiments show that the diameter B of the short-rod testpiece can be down to about one-third of the ASTM minimum size for plane strain cracking, i.e. $B \rightarrow (1/3)\,2.5\,(K_{IC}/\sigma_y)^2$, and still give K_{IC} values which tally with 'valid' CT measurements in thicker plates (Table 2.2). Furthermore by making a series of rod specimens from different locations in thicker plates, variations in K_{IC} through the plate thickness may be determined.

Fig. 2.32(a) – Short rod specimen configuration and dimensions (After Barker (1977, 1983), with permission.)

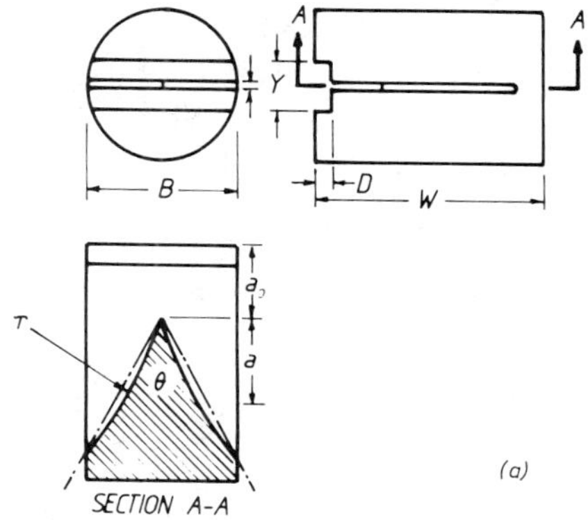

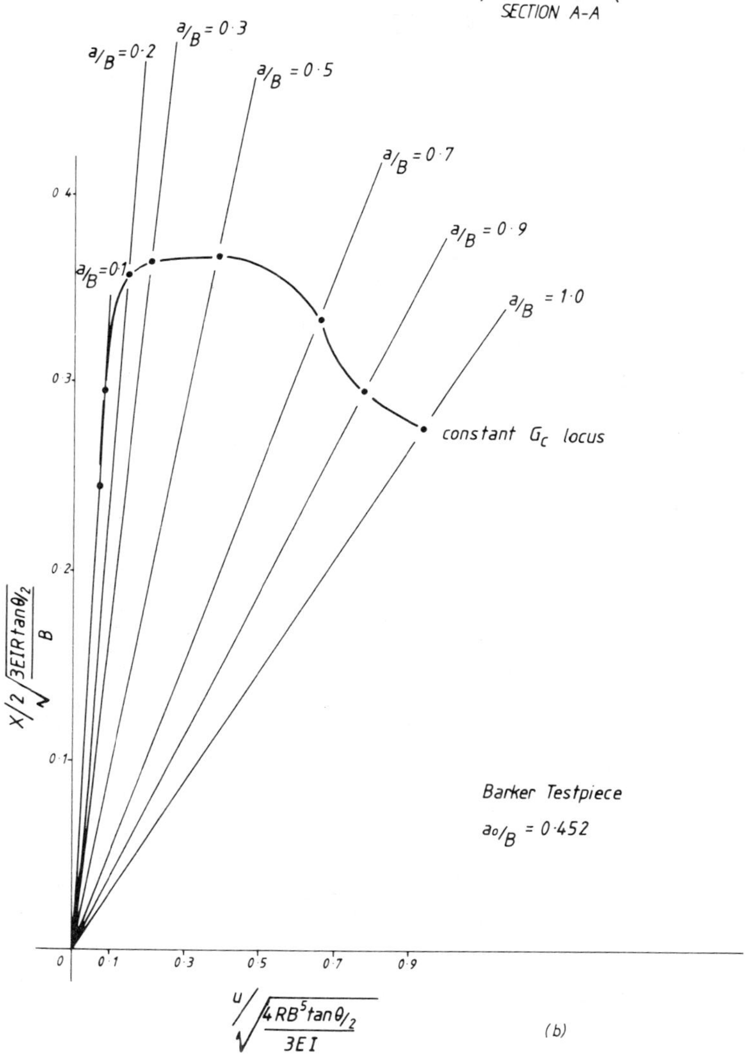

Fig. 2.32(b) – Theoretical normalised load-deflexion record for the fracture of a short rod specimen.

Table 2.2

Material type	Heat treatment	Original shape or plate thickness	Yield at 0.2% Offset (MPa)	Crack Orientation	$B_{min} = 2.5(K_{IC}/\sigma_{ys})^2$ (mm)	CT specimen thickness (mm)	Short rod diameter (mm)	K_{IC} per ASTM E 399 ($MPa.m^{1/2}$)	K_{ICSR} Short rod ($MPa.m^{1/2}$)	% Difference, $100\,K_{IC}\,(K_{ICSR} - K_{IC})$
HF-1 isothermally transformed	austenitised at 1700°F; quenched to 1150°F; held at 1150°F for 1½ h; then air-cooled	25 mm	503	T-L	8.9	22.2	22.2	30.1	28.8	$-4.4^{b,c}$
							22.2		27.4	. . .
ASTM A 470 CrMoV rotor steel	standard	1.27 m diameter, 6.73 m long	626	C-R	15.8	25.4	25.4	49.9[d]	47.8	$-4.2^{b,c}$
							25.4		51.1	. . .
ASI 4340	quenched from 1475°F, tempered at 950°F 1 h	152 by 127 by 610 mm	1070 (estimated)	L-T	13.0	32.0	25.4	77.0	81.4	5.7
D6AC	austenitised, salt quench, double temper	hollow rolled ring, 1.83-m diameter, 76-mm-thick wall	1407	L-C	17.5	31.8	25.4	118.2	111.1	−6.0
D6AC	austenitised, salt quench, double temper	hollow rolled ring, 1.83-m diameter, 76-mm-thick wall	1385	L-C	19.3	31.8	25.4	122.2	114.3	−6.5
6–4	standard anneal	irregular forging	868	T-L	23.9	31.8	25.4	85.1	69.5	−18.3
6–4	standard anneal	irregular forging	885	T-L	15.2	31.8	25.4	69.1	66.8	−3.3
2124	T851	76 mm	393	T-L	11.9	31.8	25.4	27.2	24.9	−8.5
2124	T851	140 mm	406	L-T	20.1	38.1	25.4	36.4	34.6	−4.8
7050	T73651	83 mm	456	T-L	13.5	50.8	25.4	33.5	31.5	−6.2
							25.4		30.2	-9.8^{b}
7475	T7351	44 mm	450	L-T	37.6	44.5	25.4	55.2	62.6	. . .[f]
							25.4		64.0	. . .[b,f]

(after Barker, 1977)

[b] Duplicate short rod specimens were prepared from the supplied CT specimen.

[c] Estimated 10% uncertainty in K_{ICSR} because of out-of-tolerance short rod specimen and large pop-in; thus, the K_{IC} versus K_{ICSR} comparison is less significant than for the other HF-1 short rod specimen.

[d] Average of 3 tests.

[e] Estimated 10% uncertainty in K_{ICSR} because of large crack jumps; thus, the K_{IC} versus K_{ICSR} comparison is less significant than for the other ASTM A470 short rod specimen.

[f] Comparison of K_{ICSR} with K_{IC} is invalid because toughness varied with position in the material.

Example 2.17 Derive, from first principles, the form of equation (2.38) assuming elastic fracture mechanics.

Ans. Since the testpiece in Fig. 2.32(a) is short and stubby, an expression is required for the compliance which takes into account shear deflexions, the warping of cross-sections and constraints to warping at built-in ends of the upper and lower cantilever beam arms of the testpiece. Following the method in section 39 of Timoshenko (1955), and in particular the derivation of equation (j) on p. 175 relating to rectangular beams, we find that for beams of semicircular cross-section

$$u/X = \frac{l^3}{3EI}\left[1 + \frac{0.2B^2}{l^2}\right]$$

for each arm where l is the length of the beam, i.e. $l = (a_0 + a)$. The crack area increment is $dA = 2a \tan(\theta/2)\, dl$ and $dl = da$, so $d/dA\ (u/X)$ for both arms together is $(2l^2/EI + 0.4B^2/3EI)/2a \tan(\theta/2)$. Thus in equation (2.7), $X^2 = 12\, EIRa \tan(\theta/2)/(6l^2 + 0.4B^2)$. Substituting $(a + a_0)$ for l, $B^4(9\pi^2 - 64)/1152\pi$ for I, 0.566 for tan 29½° and normalising, gives

$$X^2 = 0.048\,(ER)\,B^3\,(a/B)/[1.62 + 6(a/B)^2 + 5.4(a/B)]$$

As a increases, X increases (since equal increments, da, produce progressively bigger increments in crack area, dA), but the elastic cracking load passes through a maximum at $a/B = 0.52$ (proved by differentiation) and thereafter falls, Fig. 2.32(b). Notice the peculiar shape taken by the constant G_C locus for reversible cracking. Writing $K_{IC}^2 = ER$ and substituting $a/B = 0.52$ in the expression for cracking load gives, at X_{max}, $K_{IC} \approx 16\, X_{max}/B^{3/2}$. This corresponds reasonably well with Barker's equation (2.38) given that the u/X relation (even including shear effects) will not be quite correct owing to the tapered grooves ahead of the crack front. Note that the calibration constant depends only on the geometry of the testpiece and not on the material being tested.

Further comments on short-rod testing in the presence of remote plastic flow will be found in Chapter 4. As regards other test geometries, C-shaped sections of cylinders and other specimens cut from cylinders are discussed by, for example, Jones (1974), Kapp *et al.* (1977) and 'round' compact tension specimens by Newman (1981). The US National Materials Advisory Board report entitled 'Rapid inexpensive tests for determining fracture toughness' (1976) is of general interest in this context, as is Wilhem (1983) on standardised coding for orientations.

2.9 CRACK SHARPNESS AND COD

Conditions surrounding the initiation and propagation of cracks depend upon the bluntness of the crack tip. Blunt cracks usually require greater fracture loads than sharp cracks, and as described in Chapter 8, often produce unstable fractures because there is an excess of strain energy locked up in the blunt system at cracking. Thus, as may be expected, conversion of cracking loads (from specimens with varying degrees of bluntness into apparent K_C values, gives K_C increasing as ρ increases, where ρ is the crack tip radius, as shown in Fig. 2.33 for H-11 steel testpieces. Below some minimum ρ, however, K_C is approximately constant, and this suggests that cracks have a 'natural' radius ρ_C at which they initiate and propagate. The value of ρ_C depends upon the material and the particular micromechanism of fracture taking place. In ideal cleavage, ρ_C is of the order of the inter-atomic spacing.

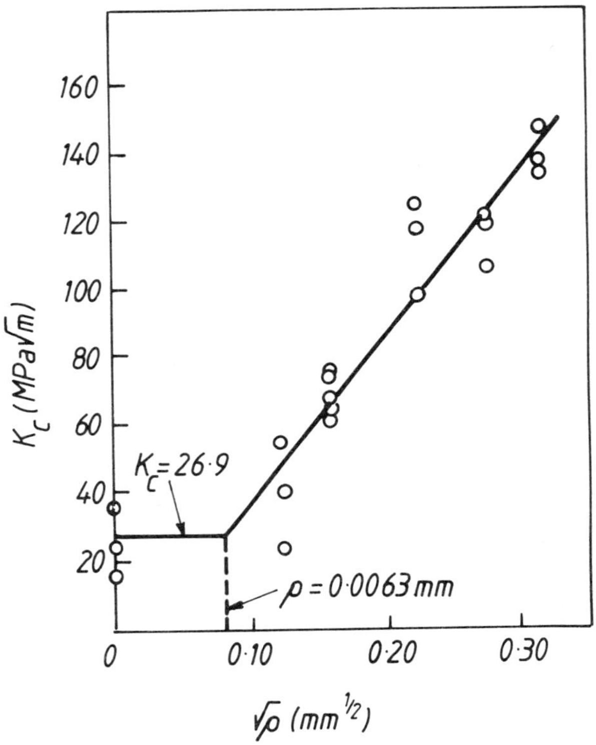

Fig. 2.33 – Variation of K_C with root radius of crack tip. (After Irwin (1964), with permission.)

In commercial metals, one type of crack initiation is associated with the splitting and debonding of inclusions ahead of the crack tip (Chapter 5), so that inclusions nearest the tip of the static starter crack or running crack have to be brought up to the critical stress or strain condition (really their 'local' K_C) at which microcracks are produced before the main crack can propagate onwards. Owing to the local crack tip process zone, local blunting occurs and the crack tip radius can never be zero as implied by an elastic continuum analysis. Strains in the blunted zone may be estimated from slip line field analysis, and Rice (1973) has shown that the most intense strains are situated a couple of crack tip radii ahead of the rounded crack front. The size of the plastic zone is proportional to ρ. Thus cracks sharper than ρ_C have high strains, but within a very small zone which may not encompass an inclusion; cracks blunter than ρ_C have lower strains distributed over a larger area, so that although many inclusions may be strained, none will be sufficiently strained to cause debonding or fracture. Consequently it seems that cracks sharper than ρ_C have to become blunted with local crack tip flow as the applied load is increased, until the crack tip zone (proportional to the blunting) has grown to encompass the requisite microstructural features that determine crack initiation. Cracks already blunter than ρ_C require higher loads to initiate fracture than those for which $\rho < \rho_C$, because the magnitudes of the strains within the crack tip zone (depending upon $1/\rho$) are more diffuse than for a sharper crack. As explained in Chapter 5, the local hydrostatic stress at the crack tip (inversely proportional to ρ) also plays a role.

The maximum tensile strain tangential to a notch contour within the plastic zone is given by Rice for an elastic/perfectly plastic material as

$$\epsilon_{max} \approx 3\epsilon_y \; [1 + K_C^2/\sigma_y^2 \rho]/4 \tag{2.39a}$$

where ϵ_y is the yield strain and σ_y the yield stress. Thus

$$K_C \approx \sigma_y \; [4\epsilon_{max}/3\epsilon_y - 1]^{1/2} \, \rho^{1/2} \tag{2.39b}$$

The experimental results in Fig. 2.33 show that K_C does vary directly with $\sqrt{\rho}$, and the slope suggests that for $\sigma_y \approx 2$ GPa, $\epsilon_{max}/\epsilon_y \approx 60$ which for $\epsilon_y \approx 0.001 \sim 0.005$ or so in metals gives $\epsilon_{max} \approx 0.06 \sim 0.3$. The larger extreme is comparable with the tensile test strain to fracture for these same materials, and has led to models of crack growth in which crack tip strain is taken to be the same as the strain to fracture in a tensile test or some fraction of it.

Once initiated, a crack tends to propagate with its own characteristic 'natural bluntness' ρ_C and sharp cracks emanating from blunt notches have crack tip radii of ρ_C. This leads to the Wells–Cottrell concept of a *critical crack opening displacement* at the crack tip during fracture (COD, $\delta_C \approx 2\,\rho_C$) which is yet another alternative to R and K_C for characteristing a material's resistance to fracture (see British Standards Institution, 1979). The use of δ_C in design is discussed in section 4.1.2.

Following arguments similar to those in section 2.7, where now crack tip forces do work by moving through δ_C before fracture can take place and this work is the source of R, it is easily shown that

$$R \approx \int_0^{\delta_C} \sigma_y(\delta)\,.\,d\delta = \sigma_y\,\delta_C \tag{2.40a}$$

Hence,

$$\delta_C \approx R/\sigma_y = \;[K_C^2 \text{ or } K_{IC}^2]/E^*\sigma_y \tag{2.40b}$$

In practice, it is found that

$$R = m\,\sigma_y\,\delta_C \tag{2.40c}$$

where $1 < m < 3$. Then

$$\delta_C = [K_C^2 \text{ or } K_{IC}^2]/mE^* \sigma_y \tag{2.40d}$$

The extent of the crack tip plastic zone is, form equations (2.36) and (2.37)

$$d_y \approx \begin{cases} (1/\pi)(K_C/\sigma_y)^2 & \text{plane stress} \\ (1/3\pi)(K_{IC}/\sigma_y)^2 & \text{plane strain} \end{cases} \tag{2.41}$$

so

$$d_y/\delta_C = \begin{cases} (1/\pi)(E^*/\sigma_y) & \text{plane stress} \\ (1/3\pi)(E^*/\sigma_y) & \text{plane strain} \end{cases} \tag{2.42}$$

As $100 < E^*/\delta_y < 1000$ for many metal alloys

$$d_y = \begin{cases} (30-300)\,\delta_C & \text{plane stress} \\ (10-100)\,\delta_C & \text{plane strain} \end{cases} \tag{2.43}$$

In other words, the crack tip plastic zone at initiation is much larger than crack tip blunting and the corresponding critical crack opening displacement. In fact, the magnitude of COD is similar to that of the process zone within the plastic zone. For many metals, $\rho_C \approx \delta_C/2 = (10-100)\mu$m (see Fig. 2.33) which is often comparable with the grain size. This returns us to the idea of crack initiation occurring when certain microstructural features become encompassed in the crack tip process (or so-called stretch zone) at a sufficiently intense elevation of strain and stress.

Example 2.18 Derive expressions for the crack tip plastic zone radius r_y and the COD δ for the inclined slip band model shown in Fig. 2.34(a).

Ans. Using Williams's pioneering 1957 formulation of the stress state around a crack tip, we have $\tau_{r\theta} = K_I/\sqrt{2\pi r}\,[\frac{1}{4}\sin\frac{\theta}{2} + \frac{1}{4}\sin\frac{3\theta}{2}] = (K_I/\sqrt{2\pi r})\,[\sin\theta \cos\theta/2]/2$ using the first term of the series approximation, for the shear stress along a ray inclined at θ to the crack direction (the set of stress components σ_{rr} $\sigma_{\theta\theta}$ and $\tau_{r\theta}$ are the versions in polar coordinates of the set σ_{xx} σ_{yy} τ_{xy} in Cartesian coordinates, of which equation (2.30) gives one member). If we consider events along the slip bands at angle θ to the direction of the crack, we might argue that they are each a mode II shear crack under its own stress intensity factor $K_{II\theta}$. If so, the shear stress $\tau_{r\theta}$ activating slip along the direction of each band is given by $\tau_{r\theta} = K_{II\theta}/\sqrt{2\pi r}$ from the mode II versions of the stress field equations. If this stress is relaxed by sliding in along the band, we have

(i) $r_y = (\pi/8)(K_{II\theta}/\tau_y)^2$ and
(ii) $\delta_\theta = K_{II\theta}^2/E^*\tau_y$

where τ_y is the yield stress in shear using the Dugdale stip-yield model given in section 4.2.3 for r_y (better than equation (2.35) here) and equation (2.40). But if

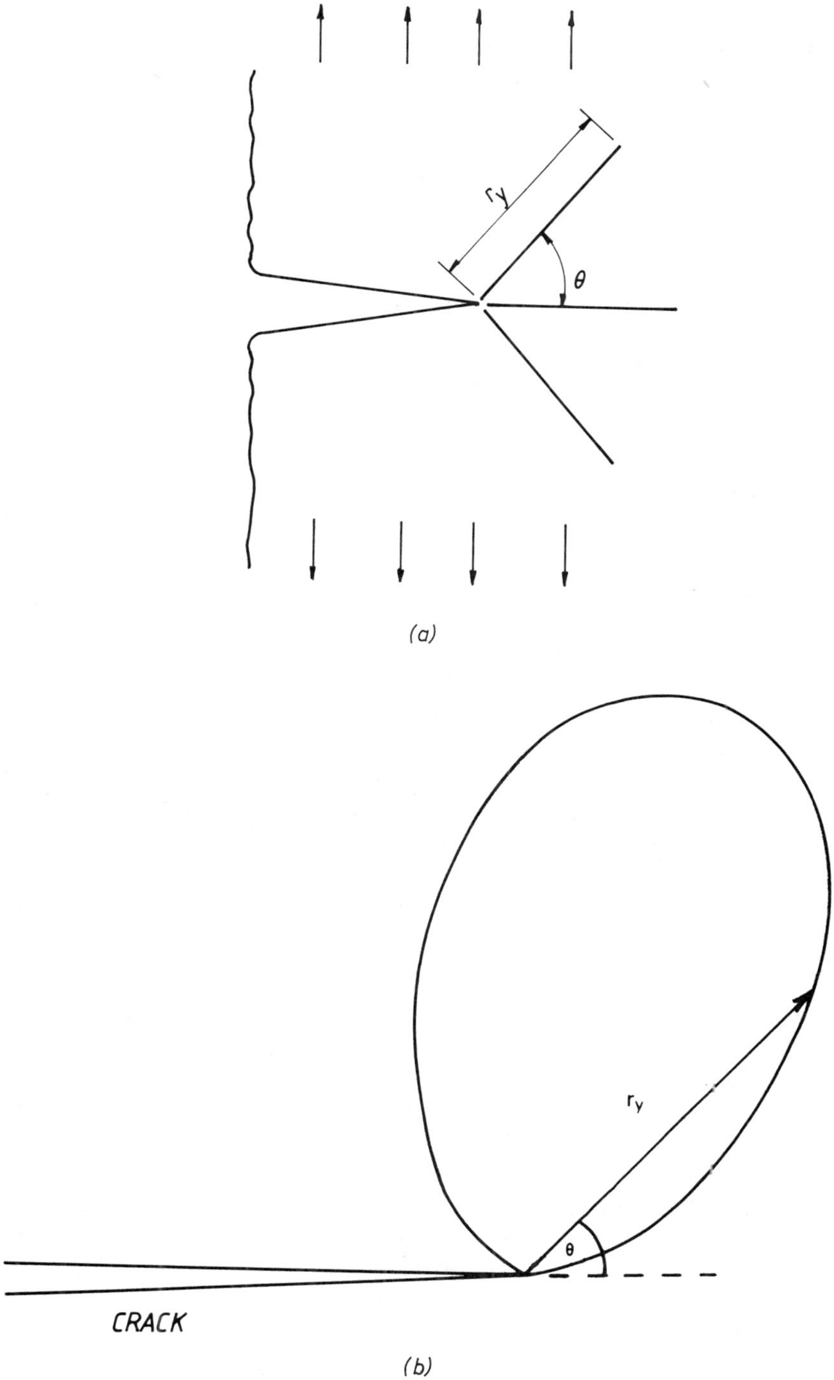

Fig. 2.34 – (a) Geometry for the inclined slip band model. (b) Plastic zone shape according to the model in (a). $r_y \propto \sin^2\theta\,(1 + \cos\theta)$.

the actual external loadings really produce a mode I stress field at the main crack tip with $\tau_{r\theta}$ as given above, comparison of the two expressions for shear stress suggests that

$$K_{II\theta} = (\sin\theta \,.\, \cos\theta/2)\, K_I/2$$

from which, substituting back for $K_{II\theta}$ gives

$$r_y \approx (\pi/64)\sin^2\theta\,(1+\cos\theta)\,K_I^2/\tau_y^2$$

and

$$\delta \approx 2\delta_\theta \sin\theta = [\sin^3\theta\,(1+\cos\theta)\,K_I^2/E^*\tau_y]/4.$$

Clearly r_y varies with θ, and differentiation with respect to θ gives $\cos\theta = 1/3$ for a maximum, i.e. $\theta = 70.6°$. At this θ, we have

$$r_y = (\pi/18)\,(K_I/\sqrt{3}\,\tau_y)^2 \approx 0.17\,(K_I/\sigma_y)^2$$

and $\delta = (16/27)\sqrt{2/3}\;K_I{}^2/E^*\sqrt{3}\;\tau_y \approx 0.48\;K_I{}^2/E\;\sigma_y$ for $\nu = 0.3$, where $\sigma_y = \sqrt{3}\;\tau_y$ is the uniaxial yield stress according to the Maxwell–von Mises yield criterion. The expressions were written in this form by Rice to compare them with numerical finite element solutions for an elastic–perfectly plastic Mises solid; the agreement is very reasonable, the coefficient for r_y being 0.152 and 0.493 for δ. That is, in equation (2.40c), $m \approx 2$. It is possible to consider the zone as being diffuse, rather than being confined to single slip bands, in which case the *shape* of the elastic–plastic boundary of the crack tip zone depends on the shape of $\sin^2\theta\,(1+\cos\theta)$, since this function determines the variation of r_y (above). Fig. 2.34(b) illustrates that the pair of 'wings' which are often found in crack tip zones is well predicted.

In experiments to determine the critical stress intensity factor, it is preferable to try and have the crack tip radii of starter cracks as sharp as ρ_C otherwise spuriously high (and non-conservative) values for K_C will be obtained. Sometimes it is adequate to sharpen the sawn or milled starter crack by means of a fine jeweller's saw or razor blade or finish the starter crack with a single point flycutter with fine radius (~25 μm), but many international standards instruct that the starter crack, of specimens subsequently to be loaded monotonically to determine K_C, be sharpended by *fatigue.* As explained in Chapter 7 the volume of irreversibly deformed material at the crack tip is less under cyclic than under monotonic loading which makes ρ small. Clearly care has to be taken in the level of fatigue loading, lest the specimen (possessing a sharp crack springing from the blunt saw-cut or milled slot) itself break in fatigue. In the case of metals, ASTM standard requirements recommend $K_F)_{max}/E < 10^{-2}\,\sqrt{\text{mm}}$ and $K_F)_{max} - K_F)_{min} > 0.9\;K_F)_{max}$ where K_F are the applied stress intensity factors in fatigue pre-cracking and E is Young's modulus. The total number of cycles to sharpen a starter crack may be of the order of 10^5 cycles.

The conventional ASTM procedures are difficult to apply to 'very' brittle solids such as ceramics because sharp cracks are not easy to introduce into the specimens and fatigue pre-cracking is not *practicable.* Available specimen sizes and geometries are sometimes also very restrictive. At present it seems that K_{IC} can be best obtained from specimens with indented pre-cracks using a pyramid hardness indenter (Lawn and Wilshaw, 1976) but residual stresses associated with the indentation must be properly accounted for (section 3.7). The double torsion (DT) and double-cantilever-beam geometries are favourite specimens for both fracture toughness and 'slow crack growth' measurements (Chapter 6). This is because of their good inherent stability for cracking (see Chapter 8) and because a sharp natural crack is produced after the first arrested 'jump' (cf. Example 8.9).

2.10 FRACTURE OF HIGHLY DEFORMABLE MATERIALS

Most fracture studies relate to the cracking of comparatively highly loaded, stiff bodies, and most of this text is concerned with such types of solid under high structural loading coefficients. What of bodies made of highly deformable materials, such as rubber and other elastomers, biological materials such as skin, gut, bladder, artery and so on? What is their specific work of fracture, how is it measured and can it be used predictively to determine critical stresses or strains at which fracture will occur?

Experience dictates that under tensile load, say, such materials can extend *reversibly* over a much wider range than metals, ceramics and stiff 'engineering' polymers. Now they might simply extend linearly over a greater range maintaining an initial Young's modulus or stiffness, but in fact their behaviour is usually highly non-linear from the outset. Fig. 1.17 contrasts the response of a typical Hookean solid used in traditional engineering with that of rubber and an animal membrane. We note that rubber has a sigmoid-shaped σ–ϵ curve which is different from the J-shaped curve of the biological material, and we recall from Chapter 1 that this difference is significant in attempts to find replacement artificial materials to substitute for biological materials in corrective operations.

In highly extensible materials, the geometry of the crack tip zone changes markedly with deformation and 'cuts' become Inglis 'ellipses' under tension (Fig. 2.35). Even if the properties in the strained state were the same as those in the unstrained state (which is usually not so, since extensive microstructural orientation takes place in the highly deformed regions around the crack or notch) we should expect differences in behaviour between stiff materials and compliant materials owing to the difference in crack tip geometry during propagation. For example, what would be the meaning of a critical COD criterion of fracture in such materials?

Again, owing to extensive deformation and the blunting of initially sharp cuts into relatively harmless notch profiles (which reduce stress and strain concentrations even in terms of large deformation theory), there is little difference between elastic strain energy density W (= Λ/volume) throughout the body and the strain energy density in 'crack tip' regions. This is a quite different situation from crack tips in stiff solids. It is difficult to transmit and concentrate energy into the very regions where crack propagation may take place. Cottrell (1964)

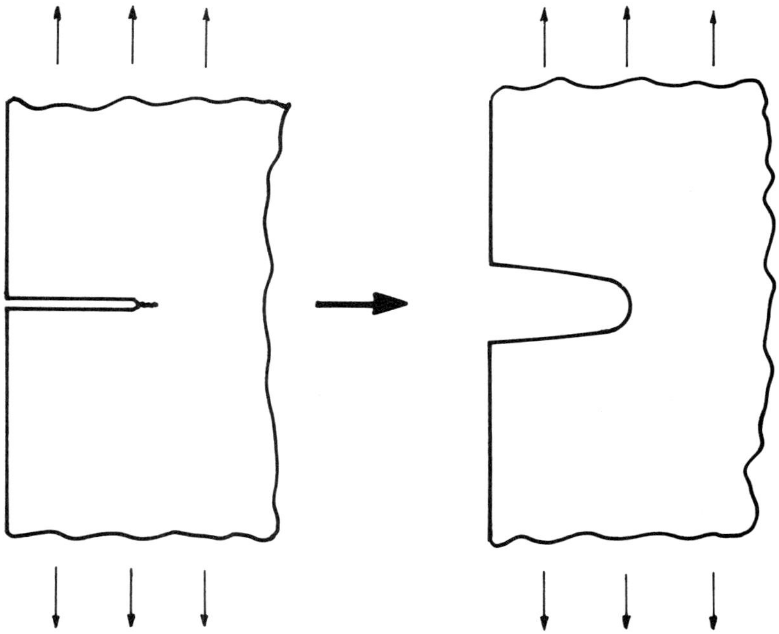

Fig. 2.35 – Crack tip geometry change of a highly extensible solid upon loading.

remarks that rubber owes its great wear resistance to elastic resilience and notch insensitivity: a motor car tyre indents without fracturing when rolled over a stone on a road and then recovers its original shape as it moves on. Again, Gordon (1979) has pointed out that, although animal membranes are remarkably difficult to tear and that therefore one might presume that the fracture toughness must necessarily be high, in fact it is typically < 10 kJ/m^2 (cf. Fig. 2.38 later) which is an order of magnitude less than that of aluminium foil, say, which does tear easily! Where rat skin, worm, insect cuticle or human arteries differ from metal sheet – and indeed from rubber sheet – is not the specific work of fracture, but rather in the shape of the stress–strain curve. The initial part of a J-shaped curve (Fig. 1.17) is almost horizontal, rather like surface tension in a liquid, and hence the material has a very low shear modulus. There is thus virtually no communication of strain energy between parts of the stressed body and it is difficult to feed energy into potential crack sites. Only in situations where membrane fracture is intended (amniotic membranes, egg shells) does Nature provide a quasi-Hookean stress–strain curve, which permits comparatively easy fracture via 'Griffith' fracture principles.

2.10.1 Graphical methods for R

The fracture resistance of *any* solid cracking elastically (irrespective of Xu non-linearities and the magnitude of u) can be determined from work areas on an Xu plot. Fig. 2.36 shows schematically a Gurney-plot in such circumstances.

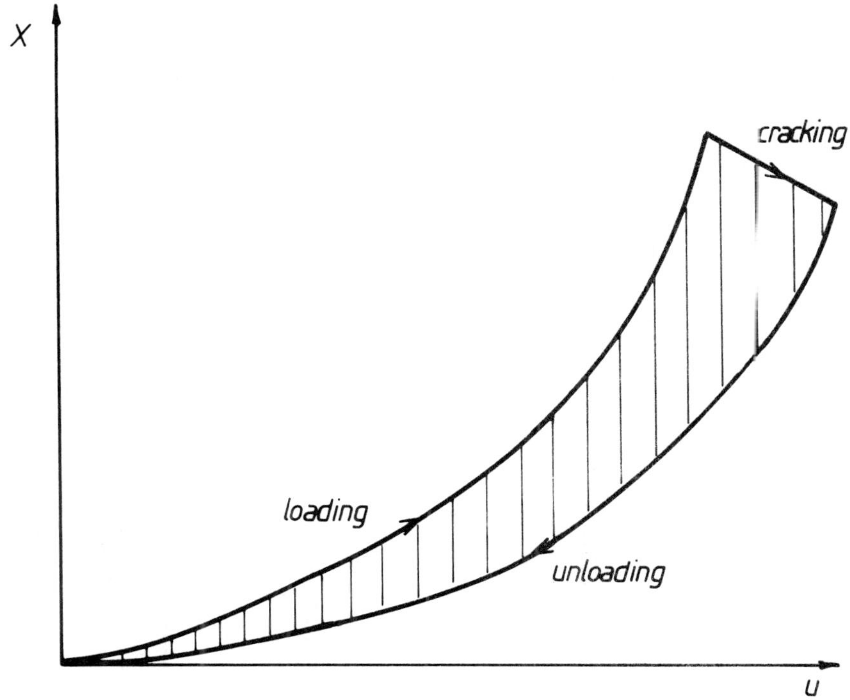

Fig. 2.36 – Fracture toughness determination for highly extensible solids using the Gurney approach.

In principle, any of the testpieces used with metals, ceramics and stiff polymers, may be employed to determine *R* for extensible materials; in practice, not all are suitable, owing to excessive distortions, and some special designs have been developed, particularly for rubbery materials (Fig. 2.37). The trousers tear test, in particular, has found favour and Fig. 2.38 gives the specific work of fracture of wet rat skin in various orientations, determined in tear experiments by Purslow (1980). Purslow's *Xu* plots (with tearing at constant load) are similar to those described in section 2.4 in connexion with tearing of paper, cloth and so on, but now there is appreciable stretching of the compliant material.

2.10.2 Algebraic formulae for cracking loads

In addition to measurement of *R* from work areas, it is desirable to have algebraic formulae for the (*X*, *u*, *R*, *A*) coordinates during cracking of extensible solids, both as a complementary method for the determination of *R* and as a means of *predicting* the onset of fracture. This is where difficulties arise, since most of the formulae available for critical cracking loads or stresses are based on small, linear, deformation theory (i.e. small strain elasticity) for Λ in equation (1.1). Even so, it was shown in section 2.2.2.2 how to deal algebraically with non-linear elastic fracture from an energy viewpoint, either where the curved *Xu* response

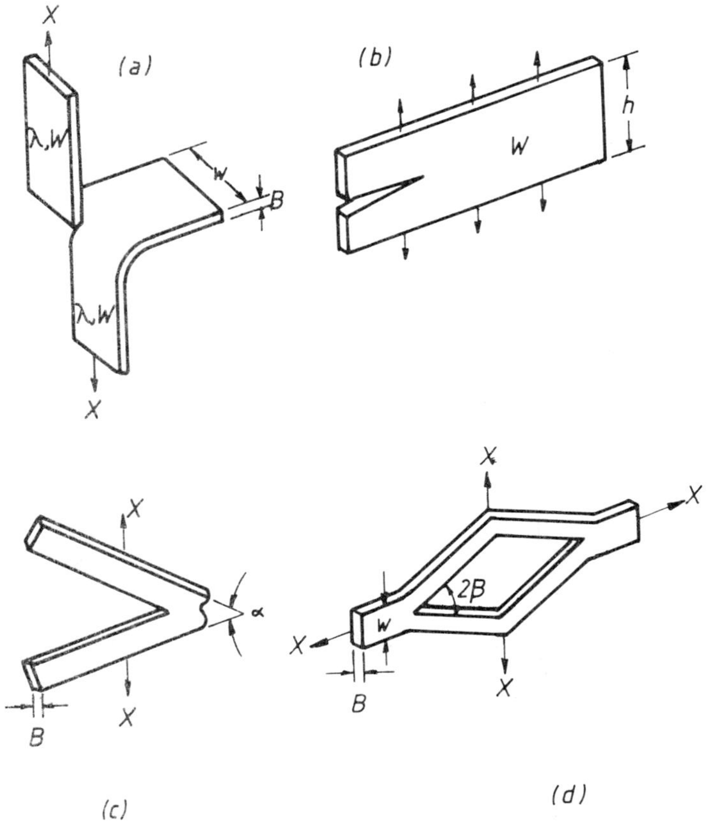

Fig. 2.37 – Various tear test pieces: (a) trousers, (b) pure shear, (c) 'angled' and (d) 'split' (After Lake and Thomas (1982), with permission.)

was available in closed-form to give Λ (e.g. DCB specimens at large rotations θ for which $u/2a = 1 - \frac{2}{\alpha}(E(k) - E(k,\theta)_1)$ and $\frac{a-\nu}{a} = \frac{\sqrt{4k^2-2}}{\alpha}$ where $\alpha = a\sqrt{X/EI}$ $= F(k) - F(k,\theta_1)$, $\sin\theta_1 = 1/k\sqrt{2}$, and E and F are elliptic integrals of the first and second kind; ν is the deflexion of the beam perpendicular to the line of action of X – Bisshopp and Drucker (1945) (see Gurney and Ngan, 1971) or where the problem could be solved using $X = k(A)$. u^n-type empirical relations, or series relations such as equation (2.15).

It was clear that the critical stress intensity factor K_C approach of linear elastic fracture mechanics, nor indeed direct application of the Griffith relation (Example 2.5), was *not* appropriate in these cases. An immediate difficulty concerns what value of E should be used when there is marked non-linearity and different sign curvatures along extended σ–ϵ plots.

The following example, suggested privately by Jeronimidis and Purslow, illustrates some of the difficulties that can arise.

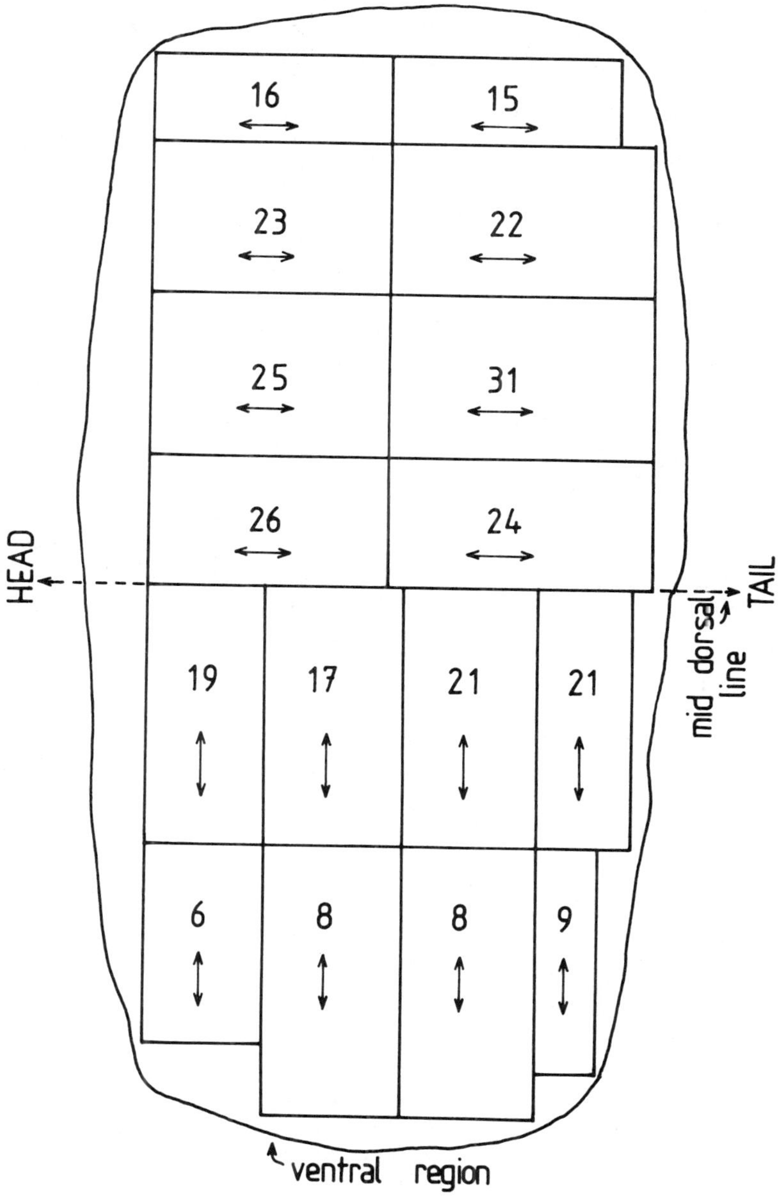

Fig. 2.38 – Directional-dependent toughness for rat skin obtained using the trousers tear test. (Courtesy P. P. Purslow.)

Example 2.19 The abdomen of a blood-sucking *Triatoma* nymph becomes distended upon feeding. Hillerton and Purslow measured the associated circumferential and longitudinal strains and found an upper limit of some 75 per cent engineering strain. From load–extension tests on longitudinal strips of abdominal tergal plate 75 per cent strain was found to correspond to a stress of 27 kPa and they found the local longitudinal modulus to be 50 MPa. Toughness for cracks propagating in the circumferential direction has a value of some 1.2 kJ/m^2. Determine the critical Griffith crack length at a stress of 27 kPa and comment on the answer in relation to the size of the insect.

Ans. The critical crack length $a = ER/\pi\sigma^2 = 50 \times 10^6 \times 1.2 \times 10^3/\pi(27 \times 10^3)^2 = 26$ m. The width of the unfed insect is only some 4 mm which is four orders of magnitude smaller! Application of formulae based on linear theory is not correct and, although there may be in fact a large 'factor of safety' in toughness of such insects at the working stress levels, it is probably not as large as suggested by these calculations. Purslow (1980) has also shown that the critical crack length in rat skin, for a 250 g mass rat suspended around its mid-region, is about 86 m which is over 40 times the circumference of the rat's belly ($\sigma \cong 18$ kPa, $R \sim 14$ kJ/m^2, local $E \approx 6$MPa). Problems 2.42 and 2.43 pose similar questions for critical crack lengths in mesogloea and pig aorta, where it will be found that direct application of the Griffith formula gives critical crack lengths which, although smaller than the size of the body, are still remarkably large. Similar questions may be asked about the fracture of rubber balloons.

While algebra based on linear deformations will not apply to highly extensible solids, the non-linear approach using, for example, $X = k(A)u^n$ should still apply, with appropriate expressions for the J-shaped curve of extensible biological materials (i.e. $n > 1$). Thus exercises such as Example 2.7 should be directly applicable to reversible extensible materials.

In the case of rubber, the sigmoid curve may be represented by a modified sinh curve, viz. $(X - X_0) = k(A)$. sinh $(u - u_0)$ where (X_0, u_0) is chosen to match the initial stiffness of the cracked body, i.e. $(\mathrm{d}X/\mathrm{d}u)_0 = k(A) \cosh u_0$. This seems to be an approach which has not been explored in the literature.

A different line of attack has been pursued over the year by Rivlin, Thomas and co-workers at the Malaysian Rubber Producers' Research Association. Indeed it is true to say that, for a time 'fracture mechanics' of rubbers was more advanced than that for many other materials. Unfortunately its separate development has led to differences in terminology,† such as the use of 'tearing energy'

† In the testpiece shown in Fig. 2.37(b), the state of strain in most of the uncracked region (for very wide specimens in which edge effects may be neglected) comprises uniform extension perpendicular to the crack balanced by uniform thinning through the thickness (with the same magnitude of strain) since the strain parallel to the crack direction is more or less zero and rubber is incompressible with Poisson's ratio $\nu = 1/2$. The cracked testpiece has been called a 'pure shear' specimen supposedly in accord with a definition in Love (1927). But the usual definition of 'pure shear' is shear with zero hydrostatic tension (equivalent to a tension and compression stress of equal magnitude at right angles) and that is *not* the state of stress in this specimen.

with symbol T for rubbers in place of specific work of fracture R (or J_C, but not really G_C since G_C is a linear property).

'Tearing' came to be used because one of the earliest and simplest-to-perform tests for the fracture resistance of rubber was the trousers tear test (cf. Figs. 2.35 and 2.36). The legs stretch considerably, but non-linearly, so the Caddell analysis (Example 2.8) cannot be applied, for the same reason that a direct Griffith or K_C approach does not apply to the behaviour of highly-extensible, non-linear materials (see Example 2.19).

The Rivlin–Thomas approach uses equations (2.11a) or (2.12a) for non-linear elastic cracking, but writes

$$(\Lambda - \Lambda_0) = ka^2 BW \tag{2.44}$$

for the energy change caused by the introduction of a crack of length a into a highly extended specimen of thickness B; Λ_0 is the strain energy of the sheet *without* the crack and Λ is the strain energy of the sheet *with* the crack, both at the same displacement, u. W is the elastic strain energy density in the specimen far removed from the crack. Relation (2.44) is based on dimensional arguments; a similar approach is the approximate solution to the Griffith problem (for small-deformation theory) which determines $(\Lambda - \Lambda_0)$ from 'stress-free' regions above and below the crack in the loaded body (see, for example, Knott (1973, p. 101 *et seq*)). Andrews' generalised theory of fracture (Chapter 1) also relates to this sort of formulation. From equation (2.12a),

$$R = -\left(\frac{\partial \Lambda}{\partial A}\right)_u = 2ka\,W \tag{2.45}$$

In general, k for rubbers depends on strain and crack length, but in some testpiece geometries only on strain. Thus, for the single edge notch (SEN) testpiece, for example, Rivlin and Thomas showed in 1953 that k is a slowly varying function of extension for sigmoid-shaped Xu curves. At small strains $k \to \pi$ which corresponds with Griffith and linear elastic fracture mechanics (since $W = \sigma^2/2E$ then, when E has a meaning). Fig. 2.39 shows k vs. engineering strain obtained by a variety of methods explained in Lindley (1972), including Lake's 1970 approximation, $k = \pi/\sqrt{\lambda}$ where λ is the 'extension ratio', i.e. $(1 + e)$ where e is the engineering strain remote from the crack.

In the case of the tear test, Rivlin and Thomas gave

$$R = \frac{2X\lambda}{B} - Ww \tag{2.46}$$

where w is the width of one of the legs, the rest of the symbols having meanings already defined. The first term in equation (2.46) relates to changes in the cut length of the specimen, the second to changes in the overall length.

In this and similar formulae for other cracked geometries, W can be calculated from Rivlin's 1948 expressions involving strain invariants given in terms of λ (see, for example, Lindley (1972) and Green and Adkins (1960)). Alternatively

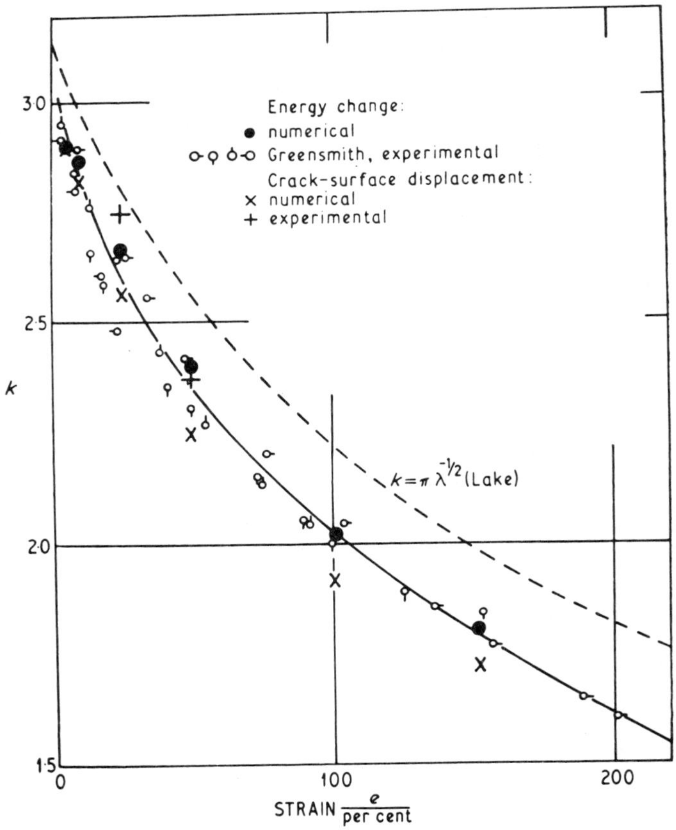

Fig. 2.39 – Variation of k in equation (2.45) with strain for simple extensible strips determined by various methods. (After Lindley (1972), with permission.)

W may be measured directly from Xu plots of un-notched samples, arguing that regions far from a crack behave as if the crack were absent. The formulae can be written in terms of deformed (rather than original) specimen dimensions (Rivlin and Thomas, 1953).

Example 2.20 Contrast the Rivlin–Thomas expression for R in the trousers tear test (equation (2.46)) with the expression for rigid legs (equation (2.17)) and the Caddell (1980) expression for linearly elastic legs given in Example 2.8.

Ans. For 'rigid' legs which do not stretch $\lambda = 1$, no strain energy is stored in the legs so $W = 0$, and relation (2.46) coincides with the simple expression $R = 2X/B$ which is equation (2.17). For linear elasticity, $W = \sigma^2/2E = X^2/2B^2w^2E$ and $\lambda = 1 + e = (1 + \sigma/E) = (1 + X/BwE)$. Hence, in equation (2.46), (remembering that there are two legs) $R = (2X/B)\,(1 + X/BwE) - 2(X^2/2B^2w^2E)w = 2X/B + X^2/wEB^2$ which is the Caddell relation.

Even in extensible rubber, the stress is often low and λ not much greater than unity, in which case $R = 2X/B$ is a good approximation.

All these calculations presume elastic displacement reversibility (recognising that, in the particular case of the trousers test, on the testing machine u remains non-zero on unloading, but that the 'rectangular' Xu work area gives genuine results for R and that, on removal from the testing machine, the torn piece will lie flat and the original untorn piece could be regained by gluing). In practice, time-dependent effects in many types of extensible material complicate interpretation of the results (Chapter 6). There may also be a further complication (different in kind from rate effects) regarding the possibility of irreversible deformation, remote from the crack region, which has nothing to do with fracture, but which will be part of the unloaded work area. This is dealt with in Chapter 4 in connexion with remote plastic flow in (comparatively stiff) metal and polymer bodies.

It is shown that if the effect of plasticity is to increase the size of the process zone, and that fracture is otherwise impossible without that increased size zone in that particular mode of deformation, then R is elevated above the level it has in the absence of plasticity. This is well-illustrated in section 4.3.2 regarding the tearing of ductile metal sheets and the relative values of R obtained by tearing and by cutting (where the process zone is smaller). A comparable example for extensible materials is the localised flow around torn edges in sheets of polyethylene, which appear as wrinkled (buckled) regions on unloading. If 'free' tearing is impossible without such deformed regions around the tear, then the associated work required for that deformation must be included with R, for that method of producing fracture. If fracture is produced in another way, by cutting for example, the specific work required may well be less since the blade limits the deformation in the fracture zone. But if fracture is accompanied by irreversibilities remote from the crack faces, the magnitude of which depends on the size and geometry of the cracked sheet, then it is *not* correct to include such specific work in the 'basic' R value. Application of equations (2.21), (2.23) or (2.24) to the tearing of extensible materials is correct *only* if *all* the dissipated work is connected with the fracture process. The general equation to be used when some of the dissipated work is irreversible flow remote from the crack faces is given in section 4.6.4.

2.10.3 Cutting of extensible materials

It is common experience (particularly in connexion with packaging materials!) that it is difficult to propagate a crack in some extensible materials by simple pulling, tearing or twisting. Probable reasons (inability to transmit energy to potential crack sites, microstructural orientation at crack tips and so on) have been highlighted in this section. Such materials, however, may often be *cut* readily, the more so with sharp knives, scissors or blades. Results of guillotining rubber sheet and biological materials have been given in section 2.4.2 and Table 2.1. George Orwell in *Down and Out in Paris and London* (1933) said that the only way to run a successful restaurant was to have sharp knives. This statement is the clue to the success of cutting as a means of separating parts of extensible materials, while tearing is more difficult. Stresses, stains and strain energy *are* now concentrated by means of the sharp tool and fracture occurs with little deformation.

Lake and Yeoh (1978), in cutting prestressed rubber sheet with razor blades, found that fracture occurred when the (rate) of total energy available (from the prestressed sheet and the work done by the blade) reached a constant value. The constant value bore the same ratio to the specific work of fracture in free tearing as the ratio of the razor blade cutting edge tip diameter to the minimum crack tip diameter in free trearing. Similar considerations apply to the cutting of ductile materials (e.g. metal cutting, Chapters 4, 5 and 10) which otherwise experience extensive deformation before fracture.

A comprehensive study of the mechanics of cutting of extensible materials has yet to be performed, but is clearly relevant to topics as diverse as surgery and food processing. The cutting of skin in surgery is an example of biaxial fracture (considered in Chapter 3), since living tissue is pre-stressed. In food processing, answers are required to questions such as: 'What is "perceived texture" in food'? 'What (scientifically) is "tough" meat?' Why, in the slicing of meat with a rotating circular blade, is it better to have a high rotational speed and low speed of feeding material into the blade? Why, in cutting bread or sponge cake, is one comparatively unsuccessful if one simply presses down with knife and does not 'saw' as well? These are neither frivolous nor trivial questions, and the answers are not likely to be trivial either, involving, as they probably do, considerations of visco-elastic transitions and so on.

PROBLEMS

2.1 A rectangular Perspex plate 600 mm × 300 mm × 6 mm is scribed into two squares in the middle of the plate by a knife leaving a uniform cut of depth 0.3 mm. What is the bending moment required to break the Perspex plate if it has a specific work of fracture of 500 J/m^2? $E = 2.5$ GPa for Perspex.

2.2 A number of identical specimens of a given metal differ only in that the starting central crack is cut to different lengths. The specimens are subjected to tensile loads and the compliance versus starting crack length is determined for each. The results are then plotted in the form of a curve of 'C' versus crack length. From these results, the *slope* of this curve at a crack length of 2.5 mm is found to be 2.2×10^{-5} (N^{-1}). A subsequent test, using a specimen having a starting crack of 2.5 mm shows that the load required to start crack propagation is 1334 N. Determine G_C if all specimens had a thickness of 9.5 mm.

2.3 How has the Griffith formula been modified for materials which show macroscopic brittle behaviour but in which there is local plastic deformation in the vicinity of the crack tip?

A sheet of maraging steel has a tensile strength of 1950 MPa. Calculate the percentage reduction in strength due to the presence in the sheet of a crack 4 mm long and perpendicular to the stressing direction. For this steel, $E = 200$ GPa, $\gamma_s = 2$J m^{-2}, and γ_p, the work of plastic deformation of each surface at the crack tip, is 2×10^4 J m^{-2}.

2.4 A cylindrical pressure vessel of diameter 3 m and length 9 m, with closed ends, is to be constructed using butt-welded steel plates which are 0.03 m thick and approximately 1 m square. It must be designed to contain a pressure P without failure by yielding or by brittle fracture. Yield occurs when the equivalent tensile stress equals the yield stress, i.e., when

$$(\sigma_1 - \sigma_2)^2 + (\sigma_2 - \sigma_3)^2 + (\sigma_3 - \sigma_1)^2 = 2\sigma_y^2.$$

The butt welds joining the plates are known to contain thumbnail (semi-circular) cracks with a maximum depth of a. For such cracks $K_I = 1.128\sigma \sqrt{(\pi a)}$ where σ is the tensile stress across a crack, and brittle failure will occur when $K_I \geqslant K_{IC}$. Three steels are available for constructing the pressure vessel. Their yield strengths and fracture toughnesses are:

Steel	σ_y(MPa)	K_{IC}(MN m$^{-3/2}$)
HY 140	965	280
T 1	690	180
HY 180	1240	180

Construct a plot of maximum pressure against crack depth a for each steel, showing the region of P and a which is safe. If non-destructive testing can detect cracks of depth $a_1 \geqslant 20$ mm (allowing welds with larger cracks than this to be repaired) which steel gives the greatest margin of safety? If a more refined technique will detect cracks of depth $a_2 \geqslant 2$ mm, which steel offers the greatest margin of safety?

2.5 A class 20 grey cast-iron has an average graphite flake size about nine times as long as a class 60 type. The strength at fracture of the former is about 140 MPa. Using arguments that would follow Griffith's analysis, what would you predict would be the strength of the class 60 iron?

2.6 A cylindrical pressure vessel, having a diameter of 6.1 m wall thickness of 25.4 mm underwent catastrophic failure when the internal pressure reached 17.5 MPa. The vessel material had an elastic modulus of 210 GPa, a yield strength of 2450 MPa and a value of G_C of 131 kJ/m^2.

(a) Show that failure would not have been expected if the Maxwell–von Mises criterion had served for design purposes.
(b) Based upon Griffith's analysis determine the size of crack that might have caused this failure.

2.7 As a cracked plate is loaded, the crack begins to increase when a load of 530 N is reached and the crack continues to lengthen under a constant load. The plate has displaced about 2.5 mm just as the crack began to run; when the crack stopped increasing in length the displacement of the plate was 5 mm. During this displacement the crack area increased by 3200 mm^2.

(a) Calculate the external work done during crack propagation.
(b) Calculate G_C, using the Gurney approach.

2.8 Consider a beam, split unsymmetrically as shown. If this material has a value of $G_C = 175$ kJ/m^2 and $E = 210$ GPa, at what load will the crack begin to lengthen?

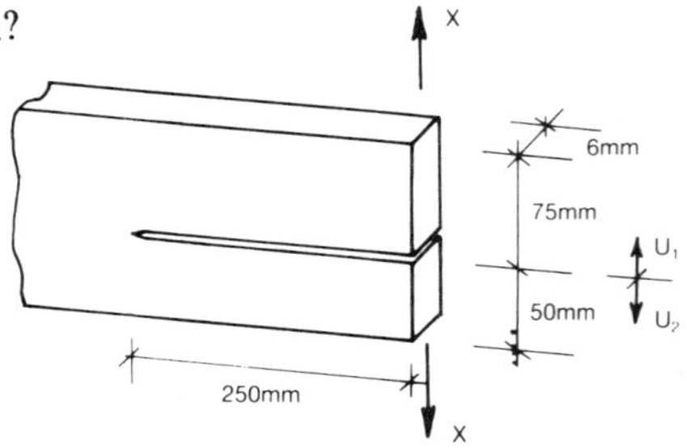

2.9 The following compliances (u/X) of 12-mm thick plates were obtained for various artifically introduced crack lengths, a. The critical loads at which the same pre-cracked plates started to fracture are also given. Calculate the specific work of fracture R

(i) from equation (2.7)

(ii) by constructing the quasi-static X,u relation for the running crack (i.e. the R locus) and evaluating R by Gurney's method.

a (mm)	u/X (mm/N $\times 10^{-6}$)	X_{crack} (kN)
50	79	2.13
75	235	1.34
100	508	1.06
125	921	0.88
150	1524	0.75
175	2286	0.66

2.10 Calculate the force X required to remove masking tape of width w, thickness t, elastic modulus E and specific fracture resistance G_C between layers of tape, when a length L has been removed. (After R. M. Caddell.)

2.11 Calculate the minimum force X required to split wood. Making some wild estimates, let the initial configuration be as shown in the sketch below, and further assume that homogeneous isotropic elastic beam theory holds (it doesn't), that the elastic modulus of the wood is 14 GN/m^2, G_C is 5 kJ/m^2, and the wedge is perfectly hard and rigid. Deflection u of a cantilever beam loaded at its end is $u = 4Xa^3/EBh^3$, where X is the load, a its length, E the elastic modulus, b the base of the cross-section and h its height. Frictional effects are to ignored.

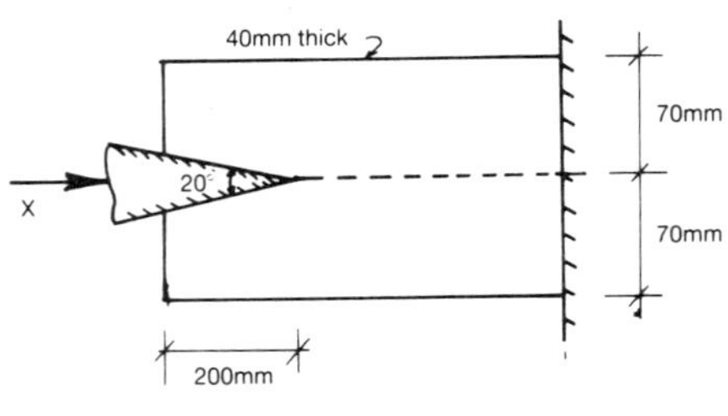

2.12 Show that the load required to peel a tapered (triangular) section of tape from a roll is given by $X = D\,(w_0 - 2\,a \tan\alpha)$ where w_0 is the width of the tape at the gripped end, 2α is the included apex angle of the triangular lamina and a the current peeled length. D is a constant equal to $EB\,\{\sqrt{1 + 2R/EB} - 1\}$ where E is Young's modulus, R the interfacial toughness, and B the thickness of the tape. Note that when $2R \gg EB$, $X \rightarrow \frac{R}{2}\{w_0 - 2\,a \tan\alpha\}$ which is analogous to equation (2.17) for the current width of tape coming off the roll.

2.13 X, u and A in Fig. 2.10 may be related empricially by $X = \sum_m C_m u^m A^{(m+1)/2}$ where C_m are constants corresponding with $m = 1, 2, 3$ etc. Integrate this expression with respect to u to obtain the non-linear strain energy function Λ. Then use $R = -\left(\frac{\partial \Lambda}{\partial A}\right)_u$ to show that $RA = \frac{1}{2}Xu$ in this case.

2.14 Following 2.13, find the Xu relation in terms of R by substituting $A = Xu/2R$ in the summation for X. If $C_1 = 3500$ $C_2 = -18{,}900$ and $C_3 = 52{,}500$ (dimensions–joules), plot constant R loci for $R = 500$ J/m^2 and 1.5 kJ/m^2 on the Xu diagram.

2.15 Analyse the trouser tear test when the legs are elastic and have unequal starting lengths.

2.16 Interfacial dislocations that are spontaneously creasted by peeling adhesive interfaces have been described by Kendall (1978). Show that the force required to peel a dislocated interface is twice that required to peel the undislocated interface. [Hint: a residual strain is left in the surface components when the dislocation is formed. Account for this in the energy rate balance for subsequent pulling.]

2.17 Show that the strain required to initiate interfacial dislocations between platelets of thickness d is given by $\epsilon = \sqrt{2R/Ed}$ where R is the interfacial work and E the Young's modulus. If this strain is high brittle platelets themselves may fracture, so there must be optimum platelet spacings in materials such as nacre (mother-of-pearl) which are caused to be strong and tough by interfacial dislocation action. If $E = 150$ GPa, $R = 0.2$ J/m^2 and the fracture strain of nacre platelets is about 0.003, determine the optimum plate thickness. (SEM pictures show that d is in fact about 500 nm.)

2.18 One method of testing for elastic reversibility in situations which are apparently displacement-irreversible is to saw-cut through the end of the crack into regions beyond the crack tip zone (either under load or off the testing machine). If displacement irreversibility is caused by effects local to the crack tip (large zones in polycarbonate, say) cutting into elastic regions beyond the (large) zone should permit the crack faces in the body to close back up. Discuss this technique and determine whether geometry-independent fracture toughness data may be determined from the X–u plots after saw cuts have been made.

2.19 Investigate non-linear versions of equations (2.21) and (2.23) for cases which elastic displacement irreversibilities, i.e. crack propagation with residual elastic displacements in bodies having stiffness at constant crack area which is a function of displacement.

2.20 Show that there is a one-to-one correspondence between the following variables in crack mechanics and classical thermodynamics: load X and pressure p; displacement u and volume V; toughness R and temperature T; crack area A and entropy S.

2.21 Show that there is a one-to-one correspondence between the following functions in crack mechanics and in classical thermodynamics: strain energy and internal energy; complementary strain energy and enthalpy; $(RA + \Lambda)$ and Helmholtz free energy; $(Xu - \Lambda - RA)$ and Gibbs availability. R is fracture toughness, A crack area, Λ strain energy, X load, u displacement.

2.22 Show that the Jacobian $J\left(\frac{Xu}{RA}\right) = 1$. [Hint: $d\Lambda$ (strain energy) is an exact differential.]

2.23 Following 2.20, determine the equivalent crack mechanics relations to the following Maxwell thermodynamic relations:

$$\partial p/\partial S)_V = -\,\partial T/\partial V)_S; \qquad \partial V/\partial S)_p = \partial T/\partial p)_S;$$

$$\partial p/\partial T)_V = \partial S/\partial V)_T; \qquad \partial V/\partial T)_p = -\,\partial S/\partial p)_T.$$

2.24 What is the difference between the stress concentration factor (K_t) and the stress intensity factor (K)?

2.25 Show that there is a relationship between stress concentration factors and stress intensity factors given by

$$K_I = \lim_{\rho \to 0} \left[\frac{\sigma_{max}}{2} \sqrt{\pi \rho} \right]$$

where ρ is notch radius and σ_{max} is the maximum stress at the root of the notch as elevated above the remote average stress by the relevant stress concentration factor for the geometry and loading in question.

2.26 A 50-mm thick standard CT testpiece of EN25T 0.3 C/2.5 Ni/0.7 Cr/ 0.6 Mo steel had a fatigued starter crack with a bowed front which measured 50.0, 50.2, 51.5, 51.7, 51.0, 51.1, 49.9 mm long at equal spaces across the thickness of the specimen. The secant load $X_Q = 258$ kN and the maximum load $X_{max} = 300$ kN in a fracture test at −9°C. Determine K_Q using the expression $K_Q = (X_Q/B\sqrt{W})\,(29.6\,(a/W)^{1/2} - 185.5\,(a/W)^{3/2} + 655.7\,(a/W)^{5/2} - 1017.0\,(a/W)^{7/2} + 638.9\,(a/W)^{9/2})$ and comment on its validity.

2.27 3-point bend specimens of a steel of yield strength 800 MPa broke at a load of 59 kN. If $a = 25$ mm, $W = 50$ mm and $S = 200$ mm, calculate K_{IC} and comment on its validity.

2.28 What, according to ASTM standards is the maximum K_{IC} value that may be determined on 50-mm thick plates of metals with 0.2 per cent proof stresses (a) 500 MPa, (b) 1 GPA?

2.29 Show from K-formulae that the dimension which mainly controls the severity of part-thickness (surface) flaws is the depth in the thickness direction.

2.30 Show that part-thickness defects at the surface are more severe than defects embedded within the thickness.

2.31 Show that the maximum inelastic cracking load in Barker's short-rod specimen occurs at $a/B = 0.52$ (see Example 2.17).

2.32 The figure shows two loading cases for an infinite half plane containing an edge crack of depth a. In case A the tensile stress is applied perpendicular to and remote from the crack plane. In case B the edge crack is loaded along its faces by a stress which increases linearly from zero at the edge

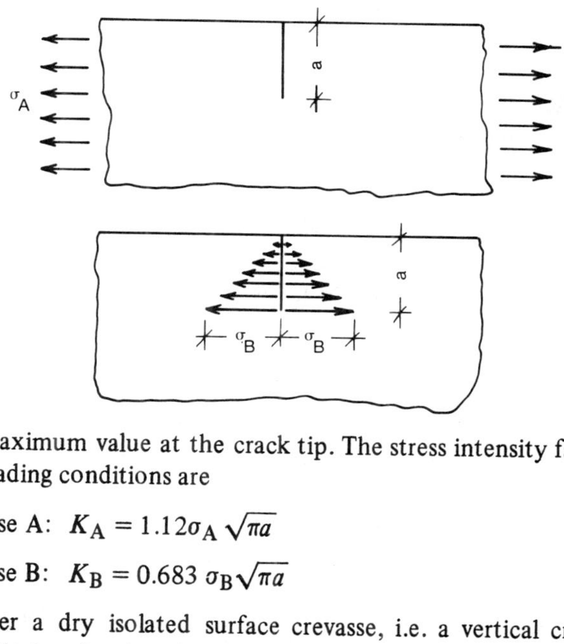

to a maximum value at the crack tip. The stress intensity factors from the two loading conditions are

$$\text{case A:}\quad K_A = 1.12\sigma_A\sqrt{\pi a}$$

$$\text{case B:}\quad K_B = 0.683\,\sigma_B\sqrt{\pi a}$$

Consider a dry isolated surface crevasse, i.e. a vertical crack, in a very thick glacier that is subjected to a uniform longitudinal tensile stress of 200 kN/m^2. Determine the depth of the crevasse given that the density of ice is 0.92×10^3 kg/m^3. (Cambridge University).

2.33 Splitting of wood around driven nails means that the sideways contact load between wood and nail is borne over only part of the nail's periphery. The amount of splitting is determined by the fracture toughness of the wood, and the contact stress is limited by the resistance of the wood to indentation. Show how these factors are interrelated, using the following heuristic modification of equation (2.32) to account for stresses applied over only part of the crack faces (see Atkins and Mai, 1975), viz.

$$p = K_{IC}/(2/\pi)\sin^{-1}(b/c)\sqrt{\pi a}\,(M_e/\sqrt{Q})$$

where $2b$ is that part of the surface crack length $2c$ over which the contact stress p acts. What is the general shape of a nail load–nail penetration plot? State what type of friction you assume between nail and wood.

2.34 Prove from first principles that the fracture parameters R and K_C are related by:

$$K_C^2 = E^*R$$

where $E^* = E$ for plane stress and $=E/(1-\nu^2)$ for plane strain. [Hint: In Equ. 2.33, take the origins of the r-coordinate for σ_{yy} and v_{yy} at opposite ends of the Δa increment.]

2.35 A long specimen of width W and thickness B has an edge crack of length a. It is subjected to simultaneous remote direct tensile stress σ and bending moment M. Determine the stress intensity factor for the combined loading, given the separate expressions (i) for tensile stress $Y(a/W) = 1.12 - 0.231(a/W) + 10.55(a/W)^2 - 21.72(a/W)^3 + 30.39(a/W)^4$ and (ii) for bending $Y(a/W) = 1.12 - 1.40(a/W) + 7.33(a/W)^2 - 13.08(a/W)^3 + 14.00(a/W)^4$ and the bending stress $\sigma = 6M/BW^2$.

2.36 Plot in the Xu plane constant R loci for the development of cone cracks of semi-angle 22° and radius r under an indenter of base radius $r_0 = 10$ mm in a block of a brittle solid of Poisson's ratio 0.25 for which $R = 4 \times 10^{-14}\ X^2/r^3$ (J/m²) and hence show that cracking occurs under increasing load. [Hint: integrate $R = (X^2/2)\ \mathrm{d}/\mathrm{d}A\ (u/X)$ to get lines of constant stiffness in terms of various (r/r_0).]

2.37 A rod of diameter d is embeded in a massive block. Show that the force required to cause a crack to propagate in mode II, at the cylindrical interface between rod and block, is given by $X^2 = \pi^2 d^3 RE/2$ where R is the interfacial toughness and E the Young modulus of the rod. (This problem is related to fracture in fibre-reinforced composites and to King Arthur's sword 'Excalibur'; see Gurney and Hunt, 1967, Atkinson *et al.*, 1982.)

2.38 Show that an average crack extension equal in magnitude to the radius of the plane strain plastic zone size is equivalent to a reduction in stiffness of 5 per cent.

2.39 Elastin is the yellow solid found in cheap meat which does not go away on boiling the meat. Does boiling affect the toughness of meat?

2.40 Meat is beaten with a hammer to 'tenderise' it. Discuss what changes in structure occur and how the toughness of meat is affected.

2.41 'The only way to run a successful restaurant is to have sharp knives' (George Orwell *Down and Out in Paris and London*). Discuss.

2.42 The maximum tensile stress, due to bending of the body column during an anemone's filter feeding, is about 39 kPa. The stiffness of *Metridium* mesogloea (the polysaccharide/collagen gel between cells) is strain-rate dependent, according to

$$E = 1.11(\dot{\epsilon})^{0.29}$$

Strain rates in mesogloea induced by wave-action are about $1.6 \times 10^{-3}\ \mathrm{s}^{-1}$

(Koehl, 1977). Show that if $R = 1.4 \text{ kJ/m}^2$, the critical crack length given by the Griffith formula is some 50 mm. Comment on the sense of this answer in relation to the size of the anemone's column (some 50 mm diameter) and on the validity of applying the Griffith formula.

2.43 The mean systolic blood pressure in a pig is 169 mm Hg (Altman and Dittmer, 1964). This corresponds to a circumferential stress in the wall of a pig aorta of 110 kPa for a specimen of average dimensions. Taking a value of 1 MPa for the longitudinal stiffness of an aorta (Patel *et al.*, 1969) and an experimental value of R in the circumferential direction of 1 kJ/m^2, show that the critical crack length in the weaker circumferential direction given by the Griffith formula is some 26 mm. Comment on the sense and validity of the answer (the circumference of the aorta in the relevant region is some 40 mm).

2.44 Practical materials are not continua on a fine scale. Discuss the assumptions in the application of continuum mechanics to practical materials highlighting the difficulties which arise when the size of a crack is comparable with the size of microstructural features.

REFERENCES

Altman, P. L. and Dittmer, D. S. (1964) *Biology Data Handbook.* Washington DC: Federation of American Societies for Experimental Biology.

Ashby, M. F. (1979) in *Fracture Mechanics: Current Status, Future Prospects.* Toronto: Pergamon.

ASTM (American Society for Testing and Materials) (1981) Document E399.

Atkins, A. G. and Mai, Y. W. (1975) *Int. J. Mech. Sci.,* **17**, 729.

Atkins, A. G. and Mai, Y. W. (1979) *J. Mater. Sci.* **14**, 2747.

Atkins, A. G. and Vincent, J. F. V. (1984) *J. Mater. Sci. Letters* **3**, 310.

Atkinson, C., Avila, J., Betz, E. and Smelser, R. E. (1982) *J. Mech. Phys. Solids* **30**, 97.

Barker, L. M. (1977) *Engr. Fract. Mech.* **9**, 361.

Bisshopp, K. E. and Drucker, D. C. (1945) *Q. J. Appl. Mech.* **III3**, 272.

Bilby, B. A., Cottrell, A. H. and Swinden, K. H. (1963) *Proc. Roy. Soc.* **A273**, 304.

British Standards Institution (1977) BS 5447. London: BSI.

British Standards Institution (1979) BS 5762. London: BSI.

Burns, S. J., Pollett, J.-C. and Chow, C.-Lun (1978) *Int. J. Fract.* **14**, 311.

Caddell, R. M. (1980) *Deformation and Fracture of Solids.* Englewood Cliffs, N.J.: Prentice-Hall.

Caddell, R. M. and Uete, H. K. (1983) *Int. J. Mech. Sci.* **25**, 87.

Chow, C. L. (1982) *J. Strain Analysis* **17**, 67.

Chow, C. L. and Lau, K. J. (1977) *J. Strain Analysis* **12**, 286.

Coker, E. G. and Scoble, W. A. (1913) *Trans. Inst. Naval Architects* **LV**, 231.

Cook, J. and Gordon, J. E. (1964) *Proc. Roy. Soc.* **A282**, 508.

Cottrell, A. H. (1964) *The Mechanical Properties of Matter.* London: Wiley.

Currey, J. D. (1977) *Proc. Roy. Soc.* **B196**, 443.
Gent, A. N. and Kim, H. J. (1978) *Rubber Chem. & Tech.*, **51**, 35.
Gordon, J. E. (1979) *Proc. Conf. Review of Progress in Quantitative NDE, LaJolla,* California: Rockwell International.
Green, A. E. and Adkins, J. E. (1960) *Large Elastic Deformations.* Oxford: Clarendon Press.
Griffith, A. A. (1921) *Phil. Trans. Roy. Soc.* **A221**, 163 (see also *Proc. 1st. Int. Congr. Appl. Mech., Delft, 1924*). University of Delft.
Gurney, C. and Ngan, K. M. (1971) *Proc. Roy. Soc.* **A325**, 207
Gurney, C. and Hunt, J. (1967) *Proc. Roy. Soc.* **A299**, 508.
Hillerton, J. E. and Purslow, P. P. (1981) Private communication.
Hodgkinson, J. M. and Williams, J. G. (1981) *J. Mater. Sci.* **16**, 50.
Inglis, C. E. (1913) *Trans. Inst. Naval Architects* **LV**, 219.
Irwin, G. R. and Kies, J. A. (1952) *Weld. J. Res. Suppl.* **17**.
Irwin, G. R. (1948) in *Fracturing of Metals,* Cleveland, Ohio: ASM.
Irwin, G. R. (1957) *J. Appl. Mech.* **24**, 361.
Irwin, G. R. (1957) *Proc. 9th Int. Congr. Appl. Mech.*, VIII, Paper 101(II), University of (Brussels), Brussels.
Irwin, G. R. (1964) *Appl. Mater. Res.* **3**(2), 65 (see also 3rd Rept. Special C'ttee Mater. Res. Stds. 1961, 883).
Isherwood, D. P. and Williams, J. G. (1978) *Engr. Fract. Mech.* **10**, 887.
Jones, A. T. (1974) *Engr. Fract. Mech.* **6**, 435.
Jones, A. T. (1974) *Engr. Fract. Mech.* **6**, 653.
Kapp, J. A., Pflegl, G. A. and Underwood, J. H. (1977) *Int. J. Fract.* **13**, 721.
Kendall, K. (1975) *J. Phys. D* **8**, 1449.
Kendall, K. (1978) *J. Phys, D.* **11**, 1519.
Kenny, P and Campbell, J. D. (1968) *Progress in Materials Science* **13**, 135.
Knott, J. F. (1973) *Fundamentals of Fracture Mechanics.* London: Butterworths.
Koehl, M. A. R. (1977) *J. Exp. Biol.* **69**, 127.
Krafft, J. M., Sullivan, A. M. and Boyle, R. W. (1961) *Proc. Symp. Crack Propagation, Cranfield, 8.* College of Aeronautics, Cranfield.
Lake, G. J. (1970) *Proc. Conf. on Yield, Deformation, and Fracture of Polymers* London (Institute of Physics and the Physical Society).
Lake, G. J. and Yeoh, O. H. (1978) *Int. J. Fract.* **14**, 509.
Lake, G. J. and Thomas, A. G. (1982) *Proc. Conf. Absorbed Specific Energy* (Eds. Sih, Gzoboly and Gillemot). Den Haag: Martinus Nijhoff.
Lawn, B. R. and Wilshaw, T. R. (1976) *Fracture of Brittle Solids.* London: Cambridge University Press.
Lindley, P. B. (1972) *J. Strain Analysis* **7**, 132.
Love, A. E. H. (1927) *Mathematical Theory of Elasticity,* 4th edition. London: Cambridge University Press.
Mai, Y. W. (1976) *Int. J. Fracture,* **12**, 107.
Mai, Y. W. (1983) *Appita,* **36**, 461.
Mostovoy, S. and Ripling, E. J. (1966) *J. Appl. Poly. Sci.* **10**, 1351.
National Academy of Sciences (1976) *Rapid Inexpensive Tests for Determining Fracture Toughness.* Washington, D.C.

Newman, J. C. (1971) NASA TND-6376.

Newman, J. C. (1981) *Int. J. Fract.* **17**, 567.

Neuber, H. (1961) *J. Appl. Mech.* **28**, 544 (see also Kerbspaunungslehre, trans. AEC – tr – *4547*, 1961, OTS, Dept. Commerce, Washington, DC).

Orowan, E. (1949) *Rept. Prog. Phys.* **12**, 214 and 185S.

Orowan, E. (1955) *Weld. J. Res. Suppl.* **20**, 157.

Orwell, G. (1933) *Down and Out in Paris and London.* London: Victor Gollancz.

Parry, G. W. and Lazzeri, L. (1969) *Engr. Fract. Mech.*, **1**, 519.

Patel, D. J., Janicki, J. S. and Carew, T. E. (1969) *Circulation Research*, **25**, 765.

Peters, R. W. and Kuhn, P. (1957) NACA Tech. Note No. 3993.

Peterson, R. E. (1974) *Stress Concentration Design Factors,* 2nd edition. New York: Wiley.

Purslow, P. P. (1980) Ph.D. dissertation, University of Reading.

Rice, J. R. (1973) Crack tip plasticity and fracture initiation, ICF-3, Munich, paper I-441.

Rivlin, R. S. (1948). *Phil. Trans. Roy. Soc.*, **241**, 379.

Rivlin, R. S. and Thomas, A. G. (1953) *J. Polymer Soc.*, **10**, 291.

Rooke, D. P. and Carwright, D. J. (1976) *Compendium of Stress Intensity Factors.* London: HMSO.

Savin, G. N. (1961) *Stress Concentration around Holes.* Oxford: Pergamon.

Sih, G. C. (1973) *Handbook of Stress Intensity Factors.* Bethlehem, Penna: Lehigh University.

Tada, H., Paris, P. and Irwin, G. R. (1973) *The Stress Analysis of Cracks Handbook.* Hellertown, Penna: Del Research Corp.

Theocaris, P. S. (1971) *Applied Optics* **10**, 2240.

Timoshenko, S. and Goodier, J. E. (1951) *Theory of Elasticity,* 2nd edition. New York: McGraw-Hill.

Timoshenko, S. (1955) *Strength of Materials,* 3rd edition. Princeton, N.J: van Nostrand.

Timoshenko, S. and Young, D. H. (1945) *Theory of Structures.* New York: McGraw-Hill.

Uete, H. K. (1981) Ph.D dissertation, University of Michigan.

Wecharantana, M. and Shah, S. P. (1980) *Cement and Concrete Research* **10**, 833.

Wessel, E. T. (1968) *Engr. Fract. Mech.* *7*, 77.

Wilhem, D. P. (1983) *Int. J. Fract.* **23**, R77.

Williams, M. L. (1957) *J. Appl. Mech.* **24**, 109.

Yap, O. F., Mai, Y. W. and Cotterell, B. (1983) *J. Mater. Sci.* **18**, 657.

3

Special topics in quasi-static elastic fracture

'I know what you're thinking about,' said Tweedledum; 'but it isn't so, nohow.'
'Contrariwise,' continued Tweedledee, 'if it was so, it might be; and it were so, it would be; but as it isn't, it ain't. That's logic.'

A number of separate topics are investigated in this chapter. They are considerations of crack velocity, fracture in compression (including hardness of brittle solids), fracture of porous bodies, cracking under multiaxial loading, crack paths, cracking of layered structures, the influence of residual stresses on fracture, and crack propagation from notches. The final section of the chapter deals with the ensuring of globally elastic fracture by the avoidance of remote yielding. What happens when remote flow is unavoidable is the subject of later chapters.

3.1 CRACK VELOCITY

If testing machine crossheads are driven at different speeds, the associated crack velocities will change. Equation (1.1) for quasi-static cracking may be written as:

$$X \frac{\mathrm{d}u}{\mathrm{d}t} = \frac{\mathrm{d}\Lambda}{\mathrm{d}t} + R \frac{\mathrm{d}A}{\mathrm{d}t} \tag{3.1}$$

where t is time.

For linear systems $\Lambda = \frac{1}{2} Xu$ and $\mathrm{d}\Lambda/\mathrm{d}t = \frac{1}{2} (X\mathrm{d}u/\mathrm{d}t + u\mathrm{d}X/\mathrm{d}t)$. Using the fluxional notation,

$$X\dot{u} = \tfrac{1}{2} (X\dot{u} + u\dot{X}) + R\dot{A}$$

where $\dot{A}$ is areal velocity. Hence,

$$2R\dot{A}/\dot{u} = X - u\dot{X}/\dot{u} = X - u\,\dot{X}\dot{A}/\dot{A}\dot{u}$$

or

$$\frac{\dot{A}}{\dot{u}} = \frac{B\dot{a}}{\dot{u}} = \frac{X}{2R + u\dfrac{dX}{dA}} \tag{3.2}$$

in a testpiece of uniform thickness.

Consider the cracking of a DCB testpiece, for which $u/X = 2a^3/3EI$, and $X^2a^2 = REIB$ (equation (2.6)). We have

$$u\frac{dX}{dA} = \left(\frac{2a^3X}{3EI}\right)\frac{1}{B}\frac{d}{da}\left(\frac{\sqrt{EIRB}}{a}\right) = -\frac{2}{3}R$$

(X decreases as A increases in the DCB specimen at constant R.) Hence in equation (3.2),

$$\frac{\dot{a}}{\dot{u}} = \frac{3X}{4BR} = \frac{3}{8a}\sqrt{\frac{Eh^3}{3R}} \tag{3.3}$$

and the crack velocity slows down as the crack propagates, and is faster the greater E, h and faster the lower R. Even though $\dot{a}$ may not be 'slow', a quasi-static analysis may be used since the kinetic energy associated with $\dot{a}$ is still small compared with the other terms in equation (1.1). Upper limiting ranges for $\dot{a}$ where this ceases to hold, are discussed in section 8.2. To help the reader, these limiting velocities are some 6 m/s in metals and 2 m/s in polymers (e.g. Gurney and Ngan, 1971).

Example 3.1 A DCB testpiece is being used to investigate the possible variation of toughness with crack velocity in PMMA (discussed in detail in Chapter 6). Gurney's segmental area approach is to be used and crack propagation is to be tracked visually. What useful range of $\dot{a}$ may be covered in a single typical-sized specimen with $E = 3$ GPa and R about 500 J/m^2?

Ans. Cracks cannot be tracked visually with much accuracy if $\dot{a} > 5$ mm/s, so let this be the fastest initial velocity permissible. Assume that $h = 15$ mm and the starting a is 100 mm. Then the $\dot{u}$ which produces $\dot{a} = 5$ mm/s comes from equation (3.3) with

$$5/\dot{u} = (3/8 \times 100 \times 10^3)(3 \times 10^9 \times 15^3 \times 10^{-9}/3 \times 500)^{1/2} \text{ or } \dot{u} = 510\ \mu\text{m/s}$$

Hence the crosshead velocity should be kept below this. If a propagates to, say, 150 mm, $\dot{a}$ slows to (100/150) of 5 mm/s $\simeq$ 3 mm/s.

Many recommended test pieces are too stubby for beam theory to apply in evaluating udX/dA for use in equation (3.2). Fracture mechanics formulae for

K for such specimens provide a means of deriving the appropriate $u\mathrm{d}X/\mathrm{d}A$. Since $K_\mathrm{C} = \sqrt{E^*R}$, we have, using equation (2.6),

$$X = \sqrt{\frac{2R}{\dfrac{\mathrm{d}}{\mathrm{d}A}\left(\dfrac{u}{X}\right)}} = K\sqrt{\frac{2}{E^*\dfrac{\mathrm{d}}{\mathrm{d}A}\left(\dfrac{u}{X}\right)}} \tag{3.4}$$

and we may determine the compliance u/X by integration and hence $u\mathrm{d}X/\mathrm{d}A$.

Consider the compact tension type of test piece for which

$$K = \frac{X}{B\sqrt{W}}\left[Z\left(\frac{a}{W}\right)\right]$$

where $Z(a/W)$ is already given in Example 2.11 in Chapter 2. Thus, from equation (3.4) we have

$$X = \left(\frac{X}{B\sqrt{W}}\right)(Z)\sqrt{\frac{2}{E^*\dfrac{\mathrm{d}}{\mathrm{d}A}\left(\dfrac{u}{X}\right)}}$$

i.e.

$$\frac{\mathrm{d}}{\mathrm{d}A}\left(\frac{u}{X}\right) = \frac{2}{E^*B^2W}Z^2 \tag{3.5}$$

Since $\mathrm{d}A = B\mathrm{d}a = BW\,\mathrm{d}(a/W)$, equation (3.5) can be integrated to give

$$\frac{u}{X} = \frac{2}{E^*B}\int Z^2\,\mathrm{d}(a/W) = \frac{2}{E^*B}(I_1) \tag{3.6}$$

where I_1 is the integral of $\int Z^2\,\mathrm{d}(a/W)$ neglecting the compliance of the crack-free body in comparison with that of the cracked body (i.e. $u/X = 0$ when $a/W = 0$). Equations (3.5) and (3.6) appear in Turner's η factor formulation relating work area to toughness (sections 4.5.2 and 8.8).

During cracking, $K_\mathrm{C} = \dfrac{X}{B\sqrt{W}}(Z)$ so that

$$X = \frac{B\sqrt{W}K_\mathrm{C}}{Z} \tag{3.7}$$

and therefore,

$$\frac{\mathrm{d}X}{\mathrm{d}A} = \frac{K_\mathrm{C}}{\sqrt{W}}\frac{\mathrm{d}}{\mathrm{d}(a/W)}(Z)^{-1} = \frac{K_\mathrm{C}}{\sqrt{W}}(D_1) \tag{3.8}$$

Note that $\mathrm{d}X/\mathrm{d}A$ is negative because during quasi-static cracking X decreases as A increases. D_1 is the crack length derivative of Z^{-1}. Consequently from equations (3.6), (3.7) and (3.8), we obtain:

$$u\frac{\mathrm{d}X}{\mathrm{d}A} = X\left(\frac{u}{X}\right)\frac{\mathrm{d}X}{\mathrm{d}A} = \frac{K_C^2}{(Z)}\ \frac{2}{E^*}\ (I_1)(D_1) \tag{3.9}$$

This expression (and others like it for other geometries) is not as cumbersome as it appears, and is readily evaluated by pocket calculators. For example, if $(a/W) = 0.5$ (the recommended starting crack size in the standard CT specimen used to determine K_{IC}), we find $u\mathrm{d}X/\mathrm{d}A = -1.24R$ (using $K_C^2 = E^*R$ to eliminate E^*). Hence in equation (3.2),

$$\frac{\dot{A}}{\dot{u}} = \frac{B\dot{a}}{\dot{u}} = \frac{X}{(2R - 1.24R)} = \frac{1.31X}{R}$$

Substituting for X from equation (3.7),

$$\frac{\dot{a}}{\dot{u}} = \frac{1.31\ W^{1/2}K_C}{(Z)\ R} = \frac{1.31}{(Z)}\sqrt{\frac{E^*W}{R}} \tag{3.10}$$

In this case, as with the DCB specimen, the crack slows down as it propagates (the function (Z) increases as a/W increases) and is faster the greater E^*, W and faster the lower R. As an example, consider a K_{IC} test on a low-alloy structural steel with, say, $K_{IC} = 60\ \mathrm{MPa}\sqrt{\mathrm{m}}$, $R = 17\ \mathrm{kJ/m^2}$, and $\sigma_y = 600$ MPa at $-50°$C (typical of A533 steels). In a 30-mm thick specimen (which satisfies the plane strain size requirement – see Example 2.11), the standard CT specimen will have $W = 2B = 60$ mm. (Z) for $a/W = 0.5$ is 9.60, so that $\dot{a}/\dot{u} \approx 118$, and cracks propagate at two orders of magnitude faster than the crosshead velocity. It will be shown in section 8.2.2 that cracking under displacement control is *stable* in the CT specimen (i.e. satisfies a quasi-static energy rate balance, equation 1.1) even though, as we show here, the crack velocity is 'fast'. 'Fast fracture' is *not* necessarily 'unstable fracture'.

3.1.1 Changes in crack speed at interfaces

A crack propagating in a layered structure may cross a number of interfaces with different elastic properties. Section 3.5.2 describes how cracks may *turn* in their path when meeting interfaces between anisotropic layers; the discussion is relevant to mechanisms of toughening in composite bodies. Consideration of what happens to crack *velocities* at such interfaces also sheds light on synergisms in toughness in composites.

Kendall (1975b) extended his study of peeling (section 2.4.1) to tapes having periodic changes in bending stiffness along the interface; changes in EI were produced either by simply thickening-up the tape or by locally reinforcing it with woven glass fibre cloth (Fig. 3.1). For a tape with uniform EI, obeying the simple relations (2.16) and (2.17), peeling takes place at constant velocity equal to the rate of displacement of the applied force (i.e. $\dot{a} = \dot{u}$). For uniform EI, the

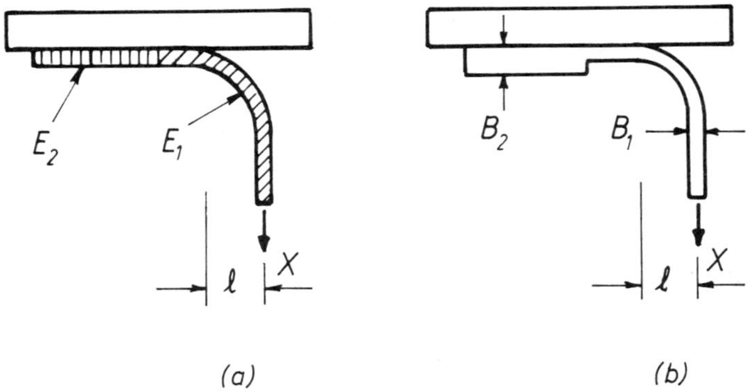

Fig. 3.1 – Peeling of tapes of varying bending stiffness. (a) Differing E_1 and E_2 and (b) change in film thickness. (After Kendall (1975b), with permission.)

peel bend shape remains steady. However, if the crack meets an interface, the peel bend shape changes and the energy rate balance is altered. The stiffer material just past the interface is reluctant to deform and consequently the movement of the force X is reduced even though the crack moves the same distance as before. As remarked by Kendall, the interface with stiffer material is seen by the crack essentially as an 'elastic gear change' into a higher ratio. At constant force, therefore, the crack is *retarded.* It may be shown (Problem 3.6) that when the crack just meets the interface between materials of stiffness E_1I_1 and E_2I_2 respectively, equation (2.17) must be modified to

$$\frac{X}{w} = \frac{E_2I_2}{E_1I_1}R \tag{3.11}$$

(using w in place of B which applies to tearing).

This means that the toughness of the system now has two components: the intrinsic material toughness reflected in the fracture energy R, and an interface toughening effect embodied in the toughening factor E_2I_2/E_1I_1 which gives a clue to the cause of extraordinary toughness of composite materials when $E_2I_2 > E_1I_1$. (Simple illustrations of the effect are found in tearing postage stamps.)

When the crack has travelled well beyond the location of the change in EI and a new steady state geometry for the peel bend has been established, peeling is controlled once more by equation (2.17) (i.e. $X/w = R$). The effect of the interfacial toughening factor E_2I_2/E_1I_1 decays, after the crack passes the place where EI changes. Kendall writes

$$X/w = [E_2I_2/E_1I_1]^{Ke}\,R \tag{3.12}$$

where Ke has the value zero before meeting an interface where EI changes, unity at the interface and zero again well beyond the interface. (The dimension of the fully-developed peel bend in the E_2I_2 region, which governs the distance at which the effect of the toughening factor is lost, is given by $l^2 = 2E_2I_2/X$,

Problem 3.7.) Fig. 3.2 shows experimental results for crack velocity at a glass-reinforced interface. Before reaching the interface the crack travels at constant speed. There is a sudden slowing, by a factor of nearly 100 at the interface, and, subsequently, the crack gradually accelerates to reach its original speed a long way past the interface. The full line shows the velocities to be expected on the basis of independent $R(\dot{a})$ data for peeling between glass sheet and rubber tape, viz. R = (const.) $\dot{a}^{0.53}$); (relations of this sort are discussed in Chapter 6). Also shown in Fig. 3.2 is the effect of the interface on a crack travelling forwards in the opposite direction, from the stiffer material into the more compliant. Here the theory predicts that the crack will be accelerated at the interface according to the broken curve. The experimental crosses demonstrate this effect. Similar results were obtained for an interface where the increase in stiffness was provided by suddenly doubling the rubber thickness.

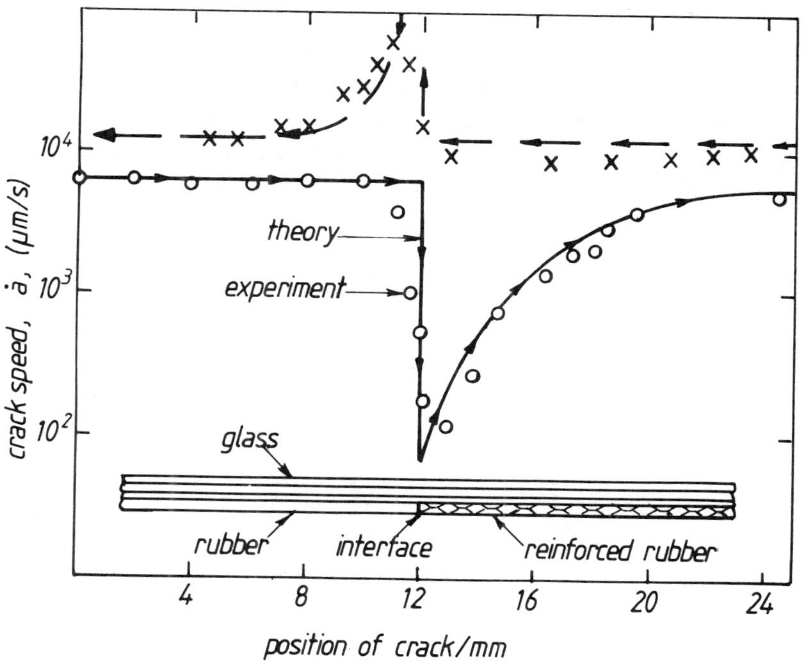

Fig. 3.2 – Effect of reinforcement interface on crack speed at constant applied load. Crack propagation direction is indicated by arrows on the theoretical curves. (After Kendall (1975b), with permission.)

If a material possesses a static threshold toughness, below which no monotonic propagation occurs, and if on meeting a stiff interface the value of the force calculated from equation (3.12) turns out to be less than the value being applied (which is adequate to give propagation in the E_1I_1 section) the crack must *arrest* in the E_2I_2 section. Crack arrest is discussed in full in section 8.6.

Kendall discusses the design of composite tapes with periodic changes in stiffness which have improved resistance to fracture over tapes with constant *EI*. Improvement comes about, not only by the possibility of arrest at stiff interfaces, but also by an increased resistance to propagation of a moving crack owing to the fluctuations in speed as the interfaces are crossed (correct design of reinforced tape will ensure that the crack, while still moving continuously, spends relatively more time going slowly than quickly, thus reducing the average crack speed (for Kendall's $R(\dot{a})$ relation) for a given applied force). However, owing to the inertia of fast-moving cracks and the influence of kinetic energy in the work rate equation at those speeds, interfaces of the sort discussed can bring *no* improvement to dynamic toughness or 'impact peel resistance'. Even so, Kendall's work is germane to crack propagation in anisotropic materials, and is relevant to the use of crack arresters (ribs and such like) in large structures.

3.2 ELASTIC FRACTURE IN COMPRESSION

The major modes of cracking of brittle materials in compression are either cone fractures (mentioned by Coulomb in the eighteenth century) or axial splitting (noted by Foppl at the end of the nineteenth century), Fig. 3.3 (taken from Kendall, 1978). In practice, fractures are often complicated to interpret in low (K_{IC}/σ_y) materials owing to specimens shattering to many pieces which is caused by crack stability difficulties discussed in Chapter 8.

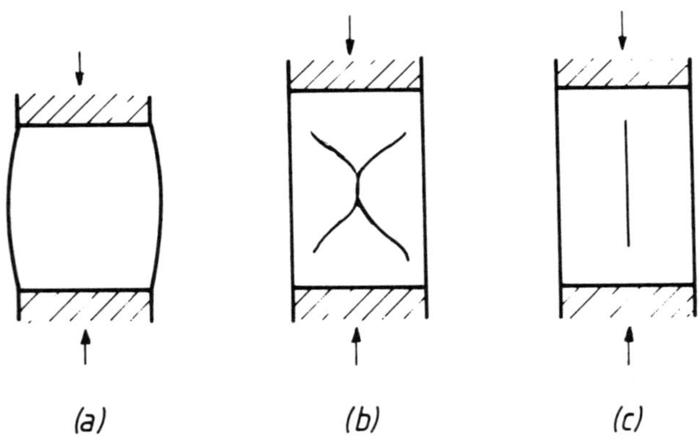

Fig. 3.3 – Compression failure: (a) plastic yielding, (b) cone fracture, (c) axial splitting. (After Kendall (1978), with permission.)

These days, experiments are performed on specimens which have starter cracks already cut in them and fracture mechanics principles applied to the results or, in supposedly flaw-free materials, as with Griffith's tensile experiments on glass, fracture initiation is attributed to inherent microflaws in the material. Historically, however, it was customary to talk in terms of a 'compressive strength', given either by the fracture load divided by the contact area between specimen and loading platens, or by the maximum stress in the body predicted by some strength-of-materials expression. Even today, the Brazilian disk test (Fig. 3.4) for example, employs compression splitting to assess the strength in compression of brittle materials used in civil engineering; related geometries used by other workers, such as geologists, are also shown in this figure. From such measurements, the 'compressive strength' is found to be an order of magnitude greater than the 'tensile strength' of the same material (e.g. Jaeger and Cook, 1976).

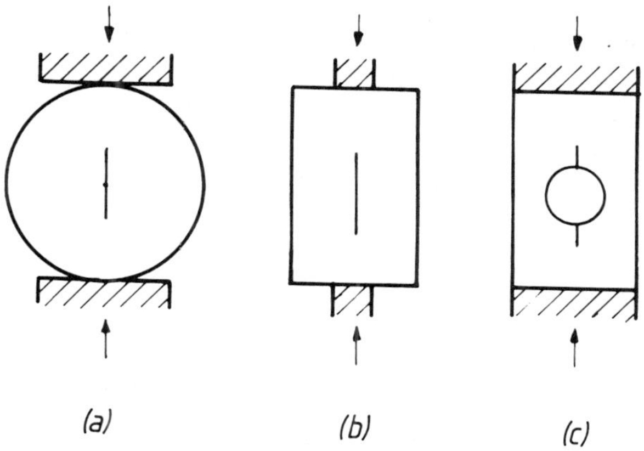

Fig. 3.4 – Compression splitting: (a) disk or Brazilian, (b) indentation and (c) circular hole. (After Kendall (1978), with permission.)

One of the problems with interpretation of brittle compression tests is that there is great variability in the results and that, in pure compression, cracks are closed up and fracture is inhibited. It is well-known that von Karman in 1911 caused marble and sandstone to deform plastically in compression by application of sufficiently high superimposed hydrostatic pressure; similar studies have been performed by Böker and by Bridgman (1935). As pointed out by Polakowski and Ripling (1966), the plasticity of brittle materials under high hydrostatic pressures accounts for the 'crackless' formation of undulations and bends in geological strata by normally brittle rocks.

Pure compression is difficult to achieve experimentally and there will always be regions of tensile stress in testpieces caused by, for example, platen friction, if not testpiece geometry itself. Thus, failures in *compression* in Brazilian and similar tests have been analysed as *tension* fractures caused by tensile stresses perpendicular to the crack plane; strengths calculated on this basis allegedly agree with tensile measurements (the Brazilian test is sometimes called an indirect tensile test).

Results on many materials for a wide range of specimen geometries show, however, that the compressive stress required for cracking is not really constant, but varies with geometry, lateral pressure, and specimen size (large specimens cracking, but small specimens plastically yielding). From what has been written so far in this book, it should not be surprising that the compressive strength of brittle solids is not constant for, if the global failure stress σ_f depends on the propagation of cracks which is governed by $\sigma_f = (1/Y)\sqrt{E^*R/\pi a}$, σ_f varies with inherent flaw size (a), and specimen geometry and load distribution (Y). The invariant controlling parameter in such tests must be R, the specific work of fracture; that, not σ_f is the constant material property over all tests.†

Kendall (1978) has presented an analysis of axial splitting in compression in terms of an energy rate balance which brings out the salient features of compressive fracture. In Fig. 3.5 a solid block of material containing an axial crack is built in at one end and compressed axially by a rigid flat platen of width w at the other. It is clear that compression is not the sole mode of deformation of the sample, since the crack effectively divides the specimen into two short struts which can bend and shear outwards under axial compression to open up the central starter crack. This tendency is greatest for small w and high platen friction.

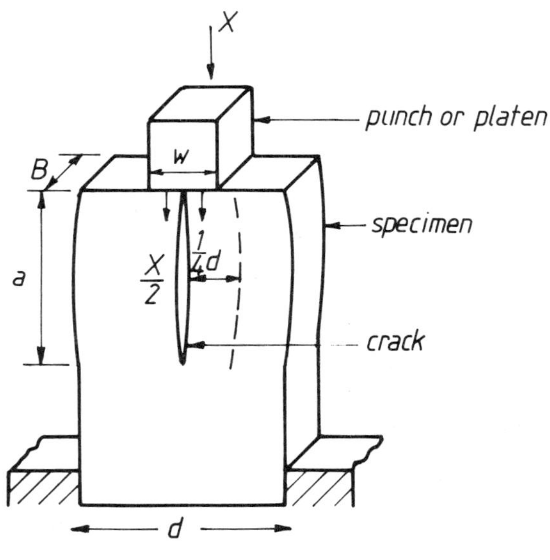

Fig. 3.5 – Geometry of the compression split. (After Kendall (1978), with permission.)

The total elastic strain energy is composed of (i) elastic compression of the struts and (ii) elastic bending of the struts. For small axial deformations prior to cracking, the strain energy produced by elastic compression remains essentially constant with respect to crack length whereas the bending strain energy increases

† Similar thoughts apply to the invariance of interfacial toughness in peeling tests of thin films of paint and so on, rather than peel 'strength'.

with sideways deflexion. Furthermore, the bending moment for small axial deformations is essentially constant along the struts, and is given by $X(d-w)/8$. From 'bending strain energy = $(M^2/2EI)$ (beam length)' we have

$$\Lambda_{\text{bending}} = 2\,(X^2\,(d-w)^2/64 \times 2\,EI)\,a$$
$$= 3\,X^2\,(d-w)^2\,a/2E\,Bd^3$$

using $I = Bd^3/96$.

In a linear elastic system $\Lambda = \Omega$ so using $R = (\partial\Omega/\partial A)_X$ we obtain

$$R = 3X^2\,(d-w)^2/2E\,B^2d^3$$

or
$$X = \frac{B}{(1-w/d)}\,(2ERd/3)^{1/2} \tag{3.13}$$

This relation demonstrates that 'compressive strength' cannot be constant and will vary with elasticity, geometry and fracture toughness. For example, as the platen is increased in width, the force needed for cracking increases as $(1-w/d)^{-1}$. In particular, the splitting force rises rapidly as $w \to d$. In practice, the force cannot rise indefinitely owing to plastic yielding under the platen, when $X = Bw\sigma_y$, with σ_y the flow stress in compression. There is, in other words, a 'deformation transition' from splitting to yielding: this sort of thing is discussed in section 3.9 as it relates to fracture toughness testing, and in detail in Chapter 10.

The foregoing treatment assumes simple axial splitting, but the more general problem concerns cracks or notches inclined to the direction of loading, Fig. 3.6. Not only are we concerned with the forces associated with crack propagation, but also the very direction of propagation. Does the crack start off in its original direction from its tip and continue in that direction, does it begin to turn, or does it start off in a direction different from the initial crack direction or even start off not from the original crack tip? These questions of path stability are discussed in section 3.5 including the observation that whereas, in *tensile* loading of an asymmetrical starter flaw, the propagating crack follows a curved path which eventually becomes *normal* to the applied load, in *compression* loading of the same crack, the path is such that it becomes *parallel* to the applied compression (Fig. 3.6; cf. Fig. 3.18 later).

In analyses of the compression failure of brittle solids, the idea that fracture occurs when the local crack tip tensile stress exceeds the 'cohesive strength' of the material, has found popular appeal. Analyses of brittle fracture under biaxial and triaxial stress systems, in terms of $\sigma_1\ \sigma_2\ \sigma_3$ 'failure loci', using this approach, may be found in McClintock and Argon (1966). Griffith himself showed that the dependence of strength on crack length given by his energy-based criterion for cracking (viz. $\sigma_{cr} \propto 1/\sqrt{a}$) was mimicked by crack-tip cohesive strength models, *if* it were assumed that the root radius of the crack is small and independent of crack length. The next example shows this.

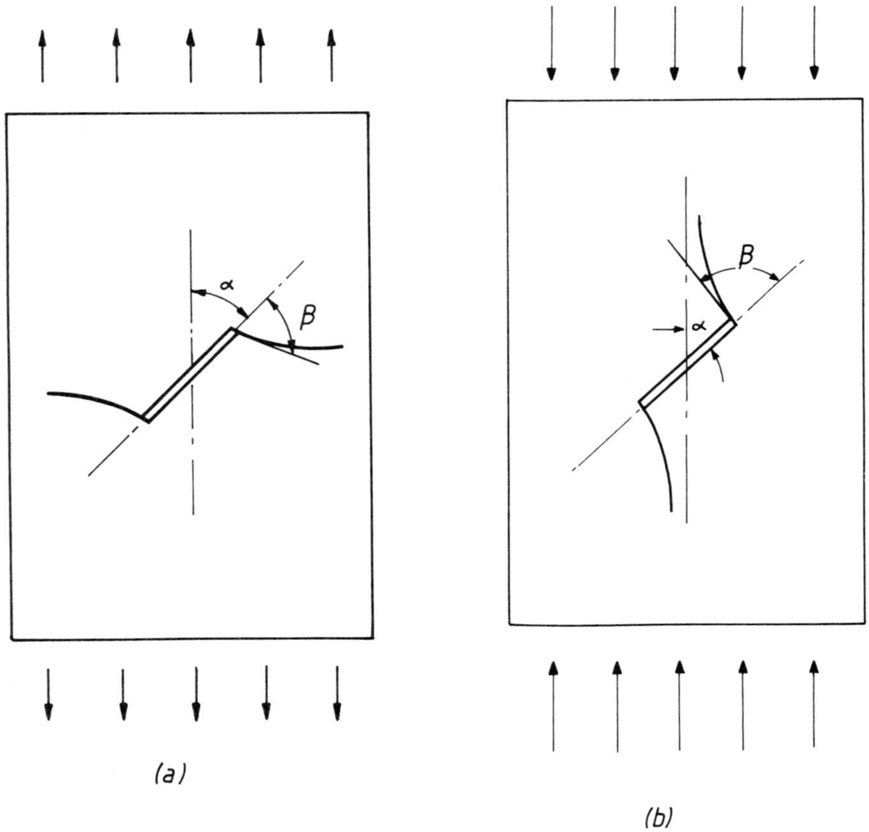

Fig. 3.6 – Fracture path of an inclined crack in brittle solids under (a) axial tensile stress and (b) compression.

Example 3.2 Show under what conditions the same σ_{cr} vs. a dependence is given by both the Griffith theory and a cohesive strength theory.

Ans. The σ_{yy} stress concentration at the end of the major axis of an elliptical slit loaded remotely by $\sigma_{applied}$ is given by the Inglis solution as $(1 + 2\,a/b) \approx (1 + 2\sqrt{a/\rho})$ where a and b are the lengths of the semi-major and semi-minor axes and ρ is the radius of curvature at the end of the major axis (section 2.5). For $\rho \ll a$ (i.e. sharp cracks) the stress concentration is given essentially by $2\sqrt{a/\rho}$ and if fracture occurs when the concentrated stress is equal to the theoretical cohesive strength σ_{th}, we have $\sigma_{applied}\ 2\sqrt{a/\rho} = \sigma_{th}$. An estimate for σ_{th} is as follows: the inter-atomic stress–strain curve may be approximated as $\sigma = \sigma_{th} \sin(2\pi x/\lambda)$ where x is the distance atoms are moved from their equilibrium spacing a_0 by the application of a load, and λ is some artificial wavelength. For small x, the crystal strain increments are dx/a_0 and in the elastic region $d\sigma \div dx/a_0 = E$. Thus since $d\sigma/dx = (2\pi/\lambda)\,\sigma_{th} \cos(2\pi x/\lambda)$ and $\cos(2\pi x/\lambda) \approx 1$ for small

(x/λ), $\sigma_{th} = \lambda\, E/2\pi a_0$. The surface energy 2γ corresponds effectively with the area under the inter-atomic stress–strain curve, i.e. $2\gamma \approx \int_0^{\lambda/2} \sigma_{th} \sin(2\pi x/\lambda)\, dx = \lambda\sigma_{th}/2\pi$. Therefore, replacing λ in $\sigma_{th} = \lambda E/2\pi a_0$ by this relation involving γ, gives $\sigma_{th} \simeq \sqrt{2E\gamma/a_0}$. Thus, from $\sigma_{th} = a_{applied} \cdot 2\sqrt{a/\rho}$, we have $\sigma_{applied} = \sqrt{E\gamma\rho/2\, a_0 a}$. (see, for example, Cottrell (1964)). Hence the fracture stress based on a cohesive strength model varies inversely with $\sqrt{a}$, exactly the same as predicted by the energy-based theory, *when cracks are sharp.*

Example 3.2 employed the stress concentration of a crack *perpendicular* to the remote stress. Cotterell (1972) remarks that the conclusion of Example 3.2 (based on cracks normal to the tension field) is *not* general for *inclined* cracks, so cohesive strength models do *not* predict the same results as energy models, cf. the cross terms in equation (3.23) following preclude a direct equivalence between energetic and maximum stress criteria (see Swedlow, 1965). Furthermore, the conditions under which the two approaches do give predictions of similar magnitudes depend on the crack sharpness, and that the degree of required sharpening is *different* for tension and compression. He goes on to compare and contrast the predictions of the different lines of attack (both for tension and compression) and the results of experiments. Interpretation of data is complicated by the fact that starter cracks introduced into model testpieces by machining, for example, can hardly be expected to possess the sharpness of natural cracks. Analysis should therefore be in terms of 'open elliptical cracks' (e.g. using the Inglis analysis for ellipses with finite root radii) rather than 'slits'. This realisation is just as germane to historical interpretations of compressive stength via cohesive strength models as to fracture mechanics analyses. Cotterell concludes that, although fractures in compression and tension from open cracks can *apparently* be explained by the two separate theories, in truth it is always when G reaches a critical value (R) that fracture occurs; the point about compression is that the solution always depends on the root radius of the slit which, in turn, depends on the machining process used to make the starter flaw. The mechanics in practice concern fracture from notches rather than from sharp cracks. In other words, in laboratory tests, it is impossible for the initiation of fracture from a machined slit in compression to occur under conditions controlled solely by the length of a crack and its orientation. Only for the sharpness of a natural crack can the theory of an infinitely sharp slit be applied and then, because the crack would *close up* in compression, the problem is no londer one of an open slit!

3.2.1 Indentation hardness of brittle solids

Hydrostatic compression exists under many indenter geometries so, owing to its suppressive effect on crack formation, plastic flow can often be produced in hardness tests on materials which would normally fracture brittlely in tension. Hardness testing is a useful method of assessing flow properties of materials when other methods fail. That fracture is suppressed under indenters is seen from relation (3.13): as w is decreased, the local stress under the punch rises, whereas calculations show that the splitting force is considerably less sensitive to changes in w; thus plastic flow is favoured over splitting. (This deformation transition is discussed further in Chapter 10.)

Indentations combined with cracks are often produced in hardness tests, however, or even simple cracking alone with no permanent impressions. There was a time when people who produced cracks around indentations used to hide them away. Nowadays the study of fractures around indentations is receiving wide attention. Since the stress and strain fields beneath indenters are complex, it is not surprising that patterns of cracking turn out to be complex, particularly when some cracks are produced during unloading and by residual stresses, and interact with those cracks which have been produced during loading.

The best-known type of crack associated with indentation is probably the Hertzian ring crack produced by spherical indenters in brittle solids which propagates down into the interior of the solid under increasing load as a stable cone crack (Fig. 3.7). In 1891 Auerbach observed that the critical (elastic) stress for the formation of a cone crack in glass beneath a ball indenter decreased with increasing indenter radius: the 'fracture strength' was *not* constant. For

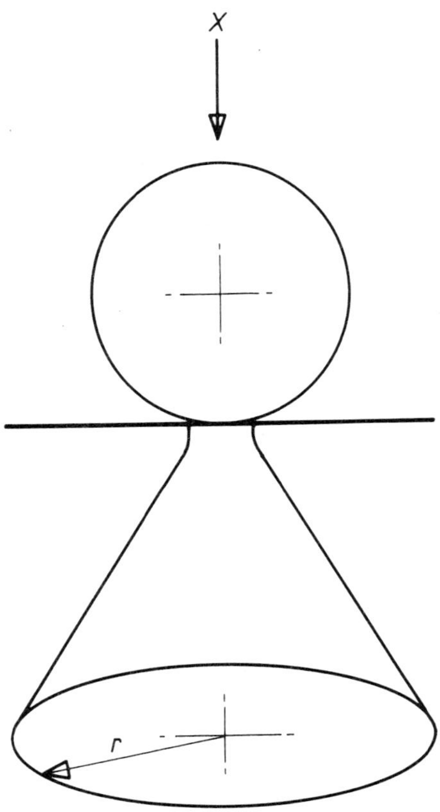

Fig. 3.7 – Cone crack developed by ball indentation.

many years this, and similar results were explained statistically on the basis of the distribution of sizes of natural flaws in the glass. However, in a classic study in 1956, Roesler showed that all the data satisfied a condition of constant strain energy just before fracture. This observation is identical with a constant R for fracture. Roesler (1956) gave the load for cone fractures as

$$ER = f(\nu)\, X^2/r^3 \tag{3.14}$$

where $f(\nu)$ is a function of Poisson's ratio and r is the base radius of the frustum (Appendix 1). At constant R, the cracking load increases with penetration, since equal vertical penetrations produce progressively larger increments of crack area

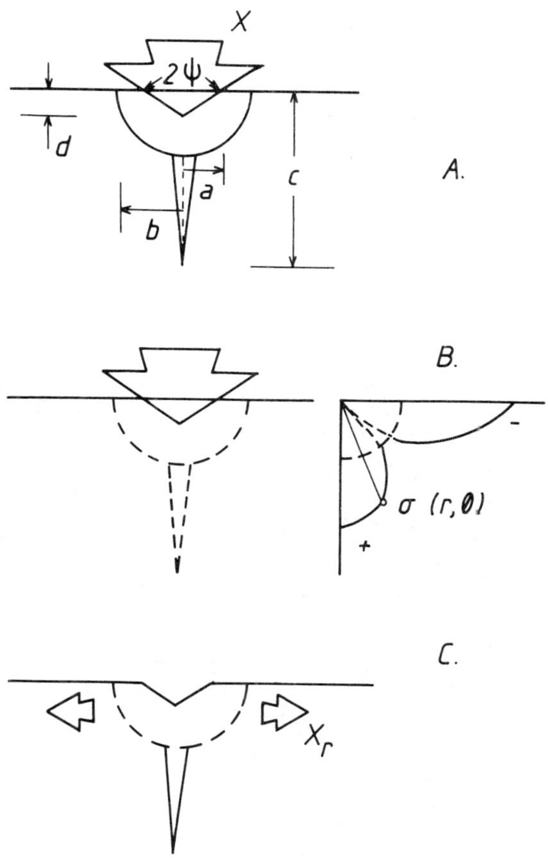

Fig. 3.8 – Median/radial crack system, showing (A) elastic/plastic configuration at full load subdivided into (B) elastic component at full load plus (C) residual component at complete unload. In Lawn *et al.*'s model the indentation load X determines the intensity of of the crack driving forces: the elastic component is characterised by the distribution of prior stresses $\sigma(r, \phi)$ normal to the crack plane (shown in (B) by stress contours in the median plane itself); the residual component is characterised by a residual centre-opening force exerted by the radially expanded plastic zone. (After Lawn *et al.*, (1980), with permission.)

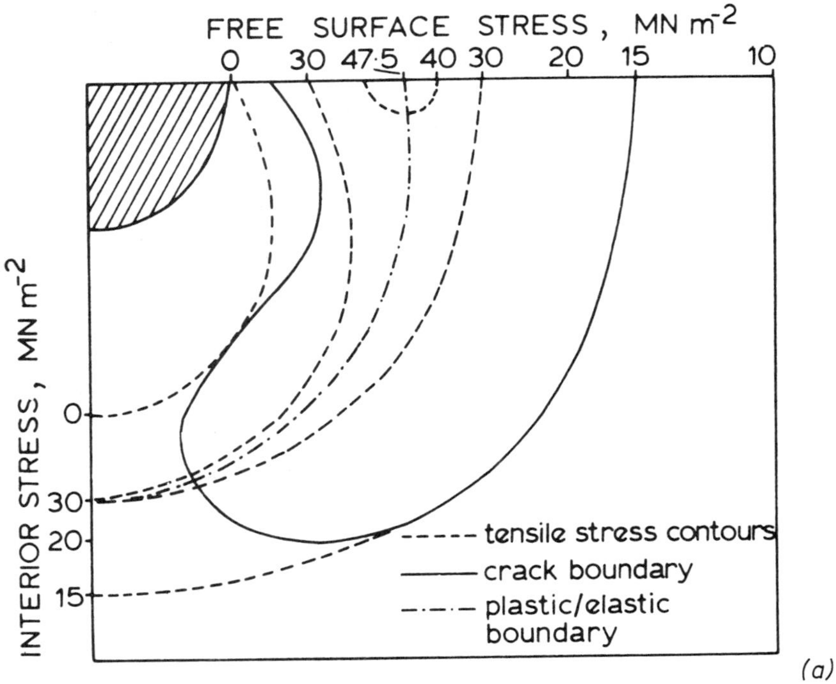

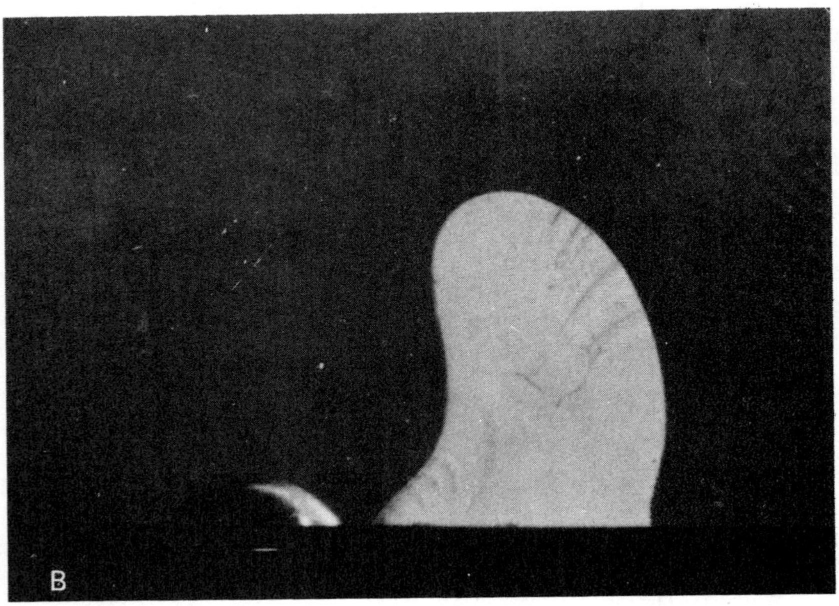

Fig. 3.9 – Formation of radial cracks for $\sigma_\theta > \sigma_r$. (a) Stress field. (b) Asymmetric fracture lobe around outside of elastoplastic zone. (After Puttick *et al.*, (1977), with permission.)

growth around the skirt of the growing cone fracture. Cracking under these conditions is stable and is easily controlled (Chapter 8).

Indentations in brittle solids with 'sharper' geometries of contact, produce a different sort of cracking, wherein fractures initiate below the indenter in the interior and subsequently spall through to the surface (Fig. 3.8). Cracks formed in these circumstances are classified as 'median' cracks (on planes containing the load axis) and 'lateral' cracks (on planes parallel to the indented surface).

With elastoplastic solids possessing higher E/σ_Y ratios, the stress system under ball indentations is radically different from the Hertzian case with low E/σ_Y: instead of high tensile *radial* stresses (which produce the ring crack) we have tensile *circumferential* stresses, which promote radial cracking (Fig. 3.9). For all types of indenter there is a tendency for surface radial cracking to predominate, either cracks from the corners of pyramidal indenters or single cracks (or perhaps pairs of cracks at opposite ends of a diameter) with spherical indenters. As discussed by Lawn *et al.* (1980), this remains true even though the *final* shape of the fully developed crack down into the body, after removal of the indenter, may often be roughly semicircular whatever the E/σ_Y ratio.

Example 3.3 Is it likely that indentation fracture can be characterised by surface measurements alone, without regard to the associated depth of fracture?

Ans. Cracks can be formed during both loading *and* unloading. Lawn and co-workers (1980) developed earlier analyses for indentation fracture using an elastoplastic expanding cavity model for hardness, central to which is their separation of the indentation field of force into 'elastic' and 'residual' components. They show that median cracks (i.e. cracks where depth is the controlling dimension) are expected to attain maximum growth during the loading half-cycle of a test, but that radial cracks (i.e. cracks in which the surface dimension is controlling) should continue growing during the unloading phase. Indeed for low E/σ_Y materials, radial cracks may not occur at all until unloading takes place; this happened in their experiments on soda-lime glass which has $E/\sigma_Y \approx 40$; with ZnS, for which $E/\sigma_Y \approx 160$, radial cracks grew throughout (for both solids, $0.7 < K_{IC} < 1.0$ $MPa\sqrt{m}$). Lawn *et al.*'s treatment (which involves addition of K_C expressions for the reversible elastic and irreversible residual components) shows that the radial crack configuration is in equilibrium throughout both loading and unloading parts of the cycle, unlike the median crack configuration. This *does* suggest that indentation fracture may be characterised from surface crack measurements alone, which is an obvious advantage for opaque materials. This conclusion bears on the construction and form of 'universal' diagrams in which radial crack data are used to determine absolute values of fracture toughness of ceramic solids, and also provides an analytical base for the types of plot constructed by Evans and Charles (1976).

The conclusions of Example 3.3 are important for fracture experiments (of all sorts, not only compressive) in brittle solids in which starter flaws have been introduced by mechanical contact in some way. The residual stress field associated with the formation of the starter flaw must be taken into account in assessing the strength of the flawed component (cf. section 3.7). Lawn *et al.*'s paper explains why fracture toughnesses, determined from experiments in which starter flaws have been introduced by making indentations (e.g. Inglestrom and Nordberg, 1974) are often consistently lower than those determined from independent methods. Similarly, the experimental 'failure strength' of such flawed bodies will be lower than that given by a fracture mechanics formula, using the correct R value, owing to the residual tensile stresses introduced by the indentation (cf. section 3.7).

The reader interested in indentation hardness is referred to a review by Atkins (1982) which contains many important references including among other papers, that of Evans (1979).

3.3 FRACTURE IN THE PRESENCE OF MANY CRACKS: POROUS BODIES

Porous bodies, such as ceramics, cements and so on, contain many holes, and therefore potential starter cracks, throughout their volume and on their surface. Again, on other solids, surface scratches can provide microflaws sites which weaken the material. For a given geometry of body, and type of applied loading, one or two of the cracks will have the right combination of length and orientation to satisfy '$K_C = \sigma_{cr}\sqrt{\pi a}\, Y$' and hence cause fracture, whereas the rest will not and will not influence cracking. Conditions of length *and* orientation have to be satisfied, so fracture will not start from the longest cracks if they are aligned towards the direction of loading, say. Within the whole population of cracks and scratches, the 'favourably orientated, critical length' crack is thought of as the controlling parameter in a 'weakest link' argument.

In the absence of information about the distribution of crack orientations, it is customary to find a starter crack length that makes experimental cracking stresses agree with $\sigma_{cr} = K_C/Y\sqrt{\pi a}$. The characteristic crack size is then often identified with microstructural features. Thus, since the strength of beer bottle glass is about 170 MPa with $E = 70$ GPa and R about 20 J/m^2, $\sigma_{cr} = \sqrt{ER/\pi a}$ gives $2a = 30$ μm for the size of Griffith flaws in this type of glass. In ceramics, the derived 'a' often match pore sizes, or grain sizes in highly dense solids.

The lower curve in Fig. 3.10 shows the results of Birchall *et al.* (1982) on cement beams in which artificial flaws of various lengths were cut with a diamond wheel in order to test the inverse crack length relation. The results fit a Griffith curve, but only for artificial flaws above 1 mm size. Smaller flaws have no effect because the cement beams contain natural flaws of comparable size – which can be seen under the microscope on polished surfaces of cement. Thus it is large natural flaws, say 0.1–1 mm in size, which restrict the bending strength of (unnotched) cements to about 10 MPa. This suggests that a fruitful line of

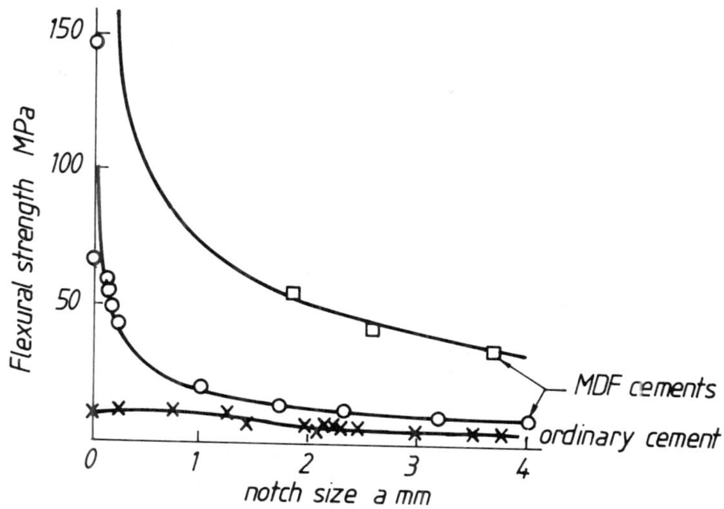

Fig. 3.10 – Bending strength results plotted against notch depth for thee cements. Curves are Griffith lines, the upper macro-deflect-free cement having lesser porosity. (After Birchall *et al.* (1982), with permission.)

attack to improve the strength of hydraulic cements would be to eliminate these large natural flaws, rather than attempt modification of the microstructure by chemical intervention. These thoughts led Birchall and his colleagues at ICI to improve the strength of cements by removing the largest natural flaws by a combination of rheological control and efficient mixing. Surface-active chemicals were added to 'lubricate' the cement grains and thereby permit better packing during moulding. The upper curves in Fig. 3.10 show the strength variation of these so-called 'macro-defect-free' (MDF) cements. For the middle curve, the strength rises to 70 MPa and only aritificial flaws below about 0.1 mm cease to have an effect. Furthermore the strength at any given flaw size is greater, owing to the higher modulus of MDF cement ($E = \sim 40$ GPa in place of ~ 20 MPa) and its higher toughness ($R = \sim 30\ \text{J/m}^2$ in place of ~ 20 J/m^2). Flexural strengths of some 70 MPa are comparable to those of sintered ceramics, but have been achieved without the need for fibrous reinforcement or high pressure compaction. The compressive strengths of MDF cements are likewise elevated, to some 200 MPa, in contrast to some 40 MPa for ordinary cements. The uppermost curve relates to cement with very low volume porosity (ca. 1%), and displays even greater improved properties.

According to $\sigma_{cr} = K_C/Y\sqrt{\pi a}$, strength falls as the characteristic flaw size increases and, it is important to note, the total number of flaws (i.e. porosity or total pore volume) is *only* relevant insofar as it determines the statistical chance of a 'favourably orientated, critical length' crack being in the region of a loaded body actually under strain. Chapter 6 discusses the theory of statistical strength of solids.

Even though fracture strength should not correlate with total pore volume measurements (except in the accidental case of the pore size distribution remaining constant over a series of samples), there is a long-held view (contained in Feret's and Abrams's empirical laws) that total porosity is the prime determinant of strength in cements. These apparently contradictory statements have been resolved by Birchall and co-workers (1982) as explored in the following example.

Example 3.4 The variation of Young's modulus and toughness of cements with porosity fit the relations

$$E = E_0\,(1-p)^3 \qquad \text{(Helmuth and Turk, 1966)}$$

and

$$R = R_0\,\exp(-dp) \qquad \text{(Rice } et\ al.\text{, 1978)}$$

where E_0 and R_0 are the values for fully dense material, p is volume porosity and d is a constant. Give an expression for the fracture stress which takes into account porosity, and show that the Feret and Abrams laws are special cases of the Griffith equation where flaw size remains constant as pore volume changes.

Ans. We may write $\sigma_{cr} = \sqrt{E_0 R_0 (1-p)^3\,\exp(-dp)/\pi a}$ as a modified Griffith formula for fracture strength which takes into account volume porosity, p. The curves in Fig. 3.10 are, in fact, plots of this relation rather than the simple $\sigma_{cr} = \sqrt{ER/\pi a}$ and the difference between the upper two curves concerns p. If the largest flaw size in a series of cement samples is constant, $\sigma_{cr}^2 \propto (1-p)^3$ exp $(-dp)$. This algebraic relation just as well fits data for strengths against the (weight water/weight cement = w/c) ratio in the mix, as does Abrams's rule given by $\sigma_{cr} = (B/D)^{w/c}$ where B and D are constants.

The modified Griffith relation in Example 3.4 explains why both compressive and bending strength display the same porosity dependence, since both types of failure depend on $\sqrt{ER}$. The modified relation should hold for other ceramics, where both the volume of pores and the size of the largest holes influence strength, although it is not clear whether such analyses have yet been attempted. The relationship should be contrasted with expressions such as

$$\sigma = \sigma_0\,\exp(-bp) \tag{3.15}$$

sometimes used for ceramics (e.g. Ryskewitch, 1953) where σ_0 is the (curve-fitted) strength at zero porosity and b is an empirical constant. Data for the strength of ordinary cements suggest $\sigma_0 \approx 20$ MPa in bending and 140 MPa in compression. This has led to the conventional wisdom that cement is only useful in compression, is intrinsically weak in tension, and must be supported by reinforcement. But, as shown by Birchall and co-workers, these values are much elevated when attention is paid to the role of flaw size in strength-porosity relations given by the generalised Griffith relationship in Example 3.4. Such considerations have permitted the manufacture of a helical tension spring from MDF cement! (See Birchall *et al.*, 1982.)

The fracture toughness of some brittle ceramics has been improved by encouraging multiple microcracking in an energy-dissipative zone at the main crack tip (Kreher, Pompe and co-workers, e.g. 1981). An expression for the dependence of the fracture toughness on the density of microcracks, the amount of residual stresses caused by thermal expansion mismatch between the ceramic matrix and small particles embedded in it and the volume fraction of these particles was derived, and used to state conditions necessary for the fracture toughness to be increased. The theory agreed well with experimental results taken from literature relating to alumina containing zirconia particles.

3.4 CRACKING UNDER MULTIAXIAL LOADING

Many of the examples in this book concern loading systems that are essentially uniaxial. But many practical situations concern fracture under biaxial or triaxial external loadings (the crack tip process zone is of course, already under a complicated state of stress and strain even in simple uniaxial load situations).

Axisymmetric problems, such as the stress intensity for a plane circular ('penny-shaped') crack inside a large body loaded by a remote uniform stress, provide examples of simple multiaxial situations. Sneddon in 1946 gave a solution for this rotationally-symmetric version of the Inglis–Griffith two-dimensional geometry, which is

$$K_I = \sigma_\infty \sqrt{\pi a}\,.\,(2/\pi) \tag{3.16}$$

where a is the radius of the crack. Owing to the axial symmetry, the crack tip stress intensity remains simple, i.e. there are no K_{II} or K_{III} components. This sort of relation has been used by Dobbs *et al.* (1973) to study laser-induced internal cracks in bodies which are, of course, isolated from the external environment.

A related geometry is that of the 'pressurised blister test', Fig. 3.11(a) used originally by Dannenburg in 1961 for assessment of the adhesion of paint. Depending on the thickness of the film, the elastic strain energy in the 'blister' may be analysed in terms of bending energy (if the film is really a 'thick plate'), stretching energy (if the film is a 'thin membrane') or a combination of the two (if the film is a 'thin plate'). By way of illustration we note the analysis of Williams (1972), in connexion with debonding of an elastic disk adhered to a rigid substrate. The deflexion $w(r)$ of a clamped plate of thickness h and radius a uniformly loaded by pressure p on one side, is given by

$$w(r) = (p/D)\,(a^2 - r^2)^2/64 \tag{3.17}$$

where D is the flexural rigidity of the plate given by $D = Eh^3/12(1-\nu^2)$.

The volume contained by the blister is

$$\begin{aligned} \mathrm{v} &= 2\pi \int_0^a rw\,dr \\ &= \frac{pa^6\pi}{192D} \end{aligned} \tag{3.18}$$

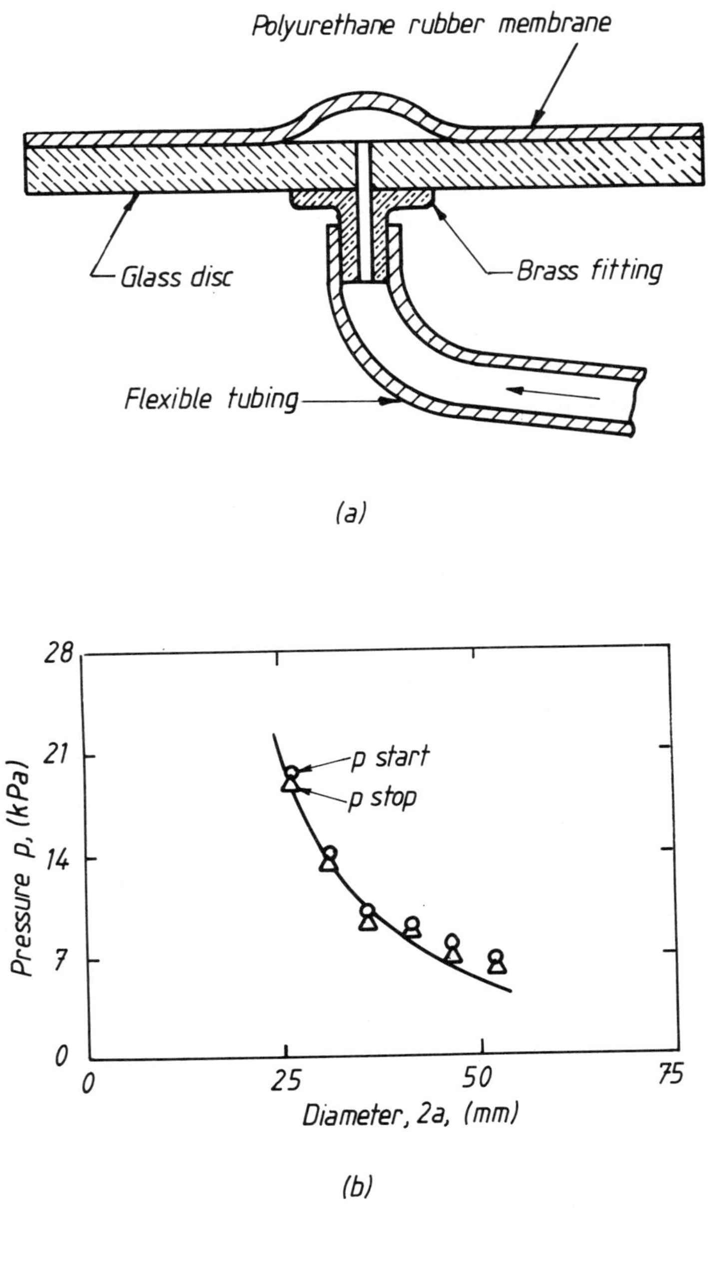

Fig. 3.11 – (a) Sketch of pressurised blister specimen. (b) Typical pressure (p) versus diameter ($2a$) data for the pressurised disk. E = 2.8 MPa, h = 1.0 mm and R = 250 J/m². (After Williams (1972), with permission.)

Therefore, in a linear elastic system, with $\Omega = \frac{1}{2}$pv, we have

$$\Omega = p^2 a^6 \pi / 384D \tag{3.19}$$

and with $A = \pi a^2$,

$$R = (\partial\Omega/\partial A)_P = \frac{3(1-\nu^2)\,p_c^2}{32(E/a)}\,(a/h)^3 \tag{3.20}$$

where p_C is the pressure that causes debonding around the periphery of the blister or 'pancake'. It follows from relation (3.20) that

$$p_C \propto 1/(2a)^2 = 1/(\text{diameter})^2$$

and Fig. 3.11(b) shows the results of Williams's experiments on ca. 1 mm thick rubber membranes (E = 2.8 MPa) bonded to glass plates. The curve corresponds with R = 250 J/m^2.

Note that for large diameter disks, the data depart from the constant R line given by equation (3.20); this is because the decrease in $(h/2a)$ below, say, 1/50 is changing the problem to one of membrane behaviour rather than plate behaviour (cf. the case of a pressurised thin-walled cylinder in section 2.2.2.2 which is discussed in terms of membrane theory). The Sneddon predictions of p are, of course, considerably larger at all a.

Williams and co-workers have extended the analysis to account for the separate effect of an intermediate adhesive layer, by representing the layer as a Winkler elastic foundation (cf. section 8.7.2.2).

An interesting application of this type of test has been to assess the relative *in vivo* efficiency of natural barnacle cement (under water) against commercial dental adhesives. Barnacles were grown over wax-filled holes in PMMA sheets, and the interfacial R determined by the blister tests. Williams (1972) reports that the adhesion shown by barnacles under water was as good as that shown by the best commercial dental cements in dry conditions!

The adhesive contact between a rigid spherical punch and an elastomeric solid has been studied by Barquins (1984) using axisymmetric fracture mechanics, since the edge of contact between indenter and elastomer appears as a crack propagating in mode I at the interface (cf. Kendall's experiments in section 2.4.1).

The following example illustrate a situation where loads in different directions may be altered independently.

***Example** 3.5* Determine the compressive force required for axial splitting of a block of a brittle solid when lateral compressive forces are applied to the body.

Ans. Kendall's energetic analysis for axial splitting (section 3.2) may be made more general to include concentrated lateral force X_L (Fig. 3.12). For simplicity, let X_L act at a distance $\frac{1}{2}a$ along the pre-cracked block. The bending moment produced by the equipollent axial forces $X/2$ is, as before, $X(d-w)/8$ on each beam and may be taken to be constant along the beam. In addition, the forces

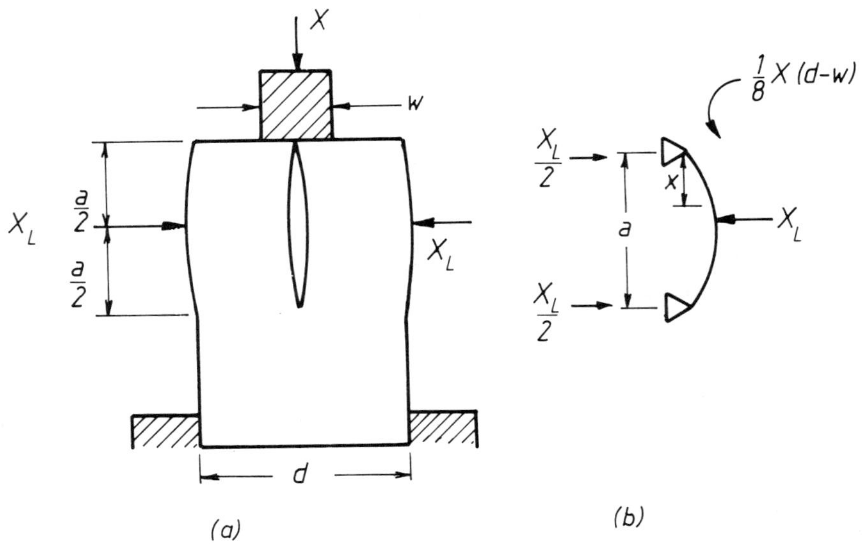

Fig. 3.12 – (a) Compression geometry with lateral force X_L inhibiting bending; (b) the forces acting on the right-hand strut. (After Kendall (1978), with permission.)

X_L acting at 'mid-span' of supposedly simply-supported beam struts (the actual end conditions are unclear) give a linearly-varying bending mement $X_L(x/2)$ of opposite sign to the moment caused by the axial force. The net bending moment at some x ($< a/2$) is $\{X(d-w)/8 - X_L\,x/2\}$ and, from '$\Lambda_{\text{bending}} = (M^2/2EI)\,dx$' for each beam we obtain for the total $\Lambda_{\text{bending}} = 2[3X^2(d-w)^2 a/4 + X_L^2\,a^3 - 3X\,X_L(d-w)\,a^2/2]/EBd^3$. Since $\Lambda \equiv \Omega$ in a linear system, application of $R = (\partial\Omega/\partial A)_X$ gives (assuming X_L is constant) a quadratic equation for X which leads to

$$X = \frac{B}{(1-w/d)}\;[2ERd/3]^{1/2} + 2\,X_L\,a/(d-w) \tag{3.21a}$$

For $X_L = 0$ this reduces to the solution given in section 3.2, and otherwise predicts the requirement of an increased axial splitting force in the presence of lateral compressive loads X_L. The relation shows that X_L causes a linear increase in X and that the increase is also proportional to a. Long cracks are therefore more inhibited than short cracks. Furthermore, small X_L have a large influence on X when the 'magnifying factor' $2a/(d-w)$ is much greater than unity. Experiments by Kendall (1978) confirm the validity of this relation.

Were X_L to represent lateral *tension* loads in Example 3.5, the components of the nett bending moment would be additive and the required X would be reduced. It may be shown that then

$$X = \frac{B}{(1-w/d)}\;[2ERd/3]^{\frac{1}{2}} - 2X_L a/(d-w) \tag{3.21b}$$

Note that no axial force is required (i.e. $X = 0$) when

$$X_L = (Bd/2a)\left(\frac{2ERd}{3}\right)^{\frac{1}{2}}$$

which is a version of the formula in Example 2.3 for DCB specimens, but loaded half-way along the arms and with the ends constrained.

Note that in the solution of Example 3.5, X_L was taken as constant, i.e. a fixed pre-applied lateral load. In contrast, many biaxial load problems relate to proportional loading, where X and X_L are related throughout the test. For example, if $X_L = kX$ in the Kendall analysis (where X_L and X are both compressive), we obtain

$$X = \frac{Bd\,(2ERd/3)^{\frac{1}{2}}}{\sqrt{(d-w)^2 - 4k(d-w)a + 4k^2a^2}} \tag{3.22}$$

Again, for $k = 0$, the relation reduces to that for simple splitting in section 3.2. X is always greater in the presence of lateral compressive X_L so long as $4k^2a^2 > 4k\,(d-w)\,a$, i.e. $k > (d-w)/a$. We remember that the use of simple bending theory implies $a \gg d$, so this theoretical condition probably has no practical significance.

If X_L is tensile, k is negative and equation (3.22) has for the denominator $\sqrt{(d-w)^2 + 4k\,(d-w)\,a + 4k^2a^2}$ and X is less.

Griffith's seminal analysis is usually presented using Inglis's solution for remote *uniaxial* loading, normal to the axis of the crack. Griffith in fact considered *biaxial* stressing, but he assumed that loadings parallel to the crack had no influence on the critical stress for fracture. Eftis and Jones (1982) explain that, over the years, Griffith's opinion in this regard has met with both disagreement and support. The difference of view persists because (i) there is disagreement as to the proper mathematical procedures for calculating the rate at which the elastic strain energy Λ of a cracked infinite sheet changes as the crack length is increased, viz. either calculate $\Delta\Lambda$ for some Δa and take the limit of $\Delta\Lambda/\Delta a$, or calculate Λ for the whole body first and then formally differentiate with respect to a and (ii) until recently there have been few biaxial fracture experiments which determined fracture loads for the Inglis geometry which would provide conclusive supporting evidence one way or the other.†

Swedlow (1965) in a re-examination of the Griffith problem remarked that consultation of a 1923 article by Wolf leads to the suspicion that the Griffith criterion does in fact depend upon the distribution of applied stresses. This may be seen by considering two stress states, one of which contains a stress singularity. If the strain energy density for each stress system is computed and these are summed, the singular behaviour is unaffected by the regular stresses. If, however, the stresses are first superposed and then the strain energy density is computed,

† Experiments on the influence of lateral loads on crack *paths* show, of course, that cracks may be prevented from turning by compressive stresses across the crack plane (section 3.5). Prediction of paths and fracture stresses are intimately connected as we shall see.

the non-singular stresses will in general affect the singular behaviour of the energy through the presence of a cross-product term. A fracture criterion computed from the latter strain energy can therefore be expected to reflect the influence of the regular stresses. In particular, while a given stress singularity will not reflect biaxial loading of a cracked plate, the singularities in the strain energy density may do so, and the fracture criterion must behave in a like manner. Eftis and Jones (1982) review the work of a number of authors who have tackled the Griffith problem, and point out that procedures involving the linear superposition of Λ (which is *not* linear in σ or ϵ) are dubious.

Swedlow (1965) extended the uniaxial Griffith solution (i.e. mathematics which acknowledge the role played by combinations of stresses) to other loadings, viz. (i) a uniformly loaded plate having a pressurised crack and (ii) the general case of remote biaxial stresses combined with remote shear stresses.† For the latter case, the generalised Griffith formula is

$$\sigma_{y\infty}^2 - (\kappa - 2)\,\sigma_{y\infty}\,\sigma_{x\infty} + (\kappa - 1)\,\tau_{xy}^2 = 16\mu\gamma/\pi a = (\sigma^*)^2 \qquad (3.23)$$

where $\kappa = (3 - \nu)/(1 + \nu)$ in plane stress and $(3 - 4\nu)$ in plane strain, with $\sigma_{y\infty}$ the remote stress perpendicular to the crack and μ the shear modulus. (Note the role played by $\sigma_{y\infty}\ \sigma_{x\infty}$ cross-product term in a discussion of the possible equivalence of maximum stress and energetic criteria of fracture in sections 3.2 and 3.5.1.) Swedlow also puts this result into a useful alternative form in which loads are orientated at an angle $-\alpha$ to the x–y axis;) the $K_I K_{II}$ polar distribution of stresses at the crack tips in these circumstances are given by Williams (1957).

Figs. 3.13 and 3.14 give plots for the particular cases of biaxial tension, and shear plus uniaxial tension, given by equation (3.23). Depending on κ (i.e. on Poisson's ratio), the transverse stress $\sigma_{x\infty}$ in the biaxial tension case may either strengthen or weaken the plate. Eftis and Jones (1982) performed plane stress biaxial tensile experiments on 7075-T6 aluminium ($\nu \simeq 0.3$, $0 < \sigma_{x\infty}/\sigma_{y\infty} < 2.0$) and PMMA ($\nu \simeq 0.4$, $-0.5 < \sigma_{x\infty}/\sigma_{y\infty} < 2.5$) and indeed demonstrated an increasing trend for $\sigma_{y\infty}$ with $\sigma_{x\infty}$ in the first case and a decreasing trend in the second case, exactly as predicted. They also showed that Radon *et al.*'s 1976 (Radon and Leevers) and 1977 (Leevers *et al.*) PMMA results agreed with their analysis. Shear may likewise have an important influence on the axial cracking stress, again depending on the magnitude of κ.

Goetz (1965) gives related solutions for remote biaxial stresses and frictional shear stresses on the crack faces, for application to brittle fracture in compression.

In all cases, the size and shape of the crack tip zones are affected by stress biaxiality (see section 7.6).

† While Swedlow in 1965 reproduced the 1924 version of the *uniaxial* Griffith formula (i.e. the one used today without Poisson's ratio under the square root – see Example 3.6 following), these generalised 1965 formulae for *combinations* of stresses are, in fact, based on the 1921 version, as are the much later, but similar, results of Eftis and Jones (1982). Swedlow (1984) has solved the biaxial problem for the proper 1924 formulation, but it is unpublished.

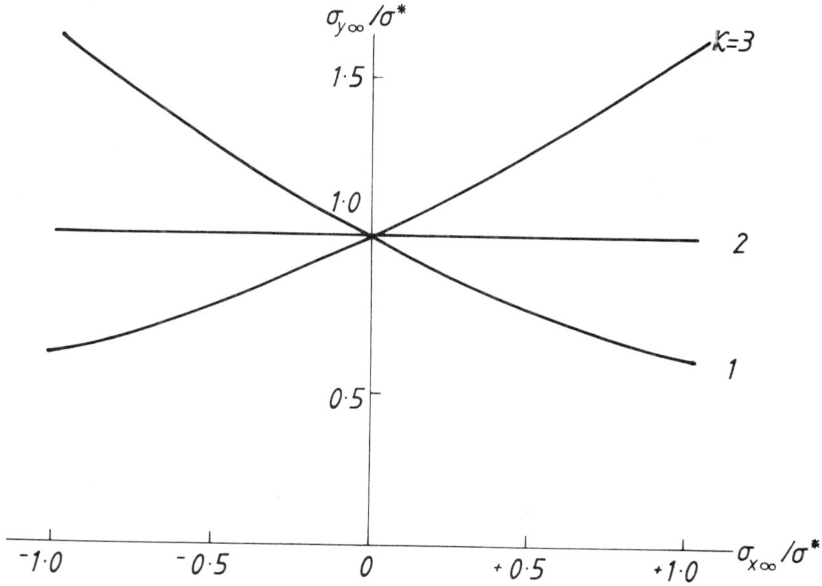

Fig. 3.13 – Interaction of remote longitudinal stress $\sigma_{y\infty}$ (perpendicular to crack) and remote transverse stress $\sigma_{x\infty}$. Influence of Poisson's ratio through κ. (After Swedlow (1965), with permission.)

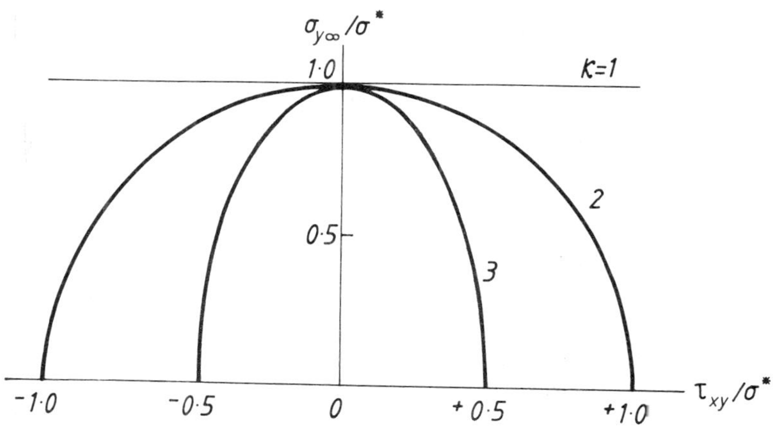

Fig. 3.14 – Interaction of remote longitudinal stress $\sigma_{y\infty}$ (perpendicular to crack) and remote shear stress τ_{xy}. Influence of Poisson's ratio through κ. (After Swedlow (1965), with permission.)

Example 3.6 At the galley-proof stage, a footnote was added at the end of Griffith's (1921) *Phil. Trans.* paper to the effect that '. . . the method of calculating the energy of a crack plate . . . requires correction. . . . The correction affects the numberical values . . . but not their order of magnitude. The main argument of the paper is therefore not impaired . . . but some reconsideration of the experimental verification of the theory is required'. Griffith later presented a revised version of the analysis in his 1924 Delft paper. Discuss the differences between the predictions of the theory and Griffith's emendations. Was he correct to make changes, or was he right in the first place?

Ans. The presence of a crack decreases the stiffness of a body and thus, under given boundary tractions, must increase the strain energy by $\Delta\Lambda$, say (cf. Example 2.5). Griffith in his 1921 *Phil. Trans.* paper using Inglis's mathematics, gave $\Delta\Lambda = (3 - \kappa)(1 + \nu)\,\pi\sigma^2 \cosh 2\alpha_0\, \Delta(a^2)/4E$ for a general elliptical cavity defined by the elliptical coordinate α_0, which for a slit with $\alpha_0 = 0$ is $\Delta\Lambda = (3 - \kappa)(1 + \nu)\,\pi\sigma^2\, \Delta(a^2)/4E$ in present notation. Using this in equations (2.11) or (2.12), he went on to give $\sigma = \sqrt{2E\gamma/(1 + \nu)\nu\,\pi a}$ for plane strain and $\sigma = \sqrt{2E\gamma/\nu\pi a}$ for plane stress, where $\gamma = R/2$ is the surface free energy (see Example 2.5). He obtained excellent agreement between his measured fracture stresses on glass rods and fibres and $\sqrt{2E\gamma/\nu\pi a}$. He came to believe, however, that Poisson's ratio ν should not be in the denominator, hence the footnote to the paper. If so, this meant that the derived values of γ were low by a factor of 3 or thereabouts (i.e. $\nu \approx 1/3$). His reason for changing his analysis is given in Griffith's 1924 Delft paper: 'A solution of this problem was given in a paper read in 1920 but in the solution there given the calculation of the strain energy was erroneous, in that the expressions used for the stresses gave values at infinity differing from the postulated uniform stress at infinity by an amount which, though infinitesimal, yet made a finite contribution to the energy when integrated round the infinite boundary. This difficulty has been overcome by slightly modifying the expressions for the stresses, so as to make this contribution to the energy vanish . . .'. He revised $\Delta\Lambda$ to $\Delta\Lambda = (1 + \kappa)(1 + \nu)\,\pi\sigma^2\, \Delta(a^2)/4E$ and obtained $\sigma = \sqrt{2E\gamma/(1 - \nu^2)\pi a}$ for plane strain and $\sigma\, \sqrt{2E\gamma/\pi a}$ for plane stress, which are the familiar usual forms quoted for Griffith, remembering that $E^* = E/(1 - \nu^2)$ for plane strain. He did not indicate in 1924 the modifications he made to his manipulation of the basic Inglis solution. Now the energy balance of the plate must include the surface free energy created by additional crack surface, the change in elastic strain energy and the transfer of energy at the edges of the plate. It is this term which is not correctly handled in Griffith's original paper according to Goetz (1965). Since the stress distribution remote from the crack approaches constant values in the Inglis problem, Griffith, it seems, took their rate of change with respect to crack length to be zero on the far boundaries (Sack in 1946 also did this). However, a subtle point arises, in that second order terms in the complicated stress distribution at regions remote from the crack make first order contributions to the total energy change, and when the mathematical operations are performed, Goetz found

$$2(\kappa + 1)\,\pi a\, \sigma^2/8\mu = 4\gamma$$

which, with $\kappa = (3 - \nu)/(1 + \nu)$ and $E = 2\mu(1 + \nu)$ for plane stress, gives the classic corrected Griffith equation, viz. $\sigma = \sqrt{2E\gamma/\pi a}$. However, Swedlow

(1984) remarks that Goetz's line of attack is a modification of the method of calculation, whereas Griffith said he had modified the stresses. Also, as Swedlow additionally comments, the problem is more complicated than anyone might have guessed! Indeed Eftis (1985) now believes that the 'Westergaard equations,' upon which much of linear elastic fracture mechanics has been based, are partially in error owing to a mistake unwittingly carried over by Westergaard from MacGregor's 1935 paper on potential functions. In this 1965 paper, Swedlow reproduced the 1924 Griffith result by considering a uniform stress to act on the crack faces in the Inglis geometry with the outer periphery free of loads; the stress system for this case differs from the usual Griffith case solely by a constant term and Swedlow points out that the distinction between the two is the same as that found between a 'first-term-of-a-series' (lowest eigenvalue) representation of stresses and a full solution. In his first paper, Griffith had argued that it was necessary to anneal his specimens (causing the fracture strength to increase) in order to remove residual stresses, and got very good agreement between the measured fracture strengths and those predicted from the surface energy measurements. Having come to believe that the 1920 analysis was in error, he then had to argue that the annealing treatment had blunted the cracks in glass and so had artifically increased the observed fracture strength. By decreasing the annealing time he obtained fracture strengths that agreed with the corrected theory.

Inspection of the expression given by Eftis and Jones (1982) for $d\Lambda/da$ in the biaxial case shows that, for given σ and a, $d\Lambda/da$ may be positive, negative, or zero depending on the values of σ_x/σ_y and κ. The $(\sigma_x/\sigma_y, \nu)$ boundaries at which $(d\Lambda/da)$ changes sign are shown in Fig. 3.15. Both pairs of loci are asymptotic to $\sigma_x/\sigma_y = \infty$ at the Poisson's ratio values (1/3 in plane stress and 1/4 in plane strain) at which σ_y in Fig. 3.14 changes from increasing with increasing σ_x to decreasing with increasing σ_x. We note that all the experiments performed by Eftis and Jones, to substantiate the biaxial Griffith expression for fracture stress, lie in the $d\Lambda/da < 0$ region of the diagram.

Eftis and Jones (1982) remarked that when $d\Lambda/da$ vanishes the Griffith hypothesis is no longer meaningful. This cannot be true: it would suggest that the application of a biaxial stress of finite magnitude should prevent fracture altogether. Swedlow (1965) also remarks that, according to equation (3.23), incompressible brittle materials under conditions of plane strain are predicted to possess infinite shear strength, cf. the horizontal line for $\kappa = 1$ in Fig. 3.14(b).

What is the explanation? Firstly, we note that the conditions under which $d\Lambda/da$ changes sign are (i) (in the tensile cases with σ_x positive) when the loading parallel to the crack σ_x is much greater than the stress σ_y applied perpendicular to the crack and (ii) (in compressive lateral cases with σ_x negative) when the magnitude of the lateral loading is much greater than σ_y. This is where the question of likely crack paths must feature in any analysis for, under these sorts of biaxial loadings, it might be expected that the crack will veer away from its starting direction and *turn*. The Eftis and Jones (and Swedlow) calculations

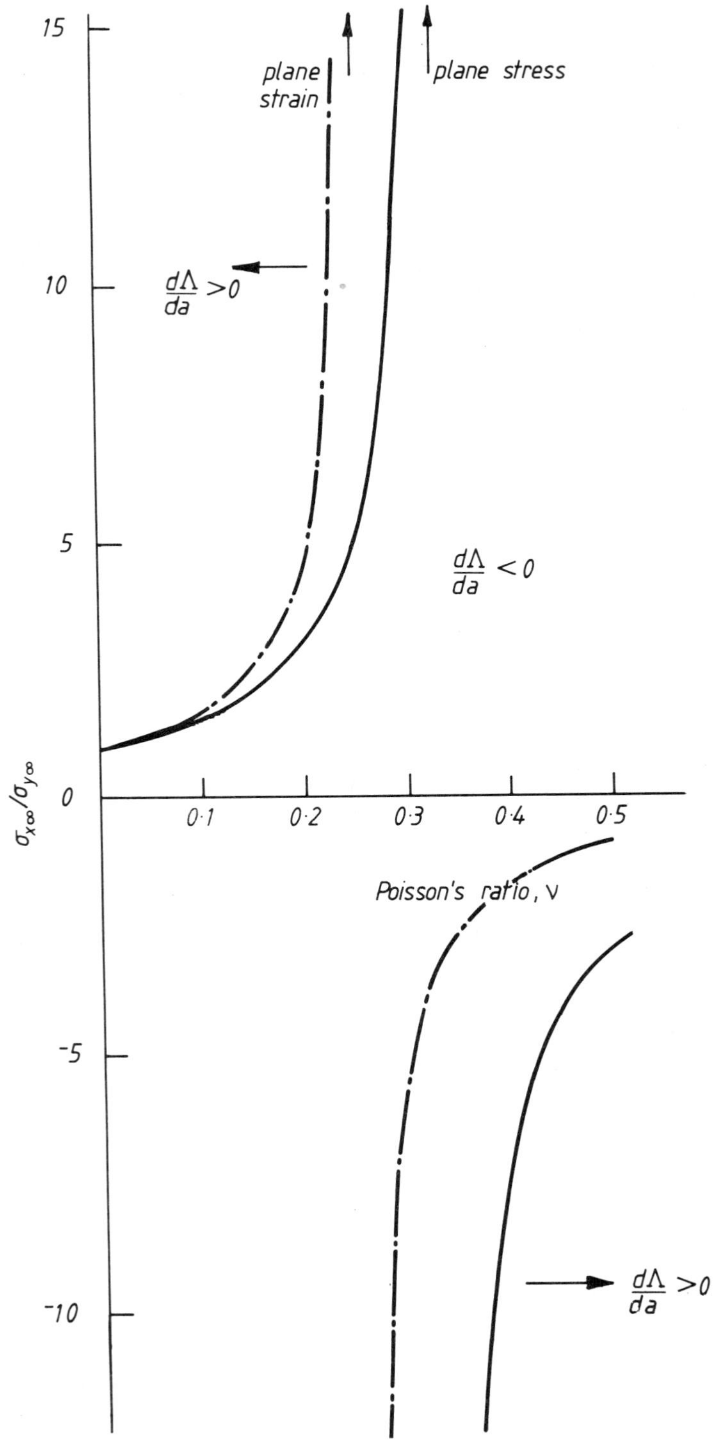

Fig. 3.15 – Relation between load biaxiality ratio and Poisson's ratio (ν) for which the strain energy release rate changes sign.

assume that the crack grows in its own starting direction (i.e. propagation is what is called 'self-similar'). When $d\Lambda/da$ calculated with this assumption changes sign, and $\partial\Omega/\partial a)_{\sigma \text{ or } X}$ does *not* satisfy the value of R, it means that it is energetically more favourable for the crack to propagate in *another direction* in which the (different) $(\partial\Omega/\partial A)$ *does* satisfy $(\partial\Omega/B\partial a)_{\sigma \text{ or } X} = R$.

Thus cracking loads depend on crack paths and vice versa. Further discussion on path 'stability' (i.e. whether the crack turns from its starting direction, or even starts off in an entirely different direction) will be found in section 3.5, where the general 'kinked' or 'branched' crack problem is discussed. It will be explained that there is an optimum direction (not always in the original line of direction of the crack) along which the crack propagates (satisfying $(\partial\Omega/\partial a)_{\sigma \text{ or } X} = R$). While there are mathematical differences between the Eftis and Jones, and Swedlow, papers (Eftis and Jones believing that the 1921 Griffith paper is, in fact, correct), both sets of calculations presume that the crack gets longer in its original orientation. The type of calculation performed for the uniaxial Griffith case by Palaniswamy and Knauss (1974) has, it seems, yet to be performed for the general case of an angled crack with a kink at the tip under general loading (see section 3.5.1).

The assumption of self-similar cracking is implicit in the procedure of adding elastic stress intensity factors, caused by different superimposed elastic loadings, to get the effective crack tip applied K. For example, if a single-edge-notch specimen is subjected to combined axial loading σ and pure bending M (Fig. 3.16) we have

$$K_{\mathrm{I_{nett}}} = \sigma\sqrt{\pi a}\,.\,Y_1\,(a/W) + \frac{6M}{BW^2}\,\sqrt{\pi a}\,.\,Y_2(a/W)$$

where Y_1 is the geometric factor for simple tension and Y_2 that for pure bending. From calibration cruves, or polynomial expansions, we have at $a/W = 0.5$, say, $Y_1 = 2.83$ and $Y_2 = 1.49$. Hence

$$K_{\mathrm{I_{nett}}} = (2.83\,\sigma + \frac{8.97M}{BW^2})\,\sqrt{\pi a}$$

Fracture should occur when $K_{\mathrm{I_{nett}}} = K_{\mathrm{IC}}$.

In other states of combined loading, stress intensities in different modes may occur simultaneously. For example, a notched round bar if twisted as well as simultaneously bent, will have both K_{I} and K_{III} components at the crack front, with K_{I} relating to crack opening on the tensile side of the bend but crack closure on the compressive side. These applied Ks have nothing to do with fracture and will disappear on unloading. Fracture occurs when the stress intensity reaches a critical value. In combined mode cracking, it is a critical value of an 'effectve' stress intensity factor which is a function of K_{I} K_{II} and K_{III}. The particular combination of Ks leading to the critical value is not altogether clear, and is the subject of research; one suggestion is that failure under monotonic biaxial loading is governed by

$$2(1 - K_{\mathrm{I}}) = 3\,K_{\mathrm{II}}^2 \qquad (3.24)$$

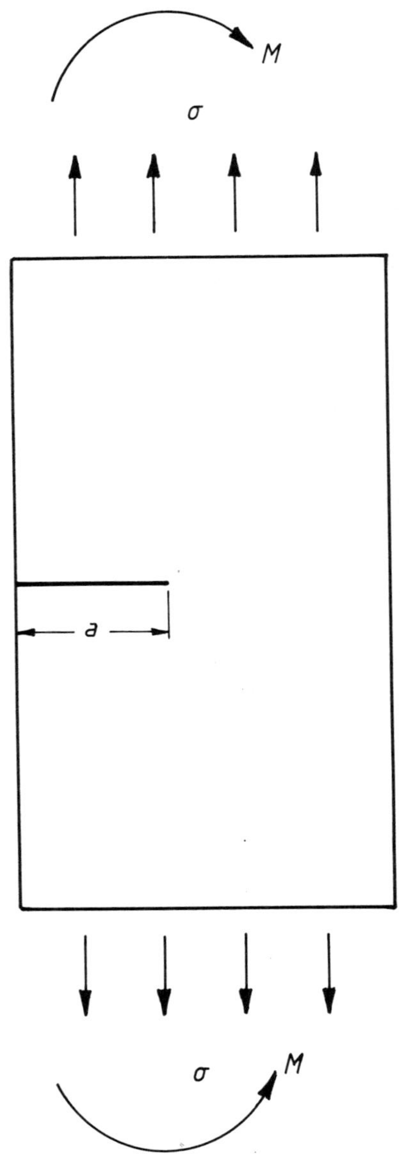

Fig. 3.16 – An edge crack plate subjected to bending and uniaxial stressing.

3.5 CRACK PATHS

Simple experiments on brittle solids seem to indicate that cracks propagate perpendicularly to the direction of maximum tensile stress. For example, a tension test of glass or cast iron results in a fracture surface perpendicular to the tensile axis of the specimen, and the paths of cone cracks in glass produced by ball indentation apparently coincide with the trajectories of maximum

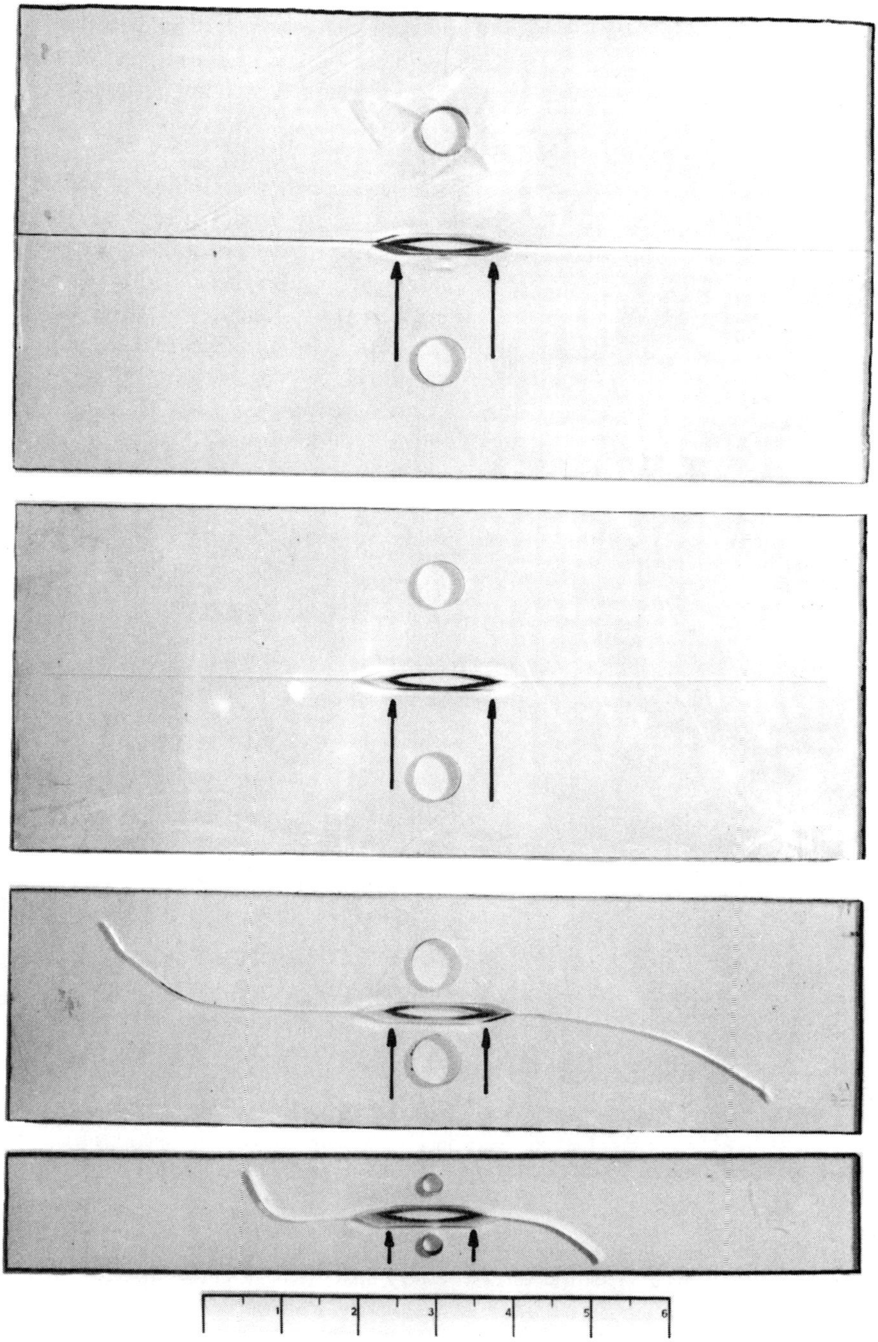

Fig. 3.17 – Effect of beam height on crack paths in the fracture of PMMA (bottom marker is in inches).

tensile stress. A stick of chalk, when twisted, breaks along helical paths where the tensile stress is greatest, but breacks across in a flat fracture when bent. With ductile solids, crack paths in plastic fracture are often more influenced by shear stresses (e.g. final cracks sometimes running around the outside of the plastic hinge in a Charpy testpiece).

Consider the series of centre-cracked panels in Fig. 3.17 loaded in tension in which the only difference is the height of the panel. For tall panels, the cracks run straight across the specimen, but as the height is reduced, the crack path becomes smoothly curved after some propagation, until at very small heights the cracks veer round immediately propagation begins with sharper curvature than before. When the starter crack is at an angle to the major axis of loading, the direction of propagation can itself be at an angle to the line of the starter crack, i.e. not a smooth continuation of the starter crack, but a kink (Fig. 3.18; cf. Fig. 3.6). Propagation follows a curved path which eventually becomes normal to the load. Similar results are found when bodies are loaded in compression, with the additional observation that when the starter crack is aligned well away from the line of the applied stress, the incremental crack initiates not at the crack tip itself but several crack tip radii away.

Fig. 3.18 – Fracture path in a plate with an inclined crack loaded uniaxially. (Courtesy Pook; Crown copyright reserved; reproduced by permission.)

Why do the cracks curve and depart from a coplanar path? The hypothesis that propagation occurs normal to the direction of maximum tensile stress may be tested in the following simple manner. Consider the DCB-type specimen in Fig. 3.19. The external load X is equipollent with the same load applied to the centre of the ligament of length $(W - a)$ in the direction of the existing crack, together with the couple of magnitude $X(W + a)/2$, as shown. Assuming, for

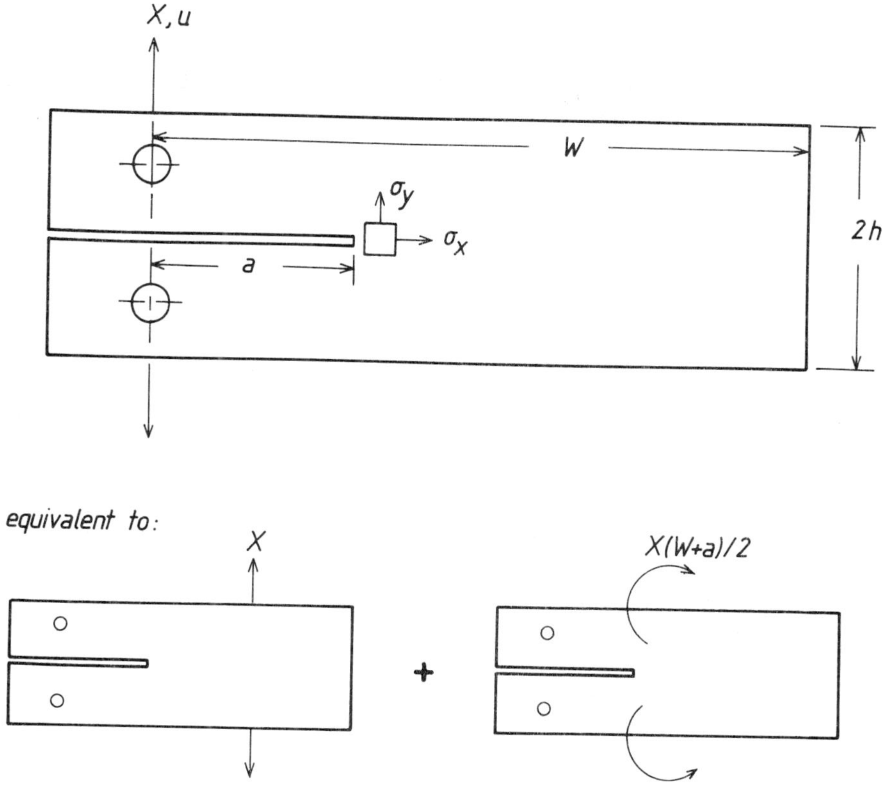

Fig. 3.19 – Equivalent loadings in a DCB specimen.

ease of calculation, that simple beam theory may be used for the normal stresses σ_x and σ_y in the ligament near the end of the crack, '$\sigma = My/I$' and we have

$$\sigma_y = \frac{6X}{BW}\left(\frac{a/W + (2/3)(1 - a/W)}{(1 - a/W)^2}\right) \tag{3.25}$$

$$\sigma_x = \frac{6X}{BW}\left(\frac{a/W}{(h/W)^2}\right) \tag{3.26}$$

Cotterell in 1970 postulated that the crack will propagate in its own direction for $\sigma_y > \sigma_x$ but curve for $\sigma_y < \sigma_x$. At the transition between straight and curved $\sigma_y = \sigma_x$, i.e.

$$\sigma_y = \sigma_x = \frac{6X}{BW}\left(\frac{2 + a/W}{3(1 - a/W)^2}\right) \tag{3.27}$$

Of course simple beam theory is not correct for all combinations of a, h and W, but even so let us plot relations (3.26) and (3.27) against a/W, in Fig. 3.20.

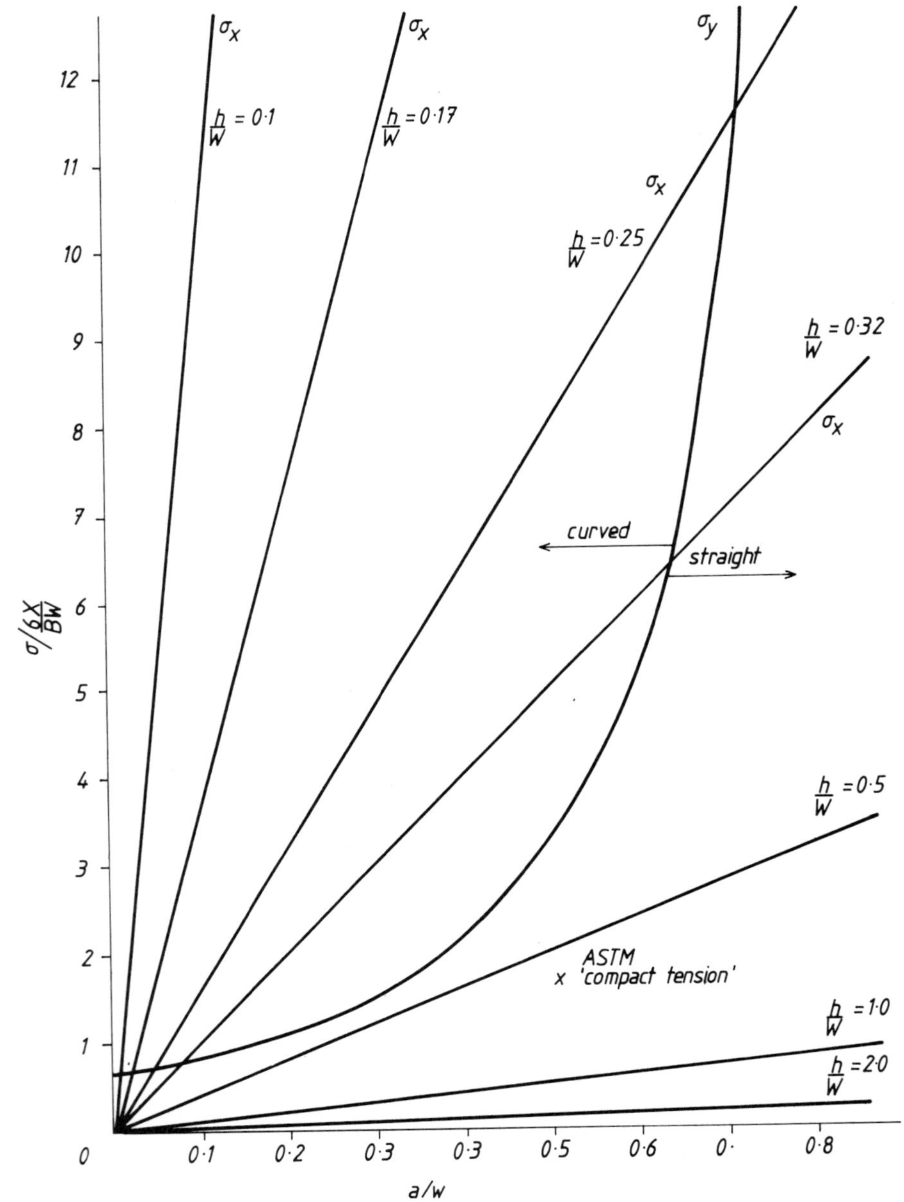

Fig. 3.20 – Variation of σ_x and σ_y with a/W for the DCB specimen in Fig. 3.19.

A family of radial lines with slope $\tan^{-1}[1/(h/W)^2]$ represents σ_x and a concave-up curve for σ_y, intersecting the ordinate at $\sigma/(6X/BW) = 2/3$, describes relation (3.27). Proportions of testpieces (i.e. a/W and h/W) may be located on this diagram to determine whether cracking is likely to continue straight or be curved. The ASTM compact tension CT specimen has $h/W = 0.6$ and $a/W = 0.5$ and is located in the 'straight' region. Fig. 3.21 shows the results of experiments on three geometries (a) in the curved region according to Fig. 3.20; (b) in the

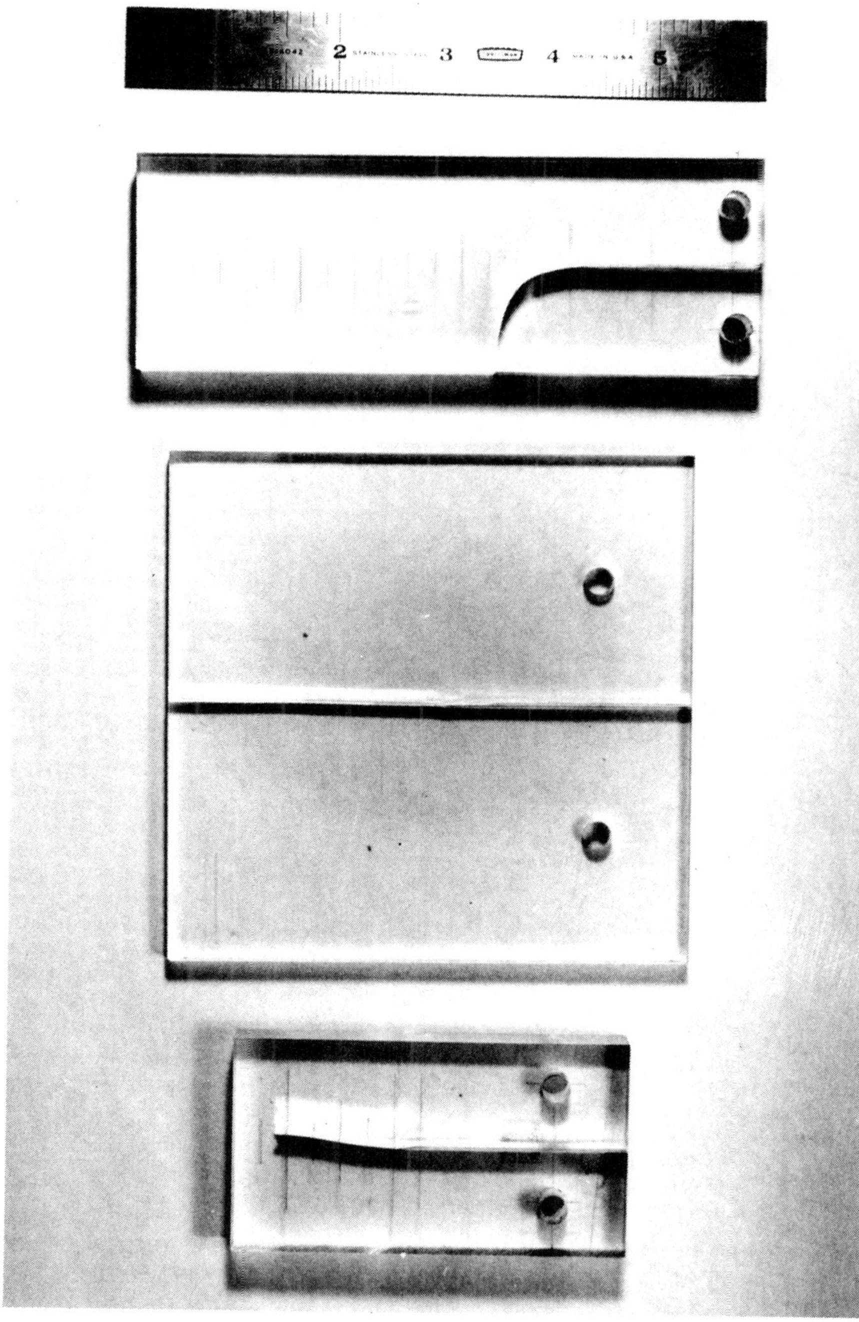

Fig. 3.21 – Fracture paths in DCB type of specimens for (top) $a/W = 0.2, h/W = 0.1$; (middle) $a/W = 0.3, h/W = 0.5$ and (bottom) $a/W = 0.41, h/W = 0.33$. Marker in inches. (After Caddell and Atkins.)

straight region; and (c) in the transition region. The simple criterion works well. Note also that in Fig. 3.21(c) the crack becomes straight again after some initial curved propagation: the starting conditions of $a/W = 0.41$, $h/W = 0.33$ are marginally in the curved region (according to beam theory); the crack begins to inflect back to a straight path when $a/W \approx 0.67$ which agrees well with the prediction where the $h/W = 0.33$ line intersects the curved/straight boundary given by relation (3.27). Cotterell investigated path stability in CT testpieces made from araldite, PMMA, glass and steel; crack deviation was small when the bending stress in the ligament was greater than that in the arms.

Example 3.7 Grooves are sometimes cut in the paths of cracks to encourage propagation in that direction. Show that straight cracks may be obtained with smaller h/W than in ungrooved DCB-type specimens.

Ans. If the net section thickness at the groove is B^*, σ_y becomes $\sigma_y^* = (B/B^*)\sigma_y$ but relation (3.26) for σ_x remains the same. The transition curve given by relation (3.24) becomes multiplied by (B/B^*) and is hence shifted upwards. As $\sigma_y^* > \sigma_y$, h/W can be smaller than in ungrooved specimens. For example, if $B/B^* = 2$ and $a/W = 0.4$, h/W may be reduced to about the 0.32 line and still give a straight path. Even so, there are numerous combinations of (h/W) and (a/W) which will give curved cracks even in the presence of grooves. Note then that when a crack, curving in the plane of the 'web' at the base of the groove, meets the full thickness of the testpiece, it enters a region which is even more favourable for curving (the criterion then being the original lower position of relation (3.27) in Fig. 3.20).

Benbow and Roesler (1957) performed experiments on specimens with central slits where compressive stresses were applied to the top and bottom external faces of the testpieces by clamps. Cracks propagated in their original direction for longer distances as the external compressive forces were increased. This may be viewed as in Example 3.7 again as moving the curved/straight boundary line upwards in Fig. 3.20 so that the σ_x stresses are reduced by the level of the applied compressive stresses and the σ_y stresses have a greater influence on propagation direction.

A more general line of attack for crack paths employs the formulae for the distribution of stresses around crack tips and looks for the directions of maximum stress. In Chapter 2 we concentrated on merely the σ_{yy} stress component as it was sufficient for the introduction of the stress intensity factor approach to LEFM. There are, however, expressions for σ_r σ_θ and $\tau_{r\theta}$ (polar coordinates at the crack tip cf. Example 2.18) and these were employed by Erdogan and Sih in 1963 for crack path prediction. Instead of merely the first-term-of-the-series approximation for stresses, Cotterell in 1966 used many more terms to determine a criterion for path stability. He found that if an excursion was made from the ideal fracture path, because of some local inhomogeneity, the path either turned back to its initial direction or deviated even further, depending on the sign of the coefficients of the second term in the full series expansion for σ_θ. Later workers (Williams and Ewing in 1972; Finnie and Saith in 1973) have employed similar arguments but investigated σ_θ at the boundary of the process zone rather than the crack tip itself, which better accords with experimental data.

When a crack turns from a linear path, crack opening ceases to be merely mode I, and mode II component appears in plate geometries (and mode III in the general case of a twisting, as well as curving, crack).† However, in a curved path, if propagation is normal to directions of principal stress, $\tau_{r\theta} = 0$ and the K_{II} shearing mode intensity factor is much reduced. Indeed, setting $K_{II} = 0$ forms the basis of an alternative method of deducing the directions of crack paths (Iida and Kobayashi, 1969; Bilby and Cardew, 1975; Cotterell and Rice, 1980; Maiti and Prasad, 1980).

While the notion of cracks propagating perpendicular to a maximum tensile stress may seem to be a useful working hypothesis, with which the resolution of simple experiments agree, the path taken by a propagating crack ought also to be connected with displacements and rates of performing work, since crack propagation is concerned with interchanges of *energy*. We remember the principle of minimum potential energy in mechanics which states that 'Of all the possible displacements that satisfy given boundary conditions, those that satisfy the equilibrium conditions make the potential energy assume a stationary value, which in stable equilibrium is a minimum'. From this statement (and the equivalent one for complementary energy) we should guess that the criterion for crack paths involves lowering the energy of the cracked body by the quickest route, i.e. that the favoured crack path absorbs toughness work at a greater rate than all other possible paths.

In seeking an energy-based criterion for the spatial path of a crack, we can perform experiments in which the crack is forced to take unnatural paths and observe the behaviour of *Xu* diagrams. The top testpiece shown in Fig. 3.22(a) is grooved and during propagation the crack follows the grooved path ($B/B^* =$ 6 mm/3 mm $= 2$) and considerations similar to those in Example 3.7 say that the crack does not climb out of the groove. All the testpieces in Fig. 3.22(a) have identical starter cracks but the lower two have the grooves cut in unnatural paths. The middle specimen has grooves which follow the sort of ogive-shape which is the path of a propagation in a *full thickness* testpiece of the same external size (cf. Fig. 3.17); when loaded and broken, it is seen that the crack chooses to run out of the groove into the full-thickness plate and follow an ogive curve (but even so, not the artificially introduced grooved ogive-path). The lowest design in Fig. 3.22(a) has grooves cut in a U-shape and again the propagating crack runs out of the groove into full-thickness plate.

---

†Such paths are often said to be 'mixed mode' fracture – but note that the term mixed mode is also used for the quite different situation of (straight or curved) fracture surfaces which have a mixutre of flat fractures and shear lips (cf. Chapter 2).

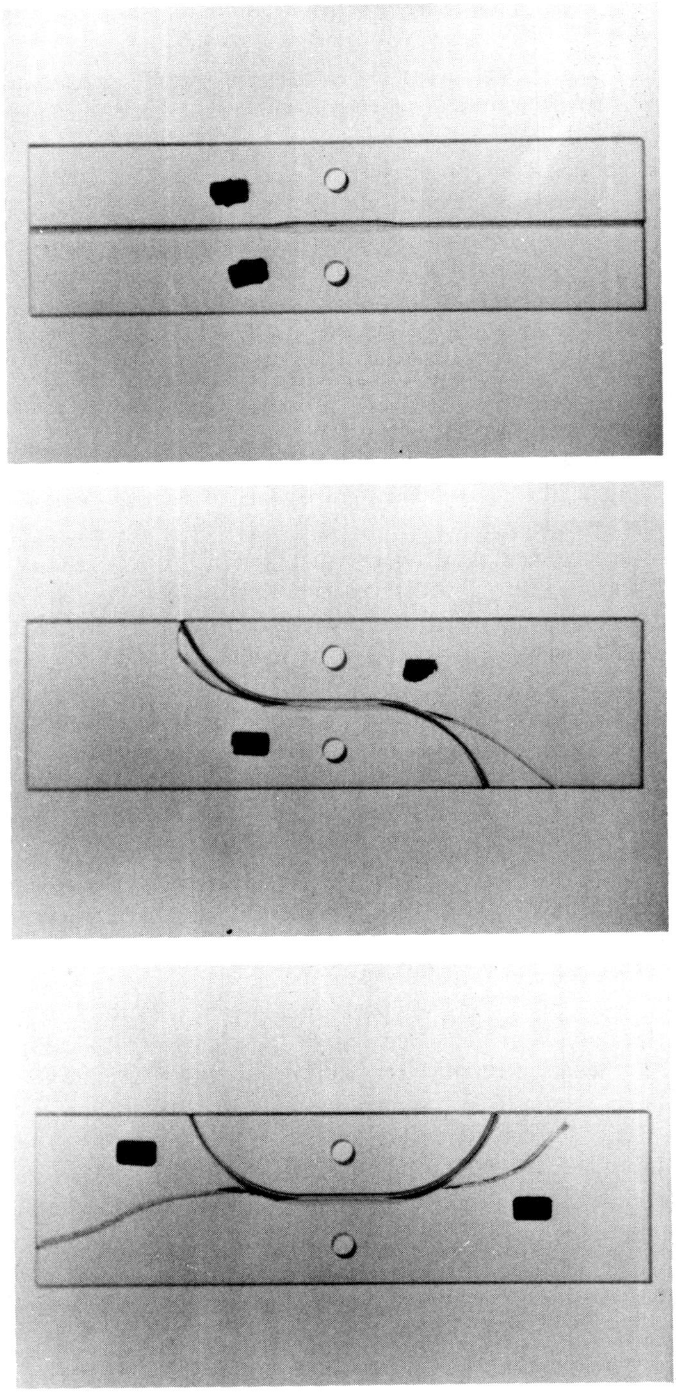

Fig. 3.22(a) – Beam-like specimens with (top) natural and (middle and bottom) unnatural crack paths.

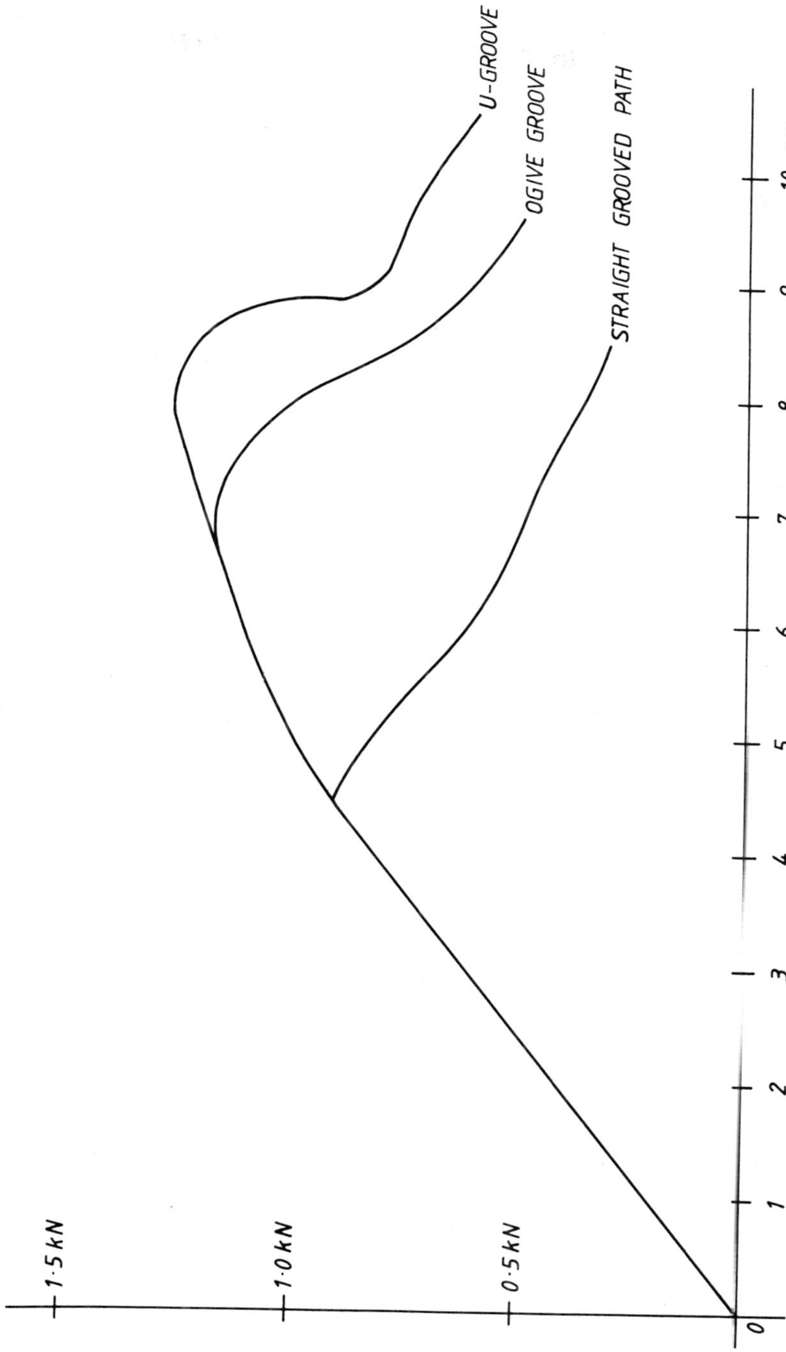

Fig. 3.22(b) – Load–deflexion traces for specimens in (a) above.

The corresponding three quasi-static Xu curves are shown in Fig. 3.22(b). Various conclusions may be drawn about X, u and Λ during propagation along the natural path in comparison with the unnatural paths, viz. (i) X is least at the same u; (ii) $\mathrm{d}X/\mathrm{d}u$ is least during propagation; (iii) increments in Λ, over the same $\mathrm{d}u$, are least for the natural path; (iv) the external work consumed for a given crack length change is least along the natural path.†

Some consideration shows that all these observations from quasi-static experiments are consistent with a criterion of maximised $R\mathrm{d}A$ along the natural path; i.e. a maximum absorption of fracture work. Equivalently (and also for unstable non-quasi-static propagation) the natural crack path corresponds with a maximum strain energy release rate $G\mathrm{d}A$ into the path of the crack. (We remember that the alternative name for the Principle of Minimum Potential Energy is the Principle of Least Work, i.e. to go quasi-statically from one equilibrium state to another (one crack length to a longer) the system will naturally do this with least expenditure of effort).

Proofs are as follows. Equation (1.1) may be recast as

$$X - u\mathrm{d}X/\mathrm{d}u = 2J\,\mathrm{d}A/\mathrm{d}u \qquad (3.28)$$

For linear elastic systems, $X = k(A)u$ where $k(A)$ is the stiffness of the cracked body. Hence, using G for J,

$$u\,\{k(A) - \mathrm{d}X/\mathrm{d}u\} = 2G\,\mathrm{d}A/\mathrm{d}u \qquad (3.29)$$

For maximum $G\mathrm{d}A/\mathrm{d}u$, $\mathrm{d}X/\mathrm{d}u$ must be least for a given displacement at fracture u. It follows that if $\mathrm{d}X/\mathrm{d}u$ is least, X must also be least. Again, if we rewrite equation (1.1) as

$$\mathrm{d}\Lambda/\mathrm{d}u = X - G\mathrm{d}A/\mathrm{d}u \qquad (3.30)$$

(where $\Lambda = \frac{1}{2}Xu$ for a linear body), it follows that $\mathrm{d}\Lambda/\mathrm{d}u$ is a minimum if X is a minimum (i.e. Λ of the system after crack propagation is least) and $G\mathrm{d}A/\mathrm{d}u$ is a maximum.

Another corollary of the maximum $G\mathrm{d}A$ criterion, not immediately obvious from Fig. 3.22(b), is that the rate of change of compliance with crack area is greatest for the natural crack. This may be seen be rewritting equation (2.6) as

$$2/X^2 = (1/G)\,\mathrm{d}/\mathrm{d}A\,(u/X) \qquad (3.31)$$

As X is least, $\mathrm{d}/\mathrm{d}A\ (u/X)$ must be a maximum.

Gurney and Hunt in 1967 pointed out that maximising $G\mathrm{d}A/\mathrm{d}u$ maximises the entropy increase of the (isothermal) testpiece, which is sensible.

---

† Similar values for R are given by *all* three Xu plots (ca. 600 J/m²). The unnatural paths envelop larger work areas than the natural path, but the unnatural crack runs in full-thickness plate and not in the grooves and requires more work input.

Example 3.8 Consider Fig. 3.17. Show that a plausible interpretation for the changing patterns of natural crack path (particularly why straight paths in tall specimens become ogive-shaped paths in smaller) is that the elastic compliance of the specimen is always being maximised as propagation proceeds.

Ans. Consider the ogive-shaped paths, in the shorter specimens of Fig. 3.17. Consider also the crack paths in Fig. 3.22(a) (do not be confused that in Fig. 3.22(a) the natural path for the top specimen is straight, owing to the grooving, whereas the natural path for an ungrooved specimen of the same size is ogive-shaped in the full thickness). Following arguments of Gurney and Hunt (1967), we need first a reason for the asymmetry of ogive path branches. If a crack followed a U-path, as in the third specimen of Fig. 3.22(a) then, just before the crack fronts reached the surface, the upper part would have the flexibility of a pin-ended beam, but the lower part would be stiffer because of the thrust developed by its acting as an arch. Equally, we may argue that a U-path is impossible because the thrusts will cause of compressive strain at the crack ends. By taking the asymmetrical path these thrusts are avoided and both parts have stiffness appropriate to pin-ended beams, as the crack approaches the surface. Next we need an arguement for the departure of the crack from a straight line in the lower specimens of Fig. 3.17. A simple, if crude, explanation is the following. Suppose that the crack has extended to a length a along the mid-length of the testpiece. Consider (i) that the crack then turns through a right angle in opposite directions at each end of the straight part and that each crack front has nearly reached the surface. The crack length is then nearly $(a + 2h)$ ($2h$ being the height of the testpiece) and the cracked test piece has flexibility appropriate to two *simply-supported* beams of length a so that its elastic compliance is $u/X = (a^2/24EI)$ where I is the second moment of area of the half depth of the test piece. Consider (ii) that the crack runs straight. When it has a length $(a + 2h)$ its elastic compliance as two *fixed-end* beams will be $u/X = [(a + 2h)^3/96EI]$. The testpiece with crack path (i) will be more flexible than that with crack path (ii) if $(a^3/24) > [(a + 2h)^3/96]$. We conclude that (i) will be more flexible than (ii) if $a > 3.4h$. Finally, we need an argument for the reversal of curvature of the crack front as it approaches the boundaries of the testpiece. This is provided by the consideration that, after the crack has turned appreciably, but before it nearly reaches the boundaries, the effective fixing moment will have become negligible, and that no more flexibility is to be gained by reduction of fixing moment. The crack, therefore, turns and the testpiece gains flexibility by increase in the span of the two effectively pin-ended beams which then comprise the testpiece. (If a testpiece is suitably waisted in plan form with the least depth at the centre, it is then found that the crack runs straight down its longitudinal centre line. This is because if h increases suitably with a, the foregoing inequality is never true, cf. Problem 3.16).

Example 3.9 Investigate why, in the compression of blocks of brittle materials, axial splitting can be a stable direction for crack propagation.

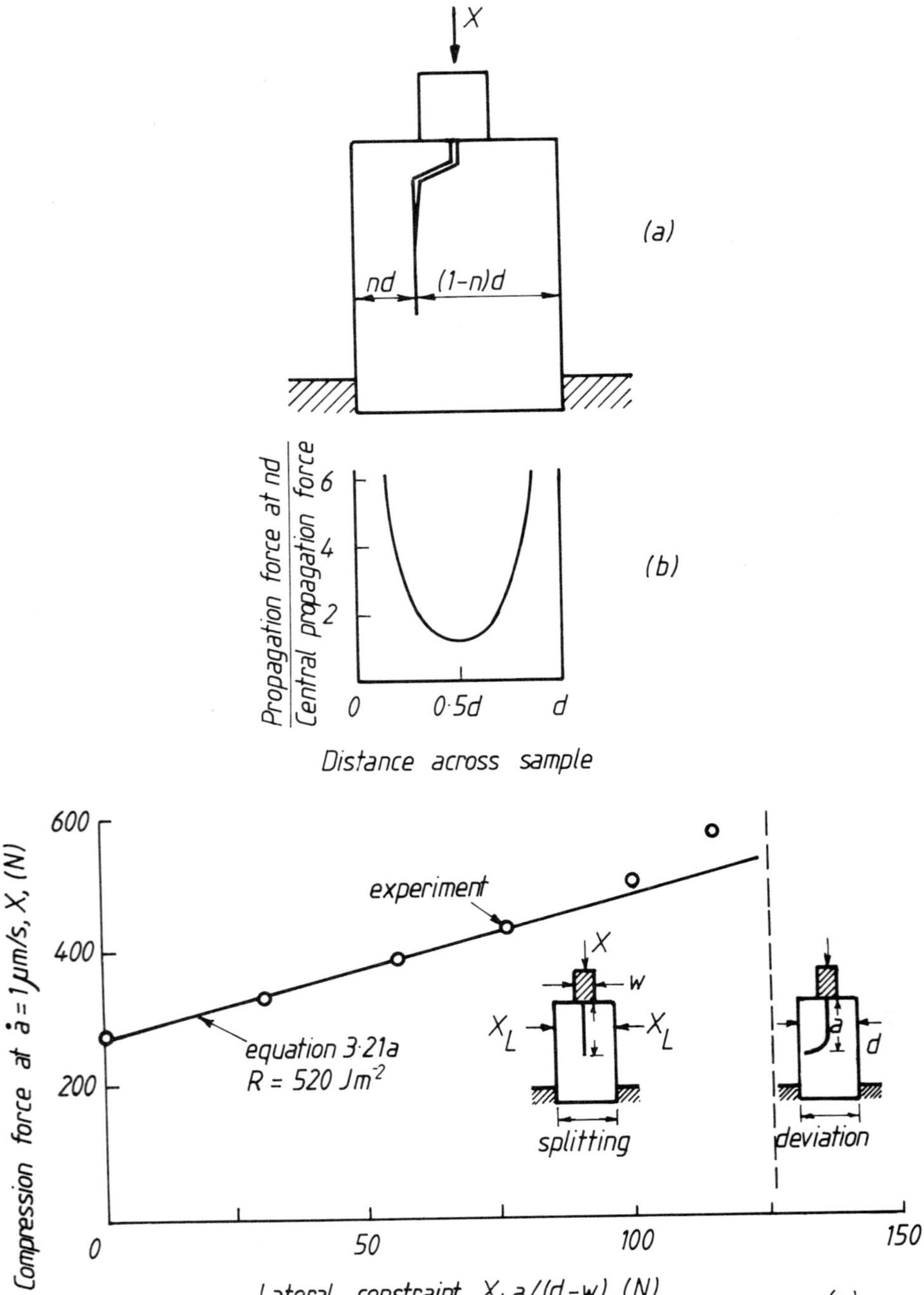

Fig. 3.23 – (a) Compression splitting off the central plane. (b) The force required for compression splitting with distance from the central plane showing that central splitting is preferred. (c) Effects of lateral force on the compression force required to cause splitting. Crack path deviation occurs at a critical lateral force shown. (All after Kendall (1978), with permission.)

Ans. Using Kendall's analysis (section 3.2), we can determine the axial force required for asymmetrical splitting of low ER/σ_y^2 materials. In Fig. 3.23(a), the bent struts forming the crack have different sizes viz. nd and $(1 - n)d$ rather than $d/2$ as in Fig. 3.5. The second moments of area are different and given by $Bn^3d^3/12$ and $B(1 - n)^3\, d^3/12$. Again, the bending moments produced by the equivalent forces $X/2$ (where X is still applied centrally) are different and given, for the left-hand strut, by $X(d\text{-}nd\text{-}w/2)/4$. Using the procedures of section 3.2 we obtain, for the total force for splitting,

$$X = \frac{B}{(1 - w/d)} \left(\frac{2ERd}{3}\right)^{\frac{1}{2}} \left(\frac{(1-n)^5 + n^5}{(1-n)^2\, n^2}\right)^{\frac{1}{2}}$$

which, as before, is independent of crack length (for $n = \frac{1}{2}$, this reduces of course to equation (3.13)). The predictions of this relation for varying n (i.e. varying positions of the crack across the sample) are shown in Fig. 3 23(b) and it is seen that the force (and hence applied work) required for compression splitting increases rapidly with distance from the central plane. Hence it is not unreasonable to presume that axial propagation will be preferred. When specimens are, in addition, subjected to lateral compressive loading (section 3.4) axial splitting is even more highly favoured, but even so crack turning, according to the principles already discussed in this section, can occur under the correct conditions in the presence of sideways compressive force as shown in Fig. 3.23(c).

3.5.1 The kinked angled crack problem

The general problem that has to be solved is shown in Fig. 3.24. A pre-existing crack of area A in a plate specimen is loaded at an angle α; incremental crack

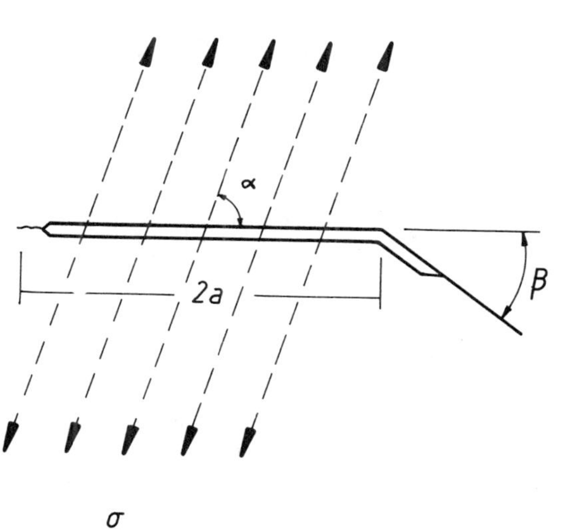

Fig. 3.24 – The general, inclined, branched crack problem.

extension occurs along the kinked segment inclined at β to the direction of the crack. We require to find how β varies with α for given loading of a given testpiece geometry. In Fig. 3.17, α is $90°$ and β is $0°$ for large h; β takes on increasing values in smoothly curving cracks at smaller h; when $\alpha \neq 90°$, β starts out with finite values at large h (Fig. 3.18) and the crack extension is discontinuous with the starter crack.

The mathematical procedure is to investigate the strain energy of the cracked body before and after the incremental propagation along the kinked path, and calculate the strain energy release rate in the usual way, i.e. $-(\partial\Lambda/\partial A)_u = G$ ($= R$ for quasi-static propagation) but now G is additionally a function of crack orientation β. In most of the examples in this book, crack propagation is taken to be 'self-similar' with the starting crack direction unless otherwise stated; and also in most examples, $\alpha = 90°$. From the maximum strain energy release rate $G\mathrm{d}A$ criterion, it follows that of all possible extensions $\mathrm{d}A(\beta)$ there should be one $\mathrm{d}A(\beta^*)$ which makes G an absolute maximum, and β^* defines the natural crack path. That the energy release rate should be a maximum is apparent from the following argument, which is complementary to the comments already provided. Let the load at fracture along the natural path be X_f, and consider a crack extension $\mathrm{d}A(\beta) \neq \mathrm{d}A(\beta^*)$ so that $G\ (X_f, \mathrm{d}A(\beta)) < G_{max}$. To produce fracture along the unnatural path, X_f has to be increased to $(X_f + \mathrm{d}X_f)$, so that then $G\ ((X_f + \mathrm{d}X_f), \mathrm{d}A(\beta)) = G_{max} = R$ in quasi-static cracking. Thus there exists a relation between the magnitude of the load at fracture and the orientation of the crack extension: deviation from $\mathrm{d}A(\beta^*)$ decreases G, while increase in X_f increases G. Therefore if G is to be constant at fracture ($\equiv R$), then minimum loading at rupture requires that $\mathrm{d}A(\beta^*)$ be such as to maximise G. These comments should be related back to the experiments in Fig. 3.22.

The computational procedures of the 'angled crack' problem are formidable. The interested reader is referred to the review by Palaniswamy and Knauss (1974), who have solved the problem for the Griffith geometry (i.e. small angled crack in large plate loaded in tension at the remote boundaries). While that geometry is different from the sorts of testpieces shown in Figs. 3.17 and 3.18, it is important to show Palaniswamy and Knauss's analytical results (Fig. 3.25), not only for their intrinsic value but also because comparison is given with the results of the simpler maximum stress criterion for crack path for the same problem.

The relation between β and α given by the stress criterion deviates only mildly from the energy analysis, virtual agreement prevailing for $\alpha > 60°$. This helps to explain the agreement between the stress and energy criteria for earlier problems (of admittedly different geometry; in such problems with non-remote boundaries the α, β relation will depend on crack length as well, as is evident in Fig. 3.17). But why should there be agreement between the two different criteria and is it likely to be a general agreement?

An explanation for the closeness of the results is given by Palaniswamy and Knauss. Consider components of stress $\sigma_\theta\ \tau_{r\theta}$ and displacement u_θ, u_r at the crack tip before (b) and after (a) the crack branches at the angle β. We have

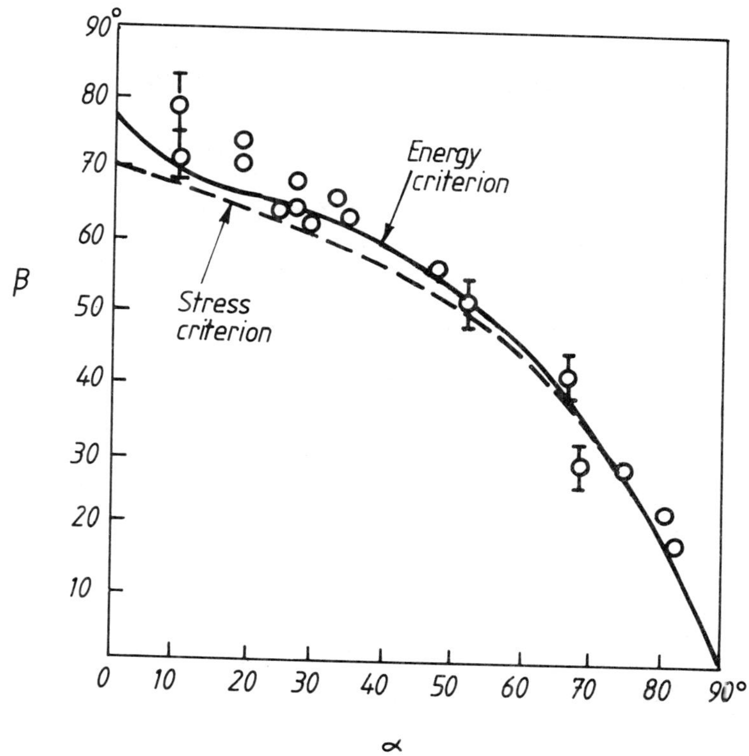

Fig. 3.25 – Experimental results on a toluene-swollen polyurethane: crack branching angle (β) as a function of load orientation (α). (After Palaniswamy and Knauss (1974), with permission.)

$$G(\beta) = \underset{\Delta A \to 0}{\mathrm{Lt}} \left(\frac{\Delta \Lambda(\beta, \alpha)}{\Delta A} \right)$$

$$= \underset{\Delta A \to 0}{\mathrm{Lt}} \left(\frac{\Lambda^a(\beta, \alpha) - \Lambda^b(\alpha)}{\Delta A} \right) \tag{3.32}$$

Using Clapeyron's theorem for strain energy in terms of σ_θ, $\tau_{r\theta}$ and u_θ, u_r and Betti's reciprocity relation on the conditions before and after crack branching we obtain,

$$G(\beta) \underset{\Delta A \to 0}{\mathrm{Lt}} \left\{ \frac{B}{2\Delta A} \int_0^{\Delta A} \left\{ \sigma_\theta^{\mathrm{b}}(\tau, \beta) \; [u_\theta^{\mathrm{a}}(\tau, \beta)] + \tau_{r\theta}^{b}(\tau, \beta) \; [u_r^{\mathrm{a}}(\tau, \beta)] \right\} \mathrm{d}A \right. \tag{3.33}$$

where [ ] denotes the jump in the displacement components across the crack and B is the plate thickness.

Now the energy criterion of fracture which requires maximisation of $G(\beta)$ clearly involves *both* terms under the integral. In contrast, as the maximum σ_θ stress criterion concerns situations where $\tau_{r\theta} = 0$ (otherwise fractures would not occur along rays from the crack tip), crack paths predicted on the basis of maximum σ_θ concern only the *first* term under the integral. If the jump in u_r^a is of the same order of magnitude or smaller than u_θ^a, then for small or vanishing $\tau_{r\theta}$ both the energy and stress criteria give (approximately) the same result. This situation corresponds with cases when β is small, which occurs (in this Griffith problem) when α is near 90°. In fact, in the limit of $\alpha \to 90°$ and $\beta \to 0$ the jump $[u_r^a(\tau,\beta)]$ vanishes identically and the stress criterion is then identical with the energy criterion.

Having decided on the optimum crack branching angle β^*, it is possible to work out the remote stresses σ_∞ that cause crack extension along β^* in terms of α. The boundary stresses given by the energy criterion for this Griffith problem are *marginally* smaller than the boundary stresses given by the stress criterion; this reflects the least X corollary of the maximum GdA criterion. We note that this means that the approximate maximum stress criterion for crack branching does not, in the general sense, predict the fracture load conservatively (although in this particular illustration, the difference is small).

Fig. 3.25 includes experimental data which show that, despite the scatter, the energy criterion seems better followed.

The reader will come across other criteria for crack paths, such as the strain energy density (S_c) theory due to Sih (1973). That postulates that fracture occurs in the direction of maximum potential energy density, which is equivalent to the radial direction from an existing crack tip along which the strain energy density reaches a minimum with respect to θ at some small distance from the tip. The small distance is a material parameter linked to the size of the process zone. A wide-ranging series of papers concerned with the S_c criterion may be found in Sih, Czoboly and Gillemot (1982).

Since strain energy, rather than stress alone, is the criterion, predictions depend on Poisson's ratio, ν. While for many cases the S_c theory works as well as the strain energy release rate GdA, or simple stress, criteria there is growing evidence that it is not true in general. Discrepancies arise in cone indentation cracks and Finnie and Weiss in 1974, for example, showed that a stress criterion better fits crack path data in cross-rolled beryllium (which has the unusual property of $\nu = 0$). Again Maiti in 1980 demonstrated wide differences in predicted crack paths in bar shearing, with the experimental data following a stress prediction rather than the S_c prediction, and Maiti and Smith (1983) have shown that the S_c prediction is inadequate for path prediction in *DCB* specimens.

3.5.2 Crack path in anisotropic solids

Fig. 3.26 shows crack paths in a variety of solids some of which are anisotropic. If materials possess planes and directions having lower resistance to crack propagation than others, it is not surprising that cracks may endeavour to turn into such planes of 'weakness'. This often poses experimental difficulties when

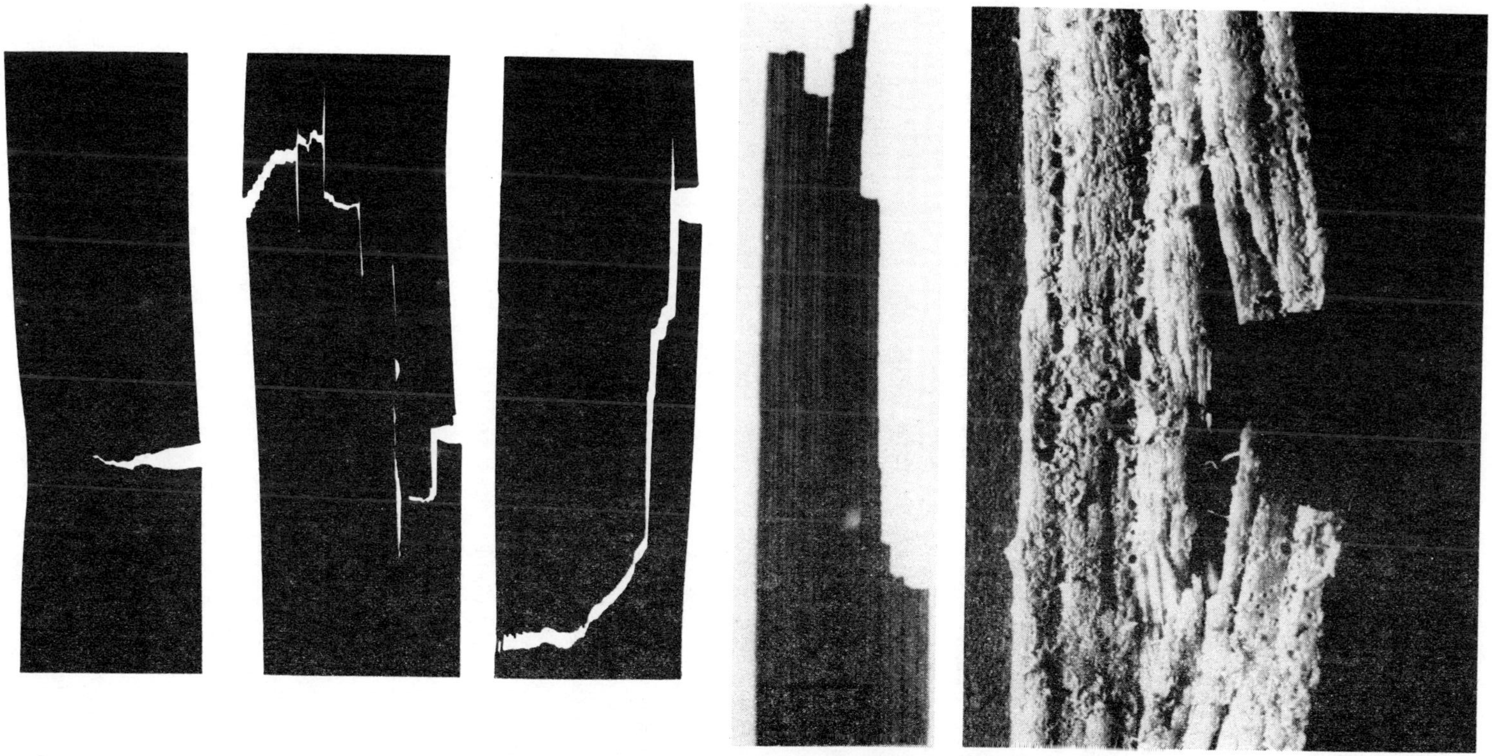

Fig. 3.26(a) – Crack paths in anisotropic solids deformed in uniaxial tension (parallel to long dimension of testpiece). Left to right: copper, copper reinforced with silica fibres, copper reinforced with tungsten wires, grass (*lolium perrene*), meat (semi-tendinosus). All except copper are notch-sensitive. (Courtesy G. A. Cooper and A. Kelly; J. A. V. Vincent; P. P. Purslow.)

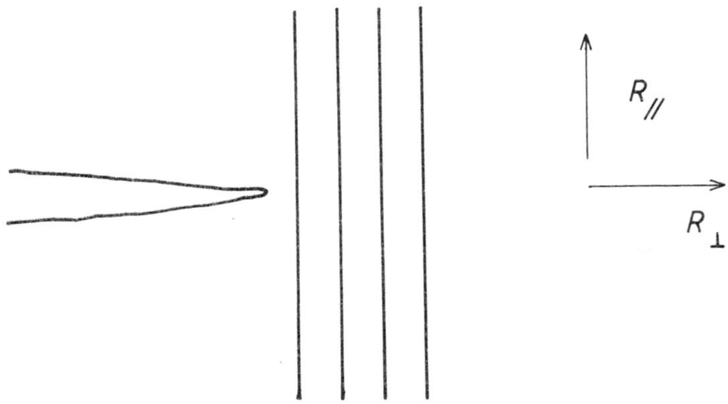

Fig. 3.26(b) – Specific works of fracture parallel and perpendicular to fibres.

attempting to measure fracture toughness in the strong directions, and can be a nuisance in, for example, woodworking.

Cook and Gordon in 1964 studied the stress components at the tip of a crack and proposed that a crack will propagate in the direction which displays the least material strength. Similarly Cooper and Kelly in 1967 argued that a transition from crack propagation parallel to reinforcing fibres to propagation transverse to the fibres occurs when the ratios of the strengths of the material in the different directions reach a specific value. As Cherry and Harrison pointed out in 1970, it is implict in these arguments that crack propagation occurs in the direction requiring the least applied load. This accords with Gurney's arguments in section 3.5, and is consonant with a maximised GdA criterion. Cherry and Harrison applied the maximum $d/dA(u/X)$ version of the criterion to crack propagation in composites. They considered the alternative cases of cracking perpendicular to aligned fibres or cracking parallel to aligned fibres (called by them splitting), Fig. 3.26b. They showed that in quasi-static deformation the two types of path have equal probabilities of occurring when

$$\frac{1}{R_{//}}\left(\frac{d(u/X)}{dA}\right)_{//} = \frac{1}{R_{\perp}}\left(\frac{d(u/X)}{dA}\right)_{\perp} \tag{3.34}$$

where the subscripts // and $\perp$ mean respectively propagation parallel to (i.e. splitting) and perpendicular to the reinforcement, Fig. 3.26(b). It follows that a necessary condition for a crack to turn from a path which cuts fibres to one which runs parallel to them, is

$$G_{//}/G_{\perp} > R_{//}/R_{\perp} \tag{3.35}$$

since $G \propto d(u/X)/dA$. (When cracks turn in isotropic solids with the same R in all directions it is to maximise GdA and $d/dA(u/X)$, cf. Example 3.8; cf. also section 3.4 on cracking under biaxial stresses.)

Harrison (1972) went on to determine experimentally $G_{/\!/}/G_{\perp}$ by considering $\partial\Lambda/\partial A$ in the presence of artifically introduced small cuts at the ends of starter cracks. Cuts ($\equiv \Delta A$) were either along the line of the original crack (for 'cracking') or perpendicular to the original crack (for 'splitting'). With the proviso that the total ΔA had magnitude comparable with the size of the crack tip zone, he found experimentally that $G_{\perp}/G_{/\!/} \approx 8$ in a 0.5 volume fraction vacuum impregnated unidirectional glass fibre reinforced epoxy resin.

While it is possible, in principle, to have solids with anisotropic R values which still, however, display isotropic elasticity, it is more common to find overall anisotropy. For example the Young's modulus of a number of hardwoods (khaya walnut, birch, ash, beech) and softwoods (spruce, pine, fir) is 10 ~ 16 GPa along the grain, but about 1 ~ 2 GPa in the trunk radial direction and 0.5 ~ 1.0 GPa in the trunk circumferential direction; similarly, typical values for uniaxial fibre reinforced composites along and perpendicular to the reinforcing filaments are 140 and 15 GPa (graphite expoxy), 42 and 12 GPa (glass/epoxy), 210 and 15 GPa (boron/epoxy), 126 and 25 GPa (graphite/aluminium), the precise values depending on volume fraction of filaments of course. In this book we do not discuss theoretical K or G expressions for anisotropic solids (we do of course concern outselves with *experiments* on such solids) but it may be expected that magnitudes of the strain energy release rate G will depend on orientation, as the elastic stiffness of the anisotropic body depends on orientation (see Williams and Birch, 1976). Thus in condition (3.35) it does not follow that cracks will *always* veer towards planes of low fracture toughness, since the ratio of the Gs has to be appropriate. This is shown up by Harrison's finite element calculations using anisotropic K and hence G expressions (e.g. Sih *et al.*, 1965). Composites reinforced with different volume fractions of glass fibres, or carbon fibres, and also wood were studied, where the ratio of orthogonal elastic moduli ranged from unity to 52, and where the shear modulus varied by a factor of over 70. The calculated $G_{\perp}/G_{/\!/}$ varied, within these same ranges, by about a factor of three (~6 to ~17), increasing with increasing anisotropy. Thus, *ceteris paribus,* a crack is more likely to propagate *across* the fibres in a material whose elastic properties are strongly anisotropic according to Harrison's calculations than in one for which the elastic properties are more nearly equal.

Example 3.10 The Cook–Gordon mechanism of crack stopping is based on weak interfaces perpendicular to line of the main crack (section 1.2.3). Isotropic elastic calculations show that cracks will be diverted into the plane of weakness when the tensile strength of the interface is about 1/5 that of the main body (cf. Fig. 2.23). What value might this ratio take in an elastically anisotropic solid?

Ans. In the light of Harrison's calculations that $G_{\perp}/G_{/\!/}$ increases as the elastic anisotropy increases and that therefore cracks are likely to remain in straight paths the more anistropic the body, it follows that the tensile strength of the plane of weakness must be less than 1/5 for crack diversion. Bishop in 1972 showed that the ratio should be 1/70 for a composite having properties appropriate to carbon-fibre reinforced bodies, i.e. $E_{//} = 240$ GPa, $E_{\perp} = 7.5$ GPa, in-plane shear modulus of orthotropic sheet = 5.8 GPa, $\nu = 0.27$.

In calculations for $G_{\perp}/G_{//}$, it must be noted that the elastic constants in anisotropic elasticity are not simply related as they are in isotropic theory by a single Poisson's ratio (i.e. shear modulus = $E/2(1 + \nu)$). In particular, the shear modulus plays an important role in $G_{\perp}/G_{//}$; quoting merely orthogonal Young's moduli does not specify an anisotropic cracking problem completely, since the shear modulus may vary independently. The answer to Example 3.10 has to be viewed in this light. Jeronimidis has observed that low values of the shear modulus, at the same ratio of inplane Young's moduli, give abnormally high R in many materials. As discussed in Chapter 1, this is not due to an intrinsically high work of fracture, rather that it is difficult to concentrate the energy in order to promote cracking. Hence the resistance of many extensible biological tissues, and woven and knitted fabrics, to fracture as described in section 2.10.

A valuable introduction to theory and experiments for crack propagation in anisotropic media will be found in Williams and Brich (1976).

Example 3.11 Why are tears in the seats of trousers always L-shaped?

Ans. There is hardly any shear resistance to woven fabrics at 45° to the warp and weft. Hence it is very difficult to concentrate energy release into that direction and fracture occurs along the stiffer directions.

In elastic fracture at high speed, and in unstable fracture, where kinetic energy and stress wave effects should be taken into account, there is an upper limit to crack speed of about $0.38c_0$ where c_0 is the longitudinal elastic wave velocity. (The upper limiting crack speed for quasi-static analyses to apply is some $10^{-3}c_0$.) A limiting dynamic crack velocity occurs because the kinetic energy cannot increase faster than energy is supplied from external work; at the limiting velocity, atomic bonds at the crack front are being strained to that limit in times of the order of one atomic vibration. In order to dump excess energy at higher rates, a single crack is encouraged to fork (branch or bifurcate) under these conditions and thereby provide more crack surface area. The phenomenon is familiar in glass. Criteria for such multiple paths are similar to those already given, but analysis of the problem is complicated, needless to say.

Crack paths under quasi-static elastoplastic conditions have been studied theoretically by Kfouri and Miller (1981). Crack paths were predicted on the basis of a maximised G^{Δ} criterion where G^{Δ} is the crack separation energy rate calculated for a *finite* amount of crack advance (unlike G in elastic fracture mechanics). Their results are not markedly different from paths predicted by the various other criteria mentioned in this section (in the case of the centre-cracked plate geometry). They say that this fact helps to explain the success of predictions obtained from linear-elastic analyses in spite of the known occurrence of irreversibilities right at the crack tip.

3.6 CRACKING OF LAYERED STRUCTURES

Fig. 3.27 shows two tensile test pieces of a tough polymer, one of which had epoxy resin smeared over its surface. Williams (1976) first showed that whereas the uncoated sample displays characteristic yielding and 'drawing' behaviour before fracture, the sample coated with the brittle epoxy resin layer fractures in an elastic brittle fashion. Clearly there must be interaction between the two layers of widely different mechanical behaviour as the component body is loaded. What are the appropriate mechanics for such combined deformation? Why is there a change in the response to loading?

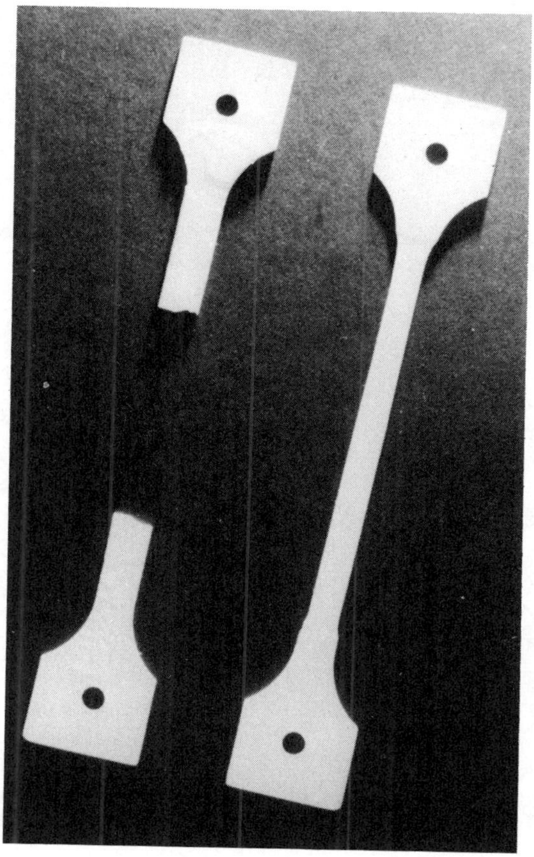

Fig. 3.27 – Tensile testpieces of polyethylene, the left coated with epoxy, the right uncoated. The coated specimen broke brittlely, the natural specimen flowed plastically and drew down (courtesy R. Dakin).

Consider two cracked bodies identical in all respects except that they possess different R (the general problem of bodies with different E and sizes is considered later). The Xu plot for crack propagation in the tougher component (by itself) is OM_1N_1O (Fig. 3.28(a)) along the R_1 locus; for the less tough component it is OM_2N_2O along R_2 over the *same* increase in crack length, defined by the common stiffness lines OM and ON. Were these two plates to be loaded in parallel through common loading pins, they would crack individually and, behaving as springs in parallel, the total load would be the sum of the separate

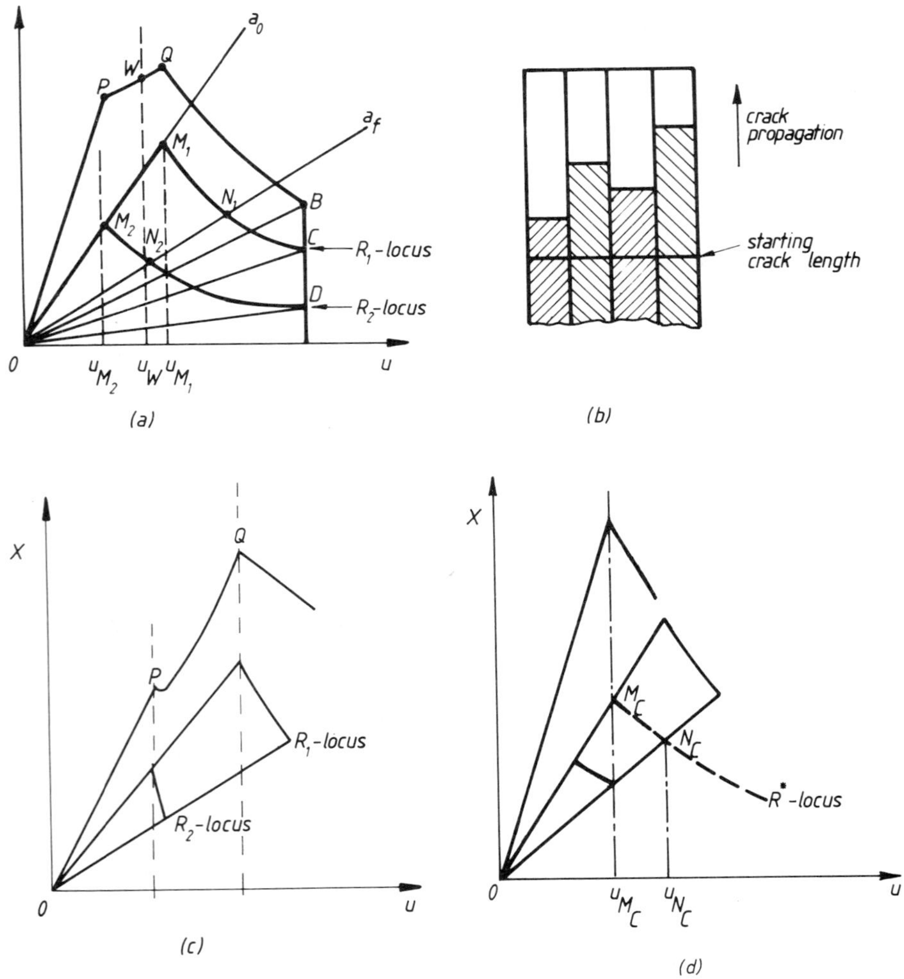

Fig. 3.28 - (a) Load–deflexion curves for cracking of a 'free' sandwich containing a tough and a weak component whose toughnesses are R_1 and R_2 respectively. (b) Crack fronts in a many-layered 'free' sandwich showing a stepped appearance with different crack lengths in each layer. (c) 'Pop-in' observed in cracking of a free sandwich. (d) Load–deflexion curves for cracking of a two-layer sandwich with perfect bonding between them. (After Atkins (1979), with permission.)

loads at various u (Fig. 3.28(a)). There is a kink in the total load diagram at P when the component with lower toughness starts to crack; between u_{M2} and u_{M1} only the less tough body is cracking, but after u_{M1} both are cracking. The crack lengths in each component are different, and are the same as each body has when cracking individually. Thus at u_W, the crack length in the layer 2 has become a_f, but in layer 1 it is still a_0. Triangular work area OPWO must be the same as OM_2N_2O, where stiffness OW is the combined stiffness of the cracked body 2 plus uncracked body 1. Later, work area OPQBO represents work of cracking in both layers from starting length a_0 to propagated lengths having stiffness lines OD (body 2) and OC (body 1), and so on. In plan view, the crack fronts in a many-layered sandwich present a stepped appearance, with different propagated lengths in each layer (Fig. 3.28(b)).

First cracking in body 2 occurs as the *total* load in the system *increases* along PQ, owing to the stiffening action of the uncracked body 1, even though of itself body 2 would crack under *decreasing* net load along M_2N_2. However, as the total load is the combination of decreasing loads along M_2N_2 and increasing loads along M_2M_1, it is possible to get a local decrease in total load at P if curvature in the R_2 locus at M_2 drops X_{M2} sufficiently (Fig. 3.28(c)). This behaviour is called 'pop-in' in slightly different circumstances (viz. in the cracking of thick plates displaying centreline 'tunnelling', section 2.8 where, in the present context, longer tunnelled crack lengths correspond with regions of low toughness which crack earlier than others (see Fig. 3.28(b)).

If the bodies were stuck together, instead of being free, they would have a composite behaviour, in between the individual responses. The action of the adhesive, solder or whatever is to transmit shear forces between the layers, the behaviour of which is thereby modified. Assume that the glue is capable of transmitting shear tractions which force identical propagated crack lengths in all layers so that there is negligible interfacial slip (experiments show that this is possible). As the Δu for each glued layer must be identical through the common loading pins, it follows that cracking in the less tough layer is suppressed (until OM_cN_cO in Fig. 3.28(d) and cracking in the tougher layer occurs earlier (again at OM_cN_cO) than in the absence of glue action (the common crosshead movement is u_{Mc} to u_{Nc}). In consequence, during cracking of glued composite bodies, the tougher component is forced to crack at a load below its free cracking load, and the less tough component at a higher load than normal. In such a constrained condition, the first material will appear to have a lower fracture toughness than R_1 and the second a higher value than R_2. If energy is conserved, the total work required will be as before for the same increases in crack length, so for quasi-static cracking

$$OM_2N_2 + OM_1N_1 = 2\,OM_cN_c$$

or

$$R^* = (R_1 + R_2)/2, \tag{3.36}$$

which is a rule of mixtures for R^* (the apparent toughness of the layered body) for the present particular problem of two equal layers. It is physically reasonable to add up the individual works of fracture required in each layer. A rule-of-mixtures approach – well-known for the modulus and 'strength' in fibre-reinforced composites – was first applied to strain energy release rates G by Bluhm in 1961 in connection with the total work of fracture in laminated plates and thick solid plates possessing low centreline toughness and high edge (shear-lip) toughness.

The rule of mixtures sets the level of X_{M1} in the following way: Writing $R = \frac{X^2}{2B}\frac{d}{da}\left(\frac{u}{X}\right)$, and noting that $\frac{d}{da}\left(\frac{u}{X}\right)$ is the same in each layer (in this particular example) we obtain

$$X_{Mc} = \sqrt{(X_{M1}^2 + X_{M2}^2)/2} \tag{3.37}$$

and for the total load on the system

$$X_T = 2X_{MC} = \sqrt{2(X_{M1}^2 + X_{M2}^2)}\ . \tag{3.38}$$

Note that $X_T > (X_{M1} + X_{M2})$ so that there is a synergism is cracking load, owing to the addition of terms in X^2 (proportional to energy) rather than terms in X alone. Likewise, since the stress intensity factor K is proportional to load alone (the crack lengths being the same in the different layers), we have

$$K_{MC} = \sqrt{(K_{M1C}^2 + K_{M2C}^2)/2} \tag{3.39}$$

for sandwiches with equal bending stiffness. We note that K_{MC} is given by $(K_{M1C} + K_{M2C})/2$ (i.e. a rule of mixtures on K) *only* when there is no glue, and no interlaminar shear tractions.

This interesting and unexpected result of a synergism on X or K has been observed in PMMA/PVC double-cantilever-beam (DCB) sandwich specimens glued together with a thin layer of cyanoacrylate adhesive as shown in Table 3.1.

Table 3.1

K_C values for PMMA, PVC and composite beam (Guild, Harris and Atkins, 1978)

PMMA	PVC	Composite	Mixture rule average	Equation (3.39)
1.09 ± 0.04†	3.56 ± 0.20†	2.68 ± 0.08‡	2.32	2.63

†9 samples were used and 13 results obtained.
‡5 samples were used and 9 results obtained.

The mean composite K_C is much higher than the mixture rule average and is very close to the synergistically higher value (statistically, the difference between the means is about nine times the standard error, so the effect is real – the effects of adhesive toughness, specimen thickness, and rate effects in the separate layers and in the composite were considered in the experiments).

For the more general problem of layers of different thicknesses and different mechanical properties, consider Fig. 3.29(a) and (b) which shows a stiff, tough

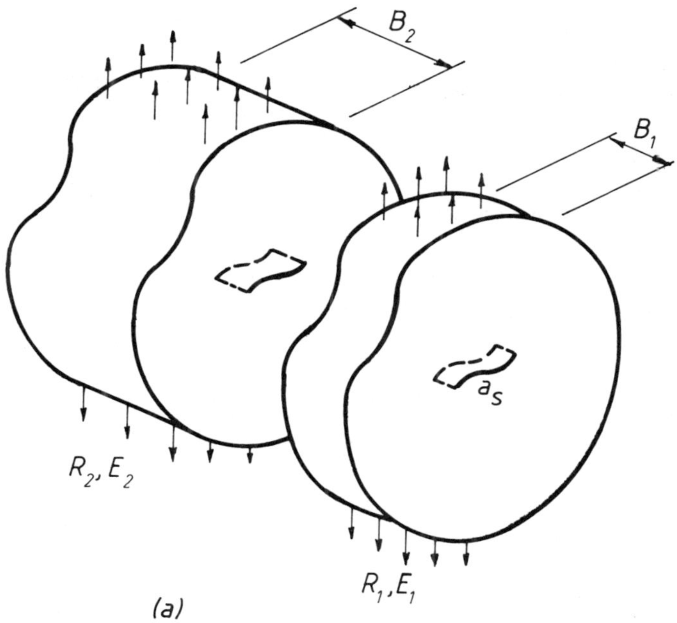

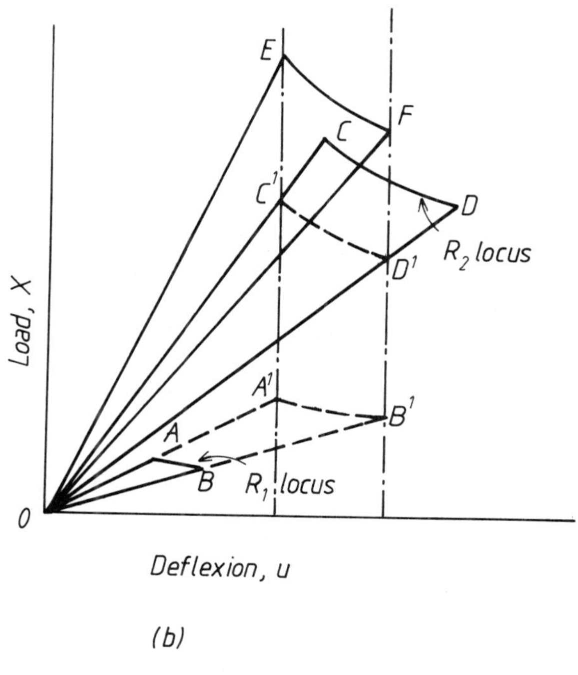

Fig. 3.29 – (a) A two-layer composite with different Young's moduli and toughnesses but same starter crack length. (b) Load–deflexion curves for cracking of the composite in (a).

layer (with stiffness OC, corresponding to crack length a_s, cracking along CD when free and unloading down DO at propagated length a_f) coupled to a compliant, less-tough layer (with stiffnesses OA for a_s and OB for a_f, cracking along AB when free); other combinations are of course possible (Atkins, 1979). When interlaminar glue tractions force a common crack front, simultaneous cracking from length a_s to length a_f in both layers occurs over the common displacement u_E to u_F instead of u_A to u_B, and u_C to u_D separately. Point E lies on the combined stiffness line (OE) of the sandwich with starter length a_s; point F lines on the combined propagated stiffness line. Points E and F lie on the toughness locus for the combined sandwich, and their location is controlled by the work area requirement,

$$\text{OABO} + \text{OCDO} = \text{OEFO}$$

i.e.

$$R_1 B_1 [a_f - a_s] + R_2 B_2 [a_f - a_s] = R^* [B_1 + B_2] [a_f - a_s] \tag{3.40}$$

R^* is thus the rule-of-mixtures toughness $(R_1 B_1 + R_2 B_2)/(B_1 + B_2)$. Proceeding as in the earlier example we obtain,

$$X_1^2 \frac{\mathrm{d}}{\mathrm{d}a}\left(\frac{u}{X}\right)_1 + X_2^2 \frac{\mathrm{d}}{\mathrm{d}a}\left(\frac{u}{X}\right)_2 = X_T^2 \frac{\mathrm{d}}{\mathrm{d}a}\left(\frac{u}{X}\right)_T \tag{3.41}$$

where X_T is the total load in the sandwich and $(u/X)_T$ is the compliance of the two-layer sandwich. Now, in general, $\frac{\mathrm{d}}{\mathrm{d}a}\left(\frac{u}{X}\right) = \frac{2Z^2}{EBW}$ (equation (3.5)), so that

$$\frac{X_1^2}{E_1 B_1} + \frac{X_2^2}{E_2 B_2} = \frac{X_T^2}{E_T (B_1 + B_2)} \tag{3.42}$$

As we are assuming no interlaminar slip, the Voigt rule-of-mixtures for E_T (based on identical strains) is appropriate where $E_T = (E_1 B_1 + E_2 B_2)/(B_1 + B_2)$. Then

$$X_T^2 = X_2^2 (1 + \alpha) \left(\frac{X_1^2}{\alpha X_2^2} + 1\right) \tag{3.43}$$

where $\alpha = E_1 B_1/E_2 B_2$. Furthermore, from general K formulae, $X_1^2/X_2^2 = (E_1 R_1 B_1^2/E_2 R_2 B_2^2) = \alpha\beta$, where $\beta = R_1 B_1/R_2 B_2$. Hence

$$X_T = \begin{Bmatrix} X_2 \sqrt{(1+\alpha)(1+\beta)} \\ X_1 \sqrt{(1+\alpha)(1+\beta)/\alpha\beta} \end{Bmatrix} \tag{3.44}$$

which is the expression for X_T along EF . . . in terms of X_2 along CD . . . or X_1 along AB . . . at the same a; if free, the separate layers would require X_2 and X_1 at that same a but at different u of course.

When $\alpha = 1$, we revert to the simple case of identical layers differing only in R, and equation (3.43) reduces the equation (3.38).

The actual loads in each layer when cracking occurs under full restraint are given by the loci A′B′ . . . and C′D′ . . . where A′ and B′ lie on OA and OB extrapolated to encompass the displacement $(u_F - u_E)$ over which the sandwich cracks. It may be shown (Problem 3.19) that

$$\left.\begin{aligned} X_A' &= X_A \sqrt{\alpha(1+\beta)/\beta(1+\alpha)} = X_E\,[\alpha/(1+\alpha)] \\ X_C' &= X_C \sqrt{(1+\beta)/(1+\alpha)} = X_E\,[1/(1+\alpha)] \end{aligned}\right\} \qquad (3.45)$$

where X_E is X_T at the start of cracking. When $\alpha = 1$, we obtain equation (3.38). As in general X_E changes during cracking so will $X_A{}'$ and $X_C{}'$.

In section 3.9.4 later it will be shown that yielding can be suppressed using layered structures. This comes about because the fracture load for the tough layer is $X_C{}'$ and not X_C (which would be greater than the yield load without the brittle layer). So long as $X_C{}'$ is less than the yield load, valid toughness K_{C_2} or R_2 for the tough solid can be obtained. Clearly, from the second equation of (3.45) $X_C{}'$ is least (and suppression of yielding is most effective) when β is least or α is a maximum. Williams (1976) and Yap *et al.* (1980) have used thin layers of epoxy resin to promote brittle cracking in ductile polymers like PVC and HIPS.

Example 3.12 Epoxy resin displays stick–slip fracture so that Gurney's work area method is difficult to apply to determine its R value (see Chapter 8). By glueing a layer of PMMA to the epoxy layer it has been found that the composite shows slow stable cracking with a common crack front giving a composite toughness fo 400 J/m^2 using Gurney's segmental area method. Assuming the thickness for the PMMA and epoxy are 6 mm and 3 mm respectively, calculate R for epoxy. Separate experiments show that R for PMMA is about 500 J/m^2.

Ans. The reason why PMMA can stabilise epoxy will be discussed in Chapter 8. Here for cracking of the PMMA/epoxy composite with no interface delamination, equation (3.36) gives

$$R_{\text{composite}}(B_{\text{PMMA}} + B_{\text{epoxy}}) = R_{\text{PMMA}}B_{\text{PMMA}} + R_{\text{epoxy}}B_{\text{epoxy}}$$

so that

$$400 \times (6+3) = 500 \times 6 + R_{\text{epoxy}} \times 3$$

therefore

$$R_{\text{epoxy}} = \frac{400 \times 9 - 500 \times 6}{3} = 200 \text{ J/m}^2.$$

Example 3.13 Obtain an expression for the apparent toughnesses at which the glued layers crack in Fig. 3.29.

Ans. We have OAB + OCD = OEF = OA'B' + OC'D' so $R_1 B_1 + R_2 B_2 = R'B_1 + R_2' B_2$ where R_1' and R_2' are the apparent toughnesses displayed by layers (1) and (2) along loci A'B' and C'D'. Thus $R_1' = R_1 + (B_2/B_1)(R_2 - R_2')$. But $R \propto X^2/B$ so $R_2' = (X_C'/X_A')^2 (B_1/B_2) R_1'$. Hence $R_1' = (R_1 + (B_2/B_1) R_2)$ $(1 + 1/\alpha^2)$ using equation (3.45). Similarly $R_2' = R_2 + (B_1 R_1 - \alpha^2 B_2 R_2)/B_2$ $(\alpha^2 + 1)$. When $E_1 = E_2$ and $B_1 = B_2$, $\alpha = 1$, and $R_1' = R_2' = 1/2\,(R_1 + R_2)$ as before. Note that more work is obtained from the frangible component (OA'B' in place of OAB) by making it deflect more than it otherwise would, and less work (OC'D' vs. OCD) is available from the tough layer. There must therefore be an *energy transfer* through the glue from layer (1) to layer (2); this true for Fig. 3.28(d) as well.

Example 3.14 Discuss the magnitudes of the shear tractions required to force a common crack front in layered structures possessing different mechanical properties in different layers.

Ans. A measure of the tractions that have to be withstood by the interface is given in Fig. 3.29 by $(X_A - X_A')$, which produces a crack tip closing effect in the less tough layer and $(X_C' - X_C)$ which produces a crack tip opening effect in the tougher component. Note that these are not necessarily equal, but they are self-equilibrating in each layer ('above' and 'below' the crack); they will change as X_E changes during propagation. The distribution of these tractions as shear stresses is not clear, but obviously they may be high enough to cause interlaminar breakdown in an adhesive. Such 'breakdown' may be brittle cracking at the glue line, or may be viscous flow in a ductile adhesive. Cracking under the action of $(X_A - X_A')$ and $(X_C' - X_C)$ is similar to cracking under the influence of residual stresses (section 3.7) where the controlling toughness may be apparently increased or decreased.

It is not always possible to find an adhesive capable of withstanding these shear tractions, so that interlaminar slip occurs and a common crack front is not achieved. Even so, there may be some coupling between the layers and the behaviour then is somewhere in-between that of free and fully constrained sandwiches, as shown in Fig. 3.30 where, for simplicity, the case of bodies cracking at constant load is illustrated. The (limited) shear tractions are capable of bringing down CD to only C''D'' and AB up to A''B'', which are the apparent toughness loci. Analysis of this type of problem is complicated as $d/da\,(u/X)$ now varies with the different crack lengths in the different layers, but it may be shown that when every layer is cracking in the general problem,

$$X_T)_{\text{free}} < X_T)_{\text{partial slip}} < X_T)_{\text{no interlaminar slip}} \tag{3.46}$$

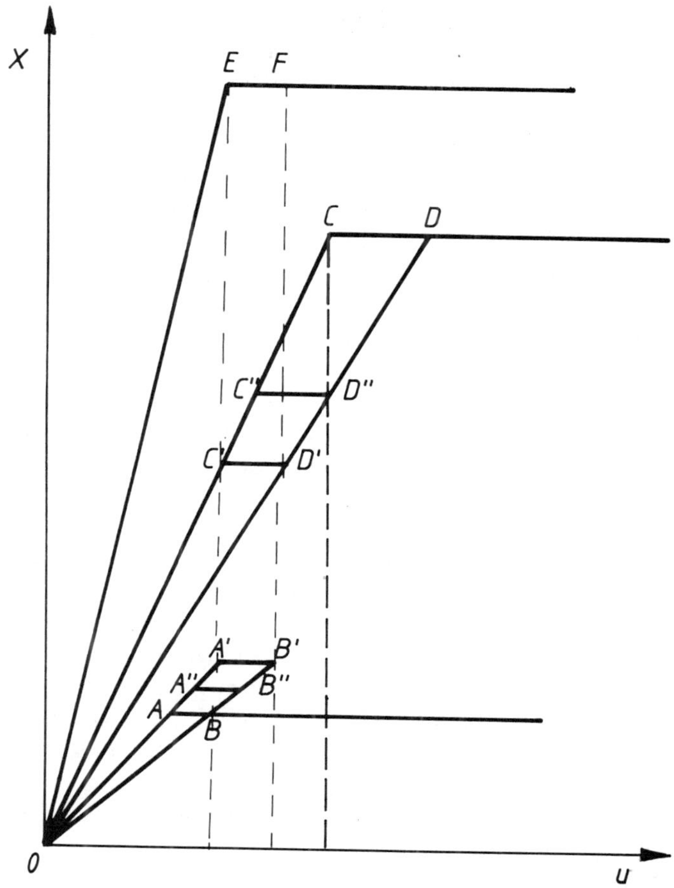

Fig. 3.30 – Load–deflexion traces for the cracking of a sandwiched composite with interlaminar slipping.

An interesting question arises for the start of cracking in this problem, viz. in the period A to A″, the only cracking work being consumed is that in layer (1) – layer (2) does not begin to crack until C″ – the work requirement for which is OAB; yet in the presence of interlaminar tractions the energy available is OA″B″ over the same crack length increment, and this is greater than OAB; (later, when layer (2) cracks there is an energy transfer to that layer to compensate for its 'early' cracking at C″ instead of C). Two explanations are possible: either (i) a local interlaminar debonding instability occurs in layer (1), followed by arrest of the crack by the action of the layers slipping over one another as propagation ensues in layer (1), followed by loading at the arrested longer length in layer (1) and the original length in layer (2), until such time as they propagate together along A″B″ and C″D″. Alternatively, (ii) crack initiation occurs still at A, but propagation stops eventually owing to the build-up in traction with interlaminar slip, and events then proceed as in (i). The local debonding described in (i) is similar to debonding in filament-reinforced composites where, in glass fibre for example, the regions become milky in colour.

Another application of the partial slip model is to cracking in thick plates which have low centreline (plane strain) toughness and high edge (plane stress) toughness, which display the phenomena of pop-in, slow crack growth and thumbnail-shaped crack fronts. Discrete layers are replaced by incremental layers, of the same stiffnesses, whose toughness varies with position. Discrete interlaminar tractions become a distribution of through-thickness shear stresses, limiting values for which occur when they reach the shear yield stress. In the general case where $A''B''$ and $C''D''$ drop down during propagation (they are shown as constant in Fig. 3.30 for simplicity of illustration), loads along $A''B''$, corresponding with interior plane strain deformation in thick plates, may locally fall more quickly than the rising loads of the uncracked outer plane stress regions, i.e. 'pop-in'. The thumbnail crack front corresponds with a smoothly curving version of Fig. 3.28(b), and the 'pop-in' load for thick plates (corresponding with P in Fig. 3.28(c)) indicates that the apparent toughness level at A'' in Fig. 3.30. The pop-in load would seem to be geometry-dependent (competition between rising loads along uncracked stiffness lines and falling R loci). Clearly there are shear irreversibilities in planes perpendicular to the crack plane, which must become an integral part of fracture work in these circumstances.

We conclude by noting that the toughness of WC–Co 'hard-metal' alloys may be predicted from the properties of the hard WC particles and the ductile cobalt matrix (Murray, 1977) via a 'mechanics of sandwiches'. The mean crack tip plastic zone size (and hence crack opening displacement) is spatially constrained by the interparticle spacing and Murray found that the constrained flow stress is about twice the theoretical shear stress. Fracture paths are predominantly through the cobalt phase and the specific work of fracture (i.e. K_{IC}) may then be related to the mean free path of the cobalt.

3.7 CRACKING IN THE PRESENCE OF RESIDUAL STRESSES

Many materials are left in a state of macroscopic stress as a result of manufacturing processes. Indeed some semi-finished components have processing cracks caused by relief of stresses set up by differential thermal straining (casting cracks, cracks around inclusions, heat treatment queching cracks and so on) and cracks unintentionally produced by mechanical processing (rolling and forging cracks, splits in sheet metal and so on). Cracks in dried mud and cracks produced in dough under the rolling pin are similar. It is well-known that when materials containing unrelieved residual stresses are slit or machined, the cut pieces distort in various ways, and the components may even spontaneously break up. The mechanics of initial crack formation, and subsequent cracking produced by external loads, must clearly be influenced by the pattern of residual stresses. Experiments show that in elastic fracture in the presence of residual stresses, loading/unloading lines do not in general pass out from the origin of the Xu coordinate system and there are residual displacements as the body 'springs' open. The body is, however, deforming elastically, so we note that 'reversibility' in cases of residual stress, means restoration of *both* displacement *and* a force system at least equipollent with the original residual stress distribution. This is given by the force necessary to restore the body to its original geometry.

At a crack tip, residual stress fields may be tensile or compressive but, as there are zero net forces and moments in a free body containing residual stresses, there must be compensatory compressive and tensile stresses elsewhere in the body. However, there will always be a finite strain energy Λ_r associated with the complete residual stress and strain distribution. Λ_r changes when the residual stresses and strains are relieved by crack propagation, and clearly $d\Lambda_r/dA$ must enter the energy rate balance, equation (1.1), and affect the value of dJ/dA or dG/dA. This will then determine the external loads required to cause crack propagation (in broad terms smaller loads than otherwise if residual energy is released, or even greater loads if residual energy is taken in). Under the correct conditions, spontaneous cracking in the absence of additional external loads is possible if $d\Lambda_r/dA$ is energy released just at the correct rate to feed the crack with fracture toughness work.

It might be expected that an opening or closing notional load (or stress intensity) could represent tensile or compressive residual stresses at a crack tip. In this way, by superposition, it would be argued that the stress intensity at the crack tip is the sum of the applied stress intensity and the notional stress intensity, and that cracking would occur when the usual critical stress intensity is reached. However, as energy considerations involve displacement as well as load, the relation may not be quite so straightforward (similar difficulties arise with stress or energy criteria for crack paths, section 3.5. This point is investigated in what follows.

A simple experimental simulation of residual stresses may be produced by glueing together strips of material which have mutually opposite curvatures (Mai, 1972). Such curved strips can be made by hot plastic bending of metal or polymer strips; at the test temperature, the strips should remain in the elastic range when flattened. Unglued portions act as starter cracks, and cracking of glued joint can take place in the presence of both tensile crack tip stress fields at the crack ends (Fig. 3.31(a)) and compressive fields (Fig. 3.31(b)) depending on the sense of the initial curvatures. In the tensile case, crack faces are opened up by the elastic curved strips wanting to regain their former shape, so that 'daylight' exists through the crack. In the compressive case, the crack faces are jammed together.

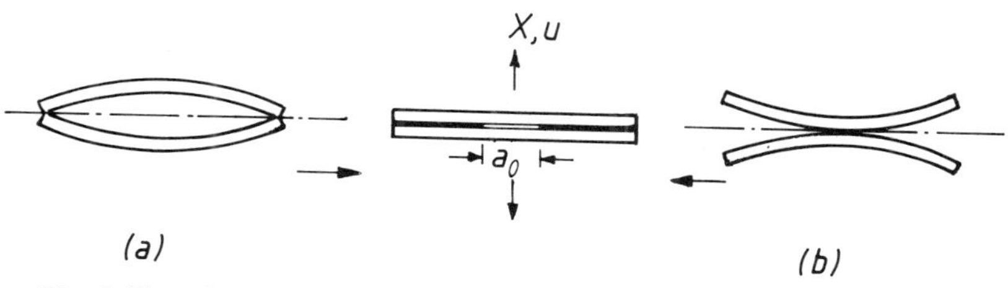

Fig. 3.31 – Splitting of an adhesive joint formed by two slender beams with initial curvature: (a) tensile residual stress case, and (b) compressive residual stress case. (After Mai (1972), with permission.)

3.7.1 Tensile residual stresses

Let us first consider the tensile case. Work has to be performed on the curved strips of Fig. 3.31(a) in order to flatten them before glueing. When the glue has set, that work is locked up as residual strain energy Λ_r. Tensile stresses and strains in the glue near the crack tip, and compressive ones remote from the crack contribute to Λ_r. If the flattening is performed by compressive point loads at the apices of the 'arches' (the shape of the initially curved strips, given by u_r), the separation of the strips at the centre will be

$$u = u_r - \left(- \frac{Xa^3}{96EI} \right) \tag{3.47}$$

where a is the current length of unflattened strip at that instant (which acts effectively as a fixed-ended beam) and X is negative being in the opposite sense to u. The strips would be completely flat when $u = 0$. Writing $u_r = \alpha a^n$ (for the starting shape of the curved strips) we have for $u = 0$ (i.e. complete initial flattening),

$$\alpha a^n = - \frac{X_0 a^3}{96EI} \tag{3.48}$$

where X_0 is the load to cause complete flattening.

If we choose $n = 3$, $X_0 = -96EI\alpha$ which is independent of a. The graphical significance of this simplification follows in Fig. 3.32 presently. For any length

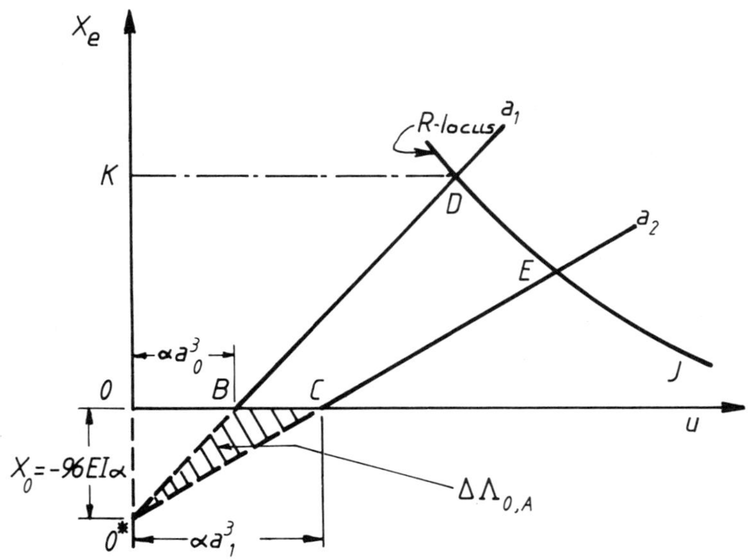

Fig. 3.32 – Diagrammatic representation of residual strain energy released during quasi-static cracking due to tensile residual stresses. (After Mai (1977), with permission.)

of strip a, initially bowed to $u_r = \alpha a^3$, the residual strain energy of two such strips glued together is

$$\Lambda_r = \tfrac{1}{2} X_0 u_r = -\tfrac{1}{2}(96EI\alpha)(\alpha a^3) \tag{3.49}$$

assuming that chordal lengths and arc lengths are the same. Now if the strips become unstuck for any reason (either spontaneously because the glue is inadequate, or because the strips are forced apart), Λ_r is *released* from the system. Since an unstuck portion (crack) of length a, which increases its length by da, creates total new crack area wda, we have

$$\mathrm{d}\Lambda_r/\mathrm{d}A = -3(96EI\alpha)(\alpha a^2)/2w \tag{3.50}$$

the negative sign indicating the release of Λ_r as A increases. The toughness of the glue has to be capable of withstanding this potential strain energy release, otherwise spontaneous springing aprt of the joint will take place. The joint will hold when its threshold toughness

$$R_0 \geqslant -\mathrm{d}\Lambda_r/\mathrm{d}A = +3(96EI\alpha)(\alpha a^2)/2w \tag{3.51}$$

for a current unglued portion of length a. Note that this requirement, increases with a which suggests that, once spontaneous delamination occurs, it does not arrest.

External loads have to be applied to the joint to break it if the glue toughness exceeds R_0. For an imposed tensile load X_e (in the same sense now as u) we have (with reference to origin O^* in Fig. 3.32):

$$X_e \mathrm{d}u = \mathrm{d}\Lambda_r + \mathrm{d}\Lambda + R\mathrm{d}A \tag{3.52}$$

where the extra work required to crack the joint is supplied by the external force and the additional energy release of the joint itself. Using Equation (3.50) for $(\mathrm{d}\Lambda_r/\mathrm{d}A)$ and $\mathrm{d}\Lambda = \tfrac{1}{2}\,\mathrm{d}[X_e(u-u_r)]$, so

$$X_e \frac{\mathrm{d}u}{\mathrm{d}A} = -\frac{(96EI\alpha)\ 3\alpha a^2}{2w} + \tfrac{1}{2}\frac{\mathrm{d}}{\mathrm{d}A}\left(X_e(u-\alpha a^3)\right) + R$$

i.e.

$$X_e \frac{\mathrm{d}u}{\mathrm{d}a} = -\ 3(96EI\alpha)\alpha^2 - X_e\,3\alpha a^2 + (u-\alpha a^3)\frac{\mathrm{d}X_e}{\mathrm{d}a} + 2Rw \tag{3.53}$$

Under the action X_e, the deflexion of the pair of glued beams is

$$u = u_r + \frac{X_e a^3}{96EI} = \alpha a^3 + \frac{X_e a^3}{96EI}$$

Thus,

$$\frac{\mathrm{d}u}{\mathrm{d}A} = \frac{1}{w}\left[3\alpha a^2 + \frac{3X_e a^2}{96EI} + \frac{a^3}{96EI}\left(\frac{\mathrm{d}X_e}{\mathrm{d}a}\right)\right] \tag{3.54}$$

Substitution back in equation (3.53) gives a differential equation, the solution of which is

$$X_e = \sqrt{\frac{2 \times 96EIwR}{3a^2}} - 96EI\alpha \tag{3.55}$$

This diminishes with a as we expect, since $|-\mathrm{d}\Lambda_r/\mathrm{d}A|$ increases with a thus requiring less external effort to break the joint.

The load X_{abs} to crack the joint in the *absence* of residual stresses is given by

$$X_{abs}^2 = \frac{2R}{\dfrac{\mathrm{d}}{\mathrm{d}A}\left(\dfrac{u}{X}\right)} = \frac{2R}{\dfrac{\mathrm{d}}{\mathrm{d}A}\left(\dfrac{a^3}{96EI}\right)} = \frac{2 \times 96EIwR}{3a^2}$$

which is the same as the first term on the r.h.s. of equation (3.55). Thus equation (3.55) becomes simply

$$X_e = X_{abs} + X_0 \tag{3.56}$$

since $X_0 = -96EI\alpha$ for $n = 3$ (equation (3.48)). Hence in this case, cracking occurs at the residual-stress-free load modified by the load X_0 associated with forcing the residual stresses on the system.

The foregoing analysis is readily understood from the work areas in Fig. 3.32. Constant crack length compliance lines fan out from the origin O*, which, as far as *external* loads X_e are concerned, is located at ($u = 0$, $X_0 = -96EI\alpha$). Remember that the lines fan out from the same origin owing to the form of $u_r = \alpha a^3$. The origin of external displacements is at B (where $u_0 = \alpha a_0^3$) since the strips are already bowed out over the unstuck length a_0. Loading/cracking/unloading experiments do not revert to their starting displacement B and have progressively increasing unloaded crack openings.

Since X_e falls with increasing a it will become zero when $a = \sqrt{wR/EI}/12\alpha$ (equation (3.55)) and $u_r = (wR/EI)^{3/2}/1728\alpha^2$. Then spontaneous cracking takes over. In Fig. 3.32 this would be when the continuation of the DEJ locus cuts the abscissa. Thereafter an adquate rate of supply of work is provided from Λ_r below that axis.

Stress corrosion cracking in the presence of residual stresses occurs for similar reasons, when the action of a hostile environment lowers R, and eventually Λ_r causes spontaneous cracking (Chapter 6).

From the analysis, it follows that area O*DE is the segmental work area for $Rw(a_1 - a_0)$, the cracking load in the absence of residual stress being the total height of the ordinate from O*. Experiments will only pick up work area BDEC, which is less than the toughness work required by the amount OBC, which represents $\mathrm{d}\Lambda_r$ which is helping crack propagation. We have

$$\begin{aligned} \text{area BDEC} &= Rw(a_1 - a_0) + (\Lambda_{0,a_1} - \Lambda_{0,a_0}) \\ &= Rw(a_1 - a_0) + \Lambda_r \end{aligned} \tag{3.57}$$

where the notation $\Lambda_{0,a}$ means the strain energy in the system at zero load and crack length a, i.e. the work areas below the abscissa. In the limit,

$$\frac{d}{dA}\,(\text{BDEC}) = R + \frac{d}{da}\,(\Lambda_r)\ \ (= R - R_0) \tag{3.58}$$

and

$$\frac{d}{dA}\,(\text{BDEC}) = \frac{\partial}{\partial a}\,(\text{OBDK})_X = R^* \tag{3.59}$$

where R^* may be thought of as the apparent fracture toughness, obtained only from those work areas above the abscissa. Hence

$$R^* = R + \frac{d}{da}\,(\Lambda_r) = (R - R_0) \tag{3.60}$$

and R^* coincides with the 'real' toughness R only when $d/dA\ (\Lambda_r) = 0$, i.e. in the absence of residual stresses. In tests to measure R, $d/dA\ (\Lambda_r)$ can be obtained graphically as indicated below in Example 3.15 using back extrapolation of unloading lines, but in analyses to predict X_e, for known R, the situation is difficult as $d/dA\ (\Lambda_r)$ will not usually be known.

Example 3.15 Show that the 'true' fracture toughness of a solid in the presence of residual stresses is obtained from the work area given by back-extrapolating loading/unloading lines to a common origin.

Ans. Back-extrapolation of lines BD and CE gives O* in Fig. 3.32. Integration of $X du = d\Lambda + R dA$ along BD gives (—area BOO* + area BDM = $\Lambda_D - \Lambda_{O^*}$). Similarly along CE, (—area COO* + area CEN = $\Lambda_E - \Lambda_{OE^*}$). Since area DENM = $\int_D^E X du$, we conclude that area O*DE = $\Lambda_D - \Lambda_E + \int_D^E X du = R(A_2 - A_1)$ over crack area increase $(A_2 - A_1)$ from the basic relation $X du = d\Lambda + R dA$. Thus R = area O*DE/$(A_2 - A_1)$.

Example 3.16 Show that equation (2.21) in Chapter 2 can be applied to the residual stress problem described above.

Ans. Equation (2.21) gives

$$R = \tfrac{1}{2} X^2 \frac{dC}{dA} + X \frac{du_r}{dA}$$

which is valid for situations where no residual strain energy is released with crack propagation. In terms of Fig. 3.32 it is modified to

$$R^* = \tfrac{1}{2} X_e^2 \frac{dC}{dA} + X_e \frac{du_r}{dA} \tag{a}$$

where R^* is the apparent fracture toughness and X_e the effective fracture load. From equation (3.52) we obtain

$$X_e \, du = d\Lambda_r + R\,dA + d\Lambda$$

and since $u = u_r + CX_e$ (where the compliance $C = a^3/96EI$),

$$X_e \, du_r + X_e^2 \, dC + X_e C \, dX_e = d\Lambda_r + (R^* + R_0)\, dA + d\Lambda \qquad \text{(b)}$$

Now $R_0 = -d\Lambda_r/dA$ and $d\Lambda = X_e C \, dX_e + \dfrac{X_e^2}{2} \, dC$, therefore we have

$$R^* = \frac{X_e^2}{2} \frac{dC}{dA} + X_e \frac{du_r}{dA} \qquad \text{(c)}$$

which is the same as equation (a) above. It can also be shown that equation (3.52) gives directly equation (2.33), i.e.

$$R^* = X_e \frac{du}{dA} - \tfrac{1}{2} X_e^2 \frac{dC}{dA} - X_e C \frac{dX_e}{dA} \qquad \text{(d)}$$

Experimental verification of the trends of this analytical treatment for the cracking of an adhesive joint in the presence of residual crack-tip tensile stresses is shown in Fig. 3.33. Table 3.2 gives values of Λ_r which are determined both from equation (3.49), i.e. $\Lambda_r = -\frac{1}{2}(\alpha a^3)\,(96EI\alpha)$, and from the graphical relation equation (3.60), i.e. $\Delta\Lambda_r$ = area BDEC $- R\Delta A$. Comparison of the results shows that fairly good agreement is obtained.

Table 3.2
Estimation of $\Delta\Lambda_r$
(Mai, 1972)

Crack area increment ΔA (cm^2)	Genuine fracture toughness R (J/m^2)	Residual strain energy released $\Delta\Lambda_r$ (mJ) Equation (3.60)	Equation (3.49)
31.24 − 24.88 = 6.36	11	−3.33	−1.00
40.13 − 31.24 = 8.89	12	−4.22	−1.57
48.77 − 40.13 = 8.64	15	−2.06	−2.16
57.66 − 48.77 = 8.89	14	−3.33	−2.60
66.55 − 57.66 = 8.89	14	−4.12	−3.25
75.31 − 66.55 = 8.76	15	−3.82	−3.82

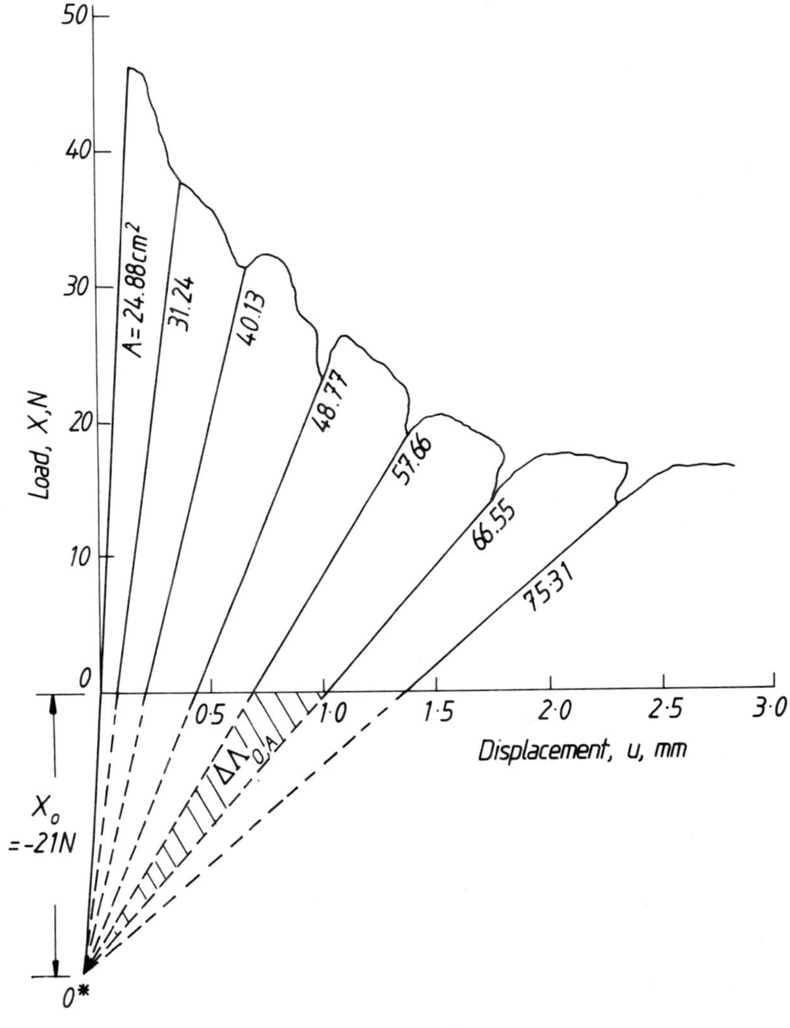

Fig. 3.33 – Estimation of residual strain energy by splitting an adhesive joint shown in Fig. 3.31(a). (After Mai (1972), with permission.)

Example 3.17 Plywood laminations sometimes fail and buckle because of swelling due to absorption of water. Severe buckling may lead to cracking along the interlaminar planes of adhesion. Investigate the problem, incorporating the moisture content as a variable, and determine the minimum fracture toughness of adhesion required to resist fracture due to swelling.

Ans. Fig. 3.34 is a two-dimensional model of a strip of plywood with the transverse grain along its length, stuck to another layer acting as an effectively rigid substrate. The adhesion has failed over a length a which constitutes a crack.

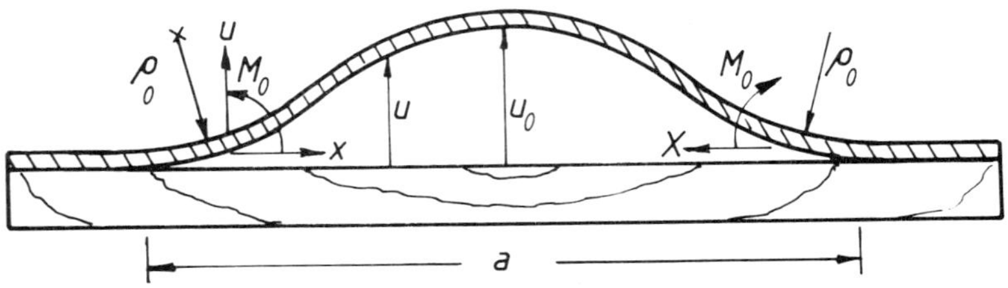

Fig. 3.34 – Splitting of a slender beam by swelling due to moisture absorption. (After Mai (1977), with permission.)

Swelling caused by water absorption lengthens the layer which bows out as shown. Simple buckling may occur without change in a; on the other hand a may increase by cracking along the interface.

There will be a buckling end-load X caused by moisture absorption, which compresses the swollen unattached strip and buckles it. There are also bending moments, M_0, over the buckle at the 'built-in' ends. There are thus two components of strain energy (i) due to compression $= X^2a/2EwB$ and (ii) due to bending or buckling $= M_0^2a/2EI$. Cracking will occur in the absence of external loads or moments when equation (3.51) is satisfied. Thus,

$$R = -\frac{d\Lambda_r}{dA} = \frac{X^2}{2Ew^2B} + \frac{M_0^2}{2EIw}, \tag{a}$$

where $dA = w da$. M_0 may be inferred from the end curvature $1/\rho_0$ of the buckled portion, via beam theory (i.e. $M_0 = EI/\rho_0$). Assuming a cosine deflexion curve for the debonded bowed section, with (u,x) coordinates as shown in Fig. 3.34, this gives

$$u = \left(\frac{u_0}{2}\right)\left(1 - \cos\frac{2\pi x}{a}\right) \text{ for } 0 \leqslant x \leqslant a/2 \tag{b}$$

and

$$\frac{1}{\rho_0} \approx \left(\frac{d^2u}{dx^2}\right)_{x=0} \cong 2u_0\left(\frac{\pi}{a}\right)^2$$

Thus

$$R = \frac{X^2}{2Ew^2B} + 2\left(\frac{EI}{w}\right)u_0^2\left(\frac{\pi}{a}\right)^4 \tag{c}$$

which gives the necessary condition for failure of adhesion in terms of X and u_0. As will be shown presently, X may be eliminated in the algebra by reference to the lengthening of the lamination (i.e. bowing) caused by moisture absorption.

Depending upon the circumstances, four events can take place upon swelling of the wood, viz. (a) cracking in the sliding mode II without buckling; (b) cracking in bending mode I before buckling; (c) buckling with no cracking; (d) buckling followed by cracking in a mixed sliding and bending mode. Different conditions apply to these cases as follows. In (a) as there is no bowing, there is no contribution from M_0 in equation (a) and $R_{II} = \frac{1}{2}X^2/Ew^2B$. Case (b) is discussed later on. In (c) the Euler buckling load is $X = 4(\pi/a)^2EI$ and the initial condition is $u = u_0 = 0$ in equation (b). (d) The initial conditions for this event are that X is the buckling load as given in case (c), but now $u = u_0$, and R is given by equation (a).

X may be eliminated by reference to the lengthening caused by moisture absorption. Experiments show that the swelling and proportional lengthening of the wood ($\Delta a/a$) caused by absorption of moisture (m, mass) follows a relation

$$\frac{\Delta a}{a} = \beta m \tag{d}$$

Thus the unattached length a of the strip, acted upon by m (which lengthens it by βma) and by X (which shortens it by Xa/BwE) has a net length $a\left(1 - \dfrac{X}{BwE} + \beta m\right)$ which must be the same as the integrated length along the deflected cosine curve which is $a\left(1 + \dfrac{\pi^2}{4}\left(\dfrac{u_0}{a}\right)^2\right)$.

Thus in case (a) above, where the compression due to X is counterbalanced by the swelling due to m and no bowing occurs, $u_0 = 0$ so $X = \beta m_{II}BwE$ and so for case (a),

$$m_{II} = \frac{1}{\beta}\sqrt{\frac{2R_{II}}{EB}} \tag{e}$$

where m_{II} is the moisture absorption necessary to cause cracking before buckling, and R_{II} is the fracture toughness of the joint in the sliding mode II.

In simple buckling, substitution of $\beta m_{\mathrm{b}} = X/BwE$ in the Euler expression gives the case (c):

$$m_{\mathrm{b}} = \frac{4}{\beta}\left(\frac{\pi}{a}\right)^2\left(\frac{I}{Bw}\right) \tag{f}$$

where m_{b} is the moisture content to cause buckling.

Where cracking occurs after this buckling, the bowed arc length at cracking $a(\pi^2/4)\,(u_0/a)^2$ is produced by moisture content m_{c} additional to m_{b}, so that for case (d)

$$\beta(m_{\mathrm{c}} - m_{\mathrm{b}}) = \left(\frac{\pi^2}{4}\right)\left(\frac{u_0}{a}\right)^2$$

Then using equation (c), we have

$$m_{\mathrm{c}} = \frac{1}{\beta}\left(\frac{RBa^2}{8EI\pi^2} + 3\left(\frac{\pi}{a}\right)^2\left(\frac{I}{Bw}\right)\right) \tag{g}$$

where R is the mixed mode toughness.

When mode I facture is caused by the bending moment which exists before the buckle actually 'pops out', i.e. case (b), only the first term in (g) is significant and we have

$$m_{\mathrm{I}} = \frac{1}{\beta}\,\frac{Rw}{8EI}\left(\frac{a}{\pi}\right)^2 \tag{h}$$

When $m_{\mathrm{II}} < m_{\mathrm{b}}$ fracture by the swelling force precedes buckling. The minimum fracture toughness of adhesion R_{II} to resist fracture due to swelling is therefore given from

$$\frac{1}{\beta}\sqrt{\frac{2R_{\mathrm{II}}}{Ew}} < \left(\frac{4}{\beta}\right)\left(\frac{\pi}{a}\right)^2\left(\frac{I}{Bw}\right)$$

i.e.

$$R_{\mathrm{II}} < 8\left(\frac{\pi}{a}\right)^4\frac{EI^2}{Bw^2} \tag{i}$$

Various other residual stress problems in glued joints are discussed by Chow and Ngan (1981) as well as Cherry and Hang (1974).

Some thermal stress problems may be tackled in the same manner as Example 3.17. For example, consider two strips, with different coefficients of expansion α_1 and α_2 which are glued together. When the temperature changes (a) cracks may propagate in mode II by shear along the interface; or (b) cracks may propagate in mode I caused by bending at the crack tip; or (c) thermal buckling may occur which may be followed by cracking. As illustrated by Problem 3.23, corresponding equations given in Example 3.17 apply with βm replaced by $\Delta T(\alpha_1 - \alpha_2)$ where ΔT is the temperature change.

3.7.2 Compressive residual stresses

Consider the mutually curved bent strip problem of Fig. 3.31(b). The analysis is as before, but uses negative α for u_r, i.e. $u_r = -\alpha a^n$ ($= -\alpha a^3$ for simplicity) with some additional modifications. Even when a starter crack has been left unglued in the middle, the surfaces are jammed together and there is no daylight visible, unlike the tensile residual stress case. Thus, if

$$u = -\alpha a^3 + \frac{Xa^3}{96EI} \tag{3.61}$$

$u = 0$ when $X_0 = +96EI\alpha$ and $u = -\alpha a^3$ when $X = 0$. Loading/unloading compliance lines fan out from O*, at $(0, X_0)$ the location of which is independent of a (Fig. 3.35), owing to the choice of $n = 3$.

Following equation (3.56) we obtain

$$X_e = X_0 + X_{abs} = 96EI\alpha + X_{abs} \tag{3.62}$$

$X_e > X_{abs}$ because of the residual compressive load. After some cracking, the residual displacements increase negatively ($u_r = -\alpha a^3$), as the compliance lines cross over at O*. The larger residual u_i reflect the bigger chordal bowing of the 'released' strips at longer a. Daylight between the strips will only be observed at positive values of u on loading, and on unloading, longer crack surfaces are

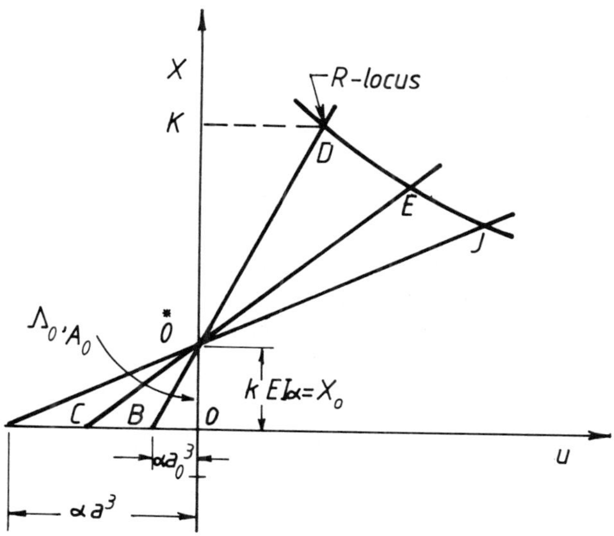

Fig. 3.35 – *X-u* relation for cracking in an adhesive joint under residual compressive stresses shown in Fig. 3.31(b).

always jammed together. The load necessary to just open them is always $+96EI\alpha$, however, independent of a again because of the choice of the cube power for a in $u_r = -\alpha a^3$.

As in the tensile problem, the residual strain energy is given by

$$\Lambda_r = \tfrac{1}{2}(96EI\alpha)(\alpha a^3)$$

and there is a release of strain energy because $(\Lambda_{0,a_2} - \Lambda_{0,a_1})$ is negative since $|-\alpha a_2^3| > |-\alpha a_1^3|$. Since α appears as α^2 in expression such as

$$\frac{\partial}{\partial A}(\mathrm{OBDK})_X = R + \frac{\mathrm{d}}{\mathrm{d}A}(\Lambda_r) = R - \tfrac{3}{2}\left(\frac{\alpha^2 a^2}{w}\right) 96EI$$

for Figs. 3.32 and 3.35 it matters not whether α is positive (tensile residual stress problem) or negative (compressive). However, the meaning of the area equivalent to OBDK (in Fig. 3.32) in the present problem (Fig. 3.35) has to be considered. Part of this crossed-over area is positive and part is negative, unlike the tensile problem where it is all positive. An interesting conclusion follows by $\mathrm{d}/\mathrm{d}A$ (O*DE) = $\partial/\partial A$ (O*KD)$_X$. Now O*KD = $\int_{X_0}^{X_e} u\mathrm{d}X$ and substitution of u from equation (3.61) gives $\partial/\partial A$ (O*KD)$_X$ = $3a^2X^2/2 \times 96EIw = R$ alone (refer back to the tensile residual case).

Thus the segmental areas of positive work centred on O* give the *actual* fracture toughness. Cracking loads are higher than if there were no residual stresses ($X_e > X_{abs}$ by $+96EI\alpha$), but the work areas give the *real* R. How then do we interpret equation (3.60), where

$$\frac{\mathrm{d}}{\mathrm{d}A}(\Lambda_r) = \mathrm{O^*BC} = -\tfrac{3}{2}\left(\frac{\alpha^2 a^2}{w}\right) 96EI?$$

In the tensile case (Fig. 3.32) OBDK is all positive, but in Fig. 3.35 work area OBDK is (O*KD − O*BO), and since $R = \partial/\partial A$ (O*KD)$_X$, $\mathrm{d}\Lambda_r/\mathrm{d}A$ must be equal to $\partial/\partial A$ (O*KD − O*BO − O*DE)$_X$. But $\partial/\partial A$ (O*KD)$_X$ = $\partial/\partial A$ (O*DE)$_X$ so the $\mathrm{d}\Lambda_r/\mathrm{d}A$ term has been 'taken out' of the work area equivalent OBDK and subsequently put back in via equation (3.60), giving actual R values for the segmental areas fanning out from O*.

3.7.3 More general residual stress problems and unloading after considerable crack propagation

If the bent strip simulation of residual stresses is retained in the form $u_r = \pm\alpha a^n$ rather than αa^3, Mai (1977) showed

$$\left.\begin{aligned} u &= \left(\frac{Xa^3}{96EI}\right) \pm \alpha a^n \\ u &= 0 \text{ when } X_0 = \mp 96EI\alpha a^{n-3} \\ u &= \pm \alpha a^n \text{ when } X = 0 \\ X_e &= X_{abs} \mp 96EI\alpha a^{n-3} \end{aligned}\right\} \qquad (3.63)$$

and it may be shown that

$$\frac{d}{dA}(\Lambda_r) = -\left(\frac{2n-3}{2}\right)\frac{96EI\alpha^2 a^{(2n-4)}}{w}\left\{1-(n-3)\alpha a^{n-2}\sqrt{64EIRw}\right\} \quad (3.64)$$

Various points arise, principal among which is that X_0 now varies with a, so that as shown in Fig. 3.36 loading/unloading lines fan out from an origin O* which wanders as a changes, unlike Figs. 3.32 and 3.25. For $n < 3$, $|X_0|$ decreases as a increases (Fig. 3.36); for $n > 3$, $|X_0|$ increases as the crack propagates. As regards $d\Lambda_r/dA$, its released value is increased for $n < 3$ and decreased for $n > 3$ at the same a. Even so, it may be shown (Mai, 1977) as in equation (3.60) that:

$$\frac{\partial}{\partial A}(\text{OBDK})_X = R + \frac{d\Lambda_r}{dA} \quad (3.65)$$

and that 'true' fracture toughness is obtained by extrapolation to O*, even though O* moves with propagation. (Example 3.15 is generally true). Michener and Burns (1983) have subsequently written on this topic.

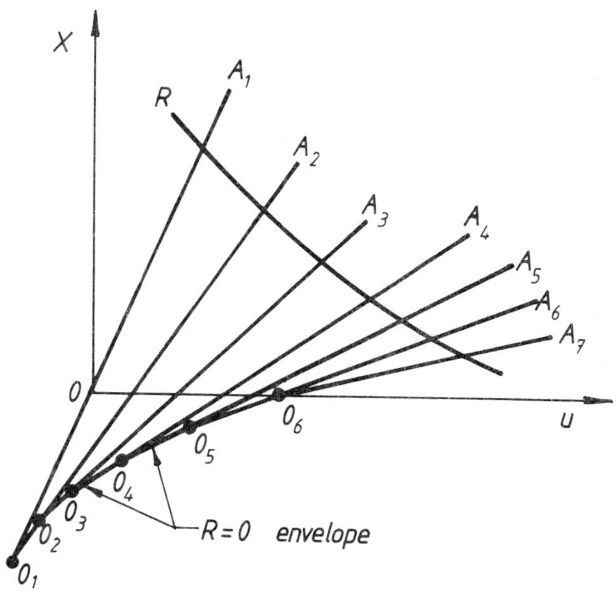

Fig. 3.36 – 'Edge of regression' showing the line $R = 0$ which envelopes the points of intersection of constant stiffness lines. (After Mai (1977), with permission.)

3.7.4 Analysis in terms of stress intensity factors

In the simple types of problem so far described it happens that

$$X_e = X_{abs} + X_0$$

i.e. the externally applied load X_e is equal to the load X_{abs} (which would apply to same situation unaffected by residual stresses) modified by X_0 (the notional load on the crack tip region caused by the residual stresses). We may write

$$X_e \sqrt{a} = X_{abs}\sqrt{a} + X_0\sqrt{a} \tag{3.66}$$

where a is the current length of the crack. Hence *in these circumstances*

$$K_e = K_{abs} + K_{residual} \tag{3.67}$$

where $K_{abs} = K_{critical}$ applying to the residual stress-free problem and $K_{residual}$ is the stress intensity associated with the residual stresses. Even when X_0 is independent of crack length (Figs. 3.32 and 3.35), $K_{residual}$ must vary with a; when X_0 varies with a (Fig. 3.36), $K_{residual} = 96EI\alpha a^{n-3}\sqrt{a}$ and only for the special case of $(n - 3 + \frac{1}{2}) = 0$, i.e. $n = 2.5$ is $K_{residual}$ constant for all a.

It is clear that rigorous analysis for cracking loads, and interpretation of Xu plots in the presence of residual stresses, is very complicated since $d\Lambda_r/dA$ or $K_{residual}$ will not usually be known. Ahmad and Kanninen (1984) discuss the difficult problems of thermal residual stresses in welds and their effect on fracture. An application to fatigue cracking in the presence of compressive residual stresses (deliberately introduced into the surface of components by shot peening in order to prolong fatigue life) is discussed in Chapter 7.

3.7.5 Determination of residual stress using *K*-formulae

K-formulae may be used to determine the magnitude and direction of (unknown) residual stresses although no crack propagation as such is involved. Instead cracks are introduced into the component and enlarged by saw-cutting, and the associated changes in stress patterns around the crack ends are measured (via photoelastic coatings, or by measuring displacements, etc.). The crack tip stresses give the current stress intensity factor K, and the procedure is repeated for a range of crack saw-cut lengths. Then use is made of Barenblatt's important observation (1962) that these stress intensity factors are the same, at the same locations, as would be calculated using the residual stress pattern existing in the *uncracked* body. The *K*-formula is then 'inverted' to give the residual stress. Essentially, by introducing the crack, the effect of the residual stress field is magnified and relatively small stresses are detectable.

Vaidyanathan and Finnie (1971) have shown how the technique may be used to determine the residual stress distribution in an electron-beam-welded aluminium plate (Fig. 3.37(a)). For a through-crack of length a introduced into the residual stress field $\sigma(x)$ to the side of the weld we have

$$K_I = \sqrt{\frac{2}{\pi a}} \int_0^a \frac{\sigma(x)\sqrt{x}\,dx}{\sqrt{a-x}} \tag{3.68}$$

for the end $x = a$ of the crack. Inversion of equation (3.68) gives $\sigma(a)$ as:

$$\sigma(a) = \frac{1}{\sqrt{2\pi a}} \frac{d}{da} \int_0^a \frac{K(x)\sqrt{x}\,dx}{\sqrt{a-x}} \tag{3.69}$$

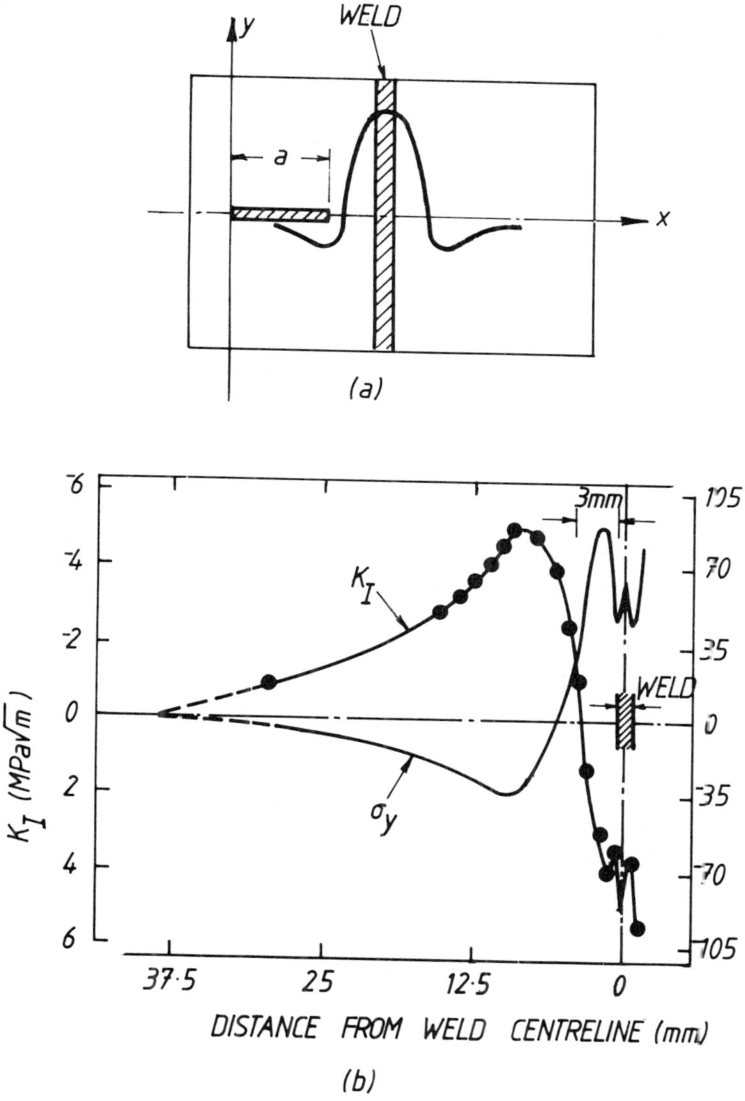

Fig. 3.37 – (a) Schematic drawing of residual stress distribution $\sigma_y(x)$ on the x-axis of a welded plate prior to the introduction of edge crack. (b) Measurement of K_I and calculated residual stress. (After Vaidyanathan and Finnie, 1971; ASME, with permission.)

(σ is presumed not to vary in thickness direction.) Fig. 3.37(b) shows the variation of K (determined from a photoelastic coating put on the welded plate) and the derived residual stress distribution. It is seen that the tensile stresses reach a maximum of around 90 MPa at 3.5 mm from the weld centre-line, drop to 43 MPa at 0.5 mm from the weld centre, and rise again to 68 MPa over the centre-line. An independent check on the stress distribution was obtained by cutting out a small strain-gauged strip parallel to the weld about 2.5 mm from the centre-line; the change in strain readings corresponded to a stress of some 72 MPa, which agrees well with Fig. 3.37(b).

3.8 CRACK PROPAGATION FROM NOTCHES

In Inglis's classic paper of 1913 he observed that the elastic stress concentrations he calculated for ellipses could be applied with reasonable accuracy to holes of other geometries, to notches and so on, using the local radius of curvature of the hole as well as its size. It might seem reasonable to employ similar arguments in fracture mechanics calculations and say that if a sharp crack is growing out of a notch, the equivalent crack length of the notch and crack is simply the sum of both sizes, that is to say a crack of length a growing out of the side of a hole of diameter 2α has an effective length of $(2\alpha + a)$. But the story is not that simple as it neglects the stress 'disturbing' field of the notch itself. Is the crack short and thus propagating in the notch stress field (which is locally different from the remote stress used in the K-formula), or is the crack long and propagating in a stress field undisturbed by the notch?

Newman in 1971, and Nisitani in 1974, have obtained K-formulae for cracks growing from elliptical holes in finite and semi-finite plates but the Y-factors are complicated, and depend on the shape of the holes as well as the sharp crack length (Fig. 2.28(b) applies to an elliptical surface crack *itself* growing). Smith and Miller (1977) have put the results into a simplified form using the concept of an 'equivalent crack length' L for the notch and crack taken together in place of the Y-factors. The proportion of the depth of notch that has to be added to the crack length a to give an equivalent crack length depends on the relative sizes of notch and crack (Fig. 3.38(a)). The notch effect is simply defined as

$$e = L - a \tag{3.70}$$

where e is the proportion of α (depth of blunt notch, say) that must be added to the sharp crack length to give the equivalent crack length. 'Equivalence' is obtained from equating the K expression for the equivalent sharp cracks of length L in a large plate i.e. ($K = \sigma\sqrt{\pi L}$) to the exact K expression for the notch-and-crack combination (i.e. $K = Y\,\sigma\sqrt{\pi(\alpha + a)}$ where α is the semi-major axis of the elliptical hole or notch). Hence

$$Y\sqrt{(\alpha + a)} = \sqrt{L}$$

or

$$\frac{e}{\alpha} = Y^2\left(1 + \frac{a}{\alpha}\right) - \frac{a}{\alpha} \tag{3.71}$$

This may be generalised to a notch of any shape by noting that the radius of curvature, of the part of the ellipse from which the sharp crack is growing, is given by $\rho = \beta^2/\alpha$. The non-dimensionality of the (a/α) parameter in equation (3.71) is preserved if we multiply (a/α) by the 'shape' parameter of the ellipse

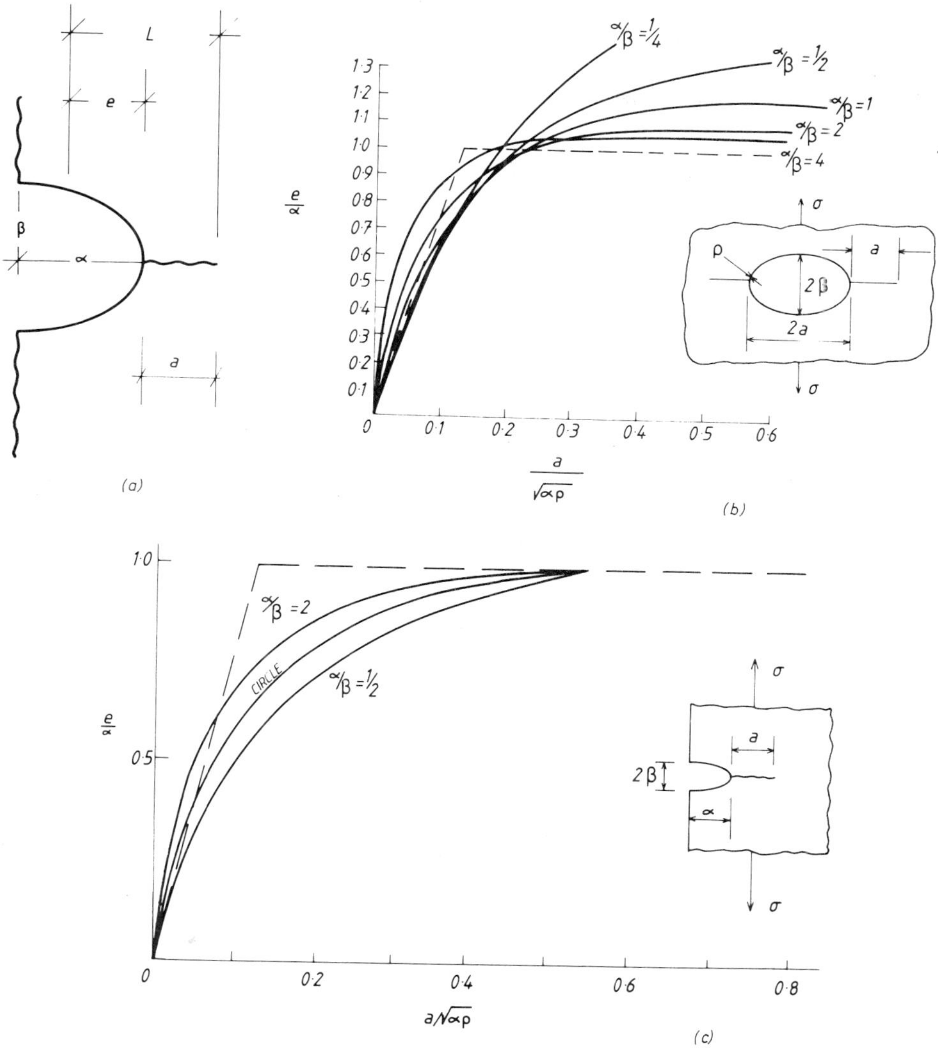

Fig. 3.38 – (a) Rationalisation of notch depth contributions to crack length for various (b) centre notch and (c) edge notch profiles. (After Smith and Miller (1977), with permission.)

α/β, (which in terms of ρ is $\alpha/\sqrt{\alpha\,\rho} = \sqrt{\alpha/\rho}$) to give $(a/\alpha)\,(\alpha/\beta) = (a/\alpha)\sqrt{\alpha/\rho} = a/\sqrt{\alpha\rho}$. This new new parameter combines the shape (α/ρ) and size α of the notch (cf. Inglis).

Numerical calculations by Smith and Miller show that (e/α) depends on $a/\sqrt{\alpha\rho}$ as illustrated in Fig. 3.38(b) and (c). At very large $a/\sqrt{\alpha\rho}$, i.e. when the sharp crack is outside the zone of influence of the notch, $e \rightarrow 1$ and the equivalent crack is (as we first postulated) simply the sharp crack length plus the size of notch. However, for very small $a/\sqrt{\alpha\rho}$, $e \rightarrow 0$ and L is simply the (very small) value of a, the size of the notch playing no part. For $0 < a/\sqrt{\alpha\rho} < 0.13$, for both central and edge notches,

$$e/\alpha \approx 7.69a/\sqrt{\alpha\rho}$$

which describes the increasing proportion of the (fixed) notch size that has to be added to the actual sharp crack length as the crack grows. The extent of influence of the notch field is given by $a/\sqrt{\alpha\rho} < 0.13$ approximately (dashed lines in Fig. 3.38).

Hence for short cracks springing from a notch

$$e = 7.69\, a\sqrt{\alpha/\rho}$$

and

$$L = (1 + 7.69\sqrt{\alpha/\rho})\, a \tag{3.72}$$

Returning to the defintion of the 'equivalent crack' (equation (3.71)), the stress intensity factor for a crack in the notch field will be

$$K = (1 + 7.69\sqrt{\alpha/\rho})^{1/2}\ \sigma\sqrt{\pi a} \tag{3.73}$$

This expression, based on the dashed line in Fig. 3.38 is an approximation for Newman's and Nisitani's results over a useful range of variables ($0.06 < \alpha/\rho < 16$).

For cracks larger than the extent of the notch field, $L = a + \alpha$ and $K = \sigma\sqrt{\pi(a + \alpha)}$ for a centrally cracked plate, or $K = 1.12\sqrt{\pi(a + \alpha)}$ for an edge notched plate (the factor 1.12 is the geometric Y correction for an edge notch, Example 2.11 – this has been included implicitly in the Smith–Miller calculations).

Crack propagation at notches is particularly important in fatigue of engineering components where failure usually begins at some stress or strain concentrating feature, and use is made in Chapter 7 of these K expressions to explain, among other things, fatigue cracks which initiate but do not propagate.

Example 3.18 Determine the size of the notch field for a circular notch of 10 mm diameter in a large plate.

Ans. The notch field is given by $0.13\sqrt{\alpha\rho}$. Now $\alpha = (10/2)$ mm and $\rho = 5$ mm, so $0.13\sqrt{\alpha\rho} = 0.65$ mm.

Example 3.19 A glass shower screen contains a mesh of steel wires in its mid-thickness. The glass is 6 mm thick and the wire diameter is 0.50 mm with a mesh size 12.5 mm square. A thermal gradient is set up across the glass plane when hot water impinges on it which gives rise to a thermal stress of some 8 MPa. Although the strength of the shower screen is measured to be some 35 MPa, fracture occurs. Explain why this happens. Assume that the critical stress intensity factor for glass is $K_{IC} = 0.75$ MPa$\sqrt{m}$.

Ans. For the thermal stress of 8 MPa, the maximum stress at edge of the circular holes containing the wire reinforcement is $3\sigma = 24$ MPa. As even this stress concentrated effect is less than 35 MPa fracture should not occur. What actually happens is that (i) there is no adhesion between glass and wire and (ii) there will be slight chipping or damage at the interface between glass and wire. Assume cracks of size a at each end of the hole ($a \ll d$), Fig. 3.39. Using equation (3.73), we have: $K_{IC} = [1 + 7.69\sqrt{(\alpha/\rho)}]^{1/2}\, \sigma\sqrt{\pi a}$ and $\rho = \alpha$ for a circle, so $0.75 \times 10^6 = 8 \times 10^6\, \sqrt{\pi a} \times (1 + 7.69)^{1/2}$. Thus $a = \left(\frac{0.75}{8}\right)^2 \left(\frac{1}{\pi}\right)\left(\frac{1}{8.69}\right) 1000$ $= 0.32$ mm. (Note that this is actually the same as saying that if $a \ll d$, $K = \sigma_{yy}\sqrt{\pi a}$, where $\sigma_{yy} = 3\sigma$. That is, $K = 3\sigma\sqrt{\pi a}$. Above we had $\sqrt{8.69} = 2.95$ instead of 3 in Smith–Miller solution when $\rho = \alpha$.)

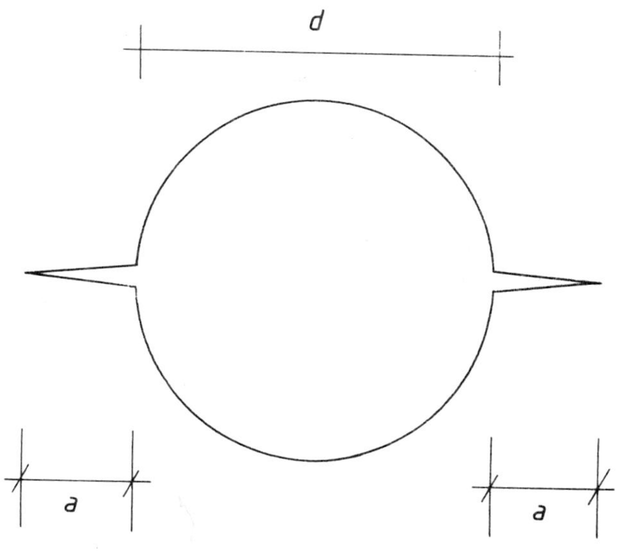

Fig. 3.39 – Edge cracks in a circular hole.

3.9 AVOIDANCE OF REMOTE YIELDING

Local plastic flow can occur at notches and disturb the crack tip stress field. While the existence of a boundary layer of irreversibly-flowed material alongside the crack is recognised in elastic fracture, overall displacement-reversibility (at least in an engineering sense) is satisfied. But in many practical fracture situations remote plastic flow other than localised at the crack tip could occur.

Depending on the magnitude of the cracking load (which depends upon material toughness and the crack size) and on the yield strength of the material, it is possible for irreversible plastic flow to occur at various regions of the cracked body, either before, during or after crack initation and propagation. If yielding occurs it may be spread generally through the body (as in forging or sheet metal forming) or be concentrated in smaller zones with parts of the body being still elastic (e.g. the plastic hinge in three-point bend tests and the narrow zones of intense plastic shear in operations such as punching or cropping). The mechanics of plastic fracture are deferred until Chapters 4 and 5, but from remarks in Chapter 1 (Figs. 1.19 *et seq.*) it is clear that unless the total work of flow and fracture can be partitioned into separate components (and the associated forces and displacements identified) it is impossible to determine unambiguous values of toughness and, vice versa, to predict when cracking will take over from flow.

Owing to the scale effect in the mechanics of fracture (whereby the stress to cause fracture in large bodies can often be less than the stress to cause fracture in similar small bodies, cf. Chapter 1 and Chapter 9 later), laboratory testpieces often yield before cracking and it is difficult to extract unambiguous K_C K_{IC} or R values from such tests. Yet these toughness values are required for the design against fracture of large structures made from the same material. The situation is exacerbated the greater the toughness-to-strength (ER/σ_y^2) ratio of the solid, and it will be appreciated therefore that it is difficult to measure the toughness of mild steel (of which so many large structures are made) in the laboratory.

If means can be found of suppressing plastic flow in the laboratory, however, the resulting elastic fracture deformations may be analysed in the usual manner and unambiguous toughness values obtained from experiments. In this way fracture toughness measurements are possible in the laboratory on small-size testpieces of ductile solids which, in the normal course of events, would yield extensively before cracking. Procedures for suppressing plastic flow are given in this section; even so, it is sometimes impossible to prevent remote yielding and the mechanics for combined flow and fracture must be employed to analyse problems.

In bodies loaded by tensile forces, shown schematically in Fig. 3.40, yielding may occur in the paths of the crack along the 'net section' or 'uncracked ligament'; or the body may yield in bending; and so on. Yielding will clearly depend upon the size of the specimen. For example, in beam-like testpieces for which engineers' bending theory may apply, the bending stress in the arm is

$$\sigma_{\text{bend}} = My/I = 6M/Bh^2 \tag{3.74}$$

where M is the bending moment on a section, h is the height of the arm and B the thickness. The bending moment to cause cracking in a slender split bending testpiece, is $M^2 = REIB = REB^2h^3/12$ (Chapter 2) so that the bending stresses produced during cracking must be

$$\sigma_{\text{bend}} = \frac{6}{Bh^2}\sqrt{REB^2h^3/12} = \sqrt{3RE/h} \tag{3.75}$$

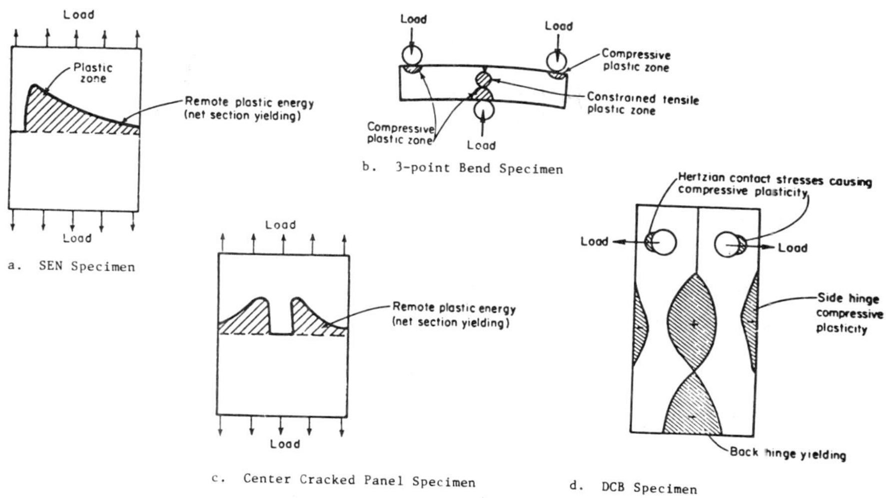

Fig. 3.40 – Yielding zones in various common testpiece geometries. (After Wilkowsky *et al.* (1979), with permission.)

Yielding will not occur provided that $\sigma_{bend} < \sigma_Y$, i.e. when

$$h > 3RE/\sigma_Y^2 = 3(K_C/\sigma_Y)^2 \qquad (3.76)$$

Similarly if simple beam theory can be applied to the crack path ligament it may be shown that ligament length $\approx h > 3(K_C/\sigma_Y)^2$ also to prevent yielding in the ligament. Experiments on practical stubby bending testpieces suggest that $h > 1.5(K_C/\sigma_Y)^2$, the difference in coefficient being associated with specimens for which simple beam theory does not apply. These restrictions on specimen *height* and *ligament* are not to be confused with the ASTM empirical minimum *thickness* limitation of $2.5(K_{IC}/\sigma_Y)^2$ for plane strain, which comes about from limiting the proportion of the fracture surface taken up by shear lips in metals to a few per cent (Example 2.15). It so happens, because of the way in which ASTM testpieces are proportioned that if the thickness requirement is met, so also are the 'height' and 'ligament width' restrictions met (e.g. in the compact tension testpiece $a/W = 0.5$ and $h/W = 0.6$ and $a = B$). However, it is perfectly possible to have globally reversible *plane stress* fractures, by choice of suitable testpiece geometries, where only height and width criteria are met.

The testpiece sizes required to avoid remote yielding in typical low strength/high toughness engineering materials can be enormous. For example, although in high strength heat-treated steel (SAE 4340 say) with $(K_{IC}/\sigma_Y) = 0.1\sqrt{m}$, the height and ligament requirement is ~25 mm, for annealed low carbon steels with $(K_{IC}/\sigma_Y) = 0.5\sqrt{m}$ it becomes about 600 mm. These are plane strain (K_{IC}/σ_Y) values, so that they also indicate thickness requirements! Gigantic compact tension testpieces 300 mm thick have been tested (Wessel, 1969), but there are obvious difficulties in measuring in the laboratory high toughnesses which are uncomplicated by remote yielding. Plane stress (K_C/σ_Y) values are at least twice as big as plane strain values, so that very large thin sheets have to be employed in the laboratory if remote yielding is to be avoided in plane stress cracking. (There is a chicken-and-egg argument here, of course, since the size requirements cannot be determined without valid toughness data, the obtaining of which is the aim of the experiments; cf. K_Q in Example 2.16).

3.9.1 Grooving

A simple way of getting around the large size requirements is to groove the testpiece. This results in two effects (i) a relative reduction in cracking load (dA is less for the same da) and (ii) a relative increase in stiffness of the rest of the testpiece compared with that along the groove. For a groove which leaves a net crack front width B^* in a specimen of thickness B, the bending requirement becomes

$$h > \frac{3RE}{\sigma_y^2} (B^*/B)^2 = 3(K_C/\sigma_y)^2 (B^*/B)^2 \tag{3.77}$$

Clearly for grooving to, say, half the depth, the height requirement is diminished by one-quarter. There are limitations to this procedure, however, because the thin crack fronts that occur with (B^*/B) very small may crack in plane stress, for which R is increased, so that the favourable reduction in h is not so marked. Again, if plane strain fractures are of interest, $B^* > 2.5\ (K_{IC}/\sigma_y)^2$ in any case, and since this value is often very high, it is often impracticable to make $B > B^*$. It is possible that very deep grooving could encourage plane strain fractures in thicknesses smaller than $2.5\ (K_{IC}/\sigma_y)^2$ because of triaxial constraint, but experiments on materials with very large (K_C/σ_y) suggest that the technique does not produce the desired effect to a marked degree, although success has been achieved with materials of moderate (K_C/σ_y). Distinction should be made between grooving to eliminate remote yielding (which always works in principle) and grooving to promote plane strain fractures (which does not always seem to work).

The effective stiffening action of grooving may be improved upon if testpieces are reinforced with separate attachments. Gurney and co-workers (1974), Weitzmann and Finnie (1972), and Seth and Page (1979) independently introduced schemes which stiffen the regions of the testpiece that are likely to yield, thereby limiting the strains in those regions to within the elastic limit; the crack elsewhere in the testpiece thus propagates in an elastic field. Alternatively the ductile testpiece may be sandwiched with a brittle component so proportioned that the whole composite breaks elastically. These different methods are discussed in the following sections. Further information on the transition between fracture and flow is contained in Chapter 10.

3.9.2 Gurney rig

Consider a bending testpiece to the arms of which are attached plates of material possessing much higher tensile strength (Fig. 3.41). The method of flitching the pieces together does not matter for the moment. The object of the exercise is to

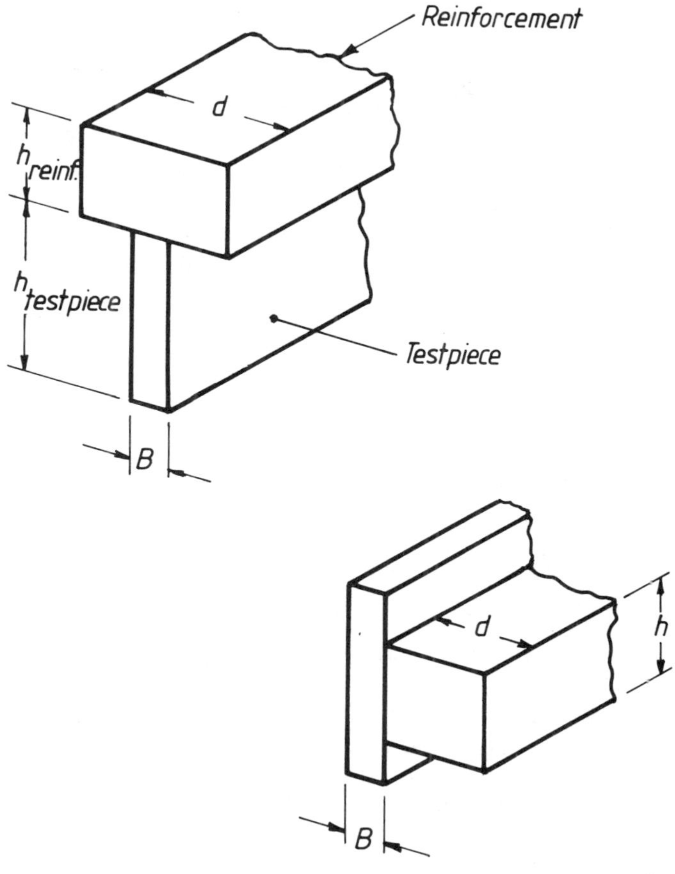

Fig. 3.41 – Reinforced bending testpieces.

prevent yielding in the testpiece (and certainly of course in the higher strength reinforcement), so for bent neutral plane radii of curvature $\rho_{testpiece}$ and $\rho_{reinforcement}$ we may write from engineers' bending theory for no yield for the outer fibre bending stresses:

$$\left.\begin{aligned}\sigma_{testpiece} &= (Ey/\rho)_{testpiece} < (\sigma_y)_{testpiece}\\ \sigma_{reinforcement} &= (Ey/\rho)_{reinforcement} < (\sigma_y)_{reinforcement}\end{aligned}\right\} \quad (3.78)$$

where the ys are the maximum distances of parts of the testpiece and reinforcement from their neutral axes. Were a low carbon steel testpiece to have a high tensile strength steel reinforcement of say four times the yield strength (but the same E), we conclude that if yielding in both is to be prevented

$$y_{testpiece}/y_{reinforcement} \leqslant 1/4 \quad (3.79)$$

assuming some common bent radius of curvature imposed by flitching. This relationship gives a rough guide to the relative 'height' dimensions ($h_{testpiece}$ and $h_{reinforcement}$) required between testpiece and reinforcement. Were $h_{reinforcement} < 4\,h_{testpiece}$, there would be the possibility of the reinforcement being strained less than its own yield strain yet at the same time producing (through the common curvature) strains in the testpiece greater than its yield strain. Even smaller $h_{reinforcement}$ would give yielding in both.

The total bending moment required for cracking the testpiece, remembering that both the testpiece and reinforcement have to be bent, is from equation (2.6), (Example 2.3):

$$M^2 \left\{ (1/EI)_{testpiece} + (1/EI)_{reinforcement} \right\} = (RB)_{testpiece} \tag{3.80}$$

where B is the thickness of the crack path in the testpiece. This follows because (if yielding is totally suppressed) there is no other dissipation than R. Because of the comparative massiveness of the reinforcement, the bending stiffness of the testpiece may usually be neglected, so that

$$M^2 \approx (RB)_{testpiece}\,(EI)_{reinforcement} \tag{3.81}$$

The bending stress in the reinforcement is given by

$$(\sigma)_{bend\ reinforcement} = 6M/dh^2 \tag{3.82}$$

where d is the width of the reinforcement. Thus, using the approximate expression for M, we have for no yielding in the reinforcement

$$(\sigma_y)^2_{reinforcement} > 3(RB)_{testpiece}\,(E/hd)_{reinforcement} \tag{3.83}$$

In this way, a minimum cross-sectional area (hd) of reinforcement is required to prevent yielding (Fig. 3.4.1). The individual values of h and d come from noting that for common curvature

$$\frac{y_{testpiece}}{y_{reinforcement}} = \frac{(\sigma_y)_{testpiece}\,(E)_{reinforcement}}{(\sigma_y)_{reinforcement}\,(E)_{testpiece}} \tag{3.84}$$

Example 3.20 Determine the half-beam depth required to prevent yielding in a DCB testpiece made from 50 mm thick 2024 aluminium plate having $R = 196$ kJ/m^2 and $\sigma_y = 98$ MPa. Investigate the feasibility of using a high tensile steel reinforcing rig to eliminate generalised yielding in a small aluminium specimen of half-beam height 10 mm (for steel $E = 210$ GPa, $\sigma_y = 1.66$ GPa).

Ans. From equation (3.76) we have

$$h \geqslant \frac{3ER}{\sigma_y^2} = \frac{3 \times 70 \times 10^9 \times 196 \times 10^3}{98^2 \times 10^{12}} = 4.2 \text{ m}$$

taking E for aluminium is 70 GPa. This is an impossibly large specimen. With reinforcement, equation (3.83) yields:

$$(hd)_{\text{reinforcement}} \geqslant \frac{3 \times 196 \times 10^3 \times 50 \times 10^3 \times 210 \times 10^9}{1.66^2 \times 10^{18}}$$

$$= 0.0023 \ \text{m}^2$$

The individual values of h and d (rather than merely their product) come from use of equation (3.84). Thus

$$\frac{h_{\text{testpiece}}}{h_{\text{reinforcement}}} = \frac{y_{\text{testpiece}}}{y_{\text{reinforcement}}} = \frac{98 \times 210}{70 \times 1.66 \times 10^3} = 0.18$$

If h for testpiece is 10 mm then h for reinforcement is 55 mm (= 10/0.18), whence d for reinforcement is 36 mm since (hd) = 0.0023 m^2. These relative values may be varied to suit given circumstances.

Although these calculations are based upon simple bending analyses the ideas are the same for reinforcement of testpecies of more complex load/deflexion behaviour. The material from which the stiffening attachments are made may be tailored to the testpiece material (e.g. high-strength aluminium alloys as reinforcements for testing polymers, but precipitation-hardened stainless steels for testing tough, strong metals).

Different methods have been used to attach the reinforcement to the testpiece, which itself can take various forms. In principle the flitched high-strength material could be bolted on to the material to be tested, but relative slippage between testpiece and reinforcement is likely on materials with very high K_C/σ_y which not only impairs the reinforcing ability of the strong material but also produces erroneous readings for displacement or rotation in the work areas for R. An arrangement where the testpiece is in the form of a long I-beam, the flanges of which slide in slots in the reinforcement has been used with great success by Gurney and co-workers (Fig. 3.42). The method of loading is by compression through rollers on the outrigger arms, which produces the propagation bending moment at the crack tip. Materials with plane stress $(K_C/\sigma_y)^2$ up to 450 mm have been reversibly cracked in such a test rig. Toughness is determined either by using Gurney's segmental work area (Fig. 3.43(a) and (b)), or by use of analytical relationships between the applied moments, crack length and toughness. These latter relationships were derived from a 'beam on elastic foundation' cracking analysis (section 8.7.2.2) of the test rig. One of the relations shows that the crack tip opening displacement during quasi-static cracking must have a critical value, δ_C given by

$$k\delta_C^2 = RB \tag{3.85}$$

where k is the stiffness of the elastic foundation (i.e. the testpiece); δ_C may be measured using clip gauges, although experimentally this is not easy.

There are many design choices available in the manufacture of the rig, but it is of prime importance to have sufficient length in the path of cracking; the right height should be at least 10 times the depth of the reinforcement ($10h$). It is found that crack propagation is quite stable, even in materials whose toughness decreases with increase with crack velocity. The geometric stability factors (gsf) (Chapter 8) for the rig depend upon size, relative values of end moments on the top and bottom pairs of outriggers (i.e. whether the top and bottom loads

Fig. 3.42 – Testing rig for high toughness/low yield strength materials.

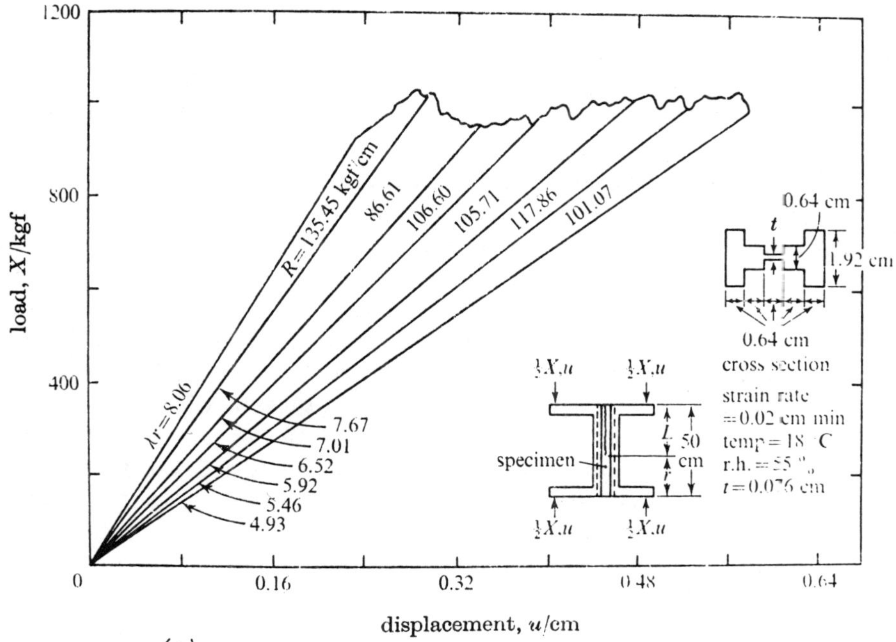

(a)

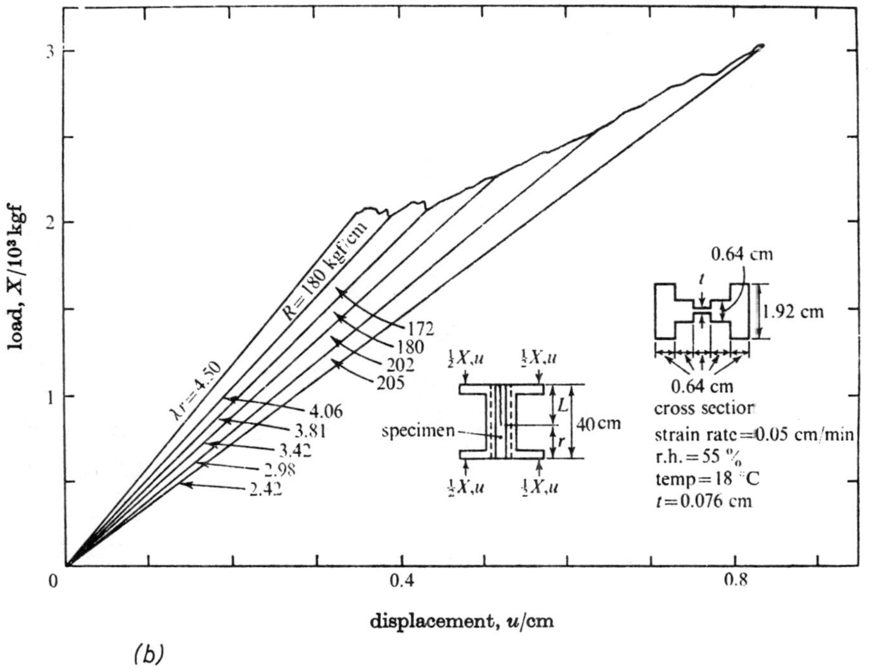

(b)

Fig. 3.43 – Cracking of (a) an aluminium alloy 7075 and (b) mild steel using the Gurney–Mai–Own rig; 1 kgf/cm ≈ 1 kJ/m². (After Gurney *et al.* (1974), with permission.)

have equal moment arms). Indeed the rig has application simply for the purpose of promoting stable cracking irrespective of the suppression of generalised yielding, as discussed in Chapter 8.

The design of a high capacity Gurney rig is discussed in Problems 3.26 *et seq*.

3.9.3 Weitzmann and Finnie, and Seth and Page rigs

Weitzmann and Finnie introduced in 1972 a reinforced testpiece similar to Gurney's, and Seth and Page (see Seth, 1979) employed clamped specimens in a study of the fracture of paper. Although the original purpose of Weitzmann and Finnie's apparatus was to give a continuous elastic cracking test which required no compliance calibration, remote yielding may also be eliminated. Their design is basically a modified 'wedge opening loaded' (WOL) long beam specimen where the wedge is no longer in contact with the split crack faces, but is replaced by a roller carriage assembly which, as it is pulled along, pushes the testpiece arms apart via tracks attached to the specimen arms (Fig. 3.44). The opening load is given by a strain-gauge load cell mounted on the carriage. In equation (1.1), $X\mathrm{d}u = R\mathrm{d}A + d(\frac{1}{2}\ Xu)$, u = constant (the interference fit of the rollers against the tracks), so $\mathrm{d}u = 0$. Then

$$R = -\tfrac{1}{2}u\left(\frac{\mathrm{d}X}{\mathrm{d}A}\right) = -\tfrac{1}{2}\left(\frac{u}{B}\right)\left(\frac{\mathrm{d}X}{\mathrm{d}a}\right)$$

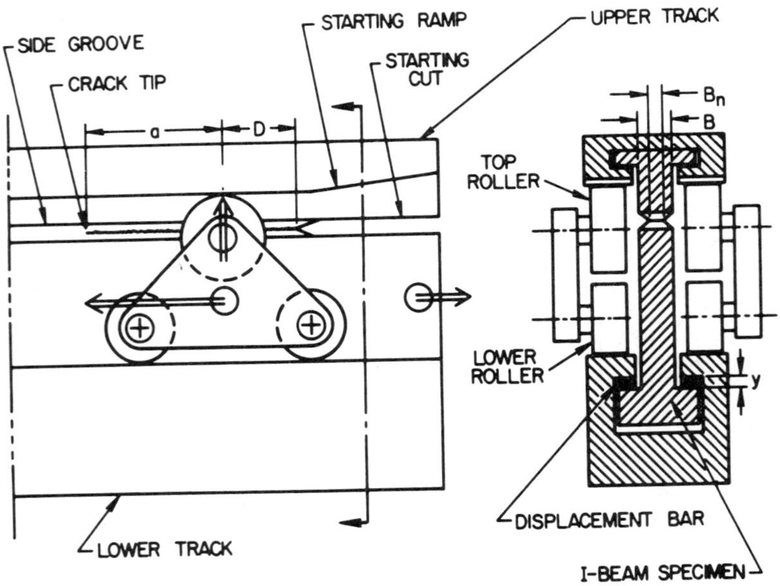

Fig. 3.44 – The Weitzmann and Finnie rig for controlled-displacement fracture testing. (Weitzman and Finnie, 1972. Copyright ASTM, 1916 Race Street, Philadelphia PA 19103. Reprinted with permission.)

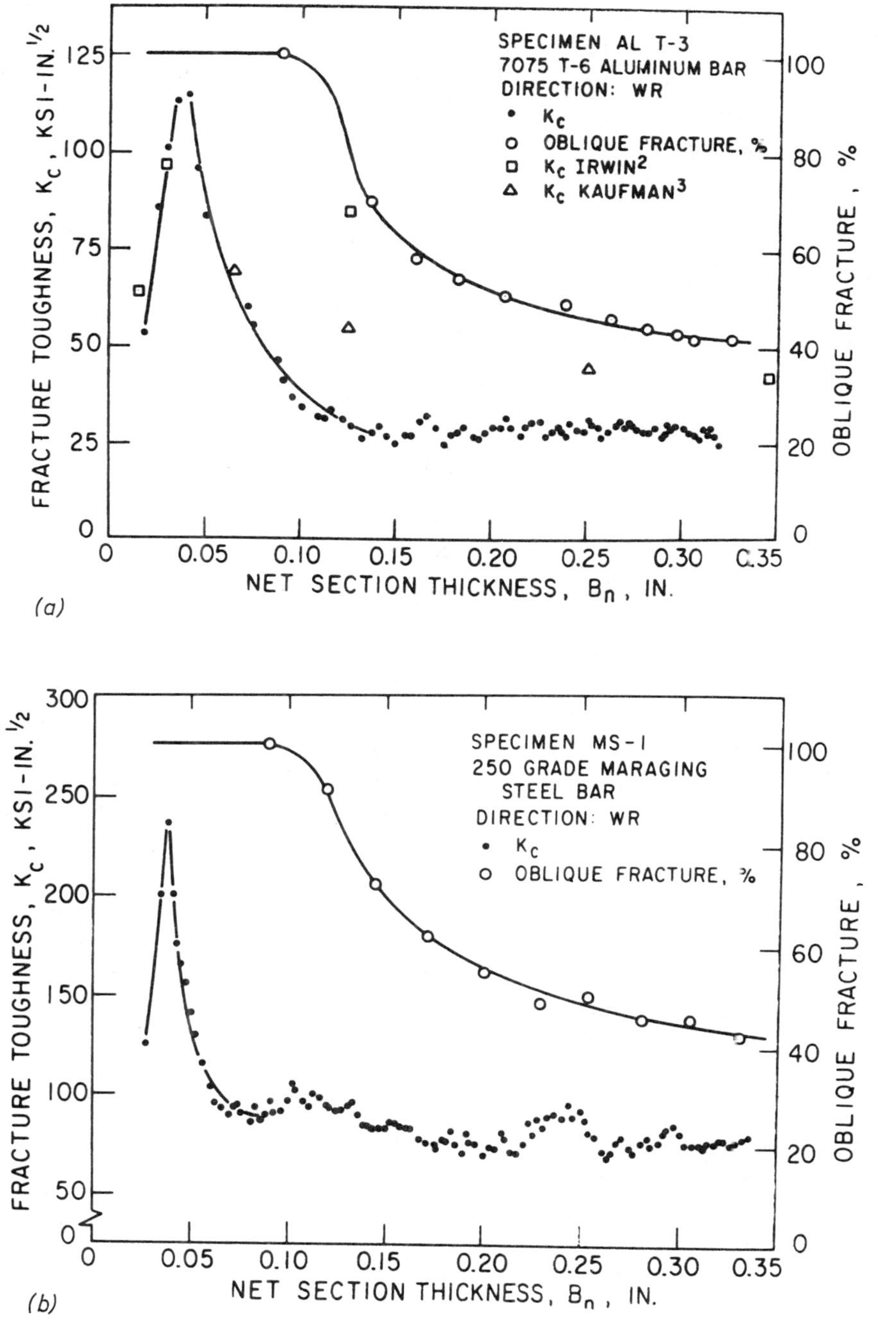

Fig. 3.45 – Variation of K_C with net section thickness (B_n) for (a) 7075-T6 aluminium and (b) 250 grade maraging steel using the test apparatus in Fig. 3.44; 1 ksi$\sqrt{\text{in}}$ ≈ 1 MPa$\sqrt{\text{m}}$ (Weitzmann and Finnie, 1972. Copyright ASTM, 1916 Race Street, Philadelphia PA 19103. Reprinted with permission.)

where $A = Ba$. The negative sign arises because X decreases as a increases. Monitoring of dX/da allows R to be calculated. dX/da amounts to dX/dL, where L is the movement of the roller carriage, because as the wedge moves along the specimen the crack front keeps pace or jumps ahead and arrests at essentially the same distance in front of the carriage for a given u. u may be varied by use of tapered displacement blocks on the roller carriage, but R should not vary with u, merely with the crack load and associated crack length. This is different from the usual WOL testpiece where u increases as the wedge is driven into the split at the same location. The elimination of a compliance calibration in the Weitzmann–Finnie apparatus is similar to Gurney's segmental area technique for R. Also, if continuous quasi-static cracking is achieved, the need for precise fatigue cracking of the starter crack is perhaps by-passed. As with Gurney's methods, initial instabilities (connected for example with blunt starter cracks) do not necessarily spoil the rest of the test (from which many estimates for R can be obtained).

The hardened load tracks not only serve to accept the line-contact of the roller force but also make possible the distribution of the opening-mode load over a large portion of the flange of the I-beam specimen. In this way the possibility of remote yielding is diminished. Calculations of the reinforcement potential of the tracks follow the treatment given earlier in this section. Checks upon reversibility can be performed by running back the roller carriage away from the crack front.

Valuable data have been generated on this apparatus, particularly the way in which R varies with thickness. For example, Figs. 3.45(a) and (b) show the behaviour of 7075-T6 aluminium alloy (with plane stress $(K_C/\sigma_Y)^2 = 28$ mm and plane strain $(K_{IC}/\sigma_Y)^2 = 6$ mm) and 250 grade maraging steel (with plane stress $(K_C/\sigma_Y)^2 = 8$ mm and plane strain $(K_{IC}/\sigma_Y)^2 = 2.5$ mm). Tapered grooves were employed in the specimen so as to pick up many thickness variations from single tests.

Note in these results that full 45° oblique fracture surfaces occur at thicknesses considerably greater than the point of peak toughness. This suggests that some criterion other than the 'meeting of the shear lips' is involved in bringing about maximum toughness; cf. sections 2.8 and 4.2.2. It also suggests that 'mixed mode' toughnesses would not, in this case, be given by some rule-of-mixtures between the peak (plane stress) and the lowest (plane strain) R values.

A final point should be made in connextion with these reinforced testpieces and with grooved testpieces in general. It is that in very tough materials the extent of the plastic zone may, in some cases, be comparable to or greater than the web dimension of the testpice. The influence of the flanges on toughness is not altogether clear. Least effect would occur with materials displaying Dugdale-type slit crack tip zones (section 4.2.3).

3.9.4 Suppression of yielding in layered bodies

It was shown in section 3.6 that the loads at which layers in a sandwich crack are modified from the loads at which they crack when free, owing to interlaminar shear tractions. This suggests that if a ductile solid is sandwiched by brittle layers, the traction-modified load in the ductile layer could be brought down below its yielding load, and the composite body then would crack elastically, from which valid toughnesses could be determined. In Fig. 3.29(b), X_C of the separate layer would be greater than the load to cause general yielding, but X_C' of the sandwich would be less.

Williams in the early 1970s first performed experiments in which ductile PVC sheet, smeared with a thin layer of epoxy resin, cracked in a brittle fashion (cf. Fig. 3.27). The behaviour of case-hardened ductile metals is similar. Fig. 3.46 shows Xu traces of pre-cracked 10 mm square section three-point bending specimens of EN3B steel with different case depths; at small case depths there is still general yielding and considerable non-linearity (and sideways necking at the root of the starter crack), but at the greatest case depth the fracture is elastic with a flat crack front.

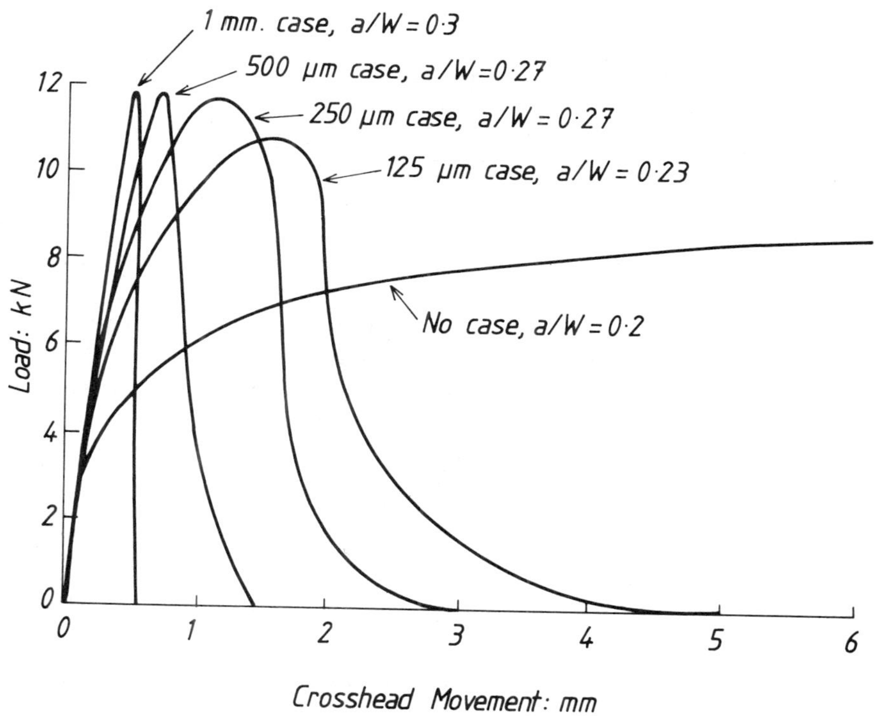

Fig. 3.46 – Load–deflexion traces for three-point bending beams of EN3B steel with varying case-hardened depths. (After Atkins (1979), with permission.)

Equation (3.42) permits the testing machine load X_T to be broken down into the loads at which the case would have cracked (X_1) and the soft ductile core would have cracked (X_2) if free. (Of course X_2 is never actually attained when free owing to yielding.) Assuming that the toughness of the case is negligible, so that $\beta = 0$, we have

$$X_T = X_2 \sqrt{1 + \alpha} \tag{3.86}$$

Furthermore $\alpha = B_1/B_2$ as the moduli of case and core are comparable. For 1 mm case depth on each flank, $B_1/B_2 = 2/8 = 0.25 = \alpha$, and from Fig. 3.46 $X_T = 12$ kN whence $X_2 = 10.7$ kN. K_{IC} for the core may be determined from the appropriate fracture mechanics formula for three-point bending, which for a span S of 50.4 mm, crack depth $a = 3$ mm, beam depth $W = 10$ mm, is $K = 3X\,S\,\sqrt{\pi a}\;Y\,(a/W)/2BW^2$. Using $X = 12$ kN, and $Y(a/W) = Y(0.3) = 1.06$, gives $K_{IC} = 104$ MPa $\sqrt{\text{m}}$ (equivalent to $R = (1-\nu^2)K_{IC}^2/E = 46$ kJ/m^2). This is an acceptable value for the 'fast fracture' toughness of a 0.15% C steel at room temperature, which is characterised (in the experiments at greatest case depths) by cleavage facets in place of the dimpled fibrous fracture found with least case depths.

PROBLEMS

3.1 For a non-linear elastic structure following $X = k(A)u^n$ prove that the crack velocity ($\dot{a}$) is related to the driving velocity ($\dot{u}$) by

$$\frac{\dot{a}}{\dot{u}} = \frac{nX}{RB(n+1) + n\dfrac{\mathrm{d}X}{\mathrm{d}a}} = \frac{-k'(n+1)}{Bk''u}$$

where the prime notations indicate crack length (a) derivatives and B is the thickness.

3.2 Show that if a crack propagates at constant velocity, a linear relation between X and u during propagation is implied.

3.3 The Gross–Srawley expressions for edge crack specimens are

$$KB\sqrt{W}/X = \frac{0.537 + 2.17(1 + a/W)/(1 - a/W)}{(1 - a/W)^{1/2}} \quad \text{for} \quad \frac{W}{H} \leqslant 2$$

$$KB\sqrt{H}/X = 3.46\,(a/H + 0.7) \qquad W/H \geqslant 2$$

Determine $u\,.\,\mathrm{d}X/\mathrm{d}A$ and hence show from equation (3.2) that

$$(\dot{a}/\dot{u})\sqrt{R/WE} = (\dot{a}/\dot{u})\,(K/E\sqrt{W}) = f\left(\frac{h}{W}, \frac{a}{W}\right)$$

3.4 Work out $\int Z^2\ \mathrm{d}(a/W) = I_1$ in equation (3.6) for the compact tension specimen, using the expression for Z given in Example 2.11.

3.5 Work out $\mathrm{d}/\mathrm{d}(a/W)\ (Z^{-1}) = D_1$ in equation (3.8) for the compact tension specimen, using the expression for Z given in Example 2.11.

3.6 Show that interfacial peeling between a substrate and a tape possessing a changing bending stiffness is governed, just as the interface is crossed, by $X/w = (E_2 I_2/E_1 I_1)R$ (cf. equation (2.17)).

(Hint. Apply equation (1.1) to conditions before and after the crack crosses the change in EI: $\Lambda = (1/2)M^2/EI$ per unti length around peel bend (Fig. 3.1) and $\mathrm{d}^2y/\mathrm{d}x^2/(1 + (\mathrm{d}y/\mathrm{d}x)^2)^{3/2} = Xy/EI$).

3.7 Show that the dimension of a fully developed peel bend in a tape of bending stiffness $E_2 I_2$ is given by $l^2 = 2E_2I_2/X$ where l is measured from the crack tip to the line of action of the peel force X.

3.8 Show that the elementary stresses used for crack path determination in tapered DCB specimens *loaded at the apex* are:

$$\sigma_x = 6X(a/W)/W \tan^2\theta . (1 - a/W)^2 \text{ and } \sigma_y = 2X(2 + a/W)/W(1 - a/W)^2$$

where the symbols are defined in Fig. 3.19.

3.9 Show that the elementary stresses used for crack path determination in tapered DCB specimens *loaded at the base* (not at the apex) are

$$\sigma_x = 6X(a/W)/W \tan^2\theta . (e/W + a/W) \text{ and } \sigma_y = 2X(2 + a/W)/W(1 - a/W)^2$$

(symbols defined in Fig. 3.19) and that straight cracks are much more readily produced than in similar specimens loaded at the apex.

3.10 Prove that a straight crack will always propagate along the mid-plane of a triangular lamina loaded at its base when the taper semi-angle is greater than 45°.

3.11 Show that in edge-crack rectangular testpieces the σ_y/σ_x ratio in the ligament near the crack tip (Fig. 3.19) becomes greater the longer the crack. Determine the profile of the specimen which maintains σ_y/σ_x constant and show that it should give straight, stable, and slowly propagating cracks in brittle materials such as glass.

3.12 Determine the magnitude of the distributed compressive load required on the top and bottom faces of an edge-crack specimen (Fig. 3.19) which should ensure straight cracking with $a/W = 0.5, h/W = 0.25$.

3.13 Consider a specimen made up of two materials of different bending stiffness E_1I_1 and E_2I_2 as shown below. Assume that the crack has spread to the interface along the mid-length of the first medium. Determine the path in which the crack will subsequently propagate, in terms of the interfacial toughness R_y and the toughness of the second medium R_x.

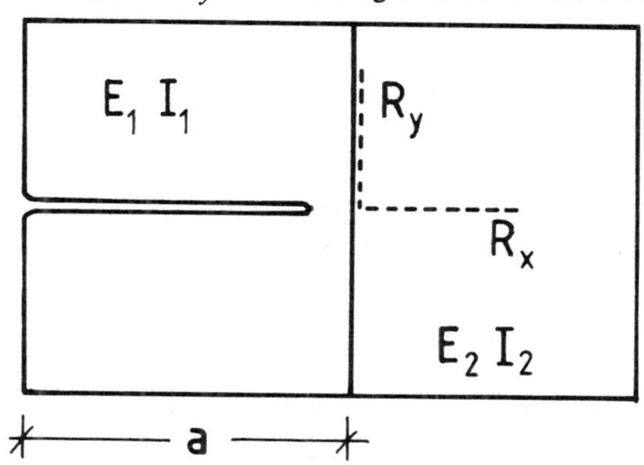

3.14 Adhesive tape is stuck on 'brown' paper used for making parcels. When peeled off slowly, there is little paper pick-up on the glued side of the transparent tape; in fast peeling the paper remains stuck to the tape. Explain.

3.15 Consider strips of PMMA joined by epoxy cement. R for expoxy is some 130 J/m^2; for PMMA it is perhaps 400 J/m^2 under similar conditions of temperature, crack speed and environment. Will cracking follow the glue line?

3.16 Consider two putative crack paths in a double-ended cantiliver beam, (where the existing straight crack has length a), the one comprising equal increments d in crack extension along the direction of the existing crack, and the other comprising equal antisymmetric increments d in crack extension perpendicular to the crack starter. Show that the beam theory of the specimen for these two possible paths are $(a + 2d)^3/96EI$ (straight) and $a^3/24EI$ (perpendicular) and that straight cracks occur when $a < 3.4d$.

3.17 Show that for the modelling of fracture in thick monolithic plates two equations for the fracture toughness (K_C) can be derived. Based on the rule-of-mixtures of K

$$K_C = K_{C1}\left(1 - \frac{2r_p}{B}\right) + 2K_{C2}\frac{r_p}{B} \tag{a}$$

and based on the rule-of-mixtures of G

$$G_C = G_{C1}\left(1 - \frac{2r_p}{B}\right) + 2G_{C2}\frac{r_p}{B} \tag{b}$$

The subscripts ($C1$) and ($C2$) denote critical K and G values for plane strain and plane stress conditions respectively. B is the monolithic plate thickness and r_p is the extent of the plane stress plastic zone on the sides. Compare the ratio of the K_C values as predicted by these two approaches.

3.18 High-impact polystyrene (HIPS) is normally a ductile and tough material that does not give valid toughness results using small size laboratory specimens. Yap *et al.* (1980) have successfully promoted brittle fracture in this material by laminating a 3-mm epoxy resin layer on the HIPS specimen. Using single-edge-notched tension specimens they obtained a combined critical stress intensity factor of 1.58 MPa$\sqrt{\text{m}}$. Determine K_C for the HIPS if its thickness is 4.7 mm and K_C for epoxy is 0.94 MPa$\sqrt{\text{m}}$. Explain why brittle fracture is possible in this example.

3.19 The load–displacement relations of cracking of a sanwich composite with different stiffness but same initial crack length are given in Fig. 3.29(b). Show that equation (3.45) gives the fracture loads for the separate layers and hence explain why brittle fracture is possible in the ductile layer by this sandwiching action.

3.20 Plastic liners in refrigerators and freezers sometimes fail during service. It is intended to see how the polyurethane foam which is an insulating material between the liner cabinet and the steel case affects the fracture behaviour of the liner/foam sandwich. Thermal stresses may be considered as the only applied loading and it may be assumed that there is perfect bonding between the two materials. The following tensile and fracture experimental results are obtained.

	Polyurethane			Plastic liner			Composite	
Temperature (°C)	E (MPa)	B (mm)	K_C (kPa$\sqrt{m}$)	E (MPa)	B (mm)	K_C (MPa$\sqrt{m}$)	E (MPa)	K_C (MPa$\sqrt{m}$)
20	12	33	26	1.75	4.8	5.95	264	3.88
0	13	33	26	2.42	4.9	6.60	243	5.07
−20	14	32	29	2.65	4.9	8.00	285	5.12
−40	15	33	28	2.70	4.9	6.20	304	4.78

The K_C values for the plastic liner are calculated based on maximum load obtained in the fracture experiments. Initiation values for this liner material are independent of temperature, $K_i = 2.50$ MPa$\sqrt{m}$. Calculate the modified toughness of the plastic liner as affected by the polyurethane layer in the composite. Comment on these results.

3.21 Because of release of residual strain energy a stiff adhesive tape sometimes unwraps by itself without any external force applied to it. Determine that the condition for self-unwrapping of the roll of adhesive tape shown below is given by $R \leqslant EI/2w\rho^2$ where w is the width of the tape. If the

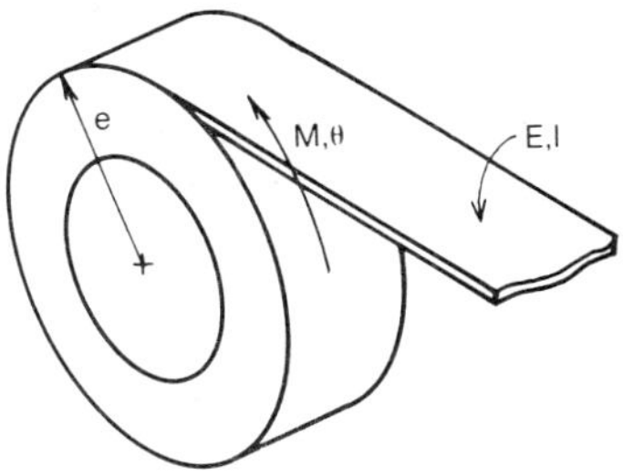

width of the tape is 10 mm and thickness 1.0 mm determined the diameter of the roll if self-unwrapping is to be avoided. Assume that $E = 500$ MPa and R (adhesive) $= 50$ J/m^2 for the adhesive tape.

3.23 Two rectangular strips of metal of thickness B and width w have the same modulus E but different coefficients of expansion α_1 and α_2. They are glued together to leave a central crack of area $a \times w$ (cf. Fig. 3.34). Show that if the temperature is changed by ΔT,

(a) fracture by axial compression alone is given by

$$X = \sqrt{2R_{II}EBw^2}$$

and

$$\Delta T_{II} = \frac{1}{(\alpha_1 - \alpha_2)} \sqrt{\frac{2R_{II}}{EB}}$$

(b) fracture by bending moment at the crack tip is given by

$$\Delta T_{\mathrm{I}} = \frac{1}{(\alpha_1 - \alpha_2)} \left(\frac{R_{\mathrm{I}}B}{8EI}\right)\left(\frac{a}{\pi}\right)^2$$

(c) thermal buckling occurs when

$$\Delta T_{\mathrm{B}} = \frac{1}{(\alpha_1 - \alpha_2)} \, 4 \left(\frac{\pi}{a}\right)^2 \left(\frac{I}{Bw}\right)$$

R_{I}, R_{II} refer to the toughness of the bond between the layers in the different modes of cracking.

3.23 Two warped twisted sheets of thickness B are elastically flattened and butt-glued together. The glue is capable of retaining the flattened shape but the glued joint is thereby under a state of internal (residual) stress. Determine the conditions for spontaneous cracking, without application of external load, if fracture of some unglued portion of the joint is triggered off.

(Hints. Assume that the original out-of-plane warping along the glue line is linear and given by $u = \alpha a$. View the glued body as an Outwater testpiece for which $u = kAX$. From the strain energy locked up in flattening the plates initially, show that $R \leqslant \alpha^2/2\,kB^2$)

3.24 Car windscreens are 'toughened' by heat treatment which produces residual compressive stresses in the outside layers and tensile stresses in the interior. Cracking from exterior damage is thereby inhibited, but if cracks do reach the interior catastrophic failure takes place. Even so, the windscreen usually shatters quietly with little energy to spare. Explain. (Residual tensile stress is about 10 MPa; R for glass is about 5–10 J/m^2.)

3.25 The liner materials for most refrigerators are made from rubber modified polymers such as high-impact polystyrenes (HIPS). To improve the gloss a biaxially stretched thin layer of polystyrene film (25–50 μm) is usually laminated on the HIPS. It has been found that this cosmetic improvement reduces the impact toughness of the HIPS and promotes brittle fracture. Explain why this may be so.

3.26 How would you design a test rig similar to the Gurney–Mai–Owen rig in testing full thickness X60 rolled pipeline steels with $(K_{\mathrm{C}}/\sigma_{\mathrm{y}})^2 = 3{,}600$ mm for use in a 1,000-kN testing machine?

3.27 A heavy-duty Gurney rig has been built in Battelle Columbus Laboratories to measure the toughness of 12 mm thick steel ($\sigma_{\mathrm{y}} > 450$ MPa). The load–deflextion diagram for a typical test is reproduced below. Comment on the results obtained. (See Wilkowsky *et al.*, 1979).

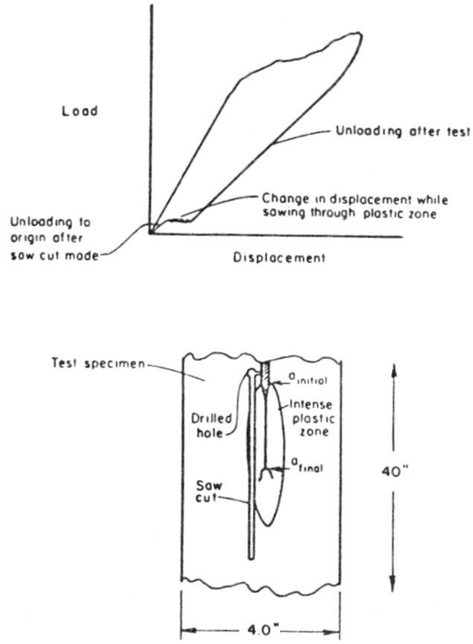

(a) Load-Displacement Crack Length Record and Illustration of Localized Crack Tip Plasticity

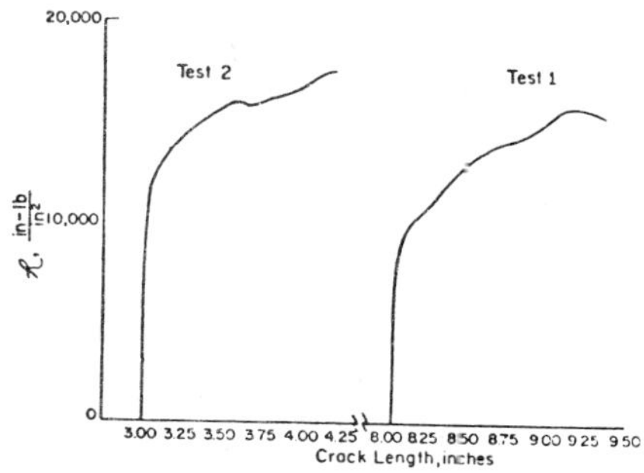

(b) R-Curve Evaluation of Steel AG-3 Using Different Initial Crack Length

3.28 Papers have very large $(K_C/\sigma_Y)^2$ ratios in the range 15–100 mm and it is therefore necessary to avoid remote yielding by building a special test rig to achieve this. Seth (*Tappi,* vol. 62, 1979, 92) has constructed reinforcement clamps to be used with rectangular cracked sheets. Comment on the success of such a method.

If you are to build a miniature Gurney rig for testing papers how would you design such a rig?

3.29 Equation (3.85) is a simple solution for the cracking of an I-beam specimen in Gurney's rig using a beam-on-elastic foundation model. Discuss possible modifications to this equation if the transverse shear deflexion and rotary inertia of the moving beams are included in the analysis. (See M. F. Kanninen, *Int. Jnl. Fracture,* vol. 10, 1974, 415–430.)

REFERENCES

Abrams, D. (1918) *Bull. Struct. Mater. Res. Lab. Lewis Inst.,* Chicago, No. 1.
Ahmad, J. and Kanninen, M. F. (1984) *Proc. Conf. Numerical Methods in Fract. Mech.* University College, Swansea.
Atkins, A. G. (1979) *Proc. ICM-3* **3**, 341.
Atkins, A. G. (1982) *Metal Sci.* **16**, 127.
Auerbach, F. (1891) *Ann. Phys. Chem.* **43**, 61.
Barenblatt, G. I. (1962) *Adv. Appl. Mech.* **7**, 55.

Barquins, M. (1984) Chapter 3 in *Adhesion 8,* K. W. Allen (ed.). London: Elsevier Applied Science Publishers.
Benbow, J. J. and Roesler, F. C. (1957) *Proc. Phys. Soc.* **B70**, 201.
Bilby, B. A. and Cardew, G. E. (1975) *Int. J. Fract.* **11**, 708.
Brichall, J. D., Howard, A. J. and Kendall, K. (1982) *J. Mater. Sci. Letters* **1**, 125.
Birchall, J. D., Howard, A. J. and Kendall, K. (1982) *Proc. British Ceramic Soc., No. 32, 'Engineering with Ceramics',* R. W. Davidge (ed.), pp. 25–32.
Bishop, S. M. (1974) *J. Phys. D. App. Phys.* **7**, 1489.
Bishop, S. M. (1972) Royal Aircraft Establishment, RAE Tech. Rept. 72026. Farnborough, UK.
Bluhm, J. I. (1961) *Proc. ASTM* **61**, 1324.
Boker, R. (see Polakowski and Ripling, 1966).
Bridgman, P. W. (1935) *Phy. Rev.* **48**, 825.
Cherry B. W. and Hang, P. T. (1974) *Proceedings of the Australian Fracture Group,* R. H. Leicester (ed.), Melbourne: Monash University, pp. 41–50.
Cherry, B. W. and Harrison, N. L. (1970) *Fibre Science and Technology* **2**, 299.
Chow, C. L. and Ngan, K. M. (1981) *J. Strain Analysis* **16**, 9.
Cook, J. and Gordon, J. E. (1964) *Proc. Roy. Soc.,* **A282**, 508.
Cooper, G. A. and Kelly, A. (1967) *J. Mech. Phy. Solids* **15**, 279.
Cotterell, B. (1966) *Int. J. Fract.* **2**, 256.
Cotterell, B. (1970) *Int. J. Fract.* **6**, 189.
Cotterell, B. (1972) *Int. J. Fract.* **8**, 195.
Cotterell, B. and Rice, J. R. (1980) *Int. J. Fract.* **16**, 155.
Cottrell, A. H. (1964) *The Mechanical Properties of Matter.* New York: John Wiley.
Dannenburg, H. (1961) *J. Appl. Polym. Sci.* **5**, 125.
Dobbs, H. S., Field, J. E. and Maitland, A. H. (1973) *Phil. Mag.* **28**, 33.
Eftis, J. (1985) Private Communication.
Eftis, J. and Jones, D. L. (1982) *Int. J. Fract.* **20**, 267.
Erdogan, F. and Sih, G. C. (1963) *Trans. ASME, J. Basic Engr.* **85**, 519.
Evans, A. G. (1979) *Fracture Mechanics Applied to Brittle Materials,* STP No. 678, Freiman, (ed.). Philadelphia: ASTM.
Evans, A. G. and Charles, E. A. (1976) *J. Am. Ceram. Soc.* **59**, 371.
Feret, R. (1897) *Bull. Soc. Encouragement Indust. Nat.* **II**, 1604.
Finnie, I. and Saith, A. (1973) *Int. J. Fract.* **9**, 484.
Finnie, I and Weiss, H. D. (1974) *Int. J. Fract.* **10**, 136.
Goetz, C. G. (1965) US Army Cold Regions Research and Engineering Laboratory Report 27021-P5-01724, Hanover, N. H.
Guild, F. J., Harris, B. and Atkins, A. G. (1978) *J. Mater. Sci.* **13**, 2295.
Gurney, C. and Hunt, J. (1967) *Proc. Roy. Soc.* **A299**, 508.
Gurney, C., Mai, Y. W. and Owen, R. C. (1974) *Proc. Roy. Soc.* **A340**, 213.
Gurney, C. and Ngan, K. M. (1971) *Proc. Roy. Soc.* **A325**, 207.
Harrison, N. L. (1972) *Fibre Science and Technology* **5**, 197.
Helmuth, R. A. and Turk, D. A. (1966). Symposium on Structure of Portland

Cement Paste and Concrete. Washington D. C. Highway Research Board Special Report No. 90, p. 135.
Iida, S. and Kobayashi, A. S. (1969) *J. Basic. Engr.* **91**, 764.
Inglestrom, N. and Nordberg, H. (1974) *Engr. Fract. Mech.* **6**, 597.
Inglis, C. E. (1913) *Trans. Inst. Naval Architects,* **LV**, 219.
Jaeger, J. C. and Cook, N. G. W. (1976) *Fundamentals of Rock Mechanics.* London: Chapman & Hall.
von Kármán, T. (1911) *Z. ver. dtsch. Ing.* **55**, 1749.
Kendall, K. (1975a) *Int. J. Fract.* **11**, 3.
Kendall, K. (1975b) *Proc. Roy. Soc.* **A341**, 409.
Kendall, K. (1978) *Proc. Roy. Soc. (London)* **A361**, 245.
Kfouri, A. P. and Miller, K. J. (1981) *Three-Dimensional Constitutive Relations* Nemat-Nasser (ed.). Amsterdam: North-Holland.
Kreher, W. and Pompe, W. (1981) *J. Mater. Sci.* **16**, 694.
Lawn, B. R., Evans, A. G. and Marshall, D. B. (1980) *J. Am. Ceram. Soc.* **63**, 574.
Lawn, B. R. and Wilshaw, T. R. (1975) *J. Mater. Sci.* **10**, 1049.
Lawn, B. R., Hockey, B. J. and Wiederhorn, S. M. (1980) *J. Mater. sci.* **15**, 1207.
Leevers, P. S., Radon, J. C. and Culver, L. E. (1976) *Polymer* **17**, 627.
MacGregor, C. W. (1935) *Trans. American Math. Soc.* **8**, 177.
Mai, Y. W. (1972) Ph.D dissertation, University of Hong Kong.
Mai, Y. W. (1977) *Int. J. Mech. Sci.* **19**, 325.
Maiti, S. K. (1980) *J. Strain Analysis* **15**, 183.
Maiti, S. K. and Prasad, K. R. S. K. (1980) *Int. J. Solids and Structures* **16**, 563.
Maiti, S. K. and Smith, R. A. (1983) *J. Mech. Phys. Solids* **31**, 389.
McClintock, F. A. and Argon, A. S. (1966) *Mechanical Behaviour of Materials* Reading, Mass: Addison-Wesley.
Michener, J. R. and Burns, S. J. (1983) *Int. J. Fract. Mech.* **23**, R45.
Murray, M. J. (1977) *Proc. Roy. Soc.* **A356**, 483.
Newman, J. C. (1971) NASA TND-6376.
Nisitani, H. (1974) *Proc. Joint US/Japan Seminar, Strength and Structure of Solids,* Syracuse University, New York.
Palaniswamy, K. and Knauss, W. G. (1974) On the problems of crack extension in brittle solids under general loading, Graduate Aeronautical Laboratories, CIT Report SM 74-8. (See also *Mechanics Today,* **4**.)
Polakowski, N. H. and Ripling, E. J. (1966) *Strength and Structure of Engineering Materials.* Englewood Cliffs, N. J: Prentice-Hall.
Puttick, K. E., Smith, L. S. A. and Miller, L. E. (1977) *J. Phys. D. (Appl. Phys.)* **10**, 617.
Radon, J. C. and Leevers, P. S. (1977) *Proc. ICF-4, Waterloo* **3**, 1113.
Rice, R. W., Freiman, S. W., Pohianka, R. C., Mecholsky, J. J. and Wu, C. Cm. (1978) *Fracture Mechanics of Ceramics,* Vol. 4. London: Plenum Press.
Roesler, F. C. (1956) *Proc. Phys. Soc.* **B69**, 981.
Ryskewitch, T. (1953) *J. Am. Ceram. Soc.* **36**, 65.
Sack, R. A. (1946) *Proc. Phys. Soc. (Lond.)* **58**, 729.
Seth, R. S. (1979) *Tappi* **62**, 92.

Sia, T. K., Mai, Y. W. and Cotterell, B. (1981) *Proc. 2nd Australian Conference in Engineering Materials,* D. J. Cook (ed.). Sydney: University of Sydney, pp. 515–529.

Sih, G. C. (1973) *Engr. Fract. Mech.* **5**, 365.

Sih, G. C., Czoboly, E. and Gillemot, F. (eds.) (1982) *Absorbed Specific Energy and/or Strain Energy Density Criterion.* Den Haag: Martinus Nijhoff.

Sih, G. C., Paris, P. C. and Irwin, G. R. (1965) *Int. J. Fract. Mech.* **1**, 189.

Smith, R. A. and Miller, K. J. (1977) *Int. J. Mech. Sci.* **19**, 11.

Sneddon, I. N. (1946) *Proc. Roy. Soc.* **A187**, 229.

Swedlow, J. L. (1965) *Int. J. Fract. Mech.* **1**, 210.

Swedlow, J. L. (1984) Private communication.

Weitzmann, R. H. and Finnie, I. (1972) *J. of Materials* **7**, 294.

Wessel, E. T. (1969) in *Practical Fracture Mechanics for Structural Steel,* Paper H. London: UKAEA/Chapman & Hall.

Westergaard, H. M. (1939) *Trans. ASME, J. Appl. Mech.* **6**, A-49.

Wilkowsky, G. M. (1979) *Proc. 6th Symposium on Pipe Line Research.* American Gas Associaton.

Wilkowsky, G. M., Maxey, W. A. and Eiber, R. J. (1979). ASME Paper 79-PVP-115.

Williams, J. G. (1976) Private communication.

Williams, J. G. and Birch, M. W. (1976) STP 601, 125. Philadelphia: ASTM.

Williams, J. G. and Ewing, P. D. (1972) *Int. Fract.* **8**, 441.

Williams, J. G. and Ewing, P. D. (1974) *Int. J. Fract. Mech.* **10**, 135.

Williams, M. L. (1957) *J. Appl. Mech.* **24**, 109.

Williams, M. L. (1972) *J. Adhesion* **4**, 307.

Wolf, K. (1923) *Zeits. für ange. Math. u. Mech.* **3**, 107.

Vaidyanathan, S. and Finnie, I. (1971) *Trans. ASME, J. Basic Engr.* **93**, 242.

Yap, O. F., Mai, Y. W. and Cotterell, (1980) *Proc. Int. Conf. on Analytical and Experimental Fracture Mechanics,* G. C. Sih and M. Mibabile (eds.). Netherlands: Sijthoff and Noordhoff, pp 919–930.

4

Combined flow and fracture

But she couldn't help thinking to herself 'What dreadful nonsense we *are* talking!'

'She can't do sums a *bit*!' the Queens said together, with great emphasis.

'Can *you* do sums?' Alice said, turning suddenly on the White Queen, for she didn't like being found fault with so much.

4.1 INTRODUCTION

What is often called 'fracture mechanics' is really the mechanics of only *elastic* fracture where, except in regions very close to the crack faces, the behaviour is reversible so that broken parts may be refitted together to regain the original component dimensions. Materials which behave in this manner are those with comparatively low toughness-to-strength (E^*R/σ_y^2) ratios. Although the utility of fracture mechanics is not restricted to 'brittle' materials (owing to the scale effect – Chapter 9 – whereby materials which behave in a ductile fashion in the small can behave in a brittle manner in the large), traditional stress-intensity factor fracture mechanics are inadequate to describe the behaviour of high (E^*R/σ_y^2) solids in the small, owing to the gross irreversibilities which accompany cracking. In the development of an analysis for *plastic* fracture, we shall be concerned both with the propagation of existing cracks (now in extensive plastic, rather than elastic, fields) and with the initiation of cracks in supposedly crack-free materials undergoing extensive plastic flow. This chapter deals with the former problem, Chapter 5 with the latter.

Consider a previously existing crack of area A in a ductile solid. On loading, the ductile solid will flow plastically – in general both prior to, and during, crack propagation. We have

$$X\mathrm{d}u = \mathrm{d}\Lambda + \mathrm{d}\Gamma \qquad (1.3\ \text{bis})$$

prior to crack propagation, and during quasi-static crack propagation

$$X\mathrm{d}u = \mathrm{d}\Lambda + \mathrm{d}\Gamma + R\mathrm{d}A \qquad (1.2\ \text{bis})$$

R is now the specific work of fracture in the presence of extensive flow.

Whether R in equation (1.2) is identical with R obtained when Γ is negligible (i.e. in an *elastic* cracking situation, but at identical temperature, crack velocity, environment) depends on circumstances. If the process zone size is still small during plastic fracture, then the two values will be similar in magnitude. Often, however, plastic fracture involves additional localised deformation near the crack faces, such as localised necking in the tensile fracture of ductile sheets or combined bending and necking in the tearing of ductile sheets (similar to burring in cut sheets), without which plastic fracture does not occur. The size of the process zone is increased therefore above the elastic value and the associated R is greater. These points are discussed in detail later; to employ equation (1.2), it is sufficient that the relevant R values are shown by experiments to be reproducible.

To help the reader who may know the literature on the subject, we note that a different line of attack to elastoplastic fracture used by many workers utilises equation (1.2) but does *not* distinguish between R and $\mathrm{d}\Gamma/\mathrm{d}A$ and lumps them together in the so-called J_R approach. Where this formulation may apply, and where it most certainly does *not* apply, is discussed in later sections. In addition, there is the 'multiplicative' approach of Andrews, mentioned in section 1.3 and discussed later in this Chapter.

$\mathrm{d}\Lambda$ may be written in terms of the elastic strain energy density W and the volume of the body V undergoing elastic deformation (which will be the whole body, of course), as in equation (2.1). A similar approach is possible for $\mathrm{d}\Gamma$ in terms of the volume of the body V^{p} ($<V$) underding plastic flow and the plastic work done per unit volume W^{p}, which is given by $\int\bar{\sigma}\mathrm{d}\bar{\epsilon}$ (the area under the effective stress-effective strain curve). Hence

$$X\mathrm{d}u = W\mathrm{d}V + W^{\mathrm{p}}\,\mathrm{d}V^{\mathrm{p}} \tag{4.1}$$

before propagation. This equation will apply both in a body already containing a crack which will subsequently propagate, and in a crack-free body in which cracks have to be initiated – really these two cases are probably the same since the tip of the pre-existing crack has to experience sufficient critical deformation to cause it to 'continuously re-initiate' (i.e. propagate). Similarly

$$X\mathrm{d}u = W\mathrm{d}V + W^{\mathrm{p}}\mathrm{d}V^{\mathrm{p}} + R\mathrm{d}A \tag{4.2}$$

during propagation. In both (4.1) and (4.2) the elastic strains within V^{p} are taken care of by V which embraces V^{p}.

V^{p} will often be less than the total volume of the body V, but will at the same time be much greater than the crack tip process zone of irreversibility which accompanies even elastic cracking (Fig. 1.19). Note that as the plastic work increment involves both the 'work density' W^{p} and the extent of flow $\mathrm{d}V^{\mathrm{p}}$, a small volume of highly deformed material (such as a shear band, Fig. 1.19(f)) may produce the same $W^{\mathrm{p}}\,\mathrm{d}V^{\mathrm{p}}$ as a much larger volume of plastically deforming material over which a smaller W^{p} is diffused, Fig. 1.19(c) or 1.19(d).

Note also that during crack propagation in an extensive plastic flow field, it does not follow that there is a relation between the size of the plastically deforming region represented by V^{p} and the crack tip events represented by the crack tip advance $\mathrm{d}A$, except insofar as the externally applied load provides the work of simultaneous flow and fracture and that minimum work arguments must apply. In some problems, such as the tearing of sheet metal, $\mathrm{d}V^{\mathrm{p}}$ and $\mathrm{d}A$ are coupled which provides great simplification (section 4.4.1), but the general solution of equation (4.2) is formidable, not least because as a crack propagates, regions of the body may unload, while others still are experiencing increasing deformation.

There is, of course, a whole range of behaviour in between ideal elastic cracking (as exemplified by the cleavage of mica, say) and cracking to the accompaniment of extensive plastic plastic flow (Fig. 1.19). This is reflected in the relative significance of the terms in equation (4.2). As we know, in elastic cracking $W^{\mathrm{p}}\,\mathrm{d}V^{\mathrm{p}}$ is zero, but in extensive plastic flow fields $W^{\mathrm{p}}\,\mathrm{d}V^{\mathrm{p}}$ may be so large as to swamp $W\mathrm{d}V$. The latter occurs in many metal-forming problems in which large plastic strains are involved, and analysis is simplified by ignoring the elastic term in equation (4.2) to give

$$X\mathrm{d}u = W^{\mathrm{p}}\,\mathrm{d}V^{\mathrm{p}} + R\mathrm{d}A \tag{4.3}$$

Problems involving extensive flow are sometimes called *large-scale yielding fracture*. We call them 'plastic fracture'.

On the other hand, $W\mathrm{d}V$ and $W^{\mathrm{p}}\,\mathrm{d}V^{\mathrm{p}}$ may have comparable magnitudes. This happens in so-called *small-scale yielding fracture*, a common example of which is laboratory toughness testing of solids with high $(E^*R/\sigma_{\mathrm{y}}^2)$ which do not satisfy the minimum size requirements for valid K_{IC} testing (i.e. are not thick enough to keep $W\mathrm{d}V \gg W^{\mathrm{p}}\mathrm{d}V^{\mathrm{p}}$). A particular treatment of equation (4.2) in these circumstances has given rise to the J-integral and related criteria for 'post-yield' cracking (section 4.5).

A major difficulty in problems of this sort has been highlighted in Fig. 1.21 regarding the partitioning of fracture work and work of remote flow ($R\mathrm{d}A$ and $W^{\mathrm{p}}\mathrm{d}V^{\mathrm{p}} = \mathrm{d}\Gamma$) in a total work area contained beneath an Xu plot; and the associated prediction of X and u separately. Indeed where fracture is produced by local tooling as in many cutting operations, the total work area contains yet another component, namely incremental work against friction at the tool–workpiece interface, $\mathrm{d}F$

$$X\mathrm{d}u = W\mathrm{d}V + W^{\mathrm{p}}\,\mathrm{d}V^{\mathrm{p}} + R\mathrm{d}A + \mathrm{d}F \tag{4.4}$$

Much of this chapter will be concerned with different lines of attack to these difficulties and includes considerations of residual strain energy which must be accounted for in *unloaded* work areas whenever elastoplastic deformation has been non-uniform.

Interest in the fields of elastoplastic and plastic fracture takes on a variety of forms, depending on the end-use to which the information is to be put. There are three basic emphases:

(i) where it is *elastic* K_{IC} or R values that are really required for elastic design of large structures in solids possessing large $K_{\mathrm{IC}}/\sigma_{\mathrm{y}}$ ratios, but where the scale effect in fracture (Chapters 1 and 9) causes the bodies to behave elasto-plastically or even fully plastically in standard laboratory-sized toughness test pieces. Interest centres on inferring elastic data from non-elastic tests, either via mechanics procedures or via correlations with other properties (e.g. K_{IC} and Charpy impact energy, section 4.3.1), for use in basic or modified elastic fracture expressions;

(ii) where in applicatons the behaviour is elastoplastic (pressure vessels made from solids with medium-to-high K_{IC}/σ_y ratios and so on) and where elastoplastic fracture criteria have to be employed in testpiece and structure for design against failure;

(iii) where behaviour is plastic and where sometimes, in contrast to (i) and (ii), fracture is *required* (as in opening a beer can).

In many examples, the possibility of propagation of already-existing cracks is investigated, where the 'starter cracks' are found by non-destructive flaw inspection techniques. In other examples, the initiation of cracks in limited or extensive flow fields within supposedly flaw-free solids is of concern and more usually applied to (iii). Such crack nucleation is discussed in Chapter 5 and although often it is equated with 'failure', subsequent propagation may be of equal concern, as in the joining of cracks in processes such as shearing, blanking and other metal cutting operations.

Of the three divisions, in (i) and (ii) workers would prefer to avoid plasticity which is seen as a nuisance and complication. Indeed, elastoplastic mechanics *are* difficult. Only in (iii) might one say there is real interest in plastic fracture, and here it is often permissible to neglect elastic contributions and simplify problems via equation (4.3). Most developments have extended elastic fracture problems into the elastoplastic range; comparatively little has been done starting at the fully plastic end of the spectrum, and then reducing the magnitude of the plastic strains. Of course, both approaches should coincide in elastoplastic problems via a unified theory.

There are really two classes of problem to be considered:

(i) where R is known independently, such as in the fracture of adhesive joints duirng which the adherands deform plastically;

(ii) where R is not usually known independently as in the general cohesive cracking of ductile solids. As already mentioned in this Chapter, whether 'R' in equation (1.2) is then the same as R obtained from elastic fracture when Γ is negligible (but otherwise under identical conditions) is a crucial question in the mechanics of combined flow and fracture and this is the subject of much research and argument (see use of J_R later).

Displacement irreversibilities u_r in the Xu plot on unloading are usually indications of remote plastic flow (Fig. 1.21). However, they may also indicate events at the crack tip in materials such as reinforced composites where fibres bridging the crack faces prevent the crack from closing up even though remote regions may still be elastic; similarly, with polycarbonate, large necked-down regions bring about the same result (cf. section 2.4.3). Such effects are more marked when the crack length is comparable with the size of the crack tip zone of irreversibility. Displacement irreversibilities in Xu plots are also possible even in particular examples of properly-elastic fracture (where the broken pieces may be refitted together). The examples quoted in section 2.4 included ripping paper or cutting it with scissors: although there is a residual displacement in the direction of the force, if the overall displacement is elastic the cut pieces can be laid flat and refitted. Cracking in the presence of residual stresses also came into this category (Chapter 3) where it was noted that the (elastic) 'forcing' required to refit the broken pieces not only regained the original size body but also the internal stress pattern.

Apart from these particular elastic cases, there is uncertainty regarding how to analyse fracture in the presence of residual displacement irreversibilities. When u_r are large, we are in the field of combined flow and fracture, but when u_r are small we are in a transitional 'grey area' between elastic and elastoplastic cracking. In some circumstances, a tolerable limit on u_r is set and problems are analysed using basic elastic fracture mechanics. The offset procedure in valid K_{IC} testing is an example and implies a small u_r consonant with a small error on the derived K_{IC} (i.e. K_Q), cf. Examples 2.14/15; in practice specimens are rarely unloaded and fracture is rapid, so the actual u_r is not measured. At other times, modifications to the elastic fracture mechanics parameters are made, e.g. the embedded crack tip in the plastic zone, leading to apparent enhanced crack length in basic analyses (Example 2.14). These dodges avoid becoming embroiled in the solution of equation (1.2). But what *is* an acceptable limit for u_r, and is u_r likely to be different in different cracked geometries? What are the implications both for (i) determination of R or K_C and for (ii) predictions of X, u and A during fracture? We shall discover that it is crucial to know whether u_r is associated wholly with remote plastic flow, partially with 'suppressed' elastic effects, or what ever else (section 4.6.4).

Example 4.1 Is it possible for Xu plots to revert to the origin on unloading but where irreversible flow, redundant to cracking, has yet taken place?

Ans. This situation is very rare, but an example will be found in Chapter 8 (Example 8.7) on the stabilisation of crack propagation by the use of adhesive tapes (Fig. 8.16). The tapes slip at the interface between tape and testpiece, and at the relatively large displacements associated with tough materials the work against viscous flow of the glue and tape is not negligible. That work is the sink for the excess energy at crack initiation which would otherwise cause unstable cracking. On unloading, the displaced tapes buckle back into loops which offer negligible resistance to the crack faces closing back up. Any residual stress set up by the tape is relieved by easy buckling. The work of viscoelastic flow has to be subtracted from the Gurney segmental work area in order to determine the true fracture toughness (Chapter 8).

4.2 ELASTOPLASTIC TOPICS FOR APPLICATION TO ELASTIC FRACTURE

In addition to the three areas covered in this section, note should be taken of the related topics of (i) corrections to crack length using the embedded crack tip concept (section (2.8); (ii) scaling techniques for inferring elastic toughness values from elastoplastic small testpieces (in Chapter 9); and (iii) methods of avoiding remote flow in fracture toughness testing (section 3.9).

4.2.1 Use of 'small' specimens for K_{IC}

The size requirement for valid K_{IC} testing, where $W^P dV^P$ is small, has been discussed in Chapters 2 and 3. For materials with $K_{IC}/\sigma_y > 0.2\sqrt{m}$ the size requirement is difficult to satisfy – and also valid testpieces sometimes become thicker than the section of plate to be used in the large-scale structure. Barker's short-rod specimen, described in section 2.8, seems to give valid K_{IC} in specimen sizes down to one-third the ASTM requirement for minimum thickness of 2.5 $(K_{IC}/\sigma_y)^2$. If the short-rod testpiece remains globally elastic, unloading lines return to the origin of displacement and the determination of R or K_{IC} values by use of algebraic formulae, or Gurney's area method, is *no different* from the methods of Chapter 2. The behaviour of Barker's testpiece is unusual insofar as cracking occurs under *increasing* load at first which, although not unknown in certain cracking geometries, is not common.

There is, of course, always a danger of remote plastic flow and unloading irreversibilities in the likes of the Barker specimen, particularly in materials with high (K_{IC}/σ_y) ratios. The degree to which LEFM assumptions are violated may be determined in experiments by a few unloading/reloading excusions near the point of maximum load. Barker (1979) gives a correction factor for equation (2.38) in terms of residual displacements, and writes

$$K_{IC} = [(1 + p)/(1 - p)]^{1/2} C . X_{max}/B^{3/2} \qquad (4.5)$$

where, in the notation of Fig. 2.20, $p = \Delta u_r/\Delta u_Q$. Since $K_{IC}^2 = E^*R$, the factor is a correction on compliance, i.e.

$$G_c = R = ½ \left(\frac{1+p}{1-p}\right) X^2 \, d/dA\,(u/X) \qquad (4.6)$$

In the derivation of these expressions, it is assumed that *all* dissipated work goes to work of fracture and not (as is really the case) some to remote flow. If all the work went to fracture, equations (2.21) or (2.23) would apply. Barker and Baratta (1980) say that excessive plastic flow precludes the obtaining of valid K_{IC} results. Thus while the short rod method produces, in small specimen sizes, essentially the same K_{IC} values as would be measured by ASTM method E399 in larger specimens – which is of value experimentally – the test method inevitably fails to give valid data in materials with very high (K_{IC}/σ_y).

The use of small specimens in J_{IC} testing is discussed in section 4.5.

4.2.2 Empirical correlations between K_C and other measurements

Notched impact tests (Charpy, Izod) have been employed historically to aid in the design and selection of materials, although they do not, of course, give values which a designer can use in calculations. It is simply a case of 'if a material with such-and-such a Charpy value (CVN) worked in a given situation last time, then another material with at least the same CVN should work just as well next time'. However, it is observed that when valid quasi-static K_{IC} or dynamic K_{Id} (see Chapter 6) data are plotted against temperature or loading rate, the curves exhibit the same shape as notched impact data (Figs 4.1a, b). It would appear that there exist empirical correlations between the two, and since it is easier to make CVN specimens and test them than perform full critical K tests, an inexpensive short-cut is suggested to get valid toughness data to use in design. In fact various design codes which are based on fracture mechanics principles incorporate auxiliary tests such as notched impact and dynamic tear (DT) specifications instead of formal K_{IC} specifications (a dynamic tear test is a sort of dynamic three-point bend test) e.g. bridge steels (US-AASHTO code); thick-walled nuclear pressure vessels (based on minimum K_{Id} but specified in terms of nil ductility temperature (NDT) and CVN values defined in terms of lateral expansion); welded ships' plates (based on K_{Id}/σ_y but specified in terms of NDT and DT impact values).

Examples of the sorts of empirical relation between CVN and K_{IC} for steels are the following. For $270 < \sigma_y < 1700$ MPa and $4 < \mathrm{CVN} < 82$J, Barsom and Rolfe (1970) give

$$K_{IC}^2/E = 0.22\,(\mathrm{CVN})^{1.5} \quad (\mathrm{kPa\text{-}m, J}) \tag{4.7}$$

Alternatively Roberts and Newton (1981) give

$$K_{IC} = 8.47\,(\mathrm{CVN})^{0.63} \quad (\mathrm{MPa}\sqrt{\mathrm{m}}, \mathrm{J}) \tag{4.8}$$

Many other empirical correlations between toughness and impact, or alternatively slow, CVN values have been given in the literature. Table 4.1 lists a number of them; note the ranges of σ_y and CVN over which they have been determined.

A possible explanation for these correlations may be found in the relation between work areas and J (Section 4.5.2).

Example 4.2. Steel bolts to be used for transmission lines are required to have CVN values of at least 20 J and tensile strength 500 MPa. The diameter and length of the bolt are 24 mm and 80 mm, respectively. If, with time, a fatigue crack develops under the head as shown in Fig. 4.2 estimate the critical length of the crack. Assume that the direct stress due to initial tightening is 120 MPa and the bending stress caused by the shear load at the line is 250 MPa.

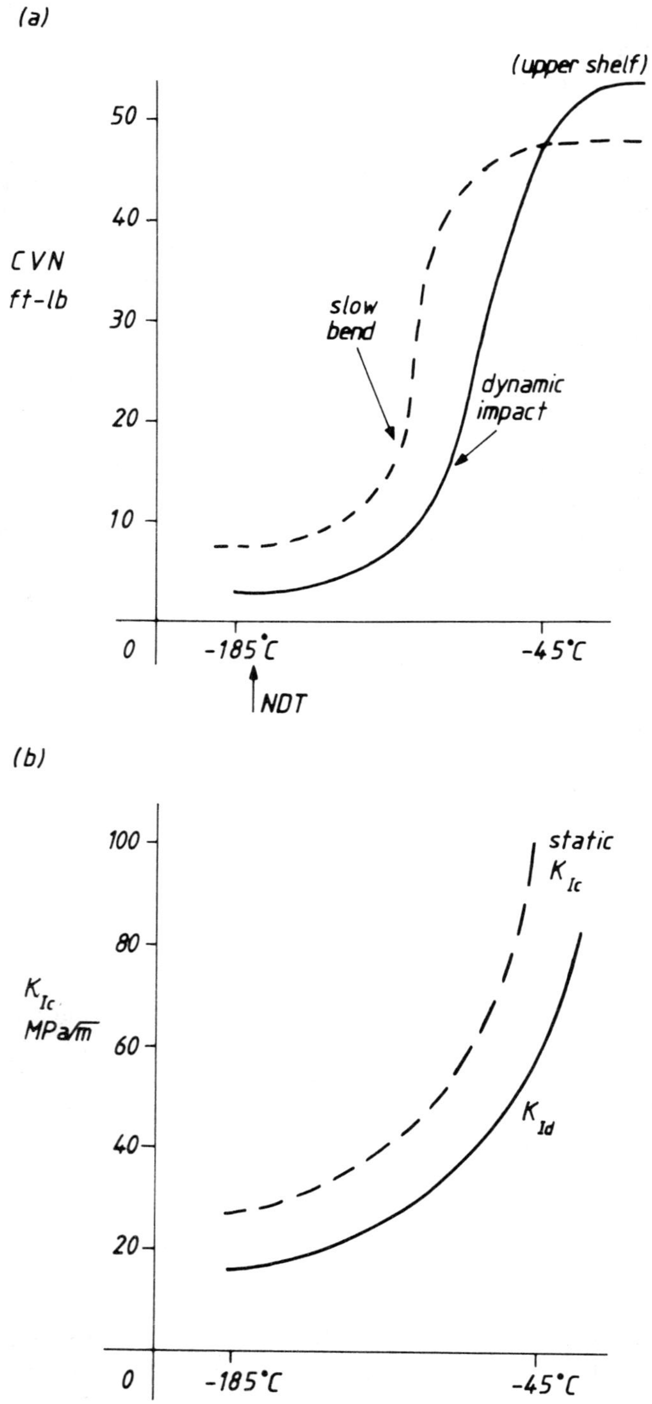

Fig. 4.1 – Variation of (a) Charpy energy absorption (CVN) and (b) fracture toughness (K_{IC}) versus temperature for a typical structural steel.

Table 4.1
Fracture toughness – Charpy energy correlations (after Roberts and Newton (1981), with permission) (SI units)

Material	Notch	Test	Temperature range	Range of Charpy results	Yield strength range (MPa)	Correlation
A517D 4147 HY 130 4130 12 Ni-5Cr-3Mo 18 Ni-8Co-3Mo	V-Notch	Impact	Upper shelf	31–121	760–1700	$\frac{K_{IC}^2}{\sigma_{ys}} = 0.64 \frac{\text{CVN}}{\sigma_{ys}} - 0.01$
High-strength Steels	V-Notch	Impact	Upper shelf	15–28.6	1610–1980	$\frac{K_{IC}^2}{\sigma_{ys}} = 0.18 \frac{\text{CVN}}{\sigma_{ys}} - 0.0011$
A517F A3202B ABS-C HY-130 18 Ni(250) Ni-Cr-Mo-V Ni-Mo-V Cr Mo V	V-Notch	Impact	Transition	4–82	270–1700	$\frac{K_{IC}^2}{E} = 0.22\,(\text{CVN})^{1.5}$
A533B A517F A542	V-Notch	Impact	Transition	7–68	410–480	$K_{IC} = 14.6\,(\text{CVN})^{0.5}$

Table 4.1 – *continued*

Fracture toughness – Charpy energy correlations (after Roberts and Newton (1981), with permission) (SI units)

Material	Notch	Test	Temperature range	Range of Charpy results	Yield strength range (MPa)	Correlation
En25 3% Ni-Cr-Mo-V	V-Notch	Impact	Transition	6–55	820–1420	$K_{IC} = 18.2 \frac{CVN^{0.5}}{A}$
HY60 Ti3 Ti6	V-Notch	Impact	Transition & Upper shelf	5–125	400–560	$K_C = 15 \frac{CVN^{0.543}}{A}$
ABS-C A305-B A517-F	V-Notch	Impact Slow bend	Transition	2.7–61	250–345	$\frac{K_{Id}^2}{E} = 0.64\ CVN$ $\frac{K_{IC}^2}{E} = 0.64\ CSB$
ABS-C A302-B A517-F A533-B	V-Notch	Impact	Transition	3–95	270–815	$K_{Id} = 15.5\ (CVN)^{0.375}$
A533-73 2.25% Cr-1Mo C-Mn-Nb-V 12MDV C4-03M	V-Notch	Impact	Transition	5–50	303–820	$K_{Id}^2 = 19\ (CVN)^{0.5}$

ABS-C A302-B A517-F	Precracked Slow bend	Impact	Transition	2.7–61	250–345	$\frac{K_{IC}^2}{E} = 0.52\ \text{PCI}$ $\frac{K_{IC}^2}{E} = 0.52\ \text{PSB}$
HY60	Precracked	Impact	Transition	16–40	400	$K_C = 25\ \frac{\text{PCI}^{0.5}}{A}$
4340 D6AC Al, Ti	Precracked	Slow bend	*	3–16	*	$K_{IC}^2 = \frac{E}{2(1-\nu)^2}\ \frac{\text{PSB}}{A}$
Steel Al Ti	Precracked	Slow bend	*	0.6–33	310–2080	$\frac{K_{IC}^2}{E} = 0.95 + 0.45\ \frac{\text{PSB}}{A}$
Ti-6Al-4V	Precracked	Instrumented Slow bend instrumented impact	*	*	*	$K_{IC} = 0.89\ \frac{\text{PSB}}{A} + 17.8$ $K_{IC} = 0.59\ \frac{\text{PCI}}{A} + 7.4$
Low–medium Strength Steels	V-Notch	Instrumented impact	Transition	*	*	$K_{Id} = 0.02\ P_f$

*Not readily available.
PSB: Precracked slow bend; PCI: Precracked Charpy impact; CVN: Charpy V-notch;
CSB: Charpy slow bend; P_f: Fracture load (kg).
A: Cross-sectional area

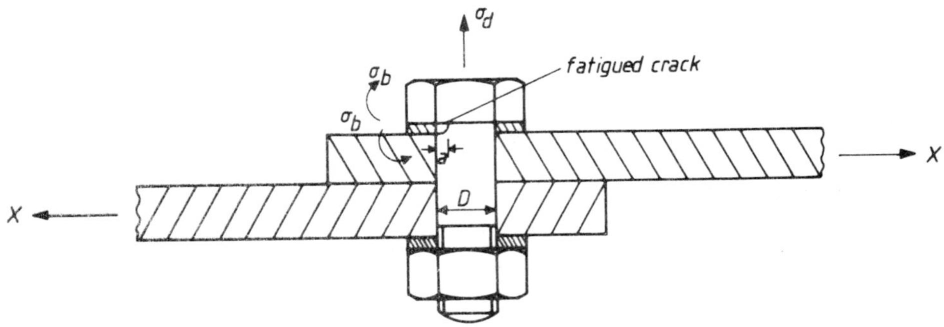

Fig. 4.2 – Schematic drawing of a bolt in a transmission line where the crack underneath the head is subjected to both tension and bending.

Ans. We are given $\sigma_d = 120$ MPa (direct stress), $\sigma_b = 250$ MPa (bending stress induced by shear load). Using the K-superposition principle (section 2.5)

$$K = \sigma_d \sqrt{\pi a}\, Y_d\,(a/D) + \sigma_b \sqrt{\pi a}\, Y_b\,(a/D).$$

If $a \ll D$ then $Y_d(a/D) = Y_b\,(a/D) = 1.12$, so the equation simplifies to $K = (\sigma_b + \sigma_d)\ 1.12\ \sqrt{\pi a}$. The fracture toughness K_{IC} value can be evaluated from either equation (4.7) or (4.8) which gives 64 MPa$\sqrt{\text{m}}$ and 55 MPa$\sqrt{\text{m}}$, respectively. Taking the more conservative estimate of 55 MPa$\sqrt{\text{m}}$ and assuming $a \ll D$, the critical crack length is

$$a_c = \frac{1}{(1.12)^2\pi}\left(\frac{55}{370}\right)^2 \times 1000\ (\text{mm}) = 5.60\ \text{mm}$$

When $a \not\ll D$, and $Y_d \neq Y_b$, a better answer is obtained from the unsimplified equation with $Y_b\,(a/D)$ and $Y_d\,(a/D)$ taken from stress intensity factor handbooks given in the references of Chapter 2.

Another empirical correlation between K_{IC} or K_C and an easily measured parameter, concerns the size of the shear lips in the path of fracture, as first postulated by Krafft, Sullivan and Boyle in 1961.

Lai and Ferguson reported in 1980 that the shear lip size B_{SL} in compact tension test pieces of a variety of metals was constant along the path of propagation and was related to valid K_{IC}/σ_y data by $B_{SL} = 0.41\,(K_{IC}/\sigma_y)^2$ (Fig. 4.3). The correlation arises through the link between crack tip plastic zone size and shear lip size (section 2.8). Note, however, that it is often argued that B_{SL} should be related to the plane *stress* K_C, but Krafft and co-workers did find better correlation between d_y and K_{IC}. Also, it should be pointed out that most polymer data seem to fit $B_{SL} \propto K_C$ (where K_C is a mixed-mode value, Yap *et al.*, 1983).

As illustrated by Example 4.2 (also Problem 4.5) these inferred K_{IC} would be used in design – particularly in the case of large structures made from materials

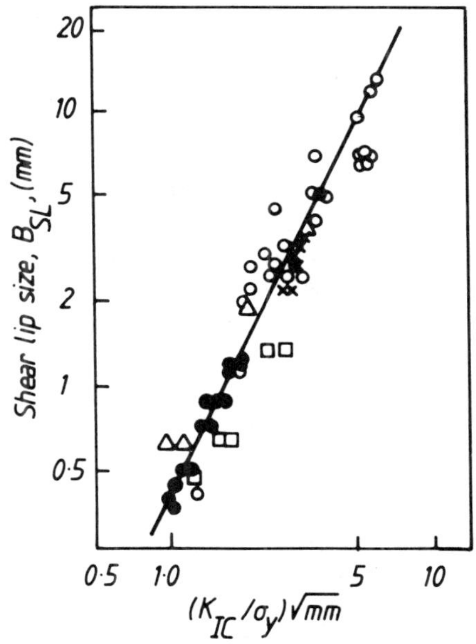

Fig. 4.3 – Plot of shear lip size (B_{SL}) against the relative plane strain fracture toughness for several metallic materials (After Lai and Ferguson (1980), with permission).

which are ductile in the laboratory. The derived K_{IC} may possibly be also employed in cracking situations where a small amount of elastoplastic deformation takes place.

4.2.3 Crack opening displacement (COD)

When valid K_{IC} measurements cannot be made owing to elastoplastic deformation, a critical value of COD at fracture δ_c has been widely used in Europe and Japan as the criterion for fracture. Subsequent application to design relies on a COD 'design curve' which gives allowable defect sizes (to be checked-for by non-destructive testing) in terms of the level of strain applied to the structure in use.

The concept of a critical crack opening displacement as an alternative to R or K_C in elastic fracture has been introduced in section 2.9. It was shown that $\delta_c \doteqdot d_y/100$ where the plastic zone size $d_y \propto (K_c/\sigma_y)^2$ and where $R \approx m\,\sigma_y\delta_c$ (where $1 < m < 3$). For materials which fracture elastically in the small (i.e. those displaying $(K/\sigma_y)^2 < 10^{-2}$ m say), δ_c is very small, less than 10^{-2} mm; for materials with greater toughness-to-strength ratios, δ_c is larger ($\delta_c \approx 0.2$ mm for mild steel in thick sections). Wells (1961) and Cottrell (1961) observed that δ_c seemed to be a material constant and could be reliably measured even in the presence of significant yielding at the crack tip. It is then argued that the same δ_c will apply for brittle fracture in large structures made from materials which show considerable ductility in the laboratory. How the same δ_c may be accommodata plastically in the small, and elastically in the large, has been discussed in section 1.5 (see also Chapters 9 and 10).

Dugdale in 1960 presented a solution for the extent of local plastic flow at the crack tip in elastoplastic deformation, which is a more refined model than Irwin's for d_y given in section 2.8. He considered the Griffith geometry of a small crack of length $2a$ in a large plate remotely loaded by σ but where the crack was extended by a distance c at each end over which a uniform yield stress σ_y acted (Fig. 4.4). The length c thus corresponds with the size of a crack tip 'line zone', and is similar to the array of crack tip dislocations which is central to the Bilby–Cottrell–Swinden 1963 model of non-linear crack tip deformation.

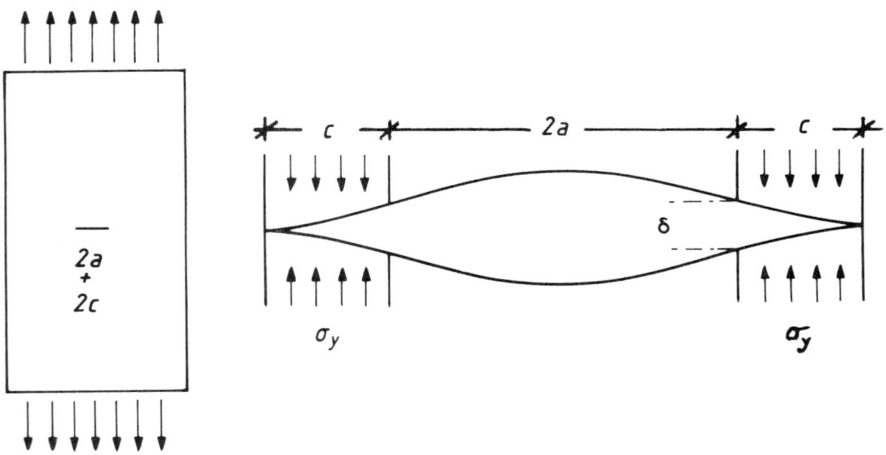

Fig. 4.4 – The Dugdale–Barenblatt fracture model.

The stress at the crack tip is no longer singular (as it is limited to σ_y) and Dugdale showed that for the separate solutions (of the remote applied stress and the imagined local stress σ_y) to be compatible, the length of the plastic region had to be

$$c = a \left[\sec \left(\frac{\pi \sigma}{2\sigma_y} \right) - 1 \right] \tag{4.9}$$

(Note that Dugdale did not need to use plasticity theory for this 'yield-strip' model, and that the extent of the plastic zone in the direction normal to the crack is not considered, cf. Example 2.15. Note also that when the work was done, Dugdale was concerned with the size of the plastic zones around slits in thin sheets and not, paradoxically, with fracture as such.) The plastic zone area proportion of the crack length c/a is 0.01 for $\sigma/\sigma_y = 0.1$ and increases to 0.4 for $\sigma/\sigma_y = 0.5$. For $\sigma/\sigma_y = 1$, $c/a \to \infty$ and the zone has spread across the whole plate. This suggests limits for σ/σ_y in the applicability of the Dugdale model since for reversible elastic behaviour (and hence the use of K-formulae) c/a should be small, cf. Example 2.14. The use of a Dugdale crack tip zone in the construction of an R-6 'interaction' diagram in elastoplastic fracture is described in Chapter 10.

Burdekin and Stone (1966) evaluated the displacement at the crack tip of the Dugdale model crack by combining the elastic solutions for remote and local stresses. They found for COD

$$\delta_c = \frac{8\sigma_y a}{\pi E} \ln \left(\sec \frac{\pi\sigma}{2\sigma_y} \right) \tag{4.10}$$

which, on retaining the first term of the series expansion for ln (sec), becomes

$$\delta_c \approx \pi\sigma^2 a/E\sigma_y = K_C^2/E\sigma_y \tag{4.11}$$

and since $K^2 = E^*G$, $G_C = R = \sigma_y\,\delta_c$ as before. Similar results may be obtained from the Bilby–Cottrell–Swinden dislocation model (1963) and Barenblatt cohesive model (1962) of crack tip zones.

Example 4.3 δ_c is sometimes measured directly (using the early idea of paddle gauges or, more recently, using rubber infiltration techniques). Alternatively, the movement of clip gauges mounted on the cracked body at a location which gives a magnified displacement u is measured, Fig. 4.5. Derive the relationship between the measured u and δ_c (z is the thickness of the knife edges and $r\,(W - a)$ locates the apparent centre of rotation beneath the crack tip).

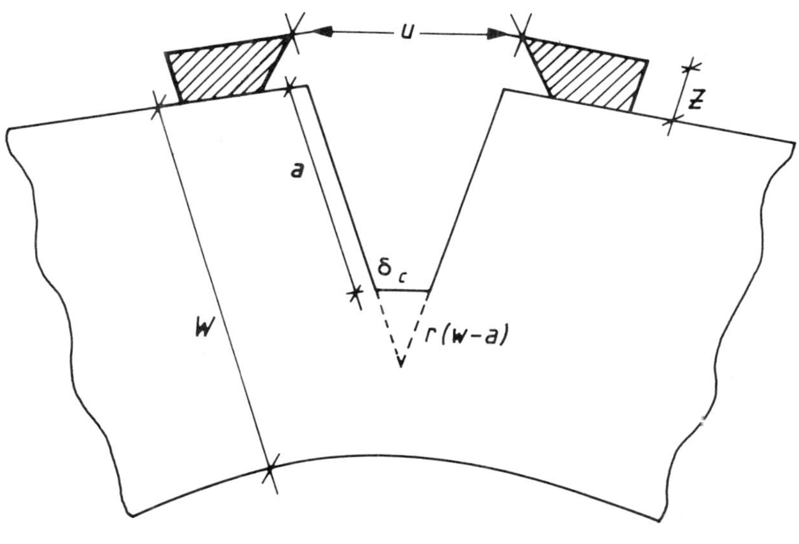

Fig. 4.5 – Notch deformation in a standard COD specimen.

Ans. From similar triangles $\delta_c = r(W-a)u/\{z + a + r(W-a)\}$. If $r = ½$ the centre of rotation coincides with the neutral axis and $\delta_c = (W-a)u/(2z + W + a)$ and if, say, $a/W = ½$ and $z/W = 1/10$, $\delta_c = 3.4u$. In the early stages of deformation, before the plastic hinge has fully developed, $r < ½$ and the centre of rotation is closer to the crack tip. British and Australian standards (see references) recommend $r = 0.4$; but there is increasing experimental evidence that r is closer to ca. 0.5 as would be predicted from slip line field solutions. The Chinese standard has $r = 0.45$; the Russian standard $r = 0.5$ (see references and papers by Lin *et al.* (1982) and Matsoukas *et al.* (1984)).

If K_{IC} were known for ductile solids, design against brittle fracture would proceed using equation (2.31) to give the design stress or critical crack size, i.e. $a_{crit} = (1/\pi)\ [K_{IC}/(Y\sigma)]^2$. If such lefm arguments are retained in elastoplastic conditions and we say that σ has become σ_y, we would argue that the allowable flaw size $a_{crit} \propto (K_{IC}/\sigma_y)^2$. But as $K_{IC}^2 = ER \propto E(\sigma_y\delta_c)$, we might expect $a_{crit} \propto E\delta_c/\sigma_y = \delta_c/\epsilon_y$ where ϵ_y is the yield strain σ_y/E. The *COD design curve* is the experimentally found correlation between a_{crit} and δ_c/ϵ_y and is shown in Fig. 4.6(a). We see that $a_{crit} = C(\delta_c/\epsilon_y)$, where C takes different values depending on the ratio of applied strain to yield strain (ϵ/ϵ_y). The curve is the result of a combination of theoretical predictions and experimental observations; note that there is a 'safety factor' of two deliberately incorporated in the COD design curve in the relation between δ_c and ϵ (see Example 10.4). As explained by Burdekin (1981) a basic requirement of toughness tests for critical COD is that they should be carried out at full thickness. These tests do not have to satisfy a criterion related to plane strain conditions, but seek to determine a toughness value relevant to the particular thickness of interest. For application to welded structures or components, it is of course necessary for tests to be carried out on material representing different regions of welded joints.

As with lefm design, the use of COD procedures concerns either the toughness (COD) necessary to prevent fracture if the likely flaw size is specified, or the NDT inspection accuracy required if a material with a given toughness must be employed. An alternative to Fig. 4.6(a) is to employ the curve in Fig. 4.6(b) given by Burdekin and Dawes (1971) which relates non-dimensional COD ($\delta_c/2\pi\epsilon_y a$) with normalised strain (ϵ/ϵ_y). Design then involves determination of ϵ/ϵ_y for the *uncracked* body from strength-of-materials, which then gives ($\delta_c/2\pi\epsilon_y a$). In this way, as illustrated in Example 4.4, the required δ_c to prevent cracking in a given location is established for some chosen assumed flaw size in that location.

Example 4.4 Fissures, about 1 mm deep, were found on the inside surface of a cylindrical chemical plant tower having diameter 1.5 m and wall-thickness 12.7 mm. The column was made of a low carbon steel having 320 MPa yield stress

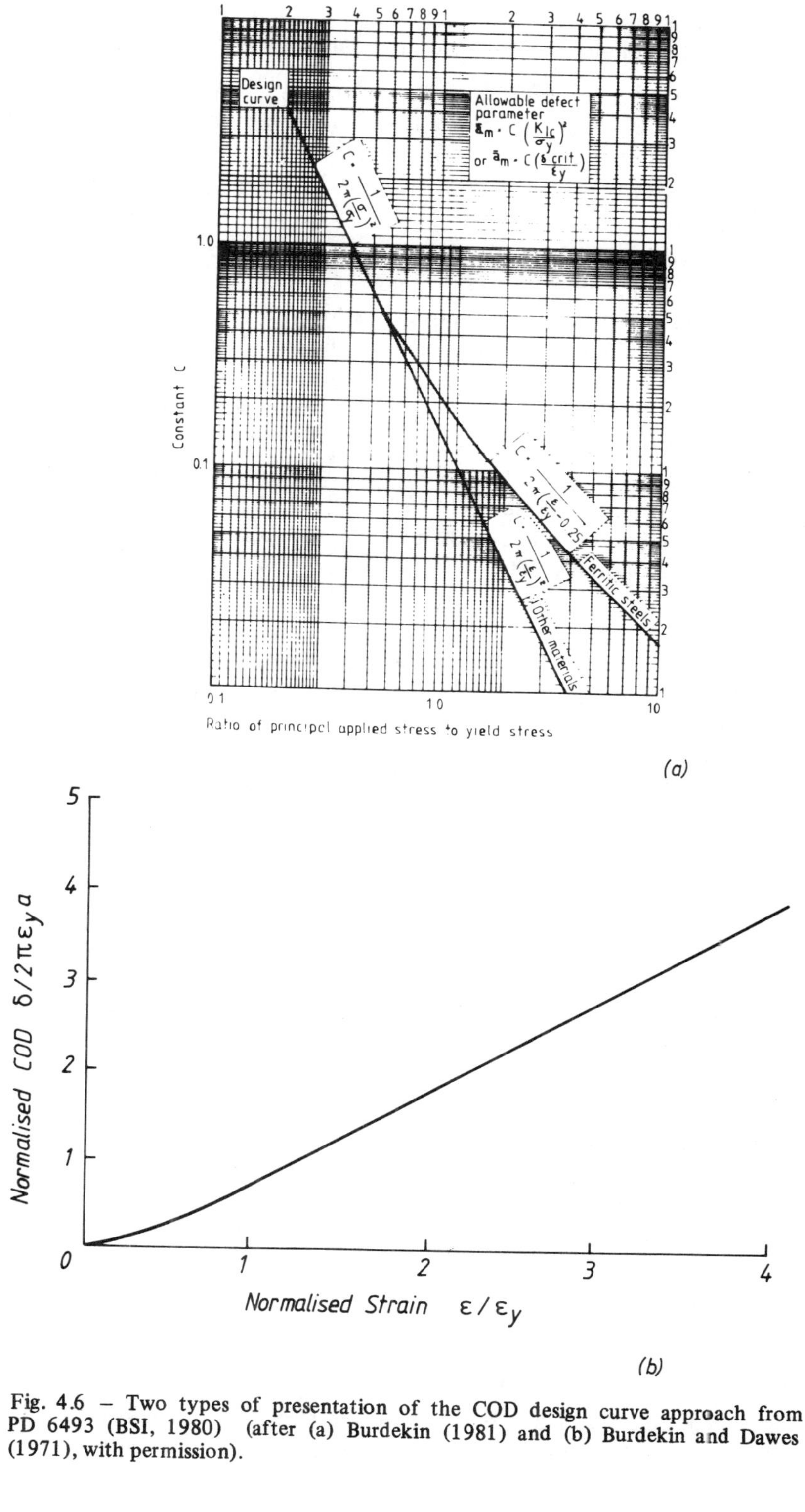

Fig. 4.6 – Two types of presentation of the COD design curve approach from PD 6493 (BSI, 1980) (after (a) Burdekin (1981) and (b) Burdekin and Dawes (1971), with permission).

and was operated at a pressure of 800 kPa. Experiments on testpieces taken from the column showed that $\delta_c \approx 0.4$ mm. Discuss the probable safety of the column.

Ans. The hoop stress in the column is $\sigma_\theta = pd/2B = 800 \times 10^3 \times 1.5/2 \times 12.7 \times 10^{-3} = 47.2$ MPa. Hence $\sigma/\sigma_y = \epsilon/\epsilon_y = 47.2/320 = 0.15$. It is not clear from the problem statement whether the defective areas are near nozzles (which might introduce a stress concentration of 3, say) or whether the vessel was stress-relieved after welding – if not, it is customary in the COD design approach to say that strains of magnitude ca. ϵ_y are present owing to residual stresses. Hence in this problem ϵ/ϵ_y is increased above 0.15 by (i) a factor of 3 to account for nozzles, manifold openings, and so on and (ii) by the addition of unity (i.e. $\epsilon/\epsilon_y = 1$) to account for residual stress effects. Hence $\epsilon/\epsilon_y \rightarrow 1.45$. This value corresponds with $(\delta_c/\pi\epsilon_y a) = 1.12$ from Fig. 4.6(b). From this, the critical crack half length is $a = 38$ mm using $\delta_c = 0.4$ mm and $\epsilon_y = \sigma_y/E = 320/210 \times 10^9 = 1.5 \times 10^{-3}$. The critical crack length of 76 mm is much larger than the 1 mm depth of the fissures (since these flaws are surface cracks it is their depth, rather than length, which is important). Despite the uncertainties in the analysis, it is concluded that the column is safe to operate (subject to periodic inspection, of course).

Example XXV in Knott and Elliott (1979) relates the application of the Welding Institute COD design curve to cylindrical pressure vessels with pipes welded into circular openings.

Slow crack growth can cause discrepanices between COD test results and the prediction of failure stresses in large components based upon those COD data. Even so, Fearnehough and co-workers concluded in 1971 that cracking in full-size burst tests of flawed pressure vessels can be predicted by COD test results on small specimens, providing the initiation of cracking in the vessels is carefully detected. That is, although failure occurred at a considerably later stage than crack initiation (in some cases the crack growth was 36% of the initial crack size), the *initiation* COD test results were very useful in predicting failure. Turner (1979) has observed that tests carried out in the UK on bend bars, pressurised spheres and cylinders seem to confirm that COD at the initiation of slow crack growth from a sharp pre-existing crack is a material property (for given temperature, strain rate and environment). Questions of initiation and propagation toughness arise in the following section.

The connexion between δ and the J-integral is discussed in section 4.5.2 *et seq.*, including the differences (during crack propagation) between δ measured at the original position of the crack tip and at the moving crack tip.

4.2.4 *R*-curves and elastic-plastic fracture

R-curves were introduced in Chapter 2 and, like the COD concept, are only of passing interest in wholly elastic fracture unless there are peculiar rate, temperature or environmental effects at the crack tip which cause R to increase with crack growth. In thin sheets, however, (where the plane stress toughness may be anything from twice to ten times the plane-strain resistance to fracture), crack tip zones are comparatively large and, as we shall see with the plastic fracture of deep edge-notched sheets (section 4.3.1), the crack may begin to propagate *before* the crack tip plastic zone is fully grown. There may also be transitions in fracture mode once the crack begins to propagate (cf. section 2.8): in particular a flat starter crack will twist into a slant fracture, and the increasing width of shear lips, plus increasing plasticity of the possibly growing crack tip plastic zone, both contribute towards a rising R curve, as already shown in Fig. 2.30 (cf. also Figs. 4.14 and 4.15).

For use in design it is customary to convert the R vs. Δa curve into a K_R vs. Δa curve, where the subscript shows that K was derived from an R-curve. In the elastic range (where there is essentially no remote yielding, and where R-curve effects arise from large crack tip zones which are still encapsulated within remote elastic fields), K_R is simply given by the relevant fracture mechanics formula, viz. $K_R = \sigma Y\sqrt{\pi a}$ where, in the early stages of crack growth (small increases in a), K_R is rising. In the elastoplastic, or even fully-plastic regime, an apparent K_R can be obtained from COD values. Since $R \approx \sigma_y \delta_c$, $K_R = \sqrt{E\sigma_y\delta_c}$, and K_R increases if R or δ_c increase. Of course in the presence of extensive plastic flow in materials with large K/σ_y, where displacement reversibility is patently not obyed, the justification of using an equivalent elastic K_R must be questioned. Even so, the K_R curve is employed by some workers to find points of instability and hence used for failure prediction.

Example 4.5 Show that increasing shear lip width fracture will contribute to a rising R-curve in a constant thickness plate.

Ans. Consider a partially propagated crack face consisting of a central region of flat fracture (plane strain) flanked by 45° shear lips (Fig. 4.7). To advance the crack length by da, the incremental total work dW_T required is the sum of the work to propagate the flat fracture area plus the plastic work done on the volume of shear lips leading to slip-off. We have

$$dW_T = \left(\frac{\partial W_T}{\partial A_{PS}}\right)_{A_{SL}} dA_{PS} + \left(\frac{\partial W_T}{\partial A_{SL}}\right)_{A_{PS}} dA_{SL}$$

with

$$A_{PS} + A_{SL} = A_T = (B_{PS} + B_{SL})\,a_T$$

where PS denotes plane strain and SL shear lip. But

$$R = dW_T/dA_T = \left(\frac{1}{B_{PS} + B_{SL}}\right)\frac{dW_T}{da_T} = \frac{B_{PS}}{(B_{PS} + B_{SL})}\left(\frac{\partial W_A}{\partial A_{PS}}\right)_{A_{SL}} + \frac{B_{SL}}{(B_{PS} + B_{SL})}\left(\frac{\partial W_T}{\partial A_{SL}}\right)_{A_{PS}}$$

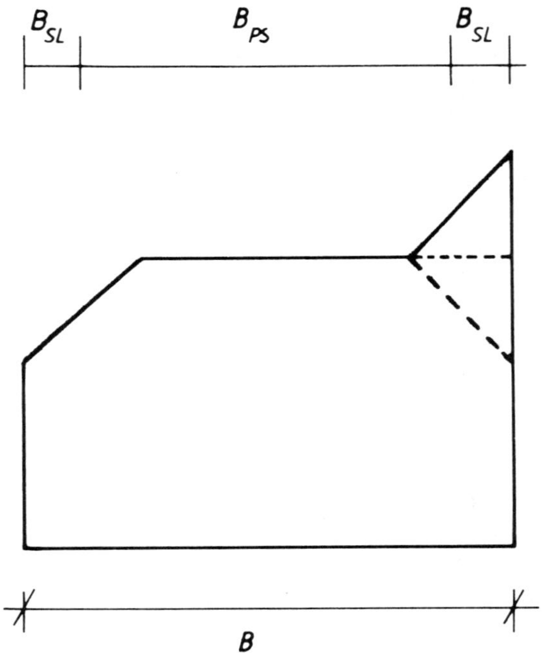

(a)

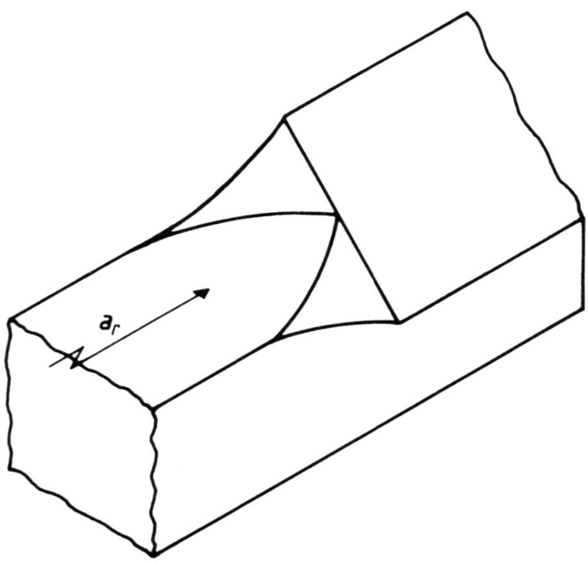

(b)

Fig. 4.7 – (a) Mixed mode fracture showing flat and slant surfaces. (b) Transition from mixed mode to fully slant fractures in the direction of crack propagation.

For

$$\left(\frac{\partial W_T}{\partial A_{SL}}\right)_{A_{PS}} > \left(\frac{\partial W_T}{\partial A_{PS}}\right)_{A_{SL}}$$

it follows that increasing B_{SL} will give rise to increasing R (cf. Fig. 2.30).

In thin sheets of metals, it is sometimes found that the R-curve continues to rise even after the shear lips span the whole thickness; clearly other effects are contributing to the rising R-curve.

4.3 INTERFACIAL ADHESIVE FRACTURES IN PRESENCE OF PLASTIC FLOW

In this class of elastoplastic fracture problem it is presumed that independent measurements of R can, if necessary, be obtained from other tests in which remote plastic flow has been eliminated. The cracking of glued interfaces forms a suitable illustration and in this section we consider what happens when plastic flow occurs in the adherands.

4.3.1 Peeling of glued soft metal strips

Fig. 4.8(a) shows the peeling of strips of soft metal which have been glued together. Let us neglect elastic effects initially, and let us consider steady tearing where the shape of the bent curves has already been established. There is simultaneous work of plastic bending and unbending as material is pulled into and out of the bend. Assume for simplicity that the bend is a semicircle or radius ρ with (work-hardened) fully plastic hinges at every point (ρ is not really constant as the bending moment given by Xz changes. The shape of the bend is given in Problem 4.9 for a work-hardening material). The average plastic strain in the bend of the sheet is $\bar{\epsilon}^P = (B/4\rho)$ and the plastic work done per unit volume as the metal is pulled into the bend is $\sigma_Y\ \bar{\epsilon}^P = \sigma_Y(B/4\rho)$. As the metal comes out of the bend in steady-state tearing, further plastic flow takes place with the strain increasing from $\bar{\epsilon}^P$ to $2\bar{\epsilon}^P$, again requiring $(\sigma_Y\ B/4\rho)$ work per unit volume, assuming for simplicity the same average flow stress. All elements passing into and out of a bend experience

$$W^P = 2\,(\sigma_Y\,.\,B/4\rho) = \sigma_Y\,B/2\rho \tag{4.12}$$

Hence, in equation (4.2) with the elastic $W\mathrm{d}V = 0$, we obtain for this problem (remembering there are two bends)

$$X\mathrm{d}u = 2\,(\sigma_Y\,.\,B/2\rho)\,B\mathrm{d}A + R\mathrm{d}A \tag{4.13}$$

during steady tearing, since $\mathrm{d}V^P = B\mathrm{d}A$ in each bend. The fracture and plastic works are coupled directly, and it is as if the fracture toughness R were increased to $(R + \sigma_Y\,B^2/\rho)$. Notice that unless B^2/ρ is constant, the augmented toughness is geometry-dependent.

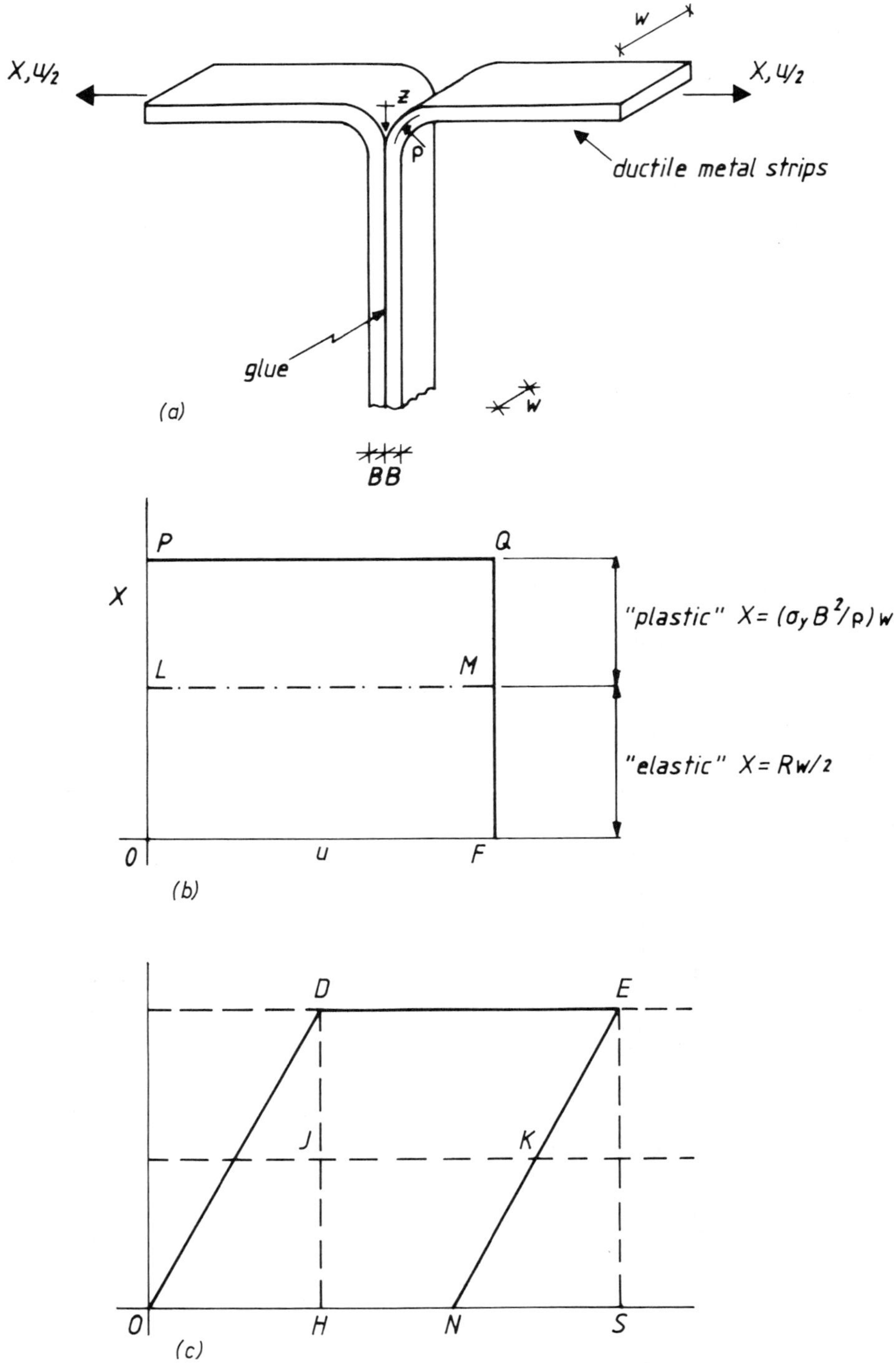

Fig. 4.8 – (a) Peeling geometry of a glued soft metal joint. (b) Load–deflexion traces for the joint in (a) when peeled. The adherands are assumed to be perfectly plastic solids. (c) Same as (b) except the adherands store elastic energies which are recovered upon unloading.

Since $dA = w\,du/2$, rearrangement of equation (4.13) gives

$$X = (\sigma_y B^2/2\rho)\, w + Rw/2 \qquad (4.14)$$

or

$$X/w = \sigma_y B^2/2\rho + R/2 \qquad (4.14)$$

The steady-state peeling load is thus constant for a given material and specimen dimensions.

The graphical representation of equation (4.14) as an Xu diagram obtained from a testing machine is shown in Fig. 4.8(b). The rectangular work area OPQF divided by the fracture area gives the 'enhanced' or 'effective' toughness $R_{aug} = (R + \sigma_y B^2/\rho)$; the displacement OF is common to both the new crack area (rigid-plastic deformation) and the associated plastic volume brought into play. In the absence of plastic flow, the load required would be simply $X = Rw/2$ as in a trousers test (section 2.4.1), and is represented by the load level along LM. The increased load (to level PQ) required in the presence of plastic flow is given by $(\sigma_y B^2/2\rho)w$.

The stiffness of the fully-developed (steady-state) elastic quadrants in Fig. 4.8 is $X/u = 4EI/\pi\rho^3$ where $I = wB^3/12$ (e.g. Timoshenko, 1955). For a given w and ρ, this is constant, so Λ is constant during propagation, $d\Lambda = 0$ in equation (1.2) and $W dV = 0$ in equation (4.2). Thus for real problems, work areas in the presence of elasticity are shown in Fig. 4.8(c), where the elastic triangular areas at the beginning (ODH) and end (NES) of peeling are identical. Parallelogram areas ODEN represent $[R + \sigma_y B^2/\rho]\, w(u_2 - u_1)/2$ where $(u_2 - u_1) = $ ON. Again, the load levels JK and DE correspond with LM and PQ in Fig. 4.8(b).

Experiments show that as w is varied in a sheet of constant thickness not only does X change but so also does the 'free standing' ρ (which changes less rapidly than w). Equation (4.14) suggests that a plot of X/w versus $1/\rho$ should be a straight line, the intercept of which gives $R/2$ of the glue and the slope of which is $\sigma_y B^2$.

After rigid-plastic peeling, the strips are flat and could be glued back together. The situation *apparently*, therefore, satisfies the acid test of reversibility. But clearly the strips have experienced fully plastic flow, and are work-hardended in comparison with their starting conditions. Although the toughness work may have been confined to the glue layer the rest of the system has *not* remained elastic, and elastic fracture mechanics analyses are inappropriate. The whole of work areas OPQF or ODEN in Fig. 4.8 have *not* gone to fracture work alone This important point is brought home if attempts are made to apply Equation (2.21) to this problem, because we end up with only the elastic $Rw/2$ component of the total load. This comment applies to a number of the problems and situations following in this chapter. It is such an important point that it is dealt with separately in section 4.6.4, in which will be found yet a further generalization of equation (2.21) to include plastic flow.

4.3.2 The soldered can lid problem

Certain containers such as some designs of sardine can have soldered lids and the action of opening such cans may be analysed by considering fracture through the solder layer (Fig. 4.9). (In other designs of container, it is the metal itself that is cut and torn when opening up; this is considered in section 4.4.)

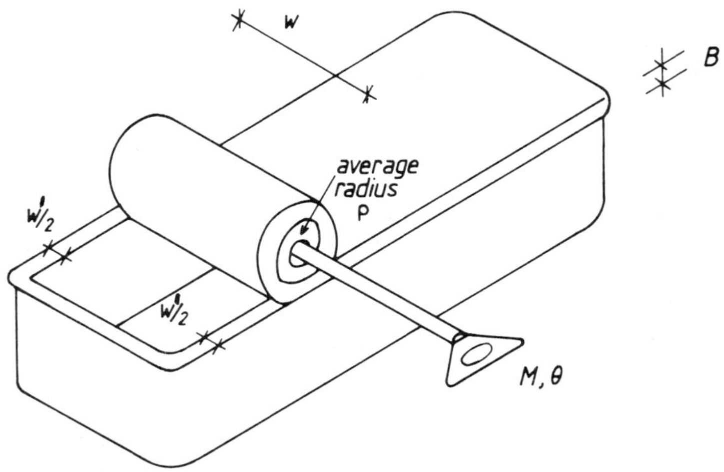

Fig. 4.9 – Opening of a soldered can lid.

The rolling-up of the lid on to the key consumes plastic work and the external incremental work done ($Md\theta$ rotating the key) has to provide that work in addition to the essential fracture work in the solder layer. Proceeding as before, and assuming an average wound-up radius ρ, $\bar{\epsilon}^P = B/4\rho$. (Unlike the glued strip tearing problem in section 4.3.1 where the strip takes up a 'free-standing' plastic curvature, the size of the key limits ρ and the plastic strain as plastic hinges are continuously collapsed on to it.) Using $dA = w'\rho d\theta$ and $dV^P = wB\rho d\theta$, equation (4.3) gives

$$Md\theta = (\sigma_Y B/4\rho)\, wB\rho d\theta + Rw'\rho d\theta$$

i.e.

$$M = \sigma_Y B^2 w/4 + Rw'\rho \qquad (4.15)$$

The moment is greater the greater the thickness of the metal and the wider the rolled-up tongue of metal. Note also that M increases as ρ increases, although in practice σ_Y falls during opening as $\bar{\epsilon}^P$ ($\propto 1/\rho$) is smaller.

4.3.3 Elastoplastic deformation of adherand

Consider the adhesive joint shown in Fig. 4.10(a) where a rectangular ductile metal beam of depth $2h$ and width w is glued to a rigid block. The beam is loaded by an end-couple which is increased until the joint fractures. Assume that the beam goes partially plastic before fracture, so that the applied moment produces uniformly along the beam an elastic core of height ξh on either side of the neutral

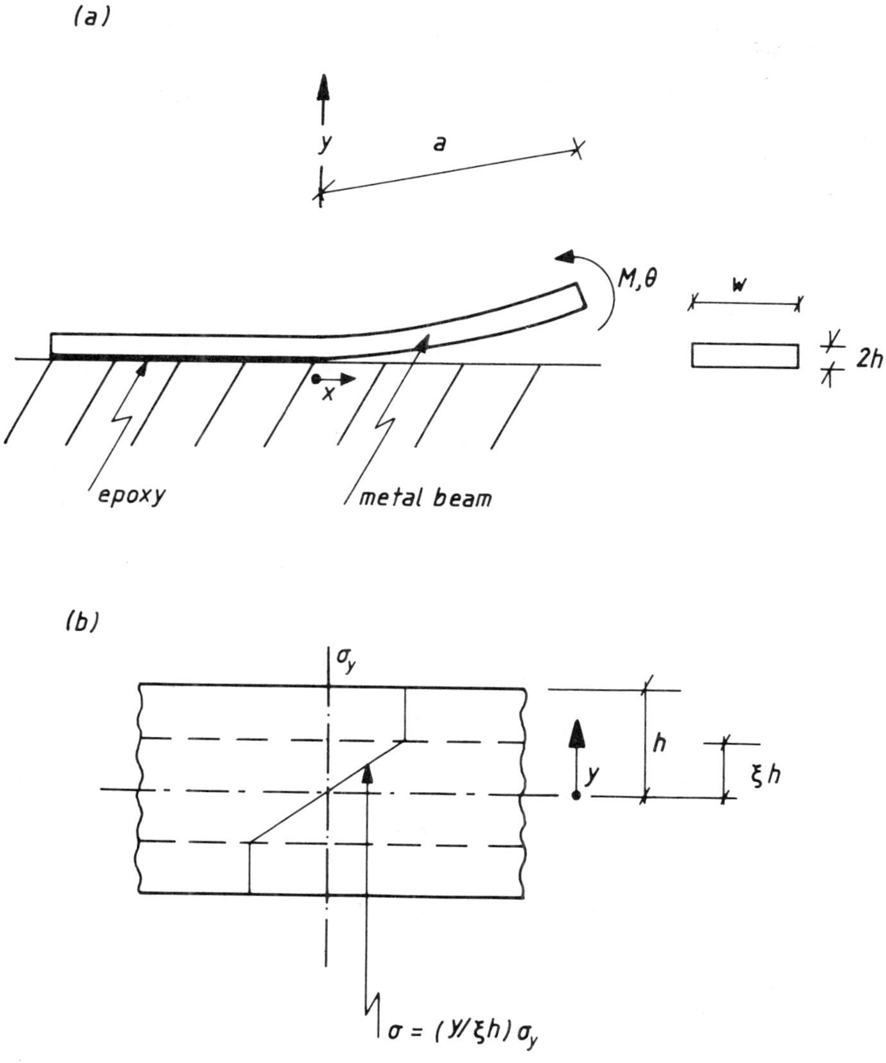

Fig. 4.10 – (a) Peeling of an elastoplastic metal–epoxy joint by an end couple. (b) Stress distribution across the beam depth showing the regions of elastic and plastic bending.

axis and plastic outer layers of height $(1-\xi)h$.† If the metal has a constant yield stress σ_Y, the stress distribution is as shown in Fig. 4.10(b), and the moment is given by

$$M = wh^2\,\sigma_Y(3-\xi^2)/3 \tag{4.16}$$

† This analysis is based on the more complicated treatment of a glued cantilever loaded by an end force given by Chang *et al.* (1972) which is referred to later.

In the elastic core the strain energy per unit volume is $(\sigma^2/2E) = (y^2\sigma_y^2/2\xi^2h^2E)$. There will also be elastic energy in the plastic regions, the density of which is constant at $(\sigma_y^2/2E)$. Thus, the total elastic strain energy is

$$\Lambda = \int W \mathrm{d}V$$

$$= \int_0^a \int_{-\xi h}^{+\xi h} (y^2\sigma_y^2/2\xi^2h^2\,E)\,w\mathrm{d}y\mathrm{d}x + 2\int_0^a \int_{\xi h}^{h} (\sigma_y^2/2E)\,w\mathrm{d}y\mathrm{d}x$$

$$= \frac{w\sigma_y^2\,\xi ha}{3E} + \frac{w\sigma_y^2\,(1-\xi)\,ha}{E}$$

$$= w\sigma_y^2\,ha\,(3-2\xi)/3E \tag{4.17}$$

In the plastic parts, the work done per unit volume is $\int \sigma_y \mathrm{d}\epsilon$ where the strain level depends on position in the beam. With the usual assumptions, the total strain is $\epsilon = y/\rho$ where ρ is the deformed radius of the beam. ρ may be obtained from conditions at the elastic/plastic interface using elastic beam theory, i.e. $1/\rho = M/EI$. Thus, with M from equation (4.16) we obtain $1/\rho = \sigma_y/E\xi h$, which is constant. In the plastic region ($|y| > \xi h$), the total strain is $\epsilon = y/\rho = \sigma_y y/E\xi h$. The plastic strain is this total strain less the yield strain σ_y/E, so we obtain

$$\Gamma = \int W^{\mathrm{P}}\,\mathrm{d}V^{\mathrm{P}}$$

$$= 2\int_0^a \int_{\xi h}^{h} \sigma_y\,(\sigma_y y/E\xi h - \sigma_y/E)\,w\mathrm{d}y\mathrm{d}x$$

$$= w\sigma_y^2 h\,(1-\xi)^2\,a/E\xi \tag{4.18}$$

If fracture along the glued interface is occurring, we have for the next increment of deformation (given by the rotation of the beam $\mathrm{d}\theta$),

$$M\mathrm{d}\theta = \mathrm{d}(w\sigma_y^2\,ha\,(3-2\xi)/3E) + \mathrm{d}(w\sigma_y^2\,h(1-\xi)^2\,a/E\xi) + Rw\mathrm{d}a \tag{4.19}$$

using equation (4.2).

This equation may be solved knowing $\mathrm{d}\theta$ in terms of $\mathrm{d}a$. From the elastic-plastic interface and the curvature of the beam ($1/\rho = \sigma_y/E\xi h$), together with $\mathrm{d}a = \rho\mathrm{d}\theta$, we have $\mathrm{d}\theta = \sigma_y\,\mathrm{d}a/E\xi h$. Then

$$M = w\xi h^2\,\sigma_y\left(\frac{(3-2\xi)}{3} + \frac{(1-\xi)^2}{\xi}\right) + \frac{RwE\xi h}{\sigma_y}$$

i.e.

$$M^2 = \frac{REIw}{2}\;\frac{(3-\xi^2)^2}{(3-2\xi)} \tag{4.20a}$$

or

$$M = wh^2\sigma_y\left((1 - \frac{3}{4}(1 - RE/\sigma_y^2 h)^2)\right) \tag{4.20b}$$

or again

$$R = (h\sigma_y^2/E)\,(1 - 2\xi/3) \tag{4.20c}$$

using equation (4.16) for σ_y (and ξ). Crack propagation occurs at constant moment, independent of the length of the beam. Note that for given geometry (w, h) and given materal properties (σ_y, E, R) there is a unique value for M during elastoplastic cracking given by equation (4.20b). ξ in equation (4.20a) and (4.20c) is a dependent variable, taking the value set by equation (4.16) for the relevant level of M.

Example 4.6 Determine the magnitudes of M predicted by equation (4.20a) for $\xi = 1$ and $\xi = 0$ and comment on an apparent paradox in their relative values.

Ans. When $\xi = 1$, and the outside fibres are just on the point of yielding, $M^2 = 2\,REIw$ (cf. the elastic solution for a DCB testpiece with *two* arms storing strain energy (here there is only one) which is $M^2 = REIw$, using w for 'thickness'). If cracking occurs just as the elastic core disappears and full plastic collapse is about to occur, $\xi = 0$ and $M^2 = 3REIw/2$ from equation (4.20a). This latter moment for $\xi = 0$ is apparently less than that for earlier in the deformation at $\xi = 1$, which seems odd. However, these M values are *not* those that a given beam geometry/material combination experiences during progressive deformation from the unloaded state, i.e. cracking at $\xi = 0$ necessarily is preceded by prior elastic deformation at $\xi = 1$ when the applied moment is *smaller* (not, as here, apparently *larger*). The M values in equations (4.20a) and (4.20b) are merely the unique values at which fracture occurs and they depend on the beam geometry/material combination; if, for example, a plastic hinge fully develops before any fracture occurs, the beam will simply curl up plastically and no fracture takes place at all. Rearrangement of equations (4.20b) and (4.20c) gives $M^2 = (h\sigma_y^2/E)\,(1 - 2\xi/3)\,(EIw/2)\,(3 - \xi^2)^2/(3 - 2\xi)$ which for $\xi = 1$ becomes $(4/9)\,w^2h^4\sigma_y^2$ using $I = 2wh^3/3$ and for $\xi = 0$ becomes $w^2h^4\sigma_y^2$. These magnitudes now have the sensible ranking. This point is part of the wide field of 'fracture transitions' dealt with in Chapter 10; further clarification of this particular glued beam problem will be found in section 4.6.

Notice that equation (4.20) may be adapted to the soldered sardine can problem, when the metal is more springy and is not so soft that it collapses tightly on to the key. We have, using B for $2h$ and w' for the width of the soldered joint (Fig. 4.9)

$$M = \frac{w\xi B^2}{4}\,\sigma_y\left(\frac{(3-2\xi)}{3} + \frac{(1-\xi)^2}{\xi}\right) + \frac{Rw'E\xi B}{2\sigma_y}$$

i.e.

$$M^2 = REIw'\,(3 - \xi^2)/2(3 - 2\xi) \tag{4.21a}$$

or

$$M = \frac{wB^2\sigma_y}{4}\left(1 - \frac{3}{4}\,(1 - \frac{2RE}{\sigma_y^2 B}\,\frac{w'}{w})\right)^2 \tag{4.21b}$$

or, again

$$R = \frac{B\sigma_y^2}{2E}\left(\frac{w}{w'}\right)(1 - 2\xi/3) \tag{4.21c}$$

as for the corresponding set of equations (4.20). These relations should be contrasted with equation (4.15) in which rigid-plastic flow and collapse on to the winding key has been assumed *ab initio*.

Example 4.7 Determine the contribution of the elastic term in equation (4.19) in relation to the incremental total energy, and compare the result with analyses (such as those leading to equations (4.14) or (4.15)) which neglect elasticity.

Ans. The incremental elastic term in equation (4.19) is $w\sigma_y^2 h(3-2\xi)/3E$. Its contribution to M, as a proportion of the total, is given by

$$w\sigma_y^2 h(3-2\xi)/3E \quad / \quad w\sigma_y^2 h(3-2\xi)/3E + w\sigma_y^2 h(1-\xi)^2/E\xi \;=\; (3\xi-2\xi^2)/(\xi^2-3\xi+3)$$

For $\xi = 1$ (the beam remains just completely elastic), the contribution is 100%; for $\xi = 0.8$, 90%; $\xi = 0.5$, 57%; $\xi = 0.2$, 21%; $\xi = 0.1$, 10%; $\xi = 0$ (plastic collapse), 0%. This shows that the elastic contribution can still be appreciable even in the presence of extensive plastic flow, and that therefore neglect of elasticity may be in error in some problems. Further considerations of these matters (including the effects of residual strain energy on unloading) will be found in section 4.6.

Chang, de Vries and Williams in 1972 solved the more complicated glued cantilever problem where the beam is loaded by an end force X rather than moment M. As the bending moment now increases along the length of the beam, the elastoplastic interface changes, i.e. ξ varies along the beam. They give the complete solution for a cantilever beam which remains elastic at the load end and is partially plastic at the root end, (Fig. 4.11). The method of solution is similar but the equation of the deflexion curve has to be determined (with varying ρ) in order to substitute for du in terms of da. Problems 4.11 and 4.14 give the solution for the equation over the whole length of the beam. Problem 4.13 gives the resulting expression for R in terms of X, σ_y and the geometry of the beam.

Experiments were performed by Chang and co-workers on beams (approximately 3 mm × 25 mm × 230 mm long) of 6061-T6 aluminium alloy (which very nearly displays an elastic–perfectly plastic $\sigma\epsilon$ curve with $E = 71$ GPa and $\sigma_y = 284$ MPa), in order to confirm their analysis. The beam length was varied and X measured at first cracking. Substitution in their ($X, a, R, \sigma_y \ldots$) relation for an elastoplastic cantilever (given in Problem 4.13) gave a constant value of R of 1.5 kJ/m^2 for the particular structural adhesive employed (3M Company 2216 B/A), which corresponds with the same value for R (at comparable crack speed, temperature and environment) obtained under fully elastic behaviour in the adherands. In this way, they demonstrated that the remote irreversible yielding (which has nothing to do with the process of fracture of the glued

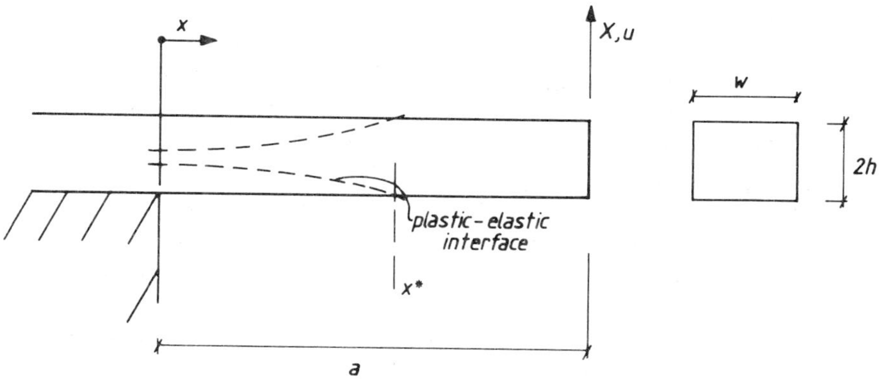

Fig. 4.11 – Peeling of an elastoplastic metal–epoxy joint by an end force X (after Chang *et al.* (1972), with permission).

joint and which arises simply because of the properties and geometry of the beam) could be accounted for and the ('true') toughness of the joint measured. Vice versa, it is possible to predict the loads and deflexions in the plastic range at which glued metal components will fail. They went on to discuss the use of the method of determine the toughness of explosively-welded metal interfaces.

If the beam material work hardens in these problems the picture is modified, of course, but the same general principles apply.

4.4 CRACKING IN MONOLITHIC BODIES IN THE PRESENCE OF EXTENSIVE REMOTE FLOW

In the glued beam type of problem in section 4.3.3, the crack path is predetermined (along the glue line) and, of particular importance, the resistance to fracture R along that path may be determined independently. For example, it could be arranged for R of the glue to be measured under 'elastic' fracture conditions (without the complications of remote yielding), such as described in Chapters 2 and 3, by choice of beam materials and geometries that do not yield. There do not, however, seem to have been many elastoplastic studies of this sort, because even when R along a glue line is known *a priori*, most crack geometries of interest are more complicated than the simple beam examples described here and the associated elastoplastic analyses are difficult although, of course, it is always open to the experimentalist to perform loading/unloading tests. But in most designs of glued joints, the adherands would be expected to behave elastically. Even so, the examples in section 4.3 are valuable in that they demonstrate the applicability of equation (1.2) for elastoplastic problems.

In elastoplastic and fully plastic cohesive crack propagation in solid pieces of one material, there is no artificial line of fracture with known crack resistance. For materials possessing moderate-to-high K/σ_y ratios, it is often very difficult or even impossible in the laboratory to produce elastic fractures, which would give measures of R, owing to the scale-effect. As we shall see, it is possible in some circumstances to make corroboratory measurements of R in different crack geometries of the same material, and partition out the remote work and recoverable elastic work to leave the fracture work. The general problem of partitioning energy components in workhardening metals is, however, formidable. In fact there have been comparatively few studies along these lines since, as explained in section 4.1, design applications usually concern only 'small-scale yielding' situations. Modified elastic fracture mechanics procedures or critical COD criteria are often employed (section 4.1) or, where 'new' criteria of elasto-plastic fracture have been proposed (particularly a critical value for the J-integral (section 4.5)), they have been based on different premises. A discussion of these points is given later, but we start this section by describing experiments on monolithic materials, which are logical extensions of the experiments in section 4.2 in the application of equation (1.2) to crack propagation in elastoplastic and fully plastic fields.

4.4.1 Fracture in ductile DEN sheets

If a deep double-edge-notched sheet yields completely before fracture, the plastic region is confined to an almost circular patch, with the ligament as a diameter (Fig. 4.12(a)). Thus in these circumstances $V^P = (\pi L^2/4)B$ and we note that V^P is less than the total volume of the body. With increase of load, necking eventually occurs in the line of the starter cracks – at first at the crack tips themselves, after which the necked-down zones work their way across the specimen to meet one another. As this occurs, so crack propagation follows the necks in the path of reduced thickness. Cracks will start to propagate, in general, before the thinned-down necks meet in the centre of the plate, as illustrated in Fig. 4.12(b).

As cracking in these conditions is impossible without preceding neck formation, it is proper to think of the whole necked region at the crack tip as the fracture process zone. The R associated with fracture in these circumstances will include contributions from work of necking as well as actual fracture since cracking is impossible without necking.

Experiments show that along the line of cracking the strains are highest at the crack tip, and as the ligament shortens as the crack propagates, these high strains are eventually experienced by all particles in the region of the fracture surface. Even though the plastic strain distribution and strain history in the circular patch is complex as the crack propagates, by the time the testpiece has fractured right across, it is possible to talk of a final average plastic strain $\bar{\epsilon}_{av}$ for the whole deformation. If the associated average flow stress is $\bar{\sigma}_{av}$, $W^P = \bar{\sigma}_{av}\,\bar{\epsilon}_{av}$ and we have the total work W_T consumed by the end of deformation (assuming rigid-plastic behaviour since the elastic contribution is small)

$$X\mathrm{d}u = W_T = (\bar{\sigma}_{av}\,\bar{\epsilon}_{av})\,(\pi L^2/4)B + RBL \tag{4.22}$$

where the right-hand side is already integrated up as we are considering the *final* conditions. Dividing by BL, we have

$$W_T/BL = w_T = (\bar{\sigma}_{av}\,\bar{\epsilon}_{av})\,(\pi/4)\,L + R \tag{4.23}$$

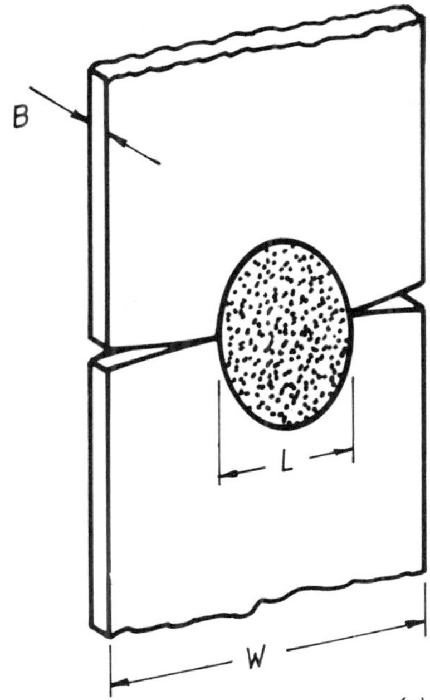

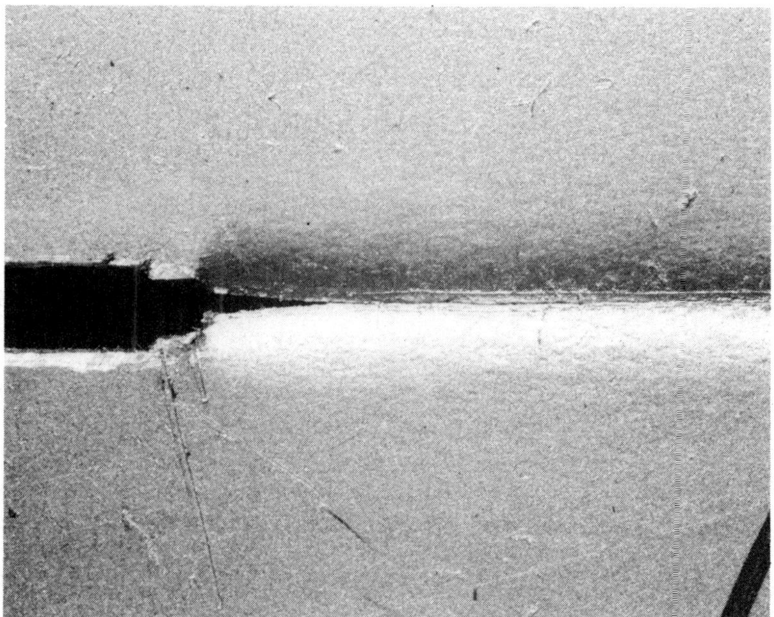

Fig. 4.12 – (a) The deep edge notched tension specimen. The shaded circular region undergoes yielding when stretched. (b) Necking at crack tip when the crack propagates.

This suggests that experimental total work areas normalised by BL should plot as a straight line against starting ligament length L, with the ordinate intercept giving R. This neat result is due to Cotterell and Reddell (1977). Plots for a 0.1% carbon steel sheet, possessing varying degrees of prestrain, are shown in Fig. 4.13. The model is seen to work very well. The effect of prestrain is to lower the slope (there is less plastic work done per unit volume in the circular patch) and, according to Fig. 4.13, all lines back-extrapolate to approximately the *same R* which, for this material, seems therefore to be independent of prior cold work. (This is not true in general, however, as we shall see, and depends upon the workhardening index – materials with greater n in $\bar{\sigma} = \sigma_0 \, \bar{\epsilon}^n$ displaying different R with degree of cold work.)

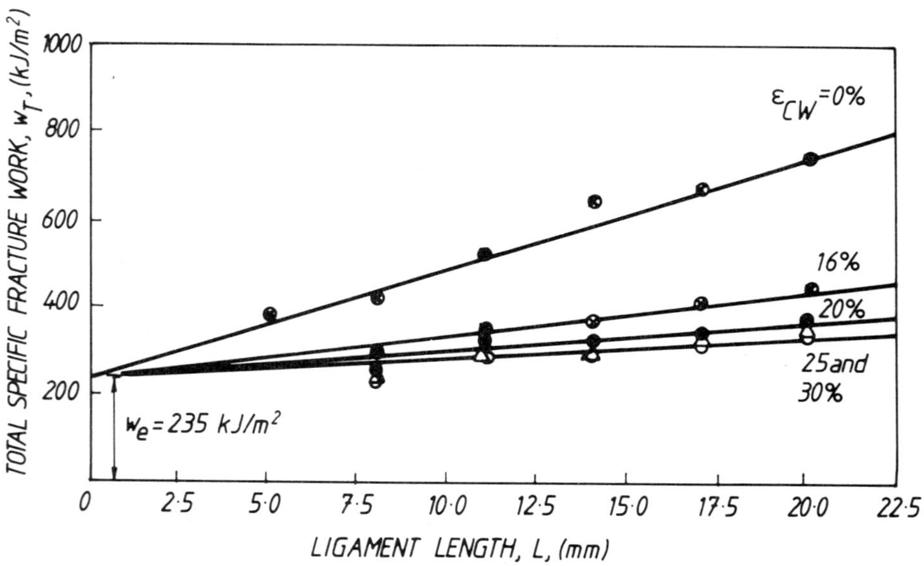

Fig. 4.13 – Plots of specific fracture work against ligament for a low carbon steel subjected to various pre-strain levels (after Mai *et al.* (1979), with permission).

As in earlier examples in this chapter, it is clear that as far as the separation of the notched panel into two parts by tensile pulling is concerned, only some of the work is actually required for fracture and the rest is remote flow which occurs concurrently owing to the geometry of the body and the properties of the material. Cotterell has partitioned w_T into the specific *essential* work of fracture w_e and specific *non-essential* plastic work w_P. In equation (4.23), $w_e \equiv R$ and $w_P \equiv (W^P \pi/4)L$. As already explained, w_e here will include work of necking over and above specific work of fracture, since fracture is impossible without necking in the presence of remote flow.

The simple linear plot of normalised total work against ligament length predicted by equation (4.23) and shown in Fig. 4.13, is similar to the relations given in equations (4.14) and (4.15) for fracture of adhesive or soldered joints when there is gross plastic flow in the adherands. However, whereas the geometry shown in Fig. 4.8 produces steady-state cracking at constant load or moment (forgetting the transient start to the process), fracture in ductile DEN sheets is a non-steady-state problem, with the load rising to a peak and falling again. Cotterell and Reddell got round this difficulty in 1977 by considering total work consumed, and equation (4.23) concerns normalised work, not normalised steady force or moment. Later, Cotterell and Mai (1982) presented an incremental analysis for the non-steady deformation of ductile DEN sheets.

Experimental evidence suggested the following assumptions for their analysis:

(i) before crack initiation, the stretch at the crack tip δ is the same as the remote elongation u of the whole sheet (i.e. the displacement of clip gauges on the specimen);

(ii) necking occurs when the longitudinal strain in the sheet reaches a critical value $\bar{\epsilon}_u = n$ (the workhardening rate in $\bar{\sigma} = \sigma_0 \bar{\epsilon}^n$); for sheets prestrained by $\bar{\epsilon}_{cw}$ the necking strain is $(n - \bar{\epsilon}_{cw})$ for $\bar{\epsilon}_{cw} < n$; for $\bar{\epsilon}_{cw} > n$ necking occurs immediately;

(iii) crack initiation occurs when $\delta = \delta_c$ and the COD at the tip of the propagating crack is constant and equal to δ_c, the shape of the fracture process zone not changing during propagation;

(iv) even in the case of prestrained samples, the stretch required to produce fracture after necking is constant and equal to $(\delta_c - e_u d)$ where e_u is the engineering strain equivalent of $\bar{\epsilon}_u$ (i.e. $e_u = \exp \bar{\epsilon}_u - 1$) and d is the height of that part of the undeformed sheet that later forms the process zone (Fig. 4.14); d is always of the order of the sheet thickness according to Hill's 1952 analysis of necking between notches in sheets.

The specific essential work of fracture, which takes place in the process zone, may be divided into two components w_{e1} and w_{e2} which relate to events preceding neck formation and to subsequent events in the necked down zone in which the crack eventually runs. Since there is uniform straining up to the strain $\bar{\epsilon}_u$ at necking and since the plastic volume that forms the necked process zone is *LBd*,

$$w_{e1} = \left(\int_0^{\bar{\epsilon}} \bar{\sigma} \mathrm{d}\bar{\epsilon} \right) (LBd)/(LB)$$

$$= d\sigma_0 \, n^{n+1}/(n+1) \tag{4.24}$$

using $\bar{\sigma} = \sigma_0 \bar{\epsilon}^n$. After necking, the deformation in the process zone is localised and non-uniform, so instead of attempting to accommodate the complex non-uniform plastic flow field, we employ the actual displacements in the neck and the functional dependence of σ and δ (which may be obtained from independent tension specimens with shallow notches which force necking normal to the applied stress). The displacement as necking commences is $e_u d = (\exp n - 1) d$ and at fracture it is δ_c, so

$$w_{e2} = \int_{(\exp n - 1)d}^{\delta_c} \sigma(\delta) \, . \, \mathrm{d}\delta \tag{4.25}$$

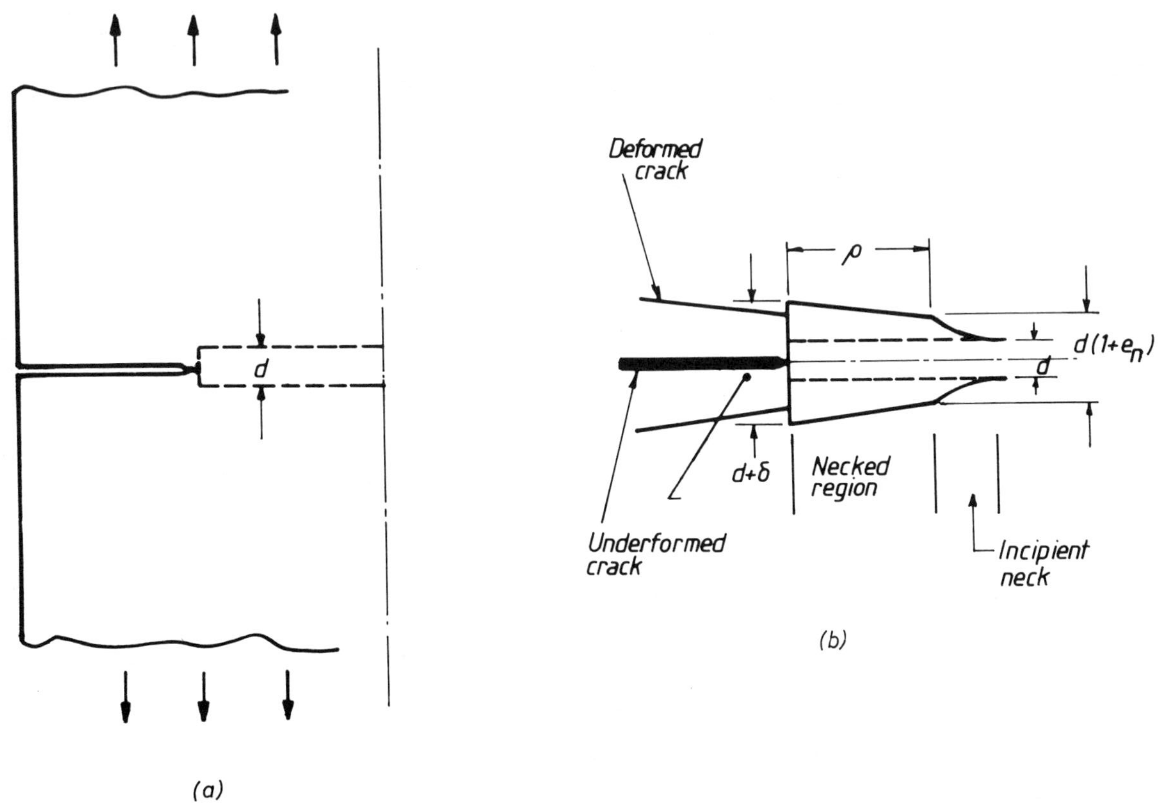

Fig. 4.14 – (a) The width of the zone in the ligament which later necks and becomes the fracture process zone (b) Details of the deformation geometry at the crack tip. Both necked and incipient neck zones are shown.

Thus

$$w_e = w_{w1} + w_{e2} = d\sigma_0 n^{n+1}/(n+1) + \int_{(\exp n-1)d}^{\delta_c} \sigma(\delta)\,.\,d\delta \qquad (4.26)$$

Example 4.8 A low carbon steel sheet of thickness 1.6 mm displays $\bar{\sigma} = 465\bar{\epsilon}^{0.10}$ MPa. Experiments show that fracture in DEN sheets occurs at $\delta_c \approx 0.73$ mm and that $\sigma(\delta)$ after necking is approximately constant and numerically equal to the UTS of 360 MPa. The necked region is about 1.2 mm across. Calculate w_e and comment on the answer.

Ans. $w_e = w_{e1} + w_{e2} = (1.2 \times 10^{-3})(465 \times 10^6)\,0.1^{1.1}/1.1 + 360\,(0.73 - 0.13)10^{-3} \approx 40 + 216 = 256$ kJ/m^2. This value compares well with the R intercept of 235 kJ/m^2 in Fig. 4.13 for annealed sheet. We observe that w_{e1} is only some 16% of w_e overall.

When the sheet possesses a prestrain of $\bar{\epsilon}_{cw}$ (less in magnitude than the annealed necking strain n) the effect is to reduce w_{e1} by changing the lower limit from zero (annealed material to which $\bar{\sigma} = \sigma_0 \bar{\epsilon}^n$ refers) to $\bar{\epsilon}_{cw}$. Since the stretch after necking remains approximately constant at $(\delta_c - e_u d)$, w_{e2} is unaffected. Thus

$$_{cw}w_e = (\sigma_0 d/(n+1))\,(n^{n+1} - \epsilon_{cw}^{n+1}) + \int_{(\exp n-1)d}^{\delta_c} \sigma(\delta)d\delta \qquad (4.27)$$

When $\bar{\epsilon}_{cw} \geqslant n$ ($\approx 10\%$ reduction of area for $n = 0.1$), $w_{e1} \to 0$ and we have

$$_{cw}w_e = \int_0^{\delta_c} \sigma(\delta)\,.\,d\delta \qquad (4.28)$$

Experiments by Mai and Cotterell (1980) on 1.02 mm thick α-brass sheet (displaying $\sigma = 625\ \epsilon^{0.32}$ MPa in the as-received condition) permit closer inspection of the influence of $\bar{\epsilon}_{cw}$ on the mechanics of crack propagation in ductile DEN sheets. The material was cold worked by (uniform) uniaxial stretching up to 35% reduction of area. As before, $\sigma(\delta)$ proved to be approximately constant at the UTS of 310 MPa; $\delta_c \approx 0.90$ mm and $d \approx 1$ mm. The stretch of the zone that forms the neck is $(\exp 0.32 - 1)\,d = 0.38$ mm just at necking. Thus, for the as-received annealed sheet, equation (4.26) gives

$$w_e = w_{e1} + w_{e2} = (10^{-3})(625 \times 10^6)\,0.32^{1.32}/1.32 + (310 \times 10^6)$$
$$(0.90 - 0.38)\,10^{-3} = 105 + 162 = 267 \text{ kJ/m}^2$$

w_{e1} is no longer small compared with w_{e2} owing to the greater uniform deformation prior to neck formation.

For a specimen with, say $\bar{\epsilon}_{cw} = 0.20$, equation (4.27) gives

$$_{0.20}w_e = [(10^{-3})(625 \times 10^3)/1.32]\ (0.32^{1.32} - 0.20^{1.32}) + (310 \times 10^6)$$
$$(0.90 - 0.32)\ 10^{-3} = 49 + 162 = \underline{211\ \text{kJ/m}^2}$$

Table 4.2 gives w_e for a variety of prestrains and compares them with experimental values obtained from Fig. 4.15. It is seen that w_{e2} remains constant at 162 kJ/m^2 independent of $\bar{\epsilon}_{cw}$, and is the value taken by all material cold-worked greater than 38% reduction of area. It is clear that for solids like α-brass showing appreciable workhardening before necking, w_e decreases with increasing prestrain. In contrast, low carbon steels and some aluminium alloys with low strains to necking show effectively constant w_e.

Table 4.2

Comparison of predicted and experimental essential work of fracture for annealed and prestrained Austral 207 (30/70) brasses (after Mai and Cotterell (1980), with permission).

		w_e (kJ/m^2)			
e_{cw}	ϵ_{cw}	experimental	predicted	w_{e1}	w_{e2}
0	0	262	267	105	162
0.05	0.05	250	258	96	162
0.10	0.10	240	244	82	162
0.15	0.14	220	231	69	162
0.20	0.18	195	217	56	162
0.25	0.22	170	202	40	162
0.35	0.30	150	162	0	162

Two further observations should be made which affect the detailed interpretation of the DEN measurements, (i) the assumption of a constant δ_c and (ii) the behaviour of specimens with very small ligament lengths.

While the assumption of a constant δ_c is acceptable for materials with reasonably high workhardening indices (where the fracture resistance, R, is practically independent of crack growth, Δa, such as observed in Lyten for which $n = 0.25$ (Cotterell and Mai, 1982)), in sheet materials with lower n values, and consequently larger necked zones, initiation occurs at a *earlier* reduction in thickness than does propagation. This means that there are *two* critical crack opening displacements, viz. δ_i for initiation and δ_p for propagation; δ_i increases to δ_p through the distance taken for the reduction in sheet thickness to reach a maximum (Fig. 4.14). R increases from an initiation value to a larger propagation value over the first stages of growth – exactly as encountered in 'R-curves' (section 4.2.4) and J_R-curves (section 4.5 later). It is the larger

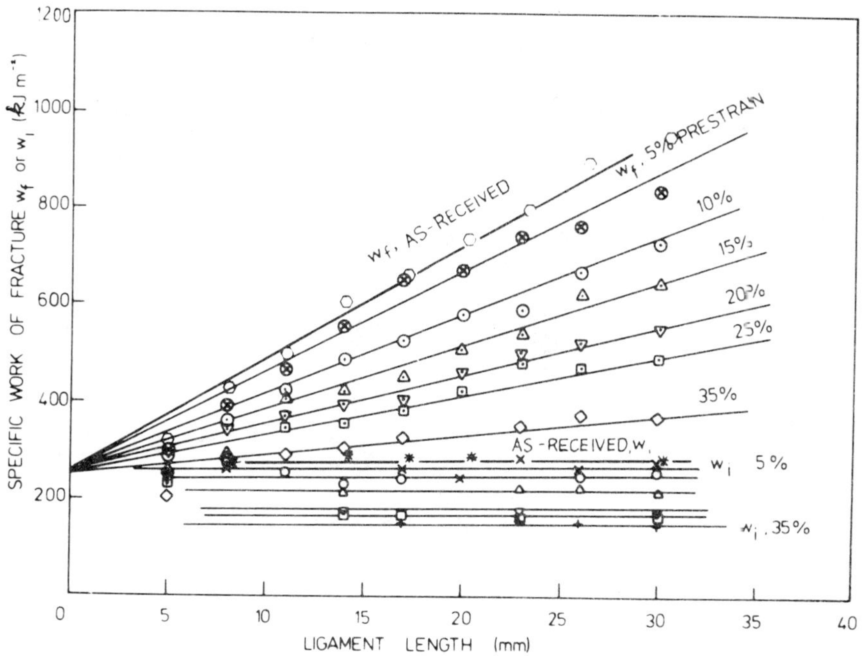

Fig. 4.15 – Plots of specific work of fracture against ligament for Austral 207 (70/30) brasses with varying amounts of prestain. (1 mJ/mm² = 1 kJ/m²) (after Mai and Cotterell (1980), with permission).

propagation R-value to which equation (4.23) back-extrapolates; the lower initiation value corresponds with the specific work of initiation $(w_i^e + w_i^p)$. Fig. 4.16 displays these features for a half-hard aluminium alloy AA2S (which has a reduction in thickness of 24% at initiation and 76% for propagation) and for a low alloy steel (with corresponding values of 12% and 36%, respectively). Initiation and propagation values on the ordinate are distinct for the aluminium alloy in Fig. 4.16(a), but are essentially the same in the steel shown in Fig. 4.16(b).

Figs. 4.16 also show that when the ligament length L is small, w_T vs. L data fall below the lines and curve down to intersect the ordinate at smaller values than R. This occurs because when $L \approx 2\rho_c$ (where ρ_c = length of necked zone (Fig. 4.14)) there will be a transition from plane stress to plane strain, and owing to the extra constraint in plane strain, the work of fracture takes on lower values.

The w_e and w_P components which make up w_T have each been broken down into further components relating to the initiation and propagation phases in Fig. 4.16. Thus there are (i) w_e^i and (ii) w_e^p which represent respectively (i) the localised irreversible work required to generate the necked-down crack tip process zone (which is a necessary prelude of initiation before propagation in ductile fracture of thin sheets) and (ii) the toughness work required to propagate the crack as the crack tip process zone moves along. Again there are (iii) w_p^i and

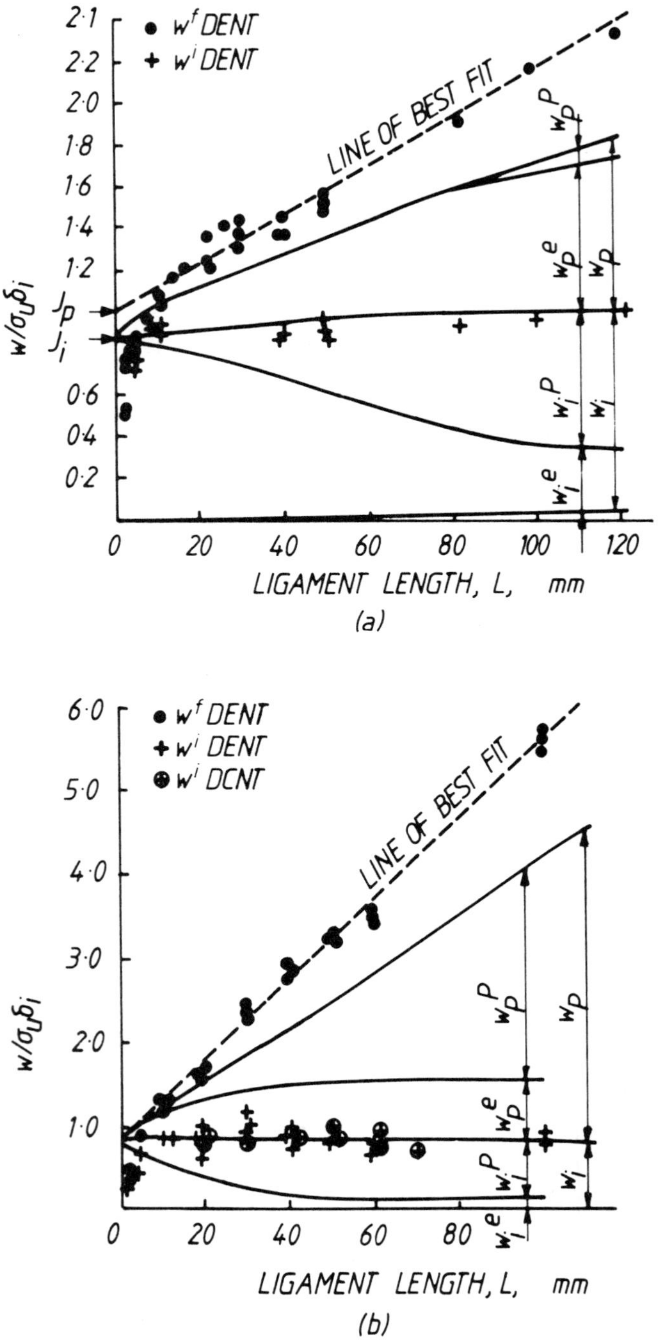

Fig. 4.16 – Specific works of initiation and fracture for (a) AA2S and (b) Lyten, showing the four work components at each ligament length (after Cotterell and Mai (1982), with permission).

(iv) w_P^p giving respectively (iii) the plastic work done in the circular patch before propagation and (iv) the remote plastic work during propagation. For futher details of partitioning of total work, the interested reader is referred to Cotterell and Mai (1982).

The fracture of DEN sheets of polymers may be analysed by the same methods. Fig. 4.17 shows results on ultra-high molecular weight polyethylene

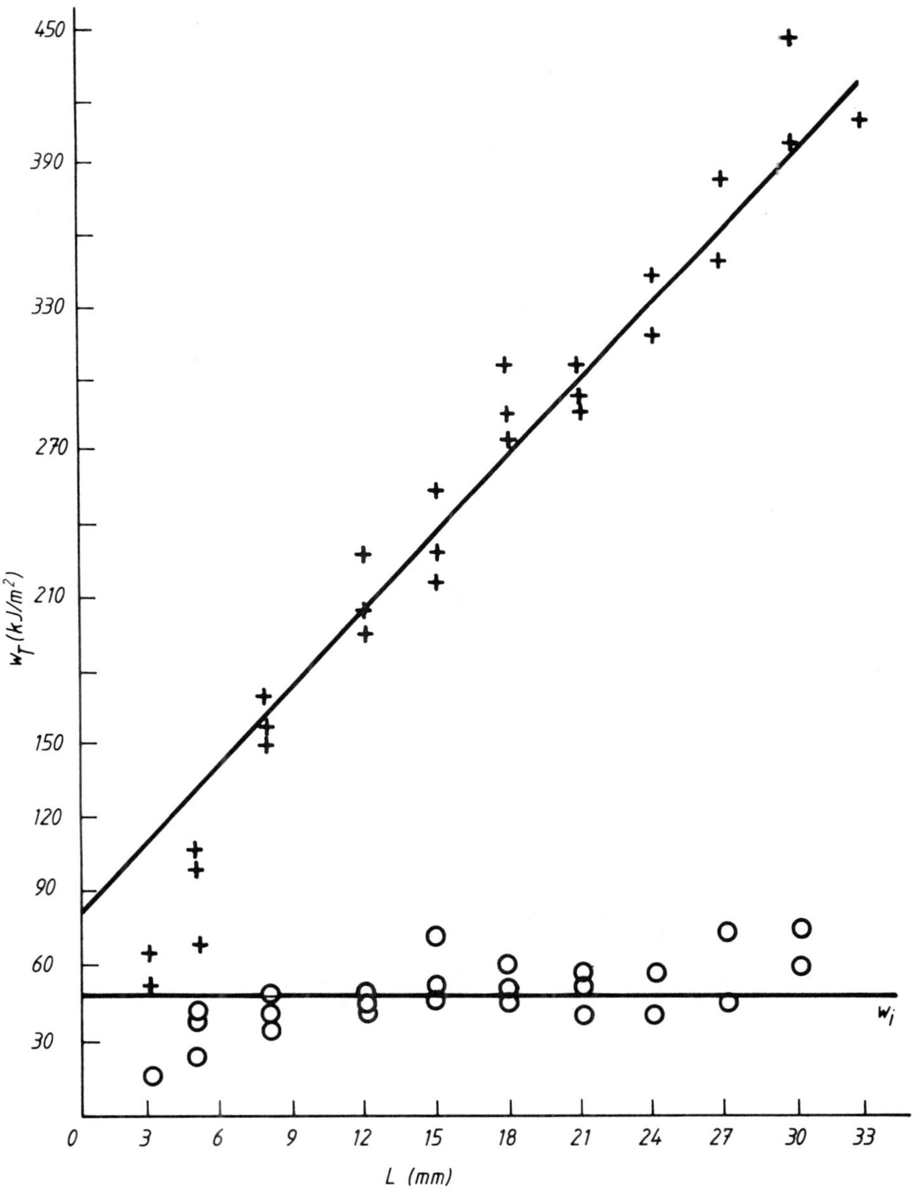

Fig. 4.17 – Plot of specific work of fracture against ligament for an ultra-high molecular weight polyethylene at 20 ± 2°C (after Mai *et al.* (1984), with permission).

(UHMWPE) obtained by Mai and Cotterell (1985) with two special essential fracture work values. One is about 50 kJ/m^2 for crack initiation and the other is ~80 kJ/m^2 for crack propagation. Results for a high impact polystyrene (HIPS) given by Yap *et al.* (1983) are shown in Fig. 4.18 with $w_e \approx 1.1$ kJ/m^2. Note that this is a plane strain toughness value, (not plane stress as for metals and UHMWPE discussed above), because the data obtained are in the plane stress–plane strain transition region. This value is in good agreement with the plane strain, R, measured independently (Yap *et al.*, 1983). Note also that for ligament

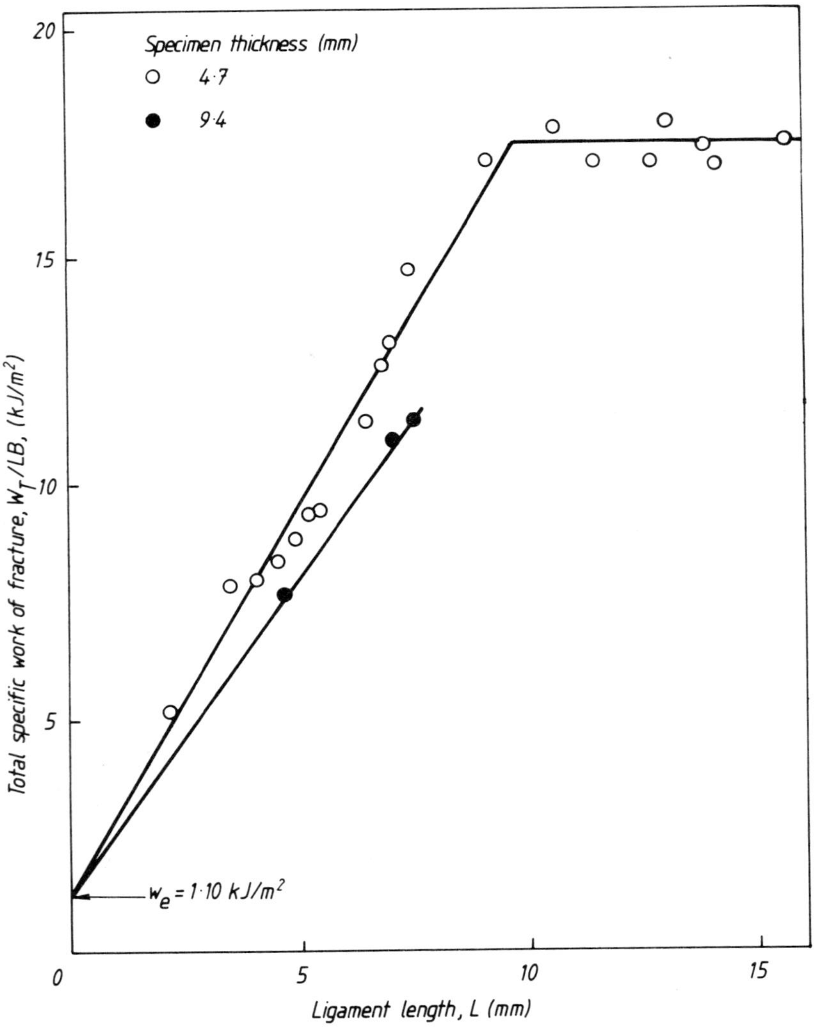

Fig. 4.18 – Plot of specific work of fracture against ligament for a high impact polystyrene (HIPS) at 20 ± 2°C (after Yap *et al.* (1983), with permission).

lengths larger than 10 mm, plane stress conditions prevail and crack initiation occurs prior to the full ligament being completely crazed. Of particular interest is that in the thicker, 9.4 mm, sheets it was impossible to produce ductile fractures for ligament lengths longer than about 8 mm since the elastic fracture stress was now lower than the plastic yield stress. This is an example of a deformation transion, the topic of Chapter 10.

Experiments may also be performed on testpieces with staggered starter cracks (Fig. 4.19(a)), which produce interesting fractures in mixed modes I and II along the inclined neck which forms between the tips of the starter cracks. For an isotropic material following the Maxwell–von Mises yield criterion, the direction of relative motion of material across the neck is at an angle $\psi = \tan^{-1}\{(\cot\theta)/4\}$ (Fig. 4.19(a)). The specific total work input w_T may thus be separated into a mode I component w_{TI} for the displacement component perpendicular to the neck and a mode II component w_{TII} for the displacement component parallel to the neck. It is easily shown that

$$R_I = w_{TI} = (1/BL)\,Xu_{rel}\sin\psi\cos\theta$$

$$\text{and } R_{II} = w_{TII} = (1/BL)Xu_{rel}\cos\psi\sin\theta$$

where u_{rel} is the displacement along direction ψ; w_T is of course $Xu_{rel}\sin(\theta+\psi)$. The ratio of R_{II}/R_I is $\tan\theta/\tan\psi$, i.e. $4\tan^2\theta$.

The manner in which the combined R (i.e. $R_I + R_{II}$), measured from testing machine crosshead load and displacement, varied with inclined ligament length is shown in Fig. 4.19(b)), and the subdivision into $(R_I + R_{II})$ is given in Fig. 4.19(c), all for low carbon steel. It appears that $R = R_I + R_{II}$ is essentially constant in this material, but this particular result is not general for materials with lower workhardening capacity. Further work by Mai and Cotterell (1983) has shown that if the workhardening exponent (n) is low and necking is intensive, such as in aluminium alloy B1200-H14 where $n \approx 0.045$, R varies with the staggered angle (θ) and decreases to a minimum at $\theta = 90°$. For materials such as Lyten and low carbon steel ($n = \sim 0.2$), where necking is moderate and n is comparatively large, R is constant being independent of θ (cf. Fig. 4.19(b) and 4.19(d)).

We note that solutions for fully-plastic and elastoplastic deformation of DEN testpieces may be evaluated from the GE/EPRI studies described in section 4.5.4. A comparison between those J_R-based analyses and the analysis here, and the particular experimental sizes used, has yet to be performed. We also note that an equation like (4.23) seems to apply to elastoplastic fracture of scaled three-point bend bars (section 9.6).

An alternative method of analysis for plane stress ductile fracture in sheets is Andrews's generalised theory of fracture in which work inputs are treated in a multiplicative, rather than additive, manner. For example, Andrews and Bhatty (1982) found that the specific essential work of fracture in 0.62 mm thick low carbon steel was 240 kJ/m² in tests where $\dot{u} \approx 10$ µm/s and 100 kJ/m² for $\dot{u} \approx 200$ µm/s; for 0.88 mm thick 18/8 stainless steel sheet at $\dot{u} \approx 40$ µm/s, it was 570 kJ/m².

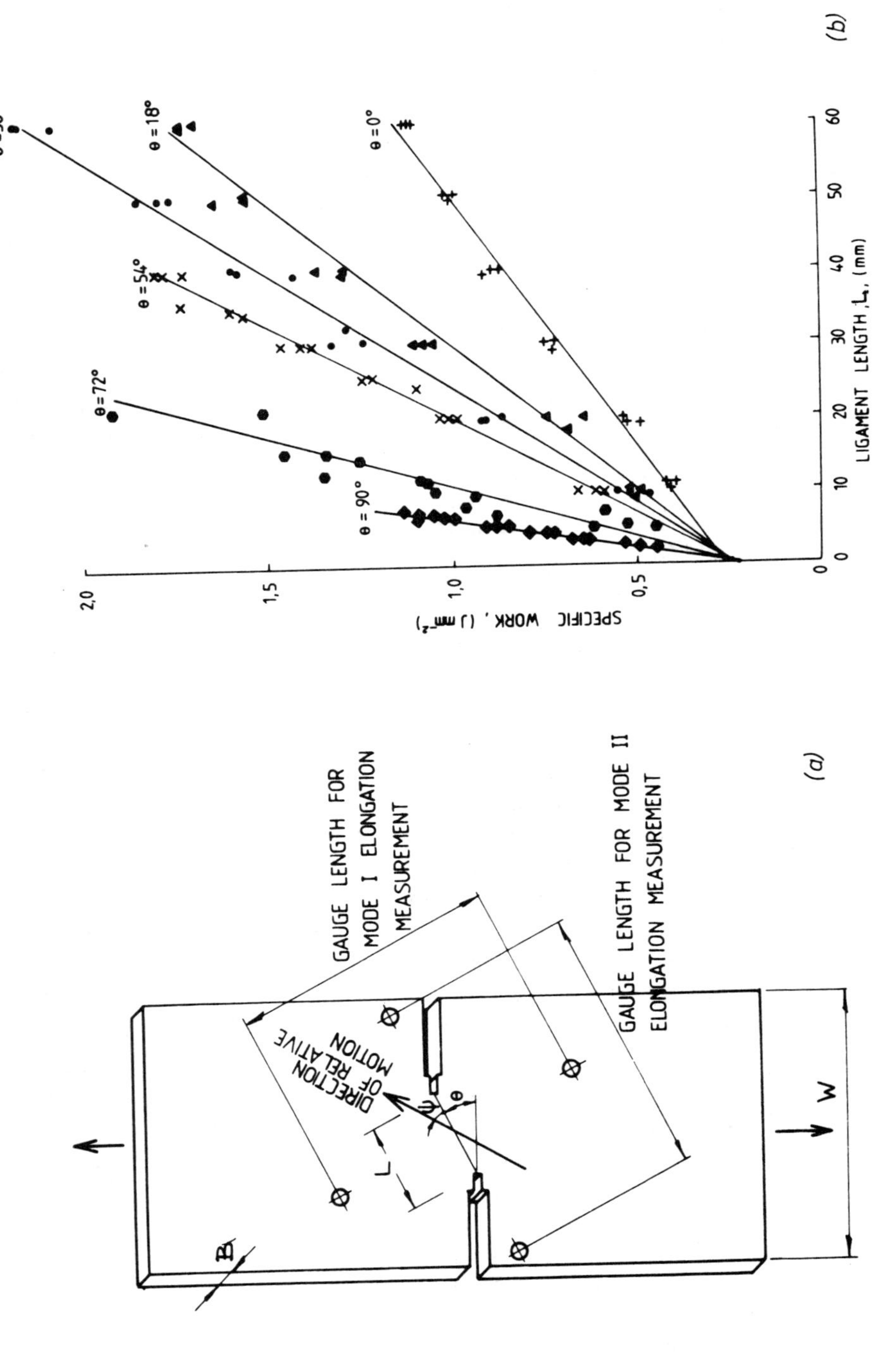
GAUGE LENGTH FOR MODE I ELONGATION MEASUREMENT
GAUGE LENGTH FOR MODE II ELONGATION MEASUREMENT
DIRECTION OF RELATIVE MOTION
ψ
θ
L
B
W
(a)
SPECIFIC WORK, (J mm⁻²)
LIGAMENT LENGTH, L, (mm)
θ = 90°
θ = 72°
θ = 54°
θ = 36°
θ = 18°
θ = 0°
0
0,5
1,0
1,5
2,0
10
20
30
40
50
60
(b)

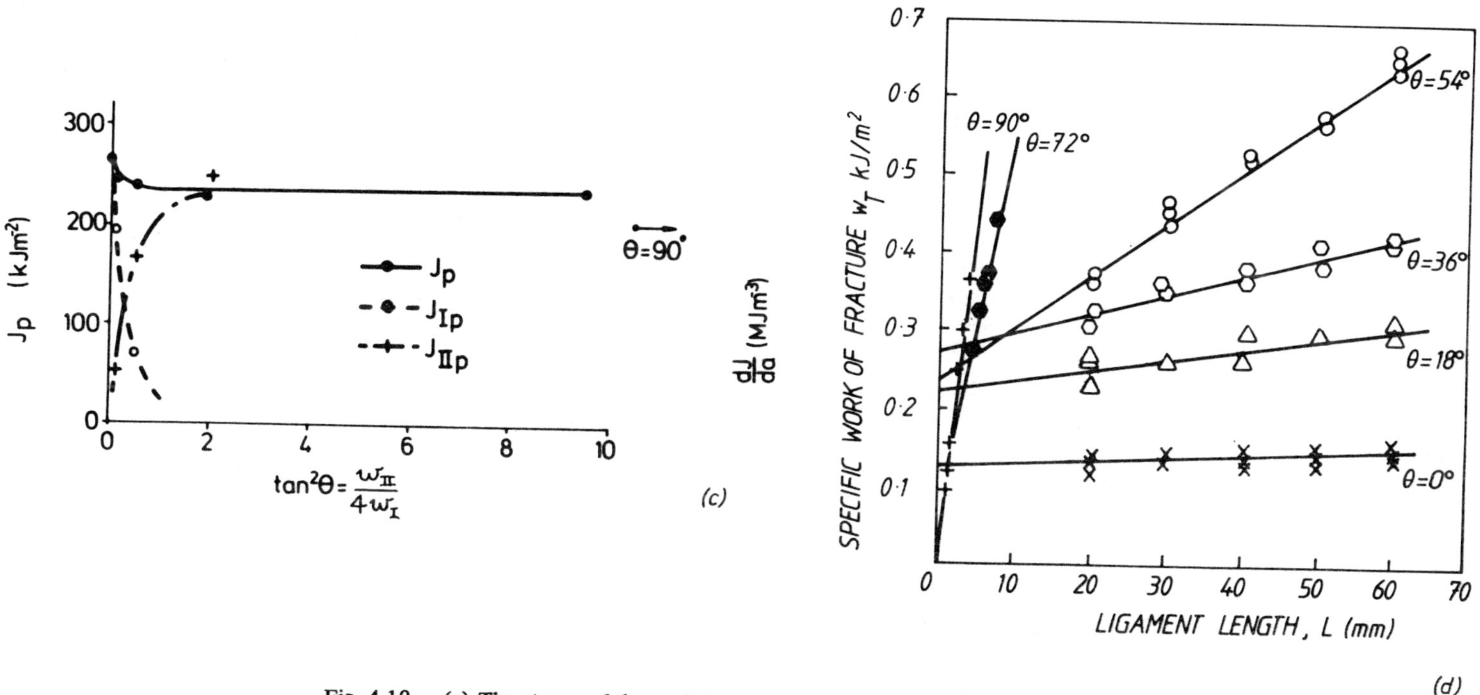

Fig. 4.19 – (a) The staggered deep edge notch tension specimen. (b) Plots of specific fracture work against ligament length for varying stagger angles (θ) for Lyten. (c) J_p (i.e. R) as a function of $\tan^2\theta$ for Lyten. (d) Plots of specific fracture work against ligament length for varying stagger angles (θ) for an aluminium alloy B1200-H14 ((a), (b), (c) after Cotterell *et al.* (1982), with permission; (d) after Mai and Cotterell, with permission).

4.4.2 Tearing of ductile metal sheets

Trouser-type tearing tests of ductile materials may be analysed by the normalised total work methods employed in section 4.4.1 (for a review of elastic trousers tests, see section 2.4.1). Fig. 4.20 shows the deformation geometry of a specimen after steady state tearing has been reached. There is virtually no deformation in the sheet until it reaches point B where the metal enters the fracture process zone BD whose width is $2s$. There, the metal which eventually flows into the two legs is twisted and eventually fractures. The torn edges are 'turned-up' (like burrs on the edges of guillotined sheets) owing to the shearing experienced in the process zone (see cross-sections XX and YY). While tearing takes place, the legs of the specimen are bent plastically to a mean radius of curvature ρ_0, but from D to F the legs undergo elastic unbending, with the bending moment reaching zero and changing sign at E (along the line of action of the pulling load). At F the legs begin to unbend plastically which is complete at G. In a non-workhardening material all the unbending would take place at G where the unbending moment is a maximum. From G to H the legs deform plastically so that the load line can adjust itself to maintain the necessary bending moments (cf. the elementary analysis for plastic bending in section 4.3.1 and 4.3.2).

Let us first examine the non-essential work associated with fracture in this trousers geometry, which concerns the work of plastic bending and unbending of the legs of the specimen. We shall eventually employ equation (4.3), so we require W^{P} relating to bending and unbending. The mean plastic strain, for a radius of curvature ρ_0, is $(B/4\rho_0)$ in a sheet of thickness B (Fig. 2.16(a)). Thus the *mean* work/volume for bending in a material following $\bar{\sigma} = \sigma_0 \bar{\epsilon}^n$ is

$$W^{\mathrm{P}}_{\mathrm{bend}} = \int_0^{\bar{\epsilon}} \bar{\sigma}\,\mathrm{d}\bar{\epsilon} = \sigma_0\, \bar{\epsilon}^{n+1}/(n+1)$$

$$= \sigma_0 B^{n+1}/(n+1)\ 4^{n+1}\ \rho_0^{n+1} \qquad (4.29)$$

For unbending, where the mean unbending strain is equal in magnitude to the bending strain and has to be added to it,

$$W^{\mathrm{P}}_{\mathrm{unbend}} = \frac{\sigma_0}{(n+1)}\left[\left(\frac{B}{2\rho_0}\right)^{n+1} - \left(\frac{B}{4\rho_0}\right)^{n+1}\right] \qquad (4.30)$$

Thus

$$W^{\mathrm{P}} = W^{\mathrm{P}}_{\mathrm{bend}} + W^{\mathrm{P}}_{\mathrm{unbend}}$$

$$= \sigma_0\ B^{n+1}/2^{n+1}\ \rho_0^{n+1}\ (n+1)$$

The incremental volume $\mathrm{d}V^{\mathrm{P}}$ passing through the process zone is

$$\mathrm{d}V^{\mathrm{P}} = 2\,wB\mathrm{d}u$$

where w is the width of each leg and u is the displacement of the load (elasticity

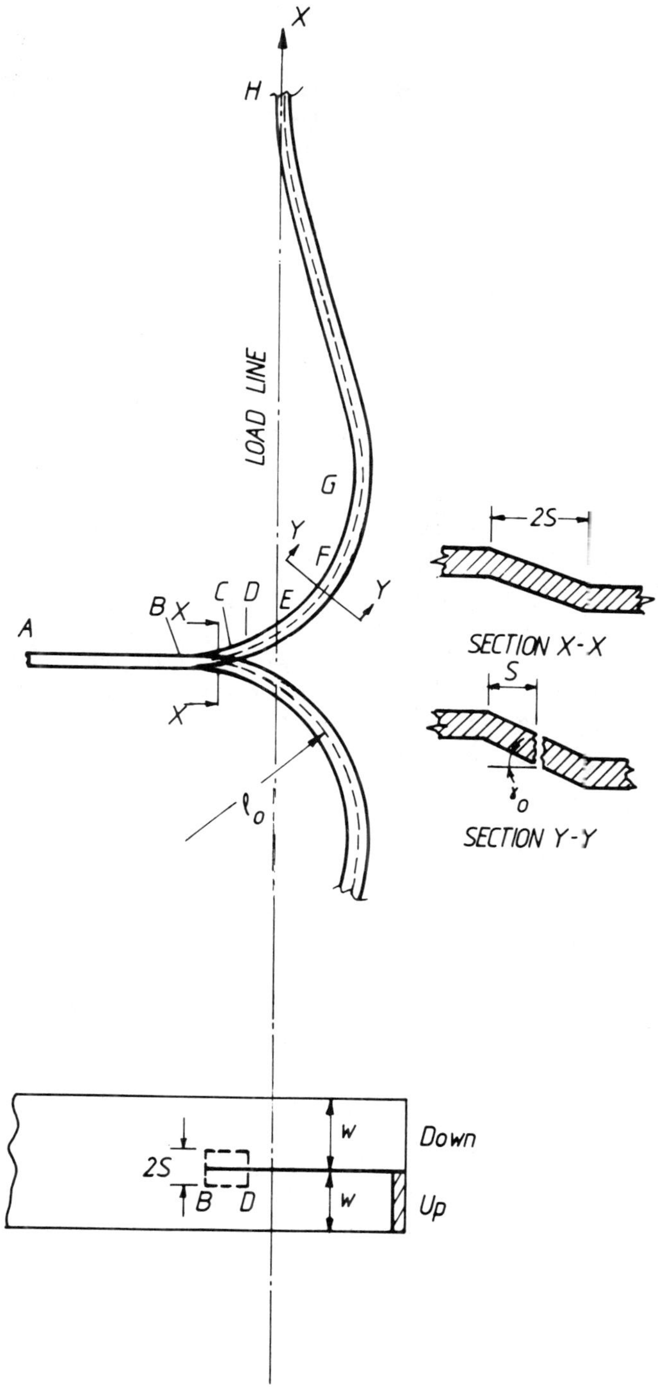

Fig. 4.20 – Steady-state geometry for tearing of a thin metal trousers specimen (after Mai and Cotterell (1984), with permission).

is neglected here). The corresponding incremental crack area generated is $dA = B\mathrm{d}u$, so in equaton (4.3)

$$2X\mathrm{d}u = \frac{\sigma_0 B^{n+1}}{2^{n+1}\rho_0^{n+1}(n+1)} 2w\, B\mathrm{d}u + R\, B\mathrm{d}u$$

$$\frac{X}{B} = \frac{\sigma_0 B^{n+1} w}{2^{n+1}\rho_0^{n+1}(n+1)} + \frac{R}{2} \tag{4.31}$$

ρ_0 is expected to be a weak function of the leg width w and it was assumed to be approximately constant for the range of w investigated by Mai and Cotterell in 1984 on 1.6 mm thick low carbon steel and 2 mm thick 5251 aluminium alloy sheets. According to equation (4.31), a plot of X/B vs. w should be a straight line of intercept $R/2$. Fig. 4.21(a) shows that the agreement is good with R values of 1040 kJ/m^2 (steel) and 600 kJ/m^2 (aluminium). The prediction of ρ_0 from the slopes of the lines also agrees well with experimental values of ca. 30 mm.† (For the steel $\bar{\sigma} = 893\bar{\epsilon}^{0.21}$ MPa; for 5251 aluminium alloy $\bar{\sigma} = 182\bar{\epsilon}^{0.05}$ MPa).

The R values are very much larger than the corresponding values of the essential work of fracture obtained with DEN tensile tests, which are some 260 kJ/m^2 and 125 kJ/m^2 respectively. The mode of deformation is, of course, different, being ideally out-of-plane mode III in the trousers test, but practically involving some mode I deformation, and there is no reason why the R value should coincide. The essential work of fracture R consists of two components, (i) the work necessary to produce out-of-plane fracture and (ii) the plastic work associated with the thin lip or burr of width s on the edges of the torn legs (which are inherent to the process if the metal is ductile). The shear zones are similar, but physically much larger, than the burrs produced on cut edges by guillotines with blunt blades (section 2.4.2) (Atkins and Mai, 1979). It turns out that most of the essential work of fracture goes into deforming the burr zones rather than producing the final tear.

A shear strain γ_0 in the lip (Fig. 4.20), corresponds with an effective strain $\gamma_0/\sqrt{3}$, so the plastic shearing work/unit volume is $\sigma_0 {\gamma_0}^{n+1}/(n+1)\, 3^{(n+1)/2}$ The incremental lip volume generated on the edge of each leg in passage through the process zone is $sB\mathrm{d}u$, so the total incremental work in forming the two lines of burred edges is

$$2sB\sigma_0 {\gamma_0}^{n+1}\, \mathrm{d}u/(n+1)3^{(n+1)/2}$$

When this work, which is local to the torn edges and which must be included in the essential work of fracture is normalised by the incremental crack area $B\mathrm{d}u$, its contribution to R is

$$w_{\mathrm{e(burr)}} = 2s\sigma_0 {\gamma_0}^{n+1}/(n+1)3^{(n+1)/2} \tag{4.32}$$

---

† Such large ρ_0 values are associated exclusively with plastic bending in the legs. For elastic bending of the legs such as occurs in the tearing of materials with very large flexibilities Gurney and Ngan (1971) showed that $\rho_0 = \sqrt{EI/RB}$ and thus ρ_0 is small (Chapter 2).

(a)

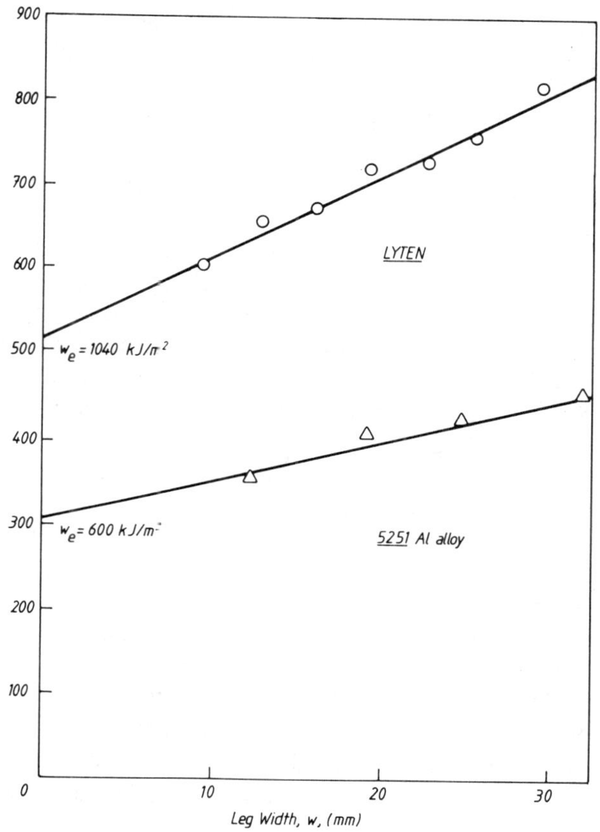

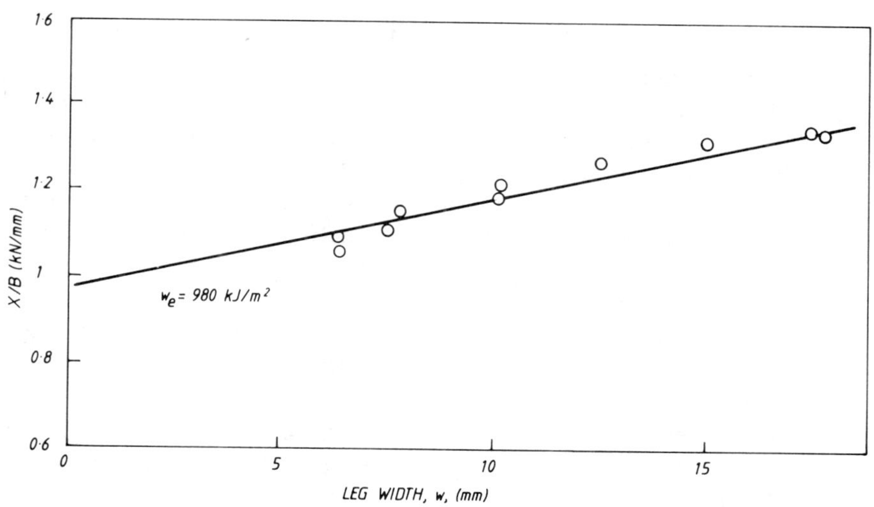

(b)

Fig. 4.21 – (a) The tearing force (X/t) plotted against the leg width (B) for Lyten and aluminium alloy 5251. (b) Tearing force (X/t) plotted against leg width (B) for a three-leg trousers specimen for Lyten (after Mai and Cotterell (1984), with permission).

It is difficult in practice to measure both s and γ_0. Roughly $s \approx 2B$, and Mai and Cotterell found $0.5 < \gamma_0 < 0.9$ for both their steel and aluminium samples. Then, $380 < w_{e(burr)} < 620$ kJ/m^2 for the aluminium alloy and $970 < w_{e(burr)} < 2{,}520$ kJ/m^2 for the low carbon steel. Clearly the formation of the burred edges is the major effect in setting the level of R.

A series of additional tests was also carried out on low carbon steel sheets using a trousers test with three legs, with the middle leg width twice that of the two outside legs. In this way, the plastic non-essential work could be made to differ from the two-legs trousers test without altering the essential fracture work. This is a test of whether the essential fracture work is truly a material property, unrelated to the remote flow in the legs, which has been the assumption of this chapter. Fig. 4.21(b) confirms the same large R value of ca. 1,000 kJ/m^2 while the associated inessential work canges markely. Owing to the change in geometry from the two-legged to three-legged trousers test, with both the essential and non-essential work being doubled, the intercept in a plot of X/B vs. w gives R not $R/2$ as in a two-legged test.

We have recently discovered that Hennephof in 1956 studied the tearing of strips of aluminium foil at various angles to the plane of the sheet. He analysed his results in terms of plastic bending (which varies with pulling angle) and burr formation, but not in terms of work of fracture. Inspection of his data using the present analysis suggests $R \approx 53$ kJ/m^2 for aluminium possessing a UTS of some 119 MPa. Hennephof comments on the crack path taken in the plane of the sheet, since a parallel starting strip eventually comes off with a triangular 'tongue' at the end. In fact Jevons (1942) recommends measurement of the length of the tongue as a test of sheet formability (see Chapter 5).

4.5 ESHELBY–RICE *J*-INTEGRAL

4.5.1 Derivation of *J*

Irreversibilities remote from the crack faces in elastoplastic and fully plastic deformation cause two effects, viz. curvature in the *Xu* diagram on loading and displacement irreversibilities on unloading (Fig. 1.21). The former is similar to conventional plastic deformation of workhardening metal, but now is in the presence of cracks; the latter, and associated distortions of the cracked body, prevent the original shape and size of the component being obtained on refitting the broken parts back together.

On the face of it, there is no difference between curvature in *the loading part* of an *Xu* diagram produced by irreversibile plastic flow and curvature produced by reversible non-linear elasticity. The physical causes are different, but the effect is the same. The difference arises on *unloading,* Fig. 1.21 (broken pieces having experienced non-linear elastic deformation *can* be refitted to give the original component).

The comparison between non-linear elasticity and plasticity during loading has an analytical basis in Hencky's deformation theory of plasticity in which *total* strains (not *increments* of strain as in proper plasticity theory) are functions of the current stress. As shown in Appendix 3 the Hencky theory is formally identical with non-linear elasticity. While plasticity based on incremental strains is the 'correct' theory, there are formidable difficulties in achieving full solutions to many problems, and the Hencky equations are often used for conventional elastoplastic flow problems in crack-free bodies. Results for some monotonic loading problems agree well with experiment, particularly if during loading the stress or strain ratios do not vary much (in 'proportional' or 'radial' loading where the ratios are constant, the two approaches coincide).

The Hencky equations may also be employed to analyse the elastoplastic deformation of *cracked bodies*. The correspondence with non-linear elastic behaviour has led a number of authors to propose that cracking in the presence of ***plastic*** flow ought to be capable of analysis using methods similar to those employed in Chapter 2 for non-linear ***elastic*** cracking. That is, expressions for the loads and deflexions during elastoplastic fracture should be derivable both from energy methods and from considerations of stresses around crack tips in non-linear elastic bodies. This approach ignores the marked differences that occur on unloading in the two cases, and considers merely monotonic loading. A discussion of the significance of the differences (i.e. displacement irreversibility in the plastic case and complete reversibility in the non-linear elastic case), and whether it is justifiable to ignore them in the practical applications of a theory of elastoplastic fracture based on Hencky total strain plasticity theory, is deferred until section 4.5.3 later. It is a crucial and contentious matter, for even in globally monotonic loading there may be *local unloading* at the crack tip.

We may recast equation (1.1) in the following way:

$$J = \Sigma\, X_i \frac{\mathrm{d}u_i}{\mathrm{d}A} - \frac{\mathrm{d}\Lambda}{\mathrm{d}A}$$

or

$$J = \Sigma\, X_i \frac{\mathrm{d}u_i}{\mathrm{d}A} - W \frac{\mathrm{d}V}{\mathrm{d}A} \tag{4.33}$$

or

$$J = -\frac{\mathrm{d}\Pi}{\mathrm{d}A} \tag{4.34}$$

where $\Lambda = \int\!\!\int W \mathrm{d}x\mathrm{d}y$, with the strain energy density $W = \int \sigma_{ij} \mathrm{d}\epsilon_{ij}$ in general for a non-linear elastic body of volume V (cf. equation (2.2) for a linear body) and where $\Pi = \int (\mathrm{d}\Lambda - X\mathrm{d}u)$ is defined as the 'potential energy' of the system.

In place of the summation of the components of the work done by the external forces $\Sigma X_i du_i$ in equation (4.33) (usually at the boundaries of the cracked body), consider conditions along some contour C around the crack tip (but outside the process zone). Under load and in equilibrium, there is strain energy within the contour, and tractions **T** and displacements **u** on the contour itself. We have for Π for the region V encapsulated by the contour

$$\begin{aligned} \Pi &= \int (\mathrm{d}\Lambda - X\mathrm{d}u) \\ &= \int_V W \mathrm{d}V - \int_C \mathbf{T}\,.\,\mathbf{u}\,.\,\mathrm{d}s \end{aligned}$$

where s is distance along the contour and contour integration is performed anticlockwise. Thus, according to equation (4.34)

$$J = -\frac{\mathrm{d}\Pi}{\mathrm{d}A} = -\frac{\mathrm{d}}{\mathrm{d}A}\left(\int_V W\mathrm{d}V - \int_C \mathbf{T}.\mathbf{u}.\mathrm{d}s\right)$$

In the two-dimensional case with unit thickness body, $\mathrm{d}V = \mathrm{d}x\mathrm{d}y$ and this becomes

$$J = -\frac{\mathrm{d}}{\mathrm{d}a}\left(\int W\mathrm{d}x\mathrm{d}y - \int_C \mathbf{T}.\mathbf{u}.\mathrm{d}s\right)$$

where $\mathrm{d}a$ is incremental crack length increase. With a fixed contour, $\mathrm{d}/\mathrm{d}a = -\mathrm{d}/\mathrm{d}x$ so that

$$J = -\frac{\mathrm{d}\Pi}{\mathrm{d}A} = \int_C W\mathrm{d}y - \int_C \mathbf{T}.\left(\frac{\partial u}{\partial x}\right)\mathrm{d}s$$

$$= \oint_C W\mathrm{d}y - \mathbf{T}\left(\frac{\partial u}{\partial x}\right)\mathrm{d}s \qquad (4.35)$$

Since equation (1.1) is an *elastic* relation (for which all deformations are path-independent) the contour integral in equation (4.35) is also *path-independent*. That is, in the likes of Figs. 2.7 and 2.8 in linear cases, or Figs. 2.9 and 2.10 in non-linear cases, a given pair of Xu coordinates corresponds with a unique G or R-locus, irrespective of the loading route followed prior to the onset of cracking.

Despite its quite different and arcane appearance, expression (4.35) is but another version of equations (2.10) and (2.11) (linear elastic cases) or (2.10a) and (2.11a) (non-linear elastic cases), i.e.

$$J \text{ or } G = \left(\frac{\partial\Omega}{\partial A}\right)_X = -\left(\frac{\partial\Lambda}{\partial A}\right)_u \qquad (2.10, 2.11 \text{ bis})$$

It may be shown, simply by graphical methods, that in the constant displacement case, the *change* in potential energy $\mathrm{d}\Pi$ is the same as the *change* in elastic strain energy $\mathrm{d}\Lambda$; in the constant load case, the *change* potential energy $\mathrm{d}\Pi$ is the same as the *change* in complementary strain energy $\mathrm{d}\Omega$. The absolute values of Π, Λ or Ω do not, of course, coincide.

Thus equation (4.35) presents an alternative way of computing J (or G and hence K) in *elastic* problems, although historically the contour integral did not emerge from studies of elastic fracture. Rather, it was first presented in the late 1960s by Rice† and taken up by workers at the Westinghouse Laboratories as a

† In 1951, in a paper on the force on an elastic singularity, Eshelby introduced, in analogy with the Maxwell tensor, the elastic energy momentum tensor, and he was much concerned with the development of this concept during his later years. In 1968 he published accounts of its application to the calculation of forces on static and moving cracks in elastic media, following this with a discussion of the equation of motion of a crack, including a proper energy balance about the moving tip. Related work on the Maxwell tensor, formulated also for elastic-plastic materials and published independently by J. R. Rice, also in 1968, has found widespread application by engineers in fracture mechanics, sometimes in a way to which Eshelby did not always subscribe. [Quoted from B. A. Bilby's obituary notice of Professor Jock Eshelby.]

criterion of *elastoplastic* fracture invoking Hencky plasticity theory which permitted the (really irreversible) plastic problem to be viewed as a (reversible) non-linear elastic problem. Rice employed the symbol J, and the path-independent contour integral in equation (4.35) is now familarly known as the 'J-integral'. We have employed the symbol J throughout this book for the energy-derived relations (2.10), (2.11) and so on, in anticipation of this nomenclature, although historically G was used for the linear elastic case and no symbol for the non-linear case (unless it was R, presuming quasi-static fracture).

Example 4.9 Using the elastic expressions for stress and displacement around a crack tip given in section 2.5, use equation (4.35) to show that $J = K_I^2/E^*$.

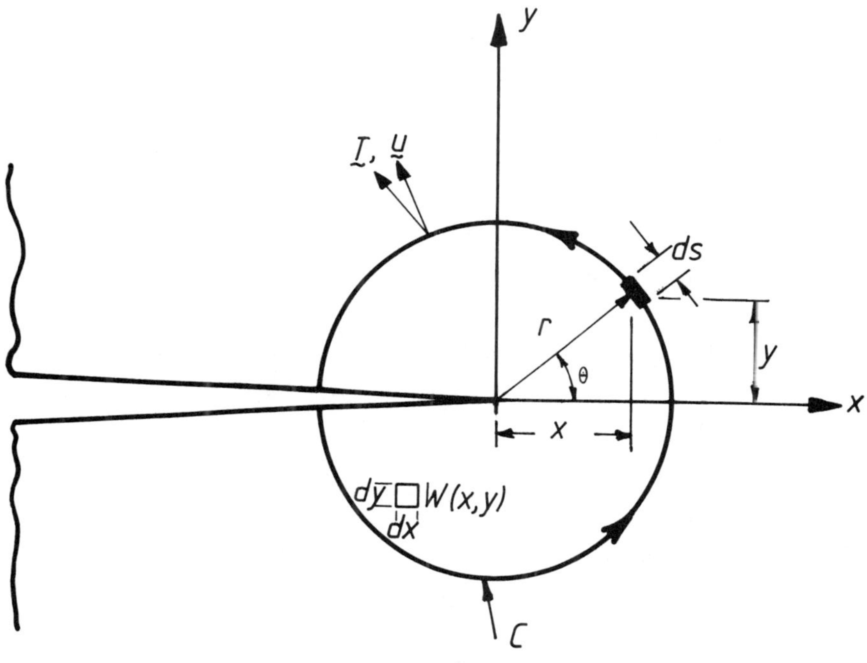

Fig. 4.22 – Continuum crack tip zone and the J-contour integral.

Ans. Consider a circular contour C of radius r. In Fig. 4.22, $y = r \sin\theta$, $x = r\cos\theta$, $ds \fallingdotseq r d\theta$. Now $\oint W dy = \int_{-\pi}^{+\pi} Wr \cos\theta \, d\theta$ and for linear behaviour

$$W = \tfrac{1}{2}\sigma_{ij}\epsilon_{ij} = \frac{(1+\nu)}{2E}\left\{(1-\nu)({\sigma_{xx}}^2 + {\sigma_{yy}}^2) - 2\nu\sigma_{xx}\sigma_{yy} + 2{\sigma_{xy}}^2\right\}$$

But

$$\sigma_{xx} = \frac{K_I}{\sqrt{2\pi r}}\;[1 - \sin(\theta/2)\sin(3\,\theta/2)]\;\cos\theta/2 \text{ etc.,}$$

so substituting into the above, we have

$$W = \frac{K_I{}^2}{2\pi r}\left(\frac{1+\nu}{E}\right)((1-2\nu+\sin^2\theta/2)\cos^2\theta/2)$$

and

$$\oint_C W\mathrm{d}y = \int_{-\pi}^{+\pi}\frac{K_I{}^2}{2\pi r}\left(\frac{1+\nu}{E}\right)(1-2\nu+\sin^2\theta/2)\cos^2(\theta/2)\,r\cos\theta\,\mathrm{d}\theta$$
$$= \frac{(1+\nu)(1-2\nu)}{E}K_I^2$$

Now

$$\oint X_i\frac{\partial u_i}{\partial x}\,\mathrm{d}s = \int_{-\pi}^{\pi}\left(T_x\frac{\partial u}{\partial x}+T_y\frac{\partial v}{\partial x}\right)r\mathrm{d}\theta$$

Now for forces on the boundary, equilibrium demands that

$$X = \sigma_{xx}\cos(n,x)+\sigma_{xy}\cos(n,y)$$
$$X_x = \sigma_{xx}\cos\theta+\sigma_{xy}\sin\theta$$

and

$$X_y = \sigma_{xy}\cos\theta+\sigma_{yy}\sin\theta$$

From the relations between σ_{xx}, σ_{xy}, σ_{yy}, etc. with K_I and θ we have

$$X_x = \frac{K_I}{\sqrt{2\pi r}}\left(\frac{3}{2}\cos\theta-\tfrac{1}{2}\right)\cos\theta/2 \text{ and } X_y = \frac{K_I}{\sqrt{2\pi r}}\left(\frac{3}{2}\sin\theta\right)\cos\theta/2$$

Also, the associated displacements are

$$u = \frac{2(1+\nu)}{E}K_I\sqrt{\frac{r}{2\pi}}(1-2\nu+\sin^2\theta/2)\cos\theta/2$$
$$v = \frac{2(1+\nu)}{E}K_I\sqrt{\frac{r}{2\pi}}(2-2\nu-\cos^2\theta/2)\sin\theta/2$$

Manipulation of these relations leads to

$$\oint \mathbf{X}_i\frac{\partial \mathbf{u}_i}{\partial x}\,\mathrm{d}s = -(1+\nu)\,\nu K_I{}^2/E$$

and so

$$J = \oint\left(W\mathrm{d}y-\mathbf{T}\frac{\partial u}{\partial x}\mathrm{d}s\right) = \frac{(1+\nu)(1-2\nu)}{E}K_I^2+\frac{\nu(1+\nu)}{E}K_I^2$$

i.e.

$$J = \frac{(1-\nu^2)}{E}K_I^2\,(=G) = K_I^2/E^*$$

Since in linear elastic fracture the stress intensity factor K is an alternative parameter to J or G, we expect that the stresses, strains and displacements around the crack tip in a *non-linear* elastic body will also be expressible in terms of some 'non-linear K' which will be related to J, and to the degree of non-linearity (given by typical non-linear stiffness expressions such as $X = k(A)u^n$ used in Chapters 2–4). This has not been done as such, but owing to the application of the J-integral to elastoplastic fracture, use is made of the Hutchinson (1968), Rice and Rosengren (1968) ('HRR') field equations for crack tip stresses and strains for a material following a non-linear *stress–strain* relation given by $(\epsilon/\epsilon_y) = \alpha\,(\sigma/\sigma_y)^N$ or, equivalently, $\sigma = \sigma_0\epsilon^n$ where $\sigma_0 = E\epsilon_y{}^{1-(1/N)}/\alpha^{1/N}$. It is possible to write

$$\sigma_{yy} \propto [EJ/\sigma_y^2 I_N r]^{1/(N+1)} \tag{4.36}$$

where r is distance ahead of the crack tip and I_N is a constant of integration which is a function of N only. The linear elastic expressions given in equation (2.21) are recovered when $N = 1$, which gives $(EJ)^{1/2} = K$ and a singularity of order $1/\sqrt{r}$.

As we shall see in the next section, to the extent that J characterises the stationary crack tip field, we might expect a critical J to define the onset of crack extension. A crucial question to be answered (particularly so in connexion with crack propagation rather than initiation) however, is how valid does in fact J describe crack tip events and/or energy interchanges? The answer must centre on the size of the crack tip process zone contrasted with the size of the region over which the HRR singularity dominates. If the fracture process zone is completely encompassed by the HRR field, (so that we have 'J-dominance' in the cracked non-linear material similar to 'K-dominance' in cracked linear materials) then a J-description is probably acceptable. If the fracture process zone is not encompassed by the HRR field, a J-description for initiation is not acceptable (and a J_R-description – see later – in propagation is not acceptable) and other formulations must be employed. These very important aspects of elastoplastic fracture are discussed in the following sections.

4.5.2 J_C as a criterion for crack initiation

The J-integral, as such, has *nothing* to do with elastic or elastoplastic fracture, exactly as K and the energy-derived J or G in elastic deformation of cracked bodies has nothing to do with fracture. But, as in elastic fracture, it is argued that fracture occurs in a body experiencing elastoplastic deformation when J reaches a critical value J_C or J_{IC} (by analogy with K_C, K_{IC} or G_C) which is characteristic of the material. Like R, J_C must be determined experimentally, and in a design application J_C would be compared with J predicted mathematically using equation (4.34) for the cracked geometry and loading system of interest (i.e. pressure vessel, structure etc. made from 'ductile' material). In many practical cases the applied J has to be evaluated using finite-element procedures, particularly in complex geometries such as cracks in manifolds and nozzles of pressure vessels (section 4.5.4).

While, in principle, J_C at fracture in laboratory testpieces may be determined using equation (4.35), it is customary instead to derive J_C from work areas on experimental *Xu* diagrams obtained from combined flow and fracture tests, empolying equation (4.34) and (in the constant displacement case) the identity between $d\Pi$ and $d\Lambda$. We must repeat that, for the moment, we are ignoring the fact that such areas will include work of irreversibile flow Γ (and will not therefore all be 'available' for fracture, which would be the case if the situation were really one of non-linear elasticity) and also ignoring the fact that the *Xu* diagram will display marked differences on unloading from the presumed non-linear reversible elastic model.

One of three methods is usually used to evaluate J_C; two require multiple specimens and one method manages with a single specimen. Thus

(i) a number of specimens with varying initial crack lenths is loaded until fracture commences and J_C is given directly by the curvilinear work sector areas divided by the crack area growth $(A_2 - A_1)$, cf. Figs. 2.10 and 2.11.

(ii) A number of specimens is loaded to the start of fracture and beyond, and the complete areas U_T beneath the different *Xu* curves out to various displacements u (*under load*) are plotted against crack area A. Since, at constant u, $d\Pi = d\Lambda = dU_T$ here (since we choose to ignore that $U_T = \Lambda + \Gamma$ really in elastoplastic fracture and not just Λ), it follows that the slopes $(\partial U_T/\partial A)_u$ coincide with $d\Pi/dA$ and are therefore the values of $-J$ at those displacements u (cf. section 2.2.2 under 'Non-linear elastic behaviour'). The derived J's are plotted against u, and J_{IC} is the value of J at the displacement at which fracture is seen to occur. The meaning of J beyond J_{IC} (so-called J_R) is discussed later in section 4.5.3. Figs. 4.23 (a) and (b) show J_{IC} data obtained by this procedure for Ni–Cr–Mo–V alloy steel three-point bend bars.

(iii) A third method (described presently) enables J to be determined from a *single Xu* plot for certain geometries of cracked body.

Regarding methods (i) and (ii), there is often difficulty in deciding the precise u at which the crack initiates and, as inspection of Fig. 4.23(b) shows, this can have a marked effect on the derived value for J_{IC}. The crack tip region often stretches and blunts before fracture. Furthermore, in thick sections, the crack may initiate in the centre of the body, slowly grow under increasing load in a 'thumb-nail' shape and may be well-developed by the time any evidence of fracture appears on the surface of the specimen. (Proof of curved crack fronts is easily obtained by inspection of completely broken specimens.) To overcome this difficulty Landes and Begley (1972) recommended a modification of the procedure outlined in (ii) above, whereby instead of plotting J versus u, J is plotted against crack extension Δa. Data tend to fall into two regions (Fig. 4.24) the point of transiton corresponding with J_{IC}. The identification of J_{IC} is aided by superimposing the straight line

$$J = 2\,\sigma_{\text{flow}}\,\Delta a \qquad (4.37)$$

(a)

(b)

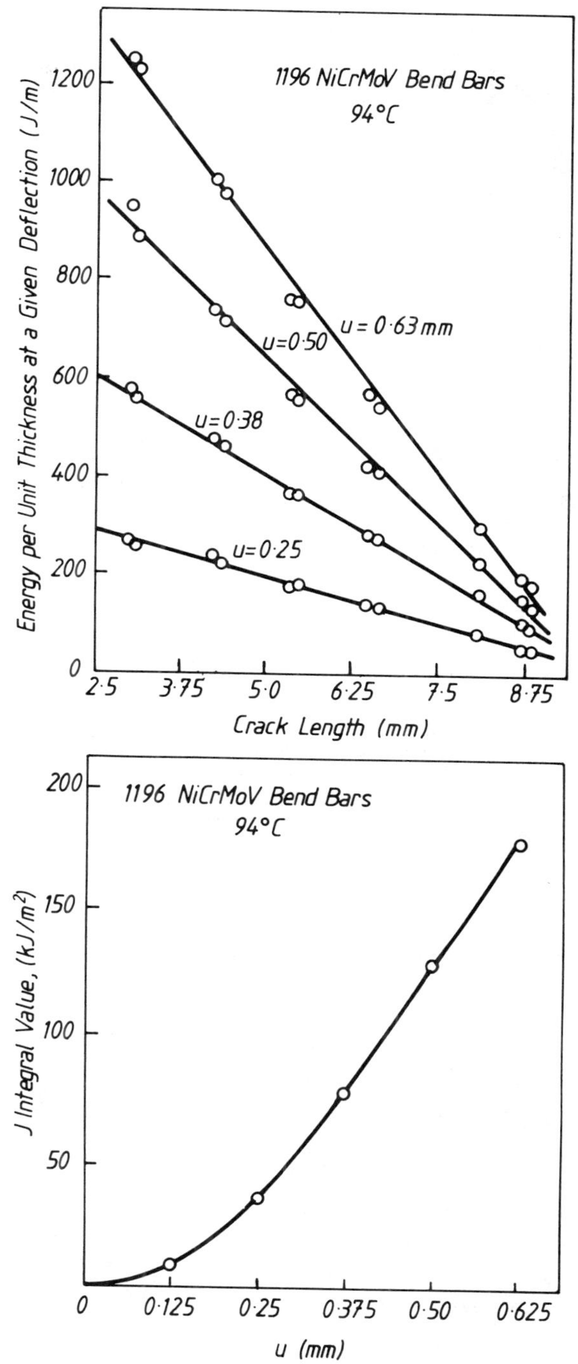

Fig. 4.23 – (a) Energy absorbed at a given deflexion versus crack length for NiCrMoV steel bars (after Landes and Begley (1974); copyright ASTM, 1916 Race Street, Philadelphia, PA 19103. Reprinted with permission.) (b) J value as a function of deflexion, NiCrMoV bend bars (after Landes and Begley (1972); copyright ASTM, 1916 Race Street, Philadelphia; PA 19103. Reprinted with permission.)

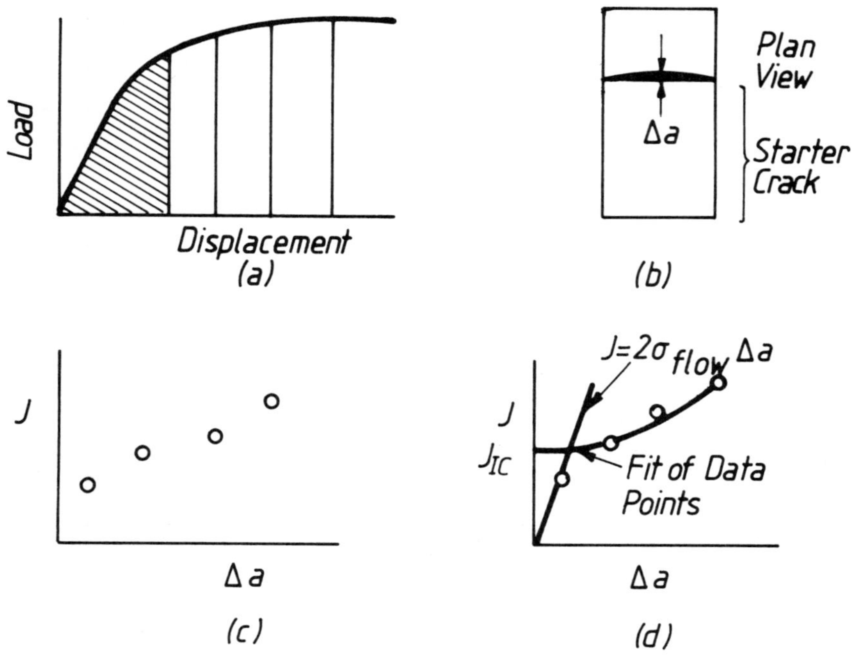

Fig. 4.24 – Procedures for J measurement: (a) Load specimens to various displacements, (b) measure crack extension, (c) calculate J for each specimen and plot J versus Δa, and (d) construct two curves for J_{IC} measurement (after Landes and Begley (1974); copyright ASTM, 1916 Race Street, Philadelphia, PA 19103. Reprinted with permission. Also see ASTM Standard E813.)

on the early data at small Δa. This is a line which represents crack blunting, where there is an increase in crack tip stretch (i.e. COD) without actual crack advance; Δa in equation (4.37) is *not* a real crack length increment, rather the notional advance that would have occurred at increasing COD had the crack remained sharp.† The relation may be derived (Problem 4.16) assuming that plastic deformation takes place at the crack tip along 45° slip lines, so that apparently $\Delta a = \delta_c/2$ and this is substituted in $J = \sigma_y \delta_c$. For materials which workharden, σ_y is replaced by some arbitrary flow stress between yield and ultimate. This seems to work acceptably well for most structural steels and aluminium alloys, but as Fig. 4.25(a) shows, for materials such as 316 stainless steel which workhardens apreciably, the blunting line is better represented by something like $J = 4\,\sigma_{flow}\Delta a$ where $\sigma_{flow} = 384$ MPa ($= (\sigma_y + \sigma_{UTS})/2$ with $\sigma_y = 211$ MPa and $\sigma_{UTS} = 550$ MPa). This is readily explained by the observation that cracktip slip occurs at some 60° to the line of crack extension in this stainless steel rather than the 45° assumed in the derivation of equation

† Strictly speaking, blunting lines should also be used in COD-Δa curves although they are not in COD standards (see Matsoukas *et al.*, 1985).

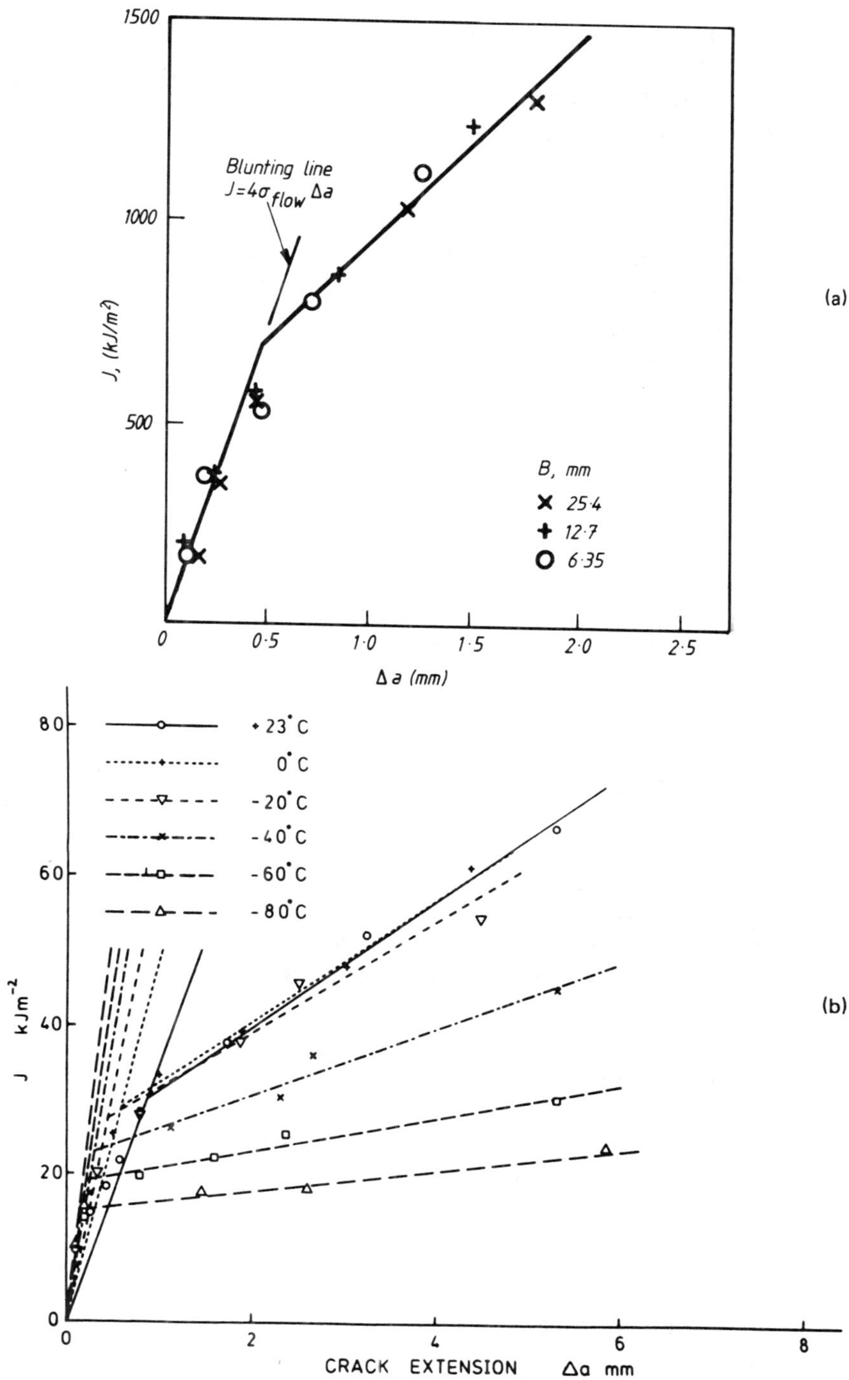

Fig. 4.25 – (a) The determination of J_{IC} for a 316 stainless steel from a J_R–Δa curve using compact tension specimens of three different thicknesses (after Hodgkinson (1980), with permission). (b) J-integral procedures applied to fracture of high density polyethylene (after Chan and Williams (1983), with permission).

(4.37). Chan and Williams (1983) have studied cracking in high density polyethelene in terms of the J-integral (Fig. 4.25(b)). They detail the experimental difficulties of determining the point of first cracking in such a tough material and also the important influence of rates.

The third, more recent, method enables J to be determined from merely a single Xu plot. The distinguishing feature of the particular testpieces which permit this is that there is only one dimension of importance, viz. the size of the uncracked ligament into which all plastic flow is confined. Examples are compact tension, three-point bend bar, edge-and central-notched plate and rod specimens, all of which need to be *deeply cracked*. The method centres on the relation between J and work areas under an Xu plot, rather than J and work rate (which is the derivative of those areas).

From

$$X\mathrm{d}u = \mathrm{d}(\textstyle\int X\mathrm{d}u) + J\mathrm{d}A$$

we have

$$J = \int (-\partial X/\partial A)_u \,\mathrm{d}u = \int (\partial u/\partial A)_X \,\mathrm{d}X \qquad \text{(2.11a bis)}$$

The total deformation u in a practical case comprises elastic and plastic components u_e and u_P, respectively. Hence

$$J = \int \left(\frac{\partial u_e}{\partial A}\right)_X \mathrm{d}X + \int \left(\frac{\partial u_P}{\partial A}\right)_X \mathrm{d}X \qquad (4.38)$$

The first term on the right-hand side is the elastic strain energy release rate (G) (from equation (2.12a)). To evaluate the second term, for the common test geometries of deeply double edge notched and central notched specimens but *not* compact tension or three-point bend specimens, Rice *et al.* (1973) noted that on dimensional grounds, $u_P = (W-a)f(X/B(W-a))$ where f is the plastic flow rule function relating generalised plastic strain (dependent on $u_P/(W-a)$) to generalised stress (dependent on $X/(W-a)B$). Then,

$$\left(\frac{\partial u_P}{\partial A}\right)_X = -\frac{1}{B(W-a)}\left[-X\left(\frac{\partial u_P}{\partial X}\right)_X + u_P\right]$$

since $\mathrm{d}A = B\mathrm{d}a$. The integral $\int (\partial u_P/\partial A)_X \,\mathrm{d}X$ may be found by integrating by parts to be $(1/B(W-a))(2\int X\mathrm{d}u_P - Xu_P)$. So

$$J = G + \frac{1}{B(W-a)}(2\textstyle\int X\mathrm{d}u_P - Xu_P) \qquad \text{(4.39a)}$$

This gives the value of J at all points along the Xu plot for CN and DENT specimens.

A graphical interpretation of eqution (4.39a) is shown in Fig. 4.26a, viz.

$$J_C = G + \frac{2A^*}{B(W-a)} \qquad \text{(4.40a)}$$

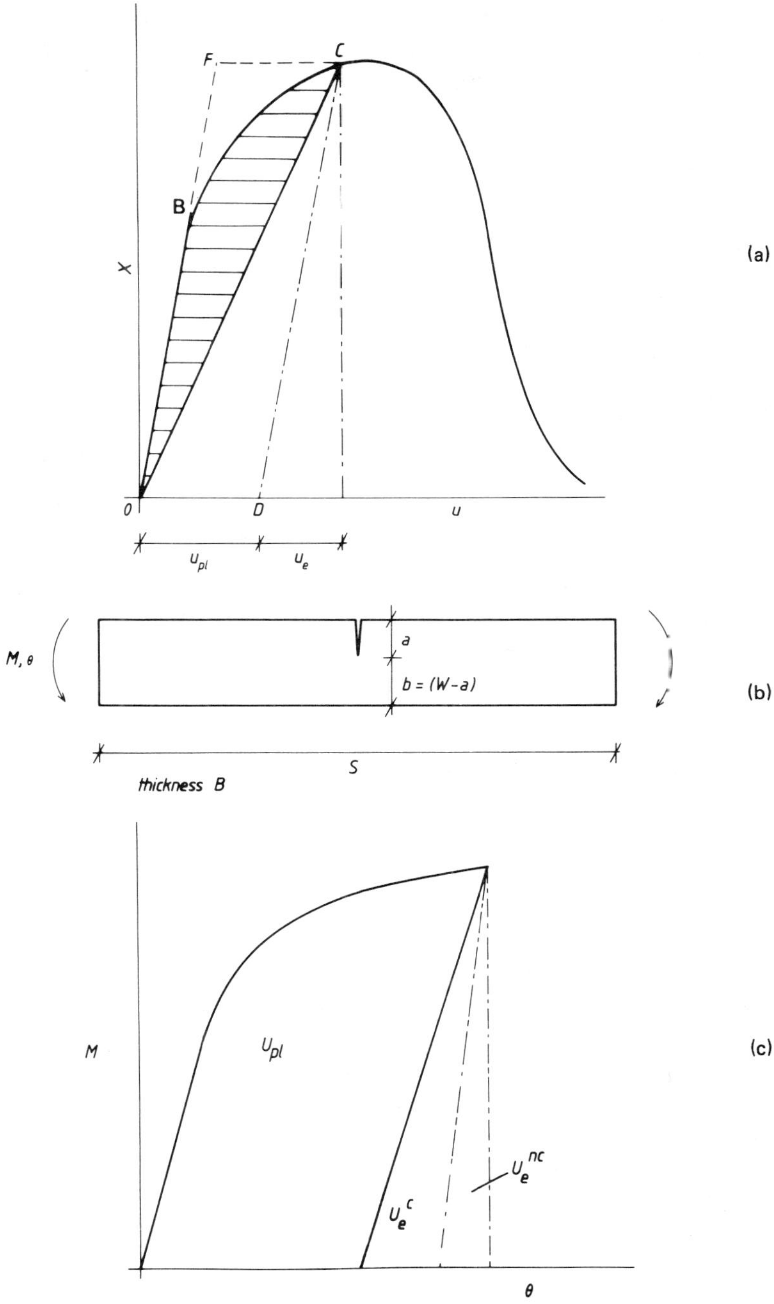

Fig. 4.26 – (a) Load–deflexion graphical interpretation of equation (4.39a) for deeply double-edged notched, and centrally notched, testpieces under load. (b) Deeply-notched three-point beam bend testpiece. (c) Subdivision of loaded work area under $M\theta$ diagram into (i) U_e^{nc} the elastic work area associated with rotation of the beam in the absence of the crack (ii) U_e^c that associated with elastic rotation in the presence of the crack and (iii) the irreversible work associated with plastic rotation of the beam.

where A^* is the segmental area bounded by the curvilinear loading line up to the point of crack initiation and a radial line (so-called secant line) returning to the orgin; ($A^* = \int X du_P - Xu_P/2$ by gemetry). We note the limiting cases of

(i) elasticity with $u_P = 0$ when $J = G$, and
(ii) rigid-perfectly plastic flow when

$$J = X_L\, u_P/B(W-a)$$

where X_L is the collapse load, if fracture and collapse happen to coincide (see Chapter 10).

For some other deeply-notched samples (CT and three-point bend), J can be related to the *total* area under the Xu plot up to initiation instead of the segmental work area of Fig. 4.26, and they are recommended in ASTM standards, such as E-813-81 for J_{IC} assessment.

The reader should be alerted that there is a deal of confusion in the literature as to whether in practical elastoplastic examples, the 'total' area should include elastic (reversible) as well as plastic (irreversible) components (the J-based theory, which appeals to Hencky total strain plasticity theory, does not make this distinction).

Consider a deeply-notched three-point bend specimen, Fig. 4.26(b). To evaluate J, we shall use the M, θ version of equation (4.38), i.e.

$$J = \int (\partial\theta/\partial A)_M\, dM$$

The elastoplastic rotation θ may be thought of being made up of (i) the rotation that would have occurred for a beam with no crack; (ii) the additional elastic rotation owing to the presence of the crack which lowers the elastic stiffness of the beam, and (iii) the rotation caused by plastic flow. Hence

$$\theta = \frac{2ML}{EI}\,(1 + f(a/L)) + g\left(\frac{M}{(W-a)^2}\right)$$

where L is the half-span. The function $f(a/L)$ gives the increased elastic rotation due to the crack and the function g gives that caused by plastic flow. Let us write $(W-a) = b$ for algebraic convenience, in which case

$$\partial/\partial A = \frac{1}{B}\,\partial/\partial a = -\frac{1}{B}\frac{\partial}{\partial b}$$

Then,

$$[\partial\theta/\partial b]_M = -\frac{2ML}{EI}\, f'\, \frac{a}{L}\cdot\left(\frac{1}{L}\right) - \frac{2M}{b^3}\, g'\left(\frac{M}{b^2}\right)$$

and

$$J = \frac{1}{B}\int_0^M \left(\frac{2M}{EI}\, f'\left(\frac{a}{L}\right) . dM + \frac{2M}{b^3}\, g'\left(\frac{M}{b^2}\right) . dM\right)$$

$$= \frac{M^2}{EIB}\, f'\,(a/L) + \frac{2}{Bb}\int_0^{\theta_P} M d\theta_{pl}$$

i.e.

$$J = G + \frac{2}{Bb} \int_0^{\theta P} M \mathrm{d}\theta_{\mathrm{pl}} \tag{4.39b}$$

For the particular deeply notched three-point geometry and size of specimen recommended in ASTM E-399, where $(S/W) = 4$ and $0.5 \leqslant a/W \leqslant 0.7$, Srawley (1976) has shown that

$$\frac{M^2}{EIB} f^1 \left(\frac{a}{L}\right) \approx \frac{2(U_e^c + U_e^{nc})}{B(W-a)}$$

and

$$\frac{2}{B(W-a)} \int_0^{\theta P} M \,\mathrm{d}\theta_{\mathrm{pl}} \approx \frac{2U_P}{B(W-a)}$$

where the total (loaded) $M\theta$ work area U_T is subdivided, as shown in Fig. 4.26(c) into (i) U_e^{nc} the elastic work area associated with rotation in the absence of the crack, (ii) U_e^c that associated with elastic rotation in the presence of the crack, and (iii) the irreversible work area associated with the plastic rotation of the beam. Substituting these component work areas for the equivalent terms of equation (4.39b) gives the interesting result *for this particular type and size specimen* that

$$J_C = 2\, U_T/B(W-a) \tag{4.40b}$$

i.e. *J* is given by twice the total work area up to crack initiation in a *single* specimen divided by the starting ligament area.†

For the CT geometry, Merkle and Corten (1974) suggest corrections to equaion (4.40b) to account for tension in the specimen and write for their simplest modification,

$$J = \left(\frac{1+\alpha}{1+\alpha^2}\right) \frac{2U_T}{B(W-a)} \tag{4.40c}$$

where α depends on $(a/(W-a))$ and can be obtained in the ASTM (E-813-81) J_{IC} standard.

A complementary viewpoint to equations (4.38) *et seq.* is provided by the work of Turner. It is shown in Chapter 8 that in *elastic* fracture, Turner gives:

$$G_C = \eta\Lambda/B(W-a) \tag{8.30 bis}$$

† Sumpter and Turner (1976) and Srawley (1976) have observed that in Rice *et al.* (1973), for three-point bend testpieces in general, it was suggested that $J = 2\,(U_T - U_e^{nc})/B(W-a)$ be used, not U_T alone. This came about because the rotation of the beam outside the plastic ligament was omitted in the analysis, which means that *G* is omitted from (the correct) equation (4.39) here. It so happens that for $(S/W) = 4$, experiments show that contribution from U_e^{nc} is negligible, in which case the 1973 Rice *et al.* expression coincides with the *special* relation given by equation (4.40b), but this relation is *not* generally valid and applies *only* for $(S/W) = 4$ and $0.5 \leqslant (a/W) \leqslant 0.7$. ASTM standards for *J* and J_R nowadays talk about U_T. Although subtraction of U_e^{nc} was part of the early result, actual unloading was *not* implied. Interpretation of unloaded *Xu* diagrams in elastoplastic fracture as they may relate to *J* and J_R is considered separately in section 4.6.5.

where $(W - a)$ is the length of uncracked ligament in the path of the crack. The factor η, which is a function of testpiece geometry, relates the elastic strain energy just at fracture to the quasi-static toughness, via G_C.

In the present elastoplastic context we write

$$J = \eta U_T/B(W-a) \tag{4.41}$$

where, still, as shown in Chapter 8,

$$\eta = (W-a)\ \frac{\partial}{\partial a}\ ((u/X)_u)/(u/X)$$
$$= -(W-a)\,(1/X)\,(\partial X/\partial a)_u \tag{8.32 bis}$$

under load. Relation (4.41) should be contrasted with relation (4.34).

In general elasoplastic deformation, η will vary with testpiece geometry and with the degree of flow (i.e. with position along the Xu diagram). Comparison of equations (8.30 bis, 4.38 and 4.41) (and remembering that the total area under the X–u curve in practical elastoplastic fracture consists of the strain energy Λ and the plastic work absorbed Γ) suggests

$$J = \left\{\begin{array}{l} J_e + J_p \\ (\eta_{el}\Lambda + \eta_{pl}\Gamma)/B(W-a) \end{array}\right\} \tag{4.42}$$

as proposed by Sumpter in 1973. Values of η_{el} are known from Chapter 8; values of η_{pl} are difficult to evaluate owing to the difficulties of predicting Xu relations during non-elastic fracture. Turner (1979) discusses this in great detail. However, should fracture coincide, more or less, with plastic collapse it is possible to determine η_{pl} for that particular position on the Xu plot. The applied load approaches the limit load X_L and

$$\eta_{pl} = -(W-a)\,(1/X_L)\,(\partial X_L/\partial a)_u$$

from equation (8.32).

In three-point bending, the fully plastic moment across the ligament is proportional to $B(W - a)^2\sigma_y$, so $X_L \propto \sigma_y\ B(W - a)^2/S$ where S is the span. $\partial X_L/\partial a \propto -2\sigma_y B(W-a)/S$. Consequently $\eta_{pl} = 2$ in this instance and remains constant during collapse since constraint to flow in deeply notched samples does not change appreciably with deformation. In *elastic* fracture of three-point bend testpieces (with $S/W = 4$ and $0.4 < a/W < 0.8$), η_{el} *also* is about 2 in value in Fig. 4.26(c). Thus in this *special* circumstance $\eta_{el} = \eta_{pl}$ and $J = 2(\Lambda + \Gamma)/B(W-a)$ $= 2U_T/B(W-a)$, as already established in equation (4.40b) by a different route.

Example 4.10 Landes, Walker and Clarke (1977) performed an extensive series of experiments on CT, three-point bend and centre-cracked tension testpieces to assess the various different methods of evaluating J (i.e., for example, equations (4.39) and (4.40) etc. given here). The material was high strength HY130 alloy steel (σ_y = 904 MPa, UTS = 957 MPa, percentage reduction in area to fracture 67 per cent). All starter cracks had blunt tip radii to eliminate crack

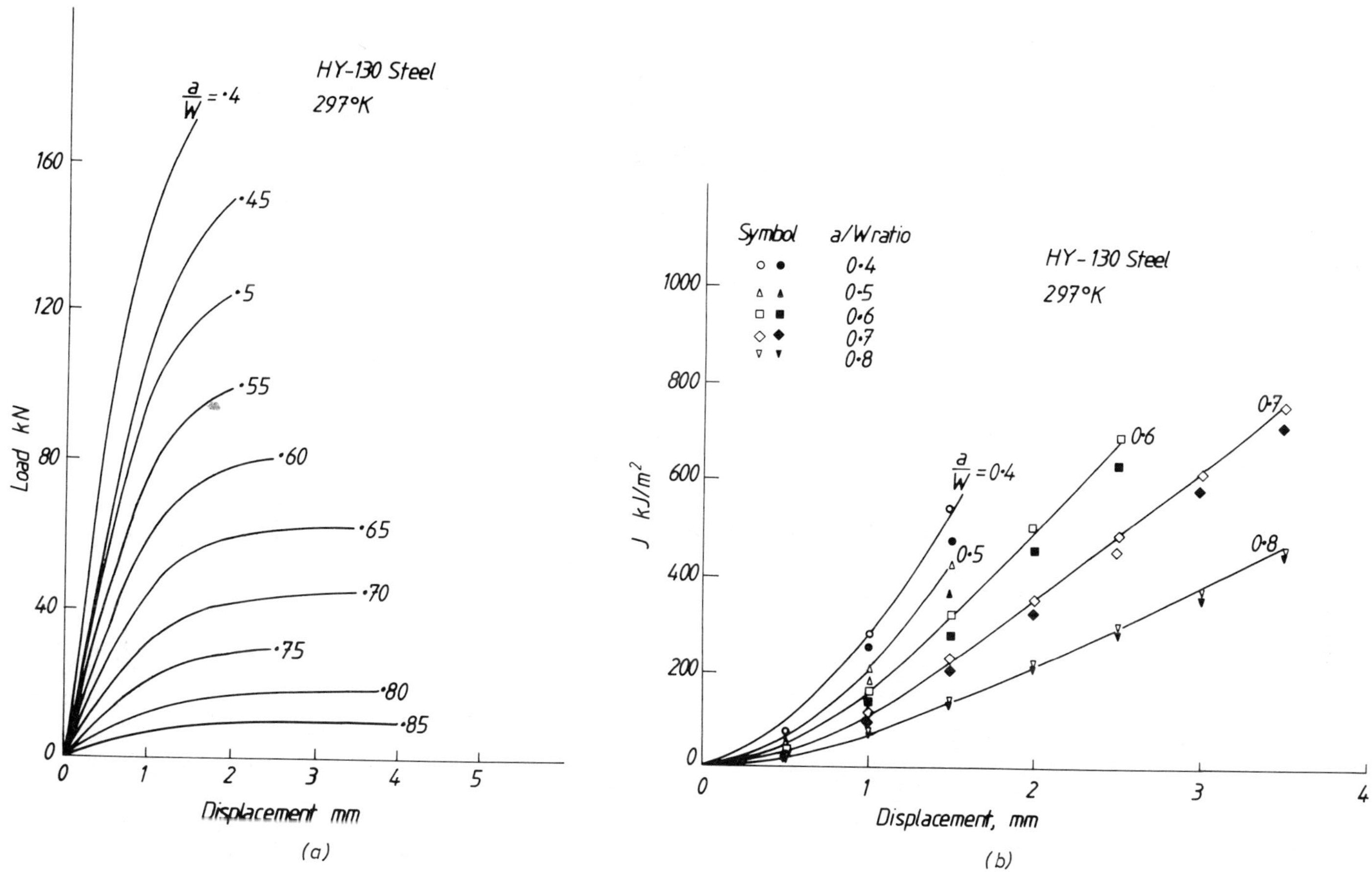

Fig. 4.27 – (a) Load–deflexion curves for compact tension testpieces, having various a/W ratios, made from HY130 alloy steel. (b) J versus displacement showing comparison of the energy rate definition of J (equation (4.34)) given as lines, and the estimated values of J using equation (4.40b) (solid points) and eqution (4.40c) (open points), all derived from the data in Fig. 4.27(a) (after Landes *et al.* (1977), with permission).

growth (subsequently proved by absence of oxidation after heat-tinting and breaking in liquid nitrogen at the conclusion of the tests). Fig. 4.27(a) gives the Xu plots obtained on compact tension testpieces with various a/W ratios. Determine J by a variety of methods.

Ans. The areas under each curve at various displacements may be calculated and the energies plotted against crack length for various displacements. J is then obtained by method (ii), equation (4.34) and the lines in Fig. 4.27(b) give Landes *et al.*'s results. They also worked out J for the CT specimen by equation (4.40b) (solid points) and equation (4.40c) (open points). We see that the Merkle–Corten expression which accounts for the tension component in the specimen agrees well with equation (4.34) but that equation (4.40b) underestimates J (cf. footnote regarding the original 1973 J estimation method given by Rice *et al.* given after equation (4.40b)). The Landes *et al.* 1977 report gives results for the other two specimen geometries and, among other points, concludes that U_T should be used for three-point bend specimens and substantiated the fact that, experimentally $\eta = 2$ for three-point bend specimens when $S/W = 4$. Method (i) using segmental areas was not employed in the report, but provides (as it should) correct values for J, as the reader may check.

In the Westinghouse study of Example 4.10, the potential energy rate definition of J (equation (4.34)), method (ii), was used as 'correct' reference values for all other Js. That is acceptable, but there are still uncertainties in the determination and application of J-based fracture criteria in that quantities determined from work areas are *called J*, but are they really the same as the formal path-independent line-integral definition (equation (4.35)) in practical solids for which Hencky total strain plasticity theory most probably does not apply during elastoplastic cracking?

Since COD measurements and J_{IC} values are nominally characterising the same event, there should be a correlation between them. Fig. 4.28(a) shows Kanazawa's data for 20 mm thick Mn steel plate (300 MPa room-temperature yield stress) obtained from a variety of testpieces (CT, three-point bend, CCP) over the range of temperature $-200°C \sim -80°C$ where some J values were calculated from work areas and others were computed via a finite-element plane stress analysis from the applied loads at fracture using the contour integral definition of J_C. In broad terms $J_C = \sigma_y \delta_c$, but Turner (1979) remarks that closer examination of the data shows that $J_C \approx 1.2\, \sigma_y \delta_c$ at low temperatures (low J) and $J_C \approx 2\, \sigma_y \delta_c$ at higher temperatures. It will be recalled that, for elastic fracture, $R \approx \sigma_y \delta_c$ (Chapter 2) and, although a unique algebraic relation between J and δ seems to exist only for a Dugdale model of crack tip plasticity (section 4.2.3), even so it appears that

$$J_C = m\, \sigma_y\, \delta_c \qquad (4.43a)$$

where $1 < m < 3$, with m being a function of the strain and extent of the plastic flow field, the overall configuration and workhardening.

If, as suggested by Rice, the crack tip opening displacement is defined, for practical purposes, as the mouth opening where lines drawn from the crack tip, at 45° above and below the centre line of the crack, intercept the crack faces, Shih (1981) shows that

$$J_C = \left(\frac{1}{d_N}\right) \sigma_y \, \delta_c \tag{4.43b}$$

where d_N is a tabulated function of σ_y/E and N, with E, σ_y and N the parameters of the Ramberg–Osgood $\sigma\epsilon$ constitutive equation, viz.

$$\epsilon/\epsilon_y = \sigma/\sigma_y + \alpha(\sigma/\sigma_y)^N \tag{4.44}$$

(cf. the non-linear relation between σ and ϵ in the HRR model, equation (4.36)). It is found that d_N varies only slightly with σ_y/E but significantly with N. For steels with $N > 5$, say, (i.e. in the Ludwik relation $\sigma = \sigma_0 \, \epsilon^n$, $n < 0.2$), $1.25 \leqslant (1/d_N) \leqslant 2.5$. For more highly workhardening solids with smaller N (bigger n), it seems that $(1/d_N) \to 5$ or so. For perfectly plastic solids, $(1/d_N) = 1$ for plane stress (in accord with the Dugdale model) but $(1/0.78) = 1.28$ for plane strain. These

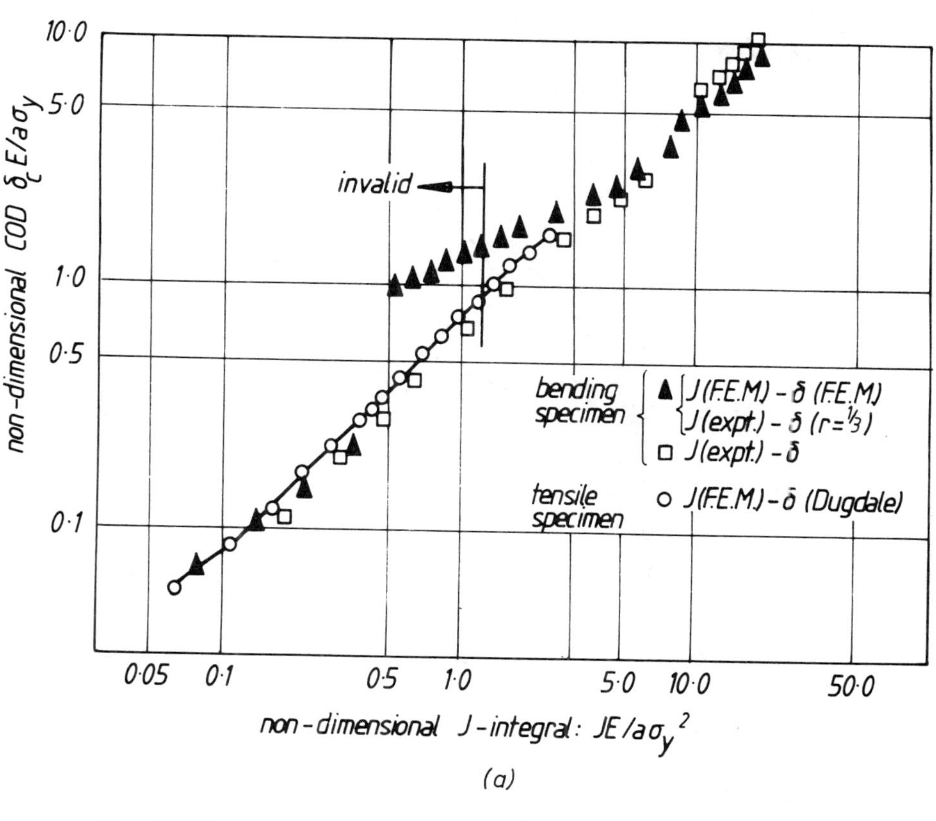

Fig. 4.28(a) – Non-dimensional plot of COD against J for a C–Mn steel (after Kanazawa (1975), with permission).

values agree with earlier work of Rice and Swalbe, Rice and Johnson and others referenced in Shih's 1981 paper (cf. equations (5.28) and (5.29) later).

Anticipating the next section regarding the use of J duing propagation, we note that $1/d_N$ is *geometry-dependent* under such circumstances.

4.5.3 J during propagation: J_R curves

The utilisation of J in this section has been as a parameter for crack tip *initiation* – when a pre-existing crack in an elastoplastic field will start to propagate.

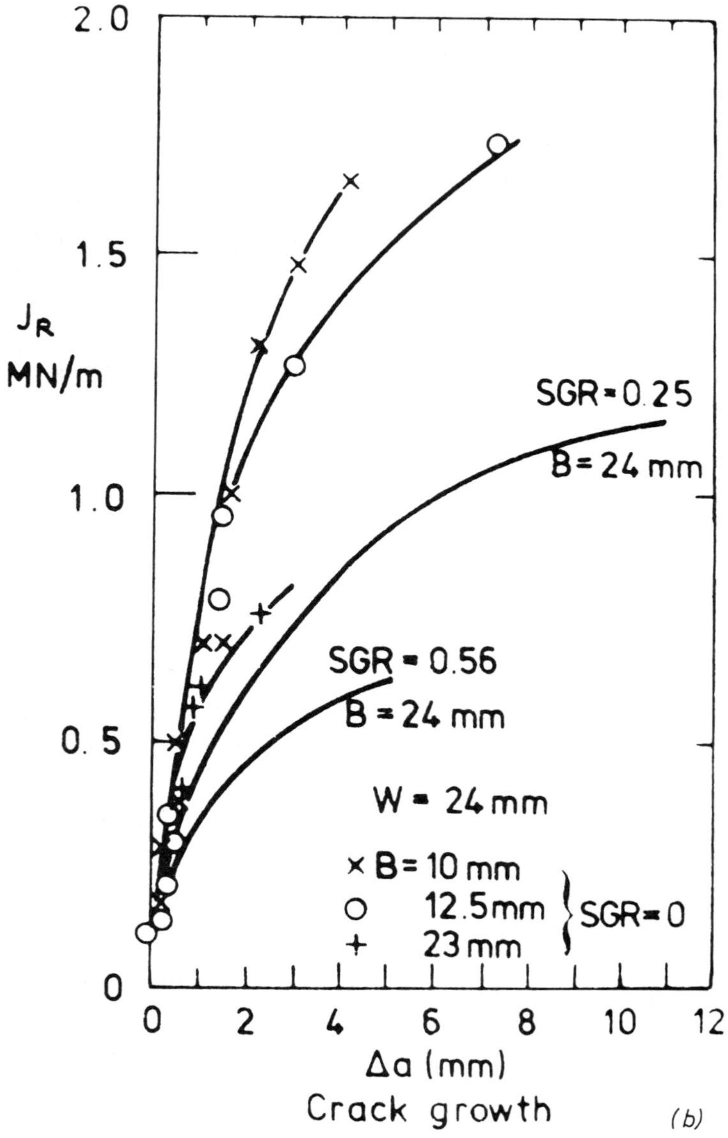

Fig. 4.28(b) – J_R–Δa curves for a normalised En32 structural steel with differing side groove (SGR) ratios (after Garwood and Turner (1977), with permission).

We have already hinted at one difficulty which can arise with even this limited use, viz. slow crack growth (which also poses problems for COD). Will different specimens produce different amounts of slow crack growth and thus affect J_C or J_{IC} values? How are values of J to be interpreted in the slow growth period and indeed when the whole crack front is propagating? Fig. 4.28(b) shows J increasing after initiation in three-point bend testpieces, which is reminiscent of R-curves in elastic fracture (section 2.3). Indeed the curve of changing J is called a $J_R - \Delta a$ curve, the slope of which is readily obtained from equation (4.41) as

$$\frac{dJ}{da} = \frac{J}{\eta}\left(\frac{\partial \eta}{\partial a}\right) + \frac{J}{U}\left(\frac{\partial U_T}{\partial a}\right) + \frac{J}{(W-a)} \tag{4.45}$$

from $J = \eta U/B(W-a)$. The η that is implied here is some weighted average of η_{el} and η_{pl} depending on the conditions; the situation is complicated if work-hardening diffuses plastic flow throughout the body (and not merely in regions adjacent to line of cracking) when, according to Turner (1979), η can drop below even η_{el} (which is usually less than η_{pl}).

Tentative test procedures for J_R, using CT and three-point bend specimens, analogous to method (iii) of section 4.5.2 for J_C, are given in Albrecht *et al.* (1982). J_R is evaluated using the special relationships, equations (4.40b) or (4.40c) where the work area goes out to points along the Xu plot at which crack growth has taken place. The increased crack lengths are inferred from compliance measurements of partially unloaded specimens. The precedure refers J_{i+1} to J_i as $a_i \rightarrow a_{i+1}$.

A crucial point to appreciate is that the increase in J_R after initiation represents an overall increase in work dissipated in the body, *not* necessarily an increase in toughness. Indeed Garwood and Turner's results on EN32 structural steel reproduced in Fig. 4.28(c) show that the COD at the tip of the advancing crack remains constant, with J (expressed in terms of the COD δ_f at the *original* position of the crack tip) increases markedly. Similar thoughts apply to crack opening *angles* measured at the starter or running positions of the crack tip. Note that these tests were performed with different thicknesses and different degrees of side-grooving, and that the results show that the increase in J is only partly due to shear lip growth at the free edges, much of the increase being also due to remote plastic flow ahead of the advancing crack. Thus, whereas *elastic* R-cruves are supposedly geometry-independent, the same is *not* likely to be true of J_R-curves, and this casts doubt on the application of J_R information to situations different from those in which they were measured.

Even so, many workers employ J-based theories to investigate crack propagation in elastoplastic deformation fields appealing to the HRR field equations for stationary crack tip stresses and strains (equation (4.36)). During quasi-static crack growth, the crack tip fields are more complicated than for stationary cracks. According to Rice *et al.* (1973) the singularity is milder (with form $\ln(1/r)$ rather than $(1/r)^{1/N+1}$ given by the HRR model). Shih *et al.* (1980) remark that analysis of these fields shows that if the HRR field increases in size

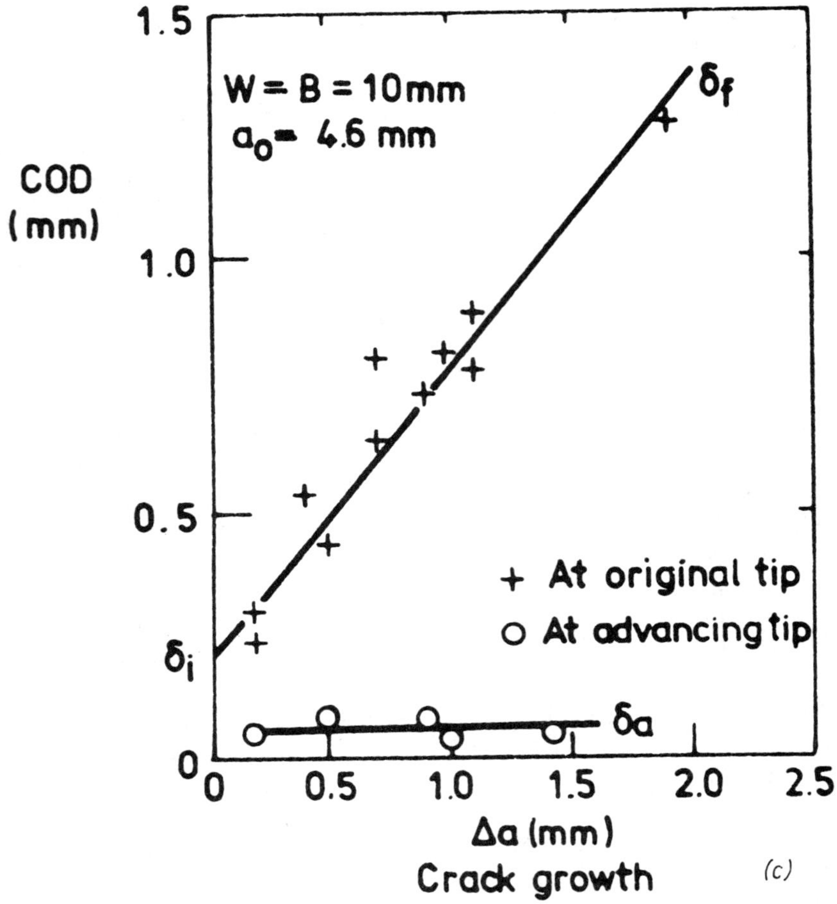

Fig. 4.28(c) – Infiltration COD at the original (fatigue) and advancing crack tip during slow growth for the steel in Fig. 4.28. Three-point bending and no side grooves (after Garwood and Turner (1978) with permission).

more rapidly than the crack advances during quasi-static propagation, then J_R (or alternatively COD) is an acceptable crack tip characterising parameter. As usually expressed, J_R-controlled growth occurs when

(i) $$\Delta a \ll (W-a) \tag{4.46a}$$

in which $(W-a)$ is the length of the remaining ligament; together with

(ii) $$\omega \equiv \frac{(W-a)}{J_{IC}}\left(\frac{dJ_R}{da}\right) \gg 1 \tag{4.46b}$$

where ω is a parameter which describes the relative magnitude of non-proportional strains (in the process zone) and proportional strains (within the assumptions of a J-formulation Hencky plasticity theory) in the plastic flow field in the rest of the body; and also

$$\text{(iii)} \qquad \rho \equiv \frac{(W-a)}{(J_{\mathrm{IC}}/\sigma_{\mathrm{y}})} > \begin{cases} 25 \text{ (in CT specimens)} \\ 200 \text{ (in CCP specimens)} \end{cases} \qquad (4.46\text{c})$$

If these conditions are not met, J_{R} is *not* an acceptable controlling parameter.

It follows that, although equations (4.40) and related expressions may be used to calculate J_{R} from experimental Xu plots, the answers will be meaningful only if the conditions given in equations (4.46) are met. Thus, for example, application of equation (4.40a) to Fig. 4.26(a) for elastoplastic crack *growth* data on DEN and CN specimens will be valid only for small crack growth. Inspection of Fig. 4.26(a) suggests that this means that areas OBCO and OCDO must be approximately equal in magnitude; otherwise large crack growth gives changed compliance and equation (4.40a) fails.

We recall the use of work areas in the Cotterell–Mai treatment of completely-fractured DEN sheets of ductile materials given in section 4.4.1. There, the specific work of fracture for initiation w_i was defined as (work area OBCD)/$B(W-a)$ using the lettering of Fig. 4.26(a). Since OBCD = (2OFCO–BFC), application of equation (4.40a) gives J_{C} or $J_i = w_i$ if *rigid-perfectly plastic* behaviour is assumed for the material (which it was in section 4.4.1).

As explained by Kanninen *et al.* (1980), a J-integral approach is unable to cope with large amounts of stable crack growth attended by large-scale plasticity, and there is pronounced specimen-dependence after some small amount of crack growth (say 10 per cent of the remaining ligament – although some finite element calculations seem to indicate that the growth can be greater than this). Kanninen *et al.* go on to say that a proper mechanics of elastoplastic fracture should delineate between the energy dissipated in direct fracture-related processes near the crack tip, and energy dissipated in geometry-dependent plastic deformation remote from the crack tip. That is the basis of equation (1.2) and the approach to problems in this book, with $R\mathrm{d}A$ being local to the process zone and $\mathrm{d}\Gamma$ applying to the remote flow. As it is conventionally viewed (particularly in the USA), the J-resistance is a far-field parameter which encompasses both forms of energy dissipation and, hence, it will increase with crack growth. In contrast, a local crack tip parameter which reflects the actual fracture process likely will not.

The distinction between the different lines of attack becomes clearer from section 4.6 below, both in which the incremental energy rate equations for initiation and propagation are contrasted with the various definitions of J up to initiation and beyond, and in which the partition of work areas beneath Xu plots is considered. Before that, however, the next section is interspersed which highlights the finite element/finite difference solutions which are available for J-controlled elastoplastic crack growth.

4.5.4 Finite element/finite difference elastoplastic fracture solutions

We have remarked that solutions for elastoplastic fracture problems present formidable computational problems. Indeed consideration of elastoplastic flow alone in many geometries is difficult. Earlier sections of this chapter demonstrate that some progress has been made with algebraic solutions in which the fracture load is identified with the plastic collapse load. But for the fracture load and collapse load to coincide is really a very special case of elastoplastic fracture (see section 4.6.1 and Chapter 10). With the advent of high-speed, large-memory computers, it has been possible to tackle many elastoplastic fracture problems analytically for the first time. While finite element/finite difference *elastic* calculations for even complicated three-dimensional problems are reasonably well founded these days (and may be used to calculate elastic stress intensity factors and so on), difficulties still exist in fully plastic and elastoplastic formulations. In solving a problem numerically, space and time must be discretised into small domains. There is a choice in selecting their characteristics, primarily the size of the elements of space and time discretisations. Simple (constant-strain) space elements can be used. However the same resolution can be achieved with more complex (high order) but fewer elements. The same applies to the time increment used for integration of the momentum equations in dynamic problems. A small timestep (explicit integration) uncouples the equation of motion. A large one (implicit integration) requires a smaller number of more complex timesteps. The simpler space elements together with smaller timesteps allows far greater generality of treatment, constitutive behaviour, etc. However, these may result in greater computing requirements for problems where the added generality is not needed, such as in elastic, small-strain calculations. The options can be described simply in each case as performing fewer but more complex calculations, or more but simpler calculations.

Nagtegaal *et al.* (1974) have pointed out that the number of degrees of freedom per space element must be greater than the number of constraints imposed by a condition of local incompressibility. Otherwise, an aritificial stiffening of the mesh occurs during plastic flow and, as a consequence, plastic collapse loads may be grossly overestimated. This applies even to infinitely refined meshes. At the other end of the spectrum, it is well known that too many degrees of freedom per element result in the appearance of unrestrained combinations of nodal displacements. This problem is known as hourglassing and is typical of meshes of constant-strain quadrilaterals in two dimensions and constant-strain hexahedrons in three dimensions.

Again, it is necessary in elastoplastic flow problems involving large deformations that incremental stresses, strains and rotations be expressed in the Jaumann (or similar) formulation. If not (and, for example, the Prandtl–Reuss equations of elementary elastoplastic flow (given in Appendix 3) are directly integrated) erroneous predictions may result, particularly if large rotations are involved. This difficulty arises particularly in large-scale plastic flow as encountered in metal-forming processes where also problems of re-zoning of the Lagrangian mesh have to be tackled as deformation proceeds to obviate wrong predictions. Even so, solutions to practical problems involving large deformations, large rotations, with full thermomechanical coupling, are now available (e.g. Marti *et al.*, 1982, 1984; Rowe *et al.*, in press). Some of the complications relating to large deformations are of less concern for the sort of elastoplastic fracture topics discussed in the present chapter, but are very important in problems of crack initation in the extensive plastic flow fields of metal forming (Chapter 5). These comments merely highlight facts well-known to the reader, namely that what the computer prints out is only as good as the formulations, assumptions and constitutive relations put in.

Finite element/finite difference methods have been applied to a variety of elastoplastic fracture problems, not only those of familiar standard testpieces but also complicated pressure vessels, manifolds and pipe fittings (e.g. Kanazawa *et al.* (1973); Sumpter and Turner (1976); Shih *et al.* and Kanninen *et al.* in EPRI Workshop Report, 1980; various papers in ASTM STP No. 803, 1983). The spur for much of this work has come from the nuclear industry. By way of illustration, we consider the GE/EPRI work of Kumar and co-workers (1981, 1982, 1983). Finite element solutions for fully plastic and elastoplastic fracture have been obtained, using both proper incremental plasticity theory and Hencky total strain plasticity (with which latter, it should be remembered, *J*-based criteria of crack initiation and propagation coincide).

As indicated above, such individual calculations (by these and other authors) require advanced finite element methods which can be expensive and time-consuming and can require experts to carry them out. However, the GE/EPRI studies incorporate the valuable proposal of interpolating the whole range of elastoplastic solutions from the already-known elastic solutions and recently obtained fully plastic solutions for the same problem, which are now tabulated in a *Plastic Fracture Handbook* (Electric Power Research Institute, 1981). The elastoplastic estimation procedure derives from the work of Shih and Hutchinson (1976). When crack initiation and propagation are *J*-controlled, we may write for *J* during elastoplastic deformation

$$J = J^{e}(a) + J^{P}(a, N) \tag{4.47}$$

where J^e is the elastic contribution and J^P is the fully plastic solution (cf. equation (4.42)). Equivalent expressions exist for COD δ_C and displacement u. J^e and J^P follow similar functional forms, viz.

$$\left.\begin{aligned} \frac{J^{e}}{\sigma_y \epsilon_y a} &= \left(\frac{X}{X_y}\right)^{2} J^{e}(a/W) \\ \text{and} \qquad \frac{J^{P}}{\alpha \sigma_y \epsilon_y a} &= \left(\frac{X}{X_y}\right)^{N+1} J^{P}(a/W, N) \end{aligned}\right\} \tag{4.48}$$

for a material with a Ramberg–Osgood elastoplastic consitutive relation (equation (4.44)). The first of equation 4.48 is simply the usual linear elastic fracture *K* expression cast in a different form, i.e. $K^2 = ER = \sigma^2 \pi a Y^2$ and division of both sides by $\sigma_y{}^2 a$, and using $\sigma_y = E\,\epsilon_y$, gives the first of equation (4.48) with $J^e(a/W) \equiv \pi\, Y^2(a/W)$. The second of equation (4.48) follows from the HRR singularly description of stress and strain fields at stationary crack tips in elasto-plastic flow (equation (4.36)), see Goldman and Hutchinson (1975). J^P is functionally similar to J^e but depends on the workhardening index N as well as crack length; tabulations of J^P for various cracked geometries and N in plane stress and plane strain form the basis of the EPRI/GE *Plastic Fracture Handbook*. (Note that in this compilation the load is load per unit thickness and that, for the elastic J, the crack length is the Irwin-modified crack length $a_e = a + \phi r_y$ where, for a workhardening solid $r_y = \dfrac{1}{2\pi}\left(\dfrac{N-1}{N+1}\right)(K_I/\sigma_y)^2$ in plane stress (6π instead of 2π for plane strain) and $\phi = \dfrac{1}{1+(X/X_y)^2}$; cf. Example 2.14).

The EPRI/GE engineering estimation line of attack for design against elastoplastic fracture does not require new finite element analyses, rather only desk-top calculations and graphical procedures to evaluate equation (4.48) from standard cases in order to construct Xu diagrams (using the expression for elastoplastic u analogous to that for elastoplastic J in equation (4.47)). Fig. 4.29 shows the comparison between calculated Xu plots in the elastoplastic fracture of a grooved compact tension testpiece and experimental results. Alternatively J–A diagrams may be constructed and elastoplastic crack stability investigated. Such procedures, using the EPRI/GE studies, are explained in Chapter 8.

Among the theoretical solutions available is that of the DEN testpiece, which will permit comparison of the Cotterell–Mai analysis and experiments with the results of Shih and co-workers, remembering, though, that many of the DEN results quoted in section 4.4.1 will *not* satisfy the condition for J_R-controlled fracture.

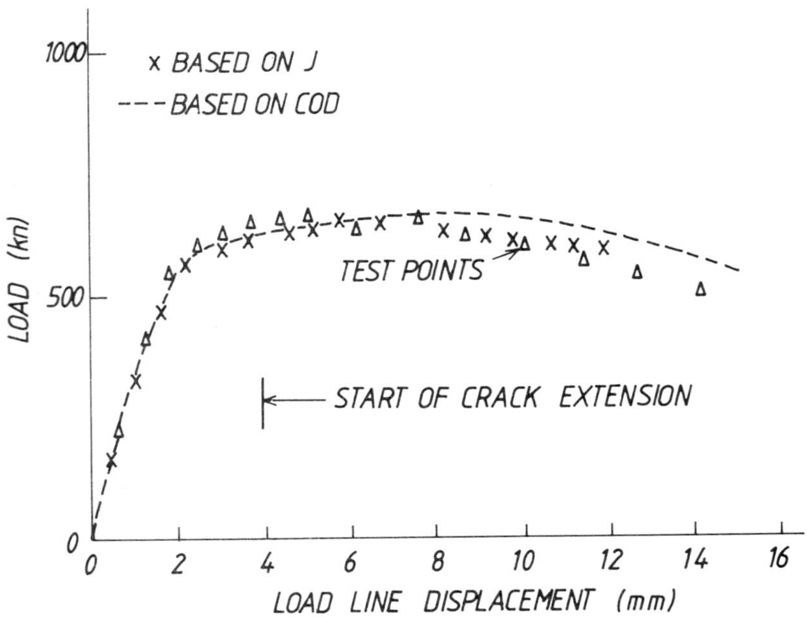

Fig. 4.29 – Theoretical and experimental load–deflexion results for compact tension specimens of A533B steel reported in the EPRI Special Study on Ductile Fracture (Shih *et al.* in EPRI Report (1980), with permission).

4.6 PARTITION OF WORK DURING COMBINED FLOW AND FRACTURE

As explained at the beginning of this chapter, (a) during initiation

$$X\mathrm{d}u = \mathrm{d}\Lambda + \mathrm{d}\Gamma \qquad (1.3\ \text{bis})$$

$$= W\mathrm{d}V + W^{\mathrm{P}}\mathrm{d}V^{\mathrm{P}} \qquad (4.1\ \text{bis})$$

and (b) during quasi-static propagation

$$X\mathrm{d}u = \mathrm{d}\Lambda + \mathrm{d}\Gamma + R\mathrm{d}A \qquad (1.2\ \text{bis})$$

$$= W\mathrm{d}V + W^{\mathrm{P}}\,\mathrm{d}V^{\mathrm{P}} + R\mathrm{d}A \qquad (4.2\ \text{bis})$$

where, throughout, changes in elastic strain energy, irreversible plastic flow and work of fracture are separated and uncoupled in general.

In terms of a J-formulation of the same events based on work areas, we have for example, both generally and for specific test geometries,

$$J \begin{cases} = -\,\mathrm{d}\Pi/\mathrm{d}A = -\,(\partial U_{\mathrm{T}}/\partial A)_u & (4.34\ \text{bis}) \\ = \eta U_{\mathrm{T}}/B(W-a) = (\eta_{\mathrm{el}}\Lambda + \eta_{\mathrm{pl}}\,\Gamma)/B(W-a) & (4.41/42\ \text{bis}) \\ = G + \dfrac{1}{B(W-a)}\,(2\int X\mathrm{d}u_{\mathrm{pl}} - Xu_{\mathrm{pl}}) & (4.39\text{a bis}) \end{cases}$$

J here may be J_{C}, J_{IC}, etc. at initiation or J_{R} at some time later during crack propagation in extensive plastic flow fields. Equations (4.41) and (4.42) or the other methods given in section 4.5.2 are employed by many authors whether the crack has begun to run or not. Immediately we see a difference between the *two* components of work input in equation (4.1) for initiation, the *three* components of work input in equation (4.2) for propagation and the *two* components in equation (4.42), for example, for both initiation *and* propagation.

For a critical J_{C} at initiation, equation (4.41) relates the total area under the Xu plot under load to the ligament area via the appropriate η factor; alternatively equation (4.34) relates the rate of change of that area w.r.t. crack area at constant u. The integrated version of equation (1.3) or (4.1) employed for the initiation phase also concerns the *same* work area, the integration taking place over complex elastic and plastic stress and strain fields. To that extent the approaches are similar and, insofar as it is the events in the process zone that actually control fracture initiation, equation (4.41) or (4.1) and so on are different attempts to link bulk parameters with microstructural events. A direct correspondence between equations (4.34) or (4.41) and (4.1) is possible in the case of void coalescence and growth models of ductile fracture initiation, as shown in section 5.8. A similar conclusion may be drawn about Andrews's generalized theory of fracture which, of course, has only one multiplicative term in his work equation corresponding with equations (1.2) and (1.3) here. A critical J_{C} at initiation seems, therefore, to be compatible with equation (4.1) (see section 4.6.2 to come) and it is probable that all the different lines of attack for the mechanics of plastic fracture are determining the same intrinsic parameter *at initiation*.

Things are different, however, during propagation. It is clear that the J-approach does not distinguish between $R\mathrm{d}A$ and $W^{\mathrm{P}}\mathrm{d}V^{\mathrm{P}}$ in equation (4.2) and all dissipated work is lumped together. When (in the case of rigid-plastic materials) $\mathrm{d}V^{\mathrm{P}}$ and $\mathrm{d}A$ are coupled (via, for example, a common remote displacement as we observed in pseudo-steady-state metal tearing problems, sections 4.3 and 4.4), this leads to an apparently enhanced specific work of fracture. The enhanced toughness must correspond with J, and insofar as it is impossible to tear the ductile sheet without also plastically bending it, more work than simply the essential $R\mathrm{d}A$ fracture work has to be performed when the sheet is torn. But clearly the level of additional work depends on w and the other term involving σ_y, B and ρ. One can obtain *any* answer one likes for the enhanced toughness simply by varying the width (w) being torn. Therefore, J_R in general cannot be geometry-independent. But, appealing to Mai and Cotterell's arguments in section 4.4.2 on 'essential' and 'non-essential' works of fracture, we would argue that the use of equation (4.2) during propagation permits 'basic' parameters (R, σ_y, geometry) to be separated in terms of the three work components of reversible elasticity, irreversible remote flow and irreversible fracture. A growing body of experiments shows this line of attack to work quite well. In contrast J_R treatments inevitably lead to a geometry-dependent cracking parameter. Of course R itself, in expressions for enhanced toughness, may be thickness-dependent but that is acceptable in the light of B-dependent fracture toughness values in *elastic* fracture. The problem with J_R-formulations is that the magnitude of the flow term in enhanced toughness expressions can be arranged to take different values, for *fixed* levels of R, by changing the proportions of the geometry of the body, i.e. B and w.

In many other problems of combined flow and fracture, $\mathrm{d}V^{\mathrm{P}}$ and $\mathrm{d}A$ are still coupled (as when plastic flow is restricted to ligaments in the line of the crack) but the problem is not steady-state, X varying with propagation (and often falling in the final stages of fracture). Enhanced values for R in such cases (i.e. J_R-curves obtained from CT, three-point bend testpieces and so on) are likely to be geometry-dependent and also crack-length-dependent. When $\mathrm{d}V^{\mathrm{P}}$ and $\mathrm{d}A$ are uncoupled altogether it is difficult to see how J during propagation can be a material property.

A re-examination of the problem given in section 4.3.3, of a glued cantilever which cracks with plastic flow in the adherand, but independently known R for the glue, is very revealing in trying to understand the mechanics of combined flow and fracture. Unlike post-yield fracture in monolithic pieces of metal, where there may be debate as to what R really is in equation (4.2), in the glued cantilever problem there can be no question regarding the fracture toughness of the adhesive and cracking along the glue line. In the following sections we discuss the partition of work areas, why cracking plus plastic flow should supervene plastic flow alone in the adherand, the correspondence between R and J_C (which can be evaluated algebraically for this problem) and path dependence. We shall learn that *residual elastic strain energy* plays an important role in the interpretation of *unloaded Xu* or *M*θ diagrams.

4.6.1 Work components for the glued cantilever problem: residual strain energy

At first, the applied bending moment will cause elastic deformation followed by plastic flow in the cantilever, all at constant crack length (Fig. 4.11(a)). In the $M\theta$ diagram there will be a departure from linearity at Y in Fig. 4.30(a) as the outer layers of the beam become plastic and thereafter ξ begins to diminish as the elastic core of the beam reduces. In the absence of fracture of the glued interface, increasing bending moment would lead to plastic collapse at C. The moment $M_y = 2\,wh^2\sigma_y/3$ and $M_c = wh^2\sigma_y$ (for a beam width w and height $2h$), and the applied moments, are both independent of the length of the beam (length of starter crack) in this particular problem. The curvature of the loaded beam is $1/\rho = \sigma_y/E\xi h$ and the corresponding rotation $\theta = a/\rho = \sigma_y a/E\xi h$ (cf. Example 4.6) with ξ and M related by equation (4.16). This permits the complete trace OYFC. . . during loading to be constructed. When ξ reaches zero at C, the rotation is unlimited since we assume an (elastic-) non-workhardening material; in practice there will be workhardening but, even so, the glue line need not decohere and the beam could simply 'roll up' plastically with no fracture whatsoever (cf. Rice's 1965 remarks that once plasticity sets in, in a rigid-plastic solid, there is no energy available to feed any crack propagation).

The stiffness of the elastoplastic beam is the same as the fully elastic beam if plane sections remain plane, i.e. $M/\theta = EI/a$ so all unloading/loading lines are parallel to the initial stiffness line OY. Superficial inspection of Fig. 4.30(a) would suggest that work areas OYFD represent irreversible plastic work Γ_F required to build up the plastic zones in the beam of constant length a to the level given by the appropriate ξ or M. But this is *not* the whole picture, for it may be shown that area OYFD contains *residual elastic strain energy* as well. The elastic strain energy at F during loading is given by equation (4.17), viz.

$$\Lambda_F = w\sigma_y^2 ha(3 - 2\xi)/3E \tag{4.17 bis}$$

The energy recovered on unloading from F is given by work area ΔFLD which is the elastic bending energy given by

$$\begin{aligned}\Delta\text{FLD} &= \tfrac{1}{2}M_B^2\, a/EI \\ &= w\sigma_y^2 ha(3 - \xi^2)^2/12E \end{aligned} \tag{4.49}$$

using equation (4.16) for M_B and $I = 2wh^3/3$. This is smaller than Λ_F so the residual elastic strain energy still contained in the unloaded beam at D must be

$$\begin{aligned}\Lambda_r &= \Lambda_F - \Delta\text{FLD} = w\sigma_y^2 ha\,((3 - 2\xi) - (3 - \xi^2)^2/4)/3E \\ &= w\sigma_y^2 ha(1 - \xi)^3\,(3 + \xi)/12E \end{aligned} \tag{4.50}$$

The pattern of residual stress and reason for its existence may be found in, for example, Timoshenko *Strength of Materials*, 3rd edn., vol II, p. 377 (Princeton, N. J., Van Nostrand, 1956).

Thus area ΔOYFD represents $(\Gamma_F + \Lambda_r)$.

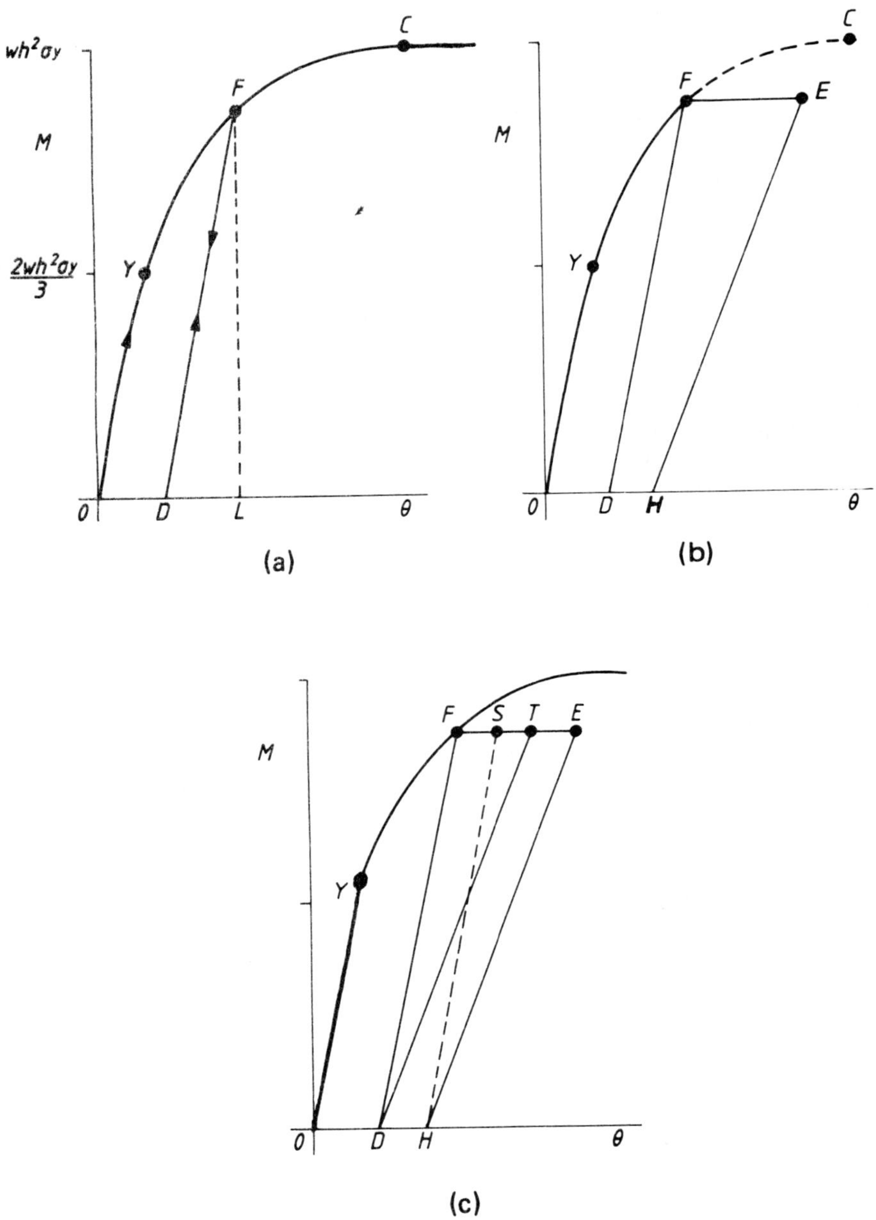
wh²σy
M
2wh²σy/3
Y
F
C
O
D
L
θ
(a)
M
Y
F
E
C
O
D
H
θ
(b)
M
Y
F
S
T
E
O
D
H
θ
(c)

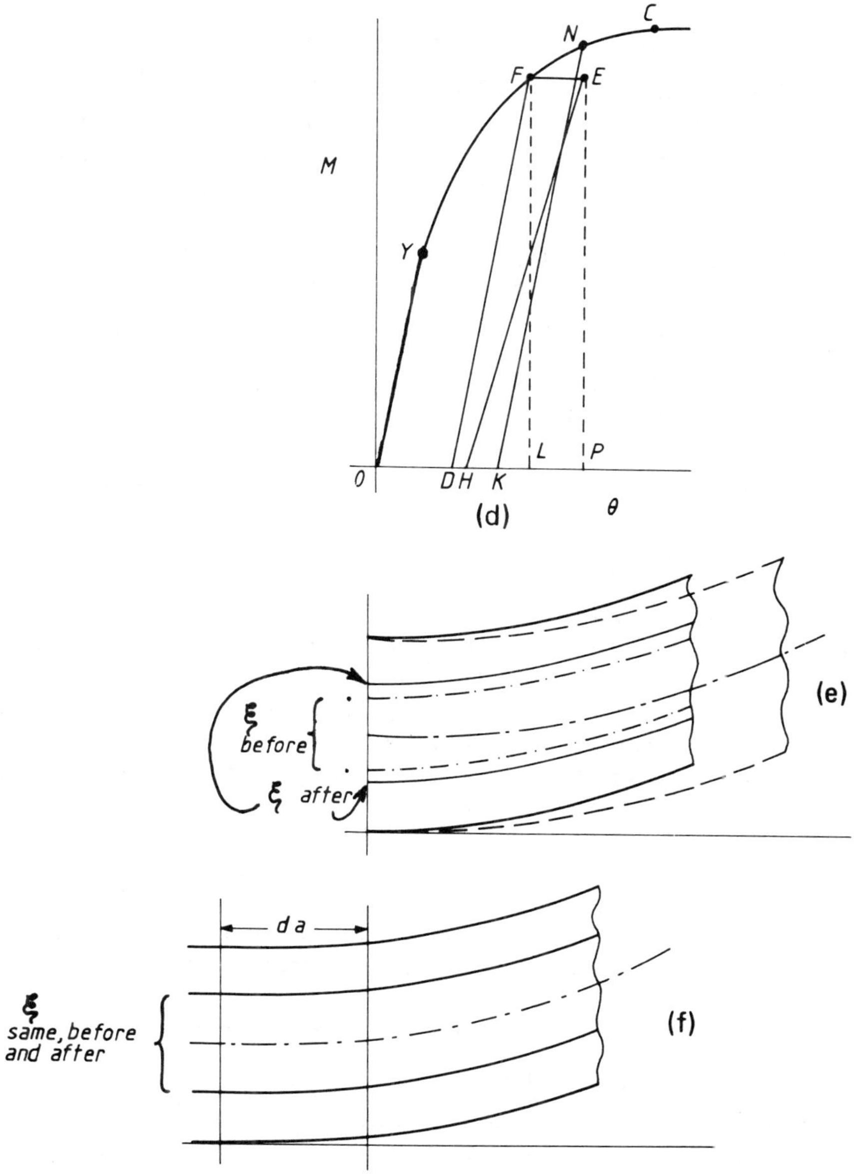

Fig. 4.30 – Moment–rotation (M–θ) curve for an elastoplastic glued beam. (a) No crack growth at F, C is plastic collapse point, (b) crack growth takes place from F to E, and (c) division of work components for the work area DFEHD in (b). (d) BYFEH is for crack growth and plastic flow and BYFNC is for plastic flow alone. (e) Deformation of the beam when no crack growth occurs. (f) Deformation of the beam after an amount of crack growth = da.

Example 4.11 What percentage of Γ_F is Λ_r, or in other words how much error is introduced if area OYFD is identified with Γ_F only?

Ans. $\Lambda_r/\Gamma_F = (1-\xi)^3\ (3+\xi)\xi/12(1-\xi)^2 = (1-\xi)(3+\xi)\xi/12$ from equations (4.50) and (4.17). For $\xi = 1$ the fraction is zero (as we should expect for the simply elastic case, but it is also zero for $\xi = 0$ in the fully plastic case as the elastic–perfectly plastic theory then says, $\Gamma \to \infty$). Calculations show that the percentage depends on ξ as follows: $\xi = 0.1$, 2%; $\xi = 0.3$, 6%; $\xi = 0.5$, 7%; $\xi = 0.7$, 6%; $\xi = 0.9$, 3%. The difference is not marked, therefore, but peaks at about 7.5% at $\xi = 0.54$ as shown by differentiation.

If fracture intervenes before plastic collapse, at F say, a loading/unloading trace looks like that in Fig. 4.30(b). The $M\theta$ history up to F is the same as in Fig. 4.30(a), but thereafter cracking occurs at constant moment given by equation (4.20a) (in terms of ξ) or equation (4.20b) absolutely. On unloading after some propagation, the stiffness of the beam has decreased since the beam has lengthened (the slope of EH is less than that of FD). As described already, work areas OYFEH represent a combination of irreversible plastic work and residual strain energy and also now, in addition, contain fracture work. It is clear that the unloaded work area OYFEH *cannot* be partitioned simply into components of irreversible work during crack initiation (supposedly area OYFD), and irreversible work of fracture and flow during crack propagation (supposedly area DFEH), without regard to the residual strain energy within these areas. The different levels of Λ_r must be accounted for in, for example, any Gurney-type measurement of R from *unloaded* $M\theta$ or Xu diagrams. Determinations of R, J or Γ made on *loaded* $M\theta$ or Xu diagrams (employing equations (4.20), (4.18) and so on in this particular problem) would be *correct*, of course. Again, if the assumption is made of rigid-plastic material behaviour (as in the Cotterell–Mai treatment of DEN sheets of ductile metals in tension, section 4.4.1; tearing tests on such sheets, section 4.4.2) there is no elastic strain energy and Λ_r does not enter the picture.

In Fig. 4.30(b), it may be shown by geometry that

$$\text{Area DFEH} = w\sigma_y^2 h\,(3-\xi^2)(4-\xi(3-\xi^2))\,(a_2-a_1)/12E\xi \quad (4.51)$$

where a_2 is the propagated crack length (giving stiffness line HE) and a_1 is the initial crack length (Problem 4.24). It may also be shown, using equation (4.18) for Γ that:

$$\text{Area DFEH} - (\Delta\Gamma + Rw(a_2-a_1)) = \Delta\Lambda_r \quad (4.52)$$

where $\Delta\Gamma = \Gamma(a_2) - \Gamma(a_1)$ and $\Delta\Lambda_r = \Lambda_r(a_2) - \Lambda_r(a_1)$; $\Lambda_r(a_2)$ is still given by equation (4.50), but using a_2 in place of a_1.

It is possible to assess the influence of Λ_r on the *combined* works of flow and fracture, i.e. on $(\Gamma + Rw(a_2-a_1))$ (Problem 4.25) but it is of more value to know what neglect of Λ_r does to values of Γ and R taken from area DFEH.

It is customary to argue (in ignorance or neglect of Λ_r) that area DFEH should be partitioned into area DFSH (plastic work associated with growth, since the stiffness does not change with merely plasticity) and ΔHSE (fracture work since crack growth alone changes the beam stiffness if plane sections remain plane). Since, in this problem, FE is horizontal, area DFSH = area DTEH and ΔHSE = ΔDFT(FT = SE) (Fig. 4.30(c)).

Thus, *geometrically* on an unloaded $M\theta$ diagram, we would identify area DFSH with Γ, i.e. area DFSH = area DFEH $- \dfrac{M^2}{2EI}(a_2 - a_1) = wh\,\sigma_y^2\,(a_2 - a_1)$ $(4(3-\xi^2) - 2\xi(3-\xi^2)^2)/(12E\xi)$ (cf. Problem 4.24). But the *actual* Γ, over the same $(a_2 - a_1)$, is given by equation (4.18) as $wh\sigma_y^2(a_2 - a_1)(1-\xi)^2/E\xi$. Thus, calling Γ* the apparent value for the plastic work obtained from area DFSH in ignorance of Λ_r, we have

$$\Gamma^* = \Gamma(4(3-\xi^2) - 2\xi(3-\xi^2))^2/12(1-\xi)^2 \tag{4.53}$$

Similarly, identifying ΔHSE $= \tfrac{1}{2}M^2(a_2 - a_1)/EI$ with $R^*w(a_2 - a_1)$, we conclude that

$$R^* = R(3-\xi^2)^2/4(3-2\xi) \tag{4.54}$$

since the 'real' R is given by equation (4.20c) as $R = (h\sigma_y^2/E)(3-2\xi)/3$.

Table 4.3 gives the ratios of Γ*/Γ and R^*/R caused by the existence of Λ_r.

Table 4.3

Ratio of Γ*/Γ and R^*/R

ξ	$\dfrac{12E\Lambda_r}{w\sigma_y^2 ha}$	Γ^*/Γ	R^*/R
1	0	∞	1.0
0.9	0.004	3.85	1.0
0.8	0.03	2.2	1.0
0.6	0.23	1.4	0.97
0.4	0.73	1.15	0.92
0.2	1.65	1.05	0.84
0	3	1.0	0.75

Notice that Λ has quite the opposite effect on Γ*/Γ and R^*/R, since Γ* approaches the correct value of Γ as ξ decreases whereas R^* deviates away from the true R for the same change in ξ. For large ξ (only just into the elastoplastic range) the *ratio* Γ*/Γ is very large since Γ is extremely small and Λ_r (although itself small) has a large influence on area DFSH. On the other hand, $R^*/R \approx 1.0$ for $\xi < 1.0$ since the system is essentially still elastic. For $\xi \to 0$, $\Gamma^*/\Gamma \to 1$ and only then, after collapse, does area DFSH (where point F is now located up at C) truly give the plastic work. Also, as $\xi \to 0$, both Λ_r and $d\Lambda_r/da$ increase (cf. equation (4.50)), and as we know from section 3.7, on cracking in the presence of residual stresses, $d\Lambda_r/da$ aids cracking and gives the appearance of the material being less tough than it really is (cf. Fig. 3.31(a)). We see in Table 4.3 that R^*/R is indeed progressively less than unity as ξ decreases. Thus, in this particular problem, if the fracture toughness of the adhesive were determined from *unloaded* work areas ΔHSE or ΔDFT just at plastic collapse in the beam, the value of R would be only 75 per cent of the true value. Values of R (or Γ) obtained from (M, θ, a) values during cracking *before unloading* (i.e. employing equations (4.20), (4.18) and so on) would, however, be *correct*.

Note, finally, the comparison between equation (4.20a) for the elastoplastic moment and $M^2 = 2REIw$ for the equivalent problem where the beam is just becoming plastic. It follows that

$$M_{\text{elpl}}^2 = M_{\text{el}}^2 (3 - \xi^2)^2/4(3 - 2\xi) \qquad (4.55)$$

Alternatively, we could write

$$M_{\text{elpl}}^2 = 2R^*EIw \qquad (4.56)$$

where $R^* = R\,(3 - \xi^2)^2/4(3 - 2\xi)$ as before. That is, the *elastoplastic* moment is given by an equation of the *elastic* type if R is replaced by the Λ_r-affected R^*, cf. Andrews (1974, 1976, 1977) 'multiplicative' approach to generalised fracture (Chapter 1). (As we have assumed that the second moment of area I is identical in both cases in this last treatment, the *only* way in which fracture can take place at different ξ experimentally in beams of the same h, is by reduction of σ_Y. This follows from the analysis in the next section.)

4.6.2 Why does cracking plus flow supervene flow alone?

In the deformation of the glued beam, there exists the possibility of increased plastic flow alone in the beam with no cracking or, alternatively, simultaneous crack propagation together with the original level of plastic flow extended over a bigger volume. Similar bifurcations in deformation are present in all elastoplastic cracking problems.

Providing (as always) that there is a mechanism at the crack tip which permits fracture to occur, it is the relative magnitudes of (incremental) work areas DFNK and DFEH (over the same $d\theta$ during loading in Fig. 4.30(d)) which determine whether merely additional plastic flow at the starter crack length occurs in the next increment of loading (i.e. the depth of the plastic zone increases), or whether a combination of plastic flow (with unchanging depth of plastic zone) and fracture occurs. In the former case, the plastic strain increases and the plastic volume increases (both because ξ decreases) so $d\Gamma = (W^P dV^P + V^P dW^P)$ increases (Fig. 4.30(e)). Since plasticity extends over a bigger region, the elastic recovery also increases (KP vs. DL) and the recovered strain energy is increased (KNP vs. DFL); unloading from N instead of F changes Λ_r since ξ increases. Alternatively, when cracking occurs, the plastic strain does not increase (since ξ remains the same so W^P is constant at the previous value) but the plastic volume increases since an additional beam volume $2hw\,da$ freed by the passage da

of the crack must sustain the same level of plasticity (defined by ξ) as existed in the beam just before propagation took place (Fig. 4.30(f)). For this case, then $d\Gamma = W^P dV^P$ only But the elastic recovery, recovered strain energy and Λ_r are all altered, *not* because unloading takes place from a higher moment (as at N) but because the unloading line from E is less steep owing to the increase in the length of the beam.

Elastic recoveries have to be accounted for in the work equations as, otherwise, cracking would always be energetically more favourable than plastic flow since work input $M d\theta$ along FE is inevitably less than $M d\theta$ over the same $d\theta$ along FN. The inclusion of elastic recoveries is a logical extension to elastoplastic fracture of 'strain energy release rate' arguments for elastic fracture in Chapter 2. This point is clearer if we write out the two work equations for the alternative incremental flows.

For combined flow and fracture we have equation (4.19), viz.

$$M d\theta = d(w\sigma_y^2 ha(3 - 2\xi)/3E) + d(w\sigma_y^2 h(1 - \xi)^2 a/E\xi) + Rw da \tag{4.19 bis}$$

in which ξ is constant but a changes (both to lengthen the crack and to increase the plastic volume of the beam).

For elastoplastic flow alone, the same equation applies, but the $Rw da$ term is omitted and now ξ changes and a is constant.
That is,

$$M d\theta = d(w_y^2 ha(3 - 2\xi)/3E) + d(w^2 \sigma_y h(1 - \xi)^2\, a/E\xi) \tag{4.57}$$

The rotation of the beam during loading is $\theta = \sigma_y a/E\xi h$, so generally

$$\begin{aligned} d\theta &= (\sigma_y/E\xi h) da - (\sigma_y a/Eh\xi^2) d\xi \\ &= (\sigma_y/E\xi h) da + (3a/2wEh^3\xi^3) dM \end{aligned} \tag{4.58}$$

using equation (4.16). In Fig. 4.30(d) the rotation during loading is produced in the one case by $(\sigma_y/E\xi h)da$ (along FE) and in the other by $-(\sigma_y a/Eh\xi^2)d\xi$ (along FN) noting that the elastic core defined by ξ decreases if M increases.

Thus cracking is favoured over flow when $(M d\theta)_{\text{cracking}} < (M d\theta)_{\text{flow}}$, i.e. when

$$\begin{aligned} &w\sigma_y^2 h(3 - 2\xi) da/3E + w\sigma_y^2 h(1 - \xi^2) da/E\xi + Rw da < \\ &-2w\sigma_y^2 ha d\xi/3E - (w\sigma_y^2 ha/E) \left(\frac{2(1 - \xi)\xi + (1 - \xi)^2}{\xi^2} \right) d\xi \end{aligned} \tag{4.59}$$

obtained by differentiating equation (4.19) w.r.t. a and equation (4.57) w.r.t. ξ. The r.h.s of inequality (4.59) simplifies to $-w\sigma_y^2 ha(3 - \xi^2)d\xi/3E\xi^2$ which becomes $+w\sigma_y h^2(3 - \xi^2)d\theta/3$ substituting for $d\xi$ in terms of $d\theta$ from (4.58) (remembering that in the case of continued plastic flow with no cracking a is constant). The l.h.s. of inequality (4.59) may be written similarly in terms of $d\theta$ for da, or equation (4.20) used directly for $(M d\theta)_{\text{cracking}}$.

We conclude that the transition from flow to cracking is favoured when†

$$REIw(3 - \xi^2)^2/2(3 - 2\xi) = (w\sigma_y h^2(3 - \xi^2)/3)^2 \qquad (4.60)$$

i.e. when

$$RE/(3 - 2\xi) = \sigma_y^2 h/3$$

or

$$h = \left(\frac{3RE}{\sigma_y^2}\right)\frac{1}{(3 - 2\xi)} \qquad (4.61)$$

(RE/σ_y^2) is a material property of the glue line and beam; h and ξ are geometric terms (linked at fracture via equation (4.61)). The relation reflects a size effect, as described in section 3.9. When $\xi = 1$ (first plasticity), $h = 3RE/\sigma_y^2$ which is the same as equation (3.76); if fracture occurs just when $\xi = 0$ (just at plastic collapse) $h = RE/\sigma_y^2$, and for h smaller the beam will simply curl up on loading with no cracking. Such 'lower' and 'upper' transitions in fracture are discussed in detail in Chapter 10. In between these two extremes in h the transition occurs at different extents of plastic flow (different ξ).

In the deviation of (4.61), we have investigated work inputs during *loading*, so that complications of different amounts of residual elastic strain energy in the unloaded system (described in section 4.6.1) do not enter the picture.

Example 4.12 Du Chang *et al.* performed experiments in 1972 on crack propagation along glue lines with the adherand in the form of end-loaded elastoplastic cantilever beams (section 4.3.3). The beams were made of 6061-T6 aluminium alloy (E = 71 GPa, σ_y = 284 MPa) and the toughness of the glue was R = 1.5 kJ/m^2. Experiments were performed with beams 3.2 mm deep. What was the extent of elastoplastic flow when fracture occurred?

Ans. Equation (4.61) may be used since, although it was derived for constant bending moment and not for end loading on the beam, it is the bending moment at the crack tip which controls fracture (cf. simple elastic fracture with $M^2 = 2REIw$ and comments in section 4.3.3). Thus

$$(3 - 2\xi) = 3RE/\sigma_y^2 h = 3 \times 1.5 \times 10^3 \times 71 \times 10^9/284^2 \times 10^{12} \times 1.6 \times 10^3 = 2.48.$$

$$\xi = (3 - 2.48)/2 = 0.26$$

That is, plastic flow in the beams extended approximately three-quarters of the way through the beam on either side of the neutral axis.

---

† The same result can, of course, be obtained simply by setting M from equation (4.20b) just to M from equation (4.16) at a common ξ since the work input is over the same θ. However, it is felt necessary in the text to go through the complete work area interactions (more particularly to include elastic effects) showing the contribution to $d\theta$ from, in the one case changes in a at constant ξ, and in the other, changes in ξ at constant a.

The reader should observe that there are *two* energy sinks when cracking occurs, (plastic flow and irreversible work of fracture), in contrast to merely the one energy-absorbing mechanism of plastic flow when cracking does not occur. Even so, the net work input can be *less* when combined flow and fracture occurs rather than flow alone and this is what governs the onset of cracking along the glue line. Minimum energy arguments such as this have wide application in problems of combined flow and fracture: the problem of cropping or bar shearing is discussed in these terms in section 5.7.

4.6.3 Correspondence between R and J_C for glued cantilever problem

If, instead of an analysis of the glued-beam problem using equation (4.2) and the concept of R as in section 4.3.3, the problem was attacked using a J-based criterion of fracture, we should proceed as follows.

For otherwise identical glued beams with different starter crack lengths, the OYFC. . . loading traces of Fig. 4.30 will be different on an $M\theta$ plot (Fig. 4.31), owing to the different starting stiffnesses. In terms of a J_C criterion for initiation, following procedure (ii) section 4.5.2, the areas under the $M\theta$ curves at various θ

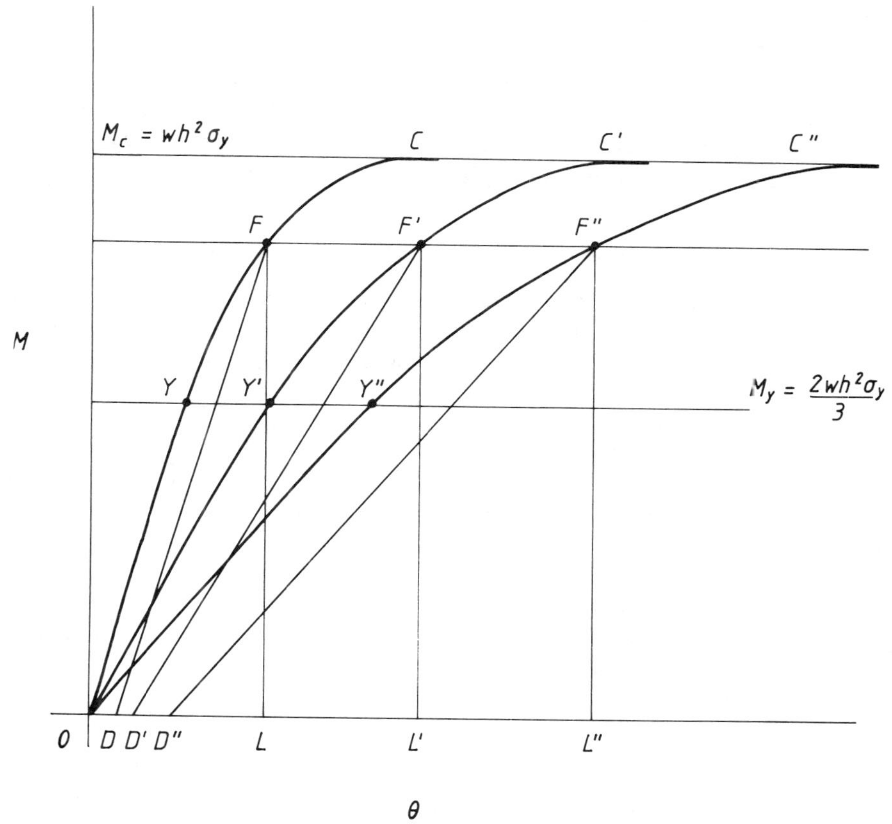

Fig. 4.31 – Evaluation of J from the M–θ diagram for the elastoplastic glued beam.

would be plotted against crack area and J determined from the slopes. J_C would be the value of J at which fracture occurred. The problem may be written out algebraically since the solution for the glued-beam problem is already known, i.e. the cracking moments at F, F′, F″ are known. (They are independent of crack length in this example, but the rotations at which they occur depend, of course, on starter crack length.)

Area OY′F′L′ in Fig. 4.3 is given by

$$U_T = \Delta OY'F'L' = w\sigma_y^2 ha(3-2\xi)/3E + w\sigma_y^2 h(1-\xi)^2 a/E\xi \qquad (4.62)$$

from equations (4.17) and (4.18) including the elastic contribution (and from what has been discovered in section 4.6.1 about Λ_T, elastic energies are best left in J formulations). In terms of θ, using $\xi = \sigma_y\, a/E\theta h$, we have

$$\Delta OY'F'L' = -w\sigma_y^2 ah/E + w\sigma_y^3 a^2/3E^2\theta + w\sigma_y h^2\theta \qquad (4.63)$$

Hence

$$-J = (\Delta OY'F'L'/w\Delta a)_{\theta\, =\text{constant}} \qquad (4.64)$$

i.e. $$J = \sigma_y^2 h/E - 2\sigma_y^3 a/3E^2\theta \qquad (4.65)$$

or $$J = (\sigma_y^2 h/E)(1-2\xi/3)$$

Equations (4.64) or (4.65) are algebraic expressions for J along paths such as OYFC in either Fig. 4.30(a) or 4.31. When the rotation becomes θ_C (at which cracking actually occurs) or equivalently ξ falls to ξ_C (the extent of the plastic zone when cracking occurs), the expressions are formulae for J_C.

Example 4.13 If fracture along the glue line in Fig. 4.10 occurs just as the outside layers of the beam become plastic, show that equation (4.65) gives a value for J_C which is identical with G_C and R from elastic theory.

Ans. In equation (4.65) there is a lower limit on θ given by $(\sigma_y a/Eh)$ at $\xi = 1$ when the outer layers of the beam are just reaching the yield strain, and if cracking is about to occur under those conditions $J_C = \sigma_y^2 h/3E$. This corresponds with the prediction of elastic fracture mechanics, i.e. $\Lambda = \int M d\theta = EI\theta^2/2a$ and $J_C = G_C = -(\partial\Lambda/\partial A)_\theta = EI\theta^2/2a^2 w = \sigma_y^2 h/3E$, when the outer layer bending stress equals σ_y. The expression also corresponds with $M^2 = 2REIw$ given in section 4.3.3 for elastic fracture with $\xi = 1$ since at first yield, $M = 2wh^2\sigma_y/3$ from equation (4.16) with $\xi = 1$. R and J_C thus coincide if the beam remains elastic.

Example 4.13 shows, as we should expect, that J_C and R are equivalent during elastic deformation of the beam. What, though, of the equivalence of J_C and R when the beam is partially plastic before cracking occurs? Comparison of equations (4.20c) and (4.65) shows that they correspond for all ξ_C through to collapse, i.e.

$$R \equiv J_C = (h\sigma_y^2/E)(1 - 2\xi_C/3) \tag{4.66}$$

$$= (h\sigma_y^2/E)\left(1 - \sqrt{\frac{4}{3}\left(1 - \frac{M_C}{wh^2\sigma_y}\right)}\right) \tag{4.67}$$

(cf. equations (4.61) and (4.20b)). Thus insertion of ξ_C or M_C (obtained experimentally at cracking under load) into these algebraic relations gives J_C which is identical to R. For example, if crack propagation were to occur just at complete plastic hinging with $\xi_C = 0, J_C = (h\sigma_y^2/E)$.

As far as graphical procedures for J_C are concerned, consideration of segmental work areas such as OY′F′F″Y″O in Fig. 4.31, given by the difference between two *loading* curves, gives relation (4.66) again, this alternative method for J_C once more confirming the equivalence of R and J_C in this problem. The result makes sense physically, in that if we had two specimens differing only in cantilever (crack) lengths a_1 and a_2, the difference in work areas on loading out to the (identical) cracking moment, i.e. area OY′F′F″Y″O, must represent the 'work equivalent' of the difference between the two cases – which is the existence of an additional length $(a_2 - a_1)$ of free-standing crack in the second case. Hence, area OY′F′F″Y″O must have a value $w(a_2 - a_1)R$ and hence R and J must coincide. While this is similar to diagrams for non-linear elastic fracture in Chapter 2 (e.g. Fig. 2.10), the similarity falls down as soon as *unloading* enters the picture. The elastoplastic glued cantilever loaded up along OYF and cracking to F′ unloads down F′D′, not down F′Y′O (Fig. 4.31), and DFF′D′ represents the sum of Γ, RA and Λ_r. Fig. 4.29(c) and Table 4.3, show that values determined from *unloaded* work areas HSE gives Λ_r-affected R values which are underestimates of the real R.

As regards propagation, it may be shown by application of equation (4.34) that $J_R = J_C = (\sigma_y^2 h/E)(1 - 2\xi/3) =$ constant throughout in this steady-state problem (Problem 4.27); see end of section 4.6.4.

4.6.4 General (X, u, R, A) relations in presence of remote flow

In Section 2.4.3 a generalisation of the standard 'compliance calibration' equation for R or G_C in elastic fracture was derived, viz.

$$R = G_C = \tfrac{1}{2} X^2 \, \mathrm{d}C/\mathrm{d}A + X \mathrm{d}u_r/\mathrm{d}A \tag{2.21 bis}$$

which took into account residual displacements u_r. Such residual displacements were *elastic* in nature and the whole of the unloaded work area went to fracture work. The general equation (and the equivalent relation (2.23)) applied, among other cases, to elastic tearing problems and elastic cracking in the presence of residual stresses (Chapter 3).

When residual displacements in *Xu* plots are caused *not* by elastic effects but by remote flow (which has been the theme of this chapter), it is *not* correct to apply equation (2.21) to the problem, even though the *Xu* plots on cursory inspection seem the same (e.g. Figs. 2.20, 4.30 and 4.31), a yet more general relationship should be used which takes into account plastic flow and even residual elastic strain energies on unloading, the importance of which has been discovered already in section 4.6.1.

To illustrate that equation (2.2.1) gives the wrong answer for cracking in the presence of remote flow, consider the problem of peeling of glued soft metal strips given in section 4.3.1. For rigid-plastic deformation, $u_r = 2A/w$ and $du_r/dA = 2/w$; $dC/dA = 0$, so equation (2.21) gives $R = 2X/w$ or $X/w = R/2$ and *not* equation (4.14), which is $X/w = (R/2 + \sigma_y B^2/2\rho)$. The plastic flow component of the 'enhanced toughness' $(R + \sigma_y B^2/\rho)$ has not been predicted at all. Similar conclusions are reached if equation (2.21) is applied to other problems where residual displacements are caused by plastic flow, e.g. the metal tearing problem of section 4.4.2.

Since equation (2.21) is a version of $Xdu = d\Lambda + RdA$ which accommodates residual displacements caused by elastic effects, we aim to derive a version of $Xdu = d\Lambda + RdA + d\Gamma$ which accommodates residual displacements caused by both plastic and ealstic effects. Consider Fig. 4.32 which may be compared with Fig. 2.20(a). Remote flow commences at Y (the body is considered to be linear-elastic for simplicity) and cracking commences at L (which is shown on a rising part of the *Xu* plot, but which just as well could be after a maximum in the plot. The stiffness of the plastically deforming body just at first cracking is the same as the original stiffness, so OY//SL but at M, with a longer crack, the slope NM is less. The total work done on the system under load at M is given by area

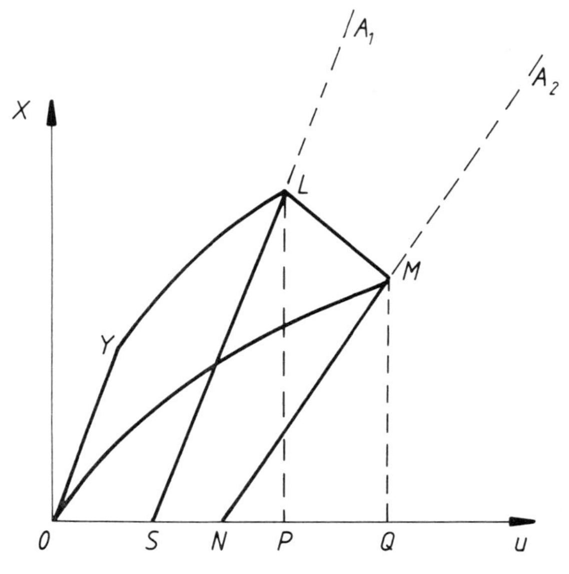

Fig. 4.32 – Generalised load–deflexion curve OYLM when crack growth occurs in the presence of combined plastic flow and residual strain energy release.

OYLMQO. The unloaded area OYLMNO represents work dissipated in fracture and flow, *together with* residual strain energy if the plastic flow has been non-uniform (which usually is the case, cf. section 4.6.1).

Consider events between L and M when the crack area increases from A_1 to A_2. Work area SLPS $= \frac{1}{2} C_{A_1} X_L^2$ is the elastic strain energy stored at L under load, where $C_{A_1} = (u/X)$ is the compliance with crack area A_1; area PLMQP $= \int X du$ is the external work done to cause *both* crack propagation *and* additional plastic flow; area NMQN is the elastic strain energy recovered $(= \frac{1}{2} C_{A_2} X_M^2)$. Let α be the proportion of the extenal work required for plastic flow i.e.

$$\alpha \int_L^M X du = (\Gamma_M - \Gamma_L)$$

Again, $(1 - \alpha) (\int_L^M X du)$ is the fraction of the external work required in the energy rate balance to help produce fracture.

We may identify an elastic strain energy release rate G in the system, given by the rate of change in *total* elastic strain energy with crack area, i.e.

$$G = \frac{(1-\alpha) \int_L^M X du + (\frac{1}{2} C_{A_2} X_M^2 - \frac{1}{2} C_{A_1} X_L^2) + (\Lambda_{r_M} - \Lambda_{r_L})}{(A_2 - A_1)}$$

where $(\Lambda_r)_{L,M}$ is the residual elastic strain energy retained in the system which must be included in the analysis since the recovered strain energies SLPS, NMQN are *less* than the elastic strain energies actually in the system under load at L and M owing to non-uniform plastic flow. In the limit, therefore, as $A_2 \to A_1$

$$G = (1-\alpha) \frac{d}{dA} (\int X du) - \left(XC \frac{dX}{dA} + \frac{X^2}{2} \frac{dC}{dA} \right) - \frac{d\Lambda_r}{dA} \qquad (4.68)$$

since $d(\frac{1}{2} CX^2) = XC\, dX + (X^2/2) dC$.

Equation (4.69) is the new generalised relation for (X, u, R, Γ, A) which accommodates irreversible as well as reversible residual displacements.

Only if α is known can G be computed. If $\alpha = 0$, there is no plasticity to cause Λ_r, and all the external work is fracture work. Then

$$R = G = X du/dA - (XC\, dX/dA + (X^2/2) dC/dA)$$

and since $u = u_r + XC$, manipulation gives

$$R = G = (X^2/2)\,(dC/dA) + X du_r/dA$$

which is equation (2.21). This is the correct equation to apply to problems such as cracking in fibre-reinforced solids where all the work is fracture work and u_r is caused by geometric interference at the crack tip.

For rigid-plastic flow, $d\Lambda$ and $d\Lambda_r = 0$ and $R = G = (1 - \alpha) \frac{d}{dA} (\int X du)$. The factors α and $(1 - \alpha)$ represent the partitioning of the total work into fracture work and plastic work, such as in the peeling of glued soft metal strips (section 4.3.1) where $(1-\alpha)\dfrac{\mathrm{d}}{\mathrm{d}A}(\int X\mathrm{d}u) \equiv R$ and $\alpha\ \mathrm{d}/\mathrm{d}A\ (\int X\mathrm{d}u) \equiv \sigma_{\mathrm{Y}}B^2/\rho$ (cf. equation (4.13)). Similar comments apply to tearing of metal strips. Example 4.14 following shows how the full relation, including Λ, Γ and Λ_{r} may be applied to the glued cantilever beam problem discussed earlier.

Example 4.14 Determine the residual rotation θ_{r} of the unloaded elastoplastic beam in Fig. 4.10(a), and employ equation (4.68) to predict the conditions for crack propagation.

Ans. In terms of (M, θ), equation (4.68) becomes

$$R = G = (1-\alpha)\ \frac{\mathrm{d}}{\mathrm{d}A}\ (\int M\mathrm{d}\theta) - \left(M(\theta/M)\ \frac{\mathrm{d}M}{\mathrm{d}A} + \frac{M^2}{2}\ \frac{\mathrm{d}(\theta/M)}{\mathrm{d}A} \right) - \frac{\mathrm{d}\Lambda_{\mathrm{r}}}{\mathrm{d}A} \qquad \text{(a)}$$

In this problem, M is independent of crack length and $\theta = \theta_{\mathrm{r}} + Ma/EI$. This results in

$$R = G = M^2/2wEI + M\mathrm{d}\theta_{\mathrm{r}}/\mathrm{d}A - \alpha\ \frac{\mathrm{d}}{\mathrm{d}A}\ \left(\int M \left\{ \mathrm{d}\theta_{\mathrm{r}} + \frac{M\mathrm{d}a}{EI} \right\} \right) - \mathrm{d}\Lambda_{\mathrm{r}}/\mathrm{d}A \qquad \text{(b)}$$

The penultimate term on the right-hand side is the component of the external work for plastic flow of the beam, which is $\mathrm{d}\Gamma/\mathrm{d}A$ and is known algebraically; again $\mathrm{d}\Lambda_{\mathrm{r}}/\mathrm{d}A$ is also known. The graphical interpretation of equation (b) is shown in Fig. 4.33.

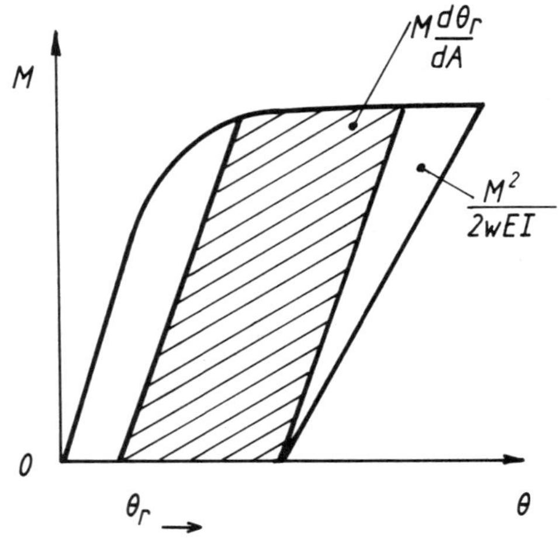

Fig. 4.33 – Division of work areas in the M–θ diagram for fracture of the elasto-plastic glued beam (see text).

The curvature of the loaded beam is $(1/\rho) = (\sigma_y/E\xi h)$ and the corresponding rotation under load is $a/\rho = \sigma_y a/E\xi h$. Hence, $\theta_r = \sigma_y a/E\xi h - Ma/EI$. Equation (4.16) relates M and σ_y, so eliminating M we obtain

$$\theta_r = (\sigma_y a/Eh)\,(1/\xi - (3-\xi^2)/2)$$

Now

$$M^2/2wEI = (\sigma_y^2 h/12E)\,(3-\xi^2)^2$$

$$M\mathrm{d}\theta_r/\mathrm{d}A = (\sigma_y^2 h/E)\,\{4(3-\xi^2) - 2\xi\,(3-\xi^2)^2\}/12\xi$$

$$\mathrm{d}\Gamma/\mathrm{d}A = (\sigma_y^2 h/E)\,(1-\xi)^2/\xi$$

and

$$\mathrm{d}\Lambda_r/\mathrm{d}A = (\sigma_y^2 h/E)\,(1-\xi)^3\,(3+\xi)/12$$

This gives, in the M,θ version of equation (4.68) stated above, $R = (\sigma_y^2 h/E)(1 - 2\xi/3)$ which coincides with equation (4.20c).

It is clear that application of equation (2.21), rather than equation (4.68), to the problem in Example 4.14 will fail to give the correct answer. It may be shown that if so, $M^2 \to REIw\, 2\xi(3-\xi^2)/(4 - 3\xi + \xi^3)$.

It is also clear that equation (4.68) gives the correct R or J_C *during propagation* and not merely from the conditions at first fracture. We have seen in section 4.6.1 that, even if Γ is accounted for in incremental total work areas during propagation, R is underestimated owing to the existence of Λ_r; most analyses, however, do not even remove Γ which, in most problems, leads to an increasing 'J_R' behaviour.

4.6.5 Interpretation of unloaded *Xu* diagrams after elastoplastic flow

Section 4.6.1 serves the valuable purpose of introducing the notion that residual elastic strain energy must be accounted for in the partitioning of *unloaded* work areas after non-uniform elastoplastic flow. Indeed it is probable that equations (1.2) and (4.2) should always be written with a residual elastic strain energy component so as properly to describe events when elastoplastic cracks propagate under decreasing load, i.e.

$$X\mathrm{d}u = \mathrm{d}\Lambda + \mathrm{d}\Gamma + R\mathrm{d}A + \mathrm{d}\Lambda_r \tag{4.69}$$

and so on.

Since the concept of J is lost as soon as unloading occurs, and since it is impossible to partition total work areas *without* unloading, it is clear that attempts to interpret unloaded *Xu* or *Mθ* diagrams in terms of J are meaningless. But is it possible to relate other quantities determined from unloaded work areas, and J determined from the work area just before unloading?

Consider Fig. 4.34(a) which depicts two loading lines for two specimens with crack area A and $(A + \Delta A)$, with unloading from the same displacement.

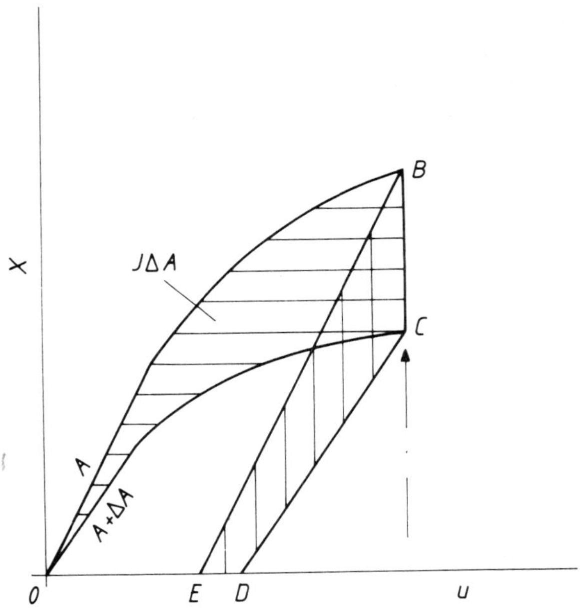

(a)

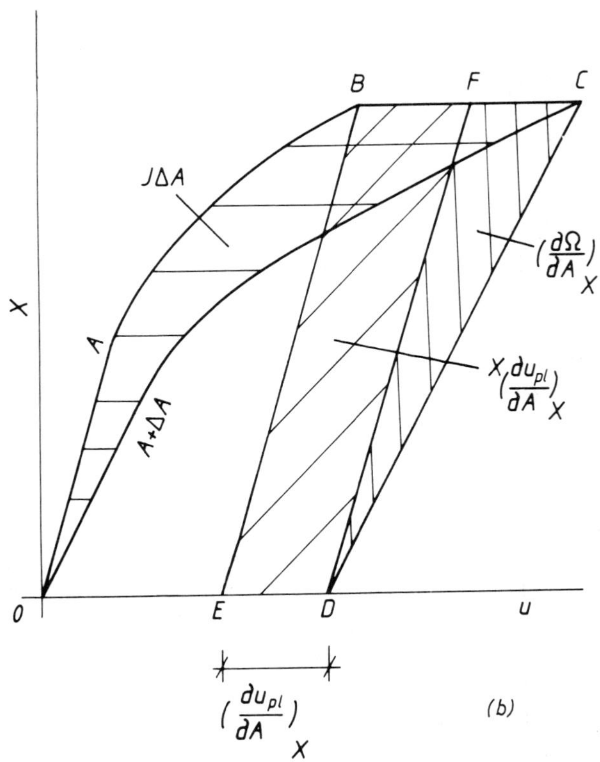

(b)

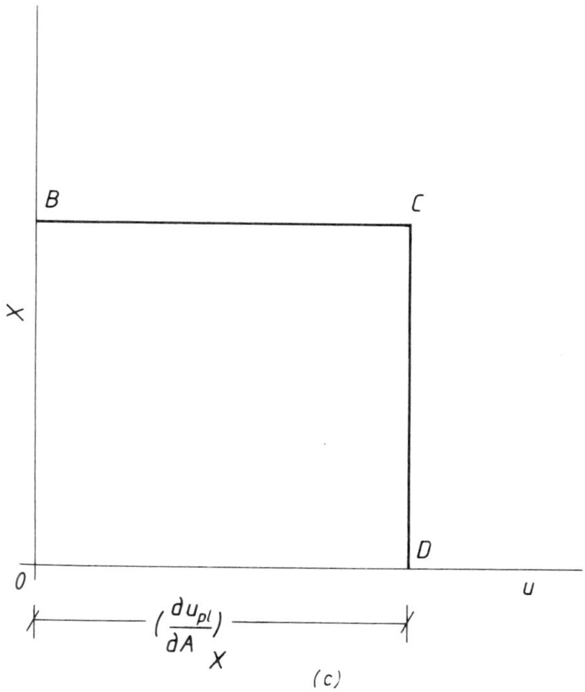

Fig. 4.34 – (a) Two elastoplastic loading lines for two specimens, having starter crack areas A and $(A + dA)$, with linear elastic unloading from a common displacement (b) Similar to (a), but with unloading from a common load. (c) The comparable behaviour of specimens displaying rigid-plasticity.

We have

$$\Delta\,\mathrm{OBC} = J\Delta A$$

i.e.

$$\frac{\Delta\mathrm{OBC}}{\Delta A} = J$$

and

$$\Delta\mathrm{EBCD} = \mathrm{R}\,\Delta A + \Delta\Gamma + \Delta\Lambda_{\mathrm{r}}$$

and

$$\frac{\Delta\mathrm{EBCD}}{\Delta A} = R + \frac{\mathrm{d}\Gamma}{\mathrm{d}A} + \frac{\mathrm{d}\Lambda_{\mathrm{r}}}{\mathrm{d}A}$$

Since, by geometry, area OBC is greater than area EBCD, the following inequality arises

$$G \leqslant R + \frac{\mathrm{d}\Gamma}{\mathrm{d}A} + \frac{\mathrm{d}\Lambda_{\mathrm{r}}}{\mathrm{d}A} \leqslant J \qquad (4.70)$$

with G as the lower limit for (real) elastic fracture.

Again, consider Fig. 4.34(b) for a similar situation to Fig. 4.34(a) but with unloading from the same load. We have

$$\frac{\Delta \mathrm{OBC}}{\Delta A} = J$$

$$\frac{\Delta \mathrm{EBCD}}{\Delta A} = R + \frac{\mathrm{d}\Gamma}{\mathrm{d}A} + \frac{\mathrm{d}\Lambda_{\mathrm{r}}}{\mathrm{d}A}$$

While the areas OBC and EBCD are not equal in Fig. 4.34(b), it is not clear this time whether $J \gtrless R + \frac{\mathrm{d}\Gamma}{\mathrm{d}A} + \frac{\mathrm{d}\Lambda_{\mathrm{r}}}{\mathrm{d}A}$. Using the complementary elastic strain energy $\Omega = Xu_{\mathrm{el}} - \Lambda$, we may rewrite this in the form

$$R + \frac{\mathrm{d}\Gamma}{\mathrm{d}A} + \frac{\mathrm{d}\Lambda_{\mathrm{r}}}{\mathrm{d}A} = \left(\frac{\partial \Omega}{\partial A}\right)_X + \left(\frac{\partial u_{\mathrm{pl}}}{\partial A}\right)_X \qquad (8.39\text{ bis})$$

so whether the rate of change with respect to crack area of the unloaded work area EBCD is greater or less than J depends on the magnitude of $X\left(\frac{\partial u_{\mathrm{pl}}}{\partial A}\right)_X$

The important point that emerges is that in neither the constant displacement case nor the constant load case does the rate of change w.r.t. crack area of the unloaded elastoplastic work area EBCD coincide with J determined (under load) from equation (4.34).

For a rigid-perfectly plastic solid, it follows that

$$J = R + \frac{\mathrm{d}\Gamma}{\mathrm{d}A}$$

being given by the rectangular work area in Fig. 4.34(c) (cf. the elementary problems in sections 4.3 and 4.4). However, this is *only* true when $\mathrm{d}\Lambda$ and $\mathrm{d}\Lambda_{\mathrm{r}}$ are zero (cf. section 4.6). The connexion between J and R is more complicated than this simple relation when elastic effects are significant as shown by the foregoing inequalities.

Turner (e.g. 1979) gives the symbol I to $(R + \mathrm{d}\Gamma/\mathrm{d}A)$ and applies it to a number of plastic fracture problems where fracture load is supposed to coincide with the plastic collapse load. As explained in Chapter 10 (and as the reader will have observed from the analysis of the elastoplastic glued cantilever beam problem in this chapter) it is a special case for the fracture and collapse loads to coincide. Even so, if the plastic flow is extensive and if elastic effects may be neglected, I is a useful parameter. In general elastoplastic flow, however, the I term should formally include the $\mathrm{d}\Lambda_{\mathrm{r}}/\mathrm{d}A$ component as well as $(R + \mathrm{d}\Gamma/\mathrm{d}A)$. Further consideration of this topic occurs in section 8.10 concerning crack stability in combined flow and fracture.

PROBLEMS

4.1 An experimental steel with a room temperature yield strength of 700 MPa is to be used for the construction of ship hulls. It is required to determine the plane strain fracture toughness of this metal at 0°C. Estimate the thickness of the specimen required if a standard Charpy V-notch impact test gives an energy absorption of 25 J and a tensile test gives a yield strength of 800 MPa at 0°C.

4.2 Show, in the Dugdale strip yield model, that if the second term of the ln(sec) series is retained as well as the first term, an expression for K_C results which is similar to Irwin's embedded crack-tip zone corrected formula (see Example 2.14), differing only in a coefficient of $\pi^2/24$ in place of 1/2.

4.3 Using the approximate expressions for the mode I *elastic* stresses near the tip of a crack given below draw approximate boundaries for the plastic enclave under conditions of plane stress and plane strain. Use the Maxwell–von Mises criterion for plastic flow. Calculate the furthermost distance of the edge of the plastic zone from the crack tip measured in the direction of the crack for both cases as a function of $(K/\sigma_Y)^2$. Compare these values with Irwin's correction factor and comment.

The energy dissipated per unit area of crack growth in plane stress on the basis of the Dugdale model is $\sigma_Y\delta$ where δ is the crack opening displacement. Use this fact to comment on the Irwin correction factor in light of the Dugdale model.

$$\sigma_{xx} = \frac{K_I}{(2\pi r)^{1/2}} \cos\frac{\theta}{2}\left[1 - \sin\frac{\theta}{2}\sin\frac{3\theta}{2}\right]$$

$$\sigma_{yy} = \frac{K_I}{(2\pi r)^{1/2}} \cos\frac{\theta}{2}\left[1 + \sin\frac{\theta}{2}\sin\frac{3\theta}{2}\right]$$

$$\tau_{xy} = \frac{K_I}{(2\pi r)^{1/2}} \sin\frac{\theta}{2}\cos\frac{\theta}{2}\cos\frac{3\theta}{2}$$

4.4 Estimate the minimum COD required to prevent cleavage fracture in 25 mm thick structural steel plate at a temperature of 253K at which temperature its yield stress is 240 MPa. The Young modulus of the steel is 210 GPa.

4.5 A large spherical welded pressure vessel is to be fabricated on site from low alloy steel plates 75 mm thick of yield point 450 MPa. It will be heat treated *in situ* after fabrication and the material after simulated heat treatment has the following Charpy V-notch energy absorptions

Temp. (°C)	−40	−20	0	20	40
Charpy V (J)	20	54	81	162	189

The vessel because of its size is to be pneumatically tested to an unfactored hoop stress of 0.9 times the yield point. The vessel has been nondestructively tested but may still possess surface cracks up to 5 mm deep and 15 mm long. The maximum stress concentration at a nozzle is 2.35. Discuss the feasibility of this project. (BSI PD 6493:1980, Fig. 4.6, will be useful for solving this problem.)

4.6 In COD-based design, welding residual stresses are accounted for in calculations by increasing the applied strain by an amount equal to the yield strain. Discuss the sense of this modification to theory.

4.7 Determine the critical length (L) of the central ligament in a double deep edge-notched tension specimen (Fig. 4.12(a)) being used for plastic fracture propagation tests at which the entire ligament necks just before crack initiation.

4.8 Show that when a material workhardens according to $\sigma = \sigma_0 \epsilon^n$ the force required for steady tearing of a ductile sheet shown below is given by $X = [\sigma_0 B^{n+2}/(n+1)]\ (w/\rho^{n+1}) + RB$.

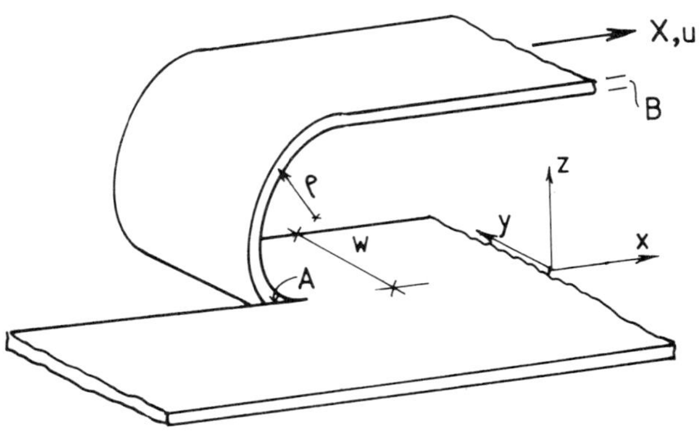

4.9 Show that the shape of the bend in steady tearing of the workhardening ductile sheet described in 4.8 is given by the solution:

$$Xz = \frac{\sigma_0 B^{n+1}}{2^{n+2}(n+1)\rho^n}$$

with $(1/\rho) = \mathrm{d}^2 z/\mathrm{d}x^2$ and $\sigma = \sigma_0 \epsilon^n$

4.10 In the guillotining of thin sheets of ductile materials, 'burrs' are sometimes formed on the cut edges caused by bluntness in the blade. Show that the occurrence of burring effectively increases the specific work of fracture to $(R + \tau_y u_0)$ where R is the specific work of fracture of the sheet in the presence of sharp tooling (no burring), τ_y is the shear yield stress of the material and u_0 is the 'height' of the burr.

4.11 Show that when cracking occurs along the glue line in Fig. 4.11, with the beam experiencing elastoplastic deformation, the toughness of the glue is given by:

$$u = [2\sqrt{w}\,\sigma_y^{3/2}/3XE]\left[\frac{2}{9X}(3\sigma_y\,wh^2 - 3Xa + 3Xx)^{3/2} - (3\sigma_y h^2 w - 3Xa)^{1/2}\,x - \frac{2}{9X}(3\sigma_y wh^2 - 3Xa)^{3/2}\right]$$

4.12 Determine the end displacement of an annealed low carbon steel beam 2 mm high and 200 mm long, loaded as in Fig. 4.10, when the beam has yielded half-way through the section ($E = 210$ GPa; $\sigma_y = 200$ MPa)

4.13 Consideration of the elastoplastic interface in Fig. 4.11 gives $d^2u/dx^2 = \sigma_y/E\xi h$. Substituting for ξ from equation (4.16) show that the deflexion curve of the elastoplastic part of the beam is given by

$$R = \frac{{\sigma_y}^2 h}{E}\left(1 - \frac{2}{\sqrt{3}}\sqrt{1 - \frac{Xa}{\sigma_y wh^2}}\right)$$

4.14 Determine the elastic deflexion curve for the right-hand portion of the beam in Fig. 4.11, and taking account of continuity at x^* (where the slopes and deflexions of plastic and elastic portions must match) show that the deflexion at the free end of the beam under X is given by:

$$u(a) = \frac{Xa^3}{2wh^3E} + \frac{2\sqrt{w}\,{\sigma_y}^{3/2}}{3XE}\left(\frac{2}{9X}(\sigma_y wh^2)^{3/2} - \frac{a}{3}(3\sigma_y wh^2 - 3Xa)^{1/2}\right)$$
$$- \frac{3X}{2wh^3E}\left(a - \frac{2}{3}\frac{\sigma_y wh^2}{X}\right)^2\left(\frac{a}{3} + \frac{1}{9}\frac{\sigma_y wh^2}{X}\right)$$
$$+ \frac{2}{3}\frac{\sigma_y wh^2}{X}\left(\frac{2\sqrt{w}\,{\sigma_y}^{3/2}}{3XE}\right)(\sqrt{\sigma_y wh^2} - \sqrt{3\sigma_y wh^2 - 3Xa})$$
$$- \frac{3X}{2Ewh^3}\left(\frac{a^2}{2} - \frac{2}{9}\frac{\sigma_y^2 w^2 h^4}{X^2}\right)$$

4.15 The load–displacement record shown overleaf was obtained from a COD test on a low carbon steel. Calculate the COD using BS 5762:1979 and estimate J_C.

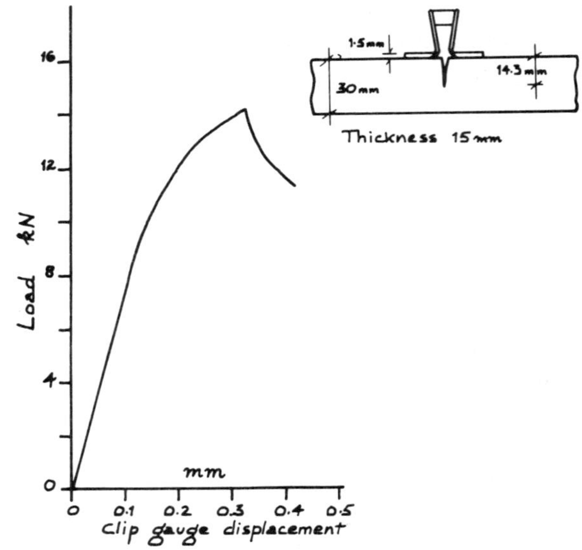

4.16 Show that Prandtl–Reuss incremental theory of plasticity reduces to Hencky's deformation theory of plasticity for the special case of proportional straining.

4.17 Show that a crack-tip blunting relation between J and apparent crack advance Δa is given by $J = 2\sigma_y \Delta a$ if crack tip plastic flow is assumed to occur by 45° slip.

4.18 Discuss the J-intergral concept making reference to the line integral and the energy approach describing the relationship between the two. How is the J integral measured experimentally for a three-point bend specimen (a) with a short crack, (b) a deep crack? What is the expression for the J integral for a deep crack in a three-point bend specimen? Derive the expression for the J integral for a deep edge notch tensile specimen in terms of the work of fracture.

4.19 Show that the J-(work area) relation for a deeply-notched round bar in tension is $J = \dfrac{1}{2\pi r^2}\,[3 \int X \mathrm{d}u - Xu]$ where r is the net section radius, and hence that under limit load conditions $\eta_{pl} = 1$.

4.20 Starting from the J-contour integral, equation (4.34) show that equation (4.26) for the specific essential work of fracture for a DEN specimen can be obtained.

4.21 The plastic constraint factor ϕ for a shallow SEN sample is given by Turner (1979) as $\phi = 1 + \ln(1 + a/W)$. Show that $\eta_{pl} \approx 3a/W$.

4.22 For notched samples in bending having $a/W < 0.2$, the plastic constraint factor is $\phi \approx 1 + \ln[1 + 2(a/W)]$. Show that $\eta_{pl} \approx 10\,a/W$.

4.23 Discuss Turner's observation that Irwin's plastic zone correction (which permits elastic fracture analyses to be used in mildly elastoplastic problems) is an estimate of J in contained yield.

4.24 Prove in Fig. 4.30 that the plastic work (Γ) is associated with area DFSH and is given by:

$$\text{area DFSH} = wh\sigma_Y{}^2(a_2 - a_1)\,[4(3 - \xi^2) - 2\xi(3 - \xi^2)^2]/(12E\xi).$$

4.25 Assess the influence of residual elastic strain energy on the combined works of flow and fracture for elastoplastic cracking of a glued cantilever (as in Fig. 4.10).

4.26 Show by geometry that work area OY′F′F″Y″O in Fig. 4.31 leads to equation (4.66) for J_C

4.27 Show that $J_R = J_C$ for the glued elastoplastic cantilever beam problem of Fig. 4.10.

REFERENCES

Ahmad, J. and Kanninen, M. F. (1984) Proceedings Conf. on Numerical Methods in Fracture Mechanics, University College Swansea.

Albrecht, P., Andrews, W. R., Gudas, J. P., Joyce, J. A., Loss, F. J , McCabe, D. E., Schmidt, D. W. and Van Der Sluys, W. A. (1982) *J. Testing and Evaluation (ASTM)* **10**, 245.

Andrews, E. H. and Bhatty, J. I. (1982) *Int. J. Fract.* **20**, 65.

ASTM Standard E813. *The Determination of J_{Ic}, a Measure of Fracture Toughness.*

Atkins, A. G. and Mai, Y. W. (1979) *J. Mater. Sci.* **14**, 2747.

Australian Standard (1983) AS.2205.7.3. *Crack Opening Displacement (COD) Test.*

Barenblatt, G. I. (1962) *Advances in Applied Mechanics,* Vol. 7. New York, Academic Press, p. 55.

Barker, L. M. (1979) *Int. J. Fracture* **15**, 515.

Barker, L. M. and Baratta, F. I. (1980) *J. Test Eval.,* **8**, 97.

Barsom, J. M. and Rolfe, S. T. (1970) *ASTM STP* **466**, 281.

Bilby, B. A. (1973) Proc. ICF-3 (Munich) XI, 1.

Bilby, B. A. (1979) Proc. 4th Tewksbury Fracture Symposium, University of Melbourne.

Bilby, B. A. (1982) *J. Mech. Phys Solids* **30**, 193.

Bilby, B. A., Cottrell, A. H. and Swinden, K. H. (1963) *Proc. Roy. Soc.* **A272**, 55.

British Standard (1979) BS 5782 *Crack Opening Displacement (COD) Testing.*

Borek, D. (1983) *Elementary Engineering Fracture Mechanics* (3rd edn). Nordhoff International.

Burdekin, F. M. (1981) *Proc. Institution of Mechanical Engineers* **195**, 73.

Burdekin, F. M. and Dawes, M. G. (1971) in *Practical Application of Fracture Mechanics to Pressure Vessel Technology,* 28. London: Institution of Mechanical Engineers.
Burdekin, F. M. and Stone, D. E. W. (1966) *J. Strain Analysis* **1**, 145.
Chan, M. K. V. and Williams, J. G. (1983) *Int. J. Fract.* **19**, 145.
Chang, M. D., de Vries, K. L. and Williams, M. L. (1972) *J. Adhesion* **4**, 221.
Chinese Standard (1980) GB 2358-80 *Crack Opening Displacement (COD) Testing Method.*
Cotterell, B. and Reddell, J. K. (1977) *Int. J. Fract.* **13**, 267.
Cotterell, B. and Mai, Y. W. (1982) in *Advances in Fracture Research* (ICF-5) D. François (ed.), Vol. 4 London: Pergamon Press, p. 1683.
Cotterell, B. and Mai, Y. W. (1984) *Int. J. Fract.* **24**, 229.
Cotterell, B., Lee, E. and Mai, Y. W. (1982) *Int. J. Fract.* **20**, 243.
Cottrell, A. H. (1961) I.S.I. Special Report No. 69, p. 281.
Dugdale, D. S. (1960) *J. Mech. Phys. Solids* **8**, 100.
Electric Power Research Institute (EPRI) (1980) Workshop Report NP-80-10-LD/WS 80-912.
Electric Power Research Institute (EPRI) (1981) *Plastic Fracture Handbook.*
Eshelby, J. D. (1951) *Phil. Trans. Roy. Soc.,* **A244**, 87.
Eshelby, J. D. (1956) *Solid State Physics,* Vol. 3. New York: Academic Press, p. 79.
Eshelby, J. D. (1968) in *Fracture Toughness.* London: Iron and Steel Institute Publication No. 121, p. 30.
Eshelby, J. D. (1970) in *Inelastic Behaviour of Solids,* M. F. Kanninen (ed.). New York: McGraw–Hill, p. 77.
Fearnehough, G. D. *et al.* (1971) in *Practical Application of Fracture Mechanics to Pressure Vessel Technology.* London: Institution of Mechanical Engineers, p. 119.
Garwood, S. J. and Turner, C. E. (1977) ICF-4 (University of Waterloo) Vol. 3, p. 279.
Garwood, S. J. and Turner, C. E. (1978) *Int. J. Fract.* **14**, R195.
Goldman, N. L. and Hutchinson, J. W. (1975) *Int. J. Solids & Struct.* **11**, 575.
Gurney, C. and Ngan, K. M. (1971) *Proc. Roy. Soc.* **A325**, 207.
Gurney, C. and Mai, Y. W. (1972) *Engr. Fract. Mech.* **4**, 853.
Hennephof, J. (1956) *J. Mech. Phys. Solids* **4**, 172.
Hill, R. (1952) *J. Mech. Phys. Solids* **1**, 19.
Hodgkinson, P. H. (now Davies) (1980) Report MAT/27, Royal Military College of Science, Shrivenham, Wilts, UK.
Hutchinson, J. W. (1968) *J. Mech. Phys. Solids* **16**, 337.
Hutchinson, J. W. and Paris, P. C. (1979) ASTM STP No. 668, pp. 37–64.
Jevons, J. D. (1942) *The Metallugy of Deep Drawing and Pressing* (2nd edn). London: Chapman & Hall.
Kanazawa, T. (1975) Int. Inst. Weld. Rep. IIW Doc. X-779.
Kanazawa, T., Machida, S., Hagiawara, Y. and Kobayashi, J. (1973) Int. Inst. Weld. Document X-702-73.

Kanninen, M. F., Abou-Sayed, I. S., Marshall, C. W. and Broek, D. (1980) EPRI Workshop Report NP-80-10-LD: WS 80-912.

Knott, J. F. and Elliott, D. (1979) Institution of Metallurgists, Monograph No. 4: Worked Examples in Fracture Mechanics, London.

Krafft, J. M., Sullivan, A. M. and Boyle, R. W. (1961) Symposium on crack propagation, College of Aeronautics, Cranfield, Vol. 1, p. 8.

Kumar, V., German, M. D. and Shih, C. F. (1981) GE/EPRI Report NP-1931.

Kumar, V., Wilkening, W. W., Andrews, W. R., German, M. D., de Lorenzi, H. G. and Mowbray, D. F. (1982) GE/EPRI Report SRD-82-048.

Kumar, V., German, M. D. and Shih, C. F. (1983) ASTM STP No. 803 I-306.

Lai, M. O. and Ferguson, W. G. (1980) *Material Science & Engineering* **45**, 183.

Landes, J. D. and Begley, J. A. (1972) ASTM STP 514, 1.

Landes, J. D. and Begley, J. A. (1974) ASTM STP 560, 170.

Landes, J. D., Walker, H. and Clarke, G. A. (1977) Westinghouse Scientific Paper 77-ID3-JINTF-P2; Pittsburgh, Penna.

Lin, I.-H., Anderson, T. L., deWit, R. and Dawes, M. G. (1982) *Int. J. Fract.* **20**, R3.

Mai, Y. W. and Cotterell, B. (1980) *J. Mater. Sci.* **15**, 2296.

Mai, Y. W. and Cotterell, B. (1983) *Proc. Int. Conference on Fracture Mechanics Technology applied to Material Evaluation and Structural Design,* G. C. Sih, N. E. Ryan and R. Jones (eds.). Netherlands: Martinus Nijhoff, p. 401.

Mai, Y. W. and Cotterell, B. (1984) *Int. J. Fracture* **24**, 229.

Mai, Y. W., Cotterell, B. and Atkins, A. G. (1979) *Fourth Tewksbury Symposium on Fracture,* D. S. Mansell and G. H. Vasey (eds.). Melbourne: Melbourne University Press.

Mai, Y. W. and Cotterell, B. (1985) To be published.

Marti, J., Kalsi, G. and Atkins, A. G. (1982) in *Numerical Methods in Industrial Forming Processes,* J. F. T. Pittman *et al.* (eds). Swansea: Pineridge Press, p. 279.

Marti, J., Goicolea, J. M., Kalsi, G. S. and Macey, G. E. (1984) Proceedings Int. Conf. on Numerical Methods on Transient and Coupled Problems, Venice.

Matsoukas, G., Cotterell, B. and Mai, Y. W. (1984) *Int. J. Fract.* **26**, R49.

Matsoukas, G., Cotterell, B. and Mai, Y. W. (1985). To appear in Engr. Fract. Mech.

McClintock, F. (1968) *Fracture: An advanced Treatise,* Vol. 3, H. Leibowitz (ed.). New York: Academic Press, pp. 47–225.

Merkle, J. G. and Corten, H. T. (1974) ASME Paper 74-PVP-33.

Nagtegaal, J. C., Parks, D. M. and Rice, J. R. (1974) *Comp. Methods in Appl. Mech. and Engr.* **4**, 153.

Paris, P. C., Tada, H., Zahoor, A. and Ernst, H. (1979) ASTM STP No. 668, 5.

Parton, V. Z. and Morozon, E. M. (1978) *Elastic-Plastic Fracture Mechanics* (in English). Moscow: Mir.

Rice, J. R. (1965) Proc. 1st. Int. Congress Fracture, Sendai, Japan, Vol. 1. 283, 309.

Rice, J. R. (1968) in *Fracture: An Advanced Treatise,* Vol. 2, H. Liebowitz (ed.). New York: Academic Press, 191.

Rice, J. R. (1968) *J. Appl. Mech.* **35,** 379.
Rice, J. R. and Rosengren, G. P. (1968) *J. Mech. Phys Solids* **16,** 1.
Rice, J. R., Paris, P. C. and Merkle, J. G. (1973) ASTM STP 536, 231 and 331.
Roberts, R. and Newton, C. (1981) Welding Research Council Bulletin, No. 265, New York.
Rowe, G. W., Sturgess, C. E. N., Hartley, P. and Pillinger, I. (in press) *Finite Element Plasticity and Metalforming Analysis.* Cambridge: Cambridge University Press.
Royal Society (1984) Report of Working group on Subcritical Crack Growth (Royal Society British National Committee on Theoretical and Applied Mechanics.)
Russian Standard (1972) in *New Method of Assessing the Resistance of Metals to Brittle Fracture,* Yu. N. Robotnova (ed.), (in English). Moscow: Mir.
Shih, C. F. (1981) *J. Mech. Phys. Solids* **29,** 305.
Shih, C. F., Andrews, W. R., de Lorenzi, H. G., German, M. D., Van Stone, R. H. and Mowbray, D. F. (1980) EPRI Workshop Report NP-80-10-LD: WS 80-912.
Shih, C. F. and Hutchinson, J. W. (1976) *Trans. ASME J. Engr. Mater. and Tech.* **H98,** 289.
Srawley, J. E. (1976) *Int. J. Fract.* **12** (3), 470.
Sumpter, J. D. G. (1973) Ph.D thesis, University of London.
Sumpter, J. D. G. and Turner, (1976) ASTM STP No. 601.
Tabor, D. (1951) *The Hardness of Metals.* Oxford: Clarendon Press.
Timoshenko, S. (1956) *Strength of Materials,* Vol. II Princeton, N. J. Van Nostrand, p. 377.
Tomkins, B. (1983) Trans. ASME Paper 83-PVP-98.
Turner, C. E. (1979) in *Post yield Fracture Mechanics,* D. G. H. Latzko (ed.). Barking, UK: Applied Science Publishers. See also Chap. 5 of 2nd edition (1984).
Turner, C. E. (1979) ASTM STP 677, 614.
Turner, C. E. (1981) *Proc. ICF-5 (Cannes),* D. Francois (ed.). Oxford: Pergamon, p. 1167.
Wells, A. A. (1961) *Symposium on Crack Propagation,* College of Aeronautics, Cranfield, Vol 1, p. 210.
Yap, O. F., Mai, Y. W. and Cotterell, B. (1983) *J. Mater. Sci.* **18,** 657.

5

Crack and craze nucleation

'Now the cleverest thing of the sort that I ever did,' [the White Knight] went on after a pause, 'was inventing a new pudding during the meat-course . . . . I don't believe that pudding ever *was* cooked! In fact, I don't believe that pudding ever *will* be cooked! And yet it was a very clever pudding to invent.'

5.1 INTRODUCTION

In Chapter 4 we discussed the mechanics of plastic fracture where cracks already existed. In the sardine can problem, for example, we were concerned with rolling up the lid and simultaneously cracking the joint. But what caused the joint to break in the first place? This is part of the much wider problem of nucleation, in supposedly flawless bodies, of microcracks in metals and crazes in polymers. Even 'flawless' bodies contain defects of one sort or another, of course (section 1.2.5).

In Section 1.2.6, on crack tip process zones, we chose to define 'nucleation' as the formation of a void or cavity from either (i) an inclusion (usually in metals, sometines in glasses or other solids containing foreign bodies), or (ii) crazed zones in polymers, or (iii) a pile up of dislocations or (iv) grain boundary junctions and so on. 'Initiation' is the 'sharpening' of these voids which is pre-requisite for crack 'propagation'. Even regions at the tips of sharp pre-existing cracks in brittle (low $ER/\sigma_y{}^2$) solids have to be nucleated and initiated, but as explained in section 1.2.5 the associated work is a minor component of the total work of fracture. Hence in Fig. 1.19 relating to elastic fracture, the irreversible work associated with nucleation and initiation is so small that loading/unloading lines along OBC, CDE cannot be separated. Elastic fracture is promoted by sharp cracks or sharp tooling (cutting edges of scissors, say) which intensify the deformation. The size of the body will satisfy a criterion such as equation (3.76) which ensures that remote yielding is avoided (see also Chapter 10), and clearly, low values of $ER/\sigma_y{}^2$ help. Even if starter cracks are blunt, or tools dull, elastic cracking still often occurs but at higher loads further up the original stiffness line and no area of irreversible work of nucleation/initiation can be identified on unloading.

In contrast, when conditions do not favour elastic cracking, deformation is not localised at the tips of pre-existing cracks, nor can even sharp tools immediately intensify the deformation into chosen regions of flawless bodies. In these circumstances, increasing amounts of irreversible deformation are required to produce eventual nucleation and initiation. On an *Xu* plot, an area of irreversible work is identifiable, before macrocracks are initiated and begin to propagate, such as area OBCF in Fig. 1.19 (which, under load, contains an elastic component), or the unloaded area OBCDE (which, although representing principally irreversible work, contains residual elastic strain energy, cf. section 4.5.1).

The magnitude of the irreversible work which precedes fracture depends on the degree of deformation (level of applied strain, say) and the volume through which it extends. Both components combine in the $W^{\mathrm{P}}\mathrm{d}V^{\mathrm{P}}$ term of equations (4.1), (4.2) etc. Small or large volumes may flow irreversibly under low or high plastic strains, depending on body geometry, value of ER/σ_{y}^2, workhardening characteristics, sharpness of any pre-existing cracks, sharpness of applied tooling and so on. There is a whole range of behaviour, with progressively increasing areas OBCDE, between ideal elastic cracking and cracking of solids with very large ER/σ_{y}^2.

This chapter is concerned with events and criteria which control crack and craze nucleation and under what circumstances, in different geometries, these criteria may be met. There are always *material* and *configurational* factors to consider. The chapter deals first with the behaviour of supposedly crack-free bodies but then addresses the problem of nucleation and initiation in bodies containing obvious starter cracks. This follows on from studies in Chapter 4, such as the use of a critical J_{C} criterion for initiation, and the partition of total work consumed in ductile fracture into 'essential' and 'inessential' parts, both before and after propagation. The connexion between nucleation/initiation criteria and events in the material ahead of a propagating crack (which relates too, of course, to cracking in standard testpieces) are investigated and this leads on to considerations of whether a one-parameter characterisation of fracture is always possible.

5.2 EXPERIMENTAL OBSERVATIONS OF CRACK NUCLEATION IN METALFORMING

In many metalforming operations limits are set by the appearance of surface or interior cracks (Fig. 5.1). Free-surface cracks occur in such processes as upsetting, bending, heading, and rolling; internal cracks in extrusion and drawing and in some forging processes; and plane-strain fracture occurs in side pressing. Thus, to achieve maximum deformation and process yields, and avoid excessive scrap, crack nucleation must be prevented in the likes of forging, stamping, rolling, pressing and extrusion. On the other hand, in processes which concern the *separation of parts,* crack nucleation, initiation and propagation is desirable.

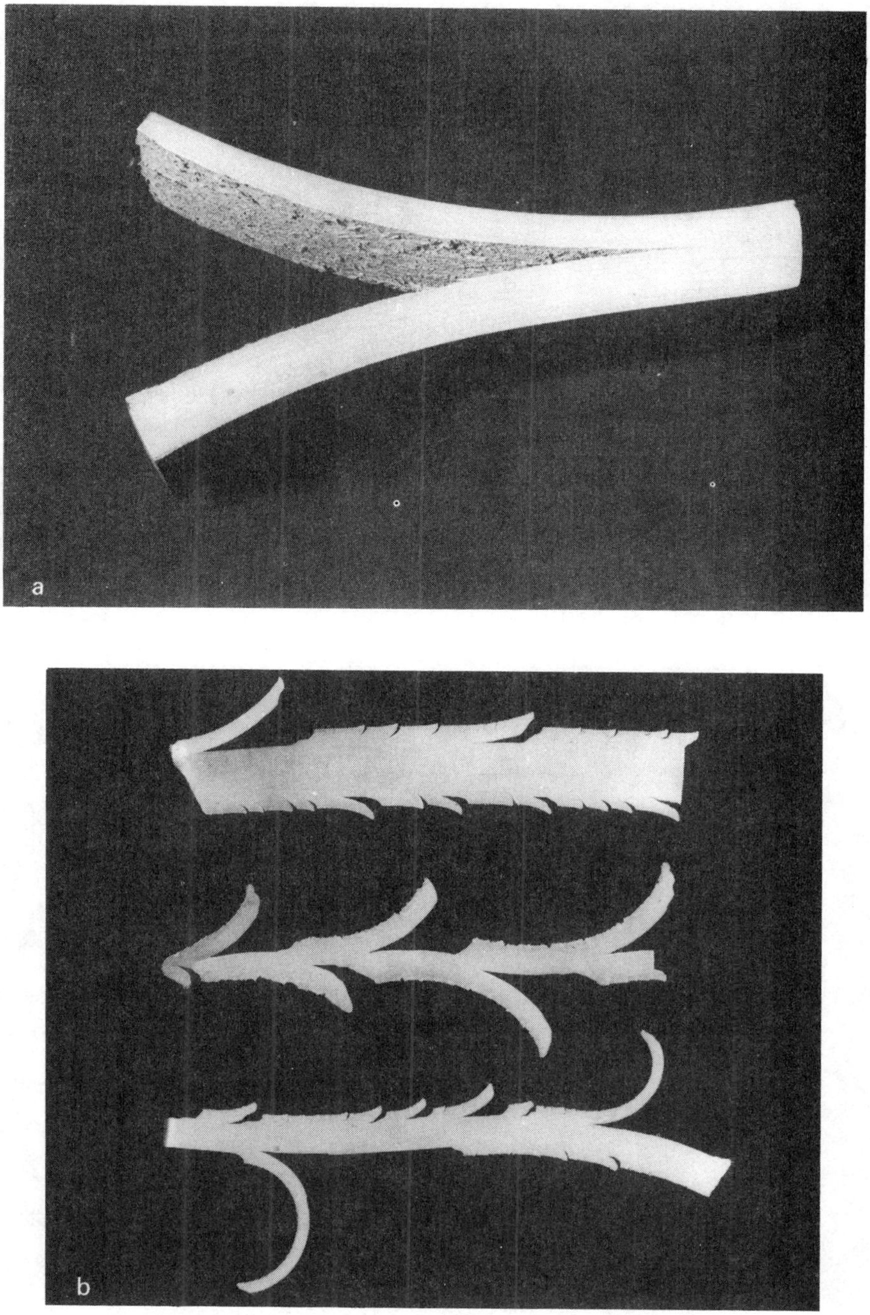

Fig. 5.1 – Surface and interior cracks in metal-forming and related processes.

(a) 'Crocodile' fracture in hot rolling Dural alloy (courtesy B. Dodd).

(b) Various forms of extensive edge cracking in cold rolling of Dural alloy (courtesy B. Dodd).

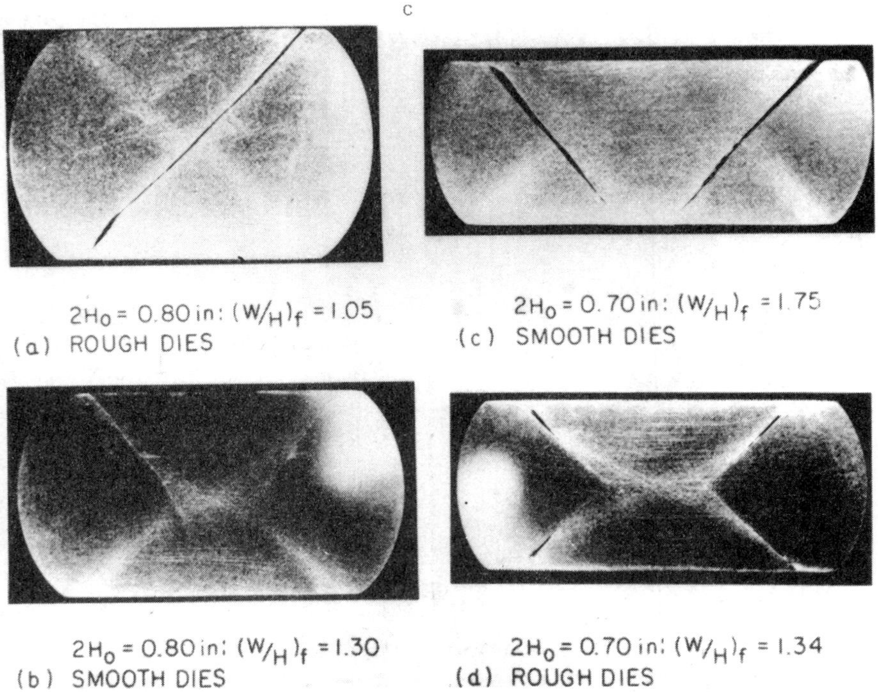

(c) Various fractures in two-dimensional 'side-pressing' of aluminium alloy, showing effects of platen lubrication and specimen aspect ratio. (After Jain and Kobayashi (1970), with permission).

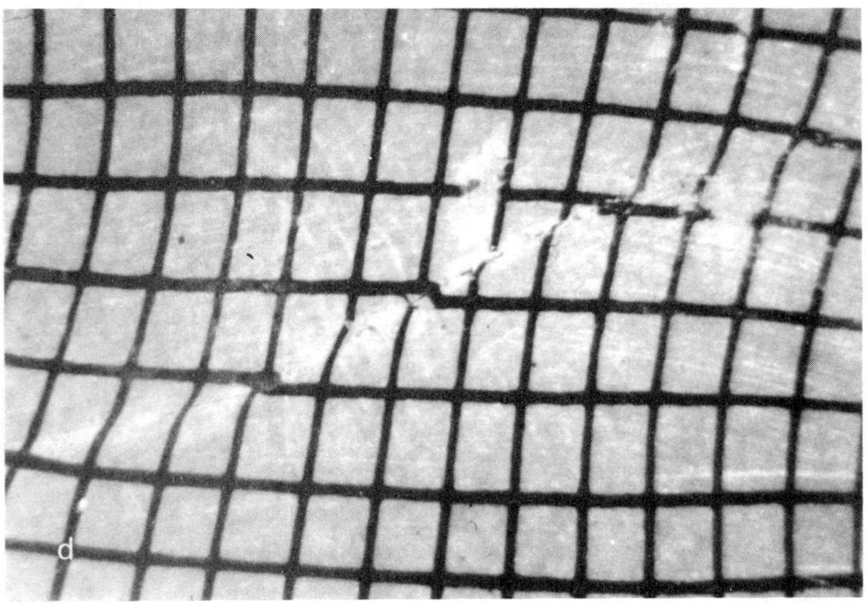

(d) Close-up of shear fracture in side-pressing of 7075-Al alloy (courtesy S. Clift).

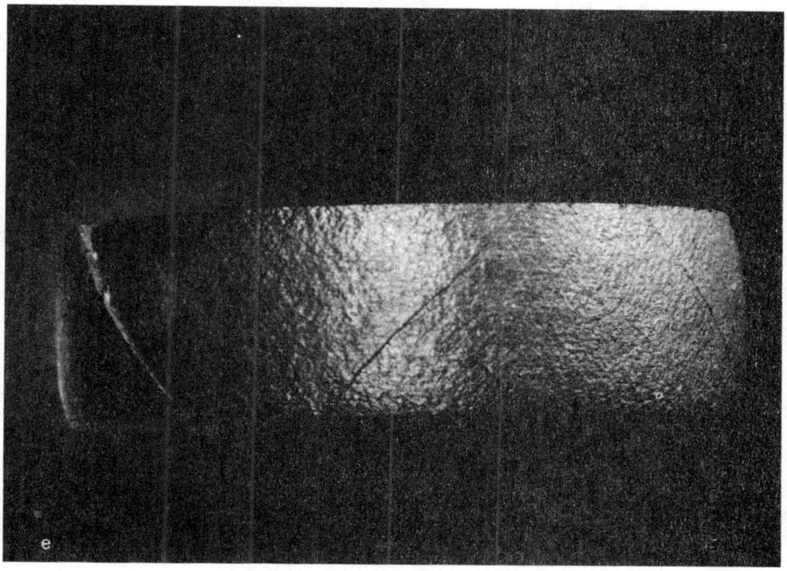

(e) Peripheral shear fractures in upset 7075-Al alloy cylinders (courtesy S. Clift).

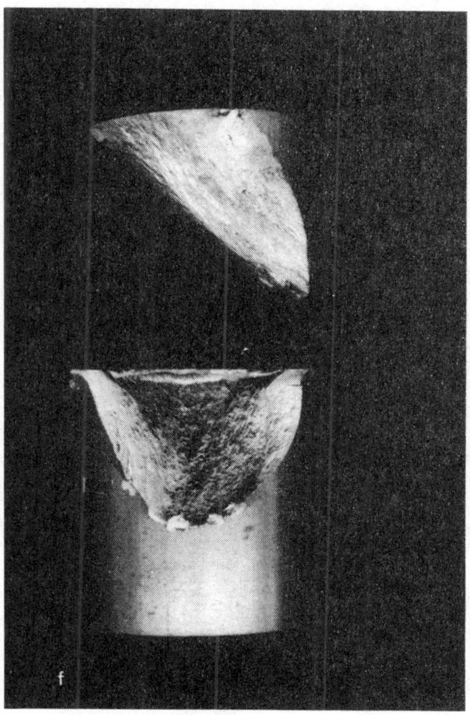

(f) Rapid fracture in upset Mg-alloy cylinders (courtesy B. Dodd).

(g) Characteristic (internal) central chevron cracking in extrusion (or rolling, as here) (courtesy W. F. Hosford).

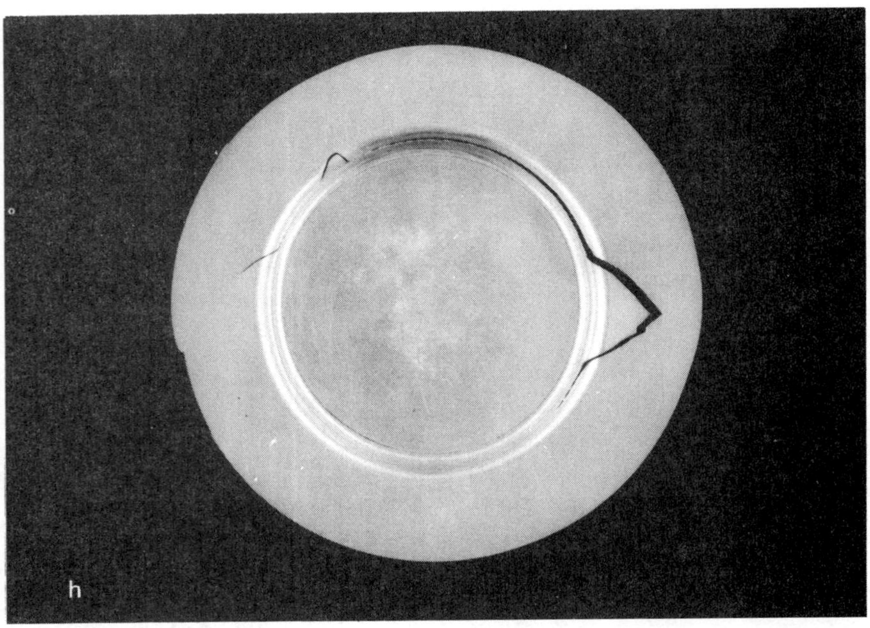

(h) Unusual shear fracture in flange or partially-drawn aluminium cup (fractures in deep drawing more often occur near the base of the wall).

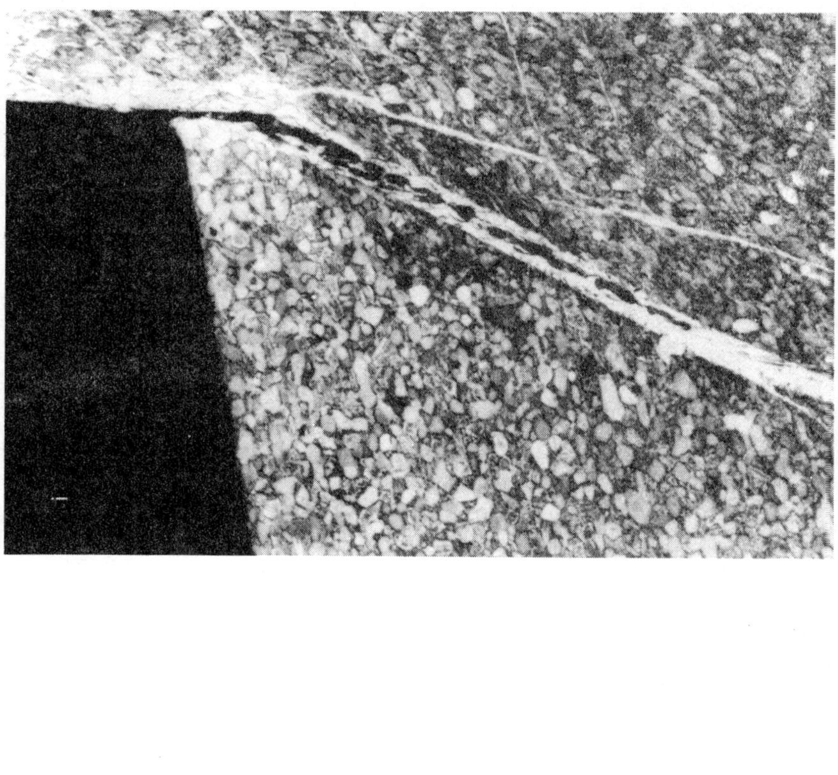

(i) Crack in adiathermal shear band formed under impact loading of Ti-6Al-4V alloy by 4.76 mm diameter plane-ended cylindrical punch at 312 m/s (courtesy S. Timothy).

(j) Cracks in turned-over head of cold chisel (courtesy J. R. Bayes).

Thus cracking is an inherent part of can-opening, blanking, cropping, guillotining, piercing, nibbling, slitting and machining. Problems in ballistics, penetration of armour, plugging of plates and so on are all related to this general theme. In all these processes, cracks nucleate within regions of deformation fields that are highly strained – either simply by extensive overall flow (as in forging, say) or by localised flow produced by the external stress-raising effect of sharp tooling (as in cropping and armour piercing).

In the period up to the nucleation and formation of macrocracks, events are controlled by equation (1.3). However, historically, in the absence of a full mechanics analysis, one approach to establish a criterion for crack initiation in ductile fracture has been to build up a body of experimental data concerning the plastic strains local to those regions of the deforming body where cracks are first observed.† Fig. 5.2 shows Kudo's and Aoi's 1967 results for upsetting, i.e. compression testing – which simulates many open-die forging operations.

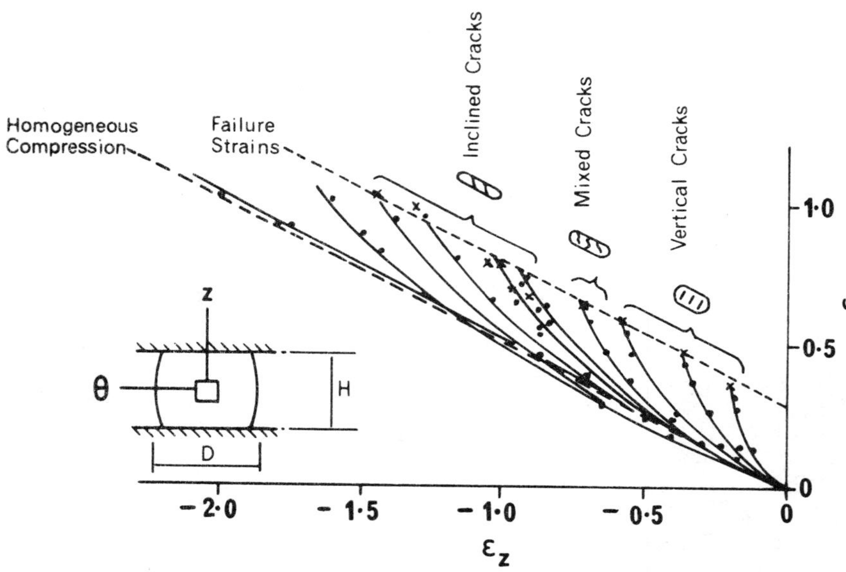

Fig. 5.2 – Hoop and axial strains in an upsetting process. The failure strain locus is also shown. Strain paths falling below homogeneous compression line produced by profiled specimens (after Kudo and Aoi (1967), with permission).

Friction at the platens produces barrelling and non-homogeneous deformation in the specimens. The severity of barrelling depends on the magnitude of the friction and aspect ratio *H/D* of the specimen, and produces the various loading

† Whether criteria based upon these strains relate to nucleation, initiation or propagation is a moot point. Clearly it depends on the level of magnification used to find the cracks (cf. section 7.2.3). In many of the sorts of experiments described here, the cracks are often about 1 mm in extent when the strains are measured, and really are already-propagated cracks.

paths. Cracks are found in the barrelled sections, and the surface strain pairs (ϵ_θ hoop tensile and ϵ_Z axial compressive), measured from grids, pop marks etc. at the site of cracking, fall on an approximately straight line 'failure locus' at $\tan^{-1}(-\frac{1}{2})$, irrespective of whether the cracks are 'vertical', 'mixed' or 'inclined'. 'Workability' of different materials, or of the same material in different conditions of prestrain or heat treatment, is indicated by the ϵ_θ intercept as shown for Ertuk's 1979 data in Fig. 5.3(a) and (b) regarding different microstructures (see also Erturk and Kazaoglu, 1982) and, in Fig. 5.3(c), Kuhn's 1976 data regarding anisotropy in hot-rolled plate; the slope of $(-\frac{1}{2})$ seems to be common for a number of different ductile solids in this type of test.

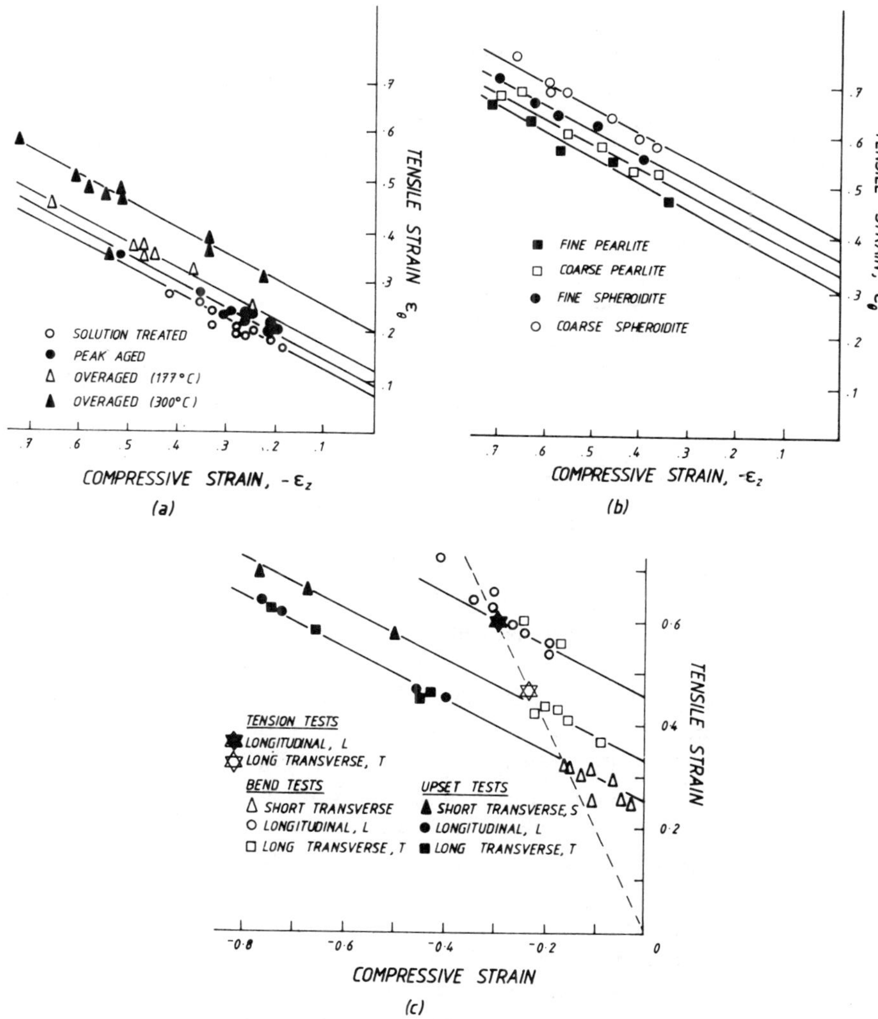

Fig. 5.3 – Hoop and axial strains in upsetting processes for (a) 2014 aluminium alloy and (b) AISI 1040 steel, regarding different microstructures, and (c) AISI 1045 steel which is hot-rolled (after Erturk (1979), and Kuhn (1976) with permission).

Kuhn and co-workers (e.g. 1976) have extensively employed this type of workability diagram for variations of solid forging (e.g. bolt heading, ring forging, bending, edge cracking in rolling, gear forging). They have modified the procedure to account for cracking at die-tool/workpiece interfaces, and for internal cracking (such as central bursting in extrusion and in hubbed-disk forging). Essentially they note that die pressure increases the capacity for the material to deform without fracture and, in effect, the height of the fracture limit line is increased. Lubrication plays an important role in this regard. In all these treatments, Kuhn argues that providing the strain path during forming does not cross the failure line (whatever the loading prehistory) no cracking will occur. He has also applied the ideas to powder metal preforms. Whether the onset of fracture is concerned simply with intersecting the 'failure locus', irrespective of prior effective stress/effective strain excursions in the plastic range, or whether fracture is actually path-dependent is discussed later.

Knowledge of the plastic strain history permits the stress components to be evaluated as described in Appendix 3. In particular, the orientation of the final crack has been compared with planes and directions of maximum stress in the quest for a criterion of ductile fracture. Thus, Kudo and Aoi (1967) showed that the observed fracture plane in simple upsets test coincided with the planes of maximum shear stress. When barrelling is slight, and the deformation tends towards homogeneous compression, the stresses local to the fracture site are compressive in the axial direction and tensile in the hoop direction. Hence, treating the surface deformation as a plane stress problem (i.e. the local radial stress is zero), leads to the conclusion that the direction of maximum shear stress is at 45° in the $z\theta$ plane, extending radially into the cyclinder. (Consider Mohr's circles, for example.) Crack orientations found under conditions of good lubrication and low aspect ratio correspond with such planes (at the left-hand side of Fig. 5.2). When barrelling is severe, however, the axial stress local to the fracture site can become *tensile* which, for tensile σ_θ and $\sigma_r = 0$ means that the planes of maximum shear are the vertical planes which are at 45° to the *radial* direction (i.e. *not* vertical and radial). The experimentally-found crack orientations under conditions of severe barrelling correspond with these planes of maximum shear stress (at the right-hand side of Fig. 5.2). Cracking in these different shear planes in monotonic fracture should be compared with related behaviour in different planes in multiaxial fatigue fracture, i.e. the type A and B cracks in section 7.6 (cf. Fig. 7.37). We shall discover too that there are microstructural reasons why localised deformation is concentrated along narrow bands in the directions of maximum shear stress in forging, and related, problems.

In sheet forming, the orthogonal in-plane plastic strains at fracture measured from grids or strain circles on hydraulically-bulged sheets, for example, also lie on a line falling from left to right, but the slope is often steeper than $\tan^{-1}(-\frac{1}{2})$, Fig. 5.4. Apart from the existence of strain data in the tension–tension strain quadrant, the diagram for sheet forming differs from that for upsetting as, in the loading paths prior to fracture, the strain ratios change markedly when localised necking occurs. In fact, in many sheet forming operations, it is the development

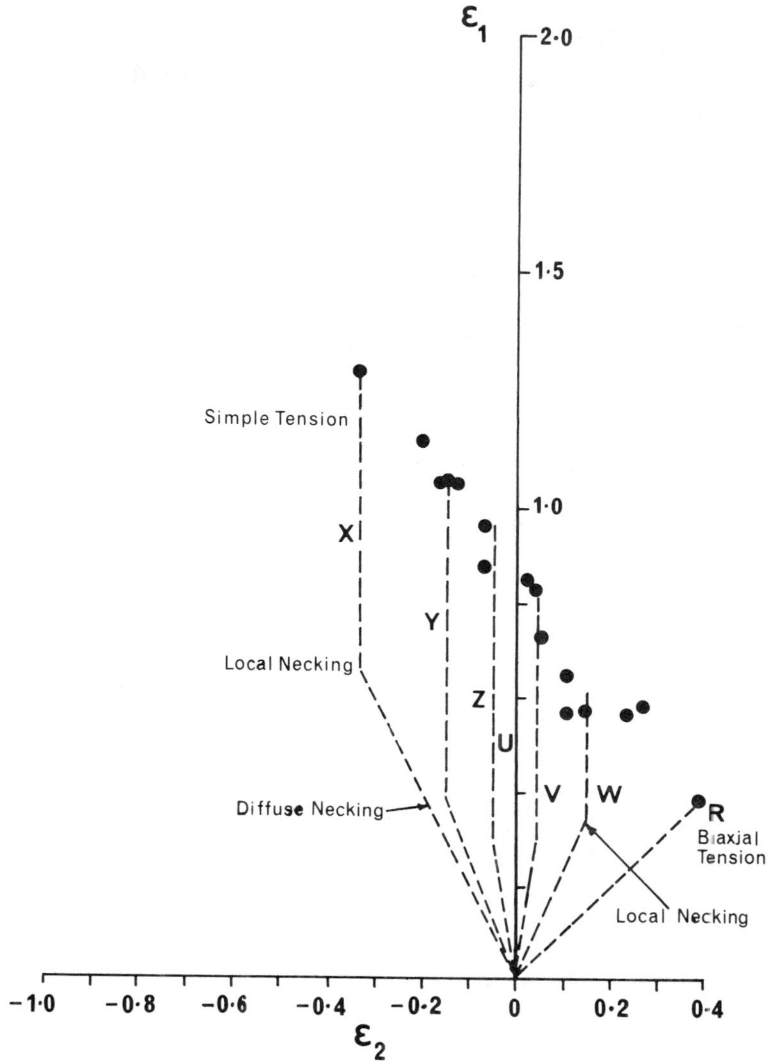

Fig. 5.4 – Orthogonal in-plane plastic strains measured on hydraulically-bulged sheets of 5154 Al–Mg alloy. Fracture strain locus is also shown (after Embury and Le Roy (1977), with permission).

of necks which effectively limits the process rather than fracture (e.g. Hosford and Caddell, 1982). Even so, there are circumstances, depending on the complex geometry of deep drawn and stretched parts, where fracture is likely to limit the attainable deformation. Examples are splits in the corners of rectangular cups and other deep-drawn components where high strain gradients occur. Consequently there has been growing interest in extending 'forming limit diagrams' to include fracture strains as well as necking strains. Although the experimental procedure of measuring grid patterns (usually circles) in the presence of a high strain gradient or a strain localisation is both tedious and inaccurate, a feature of many bulge tests is that the local neck which precedes fracture is often a plane strain neck orientated perpendicularly to the major in-plane strain; then the final portion of all strain paths in the tension–tension quadrant is approximately parallel to the major strain axis whatever the original loading path, as indicated schematically in Fig. 5.4.

In simple tension of sheet specimens which gives 2:1 strain pairs in the tension–compression quadrant, the plane strain neck is at an angle to the pulling axis (at $\tan^{-1}\sqrt{2} \approx 55°$ in the case of an isotropic sheet). On other strain paths having steeper gradients in the tension–compression quadarnt (produced in Hecker cup tests, for example) the orientation of the neck direction progressively tends towards the minor in-plane strain: under complete biaxial tension (produced, for example, by bulging over elliptical orifices) the neck coincides usually with the minor in-plane strain direction. It is along the *neck* directions that one strain increment is zero and this does not coincide with the axes of the forming limit diagram. Consequently the strain paths in the tension–compression quadrant really do *not* change direction at necking as sometimes shown in this sort

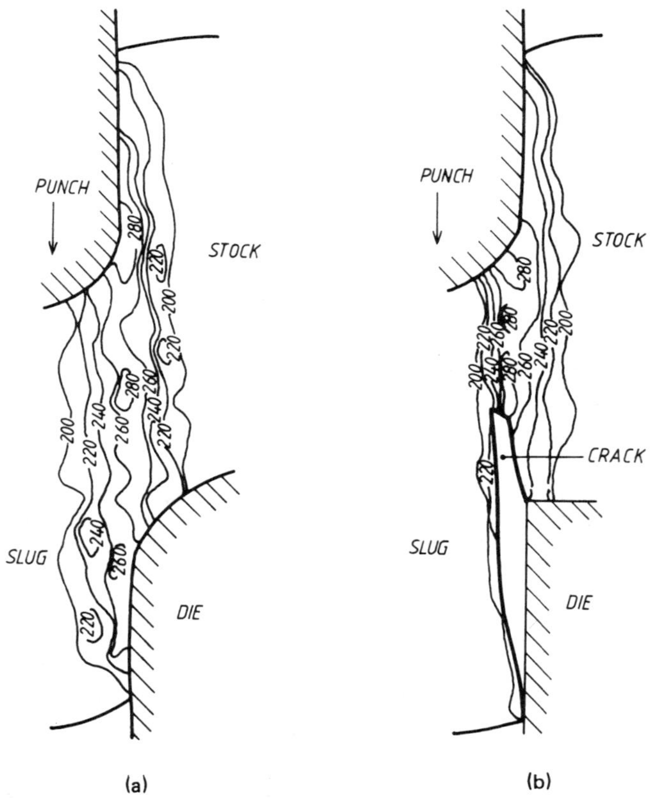

Fig. 5.5 – Cracks observed in blanking of a low carbon steel (after Noble and Oxley (1964), with permission).

of plot. Final fracture in sheets usually takes the form of through-thickness shear cracks, at some 45°, along the line of the plane strain neck.

The first appearance of cracks in the blanking of a low carbon steel plate is shown in Fig. 5.5. Here highly strained regions are produced by the *tooling* (rather than occurring 'naturally' as in the foregoing examples). The hardness contours may be thought of as strain contours and, at the sites of cracking, Noble and Oxley (1964) remarked that not only are the hardness values high, but also the hardness *gradients* are high. Fig. 5.5(b) demonstrates that tooling sharpness concentrates the deformation prior to fracture: radiusing either the die (in blanking) or punch (in piercing) also alters the path of crack propagation and hence cracking can be confined to waste material. Cracking is not only desirable in blanking and similar processes but the direction of subsequent crack propagation is important (section 5.7); in simple cropping, for example, if surface cracks are left on the ends of slugs when the cracks emanating from tool and die do not meet, they will appear as faults and blemishes in subsequent forming operations, possibly leading to later scrapping of components of high added-value.

5.3 CRITERIA FOR DUCTILE FRACTURE

Why do the experimental failure strains take a slope of $\tan^{-1}(-\frac{1}{2})$ in upsetting, or a different slope in sheet forming? What determines the value of the workability strain intercept? What is the connexion with the microstructural features?

5.3.1 Empirical criteria

Table 5.1 shows that there is a host of criteria for the onset of cracking in plastic flow fields. These include simple maximum stress or maximum strain theories, critical combinations of stress and strain, and variations on these themes; some of the criteria are entirely empirical, some are based on microstructural models. While these criteria may represent the failure data of the particular experimental conditions from which they were derived, experiments show that theories based simply upon maximum stress or maximum strain cannot be true in general. For example in Fig. 5.2 the strains at fracture depend on the straining path. While in sheet forming it is sometimes said that failure occurs at a constant through-thickness strain, (which would give a 45° failure in line in Fig. 5.4 see equation (5.10)), this does not seem to be true generally. While a tensile component of the hydrostatic stress promotes fracture (and it is important to be aware that tensile stresses can exist in regions of predominantly compressive deformation (Fig. 5.6)) the existence of a tensile stress is not a sufficient condition: clearly there is extensive deformation under tensile stresses in simple tension tests and sheet bulging tests long before fracture. Examples of crack initiation in rings and disks presented in section 5.4 provide further evidence that critical stress or strain criteria by themselves are inadequate for crack nucleation in ductile fracture.

Table 5.1
Criteria for ductile fracture in metalforming

σ_{max} maximum tensile stress	ϵ_{max} maximum strain
$\int \sigma_{max} d\bar{\epsilon}$ Cockcroft–Latham (1968)	$\int \bar{\sigma} d\bar{\epsilon}$ Gillemot (1976)
$\int \sigma_H d\bar{\epsilon}$	$(d\sigma_H/ds)_{max}$ Noble and Oxley (1964)
$\int (1 + \sigma_H/B\bar{\sigma}) d\bar{\epsilon}$ Oyane (1972)	$\frac{1}{\mathscr{H}} = \frac{1}{\mathscr{H}_0} - g(\sigma_H/\bar{\sigma} - 2/3) d\bar{\epsilon}$ Marciniak and Kuczynski (1979)
$\int \frac{2\sigma_{max} d\bar{\epsilon}}{3(\sigma_{max} - \sigma_H)}$ Brozzo *et al.* (1972)	
$\int d\bar{\epsilon}/(1 - c\sigma_H)$ Norris *et al.* (1978)	$\int (1 + 1/2L) d\bar{\epsilon}/(1 - c\sigma_H)$ Atkins (1981)

Growth of voids
$ln(1/2r) \propto \sinh(\sigma_H/\bar{\sigma}) d\bar{\epsilon} \approx (\sigma_H/\bar{\sigma}) d\bar{\epsilon}$ McClintock *et al.* (1966, 1968)
$(1 + 1/M)\sigma_1{}^2 =$ constant Ghosh (1976)
cf. also Thomason (1969); Rice and Tracey (1969); Parmar and Mellor (1978, 1980); Dillamore and Maynard (1974)

Void closure
Keife and Stahlberg (1980)

Probabilistic Approach
(size distribution of voids) van Minh, *et al.* (1974); Thompson and Hancock (1984)

ϵ, true strain. $\bar{\epsilon}$, effective strain. σ, true stress. $\bar{\sigma}$, effective stress. σ_H, hydrostatic stress. s, distance across shear band. $L =$ strain ratio $d\epsilon_1/d\epsilon_2$. $M =$ stress ratio σ_1/σ_2. $B, g, c, \mathscr{H}$, constants.
The integrated stress–strain criteria are supposed to attain critical values at fracture.

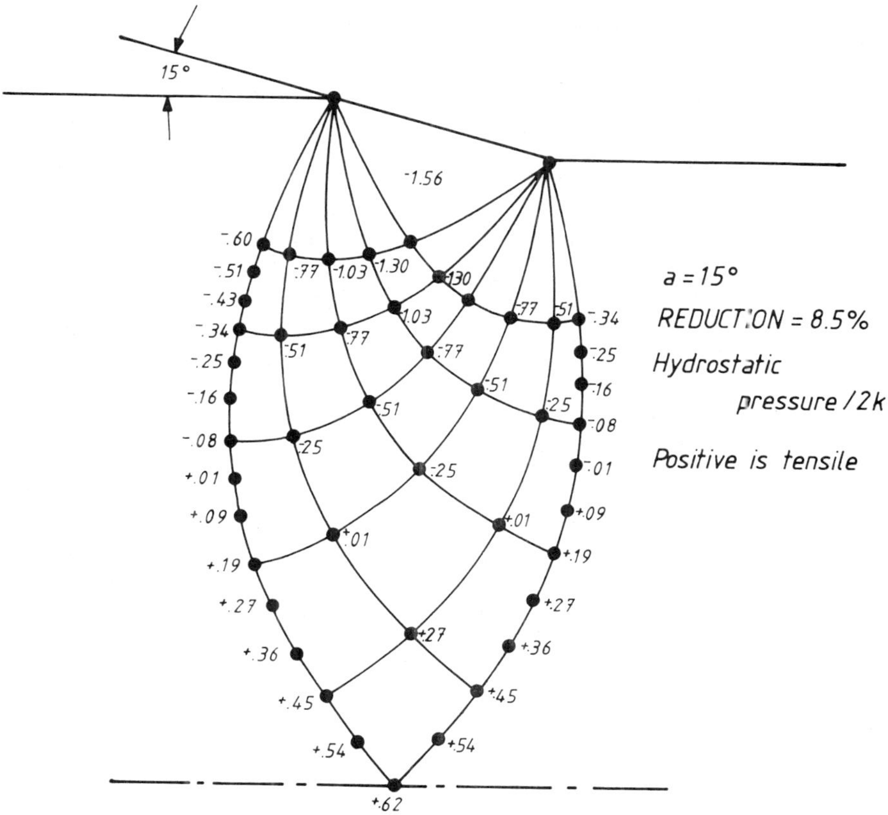

Fig. 5.6 – Distribution of hydrostatic pressure in strip for a given set of drawing conditions. $2k$ is the tensile yield stress. Hydrostatic tension near centreline of sheet (after Rogers, Leech and Coffin (1971), with permission).

Example 5.1 Plot, in $\epsilon_1\epsilon_2$ space, the predictions of some of the empirical fracture criteria of Table 5.1 for plane stress conditions (i.e. $\sigma_3 = 0$). For convenience make all curves pass through the same workability index point on the ϵ_1 axis, i.e. through the same 'plane strain–plane stress' point ϵ^0_{1f}, which is a convenient reference for the 'critical value' at fracture.

Ans. For purposes of illustration, consider the Cockcroft–Latham criterion written as $\int\sigma_1 d\bar{\epsilon} =$ constant when $\sigma_1 > \sigma_2 > \sigma_3$. From Appendix 3, using $\sigma_1 = (2L+1)\,\bar{\sigma}/\sqrt{3}\sqrt{L^2+L+1}$ and $d\bar{\epsilon} = (2/\sqrt{3})(1+1/L+1/L^2)^{\frac{1}{2}}\,d\epsilon_1$ (where $L = d\epsilon_1/d\epsilon_2 = \epsilon_1/\epsilon_2$ for radial loading paths and where $\bar{\sigma}$ and $\bar{\epsilon}$ are respectively the effective plastic stress and strain), we obtain $\int\sigma_1 d\bar{\epsilon} = \int 2(2L+1)\bar{\sigma}d\epsilon_1/3L$. For simplicity, take a rigid–perfectly plastic solid with $\bar{\sigma} =$ constant. The ordinate workability index fracture strain ϵ^0_{1f} is obtained by substituting $L = \infty$, to give for the integration constant $4\,\bar{\sigma}\,\epsilon^0_{1f}/3$. Hence for any other L which displays ϵ_{1f} we have $\epsilon_{1f}/\overset{0}{\epsilon}_{1f} = 2L/(2L + 1)$. This gives the equation of the Cockcroft–Latham failure locus for radial straining paths, i.e. $\epsilon_{1f}/\overset{0}{\epsilon}_{1f} = (2\epsilon_{1f}/\epsilon_{2f})/(2\epsilon_{1f}/\epsilon_{2f} + 1)$ or $\epsilon_{1f} = -\frac{1}{2}\epsilon_{2f} + 1$, if $\overset{0}{\epsilon}_{1f}$ is arbitarily taken as unity. This is a straight line of slope $-\frac{1}{2}$. (Fig. 5.7; cf. Figs. 5.2 and 5.3.)

Other criteria may be determined by a similar route. For example, $\int(\sigma_H/\bar{\sigma})d\bar{\epsilon} \rightarrow 2(L + 1)\epsilon_{1f}/(4L + 1)$ which corresponds with $4\epsilon_{1f}^2 + 4\epsilon_{1f}\epsilon_{2f} = 4\epsilon_{1f} + \epsilon_{2f}$ when $\overset{0}{\epsilon}_{1f} = 1$. More is said of this form of relation in section 5.3.2. If a criterion of constant effective strain $\bar{\epsilon}_f$ at fracture is taken, we obtain $\epsilon_{1f}/\overset{0}{\epsilon}_{1f} = L/\sqrt{1 + L + L^2}$, i.e. $\epsilon_{1f}^2 + \epsilon_{1f}\epsilon_{2f} + \epsilon_{2f}^2 = 1$, if $\overset{0}{\epsilon}_{1f} = 1$. Note that this corresponds with Gillemot's criterion of a critical plastic work/volume $\int\sigma d\epsilon$ in a rigid-perfectly plastic solid. In these diagrams (Fig. 5.7), non-linear failure criteria are more easily plotted using radial lines of constant L and derived ϵ_{1f} values, rather than $[\epsilon_1\,\epsilon_2]$ pairs.

We note straight away that although there is agreement between many of the criteria plotted in Fig. 5.7 in certain ranges of $\epsilon_1\,\epsilon_2$ space (particularly in the tension–tension quadrant), the criteria can diverge quite markedly elsewhere.

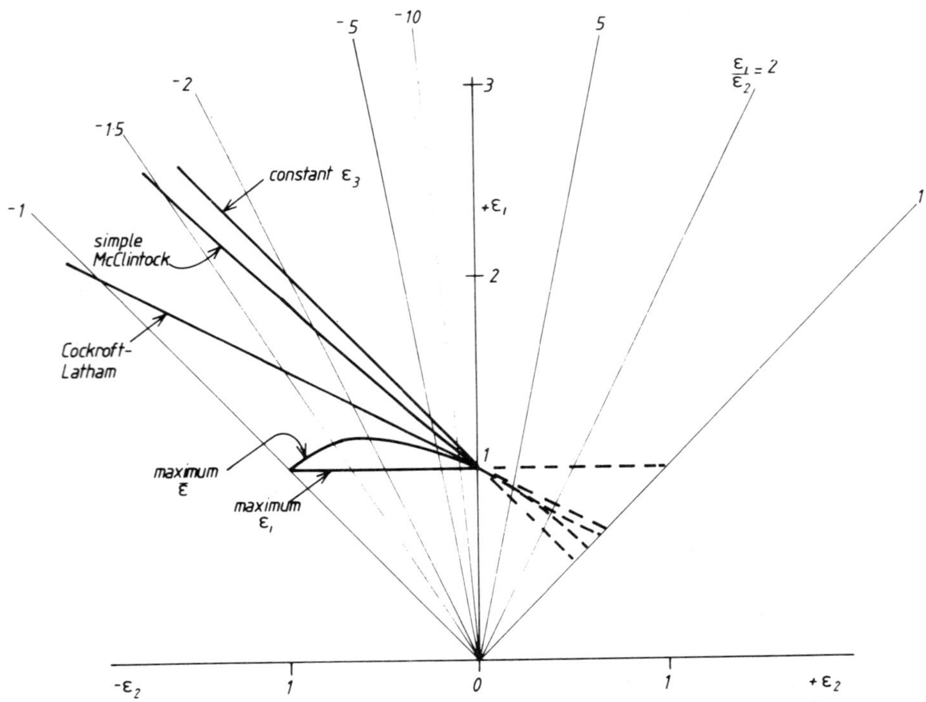

Fig. 5.7 – Empirical fracture criteria (from Table 5.1 for plane stress, $\sigma_3 = 0$) plotted in $\epsilon_1 - \epsilon_2$ space.

It is not surprising therefore that a ductile fracture criterion developed for one metal-flow process may not apply to another process having a different history of deformation (cf. Hoffmanner, 1971).

Calculations may be performed for workhardening solids using, say, $\bar{\sigma} = \sigma_0(\bar{\epsilon}_{cw} + \bar{\epsilon})^n$ to give curves similar to those shown in Fig. 5.7 (see Problem 5.1). Indeed some of the criteria mentioned in Table 5.1 can *only* be plotted for finite work hardening index n, owing to the need for a coupling between $\bar{\sigma}$ and $\bar{\epsilon}$. The Ghosh criterion is one such; a criterion of maximum tensile stress is another. The form of the ductile fracture $\epsilon_{1f}\epsilon_{2f}$ strain locus is often sensitive to the value of n, particularly in the tension–compression strain quadrant.

5.3.2 Models of void coalescence and growth

It seems obvious that the failure criterion which is most likely to succeed will be that which most closely models microstructural events. As discussed in Chapter 1, the microstructural features of crack initiation and propagation in the ductile fracture region of Ashby maps are reasonably well understood – particularly for commerical-grade metals which are metallurgically 'dirty': holes are nucleated from inclusions or hard second-phase particles and these grow and coalesce to give fracture (Fig. 1.5). The holes will both grow in size and also change their shape. The average radial growth rate of a cavity is proportional to the dilatational stress field, i.e. to the mean (hydrostatic) stress, and the shape change depends on the shear stress field (e.g. a spherical hole will become ellipsoidal).

Figs. 5.8 and 5.9 illustrate models of ductile fracture by hole growth in normal stress fields and shear stress fields respectively, which were proposed by McClintock and co-workers in the 1960s. Similar treatment is found in the work of Rice and Tracey (1969). The analyses are heuristic adapations for plastic solids of exact two-dimensional solutions for hole growth in viscous solids. Inclusions are modelled as regular arrays of cylindrical holes with known spacing-to-size ($l/2r$) ratios. Fracture is considered to occur when the voids touch each other at their element boundaries. In many problems, the holes remain relatively small over most of the deformation, so components of stress and strain may be related by incompressible fully-plastic theory (Appendix 3), e.g. Drucker (1966). Then, when holes with axes parallel to the x-axis in Fig. 5.8 touch one another, McClintock relates the applied stresses σ_x and σ_y by

$$\ln(l_y/2r) = \int_0^{\bar{\epsilon}_f} \left(\frac{\sqrt{3}}{2(1-n)} \sinh \left\{ \frac{\sqrt{3}(1-n)}{2} \frac{(\sigma_x - \sigma_y)}{\bar{\sigma}} \right\} + \frac{3}{4} \frac{(\sigma_x - \sigma_y)}{\bar{\sigma}} \right) d\bar{\epsilon} \tag{5.1}$$

where l_y is the original spacing of the holes along the y-axis, and n is the work-hardening index in $\bar{\sigma} = \sigma_0\bar{\epsilon}^n$. There are two other equations for coalescence of

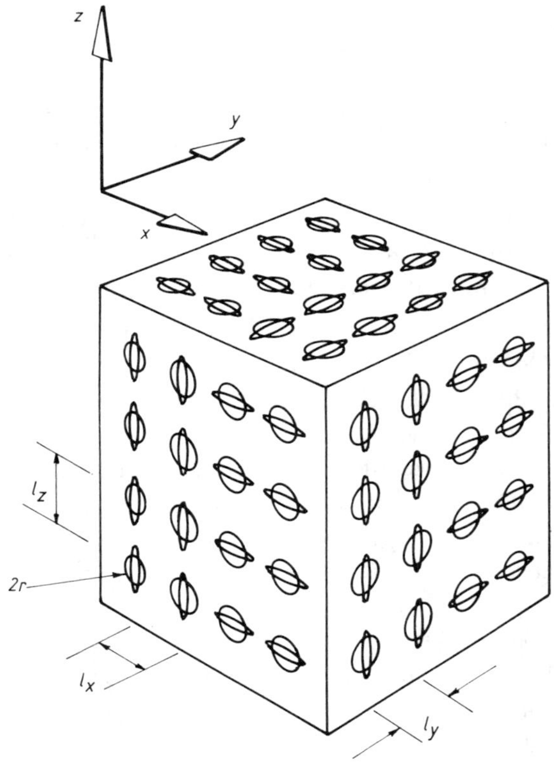

Fig. 5.8 – Model of void growth in a normal stress field (after McClintock (1968), with permission).

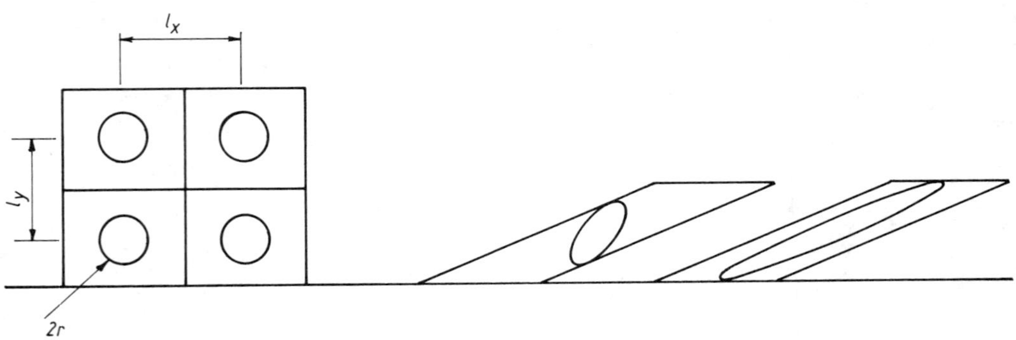

Fig. 5.9 – Model of void growth in a shear stress field (after McClintock (1966), with permission).

holes with axes parallel to the y- and z-axes, obtained by swapping the suffices correspondingly.

In shear, McClintock and co-workers give either

$$\left.\begin{aligned} \ln(l_x/2r) &= \ln\sqrt{1+\gamma^2} + \frac{\gamma}{2(1-n)} \sinh \frac{(1-n)\,\sigma_H}{\tau} \\ \text{or} \quad \ln(l_y/2r) &= \frac{\gamma}{2(1-n)} \sinh \frac{(1-n)\,\sigma_H}{\tau} \end{aligned}\right\} \quad (5.2a,b)$$

depending on whether the distorted holes touch the 'sides' or 'top and bottom' of their elemental cells (Fig. 5.9). The mean stress is σ_H $(= (\sigma_x + \sigma_y + \sigma_z)/3)$ and $\tau = \tau_0\gamma^n$.

There are related equations (not given here) for the resulting eccentricity of the elliptical holes.

The tensile relation is left in integral form, and the shear relation is the result of integration. The reason that criteria for ductile fracture are likely to be path-dependent is that a theory based on the growth of holes must depend on the history of stress, strain and rotation – unlike criteria for initial yielding or brittle fracture which require only the current state of stress. (We shall find out later, in Example 5.3 et seq., that *some* simplified versions of the McClintock criteria are path-*independent*). In both tensile and shear crack models, the right-hand sides of the relations are referred to as 'accumulated hole growth damage', which builds up to critical values set by the left-hand sides which are characteristic of the starting microstructure. Of course, practical materials do *not* consist of uniformly-spaced particles of uniform size spread through a matrix, and the critical hole growth damage at fracture is an average over a representative volume of microstructure (cf. Fig. 5.15 later). For this reason in much of what follows, the suffixes on l are dropped and some average $l/2r$ is implied. Even then there may be difficulties if there are a number of distinct populations of inclusions of different sizes (see section 5.5). The connexion between $l/2r$ and volume fracture v_f of voids (inclusions) is given by $l/2r \approx (1/v_f)^{1/3}$.

Example 5.2 An extruded steel bar has strung-out inclusions of average diameter 2 μm which are some 50 μm apart. If $\bar{\sigma} = 900\ \bar{\epsilon}^{0.2}$ MPa, at what stress should significant hole coalescence occur according to the McClintock model (a) in a uniaxial tension test where the pulling axis is perpendicular to the extrusion direction and (b) in the torsion of a thin walled tube whose axis is the same as the extrusion direction?

Ans. $l/2r = 50/2 = 25$. In (a) $\sigma_y = \sigma_z = 0$ so $\sigma_x = \bar{\sigma}$ and from equation (5.1),

$$\ln(25) = 3.22 = \int_0^{\bar{\epsilon}_f} \left(\frac{\sqrt{3}}{2(1-0.2)} \sinh\left[\frac{\sqrt{3}(1-0.2)}{2}\right] + \frac{3}{4}\right) d\bar{\epsilon} = 1.56\bar{\epsilon}_f.$$

Therefore $\bar{\epsilon}_f = 2.1$ and $\bar{\sigma} = \sigma_x = 900\,(2.1)^{0.2} = 1040$ MPa. In (b), we use equation (5.2a) as $l_y = l_x$ and the rhs reaches a critical value at smaller γ than equation (5.2b). In pure shear $\sigma_H = 0$ (the centre of the Mohr's circle sits on the origin). Therefore $\ln(50/2) = 3.22 = \ln\sqrt{1+\gamma^2}$ so $1+\gamma^2 = 2\exp(3.22)$, and $\gamma = 7$. Using $\tau = \sigma/2$ and $\gamma = 2\epsilon$ we obtain $\tau = (900/2^{1.2})\gamma^n$, assuming n is the same in shear, from which $\tau = 578$ MPa when $\gamma = 7$. (The realism of these values and *quantitative* predictions in the link between microstructure and mechanics using the McClintock model are discussed later in section 5.5).

Note that equation (5.2a) predicts fracture for $\sigma_H = 0$ (as in Example 5.2) and even when σ_H is mildly negative, owing to merely shape change of the cavity (Problem 5.2). At greater compressive σ_H, the hole growth damage is negative and has no meaning, and these models fail to predict fracture. However, in practical materials the holes are rarely 'empty' (as assumed by the analysis) and contain hard inclusions, or soft particles of lead, say, in free cutting materials. Voids therefore are prevented from closing up completely under hydrostatic compression, and fracture in shear is possible. Even in tensile stress fields, hard inclusions prevent voids from freely changing their shape (they cannot thin down in the presence of inclusions) and an alternative model for ductile fracture by hole growth, given by Ghosh in 1976, takes this into account (Problem 5.9). How holes appear in shear fields under hydrostatic compression may be readily demonstrated by rendering walls with 'dirty' plaster (i.e. containing grit).†

When σ_H and $\bar{\sigma}$ are not widely different in magnitude, there is little error in simplifying the McClintock relations by approximating $\sinh x \approx x$, to give

$$\ln(l_y/2r) = \int_0^{\bar{\epsilon}_f} \left\{ 9\sigma_H/4\bar{\sigma} + 3\,(\sigma_x - \sigma_y)/4\bar{\sigma} \right\} d\bar{\epsilon} \tag{5.3}$$

and

$$\left.\begin{aligned} \ln(l_x/2r) &= \ln\sqrt{1+\gamma^2} + \sigma_H\gamma/2\gamma \\ \ln(l_y/2r) &= \sigma_H\gamma/2\tau \end{aligned}\right\} \tag{5.4a,b}$$

(In Example 5.2(a) using $\sigma_H = \bar{\sigma}/3$, the simplified integral becomes $1.50\bar{\epsilon}$ in place of $1.56\bar{\epsilon}$ obtained previously). In this form, we see that the models are all versions of $\int(\sigma_H/\bar{\sigma})d\bar{\epsilon}$ as, in the case of shear, τ is related to $\bar{\sigma}$ by a function of the stress ratio M (Appendix 3). In proportional loading, where $M = \sigma_1/\sigma_2$ or $L = \epsilon_1/\epsilon_2$ are constant, $\sigma_H/\bar{\sigma}$ is constant and the criterion amounts to a critical effective strain $\bar{\epsilon}_f$. The strong dependency on stress triaxiality shows that the models emphasise dilatation changes in the voids rather than shape changes.

Inspection of Table 5.1 shows that many of the empirical criteria are also versions (sometimes with adjustable parameters) of $\int(\sigma_H/\bar{\sigma})\,d\bar{\epsilon}$ or merely $\int\sigma_H\,d\bar{\epsilon}$ (which, if the flow stress is constant, is the same). This comes about because

† For hole growth to occur from inclusions, the inclusion must fracture or decohere from the surrounding matrix. This process seems to be able to be described by a critical strain (Goods and Brown, 1979). The interaction of this strain level with hole growth processes is discussed by, for example, Dodd and Atkins (1983).

$\sigma_H \propto \sigma_1 \propto \sigma_2 \propto \bar{\sigma} \propto \tau$ along given linear loading paths (Appendix 3). However, not all will be as successful as each other in fitting experimental data, as the accumulated damage depends on this stress ratio $M(\sigma_1/\sigma_2)$ or the strain ratio L, i.e. although $\sigma_H \propto \sigma_1 \propto \sigma_2 \propto \bar{\sigma} \propto \tau$, the constants of proportionality depend on M and L. For example,

$$\int_0^{\bar{\epsilon}_f} (\sigma_H/\bar{\sigma})d\bar{\epsilon} \text{ becomes } \int_0^{\bar{\epsilon}_f} (1+M)d\bar{\epsilon}/3(1-M+M^2)^{\frac{1}{2}}$$

but

$$\int_0^{\bar{\epsilon}_f} (\sigma_1/\bar{\sigma})d\bar{\epsilon} = \int_0^{\bar{\epsilon}_f} Md\bar{\epsilon}/(1-M+M^2)^{\frac{1}{2}}$$

(cf. the Cockcroft–Latham criterion and so on in Fig. 5.7 and regions of agreement and disagreement between different empirical criteria).

5.4 APPLICATIONS OF VOID-GROWTH MODELS

It is possible to predict an $\epsilon_{1f}\epsilon_{2f}$ cracking locus from the various versions of the McClintock, and similar, criteria. Consider equation (5.3). Substituting for $\sigma_H/\bar{\sigma}$ and the other stresses in terms of the strain ratio L, and $d\bar{\epsilon}$ in terms of $d\epsilon_1$ (Appendix 3) we obtain, using an average $l/2r$,

$$\ln(l/2r) = \int (2 + 1/L)\ d\epsilon_1 \qquad (5.5)$$

which predicts a crack nucleation locus of

$$\epsilon_{1f} = -\epsilon_{2f}/2 + \ln(l/2r)/2 \qquad (5.6)$$

i.e. a line of slope $\tan^{-1}(-\frac{1}{2})$. This is precisely what is observed for the surface strains in barrelled upsetting tests (section 5.2).

In the plastic forming of crack-free sheet metal, slight complications arise in McClintock calculations owing to the change in strain ratio when localised necking takes place. In section 4.4, which concerns the propagation of already-existing cracks in sheets, the necked-down zone was considered to form the crack tip process zone for propagation. In bulging, the whole sheet is under a state of plane stress, the in-plane stresses being much greater than the hydraulic pressure. When necks form, experiments show that they are perpendicular to the major in-plane strain ϵ_1 and that they grow as plane strain necks, since ϵ_2 remains constant and $d\epsilon_2 = 0$. Thus, *after* necking, the situation in the necks is *both* plane stress *and* plane strain combined, $d\epsilon_2 = 0$ when $L = \infty$ and $M = 2$. Fig. 5.10(a) shows the same data as in Fig. 5.4, but plotted in $\sigma_1\sigma_2$ stress space rather than $\epsilon_1\epsilon_2$ strain space. The loading paths are labelled correspondingly. Stresses are calculated from measured strains using the usual plasticity relations given in Appendix 3. Before necking, the radial paths of constant strain ratio $\epsilon_1/\epsilon_2 = L$ in Fig. 5.4 map to radial lines of constant stress ratio $\sigma_1/\sigma_2 = M$, where $M = (2L+1)/(L+2)$, in Fig. 5.10(a). Note that the plane strain – plane stress ordinate in Fig. 5.4 having $L = \infty$ becomes the line OU_n of slope $M = 2$ in Fig. 5.10(a). The ordinate in Fig. 5.10(a) itself corresponds with the simple

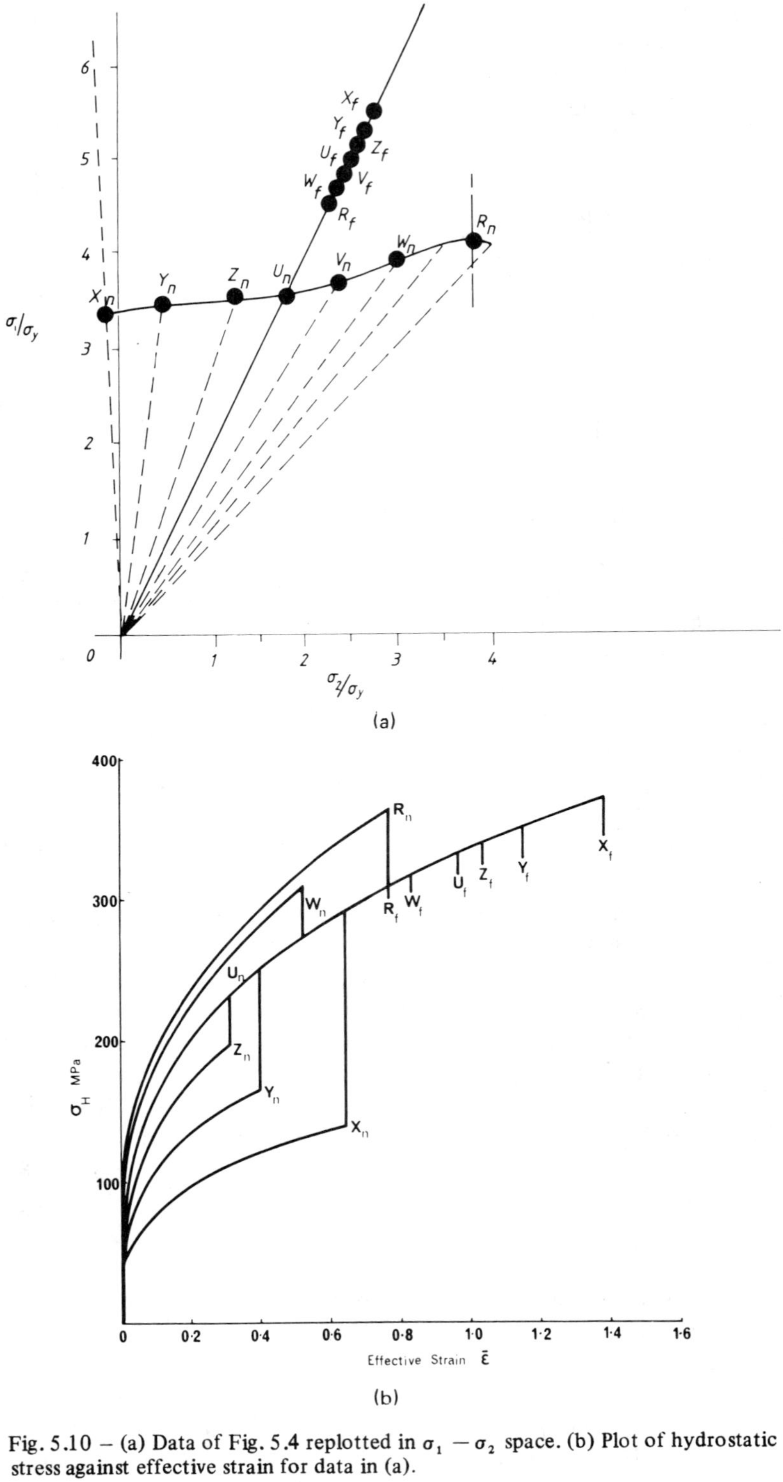

Fig. 5.10 – (a) Data of Fig. 5.4 replotted in $\sigma_1 - \sigma_2$ space. (b) Plot of hydrostatic stress against effective strain for data in (a).

tension line OX_n in Fig. 5.4 having $M = 1$ and $L = 2$.† Once necking occurs, the strain state changes and if a plane-strain neck is eventually assumed for *all* previous loading paths $L = \infty$ and $M = \sigma_1/\sigma_2 = 2$ for *all* paths during necking. Hence the separate strain paths $X_n X_f$, $Y_n Y_f$ . . . in Fig. 5.4 condense in Fig. 5.10(a) on to the single $OU_n U_f$ line is stress space.‡ The stress pairs at fracture at $X_f, Y_f \ldots R_f$ satisfy $\bar{\sigma} = \sigma_0 (\bar{\epsilon}_{neck} + \bar{\epsilon}_{after\ neck})^n$ where $\bar{\epsilon}_{after\ neck}$ occurs with $\sigma_1 = 2\sigma_2$.

The transition in stress state between, say, X_n and U_n before going on to X_f (with σ_2 increasing) or between, say, W_n and U_n before going on to W_f (with σ_2 initially decreasing and then increasing) is not known accurately (in Fig. 5.4 remember, kinks are shown in the strain paths which gloss over the experimental difficulties in determining precise changes in strain ratio).

Fig. 5.10(b) shows a plot of hydrostatic stress against effective strain for the same conditions shown in Fig. 5.4 and Fig. 5.10(a), with the loading paths again labelled correspondingly. In deformation prior to necking along paths OX_n OY_n . . . OR_n, it is seen that σ_H is greatest for greatest initial L, i.e. balanced biaxial tension (path OR_n) has greater σ_H than plane strain tension (path OU_n) and least σ_H occurs with simple tension (path OX_n). Were fracture to occur on these paths prior to necking, strains to fracture would occur at greatest σ_H; the increasing σ_H at increasing L corresponds with decreasing ϵ_f with increasing σ_H along a Kuhn upset plot. Problem 5.14 shows that this leads to the usual types of relation quoted between ϵ_f and σ_H, i.e. *increasing* σ_H produces *decreasing* ϵ_f (e.g. Kals, 1983; Atkins, 1981; Hancock and Brown, 1983). As the formation of a plane strain neck in sheet forming alters the stress state, all σ_H lie on the single plane strain OU_n line extended to greater ϵ after necking in Fig. 5.10(b). As already explained, the transitions are not really discontinuous and arise from the assumed kink in the strain path; the transition is likely to be fairly rapid even so, and data in Caddell *et al.* (1968) for example show considerable plane strain flow after necking with ϵ_2 = constant. We observe that complete $\sigma_H - \epsilon$ histories in sheet forming with necks are, for example (OX_n . . . change *up* to the plane strain curve. . . fracture at X_f) in the tension–compression quadrant and (OW_n. . . change *down* to the plane strain curve. . . fracture at W_f) in the tension–tension quadrant. Now we see that at fracture the *greatest* ϵ_f at X_f etc. coincide with the *greatest* current σ_H, and

† The reason OX_n lies slightly in the tension–compression quadrant in Fig. 5.10(a) is that we have employed an *anisotropic* yield criterion in the strain-to-stress conversion in this particular illustration. Simple tension in anisotropic yielding has L slightly different from -2, so when $L = -2$ is used in the conversion a mild σ_2 component is given and the conditions are, strictly speaking, not quite those of simple tension. For similar reasons, OR_n is slightly steeper than the $M = 1$ line it would follow for isotropic yielding. These details are not important for the interpretation of the detail in Figs. 5.10(a) and 5.10(b).

‡ Embury's and LeRoy's (1977) Fig. 9 should correspond with Fig. 5.10a here. However, they took $d\sigma_1/d\sigma_2 = 2$ rather than $\sigma_1/\sigma_2 = 2$ in the flow rules for post-neck plane strain deformation and constructed a *series* of parallel lines of slope 2 running from X_n Y_n . . . etc. to the σ_1/σ_2 stress combination which satisfied the yield criterion at the ϵ value of X_f Y_f . . . etc. This gave them a convex fracture map in stress-space, instead of the fracture $\sigma_1\sigma_2$ pairs lying on OU_n extended as in Fig. 5.10(a). If fracture occurs *prior* to neck formation, then a convex fracture locus in $\sigma_1\sigma_2$ space is acceptable.

there is a consistent trend of decreasing ϵ_f with decreasing σ_H down X_fY_f . . . R_f (the balanced biaxial last point being satisfied in this instance if an incipient plane strain neck is presumed so that R_n drops to the plane strain line just at fracture). This behaviour is unconventional and would not have occurred had the material not necked and therefore continued to follow the original loading paths (Atkins, 1981).

The behaviour in Fig. 5.10(b) reinforces the idea of some critical $\int \sigma_H d\bar{\epsilon}$ at fracture, or a version of it, and that fracture is not connected just with simple stress or strain alone. However, the simple $\int \sigma_H d\bar{\epsilon}$ criterion does not satisfactorily describe the data in Fig. 5.10, the value of the integral decreasing from about 300 MJ/m^3 along OX_nX_f to some 200 MJ/m^3 along OR_nR_f (Problem 5.5). The simplified McClintock criterion $\int(\sigma_H/\bar{\sigma})d\bar{\epsilon}$, while showing a similar decreasing trend, gives much more consistent values, viz. 0.59 for OX_nX_f through 0.54 for OU_nU_f to 0.50 for OR_nR_f, and may be adequate for predictive purposes (cf. Fig. 5.7).

A factor which has been neglected in these analyses is the effect of neck geometry on σ_H whereby σ_H will be elevated above the levels calculated (as here) merely on the basis of $\bar{\epsilon}$. It is possible to apply corrections for σ_H within plane strain necks (Bridgman, 1952, pp. 31 *et seq.*) which improves matters, but there is little experimental information on the geometry of necks under different loading histories.

The *predicted* $\epsilon_{1f}\epsilon_{2f}$ cracking locus corresponding to $\int(\sigma_H/\bar{\sigma})d\bar{\epsilon}$ may be obtained by substituting for $\sigma_H/\bar{\sigma}$ in terms of L, and $d\bar{\epsilon}$ in terms of $d\epsilon_1$, as before (e.g. equation (5.5)) to obtain

$$\ln l/2r = \int(\sigma_H/\bar{\sigma})d\bar{\epsilon} = \int 2(L+1)\, d\epsilon_1/3L \qquad (5.7)$$

In sheet forming where necks appear before fracture the integration is performed in two stages, before and after necking, so that

$$\int(\sigma_H/\bar{\sigma})d\bar{\epsilon} = \int_0^{\epsilon_{1n}} 2(L+1)d\epsilon_1/3L + \int_{\epsilon_{1n}}^{\epsilon_{1f}} 2d\epsilon_1/3 \qquad (5.8)$$

since in a plane strain neck $L = \infty$.

A valuable simplification arises in these calculations as it happens that for many forms of the McClintock criteria, the integral is (perhaps surprisingly) *path-independent,* i.e. in Fig. 5.11 the same integrated value is obtained whether the deformation proceeds along OC with strain ratio L_1 followed by CA with strain ratio of ∞, or along OA directly with strain ratio L_2, or indeed any other combinations of linear segments (including even reversed strain segments). The next Example demonstrates this unexpected result.

Example 5.3 Show that for a series of linear loading segments, $\int(\sigma_H/\sigma)d\epsilon$ is independent of path.

Ans. From the linear geometry of Fig. 5.11, $\epsilon_{1C} = (L_1/L_2)\epsilon_{1A}$. The damage built up along the direct path to A of slope L_2 is therefore $2(1/L_2 + 1)\epsilon_{1A}/3$. Along

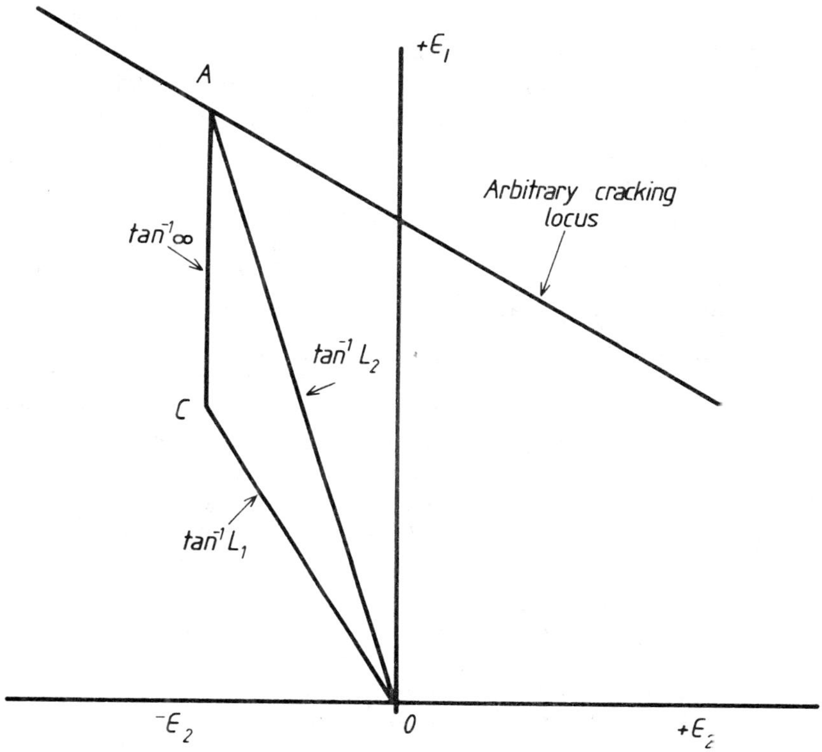

Fig. 5.11 – Path independence of equation (5.8).

the 'kinked' path OCA we have $2(L_1 + 1)\epsilon_{1C}/3L_1$ up to C and $2(\epsilon_{1A} - \epsilon_{1C})/3$ from C to A (along which $L = \infty$ in a plane strain neck). The total damage is $2\epsilon_{1C}/3L_1 + 2\epsilon_{1A}/3 = 2(1/L_2 + 1)\epsilon_{1A}/3$ which is the same.

The crack nucleation locus predicted by equation (5.8) (and Example 5.3) is

$$\ln l/2r = \text{constant} = 2(1/L + 1)\,\epsilon_1/3 \tag{5.9}$$

i.e.

$$\left.\begin{aligned} \epsilon_{1f} &= -\,\epsilon_{2f} + 3\ln(l/2r)/2 \\ &\text{or} \\ \epsilon_{1f} + \epsilon_{2f} &= -\,\epsilon_{3f} = 3\ln(l/2r)/2 \end{aligned}\right\} \tag{5.10}$$

which is a 45° line sloping down from left to right in the $\epsilon_1\ \epsilon_2$ plane, or equivalently a constant thickness strain at fracture of magnitude $3\ln(l/2r)/2$.

The McClintock cracking locus already derived from equation (5.5), i.e. $\ln(l/2r) = \int(2 + 1/L)\mathrm{d}\epsilon_1$ is also path-independent (Problem 5.12) and some (but not all) of the McClintock shear criteria, equations (5.2a, b), are also path-independent in certain shearing directions (Problem 5.13). It may be shown (Problem 5.11) that *all* criteria of the form $\int(A + B/L)\mathrm{d}\epsilon_1$, where A, B are constant, are path independent.

The stress state at fracture depends absolutely on the strain ratios in the final increments of loading, i.e. for the simple linear paths through point A considered in Fig. 5.11, on L_2 for path OA or on ∞ for path OCA, and it has already been demonstrated in Fig. 5.10 that this can mean marked differences in hydrostatic stress. Yet we have just demonstrated that ϵ_f can be essentially the same irrespective of the path. This presents a paradox to the received opinion on σ_H, ϵ_f relations. Of course we have assumed that 'damage' accumulated under a series of different L can be added directly; that must be a simplification when final fracture concerns void coalescence in different 'directions' from those in which the hole growth was built up. Research is required to clarify this matter.

The location of crack nucleation sites in expanded fully plastic thick-walled cylinders and rotating disks lends further support to a criterion of ductile fracture which must depend on *both* stress *and* strain. Tomkins (1965), in experiments concerning the fatigue life of the liners of hydrostatic extrusion presses, internally expanded thick-walled rings of a number of different ductile metals by means of driven tapered drifts. In microstructurally 'dirty' metals localised shear fractures initiated in the *mid-region* of the annulus and ensuing propagation consisted of a series of connecting zig zag shear features in a more-or-less radial direction, the outside of the thick ring being reached before the crack came in to meet the bore. Fig. 5.12(a), (b) shows the behaviour of commercial grade mild-steel and cast iron. Hydraulic tests on cylinders made from similar steel produced longitudinal splits on the *outside* surfaces while the bore remained uncracked. Close-ups of some of the 'star' nucleation sites in commercial mild steel are shown in Fig. 5.12(c).

The plastic flow of thick-walled rings may be analysed by a finite (large) strain deformation theory, with a rigid/workhardening material, due to Marin and Rimrott (1958). The maximum tensile stress in the system is the hoop stress at the outside rim; the maximum shear stress is at the bore. Similarly for strains. Yet the cracks nucleate at, and propagate from, regions *within* the annulus. Clearly theories of fracture based on maximum tensile stress or shear stress alone fail to predict the observed behaviour.

The cracks nucleate by shear as shown by the orientation of the 'stars' in Fig. 5.12(c) (we recall that the deformation field for a rigid–perfectly plastic ring is a net of logarithmic spirals e.g. Johnson *et al.* (1982)). Fig. 5.13(a) gives the distribution of γ and normalised σ_H across the radius of the ring as the radial expansion u_a increases. For convenience the same abscissa is used for corresponding points although in practice elements move radially outwards. Equations (5.2) and (5.4) involve the product of normalised hydrostatic stress and shear strain and we observe that as σ_H increases with radius, γ decreases. It is likely therefore that the metallurgical damage passes through a maximum within the

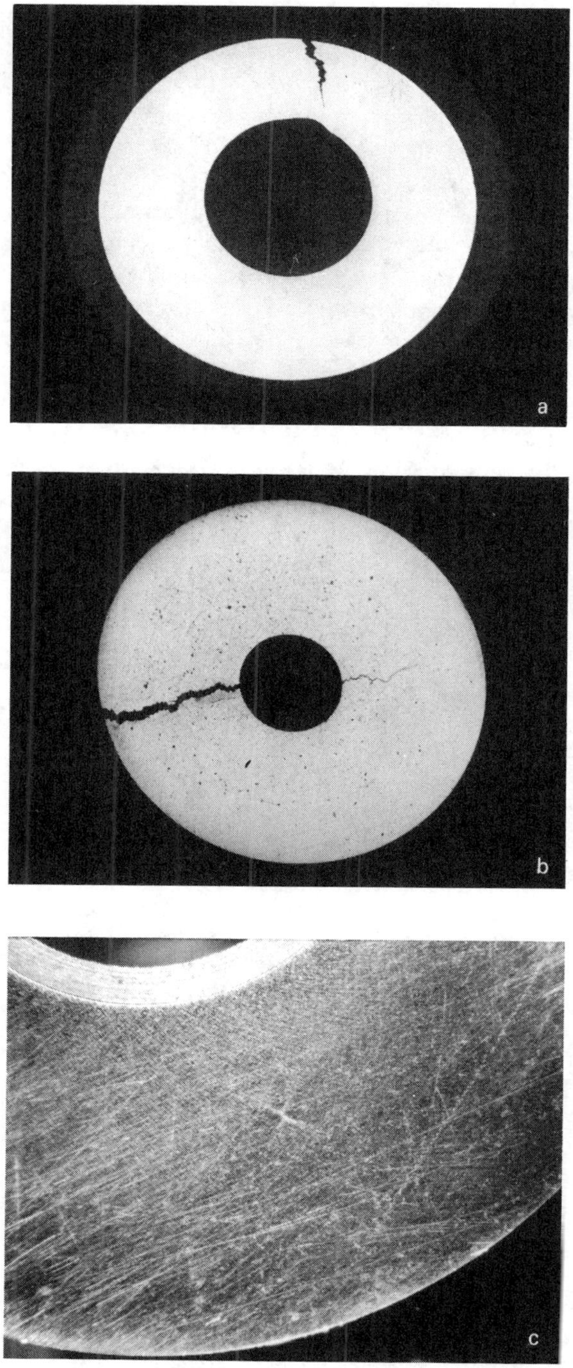

Fig. 5.12 – Cracking of expanded rings in (a) commercial grade mild steel and in (b) cast iron strained to plastic instability. In (c) close-ups of the 'star' nucleation sites in mild steel are shown (after Tomkins and Atkins (1981), with permission).

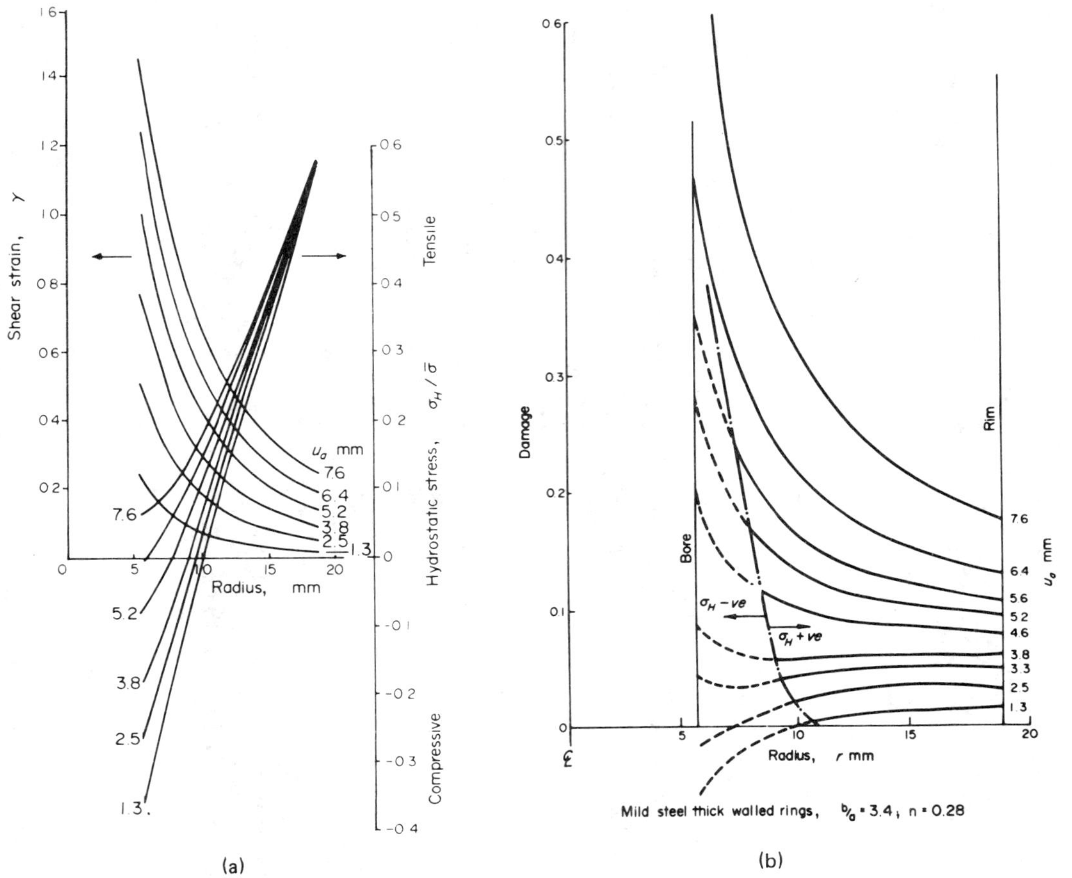

Fig. 5.13 – Normalised hydrostatic stress σ_H/σ and shear strain distributions at various bore displacements u_a for expanded rings. (b) Damage D_L as a function of radial position as u_a increases. The dashed curves to the left of the chain line are for compressive hydrostatic stresses (after Tomkins and Atkins (1981), with permission).

annulus wall. Fig. 5.13(b) shows the build-up of microvoid growth damage for mild steel as the bore displacement u_a increases. The dashed parts of the curve correspond with compressive σ_H. Up to u_a of 3 or 4 mm (which is the bore displacement corresponding with crack initiation) the curves show flat maxima over the outer two-thirds of the annulus wall. This corresponds with the location of the star nucleation sites.

The subsequent zigzag coalescence of these local shear cracks to give final fracture by a radial crack, Fig. 5.11(a), is similar to the propagated paths of cracks in upset specimens (seen by taking sections through cracked forgings parallel to the platens (Kudo, 1970), (section 5.2)).

Similar behaviour has been observed in experiments on the burst strength of rotating disks, employed to simulate turbine rotor disks. The interested reader is referred to the 1981 paper by Tomkins and Atkins for further details of the expanded ring problem and rotating disk problem, including the occurrence tensile hoop instabilities prior to fracture in 'metallurgically cleaner' materials.

Many forming problems are now being tackled by finite element/finite difference methods (see, for example, the proceedings of the 1982 Swansea meeting on numerical methods in industrial forming processes, edited by Pittman *et al.*). As hinted at in section 4.5.4, there are difficulties associated with formulation of extensive plastic flow involving large deformations and rotations but, even so, fully coupled thermomechanical problems may now be solved (see, for example, the Marti *et al.* (1984) reference in Chapter 4). As far as *fracture* following extensive flow is concerned, a number of groups is pursuing this field and applying the criteria given in this chapter to finite element/finite difference calculations. Among these groups are those of Kobayashi and co-workers (1974) and Rowe and co-workers (1985). (See S. Clift, 1984).

5.5 MECHANICS-TO-MICROSTRUCTURE LINK

The McClintock damage functions provide a mechanics-to-microstructure link, as the critical integrated value is $\ln(l/2r)$ along the failure locus. Likewise the integrated function in Oyane's criterion (Table 5.1) corresponds with the attainment of a critical volumetric strain at fracture, which may be interpreted in terms of the volume fraction of inclusions (holes). Thus, for example, the plane strain workability indices ϵ^0_{1f} of Fig. 5.3 provide a measure of $(l/2r)$ or ν_f. From equation (5.5), $\ln(l/2r) = (2 + 1/L)\epsilon_{1f} = 2\epsilon^0_{1f}$ for a path along the ϵ_1 plane strain ordinate. Hence

$$(l/2r) = \exp(2\epsilon^0_{1f}) \tag{5.11}$$

where ϵ^0_{1f} is the workability index. For the 1040 steel, the fine pearlite structure displays $\epsilon^0_{1f} = 0.3$, so $l/2r = 1.8$; the coarse spheroidite has $\epsilon^0_{1f} = 0.4$, so $l/2r = 2.2$. Inspection of micrographs in Erturk's original paper shows that these spacing-to-size ratio predictions are of the right order of magnitude, but always err on the small side (insofar as non-uniform morphology of the microstructure permits a representative $l/2r$ ratio to be assessed).

In sheet forming, the failure locus, in Fig. 5.4 for example, crosses the plain strain ordinate at $\epsilon^0_{1f} \approx 0.9$. Hence for this 5154 Al–Mg material

$$(l/2r) = \begin{cases} \exp(2 \times 0.9) \approx 6 \\ \exp 2 \times 0.9/3 \approx 2 \end{cases}$$

the second evaluation being obtained from $\int(\sigma_H/\bar{\sigma})d\bar{\epsilon} = \int(2/3)\,(1 + 1/L)d\epsilon_1$ (equation (5.7)) which predicts a 45° failure locus closer to that observed in Fig. 5.4 than the $\tan^{-1}\,(-\frac{1}{2})$ line predicted by equation (5.6). Corresponding ϵ^0_{1f} values for commercially pure aluminium sheet and alpha brass are about 1.3 and 1.0 respectively according to Caddell *et al.*'s 1968 data. These values imply $2 < l/2r < 1.13$ for annealed aluminium and $2 < l/2r < 7$ for brass, or equivalently $5 \times 10^{-4} < \nu_f < 0.12$ and $3. \times 10^{-3} < \nu_f < 0.12$ respectively using $\nu_f = (2r/l)^3$. Owing to the cube power on $(2r/l)$, the bounds on ν_f are very wide.

In simple cropping (bar-shearing) to be described in more detail in section 5.7, it is found that cracks first appear at shear strains of 20 and 1.6 respectively in lead and brass; (in Fig. 5.14, $\gamma \approx u/w$ where u is the penetration of the punch and w is the effective width of the shear band). If, for simplicity, we assume that bar shearing is a process of pure shear where $\sigma_H = 0$, equation (5.4a) gives $l/2r = \sqrt{401} \simeq 20$ for lead and $\sqrt{1 + 1.6^2} \simeq 2$ for brass. The corresponding volume fractions of inclusions are about 10^{-4} and 10^{-1} respectively.

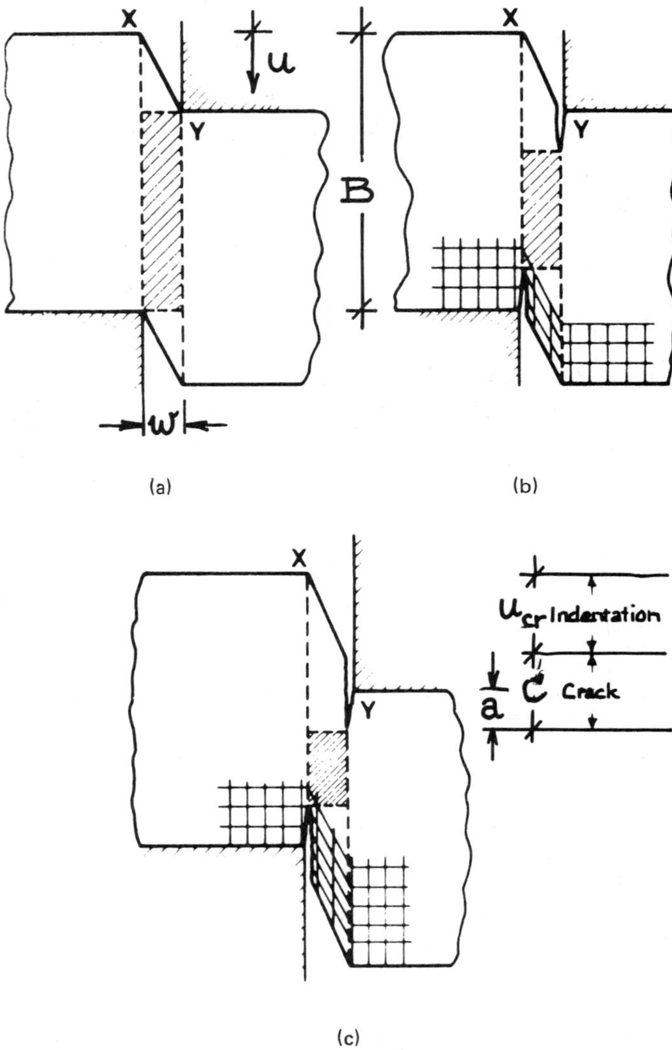

Fig. 5.14 – Single shear plane model of cropping: (a) simple flow of indentation phase; (b) flow plus cracking; and (c) flow plus cracking where tool has penetrated beyond the depth for initiation so that surface alongside the flank of the tool is part of the crack (after Atkins (1980), with permission).

Fig. 5.15 shows inclusions in the commercial grade mild steel from which the expanded rings in Fig. 5.12 were made. Fig. 5.15 shows longitudinal axial sections from the rings, the inclusions being strung out as the material had been extruded. According to Fig. 5.13(b), $\ln(l/2r)$ from the damage at crack initiation should be about 0.05, i.e. $(l/2r) = \exp(0.05) = 1.05$. This value is much too low. Fig. 5.15 shows how difficult it is to arrive at an average $l/2r$ in practical materials. The inclusions are not uniform in size and their distribution also is not uniform. Ghosh's 1976 formulation for ductile fracture by hole growth (Problem 5.9) suggests that instead of a $\ln(l/2r)$ relation between damage and microstructure, the critical hole growth damage is proportional to $(1/v_f)^{1/2n}$ where n is the workhardening index. In the case of the steel rings $n = 0.28$, so damage $\propto (1/v_f)^2$ roughly, or in other words damage $\propto (l/2r)^6$. Unfortunately it is not possible from Ghosh's analysis to give absolute values for $(l/2r)$, but comparisons with the 'cleaner' microstructure of a vacuum-melted steel that experienced a tensile hoop instability prior to fracture shows that a $(l/2r)^6$ relation is more in accord with the relative inclusion sizes and distributions as between the two steels (the vacuum-melted does not seem that clean!).

Othe workers such as Guillen-Preckler in 1967, and French and Weinrich in 1979 also find that the McClintock analysis predicts that the particles are closer together than found in practice and/or that the particles themselves are larger. We remember that the McClintock model is two-dimensional and we are

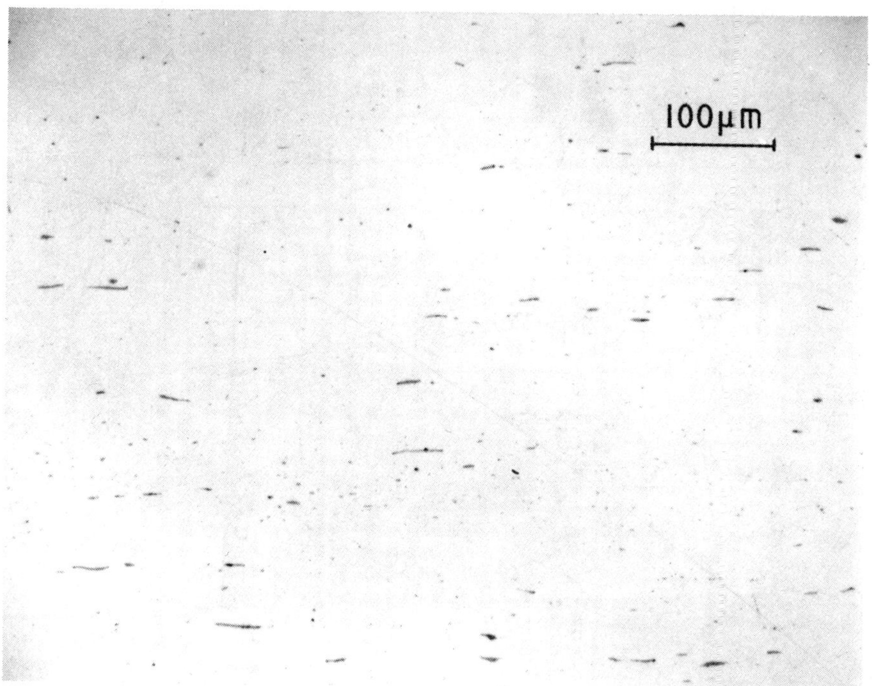

Fig. 5.15 – Inclusions in commercial grade mild steel used for rings shown in Fig. 5.12(b).

applying it often to three-dimensional problems, but even so one possible reason for the discrepancy is connected with the definition of fracture as when the expanded voids touch their cell boundaries. It could be that other events occur before the voids touch which accelerate the process of ductile fracture. The McClintock treatment assumes that the voids grow in uniform deformation fields. However, it has been shown that deformation becomes concentrated along narrow shear bands between voids (Fig. 5.16). The situation is enhanced if thermal softening occurs (as in high rate deformation which may approximate to quasi-adiabatic flow). Kobayashi and Oh (1974) from finite element calculations showed that strains along the localised deformation bands were larger than the overall strains in the bulk of the body, by a factor of 2 or 4 times, and were greater nearer the voids; also they calculated that the hydrostatic stress was

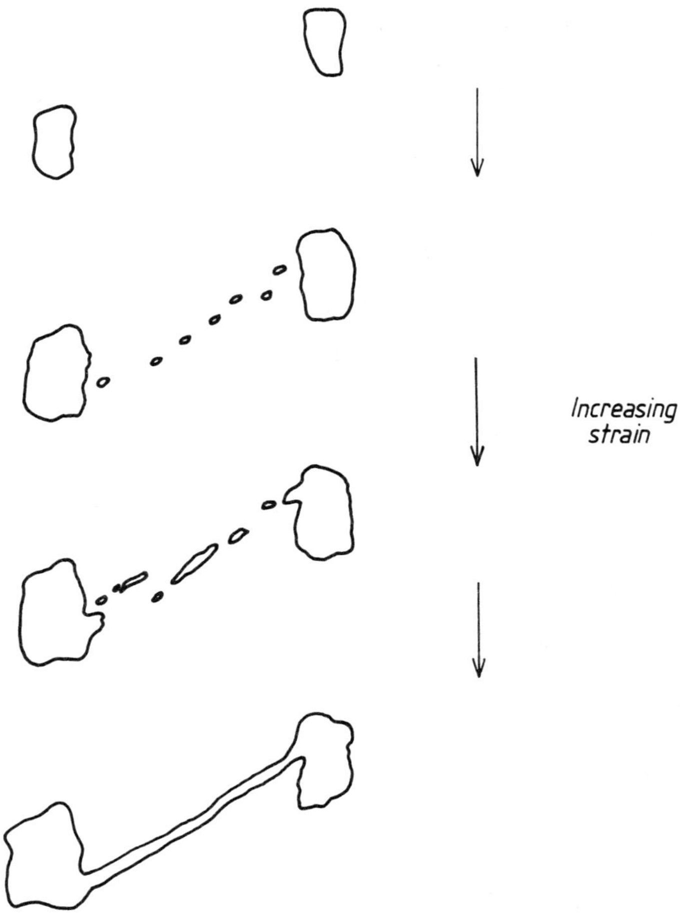

Fig. 5.16 – Flow localisation between voids (after Beachem (1975), with permission).

elevated in the shear bands. Thus, if the *local* fracture strain is $\lambda\epsilon_f$, say, instead of ϵ_f it follows that in equation (5.11), for example, the 'real' $(l/2r)$ satisfies $\exp(\lambda 2\epsilon^0_{1f})$ rather than $\exp(2\epsilon^0_{1f})$. This then brings the $(l/2r)$ calculated from workability strains more into line with values observed in microstructures.

Another important detail which bears on this problem is that, in some metal alloys, there are distinct populations of small, as well as large, inclusions. For example, in high strength steels Cox and Low (1972) found that large voids were nucleated first at manganese sulphide inclusions and, later, smaller ones at carbide particles. The first voids grew with little interaction between them. Then, at strains close to the final fracture strain, voids formed at the smaller particles, giving extensive coalescence in the form of 'void sheets' in the maximum shear stress direction. Broek (1973) and Van Stone *et al.* (1974) observed similar behaviour with high-strength aluminium alloys. In these observations, the large particles were some 10–30 μm apart, and the small particles some 300–500 nm apart.

A link between workability strains and the essential work of fracture of Chapter 4 is discussed in section 5.8.

5.6 CRAZE INITIATION IN AMORPHOUS POLYMERS

It has already been mentioned, in section 1.1, that crazes in un-crosslinked amorphous polymers are precursors to crack propagation. Crazing develops under a tensile stress only, the plane of the craze being normal to the principal stress direction. Although crazes usually are generated on the surface they can be developed internally as well, particularly when immersed in swelling agents.

Crazes in glassy polymers are not true cracks – although for many years people guessed they were, despite the fact that crazed plain tensile specimens retained appreciable strength, which true cracks would not have permitted. Rather, crazes are 'spongy' materials which are plastically deformated. Fig. 5.17 shows the detailed structure of a craze in polystyrene. Note that all spongy crazed material can occupy as much as 40–60% in void volume fraction, with the voids interconnected by fine fibrils 10 nm–20 nm in diameter. Depending on the density of the crazes the reflectivity of light also changes. This has provided a means of estimating craze densities and orientations.

Because of the fibrillar structure of the crazes, they can be distinguished easily from true cracks. Cracks have no load-carrying capacity, but crazes can sustain reasonable mechanical stresses. The mechanical behaviour of crazes differs considerably from that of the bulk polymer. Fig. 5.18 shows the cyclic stress–strain curve for a freshly formed polycarbonate craze alongside that for the bulk polymer. There are two points to note. (i) They give, for about the same maximum stress magnitude, widely different cyclic behaviour. There is no yield point and little energy loss in the stress cycle for the bulk polymer. However, the crazed material exhibits a yield at 14 MPa and presumably workhardens (as well as breaking some of the already strained fibrils) thereafter. The hysteresis loop is remarkably large. (ii) In the freshly created craze, the strains in the fibrils could be as much as 0.6. This is reasonable since crazes are formed as a

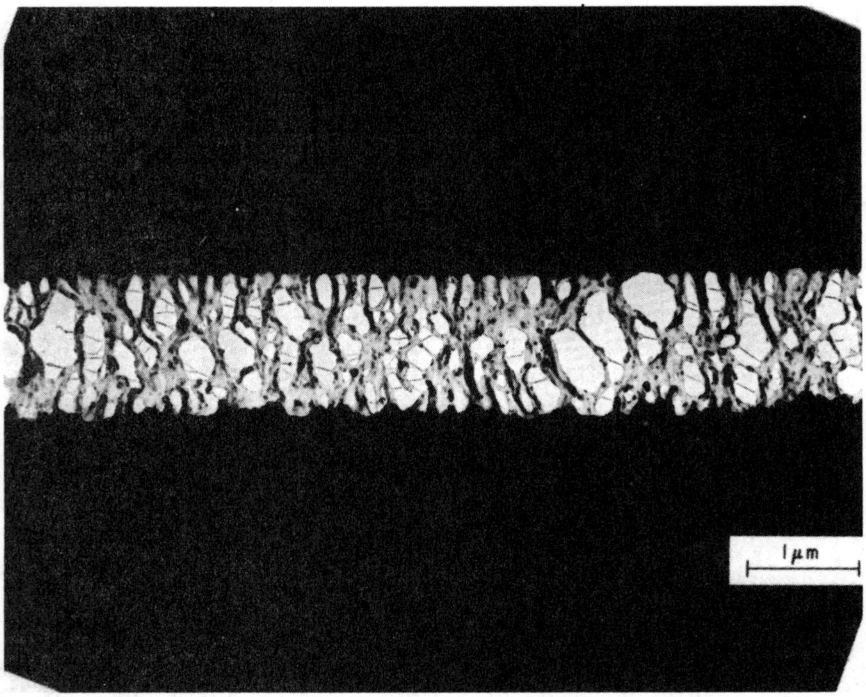

Fig. 5.17 – Transmission electronmicrograph of polystyrene craze. Stumps of broken and retracted filaments are evident as dark lumps (after Kambour and Russel (1971), with permission).

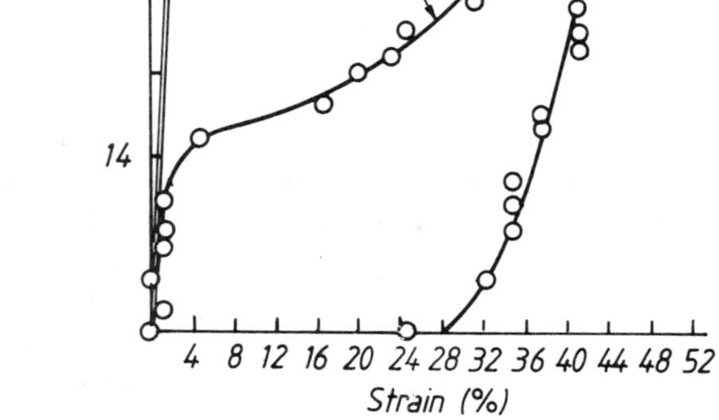

Fig. 5.18 – Cyclic stress–strain curve for completely strain-hardened craze. Strain scale calculated as though original unstressed craze had zero strain. On this scale 60% corresponds to 163% absolute strain starting from polymer of normal density (after Kambour and Kopp (1969), with permission).

result of large plastic shear deformation. When a considerable stress is applied, some of the fibrils will workharden and then break, giving a larger void content. This work must be included in the hysteresis loop.

Fig. 5.19 shows the cyclic stress–strain curve for a completely strain-hardened craze in polystyrene. Note that the large hysteresis loop originally observed in the fresh craze (Fig. 5.18) is now non-existent.

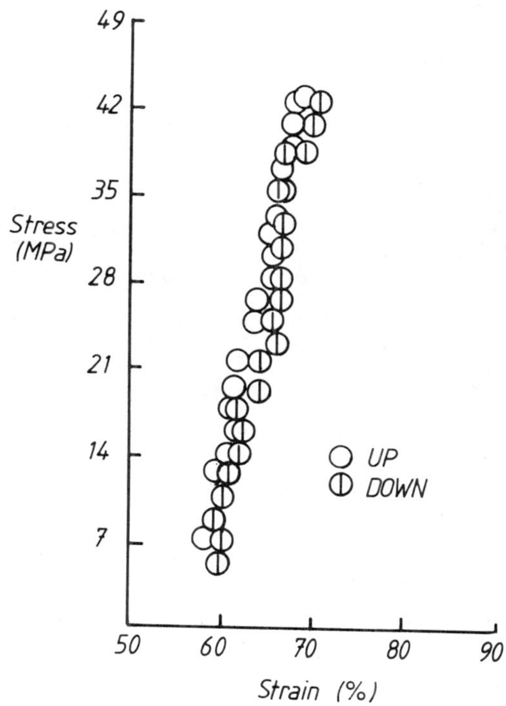

Fig. 5.19 – Cyclic stress–strain curve up to 42 MPa for bulk isotropic polycarbonate craze starting from unstressed state. Zero strain for craze is equivalent to 60% absolute strain (after Kambour and Kopp (1969), with permission).

5.6.1 Theories and mechanisms for craze initiation

How are crazes initiated and what are the thoeries for craze initiation? We shall deal first with the phenomenologically based criteria considering stresses and strains for craze initiation. The mechanisms for craze nucleation will be studied afterwards.

Since voids are observed in the crazes, it is necessary that craze initiation be involved with void formation. Any theory proposed must accommodate this effect. Consequently any model must have a dilatational component of stress to account for the volume expansion. Rather like the McClintock arguments in section 5.3.2 for hole growth (applied there to metals), Sternstein and Ongchin

(1969) suggested that a proper criterion for crazing must involve both dilatational stress (to promote the flow mobility) and a deviatoric stress for flow. Thus

$$|\sigma_1 - \sigma_1| = A(T) + B(T)/I_1 \tag{5.12}$$

where σ_1, σ_2 are the principal stresses, I_1, the first stress invariant, is $\sigma_1 + \sigma_2 + \sigma_3$ which is $3\sigma_H$, and $A(T)$ and $B(T)$ are temperature-dependent constants (σ_3 was 0 in associated experiments). It is clear from equation (5.12) that the stress bias $|\sigma_1 - \sigma_2|$ required to produce crazes decreases as the stress invariant I_1 increases. This indicates the significant role of the dilation stress in the crazing process.

Another similar criterion was proposed in 1973 by Bowden and Oxborough for crazing in PS stressed in the tension–compression quadrant. They found that craze initiation stress data were well fitted by an equation of the form:

$$\sigma_1 - \nu\sigma_2 = \frac{X(T)}{\sigma_1 + \sigma_2} + Y(T) \tag{5.13}$$

where ν is Poisson's ratio. (Note that $\sigma_1 - \nu\sigma_2 = E\epsilon_1$, so that the above indicates a critical strain criterion (as opposed to the critical stress criterion in equation (5.12)). Thus

$$\epsilon_c = \frac{X'(T)}{\sigma_H} + Y'(T) \tag{5.14}$$

where $\sigma_H = I_1/3$. When plotted out for $\sigma_3 = 0$, the biaxial-stress envelope for craze initiation has the same shape as the Sternstein–Ongchin envelope. It is evident here that the tensile strain for craze initiation decreases with increasing tensile hydrostatic stress. Apparently both the critical stress bias and the critical strain criteria work equally well. The difference is that the Sternstein–Ongchin theory is not able to explain the direction of craze orientation but the Bowden–Oxborough theory is able to do so, since $(\sigma_1 - \nu\sigma_2)$ has an intrinsic direction to it, i.e. the direction of craze strain development. Equation (5.14) is also, in theory, applicable for craze initiation under triaxial stress so that equation (5.13) is modified to:

$$\sigma_1 - \nu\sigma_2 - \nu\sigma_3 = \frac{X(T)}{\sigma_1 + \sigma_2 + \sigma_3} + Y(T) \tag{5.15}$$

The Sternstein–Ongchin theory is difficult to apply in this general triaxial stress case.

So far we have only said that equations (5.12) and (5.14) are reasonable criteria for prediction of craze initiation and that the dilatational stress given by I_1 $(= \sigma_1 + \sigma_2 + \sigma_3)$ is a requirement for craze formation. We have not yet asked *how* crazes initiate. Gent has answered this question by explaining the bifurcation of the dilatational stress acting at the tips of flaws which causes the voids to nucleate. This is sensible since crazes are usually nucleated at stress-raising flaws such as surface scratches and cracks. The stress at the tip of a sharp flaw subjected to an applied average tensile stress (σ_{av}) is given by:

$$\sigma = \left[1 + 2\left(\frac{a}{\rho}\right)^{\frac{1}{2}}\right] \sigma_{av} = k\sigma_{av} \tag{5.16}$$

where k is the stress concentration factor, a is the length or depth of the surface flaw and ρ is the tip radius (Chapter 2). For tensile test pieces k may be about 10 to 50. The dilatational stress increases the free volume of the material at the flaw tip and thus reduces its glass transition temperature (T_g) to the temperature of the test (T). Here the material localised at the flaw tip is transformed to a narrow band of rubbery material and the critical tensile stress (σ_c) required to do this is (Gent, 1970):

$$\sigma_c = [\beta(T_g - T) + p]/k \tag{5.17}$$

where β is the hydrostatic pressure coefficient of T_g, and p is any applied hydrostatic pressure. The second role of the stress is to produce cavitation in the softened thin band of material so that voids are formed spontaneously – since the hydrostatic stress required to cavitate is much less than that required to produce thin band softening. This cavitation process is similar to the way rubbers cavitate under hydrostatic tension (Gent and Lindley, 1958).

Gent's theory readily explains several aspects of craze initiation. (i) Application of external hydrostatic pressure (p) suppresses craze initiation according to equation (5.17). (ii) The craze initiation stress decreases as the test temperature increases towards the glass transition temperature of the bulk polymer. (iii) There is a marked reduction in crazing stress in the presence of organic liquids owing to the decrease in T_g caused by the environment-induced plasticisation of the material (for details see section 6.5.3). However, Gent's model has been criticised on two points. Firstly, the predicted reduction in T_g to the temperature of the test, caused by the dilatational stress, is unrealistically high. It has been shown by some researchers that the local T_g reduction is at most a few degrees and this would only produce a small decrease in the yield stress. As remarked by Andrews and Bevan (1972), it is not a necessary requirement for the yield stress to reduce to zero in order for cavitation to occur. Secondly, the microstructure for a craze predicted by the model does not agree with that observed experimentally.

A different viewpoint is taken by Andrews and Bevan (1972) regarding the conditions for cavitation. They argued that cavitation is not spontaneous in the softened material as assumed in Gent's theory but is controlled by the actual nett hydrostatic tension stress $D(= \sigma_H - p)$ necessary to enlarge the cavity. In an elastic-plastic solid, a cavity of radius r is in equilibrium with D given by:

$$D = \Psi \sigma_y + 2\gamma/r \tag{5.18}$$

where Ψ is between 2.4 and 3.3 for a wide range of glassy plastics, σ_y is the reduced tensile yield stress of the softened material and γ is the interfacial tension of the void. The work done in enlarging the cavity from a volume v_0 to a volume v by the nett hydrostatic stress is:

$$W = \int_{v_0}^{v} D\,\mathrm{d}v = 3\gamma \left(\frac{4\pi}{3}\right)^{1/3} (v^{2/3} - v_0^{2/3}) + \Psi \sigma_y (v - v_0)$$

For $v \gg v_0$ then

$$W = 4.84\, v^{2/3} \gamma + \Psi \sigma_y v \tag{5.19}$$

If there are N voids in a unit area of craze of thickness h and void fraction f then $Nv = fh$. The minimum surface work of craze formation, called R_0, is given by:

$$R_0 = 2.42\,(hf)^{2/3}\,N^{1/3}\gamma + (\Psi/2)\,\sigma_y\,fh \tag{5.20}$$

Equation (5.20) shows that the specific work of craze formation consists of two components: the first term is the surface work to nucleate the voids and the second term is that to do plastic flow work in association with void expansion. Fig. 5.20 shows experimental results for R_0 plotted against temperature for PMMA in ethanol (Andrews and Bevan, 1972). These results are in good agreement with equation (5.20). For $T > T_c$ (which may be identified with the glass transition temperature T_g) it is suggested that the stress-dilated tip region is swollen by the solvent and this reduces the local T_g value to the test temperature so that $\sigma_y = 0$. Thus, from equation (5.20) R_0 is a constant depending only on the first term. For $T < T_c$, σ_y increases with decreasing T so that the second term explains the temperature dependence of R_0.

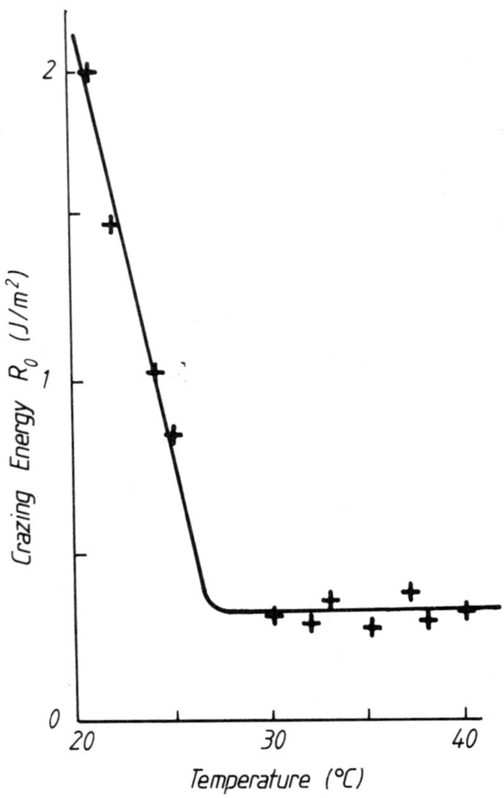

Fig. 5.20 – Work of crazing R_0 as a function of temperature for PMMA in ethanol (after Andrews and Bevan (1972), with permission).

5.6.2 Craze growth

In Gent's model for craze initiation, the craze tip is assumed to advance by repeated nucleation and expansion of isolated voids ahead of the tip. This mechanism of craze growth has been rejected by Argon and Salama (1977) because it is inconsistent with both the low stress concentration at the craze tip and the microstructure. Instead they suggest an alternative mechanism in terms of Shaffman and Taylor's meniscus instability (1958) which can give rise to craze-like structures. Details of this alternative theory will not be given here but interested readers may refer to the paper by Argon and Salama.

Craze growth in both air and organic solvents has been extensively studied by Williams and Marshall using a fracture mechanics approach (1975). Fig. 5.21 shows the craze growth curves for air-growth crazes for PMMA, polycarbonate

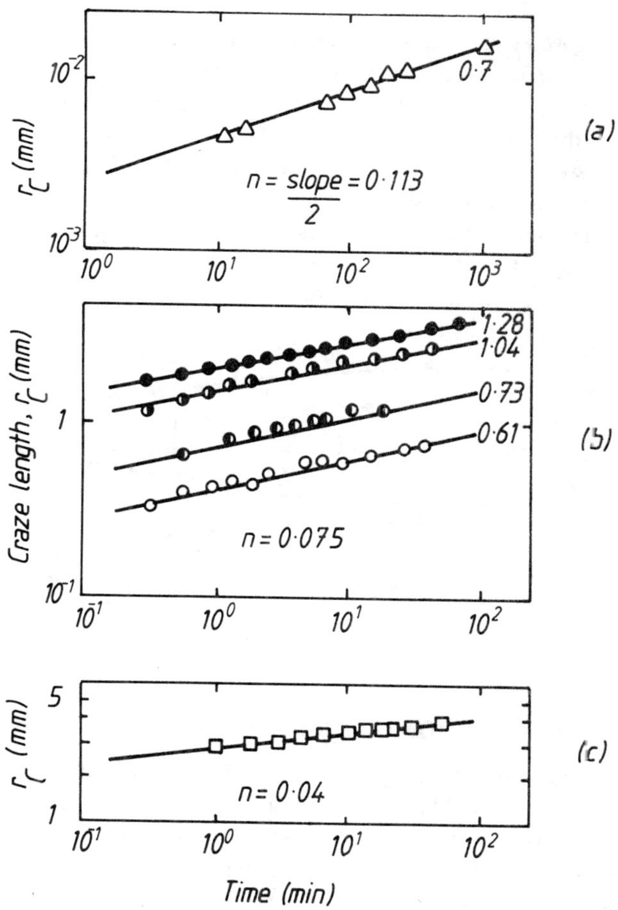

Fig. 5.21 – Craze growth curves for air-grown crazes for various polymers. (a) PMMA; (b) rubber modified polystyrene; and (c) polycarbonate. Numbers on lines are K_I values, MPa$\sqrt{m}$ (after Williams and Marshall (1975), with permission).

and rubber-modifed polystyrene obtained in single-edge-notched tension specimens. Starting from the Dugdale model the craze length (r) is given by:

$$r = \frac{\pi}{8}\left(\frac{K_I}{\sigma_c}\right)^2 \tag{5.21}$$

where K_I is the applied stress intensity factor and σ_c is the craze stress which is time-dependent according to:

$$\sigma_c = \sigma_0 t^{-n} \tag{5.22}$$

σ_0 is the unit stress and n is a constant related to the loss factor, tan δ (cf. Chapter 6). Combining equations (5.21) and (5.22) gives

$$r(t) = \frac{\pi}{8}\left(\frac{K_I}{\sigma_0}\right)^2 t^{2n} \tag{5.23}$$

The results given in Fig. 5.21 agree with the prediction of the above equation. The n values obtained from the slopes of the curve are also in very good agreement with those obtained from stress relaxation tests in which E is plotted logaritmically against time t. This shows that craze growth is relaxation-controlled.

Eventually $r(t)$ will reach a critical value (r_c) which gives a critical crack tip opening displacement (δ_C) at the craze base. (See Fig. 5.22.) Then craze breackdown occurs and allows crack propagation to proceed in a stable manner with a craze preceding its tip (see Fig. 1.11). The relationship between crack velocity and applied stress intensity factor is discussed later in section 6.3.

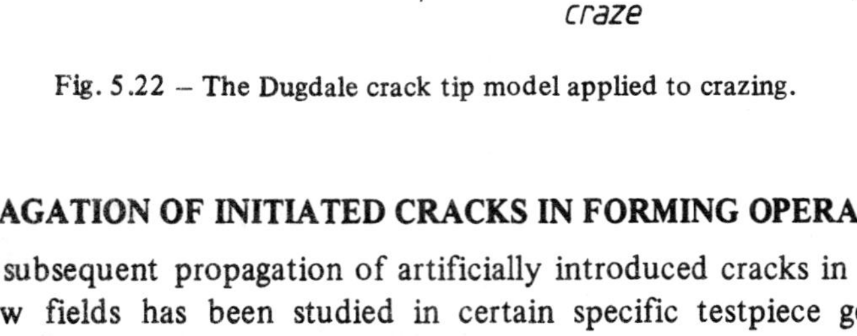

Fig. 5.22 – The Dugdale crack tip model applied to crazing.

5.7 PROPAGATION OF INITIATED CRACKS IN FORMING OPERATIONS

While the subsequent propagation of artificially introduced cracks in extensive plastic flow fields has been studied in certain specific testpiece geometries (Chapter 4), little work has been done on the subsequent growth of initiated cracks in geometries encountered in metalforming. Of course, of many processes the first appearance of cracks constitutes 'failure' and there is little interest in extensive crack propagation. (Even so, Dodd has shown that edge cracks produced in the rolling of sheet turn into the direction of rolling owing to plane stress/plane strain constraint which gives, after slitting, a reasonably wide, heavily reduced product (Fig. 5.1).) However, in processes which concern the separation of parts, the path of cracking and the forces required to effect complete separation are vitally important. For example, in shearing operations, two separate cracks emanate from the top and bottom members (puch and die sets), (Fig. 5.5). If these do not meet, final separation is by tensile tearing which can leave a rough surface displaying 'laps' and small cracks. These are sources of weakness – fatigue can originate at fastener holes manufactured by punching, and defects on the ends of cropped slugs often remain despite severe subsequent forging operations. It is clear that a better understanding of cracking in forming operations is required in order to eliminate these difficulties.

When the initiated microcracks become big enough, they propagate as macrocracks. In bar shearing the work absorbed in the absence of cracking consists of work of plastic shear and work of friction at the tool/workpiece interface. After crack initiation when there is combined flow and fracture, the work absorbed contains an additional component which is the work of fracture. In terms of a single shear-plane model of cropping, Fig. 5.14 shows the two possibilities. Minimum load and energy arguments suggest that the magnitude of such combined work must be less than that required for the alternative process of carrying on flowing with no cracking (cf. section 4.6.2). At first sight this seems odd as now *two* energy-absoring processes are taking place simultaneously, but once cracks begin to propagate the volume of material being plastically deformed is smaller than when cracks are absent. In Fig. 5.14(b), the work of cracking plus flow in the smaller plastic volume is less than work of flow alone in the larger plastic volume of Fig. 5.14(a). Such minimum work arguments must apply generally to events described by equations (1.1), (1.2), (1.3) and so on, providing of course that there is a mechanism available which enables cracking to take place, that is, which permits the RdA term to enter the work equation $\Sigma X_i du_i = d\Lambda$ for simple elasticity to give equation (1.1), or which permits RdA to enter equation (1.3) to give equation (1.2).

The minimum work hypothesis before and after crack initiation has been tested for cropping processes in metals by measuring the cumulative applied work and comparing it with predictions of a model based on plastic flow and friction alone. Fig. 5.23 shows such an analysis carried out by Atkins (1979) for annealed brass originally studied by Chang and Swift in 1950. Curve A is the cumulative work measured from the punch force punch penetration Xu diagram (shown as an inset). Curve B is the work required for work-hardening flow according to a single shear-plane model. At small u, the two curves agree, and this would seem to correspond with the indentation phase of deformation. At $u > u_{cr}$ the curves diverge, the energy absorbed being less than that required by theory. The point of divergence is likely to correspond with the onset of macrocrack propagation and indeed the value of u_{cr} given by Fig. 5.23 for brass agrees well with the value of puch penetration at which cracks seem to form according to gridded cross-sections given by Chang and Swift. It was shown that u_{cr} is quite well predicted for other materials too. Crack initiation does *not*, in general, coincide with the characteristic peak in load Xu diagrams (inset to Fig. 5.23). The peak is caused by a plastic load instability in shear (Example

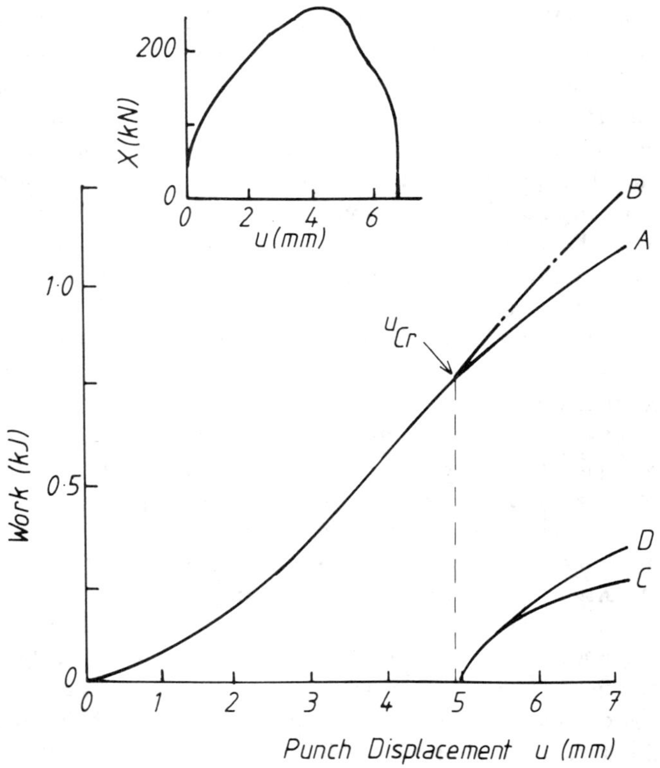

Fig. 5.23 – Cumulative work absorbed as a function of punch displacement for cropping (after Atkins (1979), with permission).

5.4 following). Beyond u_{cr} the work absorbed (curve D in Fig. 5.23) is greater than that predicted for plastic flow and friction (curve C) and it is suggested that the difference is the work of crack propagation. Values of the specific work of fracture (i.e. the fracture toughness R) obtained from the work area differences between curves C and D are 120 kJ/m^2 for aluminium and brass and 30 kJ/m^2 for lead (all results with an accuracy of about ± 20%). Although 'valid' toughnesses for such materials possessing high toughness-to-strength ratios are not known with confidence, these levels seem sensible. It should be noted that although R for lead is smaller than R for brass, the yield strength of brass is much greater than that of lead. From information in Change and Swift and these data, $R/\sigma_y \approx 2.3$ mm for lead but only 0.25 mm for brass; therefore lead is relatively more difficult to crack than brass.

Example 5.4 Show that the peak load in puch-force/punch-penetration diagrams is caused by a plastic instability and not by cracking.

Ans. The nett sheared interface is $(B-u)$ (Fig. 5.14), for unit thickness bar. The associated force is $X = \tau\,(B-u) = \tau_0\,(u/w)^n\,(B-u)$ if the workhardening shear τ is represented by $\tau = \tau_0\,\gamma^n$ with the shear strain γ in a simple shear-plane model being given by (u/w). For a plastic load instability $\mathrm{d}X = 0$, and taking w as constant, it follows that $u_{\text{peak}} = nB/(1+n)$. Good agreement has been obtained with experiments (Table 5.2). Modifications to account for tool friction and tractions normal to the shear plane are given in Problems 5.16–5.18. While macrocracking *may* coincide with the load maximum, cracking usually occurs after the peak in ductile solids; in more brittle solids cracks may appear before the peak (see Chapter 10). Bai and Johnson (1982) have extended this analysis to account for thermoplastic behaviour in the shear band and applied the theory to high rate plugging processes.

Table 5.2

Material tensile $\sigma = \sigma_0 \epsilon^n$ (10^3 lbf/in)	Thickness, B (10^{-3} in)	u_{peak} Predicted (10^{-3} in)	u_{peak} Observed (10^{-3} in)	X_{max} (10^3 lbf)	w (10^{-3} in)	$\gamma_{\text{peak}} \doteq (u_{\text{peak}}/w)$
Lead $\sigma = 2.46\epsilon^{0.18}$	500	76	60	2.0	11	6.8
Tin $\sigma = 4.03\epsilon^{0.26}$	500	103	95	2.7	34	3.0
Al(1) $\sigma = 23.7\epsilon^{0.41}$	500	146	155	12.3	78	1.8
Al(2) $\sigma = 24.42\epsilon^{0.49}$	500	165	150	14.3	58	2.8
Cu(1) $\sigma = 64.3\epsilon^{0.28}$	500	110	105	33.8	78	1.4
Brass $\sigma = 112.2\epsilon^{0.58}$	500	179	184	47.0	113	1.6
Mild steel (1) $\sigma = 103.3\epsilon^{0.22}$	500	90	95	65.2	41	2.2
Mild steel (2) $\bar{\sigma} = 113.6\epsilon^{0.24}$	500	97	95	68.1	54	1.8

The sequence of pictures in Fig. 5.24 shows the results of back polishing through an incomplete guillotining cut in a 1.7 mm thick low carbon steel (~ 0.15%C, 0.2% Mn, 133 VPN); the sections are etched in Nital in order to see the grain flow. Fig. 5.24(a) depicts the last stages of indentation phase, when the blade has penetrated about 0.6 mm, and in Fig. 5.24(b) when the blade has reached about 0.75 mm, cracks are seen to emanate from both the tip of the blade and edge of the guillotine baseplate. It appears from changes in surface profile that the cracks are initiated at some 0.67 mm penetration. Fig. 5.24(c) shows how the cracks (which run ahead of the tools) join to give full separation which from Fig. 5.24(d) seems to occur at 1.2 mm blade penetration (i.e. about 67% of the sheet thickness). That the top crack is longer than the bottom may be explained by differences in tool sharpness (cf. Fig. 5.5). An enlargement of Fig. 5.24(b) is given in Fig. 5.24(f).

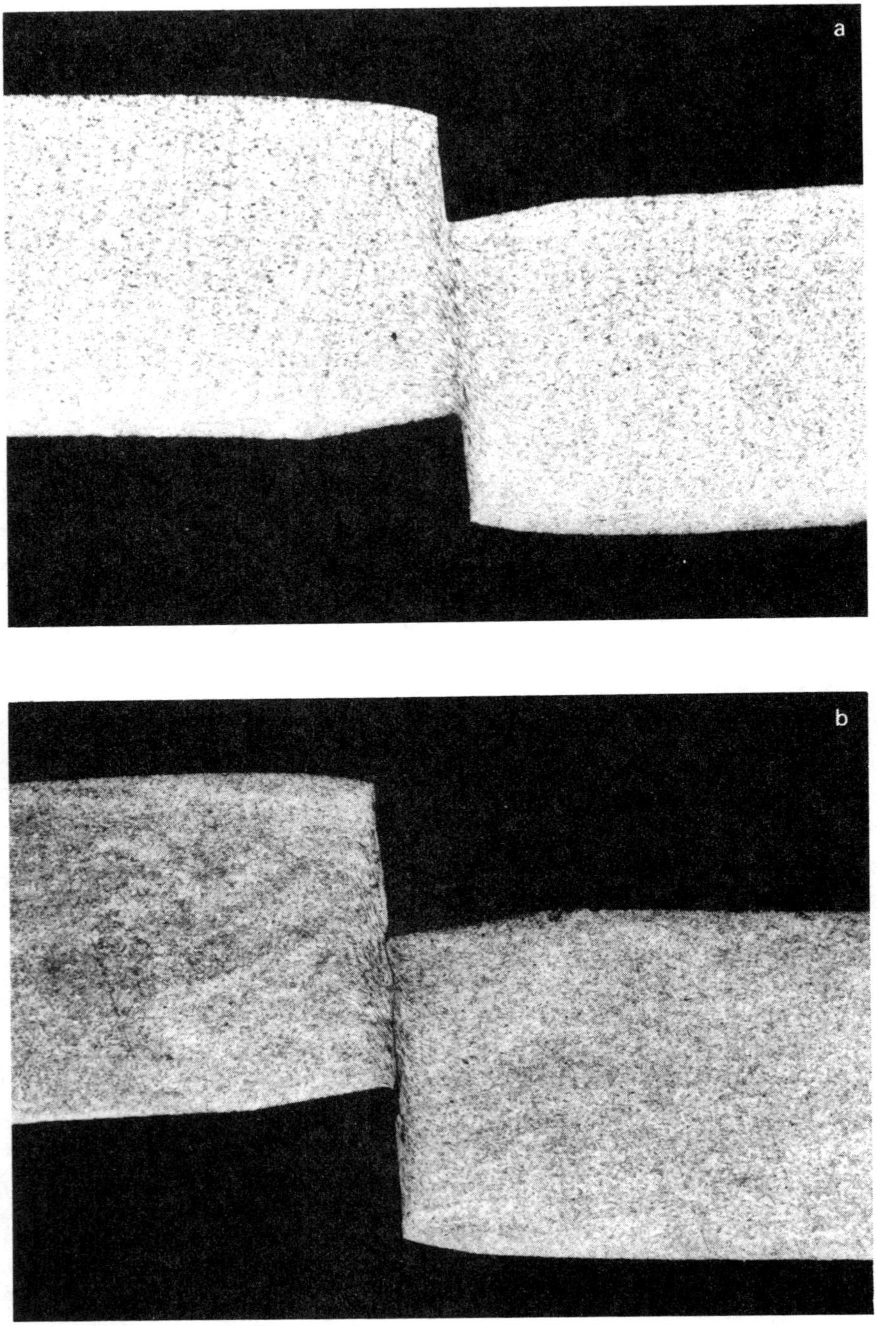

Fig. 5.24(a), (b) – Sequence of separation of guillotined surfaces shown by back-polishing and etching in solution (after Atkins (1979), with permission).

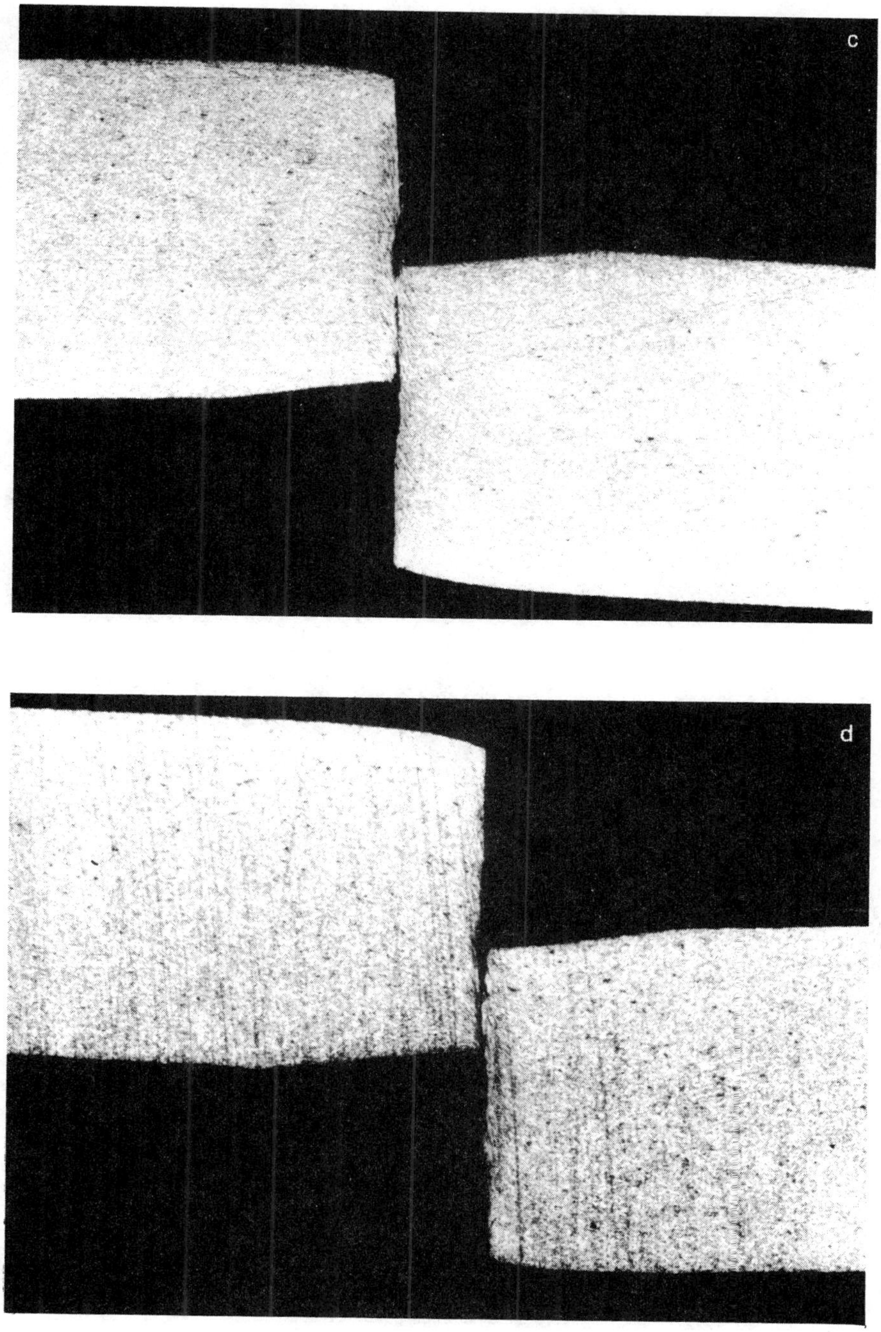

Fig. 5.24(c), (d) – Sequence of separation of guillotined surfaces shown by back-polishing and etching in solution (after Atkins (1979), with permission).

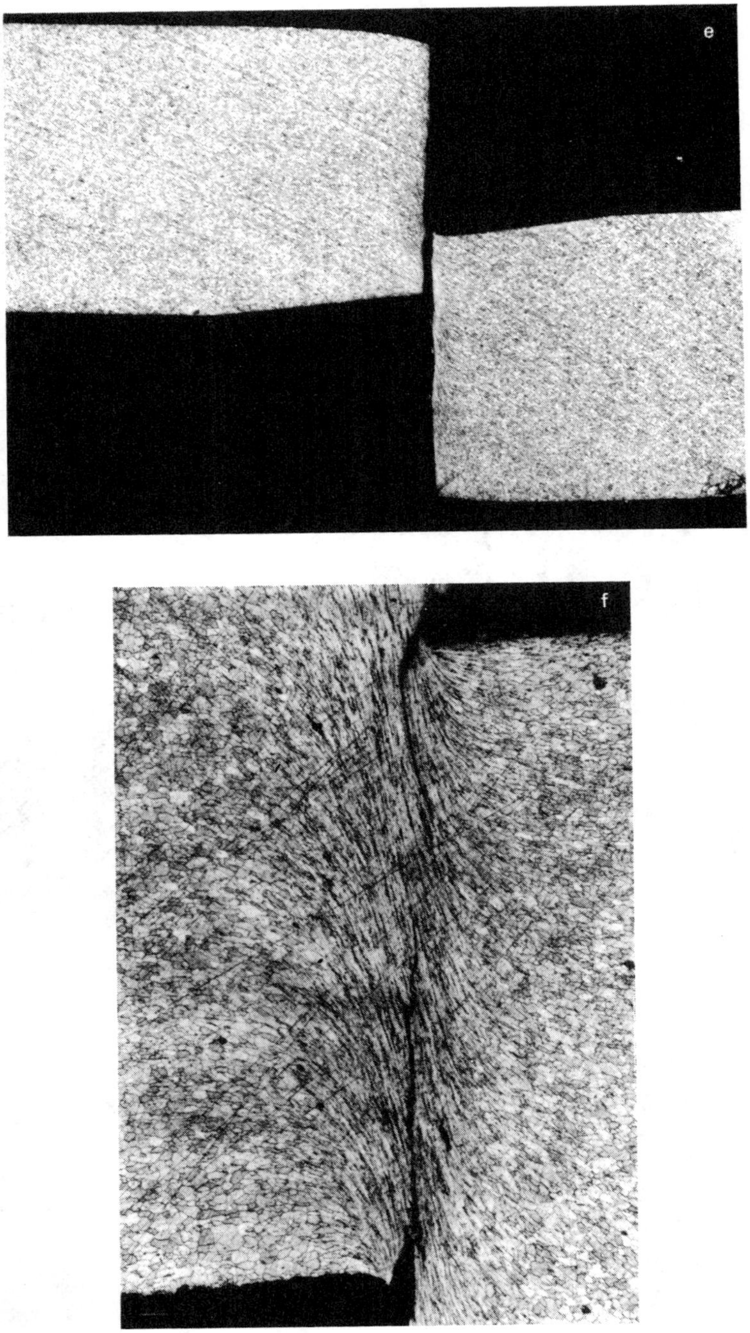

Fig. 5.24(e), (f) – Sequence of separation of guillotined surfaces shown by back-polishing and etching in solution (after Atkins (1979), with permission).

Fig. 5.25 – Side view of guillotined surface in 1.7 mm thick low carbon steel (after Atkins (1979), with permission).

On the corresponding side view of the guillotined surface (Fig. 5.25), the transition from indentation to crack is clear. In the indentation phase, tool marks are seen which are absent in the cracking region. Regions at X and Y in Fig. 5.14 are still 'attached' to one another in shear within the indentation region. Grid lines on cross-sections would be continuous if it were possible to trace them properly in the highly sheared region. Regions at X and Y cease to be attached when cracks appear, and as the blade penetrates further, obvious discontinuities will in the grid lines, particularly if a free standing crack runs ahead of the blade.

Fig. 5.25 also shows small tensile cracks, perpendicular to the motion of the blade, in the indentation zone just before full cracking. These are just discernible in Fig. 5.24 and are both reminiscent of the multiple cracks sometimes seen in cropped barstock, and related to surface finish defects in cutting operations. (Similar surface cracks may be produced by scraping a spoon over ice-cream and when rendering walls with cement). Their occurrence in the steel is influenced by the tensile pull across the shear plane, and no doubt by anisotropy in the rolled sheet. While questions of crack stability and crack path are difficult to discuss, these cracks probably initiate the main crack; the planes of the small cracks appear to be orientated along the flow lines of Fig. 5.24 and are not perpendicular to the plane of the cut.

The curved chamfer on the top surface of the sheet in Fig. 5.24 (the uppermost narrow region in the side views of Fig. 5.25) accompanies blade indentation. Unlike the sharp boundary shown in Fig. 5.14, the surface actually 'sinks in' as in hardness tests on annealed metals, and when there is clearance between blade and baseplate, the associated bending can lead to burrs in thin sheet.

The total length of crack C developed from the onset of crack propagation consists of two parts, viz. the visible crack of length a beyond the end of the punch or die and that part of the crack of length $(u - u_{cr})$ which lies alongside the flanks of the punch and die (Fig. 5.14). The second portion of the crack may or may not be visible depending on whether the material being cropped flows away from the punch and die. Discontinuities in grid lines become more pronounced as penetration proceeds and as the length $(u - u_{cr})$ of crack flanking the punch or die increases, during which the ends of the punch and die advance into regions formerly occupied by the 'free' crack of length a (Fig. 5.14(c)).

That the region flanking the punch is part of a crack seems sensible, and is readily understood, when another part of the same crack runs ahead of the tool. What, however, if the crack ahead of the tool is vanishingly small? If a punch penetration u_{cr} can be identified at which cracking starts (Fig. 5.25), the portion of the crack alongside the tool of length $(u - u_{cr})$ still exists *even though there is no visible crack ahead of the tool.* This seems to apply to lead, a guillotined edge of which is shown in Fig. 5.26. While similar to Fig. 5.25 for steel, *two* regions rather than one cracked region appear below the indentation (tool-marked) phase. The final region corresponds with a tensile tearing of the last uncut portion of the thickness (confirmed by cross-sections of partial cuts). The region immediately next to the indentation could very well be a crack which keeps pace with the movement of the cutting blade.

Fig. 5.26 – Side views of guillotined surface in 1.2 mm thick lead (after Atkins (1979), with permission).

If the observation, that 'visible' free-standing cracks are not always necessary is correct, it has wide implications in forming theory. It suggests for example that machining is continuous ductile crack formation, with the machined surface and the underside of the chip being the two surfaces of the crack (see Chapter 10). The existence of tool-tip cracks in machining is controversial and although they have been seen in some 'semi-brittle' solids, careful experiments have not revealed them in many ductile materials. Also, the tool rubs the cut surfaces, which complicates subsequent examination of machined surfaces. But possibly the idea of vanishingly small cracks which merely keep pace with the tool tip may apply for those solids in which cracks have not been observed. Parallels may be drawn between cropping or guillotining with cracks running ahead of the blade and machining with tool-tip cracks, and between cropping with cracks only alongside the flanks of the punch and machining without tool-tip cracks. The tool need not be sharp providing the neighbouring parts of the workpiece are divorced from one another for the mechanism to apply. (In simple hardness testing the relative size of the workpiece prevents piercing by the indenter, and neighbours remain neighbours within the displaced material in the plastic deformation zone.)

In a related manner, it seems that the difference between the processes of bulging or ironing and negative-rake machining or scalping Fig. 5.27 (a) and (b)

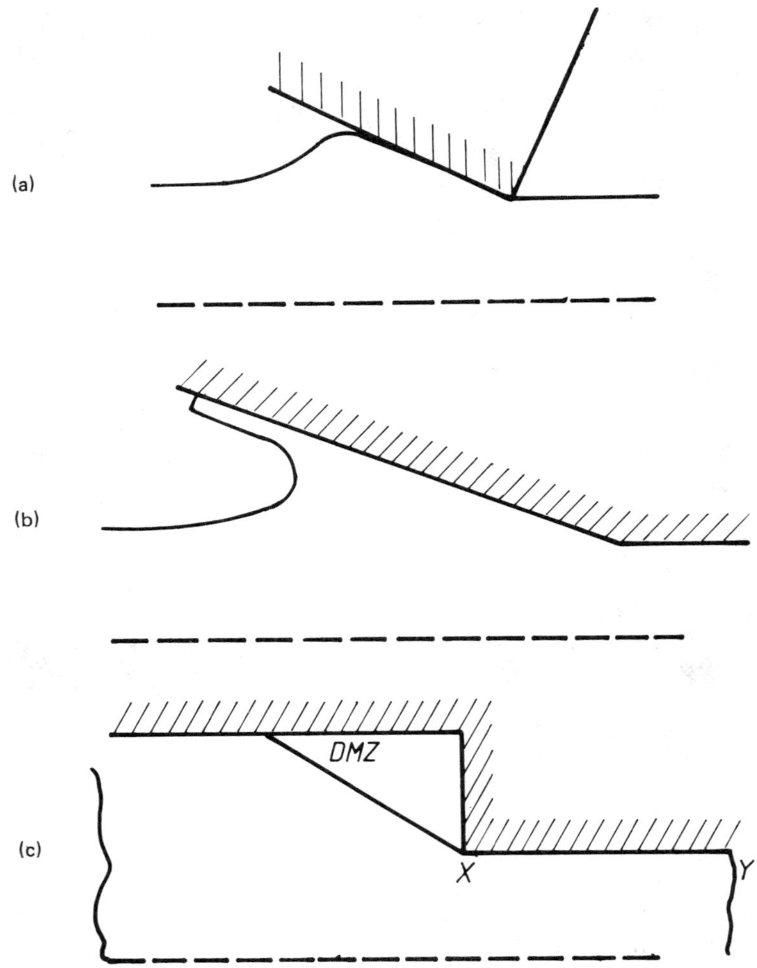

Fig. 5.27 – Difference between (a) ironing and (b) negative-rake machining processes. In (c) the boundary between the dead metal zone (DMZ) and the flow metal in extrusion may be regarded as 'cracks'.

is the introduction of a third (cracking) mechanism of deformation – rather like the difference between ocean waves and waves breaking on a shore. Again, it may not be fanciful to think that boundaries between true dead-metal zones and the rest of flow fields in extrusion are cracks in the sense described already (perhaps continuously rehealing cracks); certainly neighbours do not remain neighbours after deformation, and the characteristic discard or 'skull' left at the end of an extrusion was originally part of the front of the billet, some of which finds itself at the far end of the extrudate, i.e. regions X and Y in Fig. 5.27(c) were originally joined together, and according to simple plasticity are *still* together after extrusion.

The topic is discussed again in Chapter 10.

5.8 POSSIBLE LINK BETWEEN CRACK INITIATION AND PROPAGATION: FRACTURE IN SOLIDS WITH AND WITHOUT STARTER CRACKS

The local stress and strain conditions of material ahead of a propagating crack must be, in a general sense, similar to those which exist in loaded crack-free material prior to initiation. That is, crack propagation may be viewed as a process of continuous re-initiation along the path of cracking. This seems a particularly reasonable hypothesis for materials which crack by void growth and coalescence (cf. the remarks on natural sharpness of a crack in Fig. 2.33). It would seem that there should be a connexion between the specific essential works of steady-state crack initiation and propagation and the work required locally to nucleate a crack.

The tensile fracture of thin ductile sheets in which starter cracks have been cut is characterised by necking down in the line of the crack (section 4.4.1). Crack nucleation in flaw-free sheets also takes place within localised necks. The changes in geometry of sheets at first fracture in, for example, biaxially-bulged domes or even plain tensile sheet testpieces (where the necks are usually inclined to the axis of pulling) are similar to those at the tips of cracks which have propagated from long pre-existing cracks. Of course, the remote stress or strain ratios are different in different cases, and experiments have yet to be performed to see whether the geometry of necked-down regions developed from different starting conditions of loading are indeed the same (different neck geometries produce different σ_H elevations); again, the precise states of stress and changes in constraint within the necked regions near nucleated or propagated cracks are not known. Even so, the question arises as to whether a connexion can be made between the essential work of fracture obtained from propagation measurements (such as the Cotterell–Mai tests on sheet materials) and the local strains at which cracks nucleate in flaw-free sheets of the same material. Comparisons in terms of strains are appropriate as we remember that in uncracked sheets undergoing extensive plastic deformation it is easiest to measure strains local to the sites of crack nucleation rather than other parameters.

The specific essential work of fracture initiation J_i may be converted to a local work done/unit volume at the crack tip by dividing by the height of the necked-down process zone d (which, as explained in section 4.4.1 is often about $(2/3)B$ to B where B is the thickness of the sheet). That is, the incremental work $J_i \mathrm{d}A$ is dissipated within a volume $d.\mathrm{d}A$ at the crack tip, which therefore corresponds with a critical plastic work done/unit volume of some $J_i \mathrm{d}A/d.\mathrm{d}A = J_i/d$. If the material of the sheet workhardens according to $\bar{\sigma} = \sigma_0 \bar{\epsilon}^n$, the plastic work done/unit volume up to the crack nucleation effective strain $\bar{\epsilon}_f$ is $\sigma_0 \bar{\epsilon}_f^{(n+1)}/(n+1)$. If crack tip events are indeed the same then

$$\sigma_0 \bar{\epsilon}_f^{n+1}/(n+1) = J_i/d \tag{5.24}$$

and the effective strain at, say, the pole of a bulged sheet when cracks are just initiated is

$$(\bar{\epsilon}_f)_{\text{initiate}} \approx [J_i(n+1)/(\sigma_0 d)]^{1/(n+1)} \tag{5.25a}$$

(Note that this approach is different from, but complementary to, the method used in section 4.4.1 to derive equation (4.26); also that a critical work/volume corresponds with L. Gillemot's criterion in Table 5.1, ramifications of which are discussed by Radon and Czoboly (1982) and Rice (1968).)

In order to test equation (5.25) and compare the predicted $\bar{\epsilon}_f$ with the fracture strain pairs actually measured in biaxially stressed crack-free sheets, J_i should be that measured under the same remote loading conditions along and perpendicular to the plane strain neck in which cracks form. That is, for nucleation under biaxial stress with ratio M the necessary J_i should be determined *not* from simple uniaxial Cotterell–Mai tests, but rather from special tests in which remote stresses in the ratio M are applied along and perpendicular to the line between the starter cracks (i.e. the direction of neck formation) from the outset of the test. Such data are not yet available. The basic Cotterell–Mai test with *remote* uniaxial loads corresponds with a *local* plastic stress ratio of 2:1 in the region of the neck if the crack tips do not approach one another before necking (i.e. if a state of plane plastic strain exists in the ligament). Consequently a basis for comparison between predicted $\bar{\epsilon}_f$ and experimental data relates to bulging over an elliptical orifice which reproduces a 2:1 stress ratio before necking sets in.

Data for $\bar{\epsilon}_f$ in sheets under biaxial loading in the tension–tension strain quadrant come, in the main, from hydraulic bulge tests using elliptical orifices (cf. Fig. 5.4). A matter of debate now arises as to what J-value should be used in equation (5.25a) because, in practice, the strains measured local to the fracture sites in such bulge tests are not really the 'initiation' strains given by equation (5.25a). Instead, since the crack has actually propagated a distance which gives a crack opening sufficient to dump the hydraulic pressure (the test arrangement is clearly soft and impossssible to control), the specific work absorbed local to the fracture site is more likely to be connected with J_p rather than J_i (cf. footnote in section 5.2). If so, we have

$$\bar{\epsilon}_f \approx (J_p\,(n+1)/\sigma_0 d)^{1/(n+1)} \tag{5.25b}$$

We recall that $J_i < J_p$ usually (section 4.4.1). Only in mechanically loaded biaxial tension tests of flat sheets under displacement control is equation (5.25a) really likely to apply; it may also apply in 'Hecker' cup tests in the tension–compression strain quadrant (where strips of different width are bulged over a lubricated hemispherical punch to give different $\epsilon_1 \epsilon_2$ pairs) if the punch ram is mechanically controlled. In what follows we shall use J_p and equation (5.25b) as the basis of reasonable comparison between theory and experimental ϵ_f measured in *hydraulic bulge* tests.

Caddell and co-workers in 1968 studied the fracture of hydraulically-bulged diaphragms of annealed and cold-rolled aluminium and brass sheets under variety of biaxial in-plane strain ratios. Their data for 0.79 mm thick annealed α-brass displaying $\bar{\sigma} = 774\,\bar{\epsilon}^{0.6}$ MPa, give a necking limit and fracture limit diagram like Fig. 5.4, with the following significant strain pairs for bulging over an elliptical orifice, viz. (i) at the formation of the plane strain neck $\epsilon_{1n} = 0.30$, $\epsilon_{2n} = 0.12$

and (ii) at fracture $\epsilon_{1f} = 0.66$, $\epsilon_{2f} = 0.12$. (There was, for Caddell *et al.*'s materials, considerable neck growth and thinning within the neck before final fracture even for balanced biaxial straining over a circular orifice, unlike Fig. 5.4 where fracture occurs very soon after neck formation.) These ϵ_{1f}, ϵ_{2f} combinations may be converted into effective strains $\bar{\epsilon}_f$ using Appendix 3 (taking account of the change in strain ratio when the plane strain neck sets in, cf. equation (5.8)) to give $\bar{\epsilon}_f = 0.85$. Such $\bar{\epsilon}_f$ may then be compared with the predictions from equation (5.25b).

Cotterell–Mai tests had not been performed, of course, on the materials used by Caddell *et al.*, nor was the size d of the necked-down regions reported. Later work (Mai and Cotterell, 1980; see Chapter 4) found $J_p \approx 260$ kJ/m^2 for 1.02 mm thick Austral 207 α-brass which followed $\bar{\sigma} = 625\ \bar{\epsilon}^{0.32}$ MPa. Ignoring the different $\bar{\sigma}$–$\bar{\epsilon}$ workhardening behaviour but using the same J_p, and guessing that $d = 2B/3 = 2 \times 0.79/3 = 0.53$ mm for Caddell *et al.*'s brass, equation (5.25b) predicts $\bar{\epsilon}_f = 1.0$. It is likely that J_p for Caddell's thinner sheet is less than that for the thicker brass sheet and, if so, the predicted $\bar{\epsilon}_f$ would come more into line. Even so, the agreement is not unreasonable given that the calculations are sensitive to the valude of d which is not known, and that corrections have not been made for the elevation of σ_H in the neck.

Prestrain affects $\bar{\epsilon}_f$ since equation (5.25b) must be modified to

$$(\bar{\epsilon}_f + \bar{\epsilon}_{cw})^{n+1} - \bar{\epsilon}_{cw}^{n+1} = J_p\,(n+1)/\sigma_0 d \qquad (5.25c)$$

where $\bar{\epsilon}_{cw}$ is the equivalent prestrain from cold working the material. (We assume that the $\bar{\sigma}$–$\bar{\epsilon}$ relation for prestrained material is given by $\bar{\sigma} = \sigma_0\,(\bar{\epsilon}_{cw} + \bar{\epsilon})^n$ with σ_0 and n retaining the values obtained on the most annealed samples, rather than the less satisfactory type of relation sometimes used, i.e. $\bar{\sigma} = \sigma_0\,\bar{\epsilon}^n$, with σ_0 and n changing with degree of cold work.) For 0.79 mm thick α-brass sheet having $\bar{\epsilon}_{cw} = 0.14$, Caddell *et al.* found $\epsilon_{1n} = 0.20$, $\epsilon_{2n} = 0.08$ and $\epsilon_{1f} = 0.65$, $\epsilon_{2f} = 0.08$. This corresponds with $\bar{\epsilon}_f = 0.81$. Proceeding as before, using from Mai and Cotterell (1980) $J_p \approx 220$ kJ/m^2 for brass sheet having a similar level of prestrain, equation (5.25c) predicts $\bar{\epsilon}_f \approx 0.77$. Again the agreement is not unreasonable given the assumptions. It is clear that while the idea has some foundation, work needs to be done in which bulging tests *and* J_i or J_p measurements are determined on the *same* material.

Example 5.5 Derive from equation (5.25) an algebraic expression for $(\epsilon_{1f}\,\epsilon_{2f})$ cracking loci, and plot the result.

Ans. From Appendix 3, $\bar{\epsilon}_f^2 = (4/3)\,(\epsilon_{1f}^2 + \epsilon_{1f}\epsilon_{2f} + \epsilon_{2f}^2)$ so, in equation (5.25), if the strain ratios do not change $\epsilon_{1f}^2 + \epsilon_{1f}\,\epsilon_{2f} + \epsilon_{2f}^2 = (3/4)\,[J_p(n+1)/\sigma_0 d]^{2/(n+1)} = (\epsilon_{1f}^0)^2$ where ϵ_{1f}^0 is the workability index intercept. If J and d are strain- (or stress-) ratio independent, the rhs of the equation is constant and the relation coincides with the maximum $\bar{\epsilon}$ criterion plotted Fig. 5.7. In sheet studies, the strain ratios will *change* if necks form (Fig. 5.4) and $\bar{\epsilon}_f$ (and the $\epsilon_{1f}\,\epsilon_{2f}$ components) have to be broken down into events before necking (with the originally-applied strain ratio) and after necking (when $L = d\epsilon_1/d\epsilon_2 = \infty$) as, for related reasons, in equation (5.8). To put this down algebraically requires an expression for the $(\epsilon_1\,\epsilon_2)$ *necking* locus. This is tricky as it concerns the contentious area (at least in the tension–tension strain quadrant) of neck formation. While Hill's plane stress characteristics theory predicts necking in the tension–compression strain quadrant, it fails to do so in the tension–tension quadrant (see, for example, Hosford and Caddell, 1982). However, for annealed metal sheets, we may approximate the necking locus by $\epsilon_{1\text{neck}} = n + \epsilon_{2\text{neck}}/2$ where n is the workhardening index in $\bar{\sigma} = \sigma_0\,\bar{\epsilon}^n$. The lhs of equation (5.25) then becomes

$$\left(\frac{2}{\sqrt{3}}\right)\epsilon_{1\text{neck}}\left[1 + \frac{1}{L} + \frac{1}{L^2}\right]^{1/2} + \left(\frac{2}{\sqrt{3}}\right)\left[\epsilon_{1f} - \epsilon_{1\text{neck}}\right]$$

$$= \left(\frac{2}{\sqrt{3}}\right)\left(n + \frac{\epsilon_{2f}}{2}\right)\left[\left(1 + \frac{1}{L} + \frac{1}{L^2}\right)^{1/2} - 1\right] + \left(\frac{2}{\sqrt{3}}\right)\epsilon_{1f}$$

where L is the strain ratio prior to neck formation, and noting that $\epsilon_{2f} = \epsilon_{2\text{neck}}$.

Fig. 5.28 shows the predictions of this relation in equation (5.25) for a material with $n = 0.2$ and an (arbitrarily chosen) value of the plane stress-plane strain workability index $\epsilon_{1f}^0 = 1.4$. The treatment assumes that (J_p/d) is path independent (but see later).

Problem 5.22 asks for the equivalent equations when the sheet material is pre-strained by $\bar{\epsilon}_{cw}$.

5.8.1 *J*-void growth model connexion

Example 5.5 and equations (5.25a) and (5.25b) show that the $(\epsilon_{1f}\,\epsilon_{2f})$ cracking loci of *flawless* sheets can be expressed in terms of J_i or J_p measured in experiments on sheets containing starter cracks. But we have already shown in section 5.4 that $(\epsilon_{1f}\,\epsilon_{2f})$ cracking loci may be written in terms of a McClintock-type analysis (which relates to void growth). Consequently there should be a J_i- or J_p-to-void-growth-model connexion too.

Consider the simplest void-growth expression, i.e. $\int(\sigma_H/\bar{\sigma})d\bar{\epsilon}$. Since this is one of the cases where damage accumulation is path-independent (Example 5.3), we have

$$\left.\begin{aligned}\ln(l/2r) &= 2(L+1)\epsilon_{1f}/3L \\ &= (L+1)\bar{\epsilon}_f/\sqrt{3}\sqrt{L^2+L+1} \\ &= (\sigma_H/\bar{\sigma})\bar{\epsilon}_f\end{aligned}\right\} \qquad (5.26)$$

where L defines a *direct* strain path to the failure strain $(\epsilon_{1f}\,\epsilon_{2f})$ combination

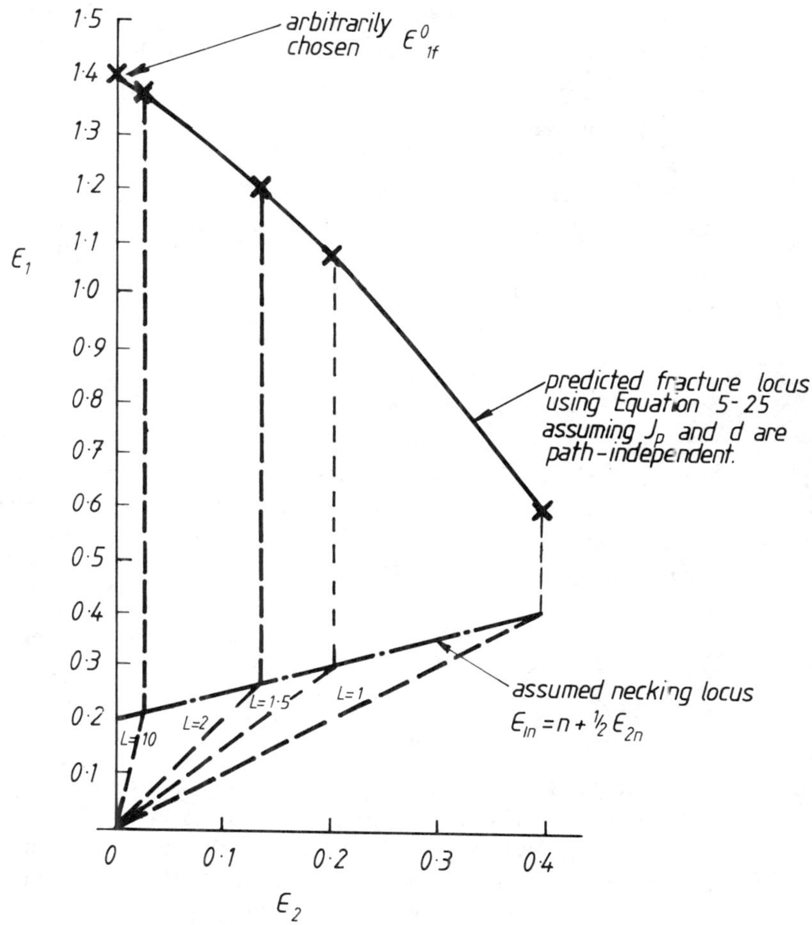

Fig. 5.28 – Fracture and necking loci in $\epsilon_1 - \epsilon_2$ space predicted by equation (5.25).

in plane stress; the actual path could be different, as in sheet forming with prior necking. But $\bar{\epsilon}_f$ is given by equation 5.25 for an annealed metal. Hence

$$\ln(l/2r) = [(L+1)/\sqrt{3}\sqrt{L^2+L+1}]\ [J_i(n+1)/\sigma_0 d]^{1/(n+1)} \tag{5.27}$$
$$= (\sigma_H/\bar{\sigma})\ [J_i(n+1)/\sigma_0 d)]^{1/(n+1)}$$

which, for a material with constant effective yield stress $\bar{\sigma}_y$ reduces to

$$\ln(l/2r) = [(L+1)J_i/\sqrt{3}\sqrt{L^2+L+1}]/(\bar{\sigma}_y d)$$
$$= (\sigma_H/\bar{\sigma}_y)\ (J_i/\bar{\sigma}_y d) \tag{5.28}$$

Example 5.6 Determine, according to equation (5.27), the average particle spacing-to-size ratio fo 0.92 mm thick annealed aluminium sheet displaying $\bar{\sigma} = 130\bar{\epsilon}^{0.23}$ MPa and $J_p = 73$ kJ/m^2.

Ans. Let us assume that J_p from simple notched sheets applies to plane strain conditions (as already explained, the sheets are already in a state of plane *stress* and 'plane strain' applies to the length of ligament between the starter cracks in Cotterell–Mai tests, which corresponds to necks in unflawed sheets occurring perpendicularly to the major in-plane strain). Along a wholly plane strain path, $L = \infty$ so in equation (5.27) using J_p in place of J_i

$$\begin{aligned}\ln(l/2r) &= (1/\sqrt{3})\,[J_p(n+1)/\sigma_0 d]^{1/(n+1)} \\ &= (1/\sqrt{3})\,(73 \times 10^3 \times 1.23/130 \times 10^6 d)^{1/1.23}\end{aligned}$$

The size d of the necked down zone is some (2/3) B to B. Then ln $(l/2r) \approx 0.63$ and 0.46 using $d = 0.62$ and 0.92 mm respectively, corresponding with $l/2r = 1.9$ and 1.6, and with $\nu_f = 14\%$ and 24% since $\nu_f \approx (2r/l)^3$. As discussed in section 5.5, the predicted $(l/2r)$ is probably smaller than that in the real microstructure, and likewise the ν_f is too large. These calculations derive from $\int(\sigma_H/\bar{\sigma})d\bar{\epsilon}$, but correspond with many of the other versions of void-growth criteria since, as already shown, the difference concerns different multiplying factors on the integration 'constant'.

In Example 5.6 we chose $L = \infty$ as that was the basis of previous comparisons between initiation and propagation data. Clearly, however, if ln$(l/2r)$ is considered to be a 'fixed' material property, the rhs of equation (5.27) must also be constant. (While $l/2r$ will have different values in different directions in a void-containing solid – which may relate to fracture by normal stresses, shear stresses and so on in different directions – equation (5.26) with constant multiplying factors will still apply overall.) (J_i/d) therefore *must depend on strain history* via L, as hinted at earlier in this section, if ln$(l/2r)$ is constant, and if the L function changes with L. Calculations show that, at least in the tension–tension strain quadrant where L is positive, the various L-functions (i.e. $(L+1)/\sqrt{L^2+L+1}$ in equation (5.27), or other similar functions deduced from equation (5.5), for example, which are all proportional to $(\sigma_H/\bar{\sigma})$) do not change markedly. Experimental investigation is required of the influence of remote plastic strain paths on the essential work of fracture. Remembering that $\sigma_H/\bar{\sigma}$ at fracture depends on neck profile (not taken into account in these equations), we recognise from the second of equations (5.27) and (5.28) that final neck profiles may very well be different for different strain paths.

Note that for given microstructural features in a given material at fixed strain ratio L, J_i should vary directly with necked zone size d (and hence sheet thickness B) according to equations (5.27) and (5.28), as long as fracture takes place by necking and the same slip-off mechanism within the neck. Aluminium sheet 0.68 mm thick gave J_p (in this case) = 50 kJ/m^2 in comparison with

$J_p = 73$ kJ/m^2 for sheet 0.92 mm thick; we observe that $(73/0.92) \approx 79$ and $50/0.68 \approx 74$.

Note also that equations (5.27) and (5.28) may be recast in the form

$$J_i\,(=R) = \frac{\sigma_0 d}{(n+1)}\left[\frac{\sqrt{3}\sqrt{L^2+L+1}}{(L+1)}\ln(l/2r)\right]^{(n+1)}$$

$$= \frac{\sigma_0 d}{(n+1)}\,[\bar{\sigma}_y/\sigma_H \ln(l/2r)]^{(n+1)} \tag{5.27a}$$

and

$$J_i\,(=R) = \left[\frac{d\sqrt{3}\sqrt{L^2+L+1}}{L+1}\ln(l/2r)\right]\bar{\sigma}_y$$

$$= (\bar{\sigma}_y/\sigma_H)\,(d\,.\,\ln(l/2r))\,\bar{\sigma}_y \tag{5.28a}$$

the second of which has the same form as the relation of COD given in sections 2.9 and 4.5.2, viz.

$$R = m\,\sigma_y\,\delta_C$$

There is correspondence clearly between

$$m \text{ and } (\sigma_y/\sigma_H) \tag{5.29}$$

and between

$$\delta_C \text{ and } (d\,.\,\ln(l/2r)) \tag{5.30}$$

in the *plane stress* examples for which all this algebra applies. The factor m represents the constraint to plastic flow which, in these plane stress cases, is sensibly represented by σ_y/σ_H which takes values of 3 for simple tension, $\sqrt{3}$ for plane strain and 2/3 for balanced biaxial tension *before* necking ensues (cf. equation (4.43b)). The connexion between δ_C and microstructure via $(d\,.\,\ln(l/2r))$ makes sense in that δ_C/d is some measure of 'crack strain' using d as the crack tip region 'gauge length' (see, for example, Backofen, 1972, Chapter 12; Havas *et al.*, 1974).

In Cotterell's and Mai's (1981) studies of the essential work of fracture, described in section 4.4.1, they used a criterion of $\delta_C \approx 0.73$ mm for crack propagation in annealed low carbon steel of thickness 1.6 mm. Other sheet materials such as brass and aluminium, in comparable thicknesses, displayed $0.5 < \delta_C < 1.0$ mm. From equation (5.30), using $d \approx B = 1.6$ mm we have $\ln(l/2r) \approx 0.73/1.6 = 0.46$ i.e. $(l/2r) \rightarrow 1.6$ for the microstructure of the low-carbon steel used by Cotterell and Mai. There are no independent data on $(l/2r)$ and while, once more, the absolute value of the prediction must be low, it does correspond well with estimates of $(l/2r)$ from McClintock analyses which have looked directly at microstructure.

From these different lines of attack, it seems safe to say that hole-growth models involving $\int(\sigma_H/\bar{\sigma})d\bar{\epsilon}$ predict the correct functional dependence of variables in crack initiation. Alternatively, the fracture criterion may be viewed as the attainment of a critical plastic work/volume given by $\int \bar{\sigma} d\bar{\epsilon}$ modified by the stress state – in particular by the hydrostatic stress. However, the value of the integrated constant (and hence the mechanics-to-microstructure link) requires further attention.

We note that there have been similar sorts of investigation into the mechanics-to-microstructure link for pre-cracked bodies in terms of predicting K_{IC} behaviour for steels from models of slip-initiated cleavage fracture (in the 'lower-shelf' region) and from models of microvoid coalescence (in the 'upper-shelf' region) (see, for example, Ritchie *et al.* (1979) which also includes relevant references). A critical fracture stress criterion is used for cleavage fracture and instead of the critical plastic work/volume line of attack used in this section, a critical effective fracture strain criterion is used (to which, of course, the McClintock void growth analysis reduces for proportional straining). A relation to predict K_{IC} in the upper shelf region is obtained by substituting for δ_C in $K_{IC}^2 = ER = m\ \sigma_y\ \delta_C$ from the Rice and Johnson (1970) solution relating the distribution of $\bar{\epsilon}$ and $\sigma_H/\bar{\sigma}$ with distance ahead of an elastoplastic blunting crack tip, to give

$$K_{IC}^2 \propto (\bar{\epsilon}_f\ .\ l_0^*\ .\ \sigma_y\ .\ E) \tag{5.31}$$

where l_0^* is some microstructurally significant size scale (similar to the 'representative volume' used in this chapter which permits continuum mathematics to be applied to microstructures which vary in fine scale). Experimental $\bar{\epsilon}_f$ and $\sigma_H/\bar{\sigma}$ are obtained from circumferentially notched round tensile testpieces with different notch acuities using Bridgman corrections, and the particular $\bar{\epsilon}_f$ used in equation (5.31) is that which corresponds with the $\sigma_H/\bar{\sigma}$ value at the distance l_0^* ahead of the crack. Clearly this approach is similar to the line of attack of this section. We note, however, the caveats in section 5.5 about the difficulties of interpretation of the characteristic microstructure size, and those in section 5.4 on the effect of strain ratio on the stress state at final fracture.

PROBLEMS

5.1 The Cockcroft–Latham criterion of ductile fracture is given by $\int_0^{\epsilon_f} \bar{\sigma}\left(\frac{\sigma^*}{\bar{\sigma}}\right) d\bar{\epsilon} = C$ where $\bar{\sigma}$, $\bar{\epsilon}$ are the effective stress and strain, $\bar{\epsilon}_f$ is the fracture strain and σ^* is the maximum tensile stress at all times in the history of a deformation process. In upsetting tests, the empirical relation between the strains local to fracture sites produced by different loading paths on the barrelled periphery of deformed specimens is $\epsilon_{\theta f} = a - \epsilon_{zf}/2$ where ϵ_θ and ϵ_z are the hoop and axial strains respectively, and a is some constant. Show that the Cockcroft–Latham criterion predicts the relation exactly for perfectly-plastic solids, but for workhardening solids only along those straining paths which satisfy $[(2 + \alpha)/2(1 + \alpha + \alpha^2)^{\frac{1}{2}}]^n = 1$, where n is the workhardening index and $\alpha = \epsilon_z/\epsilon_\theta$.

5.2 Show how externally applied hydrostatic stress may be incorporated in, say, the McClintock or Cockcroft–Latham criteria for ductile fracture.

Show that compressive hydrostatic stress shifts failure loci (as in Fig. 5.7 drawn for ambient conditions) upwards and hence improves workability, and that tensile σ_H worsens workability.

5.3 Show that the Cockcroft–Latham criterion of ductile fracture (Table 5.1) may be rewritten

$$\int \frac{\bar{\sigma}\,d\bar{\epsilon}}{(1 - 1/M + 1/M^2)^{1/2}} = \int \frac{3\sigma_H\,d\bar{\epsilon}}{(1 + 1/M)}$$

where $M = \sigma_1/\sigma_2$.

5.4 Discuss the role played by hydrostatic stress in changing a material's capacity for flow without fracture. (cf. Bridgman's studies, hydrostatic extrustion etc.)

5.5 Determine $\int \sigma_H d\bar{\epsilon}$ from the data in Fig. 5.10 and show that the values decrease from about 300 to 200 MJ/m^3 as uniaxial straining paths change to balanced biaxial straining paths.

5.6 Determine the greatest compressive hydrostatic stress at which shear cracks are still possible according to the McClintock model.

5.7 Explain why a logarithmic form is to be expected for the inclusion spacing-to-size ratio in void coalescence and growth models of ductile fracture. (Remember natural strains.)

5.8 Derive the connextion between the spacing-to-size ratio $l/2r$ of inclusions or second phase particles, and their volume fraction, i.e. $1/v_f = (l/2r)^3$.

5.9 The Ghosh criterion for ductile fracture starts with the Rice–Tracey statement that $dr \propto r_0(\sigma_H/\sigma)d\epsilon$ where r is the average radius of a void and R_0 is the starting radius. Owing to rigid inclusions which prevent thinning of the voids, the holes enlarge only in one direction. Fracture is said to occur when the product of (a) the remote maximum shear stress τ, (b) some function of the volume fraction of inclusions v_f and (c) the areal rate of increase of voids (proportional to $2\pi r_0 dr$) reaches a critical value. Show that at failure $(1 + 1/M)\sigma_1^2$ is a constant according to this criterion where $M = \sigma_1/\sigma_2$.

5.10 Show that the McClintock criterion for void growth in shear in the 13 directions given by $(\sigma_H/2\tau_{13})\gamma_{13} = (M + 1)L\epsilon_1/3(2L - 1)M$ is *not* path-independent.

5.11 Show that all void growth criteria of the form $\int\{A + B/L\}d\epsilon_1$ are path-independent. (A, B are constants, $L = d\epsilon_1/d\epsilon_2$ strain ratio.)

5.12 Show that equation (5.8) transforms to $\ln l/2r = \int(2 + 1/L)d\epsilon_1$ and that the integral is path independent.

5.13 For shear in the 12-plane, show that the damage integral is path-independent and predicts a failure locus of $\tan^{-1}(-1)$.

5.14 Write out an expression for the hydrostatic stress along the different straining paths for upsetting shown in Fig. 5.2, and demonstrate that in this case greatest effective strain to fracture occurs for least hydrostatic stress.

5.15 Explain why correlations of shearing loads in cropping and guillotining with yield strength, ultimate tensile strength and similar traditional material properties are not likely to be widely successful.

5.16 Show that in the presence of tool friction represented by some fraction f of the material shear yield stress τ, the punch penetration at peak load in cropping is given by $u_{peak} = nB/(1+n)(1-f)$ (Fig. 5.14), with $\bar{\sigma} = \sigma_0 \bar{\epsilon}^n$.

5.17 Show that when cracks appear before the peak in load in frictionless cropping, $u_{peak} = nu_f/(1+n)$ where u_f is the punch travel when the cut is complete.

5.18 Demonstrate that the magnitude of u_{peak} is unaffected by fixed axial loads, but is increased by fixed axial *stress* σ according to $nB/(1+n-2(\sigma/\bar{\sigma})^2)$ where $\bar{\sigma}$ is the effective (yield) stress and B is plate thickness.

5.19 What are crazes? What role do they play in the fracture of (i) brittle polymers like PMMA and PS and (ii) rubber-modified plastics like HIPS and ABS.

5.20 Compare the envelopes for initiation of craze yielding according to Sternstein and Ongchin's stress bias criterion and for shear yielding according to Bowden's pressure-dependent Tresca yield criterion (Bowden and Jukes, *J. Mater. Sci.* **3**, 183, 1968). Hence comment on the effects of stress state on the deformation behaviour of a typical polymer, say PMMA.

5.21 What are the envelopes for initiation of craze yielding if the pressure-dependent, anisotropic, yielding criterion of Raghava were to apply (e.g. Caddell, Raghava and Atkins, *J. Mater. Sci.* **8**, 1641, 1973).

5.22 Determine the algebraic expression for $\epsilon_{1f}\epsilon_{2f}$ derived from equation (5.25) when the sheet material is pre-strained by $\bar{\epsilon}_{cw}$ (cf. Example 5.5).

5.23 The homogeneous, annealed microstructures which are favoured for good forgeability are just those which can lead to high cutting forces and poor chip flow characteristics in machining. Conversely, the inclusion types added to steels to improve machinability tend to reduce forgeability. Discuss this conflict of requirements (reference to Chapter 10 will help in answering this question).

REFERENCES

Andrews, E. H. and Bevan, L. (1972) *Polymer* **13**, 337.

Argon, A. S. and Salama, M. M. (1977) *Phil Mag.* **35**, 1217.

Atkins, A. G. (1979) *Phil. Mag.* **43A**, 627.

Atkins, A. G. (1980) *Int. J. Mech. Sci.* **22**, 215.

Atkins, A. G. (1981) *Metal Sci.* (Feb.), 81.

Bai, Y. L. and Johnson, W. (1982) *Metals Tech.* **9**(5), 182.

Backofen, W. A. (1972) *Deformation Processing.* Reading, Mass.: Addison-Wesley.

Beachem, C. D. (1975) *Met. Trans.* **6A**, 377.

Beachem, C. D. and Yoder, G. R. (1973) *Met. Trans.* **4A**, 1145.

Bowden, P. B, and Oxborough, R. J. (1973) *Phil. Mag.* **28**, 547.

Bridgman, P. W. (1952) *Studies in Large Plastic Flow and Fracture.* New York: McGraw-Hill.

Broek, D. (1973) *Engr. Frac. Mech.* **5**, 55.

Brozzo, P., Deluca, B. and Rendina, R. (1972) *Proc. 7th Biennial Congress IDDRG, Amsterdam.*

Caddell, R. M., Duncan, J. L. and Johnson, W. (1968) *Proc. 3rd Conf. on Dimensioning, Budapest.* Hungarian Academy of Sciences, p. 385.

Caddell, R. M., Ludema, K. C. and Atkins, A. G. (1985) *Manufacturing and Process Engineering.* Englewood Cliffs, N.J.: Prentice-Hall.

Chang, T. M. and Swift, H. W. (1950) *J. Inst. Metals* **78**, 119.

Clift, S. (1984) M.Sc. dissertation, University of Birmingham.

Cockcroft, M. G. and Latham, D. J. (1968) *J. Inst. Metals* **96**, 33.

Cox, T. B. and Low, J. R. (1972) NASA Tech. Rept. Grant NGR-39-087-003.

Dillamore, I. L. and Maynard, R. A. (1974) BSC Report CDL/MT/9/74.

Dodd, B. and Atkins, A. G. (1983) *Acta Met.* **31**, 9.

Dodd, B. and Boddington, P. (1980) *J. Mech. Working Tech.* **3**, 239.

Drucker, D. C. (1966) *J. Materials* **1**, 873.

Embury, J. D. and LeRoy, G. H. (1977) *Proc. ICF-4,* Vol. 1, p. 15.

Erturk, T. (1979) *Proc. ICM-3,* Vol. 2, p. 653.

Erturk, T. and Kazazoglu, E. (1982) STP No. 753, 19. Philadelphia, ASTM.

French, I. E. and Weinrich, P. F. (1979) *Met. Trans.* **10A**, 297.

Gent, A. N. (1970) *J. Mater. Sci.* **5**, 925.

Gent, A. N. and Lindley, P. B. (1958) *Proc. Roy. Soc., London* **A249**, 195.

Ghosh, A. K. (1976) *Met. Trans.* **7A**, 523.

Gillemot, L. F. (1976) *Engr. Fract. Mech.* **8**, 239.

Goods, S. H. and Brown, L. M. (1979) *Acta Met.* **27**, 1.

Guillen-Preckler, A. (1967) *Annal. CIRP* **15**, 183.

Hancock, J. W. and Brown, D. K. (1983) *J. Mech. Phys. Solids* **31**, 1.

Havas, I., Schulze, H. D., Hagedorn, K. E. and Kochendorfer, A. (1974) *Material pruf* **16**, 349.

Hoffmanner, A. L. (1971) in *Metal Forming: Interrelationship between Theory and Practice,* A. L. Hoffmanner (ed.), Plenum Press, New York, p. 349.

Hosford, W. F. and Caddell, R. M. (1982) *Metal Forming – Metallurgy and Mechanics.* Englewood Cliffs, N.J.: Prentice-Hall.

Jain, S. C. and Kobayashi, S. (1970) *Proc. 11th MTDR.* London: Macmillan, p. 1137.

Johnson, W., Sowerby, R. and Venter, R. D. (1982) *Plane-Strain Slip-Line Fields for Metal Deformation Processes.* Oxford: Pergamon.

Kals, J. G. (1983) Private communication.

Kambour, R. P. and Kopp, R. W. (1969) *J. Poly. Sci.,* **A2** 7, 183.

Kambour, R. P. and Russell, R. R. (1971) *Polymer* **12**, 237.

Keife, H. and Stahlberg, U. (1980) *J. Mech. Working Tech.* **4**, 133.

Kivivuori, S. and Sulonen, M. (1978) *Annals. CIRP* **72**, 141.

Kobayashi, S. and Oh, S. I. (1974) AFML-TR-74-159, USAF Wright-Patterson, Ohio.

Kudo, H. (1970) Discussion to S. Kobayashi, *J. Engr. Industry (Trans. ASME)* **94B**, 391.

Kudo, H. and Aoi, K. (1967) *J. Japan Soc. Techn. Plasticity* **8**, 17.

Kuhn, H. A. (1976) Final Rept. AISI Proj. 56–350, University of Pittsburgh.

Kuhn, H. A. (1978) STP No. 647, 206, Philadelphia: ASTM.

Love, A. E. H. (1952) *The Mathematical Theory of Elasticity*, 4th ed. Cambridge: Cambridge University Press.

Marciniak, Z. and Kuczynski, K. (1979) *Int. J. Mech. Sci.* **21**, 609.

Marin, J. and Rimrott, F. P. J. (1958) Welding Res. Council Bull. Series No. 41.

McClintock, F. A. (1965) *Proc. Roy. Soc.* **A285**, 58.

McClintock, F. A. (1968) *J. Appl. Mech., Trans. ASME* **35**, 363.

McClintock, F. A., Kaplan, S. M. and Berg, C. A. (1966) *Int. J. Fract. Mech.* **2**, 614.

Noble, C. F. and Oxley, P. B. (1964) *Int. J. Prod. Res.* **2**, 265.

Norris, D. M., Reaugh, J. E., Moran, B. and Quinones, D. F. (1978) *J. Engr. Mat. & Tech.* **100**, 279.

Oh, S. I. and Kobayshi, S. (1976) USAF AFML-TR-76-61.

Oyane, M. (1972) *Bull. JSME* **15**, 1507.

Oyane, M., Sato, T., Okinoto, K. and Shima, S. (1980) *J. Mech. Working Tech.* **4**, 65.

Parmar, A. and Mellor, P. B. (1978) *Int. J. Mech. Sci.* **20**, 707.

Parmar, A. and Mellor, P. B. (1980) *Int. J. Mech. Sci.* **22**, 133.

Pittman, J. F. T., Wood, R. D., Alexander, J. M. and Zienkiewicz, O. C. (eds.) (1982) *Numerical Methods in Industrial Forming Processes.* Swansea: Pineridge Press.

Radon, J. C. and Czoboly, E. (1982) in *Proc. Conf. Absorbed Specific Energy and/or Strain Energy Density Ctriterion,* G. C. Sih, E. Czoboly and L. F. Gillemot (eds.). Den Haag: Martinus Nijhoff, p. 181.

Rice, J. R. (1968) in *Fracture,* H. Leibowitz (ed.). New York: Academic Press.

Rice, J. R. and Johnson, M. A. (1970) in *Inelastic Behaviour of Solids,* M. F. Kanninen (ed.). New York: McGraw-Hill, p. 641.

Rice, J. R. and Tracey, D. M. (1969) *J. Mech. Phys. Solids* **17**, 201.

Ritchie, R. O., Server, W. L. and Wullaert, R. A. (1979) *Proc. ICM-3,* Vol. 3, p. 489.

Rogers, H. C. (1960) *Trans. Am. Inst. Min. Metall. & Petroleum Engrs.* **218**, 498.

Rogers, H. C., Leech, A. B. and Coffin, L. E. (1971) *Metalforming: Interrelation between Theory and Practice,* A. L. Hoffmanner (ed.). New York: Plenum Press.

Rowe, G. W., Sturgess C. E. N., Hartley, P. and Pillinger, I. *Finite Element Plasticity and Metalforming Analysis* Cambridge: University Press (to be published).

Shaffman, P. G. and Taylor, G. I. (1958) *Proc. Roy. Soc., London* **A245**, 312.

Shah, S. N., Oh, S. I. and Kobayashi, S. (1976) USAF AFML-TR-76-61 (See also *Trans. ASME, J. Engr. Industry* **101**, 23, 1979)

Sternstein, S. S. and Ongchin, L. (1969) *Polym. Pre-prints, Amer. Chem. Soc.* **19**, 1117.

Thomason, P. F. (1969) *Int. J. Mech. Sci.* **11**, 187.
Thompson, R. D. and Hancock, J. W. (1984) *Int. J. Fract.* **26**, 99.
Tomkins, B. (1965) UKAEA unpublished work.
Tomkins, B. and Atkins, A. G. (1981) *Int. J. Mech. Sci.* **23**, 395.
Turner, C. E. (1979) in *Postyield Fracture Mechanics* D. G. H. Latzko (ed.). Barking, UK, Applied Science Publishers.
Van Minh, H., Sowerby, R. and Duncan, J. L. (1974) *Int. J. Mech. Sci.* **16**, 31.
Van Stone, R. H., Merchant, R. H. and Low, J. R. (1974) STP No. 556, Philadelphia: ASTM.
Waldren, N. E., Percy, M. J. and Mellor, P. B. (1965–66) *Proc. I. Mech. E.* **180**, 5/66.
Williams, J. G. and Marshall, G. P. (1975) *Proc. Roy. Soc. (London)* **A342**, 55.

6

Rate, environment and temperature effects on toughness

'No, No! The adventures first,' said the Gryphon. . . 'explanation take such a dreadful time.'

6.1 INTRODUCTION AND REPRESENTATIVE BEHAVIOUR

Resistance to cracking depends, in general, on the rate of deformation, the environment in which cracking takes place, and the temperature. The results for material toughness discussed so far in this book relate, by the large, to room temperature experiments in air in which the deformation is quasi-static, i.e. sufficiently slow that kinetic energy is not significant in equation (1.1) and where (in materials with low thermal conductivities) the irreversible work of fracture does not generate temperature increases at the crack tip which might affect the resistance to crack propagation. That fracture is affected variously by environment, speed of deformation or temperature is well-known from studies on: stress-corrosion cracking; liquid–metal and hydrogen embrittlements; the behaviour of 'silly putty'; brittle–ductile transitions with temperature; the crazing and cracking of glassy amorphous polymers by organic liquids; the stress cracking of polyamides by aqueous and alcohol solutions of inorganic metal salts; the cracking of polyolefines by alchohols and detergents; the effects of water on glasses and ceramics; ammoniacal atmospheres on brass and copper alloys; chloride solutions on aluminium alloys; water on biological materials (cutting one's nails and shaving; moisture on grass and hay); and so on. Chain-scission and cross-linking of polymers produced by ultraviolet light absorption or heating come into this category, although in this book we do not go into detail on photo-degradation and thermal degradation of polymers; (we note, however, that the mechanics of embrittled ductile polymers are well-represented by sandwich plate theory (section 3.6)). By and large, the effect of aggressive environments on the fracture behaviour of metals, polymers and ceramics is to make them 'more brittle' and, relatively, to reduce R and K_c in comparison to σ_Y. Sometimes under the action of hostile environments external loads are not required and residual stresses (from manufacturing operations) are sufficient to trigger fracture, as in spontaneous fracture of components in stress-corrosion cracking. A broad-brush treatment of these effects is given in Chapter 1.

While nowadays rate, temperature and environmental effects are usually interpreted in terms of fracture toughness parameters, much historical information is given in terms of simple stress–strain curves, tensile strains to fracture, stress/time-to-failure curves etc. As one example of the two approaches applied to the same problem, Fig. 6.1(a) shows the tensile load–deflexion curves for standard (uncracked) testpieces of α-brass in air and in a 0.1 molar mercurous nitrate solution. The two curves are identical except that in the liquid environment fracture occurs at $\epsilon_f = 1.6\%$ after modest plastic flow, compared with a plastic fracture strain in air of about 10%. When Gurney's segmental area method is used to determine the fracture toughness of α-brass in air, and as embrittled by the mercurous solution, load–displacement records like those in Fig. 6.1(b) for a double-ended beam geometry are obtained. The average R in air is about 41 kJ/m^2, and in the environment R is 3.4 kJ/m^2, which represents a reduction of over 90% in fracture resistance. In this example, the cause of embrittlement is that the amalgam between brass and mercury causes a reduction in the local cohesion of the bonds at the crack tips.

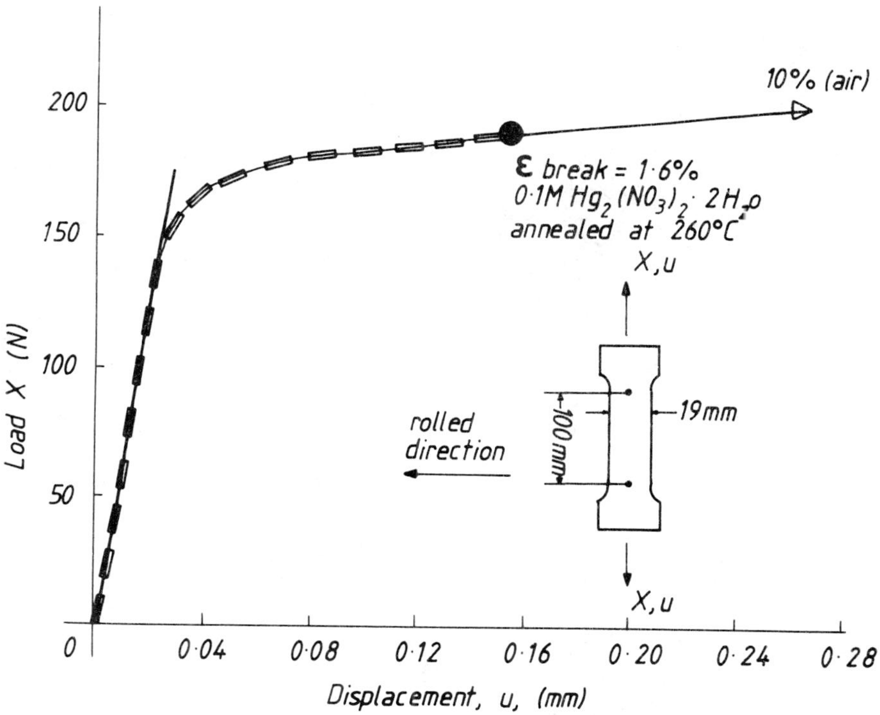

Fig. 6.1(a) – Tensile stress–strain relations for α brass tested in air and in mercurous nitrate solution.

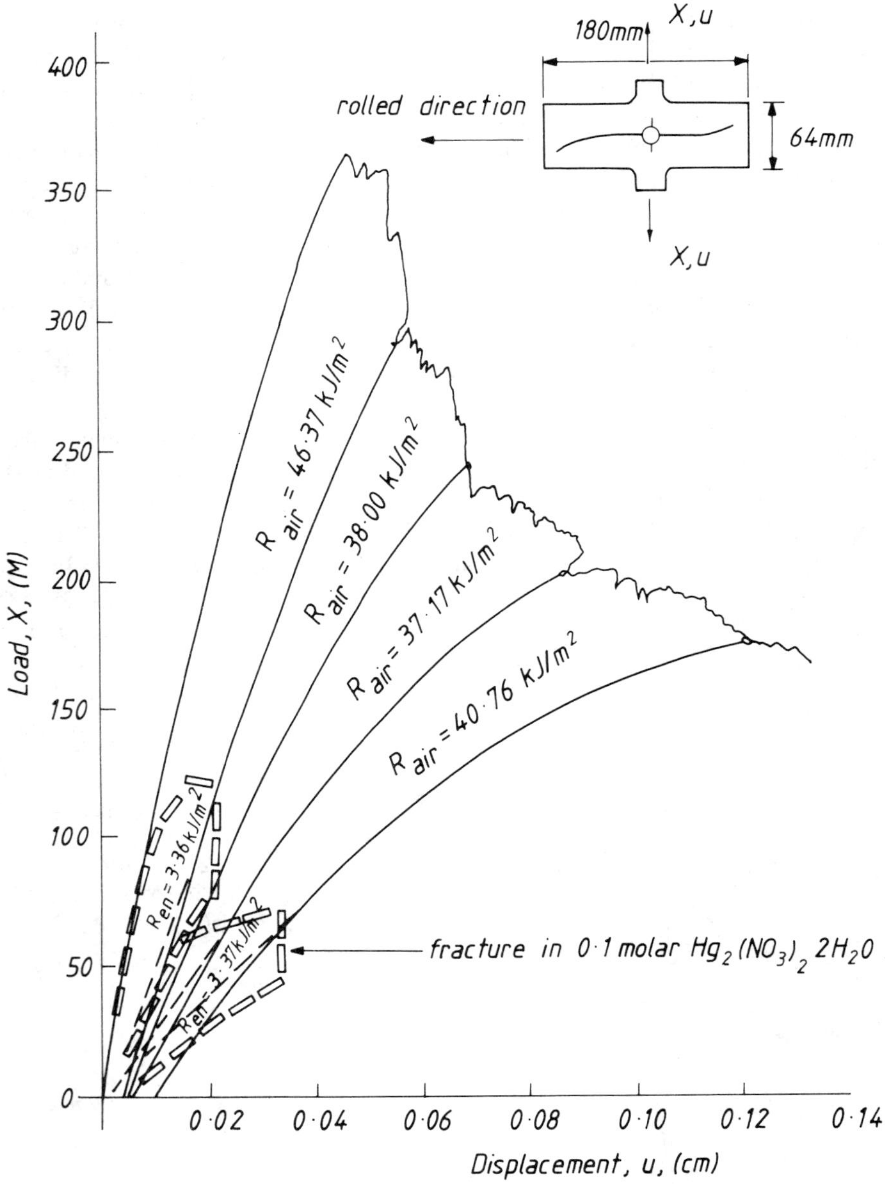

Fig. 6.1(b) – Load–deflexion diagrams for fracture of α brass in air and in mercurous nitrate solution using double-ended beam-like testpieces.

A great number of experiments have been performed in air to determine the effects of rate and temperature on specific work of fracture or critical stress intensity factor. Monotonic tests on a variety of specimens described in Chapters 2–4 have been carried out on testing machines equipped with ovens or freezing compartments, in which crack speeds were varied by driving the crosshead at different velocities. Over and above the ranges of T and $\dot{u}$ employed in these tests, experiments have been performed additionally in the presence of different liquid, gaseous or even solid hostile environments. Usually the conditions have satisfied global elasticity, but in some tests, interpretation of the data is complicated by remote plastic flow and displacement irreversibilities.

In this way, from the unique (X, u, R, A) values obtained during fracture in monotonic testing, it is possible to determine R $(\dot{a}, T)$ or R $(\dot{\epsilon}, T)$ data, in different environments (where $\dot{\epsilon}$ is the strain rate at the crack tip). Similarly it is possible to determine K_c $(\dot{a}, T)$ or K_c $(\dot{\epsilon}, T)$ data in different environments where K_c may relate either to the stationary or to the running crack. All the remarks relating to experimental technique, crack stability etc. given in Chapters 2–4 and 8 apply to these types of experiments.

In this book most of the information relates to data obtained in the quasi static range, where kinetic energy, stress wave effects and so on are not of consequence: even so, some information is given on 'dynamic' fracture toughness in this chapter.

By way of examples, Figs. 6.2 to 6.8 give representative results solids for the effects of rate, temperature and environment on crack resistance of a variety of solids. More data are provided in later Figs, e.g. Figs. 6.13–6.15, Fig. 6.17 and Fig. 6.22.

Fig. 6.2(a) shows how the toughness of A533B steel plate varies with temperature; this type of steel is used extensively in nuclear reactor pressure vessels. Toughness increases (with yield strength falling) as the temperature increases, typical of bcc metals which display a deformation transition; the so-called *nil ductility temperature* used to define the transition temperature is labelled 'NDT'. Note that all data in Fig. 6.2(a) are 'valid' K_{IC} results, i.e. the higher temperature values (which are more than three times as big as the lower temperature values) were obtained under globally elastic testing conditions. In other related methods of testing (Charpy, for example) the increase in impact energy beyond NDT in the so-called 'upper shelf' region is often linked to remote plastic flow and displacement irreversibilities (see Chapters 1 and 4).

Figs. 6.2(b) and (c) show the same trends as Fig. 6.2(a) but, additionally, highlight the fact that the 'dynamic' fracture toughness K_{Id} of low strength steel alloys (which are rate-sensitive) is lower than the static toughness (and that the transition temperature is higher) whereas $K_{Id} \approx K_{IC}$ for high strength steels such as HY-130 and 18Ni(250) maraging alloys. There is yet no standard method to determine the dynamic stress intensity factor K_{Id} other than that it is obtained at impact loading rates typically with $\dot{K} \approx 10^5$ (MPa$\sqrt{}$m/s). ASTM Task Group E24.01.06 is investigating dynamic toughness at the time of writing. Klepaczko (1984) discusses experimental methods of achieving $10^{-2} < \dot{K}_{Is} < 10^8$ MPa$\sqrt{}$m/s in the determination of the effect of stationary crack tip strain rate (effectively $\dot{K}_{Is}$ – section 6.2.2 following) on K_{IC} in metals. Rather like the problems associated with the determination of σ–ϵ properties at high strain rates (e.g. Atkins, 1970), stress wave loading becomes important in K_{IC} testing of metals beyond some 10^5 MPa$\sqrt{}$m/s and crack tip adiabatic heating in metals

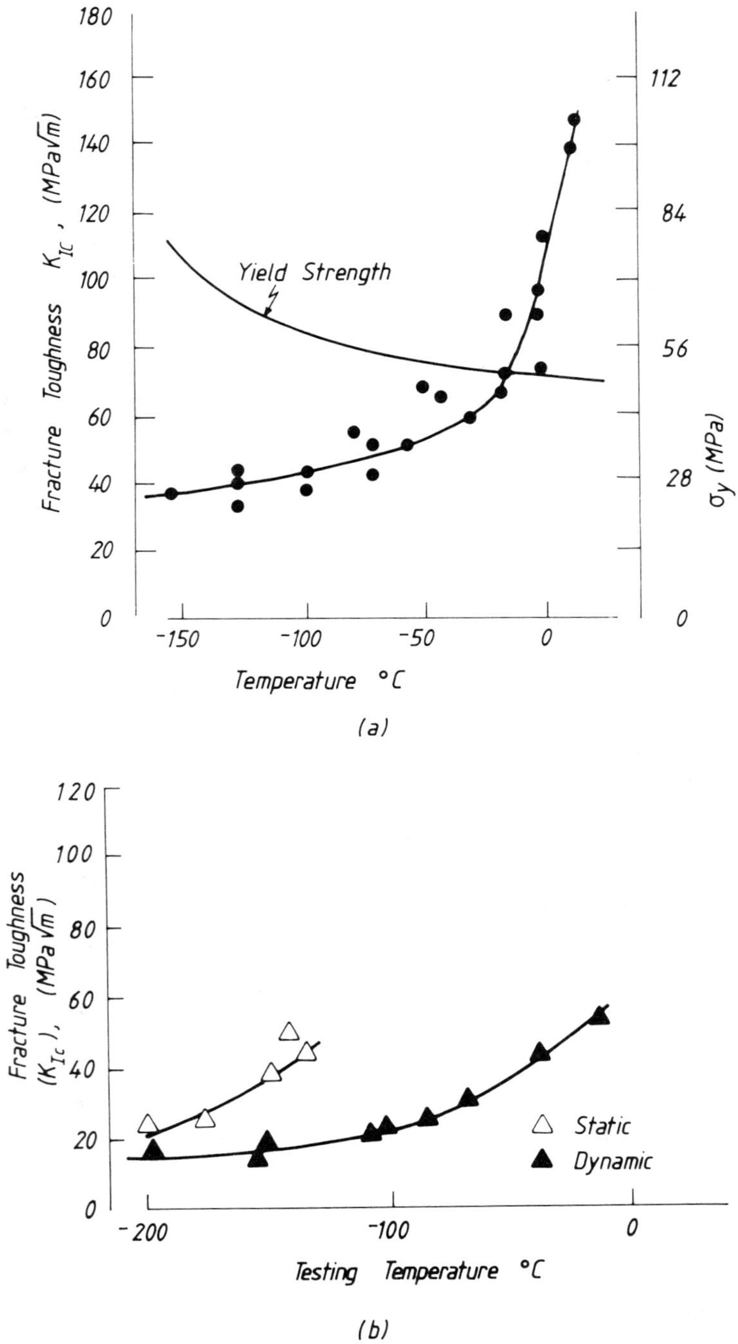

Fig. 6.2(a) – Fracture toughness at low temperature for ASTM A533, Grade B, Class 1 steel (reproduced from Seman *et al.* (1971), with permission). (b) Static and dynamic fracture toughness of ABS-C steel at low temperatures (reproduced from Shoemaker and Rolfe (1969), with permission).

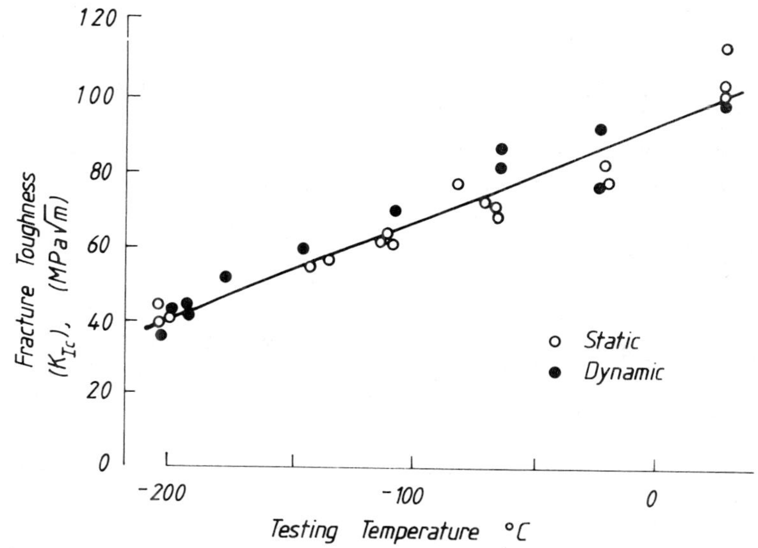

(c)

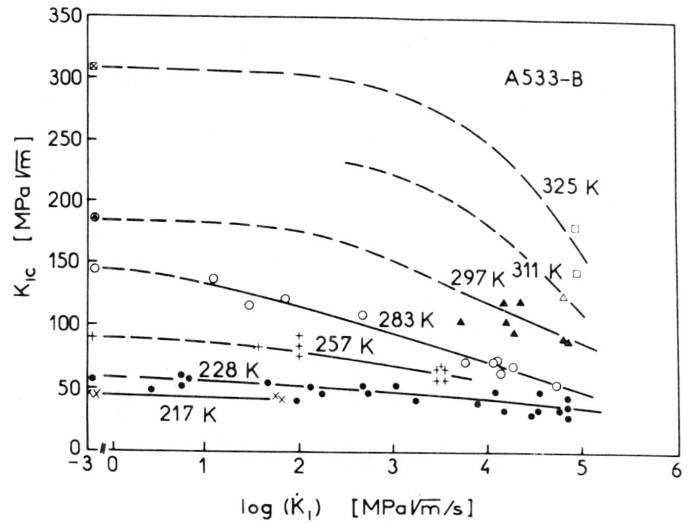

(d)

Fig. 6.2(c) – Static and dynamic fracture toughness of 18 Ni(250) maraging steel at low temperature (reproduced from Shoemaker and Rolfe (1969), with permission). (d) Variation of K_{IC} with stationary crack tip $\dot{K}$ for A533B steel at different temperature.†

---

†Data variously from Shabbits (1973), Crosley and Ripling (1975), Begley and Landes (1972), and Buchalet and Mayer (1973). Reproduced in Klepaczko (1984).

beyond some 10^3 MPa$\sqrt{}$m/s. These matters are discussed later in this chapter and in Chapter 8. Klepaczko reviews data in the literature for aluminium and titanium alloys and steels. For example, Fig. 6.2(d) shows for A533B steel that K_{IC} falls with increase of loading rate at constant temperature, the effect being more marked higher temperatures than low. However, above 373 K (in the upper shelf region) there is an abrupt change in rate sensitivity with K_{IC} increasing with increasing stationary crack $\dot{K}_{Is}$. Changes in the magnitude of the critical stress intensity factor (or equivalently R or J) with crack velocity and temperature *once the crack is propagating* are important in the topic of crack arrest discussed in Chapter 8. For high strength metal alloys at low temperatures the variation of toughness depends on the alloy base. For many aluminium alloys, K_{IC} tends to increase or remain generally constant as the test temperature is decreased, e.g. 2024 T851 has $K_{IC} \approx 25$ MPa$\sqrt{}$m between 20°C and −60°C; many of the 7000 series of aluminium alloys have $K_{IC} \approx 30$ MPa$\sqrt{}$m over the same range of temperature; others (such as 2219 T87) display K_{IC} increasing from about 35 MPa$\sqrt{}$m at 20°C to 50 MPa$\sqrt{}$m at −250°C. This favourable characteristic is attributed to the fcc crystalline structure of the aluminium alloys rather like CVN values (Chapter 1). For titanium alloys K_{IC} decreases with decreasing temperature, but the effect is influenced by the alloy content and heat treatment.

Figs 6.3(a) to (d) show results for a typical polymer, PMMA. In the higher range of temperatures, the specific work of fracture R varies linearly with crack velocity $\dot{a}$ (Fig. 6.3(a)), but at lower temperatures the relation is non-linear and displays local inversions (Fig. 6.3(b)). Fig. 6.3(c) shows that R varies in different environments, other things being equal. The data are for Outwater double torsion testpieces which crack at almost constant load (with horizontal R loci) and constant crack velocity. Fig. 6.3(d) gives the R–$\dot{a}$ results for a number of environments; (the 'air' results correspond with the room temperature line in Fig. 6.3(a)). Instead of R, stress intensity factors may be determined, giving the results in Figs. 6.3(e) and (f). While there are similar patterns to Figs. 6.3(a) and (b) etc., we remember that E* is rate- and temperature–dependent and that this should be taken into account in any comparisons between the different methods of assessing crack resistance (see section 6.3).

Toughness as a function of temperature and rate for some typical oxide and carbide ceramics is shown in Fig. 6.4(a) and (b) respectively. Some ceramics can experience a considerable increase in temperature before there is much change in crack resistance, but as Fig. 6.4(b) shows (in this case for silicon nitride) there may be considerable reduction in toughness with rate of deformation, particularly at elevated temperatures.

Materials which are difficult to crack in a controlled fashion are often those with negative $dR/d\dot{a}$ characteristics. Even when the geometric stability factor of a testpiece is negative, stable propagation is hard to achieve (see Chapter 8). Fig. 6.5 shows that epoxy resin displays negative $dR/d\dot{a}$. These data, at controlled $\dot{a}$, were obtained by stabilising crack propagation by use of adhesive tapes (see Chapter 8).

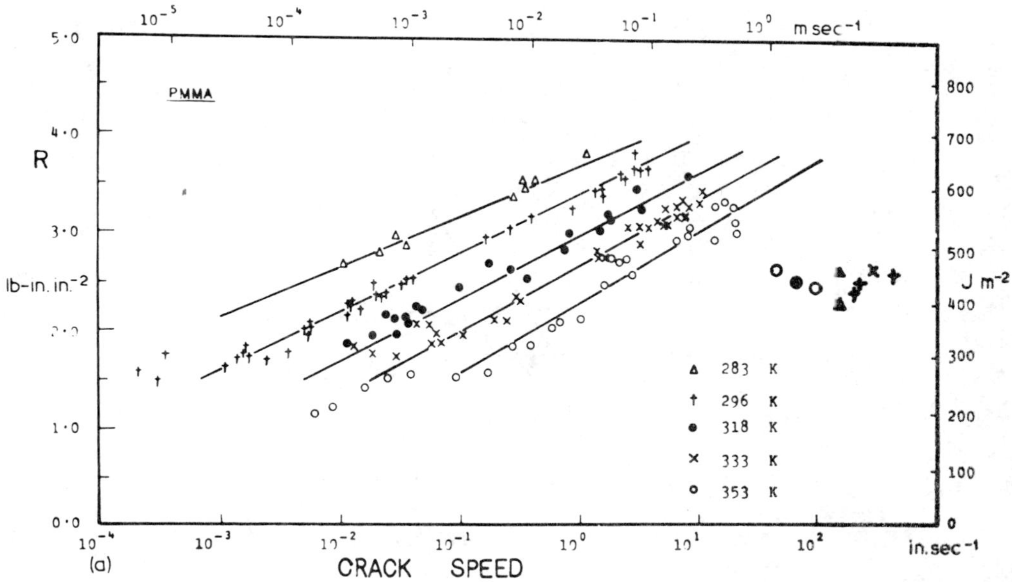

(a)

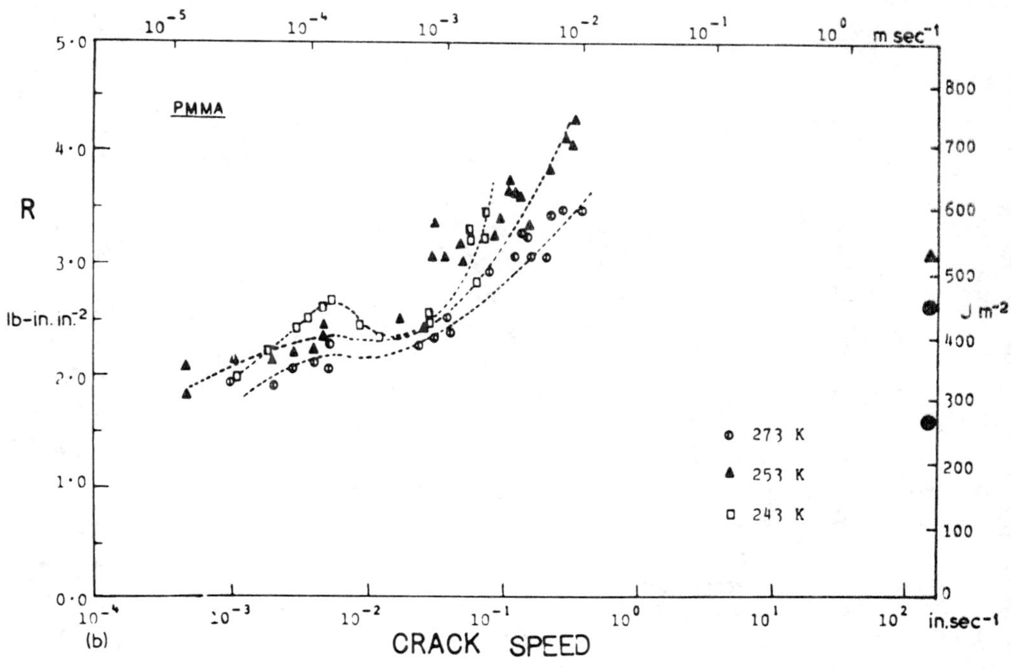

(b)

Fig. 6.3 – Experimental values of fracture toughness for PMMA plotted against crack speed, (a) for temperatures of 283 K and greater, (b) for temperatures at 273 K and lower (after Atkins *et al.* (1975), with permission).

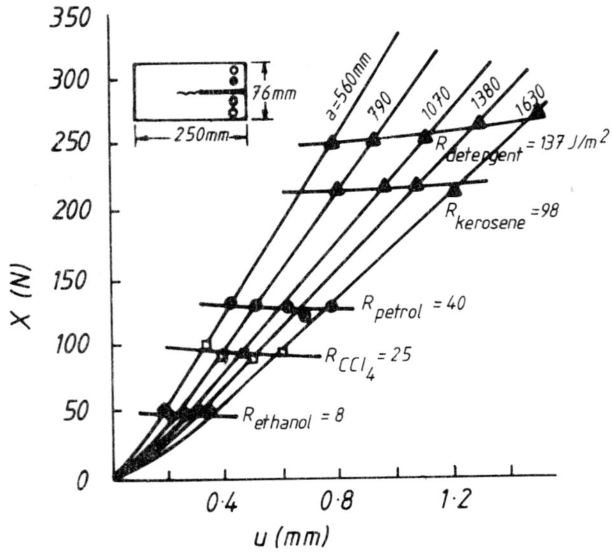

Fig. 6.3(c) – Crack initiation toughness for PMMA in various liquid environments (after Mai (1975), with permission).

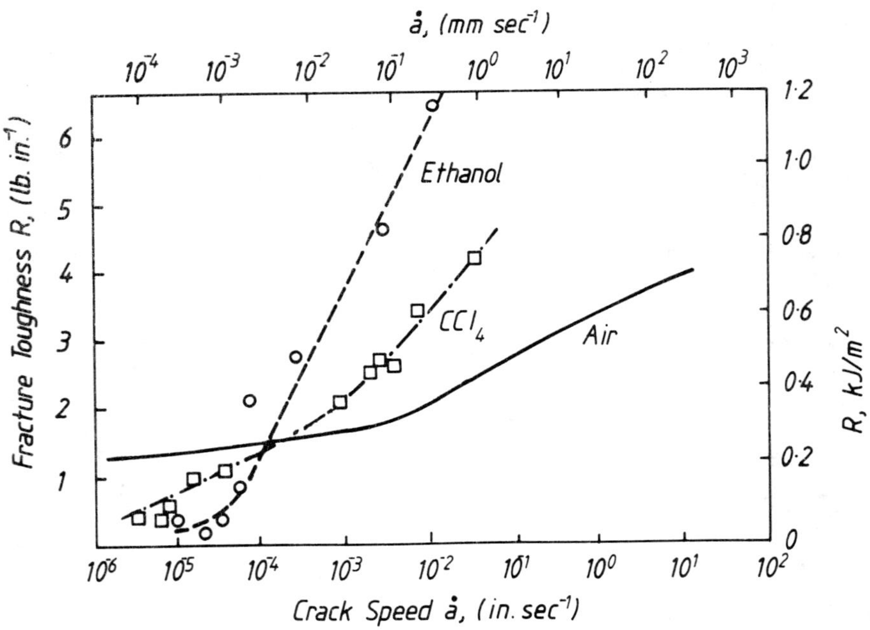

Fig. 6.3(d) – Variation of fracture toughness (R) for PMMA as a function of crack speed ($\dot{a}$) in ethanol, CCl_4 and air at room temperature (after Mai (1975), with permission).

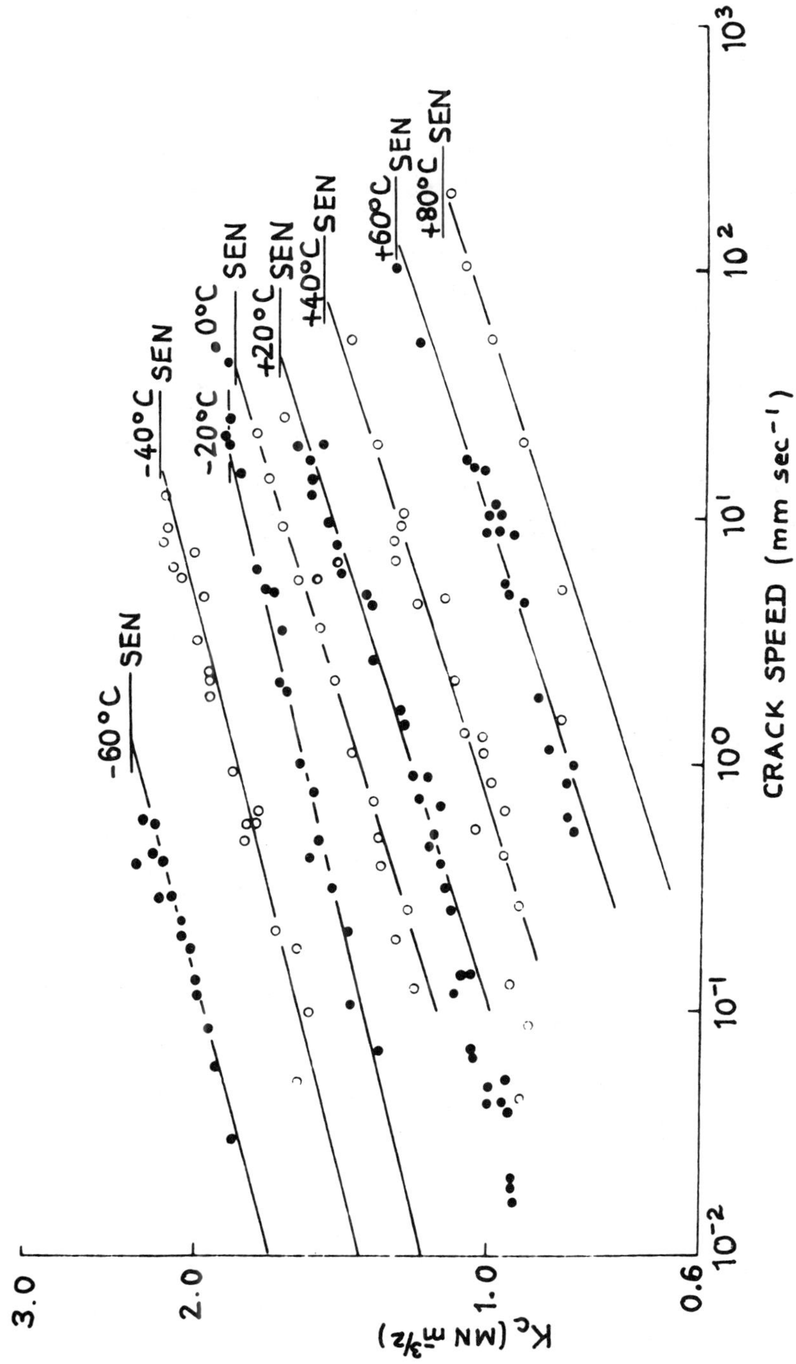

Fig. 6.3(e) – K_c as a function of crack speed for PMMA at various temperatures – stable crack growth in double torsion (DT) tests (reproduced from Marshall *et al.* (1974), with permission).

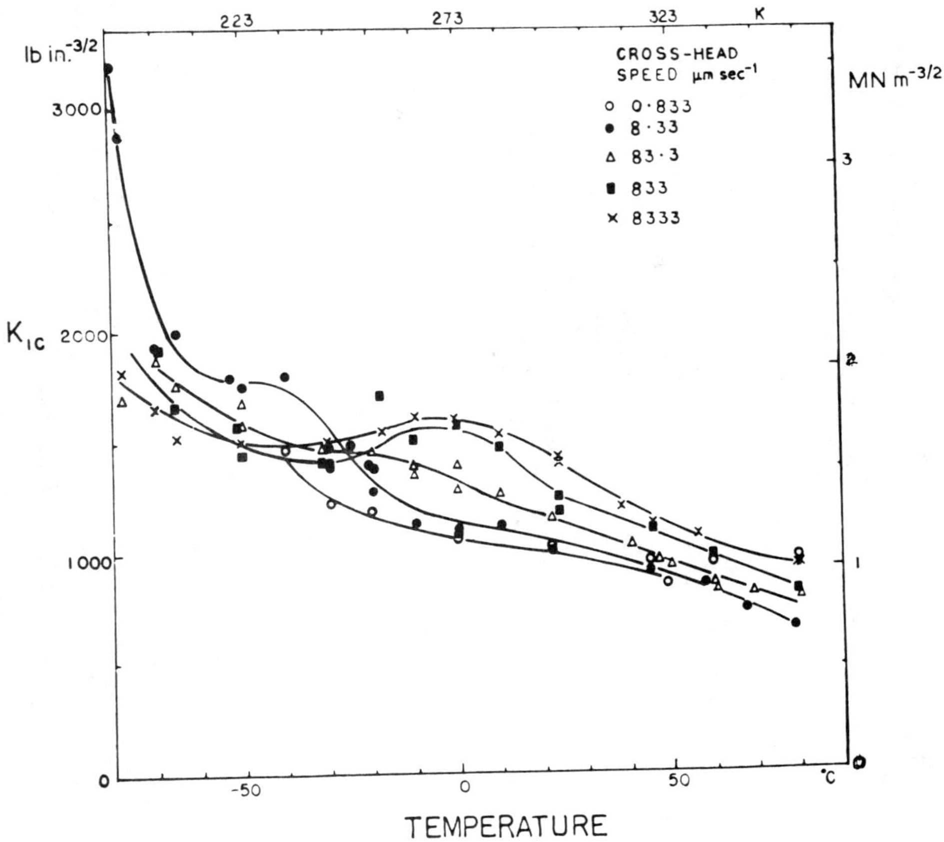

Fig. 6.3(f) – K_{IC} data plotted against temperature for various cross-head speeds for PMMA (after Atkins *et al.* (1975), with permission).

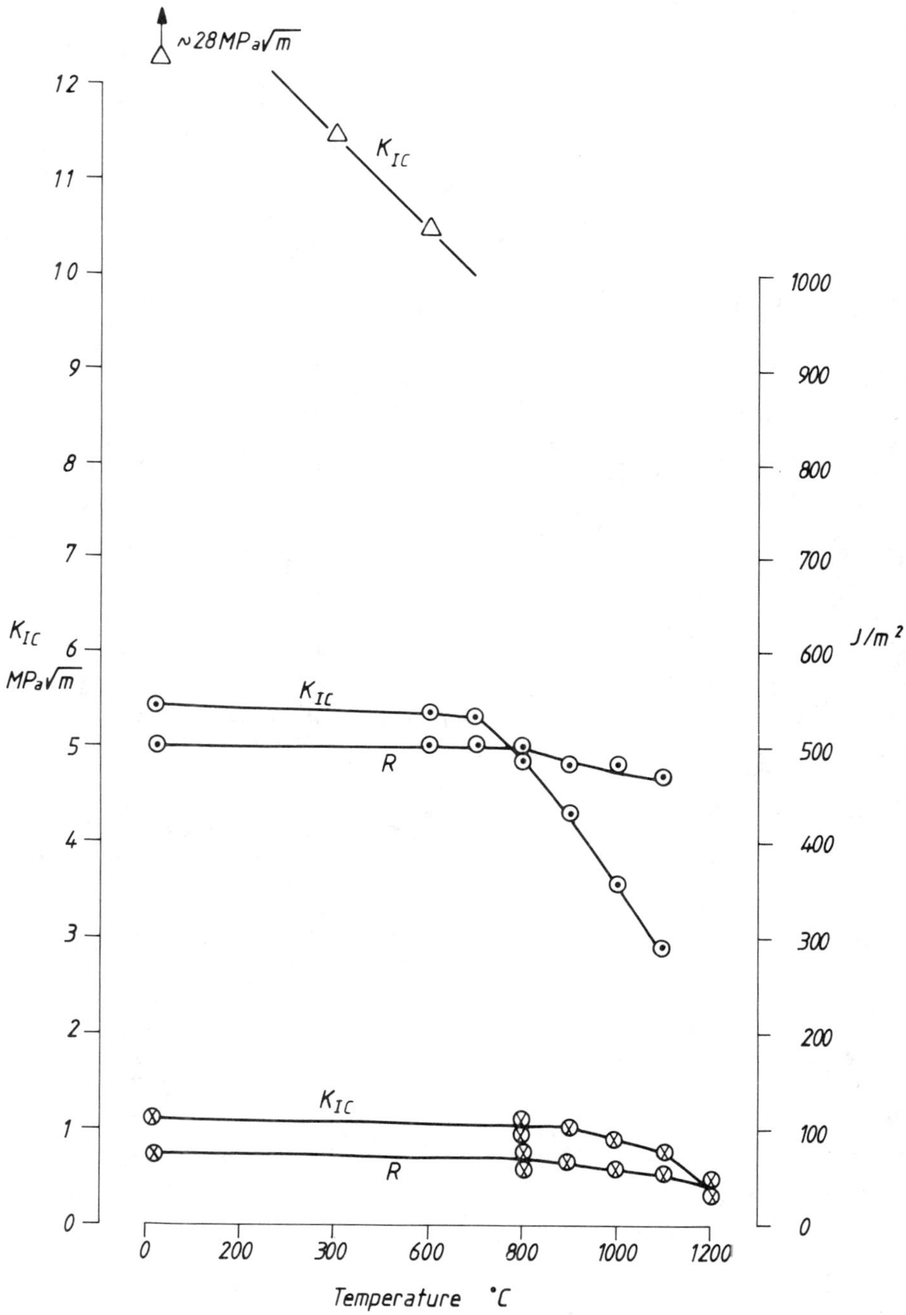

Fig. 6.4(a) – K_{IC} as a function of temperature for some tool ceramic materials: Δ GE Carboloy 370, ○ Lithium-aluminium-silicate, ○ Magnesium-aluminium-silicate (data derived from Mai and Atkins (1975)).

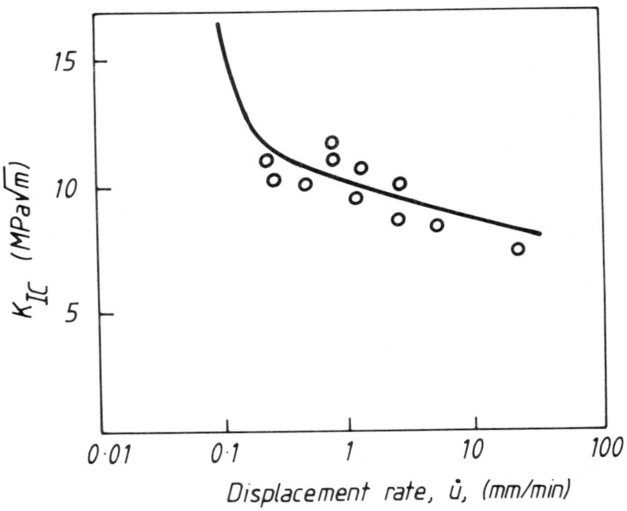

Fig. 6.4(b) – Effect of displacement rate on K_{IC} at 1225°C for Si_3N_4. (After Munz *et al.* (1979). Copyright ASTM, 1916 Race Street, Philadelphia, PA 19103. Reprinted with permission.)

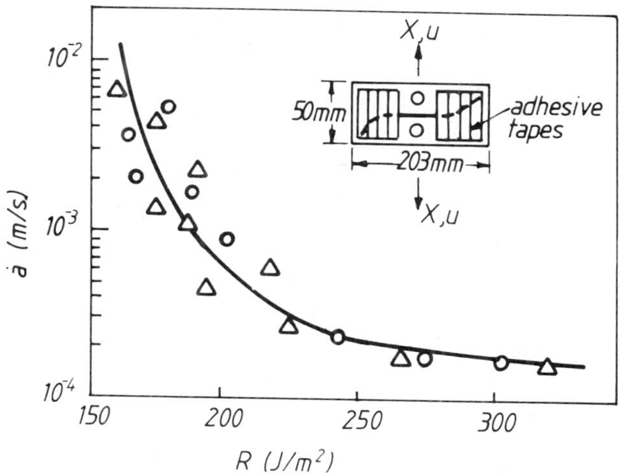

(a)

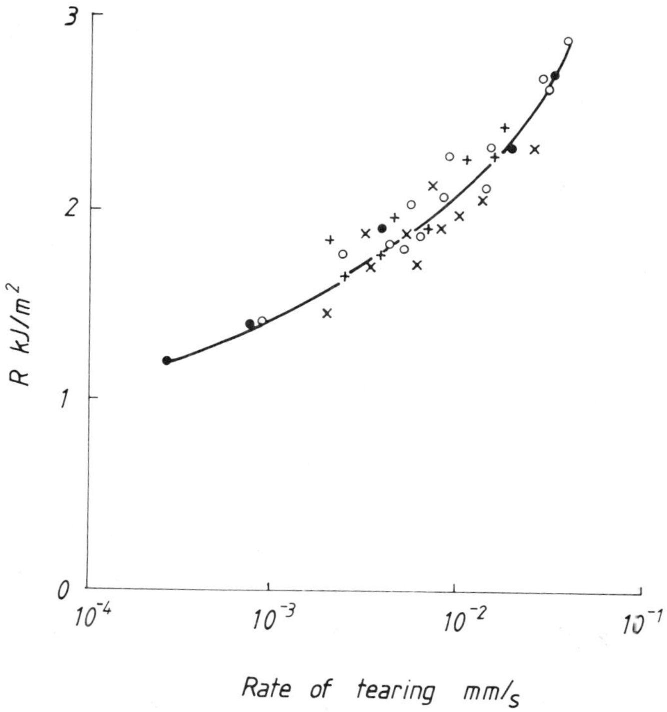

(b)

Fig. 6.5(a) – Variation of R with $\dot{a}$ for an epoxy resin (after Mai and Atkins (1975), with permission). (b) Variation of R with rate of tearing for an SBR vulcanisate using the testpieces shown in Fig. 2.37 (data from Malaysian Rubber Producers Research Association).

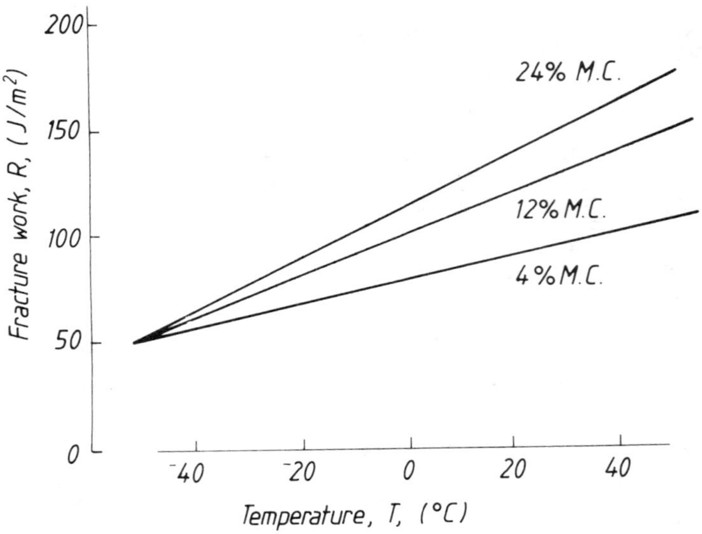

(a)

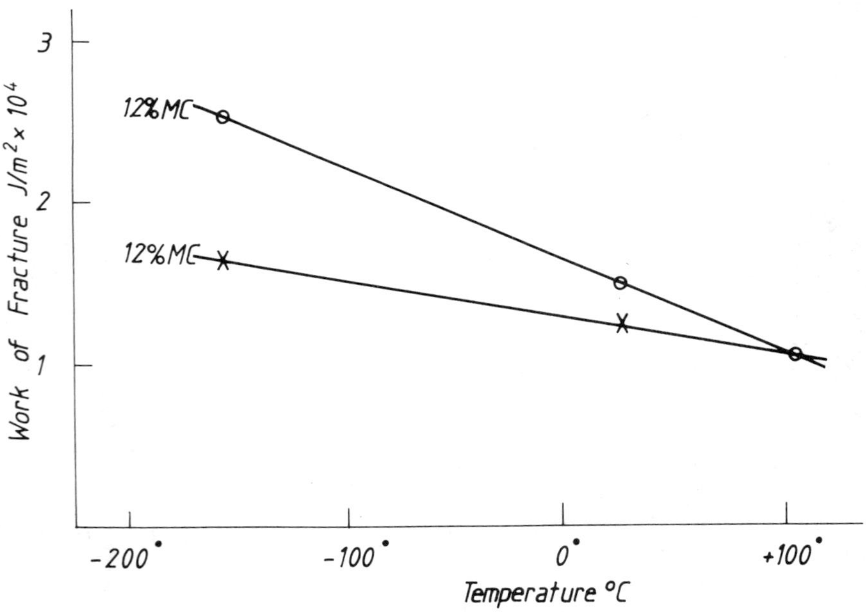

(b)

Fig. 6.6(a) – The effect of temperature on the specific fracture energy (*R*) of wood (crack propagation along the grain). (b) The effect of temperature on the fracture energy (*R*) of wood (crack propagation across the grain). (a), (b) from Jerommidis (1979), with permission.

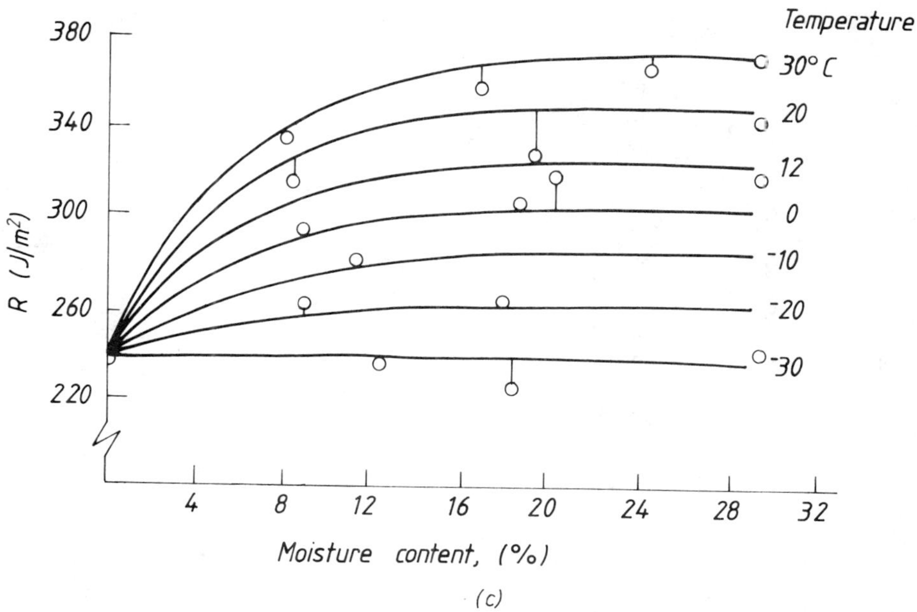

Fig. 6.6(c) – Fracture energy (R) as a function of temperature and moisture content for white pine wood. (c) from DeBaise *et al.* 1966. Copyright ASTM, 1916 Race Street, Philadelphia PA 19103. Reprinted with permission.)

The specific work of fracture of wood changes with orientation (it is well known that wood is easily split along the grain but not across the grain). Fig. 6.6(a) shows that the toughness of wood along the grain falls with falling temperature, whereas the toughness across the grain increases as the temperature falls. Young's modulus E varies with orientation in a complicated way as described in connexion with crack paths in Chapter 3 so, again, care has to be taken when making comparisons between K_{IC} in certain orientations and R in the same direction. All data in Figs 6.6(a) and (b) depend on the relative humidity: moisture has a significant effect on toughness as shown in Fig. 6.6(c) for white pine (cf. Example 3.17 on plywood laminations swelling under moisture).

Figs 6.7(a) and (b) show Castino's 1983 data for Kevlar-reinforced epoxy composites which were manufactured using the 'intermittent bonding' concept for simultaneous high stiffness, high strength and high toughness in composites (see Atkins, 1975). Depending on the interfacial coating employed to intermittently bond the composite, the specific work of fracture either increases or decreases with increase in rate or temperature.

The toughness of rat skin as a function of rate and temperature is shown in Figs 6.8(a) and (b) respectively. These results were obtained using the trousers tear test, rather than other types of test which are less amenable to such a highly extensible material (cf. section 2.10). Note that, although the material (like many other highly extensible biological tissues) is difficult to tear, its fracture toughness is not all that great really. As explained in Chapter 1, this comes about

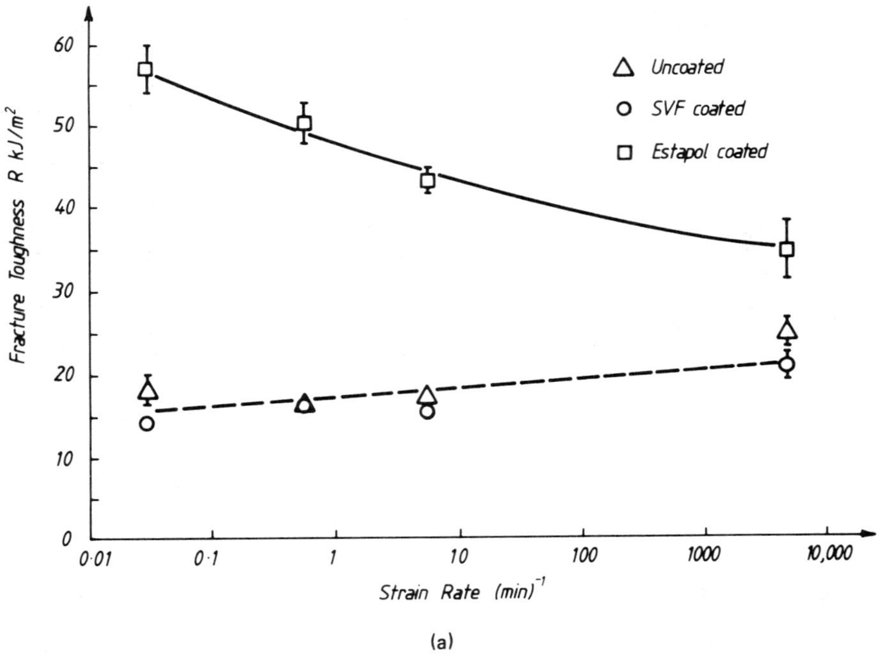

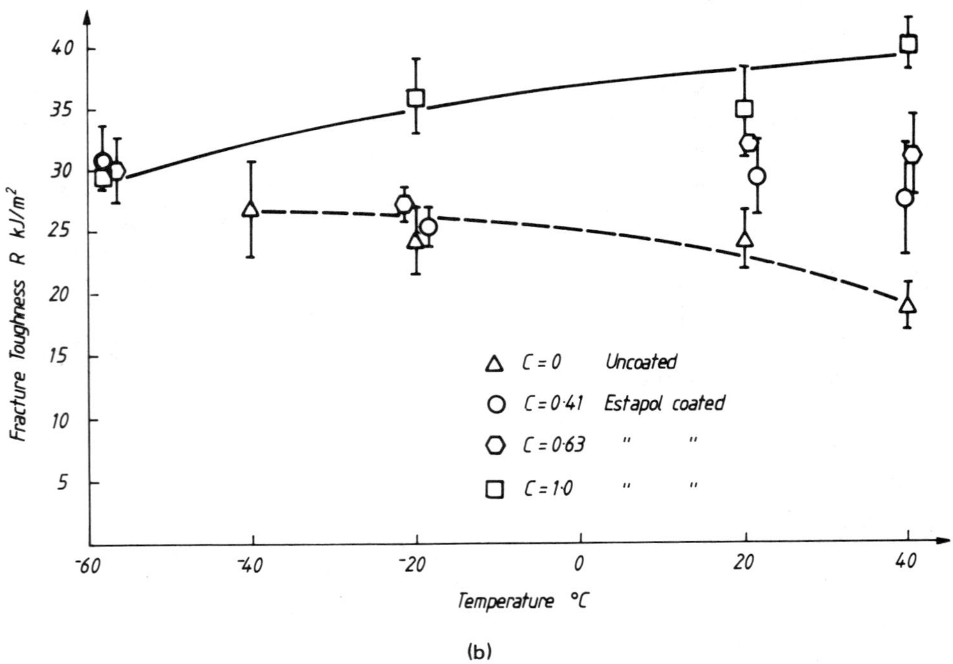

Fig. 6.7 – Effect of (a) strain rate and (b) temperature on fracture toughness of Kevlar/epoxy resin composites with and without coating fluid between fibre and matrix interface, C is ratio of per cent coated/per cent uncoated along filament (see Atkins, 1975). (Courtesy Castino, 1983).

because of the lack of good shear connexion between adjacent elements (i.e. the slope of the J-shaped stress–strain curve is very flat early in the deformation) which means that it is difficult to transmit energy into the path of fracture and therefore the material 'appears' to be tough.

6.1.1 Slow crack growth

An important point, to which the reader should be alerted, arises for materials which are capable of *creep* or *relaxation* behaviour during the conditions of rate, temperature or environment fracture testing. In these circumstances, it is possible that cracks may grow (in benign environments) under *constant load*. At first sight, this is contrary to the ideas of unique (X, u, R, A) coordinates for fracture, (unless the specimen happens to be one in which propagation does occur under

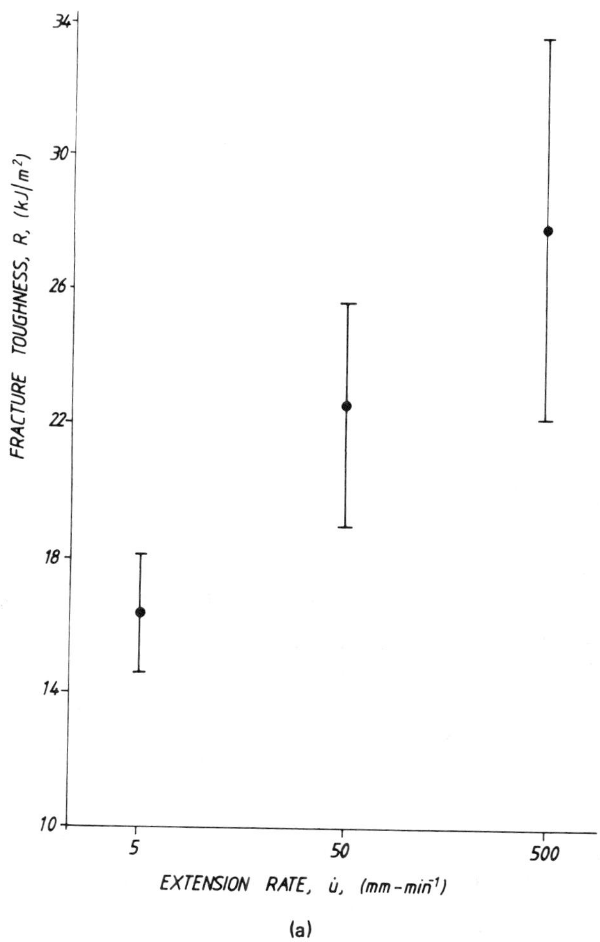

Fig. 6.8(a) – Effects of extension rate on the 'trousers test' tear resistance of wet rat skin tested at room temperature in air.

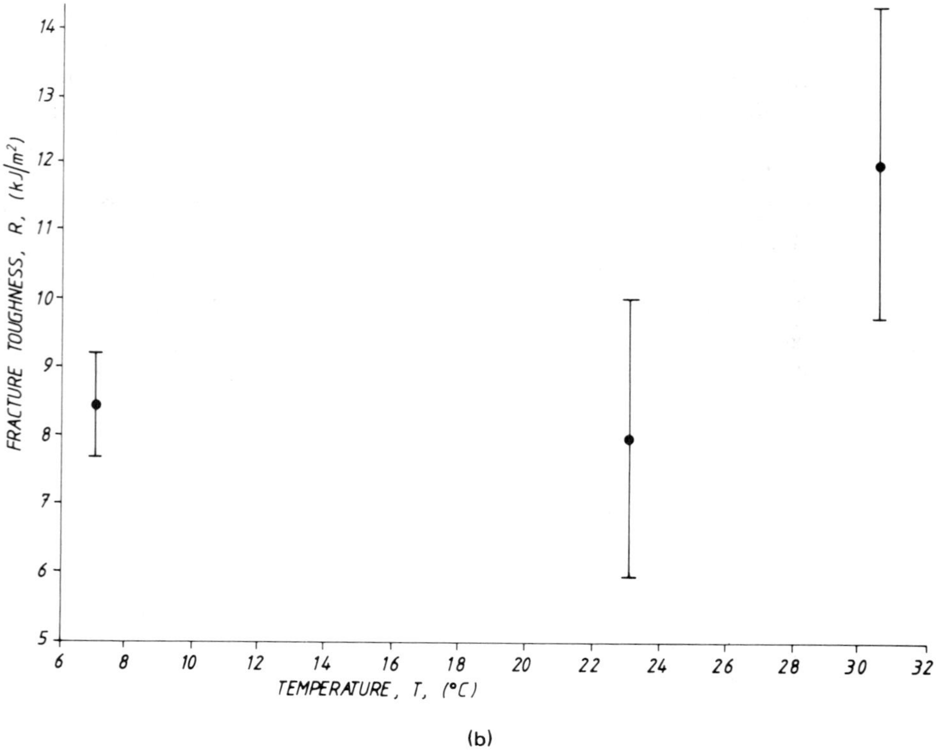

Fig. 6.8(b) – Temperature on toughness of wet rat skin in air at a tear rate 5mm/min (courtesy P. P. Purslow).

constant load). That is if, in a monotonic test, the testing machine is stopped before the load has reached the critical load for cracking (stopped before point B in Fig. 2.6) then there should be *no* cracking at all. Similarly if a displacement-controlled machine is stopped after some propagation along the R locus, cracking should *cease*. These are the predictions of time-independent mechanics of fracture. However, if the material can creep or relax during a test then†

(i) crack growth may be possible under constant load and
(ii) crack growth may occur during the loading stage of monotonic testing before the notional critical stress intensity factor is reached.

Such behaviour, if only on a limited scale (because, say, a polymer is displaying only minor viscoelastic effects) leads to a 'rounding off' of the corners of Gurney work areas; this is often seen in room temperature experiments of PMMA. On a

†Additionally, or alternatively, microstructural orientation may occur in some polymers in the highly strained crack-tip regions. E^* will increase and, in '$\sigma_{cr} \propto \sqrt{E^*R/a\pi}$' cracking will be more difficult. This sometimes occurs in the tearing of thin sheets of plastics. The highly anisotropic distribution of E^* often means that it is difficult to direct energy into the crack tip region to effect cracking (cf. results for rat skin in Fig. 6.8).

more extensive scale, the effect leads to so-called 'sub-critical crack growth' or 'slow crack growth' either before the critical stress intensity is reached in monotonic tests or in constant load tests where, according to time-independent theory, there 'should not' be cracking. Fig. 6.9 shows sub-critical crack growth in Discaloy at 650°C in both centre crack panel and compact tension testpieces.

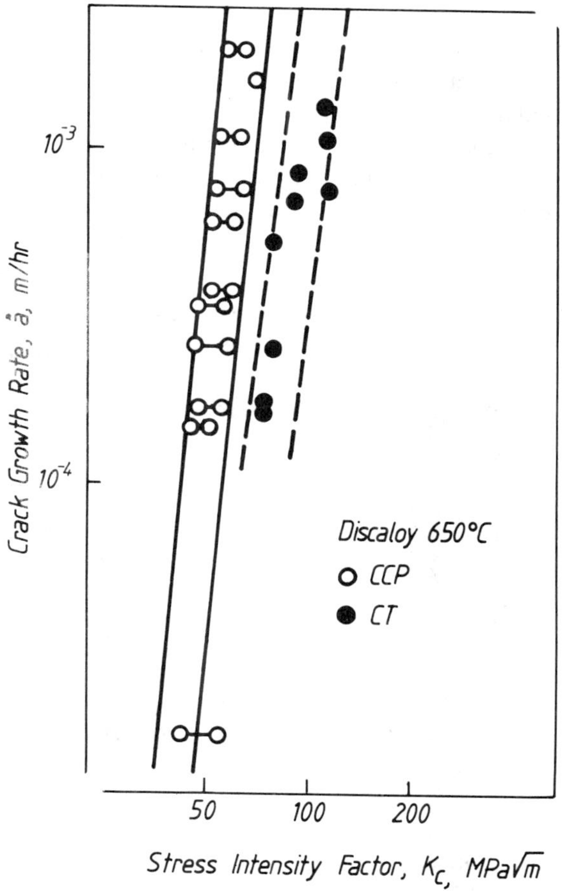

Fig. 6.9 – Crack growth rate ($\dot{a}$) versus applied stress intensity factor (K_c) for Discaloy at 650°C (after Landes and Begley (1974), with permission).

These may be environmental effects superimposed on the relaxation behaviour so that the slow crack growth produced in benign environments may be augmented in hostile environments. Of course, so-called 'static' or 'delayed' fracture under a constant load (which does not produce fracture when first applied) may occur in some materials owing solely to environmental effects: environmental-assisted slow crack growth lengthens pre-existing cracks, over a period of time under load, to the critical value at which the specimen breaks under that load. Fig. 6.10 shows the well-known delayed fracture phenomenon of glass in air. Note that *in vacuo* there is essentially no environmental effect.

The use of the term 'slow crack growth' in this general context is less than desirable, owing to its different use in connexion with rising R- or K_R-curves where the cause is quite different, viz. change of constraint at the crack tip or a screening action by the plastic crack tip zone in which the process zone is encapsulated (Chapter 4). However, we shall keep with common usuage, but the reader should be aware of the different phneomena with the same name. Again, there is a problem with symbols since, in the context of rate, temperature and environmental effects, 'K_c' is used for sub-critical or slow crack growth below K_{IC}. K_c here does *not* mean plane stress toughness; in this book we have aimed to use K_C (subscript upper case C) for plane stress toughness, K_c (subscript lower case c) for sub-critical growth. Again, δ_c is used for sub-critical COD.

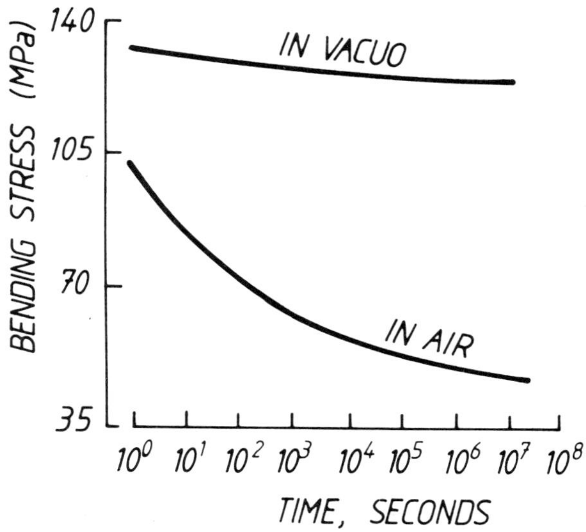

Fig. 6.10 – Results of testing glass in vacuum and in air (after Gurney (1952), with permission).

In general, slow crack growth at constant temperature is a manifestation of both time *and* environmental effects. In this sense, there is no distinction between monotonic and static load tests. A large body of data on a variety of materials displaying slow growth shows that the *same* K_c–$\dot{a}$ relations are obtained irrespective of whether the stress-intensity factor/crack velocity data are observed in dead-weight tests or in tests where the crosshead is driven at a monotonically controlled rate. In fact, in Fig. 6.3, some data are from monotonic tests and some from static load tests, and Fig. 6.11 shows K_c–$\dot{a}$ data for Scots pine obtained by both methods. The only difference between test method relates to

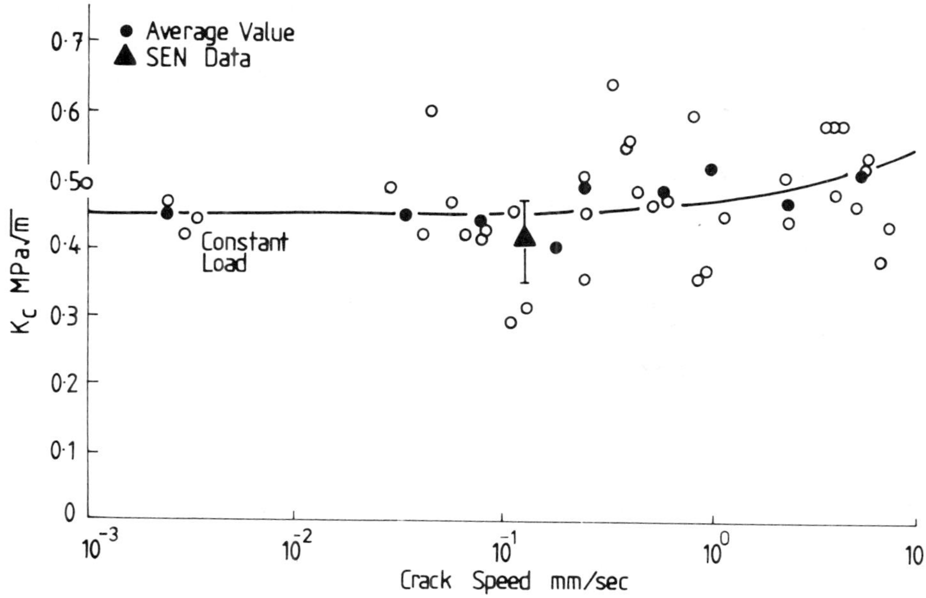

Fig. 6.11 – Fracture toughness as a function of crack speed for Scots pine (after Ewing and Williams (1979), with permission).

the range $\dot{a}$ usually encountered; in monotonic tests, $\dot{a}$ is usually greater. Thus in Fig. 6.12, which gives schematically the shape of K_c–$\dot{a}$ relations often found experimentally over the widest range of velocities, data in the upper (right-hand) part of the diagram are determined from monotonic tests; dead-weight tests, in principle, give results over the whole range but as they are difficult to control in the high K_c region, they tend to be restricted to the lower (left-hand) part of the diagram. Crack growth behaviour at very low speeds ($< 10^{-2}$ mm/s, say) presents particular experimental difficulties: it takes 10–15 minutes to produce about 1 mm of crack growth and there may in some materials be local 'stick–slip' behaviour (i.e. crack growth and subsequent arrest) or crack-tip blunting which also causes arrest – determination of such low crack velocities in these circumstances is not always accurate.

The notional three-region K_c–$\dot{a}$ diagram drawn in Fig. 6.12 is suggested by parts of a number of diagrams already given in this chapter – particularly where environment is a major contributor to the behaviour, e.g. Fig. 6.3(d). Figs. 6.13–6.15 show further evidence of this type of response. Fig. 6.13 relates to soda-lime glass and the influence, on slow crack growth, of relative humidity (Fig. 6.13(a)) and of alcohols saturated with water (Fig. 6.13(b)); and a part of Fig. 6.13(b) in finer detail relates to the effect of the concentration of water in butyl alcohol (Fig. 6.13(c)) – similar results were reported for ethyl alcohol. Fig. 6.14(a)–(c) gives representative results for adhesive joints in CCl_4, liquid water and moist air. Note that air at different relative humidities causes slower crack growth rates than cracking in liquid water. Fig. 6.15(a) and (b) concern high density polyethylene. The effects of distilled water and 'Comprox' 2% detergent solution on K_c–$\dot{a}$ relations are shown in Fig. 6.15(a), and the effect of different temperatures on crack growth in 'Comprox' in Fig. 6.15(b). Similar results for temperature on growth in distilled water were found by Chan and Williams in this 1983 series of experiments. No results are available for crack growth in air owing to the HDPE behaving in a ductile fashion: the environment has thus caused a 'deformation transition' (Chapter 10). Yet further data on three-region K_c–$\dot{a}$ diagrams will be found elsewhere in this chapter, especially in Fig. 6.22(a)–(g), and discussed in detail in section 6.4.

The end points on Fig. 6.12 have been labelled (i) $K_c = K_0$ which indicates a threshold (which varies with environment) below which crack growth does not occur (cf. section 7.2.2 regarding thresholds in fatigue) and (ii) $K_c = K_{IC}$. K_{IC} here is used in the sense of a material property that is temperature- and

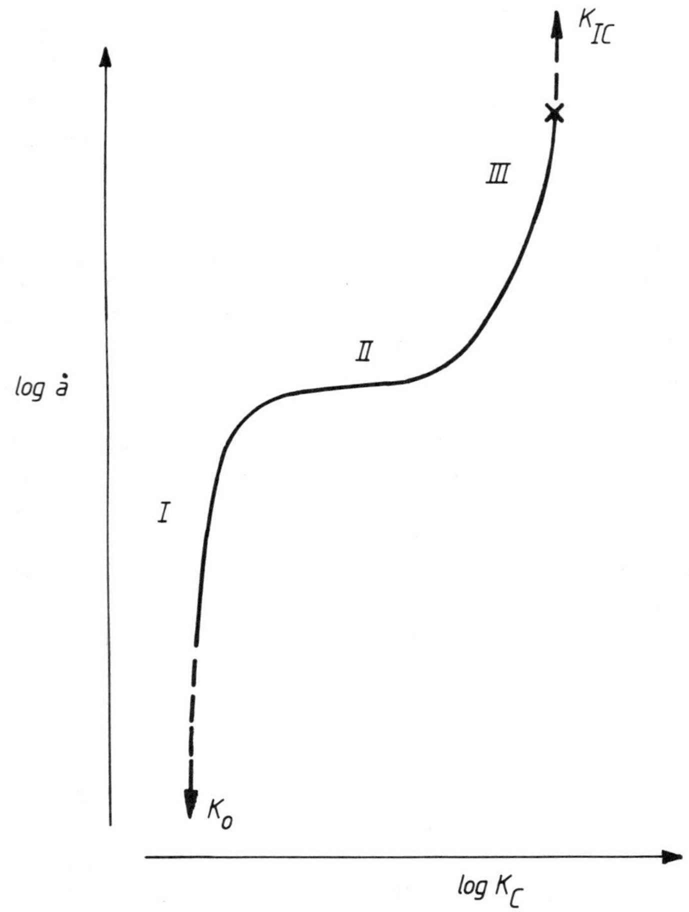

Fig. 6.12 – Schematic drawing showing the three stages of slow crack growth.

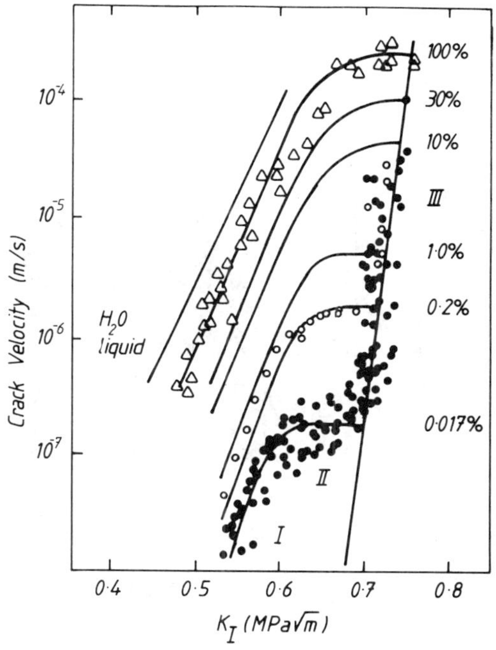

(a)

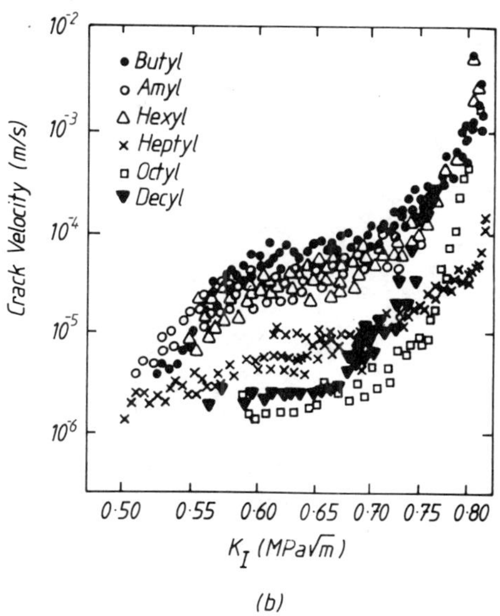

(b)

Fig. 6.13(a) – Effect of relative humidity on crack growth in soda-lime-silica glass (after Weiderhorn (1964), with permission). (b) Crack growth data on soda-lime-silica glass collected by the double-torsion technique in alcohols. With the exception of octyl alcohol, all alcohols were fully saturated. The octyl alcohol was ~50% saturated (after Wiederhorn *et al.* (1982), with permission.)

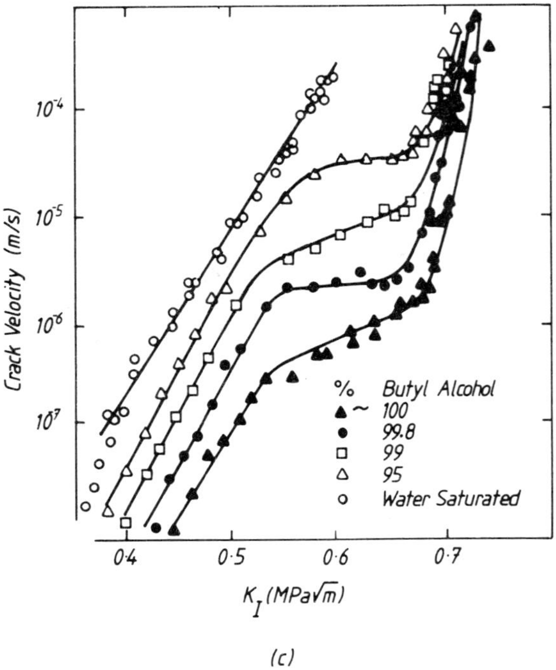

Fig. 6.13(c) – Crack propagation in octyl alcohol containing water. The concentration of water is expressed in volume % (after Wiederhorn *et al.* (1982), with permission).

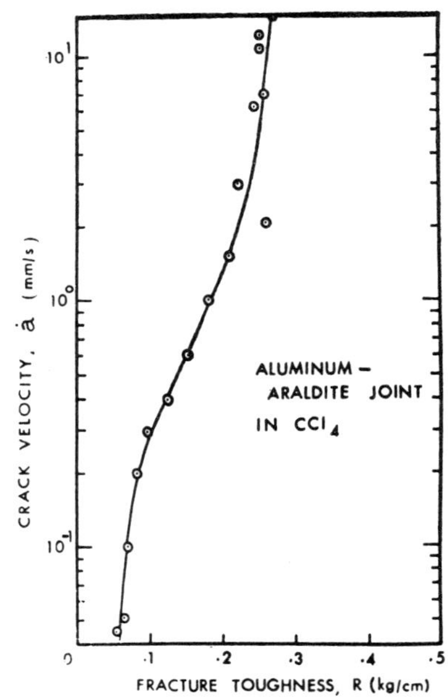

Fig. 6.14(a) – R–$\dot{a}$ relation for an aluminium/Araldite joint in carbon tetrachloride (after Mai (1975), with permission).

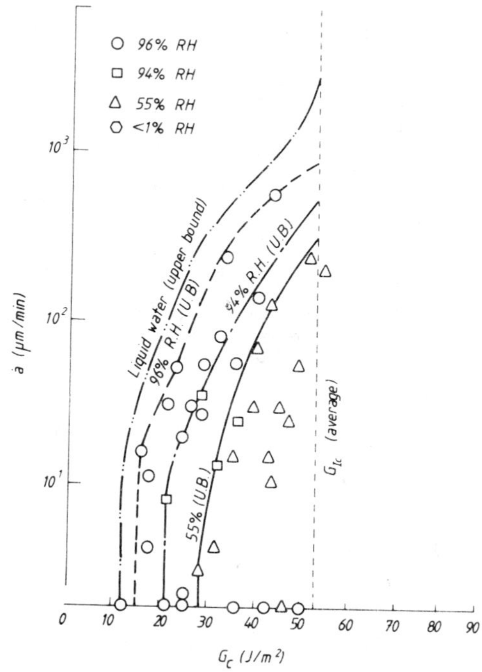

Fig. 6.14(b) – Crack growth rate ($\dot{a}$) of 12.5/270 adhesive joints in liquid water as a function of R.

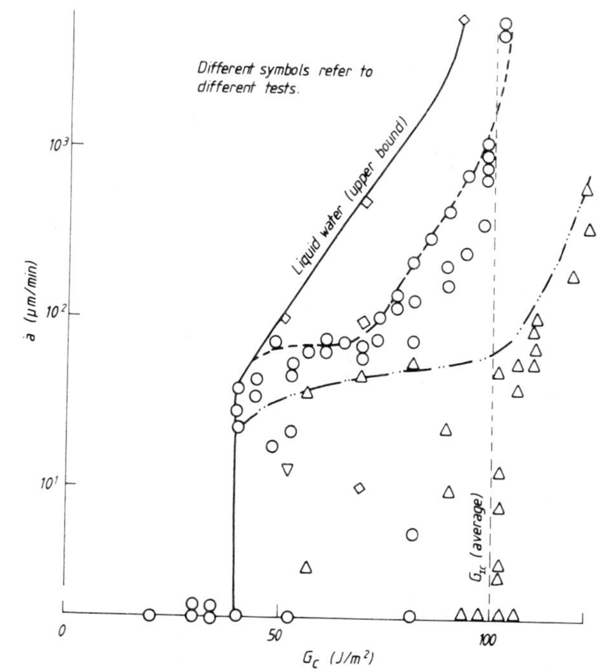

Fig. 6.14(c) – Crack growth rate ($\dot{a}$) of 10/108 adhesive joints in various relative humidities as a function of R. A/B stands for A pph of TEPA, B for post cure temperature in degrees Fahrenheit (after Mostovoy and Ripling (1969), with permission).

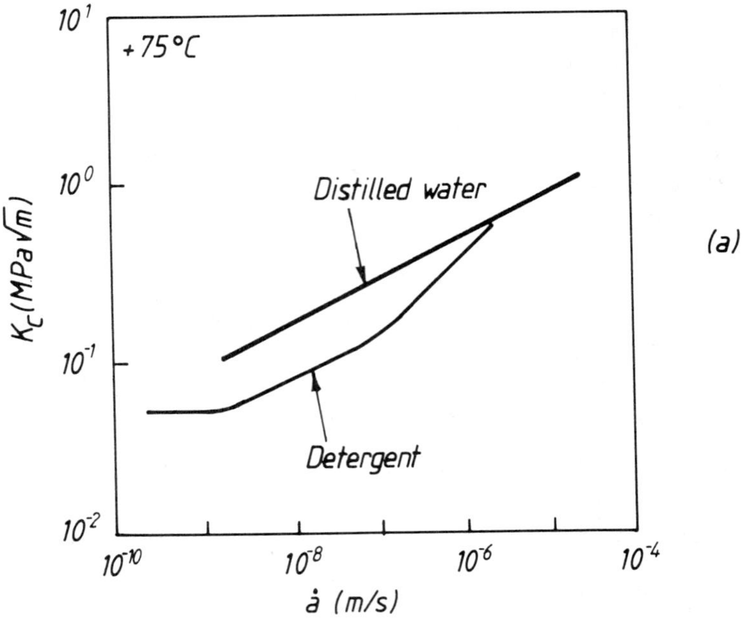

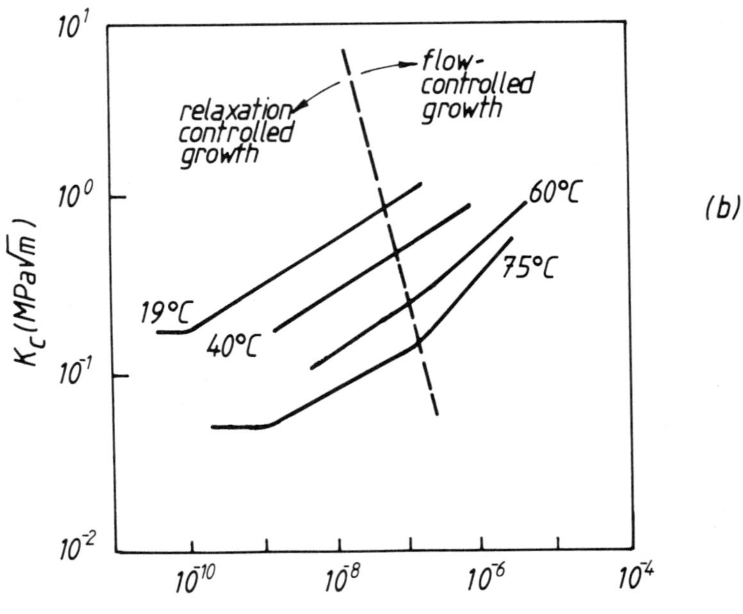

Fig. 6.15(a) – K_c–$\dot{a}$ curves for HDPE in two different environments at +75°C (after Chan and Williams (1983), with permission). (b) K_c–$\dot{a}$ curves for HDPE in Comprox 2% at different tempratures (after Chan and Williams (1983), with permission).

environment-dependent, but *not* crack speed-dependent. When, as in polymers, it is relaxation that causes slow crack growth and hence these K_c–$\dot{a}$ relationships, K_{IC} at the highest crack speed may be thought of as the rate-independent *unrelaxed* value of critical stress intensity factor. It corresponds to the isothermal-adiabatic transition point in the K_c–$\dot{a}$ plot (see Chapter 8). All other K_c are less than K_{IC} owing to creep and relaxation. Likewise there will be an R corresponding with K_{IC}^2/E^* where E^* is the unrelaxed or dynamic modulus, all other R are velocity-dependent owing to creep and relaxation effects (cf. the results in Fig. 6.3).

In Chapter 8 it will be explained that, in SEN testpieces of PMMA, some *stable* crack growth occurs in the early states of propagation, even though the testpiece has a positive geometric stability factor and therefore should display unstable behaviour. What is observed experimentally is bound up with slow crack growth (in which $dK_c/d\dot{a}$ is positive) before K_{IC} is reached (Fig. 6.16). In this specimen, K_{IC} then coincides with instability; in monotonic tests using stable testpieces, K_{IC} would apply for controlled continuous cracking at a crack speed which would be sufficiently fast to assure unrelaxed behaviour, but slow enough to remain in the quasi-static range. Depending on the degree of viscoelasticity being displayed at the test temperature, it may not always be possible for the crack velocity to satisfy both requirements, in which case the highest K_c determined in controlled quasi-static *stable* tests may not reach K_{IC} observed in *unstable* tests.

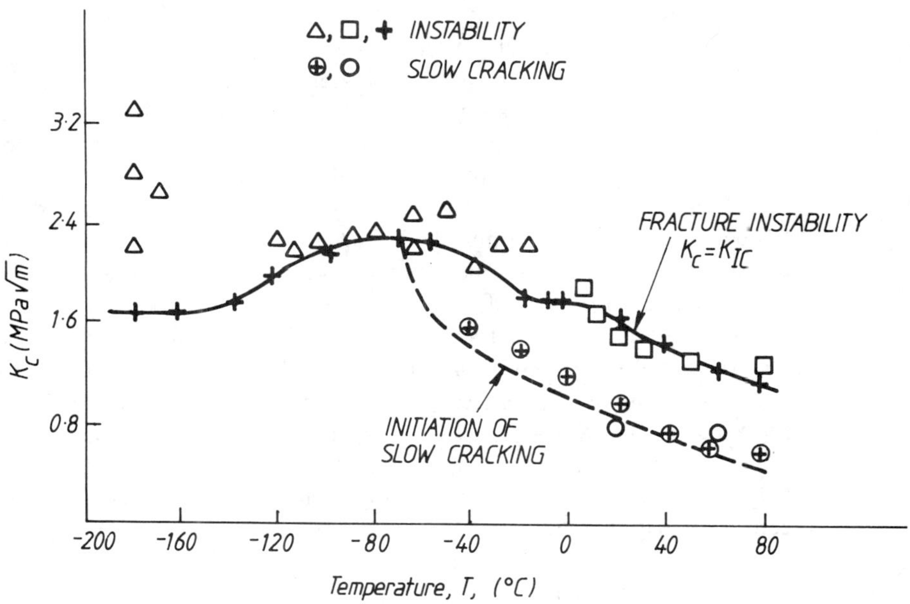

Fig. 6.16 – Variation of K_c values as a function of temperature for PMMA. + SEN instability; ▲ cleavage instability; rest of data are slow crack growth at different velocities (after Marshall *et al.* (1974), with permission.)

6.1.2 Expected effects of rate, temperature and environment

Before delving into the detail of K_c ($\dot{a}$,T, environment) and R ($\dot{a}$,T, environment) results, the reasons for them and applications of them, the following explanatory notes should help the reader understand the differences between the time-independent analyses of Chapters 2–4 and the results contained in this chapter, particulary in connexion with slow crack growth, 'monotonic' versus 'static' tests and so on. While the following discussion anticipates topics which are dealt with in detail in later sections of this chapter, we feel it is essential to make this introduction at this stage.

(a) *Rate effects*

When monotonic tests are performed at varying crosshead speeds to determine R for polymers using the Gurney method, and R is found to be a function of crack velocity (e.g. Figs. 6.3(a) and (b)), this is in fact a reflexion of slow growth. Conversion of R into K_c does not give K_{IC}. K_{IC} represents an unrelaxed value independent of crosshead velocity as shown by Marshall *et al.* (1974) for PMMA and other polymers (Fig. 6.15). The *same* R ($\dot{a}$) or K_c ($\dot{a}$) results are given by both monotonic and static load tests.

In glasses and ceramics at low homologous temperatures, K_{IC} is generally independent of crosshead speed. But tests for glasses in moist air or liquid water using Gurney's technique give a wide range of R values varying with crack velocity, which again result from slow crack growth caused by environment (relative humidity of air, for example) (see Mai and Gurney, 1975). Also, tests for ceramics such as Si_3N_4, Fig. 6.4(b), at elevated temperatures show that K_{IC} is a function of crosshead speed of the testing machine. This may be regarded as a reflexion of the relaxation effect at this high temperature. There is considerable slow-crack growth before fracture instability and its magnitude decreases as the loading rate is increased. It must be noted that this behaviour is different from that observed in brittle polymers such as PMMA which shows that K_{IC} is independent of crosshead speed (see Marshall *et al.*, 1974).

Many metals do not display time-dependent effects below half-homologous temperatures and R determined by Gurney's method is then independent of crosshead speed in monotonic tests: it *does* now convert to K_{IC} measured by other methods. However, if very high rates (over and above those available in the usual laboratory testing machine) are used, then dynamic effects can cause K_{IC} (now called K_{Id}) to be different from the quasi-static value. For ductile steels, for example, K_{Id} increases (depending on the yield strength level), but for $\sigma_y > 1$ GPa, $K_{Id} = K_{IC}$ (Fig. 6.2). Slow crack growth is not expected in steels at low homologous temperatures unless there is environmental assistance (Fig. 6.17); it will occur at sufficiently high temperatures when creep effects come into play (Fig. 6.9).

(b) *Temperature effects*

K_{IC} is definitely a function of temperature for all types of materials. Again, it is argued that K_{IC} is velocity-independent at a given temperature and environment, and that R or K_c variations with (i) stationary crack tip strain rate during

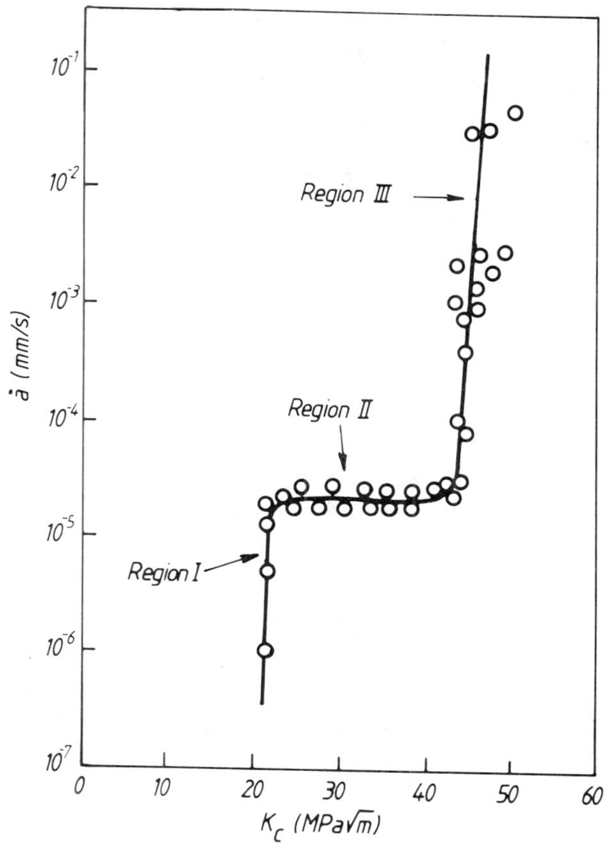

Fig. 6.17 – Effect of applied stress intensity factor (K_c) upon crack velocity ($\dot{a}$) for high strength quenched and tempered steel (AFC 77) in distilled water. (After Speidel, (1975); reproduced from Parkins (1975); reprinted by permission of the Council of the Institution of Mechanical Engineers from the *Journal of Strain Analysis.*)

initiation, or (ii) moving crack tip strain rate or crack velocity during propagation (if they occur in the quasi-static range for some types of solid) reflect slow crack growth due to relaxation phenomena. Similar $R(\dot{a},env)$ and $K_c(\dot{a},env)$ results at a given temperature are shown by both monotonic and static load tests. $K_c(\dot{a})$ values do not coincide with K_{IC} unless relaxation and creep are absent. This occurs when $\dot{a} \rightarrow$ '∞', $K_c \rightarrow K_{IC}$.

(c) *Environmental effects*

Both static load and monotonic load tests should give the same $R(\dot{a})$ or $K_c(\dot{a})$ data for slow crack growth.

Should the environmental species fail to keep up with the moving crack tip, K_c values similar to that in air will control crack propagation independent of the environment. This occurs with glasses and ceramics in moist air in the highest $\dot{a}$ part of the three-region K_c diagram (called region III in Fig. 6.12). It is in region I where the effect of environment is most keenly felt.

If, as with some polymers, the crack tip is severely crazed by the environment K_{IC} will be larger than K_{IC} in air (e.g. PMMA in organic solvents). In this case K_{IC} may also depend on crosshead speed since the amount of crazing formed at the tip can vary with crack velocity (see Mai, 1975).

In liquid–metal embrittlement K_{IC} is definitely less than that in air. This is due to the cohesive bond strength reduction at the strained crack tip. In polymers, if a solvent causes a dissolution (i.e. a chemical effect) and not crazing (i.e. a physical effect), it is possible that K_{IC} can also be reduced when compared to K_{IC} in air.

In brief, one may say that environmental effects usually cause slow crack growth; K_{IC} may or may not be affected.

Environmental effects on cracking under *cyclic loads* also occur. This is discussed fully in Chapter 7, where the apparent similarities between (static) delayed fracture and (cyclic) fatigue behaviour, and between stress corrosion cracking and stress corrosion fatigue are highlighted.

6.2 CRACK VELOCITY

Expressions for the relation between crack velocity $\dot{a}$ and crosshead velocity $\dot{u}$ have been given in section 3.1. Example 3.1 discussed how the useful range of $\dot{a}$, which permitted visual tracking of crack propagation in DCB testpieces, could be established along with the limiting upper values for $\dot{u}$, using equation (3.2)

$$\frac{\dot{A}}{\dot{u}} = \frac{B\dot{a}}{\dot{u}} = \frac{X}{2R + u\mathrm{d}X/\mathrm{d}A} \qquad (3.2\ \text{bis})$$

It was shown in section 3.3 that $\dot{a}$ could be orders of magnitude greater than $\dot{u}$ (e.g. in standard CT steel specimens with constant $R, \dot{a} \approx 100\dot{u}$) so that care is clearly required in the choice and design of test piece and experimental conditions when attempts are made to determine how R, K_c, δ_C and so on vary with $\dot{a}$. Low $\dot{a}$ can be achieved on displacement-controlled testing machines simply by reducing $\dot{u}$. In soft machines, relation (3.2) still applies but the control is now of X, and unless the specimen/geometry combination is such as to demand a positive $\mathrm{d}X/\mathrm{d}A$ during cracking (Chapter 8), control of cracking can be difficult. If X is constant, as in 'static load' tests, $\dot{a}$ will be related to the movement of the point of application of X (i.e. $\dot{u}$) by $\dot{a} = \dot{u}X/2RB$ from equation (3.2), providing that the propagation is stable.

Example 6.1 An Outwater testpiece of glass has the following dimensions: $W = 30$ mm, $W_m = 10$ mm, $B = 2$ mm (see crack geometry number 20 in Appendix 1). For the conditions of an experiment in moist air it is known that $\dot{a} = 7.3 \times 10^{-6} \exp(1.65R)$ where $\dot{a}$ is in mm/s and R is in J/m^2. Under a static load of 19 N, what will be $\dot{a}$ and $\dot{u}$?

Ans. Outwater testpieces crack at constant load, the expression for which is given at number 20 in Appendix 1. For $X = 19$ N, and the given proportions of the specimen, we find $R \approx 4$ J/m^2. Hence $\dot{a} = 4.3$ nm/s, and using equation (3.2) with X constant, $\dot{u} = 3.5$ pm/s. Thus the static load moves at about (1/1,000)th the speed of the crack (and is very slow indeed).

6.2.1 Crack tip strain rate

The strain ϵ perpendicular to the direction of a stationary starter crack is given approximately by $\sigma_{yy}/E = K/(E\sqrt{2\pi r})$. As a cracked body is loaded, the applied K increases according to $K = Y\,\sigma\sqrt{\pi a}$, where σ is the applied stress (not the local crack tip stress which is σ_{yy}). Thus, even in a *stationary* crack before propagation occurs, there will be a strain rate $\dot{\epsilon}_s$ in the crack tip zone (at fixed r) of magnitude

$$\dot{\epsilon}_s = \mathrm{d}/\mathrm{d}t\ (Y\sigma\sqrt{\pi a}/E\sqrt{2\pi r}) = \epsilon\left(\frac{\dot{K}}{K} - \frac{\dot{E}}{E}\right) \approx \epsilon\left(\frac{\dot{K}}{K}\right) = \epsilon\left(\frac{\dot{\sigma}}{\sigma}\right) \tag{6.1}$$

since for a fixed crack length $\dot{K} = Y\,\dot{\sigma}\sqrt{\pi a}$ and $\dot{E}$ can be neglected for a stationary crack. For example, in ASTM E399-81 for K_{IC} testing, a stress intensity rate $\dot{K}$ in the range $0.55 \sim 2.75$ MPa$\sqrt{\text{m}}$/s is specified. Let K rise to 50 MPa$\sqrt{\text{m}}$, say, before fracture occurs in a given testpiece. Therefore, at the elastoplastic boundary of the crack tip zone where $\epsilon \approx \epsilon_y = 0.002$ say, $\dot{\epsilon}_s \approx (0.002)\ \{(0.55 \sim 2.75)/50\} \approx (2 \sim 10)10^{-5}\,\text{s}^{-1}$.

When crack propagation ensues, there will be two additional contributions to the strain rate at some fixed point in the body as the crack front advances towards that point and eventually subsumes it: (i) owing to the $\dot{a}$ component of $\dot{K}$ and (ii) owing to the $\dot{r}$ contribution. Then we have for the *moving* crack the strain rate ($\dot{\epsilon}_m$) given by

$$\dot{\epsilon}_m = \epsilon\left(\frac{\dot{K}}{K} - \frac{\dot{E}}{E}\right) - \pi\epsilon^3\left(\frac{E}{K}\right)^2\dot{r} \tag{6.2}$$

But $\dot{r} = -\dot{a}$ (the distance between the moving crack front and the fixed point ahead of it decreases as the crack length increases), and now

$$\dot{K} = (Y\sqrt{\pi a})\,\dot{\sigma} + \sigma\,[\mathrm{d}/\mathrm{d}a\,(Y\sqrt{\pi a})]\ \dot{a}$$

Hence

$$\dot{\epsilon}_m = \epsilon\left(\frac{\dot{\sigma}}{\sigma} - \frac{\dot{E}}{E}\right) + \left(\frac{\epsilon}{Y\sqrt{\pi a}}\,\frac{\mathrm{d}}{\mathrm{d}a}\,(Y\sqrt{\pi a}) + \pi\left[\frac{E}{K}\right]^2\epsilon^3\right)\dot{a} \tag{6.3}$$

Calculations show that only the second term with the second bracket in this expression is significant for a moving crack so, using $\epsilon = \epsilon_y$ at the crack tip zone.

$$\dot{\epsilon}_m \approx \pi\left(\frac{E}{K_c}\right)^2\epsilon_y^3\,\dot{a}$$

$$\approx \pi\,(E/R)\ \epsilon_y^3\ \dot{a} \tag{6.4}$$

For example, taking a CT specimen of an aerospace metal for which $E/K_c \cong 3 \times 10^3 \sqrt{m}$ (i.e. $E = 210$ GPa and $K_c = \text{MPa}\sqrt{m}$), it was shown in section 3.3 that $\dot{a}/\dot{u} \approx 118$ for $a/W = 0.5$, so with $\dot{u} \approx 2.5 \times 10^{-5}$ m/s (equivalent to the loading rates mentioned above) we find $\dot{a} \approx 2.95$ mm/s so $\dot{\epsilon}_m \cong \pi(3 \times 10^3)^2 \times (0.002)^3 \times (2.95 \times 10^{-3}) = 66.7 \times 10^{-5}\,\text{s}^{-1}$ which should be contrasted with $\dot{\epsilon}_s \approx (2 \sim 10)10^{-5}\,\text{s}^{-1}$. The strain rate at the tip of a moving crack is thus at least an order of magnitude greater than that of the stationary crack loaded at the same rate.

Notice that the mathematics imply a discontinuity in $\dot{\epsilon}$ when the cracks starts to move. This is obviously bound up in the omission of crack accelerations from rest – which nevertheless are often rapid. Indeed, in unstable crack propagation (discussed in detail in Chapter 8) the specimen breaks 'with a bang' and the strain rate is very high. It is difficult in such cases to measure $\dot{a}$ in the early stages of crack growth when experiments are being performed to investigate rate-dependence in K_c or R. However, Williams and co-workers (Williams *et al.*, 1968; Marshall *et al.*, 1973) have successfully employed the concept of an *apparent crack speed* $\dot{a}_{app}$ under these circumstances, arguing that there must really be continuity in $\dot{\epsilon}$ in the transition from a stationary to a moving crack. They write at the transition, using the significant terms in $\dot{\epsilon}_s$ and $\dot{\epsilon}_m$,

$$\epsilon_y\,(\dot{\sigma}/\sigma) = \pi(E/K_c)^2\,\epsilon_y^3\,\dot{a}_{app}$$

Thus, writing t_f for the time to reach the cracking load

$$\dot{a}_{app} = \frac{2r_y}{t_f} \tag{6.5}$$

since $\dot{\sigma}/\sigma \approx t_f$ and $\epsilon_y = K_c/E\sqrt{2\pi r_y}$.
We have therefore

$$\dot{a}_{app} = \frac{1}{\pi t_f}\left(\frac{K_c}{\sigma_y}\right)^2 \tag{6.6a}$$

for plane stress, and

$$\dot{a}_{app} = \frac{1}{2\sqrt{2}\,\pi t_f}\left(\frac{K_{IC}}{\sigma_y}\right)^2 \approx \frac{1}{3\pi t_f}\left(\frac{K_{IC}}{\sigma_y}\right)^2 \tag{6.6b}$$

for plane strain, where σ_y is the yield strength. Good agreement with crack speeds measured by high-speed photography have been reported for PMMA and PS by Williams and his co-workers (Marshall *et al.*, 1973; 1974).

Example 6.2 Irwin (*Appl. Mater. Research,* **3**, 1964, p. 65) gives an expression for crack tip strain rate that has a coefficient of 2 in front of the $\dot{K}/K$ term in equation (6.2). Explain.

Ans. Irwin used the plastic zone size correction factor (embedded crack tip model in Example 2.9, see Fig. 6.18) on the crack length a, whereas in the Williams analysis leading to equation (6.2), r is measured from the actual crack tip which is relevant for cracking in many glassy plastics. In Fig. 6.18 we have $\sigma_y = K/\sqrt{2\pi r}$ where r is measured (in the Irwin treatment) from within the crack tip zone, ahead of the actual crack tip. Now $\dot{\sigma}_y = (\dot{K}/K - \dot{r}/2r)\sigma_y$. For negligible time-dependence of σ_y it follows that $\dot{K}/K = \dot{r}_y/2r_y$ at $r = r_y$, the radius of the crack tip zone. Distance r from the embedded crack tip is related to distance from the actual tip by $r = x - (r_y + a)$ where a is the real crack length, so $\dot{r} = -(\dot{r}_y + \dot{a})$. Thus at $r = r_y$, $\dot{\sigma}_y/\sigma_y = (\dot{K}/K + \dot{a}/2r_y + \dot{r}_y/2r_y) = \dot{S}/\sigma_y$ which is an expression for the characteristic stress rate $\dot{S}$ in the fracture process zone. Replacing $\dot{r}_y/2r_y$ in terms of $\dot{K}/K$ gives $\dot{S}/\sigma_y = (2\dot{K}/K + \dot{a}/2r_y)$ But $\dot{S}/\sigma_y =$ $(\dot{E}/E + \dot{\epsilon}/\epsilon_y)$. Consequently $\dot{\epsilon} = \epsilon_y\,(2\,\dot{K}/K - \dot{E}/E) + \dfrac{\epsilon_y \dot{a}}{2r_y}$ and since $1/2r_y =$ $\pi\,(E/K)^2\,\epsilon_y^2$, $\dot{\epsilon} = \epsilon_y\,(2\dot{K}/K - \dot{E}/E) - \pi\epsilon^3\,(E/K)^2\dot{r}$ which is the same as equation (6.2) but with a factor of 2 in front of $\dot{K}/K$. The factor 2 also appears in equation (6.5), so that the relation becomes $\dot{a}_{app} = (4r_y/t_f)$ with consequent doubling of $\dot{a}_{app}$ in equations (6.6(a)) and (6.6(b)).

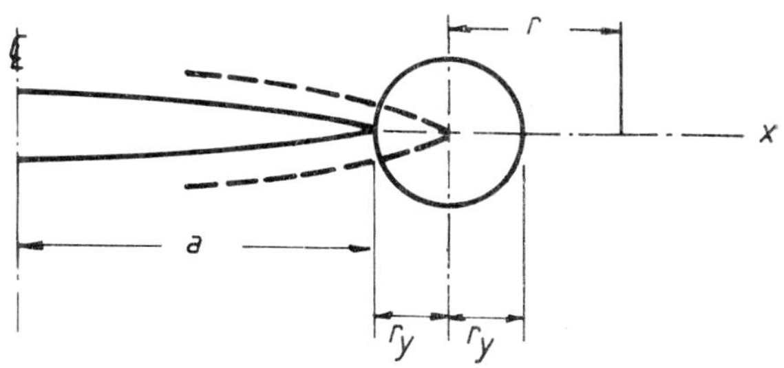

Fig. 6.18 – Crack tip plastic zone geometry and the embedded crack tip model.

Note that in solids with rate-dependent mechanical properties E, K_c and ϵ_y in equation (6.4) may all change with speed of deformation (possibly in different ways). Information given later in this chapter will permit these dependencies to be evaluated.

6.2.2 Local heating at moving crack tip

In measurements to determine the variation of toughness parameters with temperature, the whole testpiece is heated or cooled to different temperatures and standard K_{IC}, R, δ_c etc. tests performed. However, over and above changes in the bulk temperature of the specimen, it is possible for the toughness work performed in the process zone during propagation to heat up the crack tip region. The local increase in temperature ΔT depends of course on the thermal conductivity of the testpiece material, the size of the crack tip zone and the level of irreversible work dissipated. As a guide to the reader, calculations given later show that for quasi-static cracking in metals ($\dot{a} < 10$ m/s) $\Delta T \approx$ a few degrees Kelvin, but for tough polymers with low thermal conductivities $\Delta T \approx$ tens of degrees Kelvin. During dynamic cracking, ΔT can be appreciably larger. For example, Fuller *et al.* (1975) using an infrared detector, found $\Delta T \approx 450$ K in a 580 m/s crack running in PMMA and calculations by Weichart and Schönert (1978) predict $\Delta T > 1{,}000$ K in glass when $\dot{a} \sim 1{,}000$ m/s.

In quasi-static testing, ΔT may often be ignored in comparison with the bulk temperature of the specimen. However, if toughness is temperature-and rate-dependent, and in particular if toughness falls with increasing temperature, it is possible for crack instabilities to occur in quasi-static tests, whereby the local heating in the path of the crack produces $\mathrm{d}R/\mathrm{d}A$-negative effects (section 8.6.3). Alternatively, if the temperature rise is close to the glass transition or softening temperature in polymers, the zone around the crack tip will become locally very ductile and possibly bigger in size . In these circumstances R, instead of decreasing, will increase owing to additional energy required to initiate fracture.

The heating effects at the moving crack tip can be determined by considering an energy balance for the simple case of a circular plastic zone ($2r_\mathrm{y}$) at the crack tip (Williams, 1965). It is assumed that the temperature increase above ambient, $\Delta T = T - T_0$, is equal to the difference between the irreversible work of heating in the dissipative circular zone and the heat conducted across the control boundary. Thus, we have

$$(\pi r_\mathrm{y}^2)\ \bar{\rho}\ s\ \frac{\mathrm{d}\Delta T}{\mathrm{d}t} = \pi r_\mathrm{y}^2\ \sigma_\mathrm{y}\dot{\epsilon} - (2\pi r_\mathrm{y})\ \frac{k\Delta T}{r_\mathrm{y}} \tag{6.7}$$

where $\bar{\rho}$ is density, s is specific heat and k is the thermal conductivity. The crack tip strain rate ($\dot{\epsilon}$) is related to the apparent crack velocity ($\dot{a}$) by equation (6.4). Simplifying equation (6.7), the governing differential equation becomes

$$\frac{\mathrm{d}\Delta T}{\mathrm{d}t} + \frac{2k\Delta T}{\bar{\rho} s r_\mathrm{y}^2} - \frac{\sigma_\mathrm{y}\dot{\epsilon}}{\bar{\rho} s} = 0 \tag{6.8}$$

with solution

$$\Delta T = \frac{\sigma_\mathrm{y}\dot{\epsilon}}{2k}\ r_\mathrm{y}^2 \left(1 - \exp\left[-\frac{2kt}{\bar{\rho} s r_\mathrm{y}^2}\right]\right) \tag{6.9}$$

The crack tip temperature rise may be written in terms of the apparent crack speed in place of strain rate, using $\dot{\epsilon} = \pi\epsilon_\mathrm{y}^2\,\dot{a}/\delta_\mathrm{c}$, equation (6.4) and $R = E\epsilon_\mathrm{y}\delta_\mathrm{c}$. We obtain

$$\Delta T = \left(\frac{\sigma_\mathrm{y}\pi r_\mathrm{y}^2\epsilon_\mathrm{y}^2}{2k\delta_\mathrm{c}}\right)\left(1 - \exp\left[-\frac{2kt}{\bar{\rho} s r_\mathrm{y}^2}\right]\right)\dot{a} \tag{6.10}$$

Example 6.3 Determine the temperature rise at the tip of a running crack in an aluminium alloy for which $\sigma_y = 550$ MPa, $K_{IC} = 27$ MPa$\sqrt{m}$, $\bar{\rho} = 2.82 \times 10^3$ kg/m^3, k = 150 Wm^{-1} °C^{-1}, s = 0.965 kJ/kg at ambient temperature and E = 70 GPa.

Ans.

$$\epsilon_y = \sigma_y/E = 550/70{,}000 = 7.86 \times 10^{-3}$$

$$\delta_c = \frac{K_{IC}^2}{E\sigma_y} = \frac{27 \times 27}{70 \times 550} \text{ (mm)} = 0.019 \text{ mm}$$

and

$$r_y = \frac{1}{6\pi}\left(\frac{K_{IC}}{\sigma_y}\right)^2 = \frac{27 \times 27 \times 10^3}{6\pi \times 550 \times 550} = 0.13 \text{ mm}$$

Thus equation (6.10) is reduced to

$$\Delta T = 0.32\ (1 - \exp(-6523t))\ \dot{a}$$

The second term inside the bracket is negligible compared with unity when t takes values of typical tests. Then, for $\dot{a}$ within the quasi-static range, (say $\dot{a} < 10$ m/s, Chapter 4), ΔT at the crack tip region is only a few degrees at most.

Neglect of the exponential term in relation (6.10) is tantamount to assuming that the crack tip heating occurs under adiabatic conditions, with $k = 0$ in equations (6.7) and (6.8). Then $\Delta T = \sigma_y \epsilon_y / \bar{\rho} s$. An alternative estimate for ΔT under adiabatic conditions, based on a Dugdale strip yield crack tip zone rather than a circular patch, is given in section 8.6.3. Example 6.3 shows that K_{IC} is not expected to vary with crack velocity ($\dot{a}$) for most metals and alloys inside the quasi-static crack velocity regime. However, for dynamic crack propagation, with $\dot{a}$ of the order of 10^3 m/s, ΔT is substantial, i.e. of the order of 320°C, and K_{Id} may be affected.

For polymers, which are more rate- and temperature-sensitive than metals and alloys, σ_y (which now for glassy amorphous polymers becomes the crazing stress σ_c) in the governing differential relation (6.8) is both temperature- and rate-dependent. For PMMA, a reasonable representation is (Williams, 1965):

$$\sigma_c = \epsilon_y\, E \left[1 - \left(\frac{\Delta T}{\Delta T^*}\right)\right] \tag{6.11}$$

where E is the modulus at the appropriate rate and temperature T_0 and ΔT^* is the temperature rise above T_0 (i.e. $\Delta T^* = T^* - T_0$) for a 'zero' modulus. For PMMA $T^* \cong T_g \simeq 115$°C, the glass transition temperature. Solving equations (6.8) and (6.11) we obtain

$$\Delta T = \left(\frac{T_1}{1 + T_1/\Delta T^*}\right)\left(1 - \exp\left\{-\left[1 + \frac{T_1}{\Delta T^*}\right]\frac{2kt}{\bar{\rho} s r_y^2}\right\}\right) \tag{6.12}$$

where

$$T_1 = \frac{\dot{\epsilon}\epsilon_y E r_y^2}{2k} = \left(\frac{\pi E \epsilon_y^3 r_y^2}{2k\delta_c}\right)\dot{a}$$

Appropriate values for PMMA are: $\delta_c = 1.6\ \mu$m, $\epsilon_y = 0.06$, $s = 1.47$ kJ/kg at room temperature (assuming that this is independent of temperature at present), $k = 0.193$ W/m°C, $\bar{\rho} = 1.2 \times 10^3$ kg/m^3 and $r_y = (1/6\pi)(K_{IC}/\sigma_c)^2 = 12 \times 10^{-3}$ mm (taking $K_{IC} = 0.90$ MPa$\sqrt{\text{m}}$ and the crazing stress $\sigma_c = 58$ MPa). Calculations show that $2k/\bar{\rho}r_y^2 \cong 1.52 \times 10^3$, so it is therefore justified to ignore the exponential term once more, which simplifies the expression for the temperature rise ΔT to:

$$\Delta T = T_1/[1 + T_1/\Delta T^*] \tag{6.13}$$

For $T_0 = 20$°C, $\Delta T^* = 95$°C and using an expression for the rate-dependent E given by $E = 689\ (0.95 + 3.98\ \dot{\epsilon}^{0.109})$ MPa, we obtain

$$T_1 = \frac{\pi\epsilon_y^3 r_y^2}{2k\delta_c}\ (689 \times 10^6\ (0.95 + 3.98\dot{\epsilon}^{0.109}))\ \dot{a}$$

i.e. $T_1 = 108\ (0.95 + 10.45\dot{a}^{0.109})\dot{a}$.

The critical stress intensity factor K_c and fracture toughness R or G_c as functions of $\dot{a}$ and T are then given by:

$$K_c = (\epsilon_y\delta_c)^{\frac{1}{2}}\ E\left(1 - \frac{\Delta T}{\Delta T^*}\right)^{\frac{1}{2}} \tag{6.14}$$

i.e.

$$K_c = 0.213\ (0.95 + 10.45\dot{a}^{0.109})\left(1 - \frac{\Delta T}{95}\right)^{\frac{1}{2}}$$

and

$$G_c = R = \epsilon_y E\delta_c\left(1 - \frac{\Delta T}{\Delta T^*}\right)$$

i.e.

$$G_c = R = 66\left(1 - \frac{\Delta T}{\Delta T^*}\right)(0.95 + 10.45\dot{a}^{0.109}) \tag{6.15}$$

Table 6.1 shows how both K_c and G_c vary with $\dot{a}$ for PMMA. It is of interest to note that ΔT increases with $\dot{a}$, and above 10^{-2} m/s the temperature rise is appreciably larger. At low crack speeds ($< 10^{-2}$ m/s) the heat generated has enough time to be conducted away so that an isothermal condition is retained. However, at higher crack speeds ($> 10^{-2}$ m/s) there is less time for heat conduction (and the polymer is a bad conductor) so that there is a quasi-adiabatic rise. The local softening of crack tip material leads to a drop in E and K_c, and fracture instabilities may result (section 8.6.3), cf. in Table 6.1, $(dR/d\dot{a})$ is negative for $\dot{a} \geqslant 10^{-1}$ m/s.

Table 6.1
Variation of K_c, G_c and ΔT with $\dot{a}$

$\dot{a}$ (m/s)	ΔT (°C)	K_c (MPa$\sqrt{m}$)	R (J/m^2)
10^{-3}	9.63	1.19	384
10^{-2}	7.30	1.49	443
10^{-1}	48	1.36	300
10^{0}	88	0.66	56
10	94	0.32	10

6.3 RELAXATION AND CREEP EFFECTS ON TOUGHNESS

It is well-known that, phenomenologically, viscoelastic time-dependent flow in polymers, and creep flow in metals, are similar although, of course, the microstructural causes are different. Indeed similar spring-and-dashpot models simulate this sort of behaviour (see Caddell, 1980). It might be expected that time-dependent fracture in the two cases could also be similarly described from a mechanics point of view.

Consider first the variation of Young's modulus E with strain rate and temperature, such as in the case of polymers displaying viscoelasticity. Since

$$K_c^2 = E^*R \approx E^*\sigma_y\delta_c = (E^*)^2\epsilon_y\delta_c \tag{2.34 bis}$$

it is clear that K_c may vary with strain rate, crack velocity and temperature, *even if* R, ϵ_y or δ_c were *constant.* These latter may change too, of course, with rate and temperature (owing perhaps to microstructural factors), to give more complicated K_c–$\dot{a}$ or K_c–T relations.

The time-dependence of elastic modulus (when creep and relaxation are important) may be written

$$E = E_0\, t^{-n} \tag{6.16}$$

where E_0 is the unit modulus, t is time and n a constant related to tan δ, the loss factor of the solid which may be obtained from creep experiments. Similarly, the time-dependence of flow stress σ_y may be written

$$\sigma_y = \sigma_0\, t^{-m} \tag{6.17}$$

For many glassy polymers, $n \approx m \approx 0.1$; for many metals $m \approx 0.1 \sim 0.2$.

Temperature dependence is given by the Arrhenius equation which relates the time at temperature T to equivalent time t^* at the reference original ambient temperature, T_0. Thus,

$$t = t^* \exp\left\{-\frac{Q}{\bar{k}}\left(\frac{1}{T}-\frac{1}{T_0}\right)\right\}$$

where Q is the activation energy for the particular relaxation process and $\bar{k}$ is the Boltzmann constant. Thus (dropping the difference between plane strain and plane stress so that E represents E^*) we have

$$K_c = \sqrt{E^*R} = \sqrt{E\,\sigma_y\,\delta_c}$$

$$= \sqrt{E_0\,\sigma_0\,\delta_c}\quad t^{-(n+m)/2} \tag{6.19}$$

Experiment shows that ϵ_y is usually constant (particularly for polymers), independent of t and T, and since

$$E/\sigma_y = 1/\epsilon_y = (E_0/\sigma_0)\,t^{-n+m}$$

it follows that $n \approx m$ if ϵ_y is independent of t.

From equation (6.19), substituting $E\epsilon_y$ for σ_y,

$$K_c = E_0\sqrt{(\epsilon_y\delta_c)}\;t^{-n} \tag{6.20}$$

Using the Arrhenius time–temperature equivalence, equation (6.18), we obtain from equation (6.20), by substituting for t,

$$K_c = E_0\,(\epsilon_y\delta_c)^{\frac{1}{2}}\,(t^*)^{(-n)}\exp\left\{\frac{nQ}{\bar{k}}\left(\frac{1}{T}-\frac{1}{T_0}\right)\right\} \tag{6.21}$$

t^* is determined from the crack speed $\dot{a}$ and the length of the crack tip zone or craze zone Δ when δ reaches δ_c. Therefore, using the Dugdale model for the crack tip craze zone, we have

$$t^* = \frac{\Delta}{\dot{a}} = \frac{\pi}{8}\left(\frac{K_c}{\sigma_y}\right)^2\frac{1}{\dot{a}} = \frac{\pi}{8}\left(\frac{\delta_c}{\epsilon_y}\right)\frac{1}{\dot{a}} \tag{6.22}$$

Combining equations (6.21) and (6.22) gives the rate-temperature-dependence of K_c for polymeric materials as:

$$K_c = E_0\,(\epsilon_y\delta_c)^{\frac{1}{2}}\left(\frac{\pi}{8}\,\frac{\delta_c}{\epsilon_y}\right)^{-n}\dot{a}^{\,n}\exp\left\{\frac{nQ}{\bar{k}}\left[\frac{1}{T}-\frac{1}{T_0}\right]\right\} \tag{6.23}$$

Equation (6.23) predicts that, at constant temperature, log K_c should vary linearly with log $\dot{a}$, the slopes of the lines giving n. The variation of the intercepts with temperature leads to values for the activation energy. As an alternative, K_c should vary semi-logarithmically with $(1/T)$ at constant $\dot{a}$. Fig. 6.3(e) shows data for PMMA agreeing very well with equation (6.23). Within the temperature range shown, $n \approx 0.06$ and $Q \approx 84$ kJ/mole in good agreement with β-viscoelastic relaxation processes in PMMA.

We note that at constant temperature T_0 equation (6.23) reduces to

$$\dot{a} = \left(\frac{\pi}{8}\right)\left(\frac{\delta_c}{\epsilon_y}\right)(E_0^2\epsilon_y\delta_c)^{-1/2n}\,K_c^{\;1/n} \tag{6.24}$$

which may be expressed more simply as

$$\dot{a} = D K_c^{n^*} \tag{6.25}$$

where D and n^* are constant for a given temperature. Thus, the data for $\dot{a}$ vs. K_c in Fig. 6.9 for Discaloy at 650°C obey equation (6.25) with $n^* \approx 10$ or $n = 0.1$. We shall show in section 6.4 that relation (6.25) is useful for describing K_c–$\dot{a}$ behaviour for a variety of materials (not only polymers for which equation (6.24), was derived) when environmental effects come into play; D and n^* may then change with environment and also change over different ranges of $\dot{a}$ and T.

Example 6.4 Derive an equation showing the K_c–$\dot{a}$ relationship for a brittle polymer (PMMA) if the strain-rate dependent modulus determined from creep tests at 20°C gives: $E = 689\,(0.95 + 3.98\dot{\epsilon}^{0.109})$ (MPa).

Ans. Since the crack tip is a highly strained region, the strain-rate dependent modulus is determined at a high strain, viz. $\epsilon = 0.035$, of the creep test (estimated by Williams, 1977). Now the criterion for fracture is

$$G_c = \sigma_y \delta_c = E\,\epsilon_y \delta_c \tag{a}$$

where δ_c is the critical crack tip opening displacement. Replacing G_c by K_c equation (a) becomes

$$K_c = \sqrt{\epsilon_y \delta_c}\, E \tag{b}$$

Now

$$E = 689(0.95 + 3.98\dot{\epsilon}^{\,0.109}) \tag{c}$$

and from equation (6.4) and $R = E\epsilon_y\delta_c$ the apparent crack velocity is given by

$$\dot{a} = \frac{\dot{\epsilon}\delta_c}{\pi\epsilon_y^2} \tag{d}$$

Combining equations (b), (c) and (d) we obtain

$$K_c = \sqrt{\epsilon_y\delta_c}\left\{689\,\left[0.95 + \left(\frac{\pi}{\delta_c}\right)^{0.109} \epsilon_y^{0.218}\; 3.98\,\dot{a}^{\,0.109}\right]\right\} \tag{e}$$

Since ϵ_y, δ_c are constants within the errors of experiments it can be said that

$$K_c \propto \dot{a}^{0.109}$$

in which form it fits with $\dot{a} = DK_c^{n^*}$.

Figs. 6.3(a), (b) show results obtained by Atkins *et al.* in 1975 on PMMA over a wide range of crack velocities and temperatures. It is of interest to note that while there are consistent increases in K_c and R as the temperature drops for all crack speeds, there are local inversions and peaks in the values and that these do not always coincide, since E^* in $K_c^2 = E^*R$ is itself time- and temperature-dependent and is playing a role. Lee's $R(\dot{a},T)$ results were analysed in a different way from the Williams approach. Using the ideas of thermally activated processes, a 'toughness-biased' Ree–Eyring rate relation well described the data, particularly in the higher crack speed and higher temperature ranges, i.e.

$$\dot{a} = A \exp\left(-\frac{(Q - \lambda R)}{kT}\right) \tag{6.26}$$

where $\dot{a}$ is the crack velocity and A and λ are some constants (Atkins *et al.*, 1975). $Q \approx 86$ kJ/mole from these experiments which is in excellent agreement with the value obtained by Marshall *et al.* (1974) using equation (6.23). λ has the dimensions of area, and from a graph of R/T vs $\ln \dot{a}$, Lee found $\lambda \cong 53.4$ m^3/mole $\approx 9 \times 10^{-23}$ m^2 per molecule, using Avagadro's number. From their studies of fracture in glass, Kies and Clark (1969) suggest that $\lambda = \beta a_0^2$ where β is the fraction of R actually used to break bonds, and a_0 is the bond or interchain spacing. $(1 - \beta)$ performs plastic work ('lattice trapping' in glass, according to Lawn *et al.*, 1980), or in this polymer case, craze work. $\beta \approx 0.02\%$ using $a_0 = 0.7$ nm, the interchain spacing in PMMA, so most work goes to crazing. Then $\beta R = 0.1$ J/m^2 for $R \cong 500$ J/m^2 and this is the order of magnitude of the surface free energy of PMMA as we should expect.

In the case of creep fracture of metals at high homologous temperatures, K_c and $\dot{a} = DK_c^{n^*}$ are no longer useful methods of analysis. Instead we note that the flow stress is more sensitive to rate effects than is the modulus, so in $J = R \approx \sigma_y \delta_c$, R will be time-dependent owing mainly to relaxation in σ_y. Again, owing to relaxation and the fact that the (X, u, R, A) coordinates of the fracture event are now themselves time-dependent, the governing work rate equation takes the form

$$X\dot{u} = \mathrm{d}\dot{\Lambda} + \dot{\overline{(J\mathrm{d}A)}} + \mathrm{d}\dot{\Gamma} \tag{1.4 bis}$$

and parameters such as stiffness and compliance are now time-dependent.

The time dependence of flow properties may, for example, be given by secondary (power law) creep, where

$$(\dot{\epsilon}/\dot{\epsilon}_0) = \alpha\,(\sigma/\sigma_0)^M \tag{6.27}$$

Now equation (4.36) for crack tip stresses in a J-dominated stress field, in a cracked body displaying non-linear (elastic) Xu behaviour, is based on the rate-independent Ramberg–Osgood or Ludwik expression relating stress and strain, viz.

$$(\epsilon/\epsilon_y) = \alpha\,(\sigma/\sigma_y)^N$$

It follows that a definition for a time-dependent J-integral is available if strain *rates* and displacement *rates* are used in equation (4.34) (Begley and Landes, 1972). That is,

$$J(t) = C^* = \oint (W^* \,\mathrm{d}y - T_i \frac{\partial \dot{u}_i}{\partial x} \,\mathrm{d}s) \tag{6.28}$$

where

$$W(t) = W^* = \int_0^{\dot{\epsilon}_{mn}} \sigma_{ij} \,\mathrm{d}\dot{\epsilon}_{ij} \tag{6.29}$$

C^* has units $\mathrm{J/m^2\,s}$ (or in practice $\mathrm{kJ/m^2\,h}$ for creep-resistant metal alloys).

By the same token that a critical J_c criterion is sometimes used for fracture initiation in rate-independent fracture studies (particularly in non-linear cases), it has been suggested that C^* should correlate with crack growth rates in creep fracture. Goldman and Hutchinson in 1975 showed, for example, that displacements of cracked bodies following relation (6.27) are propotional to C^*/X, and this suggests that in constant load tests, $\mathrm{d}a/\mathrm{d}t$ should be proportional to C^*. Note that all the admonitions about not allowing unloading which were raised in section 4.5 in relation to J, apply to C^* also. Thus correlations between $\mathrm{d}a/\mathrm{d}t$ and C^* assume that even though unloading of the material occurs on crack advance, thus invalidating stress–strain descriptions in terms of C^* just at the tip, C^* will still be characterising the creep cavitation conditions further ahead. A plot of $\mathrm{d}a/\mathrm{d}t$ against C^* is shown in Fig. 6.19 for data on 1 Cr–Mo–V steel and Discaloy collated from a number of sources by Harper and Ellison (1977) (cf. Fig. 6.9). C^* seems to correlate crack growth data from various specimen geometries, but as remarked by Tomkins (1981) more work is needed to demonstrate that C^* is a general correlating parameter for creep crack growth. The change in slope at lower growth rates in Fig. 6.19 is probably due to stress redistribution during early parts of the test, but a threshold for creep analogous to that for fatigue must exist (see Chapter 7). Such a true threshold would be

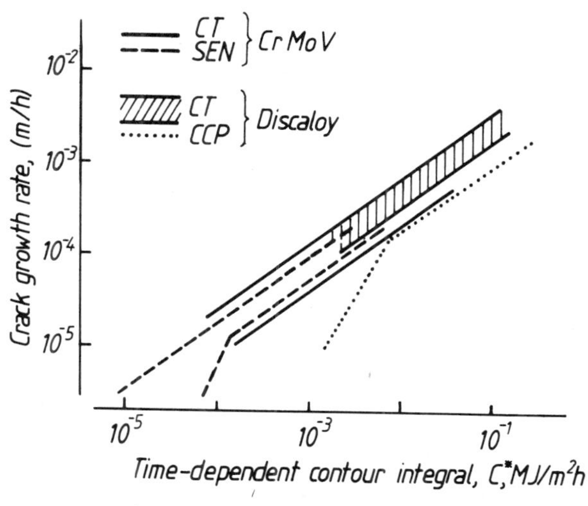

Fig. 6.19 – Summary of results of creep crack growth against C^* for 1 Cr–Mo–V steel at 565°C and Discaloy at 650°C. (After Harper and Ellison, 1977); reprinted by permission of the Council of the Institution of Mechanical Engineers from the *Journal of Strain Analysis*.)

related to either the development of creep fracture damage ahead of the crack or the achievement of a large enough crack tip opening increment to achieve void linkage. In either case, it is effectively a threshold determined by creep deformation.

In addition to analytical and numerical computations for J, methods of evaluating J based upon work areas U_T were given in Chapter 4. The following example shows how C^* may be evaluated by a method analogous to procedure (ii) in section 4.4.2, using areas U_T^* for energy rates.

Example 6.5 Creep fracture experiments were conducted by Landes and Begley (1974) in a Ni–Cr–Mo superalloy over a range of displacement rates ($\dot{u}$) at 1,200°F. After steady state creep crack growth was established, the following results were obtained.

Displacement rate ($\dot{u}$)	Load X	Crack length (a)
(mm/hr)	(MN)	(mm)
0.39	0.177	25
	0.162	30.5
	0.143	33
	0.126	35.6
	0.110	38.1
0.12	0.162	25
	0.143	30.5
	0.126	33
	0.110	35.6
	0.096	38.1
0.025	0.152	25
	0.140	30.5
	0.120	33
	0.105	35.6
	0.087	38.1
0.0033	0.146	25
	0.127	30.5
	0.105	33
	0.090	35.6
	0.070	38.1

Show how C^* varies with displacement rate ($\dot{u}$).

Ans. Using the raw data given it is possible to construct Fig. 6.20(a) showing the variation of load X with crack length u for different displacement rates,

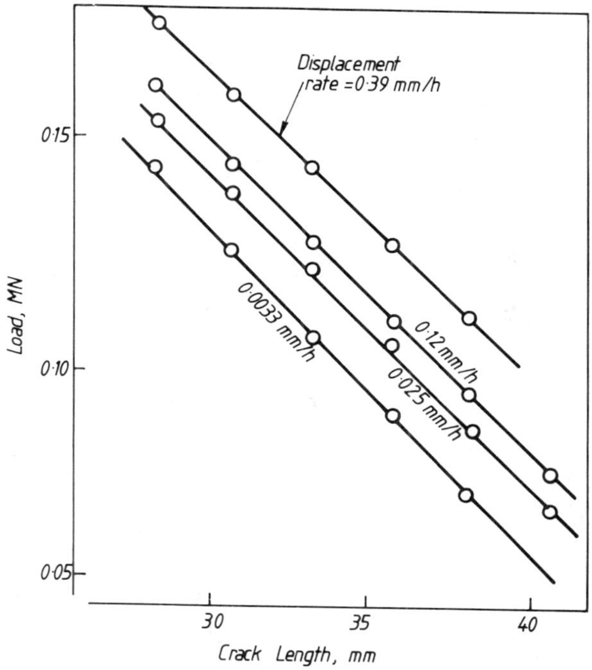

(a)

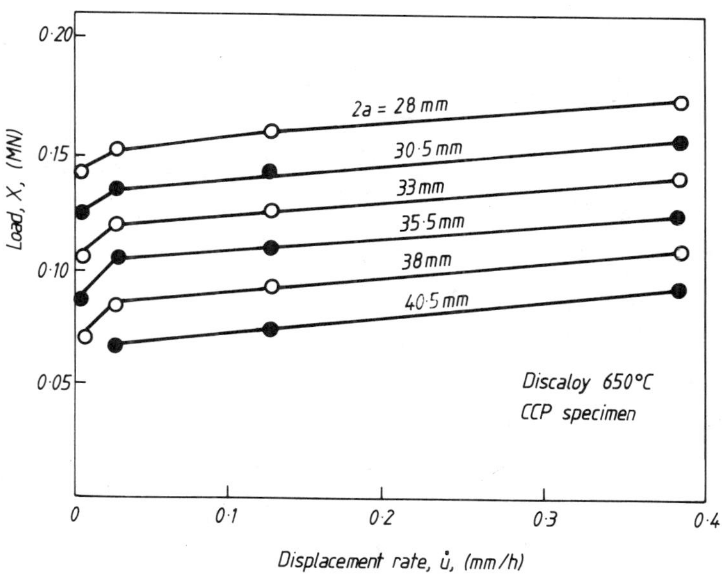

(b)

Fig. 6.20(a) – Load versus crack length for various displacement rates (Discaloy at 1,200°F). (b) Load versus displacement rates for various crack lengths.

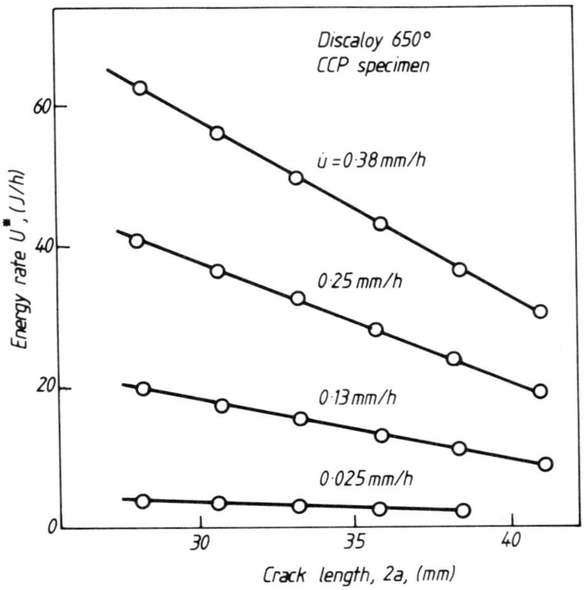

(c)

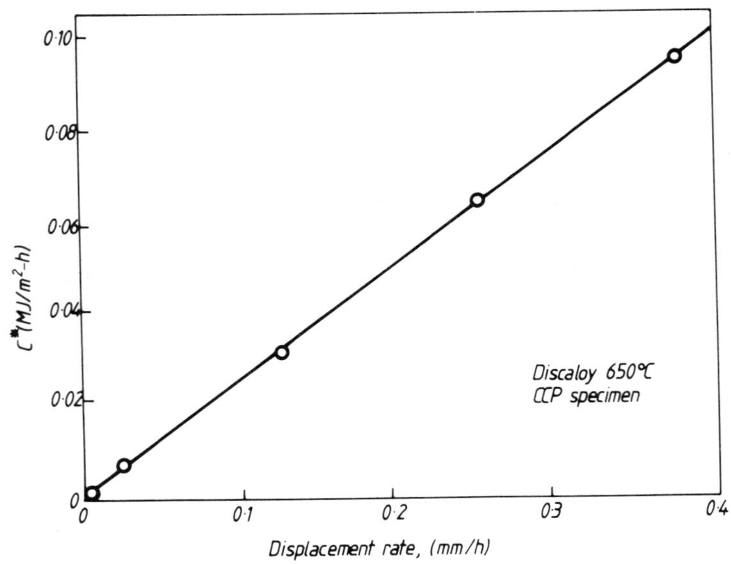

(d)

Fig. 6.20(c) – Energy rate versus crack length for various displacement rates. (d) C^* versus displacement rate. (After Landes and Begley (1974), with permission.)

$\dot{u}$. In order to construct a diagram for energy rate U_T^* vs. a (Fig. 6.20(c)) it is necessary to obtain an intermediate plot of X vs. $\dot{u}$ using Fig. 6.20(a) which is shown as Fig. 6.20(b). Differentiation of U_T^* with a and division by the speciment thickness B gives C^*, i.e. $C^* = -\partial U_T^*/B\partial a$. Note that in this particular example C^* is independent of a (Fig. 6.20(d)).

Another method of determining J employed Turner's η factor relating U_T to ligament length (section 4.5.2). There are corresponding methods for C^* using U_T^*, and η's which now depend on creep index M. The procedures for determining creep-dependent η's are analogous to the methods previously employed for time-independent η (such as equations (4.39), (4.42)) but employ displacement rates. Webster and Smith (1983) have reviewed solutions procedures for C^* as a creep growth parameter derived from U_T^*.

When deformations become extensive, as in ductile creeping solids at high stress levels, crack tip displacements could become very large, and a 'crack' would become a blunt notch. Crack growth would still occur of course by, for example, diffusional transport of matter (called 'unsintering' by Ashby and Raj (1975)) causing cavity growth, but descriptions of the crack tip stress in terms of J-dominated singular fields lose their meaning. Rather, stress redistributions in uncracked sections result in more uniform stress states, and it has been found that a simple reference stress σ_d for material ahead of the crack can adequately describe the long-term history of creep-ductile solids in terms of stress rupture plots (section 6.6). Tomkins (1981) points out that this approach suggests tht crack *growth* itself can be ignored in engineering applications and that structural failure is determined by the effect of the crack in providing a *local high creep stress* condition. Of course the overall external displacements associated with crack advance in ductile materials can be considerable, and it may not be possible to accomodate them in structures. Structural constraints may then inhibit creep crack growth. There is clearly an analogy between creep fracture and monotonic plastic (ductile) fracture in terms of accommodation of displacements: While both, in metals, involve void nucleation and growth ahead of the crack tip, the difference between the two relates to the mechanisms of crack advance – being grain boundary cavitation at high homologous temperatures but cavitation at second phase particles at low temperatures.

It is not clear yet whether methods of analysis employed, for example, in creep cracking of polymers can be applied to creep cracking of metals, and vice versa. It seems that the fields have developed separately, but one might expect that behaviour in terms of bulk parameters should be capable of similar descriptions even though the micromechanisms of crack advance may be different. There will be detail differences, no doubt, since metals tend to be linear in time but non-linear in stress (in secondary creep) whereas polymers are non-linear with time and some (at least over small ranges) linear in stress.

6.4 SLOW CRACK GROWTH: ENVIRONMENT-ASSISTED DELAYED FRACTURE AND STRESS CORROSION CRACKING

In delayed fracture under static loads, the hostile environment attacks the stretched bonds at the crack tip and gradually lengthens sub-critical flaws until one of them reaches the critical length at which $K_c \to K_{IC}$ whereupon fracture occurs. For most solids there is a lower level of loading, corresponding to some threshold K_0, below which growth rates are negligible. An upper level of loading of course relates to the critical stress intensity factor K_{IC} for fracture under monotonically increasing loading. Both points are illustrated in Fig. 6.12. In between these limits, crack growth can occur in many solids under constant sustained loads.

Historically this sort of behaviour (particularly stress-corrosion cracking of metals) was characterised by plotting stress versus time-to-failure curves for given testpieces (usually smooth, uncracked specimens) as shown in Fig. 6.10 for glass in air. The traditional appoach is clearly inadequate in view of defects known to pre-exist in structures, and events taking place at such crack tips, and it is now common practice to investigate the effect of stress intensity factor (rather than stress alone) on crack growth in hostile environments. As a crack lengthens with time the crack velocity can be determined and related to the current stress intensity factor at the crack given by $\sigma\sqrt{\pi a}.Y$; while σ remains constant under dead load, K_{applied} will increase during the test since a increases (and Y changes too). Logarithmic plots of crack velocity da/dt versus applied K_c are employed to represent the results, as shown typically in Fig. 6.13(a) for static stress corrosion data. These results (and others on delayed fracture of alumina, Fig. 6.22(b), and steel, Fig. 6.17) follow well-established patterns, namely that there exists a static threshold stress intensity factor (called K_{ISCC}) below which stress corrosion rates are negligible,† followed by a region (region I) in which da/dt increases rapidly with K_c up to a plateau region where da/dt is either only mildly dependent on K_c or even virtually independent of K_c (region II), followed by region III just below K_{IC} where da/dt is once again highly dependent on applied K_c and in which accelerated crack growth occurs leading to eventual failure. Region I concerns environment-induced relaxation (such as polyethylene in detergent solution) or reaction rate-controlled crack growth (such as glasses in moist air) where the threshold K_{ISCC} is the most important parameter; region II relates to crack growth controlled by the diffusion or mass-transport of environmental species to the crack tip (and is not very sensitive to the applied K_c); and region III generally concerns crack growth controlled by mechanical rather than environment effects since the crack tip runs ahead of the environment owing to either absorption or adsorption. K_{IC} represents the most important parameter in this final crack growth region. Wiederhorn *et al.* (1982) deal thoroughly with the effects of water and organic liquids on crack growth in soda-lime/silica glass, particularly in connexion with region III.

†In fatigue and stress-corrosion fatigue, there is a threshold ΔK_0 in the *range* of applied cyclic stress intensity below which crack growth rates are neglible. See Chapter 7. Usually $K_{ISCC} > \Delta K_0$ except at very low cyclic frequences when they are roughly equal. Note Fig. 7.49.

There are difficulties in measuring K_{ISCC}, as the value depends on the cut-off time of the experiment (Wei *et al.*, 1972). There is no standard practice to determine K_{ISCC}, but the fatigue-cracked cantilever-beam test specimen (rather like an Izod specimen under dead load, Fig. 6.21(a)) is often used in the time-to-failure tests. K_{IC} is first determined for the material in air, and then successively lower initial K_i values are applied and the successively longer times to failure t_f are obtained in the presence of the environment. The highest plane-strain

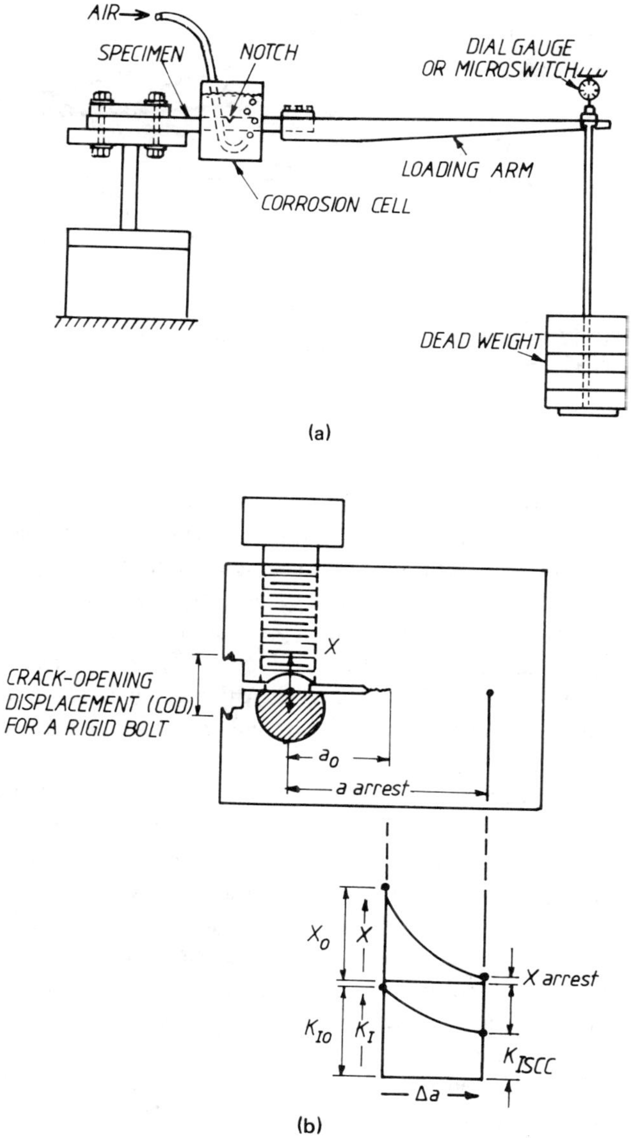

Fig. 6.21(a) – Schematic drawing of fatigue-cracked cantilever-beam test specimen and fixtures. (b) The wedge opening loaded (WOL) specimen.

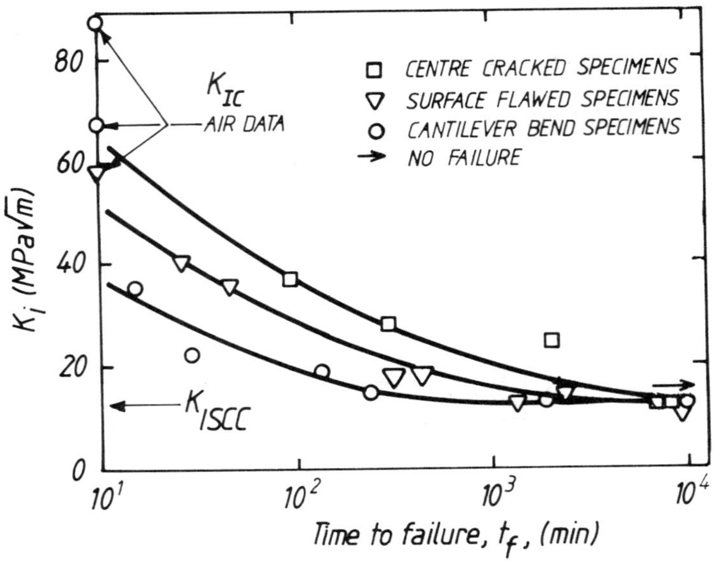

(c)

Fig. 6.21(c) – Influence of specimen geometry on the time to failure for an AISI 4340 steel. (After Rolfe and Barsom (1977), with permission.)

K_i value at which crack growth does *not* occur after a long time (e.g. > 1,000 h for steels) is taken to be K_{ISCC}. In this specimen K increases as crack extends under dead weight loading and many specimens are required to determine K_{ISCC}. A more useful specimen geometry is the wedge-opening (WOL) testpiece for which K decreases as crack extends (Fig. 6.21(b)). This means that only a single specimen is required for K_{ISCC} evaluation, for the crack will eventually stop propagating as $K < K_{ISCC}$. Fig. 6.21(c) shows results from a variety of testpiece geometries for 4340 steel. Although there are separate curves for each geometry (a consequence of different geometric effects due to integration), K_{ISCC} is independent of specimen configuration.

A three-region K_c–$\dot{a}$ diagram is common for many diverse types of material/environment combinations ranging from glasses and ceramics in water environments, steels in gaseous hydrogen to polymers in organic liquids. Figs. 6.22(a)–(g) give some typical examples of the K_c–$\dot{a}$ characteristics for a variety of materials. Note that solids, which usually behave in a ductile fashion in the absence of hostile environments, often do not display a region III. This is exemplified by stress-corrosion cracking of Al alloys in Fig. 6.22(g), detergent cracking of polyethylenes in Fig. 6.22(d), hydrogen embrittlement of steels in Fig. 6.22(c) and the fracture of ductile polymers in the presence of alcohols shown in Fig. 6.22(e). The absence of region III in these examples arises because when the sub-critical applied K_c (at which slow crack growth is occurring), approaches K_{IC} in value, the crack velocity is faster than the environment can keep up with the crack tip, so that the usual ductile mechanical fracture occurs.

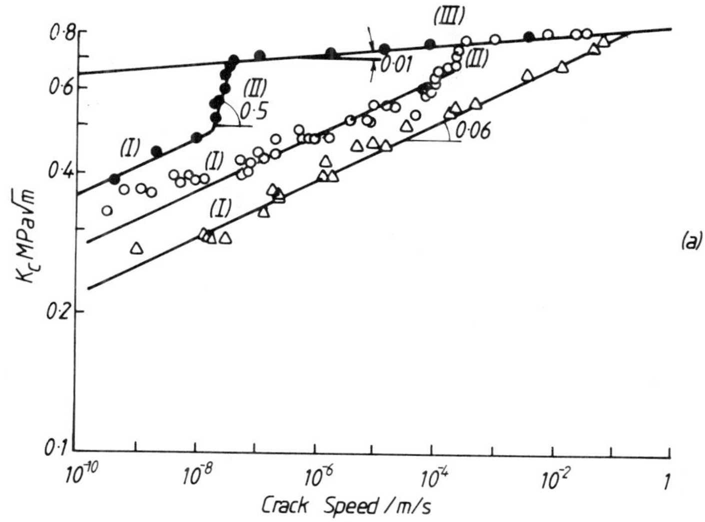

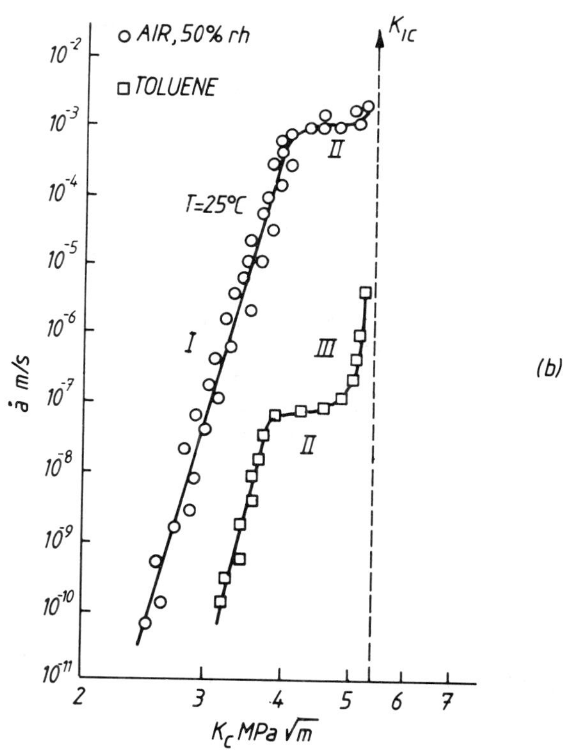

Fig. 6.22(a) – Crack growth data for inorganic glass in various environments (quoted by Williams and Marshall (1975), reproduced with permission). (b) Variations of $\dot{a}$ with K_C for polycrystalline alumina using double torsion specimen (after Williams and Evans, 1973).

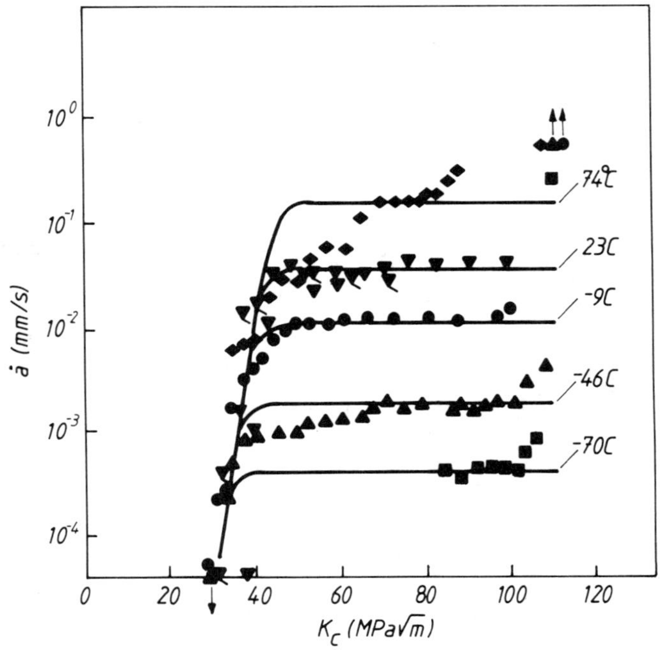

Fig. 6.22(c) – Crack growth rate ($\dot{a}$) as a function of applied K_c for a Ti–5Al–2.5 Sn alloy at different temperatures in hydrogen at 0.9 atmosphere pressure (after Williams (1973), with permission).

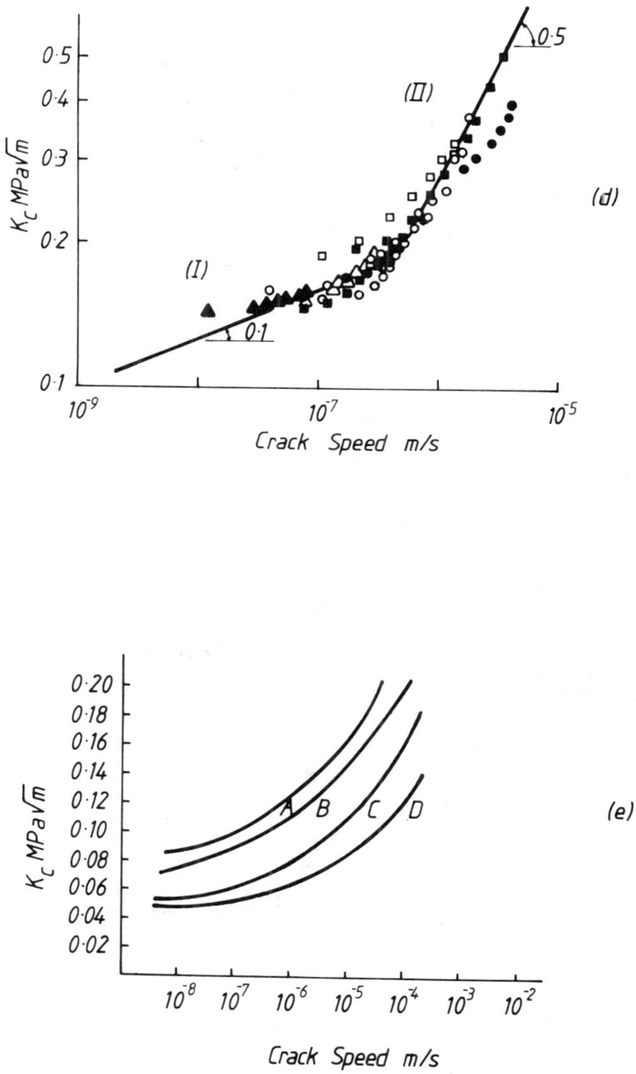

Fig. 6.22(d) – Crack growth data for HDPE in a detergent environment at 60°C. (e) Environmental crack propagation kinetics in polyethlene A = PE with melt flow index (MFI) = 7 in methanol; B:PE of MFI = 7 in ethanol; C:PE of MFI = 20 in methanol; D:PE of MFI = 20 in ethanol (after Marshall *et al.* (1970), with permission).

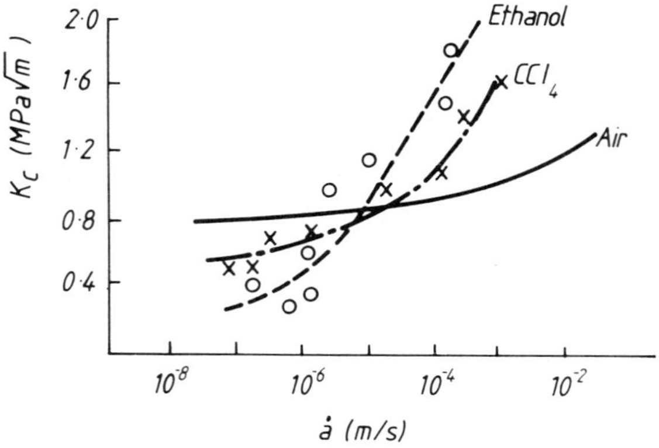

(f)

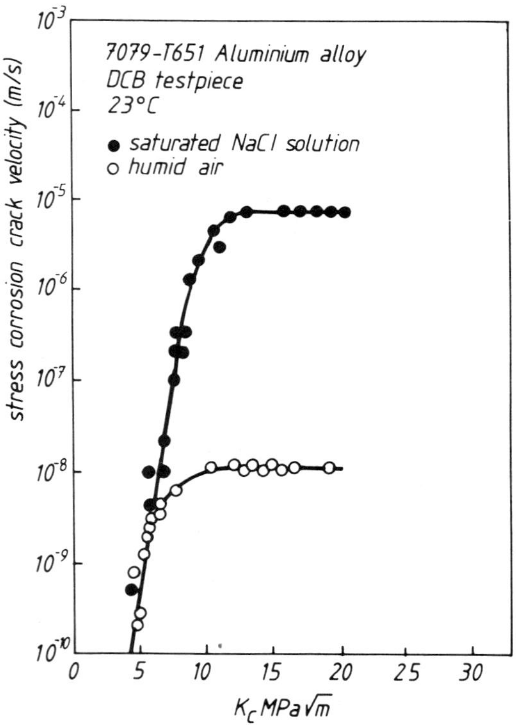

(g)

Fig. 22(f) – Environmental crack propagation kinetics in PMMA in air, in ethanol and in CCl_4 (after Mai (1975), with permission). (g) K_c–$\dot{a}$ relation for a high strength aluminium alloy in air and in aqueous NaCl solution (after Speidel (1979), reproduced with permission).

It may be noted that for the PMMA/organic solvent system in Fig. 6.22(f), at low crack speeds ($< 10^{-5}$ m/s) the crack growth resistance is lower in the solvents than in air, but at high crack speeds ($> 10^{-5}$ m/s) larger K_c values are needed to drive the crack at a given speed. These unusual results are due to the fact that, at high crack speeds, multiple crazes are developed at the crack tip whereas a single craze exists at low speeds. The absence of region III crack growth for this example is because there is fracture instability due to the crack growing away from the heavily crazed region so that $dR/dA < 0$ (Chapter 8).

Not all ceramics are subject to slow crack growth in the presence of liquid environments at room temperature. For example, hot pressed silicon nitride does not exhibit slow crack growth in water at low temperatures; however, at high temperatures (and in the absence of water) say above 1,000°C, slow crack growth does occur as shown in Fig. 6.23, owing to grain boundary flow. There also seems to be some rate effects on K_{IC} for these ceramics (see Fig. 6.4(b)).

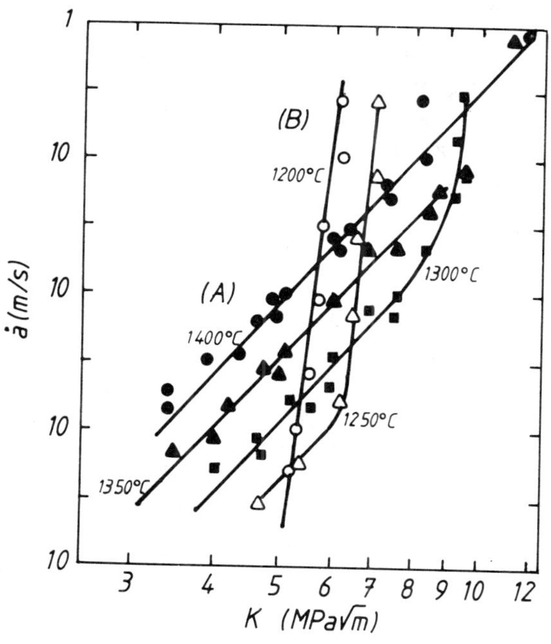

Fig. 6.23 – Crack propagation in hot pressed silicon nitride (after Evans and Wiederhorn (1974), with permission).

The three regions of K_c–$\dot{a}$ diagrams may be represented algebraically by equations of the form

$$\dot{a} = DK_c^{n^*} \tag{6.25 bis}$$

with D and n^* taking different values in different regions. For example, Ewing and Williams showed in 1979 that

$$\dot{a} = DK_c^{50} \quad \text{(m/s, MPa}\sqrt{\text{m}}\text{)}$$

where $\log_{10}D = (-317 \pm 20)$ for Scots Pine with slow crack growth along the grain. Tapered DCB specimens with $0.05 < \dot{u} < 5$ mm/min were used, and other experiments using load-relaxation methods on double-torsion specimens give the same relationship. Moisture is known to have a pronounced effect on the fracture toughness G_c or R and hence on these slow crack growth results in timber (see Fig. 6.6(c)).

Since these $\dot{a}$–K_c relations are empirical, they sometimes appear in other algebraic forms. For example, experiments by Kies and Clark (1969) on commercial soda glass in moist air at room temperature suggested that, over a limited range of variables, the raw data instead of being presented in the form of equation (6.25) can be represented by:

$$\dot{a} = D^* \exp \beta R$$

Wiederhorn has suggested that yet another form:

$$\dot{a} = D' \exp \beta' K_c$$

may well fit the same data. Despite the ability of these different algebraic equations to describe the same slow crack growth in glasses and ceramics (and in other materials) they are sensitive to lifetime predictions as a function of proof/applied stress ratio (see section 6.6 and Problem 6.16).

Example 6.6 Relation (6.25) is dimensionally awkward. Suggest a more flexible equation.

Ans. Minnear and Bradt suggested in 1975 that for ceramics and glasses one should write $\dot{a} = B\,(K_c/P)^x$. B has the dimensions of crack velocity and P the dimensions of stress intensity factor. Depending on the region to which the equation is applied, B and P represent either $(\dot{a}_0, K_0)$ in region I or $(\dot{a}_{cr}, K_{IC})$ in region III. Here, K_0 is the threshold stress intensity factor below which no crack growth occurs, and $\dot{a}_0$ and $\dot{a}_{cr}$ are the limiting velocities associated with the particular K_0 and K_{IC}.

A theoretical justification for the empirical slow crack growth equation used for environment-affected K_c–$\dot{a}$ diagrams (regions I and III) follows from the K_c–$\dot{a}$–T relations (based on viscoelastic and Arrhenius rate and temperature analyses) given for *environment-free* behaviour of polymers in section 6.3. In air, at constant temperature, assuming that slow crack growth is a manifestation of time-dependent Young's modulus and yield or craze strength, and employing a critical δ_c criterion, Williams and co-workers show that

$$\dot{a} = \left(\frac{\pi}{8}\,\frac{\delta_c}{\epsilon_y}\right)(E_0^2\,\epsilon_y\,\delta_c)^{-1/(2n)}\,K_c^{1/n} \qquad (6.24\text{ bis})$$

where the yield strain $\epsilon_y = \sigma_y/E$ is independent of time and temperature, and n is the exponent of the power law for creep modulus, i.e. $E = E_0\,t^{-n}$. Equation

(6.24) is expected to hold particularly for region III of the K_c–$\dot{a}$ diagram, not only for polymers for which it was originally derived but aslo for a variety of materials whose slow crack growth is not environment-assisted but is *mechanically* controlled owing to *plastic flow relaxation* at the crack tip, e.g. regions A and B for silicon nitride at elevated temperatures in Fig. 6.23. However, it has been queried by Lawn *et al.* (1980) that equation (6.24) is not valid to describe slow crack growth in glasses and ceramics at low temperatures, although Williams and Marshall (1975) do apply it to glasses. The argument given by Lawn is that fracture of glass is not controlled by micro-plasticity and relaxation processes at the crack tip, but is controlled by the bond rupture theory.

In the presence of liquid environments, Williams (1977) argued that the craze stress at the crack tip will be *reduced* by a factor α in region I so that equation (6.24) becomes:

$$\dot{a} = \left(\frac{\pi}{8}\right)\left(\frac{\delta_c}{\alpha\epsilon_y}\right)(E_0^2\,\epsilon_y\,\delta_c\alpha)^{-1/(2n)}\,K_c^{\,1/n} \tag{6.30}$$

in which n may increase owing to the absorption of the liquid environment. We see that D in $\dot{a} = DK_c^{n^*}$ for region I behaviour, corresponds with

$$D_I = \left(\frac{\pi}{8}\;\frac{\delta_c}{\alpha\epsilon_y}\right)(E_0^2\,\epsilon_y\,\delta_c\,\alpha)^{-n_I^*/2} \tag{6.31}$$

since $1/n \equiv n^*$. Since region III is comparatively unaffected by hostile environments (but still affected by *relaxation* phenomena), $\alpha = 1$ in the relation for D and

$$D_{III} = \left(\frac{\pi}{8}\right)\left(\frac{\delta_c}{\epsilon_y}\right)(E_0^2\,\epsilon_y\,\delta_c)^{-n_{III}^*/2} \tag{6.32}$$

The limiting velocities $\dot{a}_0$ and $\dot{a}_{cr}$ (see Example 6.6) are given by

$$\dot{a}_0 = \left(\frac{\pi}{8}\right)\left(\frac{\delta_c}{\alpha\epsilon_y}\right)\left(\frac{K_0^2}{E_0^2\,\epsilon_y\,\delta_c\,\alpha}\right)^{1/(2n)}$$

and

$$\dot{a}_{cr} = \left(\frac{\pi}{8}\right)\left(\frac{\delta_c}{\epsilon_y}\right)\left(\frac{K_{IC}^2}{E_0^2\,\epsilon_y\,\delta_c}\right)^{1/(2n)} \tag{6.33}$$

It is now possible to re-write K_c–$\dot{a}$ relations in terms of $\dot{a}_0$ or $\dot{a}_{cr}$ depending on the region of the diagram. That is

$$\dot{a} = \dot{a}_0\left(\frac{K_c}{K_0}\right)^{1/n_I}$$

in region I and

$$\dot{a} = \dot{a}_{cr}\left(\frac{K_c}{K_{IC}}\right)^{1/n_{III}} \tag{6.34}$$

for region III. Owing to the effect of environment in increasing n in region I, the slopes of K_C–$\dot{a}$ diagrams in regions I and III are different. Minnear and Bradt's empirical relation in Example 6.6 coincides with equation (6.34) which is *only theoretically* correct for those materials whose slow crack growth is caused by crack tip *micro-plasticity* and *relaxation* processes, whether or not this is enhanced by environmental effects. Although Minnear and Bradt show that equation (6.34) can be used for ceramics and glasses, this agreement is fortuitous since, as mentioned earlier, fracture of glasses and ceramics is by a sequential bond rupture process (Lawn *et al.*, 1980).

6.5 MECHANISMS OF ENVIRONMENT-ASSISTED FRACTURE

In the presence of hostile environments, the mechanics of quasi-static elastic fracture will be governed in general by equation (1.6) viz.

$$X\mathrm{d}u = \mathrm{d}\Lambda + R\mathrm{d}A + \mathrm{d}\Psi \qquad (1.6 \text{ bis})$$

where $\mathrm{d}\Psi$ is a term which embraces chemical energy, electrochemical energy, and so on.

$\mathrm{d}\Psi$ acts rather like residual strain energy $\mathrm{d}\Lambda_r$ which can promote, or suppress, fracture in comparison with the residual-free case (section 3.7 etc.); the parallel here is the promotion of fracture owing to the release of chemical energy and so on in the process of fracture. Be aware though that, unlike cracking in the presence of residual stresses where the inherent toughness R is *unaffected* by Λ_r, the environment *may* affect R *as well as* provide an additional driving force for cracking through $\mathrm{d}\Psi$. Thus, even if there is no chemical activity and $\mathrm{d}\Psi = 0$, it is still possible for environments to influence fracture via effects on R (this is particularly true for physical interactions such as the cracking of polymers in organic solvents – excluding those cases where chemical dissolution occurs).

6.5.1 Changes in *R*

In simplest terms, if R is identified with the surface free energy 2γ, it is clear that if the environment wets the solid, then γ should be replaced in equation (1.1) by the interfacial surface energy γ^* which is given by

$$\gamma^* = \gamma - \gamma_{\mathrm{env}} \cos\theta \qquad (6.35)$$

where γ_{env} is the surface energy of the environment and θ is the contact angle. Clearly $\gamma^* < \gamma$, and σ_{cr} in a Griffith-type relation must be smaller than in the absence of the hostile environment.

Although this seems a reasonable approach for ideal elastic fracture, specific work of fracture in practical cracking relates to $R = 2(\gamma + \gamma_p)$ according to Orowan and Irwin; unless γ_p (which is two or three orders of magnitude greater than γ^*) is affected by the environment to the same degree as γ, simple reductions in γ^* alone will not be able to explain environment-assisted fracture. However, a complementary viewpoint recalls the Griffith-type expression for atomic bond rupture at the crack tip: as shown in Example 3.2 integration of the interatomic force curve leads to

$$\sigma_{cr} = \sqrt{\frac{E(\rho\gamma/a_0)}{a}}$$

where ρ is the radius of curvature at the crack tip and a_0 is the interatomic spacing. For an atomically sharp crack $\rho = a_0$, but local crack tip irreversibilities in practical materials produce plastic relaxation at the crack tip and $\rho \gg a_0$. Thus, viewing R as $(\rho/a_0)\gamma$, (which is an alternative viewpoint to the Orowan–Irwin argument that $R = 2(\gamma + \gamma_p)$) it is clear why $R \gg 2\gamma$ in the presence of crack tip flow and why environments can alter values of R.

In the presence of absorbed environmental species, R and hence σ_{cr} can be decreased (i) by the lowering of γ (i.e. a local reduction in the free energy of the material/environment system) and/or (ii) by the inhibition of plastic relaxation at the crack tip so that (ρ/a_0) is smaller in the presence of the environment. Depending on the environment/material system, one or both of these two quantities come into play to given embrittlement of the solid. For example in the mercury embrittlement of polycrystalline aluminium, γ is reduced but the plastic work multiplier is probably not. In hydrogen embrittlement of steel, the hydrogen atoms accumulate within a few atomic distances from the crack tip and both the effective γ and (ρ/a_0) are altered (e.g. Howard, 1979).

6.5.2 Electrochemical Effects

An important example when $d\Psi$ plays a part in environment-affected fracture is *stress corrosion cracking* of metals and alloys. In aqueous environments, electrochemical activity at the crack tip can maintain a sharp crack profile while producing an increment of crack advance in conjunction with an applied tensile stress. During crack advance a critical balance between active and passive behaviour is maintained. Under tensile load, new material surface is exposed at the crack tip by the decohesion process of Fig. 1.7, and in an active environment this region becomes anodic relative to the adjacent crack flanks, which act cathodically. Preferential anodic dissolution mechanisms at the crack tip depend on the existence of galvanic cells set up between grains and/or precipitates. As the crack advances, crack tip material becomes crack flank material and may change from an active to a passive state (Tomkins, 1981). For crack growth to be maintained, there must be an adequate renewal both of fresh surface at the crack tip and of the controlling environment, Passivation of the crack sides while the crack tip region remains active is essential to maintain sharpness and growth in dissolution-controlled stress corrosion cracking, otherwise the crack will spread laterally to become a pit or other blunt feature. This explains why not all environments that are corrosive towards a particular metal necessarily promote stress corrosion cracking.

In these circumstances, $d\Psi$ in equation (1.6) is the electrochemical energy at the crack tip associated with the anodic process.

West (1973) gives

$$\mathrm{d}\Psi = -\frac{zF\rho\delta^*}{M}\,\eta\,\mathrm{d}A = -B\,\frac{zF\rho\delta^*\eta\mathrm{d}a}{M} \tag{6.36}$$

where da is the crack growth created by the anodic dissolution process, δ^* is approximately equal to the critical crack tip opening displacement δ_c for a brittle crack, F is the Faraday constant (96.5 kilocoulomb/mole), $\bar{\rho}$ is the density, z is valency of the solvated ions, M is the molecular mass of the metal and η is the anodic over-potential. The negative sign indicates a *release* of energy, and suggests from equation (1.6) that fracture is possible (in principle) in the *absence* of external loads when $X\mathrm{d}u = 0$ (cf. spontaneous fracture in the presence of residual stresses), and even in the absence also of elastic strain energy ($\mathrm{d}\Lambda = 0$).

We have

$$X\mathrm{d}u = \mathrm{d}\Lambda + R_{\mathrm{env}}\mathrm{d}A - BzF\rho\delta^*\eta\mathrm{d}a/M \tag{6.37}$$

where R_{env} is the fracture toughness during cracking as affected by the environment (it is assumed that the effects of environment and anodic processes are uncoupled in the equation). Compare equation (6.37) with equation (3.52) for residual stress cracking where the value of R is unaffected by Λ_r; here R may be affected by the environment as well as having dΨ help fracture propagation. Equation (6.37) corresponds with

$$-\left(\frac{\partial\Lambda}{\partial A}\right)_{\mathrm{u}} = R_{\mathrm{env}} - zF\rho\delta^*\eta/M \tag{6.38}$$

in which, during *stable* stress corrosion cracking (see Chapter 8), $zF\rho\delta^*\eta/M$ is the electrochemical energy release rate during the dissolution/crack advance process.

The whole of the right-hand side of Equation (6.38) represents an 'effective' or 'apparent' toughness for quasi-static elastic fracture. This is the quantity which a Gurney-type experimental approach would measure, as in Fig. 6.1(b). Only when the Ψ-term is independently determined can the two components be separated. We shall see presently that the corresponding K_c values during stress corrosion cracking are also 'effective' values.

Were $\mathrm{d}\Psi = 0$, $-(\partial\Lambda/\partial A)_{\mathrm{u}} = R_{\mathrm{env}}$ which is the actual toughness as affected only by a particular environment. $\mathrm{d}\Psi = 0$ occurs in *cathodic* dissolution processes such as hydrogen diffusion in metals leading to hydrogen embrittlement cracking.

Relation (6.38) applies to all stages of crack propagation and may therefore be modified to take account of situations where crack extension is discontinuous and involves consecutive 'mechanical' and 'chemical' stages. Vermilyea (1972) has proposed a model in which, following a critical creep strain increment, film breakdown at the crack tip allows anodic dissolution to proceed until, after an interval, the surface repassivates. The cycle is then repeated. The interested reader should consult West's 1973 paper.

The strain energy release rate in equation (6.38) corresponds with a critical stress intensity factor $(K_c)^*_{env}$ which in general is affected both by the environment and by the release of electrochemical energy. Thus

$$-\left(\frac{\partial \Lambda}{\partial A}\right)_u = (K_c^2)^*_{env}/E^* \tag{6.39}$$

(the asterisk on E here meaning plane stress E or plane strain $E/(1-\nu^2)$ – *not* an environment-affected E), and

$$(K_c^2)^*_{env} = E^* R_{env} - \frac{E^* zF\rho\delta^*\eta}{M} \tag{6.40}$$

Subcritical crack growth will occur, according to relation (6.40), when the applied K is increased above $(K_c)^*_{env}$. Alternatively, at a given applied K, crack growth will occur if R_{env} falls enough and/or the anodic over-potential η increases enough.

Spontaneous cracking caused by the anodic overpotential alone occurs when

$$R_{env} = zF\rho\delta^*\eta/M$$

Spontaneous intercrystalline stress corrosion cracking ('season' cracking in brass alloys) is caused by the additional contribution of relase of residual strain energy Λ_r in equation (6.40). In the usual form of stress corrosion cracking (under load) the smallest value of $(K_c)^*_{env}$ at which cracking occurs is the threshold K_{ISCC}, i.e.

$$K_{ISCC} = \sqrt{E^* \left(R_{env} - \frac{zF\rho\delta^*\eta}{M}\right)_{min}} \tag{6.41}$$

This is usually determined as if residual stresses did not exist.

Example 6.7 Explain how stress corrosion is influenced by crack tip creep strain rate effects in terms of the energy balance concept.

Ans. The fracture mechanism in stress corrosion cracking of metals involves an active/passive cell, in which the bulk of the surface is chemically passive while the tip itself advances by anodic dissolution from the bare metal. In order that electrochemical energy can contribute to the fracture process it is necessary that the local crack tip creep strain rate ($\dot{\epsilon}$) is sufficient both to rupture any passivating film that may form and to ensure that repassivation cannot occur on the yielding tip surface. It has been suggested by West (1973) that the Ψ term in equation (6.37) should be some function f given by $f(\eta(1-\delta^*/\dot{\epsilon}\tau))$ where τ is the repassivation time, i.e. the time required to reform a passive film once it has been disrupted by yielding. $\dot{\epsilon}$ depends on K, and τ depends on the electrochemistry, i.e. electrode potential, anion activity, pH and metal composition. Thus equation (6.40) becomes:

$$(K_c^2)^*_{env} = E^* R_{env} - E^* f\,(\eta\,(1-\delta^*/\dot{\epsilon}\tau))$$

At the threshold $(K_c)^*_{env} = K_{ISCC}$, there is a minimum $\dot{\epsilon}$ in order to satisfy the above equation. Note that for anodic dissolution to occur $\dot{\epsilon} > \delta^*/\tau$. Plots of da/dt vs. $\dot{\epsilon}$ in stress corrosion cracking are similar to plots of da/dt vs. K_c (e.g. Parkins, 1979).

The rate at which a crack propagates is determined by the rate of dissolution and Faraday's second law applied to the corrosion-controlled growth in stage II of an $\dot{a}$ vs. K_c or $\dot{a}$ vs. $\dot{\epsilon}$ diagram (i.e. the plateau level) gives

$$\dot{a} = \frac{IM}{zF\rho} \tag{6.42}$$

where I is the anodic current areal density (A/m^2). The possibility of passivating films will of course affect current density. Parkins (1975) showed that in order to maintain I close to the film-free value, no film should form at the crack tip or, if it does, the film must be repetitively broken.

Example 6.8 Estimate the theoretical crack propagation rate ($\dot{a}$) and the electrochemical energy (Ψ) released in the stress corrosion cracking of an aluminium alloy (Al–7Mg) in sodium chloride solution assuming the anodic dissolution model to hold.

Ans. For this environment/material system, we have for aluminium $z = 3$, $M = 26.98 \times 10^{-3}$ kg/mole, $\bar{\rho} = 2.7 \times 10^3$, $F = 96.5$ kilocoulomb/mole, $\delta^* \approx 10^{-6}$ m. Thus from equation (6.42) we have

$$\dot{a} = \frac{IM}{zF\rho} = \frac{26.98 \times 10^{-3}\, I}{3 \times 96.5 \times 10^3 \times 2.7 \times 10^3} = 34.52 \times 10 I^{-12}\,(\text{m/s})$$

I is in A/m^2 and it changes with the electrode potential at the crack tip. For $10 < I < 10^4$ A/m^2, $35 \times 10^{-8} < \dot{a} < 35 \times 10^{-5}$ mm/s in this case.

The electrochemical energy contribution to equation (6.38) is given by

$$-\frac{1}{B}\frac{\partial\Psi}{\partial a} = \frac{zF\bar{\rho}\delta^*\eta}{M} = \frac{3 \times 96.5 \times 10^3 \times 2.7 \times 10^3 \times 10^{-6}\,\eta}{26.98 \times 10^{-3}}$$

$$= 29\eta\,(\text{kJ/m}^2)$$

η is at least of the order of several hundred millivolts so that the electrochemical energy release is of the order of several kJ/m^2 (recall that bond strengths are some MJ/mole, i.e. per 10^{23} atoms).

We may use the results of Example 6.8 to check the order of magnitude estimates for K_{ISCC} for aluminium alloys according to the anodic dissolution model given in equation (6.38). We have $zF\bar{\rho}\delta^*\eta/M = 29\,\eta$ kJ/m². According to Ford (1979), the maximum value for over-potential for bare surface dissolution is $\eta \approx 0.5$ volt before liquid dissolution and/or film control reduces the effect. Let us take $\eta = 0.4$ volt, whence $d\Psi/dA \approx 12$ kJ/m² which is not much smaller than the specific work of fracture of many aluminium alloys. For $K_{ISCC} \approx 7$ MPa$\sqrt{m}$, it follows that $R_{env} \approx 13$ kJ/m² according to equation (6.41). While the results are sensible, it is clear that the calculations are very sensitive to small changes in the controlling parameters.

The interested reader is referred to Ford's 1979 paper for further information. Ford's line of attack is somewhat different from that presented here. Also, see West (1982) and related papers at that meeting.

A J-based theory for predicting stress corrosion crack growth rates when linear elastic fracture mechanics conditions do not apply has been presented by Smith (1983).

6.5.3 Plasticisation Theories

In amorphous polymers, while surface energy reduction mechanisms are candidate explanations for premature environment-affected fracture, often they are insufficient to explain actual results in polymers (Example 6.9 following). The *plasticisation* theory suggests that absorption of the environmental species into the polymer material at the crack tip causes limited plasticisation which in turn lowers the glass transition temperature (T_g) of the polymer to a certain limited degree. Application of a sufficient external stress will promote flow of glassy material in the stress direction. The greater the suppression of T_g the easier the flow. Similar remarks apply to biological materials.

As the fracture toughness of glassy polymers is controlled mainly by the craze stress (σ_c) and the associated critical opening displacement at the crack tip (δ_c) it is useful to compare the difference between an air craze and a solvent craze. The profile of a typical craze has been given in Figs. 1.1 and 5.22 which has a thickness tapering down to very small dimensions at its tip. (For conditions governing craze nucleation, see Chapter 5.) As the craze lengthens so does the thickness and when it reaches a critical value δ_c at the craze base, fracture commences and the crack tip advances through the craze. The microstructure of the craze is determined by the mechanism of craze growth which can take two forms: (i) the new fibrillar material is 'drawn' from the bulk unorientated polymer at the craze/bulk polymer interface; and (2) a fibril creep mechanism in which craze thickening occurs by further stretching of the existing fibrils. For air crazes the surface drawing mechanism prevails but for solvent crazes there is evidence that the fibril creeping mechanism predominates. The competing effects give rise to different fracture morphologies of crazes in air and in an organic solvent (n-heptane in this illustration) as shown in Figs. 6.24(a) and (b). The n-heptane craze is lower in fibril volume fraction and coarser in structure compared to the air craze. It seems that the environment has swollen and plasticised both the craze fibrils and the layer of bulk polymer contiguous to the

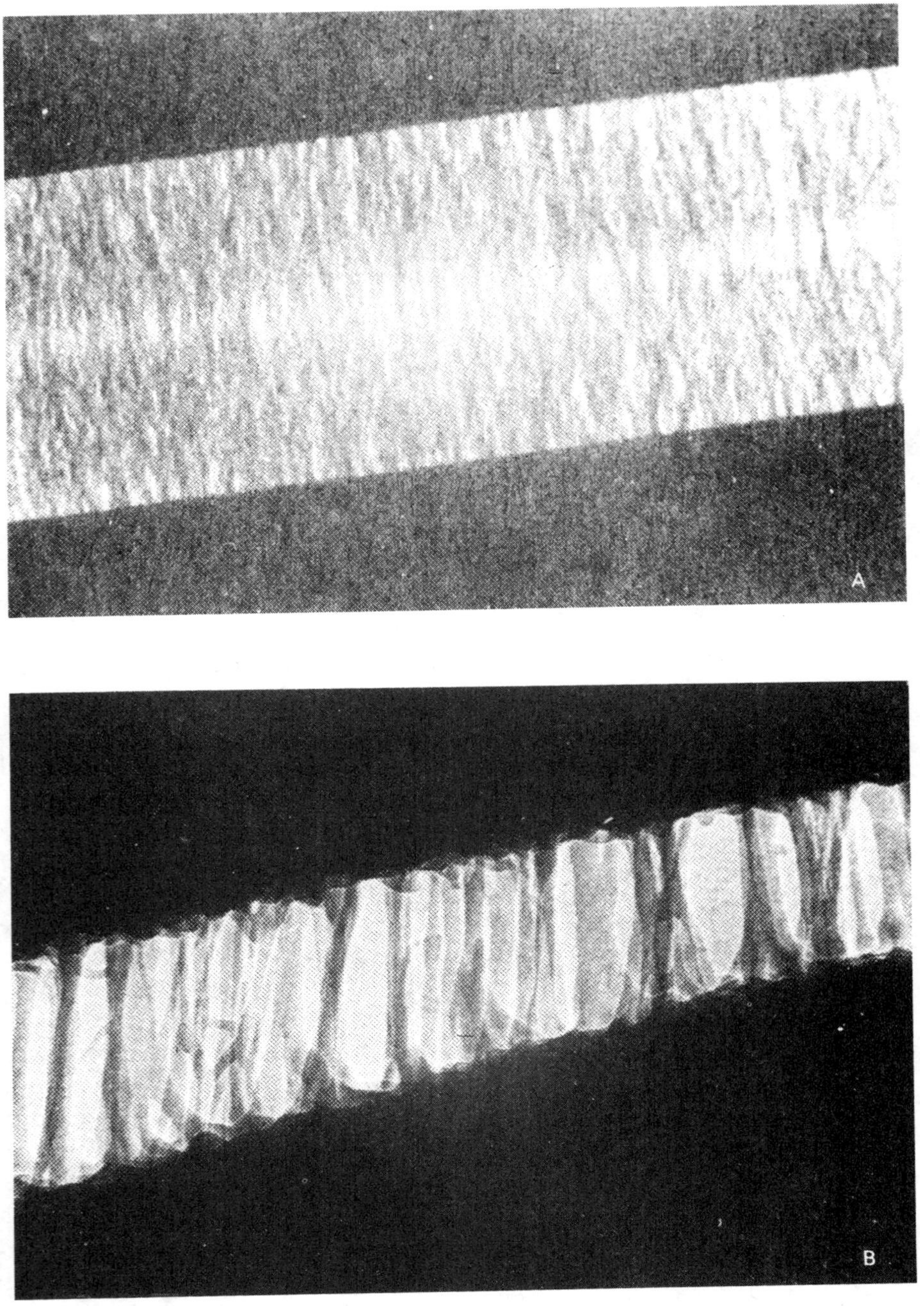

Fig. 6.24 – Transmission electron micrographs of typical crazes in films ~7000 Å thick under the stress at which the crazes originally grew. (a) Microstructure of an air craze (3 Å across) in polystyrene, and (b) microstructure of an *n*-heptane craze (1.5 Å across) in polystyrene (after Kramer (1979), with permission).

craze surface. Plasticisation lower the glass transition temperature of the polymer (and for the *n*-heptane/polystyrene system T_g is reduced to 6°C) and it seems that the larger the reduction in T_g, the coarser the craze structure.

Example 6.9 Show that the plasticisation theory and not the surface energy reduction mechanism is responsible for environment-induced embrittlement of glassy plastics.

Ans. Consider a typical glassy polymer (polystyrene) which is internally pre-plasticised with dichlorobenzene and then stressed in air. T_g varies with the degree of plasticisation and it is found that the critical strain for crazing or cracking ϵ_c decreases with increasing T_g as shown in Fig. 6.25. If the same polystyrene specimens without pre-plasticisation are now stressed in liquids of limited swelling power, it is found that the ϵ_c data, when plotted against T_g of

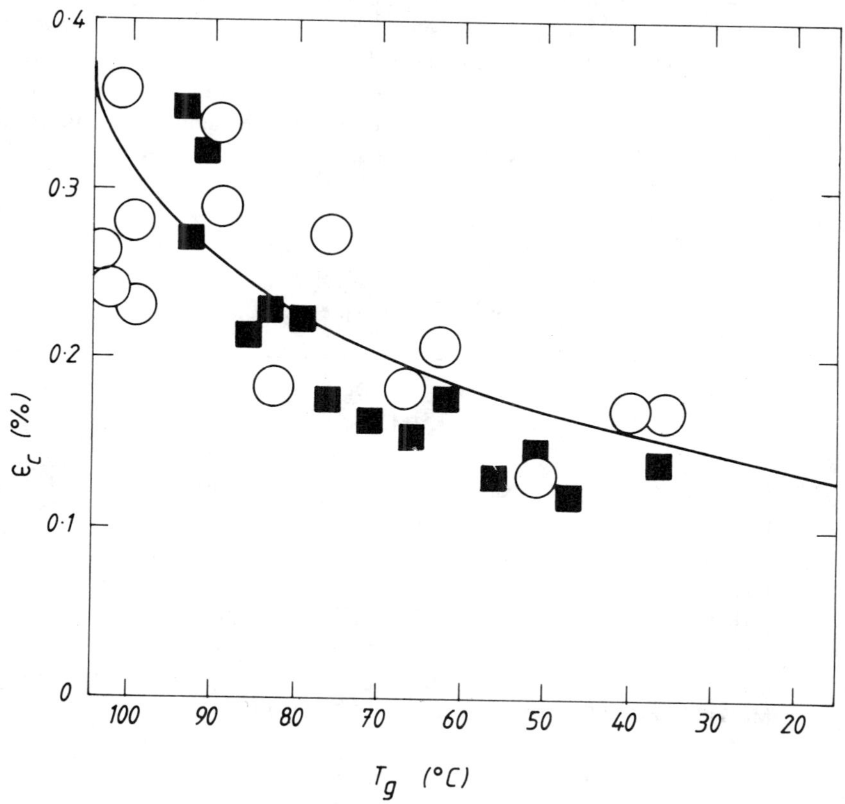

Fig. 6.25 – Critical strain (ϵ_c) for crazing or cracking of polystyrene versus T_g: ○ pure polymer stressed in liquid of limited swelling power versus T_g of film swollen to equilibrium in same agent; ■ dichlorobenzene-plasticised polymer stressed in air versus T_g of same specimen (after Kambour (1973), with permission).

the equilibrium swollen film in the same agent, fit the *same* curve as for the pre-plasticised material. These results obtained from two series of experiments confirm the validity of the plasticasation theory of environmental stress cracking for glassy polymers.

Example 6.9 shows that the plasticisation theory can explain the low stress cracking of polymers immersed in liquid environments. Although there is good correlation of craze initiation strain ϵ_c with the glass transition temperature of the plasticised fibrils T_g, it cannot be readily used to predict the reduction of ϵ_c unless T_g is measured first. But T_g measurement is tedious if the polymer has a low swelling rate. Consequently attempts have been made to use alternative parameters to characterise environment-affected mechanical behaviour of polymers. The degree of plasticisation (and lowering of T_g) is, to a first approximation, dependent directly on the difference of the *solubility parameters* of the solvent (δ_s) and the polymer (δ_p). The solubility parameter derives from the thermodynamic theory of mixing of long chain molecules and is defined as the square root of the cohesive energy density of the liquid or polymer; its units are $(J/m^3)^{1/2}$. Tables 6.2 and 6.3 give representative values of δ_s and δ_p for some common organic solvents and polymers.

Table 6.2
Solubility parameters of liquids

Liquid	δ_s $(MJ/m^3)^{1/2}$	Liquid	δ_s $(MJ/M^3)^{1/2}$
Dodecamethyl pentasiloxane	11.0	*n*-Octanol	21.1
Hexamethyl disoloxane	12.3	Isobutanol	21.5
n-Pentane	14.5	Acrylonitrile	21.5
n-Hexane	15.0	*n*-Pentanol	22.4
n-Heptane	15.2	*n*-Butanol	23.4
n-Octane	15.6	Isopropanol	23.6
n-Decane	15.8	*n*-Propanol	24.6
Methyl isobutyl ketone	17.2	Ethanol	26.1
Carbon tetrachloride	17.6	Propylene glycol	28.1
Methyl *n*-propyl ketone	17.8	Methanol	29.7
Styrene	19.1	Glycerol	33.8
Acetone	19.7	Formamide	39.4
n-Dodecanol	20.1	Water	51.3

Table 6.3
Estimated solubility parameters of polymers

Polymer	δ_p (MJ/m^3)$^{1/2}$
Polybutadiene	17.2
Poly(2,6-dimethyl-1,4-phenylene oxide)	17.6
Poly(butadiene-co-styrene) (75/25)	17.9
Polystyrene	18.7
Poly(vinyl chloride)	19.5
Poly(methyl methacrylate)	19.5
Poly(butadiene-co-acrylonitrile) (75/25)	19.5
Poly(styrene-co-acrylonitrile) (75/25)	19.7
Polycarbonate (Bisphenol A)	20.1
Polysuphone (Bisphenol A)	22.1
Polyacrylonitrile	31.6

For further information see Immergut and Brandrup (1974).

A comprehensive treatment of the thermodynamics of mixing is beyond the scope of this book, and the interested reader is referred to the review by Barton (1975) and to Mai (1985). However, an outline of the general line of attack which explains the role of solubility parameters is as follows.

Theory requires that the free energy of mixing be zero or negative (i.e. energy released) if liquid and polymer are completely miscible. For non-polar polymers interacting with non-polar liquids, this statement may be expressed as

$$[\ln(1-\phi_p) + \left(1 - \frac{1}{m}\right)\phi_p + \chi_s\phi_p^2] + \frac{V_s(\delta_s - \delta_p)^2}{\bar{k}T}\phi_p^2 = 0. \quad (6.43)$$

where the subscripts s and p refer to solvent and polymer respectively and where the symbols have these meanings: ϕ is volume fraction; m is number of elements per polymer molecule; χ is a semi-empirical interaction parameter between liquid and polymer, which contains both entropy and enthalpy condsiderations; V_s is molar volume; $\bar{k}$ is Boltzmann's constant; T is absolute temperature. The first term of the relation is the entropy contribution, and the second is that of the enthalpy. Equation (6.43) may be solved for the equilibrium volume fraction of the solvent in the polymer, i.e. $\phi_s = (1 - \phi_p)$. The lowering of T_g is directly dependent on ϕ_s and results show that the critical strain for craze initiation (ϵ_c) likewise depends on it, e.g. Fig. 6.25.

In many engineering applications it is possible to simplify equation (6.43) by considering only the second (enthalpy) term and ignoring the entropy term, in which case solubility is directly proportional to $|\delta_p - \delta_s|$. When δ_s is close or equal to δ_p dissolution of the polymer in the solvent occurs; as the difference

$|\delta_s - \delta_p|$ increases, crazing and cracking will take place. There are good correlations between δ_s and ϵ_c (for craze initiation) for a range of glassy polymers like PMMA, PS, PVC etc. in non-polar liquids (Vincent and Raha, 1972; Mai, 1976). Fig. 6.26 shows how the critical initiation stress intensity factors of plain and glass-filled PS vary with δ_s; toughness is least when $\delta_s = \delta_p$. The toughening is due to crack tip blunting produced by environment-induced clusters of crazes.

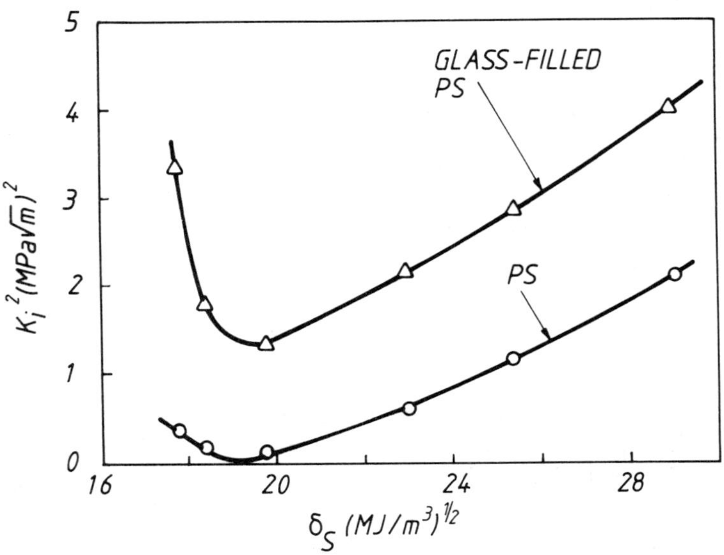

Fig. 6.26 – Correlation between K_I^2 and solvent solubility parameter, δ_s, for glass-filled and plain polystyrene (data from Mai and Atkins (1976)).

The trends in Fig. 6.26 seem to be followed by many polymer-solvent systems. It should be pointed out, however, that modifications to equation (6.43) have to be made in order to deal with polar/polar, or non-polar/polar systems. Similarly, hydrogen bonding effects must also sometimes be considered. Useful references, in addition to those already mentioned, are Henry (1974) and Jacques and Wyzgoski (1979). Solvent molecules size may also be important. For example, even with some non-polar/non-polar systems (such as polycarbonate PC and a series of aliphatic hydrocarbons), simple correlations of ϵ_c with $|\delta_s - \delta_p|$ are not fully successful and only when molar volume V_s is incorporated do the results make sense, i.e. ϵ_c correlates with $V_s\,(\delta_s - \delta_p)^2$ from the simplified version of equation (6.43). Physically this means that the larger the chain length (and hence the molar volume or size of the solvent) the more difficult it is to penetrate the polymer and this leads to large ϵ_c values. The rate-controlling process for craze growth is fluid flow through the porous craze structure, for which two possible sources of driving force have been suggested (a) atmospheric pressure as the craze voids are under a vacuum or (b) capillary pressure in the craze (see Problem 6.12).

Swelling of polymers should be increased by application of stress, and Gent (1970) has proposed an equation to estimate ϕ_p from the Flory–Huggins–Treloar treatment of polymer–liquid equilibrium (e.g. Treloar, 1958). These effects must play a role at the stressed tips of cracks and no doubt will influence environment-affected toughness, crack velocities relative to rates of growth of environment-affected zones, crack stability and so on.

As shown later, in Chapter 7, the parameters influencing environment-affected *fatigue* crack propagation in polymers can be expressed in terms of solubility parameters, and an interpretation of both abrasive and adhesive wear of polymers using solubility parameters is given in Chapter 10.

6.5.4 Liquid and Solid Metal Embrittlement

As far as we are aware, work in this field has been performed in terms of traditional parameters such as strains to fracture and so on, giving results which correspond to Fig. 6.1(a). Little information seems to be available on the liquid and solid metal embrittlement equivalents of Fig. 6.1(b) in terms of fracture toughness parameters. An exception is the investigation by Old (1978) of slow crack growth in single crystals of aluminium embrittled by liquid mercury which is raised in Problem 6.20.

6.6 STRUCTURAL LIFE PREDICTIONS USING K_c–$\dot{a}$ DIAGRAMS

In sections 6.3 and 6.4 we have shown how many engineering materials of structural importance can experience slow crack growth under static loads lower than the critical 'fracture mechanics' stresses given by $K_{IC}/\sqrt{\pi a}.Y$.

The lifetime for a statically loaded structural component consists of crack initiation (or incubation) and crack propagation periods. Most structures will contain some sort of flaw or defect which may be inherent to the material or be introduced during fabrication or manufacture. Life prediction relates to calculating failure times from integration of equation (6.35) for example. For most materials, region III of the K_c–$\dot{a}$ diagram does not make a substantial contribution to the failure time, since crack velocities here are relatively high and that part of the integrated life is therefore small. Depending on the environment, region II is also often negligible. Consequently integration is usually performed over region I only.

If $\dot{a} = DK_c^{n*}$, and $K_c = \sigma_a\sqrt{\pi a}Y$ where σ_a is the (constant) applied stress, we have

$$\int_0^{t_f} dt = \frac{2}{\sigma_a^2 \pi Y^2} \int_{K_i}^{K_t} \frac{K_c dK_c}{\dot{a}} \tag{6.44}$$

where K_i is the initial applied K_c and K_t is the K_c value separating regions I and II. Y is generally dependent on crack size but we shall show that by taking this to be approximately constant with value $Y(a_i)$ in the integration, the error is negligible. We obtain therefore:

$$t_f = \frac{2(K_i^{2-n*} - K_t^{2-n*})}{\sigma_a^2 Y^2 (n* - 2)\pi D} \tag{6.45}$$

For many solids, $n^* \geqslant 10$, in which case $K_i \leqslant 0.9\,K_t$ and equation (6.45) reduces to

$$t_f = \frac{2K_i^{2-n^*}}{D(n^*-2)\,\sigma_a^2\,Y^2\,\pi} = \frac{2(\sqrt{\pi}\sigma_a)^{-n^*}\,Y^{-n^*}\,a_i^{\,1-n^*/2}}{D(n^*-2)} \tag{6.46}$$

It is clear from this equation that t_f is determined mainly by the initial flaw size and the constant stress applied. The assumption of a constant Y at a_i is therefore justified. This equation also suggests a simple method to evaluate the constants D and n^* in the slow crack growth equation. By plotting t_f against σ_a in a log–log plot the slope gives $-n^*$ (which is usually obtained in a creep or a relaxation test, cf. section 7.2). The time-to-failure equation then allows the constant D to be determined since n^* is now known.

The time-to-failure is determined by equation (6.46) if the initial applied K_i $(= \sigma_a\ Y\sqrt{\pi a_i})$ and the K_c–$\dot{a}$ relation for the temperature and evironment of interest are known; (with the assumptions used in the derivation of equation (6.46) K_{IC} does not affect lifetime evaluations).

For a given material or structural component the initial flaw size is constant. It can be shown from equation (6.46) that the ratio of the lifetimes at two sub-critical applied stresses σ_{a1} and σ_{a2} is given by:

$$\frac{\sigma_{a1}}{\sigma_{a2}} = \frac{t_{f2}}{t_{f1}} \tag{6.47}$$

Example 6.10 The region I slow graph data for PMMA at 18°C in air are described by $\dot{a} = DK_c^{n^*}$, where $\log_{10} D = -155.34$ (m/s) and $n^* = 25$. Single-edge-notched tension specimens of 25 mm width contain initial flaws varying between 9 mm and 13 mm. They are subjected to a constant applied stress of 2.7 MPa. Calculate the expected lifetimes as a function of applied K_i.

Ans. For this specimen geometry $Y \cong 2.3$ for the range of initial crack lengths chosen (Appendix 1). Since n^*, D and σ_a (= 2.7 MPa) are known we can use equation (6.46) to estimate t_f as a function of applied K_i $(= \sigma_a Y\sqrt{\pi a_i})$. As a sample calculation, consider $K_i = 10^6$ Pa$\sqrt{\text{m}}$; we have:

$$t_f = \frac{2 \times 10^{6(2-25)}}{2.7^2 \times 10^{12} \times 2.3^2\pi \times 10^{-155.34}(25-2)} = 163 \text{ seconds}$$

Calculations for a range of K_i are collected in Fig. 6.27 where also are shown experimental data. There is a high scatter but all the points lie on or to the right of the predicted curve. It seems that the theoretical prediction based on equation (6.46) is a conservative fracture-safe design and this is probably due to the neglect of the crack initiation period in which the crack tip opening displacement develops as the craze grows to the critical value for propagation. In experiments also, poor notching at the crack tip (which produces craze bundles) will also contribute to conservative lifetime predictions, as the blunted crack needs time to re-initiate to a sharp crack.

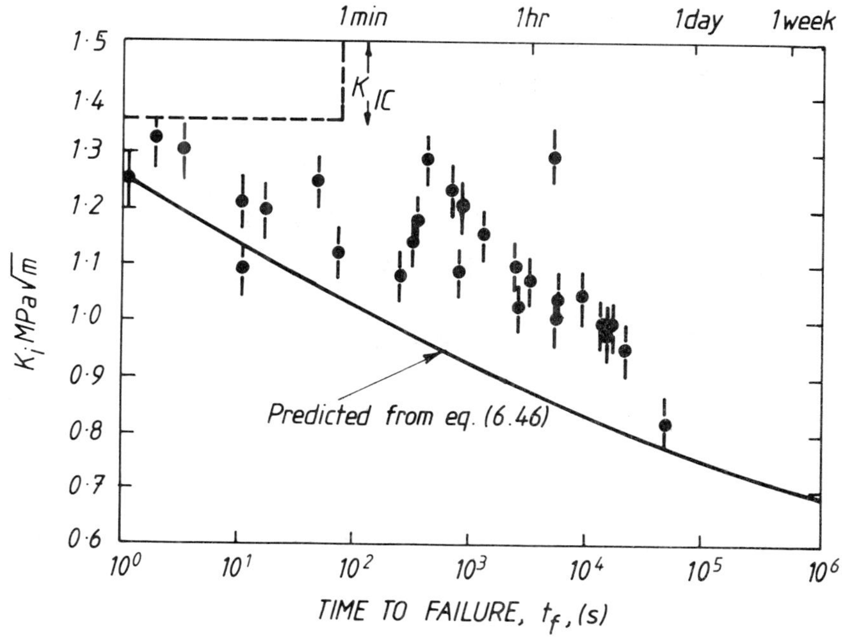

Fig. 6.27 – Experimental time-to-failure data for single edge cracked PMMA tensile specimens as a function of initial stress intensity factor (after Beaumont and Young (1975), with permission).

Example 6.11 The strength/time-to-failure data for unnotched PMMA specimens as a function of temperature are shown in Fig. 6.28. Discuss how these data can be analysed in terms of fracture mechanics.

Ans. If we assume the unnotched specimens to possess inherent flaws of size a_i, we can use equation (6.46) to determine the relationship between t_f and σ_a, i.e.

$$t_f = \frac{2(\sqrt{\pi}\sigma_a)^{-n^*} a_i^{1-(n^*/2)} Y^{-n^*}}{D(n^* - 2)} \tag{a}$$

where D and n^* are determined from K_c–$\dot{a}$ diagrams at different temperatures. a_i may be estimated as follows: consider the 18°C data at $\sigma_a = 67$ MPa and $t_f = 1$ s. Assume $Y \approx 1$ for very small values of a_i and that $K_t = K_{IC} = 1.62$ MPa$\sqrt{\text{m}}$. We can calculate K_i from equation (6.45) which in turn gives

$$a_i = \frac{1}{\pi}\left(\frac{K_i}{\sigma_a}\right)^2 \cong 0.073 \text{ mm}$$

Equation (a) now can be used to obtain a $\sigma_a - t_f$ curve for this temperature. The same procedures can be applied for other temperatures. Equation (6.47) may also be used to simplify lifetime estimation procedures if, for instance, the result at $t_f = 1$ s and $\sigma_a = 67$ MPa is used as a reference basis at 18°C and similarly for other temperatures.

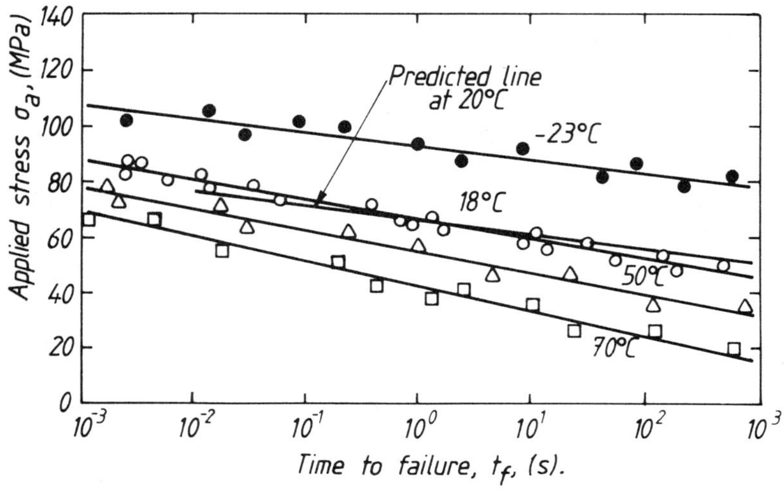

Fig. 6.28 – A strength–time diagram for unnotched PMMA sheet as a function of temperature (after Beaumont and Young (1975), with permission).

In the sorts of calculations presented here dealing with integration of $\dot{a} = D\,K_c^{n*}$ for crack growth, no threshold K_0 was considered and crack growth took place immediately from initial flaw size a_i. In the existence of a threshold (K_{ISCC} in stress corrosion cracking, sections 6.4 and 6.5 – and indeed also in the existence of a threshold range ΔK_0 in fatigue, Chapter 7) cracking should not occur for applied stress intensities lower than the threshold. One design philosophy is therefore to ensure that operating stresses give stress intensities below the threshold, in which case environment-assisted cracking should not be a problem. That is, there will be an upper 'tolerable size' for defects at the working stress level. But this necessitates knowledge of inherent flaw sizes, and there is a lower limit for reliable nondestructive crack inspection methods (say, about 1 mm). Since K_{ISCC} can be quite low for engineering alloys, (<5 MPa$\sqrt{\text{m}}$), the size of tolerable defect ($a = K^2_{ISCC}/\sigma^2 Y^2 \pi$) will be quite small unless σ is prohibitively low. It is quite possible, therefore, for the tolerable defect size to be smaller than the smallest defect that can be determined by existing NDE techniques. This poses a dilemma for the designer, as illustrated in Fig. 6.29 in which Ford (1975) shows how design stress levels and tolerable defect sizes relate to the value of K for stress corrosion cracking and stress corrosion fatigue cracking in steam turbine disk/rotor steels in water at 110°C. As Tomkins (1981) remarks, however, things are not quite so hopeless as they appear: in the

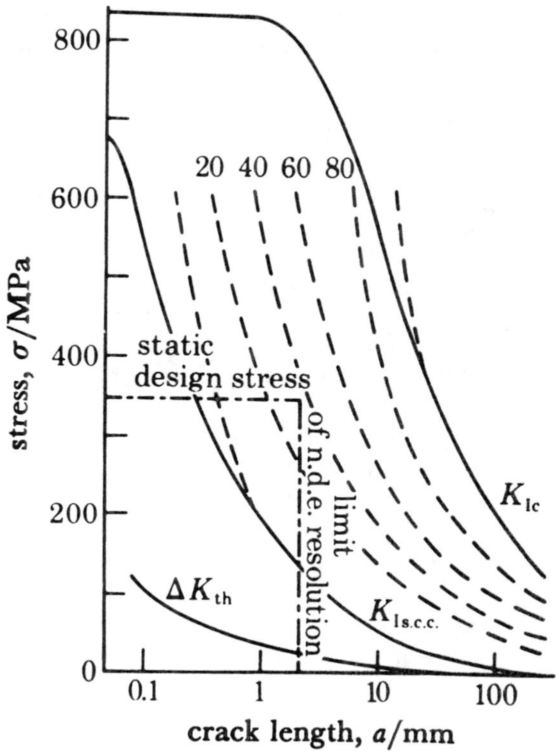

Fig. 6.29 – Stress–crack length diagram for steam turbine disk/rotor Ni–Cr–Mo–V and Cr–Mo steels in water at 110°C (after Ford (1975); taken from Tomkins (1981), with permission).

first instance, many pre-existing defects are not sharp cracks and the life of many components and structures is bound up in the time or cycles taken to initiate sharp cracks from defects. Secondly, static sub-critical crack growth (and fatigue crack growth – see Chapter 7) can in any case be predicted as already demonstrated in this section, and linked to in-service inspection.

6.6.1 Proof Testing and Statistical Variation of Structural Lives

Proof testing is a method whereby individual samples, in a batch that is not up to specifications regarding performance requirements, can be rejected before entering service. Survival of the proof test guarantees that the stress intensity at the crack tip of the most serious flaw does not exceed K_{IC}, otherwise failure would have occurred. Thus we have

$$K_{Ip} = \sigma_p Y \sqrt{\pi a_p} < K_{IC} \tag{6.48}$$

where σ_p is the proof stress, a_p is the propagated flaw size after proof testing (if it increases during the test) and K_{Ip} is the stress intensity factor on the surviving specimens at the instant σ_p is removed. In subsequent service under an applied stress σ_a less than σ_p the applied stress intensity factor K_i is

$$K_i = \sigma_a Y\sqrt{\pi a_p} \leqslant \frac{\sigma_a}{\sigma_p} K_{IC} \tag{6.49}$$

Substituting the maximum K_i value in equation (6.46), the minimum time-to-failure (t_f^*) at σ_a is:

$$t_f^* = 2\left(\frac{\sigma_p}{\sigma_a}\right)^{n^*-2} \bigg/ (n^*-2)\, D\pi\sigma_a^2 Y^2 K_{IC}^{n^*-2} \tag{6.50}$$

This same equation can be obtained if, during proof testing, a_i remains constant. It is thus clear that for a given material system t_f^* is dependent on the proof stress/applied stress ratio. Equation (6.50) may be represented in graphical form as in Fig. 6.30 for soda-lime glass in water where plots of t_f^* versus σ_a give a series of straight lines on logarithmic paper with a slope −2. This diagram gives the level of proof stressing required to guarantee no failures within a specified time

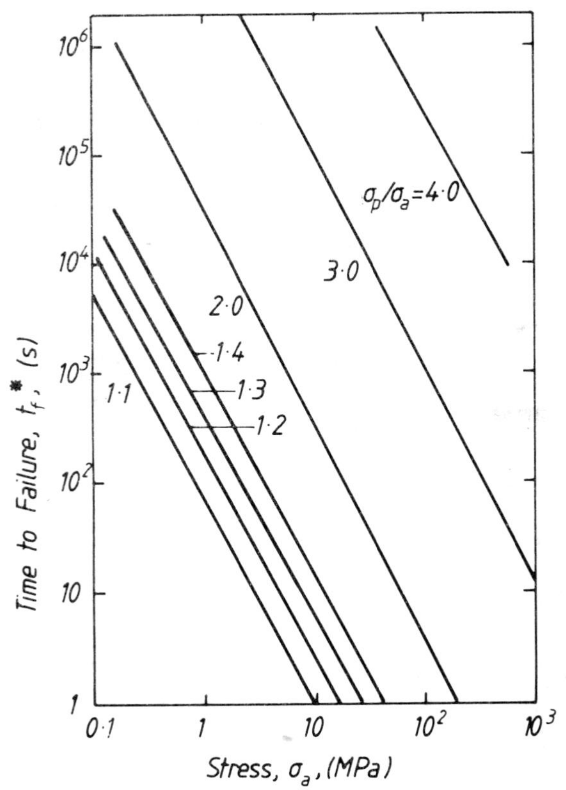

Fig. 6.30 – A proof test diagrams for soda-lime glass in water (after Evans and Weiderhorn (1974), with permission).

at the operating stress. For example if it is required that the soda-lime glass sustain an applied stress of 100 MPa for a time-to-failure of 10^5 s, the proof test stress should be 400 MPa (since $\sigma_p/\sigma_a = 4$ for these performance requirements).

Metals and polymers have more-or-less deterministic strengths and times to failure. For many brittle materials such as glasses and ceramics, however, the minimum time to failure as predicted by proof testing may be very conservative owing to the wide distribution of strengths. It is best therefore to consider the probability of survival or failure in lifetime predictions for these brittle solids.

The variation of strength within a given volume V of brittle material can be described by a Weibull distribution which, for the probability of survival $P_s(V)$ under a stress σ gives

$$P_s(V) = \exp\left[-V\left(\frac{(\sigma-\sigma_u)}{\sigma_0'}\right)^{m*}\right] \tag{6.51}$$

where σ_u is the stress below which there is zero probability of fracture (and is conveniently taken as zero), σ_0' is a normalising parameter of no physical significance and m^* is the Weibull modulus which is a measure of the strength variation – the higher m^* the less variable the strength (see, for example, Felbeck and Atkins, 1984). Values of $5 \leqslant m^* \leqslant 20$ are common for glasses and ceramics. For the purpose of plotting data it is usual to take logarithms of equation (6.51) twice and rearrange as:

$$\ln\ln\left(\frac{1}{1-P_f}\right) = \ln V + m^*\ln(\sigma-\sigma_u) - m^*\ln\sigma_0' \tag{6.52}$$

where P_f (= $1 - P_s$) is the failure probability corresponding to σ. Equation (6.46) which is derived without considering proof testing can be combined with this equation to obtain a time to failure, t_o, on a probability basis. Thus,

$$t_o = \frac{2K_i^{2-n^*}}{D\pi\sigma_a^2(n^*-2)Y^2} = \frac{2\sigma^{n^*-2}}{D\pi\sigma_a^{n^*}Y^2(n^*-2)K_{IC}^{n^*-2}} \tag{6.53}$$

where $K_{IC} = \sigma Y\sqrt{\pi a_i}$ and $K_i = \sigma_a Y\sqrt{\pi a_i}$. Taking the logarithm of this equation and substituting for σ (with $\sigma_u = 0$) we obtain:

$$\ln t_o = \frac{(n^*-2)}{m^*}\ln\ln\left(\frac{1}{1-P_f}\right) - (n^*-2)\left(\frac{\ln V}{m^*} - \ln\sigma_0'\right) - n^*\ln\sigma_a + \ln\left(\frac{2}{DY^2\pi(n^*-2)K_{IC}^{n^*-2}}\right) \tag{6.54}$$

For probability of failure $P_f < 0.1$, this reduces to:

$$\ln t_o = \frac{(n^*-2)}{m^*}\ln P_f - (n^*-2)\left(\ln\left[\frac{V^{1/m^*}}{\sigma_0'}\right]\right) - n^*\ln\sigma_a + \ln\left(\frac{2}{DY^2\pi(n^*-2)K_{IC}^{n^*-2}}\right) \tag{6.55}$$

The values of time-to-failure without proof testing can thus be plotted on the proof stress diagram (e.g. Fig. 6.30) for various P_f as a series of straight lines with slope $-n^*$, as shown in Fig. 6.31 for soda-lime glass immersed in water under a flexural load. The practical significance of this diagram is that, if for example it is required to operate a glass structural part in a water environment for 10^6 s at a stress of 5 MPa, application of a proof stress ratio of 3.15 (σ_p = 15.75 MPa) would give a probability of failure of approximately 4 per cent.

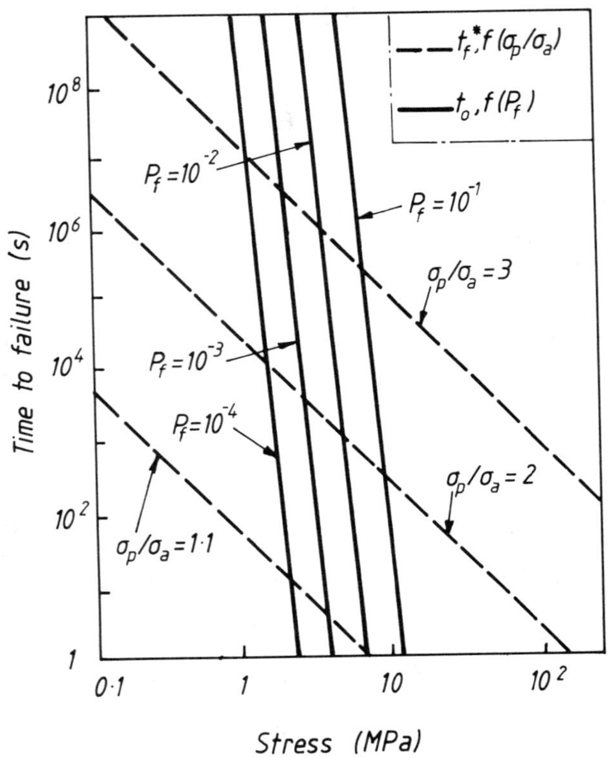

Fig. 6.31 – A complete time-to-failure diagram for ground soda-lime glass in water under a flexural load (after Evans and Wiederhorn (1974), with permission).

Example 6.12 The following experimental data were obtained for a self-bonded silicon carbide with a volume of material equal to 115 mm^3 subject to a uniform tensile stress:

σ (MPa)	221–240	241–260	261–280	281–300	301–320	321–340
No. of tests	1	2	1	5	4	4
σ (MPa)	341–360	361–380	381–400			
No. of tests	1	2	1			

Calculate the mean strength of specimens that are ten times as large, i.e. $V = 1{,}150\ \text{mm}^3$. What would be the mean three-point bend strength of these specimens? Assume the Weibull modulus to be approximately equal to 10. (Taken from Davidge (1979), with permission.)

Ans. The mean strength for the small specimen can be calculated from the given data and this is approximately 309 MPa. Consider first the case when two specimens of volumes V_1 and V_2 are subjected to a uniform stress field. The associated stresses σ_1 and σ_2 for the same probability of survival are related by Weibull's equation (6.51). For $\sigma_u = 0$, we have

$$\frac{\sigma_1}{\sigma_2} = \left(\frac{V_2}{V_1}\right)^{1/m^*} \tag{a}$$

Now with $\sigma_1 = 309$ MPa, $V_2/V_1 = 10$ and $m^* = 10$, this gives $\sigma_2 = 245$ MPa. Equation (a) implies that the strength variation of brittle materials is a manifestation of size effects.† Care must therefore be exercised when designing large components using experimental strength data derived from small testpieces. (Note that in equation (a) as $m^* \to \infty$, $\sigma_1 \to \sigma_2$, and all specimens irrespective of size break at the same stress with no variability.)

The stress distribution in the specimen will be different in three-point bending. In tension the stress is uniform over the cross-section but in elastic bending the stress is maximum at the beam surface and is zero at the neutral axis.

Owing to the linear change in bending moment from ends to mid-span in three-point bending, there is also a linear variation of stress along the length of the beam for a fixed depth distance into the beam. Thus,

$$\sigma = \sigma_b \left(1 - \frac{2D'}{D}\right)\left(1 - \frac{2L'}{L}\right) \tag{b}$$

(Fig. 6.32). In order to account for stress variation within the volume of the material the survival probability given by equation (b) can be expressed in an integral form as:

$$P_s = \exp\left(-\int_V \left[\frac{\sigma}{\sigma_0'}\right]^{m^*} dV\right) \tag{c}$$

---

†Note that there are size effects in fracture for *all* solids, even those which behave in an extremely ductile fashion in small samples but which behave in a brittle manner in large components. As explained in Chapter 9, the causes are *not* statistical, rather simply mechanical, being connected with the differences which arise when quantities scale in different ways, viz. energy available for fracture which scales with the *volume* of the cracked body, and work required for fracture which scales with the *area* of the fracture plane.

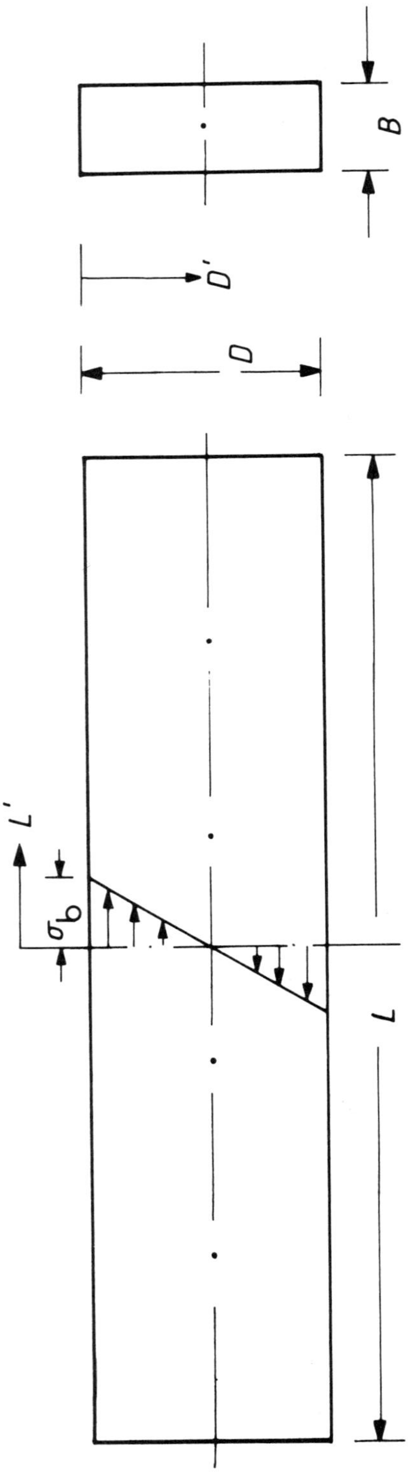

Fig. 6.32 – Stress distribution in beam of rectangular cross-section subjected to bending.

for $\sigma_u = 0$. For the uniform stress case the integral is

$$P_s = \exp\left(-\int_{BDL}\left[\frac{\sigma_t}{\sigma_0'}\right]^{m^*} dV\right) = \exp\left(-\left[\frac{\sigma_t}{\sigma_0'}\right]^{m^*}\right)BDL \qquad \text{(d)}$$

For the three-point bending case, we have

$$P_s = \exp\left(-\left[\frac{\sigma_b}{\sigma_0'}\right]^{m^*} (2B) \int_0^{D/2}\left[1-\frac{2D'}{D}\right]^{m^*} dD' \times \int_0^{L/2}\left[1-\frac{2L'}{L}\right]^{m^*} dL'\right)$$

$$= \exp\left(-\left[\frac{\sigma_b}{\sigma_0'}\right]^{m^*} \frac{BDL}{2(m^*+1)^2}\right) \qquad \text{(e)}$$

From equations (d) and (e) it can be easily shown that for the same P_s,

$$\frac{\sigma_b}{\sigma_t} = [2(m^*+1)^2]^{1/m^*} \qquad \text{(f)}$$

which for $m^* = 10$, $\sigma_b = 1.73\sigma_t$. This shows why the bend strengths of brittle materials like ceramics, glasses, cements and concretes are considerably larger than the tensile strengths. For pure bending, the second integral of equation (e) does not need to be evaluated since there is no variation of stress along the length of the beam. Thus, the pure bend strength to tensile strength ratio is given by $(2(m^*+1))^{1/m^*}$ which for the same m^* (= 10) gives a value of 1.36. Thus pure bend strengths of ceramics and similar brittle materials are usually lower than three-point bend strengths.

It may be pointed out here that the discussion so far does not distinguish between surface and interior flaws. Clearly the surface flaws are under much larger stresses and they will be first initiated when the local K initiation value is reached. Crack coalescence and slow crack growth will occur before final fracture instability which determines the quoted tensile or bend strengths.

6.7 THERMAL STRESSES AND SHOCK

Bodies which are fully restrained and experience a *uniform* change in temperature ΔT suffer a *uniform* thermal strain $\epsilon_{th} = \alpha\,\Delta T$ and elastic thermal stress $\sigma_{th} = E\,\alpha\,\Delta T$ or plastic thermal stress of $\sigma_{th} = \sigma_0\,(\alpha\,\Delta T)^n$ or some such; α is the coefficient of thermal expansion. It is possible therefore to produce fracture by uniform heating or cooling if (in the elastic range) $E\alpha\Delta T$ satisfies the 'fracture mechanics' critical stress for fracture, $\sigma_{cr} = K_C/\sqrt{\pi a}\,.\,Y$. For many ceramics, the ΔT required to produce fracture is some 100–200°C (using $\alpha \approx$ (4 to 10) $\times$ 10^{-6} °C^{-1} and $\sigma_{cr}/E \approx 10^{-5} \sim 10^{-3}$). In comparison, brittle metals that fail with little if any plastic flow (such as tool steels) require greater uniform ΔT (say, 400°C) since their σ_{cr}/E is larger, even though α is also greater.

When the change in temperature is not uniform through the body, the thermal stress is given by $\sigma_{th} = M\, E\alpha\, \Delta T$ where M depends on the thermal conductivity, k, interfacial convective heat transfer coefficient h, and geometry of body. The rate of change of ΔT may be slow, or now fast as in *thermal shock,* where ΔT occurs suddenly at the boundaries of the body. Fracture will be produced by ΔT when $\sigma_{th} = \sigma_{cr}$, so the magnitude of ΔT required to cause fracture depends on $\sigma_{cr}/ME\alpha$ or $K_C/ME\alpha\sqrt{a}$. Since, in general, M varies inversely with thermal conductivity k in a complicated fashion, ΔT depends on $k\sigma_{cr}/E\alpha$ or $k\,K_C/E\alpha\sqrt{a}$. As a guide, $k\sigma_{cr}/E\alpha$ is less than 1 kW/m for glasses, less than 10 kW/m for ceramics, but about 100 kW/m for refractory metals (see Felbeck and Atkins, 1984).

Instead of completely breaking a solid in one ΔT excursion, many small cracks may be initiated which, even if they propagate some distance, are found to arrest. These cracks (called 'thermal checking' which are similar to networks of cracks found on glazed pottery, old varnish on paintings, dried out mud in ponds, the basalt rocks on the Giants Causeway in Ireland), degrade the strength of the body since σ_{cr} is lower in the presence of longer values for a. The effect is more marked in porous bodies owing to the large number of potential crack sites (section 3.3). The effect may also be progressive with increasing number of shocks. It will be shown that the arrest of such small cracks in these circumstances, or in other words the 'extent of damage' caused by ΔT, depends inversely on $(ER/\sigma_{cr}^2) = (K_{IC}/\sigma_{cr})^2$ where R and K_{IC} are *dynamic* values over the range of ΔT.

Thus *two* parameters must be used to rank resistance to thermal shock, the one controlling crack initiation, the other crack propagation. Both should correlate with empricial tests which establish (i) critical quenching temperatures at which surface cracks are initiated and (ii) the number of further shocks that can be withstood before complete rupture. Fig. 6.33 shows the retained room-temperature bend strengths after ten thermal cycles as a function of shock temperature interval for (a) a type of titanium carbide cutting tool material and (b) a cordierite (lithium–aluminium–silicate) ceramic used in turbines. In Fig. 6.33 (a) a critical quenching temperature interval of about 400°C is shown; beyond this there is a marked reduction in residual bend strength, which remains relatively constant thereafter. The unshocked high-temperature strength of the carbide (open points) remains constant at the room-temperature value to temperatures beyond the critical temperature. The behaviour shown in Fig. 6.33(b) is different: the after-shock strength gradually decreases as the shock-temperature interval increases. The contrasting patterns are bound up in differences in the ease of crack initiation and growth at high temperatures; that is, the parameters $k\sigma_{cr}/(E\alpha)$ and K_C/σ_{cr} change their relative ranking with temperature, since k, σ_{cr}, E, and α alter much more in some materials than in others.

Hasselman's theory of thermal shock given in 1969 explains the behaviour shown in Fig. 6.33(a). Assume that a body containing many penny-shaped microcracks with numerical density N and radius r_i undergoes uniform cooling

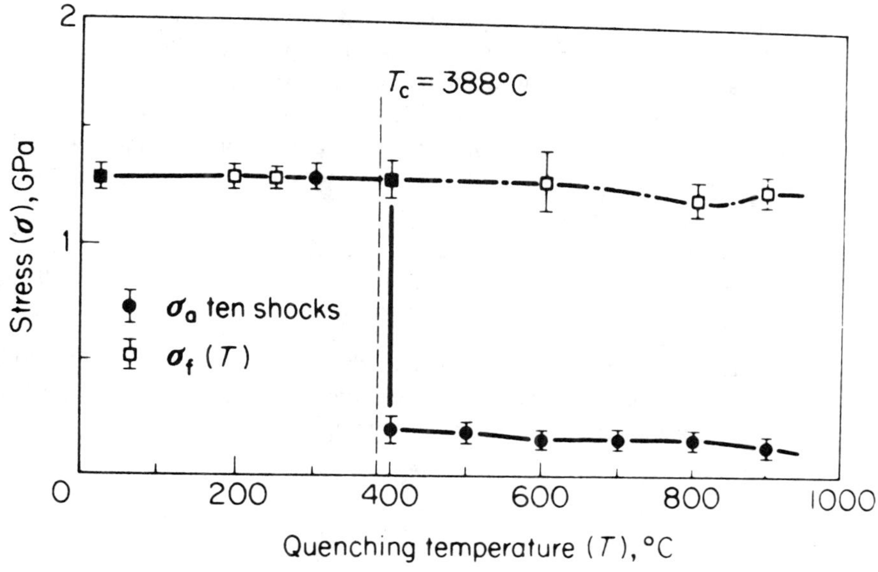

(a)

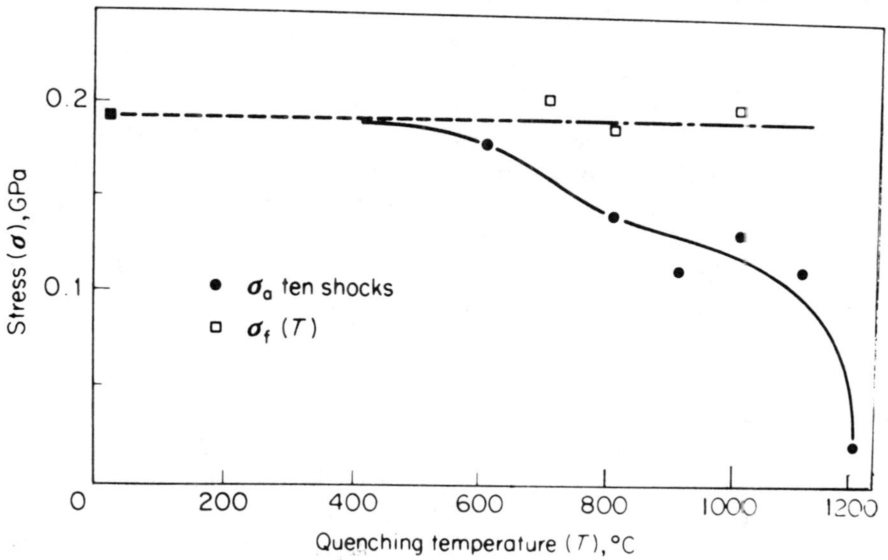

(b)

Fig. 6.33 – Retained room-temperature bend strengths after ten thermal cycles as a function of shock temperature interval for (a) a type of TiC cutting tool material, (b) a cordierite (lithium–aluminium–silicate) turbine ceramic (after Mai and Atkins (1975), with permission).

with the external surface rigidly constrained so that there is a state of triaxial tensile stress. This represents the worst possible situation since the body is subjected to uniform maximum stress. In a rapid quenching or shock experiment the stresses in the body are not uniform. By considering the potential energy change of the body at crack initiation, Hasselman proved the existence of a *critical shock temperature difference* (ΔT_c) required to initiate all N cracks, given by

$$\Delta T_c \geqslant \frac{\pi}{2E_0} \sqrt{\frac{E_0 R}{\pi r_i(1-\nu^2)}} \; \frac{(1-2\nu)}{\alpha} \left(1 + \frac{16}{9} \; \frac{r_i^3(1-\nu^2)N}{(1-2\nu)}\right) \tag{6.56}$$

where E_0 is the Young's modulus of the crack-free material. The critical stress (σ_{cr}) to cause crack initiation in an unnotched sample (three-point bend plain bar, for example) containing inherent microflaws of size r_i, is

$$\sigma_{cr} = \frac{\pi}{2} \sqrt{\frac{E_0 R}{\pi r_i(1-\nu^2)}} \tag{6.57}$$

when the flaw sizes are appropriate enough to allow continuum mechanics to be applied. R or K_C would be determined separately from tests involving specimens with deliberately introduced (long) starter cracks. Combining equations (6.56) and (6.57) we obtain

$$\Delta T_c \geqslant \frac{\sigma_{cr}}{E_0 \alpha} (1-2\nu) \left(1 + \frac{16 r_i^2 (1-\nu^2)N}{9(1-2\nu)}\right) \tag{6.58}$$

It is seen therefore that ΔT_c for crack initiation is proportional to the parameter already identified, namely ($\sigma_{cr}/E_0\alpha$). Relations for ΔT_c may be couched either in terms of σ_{cr} or K_C and r_i, as illustrated by equations (6.58) and (6.56) respectively; for historical reasons, relations involving σ_{cr} are more frequently seen in the literature. When r_i is small or when $N = 0$ (for crack-free bodies) and dropping the subscript for E_0, equation (6.58) is reduced to:

$$\Delta T_c \geqslant \frac{\sigma_{cr}}{E\alpha} (1-2\nu). \tag{6.59}$$

The sharp drop in retained strength subsequent to thermal shocks greater than about 400°C in Fig. 6.33(a) does not agree quantitatively with Hasselman's theory: equation (6.59) predicts $\Delta T_c \approx 200$°C in those circumstances. This discrepancy is not unexpected since the assumptions made in deriving equation (6.59) are not fully obeyed by the conditions of the actual experiment. For example, the stresses set up by thermal shock must depend on the heat transfer coefficient (h), the thermal conductivity (k) and the size of the body ($B/2$), i.e. on the Biot modulus ($Bh/2k$). Here B is either the thickness of a strip or the diameter of a rod-like specimen (which is assumed to be infinite in the longitudinal direction). Equation (6.59) corresponds only to the case when the Biot modulus is very *large* (Manson, 1953). For *small* Biot modulus situations the critical thermal shock temperature for rectangular beams is given by (Manson, 1953):

$$\Delta T_c \geqslant \frac{6.50}{Bh}\left(\frac{k\sigma_{cr}}{E\alpha}\right)(1-\nu) \tag{6.59a}$$

where now ΔT_c is proportional to $(k\sigma_{cr}/E\alpha)$ already identified. The term $(1-\nu)$ indicates that plane stress conditions are assumed. Equation (6.59a) now predicts $\Delta T_c \approx 388°C$ which agrees well with the experimental value of ~400°C (Mai and Atkins, 1975). Manson's complete equation is given in Example 6.14 following.

Consider now the events subsequent to crack initiation. This may involve either catastrophic crack propagation resulting in complete rupture of the specimen, or may involve *crack arrest* after some propagation (Chapter 8). If all N cracks grow from r_i to an arrest radius r_f, it may be shown that (Hasselman, 1969):

$$r_f = \frac{1}{2\sqrt{2}}\left(\frac{3(1-2\nu)}{Nr_i(1-\nu^2)}\right)^{1/2} \quad \text{for } r_f \gg r_i \tag{6.60}$$

Equations (6.57) and (6.60) together yield

$$r_f^2 = \left(\frac{3}{2\pi N}\right)\left(\frac{\sigma^2{}_{cr}}{ER}\right)(1-2\nu)$$

Thus,

$$r_f = \sqrt{\frac{3(1-2\nu)}{2\pi N(1-\nu^2)}}\left(\frac{K_{IC}}{\sigma_{cr}}\right)^{-1} \tag{6.61}$$

and is an indication of the 'damage' or extent of crack propagation (cf. Example 6.15 following). In Fig. 6.33(a) the much lower retained strengths, for bodies shocked more than the ΔT of some 400°C, are caused by all the inherent cracks of radius r_i having propagated to r_f. These enlarged cracks are 'sub-critical' until very much larger ΔT are applied, which explains why the retained strength remains constant for so long after $\Delta T = 400°C$ and why those results are not expected to decrease until quenching takes place from much higher temperatures.

The strength retained by the body (σ_a) subjected to the critical thermal shock as described by Hasselman can be obtained from

$$\sigma_a = \frac{\pi}{2}\sqrt{\frac{ER}{\pi r_f(1-\nu^2)}} \tag{6.62}$$

where r_f is given by equation (6.60). Combining equations (6.61) and (6.62) we obtain the fractional strength retained as:

$$\frac{\sigma_a}{\sigma_{cr}} = \left(\frac{\pi^3}{24}\,\frac{(1-\nu^2)}{(1-2\nu)}\,\mathrm{N}\right)^{1/4}\left(\frac{K_{IC}}{\sigma_{cr}}\right)^{3/2} \tag{6.63}$$

cf. Example 6.15 following.

Thus, materials with small r_f have good thermal shock damage resistance and good retained after-shock strengths. From both equations (6.61) and (6.62) it follows that a large (K_{IC}/σ_{cr}) parameter is needed. Clearly the bigger the magnitudes of *both* the two shock parameters, i.e. $k\sigma_{cr}/E\alpha$ and K_{IC}/σ_{cr} the better will be the resistance to crack initiation *and* damage. There is however an incompatibility here in that high σ_{cr} means high crack initiation resistance but poor crack damage resistance. There is therefore a need to improve K_{IC} of ceramics in order to improve this latter parameter for crack propagation resistance.

Since σ_c for brittle ceramics is size-dependent (Chapter 9), in that σ_{cr} is larger for a smaller size, equations (6.59) and (6.63) will predict size-dependent ΔT_c and σ_a/σ_{cr}. These predictions are supported by the results for TiC shown in Fig. 6.34. Here size *A* is less than size *B* which is less than size *C*.

A variety of ceramic materials follow the Hasselman model and display a well-defined critical ΔT_c as illustrated in Fig. 6.33(a) and 6.34. Others, however, as illustrated by Fig. 6.33(b) show a gradual reduction in after-shock retained

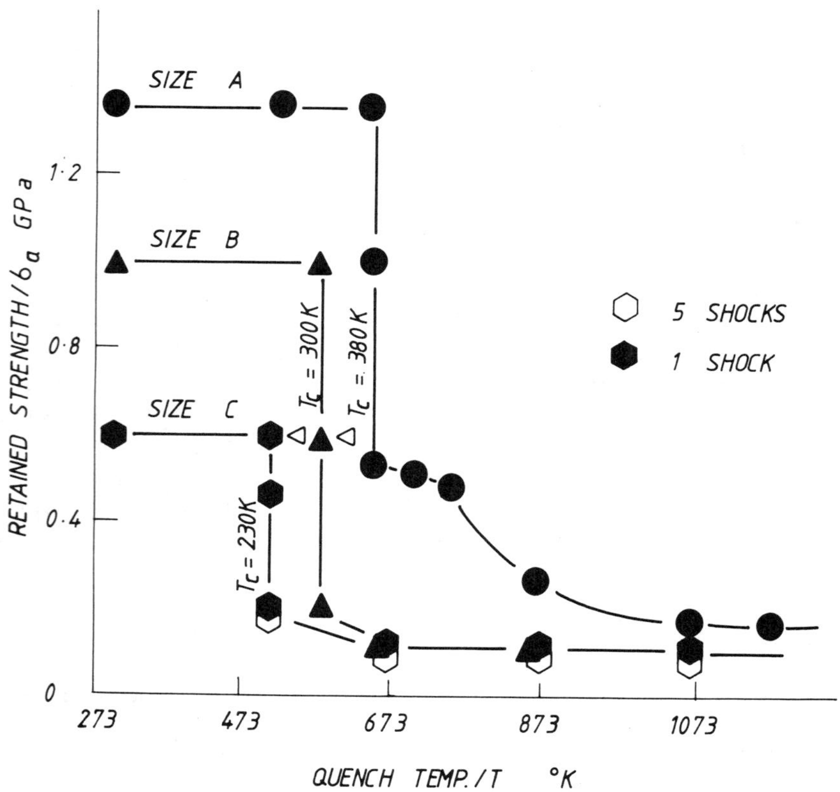

Fig. 6.34 – Size effect on critical shock temperature (ΔT_c) and retained strength (σ_a) for a titanium crabide. Size A is smaller than size B which is smaller than size C (after Mai (1977), with permission).

strength at increasing ΔT. As explained by Mai and Atkins in 1975 this arises because the two parameters $k\sigma_{cr}/E\alpha$ and K_C/σ_{cr} change their relative ranking with temperature, whereas ΔT_c given by equation (6.59) is often evaluated with data for k, σ_{cr}, E, α and K_C which strictly relate to room temperature.

Ceramic materials like alumina, oxides, silicon nitrides and carbides are among the most commonly used in cutting and turbine operations. They retain the room temperature fracture toughness even up to very high temperatures. The strength drops, however, at these elevated temperatures but this makes the (K_{IC}/σ_{cr}) large so that the material is more resistant to thermal-shock-induced crack propagation. Hot-pressed silicon carbides for example, have K_{IC} values approximately constant up to ~1,000°C and thereafter K_{IC} drops monotonically with further increase in T, Fig. 6.35. Sintered SiC does not show this drop in K_{IC} at elevated temperatures (Evans and Lange, 1975). Indeed a slight increase of K_{IC} from 1,000°C to 1,500°C has been observed by some authors. It has been argued that this increase is a result of crack tip blunting due to oxidation attack. Similarly, for hot-pressed silicon nitride, K_{IC} remains stationary at ~5 MPa$\sqrt{m}$ up to 1,000°C and then there is a sharp rise to 12 MPa$\sqrt{m}$ at ~ 1,600°C, Fig. 6.35 (cf. Fig. 6.4(a)).

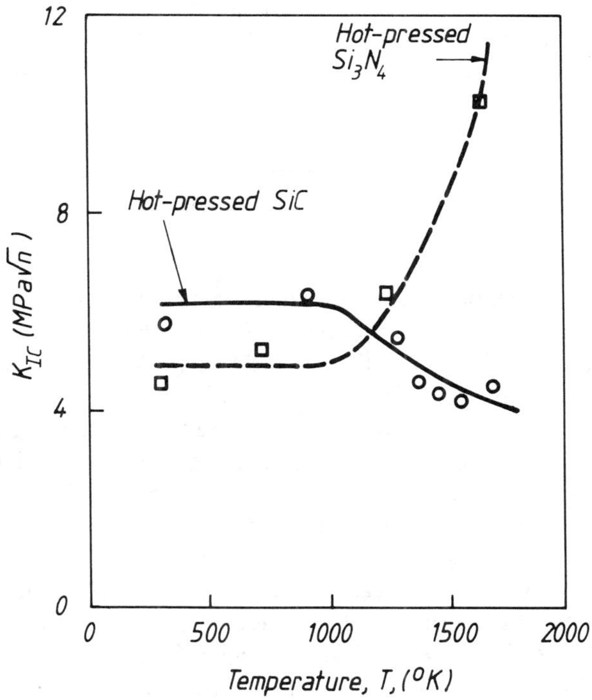

Fig. 6.35 – Temperature variation of K_{IC} for hot-pressed SiC and Si_3N_4 (after Henshall *et al.* (1977), with permission).

Example 6.13 Compare the thermal shock resistance of two carbide ceramics, WC and TiC, for use as cutting tools. The following mechanical and physical properties apply:

Property	TiC	WC
k (J/s°C m)	17	46
σ_{cr} (MPa)	500	1680
K_{IC}(MPa$\sqrt{m}$)	7.5	13
E(GPa)	446	560
ν	0.20	0.23
α (°C^{-1})	2.40×10^{-6}	5.70×10^{-6}
$\beta = Bh/k$	16.20	5.9

The empirical assessment of thermal shock resistance, based upon number of ΔT excursions required to produce complete failure in bars 6.5 mm × 6.5 mm × 38 mm, was:

Material	Thermal cycles (to temperature °C) completed before failure						
	260°	300°	400°	600°	800°	1,000°	N_f
TiC	10	10	10(5)†	10	10	4	54
WC	10	10	10	10(1)	10	–	50

†Parentheses indicate cycles required for initiation of surface cracks.

Ans. The thermal shock resistance parameters are given by $(k\sigma_{cr}/E\alpha)$ and (K_{IC}/σ_{cr}). For TiC at room temperature these values are: 7.85 (kJ/ms) and 122 $\sqrt{\mu m}$; and for WC they are: 24.36 (kJ/ms) and 88$\sqrt{\mu m}$. It is expected therefore that TiC offers better resistance to crack propagation but poorer resistance to crack initiation than WC. The theoretical ΔT_c from equation (6.59a) are 150°C (TiC) and 440°C (WC) (see Example 6.14 following). These predictions agree approximately with experiment, which shows that ΔT_c are 230°C and 360°C respectively. Thus WC can withstand more cycles before crack initiation takes place because its $(k\sigma_{cr}/E\alpha)$ parameter is larger, but the total number of cycles to failure is less because its (K_{IC}/σ_{cr}) parameter is lower.

It may be added here that experiments have shown that both WC and TiC have larger K_{IC} values at elevated temperatures above 800°C. This is because of crack tip blunting resulting from plastic flow in the metal binders, e.g. the cobalt phase in WC, and also multiple cracking caused by separation of grain boundaries (see Mai, 1976).

Example 6.14 The fracture toughness for a certain grade of tungsten carbide cutting tool is about 13 MPa$\sqrt{\text{m}}$ for the temperature range 25°C to 800°C. Estimate the size of the inherent cracks if the carbide has a flexural strength of 1,680 MPa. If the tool is to endure thermal shock during cutting estimate the critical shock temperature at which the tool would have lost a considerable portion of its original strength. What would be the residual bend strength of the tool after the critical shock? Calculate also the final size of the inherent cracks after the critical thermal shock. Other physical properties of tungsten carbide are those given in Example 6.13.

Ans. The inherent crack size of the tungsten carbide tool is given by:

$$\sigma_{\text{cr}} Y \sqrt{\pi a_{\text{i}}} = K_{\text{IC}}$$

It may be assumed that the worst crack is a semi-circular one that is situated at the free surface of the beam under flexural bending which gives $\sigma_{\text{c}} = 1{,}680$ MPa. Here $Y \approx 0.70$ assuming a_{i} to be very much smaller than the beam depth. Thus, we have:

$$1{,}680 \times 0.7 \times \sqrt{\pi a_{\text{i}}} = 13$$

$$a_{\text{i}} = 39\ \mu\text{m}$$

The critical shock temperature can be approximately estimated by equation (6.59a) which is valid for a rectangular beam geometry. It must be noted that the equation will be different for a non-rectangular cutting tool geometry (see Example 6.15 following). Nevertheless useful information can be obtained using the above equation. Let us assume that the Biot modulus $(Bh/2k)$ for the carbide beam is 3.0. Then

$$\Delta T_{\text{c}} = \frac{6.50}{6} \left(\frac{1680 \times 10^3 \times (1 - 0.23)}{560 \times 5.70} \right)$$

$$= 440°\text{C}$$

Now to estimate the final flaw size (a_{f}) after the critical shock (assuming that all cracks propagated during the shock) we need to use equation (6.61), in which N is the *volume* density of the cracks and is difficult to estimate. Instead it has to be evaluated by measuring the *surface* crack density N_{s} and then calculating N using the relationship: $N \approx N_{\text{s}}^{3/2}$. Here N_{s} is measured at about 6×10^6 cracks/m² giving $N \approx 1.5 \times 10^9$ cracks/m³. Thus,

$$a_{\text{f}} = \sqrt{\frac{3(1 - 0.46)}{2 \times 1.5 \times 10^9 \times (1 - 0.23^2)\pi}} \left(\frac{13}{1680} \right)^{-1} \approx 1.74\ \text{mm}$$

The strength retained (σ_{a}) after the shock can be obtained in two ways. Either

$$\sigma_{\text{a}}\ Y \sqrt{\pi a_{\text{f}}} = K_{\text{IC}}$$

and if $a_f \ll$ depth of the beam then $Y \approx 0.70$ still, in which case

$$\sigma_a = \frac{13}{0.70\sqrt{\pi \times 1.73 \times 10^{-3}}} = 251 \text{ MPa}$$

Alternatively if equation (6.63) is used to predict σ_a we have:

$$\frac{\sigma_a}{\sigma_{cr}} = \left(\frac{\pi^3}{24} \times \left[\frac{1-0.23^2}{1-0.46}\right] \times 15 \times 10^8\right)^{1/4} \left(\frac{13}{1680}\right)^{3/2}$$

i.e. $\sigma_a = 268$ MPa.

Note. ΔT_c is calculated using equation (6.59a) which holds for *small* Biot modulus β. The complete equation is (Manson, 1953):

$$\Delta T_c \geqslant \left(\frac{\sigma_{cr}}{E\alpha}\right)(1-\nu)\left(1.50 + \frac{3.25}{\beta} - 0.5 \exp \frac{-16}{\beta}\right)$$

i.e. $\Delta T_c \geqslant 1{,}046°\text{C}$ for $\beta = 3.0$ which is obviously much larger than the experimental value of $\sim 360°\text{C}$. If we consider the suggestion made by Davidge and Tappin (1967) that σ_{cr} should be one-third the flexural strength in order to account for size and environment effects then $\Delta T_c \approx 350°\text{C}$.

Example 6.15 A cylindrical polycrystalline alumina rod of diameter $2b = 5$ mm is subjected to thermal shock experiments. Calculate the critical shock temperature and the strength retained after the shock. The physical and mechanical properties for the alumina are given as

heat transfer coefficient (h)	=	$55 - 100$ kW/°Cm^2
thermal expansion coefficient (α)	=	7×10^{-6} °C^{-1}
tensile strength (σ_c)	=	248 MPa
Young's modulus (E)	=	310 GPa
thermal conductivity (h)	=	16.8 W/m°C
Poisson's ratio	=	0.25

Ans. Equation (6.59a) cannot be used to calculate ΔT_c for this cylindrical specimen geometry. The correct equation to be used here is given by:

$$\Delta T_c = \frac{\sigma_{cr}(1-\nu)}{\alpha E H} \qquad \text{(a)}$$

where H is a function of the Biot modulus ($\beta = bh/k$) as given by Glenny and Royston (1958) for circular cylinders. The bh/k values here are: $8 \sim 15$ for the range of h values given. Thus $H = 0.43 \sim 0.49$ from Glenny and Royston. Thus

$$\Delta T_c = \frac{248 \times (1-0.25)}{7 \times 10^{-6} \times 310 \times 10^3 \times (0.43 \sim 0.49)}$$

$$= 200°\text{C} \sim 175°\text{C}.$$

Again equation (6.63) is not valid for calculating the retained strength because it is based on the crack popagating from r_i to r_f under *uniform* cooling and not *thermal shock* conditions. Since it is not easy to calculate r_f for cracks growing in a non-uniform stress field under thermal shock, an approximate solution is to assume that all the elastic energy at fracture initiation is transformed into fracture energy (see Hasselman, 1970). This assumption is reasonable if crack propagation is extensive and many cracks are nucleated.

Thus, the theoretical elastic energy (Λ) at fracture for the rod with a parabolic temperature gradient (appropriate to thermal shock and $\nu = 0.25$) is (Hasselman, 1970):

$$\Lambda = (0.57\, \sigma_{cr}^2\, b^2/E)l \tag{b}$$

where l is the length of the rod.

If there are N_s cracks/unit area nucleated on the surface, the total number of cracks $N = 2\pi bl\, N_s$. Assuming that $r_i \ll r_f$ and that the inherent cracks grow to semi-circles on the surface then the fracture energy required is:

$$\Lambda_f = N\left(\frac{\pi r_f^2}{2}\right) R = 2\pi b\, l\, N_s\left(\frac{\pi r_f^2}{2}\right) R = \pi^2 b\, l\, r_f^{\,2}\, N_s R \tag{c}$$

where R is the fracture energy per unit area of crack growth. By equating $\Lambda = \Lambda_f$ (strictly it should be rates of change of Λ) we obtain the size of the final propagated cracks. Thus,

$$r_f = \left(\frac{0.57\sigma_c^2 b}{\pi^2\, E\, N_s R}\right)^{1/2} = \left(\frac{0.57b}{\pi^2\, N_s\,(1-\nu^2)}\right)^{1/2} \left(\frac{K_{IC}}{\sigma_{cr}}\right)^{-1} \tag{d}$$

(Note that equation (d) above is similar in form to equation (6.61) which is obtained for *uniform cooling* conditions.)

The strength retained after thermal shock (σ_a) can be obtained by considering the propagation of the semi-circular cracks. Thus, we have from equation (6.62) substituting r_f from (d) above:

$$\frac{\sigma_a}{\sigma_{cr}} = \frac{\pi}{2}\left(\frac{N_s(1-\nu^2)}{0.57b}\right)^{1/4}\left(\frac{K_{IC}}{\sigma_{cr}}\right)^{3/2} \tag{e}$$

cf. equation (6.63). Equation (e) shows that $\sigma_a \propto b^{-1/4}$ i.e. there is a size effect on strength retained, the bigger the rod the less the retained strength.

N_s is estimated here to be about $10^6/\text{m}^2$ and $K_{IC} \approx 4$ MPa$\sqrt{\text{m}}$. Thus equation (e) gives:

$$\frac{\sigma_a}{\sigma_{cr}} = \frac{\pi}{2}\left(\frac{10^6 \times (1-0.25^2)}{0.57 \times 2.5 \times 10^{-3}}\right)^{1/4}\left(\frac{4}{248}\right)^{3/2}$$

$$= 0.52$$

i.e. $\sigma_a = 128$ MPa.

PROBLEMS

6.1 Certain steels, at their NDT temperature, are thought to have a fracture toughness of 125 kJ/m^2. If the yield strength is 250 MPa at NDT temperature, what is the 'starter-crack' size if a very large steel plate fractures at one-half the yield strength?

6.2 For a certain Cr–Mo–V rotor steel the yield strength and critical stress intensity factor are given below. If the smallest dectectable defect by nondestructuve testing is 1 mm, at what temperature would the fracture strength fall to 75% of the yield strength?

Temp. (°C)	Yield strength (MPa)	Fracture toughness (MPa$\sqrt{m}$)
100	570	122
50	620	58
0	680	35
−50	750	28
−100	800	22

6.3 The ductile-to-brittle transition behaviour of polymers is affected by (a) temperature, (b) strain rate, (c) environment and (d) notch geometry. Discuss and give examples to support your answer.

6.4 Show that the stress rate at the plastic/elastic boundary at the tip of a stationary crack is given by $\dot{\sigma} = \dot{\sigma}_y \dot{K}/K$ $(= \sigma_y/t$ for a uniform loading rate, where t is time). Hence show that (i) points at different distances from the crack tip suffer different ranges of stress rate as the stress rises and (ii) points nearer the crack tip suffer greater rates and this occurs earlier in the test.

6.5 If a high stress rate is conducive to crack propagation in rate-sensitive materials, it is clear that variation in $\dot{\sigma}$ and K act in opposite senses during the test, i.e. early in the test $\dot{\sigma}$ is high, but K is low – and vice versa later in the test. Discuss this in relation to the critical condition at which cracking may eventually proceed.

6.6 Show that, to achieve a constant stress rate at the plastic/elastic boundary at the tip of a stationary crack, the loading rate must satisfy $K = K_0 \exp(\dot{\sigma}t/\sigma_y)$ where K_0 is some reference value. Show that this means an accelerating load application.

6.7 In the hatching of bird eggs, hen and duck chicks chip away at the shell (after initial 'pipping') until the damage extends for some two-thirds of the circumference of the shell. In contrast, quail and pigeon chicks chip virtually the whole way around the egg before pushing off the cap. Discuss this behaviour with regard to likely fracture properties of the different eggshells. [see Bond *et al.*, (1980)].

6.8 Illustrated is a double torson specimen which is commonly used for obtaining $(K_{\dot{c}} - \dot{a})$ data for materials that exhibit slow crack growth. If

W_m is the distance between the moment arms, show from first principles that the strain energy release rate (G) is given by:

$$G = \frac{3X^2 W_m^2}{2WB^2\mu}$$

where μ is the shear modulus. Hence prove that the stress intensity factor is:

$$K = XW_m \left(\frac{3(1+\nu)}{WB^4} \right)^{1/2}$$

where ν is Poisson's ratio.

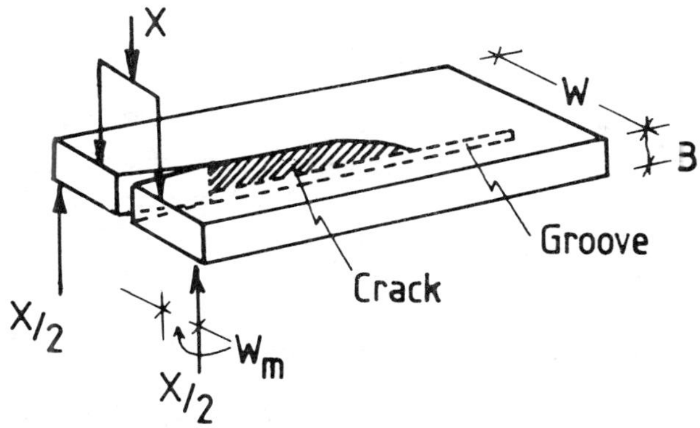

6.9 Referring to Problem 6.8, show that the power law exponent (n^*) of the slow crack growth equation: $\dot{a} = DK_c^{n^*}$ can be obtained from:

$$n = \log(\dot{u}_1/u_2)/\log(X_1/X_2) - 1$$

where (X_1, X_2) are the fracture loads at corresponding crosshead displacement rates ($\dot{u}_1$, $\dot{u}_2$).

6.10 Titanium alloy jet compressor disks in contact with cadmium plated bolts are known to fail at temperatures above the melting temperature of cadmium. Explain why and how would you avoid this problem.

6.11 Explain the modifications necessary to the basic energy rate balance (equation (1.1)) to account for 'vacancy creep'.

6.12 Assuming that a craze can be approximated by a set of parallel cylindrical pipes lying in the craze plane whose axes are parallel to the crack propagation direction show that the rate of craze growth ($\dot{\Delta}$) is given by:

$$\dot{\Delta} = \gamma_{LV}\, d \cos\theta / 8\Delta\eta$$

where γ_{LV} is the liquid–vapour surface tension, θ is the contact angle, d is the pipe diameter and η is the kinematic viscosity.

6.13 The Dugdale model has been used to describe craze growth in a viscoelastic polymeric material whose craze stress (σ_c) has a time dependence given by $\sigma_c = \sigma_0 t^{-m}$ where σ_0 is a unit time stress. Show that the craze growth rate is given by:

$$\dot{\Delta} = \frac{m\pi}{4}\left(\frac{K}{\sigma_0}\right)^2 t^{(2m-1)}$$

State any assumptions you have made in obtaining this equation.

6.14 Environmental crazing normally leads to decrease in tensile ductility. Explain why crazes formed in gaseous environments such as N_2, O_2, CO_2 etc. at low temperatures have greater ductility than in vacuum.

6.15 The rate of crack growth in brass exposed to ammonium sulphate solution is given by:

$$\dot{a} = DK_c^{n^*}$$

where $\dot{a}$ is in mm/yr, $D = 23.9$ and $n^* = 2$. K is in MPa$\sqrt{\text{m}}$. It is proposed to use the brass for piping in an ammonium sulphate plant and the pipes must withstand a pressure-induced circumferential tensile stress of 85 MPa. Experience suggests that longitudinal scratches 0.02–0.05 mm deep are likely to occur on the inner surface of the pipes. Assuming that $K_{IC} = 80$ MPa$\sqrt{\text{m}}$ estimate the lifetime of the pipe once the solution started to flow through it. (Take $K_c = \sigma\sqrt{\pi a}$). How would you protect the inside of the pipe against chemical corrosion?

6.16 The slow crack growth data in glass as shown in the following figures can be described by either of the equations below:

$$\dot{a} = DK_c^{n^*} \text{ and}$$

$$\dot{a} = C \exp\,[\beta K_c]$$

Show the sensitivity of these equations in lifetime prediction as a function of proof stress/applied stress ratio. Assume $K_{IC} = 0.75$ MPa$\sqrt{\text{m}}$ for the glass samples.

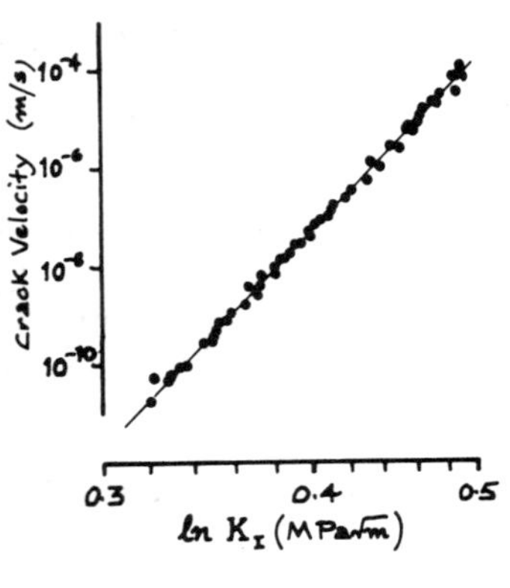

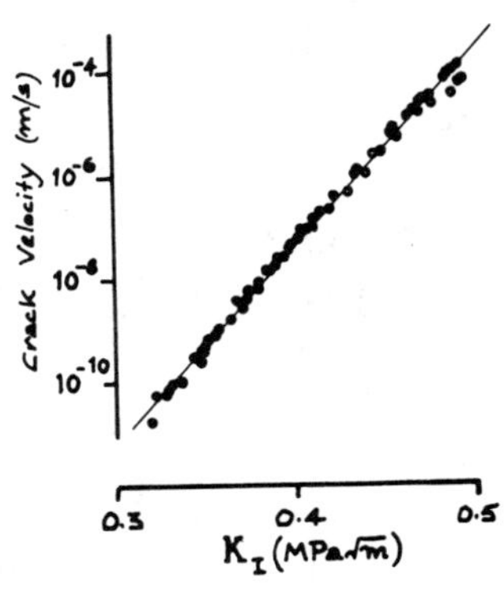

6.17 Show that the constants D and n^* in $\dot{a} = DK_c^{n^*}$ may be obtained from a plot of log σ_f vs. log $\dot{\epsilon}$, where σ_f is the fracture stress at strain rate $\dot{\epsilon}$. (Hint: assume $\dot{\sigma} \propto \dot{\epsilon}$ and show by integration that

$$\frac{\sigma_f^{n^*+1}}{n^*+1} = \frac{\dot{\sigma}}{D\sigma^{n^*}\, Y^{n^*}\, \pi^{n^*/2}} \left(\frac{2}{2-n^*}\right)\left(a_c^{1-(n^*/2)} - a_i^{1-(n^*/2)}\right)$$

where a_i is initial crack length and a_c that at fracture, and simplify this expression for large n^*.)

6.18 Explain why the suspectibility of polycrystalline zinc to embrittlement by liquid mercury is markedly influenced by alloying with small amounts of gold or copper.

6.19 In some embrittling systems, such as aluminium in liquid mercury plus gallium solutions, as the temperature is increased there is a sharp transition from brittle to ductile behaviour at a characteristic temperature, T_c (Preece and Westwood, 1969). Above T_c the fracture strain is the same as that in air; below T_c it is relatively constant. Discuss the effect of temperature on liquid-metal embrittlement in the light of these results.

6.20 Liquid-metal embrittlement usually produces spontaneous fast cracking (~ 80 mm/s depending on testpiece geometry and crack stability). There seem to be, in consequence, no relationships available in the literature between crack velocity and toughness during liquid-metal-embrittled fracture. However, Old (1978) has shown that slow crack growth can occur in single crystals of aluminium embrittled by liquid mercury. The cracking process is reported to have been accompanied by extensive plastic deformation even though the fracture itself is cleavage in nature. The illustration gives the specific fracture energy versus crack velocity relation. Discuss the magnitude of these fracture toughness values in relation to the specific work required for cleavage, and the general effect of crack velocity on liquid-metal-embrittled toughness values.

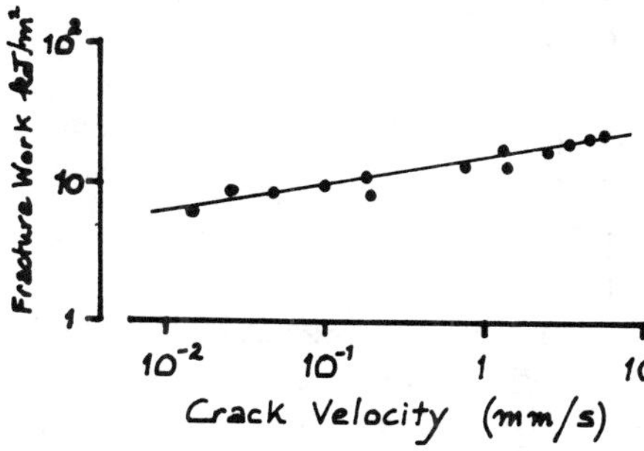

REFERENCES

Ashby, M. F. and Raj, R. (1975) *The Mechanics and Physics of Fracture.* Metals Society/Institute of Physics, p. 148.

Atkins, A. G. (1970) *J. Strain Analysis* **5**, 269.

Atkins, A. G. (1975) *J. Mater. Sci.* **10**, 819.

Atkins, A. G., Lee, C. S. and Caddell, R. M. (1975) *J. Mater. Sci.* **10**, 1381.

Barton, A. F. M. (1975) *Chemical Reviews* **75**, 731.

Beaumont, P. W. R. and Young, R. J. (1975) *J. Mater. Sci.* **10**, 1334.

Begley, J. and Landes, J. D. (1972) ASTM STP No. 514, 1. Philadelphia: ASTM.

Bond, G. M., Scott, V. D., Cooke, R. G. and Board, R. G. (1980) *The Mechanical Properties of Biological Materials,* p. 459. SEB Symposium Cambridge: Cambridge University Press.

Buchalet, C. and Mayer, T. R. (1973) ASTM STP No. 536, 281. Philadelphia: ASTM.

Burrel, H. and Immergut, E. H., (1964) *Polymer Handbook.* New York: Interscience.

Caddell, R. M. (1980) *Deformation and Fracture of Solids.* Englewood Cliffs, N. J: Prentice-Hall.

Castino, F. (1983) M. Eng. Sci. thesis, Department of Mechanical Engineering, Sydney University, NSW, Australia.

Chan, M. K. V. and Williams, J. G. (1983) *Polymer* **24**, 234.

Crosley, P. B. and Ripling, E. J. (1975) Rept. 752 MRL for EPRI on contract RP 303-1.

Davidge, R. W. (1979) *Mechanical Behaviour of Ceramics.* Cambridge: Cambridge University Press.

Davidge, R. W. and Tappin, G. (1967) *Trans. Brit. Ceram. Soc.* **66**, 405.

DeBaise, G. R., Porter, A. W. and Pentoney, R. E. (1966) *Mater. Res. Stds.* **6** (10), 493.

Evans, A. G. and Lange, F. F. (1975) *J. Mater. Sci.* **10**, 1659.

Evans, A. G. and Wiederhorn, S. M. (1974) *Int. J. Fract.* **10**, 379.

Evans, A. G. and Wiederhorn, S. M. (1974) *J. Mater. Sci.* **9**, 270.

Ewing, P. D. and Williams, J. G. (1979) *Proc. ICM-3 Cambridge* **3**, 293.

Felbeck, D. K. and Atkins, A. G. (1984). *Strength and Fracture of Engineering Solids,* Englewood Cliff, N. J: Prentice-Hall.

Ford, F. P. (1975) CERL Report LM/Mat/075. (CEGB, Leatherhead, UK.)

Ford, F. P. (1979) *Proc. ICM-3 Cambridge* **2**, 431.

Fuller, K. N. G., Fox, P. G. and Field, J. E. (1975) *Proc. Roy. Soc.* **A314**, 357.

Gent, A. N. (1970) *J. Mater. Sci.* **5**, 925.

Glenny, I. and Royston, H. (1958) *Trans. Brit. Ceram. Soc.* **57**, 645.

Goldman, N. L. and Hutchinson, J. W. (1975) *Int. J. Solids Struct.* **11**, 575.

Gurney, C. (1952) Institute of Metals Monograph No. 13, pp. 145–160.

Harper, M. P. and Ellison, E. G. (1977) *J. Strain Analysis* **12**, 167.

Hasselman, D. P. H. (1969) *J. Amer. Ceram. Soc.* **52**, 600.

Hasselman, D. P. H. (1970) *J. Amer. Ceram. Soc.* **53**, 490.

Henry, L. F. (1974) *Polymer Engineering and Science* **14**, 167.

Henshall, J. L., Rowcliffe, D. J. and Edington, J. W. (1977) *Proc. ICF-4, Waterloo* **3B**, 875.

Hertzberg, R. W. (1983) *Deformation and Fracture Mechanics of Engineering Materials,* 2nd edn. Chichester, John Wiley.

Howard, I. C. (1979) *Proc. ICM-3 Cambridge,* **2**, 463.

Immergut, E. H. and Brandrup, P. (1974) *Polymer Handbook,* 2nd edn. New York: Interscience.

Jacques, C. H. M. and Wyzgoski, M. G. (1979) *J. Appl. Polym. Sci.*, **23**, 1153.

Jeronimidis, G. (1979) *Proc. ICM-3 Cambridge* **3**, 329.

Kambour, R. P. (1973) *J. Polym. Sci. D. Macromolecular Reviews* **7**, 1.

Kies, J. A. and Clark, A. B. J. (1969) *Fracture* London: Chapman & Hall, p. 483.

Klepaczko, J. R. (1984) *Theor. Appl. Fract. Mech.,* **1**, 181.

Kramer, E. J. (1979) In *Developments in Polymer Fracture,* Vol. 1, E. H. Andrews (ed.). Barking: Applied Science Publishers, Chapter 3.

Landes, J. D. and Begley, J. A. (1974) Sci. Paper 74-1E7-FESGT-P1, Westinghouse Labs, Pittsburgh, Penna.

Lawn, B. R., Hockey, B. J. and Wiederhorn, S. M. (1980) *J. Mater. Sci.* **15**, 1207.

Mai, Y. W. (1975) *J. Mater. Sci.* **10**, 943.

Mai, Y. W. (1976) *J. Amer. Ceramic Soc.* **59**, 491.

Mai, Y. W. (1976) *J. Mater. Sci.,* **11**, 303.

Mai, Y. W. (1977) Proc. Fracture Mechanics and Technology, G. C. Sih and C. L. Chow (eds.). Netherlands: Sijthoff and Noordhoff, Vol. 1, pp. 405–419.

Mai, Y. W. (1985) *J. Mater. Sci.* (to be published).

Mai, Y. W. and Atkins, A. G. (1975) *Amer. Ceram. Soc. Bull.* **54**, 593.

Mai, Y. W. and Atkins, A. G. (1975) *J. Mater Sci.* **10**, 1904.

Mai, Y. W. and Atkins, A. G. (1975) *J. Mater. Sci.* **10**, 2000.

Mai, Y. W. and Atkins, A. G. (1976) *J. Mater. Sci.* **11**, 677.

Mai, Y. W. and Castino, F. (1984) *J. Mater. Sci.* **19**, 1638.

Mai, Y. W. and Gurney, C. (1975) *Physics and Chemistry of Glasses* **16**, 70.

Manson, S. S. (1953) NACA Technical Note TN 2933.

Marshall, G. P., Coutts, L. H. and Williams, J. G. (1970) *Plastics and Polymers* April, 95.

Marshall, G. P., Coutts, L. H. and Williams, J. G. (1974) *J. Mater. Sci.* **9**, 1409.

Marshall, G. P., Culver, L. E. and Williams, J. G. (1973) *Int. J. Fract.* **9**, 295.

Minnear, W. P. and Bradt, R. C. (1975) *J. Amer. Ceram. Soc.* **58**, 345.

Mostovoy, S. and Ripling, E. J. (1969) *J. Appl. Poly. Sci.* **13**, 1081.

Munz, D., Himsolt, G. and Eschweiter, J. (1981) ASTM STP No. 745, 69. Philadelphia: ASTM.

Old, C. F. (1978) *Proc. Conf. Mechanisms of Environmental sensitive Cracking of Material,* Swann, P. R., Ford, F. P. and Westwood, A. R. C. (eds.). Guildford: University of Surrey, p. 191.

Parkins, R. N. (1975) *J. Strain Analysis* **10**, 251.

Parkins, R. N. (1979) ASTM STP 665, 5. Philadelphia: ASTM.

Preece, C. M. and Westwood, A. R. C. (1969) *Proc. ICF-2 Brighton,* Pratt, P. (ed.). London: Chapman & Hall.

Purslow, P. P. (1980) Ph.D dissertation, University of Reading.

Rolfe, S. T. and Barsom, J. M. (1977) *Fracture and Fatigue Control in Structures.* Englewood Cliffs, N. J.: Prentice-Hall.

Seman, D. J., Kallenberg, G. P. and Towner, R. J. (1971) Rept. WAPD-TM-895, Bettis Atomic Power Lab., Pittsburgh, Penna.

Shabbits, W. O. (1973) Westinghouse Report WCAP-7623, Pittsburgh, Penna.

Shoemaker, A. K. and Rolfe, S. T. (1969) *J. Basic Engr. Trans. ASME,* September, 512.

Smith, E. (1983) *Int. J. Fract.* **23,** 213.

Speidel, L. M. (1979) *Stress Corrosion Cracking.* NATO series.

Tomkins, B. (1981) *Phil. Trans. Roy. Soc.* **A299,** 31.

Treloar, L. R. G. (1958) *Physics of Rubber Elasticity.* Oxford: Oxford University Press.

Vermilyea, D. A. (1972) *J. Electrochem. Soc.* **119,** 405.

Vincent, P. I. and Raha, S. (1972) *Polymer* **13,** 283.

Webster, G. A. and Smith, D. J. (1983) *Int. J. Fract.* **23,** R125.

Wei, R. P., Novak, S. R. and Williams, D. P. (1972) *Materials Research and Standards,* **12,** 25.

Weichart, R. and Schönert, K. (1978) *J. Mech. Phys. Solids* **26,** 151.

West, J. M. (1973) *Metal Science Journal* 7, 169.

West, J. M. (1982) *Spring Residential Conference 'Hydrogen in Steel'.* London: Institution of Metallurgists.

Wiederhorn, S. M., Freiman, S. W., Fuller, E. R. and Simmons, C. J. (1982) *J. Mater. Sci.* **17,** 3460.

Williams, J. G. (1965) *Applied Materials Research* **4,** 104.

Williams, J. G. (1977) *Advances in Polymer Sci.* **27,** 65. Berlin: Springer-Verlag.

Williams, J. G. and Marshall, G. P. (1975) *Proc. Roy. Soc.* **A342,** 55.

Williams, J. G., Radon, J. C. and Turner, C. E. (1968). *Polymer Eng. and Sci.* **8,** 130.

7

Fracture under repeated loading

'Make a remark,' said the Red Queen: 'it's ridiculous to leave all the conversation to the pudding!'

It is well known that under conditions of repeated, fluctuating or interrupted loading, many materials fracture at stresses which are both below the stress at which failure would occur were the loading monotonic, and often at stresses below the yield stress of ductile materials. The phenomenon is called fatigue. The greater the number of cycles, the lower the fracture stress, which gives the characteristic falling shape to traditional stress/number-of-cycles-to-failure (S–N) curves (Fig. 7.1(a)). Some materials (particularly ferrous and titanium alloys) show an *endurance limit* or *fatigue limit* beyond about 10^6 cycles, after which the fatigue strength falls no further. Empirically it has been found from Wohler-type rotating bending tests that the endurance limit of many commercial steels is numerically equal to about one-half the ultimate tensile stress (for steels where UTS $\leqslant$ 1.4 GPa). Most other metals, polymers and composites show no fatigue limit, but for design purposes the stress for a life of 10^7 or 10^8 cycles is often taken and called the *fatigue strength* to distinguish it from fatigue limit.

Some materials show no fatigue effects. For example, glass tested under vacuum is essentially not susceptible to fatigue. However, glass tested in air shows the phenomenon of *delayed fracture* whereby testpieces, loaded statically to stresses lower than the expected Griffith fracture stress, *do* break after some time – the higher the applied stress, the shorter the time to failure (Fig. 7.1(b)), (not unexpectedly, the effect is also called a 'creep' failure although in fact, unlike metallic creep, overall deformations are not excessive). Since the number-of-cycles-to-failure in a cyclic fatigue test also is a measure of the time to failure ($N \times$ repeat period = time), and since the two different curves in Fig. 7.1 (a) and (b) have similar shapes, delayed fracture is also unfortunately called *static*

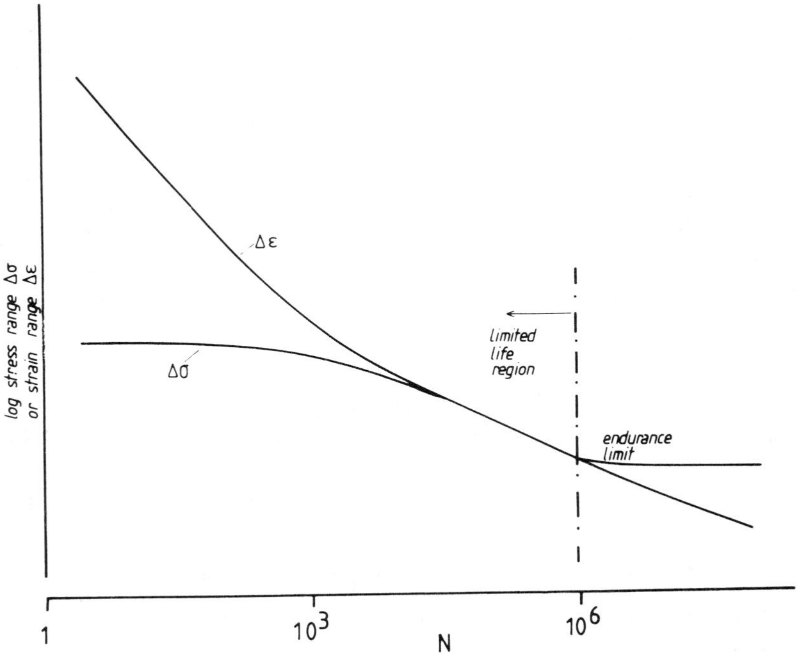

(a)

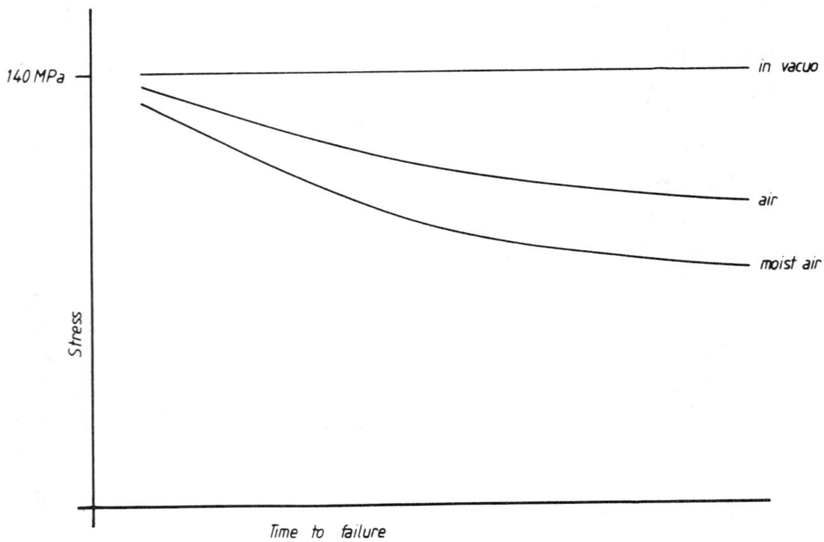

(b)

Fig. 7.1 – (a) Schematic $\Delta\sigma$ vs. N (S–N) and $\Delta\epsilon$ vs. N (ϵ–N) curves for materials under constant amplitude load and strain cycling in a relatively non-aggressive environment. Materials vary as to whether a distinct limiting stress (fatigue limit or endurance limit) exists: one does in common bcc metals. (b) Delayed fracture in glass showing stress-time to failure relationship.

fatigue, even though cyclic loads are *not* involved, and play no part in the cause of fracture (cf. Fig. 7.48 as well). Delayed fracture is an environmental effect rather like hydrogen embrittlement (Chapter 6): moisture (or other hostile environments) attacks the stretched bonds at the crack tip and gradually lengthens sub-critical Griffith flaws until one of them reaches the critical length for the particular applied stress level.

Even though delayed fracture has nothing to do with fatigue as it is usually understood, we consider the propagation of cracks in glass under repeated loading in the first section of this chapter. The important reason is that crack growth rates under repeated loading in materials which do not display fatigue effects, can be predicted directly from (velocity-dependent) *monotonic* toughness information obtained in the same environment. In contrast, it is *not* possible directly to tie together monotonic crack-resistance information with the behaviour of most materials in cyclic fatigue, and other lines of attack have to be employed. That subject forms the bulk of this chapter, but in section 7.8, on environmental corrosion fatigue, a possible connexion between the two is discussed.

7.1 CRACK GROWTH UNDER REPEATED LOADING IN MINERAL GLASS

Experiments on commerical soda glass in moist air at room temperature show that over a limited range of variables

$$\dot{a} = 7.3 \times 10^{-6} \exp(1.65G_c) \tag{7.1}$$

where $\dot{a}$ is in mm/s and G_c is in J/m^2. Such data have been generated by, for example, Kies and Clark on Outwater testpieces subjected to different crosshead velocities; (these testpieces crack under constant X, and hence constant crack velocity since $\dot{a} = (X/2RB)\dot{u}$; and $R \propto X^2$ – see Chapters 2 and 6).

Consider what happens if in a test the load flutuates. Is it possible to predict the extent of crack growth from monotonic $G_c(\dot{a})$ data? Since we know that in Outwater testpieces a certain load level gives a certain crack velocity, then if we know how long the load remains at that level, we can determine the growth of the crack over that time. Consider events in a time interval $\mathrm{d}t$ when the load is $X(t)$. $X(t)$ may be identified with a $G_c(t)$ for the same time interval, which in turn can be identified with crack velocity $\dot{a}$. Hence the crack growth in the interval becomes

$$\mathrm{d}a = \dot{a}\,\mathrm{d}t \tag{7.2}$$

(where $\dot{a}$ comes from G_c which came from X).

Consider the load to be cycled between predetermined extremes. For cyclic loading of approximately triangular waveform (such as obtained on many standard testing machines) we have

$$X(t) = \frac{2(X_m - X_0)}{T}\,t + X_0 \tag{7.3}$$

for the time interval $0 < t < T/2$, where T is the period, and X_m and X_0 the maximum and minimum values of the load (Fig. 7.2). When $X_0 = 0$, $X(t) = 2X_m t/T$.

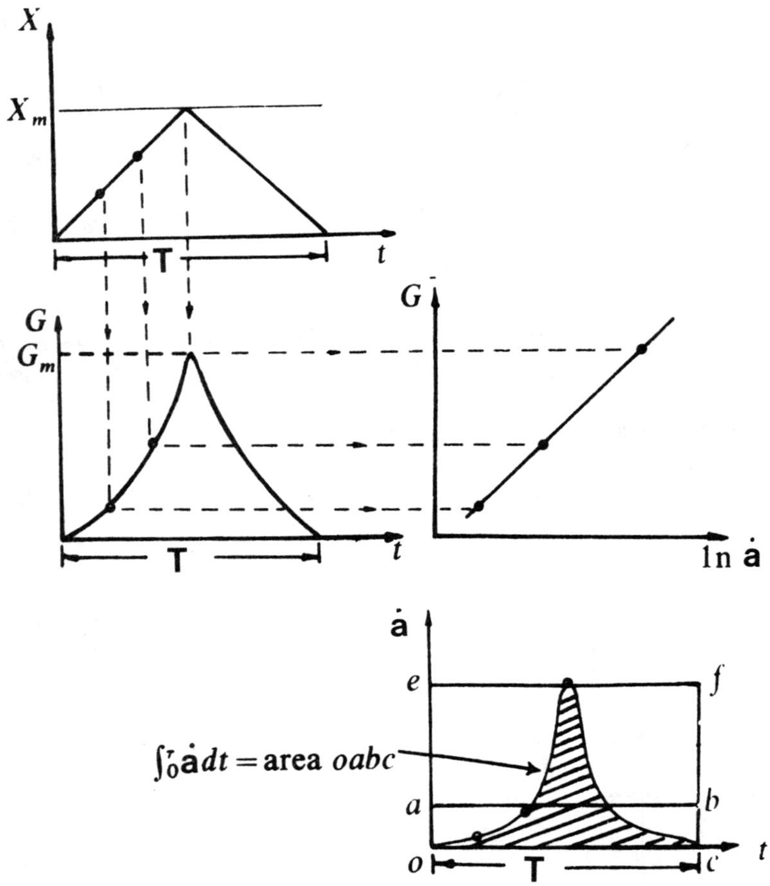

Fig. 7.2 – Illustration of graphical integration of equation (7.4) to obtain crack velocity variation, and distance propagated, per cycle (from Mai and Gurney (1975), with permission).

When cyclic fatigue effects are absent, the crack growth Δa in a cycle is given by

$$\Delta a = 2\int_0^{T/2} \dot{a}(t)\mathrm{d}t \tag{7.4}$$

Note that Δa/cycle is different from the crack speed $\dot{a}$ which is the crack increment/unit time. As $\dot{a}(t)$ is defined completely by X, via a relation like (7.1), and X itself is defined by a relation like (7.3), equation (7.4) may be integrated graphically or in closed form to obtain Δa, and hence $\Delta a/\Delta N$ the crack growth per cycle. Fig. 7.2 shows the procedures for graphical integration, which may be summarised as follows. (i) Construct the $G_c(t)$ spectrum from the load $X(t)$ spectrum using X^2 = (constant) G_c where (constant) is known for the particular size Outwater testpiece. (ii) From the independent G_c–$\dot{a}$ curve determined under constant loads, the crack velocities corresponding to the $G_c(t)$ values are located, and are then plotted as a time function. The area under this curve represents the crack growth per cycle, $\Delta a/\Delta N$, and hence is a graphical solution to the integral shown in equation (7.4).

Clearly cyclic crack growth depends largely on the load waveform: smaller load amplitude gives a smaller G range and smaller range of crack velocity, so that the rate of crack extension is smaller. Fig. 7.3 shows $\Delta a/\Delta N$ predicted graphically from constant load $G_c(\dot{a})$ data, together with experimental results, for a variety of zero-to-X_m repeated triangular load histories. The correlation is good and supports the idea that glasses are not susceptible to cyclic fatigue damage. The physical reason behind this phenomenon is that, at room temperature, glasses and indeed some ceramics such as porcelain, quartz and alumina, do not exhibit significant *plastic deformation* which, as shown by Evans and Fuller in 1974, is necessary for fatigue enhancement. However, at higher temperatures, say 400°C for alumina, where plastic deformation accompanies crack propagation, an enhanced fatigue effect of cyclic loading may be observed.

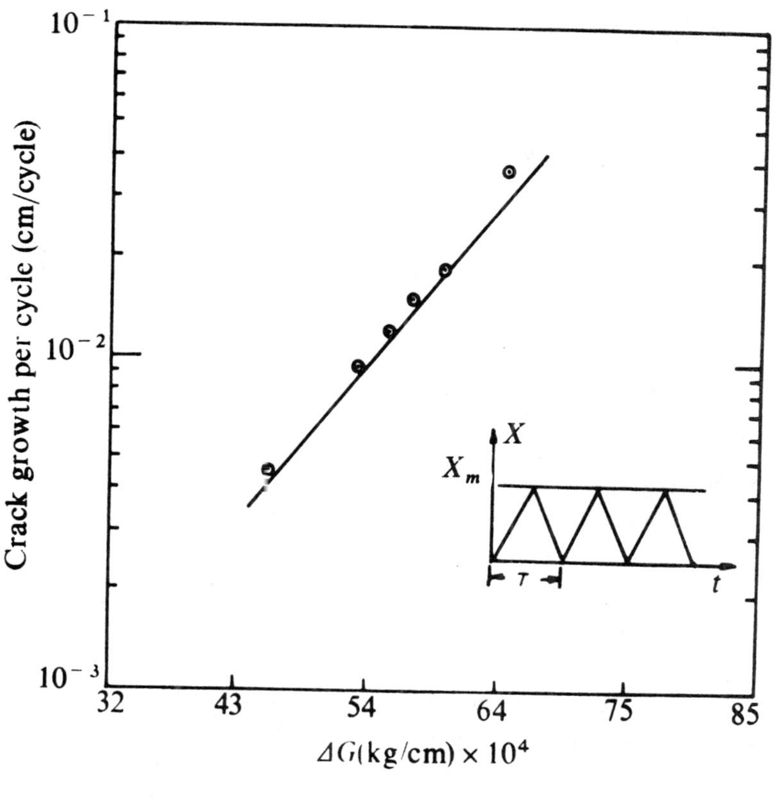

Fig. 7.3 – Growth of cracks under cyclic loading for soda-lime glass in distilled water. Line is theory; points are experimental results. 1 kgf/cm ≈1 kJ/m² (from Mai and Gurney (1975), with permission).

The experimental data and graphically-integrated prediction in Fig. 7.3 are well-represented by $da/dN = D(\Delta G)^5$ where D is some constant and ΔG is the *range* of G during cyclic loading i.e. $(G_{max} - G_{min})$ where, in this instance, $G_{min} = 0$. Since $G \propto K^2$ the relation may also be written $da/dN = C(\Delta K)^{10}$ with C another constant and ΔK the range of K experienced. We shall see later that similar relations appear in connexion with high cycle fatigue crack growth rates in metals and polymers which give log–log correlations between da/dN and ΔK (section 7.2.1). On the other hand, equation (7.1) for $\dot{a}$ rather than da/dN corresponds with a semi-logarithmic plot between $\dot{a}$ and G, a form which may be shown to be expected from chemical rate theory (Chapter 6) and which is particularly relevant for slow crack growth in glasses and ceramics.

Example 7.1 The slow crack growth equation in glass can be expressed as: $da/dt = DK_c^{n*}$ where D and $n*$ are numerical constants. Compare the times-to-failure for a glass structural component under static and cyclic loadings. You may assume that the glass contains an initial flaw size (a_i) and that $K_c = Z\sigma a^{1/2}$, Z being the geometry correction factor for the particular cracked component.

Ans. (a) Constant loading case.

$$K_c = Z\sigma a^{1/2}$$

$$da/dt = DZ^{n*}\sigma^{n*}a^{n*/2}$$

Thus

$$\frac{2}{(n*-2)}\left(\frac{1}{DZ^{n*}}\right)\times\left(\left[\frac{1}{a_i}\right]^{(n*-2)/2}-\left[\frac{1}{a_c}\right]^{(n*-2)/2}\right)=$$

$$= \int_0^t [\sigma(t)]^{n*}\,dt \qquad \text{(a)}$$

where a_c is the critical crack length for final fracture.

Now $\sigma(t) = \sigma_s$ the static load; $K_i = Z\sigma_s a_i^{1/2}$ and $K_c = Z\sigma_s a_c^{1/2}$. Therefore,

$$t_s = \left(\left[\frac{2}{n*-2}\right]\times\frac{1}{DZ^2\sigma_s^2 K_i^{(n*-2)}}\right)\left(1-\left[\frac{K_i}{K_c}\right]^{(n*-2)}\right)$$

For glasses and ceramics, $n*$ is very large so that $(K_i/K_c)^{(n*-2)}$ can be neglected, i.e. the time to failure under static loading is

$$t_s = [2/(n*-2)\ DZ^2\sigma_s^2 K_i^{(n*-2)}].$$

(b) Cyclic loading case. Assume t_c, the cyclic time to failure, can be expressed as an integer number of cycles NT plus a fraction of a cycle δT. Then $t_c = (N+\delta)T$ and

$$\int_0^{t_c=(N+\delta)T} \sigma(t)^{n^*} \mathrm{d}t = \sigma_m^{n^*} g(n^*,\zeta) t_c \times \left(N + \frac{\int_0^{\delta T} \sigma^{n^*} \mathrm{d}t}{\int_0^{T} \sigma^{n^*} \mathrm{d}t}\right) \Bigg/ (N+\delta)$$

$$\cong \sigma_m^{n^*}\, g\,(n^*,\zeta) t_c \quad (N \gg 1) \qquad \text{(b)}$$

where

$$g(n^*,\zeta) = \frac{1}{T} \int_0^T [\sigma(t)/\sigma_m]^{n^*} \mathrm{d}t$$

and σ_m is the mean stress.

Thus, the ratio of cyclic-to-static times-to-failure is obtained by comparing relations (a) and (b), i.e.

$$\frac{t_c}{t_s} \approx g^{-1}\,(n^*,\zeta) \times \left(\frac{\sigma_s}{\sigma_m}\right)^{n^*}$$

(Note. Since comparison is made at *maximum* stress, then $\sigma_s = \sigma_m + \frac{1}{2}\,\sigma_0$). We observe that $t_c = t_s$ if $\zeta = 0$, otherwise $t_c \gtreqless t_s$ depending on the expression $[g^{-1}(n,\zeta) \times (\sigma_s/\sigma_m)^{n^*}]$ and particularly the value of n^*. Some numerical results have been obtained by Evans and Fuller (1974).

Integration in closed form for $\Delta a/\Delta N$ is possible for some waveforms. We have $\dot{a} = C_1 \exp\,(C_2 G_c)$ as the general expression for relation (7.1), so if in a constant rate-of-change of compliance testpiece $G_c = C_3\, X^2$,

$$\dot{a} = C_1 \exp\,\{C_2\, C_3\, X^2\,(t)\}$$

and integration of this over period T gives the crack growth increment per cycle. The integral involves $\int \exp\,\{C_2\, C_3\, X^2\,(t)\}\,\mathrm{d}t$. By calling $\sqrt{C_2 C_3} X = \xi$, the integral may, for some waveforms, be transformed into a version of Dawson's integral, tabulated as $\mathrm{e}^{-x^2} \int_0^x \mathrm{e}^{y^2}\,\mathrm{d}y\ (x \leqslant 2)$ or $x\mathrm{e}^{-x^2} \int_0^x \mathrm{e}^{y^2}\,\mathrm{d}y\ (x > 2)$ in the National Bureau of Standards, *Handbook of Mathematical Functions* (1964). In the case of a triangular waveform $\mathrm{d}\xi = \sqrt{C_2 C_3}\, k\mathrm{d}t$, where $k = 2(X_m - X_0)/T$ from relation (7.3). For $t = 0$, $\xi = \sqrt{C_2 C_3} X_0$ and for $t = T/2$, $\xi = \sqrt{C_2 C_3} X_m$. So, crack growth per half cycle is given by

$$a/\text{half cycle} = \frac{C_1}{\sqrt{C_2 C_3}\ k} \int_{\sqrt{C_2 C_3} X_0}^{\sqrt{C_2 C_3} X_m} \exp\, \xi^2\, \mathrm{d}\xi \qquad (7.5)$$

Dawson's integral is continuous, so the limits in equation (7.5) may be altered to

$$\int_{\sqrt{C_2 C_3} X_0}^{\sqrt{C_2 C_3} X_m} \equiv \int_0^{\sqrt{C_2 C_3} X_m} - \int_0^{\sqrt{C_2 C_3} X_0}$$

In this way the crack growth per cycle $\Delta a/\Delta N$ is seen to depend on the mean load, load amplitude and the frequency of cycling.

This algebra may be applied to the problem already discussed graphically in Figs 7.2 and 7.3. We have $C_3 = 22 \times 10^{-4}$ from calibration of the Outwater testpiece and $C_2 = 1.65$ from relation (7.1) so $\sqrt{C_2 C_3} = 0.06$. For $X_0 = 0$, $X_m = 50$ N and $T = 20$ s, $k = 5$ Ns. Integrating equation (7.5) with these values of C gives a crack growth of approximately 43 μm/half cycle, i.e. $\Delta a/\Delta N =$ 86 μm/cycle. This agrees with the value calculated by graphical means in Fig. 7.3. The average crack velocity per cycle $\dot{a}_c$ is 43 μm/10s = 4.3 μm/s.

Example 7.2 Show that the average crack velocity in a cycle for a triangular waveform of constant load amplitude and constant mean load is independent of frequency for a material following relation (7.1).

Ans. Inspection of equation (7.5) shows that the crack growth in a half, triangular cycle increases as k decreases, i.e. as the period increases. But as the decrease in k is exactly counterbalanced by the increase in period, the *average crack velocity* in a cycle $\dot{a}_c$ is *independent* of period frequency. Note, however, that $\mathrm{d}a/\mathrm{d}N$ *does* depend on period or frequency. (see Example 7.15).

For varying load amplitude, at constant mean load and constant frequency, $\dot{a}$ increases with amplitude. For example, at a constant mean load of 30 N constant period of 20 seconds, cycling the Outwater glass specimen between (a) 20 and 40 N gives $\dot{a}_c$ = 0.46 μm/s; (b) 15 and 45 N gives 1.25 μm/s; and (c) 10 and 50 gives 5.9 μm/s. Note that this last value is bigger than that for cycling between zero and 50 N with the same period (which was 4.3 μm/s) because both k and the mean load are different.

Again, for varying mean stress but constant amplitude and frequency, the average crack velocity increases with mean stress. For example, cycling between (a) zero and 10 N with a period of 20 seconds gives $\dot{a}_c$ = 8.15 nm/s, (b) 10 and 20 N gives 16 nm/s; (c) 20 and 30 N gives 85 nm/s; (d) 30 and 40 N gives 834 nm/s; and (e) 40 and 50 N gives 1420 nm/s.

It is of interest to compare these average crack velocities in fatigue with the corresponding crack velocities under monotonic loadings to the same X_m. It is found that under cyclic conditions there are remarkable reductions of average crack speed compared with constant loadings. (Note that this effect is confined to materials like glasses and ceramics at room temperature which suffer no crack tip damage in fatigue; the effect is not necessarily true in metals and polymers which, as discussed later in this chapter, are sensitive to major cyclic effects). Fig. 7.4 shows the crack growth rate (per unit time) under cycle and constant loadings of the same maximum X_m and in the same distilled water environment as in Fig. 7.3. The data were obtained from the same glass testpiece which was subjected to successive periods of time-independent and cyclic loadings. For the constant loading periods, crack speed was calculated as the crack growth per unit time, and for cyclic loading periods crack speed was taken as the crack increment per cycle averaged over the periodic time T. It is seen that under triangular waveform cyclic load the crack speed is about (1/10) of that under constant load. The reason is clear from Fig. 7.2: cyclic crack speed is given by the level of ab, but crack speed under constant loading at the same X_m is represented by ef.

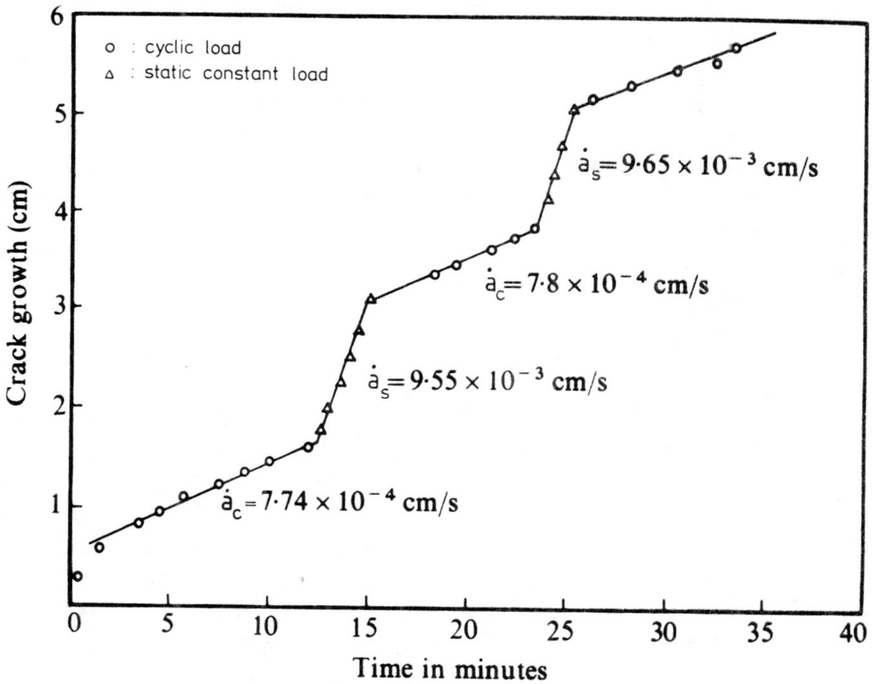

Fig. 7.4 – Comparison of crack speeds under constant and cyclic loading for double-torsion glass plate specimen in water $X_m = 52$ N and $X_0 = 2$ N (Δ constant load; o cyclic load) (from Mai and Gurney (1975) with permission).

For simplicity, the analyses for crack growth rates in this section have concerned testpieces which crack quasi-statically under *constant load.* The treatment can be extended to situations where the fracture load changes as the crack lengthens (e.g. DCB testpeices where the load drops at constant toughness) but the mathematics become very involved and the crack growth per cycle is dependent on crack length. The graphical approach is also possible but now $G_c = X^2a^2/EIB$ (for the DCB geometry) and G_c may only be obtained if X and

a are known at any given t. In this way, crack growth per cycle can be plotted as a function of crack length.

7.2 ELASTIC (HIGH CYCLE) FATIGUE IN METALS AND POLYMERS

In the case of glass under repeated loading in section 7.1, the mechanics of the Outwater specimen concerned the propagation of long pre-existing cracks of a size sufficient to cause measurable changes in compliance. For the behaviour of glass in which deliberate starter flaws have *not* been introduced, the natural Griffith flaws are the stater cracks and these eventually grow into longer cracks. The cracked body, whatever the crack length, always behaves elastically.

In the fatigue of metals and polymers, the behaviour of both short and long carcks may too be treated separately. In section 7.2.1 we consider the behaviour of long cracks which may have been deliberately introduced in testpieces, or may be large fabrication flaws in components, or they may have grown from much smaller cracks. Often, depending on the geometry of the component, these cracks will produce measurable changes in compliance. The quantitative boundary between long and short cracks is discussed in section 7.2.3, but as a guide a short crack is about the same size as the grain size in crystalline solids or largest microstructural feature (such as inclusions); about 100 μm is representative of many crystalline and amorphous materials. In the case of large single crystals used for experimental purposes, the boundary between short and long is much greater and could even be about 300 mm. Unlike glass, most metals and polymers that display fatigue do not obviously possess naturally occurring short flaws, and the question has to be asked as to where short fatigue cracks come from in supposedly crack-free materials. (A similar question regarding the nucleation of cracks in the same materials during *monotonic* ductile fracture was addressed in Chapter 5).

The characteristic striations on fatigue fracture surfaces (beach-clamshell-, or compression-marks) show that failure is caused by progressive crack advancement as the number of cycles increases, with the crack tip being successively blunted and re-sharpened. Eventually the grown crack becomes long enough to permit monotonic failure across the remaining net section. The final rupture may be elastic or plastic, depending on the material and circumstances, i.e. the grown crack either (i) satisfies the monotonic elastic fracture condition $K_c = \sigma Y\sqrt{\pi a}$ or (ii) leaves a net section small enough to permit plastic collapse and eventual ductile fracture in the remaining section.

Forsyth in 1961 classified fatigue crack propagation in metals into three stages. Stage I concerns the early growth of fatigue cracks, and being controlled by microstructure and surface finish is discussed in detail in books on the science of materials. Stage II is the period in which fatigue cracks grow to finite size, being controlled by progressively larger-scale effects than Stage I crack formation. Stage III is the final stage of fatigue and really concerns monotonic failure rather than cyclic behaviour. The important question of 'how short is a short crack' which is posed in section 7.2.3 concerns the boundary between Stages I and II where non-uniform microstructural features and crack lengths have comparable sizes. The application of continum mechanics to Stage I is doubtful, but stages II and III may be described in that way.†

Example 7.3 Discuss the range of da/dN found in fatigue. Do fracture surface striations correspond with da/dN for a single cycle?

Ans. Micromechanics of fatigue crack growth produce average growth steps per cycle over a very wide range. The existence of an endurance limit in some materials suggests that putative crack 'growth' rates must then be 'less' than atomic dimensions, i.e. $< 10^{-10}$ m/cycle whereas on approaching final fracture, fatigue crack growth rates of millimetres per cycle have been observed. Tomkins (1980) remarks that it is clear that a single micromechanism is not responsible for growth over such a wide range. The low crack growth rates in Stage I involve a scale of crack tip deformation well below the grain size of the material and thus, in metals, they are sensitive to local crystallography and dislocation structures. The lower limit to growth seems to be related to an average crack advance of one lattic spacing per cycle (Example 7.7). Crack advance in Stage I is by shear, i.e. 'mode II' (care is required over possible confusion of roman numerals here). In the intermediate growth rate region (Stage II, about 10^{-8} to 10^{-5} m/cycle) crack tip deformation can spread over several grains in a polycrystal giving plastic continuum behaviour with consequent continuum rather than crystallographic control of crack advance. That is, crack advance can occur by Cottrell's shear decohesion model (Fig. 7.7 and Example 7.4) with both slip lines operating. Crack advance is usually perpendicular to the maximum applied tensile stress, and is therefore 'mode I'. Finally, at crack growth rates beyond about 10^{-5} m/cycle, Stage III crack advance increasingly involves material inhomogeneities ahead of the crack tip (inclusions and second-phase particles) in deformation and in the nucleation of fracture (cf. Chapters 1 and 5). Crack growth becomes discontinuous involving the linkage of the resulting microvoids ahead of cracks (Fig. 1.5). This is sometimes called the 'tearing' region of crack advance (see, for example, Ashby and Tomkins (1979)).

The number of striations observed can depend on the magnification used to view fatigue fracture surfaces. The finest striations in Stage II metal fatigue are

---

†The reader should note the use of roman numberals for Forsyth's three *stages* of fatigue crack propagation. Fatigue crack propagation can, of course, occur in principle in any of the three *modes* of crack opening, for which roman subscripts are also employed (Fig. 2.26). Thus, for example, we may talk of the Stage II fatigue of cracks propagating in mode III and so on. In section 7.6, later, which deals with fatigue under multiaxial loading, we discuss the different planes on which the different types of crack may initiate and propagate. Upper-case roman letters (A,B) are used in this case. Most information is known about cracks fatiguing in mode I and nearly all the data in this text relate to mode I (tensile) fatigue cracks. However, it is known qualitatively that rates of crack propagation in the different modes are more-or-less related, so that if there are differences in the resistance of different solids to fatigue crack propagation in mode I, say, similar differences are likely to exist in modes II and III.

sub-micrometre in size. At high growth rates in Stage II, coarser striations are observed, often with finer striations lying in between (Fig. 7.5(c)). Striations are the result of the operation of local crack tip slip line fields during the opening and closing of the crack (Fig. 7.7, later), and although *sometimes* one striation spacing does represent one increment of crack advance in one cycle, this is *not* a general observation. Note also that fatigue can occur without leaving striations, that striations can be produced by interrupted stress corrosion cracking, and furthermore that striations may be left on the fracture surface of some polymers failing under *monotonic* loading.

Figs. 7.5 (which include information and ideas yet to be presented) summarises this answer in schematic form. The reason for the particular form of plot in Fig. 7.5(b) is explained presently.

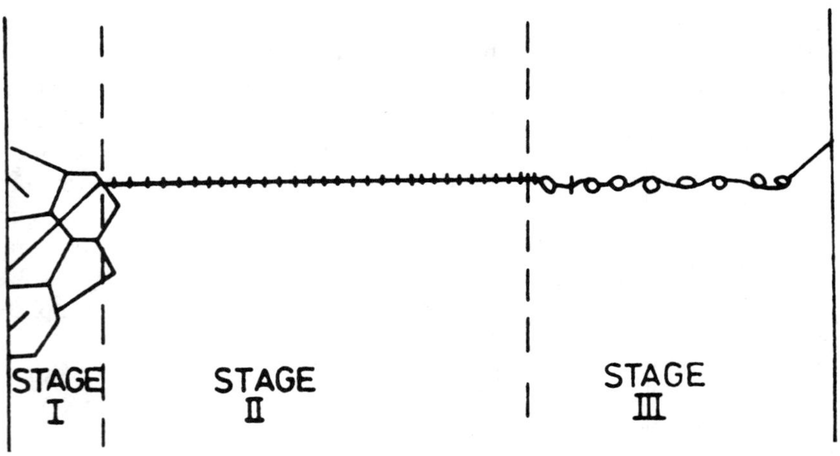

Fig. 7.5(a) – The three stages of fatigue crack propagation in a metal.

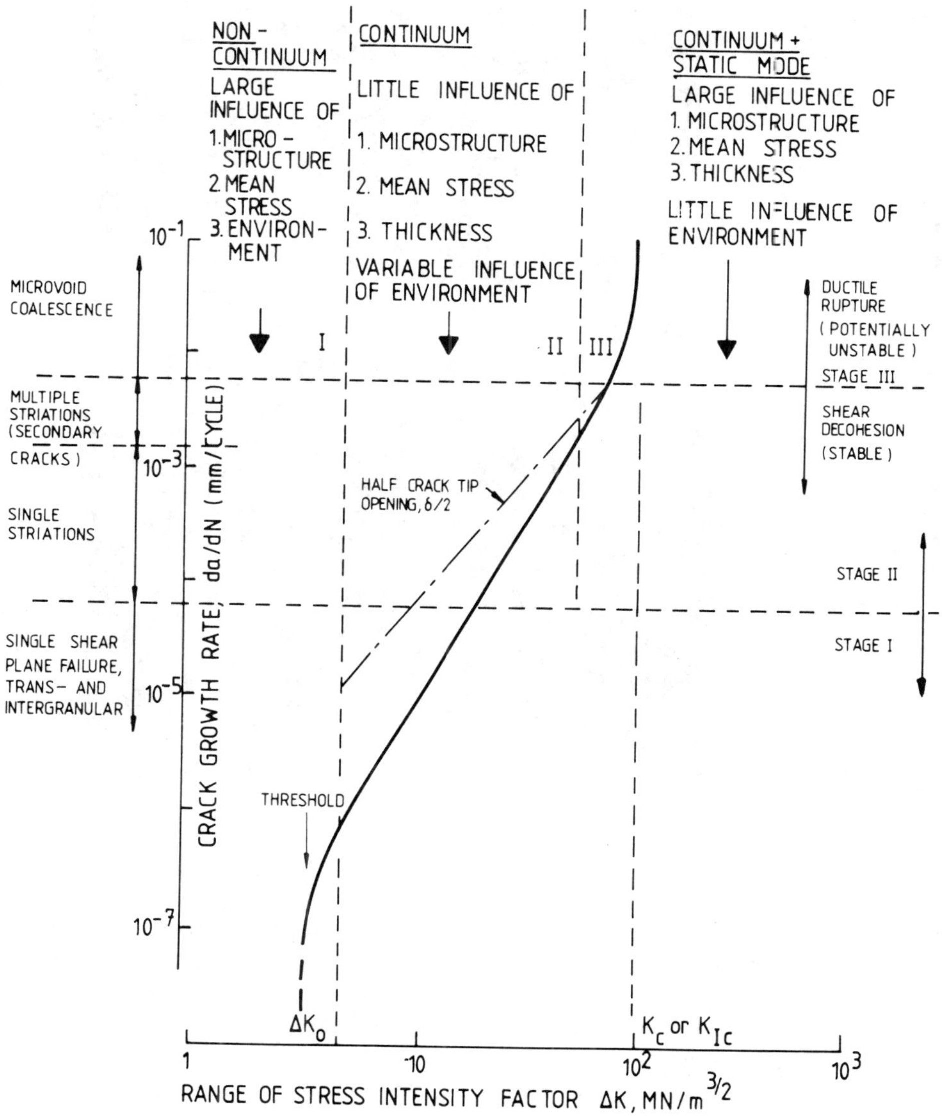

Fig. 7.5(b) – Schematic drawing showing the various features of a fatigue crack growth curve (suggested by Lindley *et al.* (1975) and Tomkins and Wareing (1977)).

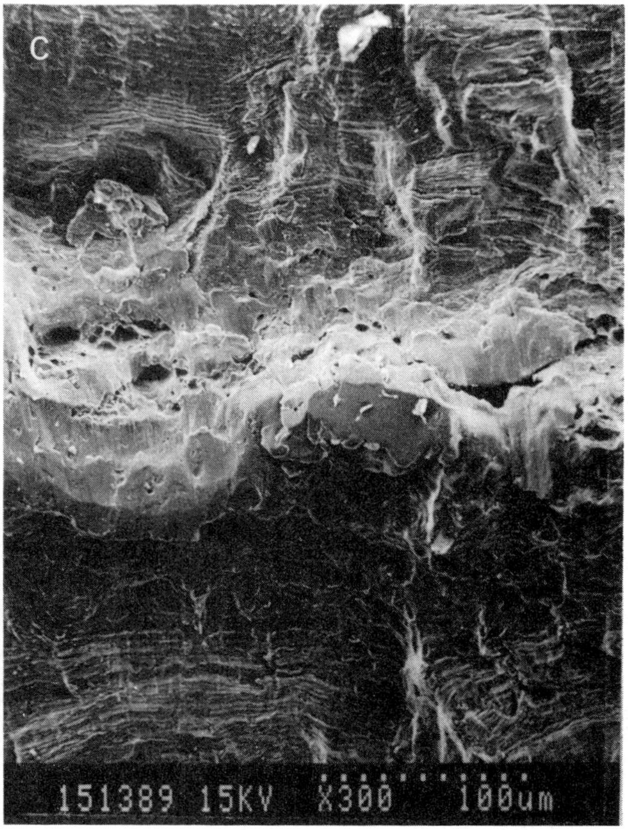

Fig. 7.5(c) – Fine striations within coarse striations at high growth rates in Stage II fatigue (courtesy D. K. Felbeck and W. Durrant).

In fatigue at low stress levels, the Stage I period of crack incubation dominates the lifetime; in tests lasting say some 10^6 cycles, Stage II may not occur until three-quarters of the total life has passed by and the mathematics of fatigue crack growth apply then only to the final quarter of the component's life. In so-called low-cycle fatigue, on the other hand, where failure occurs in a few thousand cycles, extensive regions of the body experience plastic flow each cycle (being loaded beyond the yield points in tension and compression). The Stage I incubation period is negligible in such high strain amplitude fatigue and the mechanics of crack growth are different from those where fatigue occurs at stresses below the yield point (see section 7.4). Note that, while Stage III is always a rapid process in cycle terms (and does not contribute much to lifetime), it may occupy a significant portion of the fracture *surface*.

An important realisation, which is central to the mathematical treatment of fatigue crack growth rates, is that Stage II propagation takes place under essentially globally *elastic* conditions – even in materials which would be ductile in a monotonic test to fracture. This is true even if Stage III is ductile (which would not permit refitting the broken parts). We therefore expect that *elastic* fracture mechanics methods may be useful in analysing Stage II fatigue crack growth rates at fatigue stresses below yield in materials which in a *monotonic* test would suffer extensive *plastic flow* before failure.

7.2.1 Long cracks

Fig. 7.6 represents fatigue crack growth on an *Xu* plot, where the load is being cycled between zero and X_m in a body having an initial stiffness line OB. A single monotonic loading up to X_m would not cause crack growth at all, since $X_m < X_B$ where B is the intersection with the monotonic constant G locus. But the point about fatigue is that numerous *repeated* loadings to the lower load X_m *do* cause crack growth. The cyclic *Xu* paths follow OD, OE . . . OF . . . in the later stages of fatigue crack propagation when measurable changes in stiffness may be detected. Earlier, when the fatigue crack is much shorter, the changes in stiffness are not marked and the stiffness lines are bunched together in the early stages of growth. If final fracture is elastic, point P is reached in Fig. 7.6(a), X_m being sufficient to cause the Stage III cracking over the remaining ligament in the path of the crack along the G locus. On the other hand, if plastic flow accompanies final fracture, point Y is reached in Fig. 7.6(b) where X_Y is related to the flow stress for constrained yielding on the remaining ligament. The dashed-line trace of all X_Y for flow can be drawn by locating X_Y on all stiffness lines corresponding to final ligament sizes. In Fig. 7.6(a) the X_Y trace is 'outside' the constant G locus for elastic cracking at load levels round about X_m. The X_Y yielding line in Fig. 7.6(b) is shown wholly 'within' the G locus, but it could cross the locus (depending on the geometry of the body) at both short and/or long A, giving, say, ductile monotonic behaviour at short crack lengths, but elastic cracking at long and vice versa (cf. Fig. 10.21 later).

Figs. 7.6(a) and (b) show situations in which monotonic propagation would be stable. Fig. 7.6(c) shows fatigue crack growth in specimens with unfavourable geometric stability factors (see Chapter 8): After prior cyclic growth along C, D, E. . . under zero-to-X_m loading, *unstable* fracture occurs at P.

Consider Fig. 7.6 from the point of view of incremental crack growth and energy rate balances where the difference in crack lengths between stiffness lines ODV and OEW is Δa. When steady-state Stage II fatigue growth is in progress, the work expended in fatigue for incremental crack growth Δa is area ODE. Monotonic crack growth over the identical change in crack length would require greater work, given by area OVW (= RBda). It is clear that in fatigue the resistance to crack propagation is lowered (the crack tip region has been 'damaged') and that this must be caused in some way by the cumulative effect of all the previous cycles of repeated loading. In the early stages of fatigue the crack tip process zone becomes progressively workhardened with continued cycling up to a saturation level (even though no macroscopic changes are occurring in the overall state of strain of the body) and eventually elastic cracking is possible within the highly localised process zone. Crack advance is limited by the ductile regions beyond the immediate vincinity of the crack front (rather like, but in

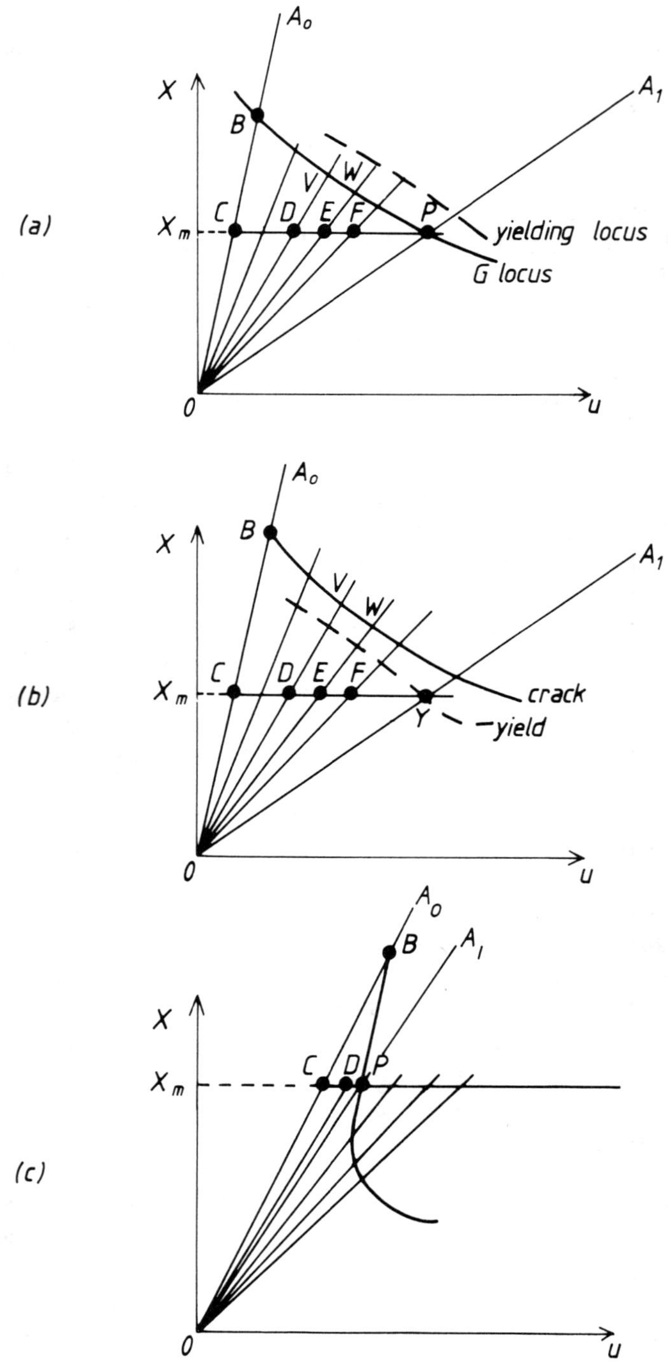

Fig. 7.6 – Interpretation of fatigue crack growth in terms of a load–deflexion X–u diagram. (a) Yield locus is above constant G locus; (b) yield locus is below constant G locus; and (c) unstable fatigue crack growth.

different circumstances, Fig. 8.10) so the crack arrests and cyclic workhardening recommences before yet another very small crack advance. These considerations are supported by the observation that fatigue cracks spread *stably* in soft (load-controlled) testing machines whereas they would be *unstable* in a *monotonic* test in the same machine (Chapter 8).

In steady-state fatigue crack growth, the crack tip damage previously accumulated in earlier cycles enables crack growth to occur at X_m instead of X_B, X_V or X_W etc. and in each cycle of propagation, events at the crack tip are such as to 'top-up' the damage to a critical level which permits cyclic growth to occur.

There are various models for fatigue crack tip damage. They differ in the parameter which is assumed progressively to reach some 'failure' condition (stress, displacement, plastic strain reversal etc.). The earliest is that of Head in 1953 who assumed that crack tip cyclic hardening led to a critical crack tip fracture *stress*. Fatigue crack growth rate da/dN was found to be proportional to the product of $\sigma^2 a$ where σ is the applied stress and a the crack length, i.e. to K^2 (although it was not expressed in terms of K in those days). Tomkins (1973) discussed a number of models, proposed by different authors, applicable both to low cycle/high strain fatigue and to high cycle/low stress fatigue, from all of which it seems that crack growth rates in fatigue, i.e. da/dN, are related to some function of the applied stress intensity factor, usually to K^2 or K^4, which is predicted also by Liu (1963) who postulated that da/dN depended directly on the size of the plastic zone r_p at the crack tip, and by Paris and Rice (1963) who argued that da/dN varied as ${r_p}^2$ (which is proportional to the volume of the plastic zone per unit length of the crack front).

Another approach, by Cotterell in 1965, starts with the premise that the value of toughness depends on da/dN and that R changes as cycling proceeds rather like the R-curve concept of R changing with a (Chapter 2).† For small crack increments and small crack tip zones he writes the growth-rate dependent R as a Maclaurin expansion, viz.

$$R(da/dN) = R(0) + (da/dN)\,R'(0) + \tfrac{1}{2}\,(da/dN)^2\;R''(0) + \ldots$$

Assuming at zero da/dN, $R(0) = 0$ and neglecting higher order terms of da/dN

$$R = (da/dN) \times R'(0) = G_C(a)$$

during quasi-static crack spreading. Hence

$$da/dN = G_C(a)/R'(0) = {K_C}^2/ER'(0)$$

i.e. in fatigue $da/dN \propto K_C^2$.

---

†It may be true to say that *crack tip plasticity* gives rise to R-curve effects which according to Cotterell explains crack growth under cyclic conditions. Thus materials which, in most environments, have a step-function R-curve (Fig. 2.15(a)) are not likely to experience fatigue damage; glass and ceramics at room temperature fit this category (cf. Evans and Fuller (1974), section 7.1). On the other hand, materials which have marked R-curves (such as metals) display pronounced fatigue effects (see section 7.8).

Plumbridge (1972) and Irving and McCartney (1977) provide valuable reviews of models for fatigue crack growth rates and the latter authors introduced a general model based upon the attainment of a critical work rate in the crack tip process zone. Again da/dN depended on some function of K.

Thus, from all the models da/dN is suggested to depend *both* on applied stress *and* on current crack length via $K = \sigma Y\sqrt{\pi a}$, and not only on stress level alone as in the traditional S–N curve approach.

Example 7.4 Describe crack advance in fatigue in terms of the shear decohesion or Bilby–Cottrell–Swinden type mechanisms of crack tip flow. What is the connexion between crack advance and crack opening displacement according to this approach?

Ans. Fig. 7.7 (after Tomkins) uses the shear decohesion mechanism to portray crack advance in Stage II fatigue. The figure uses simultaneous shear band operation to describe the crack opening process. Alternate shear band operation would lead to a lack of symmetry in striations, which is contrary to observations in homogeneous polycrystalline material. At low growth rates, e.g. 10^{-8} m/cycle one fine striation is creased over a number of cycles (Fig. 7.7(a)). During each cycle the maximum crack tip opening is achieved (of the order of 10^{-7} m/cycle) and crack growth can be continuous if the crack tip shear planes accommodate a new portion of crack surface during each cycle. Over N cycles the accumulated surface is enough to create a new striation on unloading and the process is repeated. Thus, the overall crack tip opening δ is achieved by shearing on several planes in the crack tip region, the proportion of shearing on each plane varying from cycle to cycle. The larger striation reflects the overall opening although it is not always distinct. At high growth rates, above about 10^{-6} m/cycle, one crack opening increment can involve the creation of several fine striations per cycle (Fig. 7.7(b)). Several new crack tip shear planes come into operation during one load increment. The spacing of these planes determines the intermediate fine striation spacing and this can vary with material, often being of the order of 0.5 μm (cf. Example 7.3).

In the shear decohesion or sliding-off model with slip planes at 45°, it is clear that the crack advance step Δa is equal to half the crack tip opening $\delta/2$ and therefore one might expect good correlation between da/dN and $\delta/2$ – at least up to the onset of Stage III. As discussed later, it turns out in practice that $da/dN < \delta/2$, particularly in alloys which workharden, because crack opening can then be accomodated by significant deformation back along the crack flanks in the tip region, as well as by the creation of new crack surface. Only towards the end of Stage II, when approaching Stage III, does $da/dN \approx \delta/2$ in such alloys. Equivalently we expect da/dN to approach $\delta/2$ in high strength alloys which have low work hardening capabilities and also in all alloys at high homologous temperatures. In these circumstances, coarse striations on fracture surfaces probably do correspond with da/dN and $\delta/2$. McClintock (1967) has shown that

$\delta/2$ is an upper limit for da/dN by the shear decohesion model; Kfouri and Miller (1976) have shown that δ in fatigue is smaller than δ in static tests.

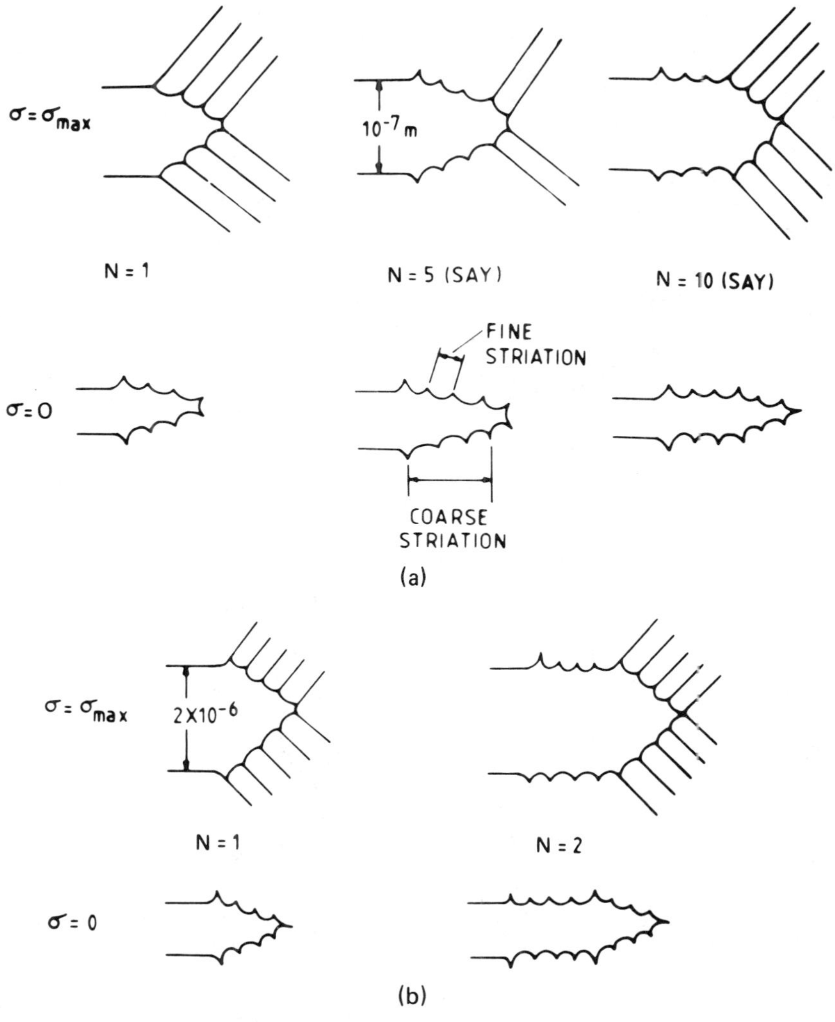

Fig. 7.7 – Crack opening and crack growth. (a) Low crack growth rate, $\sim 10^{-8}$ m/cycle; (b) high crack growth rate, $\sim 10^{-6}$ m/cycle. (After Tomkins (1980), reproduced with permission).

Since $\delta \propto K^2$, Example 7.4 reinforces the notion that da/dN depends on K, and all these models explain the pioneering study by Paris, Gomez and Anderson (reported in 1961) which showed that high cycle fatigue crack growth rates could be correlated with applied elastic stress intensity factors. Fig. 7.8(a) shows some of their early experimental results. Since that time a vast body of supporting

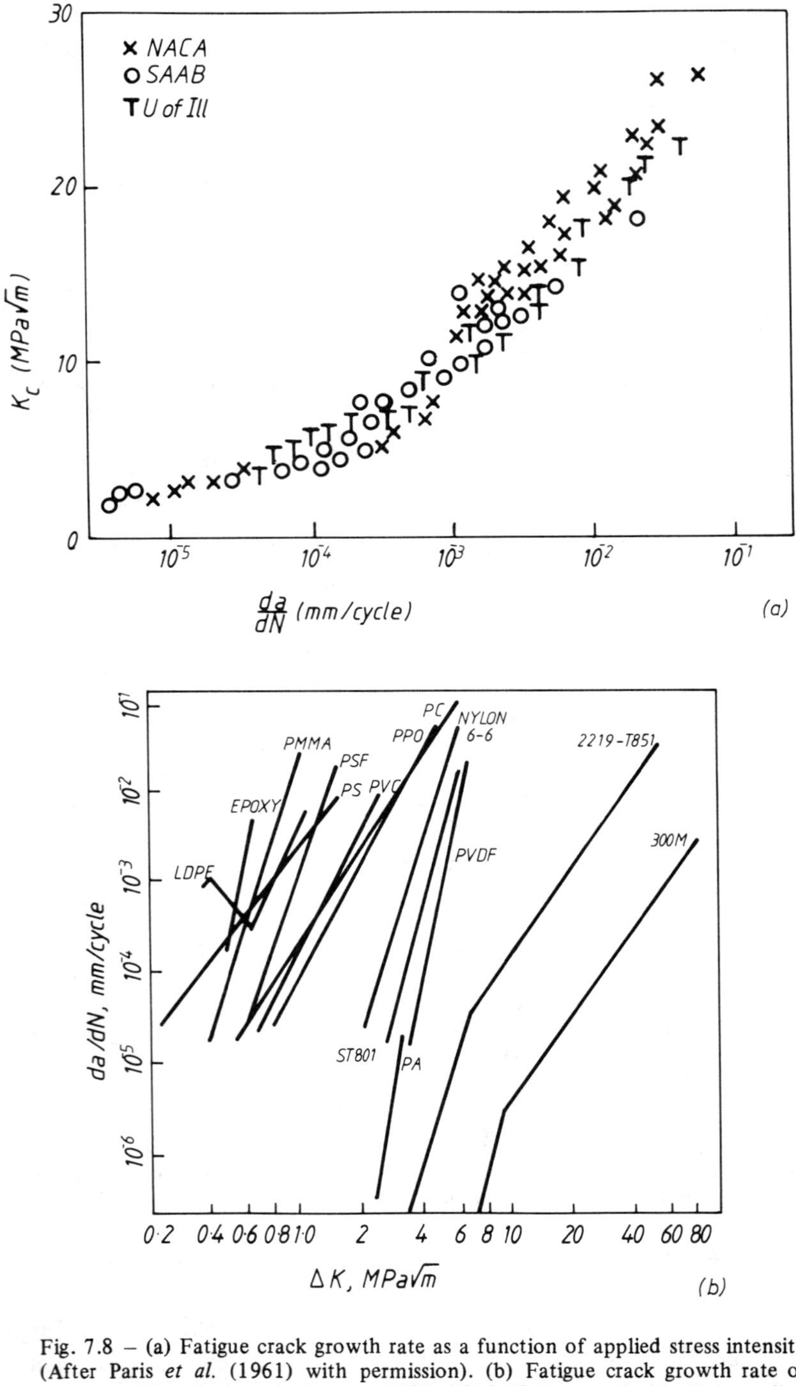

Fig. 7.8 – (a) Fatigue crack growth rate as a function of applied stress intensity (After Paris *et al.* (1961) with permission). (b) Fatigue crack growth rate of polymers (Hertzberg and Manson, 1980) 2219-T851 wrought aluminium alloy (Bucci, 1979) and 300 M structural steel (Ritchie, 1977). Note that growth rates in polymers can depend markedly on relative humidity, environment, frequency of cycling and so on.

data has been generated, over a range of applied loads, different mean stresses etc. and it is now customary to write

$$\mathrm{d}a/\mathrm{d}N = C(\Delta K)^n \tag{7.6}$$

where C is a constant and ΔK is the *range* of applied stress intensity, i.e. for cycling between stress levels σ_{min} and σ_{max} which results in $\Delta\sigma = \sigma_{max} - \sigma_{min}$,

$$\Delta K = (K_{max} - K_{min}) = (\sigma_{max} - \sigma_{min})\, Y\sqrt{\pi a} \tag{7.7}$$

When σ_{min} is compressive, it is set at zero in equation (7.7) (but see comments in section 7.5.2 on 'crack closure'). Notice that ΔK increases as the crack grows (even though $\Delta\sigma$ may remain the same) owing to the $\sqrt{\pi a}$ factor. For this reason, some experimentalists use constant rate-of-change-of-compliance testpieces, which give constant ΔK as the crack grows, particularly in stress corrosion fatigue (section 7.10). The constant C and exponent n in the Paris relation (7.6) are determined experimentally by plotting log $(\mathrm{d}a/\mathrm{d}N)$ versus log ΔK (Figs. 7.8(b) and 7.5(b) earlier). It must be emphasised that, although single lines are drawn in such plots, there is in practice considerable scatter in the data. An excellent discussion on the statistical treatment of fatigue data is given by Clark and Hudak (1975).

As explained by Knott in 1973, the complete growth rate curve on logarithmic coordinates is not a single straight line, with a single value for n, but is composed of three regions similar to Forsyth's subdivision of fatigue, as shown in Fig. 7.5(b). (The curve is reminiscent of a creep curve with primary (transient), secondary (steady-state), and tertiary regions). Early analyses of data attempted to draw a single line through all the points, which gave $n \approx 4$. Closer examination of data shows, however, that for many materials n values in the Stage II region lie in the range $2 < n < 3$. (cf. the various models for $\mathrm{d}a/\mathrm{d}N$ just discussed predict n to be 2 or 4). Even so, there is evidence that in high strength/low toughness steels, the value of n in Stage II can be higher (up to about $n = 10$). This comes about because monotonic modes of fracture occur on a fine scale in the more brittle components of the microstructure. These enhance the crack advance in some cycles giving greater $\mathrm{d}a/\mathrm{d}N$ and higher n. For most steels, n appears to have a distribution of values with maxima around 2.25 for pearlitic microstructures or 3 for ferrite/pearlite structures.

Example 7.5 Mukherjee (1972) reported a series of fatigue crack propagation experiments using wedge-opening-load (WOL) specimens of 5083 aluminium alloy subjected to constant amplitude sinusoidal loading with $X_{max} = 10.4$ kN and $X_{min} = 0$. The crack growth was measured ultrasonically and Fig. 7.9(a) shows crack length data versus number of cycles. Determine the constants in the Paris equation. (In the WOL testpiece $K = (X\sqrt{\pi a}/BW)Y$ where $Y\sqrt{\pi} = 7.8 + 37.37\,(a/W) - 117.00\,(a/W)^2 + 141.87\,(a/W)^3$) ($B = 45$ mm; $W = 65$ mm.)

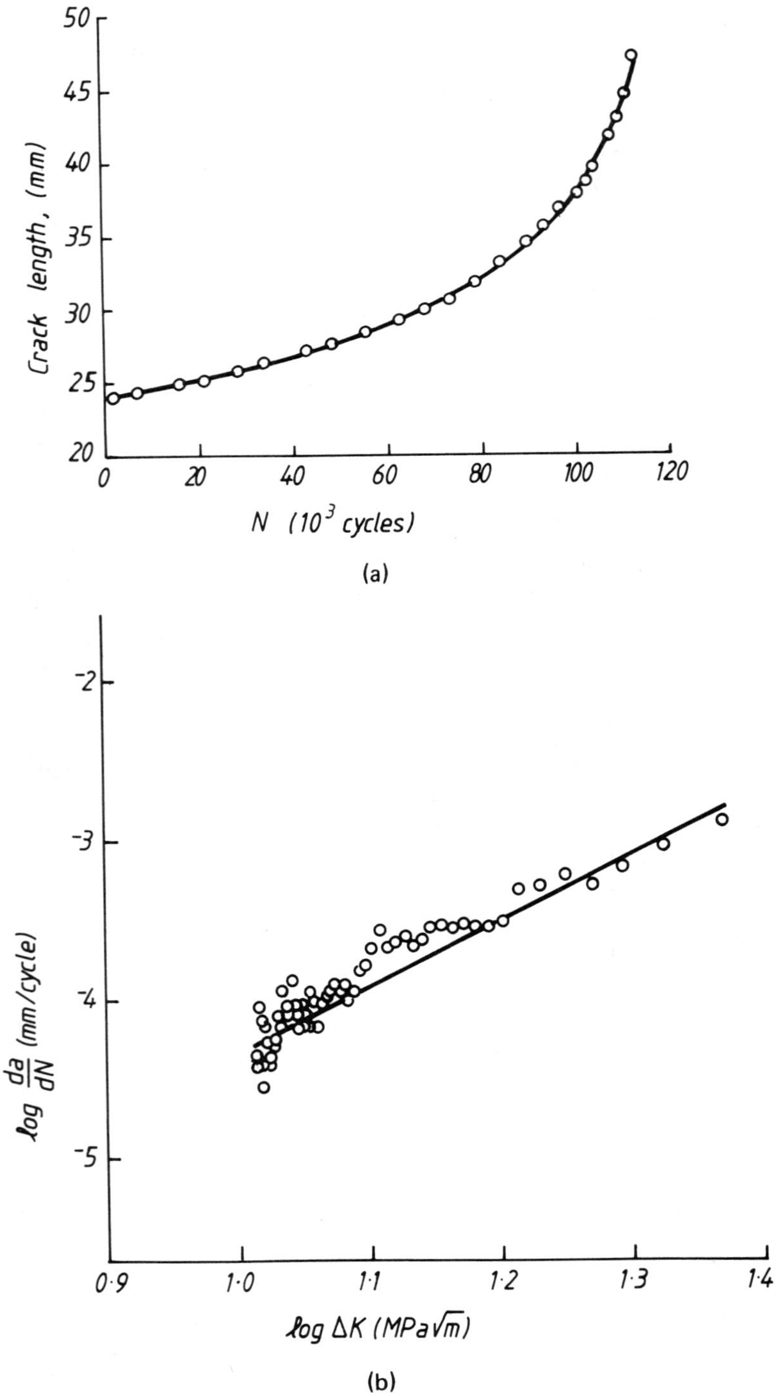

Fig. 7.9 – (a) Crack growth versus cycles elapsed for 5083 aluminium alloy. (b) Plot of da/dN against ΔK for data in (a) (both after Mukherjee (1972), with permission).

Ans. In Mukherjee's note, crack lengths were determined at equal intervals of 1800 cycles and da/dN calculated using the method of finite difference, i.e. the average growth rate at crack length, $(a_1 + a_2)/2$ is given by $(a_2 - a_1)/(N_2 - N_1)$. Fig. 7.9(b) shows $\log(da/dN)$ plotted against $\log(\Delta K)$. The equation of the best straight line through the experimental points (method of least squares) is given by $(da/dN) = 4.37 \times 10^{-34} (\Delta K)^{4.15}$ mm/cycle with ΔK in Pa$\sqrt{\text{m}}$.

The data in Example 7.5 may be analysed in a different, but equally valid, way which is to fit a polynomial through the experimental points. The curve in Fig. 7.9(a) is in fact such a line. Fatigue crack growth rates may be determined by differentiating the polynomial. Fig. 7.10 shows the log da/dN versus log ΔK results obtained in this way, differentiating the polynomial at intervals of 1,800 cycles for consistency with the previous method of analysis. The data are smoother than those in Fig. 7.9(b) owing to the process of fitting the polynomial. Mukherjee (1972) used a fifth order polynomial. Yap (1982) finds that instabilities often arise with high order polynomials and that a third order polynomial using an incremental method is better; Mukherjee's data are re-analysed that way in Problem 7.15. The best straight line through the points in Fig. 7.10 is $da/dN = 1.93 \times 10^{-30} (\Delta K)^{3.64}$ mm/cycle for ΔK in Pa$\sqrt{\text{m}}$.

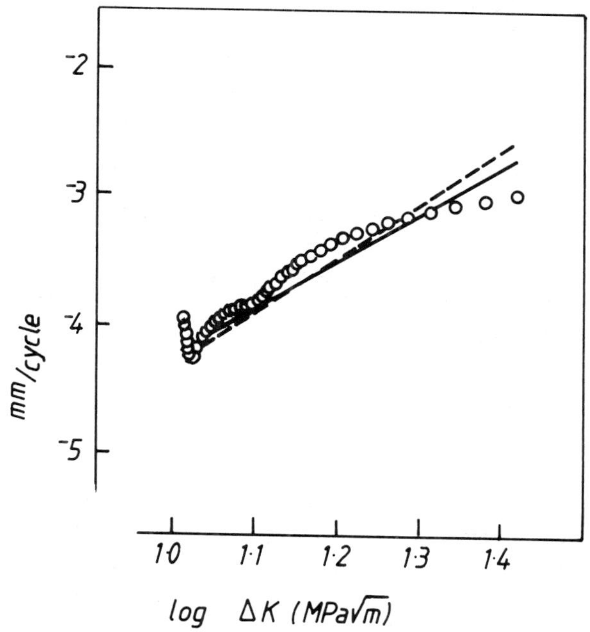

Fig. 7.10 – Fatigue crack growth equation obtained by finite difference and polynomial curve fitting methods (after Mukherjee (1972), with permission).

Notice that, although derived from the same raw data, the exponents of the Paris equation are different. This points immediately to the care which has to be taken in comparing crack growth data from different sources, which may employ different analytical methods. A basis of comparison is possible however in terms of testpiece life rather than merely the exponent of the Paris equation (Example 7.6 later).

Given an expression for $\mathrm{d}a/\mathrm{d}N$, the number of cycles to reach a specified crack length may be found by integration – in particular, the number of cycles N_f taken for the crack to grow to the critical length at which it finally fails. We have from relation (7.6)

$$\int_{a_i}^{a_f} \mathrm{d}a = \int (\mathrm{d}a/\mathrm{d}N)\mathrm{d}N = C \int (\Delta K)^n \, \mathrm{d}N \tag{7.8}$$

or equivalently

$$\int_0^{N_f} \mathrm{d}N = (1/C) \int \mathrm{d}a/(\Delta K)^n \tag{7.9}$$

where a_i and a_f are the initial and final lengths. Although the subscript 'f' on N and a usually is applied to final monotonic fracture, it could relate simply to some intermediate crack length longer than a_i and to the number of cycles to reach that length.

If $K = \sigma Y \sqrt{\pi a}$, integration of equation (7.9) gives

$$N_f = \frac{1}{C(\Delta\sigma)^n \, \pi^{n/2}} \int_{a_i}^{a_f} \frac{\mathrm{d}a}{(Y\sqrt{a})^n} \tag{7.10}$$

Y is a function of a and therefore must be left under the integral. In some cases, however, Y does not change markedly with a and we may write

$$N_f = \frac{[(1/a_i)^{(n/2)-1}] - [(1/a_f)^{(n/2)-1}]}{C(\Delta\sigma)^n \, \pi^{n/2} \, (n/2 - 1) \, Y^n} \tag{7.11}$$

where Y is understood to be either a representative value of Y between a_i and a_f, or the value $Y(a_i)$ as employed in Chapter 6 for time-to-failure evaluations. The first term in the numerator gives the life that would result were the crack able to grow to infinite length; the second term represents a life reduction because the critical crack length is, in fact, not infinite.

As an illustration consider an aircraft flight envelope which between take-off and landing produces a series of repeated gust loadings with $\sigma_{max} = 200$ MPa and $\sigma_{min} = 50$ MPa. An aluminium alloy component in the aircraft (whose monotonic R is 15 kJ/m^2) has a crack of length 4 mm, and experiments show that for this alloy $\mathrm{d}a/\mathrm{d}N = 5 \times 10^{-11} (\Delta K)^3$ m/cycle (with ΔK in MPa$\sqrt{\text{m}}$). We require to know how long it will take for the crack to propagate to the critical length. Using $K_C = \sqrt{EG_C} = \sigma\sqrt{\pi a}$, we have

$$a_f = K_C^2/\sigma_{max}^2\pi = 70 \times 10^9 \times 15 \times 10^3/\pi(200 \times 10^6)^2 = 8.4 \text{ mm}.$$

Hence

$$\int_0^{N_f} dN = \int_{4 \times 10^{-3}\,\text{m}}^{8.4 \times 10^{-3}\,\text{m}} da/5 \times 10^{-11}\,(200-50)^3\,(\pi a)^{1.5}$$

or

$$N_f = 2{,}128\left(\frac{1}{\sqrt{0.004}} - \frac{1}{\sqrt{0.0084}}\right) = 2128\;[15.81 - 10.91]$$

$$= 10{,}428 \text{ cycles}$$

If there are, say, ten gusts per flight and two flights per day, 10,428 cycles equates to about a year and a half.

This sort of calculation done for airframes can be applied to cyclic loading of various components and structures such as bridges, cranes, ground vehicles, pipework, pressure vessels and so on. In the case of offshore structures, for example, the principal source of cyclic loading is wave action. Such structures are designed typically for lives of 20 to 50 years and annually may experience 5 million cycles of wave loading (the period of waves in the North Sea is about 6 seconds).

When $a_f \gg a_i$ (which often is the case), equation (7.11) simplifies to

$$N_f \approx \left[1/a_i^{(n/2)-1}\,C(\Delta\sigma)^n\,\pi^{n/2}\,(n/2-1)Y^n\right] \tag{7.12}$$

which may be recast in terms of ΔK_i (the stress intensity range at the starting crack length) as follows:

$$N_f \approx [a_i/C(\Delta K_i)^n\,(n/2-1)] \tag{7.13}$$

Notice that when $n = 2$ in equations (7.11)–(7.13), an infinite fatigue life is predicted. However, it may be shown that (Problem 7.10) under these circumstances N_f is given by

$$N_f = [a_i/C\,\Delta K_i^2]\,\ln(a_f/a_i) \tag{7.14}$$

Calculations employing relations (7.8) *et seq.* involve a starting crack length (the lower limit of integration cannot be zero as the relation then gives an infinite number of cycles to failure). In practical components, flaws (the length of which become the lower limit of the integral) have to be found by non-destructive inspection (NDI) methods; inspections may be performed at the time of manufacture as part of quality control procedures, or at regular intervals during the use of a component. NDI is costly and can be very difficult when flaws are small. Note that surface crack detection by dye penetrant is not quantitative and while other methods (such as a.c. electric potential, ultrasonics or eddy currents) can identify the size (and shape in some cases) of cracks down to, say, 0.1 mm *on smooth machined surfaces,* on rough as-welded surfaces, say, the limit of detection would be at least an order of magnitude greater.

Where fatigue cracks, which are discovered during the life of a component, actually come from in supposedly crack-free materials is another question, which has been discussed in Chapter 5 in relation to fabricated components and processing details.

By assuming initial defect sizes for the lower limit of the integral, and hence initial K values, component lives may be predicted as shown above. This is particularly appropriate for the fatigue life of many welded details at low stress levels where it is found that the inherent tiny intrusion defects associated with a weld (e.g. slag intrusions ~ 0.5 mm deep at toes of fillet welds) act as initiation regions for fatigue, so that Stage I is negligible and virtually all the life is occupied in propagating the crack from the initial to final size. Similar considerations apply for the fatigue of cracks around, for example fastener holes which result from manufacturing processes (drilling tears, burrs etc., cf. Chapter 5. It is clear from equations (7.8) and (7.9) that the slope of a traditional S–N curve for a

Table 7.1

Representative C and n in $\mathrm{d}a/\mathrm{d}N = C(\Delta K)^n$ ($\mathrm{d}a/\mathrm{d}N$ in mm/cycle; ΔK in $\mathrm{MPa}\sqrt{\mathrm{m}}$)

Note. Actual values will usually depend on pre-strain, heat treated condition, mean stress, rate of cycling, temperature, environment, etc.

	C	n
Aluminium (commercially pure)	4.56×10^{-11}	2.9
Al/4.5%–Cu alloy (L71)	3.9×10^{-11}	3.7
Al/5%–Mg alloy	28×10^{-11}	2.7
Al/5.5%–Zn alloy (DTD 687 A)	12.6×10^{-11}	3.7
Copper	0.34×10^{-11}	3.9
Mild steel	0.42×10^{-11}	3.0
18/8 austenitic stainless steel	0.33×10^{-11}	3.1
Titanium	0.69×10^{-11}	4.4
Nylon 6-6	1.5×10^{-8}	3.2
PMMA	2.3×10^{-5}	4.7
Polycarbonate	9.5×10^{-5}	2.6
PPO	4.7×10^{-6}	3.0
Polystyrene	3.1×10^{-5}	2.2
Polysulphone	6.3×10^{-7}	5.0
PVC	2.9×10^{-6}	1.5
Westerly granite	2×10^{-7}	12

Sources. Pook (1975); Hertzberg and Manson (1980); Kim and Mubeen (1981); Hudson and Seward (1978, 1982); authors' own measurements. See Appendix 2.

cracked component, when plotted on logarithmic scales, is $(-1/n)$ where n is the index of the Paris equation (cf. Problem 7.16 and Example 7.9).

Table 7.1 gives representative values of n and C for a number of metals. We observe that the fatigue crack growth properties of steels in Stage II are broadly independent of chemistry, yield stress and mean stress (see section 7.3).

The selection of materials for design against failure which concerns the integrated equations (7.11) *et seq.* is discussed further in section 7.11.

Example 7.6 Show that estimations of crack life provide a more satisfactory comparison between the two versions of the Paris equation for the fatigue data in Fig. 7.9(a) rather than the exponents alone. (The experimental crack life for the WOL specimens was 114,200 cycles after the crack had grown from 24.1 to 48.3 mm.)

Ans. We have (i) $da/dN = 4.37 \times 10^{-34} (\Delta K)^{4.15}$ mm/cycle based on finite difference methods, and (ii) $da/dN = 1.93 \times 10^{-30} (\Delta K)^{3.64}$ mm/cycle based on polynomial curve-fitting. Both expressions may be used in equation (7.10), with the K-formula for the WOL geometry as given in Example 7.5. Numerical integration will be required since $Y(a/W)$ changes appreciably over the crack length change, viz. $Y\sqrt{\pi}$ for $a = 24.1$ mm is ~ 12.8, $Y\sqrt{\pi}$ for $a = 48.3$ mm is ~ 28.9. This prevents the use of equation (7.11). In Mukherjee's 1972 note he performed the numerical integration in a complementary way, by considering finite ΔN intervals and writing for the jth interval, $a_{j+1} = a_j + (C\Delta K^n)_j \; \Delta N$ where Y in ΔK was $Y(a_j/W)$. The first value for a_j is 24.1 mm, and the last is 48.1 mm. N_f is obtained by summing all the ΔN required to reach the final propagated length of 48.1 mm. Mukherjee used a variety of values of ΔN to check the sensitivity of the calculations to the step-size. Table 7.2 shows the crack life calculated in this way for various ΔN. The various predictions of life from both equations are within 4% of the experimental crack life of 114,200 cycles, even though the C and n values in the Paris equation are different.

Table 7.2

Calculated crack life using equations (i) and (ii) for da/dN

Intervals of numerical integration ΔN (cycles)	Crack life using (i) (cycles)	Crack life using (ii) (cycles)
500	114,500	110,000
1,000	115,500	111,000
1,500	116,200	111,500
2,000	117,000	112,200

Of course, where N_f turns out not a multiple of ΔN, the last step in the integration takes an appropriately smaller ΔN value.

ΔK changes during propagation owing both to the direct influence of increasing a and to increasing $Y(a/W)$. It is worth noting the discrepancies in predicted N_f which arise if $Y(a/W)$ is assumed constant. With $X_{max} = 10.4$ kN and $X_{min} = 0$ in the Mukherjee experiments, $\Delta K = 7.1$ MPa$\sqrt{m}$ at the starting crack length of 24.1 mm (where $Y\sqrt{\pi} \approx 12.8$) and is 22.7 MPa$\sqrt{m}$ at the final length of 48.3 mm (where $Y\sqrt{\pi} \approx 28.9$). As ΔK depends on a in a non-linear fashion, the mean ΔK of $(7.1 + 22.7)/2 \approx 15$ MPa$\sqrt{m}$ is *not* ΔK at the mean crack length of $(24.1 + 48.3)/2 = 36.2$ mm, which is 11.4 MPa$\sqrt{m}$ (with $Y\sqrt{\pi} = 16.81$). For purposes of illustration, consider a mean $Y\sqrt{\pi}$ of $(12.8 + 28.9)/2 \approx 21$ in equation (7.11). We have, for $da/dN = 4.37 \times 10^{-34} (\Delta K)^{4.15}$ mm/cycle with ΔK in Pa$\sqrt{m}$, and observing that $\Delta\sigma = (X_{max} - X_{min})/BW$,

$$N_f = \frac{\left[\dfrac{1}{(24.1 \times 10^{-3})^{1.075}} - \dfrac{1}{(48.3 \times 10^{-3})^{1.075}}\right]}{4.37 \times 10^{-37}\,(10.4 \times 10^3/45 \times 65 \times 10^{-6})^{4.15}\,(1.075)\,(21)^{4.15}}$$

$$= 130{,}000 \text{ cycles.}$$

Owing to the raising of quantities to different powers, it is easiest to write everything in basic units, so here mm/cycle has been changed to m/cycle by writing 10^{-37} in place of 10^{-34}, and the crack lengths written as meters.

Had the $Y\sqrt{\pi}$ value of 16.81, which applies to the mean crack length, been used in equation (7.11), $N_f \rightarrow 327{,}000$ which considerably overestimates the experimental crack life of *circa* 114,000 cycles. These calculations show how much care has to be taken in the integration of the Paris equation when Y changes, both as regards assumptions made and the units used.

7.2.2 Threshold ΔK_0

There is some difficulty associated with the da/dN relation in the Stage I region of Fig. 7.5(b). The implication of the graph (but not of the Paris equation (7.6)) is that there is a 'threshold' ΔK_0 below which crack growth never occurs. For many metals the apparent threshold ΔK_0 is about 1–5 MPa$\sqrt{m}$, the precise value depending on mean stress in the cycle (Table 7.3), (see section 7.3). Measurement of crack length in fatigue is not easy and if there is negligible crack advance over a number of cycles, $da/dN = 0$, so $\log(da/dN) = -\infty$, and data points for extremely low (da/dN) are difficult to locate on the graph.

As K involves both stress and crack length, low ΔK is produced either (i) by low $\Delta\sigma$ when cracks are long or (ii) by high $\Delta\sigma$ when cracks are short. The reasons for a threshold ΔK (if one exists) should *not* be expected to be the same in both cases.

With long cracks, the origin of the apparent threshold seems to be bound up with

(a) the smallness of the crack tip plastic zone associated with the small $\Delta\sigma$ which may not encompass all the microstructural features required for fracture (e.g. with $\sigma_{min} = 0$, and $K_{max} = K_0$, $d_y = (1/6\pi)(K_0/\sigma_y)^2 = 1$–$30$ μm for $1 < K_0 < 5$ MPa$\sqrt{m}$ and cyclic $\sigma_y = 210$ MPa); and

Table 7.3

Values of ΔK_0 for various materials

Material	Tensile strength (MPa)	Stress ratio, r	ΔK_0 (MPa$\sqrt{m}$)
Ferrous			
Mild steel	430	−1	6.4
		0.13	6.6
		0.35	5.2
		0.49	4.3
		0.64	3.2
		0.75	3.8
Mild steel at 300°C	480	−1	7.1
		0.23	6.0
		0.33	5.8
Mild steel in brine	430	−1	~2.0
		0.64	1.15
Mild steel in brine with cathodic protection	430	0.64	3.9
Mild steel in tap water or SAE30 oil	430	−1	7.3
Low alloy steel	835	−1	6.3
	680	0	6.6
		0.33	5.1
		0.50	4.4
		0.64	3.3
		0.75	2.5
NiCrMoV steel at 300°C	560	−1	7.1
		0.23	5.0
		0.33	5.4
		0.64	4.9
Maraging steel	2010	0.67	2.7
18/8 Austenitic steel	685	−1	6.0
	665	0	6.0
		0.33	5.9
		0.62	4.6
		0.74	4.1
Grey cast iron	255	0	7.0
		0.50	4.5

Table 7.3 – *continued*

Values of ΔK_0 for various materials

Material	Tensile strength (MPa)	Stress ratio, r	ΔK_0 (MPa$\sqrt{\text{m}}$)
Non-ferrous			
Aluminium	77	−1	1.0
		0	1.7
		0.33	1.4
		0.53	1.2
L65 Aluminium alloy	450	−1	2.1
(4.5% Cu)	495	0	2.1
		0.33	1.7
		0.50	1.5
		0.67	1.2
L65 Aluminium alloy (4.5% Cu) in brine		0.50	1.15
ZW1 Magnesium alloy	250	0	0.83
(0.6% Zr)		0.67	0.66
AM503 Magnesium alloy	165	0	0.99
(1.6% Mn)		0.67	0.77
Copper	225	−1	2.7
	215	0	2.5
		0.33	1.8
Titanium	540	0.60	2.2
Nickel	455	−1	5.9
	430	0	7.9
		0.33	6.5
		0.57	5.2
		0.71	3.6
Monel	525	−1	5.6
		0	7.0
		0.33	6.5
		0.50	5.2
		0.67	3.6
Iconel	655	−1	6.4
	650	0	7.1
		0.57	4.7
		0.71	4.0

Reproduced with permission, from Pook, P. L. (1975) *J. Strain Analysis* **10** (4), 242.

(b) the possible need for an incubation number of cycles in as-received samples at low $\Delta\sigma$ levels, during which the local crack tip saturation hardening (which has to be the precursor of crack advance) occurs by reversed plastic flow along slip bands (cf. Example 7.4); at higher levels of applied stress and thus higher ΔK these events occur more rapidly.

These matters are discussed further in section 7.5 later in terms of Elber's 'crack-closure' concept.

Other factors come into play when cracks are small and the applied ΔK is low, and are discussed in Section 7.2.3 following.

***Example** 7.7* Show that a plausible interpretation of ΔK_0 is that it is the smallest value required to produce crack advance equal to a lattice spacing.

Ans. If the applied ΔK is insufficient to create an element of new surface anywhere along the crack front equal to the lattice spacing, the crack must remain dormant. Taking 300 pm for the lattice spacing and assuming that the Paris law applies to low growth rates, we have $da/dN = 300 \times 10^{-12} = C(\Delta K_0)^n$, so $\Delta K_0 = [300 \times 10^{-12}/C]^{1/n}$. For mild steels da/dN (mm/cycle) $\approx 3 \times 10^{-9} \times (\Delta K)^3$, say, where ΔK is in MPa$\sqrt{\text{m}}$. Hence $\Delta K_0 = [300 \times 10^{-9}/3 \times 10^{-9}]^{1/3} \approx 5$ MPa$\sqrt{\text{m}}$. Again for aluminiums with lattice spacing 400 pm and $da/dN \approx 50 \times 10^{-9}\ (\Delta K)^3$, $\Delta K_0 = [400 \times 10^{-9}/50 \times 10^{-9}]^{1/3} \approx 2$ MPa$\sqrt{\text{m}}$. Both these values accord with measured ΔK_0.

The incorporation of the threshold ΔK_0 concept on an Xu diagram is interesting. For $X_{min} = 0$, ΔK_0 is represented by a G_0 locus of magnitude (K_0^2/E^*) the shape of which of course depends on the geometry of the testpiece. Fig. 7.11(a) shows the effect of a superimposed G_0 locus on the situation already given in Fig. 7.6. On unloading from, say, D and reloading, cyclic growth does not commence until point L on the G_0 locus instead of immediately. The same information (with the same lettering) is shown as a GA (RA) diagram in Fig. 7.11(b). Lines of constant X are radial from the origin in this figure owing to the assumed testpiece geometry so crack growth between zero load and X_{max} is represented by A_0 C . . . $(A_0 + \Delta A)$D. . . E. . . etc. until the crack area reaches A at P, when the crack spreads quasi-statically following the constant G (R) monotonic locus.

Similar arguments may be applied to constant cyclic *displacement* range fatigue, with $\Delta u = u_{max} - u_{min}$. The crack growth sequence is illustrated in Fig. 7.12(a) and (b). In the first excursion to u_m (when the crack size is the starter area A_0) the load reaches (in this instance) the monotonic crack spreading load at B and BC represents propagation to crack area A_1. At C the displacement has reached its maximum extent and the body unloads down CO. Thereafter the growth rate decreases each cycle until eventually point F is reached which falls on the threshold G_0 locus at crack area A_n. At F, fatigue crack propagation is *arrested*.

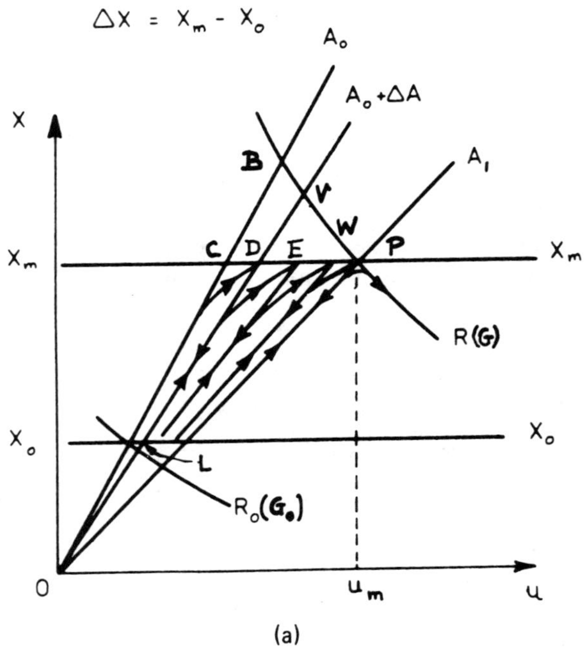

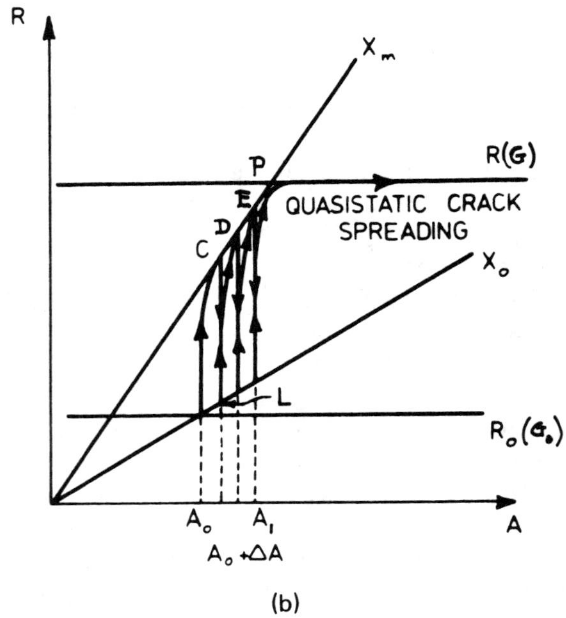

Fig. 7.11 – Interpretation of constant load range fatigue crack growth in terms of (a) X–u diagram and (b) R–A diagram when there is a fatigue threshold G_0 (or R_0).

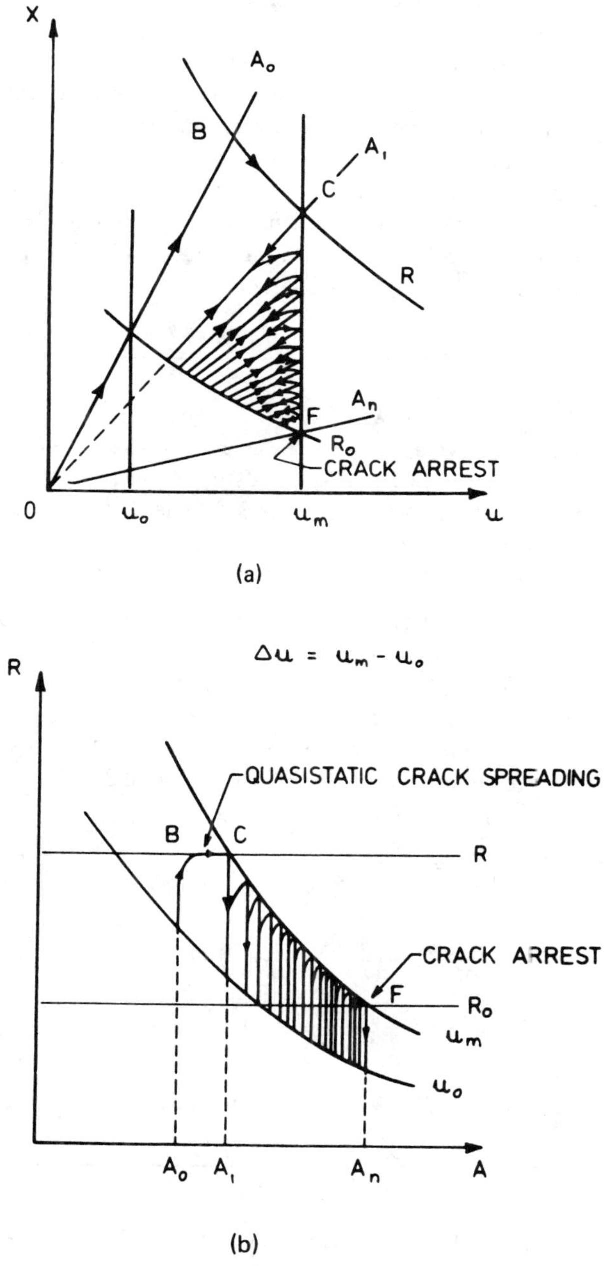

Fig. 7.12 – Interpretation of fatigue crack growth under constant cyclic displacement range in terms of (a) X–u diagram and (b) *R*–*A* diagram.

Note that in Figs. 7.11 and 7.12 the cyclic loops shown are for a *number* of cycles, and are exaggerations if presumed to apply to a single cycle.

Fatigue arrest is possible under *load*-control for some cracked geometries, especially when monotonic propagation takes place under *increasing* load – the constant G loci sloping upwards. This occurs in linear elastic cases (ball indentation, cracked ring in compression and so on, Fig. 2.6) or in non-linear elastic cases (at large displacements where membrane stresses stiffen bodies, as in Figs. 2.10 and 2.11 and in the Outwater double-torsion testpiece at large displacements). In both cases, fatigue cracks grow into decreasing K-fields and da/dN must decrease. If the decreasing K field drops below the threshold value, arrest occurs (see Example 7.8 following).

Example 7.8 The K-formula for a right-angle joint with a flaw Fig. 7.13(a) (as may apply to a welded T-joint) is given by Tada *et al.* (1973) as $K_I = 1.6 \times X/B\sqrt{\pi a}$ where X is a line load applied to the faces of the crack (their formula (8.9)). Show that in the presence of a fatigue threshold ΔK_0, cyclic loading between zero and the loads X_1 and X_2 ($X_2 > X_1$), leads to crack *arrest* and that the number of cycles to arrest is greater at the higher load. (B is plate thickness.)

Ans. Fig. 7.13(a) shows the geometry and Fig. 7.13(b) the Xu plot with the monotonic G_{IC} locus and threshold G_0 locus superimposed. Presume that $X_2 < K_{IC}B\sqrt{\pi a}/1.6$ so that no monotonic growth occurs initially. It is clear that the levels of X_2 and X_1 eventually intersect the G_0 locus so that arrest must occur at points P and Q respectively where $A_P > A_Q$. To determine the number of cycles to reach P and Q, let us assume that $da/dN = C(\Delta K)^2 = C(1.6)^2 X^2/B^2 \pi a$ for zero minimum load. Thus

$$N = \int_{a_i}^{a_{\text{arrest}}} (2.56\, CX^2/B^2\pi)^{-1} a\, da$$

where a_i is the initial flaw size. This becomes

$$N = \frac{a^2_{\text{arrest}}}{2}\left(\frac{2.56\, CX^2}{B^2\pi}\right)^{-1}\left[1 - (a_i/a_{\text{arrest}})^2\right] = \frac{a_{\text{arrest}}}{2\, CK_0^2}\left[1 - (K_0/K_i)^4\right].$$

For the same K_0, $N|_{X_2} > N|_{X_1}$ because $a_{\text{arrest}}|_{X_1} < a_{\text{arrest}}|_{X_2}$ and $K_i|_{X_2} > K_i|_{X_1}$.

Note that the N in Example 7.8 are *cycles to arrest*, not ‘fatigue life’ in the usual sense. In fact, not only is da/dN decreasing as crack length increases, under these circumstances, but the critical crack length for monotonic failure is bigger the higher the level of X (the formula is unusual in that K decreases as a increases). In the da/dN vs. ΔK plot, we are running down from right to left rather than the more common reverse direction.

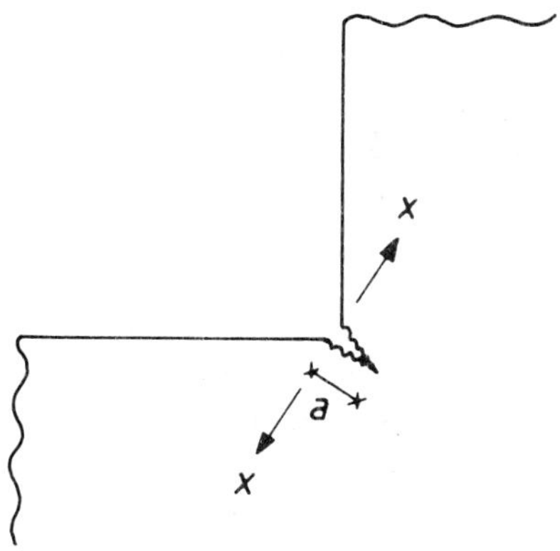

(a)

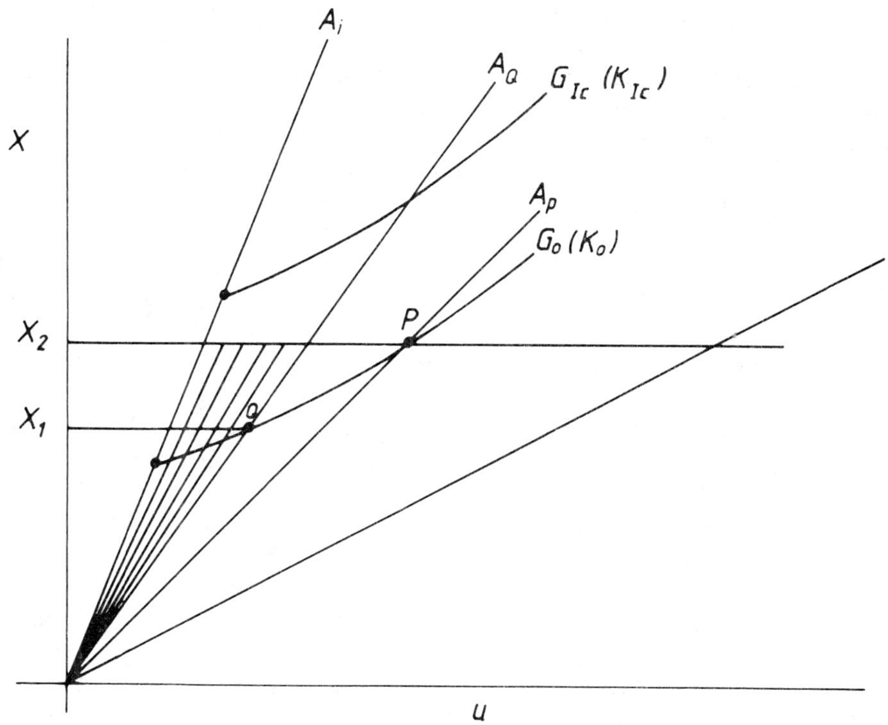

(b)

Fig. 7.13 – (a) Fatigue crack growth in a right-angle joint subjected to constant load range. (b) X–u plot with constant G-loci for the joint in (a). Note that for a constant load range crack *arrest* is predicted.

7.2.3 Short cracks

Practical materials are not continua on a fine scale, and different regions of microstructure have different properties. In metals, adjacent grains may be brittle or ductile and the grain boundaries separating them may possess yet different characteristics. Adjacent grains even with similar properties may be anisotropic with respect to each other. Even within grains, eutectic and eutectoid structures consist of quite dissimilar materials, e.g. ductile ferrite and brittle cementite in pearlite.

The application of continuum mechanics to practical materials presumes that local variations in polycrystalline solids are averaged out and that it is possible to talk in terms of homogeneous mechanical properties such as elastic moduli, yield stress and so on. This is possible when the variations in properties occur over merely a fine scale (when the grain size is small, when precipitates or inclusions are small and uniformly distributed and so on) and/or the body is large enough, even if the microstructure is coarse. In the case of cracked bodies this means that continuum mechanics may be applied to bodies containing cracks which are long enough to be unaffected by local variations in properties (i.e. even though the crack tip may be travelling through microstructure which varies on a fine scale, for practical purposes the mechanics of fracture are adequately described by a global energy balance using homogeneous mechanical properties for the body as a whole).

However, when the size of a crack is comparable with the size of microstructural features, difficulties arise. This may be put on a quantitative footing if we remind ourselves of the assumptions inherent in the application of the familiar $K = \sigma Y\sqrt{\pi a}$ formula. It is only the first term approximation of exact stress fields at the tips of cracks, and Fig. 2.24 shows that at distances greater than $r/a \approx 0.1$ errors greater than 10 per cent arise using the approximation. Thus, for confident application of the elastic $K = \sigma Y\sqrt{\pi a}$ expression, microstructural variations should certainly be encompassed within the $r \approx 0.1a$ 'K-dominant' region. If we take the size of the largest feature in a microstructure to be the grain size (of typical extent 10 μm, say), a crack of length (10/0.1) μm = 0.1 mm is required before K-formulae are considered valid. This size will vary of course with different assumptions about microstructural features.

This, however, is only part of the answer for irreversible deformation, crazing and so on occurs at crack tips in many practical materials. As already explained in Chapter 2, it is argued that *elastic* K-formulae are still able to describe crack tip conditions, providing that the plastic zone is limited in size compared with the near crack tip stress field, i.e. d_y being much smaller than $r \approx 0.1a$. Limiting the plastic zone size to, say, 1/5 or even less of the near tip elastic field then results in the condition that the encapsulated crack tip plastic zone must be limited to (1/5) (0.1) = 1/50 of the crack length or smaller; this leads to the well-known criterion for 'valid' plane strain monotonic toughness testing (Example 2.15).

When specimens with *long* starter cracks are fatigued, the applied stresses, being necessarily lower than those that caused monotonic fracture, produce smaller d_y and hence no plastic zone size difficulties; indeed thinner specimens may be used according to $B > 2.5\,(K/\sigma_y)^2$ where K is now the maximum K in the fatigue cycle rather than K_{IC}. When starting with *short* cracks in fatigue, however, microstructural features may influence the behaviour. Consider a growing small fatigue crack. Instead of using the expression for d_y to calculate the associated plastic zone size, Smith (1977) argued that the *minimum* structural feature in the material associated with the crack tip zone would be a subgrain slip band, ~ 500 nm in size. Now if the crack size is to be 50 times larger than this feature, the minimum crack size amenable to fracture mechanics treatment in fatigue becomes surprisingly large, about 25 μm.

Furthermore, consider the case of a fatigue crack on the point of growing, i.e. subjected to the threshold stress intensity factor range, ΔK_0. As the crack size to which this threshold level is applied becomes smaller, a larger applied stress is required to maintain a constant ΔK_0. If, however, this stress exceeds the plain fatigue limit σ_0 of the specimen, given by a traditional *S–N* curve, crack growth is bound to occur by definition (the starter crack is now comparable with normal surface imperfections in preparated specimens, i.e. ~ 50 nm (fine lapped surfaces, ~ 5 μm (ground), ~ 50 μm (machined)).

The effect is well shown on a plot of $(\sigma_{max} - \sigma_{min})$ versus a, i.e. the stress range necessary to produce fatigue failures versus crack length. At the threshold, $\Delta K_0 = Y\Delta\sigma\sqrt{\pi a}$, we obtain (for $\sigma_{min} = 0$)

$$\log \sigma_{max} = -\tfrac{1}{2}\log(a) + \log(\Delta K_0/Y\sqrt{\pi}) \tag{7.15}$$

which is the right-hand line in Fig. 7.14(a), reproduced from Smith and Miller (1978) using data from Kitagawa and Takahashi (1976) for HY 80 steel. Since the maximum stress cannot exceed the plain fatigue limit σ_0 (515 MPa in this case), there is a cut-off point at a crack length a_0 given by

$$a_0 = (1/\pi)\,(\Delta K_0/Y\sigma_0)^2 \tag{7.16}$$

which is 0.065 mm in Fig. 7.14(a). Similar results for mild steel are shown in Fig. 7.14(b), where $a_0 \approx 0.26$ mm.

There is a concomitant *apparent decrease* in ΔK_0 with decreasing crack length if calculated from $\Delta K_0 = Y\Delta\sigma\sqrt{\pi a}$ when $a < a_0$ since σ_{max} in $\Delta\sigma$ has reached a plateau given by the conventional fatigue limit, and $\log(\Delta K_0) \propto \frac{1}{2}\log(a)$ only. Fig. 7.15, from Smith (1977), illustrates the effect for mild steel centre-cracked plates with $\sigma_0 = 210$ MPa, $\Delta K_0 = 6$ MPa$\sqrt{\text{m}}$ and $\sigma_{min} = 0$; these are the same data as for Fig. 7.14(b). The constant σ_0 part of Figs. 7.14 corresponds with the apparently changing ΔK_0 part of Fig. 7.15, and the falling parts of Figs. 7.14 corresponds with the constant ΔK_0 part of Fig. 7.15, the lines intersecting at the appropriate value of a_0 for the material in question. Some representative values of a_0 for other metals are: grey cast iron 900 μm ($\Delta K_0 = 7$ MPa$\sqrt{\text{m}}$, $\sigma_0 = 130$ MPa); 100 μm 18/8 stainless steel (6 MPa$\sqrt{\text{m}}$ and 330 MPa); 30 μm for some for some aluminium alloys (*ca.* 2 MPa$\sqrt{\text{m}}$ and *ca.* 190 MPa); 300 μm commercially pure aluminium (1.5 MPa$\sqrt{\text{m}}$ and 46 MPa); 560 μm commercially pure copper (2.6 MPa$\sqrt{\text{m}}$ and 62 MPa); 13 μm titanium (2.2 MPa$\sqrt{\text{m}}$ and 340 MPa). Insofar as ΔK_0 and σ_0 change with stress ratio, so does a_0.

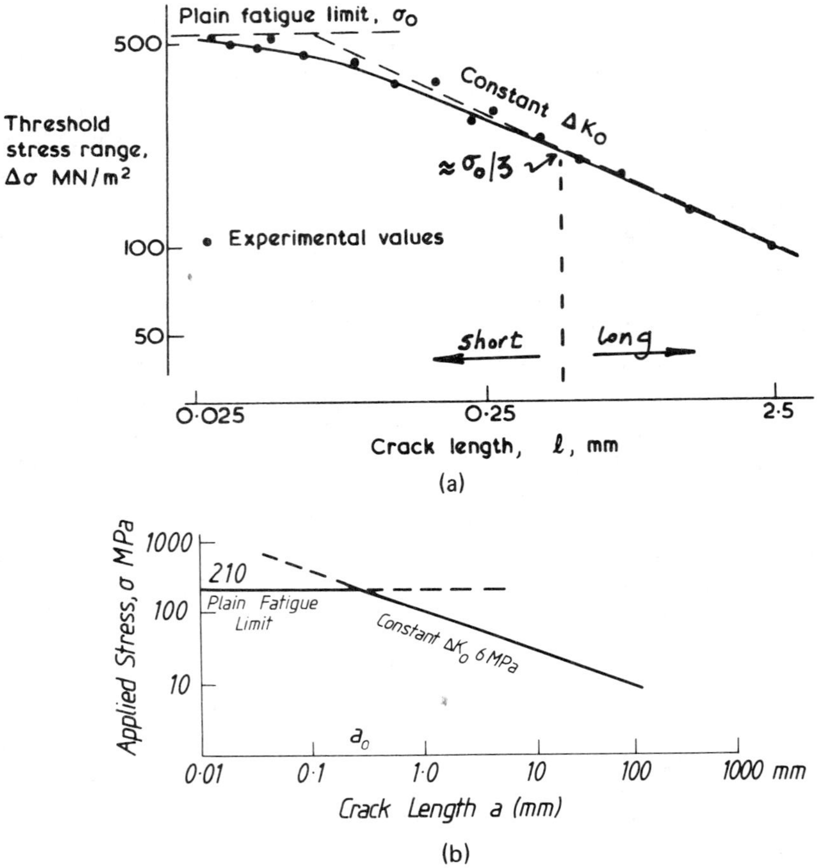

Fig. 7.14 – Effect of short cracks on the threshold stress intensity factor for (a) HY 80 steel and (b) mild steel (reproduced from Smith and Miller (1978), with permission).

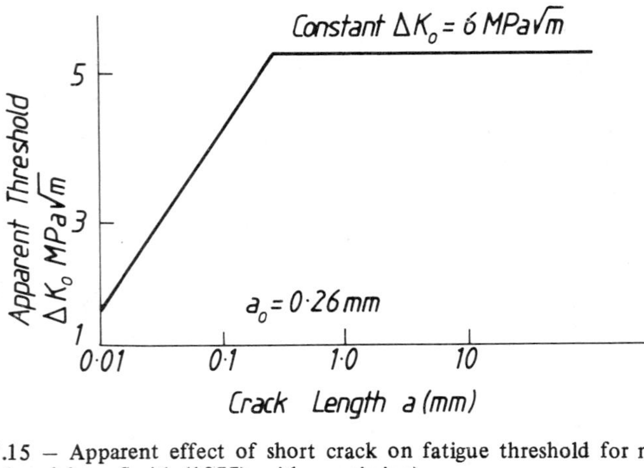

Fig. 7.15 – Apparent effect of short crack on fatigue threshold for mild steel (reproduced from Smith (1977), with permission).

The actual data for short cracks in Fig. 7.14 fall on a curve between the two straight-line regimes, and the two points at which the transition curve blends into the straight lines are called a_1 and a_2: a_2 is important as it represents the crack length below which the Paris law and predictions from it become non-conservative, the material ceasing to behave as a continuum with 'averaged out' properties; a_1 is the point below which the actual value of crack length has no effect on fatigue strength. Hence a body containing *no* cracks is just as bad (from the fatigue point of view) as a specimen containing cracks of length a_1. It is therefore necessary to postulate that it must be very easy for very short cracks to propagate to length a_1, and that in polished specimens fatigued above σ_0, cracks must appear and grow to length a_1 in a very small number of cycles.

Experimentally, propagation data are usually obtained from constant load range tests with ΔK changing owing to the increase in a. For tests in which $a < a_1$ early growth is indeed found to be faster than would be predicted by the Paris relation at the same ΔK. As seen in Fig. 7.16, short cracks (< 400 μm) grow at ΔKs *lower* than the ΔK_0 threshold established from experiments on long cracks subjected to low $\Delta\sigma$. As $a \to a_2$ the anomalous growth rate curve blends into the long crack cuve (at $a_2 \approx 400$ μm for the nickel–aluminium bronze studied by Taylor and Knott (1981) in Fig. 7.16). The location of a_2 seems to occur for some metals when $\sigma \approx \sigma_0/3$. Perhaps it is not surprising that short cracks propagate more rapidly than long cracks. After all smooth fatigue testpeices supposedly contain no cracks, but obviously fail when $\Delta\sigma > \sigma_0$ even though $\Delta K = 0$ if $a = 0$!

The magnitude of fast growth rates at small a depends very much on the details of microstructure. A crack starting within an 'easy' grain may propagate quickly, but slow up or even stop on reaching the grain boundary or other micro-structural feature; non-propagating cracks which arrested after early high growth rates are marked by downward-pointing arrows on Fig. 7.16. It is probable that a_1 should correspond with the grain size, or other major structural feature (according to Taylor and Knott, to the spacing of martensite laths in quenched and tempered steels, or to inclusion size when inclusions are important). Unfortunately there is not much information available for a_1, but Fig. 7.17 shows that a_2 (for which there are more data) correlates extremely well with 'structure size' d for a variety of metals. Taylor and Knot show that $a_2 \approx 10d$ (the 45° line of log paper going through $a_2 = 10$ μm when $d = 1$ μm). They further suggest that $a_2 \approx 4a_0$.

Miller (1985) discusses in detail the short crack growth phase with appropriate growth laws for cracks smaller than the grain size (cf. section 7.4), thus enabling S–N curves to be constructed over stages I, II and III, including the initiation as well as propagation periods. Miller argues that in practical engineering components, and many testpieces, fatigue cracks initiate immediately from surface notches, grain boundaries and triple points (as opposed to the explanation of initiation in terms of intrusion/extrusion persistent slip bands appropriate to specially prepared research samples). It follows that fatigue limits are then associated with non-propagating cracks rather than with the inability of a material to initiate a crack (cf. section 7.7; Eastabrook (1984a)).

The variation in growth rate for cracks shorter than a_2 implies that it is very difficult to make accurate predictions of their growth for design purposes. Taylor and Knott remark that the only reasonable design philosophy at present is to accept the fact that cracks shorter than a_2 will grow anomalously rapidly, and so work from the assumption that cracks of length a_2 are already present in the material as inherent defects. In many cases (e.g. the high-strength steels), a_2 is very small (less than 100 μm) and the structure will probably contain other (larger) defects such as inclusions (see section 1.2.5).

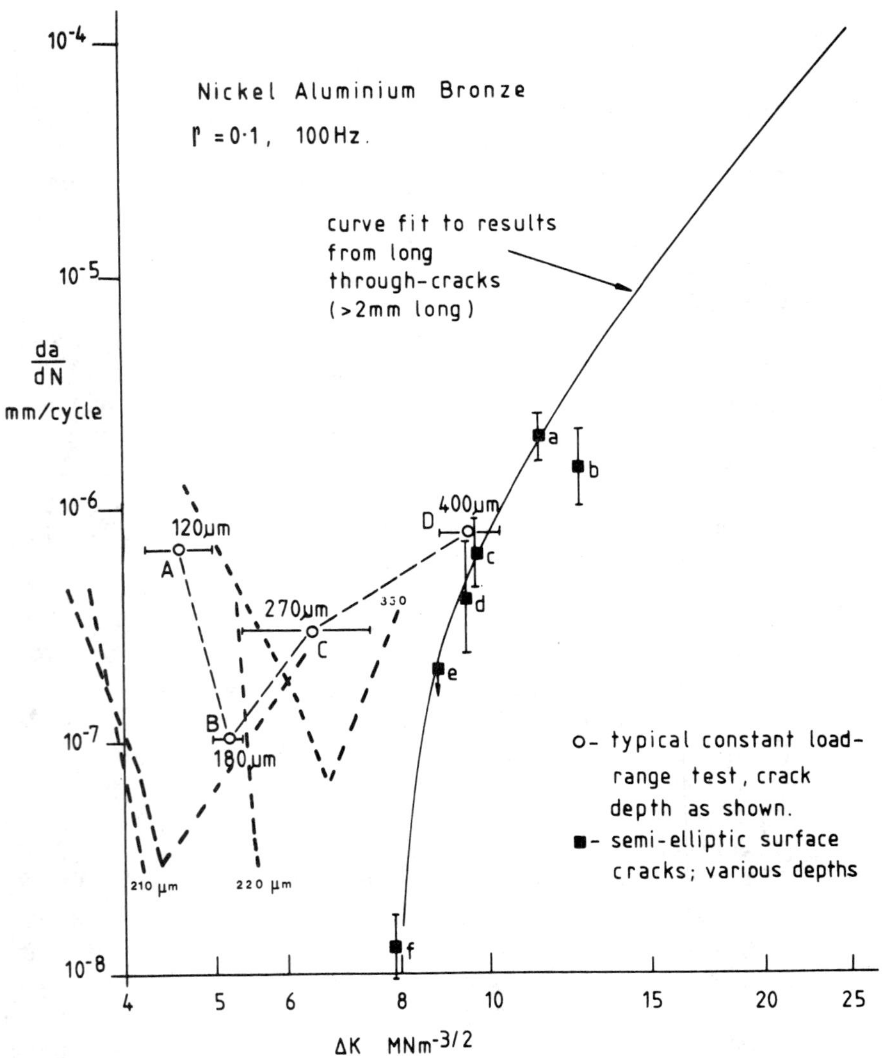

Fig. 7.16 – Fatigue crack growth as a function of applied stress intensity factor range for a nickel aluminium bronze with short cracks (reproduced from Taylor and Knott (1981), with permission).

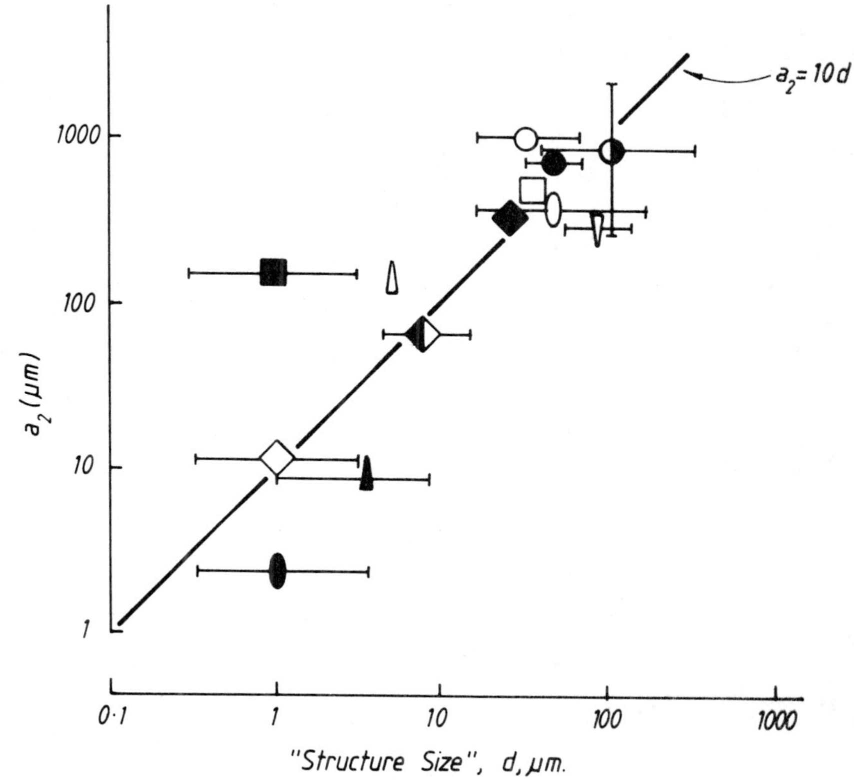

Fig. 7.17 – Correlation between characteristic crack size (a_2) and structure size (d) (reproduced from Taylor and Knott (1981), with permission).

Anomalous growth rates are also found when short cracks grow from notches (section 7.7), although there the effect is caused by plasticity at the notch root, rather than local non-homogeneities in the material continuum.

Example 7.9 Fig. 7.18(a) shows an S–N curve for cast ship propeller bronze obtained in four-point bending. The fatigue limit is ill-defined at some 124 ± 60 MPa. Pook (1974) reported that examination of the fracture surfaces showed that failure had originated at irregular crack-like flaws varying in size from 0.32 to 3.05 mm. Instead of an S–N curve, what is a better way of analysing the data?

Ans. Pook evaluated the initial ΔKs appropriate to each starter flaw and plotted them, rather than stress, against N (Fig. 7.18(b)). The scatter is much reduced and there is a reasonably well-defined fatigue limit at $\Delta K \approx 4$ MPa$\sqrt{\text{m}}$. This value turns out to correspond with the threshold ΔK_0 for the material. Pook concluded that the pre-existing flaws were controlling the fatigue behaviour

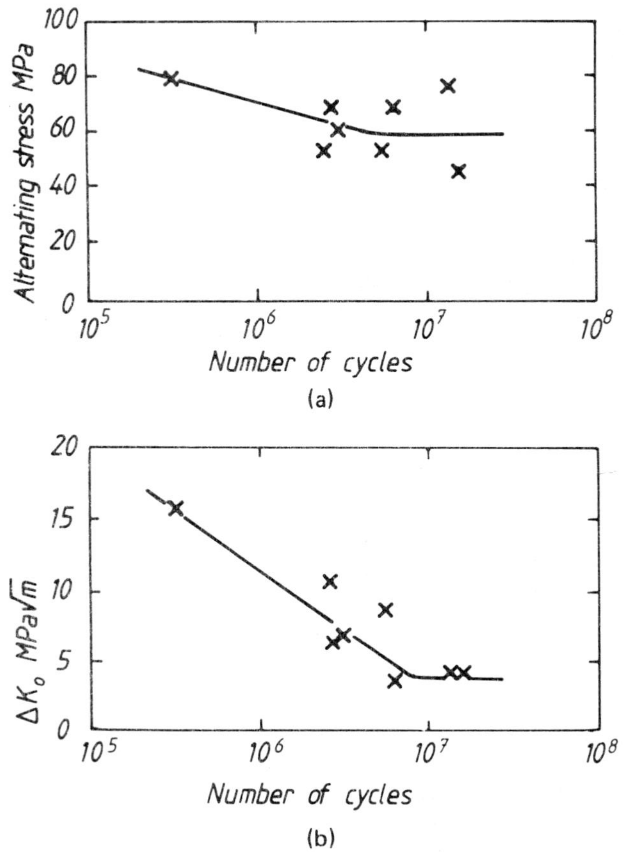

Fig. 7.18 – (a) S–N curves for case bronze bars tested in four-point bending. (b) Data in (a) plotted as initial ΔK versus N (both reproduced from Pook (1974), with permission).

and failure in a given specimen took place at the most severe flaw (with the highest ΔK). This problem illustrates that cracked components, made of materials displaying a threshold ΔK_0, will normally have a definite fatigue limit. Problem 7.16 shows that the slope of the S–N curve on log coordinates in these circumstances is $-(1/n)$ (cf. equation (7.13)). In Fig. 7.18(b), the slope is about -0.35, in which case $n = 2.9$ which is a sensible value for the Paris index. Note that since n is much higher in the transition region between the threshold and stage II, flatter S–N curves are predicted for long lives from this region.

7.3 EFFECT OF MEAN STRESS: ΔG RATHER THAN ΔK

The Paris equation relates crack growth to change in stress intensity only and, unless they are incorporated in the 'constant' C, possible effects of mean stress, frequency and so on are presumably of second order influence. That is, the same $\Delta\sigma$ in $\Delta K = \Delta\sigma Y\sqrt{\pi a}$ could be produced by the difference between two high stresses, or two low stresses; and the same ΔK could be produced by different combinations of $\Delta\sigma$ and $Y\sqrt{a}$. For example, σ_{max} and σ_{min} might be 180 and 130 MPa, or 150 and 100 MPa, both giving $\Delta\sigma = 50$ MPa and, at the same crack length, the same ΔK and, according to the Paris equation, the same da/dN, even though the stress states are different.

However, it has long been known from S–N data that a tensile mean stress reduces the fatigue limit stress range and that a compressive mean stress increases it (cf. use of Goodman-type diagrams in traditional design). This is particularly so for aluminium alloys (although Stage II crack growth rates in steels are virtually independent of stress ratio – at least for tensile mean stresses). One modification (there are others, cf. Example 7.10 following) of the Paris equation to accommodate the effect of mean stress in, for example, aluminium alloys and PMMA is the Foreman equation, viz.

$$da/dN = C(\Delta K)^n / \left[(1-r)K_C - \Delta K\right]^p \tag{7.17}$$

where $r = \sigma_{min}/\sigma_{max} = K_{min}/K_{max}$ and $p = 1$ for plane stress or $p = \frac{1}{2}$ for plane strain.† Note that $da/dN \to \infty$ when $K_{max} \to K_C$. (Other da/dN relations have been given which involve K_C – see, for example, Heald *et al.* (1972))

In order to normalise mean stress effects on fatigue crack growth it has been suggested that ΔG can be used in place of ΔK. Thus the fatigue crack propagation equation becomes:

$$\frac{da}{dN} = \bar{C}(\Delta G)^m \tag{7.18}$$

(In fact, fatigue crack propagation in rubbery materials had been described in this way – using the tearing energy in place of G (see section 2.10) – for a number of years before the Paris equation came into wide use for metals; see Fig. 7.20 and later discussion.)

Fig. 7.19(a) shows fatigue crack growth data for a commercial epoxy resin, obtained on tapered DCB (essentially constant rate of change-of-compliance) specimens and on CCP specimens, plotted in the traditional manner for the Paris equation. Separate lines are produced for each mean stress, in which the slopes are the same (ca. 10, which is comparatively large), but the increase in da/dN at increased mean stress are shown up in increased C. Fig. 7.19(b) shows the same data plotted in terms of ΔG. They now fit a single line, within experimental error, with a slope of 5.5 which, it is noted, is about one-half of that of the da/dN vs. ΔK plots.

---

† The symbol r is used here for K_{min}/K_{max} which has the same meaning as the stress ratio in conventional S–N terminology. Often the symbol R is used instead. For obvious reasons, owing to the wider use of R for fracture toughness in this book, we have changed the symbol for stress intensity ratio.

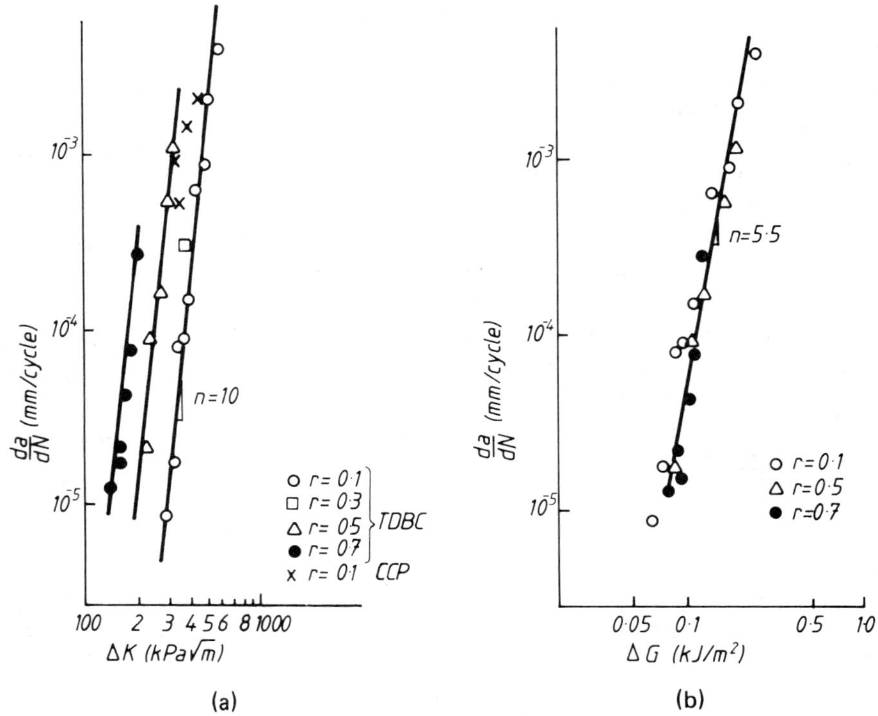

Fig. 7.19 – Fatigue crack growth in an epoxy polymer for which (a) da/dN is plotted against ΔK, and (b) da/dN is plotted against ΔG (After Sutton (1974), with permission).

Barsom (1971a) has successfully used equation (7.18) for aerospace metals and Mostovoy and Ripling (1975) apply it for adhesive joints. Also, Arad *et al.* (1972) used a modified version of (7.18) for PMMA, PC and Nylon 6.6. Although mean stress effects on fatigue crack growth in many solids can be rationalised in this way, there are some materials that do not obey this rule, e.g. PS (Mai and Williams, 1979).

Since, in equation (7.18), $E^*\Delta G = \Delta K^2$ rearrangement yields:

$$\frac{da}{dN} = (\bar{C}/E^*)\,[(K_{max} + K_{min})\,(K_{max} - K_{min})]^m$$

$$= (2^m\bar{C}/E^*)\,K_{mean}^m\,(\Delta K)^m \tag{7.19}$$

This implies that the dependence of da/dN on mean stress is to the same power (m) as the dependence on ΔG. Note that when $K_{min} = 0$ or $\sigma_{min} = 0$ equation (7.19) becomes:

$$\frac{da}{dN} = \frac{\bar{C}}{E^*}\,(\Delta K)^{2m} = C(\Delta K)^n \tag{7.20}$$

i.e. $m \approx n/2$ when $r = 0$. In Fig. 7.19(b) the slope is about one-half of that in Fig. 7.19(a). It is when the experimentally determined exponents of ΔK and K_{mean} in equation (7.19) are equal (or nearly so) that equation (7.18) may be used to normalise mean stress effects on fatigue crack growth rates. Thus Mukherjee and Burns (1971) have shown that for PMMA (where there are strong mean stress effects)

$$\frac{\mathrm{d}a}{\mathrm{d}N} = \text{constant}\ K_{\text{mean}}^{2.13}\ \Delta K^{2.39} \approx \text{constant}\ (\Delta G)^{2.26}$$

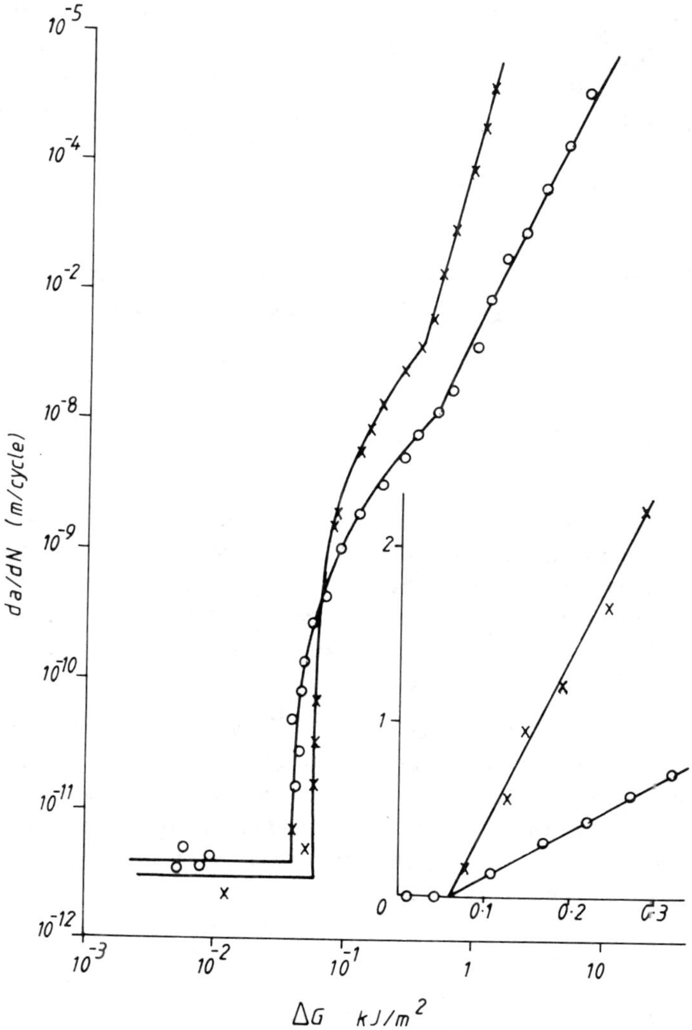

Fig. 7.20 – Fatigue crack growth rate $\mathrm{d}a/\mathrm{d}N$ plotted against ΔG for vulcanizates of natural rubber and styrene-butadiene rubber. Inset: region near G_0 plotted on linear scales (after Lake and Lindley (1964), with permission).

and Arad *et al.* (172) were able to normalise their data on PMMA using ΔG in the log–log plot. When the exponents of ΔK and K_{mean} are different, equation (7.18) cannot be used; this is the situation reported for PS by Mai and Williams in 1979.

Fig. 7.20 shows results for the fatigue of vulcanisates of natural rubber and SBR. Relation (7.18) is being followed over most of the range of da/dN with $m \approx 2$ for natural rubber and $m \approx 4$ for SBR. At low ΔG, the power law breaks down and for many elastomers there is a linear region with an intercept on the abscissa (shown as an inset to Fig. 7.20). This corresponds with a threhold for initiation of 'mechanical' fatigue crack growth. The constant growth, independent of ΔG, shown below the threshold is entirely attributable in the case of rubbers to ozone effects in the environment (e.g. Lake and Lindley, 1964).

In regard to plotting da/dN against ΔK or ΔG, we note that fatigue growth data are often plotted against $\Delta K/E$ to rationalise experimental results with appeal being made to strain-controlled (COD) processes influencing fatigue.

Example 7.10 A rotating component of a machine is made of 5Ni–Cr–Mo–V steel for which $K_{IC} = 54$ MPa$\sqrt{m}$ and cracks grow under fatigue loading according to

$$\frac{da}{dN} \text{ (mm/cycle)} = 7.72 \times 10^{-8}\ \Delta K_{eff}^{2.26} \text{ (MPa}\sqrt{m}\text{)}$$

where $\Delta K_{eff} = [(1 - br)/(1 - r)]\ \Delta K$ which is an equation which normalises the fatigue growth data to account for stress ratio r (i.e. mean stress) effects. The component is subject to an alternating stress range of $\Delta\sigma = 180$ MPa. It is found to contain cracks of length 0.2 mm. Given that $\Delta K = \Delta\sigma\sqrt{\pi a}$ calculate the number of cycles of loading to produce failure for stress ratios of 0 and -1.0. Take $b = 0.219$ for $r < 0$ and $b = 1$ for $r \geqslant 0$.

Ans. For $r < 0$ the component is partly subject to compression during cycling. We assume that compression causes the crack to close up and this does not contribute to fatigue crack growth, which is one reason why an effective stress intensity factor range has to be used. The equation given also normalises stress ratio effects on fatigue crack growth.

Failure will occur when

$$\sigma_{max}\sqrt{\pi a} = 54 \text{ MPa}\sqrt{m}$$

As $\sigma_{max} - \sigma_{min} = \sigma_{max}(1 - r) = \Delta\sigma$,

$$\sigma_{max} = \frac{\Delta\sigma}{1 - r}$$

Therefore,

$$a_c = \frac{1}{\pi}\left(\frac{54}{180}\right)^2 \quad \text{for } r = 0$$

$$= 0.0286 \text{ m}$$

and for $r_c = -1$,

$$a_c = \frac{1}{\pi}\left(\frac{54}{180}\right)^2 (1-r)^2 = \frac{1}{\pi}\left(\frac{54}{180}\right)^2 [1-(-1)]^2$$

$$= 0.110 \text{ m}$$

The fatigue life can be estimated as follows:

$$\frac{da}{dN} = 7.72 \times 10^{-8}\left[\frac{(1-br)}{(1-r)}\,\Delta\sigma\sqrt{\pi a}\right]^{2.26} \text{(mm/cycle, MPa}\sqrt{\text{m}}\text{)}$$

For $r = 0, b = 1$ and

$$\frac{da}{dN} = 7.72 \times 10^{-11} (\Delta\sigma)^{2.26} \pi^{1.13} a^{1.13} \text{ (m/cycle)}$$

$$= 3.52 \times 10^{-5} a^{1.13}$$

$$N_f = \int_{a_i}^{a_c} \frac{a^{-1.13} da}{35.18 \times 10^{-6}} = 2.84 \times 10^4 \int_{0.0002}^{0.0286} a^{-1.13} da$$

$$N_f = 2.84 \times 10^4 \left[-\frac{a^{-0.13}}{0.13}\right]_{0.0002}^{0.0286}$$

$$= \frac{2.84 \times 10^4}{0.13}\left[(0.0002)^{-0.13} - (0.0286)^{-0.13}\right]$$

$$N_f = 3.14 \times 10^5 \text{ cycles}$$

For $r = -1.0, b = 0.219$ and

$$\frac{da}{dN} = 7.72 \times 10^{-11} \times \left(\frac{1+0.219}{1+1}\right)^{2.26} (\Delta\sigma)^{2.26} \pi^{1.13} a^{1.13} \text{ (m/cycle)}$$

$$= 11.49 \times 10^{-6} a^{1.13} \text{ (m/cycle)}$$

Therefore

$$N_f = 1.13 \times 10^6 \text{ cycles}$$

Mean stress influences threshold ΔK_0 values, as shown in Table 7.3. ΔK_0 is lower the higher the mean stress (i.e. the higher $r = \sigma_{min}/\sigma_{max}$). This is thought to be caused by the crack closure stress (see Example 7.14 following) which is approximately constant for all r for a given material thickness.

All the Xu diagrams connected with sections 7.2.2 and 7.2.3 may be modified for non-zero X_{min}. If K_{min} produced by X_{min} is greater than K_0, then propagation in fatigue will ensue. Note that ΔK_0 does *not* mean X_{min} must be zero.

7.4 FATIGUE IN THE PLASTIC RANGE (LOW CYCLE FATIGUE)

Most fatigue studies relate to 'high cycle' fatigue where the bulk stress and strains during crack propagation under cyclic loading are within the elastic range. As we have already pointed out, materials with large toughness-to-strength ratios which, in *monotonic* tests, display extensive plastic flow before fracture, display elastic crack propagation in fatigue. Even so, 'low cycle' fatigue is of interest, wherein fatigue crack propagation occurs within plastic stress and strain fields caused by cycling over stress amplitudes exceeding the yield stress or strain amplitudes exceeding the yield strain. Fracture of a paper clip by repeated bending and unbending is a familiar example of low cycle fatigue, but whole cross-sections of components need not necessarily experience relatively large deformations, since high strains can be induced at stress concentrators such as notches with the rest of the body remaining at comparatively small strains. The design of gas turbine disks and pressure vessels are examples of what are essentially low-cycle fatigue problems.

7.4.1 Cyclic workhardening

It is recognised that deformation in the crack tip zone, even in elastic fracture, is intense and involves high irreversible strains. Thus in elastic fatigue it is likely that reversed plastic flow occurs in the contained crack tip zone. This must occur on a wider scale in low cycle plastic fatigue. Thus, a hysteresis $\sigma\epsilon$ loop (Fig. 7.21) develops early in fatigue which, in metals, must represent a stable dislocation configuration (often a sub-grain structure). For many materials, the relationship between the range of cyclic stress $\Delta\sigma$ and the range of plastic component of cyclic strain $\Delta\epsilon_p$ is given by

$$\Delta\sigma = \sigma_0' \, \Delta\epsilon_p^{n'} \tag{7.21}$$

This is obviously a direct parallel of the Ludwik $\sigma\epsilon$ relation for monotonic plastic flow, i.e. $\sigma = \sigma_0 \epsilon_p^n$. (Do not confuse this workhardening n with the Paris index.) While, for the same material, there are many similarities between monotonic and cyclic behaviour, some solids workharden more, or less, in cyclic than monotonic deformation and there may be higher, or lower, yield stresses in cyclic than monotonic tests. Thus neither σ_0 and σ_0', nor n and n' will necessarily be the same. Clearly this affects, for example, predictions of crack tip zone size d_y (equation (2.36)) and δ_C (equation (2.40)) in fatigue, owing to the possible differences in σ_y from static values. The situation is complicated by the fact that high strength or prior-cold-worked materials will *soften* upon plastic cycling. Again, precipitation-hardening alloys harden at first, then soften.

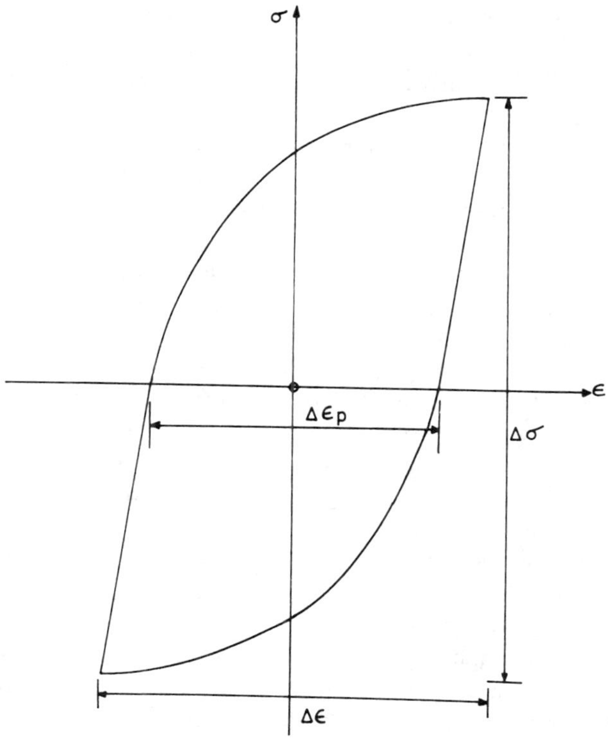

Fig. 7.21 – Hysteresis loop for fatigue in the plastic strain range. When 'ratchetting' occurs, the loop moves progressively sideways and the plastic increments per cycle are cumulative.

It is clear that the definition of a unique cyclic stress–strain response is difficult in some materials. Even so, the concept of a $\Delta\sigma$–$\Delta\epsilon$ hysteresis loop is valuable, particularly for prediction of $\mathrm{d}a/\mathrm{d}N$ in low cycle fatigue. When cyclic $\Delta\sigma$–$\Delta\epsilon$ curves are plotted (as in Fig. 7.21), the $\Delta\sigma$ and $\Delta\epsilon$ values usually relate to the stable, 'saturated' condition which applies to many alloys (e.g. 20 Cr/25 Ni stainless steels) after initial rapid hardening (i.e. after intial 'expansion' of the hysteresis loop to the size given in the diagram). Further information on cyclic stress–strain curves may be found in Tomkins (1980).

7.4.2 $\mathrm{d}a/\mathrm{d}N$ in low cycle fatigue

Although K_C is no longer an appropriate parameter for describing crack tip events in low cycle fatigue, ideas of critical absorbed energy in the process zone may still apply. Thus, instead of the Paris equation in terms of ΔK, we might expect relation (7.18) in terms of ΔJ to apply where ΔJ is now energy absorbed in the presence of remote plastic flow (cf. section 4.5). That is,

$$\frac{\mathrm{d}a}{\mathrm{d}N} = \bar{C}\,\Delta J^m \tag{7.18a}$$

ΔJ could perhaps be identified with Kfouri and Miller's G^{Δ} parameter (1976, 1981), which is the work done per unit crack extension over a macroscopically small but finite distance Δa and which gives the energy available for crack extension in both elastic and elastoplastic states. Finite element calculations for G^{Δ} have been given for a number of different loading conditions, and it is worth noting that the calculations account for unloading in the wake of the crack advance. Alternatively, a cumulative work/unit volume criterion at the crack tip might be used and this is the basis of L. Gillemot's W^{P} approach to both monotonic and fatigue fracture, where the work integral is taken over a small volume representative of the microstructure of the crack tip region (see Problem 7.20; also L. Gillemot and G. Sinay, 1958; F. Gillemot, 1973/4).

Rather like the arguments for the COD design curve (section 4.2.3), let us suppose that ΔK in elastoplastic fracture – if it has any meaning – is given by $\Delta K = \sigma_y \sqrt{\pi a}$ in zero-to-yield stress fatigue cycling. Therefore, cycling controlled by ΔG or ΔJ perhaps gives growth rates $\mathrm{d}a/\mathrm{d}N$ dependent on $\Delta(K^2/E) \propto (\sigma_y^2 a/E) = (\epsilon_y^2\, a)E$. Such sorts of relation have been found, as given later in this section; but a sounder approach for $\mathrm{d}a/\mathrm{d}N$ in low cyclic fatigue appeals to the connexion between $\mathrm{d}a/\mathrm{d}N$ and $\delta/2$ mentioned in Example 7.4. This theme is developed in what follows and since δ is related to J, we eventually return to a $\mathrm{d}a/\mathrm{d}N \sim \Delta J$ connexion.

Fractography of metals seems to indicate that crack advance mechanisms operate in a similar way irrespective of whether the situation is low or high cycle fatigue. Thus, it would appear that the demand for a given crack opening by either a high, or a low, applied stress is satisfied in essentially the same way at the crack tip. The difference arises in the definition of δ in terms of the applied stress–strain field and current crack length, with consequent influence on crack growth rate.

When major parts of a body undergo plastic flow, there is effectively a reduction in loading stiffness in the sense that a given $\Delta\sigma$ gives a much bigger $\Delta\epsilon$ in the plastic range than in the elastic (of course the body is not really displacement reversible as in elastic fracture). Owing to this, there results a great increase in the COD. Using an extension of the Bilby–Cottrell–Swinden model of crack tip flow (cf. the decohesion model of Fig. 1.7) to an elastoplastic workhardening solid, Wareing *et al.* (1973) gave expressions for the appropriate δ. Since we are concerned here with fatigue we state the result using the symbols of cyclic stress–strain response $\Delta\sigma = \sigma_0' \,\Delta\epsilon^{n'}$ rather than the monotonic $\sigma = \sigma_0 \epsilon^n$. For

$$\Delta\epsilon = \Delta\epsilon_{\text{el}} + \Delta\epsilon_{\text{p}} = \frac{\Delta\sigma}{E} + \left(\frac{\Delta\sigma}{\sigma_0'}\right)^{1/n'} \tag{7.22}$$

and for n' small with $(\Delta\sigma/T) > 0.5$, they showed

$$\delta \approx 2\,a \ln \sec\left(\frac{\pi\Delta\sigma}{2T}\right)\left(\frac{2T}{\pi E} + \frac{2\Delta\epsilon_{\text{p}}}{(1+n')}\right) \tag{7.23}$$

where $T = (\sigma_0'/2)\ (n')^{n'}$ is the cyclic analogue of the monotonic true stress equivalent ($\sigma_0 n^n$) of the 'ultimate tensile strength'. (In some applications T, as a representative flow stress for a workhardening solid, is written as the average of the monotonic σ_y and $\sigma_{fracture}$; cf. use of mean of σ_y and *UTS* in COD and *J* relations (Chapter 4).) The ln sec term is similar to that found in the expression for δ relating to Dugdale crack tip zones (section 4.1.3). At lower ($\Delta\sigma/T$), the approximate relation reduces to

$$\delta = \delta_{el} + \delta_p \approx \frac{\pi\Delta\sigma^2 a}{TE} + \frac{\pi\Delta\sigma\Delta\epsilon_p a}{T(1+n')} \qquad (7.24)$$

and we see 'elastic and 'plastic' components of δ.

The dominant role of strain, in defining δ in post-yield cases, is clear from these expressions. The Goldman and Hutchinson (1975) solution gives similar expressions for δ in finite-width centre-cracked plates.

If crack advance Δa in fatigue is linked to $\delta/2$, these expressions suggest

$$\mathrm{d}a/\mathrm{d}N \propto \frac{\pi\Delta\sigma^2 a}{2TE} + \frac{\pi\Delta\sigma\Delta\epsilon_p a}{2T(1+n')} \qquad (7.25)$$

The first term clearly corresponds to a Paris-type law with a K^2-dependency. The second term suggests that in low cycle fatigue, where ϵ_p predominates, $\mathrm{d}a/\mathrm{d}N$ varies directly as $(\Delta\sigma/T)\ (\Delta\epsilon_p)a$.

A number of authors (e.g. Solomon in 1972, Wareing *et al.* in 1973, Ibrahim and Miller in 1980) have found that

$$\frac{\mathrm{d}a}{\mathrm{d}N} = \bar{\bar{C}}\ (\Delta\epsilon_p)^{\alpha} a \qquad (7.26)$$

in low cycle fatigue where $\bar{\bar{C}}$ and α are material constants, with $1 < \alpha < 2$. Table 7.4 (taken from Tomkins *et al.*, 1979) gives α values for a variety of steels. Note that all $\alpha > 1$, the value that would agree with relation (7.25). Also experiments show that $\bar{\bar{C}}$ varies as $(\Delta\sigma/T)^2$ rather than $(\Delta\sigma/T)$ (see Tomkins, 1980); this accords with Tomkins well-known 1968 model for $\mathrm{d}a/\mathrm{d}N$, although

Table 7.4

Steel	0.2% proof stress (MPa)	*UTS*	n'	T(MPa)	α	$(1 + 2n')$
EN 1A	241	483	0.20	575	1.47	1.40
Ducol W30	545	690	0.095	710	1.18	1.19
304	276	630	0.15	793	1.23	1.30
20/25 Nb (750°C)[a]	–	–	–	–	1.53	1.54
316 (625°C)[a]	–	–	–	–	1.30	1.48
1018	–	–	–	–	1.86	1.52
'Mild steel'	–	–	–	–	2.00	–

[a]Refers to test temperature (°C).

that too implied $\alpha = 1$. Even so, Fig. 7.22 shows that when the plastic component of relations (7.24) and (7.25) is dominant in determining low cycle fatigue crack growth (as at least in low strength alloys), da/dN corresponds well with $\delta_p/2$.

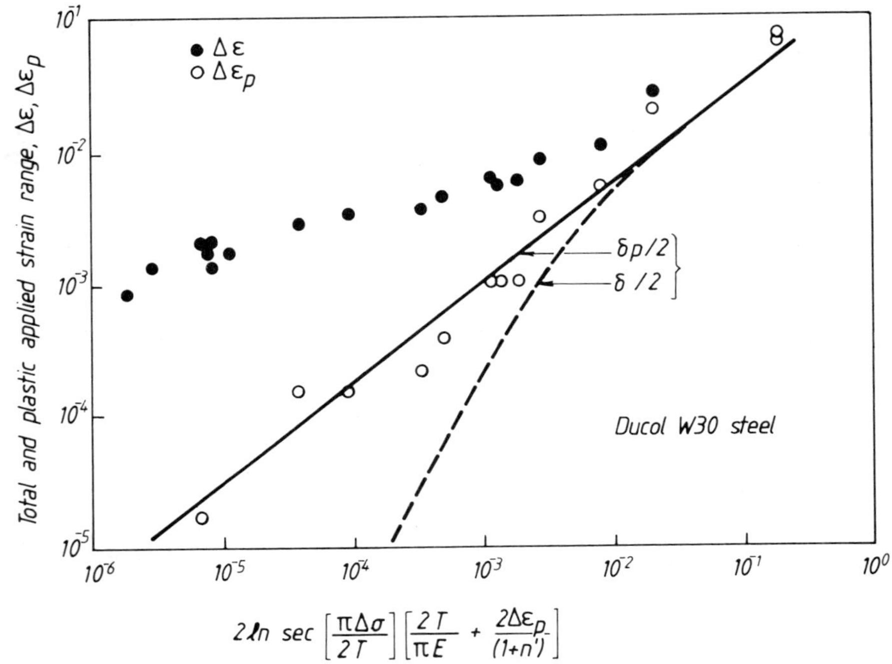

Fig. 7.22 – Applied strain range plotted against crack growth constant for a Ducol W30 steel.

For low cycle torsional fatigue, McClintock and Ritchie (1981) introduced a new 'plastic strain range' parameter to correlate mode III crack growth rates. The parameter is written in terms of the local plastic strain at a distance r ahead of the crack tip (cf. the definition of Kfouri and Miller's G^{Δ} parameter), and is defined as the value of $(r\Delta\epsilon_p)$ as $r \to 0$. Modelling mode III advance in rotor steels by initiation of voids formed at inclusions directly ahead of the crack tip, together with their subsequent coalescence in mode II shear parallel to the main crack front, and using the new parameters, Nayeb-Hashemi *et al.* (1983) found excellent correlation with experimental da/dN (which, in all cases, were a small fraction of the crack tip displacements per cycle). We note that the plastic strain range parameter is different from McEvily's (1970) definition of strain intensity, viz. $\Delta\epsilon_p\sqrt{a}$, which would appear in equation (7.26) were $\alpha = 2$.

Example 7.11 Provide a version of equation (7.26) involving stress range $\Delta\sigma$, if the material follows the cyclic relationship $\Delta\sigma = \sigma_0' \, \Delta\epsilon_p^{n'}$. Contrast the dependence on stress in low cycle fatigue with that in high cycle fatigue.

Ans. $\Delta\epsilon_p = (\Delta\sigma/\sigma')^{1/n'}$ so

$$\frac{da}{dN} = C_1 \left(\frac{\Delta\sigma}{\sigma_0'}\right)^{\alpha/n'} a$$

Since, often, $0.1 < n' < 0.3$, α/n' is of order 10 and $(da/dN) \propto \Delta\sigma^{10}$. The greater sensitivity to $\Delta\sigma$ than to $\Delta\epsilon_p$ reflects simply that $\sigma\epsilon$ plots curve over towards the ϵ axis, and that the same $\Delta\epsilon$ corresponds with smaller $\Delta\sigma$ as workhardening occurs. The Paris equation for elastic high cycle fatigue gives $(da/dN) \propto (\Delta K)^n \propto (\Delta\sigma)^n$ and $2 < n < 4$ usually. In addition to the different stress dependence, we note that the $[\Delta\sigma^2 a]$ link between stress and crack size is lost in low cycle fatigue.

Example 7.12 Show that the Coffin–Manson law relating $\Delta\epsilon_p$ and N_f in the limited-life fatigue regime (i.e. $\Delta\epsilon_p\sqrt{N}$ = constant) may be transposed from the relations for da/dN in low cycle fatigue.

Ans. Owing to the uncertainty of the power dependence of σ in the plastic component of equation (7.25) let us take $da/dN \propto \Delta\sigma^x \Delta\epsilon_p^\alpha a$. Using $\Delta\sigma = \sigma_0' (\Delta\epsilon_p)^n$ we obtain $da/a \propto \Delta\epsilon_p{}^{xn' + \alpha}\, dN$. Integrating between starter crack length and final crack length clearly results in $(\Delta\epsilon_p)\, N^{1/(xn'+\alpha)}$ = constant. The orginal relation given by Coffin in 1954 was $(\Delta\epsilon_p)N^{1/2}$ = constant, although experimental exponents on N range from about 0.1 for magnesium and some stainless steels to 0.9 for other stainless steels, nickel alloys and aluminium alloys. Tomkins (1968) shows excellent correlation between the experimental exponents and $1/(1 + 2n')$, i.e. here $\alpha = 1$ and $x = 2$. Tavernelli and Coffin (1962) pointed out that the Coffin–Manson law could be used in the high cycle range, by including the elastic strain at the fatigue limit. This corresponds with using elastic and plastic

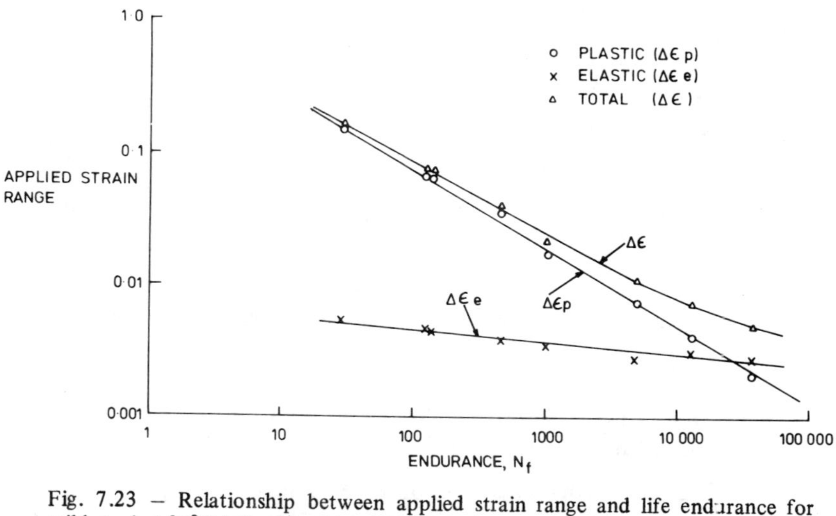

Fig. 7.23 – Relationship between applied strain range and life endurance for mild steel at 25°C (after Tomkins (1980), with permission).

components in da/dN before integrating as in equation (7.25). It also is related to plots of $\Delta\epsilon_{total}$ vs. N (the strain equivalent of S–N fatigue curves) being broken down into $\Delta\epsilon_{el}$ vs. N and $\Delta\epsilon_p$ vs. N components, with $\Delta\epsilon_{el}$ being important at high N and $\Delta\epsilon_p$ at low cycles to failure (Fig. 7.23).

Since δ is related to J, via $J = m\ \sigma_y\ \delta$ (equation (2.39a), and if da/dN is related to δ, it follows that da/dN is related to J (or rather ΔJ), as was suggested qualitatively early in this section. If we consider T as σ_y in equation (7.24) we have

$$\Delta J = \frac{\pi m\ \Delta\sigma^2 a}{E} + \frac{\pi m\ \Delta\sigma\ \Delta\epsilon_p a}{(1 + n')} \tag{7.27}$$

and Dowling (1976), for one, has employed da/dN vs. ΔJ plots for fatigue of 304 stainless steel over a wide range of growth rates. Fig. 7.24 shows comparable data for Ducol W30 pressure vessel steel.

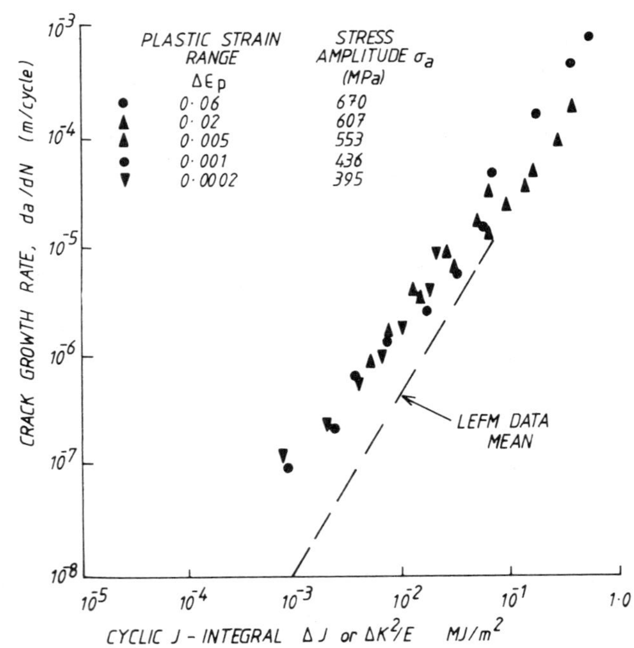

Fig. 7.24 – da/dN versus ΔJ for a pressure vessel steel (Ducol W30).

7.5 CUMULATIVE DAMAGE AND FATIGUE CRACK PROPAGATION UNDER RANDOM LOADING

For design, it is necessary to be able to take into account the influence of continual variations in stress or strain amplitude in order to be able to assess the life of the component. In practice, the fluctuating loads on components are rarely the sort of sinusoidal or triangular, constant frequency, constant amplitude, load patterns which are produced in the laboratory on basic types of fatigue testing machine. Loads in service usually have variable amplitude and are at least continually varying or more probably random in nature as, for example, for ships during buffetting at sea, aircraft gusts in flight, vibration loads caused by out-of-balance machinery, in generator rotors particularly during the start–stop operations, and so on. The problem is really rather difficult because many load histories are statistically non-stationary.

7.5.1 Palmgren–Miner rule

A body undergoing cyclic stressing at stress level σ will have a life (N_f). If that body experiences a sequence of regular cyclic loads of different stress amplitudes σ_i, the fatigue life is given by

$$\sum_i N_i/(N/_f)_i = 1 \tag{7.28}$$

where $(N_f)_i$ is the life of the structure at stress σ_i, and N_i is the number of cycles ($< (N_f)_i$) actually spent at stress level σ_i. This well-known linear damage law was proposed independently by Palmgren in 1924 and Miner in 1945 for fatigue under a variety of successive load histories.

The Palmgren–Miner rule may be 'proved' by integration of the relevant da/dN expressions for high cycle fatigue under different $\Delta\sigma$ (i.e. the Paris law, Problem 7.29) and for low cycle fatigue (Example 7.13).

Example 7.13 Determine a damage law for low cycle fatigue involving a series of stress amplitudes $\Delta\sigma_i$.

Ans. Let us use equation (7.26) in the form $da/dN = Ba = B_i\,\Delta\sigma_i^x\,a$ (Example 7.12). At each separate $\Delta\sigma_i$ the life $(N_f)_i$ is given from

$$\int_{a_0}^{a_f} \frac{da}{a} = \ln \frac{a_f}{a_0} = B_i\,\Delta\sigma_i^x\,(N_f)_i = \text{constant}$$

for all stress levels if a_0 and a_f are the same for all (according to Manson (1965) a_f is nearly constant for all $\Delta\sigma$ levels). If cycling at $\Delta\sigma_i$ goes on for only N_i cycles ($< (N_f)_i$) we have

$$\int_{a_0}^{a_1} \frac{da}{a} = \ln \frac{a_1}{a_0} = B_1\,\Delta\sigma_1^x\,N_1$$

for N_1 cycles at $\Delta\sigma_1$, followed by

$$\int_{a_1}^{a_2} \frac{da}{a} = \ln \frac{a_2}{a_1} = B_2\,\Delta\sigma_2^x\,N_2$$

for N_2 cycles at $\Delta\sigma_2$ and so on. Hence

$$N_1/(N_f)_1 = \left(\ln \frac{a_1}{a_0}\right)/\left(\ln \frac{a_f}{a_0}\right)$$

It follows that

$$\left\{N_1/(N_f)_1 + N_2/(N_f)_2 + \ldots\right\} = \left\{\ln \frac{a_1}{a_0} + \ln \frac{a_2}{a_1} + \ldots\right\} \Big/ \ln \frac{a_f}{a_0}$$

i.e. at failure, when $a_1 = a_f$,

$$\sum_i \frac{N_i}{(N_f)_i} = 1.$$

The Palmgren–Miner rule is therefore implicitly concerned with fatigue crack *propagation*. Yet we recall that in high cycle fatigue, crack *initiation* occupies the major part of the life of a component. Thus the comparative success of the rule, in conjunction with traditional S–N curves, is in a sense remarkable since N_f involves both initiation *and* growth.

We may anticipate the trends of deviations from the rule: if fatigue at low $\Delta\sigma$ is preceded by fatigue at high $\Delta\sigma$, we expect the final N_f to be *smaller* than if the low $\Delta\sigma$ had been in operation throughout. The argument is that the initial high $\Delta\sigma$ initiates a crack in *fewer* cycles than it would take the low $\Delta\sigma$ to do so, and thus when the low $\Delta\sigma$ comes into play propagation ensues straight away. Research indeed shows that $\Sigma\, N/N_f < 1$ when cycling sequence is from high-to-low stress levels, and $\Sigma\, N/N_f > 1$ when the cycling sequence is from low-to-high (Fig. 7.25).

Consider Fig. 7.26, in which two-stage cyclic loading at two different strain levels ϵ_1 and ϵ_2 is presented. Let the lower strain be applied first, for a number of cycles corresponding to some proportion x of the life at strain level 1, i.e. for $x(N_f)_1$ cycles. Upon changing to the higher strain level 2, the expected life (according to Palmgren–Miner) is $y(N_f)_2$ where $x + y = 1$, i.e. on the 45° line in the likes of Fig. 7.25 which is a plot of y vs. x. However, if the initiation phase has been exceeded at level 1 a fatigue crack will exist of a size equal to a_x, say, which will immediately have the growth rate appropriate to the higher level of cyclic strain, and there will be no initiation phase at level (2). Similarly, if higher strain cycling occurs first for $y(N_f)_2$ cycles and initiation is achieved giving crack length a_y, it follows that upon changing to the lower strain level there will be no initiation phase and crack growth will proceed immediately at the rate appropriate with strain level 1. Both these sequences will produce deviations in the Palmgren–Miner $x + y = 1$ relation.

Ibrahim and Miller (1980) have tackled the prediction of deviations from the Palmgren–Miner rule in the following way. They employed relation (7.26) for da/dN in the propagation phase at the different ϵ levels. Then, if $(a_i)_1$ is the

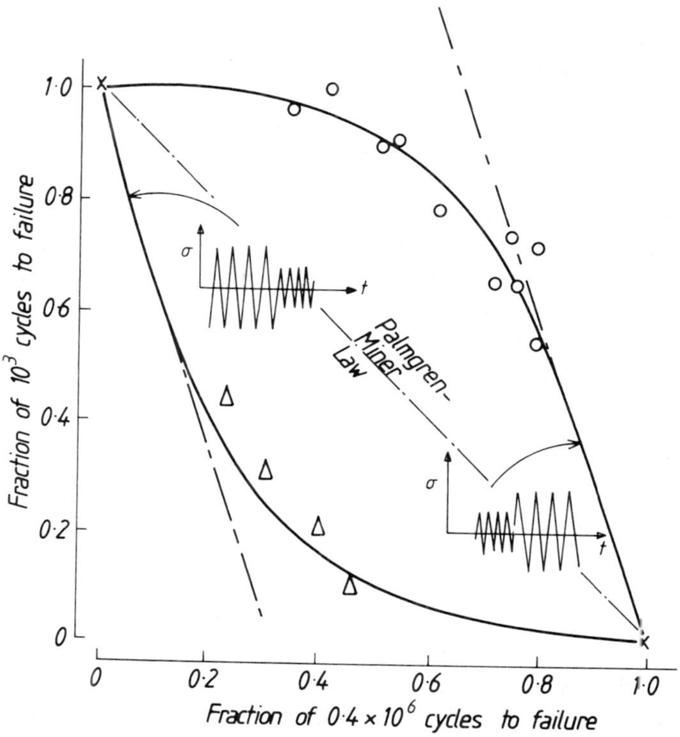

Fig. 7.25 – Accumulative damage showing deviations from the Palmgren–Miner law (courtesy K. J. Pascoe).

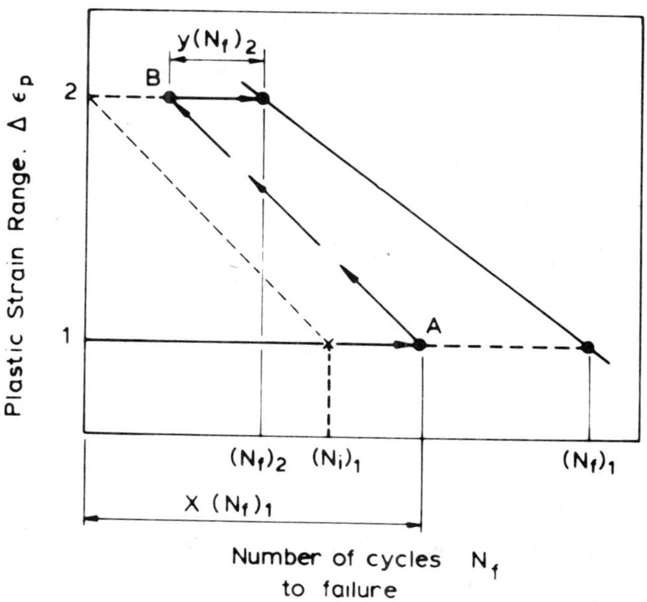

Fig. 7.26 – Sequence of two-stage cyclic loading presented on a schematic $\Delta\epsilon_p$–N_f curve (reproduced from Ibrahim and Miller (1980), with permission).

crack size corresponding to initiation at strain level ϵ_1, integration for crack growth at the ϵ_1 level gives

$$\ln\left[\frac{a_x}{(a_i)_1}\right] = \bar{\bar{C}}\,(\epsilon_1)^\alpha\,[x\,(N_f)_1 - (N_i)_1] \tag{7.29a}$$

where $(N_i)_1$ is the number of cycles at the ϵ_1 level required to initiate the crack. At the subsequent ϵ_2 level, where a_x at point A in Fig. 7.26 becomes the starter crack for continued propagation from point B (there will not be any initiation phase at the ϵ_2 level) we have

$$\ln\left[\frac{a_f}{a_x}\right] = \bar{\bar{C}}\,(\epsilon_2)^\alpha\,[y\,(N_f)_2] \tag{7.29b}$$

where $y(N_f)_2$ is the fraction of the lifetime remaining at the ϵ_2 level, and a_f is the critical crack length for final fracture. Addition of the two integrated expressions gives

$$\ln\left[\frac{a_f}{(a_i)_1}\right] = \bar{\bar{C}}\,(\epsilon_1)^\alpha\,[x\,(N_f)_1 - (N_i)_1] + \bar{\bar{C}}\,(\epsilon_2)^\alpha\,[y\,(N_f)_2] \tag{7.30}$$

If, instead of successively cycling at two (different) strain levels, fatiguing is carried out only at the initial strain level ϵ_1 from crack size $(a_i)_1$ to a_f, we have

$$\ln\left[\frac{a_f}{(a_i)_1}\right] = \bar{\bar{C}}\,(\epsilon_1)^\alpha\,[(N_f)_1 - (N_i)_1] \tag{7.31}$$

Relations (7.30) and (7.31) may be equated to eliminate $(N_i)_1$ to give

$$y = \left[\frac{\epsilon_1}{\epsilon_2}\right]^\alpha \left[\frac{(N_f)_1}{(N_f)_2}\right](1 - x) \tag{7.32}$$

for the sequence low ϵ to high ϵ.

In the other case of cycling first at the higher level followed by cycling at the lower, the relationship between y and x becomes, *mutatis mutandis,*

$$x = \left[\frac{\epsilon_2}{\epsilon_1}\right]^\alpha \left[\frac{(N_f)_2}{(N_f)_1}\right](1 - y)$$

i.e. (7.33)

$$y = 1 - \left[\frac{\epsilon_1}{\epsilon_2}\right]^\alpha \left[\frac{(N_f)_1}{(N_f)_2}\right] x$$

Both relationships predict straight lines of slope $-\ (\epsilon_1/\epsilon_2)^\alpha (N_f)_1/(N_f)_2$ passing through $x = 1$ in the case of low-to-high strain cycling and through $y = 1$ in the case of high-to-low. The lines describe the tangential end regions of the like of Fig. 7.25. The points where the experimental results depart from the predictions, and curve back towards $y = 1$ and $x = 1$ respectively, are related to the initiation periods at the respective starting cyclic strain levels, since equations (7.32) and (7.33) describe propagation events.

Ibrahim and Miller analysed Zacharia's (1974) data on low-to-high cycle torsional fatigue of mild steel tubes in the manner indicated by the above. It was found that $\alpha \approx 2$ best fitted the data. The points of departure from the analysis indicated the demarcation between initiation and propagation phases and enabled them to establish the size of the just-initiated cracks. It was found that both a_i and N_i (the number of cycles required for initiation) depended both on the cyclic stress condition and the surface condition of the testpiece. As an example, $a_i \approx 2$ μm at about 77 per cent of a total life of 400,000 cycles in a sample cyclically strained through a plastic (shear) strain range of 0.24. The grain size of the material was some 38 μm and it is sensible that the size of the initiated crack emanating from a smooth clean specimen should be much less than the grain size, unless, of course, the microstructure contains inclusions comparable in size to grains. Ibrahim and Miller quote the work of a number of authors which confirm this viewpoint. All the results indicate that high cycle fatigue lifetime is dominated by initiation and growth of very small cracks.

7.5.2 Simple and complex variable amplitude cyclic loading

The Palmgren–Miner rule concerns a sequence of regular loads of constant amplitude and must be a great approximation to flutuating loads in practice. It is useful to distinguish between simple and complex variable amplitude cyclic loads, Fig. 7.27(a), (b), (c). 'Simple' is used in the sense that there are only single or multiple tensile overloads superimposed upon constant-amplitude cyclic loads. In some cases the overloads can be compressive. 'Complex' refers to multi-value cyclic load fluctuations with variable amplitude and frequency. Various methods of establishing equivalent simple cyclic spectra for practical complex spectra have been proposed. Two averages have to be considered in the main, viz. (i) the method of counting the number of cycles of different stress ranges and (ii) the way in which cyclic amplitude is averaged. Pook (1975) argues that the most logically defensible and sophisticated method for counting cycles is the 'rainflow' procedure. Complex cyclic loads can be represented by some types of probability density curves either to characterise the stress history of the structre (e.g. Klippstein and Schilling (1976) employed Rayleigh and Erlang curves for stress spectra in short-span bridges) and/or drive fatigue testing machines in order to determine da/dN under complex cyclic loading. Fatigue crack growth behaviour and the prediction of fatigue life under random loading, whether simple or complex, is a difficult subject and in this book only a brief introduction is given.

7.5.3 Overload effects on fatigue crack growth

It is now well-established that tensile overloads introduced in a constant-amplitude cyclic load test cause delay in fatigue crack growth. (von Euw *et al.*, 1972; Probst and Hillberry, 1973). As shown in Fig. 7.28, the crack growth rate decreases to a minimum value after the overload is applied. This result gives rise

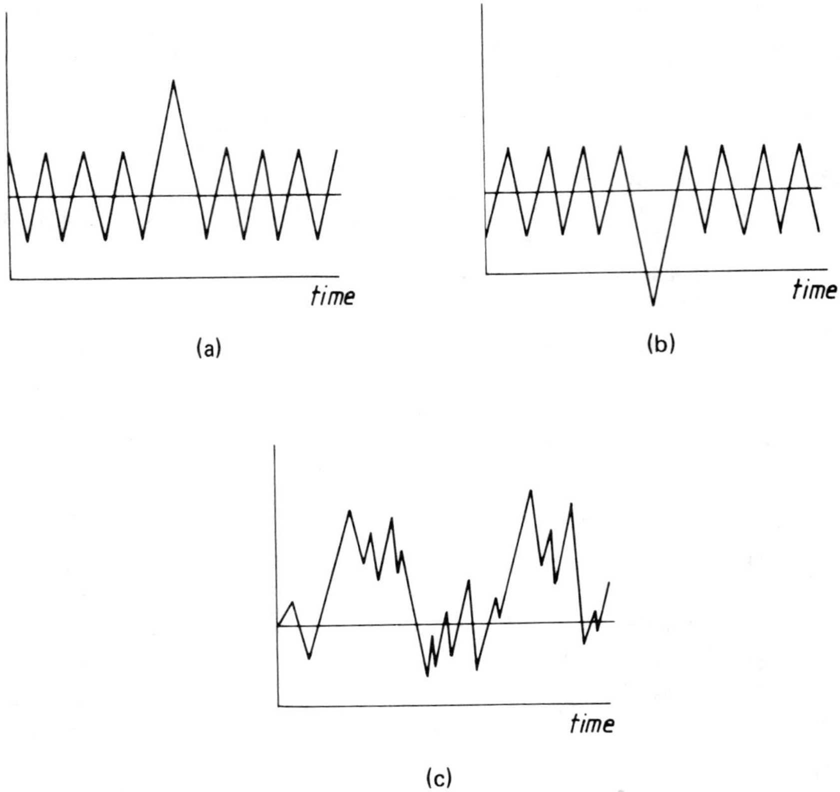

Fig. 7.27 – (a) Simple single tensile overload. (b) Simple single compressive overload. (c) Complex spectrum loading.

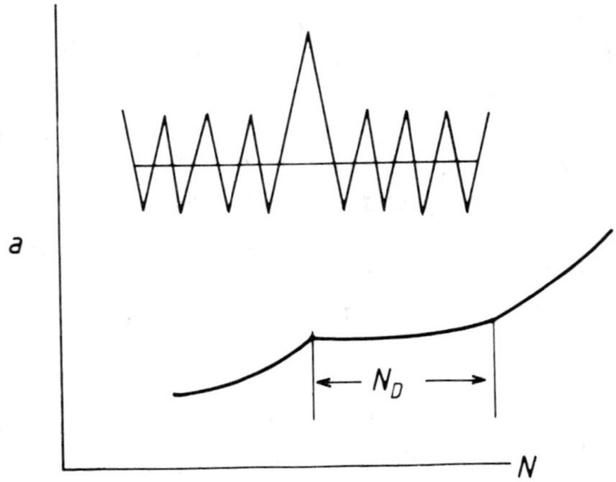

Fig. 7.28 – Fatigue crack growth retardation owing to tensile overloads.

to fatigue life enhancement equal to the period of fatigue crack growth delay (N_D). The prediction of N_D due to single or mulitple tensile overloads is difficult as it depends on a wide range of variables such as the constant-amplitude stress range, the overload magnitude, the stress ratio and the number of overloads. Such information must be available before any life prediction can be made. For example Mai (1979) and others have shown that for a few polymeric materials such as polystyrene and polycarbonate, N_D is proportional to the square of the overload ratio, all other factors being equal (the overload ratio is the ratio of the overload K to the base K).

Qualitative and semi-empirical models have been proposed to explain the crack growth delay phenomenon. Three such models are

(i) *The crack tip blunting concept* (Christensen, 1959) which suggests that, after the overload is applied, the crack tip becomes blunted as a consequence of localised plastic flow. It is necessary to resharpen the blunted crack tip before the crack can grow at rates determined by its original ΔK (Fig. 7.29(a)).

(ii) *Residual compressive stresses* developed *ahead* of the crack tip after the tensile overload (Wheeler, 1972) tend to close it and reduce the effective stress intensity factor range so that crack growth is delayed (Fig. 7.29(b)). Only when the crack tip has gone past this residual stress region will the crack resume its original crack growth rate.

(iii) The *crack closure* model proposed by Elber in 1971 assumes that residual stresses are developed at the *wake* of the crack tip when the crack faces close up so that the crack tip will not open up until this stress is overcome (Fig. 7.29(c)).

It is not possible to predict which of the models gives the correct solution to the fatigue crack growth retardation phenomenon as this depends on the material itself and a range of other variables. It seems that for metals and alloys the residual compressive stresses and associated crack closure models are commonly accepted explanations. For polymers, it seems that in PMMA the mechanism of fatigue retardation is due to crack closure, and in PC and PS it is due to crack tip blunting. For details, see Mai (1979) and Grandt and co-workers (Pitoniak *et al.*, 1974 and Banasiak *et al.*, 1977); see also Eastabrook (1984b).

Compressive overloads are not likely to build up residual stresses because the crack closes in compression. However, they are detrimental in fatigue life since they annihilate the beneficial effects of tensile overloads. There also seems to be a sequence effect of the overload cycles and much effort is needed to clarify this aspect of fatigue crack growth behaviour.

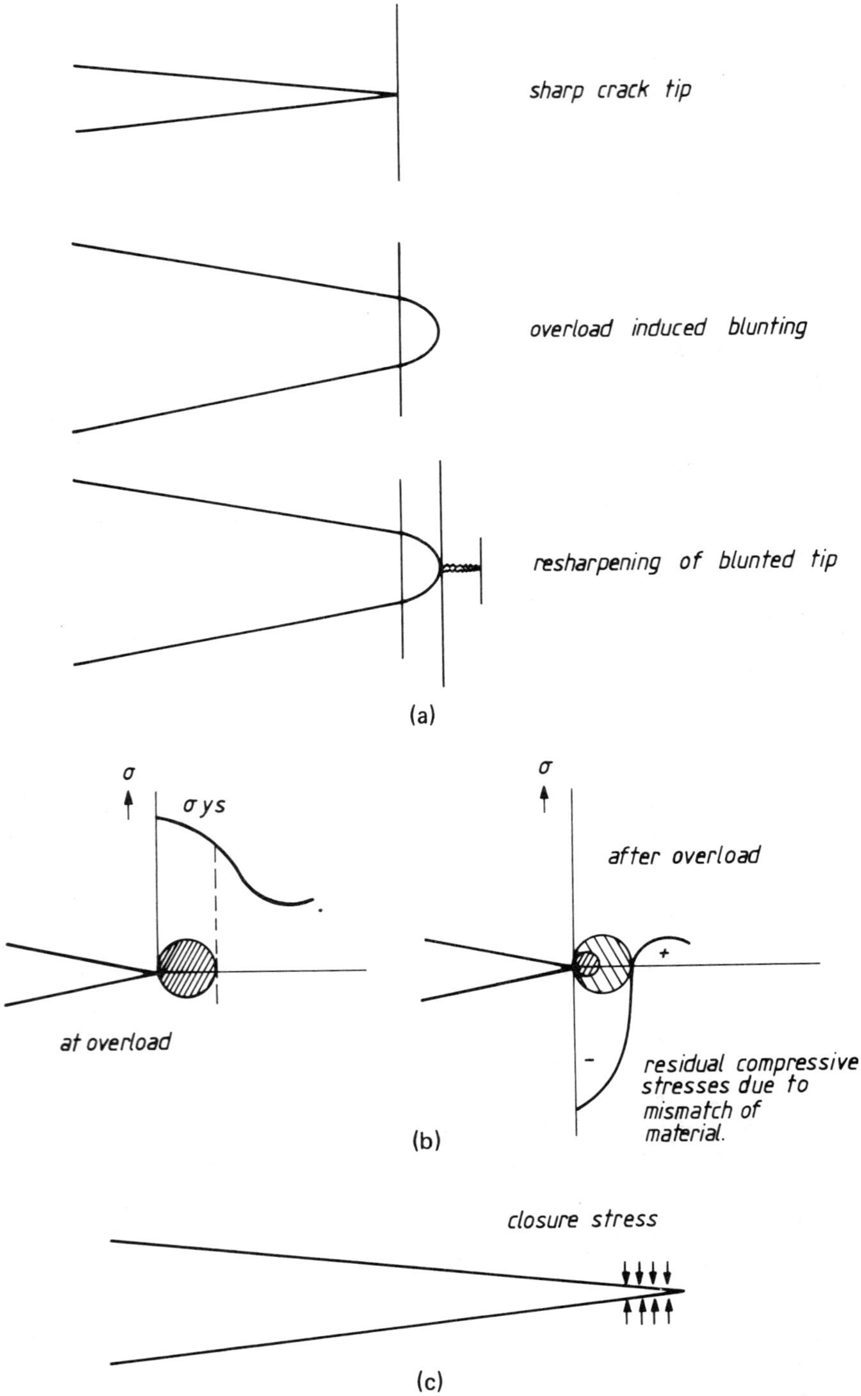

Fig. 7.29 – (a) Fatigue retardation owing to overload–crack tip blunting concept. (b) Residual compressive stress at crack tip as a result of tensile overload. (c) Residual or closure stresses at wake of crack tip after overload.

Example 7.14 The following results were obtained for fatigue tests on 2024-T3 aluminium:

Frequency (Hz)	r	K_{cl} (MPa$\sqrt{m}$)	ΔK_0 (MPa$\sqrt{m}$)
300	0.53	1.91	1.72
580	0.76	2.67	0.83

where r is the mean stress ratio, K_{cl} is the closure stress intensity owing to residual stresses and ΔK_0 is the threshold stress intensity range necessary to produce crack growth. Using the crack closure concept, explain why ΔK_0 depends on r.

Ans. Let us assume that the stress intensity needed to open a crack is constant. There must then be a portion of the stress intensity range above K_{cl}, say ΔK_{0p}, to produce crack growth at the threshold (Fig. 7.30(a)). If ΔK_{0p} is constant with changes in load ratio, two distinct cases arise.

(a) For high load ratio where $K_{min} > K_{cl}$, $\Delta K_0 = \Delta K_{0p}$ = constant. In terms of K_{max}, this is given by $K_{max} = \dfrac{\Delta K_0}{1-r} = \dfrac{\Delta K_{0p}}{1-r}$, where $r = K_{min}/K_{max}$.

(b) For low load ratio where $K_{min} < K_{cl}$ (the case shown in Fig. 7.30(a)), $K_{max} = K_{cl} + \Delta K_{0p}$ = constant. Now

$$\Delta K_0 = K_{max} - K_{min} = K_{max}(1-r) = (K_{cl} + \Delta K_{0p})(1-r).$$

These results may be plotted graphically as shown in Figs. 7.30 (b) and (c). The two sets of experimental results lie on either side of the divide given by $K_{min} = K_{cl}$ and $r = r_{cl}$, and the results fit the crack closure concept.

For the situation described in Example 7.14, it should be noted that r_{cl} cannot be predicted theoretically and it is obtained by fitting experimental data. Also it has been observed that r_{cl}, K_{cl} and ΔK_{0p} are frequency-dependent (Schmidt and Paris, 1973).

Schmidt and Paris suggested a very simple technique to measure K_{cl} directly. If K_{cl} caused by residual closure stress for the compact tension specimen (Fig. 7.31(a)) is required, for example, three strain gauges may be bonded at the locations shown. Compressive load–displacement records (displacement being converted from the strain measurements) appear as in Fig. 7.31(b) for a specimen which has $a = 38.5$ mm, $B = 2.4$ mm and $W = 63.5$ mm. It is easier to locate the closure load (X_{cl}) if X is plotted against (u/X) as shown in Fig. 7.31(c). These results indicate an average closure load of some 118 N. Thus K_{cl} is given by

$$K_{cl} = \frac{X_{cl}a^{1/2}}{BW} Z(a/W) = \frac{118 \times \sqrt{0.0385} \times 18.05}{63.5 \times 2.4 \times 10^{-6}} = 2.74 \text{ MPa}\sqrt{m}$$

since $Z(a/W)$ for a CT specimen with $a/W = 38.5/63.5 = 0.61$ is 18.05.

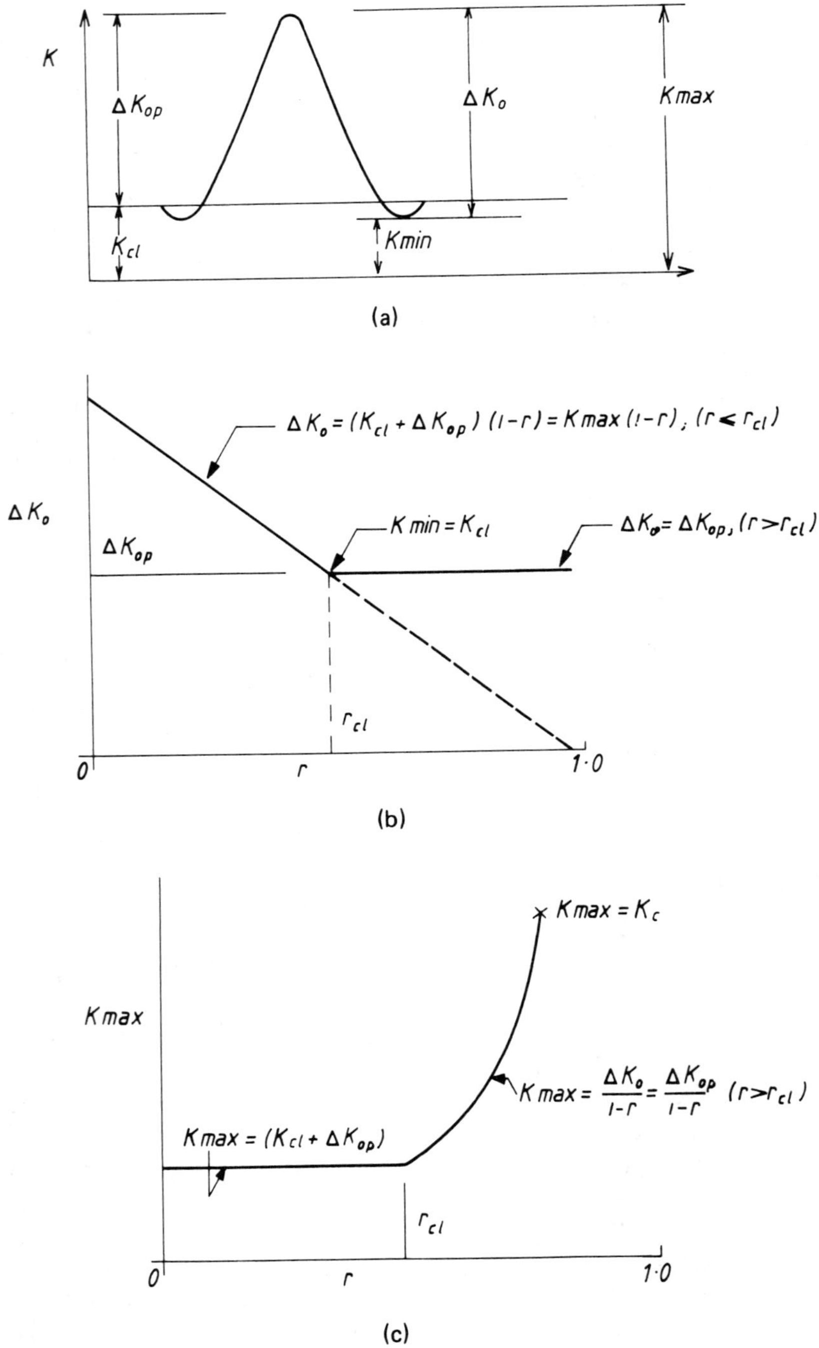

Fig. 7.30 – (a) Stress intensity parameters at threshold; (b) and (c) general form of threshold curves on basis of crack closure. (From Schmidt and Paris, 1973. Copyright ASTM, 1916 Race Street, Philadelphia, PA 19103. Reprinted with permission.)

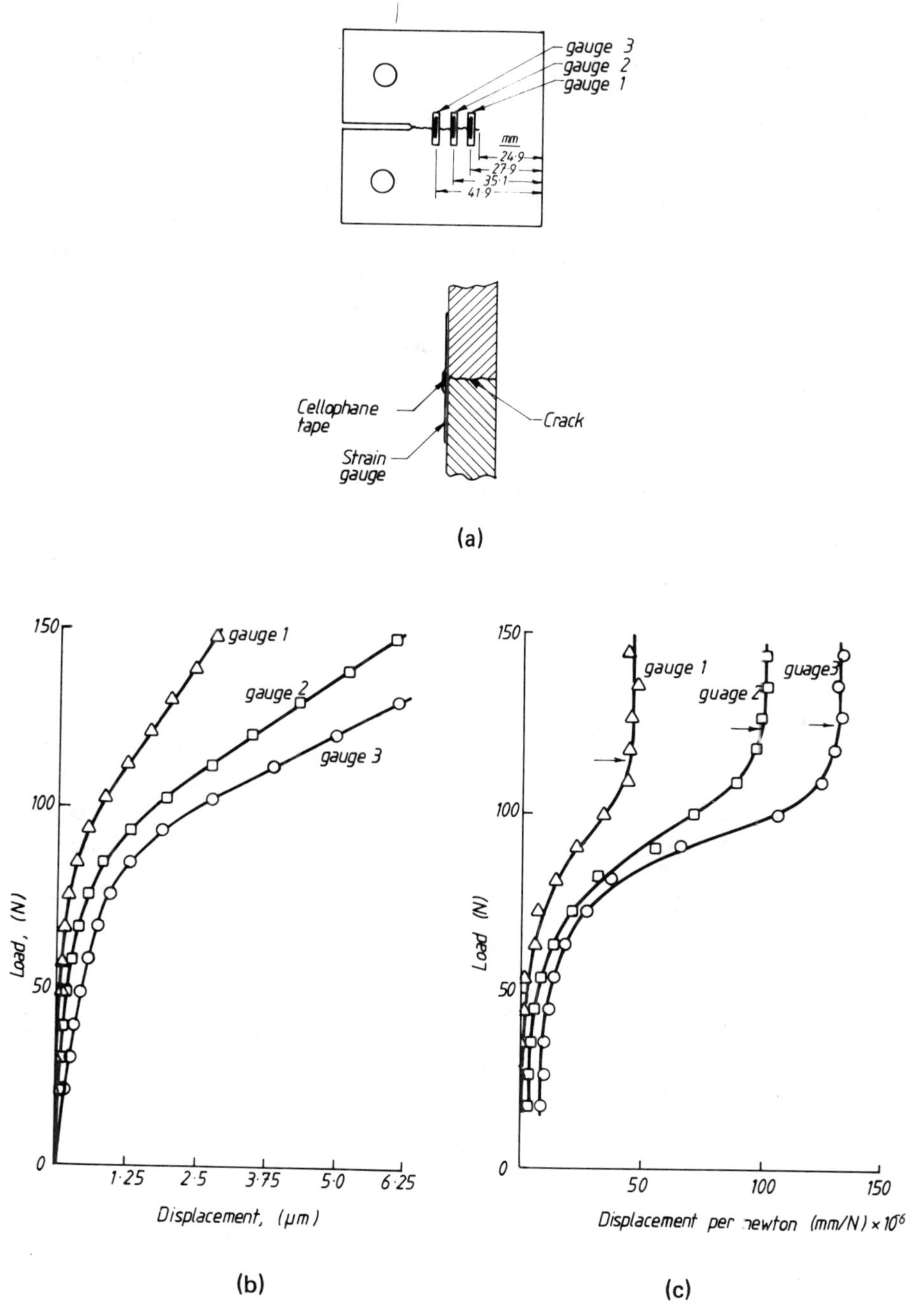

Fig. 7.31 – (a) Crack closure measurement using three strain gauges. (b) Load–displacement record obtained for strain gauges in (a). (c) Load versus reciprocal slope of (b) plotted to obtain closure load. (From Schmidt and Paris, 1973. Copyright ASTM, 1916 Race Street, Philadelphia, PA 19103. Reprinted with permission.)

7.5.4 Prediction of fatigue life due to variable amplitude loading

Fatigue life prediction of structures subjected to variable amplitude fluctuating loads is complicated by interaction effects which give rise to fatigue crack retardation. A simple solution which is conservative is to add N_D (the period of fatigue growth delay) to the cycles obtained in constant amplitude loadings. Here N_D has to be predetermined by experiments. This is very time-consuming. Barsom in 1976 proposed a simpler model which relates the average da/dN to an effective ΔK_{eff} that is characteristic of the probability density curve for the particular load spectrum (cf. the use of an effective ΔK in another circumstance in Example 7.9). He suggested that a suitable parameter is the root-mean-square value of the load distribution so that $\Delta K_{eff} = \Delta K_{rms}$. He showed that within experimental scatter, for a wide range of bridge steels subjected to certain random load distributions, the average fatigue crack growth rate is given by:

$$da/dN = C(\Delta K_{rms})^n \qquad (7.34)$$

where C and n are material constants and

$$\Delta K_{rms} = \left[\sum_{i=1}^{m} \frac{(\Delta K_i)^2}{m} \right]^{1/2} \qquad (7.35)$$

m is the number of load excusions within the load spectrum. Note that for constant-amplitude cyclic load fluctuations $\Delta K_{rms} = \Delta K$ so that C and n should be the same constants as those in equation (7.6). If so, da/dN under variable-amplitude random sequence load spectra can be predicted from constant-amplitude data by equation (7.6) with the same minimum load and ΔK replaced by ΔK_{rms}.

Example 7.15 A structure component made of a certain martensitic steel ($K_{IC} \approx 150$ MPa$\sqrt{m}$) is known to contain a fabrication crack of dimension 10 mm. The fatigue crack growth equation for the metal is given by:

$$da/dN \text{ (m/cycle)} = 1.68 \times 10^{-10} (\Delta K)^{2.25} \text{ (MPa}\sqrt{m}\text{)}.$$

Compare the lifetime of the component when it is subjected to (a) a sinusoidally varying stress having $\sigma_{min} = 0$ and $\sigma_{max} = 150$ MPa, and (b) the series of cyclic stress sequences shown in Fig. 7.32 in which σ_{max} rises from zero to 150 MPa every 500 cycles. Assume $K = \sigma\sqrt{\pi a}$.

Ans. The critical crack length,

$$a_c = \frac{1}{\pi}\left(\frac{K_{IC}}{\sigma_{max}}\right)^2 = \frac{1}{\pi}\left(\frac{150}{150}\right)^2 = 0.32 \text{ m}$$

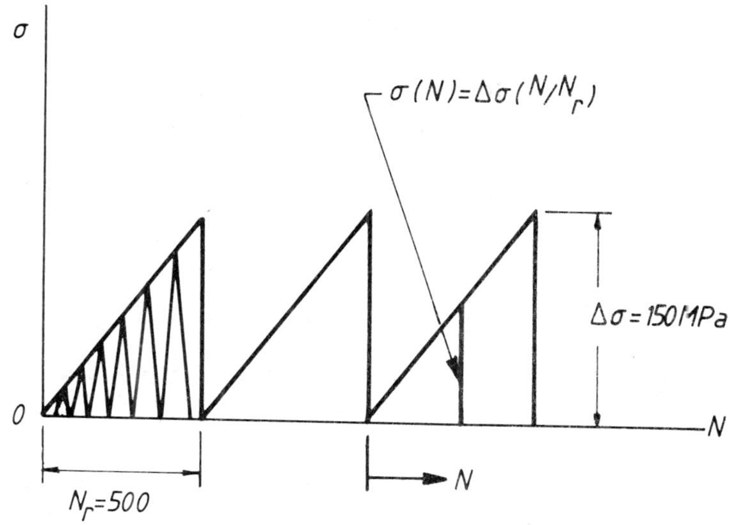

Fig. 7.32 – Linear varying (ramp) cyclic stress profile.

(a) For the constant σ_{max} case,

$$(N_f)_1 = \int_{a_i}^{a_c} \frac{da}{1.68 \times 10^{-10}(\Delta K)^{2.25}} = \frac{10^{10}}{1.68} \int_{a_i}^{a_c} \frac{da}{\Delta\sigma^{2.25}\, \pi^{1.125}\, a^{1.125}}$$

$$= \frac{10^{10}}{1.68} \left(\frac{1}{150^{2.25}\, \pi^{1.125}} \right) \left[-\frac{1}{0.125} \quad a^{-0.125} \right]_{a_i}^{a_c}$$

i.e.

$$(N_f)_1 = \frac{10^{10}}{1.68} \left(\frac{0.125^{-1}}{150^{2.25}\, \pi^{1.125}} \right) \left[-\frac{1}{0.320^{0.125}} + \frac{1}{0.005^{0.125}} \right]$$

$$= 1.3 \times 10^5 \text{ cycles}$$

(b) For the repeated linear varying case, the root-mean-square value ΔK_{rms} for each stress block is given by equation (7.35) as

$$\Delta K^2_{rms} = \frac{1}{N_r} \int_0^{N_r} (\Delta\sigma\sqrt{\pi a})^2 \left(\frac{N}{N_r} \right)^2 dN$$

$$= \frac{(\Delta\sigma\sqrt{\pi a})^2}{N_r} \left[\frac{N^3}{3\, N_r^2} \right]_0^{N_r} = \frac{1}{3} \quad (\Delta\sigma\sqrt{\pi a})^2$$

The fatigue crack growth equation is the same as for constant amplitude stress variation except ΔK is replaced by ΔK_{rms} so that

$$\frac{da}{dN} = 1.68 \times 10^{-10} (\Delta K_{rms})^{2.25}$$

i.e.

$$(N_f)_2 = \frac{10^{10}\sqrt{3}}{1.68} \int_{0.005}^{0.32} \frac{da}{(150)^{2.25}\, \pi^{1.125}\, a^{1.125}}$$

$$= \sqrt{3}\,(N_f)_1 = 2.24 \times 10^5 \text{ cycles}$$

Hence the fatigue life in case (b) is $\sqrt{3}$ times as long as in case (a).

Case (b) in Example 7.15 is not of course in true random sequence, rather it is only an ordered sequence. Barsom (1973) has shown that the fatigue crack growth equation (7.34) actually fits very well experimental data from a wide range of load sequences. The relation is independent of certain repetitive load sequences and some load distribution functions.

Fatigue crack retardation does not seem to be significant in random fatigue tests.

7.6 ELASTIC AND PLASTIC FATIGUE UNDER MULTIAXIAL LOADING

Analysis of engineering components and structures most often involves in biaxial or multiaxial stress field, – propeller shafts and crank shafts, for example, operate under varying levels of combined torsion and bending. Other components under multiaxial stressing are bends in pipework, cooling holes in gas turbine blades, nozzles and manifolds in pressure vessels and so on. Even bodies stressed simply have in fact varying degrees of multiaxial strain if there are changes in cross-section or other geometric discontinuities. Miller (1977) points out that a designer needs to know the answers to a number of questions in multiaxial stressing situations, such as (a) where is a fatigue failure likely to start (which is not the same question as where will the first fatigue crack initiate), (b) on what plane is a crack likely to grow; (c) in which direction on that plane will a crack propagate; and finally, (d) how fast will a crack grow. Unfortunately, the testing of specimens in a research laboratory frequently is only concerned with the latter question, and much traditional work does not ask even this question and is only involved with plotting *S*/*N*-type data.

Models of fatigue growth rate often relate da/dN to the size of the crack tip plastic zone, which depends on (K/σ_y), where σ_y is the cyclic yield stress. Since the size and shape of crack tip zones are affected by loads applied laterally as well as perpendicular to the crack plane (e.g. Liebowitz *et al.*, 1978, Fig. 7.34 here) we should therefore expect that da/dN in fatigue is affected by multiaxial stressing. This applies equally to *uniaxial* stressing of *inclined* cracks, as to situations where the crack is aligned along one of three principal axes of loading.

We note that there are two types of problem to be considered, viz. (i) where cycling occurs along only some axes of loading and the loads along the other axis or axes are constant and (ii) where the loads along all axes may fluctuate with time.

Fig. 7.33 shows Miller's (1977) high cycle results on cruciform specimens of RR58 aluminium alloy, where $\lambda = \sigma_{x\infty}/\sigma_{y\infty}$ ($\lambda = +1$ for equibiaxial tension, $\lambda = 0$ for uniaxial tension, and $\lambda = -1$ for balanced tension and compression, i.e. shear at 45° to the crack). In all tests, $\sigma_{y\infty}$ fluctuated and $\sigma_{x\infty}$ either also fluctuated, or was constant; λ_c refers to tests in which $\sigma_{x\infty}$ also was cyclic but λ_m means tests in which $\sigma_{x\infty}$ was constant. All data follow Paris-type relations with similar *n* values for all, but with *C* values dependent on λ; $C_{\lambda=+1} < C_{\lambda=0} < C_{\lambda=-1}$. It is clear that da/dN is always faster at the same *nominal* ΔK for $\lambda = -1$ than for $\lambda = 0$, and that da/dN is slowest for $\lambda = +1$. Similarly for some fixed da/dN, cracks under $\lambda = -1$ behave as if they were shorter than their actual length, and cracks under $\lambda = +1$ behave as if they were longer. In a valuable review paper, Smith and Pascoe (1983) show that most results in the literature for biaxial fatigue of metals and polymers fit a pattern of decreasing da/dN with increasing λ, though they pointed out that this is by no means conclusive; see Beck (1973).

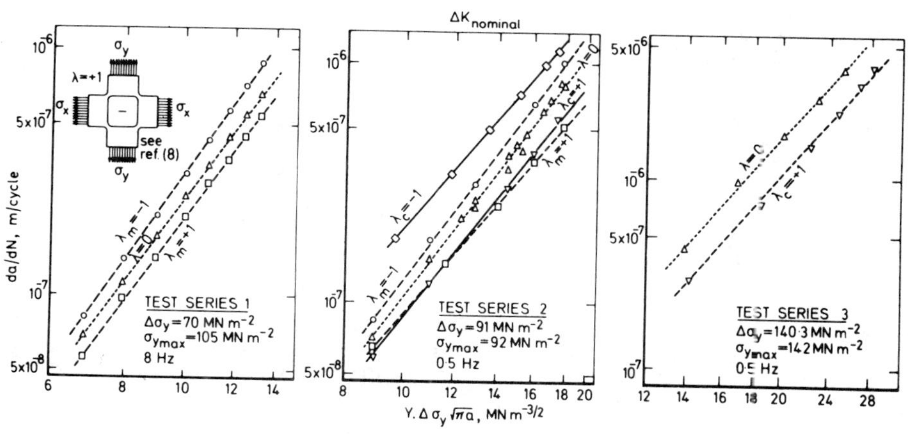

Fig. 7.33 – Biaxial fatigue results for a RR58 aluminium alloy (reproduced from Miller (1977), with permission).

It is important to note that in Fig. 7.33, the *same* expression $\Delta K = \Delta\sigma\sqrt{\pi a}.Y$ was used for *all* λ tests with a constant value for *Y* determined from a series of calibration tests for the cruciform specimen. That is why we have called the ΔK values 'nominal' since, according to the discussion of the general biaxial Griffith problem in section 3.4, *Y* changes with λ. Fatigue cracks propagate

Table 7.5

Comparison of predicted and experimental crack-growth parameters in biaxially stressed plates of RR58 alloy.

	Experimental values			Theoretical Values (base data at $\lambda = 0$)						
				Biaxial Griffith		$C(\times 10^{-11})$		Miller $a_{equivalent}$		
	λ	n	$C(\times 10^{-11})$	λ	n	$\nu = 1/3$	$\nu = 0.3$	λ	n	$C(\times 10^{-11})$
Test series No. 1										
$\Delta\sigma_y = 70$ MPa	$\lambda_m = -1$	3.616	7.05	−1	3.475	16.63	12.31	−1	3.475	10.87
$\sigma_{ymax} = 105$ MPa	$\lambda = 0$	3.475	8.34							
8 Hz	$\lambda_m = +1$	3.445	7.34	+1	3.475	5.07	6.09	+1	3.475	7.21
Test series No. 2										
$\Delta\sigma_y = 91$ MPa	$\lambda_c = -1$	2.993	2.01							
$\sigma_{ymax} = 92$ MPa	$\lambda_m = -1$	3.430	4.61	−1	3.377	8.39	6.27	−1	3.377	6.29
0.5 Hz	$\lambda = 0$	3.377	4.29							
	$\lambda_m = +1$	2.985	9.07	+1	3.377	2.65	3.16	+1	3.377	3.43
	$\lambda_c = +1$	3.249	4.70							
Test series No. 3										
$\Delta\sigma_y = 140.3$ MPa	$\lambda = 0$	3.632	3.06							
$\sigma_{ymax} = 142$ MPa	$\lambda_m = +1$	3.714	1.53	+1	3.632	1.82	2.26	+1	3.632	2.00
0.5 Hz										

Values of C in MPa$\sqrt{\text{m}}$

in plane strain for all practical purposes, so we have for the stress intensity range ΔK from equation (3.23) (with $\tau_{xy} = 0$ and $\kappa = (3 - 4\nu)$)

$$\Delta K = \frac{\Delta\sigma\sqrt{\pi a}}{2} \sqrt{\{(4\nu - 1)\,\lambda + 1\}\ (1 + \nu)} \qquad (7.36)$$

Thus from horizontal intercepts, at constant $\mathrm{d}a/\mathrm{d}N$, of a series of Paris plots whose general equation is

$$(\mathrm{d}a/\mathrm{d}N)_\lambda \;=\; C_\lambda\,(\Delta K_\lambda)^n$$

we have

$$C_\lambda = (C_{\lambda=0})\,(\Delta K_{\lambda=0}/\Delta K_\lambda)^n$$
$$= (C_{\lambda=0})\,(1/((4\nu - 1)\,\lambda + 1))^{n/2} \qquad (7.37)$$

and these C_λ apply then to Miller's plots in which ΔK was always the nominal value, dependent only on crack length and not on λ as well. For $\nu = 1/3$,

$$C_\lambda = (C_{\lambda=0})\,(1/(\lambda/3 + 1))^{n/2}.$$

In Table 7.5, values obtained in this way are given for $C_{\lambda=+1}$ and for $C_{\lambda=-1}$, and compared with the directly-determined experimental values.

Miller attacked the analysis in a different, but complementary way, arguing in terms of 'uniaxial-equivalent' crack lengths. Insofar as $(\mathrm{d}a/\mathrm{d}N)$ depends on the size of the crack tip zones, the length of the crack under uniaxial loading, which is equivalent (in terms of ΔK or $\mathrm{d}a/\mathrm{d}N$) to a crack under biaxial loading, is that which has the same extent of crack-tip plasticity. Miller quotes Kfouri's finite element calculations for the size of the plastic shear ears at the tip of a crack in a monotonically-loaded elastoplastic field as affected by different forms of (static) biaxial loading (Fig. 7.34). That is, equibiaxial tension gives the smallest

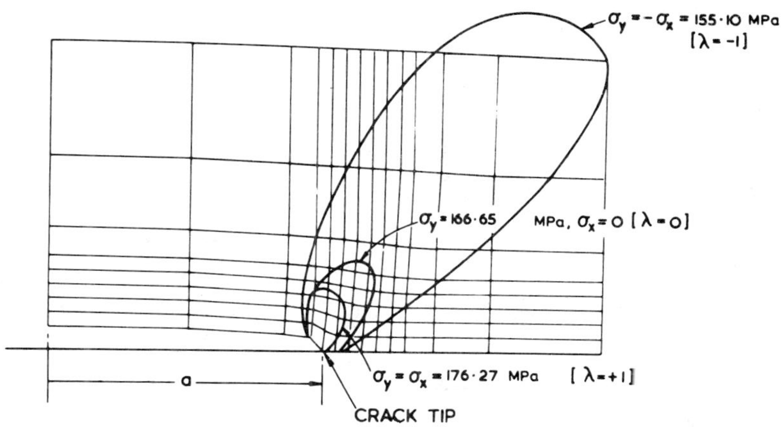

Fig. 7.34 – Effect of stress state on crack tip plastic zone size (after Miller and Kfouri, (1974), with permission).

size ears and equal tension and compression gives the largest. Then, since da/dN is a function of plastic zone size (or equivalently of COD), it follows that da/dN is greatest for $\lambda = -1$ and least for $\lambda = +1$. The *same* da/dN would be produced by 'equivalent' uniaxial cracks of length a_{equiv} which will be (a) longer than the actual crack length when $\lambda > 0$ and (b) shorter for $\lambda < 0$ (cf. previous arguments using horizontal intercepts with the biaxial Griffith solution, in which the equivalent crack length must be $a_{\text{equiv}} = \{(4\nu - 1)\,\lambda + 1\}(1 + \nu)\,a$.

Example 7.16 From the $\lambda = 0$ values, estimate values of C and n for the tests in Fig. 7.33 having $\lambda = +1$ and $\lambda = -1$, employing Miller's concept of 'equivalent uniaxial crack length' for biaxial fatigue and Kfouri's elastoplastic solutions for crack tip zone size.

Ans. Table 7.6 gives theoretical values of $\Delta K_\lambda/\Delta K_{\lambda=0}$ $(=\sqrt{a/a_{\text{equiv}}})$ and plastic zone size ratio $(d_y)_\lambda/(d_y)_{\lambda=0}$ deduced from Kfouri's calculations for $\lambda = +1, 0$

Table 7.6

Comparison of biaxial and uniaxial plastic zone sizes and stress intensities for the three test series

Test series	$\sigma_y/\sigma_{\text{yield}}$	$(d_y)_\lambda/(d_y)_{\lambda=0}$		$\Delta K_\lambda/\Delta K_{\lambda=0}$	
		$\lambda = 1$	$\lambda = -1$	$\lambda = 1$	$\lambda = -1$
1	0.154	0.92	1.165	1.042	0.927
2	0.20	0.877	1.255	1.070	0.893
3	0.308	0.790	1.72	1.125	0.763

and -1. Note that the values depend on the ratio of applied stress to yield stress $\sigma_{y\infty}/\sigma_y$ owing to the fact that Kfouri's calculations are for an elastoplastic solid. We note that the Swedlow/Eftis and Jones expression for $\Delta K_\lambda/\Delta K_{\lambda=0}$ is $[(4\nu - 1)\,\lambda + 1]^{1/2}$ and is independent of σ_y. For reference purposes, we note that if $\nu = 1/3$, the Swedlow/Eftis–Jones solution gives $\Delta K_\lambda/\Delta K_{\lambda=0} = 1.15$. for $\lambda = +1$ and 0.82 for $\lambda = -1$ (cf. Table 7.6). Using

$$C_\lambda = C_{\lambda=0}\left[(d_y)_\lambda/(d_y)_{\lambda=0}\right]^{n/2} = (\Delta K_{\lambda=0}/\Delta K_\lambda)^n$$

we obtain the values for C_λ given in the extreme right-hand columns of Table 7.5.

While the plotting of da/dN vs. ΔK on log-log scales for uniaxial fatigue data almost always results in a good straight line fit so long as ΔK has values neither near the threshold ΔK_0 nor near the value for monotonic fracture, the fact remains that unless great care is exercised in determining da/dN there will be marked scatter in the plot (cf. Example 7.5 and text following, and Smith (1973)). Indeed, when presenting data for similar types of materials, it is cutomary to represent the Paris 'line' as a scatterband and to place emphasis on the conservative side of the range. For example, for a variety of data on high yield strength structural steels from a variety of sources, Barsom (1971b) quotes for the conservative growth rate

$$da/dN = 1.35 \times 10^{-10} \, (\Delta K)^{2.25} \text{ (m/cycle, MPa}\sqrt{\text{m}}\text{)}$$

with the scatterband shown in Fig. 7.35.

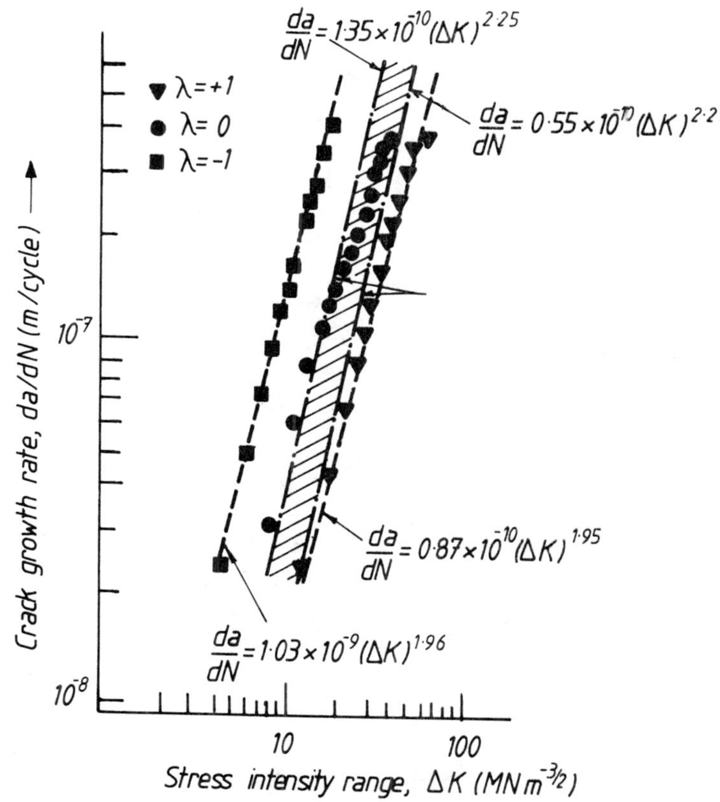

Fig. 7.35 – Fatigue crack growth rates compared for uniaxial and biaxial fatigue of HY100 steel. (reproduced from Smith and Pascoe (1982), with permission).

It is clear that if the affects of biaxiality are actual, but small in some cases, it is going to be difficult to distinguish between the different cases on log-log plots. This assertation was proved by Tanaka *et al.* (1979) and Hoshide *et al.* (1981) who found that although *linear* plots of a vs. N showed small, but discernible separations between each biaxial stress state they studied, their log da/dN vs. log ΔK curves of the same data were intermingled and could not be separated. On the other hand, Fig. 7.35 shows that Smith and Pascoe's 1982 results for HY 100 steel undergoing biaxial stressing (with $\lambda = -1$) fall distinctly to the left of Barsom's uniaxial scatterband while the data for $\lambda = +1$ appear to the right-hand side. At the same nominal ΔK, changing λ from $+1$ to -1 increases the crack growth rate by a factor of ten.

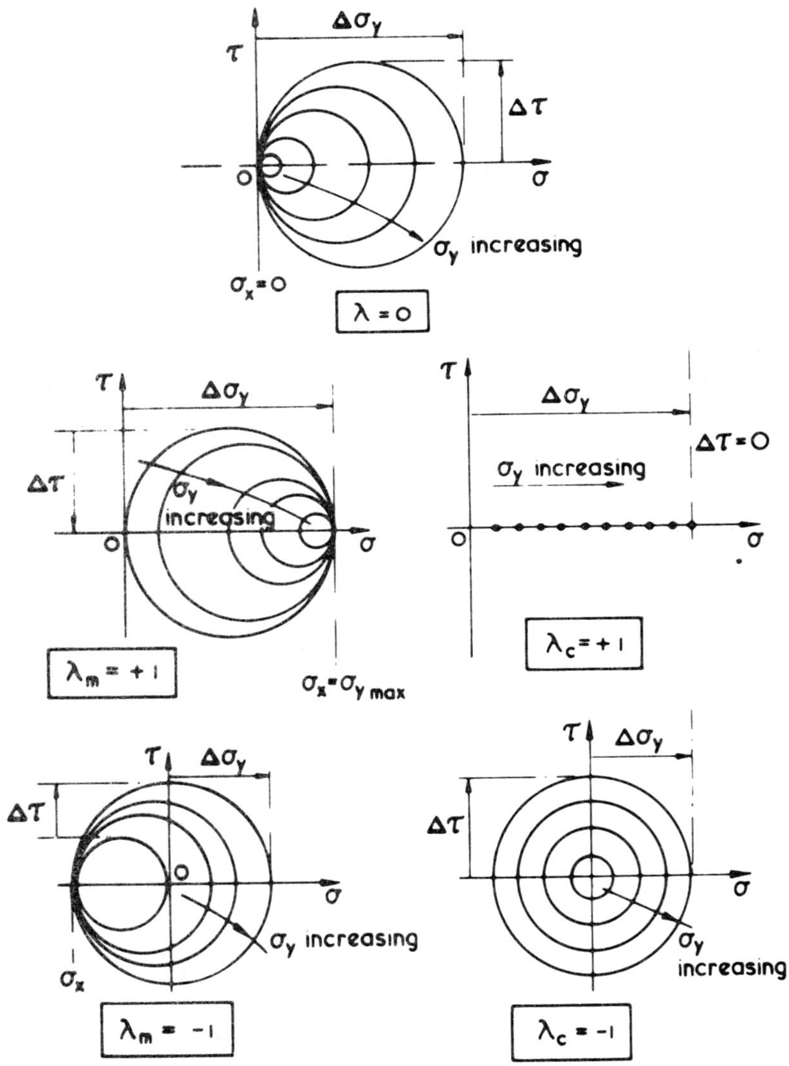

Fig. 7.36 – Mohr's circles of stresses in the plane of the plate for different λ states (After Miller (1977), with permission).

While the foregoing treatment using C_λ describes the general trend of data in Fig. 7.33, some details of behaviour remain unexplained, such as the difference between the λ_m and λ_c results particularly for $\lambda < 0$, where λ_c gives faster da/dN than λ_m. Fig. 7.36 illustrates the variations in bulk stress that occur in one cycle of loading as given by the changing sizes of Mohr's circles. Even though $\sigma_{y\infty}$ varies from zero to the same maximum value in all cases, all the Mohr's circles are different. Miller (1977) explains that to differentiate between the various tests it is necessary to invoke two parameters such as (i) $\Delta\tau$, the maximum range of shear stress in the plane of the plate for one cycle of loading, and (ii) $\Delta\sigma_n$, the range of normal stress on the plane of maximum shear (cf. McDiarmid, 1972). The first parameter defines the size variation of the circle and the second defines the range of movement of the circle. Table 7.7 summarizes this information. For the smaller crack-growth rates under the $\lambda = +1$ state it is seen that $\Delta\tau$ is zero for $\lambda_c = +1$ or negative for $\lambda_m = +1$, i.e. it will not assist crack-opening displacement. Note that $\Delta\sigma_n$ is also different in the two cases. Conversely, in the shear loading mode the maximum growth rate may be assumed to occur for the $\lambda_c = -1$ type test since $\Delta\tau$, and hence COD, is maximised while for the $\lambda_m = -1$ type test the range of shear stress in the plane of the plate is reduced and σ_n is either negative or zero.

It is not only the magnitude of the shear stress that is important, but also the plane in which it acts, and the mean stress across that plane depends on these influences. Since fatigue cracks in Stage I (Forsyth's (1961) nomenclature) are shear cracks, Brown and Miller pointed out in 1973 that two forms of Stage I fatigue cracking are suggested depending on the $\gamma_{max} = (\epsilon_1 - \epsilon_3)/2$ and $\epsilon_n = (\epsilon_1 + \epsilon_3)/2$ values, where γ_{max} and ϵ_n are the strain counterparts of maximum shear stress and normal stress, across the τ_{max} plane, with ϵ_1 and ϵ_3 being the (algebraically) greatest and least principal strains in the system ($\epsilon_1 > \epsilon_2 > \epsilon_3$). For strain ratios ($\epsilon_3/\epsilon_1$) between $-\frac{1}{2}$ and -1, both ϵ_1 and ϵ_3 lie in the surface of the specimen, with the plane of maximum shear perpendicular to the surface; for $+1 > \epsilon_3/\epsilon_1 > -\frac{1}{2}$, either ϵ_1 or ϵ_3 is perpendicular to the surface and the plane of maximum shear lies at 45° to the surface. That is, shear cracks may either run along the surface (called Type IA) or penetrate the body at 45° to the surface principal plane (called type IB), see Fig. 7.37. Since Stage II cracks propagate in mode I under tensile stresses, they start to run along the planes that bisect the planes of maximum shear. The two types of stage I crack give rise to two (different) types of Stage II crack, called types IIA and IIB, also shown in Fig. 7.37. Type B cracking systems are the more dangerous, since the crack is propagating away from the surface and reducing the net section. The data in Fig. 7.33 concern Stage II, type B cracks propagating in mode I (note the general comparison with the different patterns of peripheral cracking in barrelled upset tests (section 5.2)).

It is of interest that the components of normal stress and in-plane shear stress acting in the fibre plane are important parameters in the fatigue of filament-reinforced composites (Smith and Pascoe, 1977).

Low cycle and high cycle biaxial fatigue experiments are performed on a variety of testpiece geometries (precracked cruciform specimens loaded in-plane,

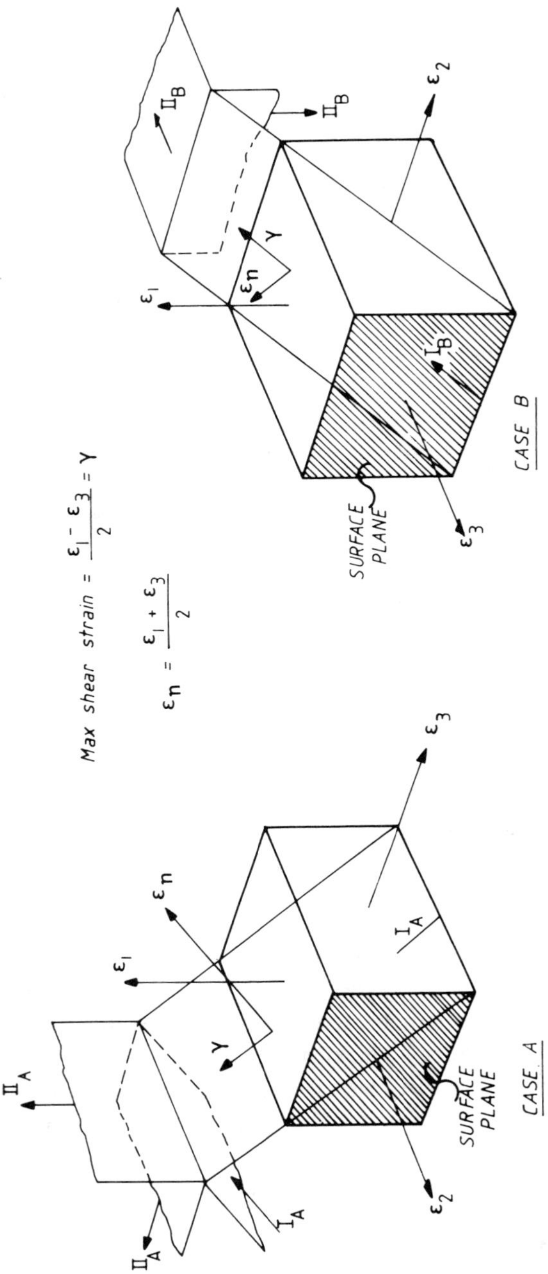

Fig. 7.37 – Types A and B fatigue crack growth for stages I and II according to Brown and Miller (1973).

precracked types twisted and pulled, and so on) in order to achieve different stress or strain ratios. Smith and Pascoe (1983) review test methods. The results may be presented in a variety of ways. Traditional Coffin–Manson plots of log (strain range) vs. log N_f may be used, with separate lines for each different state of strain. Plots of contours of equal life, using axes which represent ranges of strain in different directions, may be employed and these well illustrate effects such as the changes in life, for the same strain range in one direction, as the strain ratio changes. There are two versions of this type of plot depending on the axes used, viz. (i) using the ranges of the two principal surface strains (Pascoe and de Villiers, 1967) and (ii) using the ranges of the maximum shear strain and maximum strain normal to the plane of maximum shear (Brown and Miller, 1973).

Let us first consider the Brown and Miller approach in which $\Delta\gamma_{max} = \Delta(\epsilon_1 - \epsilon_3)/2$ is employed for the abscissa and $\Delta\epsilon_n = \Delta(\epsilon_1 + \epsilon_3)/2$ for the ordinate; (the $\Delta\gamma_{max} - \Delta\epsilon_n$ plane is sometimes called the Γ plane). Clearly all results from given 'standard' tests must lie on radial lines emanating from the origin defined by the relevant biaxial strain amplitude ratio (or stress amplitude ratio) imposed by the testing machine. Thus, in low-cycle fatigue in the plastic range, for which $\nu = ½$, simple tension has $\Delta\epsilon_2 = -½\,\Delta\epsilon_1$, and therefore a strain ratio $\Delta\epsilon_3/\Delta\epsilon_1 = -½$; $\Delta\gamma_{max} = \Delta\epsilon_1(1 + ½)/2$ and $\Delta\epsilon_n = (\Delta\epsilon_1/2)/2$ and hence uniaxial data lie on a line of slope $(\Delta\epsilon_1/4) \div (3\Delta\epsilon_1/4) = +1/3$ radiating from the origin. In torsion, $\Delta\epsilon_1 = -\Delta\epsilon_3$ and data lie on a line of slope zero, i.e. the abscissa of the Γ plane; plane strain data also lie along the abscissa; equibiaxial tension has $\Delta\epsilon_1 = \Delta\epsilon_2$, $\Delta\epsilon_3 = -(\Delta\epsilon_1 + \Delta\epsilon_2) = -2\,\Delta\epsilon_1$ and the slope in the Γ-plane is $-1/3$, and so on.

Fig. 7.38(a) shows a Γ-plane plot for QT 35 steel. The form of a constant life plot is that the greatest $\Delta\gamma_{max}$ can be withstood when $\epsilon_n = 0$ (i.e. torsion and plane strain crases), but that as ϵ_n increases, the range of $\Delta\gamma_{max}$ decreases, until the uniaxial tension line is reached. Clearly, for the same $\Delta\gamma_{max}$, a material has a shorter life as the strain ratio is increased within this region. Thereafter, as the strain ratio increases to positive values, and the equivalent Γ-plane slopes go negative, there does not seem to be much change in fatigue life expressed by the range of permissible $\Delta\gamma_{max}$, i.e. the plot seems to be more-or-less vertical between uniaxial tension and equibiaxial tension. We note that this latter range of strain ratios gives rise to type B cracking, whereas the range in which the life is sensitive to strain ratio corresponds with type A cracking (Fig. 7.37).

In Fig. 7.38(a), owing to the shape of the constant N_f loci and the fact that they intersect, different fatigue lives may exist for identically strained systems. For example, the contour for $N = 500$ on the 'vertical' part of its locus, intersects the contour for $N = 1{,}000$ during the latter's 'curved' part of its locus. This is puzzling and suggests that even a two-parameter description of fatigue is inadequate. However, we must remember the changes in orientation of the crack plane (types A and B) with change in strain ratio and that the modes of deformation along loci of constant N_f are then different. This complication is absent in the Pascoe presentation, as shown in Fig. 7.38(b) for the same QT 35 steel data. The two parts of the Miller plot which cross, and the corresponding

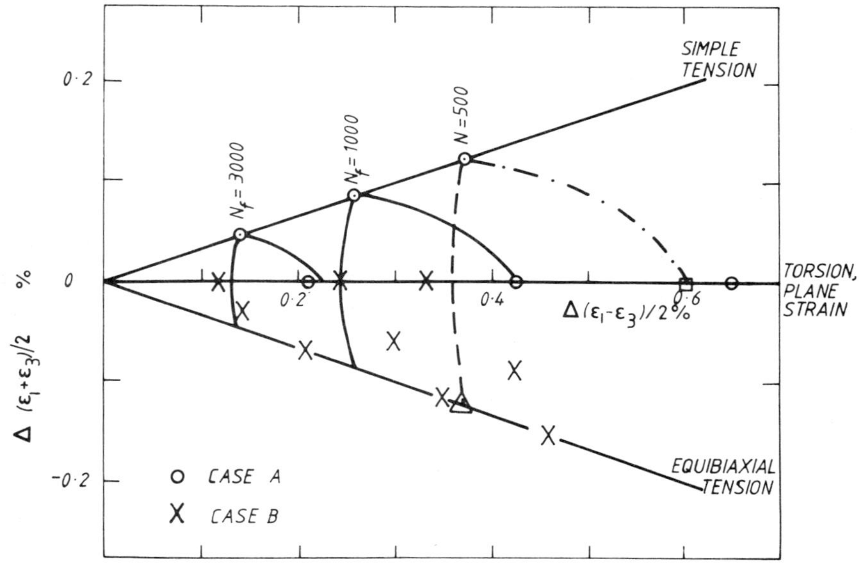

(a)

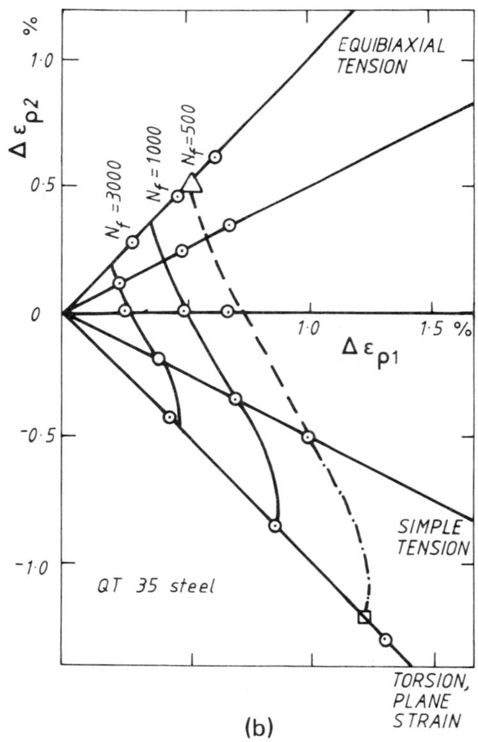

(b)

Fig. 7.38 – Plastic strain/endurance data for QT 35 steel fitted according to (a) Brown and Miller's method and (b) Pascoe and de Villiers' method (courtesy K. J. Pascoe).

parts of the Pascoe plot which do not cross, are identified by the same line markings.

Tanaka (1974) and Smith (1980) have studied the fatigue growth of inclined cracks loaded uniaxially in either tension or compressions to give mixed mode (I and II) propagation. The paths taken are the particular interest and relate to the question of paths in monotonic fracture (section 3.5).

7.7 FATIGUE FROM NOTCHES

Fatigue failures invariably start from some form of geometric discontinuity (change of section, holes, threads etc.), and many standard S–N fatigue curves have been determined from specimens with various notch 'features' which are felt to be important in given designs. This leads to the traditional strength reduction factor K_f which is the ratio of the fatigue limits of the notched specimen to an equivalent plain specimen. In turn this leads to a variety of other notch sensitivity indices (Neuber-, Stowell-, notch sensitivity factors and so on) which are used in design calculations to ensure an infinite life. Details may be found in many books on design.

Smith and Miller (1977) pointed out that in the finite life regime (i.e. working stresses higher than the endurance limit) the use of such factors can be dangerous, since the historical approach links failure with bulk stress or strain parameters and not to the behaviour of the fatigue crack itself and its local stress field. Above a certain bulk stress level, notch geometries will certainly create localised plastic flow and a crack may be initiated. But complete failure will result only if the crack is long enough, and/or the stress high enough, for it to generate its own crack tip plasticity when it enters the lower elastic stress fields away from the notch field. Long cracks will not propagate in fatigue if $\Delta K < \Delta K_0$. Bulk parameters do not take into account the various regimes of fatigue, the duration of which, as already explained, depends on stress level (i.e. at low stress levels perhaps 90 per cent of the life is concerned with initiation in smooth components whereas at high stresses, 90 per cent of the lifetime may be in Stage II propagation).

The equivalent crack length (L) concept introduced by Smith and Miller for crack propagation from notches, discussed in section 3.8 in connexion with monotonic loading, is a useful vehicle with which to discuss fatigue crack propagation from notches. Since in fatigue $\mathrm{d}a/\mathrm{d}N = C(\Delta K)^n$, the equating of K expressions to determine L is the same as saying that the real fatigue crack growing from the notch, and the equivalent sharp crack of length L, both propagate at the same rate. Smith and Miller performed experiments on centrally notched plates, using the e correction for notch size (section 3.8), and obtained excellent agreement with fatigue crack growth rates in unnotched specimens with sharp cracks equivalent of the length L. They also made life predictions for several edge-notched specimens, using the appropriate $(\mathrm{d}a/\mathrm{d}N) = C(\Delta K)^n$ equation, the constants in which were determined from sharp edge-cracked testpieces. Calculations for the notched testpieces were split into two parts, viz. the number of cycles required to propagate the crack from the starter crack length a_0 to the full extent of the notch field $0.13\sqrt{\alpha\rho}$, during which $\Delta K = \Delta\,(1.12\,(1 + 7.69\sqrt{\alpha\rho})^{1/2}\ \sigma\sqrt{\pi a})$; and the number of cycles required to extend the crack further to an arbitrary length (60 mm in Table 7.8), during which $\Delta K = \Delta\,(1.12\ \sigma\sqrt{\pi(a + \alpha)})$ (cf. section 3.8). Experiments were performed both with notches of constant depth and varying root radius and with notches of varying depth and constant root radius. Table 7.8 shows that very close agreement was obtained between the experimental and theoretical propagation lives for a wide variety of notch geometries. Since the fatigue lifetimes is dominated by growth rates at short crack lengths, the validity of the equivalent crack length concept is well supported.

Table 7.7

State	$\tau_{min} \to \tau_{max}$	$\Delta\tau$	$\sigma_{n\,min} \to \sigma_{n\,max}$
$\lambda_c = -1$	$0 \to \sigma_y$	σ_y	$0 \to 0$
$\lambda_m = -1$	$\frac{1}{2}\sigma_y \to \sigma_y$	$\frac{1}{2}\sigma_y$	$-\frac{1}{2}\sigma_y \to 0$
$\lambda = 0$	$0 \to \frac{1}{2}\sigma_y$	$\frac{1}{2}\sigma_y$	$0 \to \frac{1}{2}\sigma_y$
$\lambda_c = +1$	$0 \to 0$	0	$0 \to \sigma_y$
$\lambda_m = +1$	$-\frac{1}{2}\sigma_y \to 0$	$-\frac{1}{2}\sigma_y$	$\frac{1}{2}\sigma_y \to \sigma_y$

Example 7.17 In the historical approach to fatigue, the extent to which a notch reduces the endurance limit (in comparison with the fatigue limit σ_0 of a plain testpiece) is called the *fatigue stress concentration factor,* or the *fatigue strength reduction factor,* and is given the symbol K_f. Experiments have shown that $K_f < K_T$ (the theoretical stress concentration factor), with the difference increasing as K_T increases. (Details depend on material, microstructure and so on.) Give a possible interpretation of K_f in the light of the mechanics of fatigue crack propagation from notches just discussed.

Ans. If K_T were the proper factor, the working stress at the notch would be $K_T\sigma$ and the endurance limit of a notched component would be simply σ_0/K_T. But experiments show this is not correct. However, consider the stress intensity factor obtained when using $K_f\sigma$ for the working stress, i.e.

$$K = K_f\ \sigma\sqrt{\pi a} = (1 + 7.69\sqrt{\alpha/\rho})^{1/2}\ \sigma\sqrt{\pi a}$$

and compare it with what we would write down for K (and hence ΔK) if K_T were the correct factor, viz.

$$K = K_T\ \sigma\sqrt{\pi a}$$

Hence it is likely that K_f is more to do with $(1 + 7.69\sqrt{\alpha/\rho})^{1/2}$ than with K_T. K_T for elliptic notches is usually taken as $(1 + 2\sqrt{\alpha/\rho})$ and Fig. 7.39 compares the two factors. Clearly for notches less severe than a circle ($\sqrt{\alpha/\rho} < 1$) there is a little difference, but for sharper notches $(1 + 7.69\sqrt{\alpha/\rho})^{1/2} < 1 + 2\sqrt{\alpha/\rho}$

Table 7.8

Summary of experimental and predicted fatigue lives for edge notched plates

ρ mm	$\sqrt{\alpha/\rho}$	a_0 (mm)	l_{on}† (mm)	N_1 Cycles to grow from a_0 to l_{on}	N_2 Cycles to grow from l_{on} to 60 mm	$N_1 + N_2$	N_{Expt}	$\frac{N_{Expt}}{N_1 + N_2}$
(a) Various radii; constant depth (32 mm)								
0	∞	0	0	0	34,000	34,000	32,000	0.94
2	4	0.1	1.04	31,000	29,000	60,000	61,000	1.02
4	2.83	0.2	1.47	51,000	27,000	78,000	123,000	1.58
8	2	0.25	2.08	77,000	25,000	102,000	88,000	0.86
16	1.41	0.33	2.94	109,000	22,000	131,000	107,000	0.82
32	1	0.46	4.16	151,000	19,000	170,000	165,000	0.97
64	0.707	0.70	5.88	212,000	15,000	229,000	205,000	0.90
(b) Various depth; constant radii (16 mm)								
8	0.707	0.2	1.47	120,000	470,000	590,000	453,000	0.77
16	1	0.3	2.08	90,000	148,000	238,000	192,000	0.81
32	1.414	0.46	4.16	109,000	22,000	131,000	107,000	0.82

†$l_{on} = 0.13\sqrt{\alpha\rho}$.

which agrees with experimental trends. Smith and Miller (1977) have observed that while $K = K_T \sigma\sqrt{\pi a}$ has no physical validity once a fatigue crack exists, $K = K_f \sigma\sqrt{\pi a}$ retains its meaning, in that account is taken of the crack growing in the notch field.

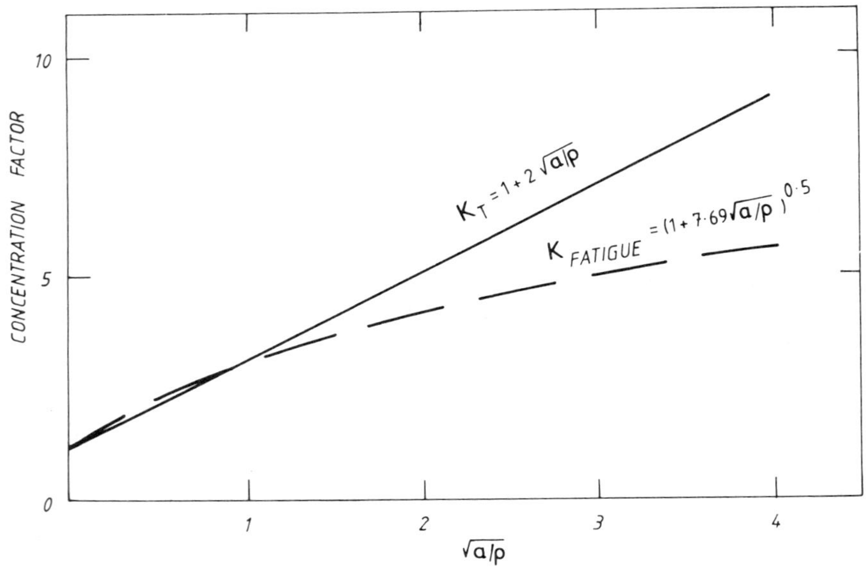

Fig. 7.39 – Comparison of stress concentration fractor K_T and the 'fatigue concentration factor', K_f (after Smith and Miller (1977), with permission).

The interaction of notch plasticity fields and sharp crack tip fields provides an explanation for non-propagating (but initiated) cracks in fatigue – particularly at very sharp notches. The initiation and early growth of a crack is controlled by the plastic strain field of the notch which diminishes rapidly (see Fig. 7.40). It follows that the early growth rate decreases with increasing crack length and will eventually stop unless the threshold ΔK_0 is exceeded. However, the crack generates its own crack tip plasticity which increases rapidly with crack length, according to $K = K_f \sigma\sqrt{\pi a} = (1 + 7.69\sqrt{\alpha/\rho})^{1/2}\sigma\sqrt{\pi a}$. Once this crack tip plasticity exceeds the amount corresponding to ΔK_0, then the crack can continue to propagate under 'fracture mechanics' control. Experimental confirmation of an initially decreasing growth rate with increasing crack length has been reported by a number of workers. The overall crack growth rate is thus seen to be the sum of two separate plasticity effects. Their relative contributions, and the minimum crack growth rate, will depend on the level of the bulk applied stress (cf. section 7.2.3).

The first stress level of interest is that required to cause crack initiation. This might be expressed in simple form as σ_0/K_T. However, Smith and Miller

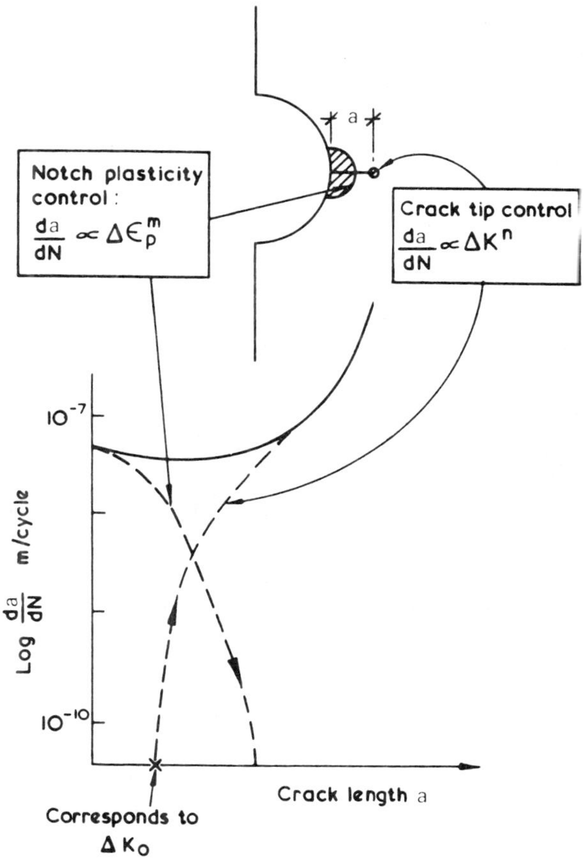

Fig. 7.40 – A two-process, crack growth schematic (after Smith and Miller (1978), with permission).

showed in 1978 that the somewhat higher experimentally obtained values are better satisfied by the fatigue concentration term given in Example 7.17, viz.

$$K_{\text{fatigue}} = (1 + 7.69\sqrt{\alpha/\rho})^{0.5} \tag{7.38}$$

For relatively mild notches, once the crack has initiated, the bulk stress level is sufficiently high for the crack to propagate to failure. For sharper notches however, the possibility of crack arrest remains to be explained. In these latter cases the fatigue limit of practical interest is based not on initiation, but on the higher stress level needed to cause propagation.

Consider a notch having parameters K_T, ρ and α, containing a crack of length a that has just ceased to propagate owing to notch plasticity, but which can just begin to propagate owing to crack tip plasticity defined by ΔK_0. Now once the fatigue crack begins to grow at the notch root, its stress intensity factor increases very rapidly to a value equivalent to that of a crack of length

($a + \alpha$) (Fig. 3.38). Furthermore, for small crack lengths at sharp notches, a is very much smaller than α. Hence, to a first approximation, the stress intensity factor at the minimum crack growth rate at an edge notch is given by $1.12\,\sigma\sqrt{\pi\alpha}$. At this condition the crack begins to grow under crack tip plasticity control, and so $1.12\sigma\sqrt{\pi\alpha} = \Delta K_0$, which, on re-arrangement, gives the minimum nominal stress level required to cause complete failure, i.e.

$$\sigma = \frac{0.5\Delta K_0}{\sqrt{\alpha}} \tag{7.39}$$

The consequence of these solutions is shown in Fig. 7.41. Here the plot of applied stress against concentration factor is divided into three regions. Namely, a zone in which no crack will initiate, a zone in which a crack will initiate but not propagate, and thirdly a zone in which a crack will initiate and then propagate to failure.

The boundaries in Fig. 7.41, i.e. the plots of $\sigma = 0.5\ \Delta K_0/\sqrt{\alpha}$ and $\sigma = \sigma_0/K_f$ relate to mild steel having $\sigma_0 = 260$ MPa, $\Delta K_0 = 6.5$ MPa$\sqrt{\text{m}}$. Experimental results conducted by Frost in 1959 on 1.3 mm deep notches in cylindrical specimens of the same material are superimposed on Fig. 7.41. A stress level for complete failure of 90 MPa is given by $0.5\ (6.5)/\sqrt{0.0013}$ which agrees well with the data.

The interested reader is referred to the book by Frost *et al.* (1974) and to the papers of Smith and Miller in which further references will be found. In the case of polymers, Constable *et al.* (1970) have discussed fatigue from notches.

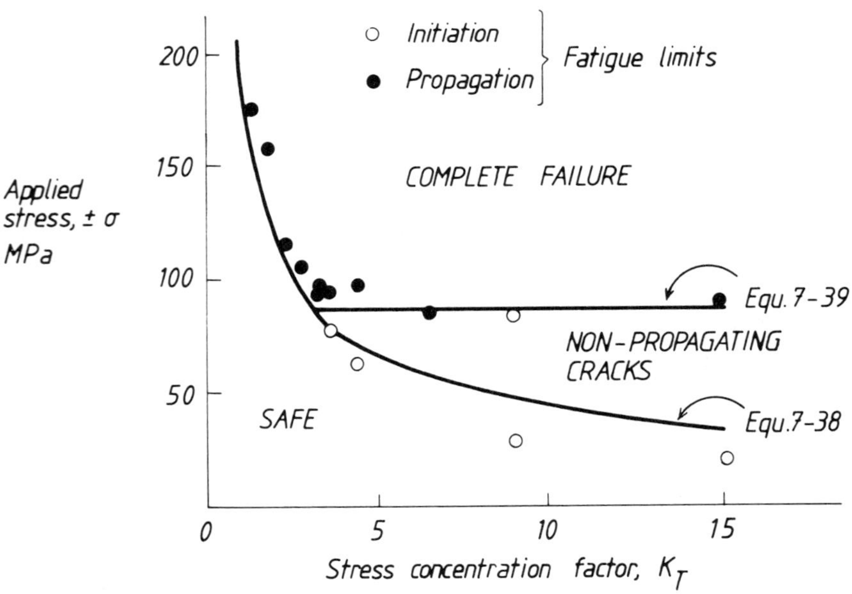

Fig. 7.41 – Comparison of predictions with experimental results for 1.3 mm deep notches in mild steel (after Smith and Miller (1978), with permission).

7.8 EFFECT OF RESIDUAL STRESSES ON FATIGUE

Residual stresses may increase or decrease the loads required for monotonic elastic fracture as explained in section 3.2.4. The same is true for crack propagation in fatigue, and it is well-known that residual compressive stresses at the surface of components prolong fatigue life. Analyses based merely on external loadings (performed in ignorance perhaps of the presence of internal stresses set up by manufacturing processes) can fail to predict the observed behaviour of components. For example, the F-111 Recovery Programme at NASA on the fracture and cyclic crack growth properties of shot-peened, surface-flawed D6AC steel revealed (a) that elastic fracture mechanics did not predict the monotonic fracture strength for short cracks and (b) that da/dN was not correlated well by the Paris law, separate relations being produced between ΔK and da/dN for different absolute values of cyclic stress.

Elber showed in 1974 that surface residual stresses caused these anomalies. Monotonic and fatigue experiments were carried out both on as-peened specimens (containing a distribution of near-surface residual stresses subsequently modelled by the piecewise linear distribution shown in Fig. 7.42(a) in which h is a convenient scaling parameter for the extent of the stress field) and on as-peened specimens which were prior-precompressed 0.1% in order to eliminate (after unloading before testing) the residual stress pattern (Fig. 7.42(b)).

Fig. 7.43 shows that at *short* crack lengths, in the presence of residual stresses, *monotonic* fracture strengths (solid circles) were higher than predicted by considering the external loading only, using $K_{IC} = 48$ MPa$\sqrt{m}$; at longer starter crack lengths the static strengths were adequately predicted. In *fatigue*, independent tests from residual stress-free material gave

$$da/dN = 1.87 \times 10^{-12} \, (\Delta K)^{2.72} \text{ m/cycle}$$

(ΔK in MPa$\sqrt{m}$) but as-shot-peened material showed slower rates of growth than this at short crack lengths and faster rates at longer crack lengths (Fig. 7.44). Furthermore, at lower values of applied stress, da/dN was higher at the same ΔK. Data from testpieces precompressed after shot-peening to remove the residual stresses, also shown in Fig. 7.44, *did* fit the basic Paris relation.

Elber analysed the behaviour by arguing that the effective crack front stress intensity is the sum of the stress intensity produced by the residual stresses $K_{residual}$ and the stress intensity produced by the external loading $K_{applied}$ (cf. comments in Chapter 3 regarding addition of K or G ($\propto K^2$)). The residual stress intensity was determined from the distribution shown in Fig. 7.42(b), (i) using the Benthem–Koiter formula given in Fig. 7.45 for $a < h$ (h defined in Fig. 7.42(a)) and (ii) assuming integrated point load expressions given in Fig. 7.45 for $a > h$. The starter cracks in the tests described in Figs. 7.43 and 7.44 involved thumbnail surface cracks, so the edge crack formulae were corrected by Newman's surface flaw shape factor (Chapter 2). The theoretical line of *monotonic*

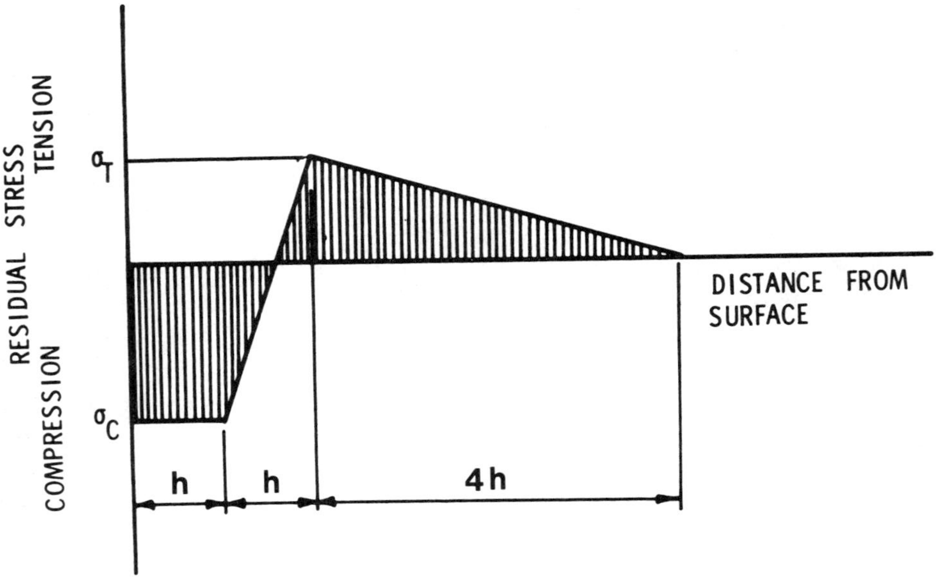

(a)

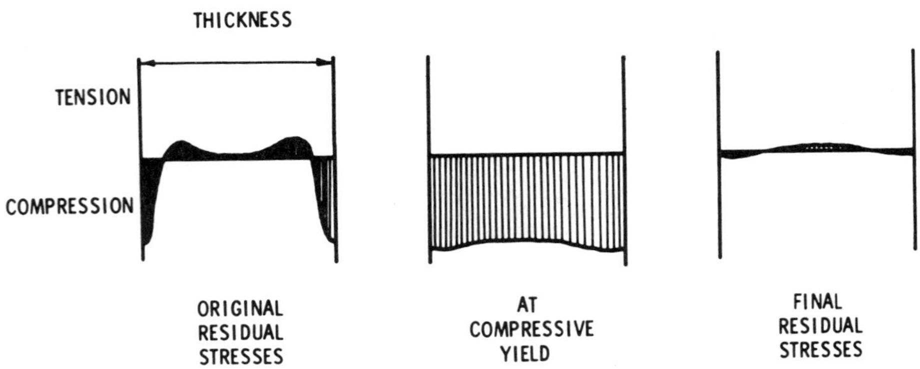

(b)

Fig. 7.42 – (a) Estimated residual stress distribution. (b) Effect of precompression on residual stresses. (Both from Elber, 1974. Copyright ASTM, 1916 Race Street, Philadelphia, PA 19103. Reprinted with permission.)

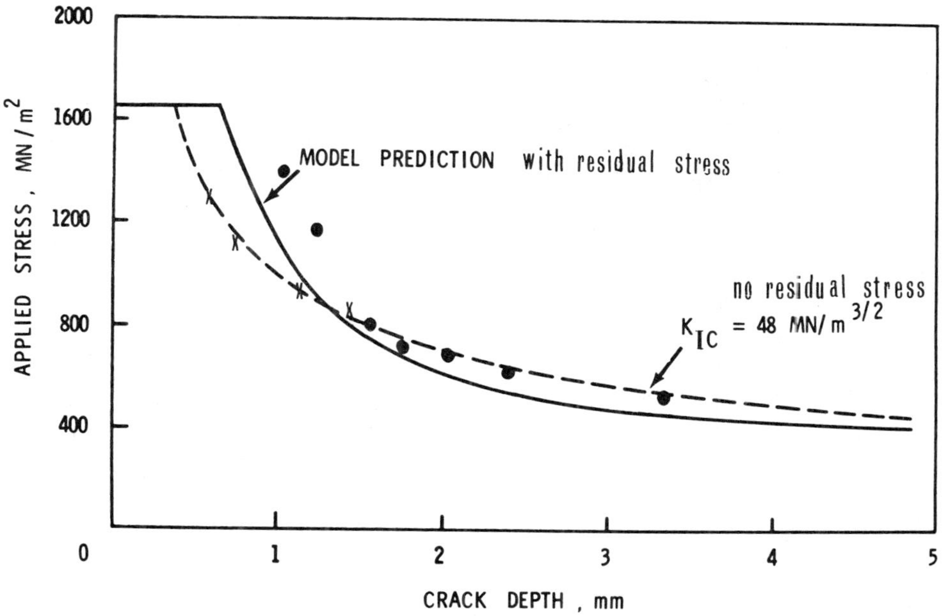

Fig. 7.43 – Monotonic fracture strength prediction. (From Elber, 1974. Copyright ASTM, 1916 Race Street, Philadelphia, PA 19103. Reprinted with permission.)

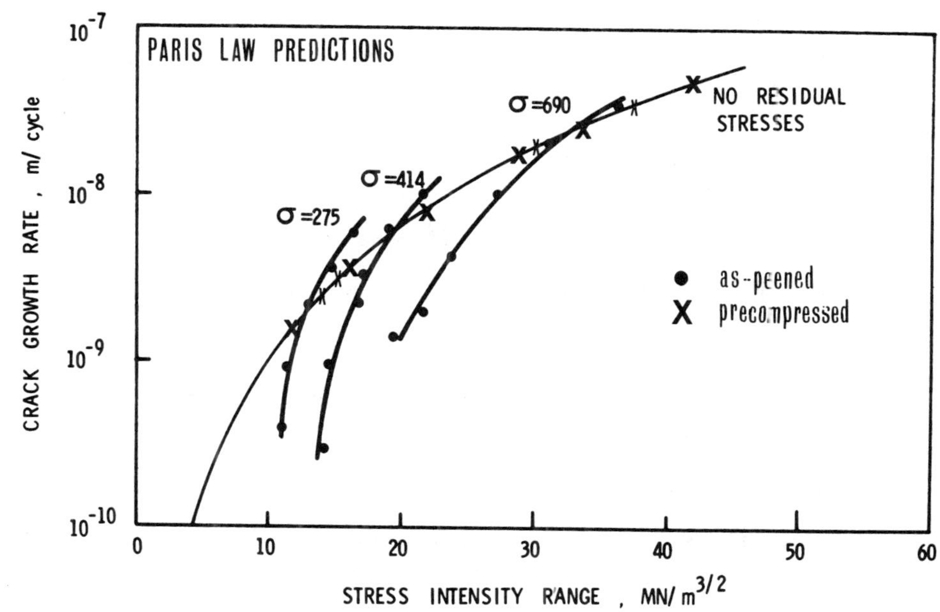

Fig. 7.44 – Effect of residual stress on fatigue crack growth rate. (From Elber, 1974. Copyright ASTM, 1916 Race Street, Philadelphia, PA 19103. Reprinted with permission.)

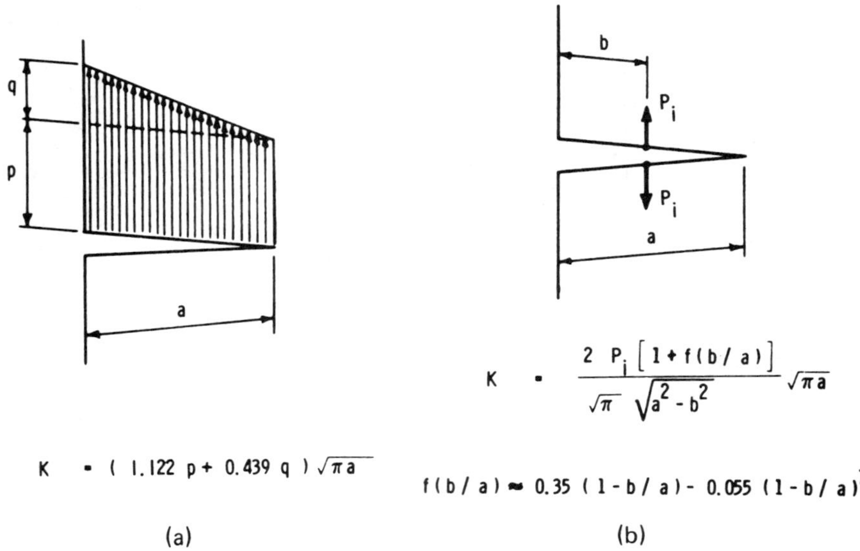

Fig. 7.45 – Stress intensity solutions for edge cracks: (a) distributed stress and (b) concentrated force. Copyright ASTM, 1916 Race Street, Philadelphia, PA 19103. Reprinted with permission.

strength in the presence of residual stress in Fig. 7.43 was obtained using $h = 0.5$ mm, $\sigma_c = 315$ MPa, $\sigma_T = 240$ MPa (Fig. 7.42(a)) and $K_{IC} = 48$ MPa$\sqrt{\text{m}}$, in

$$K_{IC} = K_{\text{applied}} + K_{\text{residual}} = 1.12\,\sigma \sqrt{\frac{\pi a}{Q}} + K_{\text{residual}} \qquad (7.40)$$

Similarly, the three separate theoretical da/dN curves in Fig. 7.44 were obtained from

$$da/dN = 1.87 \times 10^{-12}\,(K_{\text{applied}} + K_{\text{residual}})^{2.72} \qquad (7.41)$$

Both monotonic and fluctuating responses are well-predicted.

Example 7.18 Determine the residual stress intensity when $a = h/2$ and $a = h$ from the Benthem–Koiter formula in Fig. 7.45 when $h = 0.5$ mm and $\sigma_c = 315$ MPa compressive, and compare the predicted monotonic fracture stresses in the presence and absence of residual stresses ($K_{IC} = 48$ MPa$\sqrt{\text{m}}$).

Ans. The Benthem–Koiter K-formula given in Fig. 7.45 applies generally to a linear increasing stress over the crack face shown in the figure, where p is the constant stress 'base' and q the increase of stress. We are told that $a \leqslant h$ in both cases of interest, so, according to the shape of the residual stress distribution in Fig. 7.42(a), the stress on the crack faces is constant (not linearly changing). In this problem therefore $q = 0$ and $p = -315$ MPa, and $K_{\text{residual}} = (1.22p + 0.439q) \times$

$\sqrt{\pi a} = -1.122(315)\sqrt{\pi a}$ only. For $a = h/2$ mm, $K_{\text{residual}} = -9.9$ MPa$\sqrt{\text{m}}$ and -14 MPa$\sqrt{\text{m}}$ for $a = h = 0.5$ mm. Note that both answers are negative as the stress σ_c is compressive. For simple edge cracks $Q = 1$ in equation (7.40), so $\sigma = (K_{IC} - K_{\text{residual}})/1.122\sqrt{\pi a} = (48 + 9.9)/1.12\sqrt{\pi \times 0.25 \times 10^{-3}} = 1{,}841$ MPa; with $K_{\text{residual}} = 0$, $\sigma = 1{,}526$ MPa. Similarly for $a = 0.5$ mm, $\sigma = 1{,}394$ MPa in the presence of residual stresses and 1,079 MPa without (cf. the constant K-curve in Fig. 7.43 is for surface flaws, so, although the relative values for σ and a are still the same, the absolute values do not coincide). Note also that at these very short crack lengths, some of the predicted strengths are greater than the yield strength (1,450 MPa) and UTS (1,600 MPa).

Calculations show that K_{residual} is negative for $a < 1.33$ mm (for $h = 0.5$ mm, $\sigma_c = 315$ MPa, $\sigma_T = 240$ MPa in Fig. 7.42(a) (Problem 7.22), and positive for $a > 1.33$ mm. In Fig. 7.44 the fatigue crack growth rates for the as-peened bodies are sometimes slower than predicted by the Paris law at a given ΔK, and sometimes faster than da/dN for the residual stress-free material (the da/dN curves cross-over the 'master' Paris law curve). The cross-over points correspond with $a = 1.3$ mm at the appropriate applied stress levels (275, 414 MPa etc.) and growth rates are slower to the left of the cross-over point owing to compressive K_{residual} and faster to the right where K_{residual} is tensile.

In regard to the effect of residual stresses on monotonic crack propagation discussed in Chapter 3, we note that in Elber's problem compressive residual stresses at the surface of D6AC steel bodies produced an apparently *higher* fracture toughness for shallow cracks than stress-free material, and tensile residual stresses below the shot-peened layers gave apparently *lower* toughnesses for deep cracks (Fig. 7.43). This accords with the general conclusions of section 3.7.

7.9 FATIGUE AT HIGH HOMOLOGOUS TEMPERATURES: CREEP FATIGUE

As the homologous temperature is increased, fatigue crack growth rates increase, with the constant C in the Paris equation increasing and the exponent n decreasing. Typical behaviour of superalloys is shown in Fig. 7.46. However, at 704°C there is an increase in n and a marked acceleration in crack growth rate for Inconel X-750. The general increase in da/dN with temperature at constant ΔK comes about through a number of factors, e.g. a reduction in cyclic work-hardening exponent n' and in cyclic stress, a gradual involvement of relaxation and creep effects in the *flow* of the material at the crack tip (reducing σ_y and increasing COD δ and hence da/dN), and eventually creep fracture effects when crack advance becomes discontinuous owing to cavity formation ahead of the main crack tip. Under slow cycling, increments of creep crack growth can occur cycle by cycle, particularly when the crack opening displacement is large enough to make cavities join together. Large δ are most often achieved

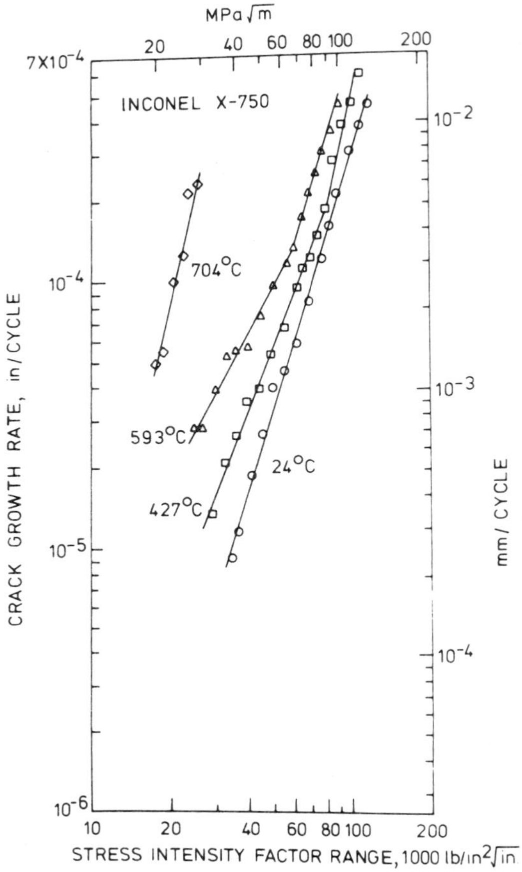

Fig. 7.46 – Fatigue crack growth rates in Inconel X-750 as a function of stress intensity factor range (cycling frequency = 0.17 Hz) (after Shahinian and Sadananda (1981), with permission).

in high strain fatigue, and indeed it was there that Solomon and Coffin (1972) first observed creep fatigue growth, but it can occur in low stress crack growth too.

Fatigue combined with creep (creep crack growth being represented in metals by the likes of Fig. 6.9 and in polymers and other solids by the various $\dot{a}$–K_c relations, given in Chapter 6, which apply at high homologous temperatures), is a complicted field. Changes in metallurgical structure, ageing, and oxidation, and changes in micromechanisms all serve to make it difficult to present a simple picture. A valuable review may be found in Ashby and Tomkins (1979) and Tomkins (1980).

When creep and relaxation are important, it is clear that the frequency of cycling will have a pronounced effect on da/dN. Furthermore, in practice, cyclic load–time–strain histories may have dwell periods (at fixed load or strain) interspersed. Creep initiation then may occur at the crack tip or even in the bulk of the body. This leads to rapid discontinuous crack advance, as shown in Fig. 7.47 for 316 stainless steel at 625°C. Transgranular crack growth is observed for both the sawtooth and dwell cycles at a rate of $da/dN \approx \delta/2$ at low ΔK, but at higher ΔK, δ is large enough for cavity linkage and da/dN both accelerates and follows an intergranular path for the tensile dwell cycle case. There is apparently no change for the sawtooth cycle case in which the da/dN vs. ΔK relationship is the same for all ΔK levels.

Creep, relaxation and frequency effects are well-known in polymer fatigue (e.g. Hertzberg and Manson, 1980); some further related comments will be found in the next section.

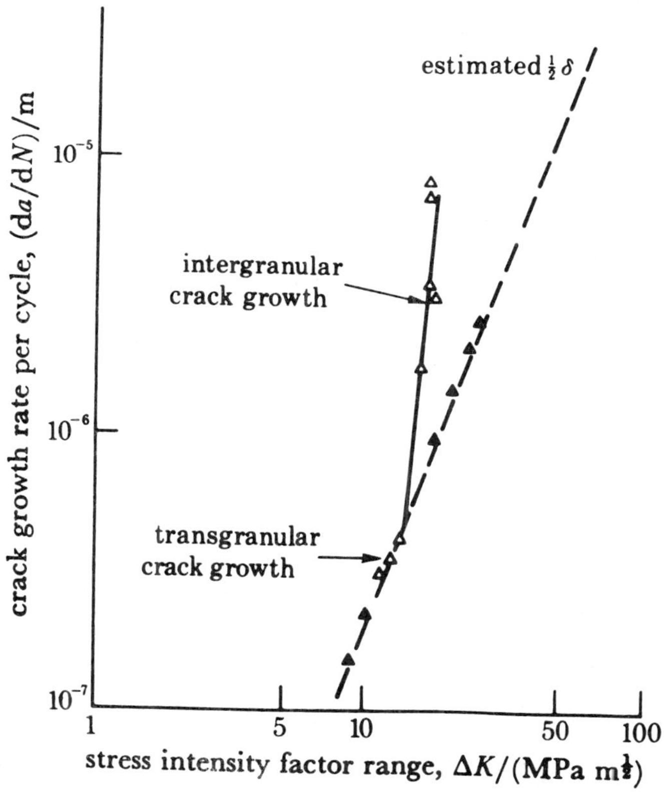

Fig. 7.47 – Effect of creep on fatigue crack growth in type 316 stainless steel at 625°C, r = 0.6. Δ sawtooth cycle, freq. = 0.01 Hz; ▲ tensile dwell cycle, freq. = 0.0001 Hz. (After Lloyd and Wareing, 1979; reproduced from Tomkins, 1981, with permission.)

7.10 STRESS CORROSION FATIGUE: CREEP STRESS CORROSION FATIGUE

In general, fatigue crack growth characteristics in hostile environments are similar to those in air shown in Figs. 7.8, 7.9 and so on, but at the same ΔK, da/dN is faster in corrosive environments. The numerical constant C in equation

(7.6) now depends on the environment as does the value of the threshold stress intensity factor (Table 7.1). Mean stress (section 7.3) and frequency may also affect da/dN (e.g. temperature rises and thermal softening may occur in polymers under high frequency loading). Calculations for number of cycles to failure proceed as before. Such procedures are important in the design of, for example, power plants where wet steam in the turbines and hydrogen used for cooling the rotor are aggressive environments (corrosion fretting fatigue is a common cause of initiating cracks in rotor bodies). Fig. 7.48 shows stress corrosion fatigue data plotted in the traditional way. Since K_{IC} values, as well as the parameters in Paris-type laws of fatigue crack growth, care affected by environment, the Stage III 'end-points' of da/dN vs. ΔK plots can be different as between environments and air – in hydrogen, for example by a factor of perhaps two.

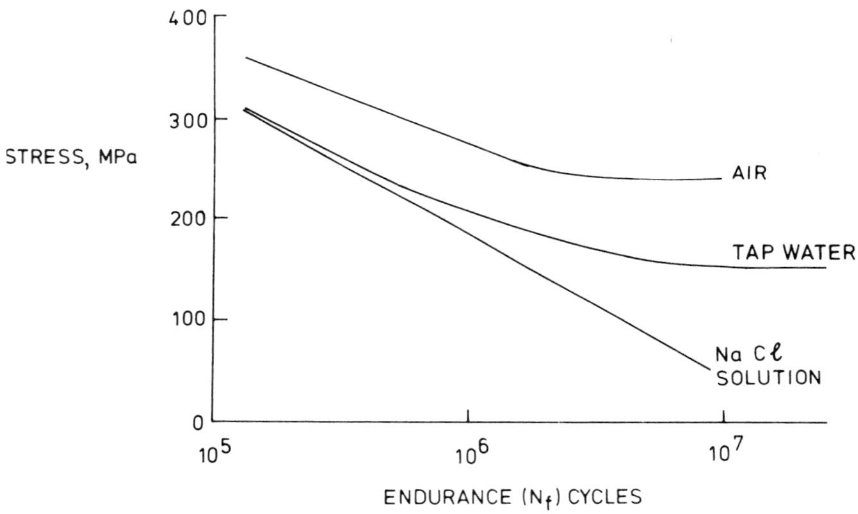

Fig. 7.48 – Effect of aqueous environment on the S–N curve of mild steel (after Duquette (1972), with permission.)

Section 7.1 highlighted the apparent similarities between delayed fracture in glasses and ceramics and conventional fatigue behaviour, in terms of the similar shapes of stress/time-to-failure and stress/number-of-cycles-to-failure curves, shown in Figs. 7.1 (a) and (b). There is a related connexion between stress corrosion data (which, as explained in Chapter 6, concerns times to failure under fixed stresses and is a delayed fracture phenomenon) and stress corrosion fatigue data. Whereas crack sharpness and crack advance in static stress corrosion is achieved through electrochemistry (section 6.5.2), load cycling itself helps to maintain sharpness so that stress corrosion fatigue usually occurs in a far greater range of environment than static stress corrosion.

Fig. 7.49 shows both *static* stress corrosion and *fatigue* stress corrosion data (in brine) for crack growth in 'Hiduminium RR58' Al–Cu–Mg–Fe–Ni alloy.

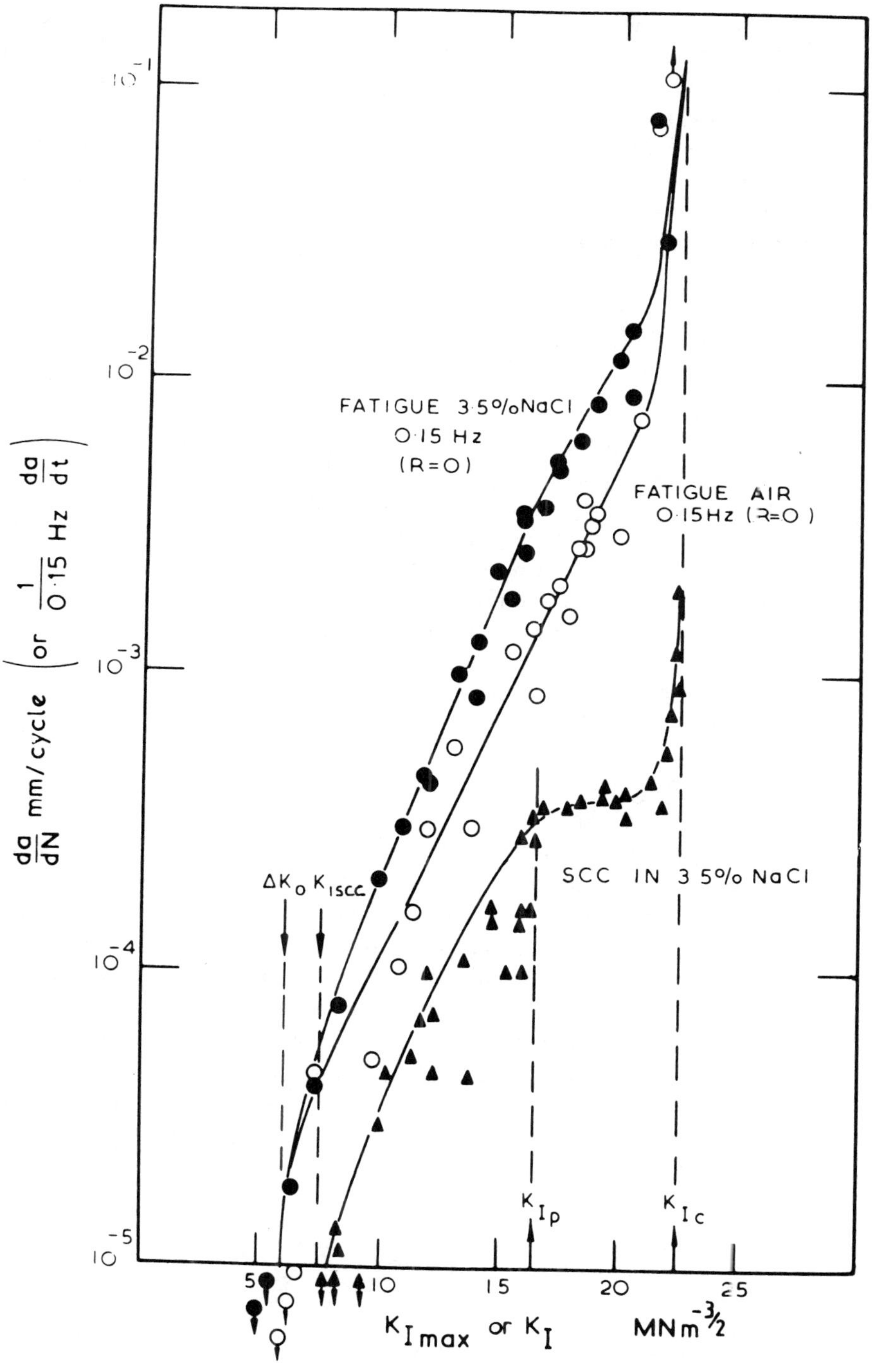

Fig. 7.49 – Effect of sodium chloride solution on fatigue crack growth in a RR58 aluminium alloy (reproduced from Rhodes and Radon (1979), with permission).

The fatigue results show faster crack growth rates in brine than in air at the same ΔK with (in this instance) similar values of the fatigue threshold $\Delta K_0 \approx 6$ MPa$\sqrt{\text{m}}$ in brine and air. Note that $\Delta K_0 < K_{ISCC}$ – in other materials often by a much wider margin than the ca. 6 to 7 MPa$\sqrt{\text{m}}$ difference of Fig. 7.49.

For glass and ceramics materials exhibiting delayed static fracture effects, it was shown in section 7.1 that crack growth rates under repeated loading (where σ and hence K varied periodically) could be predicted from velocity-dependent monotonic toughness data using

$$\mathrm{d}a/\mathrm{d}N = \int_0^T \dot{a}\ (K_c)\,\mathrm{d}t \qquad (7.42)$$

where T was the period of cycling (cf. equation (7.4) and so on). This is possible since the rate of growth of cracks is determined *solely* by velocity-dependent toughness which is an environmentally controlled process; in an inert atmosphere (under vacuum or in argon, say) there is no velocity-dependency and *no* fatigue effect at all. In metals which experience stress corrosion cracking (which is a type of delayed static fracture) it is possible to perform similar calculations using equation (7.42) for growth rates under repeated loading where $\dot{a}(K)$ is known for different environments, but in this case the predicted $\mathrm{d}a/\mathrm{d}N$ *underestimate* experimental fatigue values. Why is this if the mechanics of the processes are apparently similar? The answer lies in the behaviour of the different solids in *inert* environments. Whereas in the one case glasses and ceramics show *zero* $\mathrm{d}a/\mathrm{d}N$, in the other case metals inherently display characteristic *finite* $\mathrm{d}a/\mathrm{d}N$ under inert conditions, and this factor is inevitably absent in the integration of equation (7.42). In this context, it should be emphasised that air *cannot* be treated as a non-aggressive environment, even though it is often used as a basis of comparison for other results. Indeed many environments formerly considered non-corrosive are now realised actually to be hostile.

A great number of lengthy, tedious experiments is required to build up the data required to determine the constant and exponent in the Paris equation under many different environments. With the intention instead of predicting theoretically the $\mathrm{d}a/\mathrm{d}N$ vs. ΔK relationship, Wei and Landes proposed in 1969 a linear superposition model for fatigue crack growth rates in corrosive environments making use of the above observations regarding $\mathrm{d}a/\mathrm{d}N$ in inert and hostile atmospheres. They postulated that

$$(\mathrm{d}a/\mathrm{d}N)_{\text{total}} = (\mathrm{d}a/\mathrm{d}N)_{\text{inert}} + (\mathrm{d}a/\mathrm{d}N)_{\text{env}} \qquad (7.43)$$

where $(\mathrm{d}a/\mathrm{d}N)_{\text{inert}}$ has a characteristic value for the solid (independent of mean stress and frequency) and where $(\mathrm{d}a/\mathrm{d}N)_{\text{env}}$ is given by equation (7.42).

In the delayed fracture of glass $(\mathrm{d}a/\mathrm{d}N)_{\text{inert}} = 0$, hence leading to all the calculations in section 7.1. The success of equation (7.2) in predicting crack growth under cyclic loading for glasses has already been shown in Fig. 7.3 and will not be discussed further. While ceramics and glasses are capable of microplastic flow they do not seem to suffer from enhanced cyclic fatigue damage at room temperature and $(da/dN)_{inert}$ is insignificant when compared with $(da/dN)_{total}$. However, at higher temperatures, increased plastic flow is possible and as already remarked $(da/dN)_{inert}$ may play an increasing role (cf. the implication of Cotterell's R-curve model of fatigue crack growth in section 7.2.1 with extent of plasticity).

In the case of metals, $(da/dN)_{inert}$ is not zero. Fig. 7.50(a) shows the creep growth data for an H-11 tool steel. It seems that both humid argon and distilled water have similar effects on the creep crack growth rate. Fig. 7.50(b) shows the predicted (da/dN) vs. K_{max} curves in humid air for the same steel at varying cyclic frequencies. The experimental data obtained at 300 cycles per minute appear to agree reasonably well with theoretical predictions. These curves also predict the correct trend for the frequency effects. The same conclusions can be drawn for an 18Ni(250) maraging steel fatigue tested in a dehumidified hydrogen environment.

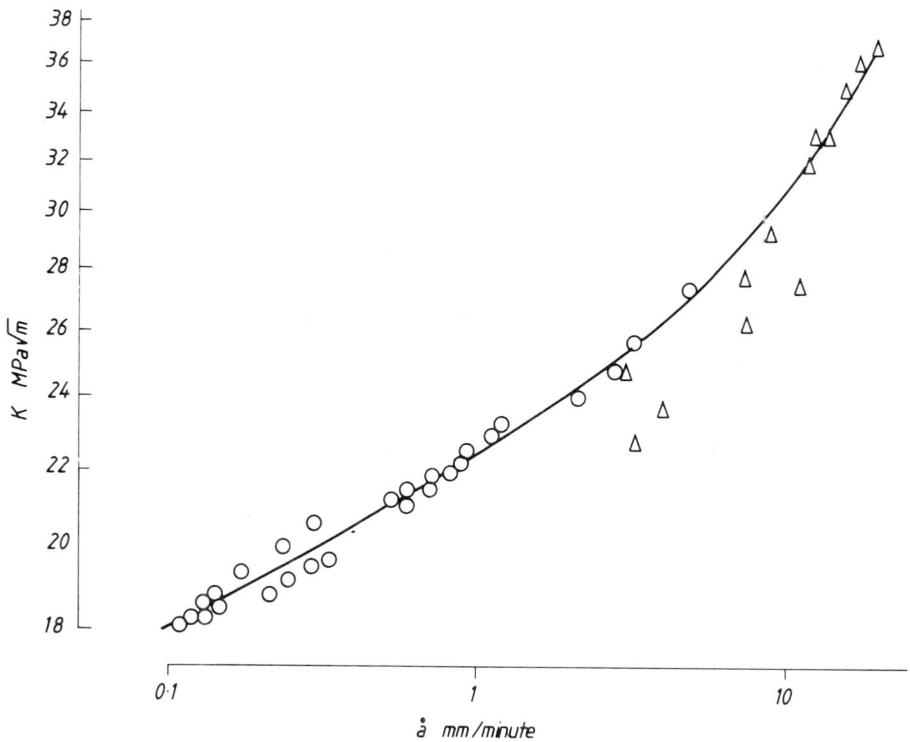

Fig. 7.50(a) – Sustained crack growth in H-11 steel in humid argon (○) and distilled water (△) (after Wei and Landes (1969).

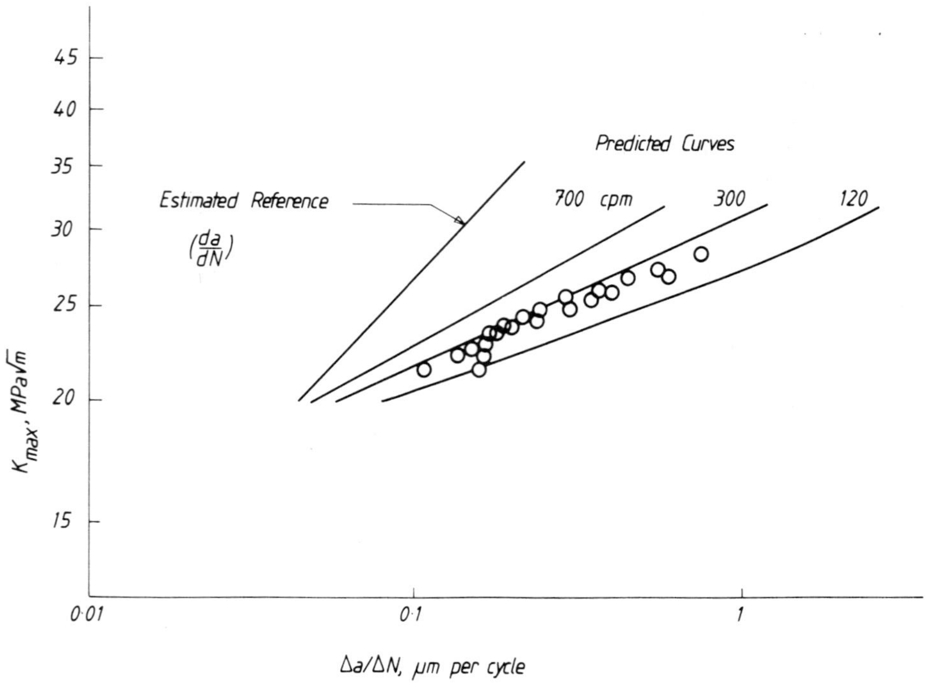

Fig. 7.50(b) – Predicted fatigue crack growth rate in humid air using equation (7.43) and data in (a). (Copyright ASTM, 1916 Race Street, Philadelphia, PA 19103. Reprinted with permission.)

Example 7.19 Show that the environmentally induced component of the total fatigue crack growth rate is proportional to the period of cycling.

Ans. Using $\dot{a} = DK_c^{n^*}$, equation (7.2) may be rewritten

$$(\mathrm{d}a/\mathrm{d}N)_{\mathrm{env}} = \int_0^T D\,[K_c(t)]^{n^*}\,\mathrm{d}t = DK_{c\,\max}^{n^*} \int_0^T \left(\frac{\sigma(t)}{\sigma_{\max}}\right)^{n^*} \mathrm{d}t$$

$$= DK_{c\max}^{n^*}\; T\left(\frac{1}{T}\int_0^T \left[\frac{\sigma(t)}{\sigma_{\max}}\right]^{n^*} \mathrm{d}t\right)$$

The quantity in the square brackets is a function of n^* and $r\,(= \sigma_{\min}/\sigma_{\max})$ and the shape of the load waveform and is simply a dimensionless number. For example

$$\frac{1}{T}\int_0^T \left(\frac{\sigma(t)}{\sigma_{\max}}\right)^{n^*} \mathrm{d}t = \tfrac{1}{2}$$

for a square waveform with $r = 0$; it is

$$\frac{1}{(n^* + 1)}$$

for a triangular waveform with $r = 0$; it is

$$\int_0^{2\pi} \sin^{n^*} x \mathrm{d}x$$

for a sinusoidal waveform with $r = -1$ etc. (see Problem 7.3). This, in all cases, $\mathrm{d}a/\mathrm{d}N \propto T$, and the faster the cycling frequency, the lower the $(\mathrm{d}a/\mathrm{d}N)_{\mathrm{env}}$. Note the difference between this problem and example 7.2. There, the average crack velocity is independent of T, but $\mathrm{d}a/\mathrm{d}N$ (as here) is proportional to T. Care should be taken to distinguish between $\mathrm{d}a/\mathrm{d}t$ and $\mathrm{d}a/\mathrm{d}N$ in this type of problem.

The prediction of fatigue crack growth rates for polymers in aggressive environments is more complicated than for glasses, ceramics and metals at low homologous temperatures. In polymers there is a very pronounced viscoelastic effect (rather like creep in metals at high homologous temperatures) and it is known that fatigue crack growth occurs at very low stress intensity factors for which no creep crack growth would have taken place under sustained loads. This means that the threshold stress intensity factor in creep is much higher than in fatigue (see Chapter 6). Fig. 7.51 shows the $\dot{a}$–(K_c) relationship† for PMMA at 21° in air (which may be regarded as a corrosion environment). Below a threshold given by $K_c = 0.80$ MPa$\sqrt{\mathrm{m}}$ creep crack growth is insignificant. Two equations are needed fully to describe the experimental data as shown in the inset of the figure. To obtain $(\mathrm{d}a/\mathrm{d}N)_{\mathrm{env}}$ a piecewise integration procedure is necessary for equation (7.42), viz.

$$\left(\frac{\mathrm{d}a}{\mathrm{d}N}\right)_{\mathrm{env}} = \int_0^{T_1} D_1 K_c(t)^{n_1^*} \mathrm{d}t + \int_{T_1}^{T_2} D_2 K_c(t)^{n_2^*} \mathrm{d}t + \ldots$$

$$+ \int_{T_{i-1}}^{T_i} D_i K_c(t)^{n_i^*} \mathrm{d}t + \ldots \qquad (7.44)$$

and

$$T = \sum_{i=1} T_i$$

Fig. 7.52 shows the computed creep component $(\mathrm{d}a/\mathrm{d}N)_{\mathrm{env}}$ for PMMA together with the experimental crack growth data obtained by Radon and Culver in 1975 under a mean stress intensity of 0.75 MPa$\sqrt{\mathrm{m}}$. As data for $(\mathrm{d}a/\mathrm{d}N)_{\mathrm{inert}}$ are unavailable, $(\mathrm{d}a/\mathrm{d}N)_{\mathrm{inert}}$ may be obtained by difference of

---

† The constants in the $\dot{a}$–(K_C) relation given in Example 6.4 are slightly different owing to slightly different experimental conditions and materials.

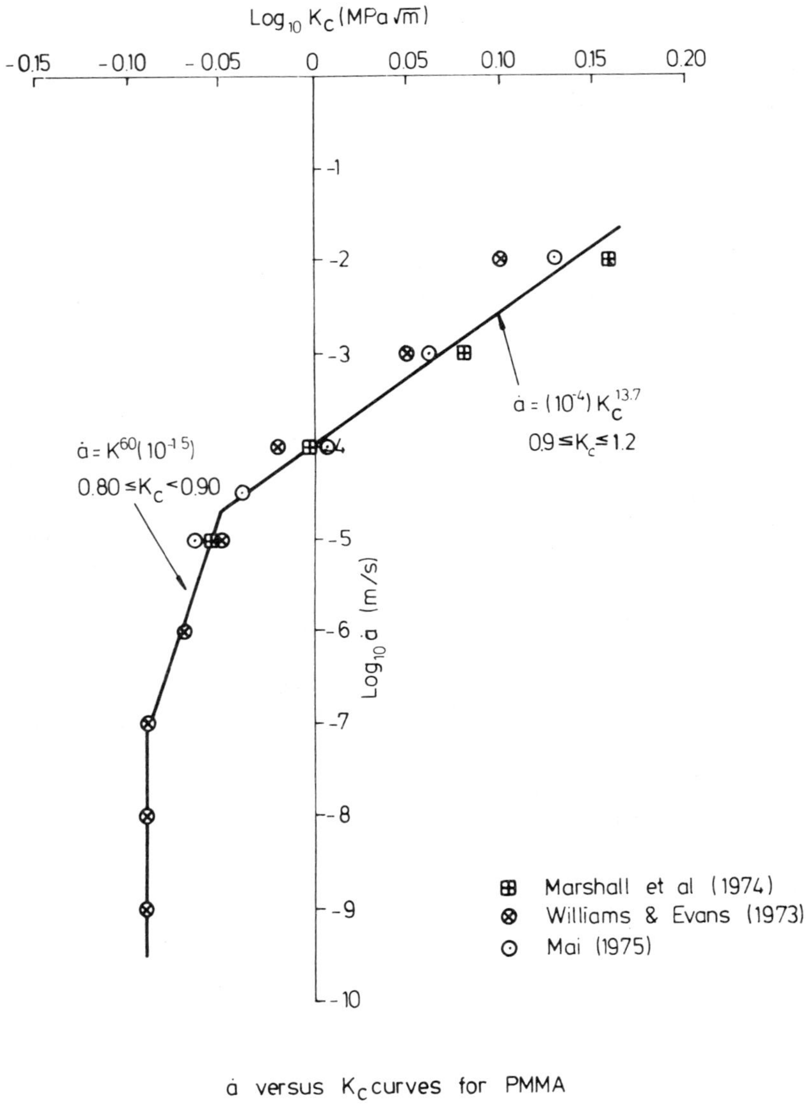

Fig. 7.51 – K_c–a relationship for PMMA; data from various authors.

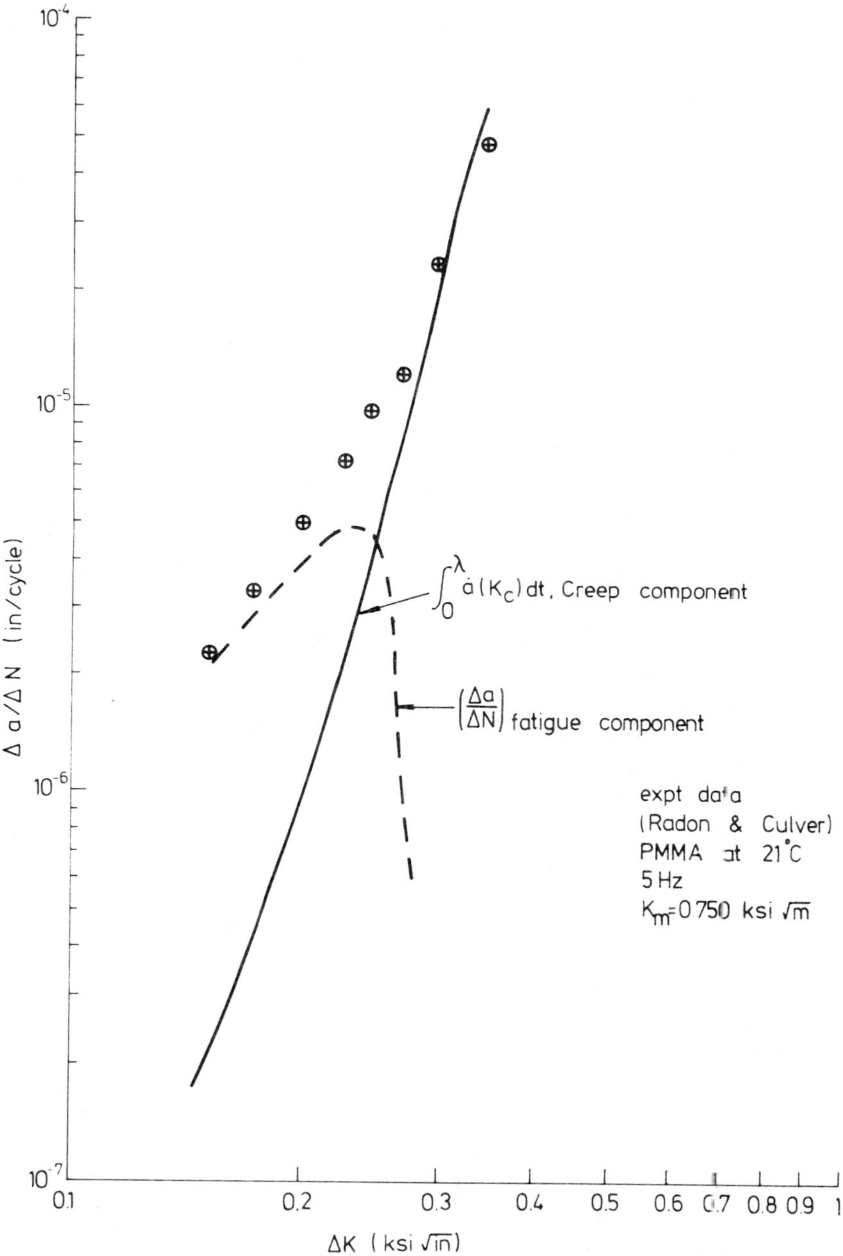

Fig. 7.52 – Comparison of creep and fatigue components of fatigue crack growth rate with experimental results.

$(da/dN)_{expt}$ and $(da/dN)_{env}$, on the assumption that the linear superposition model equation (7.43) applied (the dashed line in Fig. 7.52). It is clear that for low ΔK values the fatigue crack growth rate is dominated by the fatigue component $(da/dN)_{inert}$ and that for high ΔK values this is purely controlled by the creep crack growth component $(da/dN)_{env}$. In the mid-range ΔK values the two components are about equal. It seems that for polymers the linear superposition model only works at the high ΔK levels where $(da/dN)_{env}$ predominates. For low ΔK levels with K_{max} values less than the threshold K_o in creep crack growth, the model is inadequate since $(da/dN)_{inert}$ is not known and yet it is solely responsible for the total fatigue crack growth. This is further complicated by the fact that, unlike metals $(da/dN)_{env}$ is dependent on both frequency and mean stress. Even with metals, particularly low strength structural steels not normally susceptible to stress corrosion cracking, dramatic enhancement of da/dN can occur due to corrosion, which casts doubt on a linear superposition of fatigue and S.C.C. data.

It may be concluded that, in general, fatigue crack growth rates in polymers *cannot* be usefully determined by the linear superposition principle although there has been some success in the case of low density polyethylene fatigued in methanol as shown by El-Hakeem and Culver (1978). The fatigue component $(da/dN)_{inert}$ is not usually available and even if it is the strain rate sensitivity of the K_c threshold in creep must be accounted for in the fatigue crack growth data. Fatigue in polymers including mean stress and frequency effects may be better described in a model proposed by Williams (1979). This is based on a critical opening displacement being achieved at the crack tip. Both environmental and temperature effects can also be accommodated by this model (see Mai and Williams, 1979). We note also that the manner in which C and n in the Paris (or related) relations change with environment can be correlated with the solvent solubility parameter, used in section 6.5.3 for the variation of quasi-static K_c values with environment. Further information will be found in section 10.4, Fig. 10.17, where the results are applied to the adhesive/fatigue wear of polymers in different environments.

We conclude this section by noting remarks of Tomkins (1981) where he pointed out that, from an engineering viewpoint (particularly for metals), corrosion fatigue presents a problem in that it erodes traditional methods of fatigue data collection because it is most damaging at low cyclic frequencies. For dissolution-enhanced crack growth, the crack tip opening represents an upper limit below K_{ISCC}, but for hydrogen-assisted growth in metals it is difficult to define the limit to growth. Also for hydrogen, Wei (1979) has shown that long dwell periods between cycles can be very damaging on account of the incubation time that they provide for cracking ahead of the tip. Finally, however, because crack tip profiles can be easily modified by corrosion, below K_{ISCC} corrosion fatigue crack growth is often difficult to maintain under load variations.

7.11 SELECTION OF MATERIALS FOR DESIGN AGAINST FATIGUE

Three widely used philosophies for design are the 'safe-life', 'damage-tolerance' and 'fail-safe' methods, details of which were ably summarised by the subcritical crack growth working party of the Royal Society (1984).

The 'safe-life' philosophy is the old-established design method using S–N curves in which a statistical mean-life is determined for the expected service load spectrum. To account for scatter in experimental data and also for load interaction effects, factors of safety of 3 to 5 are employed. Owing to the uncertainties which such factors attempt to cover, and also to introduce some measure of the developing progress towards failure, the other two philosophies have come into being. The 'damage tolerant' method takes into account initial imperfections, crack growth rates and the conditions of final fracture, using equations (7.11) to (7.14) to demonstrate that cracks should not grow to their critical length within the design life, (or at least should be capable of ready detection within the design life). The 'fail-safe' method relies on redundancies in a structure such that even if failure of one component should occur, it does not lead to failure of the whole structure; this is particularly important in giving some tolerance to flaws and/or accidental damage in regions of structures difficult of access for NDI. The 'leak-before-burst' method of pressure vessel design (section 2.6) is a slightly different aspect of the fail-safe approach. The three methodologies are seldom used in isolation nowadays, and inherent in some design codes are aspects of more than one approach.

As highlighted in the Royal Society report, the exponential nature of fatigue crack growth rates means that NDI strategy has to be planned with great care. That is, there may be a danger that a growing crack, undetected at one inspection, will grow to failure before the next inspection.

Modern fatigue crack growth data for flawed components may be combined in a three-dimensional plot of applied stress, number of cycles and initial flaw size, as shown schematically in Fig. 7.53 taken from Elber and Davidson (1973). The surface represents the locus of points for which a crack will propagate from an initial length a_i to failure in N cycles at stress σ.

Under LEFM control, the number of cycles to failure N_f is given by equation (7.11) or its simplified versions (7.12) and so on, and the equation describes the operating failure surface in Fig. 7.53 for regions 2 and 4 (in region 4 the first term approximation of equation (7.11) is adequate i.e. equation (7.12)). Regions 1 and 3 concern crack growth from very small initial flaw sizes and the Paris equation will not apply (cf. section 7.2.3). In Region 1, which concerns operating conditions requiring short lives and short initial flaw sizes, yield strength and ultimate strength are the dominant material properties (but cf. section 7.4). Region 2 represents design operating conditions requiring only short lives and, consequently, relatively large initial flaw sizes. In this region, the quasi-static fracture toughness is the dominant material property. Region 3 represents design conditions requiring long lives and short initial flaw sizes. In this region, the traditional fatigue properties (S–N data) are the dominant properties. Region 4, the largest region, represents conditions requiring long lives and long initial flaw sizes; (we note that most flaws which may be detected by non-destructive inspection are 'long' in this sense). In this region the elastic crack-growth properties are dominant.

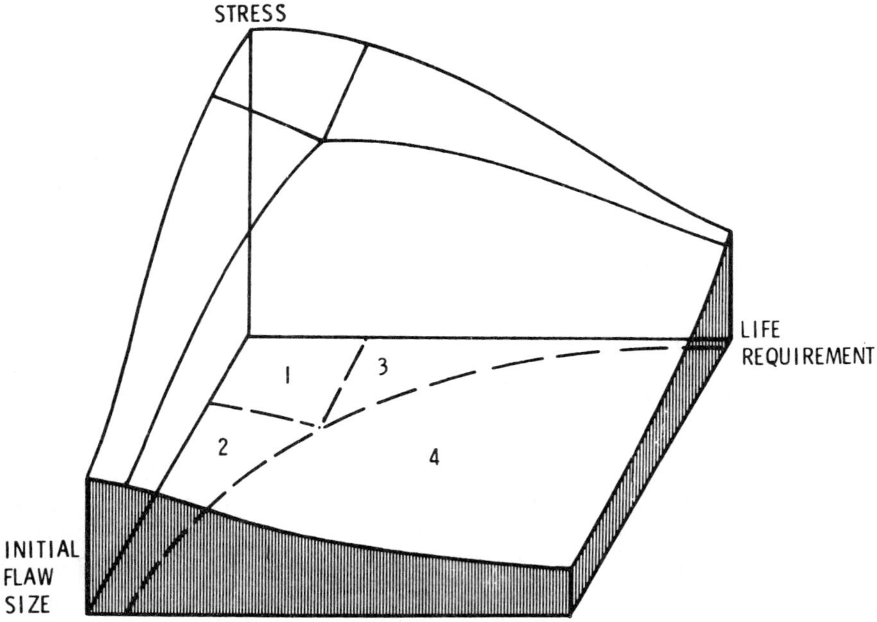

Fig. 7.53 – Regions defining dominant material properties (reproduced from Elber and Davidson (1973), with permission).

Materials may be compared under different operating conditions using the operating surface concept of Fig. 7.53. The following example is taken from Elber and Davidson (1973). It was required to design a minimum weight tension member subjected to repeated loading and containing a given initial flaw, from either D6AC steel, Ti-6A1-4V annealed titanium alloy or 2024-T3 aluminium. In Fig. 7.54(a) the operating surfaces for the three materials are plotted in such a way that only a part of the surface for any one material is visible. The part of the surface shown covers the region in which the material has a higher usable stress-to-density ratio than the other materials.

On a stress-to-density basis, each of the materials has one region in which it is superior to the others. In the small-flaw region, D6AC is superior because of its high strength and its high fatigue properties. The titanium alloy has a high strength and a relatively high toughness, but has a relatively low crack-growth resistance; its region of superiority is limited to medium crack lengths and relatively short lives. For long initial crack lengths and long life requirements, the 2024-T3 aluminium alloy is superior to the other two materials because of its higher crack-growth resistance.

Fig. 7.54(b) shows the stress contour plot of the combined operating surface, i.e. the projection of the operating surface on the a_0–N plane. The plot permits an assessment to be made to determine the sensitivity of stress levels or material choice to changes in the operating parameters of life requirement and initial flaw size. As an example, assume that point A in Fig. 7.54(b)

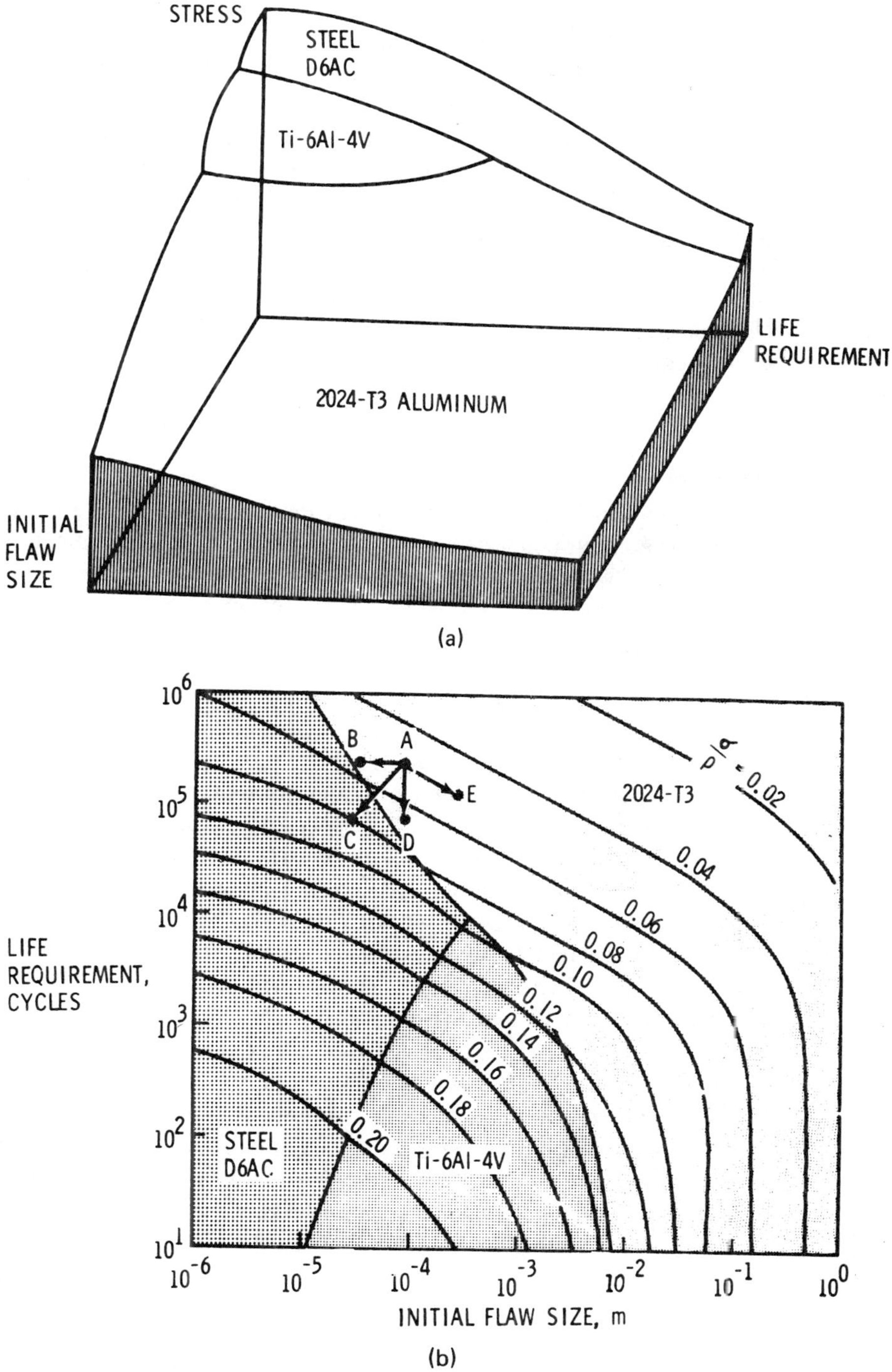

Fig. 7.54 – (a) Comparative operating surface for three aerospace materials. (b) Compound material characterisation plot for material selection. (both reproduced from Elber and Davidson (1973), with permission).

represents a set of particular design conditions for the initial flaw size and the life requirement; (the initial flaw size is determined by the sensitivity of the inspection method). These conditions then define 2024-T3 as having the least stress-to-density ratio for the (life and initial-flaw-size) coordinates of point A. To improve the available stress-to-density ratio, the design conditions must be changed by going toward point B (improve flaw inspection sensitivity), going toward point C (improve flaw inspection sensitivity and sacrifice life requirement), or going toward point D (sacrifice life requirement). For the operating conditions at point C, an additional increase in stress-to-density ratio can be gained by changing from the aluminium to the steel. Another possible trade-off – without a gain in stress-to-density ratio – is to go toward point E, in which case life requirement is sacrificed for a less severe inspection requirement.

PROBLEMS

7.1 Gurney and Pearson in 1948 showed that soda glass rods, subjected in air to given four-point bending actions, fractured in approximately equal times, whether or not the rods were rotated relative to the plane of the bending moment, and whether or not the speed of cycling was altered. This suggests that in delayed fracture ('static fatigue') only elapsed time is important. Yet if, as shown in section 7.1, crack growth velocities in cyclic loading are smaller than in static loading at the same maximum load, why did Gurney and Pearson apparently find similar times to fracture in their tests?

7.2 For a certain electrical grade of porcelain the slow crack growth equation is given by: $\dot{a}$ (m/s) $= 10^{-49}\ K_c^{30}$ (MPa$\sqrt{\text{m}}$). Experiments under repeated loading with a triangular waveform at 1 Hz gave the following cyclic crack growth rates at various ΔK: 10^{-8} m/cycle when $\Delta K = 882$ kPa$\sqrt{\text{m}}$; 10^{-6} m/cycle for 893 kPa$\sqrt{\text{m}}$ and 10^{-4} m/cycle for 912 kPa$\sqrt{\text{m}}$. Show that these results are well-predicted by equation (7.4).

7.3 Show that the environmentally induced component of total fatigue crack growth rate is: (i) $da/dN = DK_{cmax}^{n^*}\ T/2$ for a square waveform, (ii) $da/dN = (DK_{cmax}^{n^*}\ T/2\pi)\int_0^{2\pi}\ [(1 + \sin\theta)/2]^{n^*} d\theta$ for a sinusoidal waveform, both with r $(= \sigma_{min}/\sigma_{max}) = 0$ and with $\dot{a} = DK_c^{n^*}$ [Hint: refer to Examples 7.1 and 7.19].

7.4 The total energy absorbed in fatigue fracture is less than that in monotonic fracture over the same crack propagation area. Failure by fatigue may therefore be argued to be a 'more efficient' process than fracture by monotonic loading. Discuss this statement in terms of the size of the crack tip plastic zone, residual stresses developed in cyclic loading, the smoothness of the macroscopic fracture surfaces in fatigue compared with surface produced in monotonic fracture and so on.

7.5 Sketch the *Xu* and *RA* curves incorporating a threshold resistance value for fatigue crack propagation between loads X_{max} and X_{min} (i.e. curves similar to Figs. 7.11 and 7.12, but with $X_{min} \neq 0$).

7.6 Discuss how fatigue propagation under a constant cyclic *displacement* range may *not* be arrested in certain testpiece geometries.

7.7 Discuss the probable origin of values of n outside the commonly accepted range of $n \leqslant 2 \leqslant 4$ in the Paris equation.

7.8 Is it justified to apply expressions for crack tip stress derived from *static* analyses to fatigue problems?

7.9 Determine the equation for a stress-number-of-cycles-to-failure curve assuming an initial value for K using the Paris equation.

7.10 Show that the number of cycles to failure N_f in fatigue for a material following $da/dN = C(\Delta K)^2$ is $N_f = [a_i/C\Delta K_i^2] \ln(a_f/a_i)$ where a_i and a_f are the initial and final crack lengths and ΔK_i is the initial stress intensity factor range.

7.11 Repeat the calculations of Example 7.10 for $r = +0.5$ and $r = -0.5$.

7.12 A set of results of fatigue crack growth for 2024–T3 aluminium alloy centre cracked panels at various stress levels is shown below. Predict N_f and compare with the experimental data ($K_c = 110$ MPa$\sqrt{\text{m}}$, $da/dN = 1.2 \times 10^{-11} (\Delta K)^{3.64}$ mm/cycle with ΔK in MPa$\sqrt{\text{m}}$). N_f may be cycles to fracture or cycles to some predetermined crack length.

a_0 (mm)	a_f (mm)	$\Delta\sigma$ (MPa)	N_f (cycles)
2.54	15.24	205	4,850
		136	23,500
		103	58,500
		66	228,000
		53	910,000
5.08	17.78	205	2,200
5.08	22.86	136	16,300
5.08	30.48	103	47,500
5.08	40.64	66	146,000
5.08	35.56	53	420,000

(W. Elber and J. R. Davidson, NASA TN D-7221 1973).

7.13 It is sometimes possible to fit the same data to apparently contradictory 'laws' of fatigue crack growth rate. Discuss.

7.14 Show that if $n = 4$ and $Y = 1$ in equation (7.11) half the Stage II and III fatigue life of a structure is taken up in doubling the starter crack length, 75% in quadrupling the length, and so on. Modify these predictions for different bodies in which Y changes as the crack grows.

7.15 Reanalyse the da/dN vs. ΔK data in Fig. 7.9(a) using the following method: (a) fit a third order polynomial to seven data points; (b) take the differential at point 4; (c) advance by one data point, i.e. points 2 to 8, (d) obtain a new third order polynomial to fit these points; (e) obtain da/dN at point 5; and so on.

7.16 Show that if a material displays a threshold ΔK_0 in fatigue, and if there are pre-existing flaws present, (i) the *S–N* curve will display an endurance limit and (ii) the slope of the *S–N* curve on log coordinates is $-(1/n)$ where n is the index of the Paris equation.

Describe how the threshold ΔK_0 may be determined from an *S–N* curve for a *cracked* specimen with lives (N) plotted against the initial value of the applied ΔK.

7.17 In inspection procedures for the detection of cracks, it can often happen that a crack is not formed until something like three-quarters of the total fatigue life of a structure has elaspsed. Consider the question of rate of crack growth versus inspection requirements and discuss how plans for inspection should be drawn up.

7.18 Comparison of striation markings in the fatigue of alloys containing brittle intermetallic particles reveals similar growth rates in the metallic matrix at given ΔK. Yet overall growth rates (including fractures through the inclusions) can be anomalously fast, the effect being enhanced the greater the volume fraction of inclusions. Explain.

7.19 In many steels, show that for design purposes cracks less than about 100 μm in size can be analysed in terms of fatigue data from plain specimens and that fracture mechanics are required only for longer cracks.

7.20 F. Gillemot (son of L. Gillemot) gave in 1973/4 for fatigue crack growth rate $\mathrm{d}a/\mathrm{d}N$

$$\frac{\mathrm{d}a}{\mathrm{d}N} \propto \Delta K^2 \; \Delta\epsilon/\sigma_y \; W_c$$

where $\Delta\epsilon$ is the strain-range within a cycle, W_c is the work/unit volume to fracture obtained from a plain tension testpiece, ΔK is the stress intensity range and σ_y the yield stress. Show that in materials with low K_{IC}/σ_y ratios, where $\epsilon_f \cong \delta_C/L_0$ (where L_0 is a characteristic size of the process zone), this equation reduces to $\mathrm{d}a/\mathrm{d}N \propto (K/\sigma_y)^2$ and thus is a function of crack tip plastic zone size.

7.21 In fatigue testing of thin sheets at stress levels $\sigma_m \pm \sigma_a$, most available data relate to $\sigma_m > \sigma_a$. What difficulty in experimentation causes this state of affairs?

7.22 For the pattern of residual stresses shown in Fig. 7.42(a) (with $h = 0.5$ mm, $\sigma_C = 314$ MPa, $\sigma_T = 240$ MPa), show that the residual stress intensity is negative for $a < 1.3$ mm.

7.23 Compare growth rates in edge-crack testpieces of D6AC steel both in the presence and absence of residual stresses using data in Problem 7.22 and $K_{IC} = 48$ MPa$\sqrt{\text{m}}$.

7.24 Explain why near-surface residual stresses can give a solid with a shallow crack an apparent fracture toughness which is greater than the toughness of the stress-free material.

7.25 Why, in fatigue, is the effect of loading frequency likely to be of more importance in polymers than metals?

7.26 Explain why, in the laboratory, strain control is preferred in low cycle fatigue tests rather than stress control.

7.27 A feature of fatigue crack propagation in thin sheets is that crack growth initially in mode I transforms to growth on 45° planes through the thickness (i.e. the crack twists). Discuss possible reasons for this transition. Might it play a role in crack closure problems?

7.28 Explain why the use of high strength steels in welded structures often offers little advantage in terms of fatigue performance. [Hint: C and n in $da/dN = C(\Delta K)^n$ do not vary substantially between steels; in welded structures virtually the whole life is spent in Stage II propagation.]

7.29 The fatigue life of components experiencing complex stress histories is often assessed using the Palmgren–Miner law viz. $\sum_i N_i/(N_f)_i = 1$ where N_i is the number of cycles actually run at stress σ_i and $(N_f)_i$ is the number of cycles to fracture at stress σ_i. Show that this empirical relation agrees with crack rate growth laws of the Paris type. Explain why the relation is often satisfactory for the prediction of allowable stress for given life, but less satisfactory for the prediction of life for given stress.

7.30 CNC machine tools, using carbide cutters and greater feeds, speeds and depths of cut than traditional machine tools, tend to produce finished surfaces with smaller levels of residual compressive stress. Discuss the consequences for design against fatigue.

REFERENCES

Arad, S., Radon, J. C. and Culver, L. E. (1972) *Engn. Fract. Mech.* **4**, 511.

Ashby, M. F. and Tomkins, B. (1979) *Proc. ICM-3* **1**, 47.

Banasiak, D. H., Grandt, A. F. and Montulli, L. J. (1977) *J. Appl. Poly. Sci.* **11**, 1297.

Barsom, J. M. (1971a) ASTM STP No. 486, 1.

Barsom, J. M. (1971b) *Trans. ASME, J. Engr. Ind.* **93**, 1190.

Barsom, J. M. (1973) ASTM STP No. 536, 147.

Barsom, J. M. (1976) ASTM STP No. 595, 217.

Beck, E. J. (1973) Fatigue Growth Behaviour in Stiffened and Unstiffened Panels Loaded in Biaxial Tension. NASA 9-12439 MCR-73-61 Martin Marietta Corp.

Brown, M. W. and Miller, K. J. (1973) *Proc. Inst. Mech. Engr.* **187**, 745.

Bucci, R. (1979) *Engr. Fract. Mech.* **12**, 407.

Burdekin, F. M. (1981) *Proc. Inst. Mech. Engr.* **195**, 73.

Christensen, R. M. (1959) *Metal Fatigue*, McGraw-Hill, New York, p. 376.

Clark, W. G. Jr., and Hudak, S. J. Jr. (1975) *J. of Testing and Evaluation* **3**, 454.

Coffin, L. F. (1954) *Trans. ASME, J. Basic Engr.* **76**, 931.

Constable, I., Culver, L. E. and Williams, J. G. (1970) *Int. J. Fract.* **6**, 279.

Cotterell, B. (1965) *Trans. ASME, J. Basic Engr.* **87**, 230.

Dowling, N. E. (1976) Westinghouse Report 76-IE7-PALFA-P2.

Duquette, D. J. (1972) *J. Corr. Fatigue* NACE-2.
Eastabrook, J. N. (1984a) *Int. J. Fract.* **24,** R43.
Eastabrook, J. N. (1984b) Proc. 5th European Conf. on Fracture, Lisbon, p. 1037.
Eftis, J. and Jones, D. L. (1982) *Int. J. Fracture* **20,** 267.
Elber, W. (1971) ASTM STP No. 486, 230.
Elber, W. (1974) ASTM STP No. 559, 45.
Elber, W. and Davidson, J. R. (1973) NASA TND-7221.
El-Hakeem, H. A. and Culver, L. E. (1978) *J. Appl. Poly. Sci.* **22,** 2691.
Evans, A. G. and Fuller, E. R. (1974) *Metallurgical Transactions, A* **5,** 27.
Foreman, R. G., Kearney, V. E. and Engle, R. M. (1967) *Trans. ASME, J. Basic Engr.,* **89,** 459.
Forsyth, P. J. E. (1961) *Proc. Crack Propagation Symposium, Cranfield, UK,* p. 76.
Frost, N. E. (1959) *Proc. Inst. Mech. Engr.* **173,** 811.
Frost, N. E., Marsh, K. J. and Pook, L. P. (1974) *Metal Fatigue,* Oxford University Press, London.
Gillemot, L. and Sinay, G. (1958) *Acta Technica Acad. Sci., Hung.* **XXII,** 149.
Gillemot, F. (1973/4) *Int. Conf. on Creep and Fatigue in Elevated Temperature Applications,* Philadelphia/Sheffield.
Goldman, N. L. and Hutchinson, J. W. (1975) *Int. J. Solids Structures* **11,** 575.
Gurney, C. and Pearson, S. (1948) *Proc. Roy. Soc.* A192, 537.
Head, A. K. (1953) *Phil. Mag.* **44,** 925.
Heald, P. T., Lindley, T. C. and Richards, C. E. (1972) *Mater. Sci. Engr.* **10,** 235.
Hertzberg, R. W. and Manson, J. A. (1980) *Fatigue of Engineering Plastics,* Academic Press, New York.
Heywood, R. B. (1962) *Designing against Fatigue,* Chapman & Hall, London.
Hoshide, T., Tanaka, K. and Yamada, A. (1981) *Fatigue Engr. Mater. & Struct.* **4,** 355.
Hudson, C. M. and Seward, S. K. (1978) *Int. J. Fract* **14,** R151.
Hudson, C. M. and Seward, S. K. (1982) *Int. J. Fract* **20,** R57.
Ibrahim, M. F. E. and Miller, K. J. (1980) *Fatigue Engr. Mater. & Struct.,* **2,** 351.
Irving, P. E. and McCartney, L. N. (1977) *Metal Science,* **11,** 351.
Kfouri, A. P. and Miller, K. J. (1976) *Proc. Inst. Mech. Engr.* **190,** 571.
Kfouri, A. P. and Miller, K. J. (1981) in *Three-dimensional Constitutive Relations and Ductile Fracture,* Nemat-Nasser (ed.), North-Holland, Amsterdam, p. 81.
Kies, J. A. and Clark, A. B. J. (1969) *Fracture* Chapman & Hall, London, p. 483.
Kim, K. and Mubeen, A. (1981) ASTM STP No. 745. ASTM, Philadelphia.
Kitagawa, K. and Takahashi, S. (1976) *Proc. ICM-2, Boston, Mass., USA,* p. 627.
Klippstein, K. H. and Schilling, C. G. (1976) ASTM STP No. 595, 203.
Knott, J. F. (1973) *Fundamentals of Fracture Mechanics,* Butterworth, London.
Lake, G. J. and Lindley, P. B. (1964) *J. Appl. Polym. Sci.,* **8,** 707.
Liebowitz, H., Lee, J. D. and Eftis, J. (1978) *Engr. Fract. Mech.* **10,** 315.
Lindley, T. C., Richards, C. E. and Ritchie, R. O. (1975) *Mechanics and Physics of Fracture (Cambridge),* Institute of Physics/Metal Society.

Little, R. E. and Jebe, E. J. (1975) *Statistical Design of Fatigue Experiments,* Applied Science Publishers, London.

Liu, H. W. (1963) *Trans. ASME, J. Basic Engr.,* **85**, 116.

Lloyd, G. J. and Wareing, J. (1979) *J. Engr. Mater. and Tech.* **101**, 275.

Mai, Y. W. (1978) *Int. J. Mech. Sci.* **20**, 293.

Mai, Y. W. (1979) *Int. J. Fracture,* **15**, R103.

Mai, Y. W. and Gurney, C. (1975) *Phys. and Chem. of Glasses,* **16**, 70.

Mai, Y. W. and Williams, J. G. (1979) *J. Mater. Sci.* **14**, 1933.

Manson, S. S. (1965) *Proc. Int. Conf. Fract., Sendai, Japan,* p. 1387.

McClintock, F. A. (1967) ASTM STP No. 415, 170.

McClintock, F. A. and Ritchie, R. O. (1981) in *Mechanics of Fatigue,* T. Mura (ed.), AMD vol. 47, ASME, New York.

McDiarmid, D. L. (1972) Ph.D. Thesis, City University, London.

McEvily, A. J. (1970) in *Proc. USAF Conf. Fatigue & Fracture of Aircraft Structures and Materials,* AFFDLTR-70-1144, p. 451.

Miller, K. J. (1977) *Metal Science* **11**, 432.

Miller, K. J. (1985) in *Fundamentals of Deformation and Fracture* (Eshelby Memorial Symposium). Cambridge: Cambridge University Press, p. 477.

Miller, K. J. and Kfouri, A. P. (1974) *Int. J. Fract.* **10**, 393.

Miner, M. A. (1945) *J. Appl. Mech.* **12**, A159.

Mostovoy, S. and Ripling, E. J. (1975) *Poly. Sci. Technol.* **9B**, 513.

Mukherjee, J. (1972) *Int. J. Fract.* **8**, 449.

Mukherjee, B. and Burns, D. J. (1971) *Exp. Mech.* **11**, 433.

Mukherjee, B. and Burns, D. J. (1975) Symposium on Statistical Aspects of Fatigue Testing, Warwick University.

National Bureau of Standards (1964) *Handbook of Mathematical Functions: Applied Mathematics,* Series No. 55, US Govt. Printing Office, Washington, DC.

Nayeb-Hashemi, H., McClintock, F. A. and Ritchie, R. O. (1983) *Int. J. Fract.,* **23**, 163.

Palmgren, A. Z. (1924) *Z. Ver. dt. Ing.* **68**, 339.

Paris, P. C. (1963) *Trans. ASME, J. Basic Engr.,* **85**, 528.

Paris, P. C., Gomez, M. P. and Anderson, W. E. (1961) *The Trend in Engineering,* University of Washington, Seattle, **13**, 9.

Pascoe, K. J. and de Villiers, J. W. R. (1967) *J. Strain Analysis* **2**, 117.

Pitoniak, F. J., Grandt, A. F., Montulli, L. J. and Pachman, P. F. (1974) *Engr. Fract. Mech.* **6**, 663.

Plumbridge, W. J. (1972) *J. Mater. Sci.* **7**, 939.

Pook, L. P. (1974) *Welding Research Int.* **4**, 1.

Pook, L. P. (1974) *Trans. N-E Coast Inst. Engrs. & Shipbuilders* **90**, 77.

Pook, L. P. (1975) *J. Strain Analysis,* **10**, 242.

Probst, E. P. and Hillberry, B. M. (1973) AIAA Paper No. 73-325 presented at AIAA Dynamic Specialists Conference, Williamsburg, Virginia, 19–20 March.

Puttick, K. E. and Atkins, A. G. (1983) *Int. J. Fract.* **23**, R51.

Radon, J. C. and Culver, L. E. (1975) *Polymer Engr. & Sci.* **15**, 507.

Rice, J. R. (1967) ASTM STP No. 415, 247.
Ritchie, R. O. (1977) *Met. Trans.* **8A,** 1131.
Rhodes, D. and Radon, J. C. (1979) *Fatigue of Engr. Mater. & Struct.* **1,** 383.
Royal Society (1984) Report of Working Group on Subcritical Crack Growth (Royal Society British National Committee on Theoretical and Applied Mechanics).
Schigve, J. (1974) *Engr. Fract. Mech.* **6,** 245.
Schmidt, R. A. and Paris, P. C. (1973) ASTM STP No. 536, 79.
Shahinian, P. and Sadananda, K. (1981) *Metal Sci.* **15**(10), 425.
Smith, E. W. and Pascoe, K. J. (1977) *Composites,* **8,** 237.
Smith, E. W. and Pascoe, K. J. (1982) To be published in ASTM STP.
Smith, E. W. and Pascoe, K. J. (1983) *Fatigue of Engr. Mater. and Struct.,* **6,** 201.
Smith, R. A. (1973) *Int. J. Fract.* **9,** 352.
Smith, R. A. (1977) *Int. J. Fract.* **13,** 717.
Smith, R. A. (1980) Fundamentals of Tribology, p. 605, Cambridge, Mass: MIT Press.
Smith, R. A. and Miller, K. J. (1977) *Int. J. Mech. Sci.* **19,** 11.
Smith, R. A. and Miller, K. J. (1978) *Int. J. Mech. Sci.* **20,** 201.
Solomon, H. O. and Coffin, L. F. (1972) ASTM STP No. 520, 112.
Sutton, S. A. (1974) *Engr. Fract. Mech.,* **6,** 587.
Tada, H., Paris, P. C. and Irwin, G. R. (1973) *The Stress Analysis of Cracks Handbook,* Del. Research Corp., Pennsylvania.
Tanaka, K. (1974) *Engr. Fract. Mech.* **6,** 361.
Tanaka, K. (1974) *Engr. Fract. Mech.* **6,** 493.
Tanaka, K., Hoshide, R., Yamada, A. and Taira, S. (1979) *Fatigue Engr. Mater. & Struct.* **2,** 181.
Tavernelli, J. F. and Coffin, L. F. (1962) *Trans. ASME, J. Basic Engr.* **84**D 533.
Taylor, D. and Knott, J. F. (1981) *Fatigue Engr. Mater. & Struct.* **4,** 147.
Taylor, D. and Knott, J. F. (1981) Euromech. Colloquium on Short Cracks, Sheffield University.
Tomkins, B. (1968) *Phil. Mag.* **18,** 1041.
Tomkins, B. (1973) *Proc. Conf. on Mechanics and Mechanisms of Crack Growth,* Churchill College, Cambridge.
Tomkins, B. (1975) *Trans. ASME,* (series H), 289.
Tomkins, B. (1980) *Metal Sci.* **14,** 408.
Tomkins, B. (1980) UKAEA Northern Division Report, ND-R-415(S).
Tomkins, B. (1981) *Phil. Trans. Roy. Soc.* A299, 31.
Tomkins, B. and Wareing, J. (1977) *J. Metal Sci.* **11,** 414.
Tomkins, B., Wareing, J. and Sumner, G. (1979) Proceedings of International Symposium on Low Cycle Fatigue Strength and Elastoplastic Behaviour of Materials, Stuttgart, K. T. Rie and E. Haibach (eds.), p. 495.
USAF (1972) *Damage Tolerant Design Handbook,* Battelle Laboratories, Columbus, Ohio.
US DoD (1977) *Aerospace Structural Metals Handbook,* Battelle Laboratories, Columbus, Ohio.

von Euw, E. F. J., Hertzberg, R. W. and Roberts, R. (1972) ASTM STP No. 513,230.

Wareing, J., Tomkins, B. and Sumner, G. (1973) *Proc. ICF-3,* p. 17.

Wei, R. P. (1979) ASTM STP No. 675,816.

Wei, R. P. and Landes, J. D. (1969) *Mater. Research and Standards,* **9**, 25.

Wheeler, O. E. (1972) *Trans. ASME, J. Basic Engr.* **94**, 181.

Williams, J. G. (1977) *J. Mater. Sci.* **15**, 2525.

Yap, O. F. (1982) Ph.D. Thesis, University of Sydney, Australia.

Zacharia, K. P. (1974) Ph.D. Dissertation, University of Cambridge.

8

Energetic stability

'You don't know much', said the Duchess; 'and that's a fact.'

8.1 INTRODUCTION

In this chapter we consider the conditions under which crack propagation is not catastrophic, where cracks may be stopped and started in a controlled fashion during elastic fracture, plastic fracture and fatigue fracture. This is the field of energetic stability of crack propagation; the word stability is also used in connexion with crack paths and whether a crack deviates from its initial orientation (*directional stability*). That topic has been discussed in section 3.5. Linked to the study of energetic stability is the topic of *crack arrest* where a crack which is initially unstable subsequently stops propagating.

Cracks may run slowly in a controllable fashion or fast and catastrophically. Whether cracking is well-behaved or, on the other hand, 'goes bang', depends upon both the geometry of the cracked structure, the method of loading and the type of material. Thus the same material may display both slow and fast cracks, depending upon the shape and size of the testpieces; and the same shape and size testpiece, under the same loading conditions, may involve slow or fast cracking depending upon the material from which it is made. A glass rod or sheet, having its natural Griffith damage cracks, breaks catastrophically with a bang but a window pane with an adequately long starter crack (in the form of a DCB specimen) will crack progressively in a controlled manner. Laboratory-size testpieces of low carbon steel yield before cracking and Charpy specimens sometimes do not break in two, yet large ships, pipelines and so on made from the same material experience fast brittle fracture. A lath of wet wood splits slowly but a cast epoxy resin of a similar size would crack catastrophically.

We have to be very careful in the use of the words *slow, fast, stable,* and *unstable* in connexion with fracture. A variety of different types of Xu plot may be realised experimentally, as shown in Fig. 8.1, depending on the testpiece geometry, the material, and the characteristics of the testing machine. The Xu trace in Fig. 8.1(a) is found with single- or double-edge-notch specimens (numbers 7 and 8 in Appendix 1), or centre-cracked panels (number 6); that in Fig. 8.1(b) is found with DCB specimens (number 2); Fig. 8.1(c) represents peeling tests

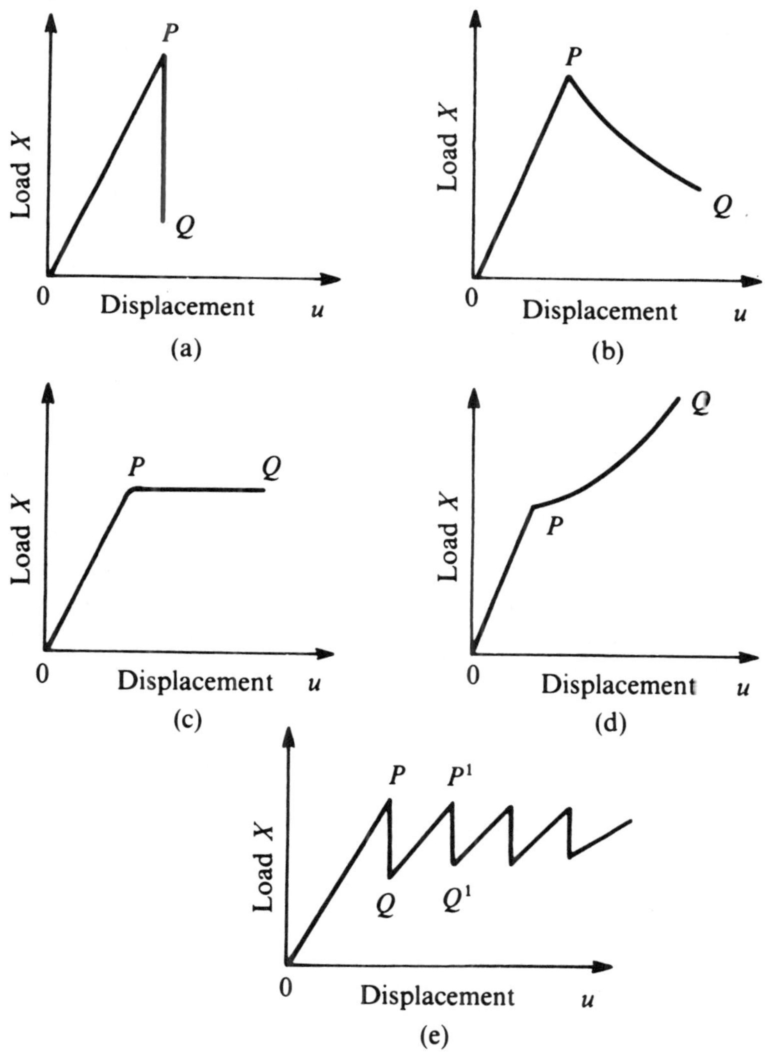

Fig. 8.1 – Typical load–displacement diagrams for: (a) unstable cracking, (b) stable cracking with decreasing load, (c) stable cracking with constant load, (d) stable cracking with increasing load, and (e) 'stick-slip' cracking. (From Mai and Atkins (1980). Reprinted by permission of the Council of the Institution of Mechanical Engineers from *J. Strain Analysis.*)

(number 12), Outwater double torsion testpieces or contoured DCB testpieces (number 10); Fig. 8.1(d) is found with ball indentations (number 13) and so on. The stick–slip sawtooth-like *Xu* type of plot shown in Fig. 8.1(e), where the crack jumps ahead in a series of jerks, interspersed with a series of crack arrests, is found with materials such as polyesters and epoxy resin whose crack tips blunt and resharpen, as well as in materials like low carbon steels whose toughness decreases as the velocity increases. How should these types of crack propagation be classified? Experiments which display precipitous load-drops, in which cracking cannot be visually tracked, are usually 'unstable'. However, this is too vague and qualitative a classification, as we shall see.

The global energy line of attack to cracking provides a means of distinguishing formally between stable and unstable crack propagation. It revolves around the amount of energy *absorbed* in cracking (to create new surfaces, do plastic work at crack tips in metals, do shear and craze yielding work at crack tips in polymers etc., do remote irreversible work in the case of ductile (plastic) fracture) compared with the amount of energy (from stored elastic strain energy, external work or chemical energy) *available* in the system to 'feed' the crack. Stable cracking occurs when the *rate* of release of energy into the crack is in equilibrium with the *rate* at which toughness work is done. If the rate of energy available for cracking is less than that required for propagation, the crack cannot move. On the other hand, if more energy is released into the crack than required for propagation, the situation becomes unstable unless there are subsidiary means of dumping the excess energy. Excess energy accelerates the crack quickly through the body and considerable kinetic energy may be generated. We define here in this book that 'continuous quasi-static cracking is stable'.

In broad terms, the amount of energy available for cracking depends on the shape and size of the cracked body and the size and orientation of the crack relative to the loading system; the amount of energy required for propagation depends upon the fracture toughness characteristics of the material. Thus, as we shall see, two factors always enter considerations of crack stability: (i) geometrical, and (ii) material. The geometry effect leads to so-called geometric stability factors (gsf) for testpieces, and the material behaviour rleates to the variations in R as the crack advances along its path. We shall be concerned with situations where too much energy is being released into the crack from the cracked structure, or where too little energy is being absorbed by the material.

8.2 CRACK STABILITY IN ELASTIC FRACTURE

8.2.1 Graphical interpretation of crack stability in elastic fracture

During quasi-static stable cracking, the variables X, u, A and J (and R) must always satisfy equation (1.1) and all its derivative relations (2.7)–(2.12), where the rate of supply of energy into crack propagation is exactly balanced by the rate of absorption of energy as work of cracking. If, for some reason, the variables are prevented from obeying equation (1.1) by, for example, constraints on their values, crack instabilities may result. Such constraints can arise from methods of loading, in particular the characteristics of testing machines.

Ideally, when determining R, it would be useful to have crack speed as the independent variable, with load and displacement being simultaneously varied to suit. Such conditions are difficult to achieve experimentally, and in practice traditional types of testing machine are used, i.e. the hard, stiff, or displacement-controlled testing machine, and the soft, load-controlled machine. In the displacement-controlled machine, increments of crosshead displacement (du in equation (1.1)) are always *positive* (i.e. the crosshead keeps on driving down, and does not reverse direction). In such circumstances, the operator has no control over the load and accepts whatever happens to it: it may increase, decrease or fluctuate as the test situation dictates. In the load-controlled machine, on the other hand, it is the increment of load dX that is always positive, and there is no control over crosshead displacement. In Fig. 8.1, $du > 0$ in (b), (c), (d) and parts of (e); $dX > 0$ in (d) and parts of (e). It should be clear that there must be occasions where satisfaction of equation (1.1), for a particular geometry of testpiece and for a particular toughness characteristic in the tested material, will require certain changes in u or X, which will not be compatible with the limitations imposed by the loading system. Thus do instabilities arise.

The following analogy with an ice skater demonstrates the influence of external constraints on stability. Consider the following three possibilities:

(i) a skater who falls through ice that has cracked because of his weight,
(ii) a skater who happens to be under an overhanging tree, and who can hang on to a bough to prevent himself dropping through the ice, even though the ice sheet may be breaking,
(iii) the skater who not only can grab on to a tree as soon as he sees cracks appear but also haul himself up off the ice sheet.

In all cases the strain energy in the ice sheets that helps to feed the cracks is caused by the deflection of the ice under the skater's weight. In the first case, the cracking is catastrophic and unstable because the descent of the skater's body provides *more* energy to the ice sheet than is required for balanced stable cracking (i.e. to satisfy power equilibrium the load should fall once the cracks start off, but the load cannot diminish). In the second case, the skater's body is prevented from dropping, and if he grabs onto the tree as soon as he sees cracks initiating he may prevent further cracking if his body is locked at a fixed height and can supply no further energy to the system. In the third case, if he quickly lifts himself off the ice sheet upon sight of the first cracks, he may prevent catastrophic cracking altogether in that he removes energy from the ice sheet by changing the way his weight is supported (i.e. by his feet and hence the ice, or by his arms and hence the tree).

All testing machines themselves deflect under load (so, in practice, would the bough of the tree), so that the displacement of the crosshead is usually an overestimate of the stretch in a tension testpiece or the reduction in height of a compression testpiece. This is, of course, why extensometers or clip gauges are used to measure specimen deflexions. During a test, the more a testing machine deflects, the more it stores strain energy in its frame. In cracking experiments, this energy is available to feed a crack once started. If the specimen geometry is such that it is capable itself from its own deflected strain energy adequately to feed the propagating crack, the extra energy released from the machine may tip the balance and cause the crack to become unstable and accelerate. Clearly soft machines are more likely to produce instabilities than hard machines because they contain more energy at a given load (as they have greater complicances), Fig. 8.2. Cracking experiments are best performed in the hardest machines available.

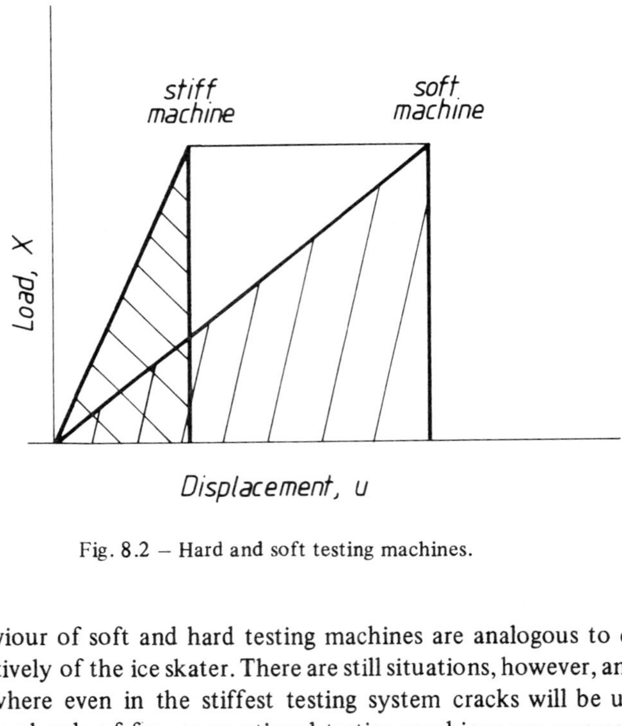

Fig. 8.2 – Hard and soft testing machines.

The behaviour of soft and hard testing machines are analogous to cases (i) and (ii) respectively of the ice skater. There are still situations, however, analogous to case (iii), where even in the stiffest testing system cracks will be unstable, because the crossheads of few conventional testing machines can reverse rapidly during a test, so that the machine is incapable of reducing the total energy in the cracked system as the skater does in case (iii).

Let us see on an Xu plot what stability means in general terms. The thick line in Fig. 8.3 represents a notional X–u relation during cracking for some testpiece material and geometry. For simplicity it can be considered as a constant J locus, where the X, u, A coordinates always satisfy equation (1.1). The specimen geometry would determine the location of the radial stiffness lines at constant crack area, and the values of X (and u) on those lines that go along with chosen J values would be determined from relationships such as equation (2.7)–(2.12). Every segmental area represents $J\mathrm{d}A = R\mathrm{d}A$ for quasi-static crack propagation. Exercises to construct such loci have been given in Chapter 2. Remember that such a curve is constructed from *theory* from the solution of equation (1.1); it does *not* follow that *all* the curve could be traced out in practical cracking tests. For prescribed external constraints, such as $\mathrm{d}u/u > 0$ (stiff

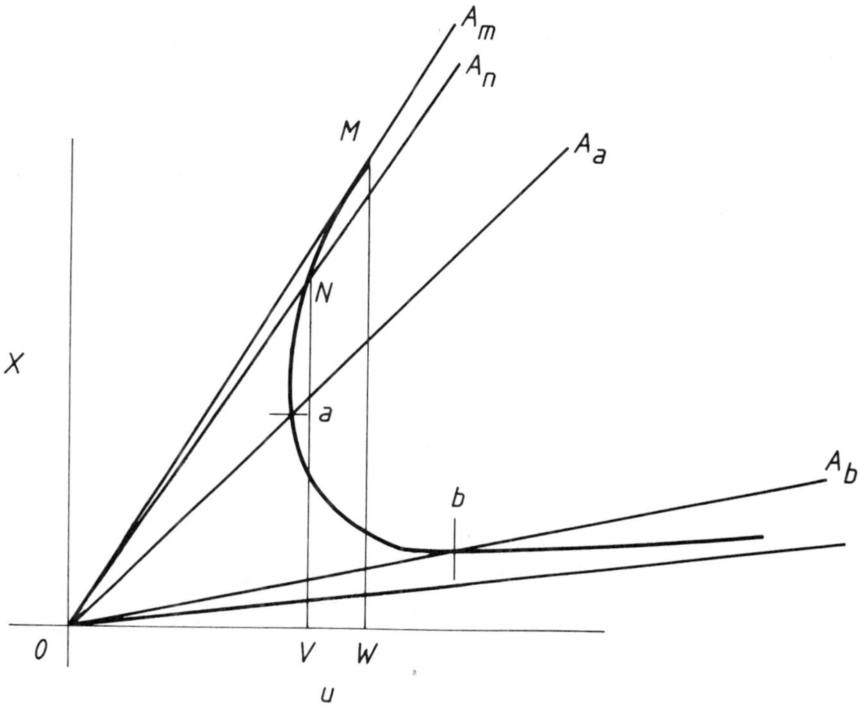

Fig. 8.3 – Features of a general R-locus.

testing machine) or $dX/X > 0$ (soft machine), it is a simple matter to see in Fig. 8.3 what parts (if any) of an R-locus can be followed subject to those constraints. Clearly, $du/u > 0$ is defined on the diagram by the vertical tangency at point 'a'; $dX/X > 0$ occurs at the tangency point 'b'. For both points there are corresponding crack areas, A_a and A_b, i.e. those having the particular stiffness lines that happen to go through points a and b. Since crack propagation means growth of crack area ($dA > 0$), the R-locus can only be followed for cracks longer than A_a (for $du/u > 0$) and longer than A_b (for $dX/X > 0$). In other words, cracks will be unstable in stiff machines for all starter cracks shorter than A_a; in soft testing machines, only cracks longer than A_b will be stable. Notice straight away that since $A_b > A_a$, stability is easier to achieve in hard machines than soft. The precise values of A_a and A_b, and how far apart they may be, and indeed whether both exist in real terms for a given specimen, depend upon the way in which the testpiece compliances and stiffness change with crack area (i.e. $(d/dA)(u/X)$ etc.) which leads to the 'geometric stability factors' already mentioned, and to be discussed in detail in section 8.2.2.

Before arriving at analytical expressions for A_a and A_b, we should ask what is going on at crack lengths shorter than A_a (for $du/u > 0$) and shorter than A_b (for $dX/X > 0$), if practical cracking is unstable yet the J-locus evidently must be obeying the quasi-static energy rate balance. Consider an increment of cracking with starter cracks shorter than A_a in Fig. 8.3. Just before crack propagation at M, the strain energy in the testpiece caused by loading along OM (which is the stiffness appropriate to crack starter area A_m) is given by the area OMW. Let us assume, for the moment, that we *can* follow the 'backwards-pointing' J-locus, and that, after so doing for a while, we unload the testpiece at N when the crack area has increased to A_n. We have expended area OMN in doing fracture work $(= R(A_n - A_m))$, and after unloading we recover area ONV. Clearly work given by the area MNVW is excess to that required and, unless it is removed from the system, it can only serve to make the situation unstable. That is, there is *so much* strain energy present in the system at crack initiation that the testpiece can afford to *unload*, giving back work (MNVW) to the loading device, yet still have enough energy to feed the crack at the appropriate rate. The excess work could be removed in principle, by the crosshead reversing at the instant of first cracking (M), so that further increments in du were *negative* (opposite direction from before), thus taking energy out of the testpiece and putting it back into the testing machine; in equation (1.1) $X\mathrm{d}u$ is then negative. Since testing machine screws do not normally respond in this way in a test, the situation is unstable and the testpiece goes 'bang' with an instantaneous vertical drop in load.† Likewise, for cracks shorter than A_b in load-controlled machines, it is impossible to have stable cracking because what the testpiece wants to do (i.e. crack with $\mathrm{d}X < 0$) is not what the testing system will allow it to do. Thus, according to this explanation of stability, the cases in Fig. 8.1 (a) and (e) are unstable in a displacement-controlled machine and all the rest are stable; whereas only Fig. 8.1(d) would be stable in a load-controlled machine.

8.2.2 Geometric stability factors in elastic fracture

We have for a linear system

$$X^2 = 2G\Big/\mathrm{d}/\mathrm{d}A\ (\dot{u}/X) \qquad \text{(2.6 bis)}$$

and for equilibrium fracture $G = R$ so that

$$X^2 = 2R\Big/\mathrm{d}/\mathrm{d}A\ (u/X) \qquad \text{(2.7 bis)}$$

Recognising that R may change during propagation (Chapter 6), the variation of R with incremental input X and output A is obtained by differentiating relation (2.7) to give

$$\frac{2}{X}\left(\frac{\mathrm{d}X}{\mathrm{d}A}\right) = \frac{1}{R}\,\frac{\mathrm{d}R}{\mathrm{d}A} - \frac{\mathrm{d}^2/\mathrm{d}A^2\,(u/X)}{\mathrm{d}/\mathrm{d}A\,(u/X)}$$

† Special control circuitry for crosshead displacements can permit the unstable part of R-loci to be followed, but it is never done in regular testing (Mamaros and Jones, 1974).

For stability in a soft testing machine, $dX/X > 0$ so

$$\frac{1}{R}\frac{dR}{dA} \geqslant \frac{d^2/dA^2\,(u/X)}{d/dA\,(u/X)} = \frac{d^2/dA^2\,(X/u)}{d/dA\,(X/u)} - 2\,\frac{d/dA\,(X/u)}{(X/u)} \tag{8.1}$$

by manipulation.

Again, from $u^2 = 2G/d/dA\,(X/u)$ we obtain for $du/u > 0$,

$$\frac{1}{R}\frac{dR}{dA} \geqslant \frac{d^2/dA^2\,(X/u)}{d/dA\,(X/u)} = \frac{d^2/dA^2\,(u/X)}{d/dA\,(u/X)} - 2\,\frac{d/dA\,(u/X)}{(u/X)} \tag{8.2}$$

(note $d/dA\ (X/u)$ is negative as the stiffness decreases with increasing crack length). Equations (8.1) and (8.2) were given by Gurney and Hunt in 1967.

Hence to achieve quasi-static stable crack propagation, the *material* property dR/dA at the crack tip (l.h.s. of the inequalities) must be greater than certain values characteristic of the *geometry* of the testpiece (r.h.s. of the inequalities) which are called the geometry stability factors (gsf) of the testpiece. We may write relations (8.1) and (8.2)

$$\frac{1}{R}\frac{dR}{dA} \geqslant \frac{n_X}{A} \text{ or } \frac{n_u}{A} \tag{8.3a}$$

where n_X, n_u are simply numbers characteristic of the testpiece. In plate structures of constant thickness, these relations become

$$\frac{1}{R}\frac{dR}{da} > \frac{n_X}{a} \text{ or } \frac{n_u}{a} \tag{8.3b}$$

with n_X and n_u the same. For example, in a DCB specimen $X/u = 3EI/2a^3$ so $d/dA\ (X/u) = -9EI/2a^4B$ and $d^2/dA^2\ (X/u) = 36EI/2a^5B^2$. Hence, in equation (8.2) for stability in a stiff testing machine.

$$\frac{1}{R}\frac{dR}{dA} > -\frac{4}{A}$$

For constant R, this becomes $0 > -4/A$ and we conclude that stability should be easily attained in DCB specimens, which experiments on many solids where R does not change markedly, confirm.

Note that even testpieces with good gsf can have instabilities should, for example, the crack front be blunt (which causes an excessive starter energy in the system), or if (even with a sharp crack) $\dot{u}$ is so high as to take $\dot{a}$ out of the quasi-static range.† Again, if the material requires less work to spread cracks at high velocity than at lower i.e. ($dR/d\dot{a}$ negative), instability may result once cracking is initiated. On the other hand $dR/d\dot{a}$-positive solids encourage stable cracking unless the specimen gsf is particularly bad.

---

† The limiting velocities of cracks in quasi-static cracking (below which kinetic energy dH is negligible in equation (1.6)) have been estimated by Gurney and Ngan (1971) as some 6 m/s in metals and 2 m/s for polymers.

Example 8.1 Show that stability in DCB specimens is difficult under load-controlled conditions.

Ans. Using the second of relation (8.1) we have

$$(1/R)\mathrm{d}R/\mathrm{d}A > -4/A - \frac{2\,(-9EI/2a^4B)}{3EI/2a^3} = -4/A + 6/A = 2/A$$

This time the gsf is positive, so for R constant we require $0 > 2/A$ which cannot be achieved. Remember that in Fig. 8.1(b), dX is always decreasing which does not admit stability in a soft system; du is increasing in the DCB geometry which does give stability in a hard system.

The parameters n_u and n_X in equation (8.3) may be determined (i) experimentally from compliance or stiffness calibrations as a function of crack area or length, (ii) directly from simple strength-of-materials where applicable (as in Example 8.1 above) or (iii) by means of integration of stress intensity factor expressions which gives (X/u) or (u/X) and their derivatives. The procedures in (iii) are similar to those employed for crack velocity predictions in section 3.1 and are as follows.

If we equate G given by K^2/E^* to G from relation (2.6) we obtain

$$K^2 = E^*G = \left(\frac{E^*X^2}{2B}\right)(\mathrm{d}/\mathrm{d}a\,(u/X)) = Y^2\sigma^2\pi a \qquad \text{(3.4 bis)}$$

with Y the geometry function dependent on the shape of the testpiece. Hence the stability conditions (8.1) and (8.2) may be written for $\mathrm{d}u/u > 0$

$$\frac{1}{R}\frac{\mathrm{d}R}{\mathrm{d}a} \geqslant \frac{1}{a}\left(1 + \frac{2aY'}{Y}\right) - \frac{2Y^2a}{\int_0^a Y^2 a\,\mathrm{d}a} \qquad (8.4)$$

and for $\mathrm{d}X/X > 0$

$$\frac{1}{R}\frac{\mathrm{d}R}{\mathrm{d}a} \geqslant \frac{1}{a}\left(1 + \frac{2aY'}{Y}\right) \qquad (8.5)$$

where in both expressions $Y' = \mathrm{d}Y/\mathrm{d}a$.

By way of example, consider the ASTM compact tension specimen (number 19 in Appendix 1, with $2H = W$ and $a = W/2$) already discussed in Example 2.11, where Y is given there for a range of (a/W). Differentiating Y with respect to a and substituting into inequality (8.5) gives the stability condition in a soft load-controlled machine. Thus, in the recommended compact tension geometry with $a/W = 0.5$, $Y = 7.66$ and $Y'a = 7.59$. Hence relation (8.5) gives

$$\frac{1}{R}\frac{\mathrm{d}R}{\mathrm{d}a} > \frac{n_X}{a} = \frac{2.98}{a} \qquad (8.6)$$

Thus in a load controlled situation, unless $\mathrm{d}R/\mathrm{d}a$ is sufficiently positive, stable cracking in a CT testpiece is difficult to achieve. However, under displacement control n_u is -4.3, and since the factor is negative, stability should be achieved for R constant and even if $\mathrm{d}R/\mathrm{d}a$ is mildly negative. Note the conclusion at the end of section 3.1 that $\dot{a}/\dot{u} \approx 118$ for the CT geometry even though the crack propagation is stable under displacement-control. Thus 'fast fracture' is *not* necessarily 'unstable fracture'.

Appendix 1 (taken from Mai and Atkins, 1980) is a compilation of gsf for a wide range of testpiece geometries. Gsf for other geometries may be worked out along the lines described above from relevant K-formulae.

Relations (8.1) and (8.2) may be rewritten and generalised to non-linear (reversible) cases in the following way. Remember that in quasi-static cracking, increments in the critical energy release $J_C\mathrm{d}A$ are balanced exactly by the increments in specific work of fracture $R\mathrm{d}A$, and loci of constant J_C are constant R-loci on Xu diagrams.

Consider that, in general, the material fracture toughness varies so that values obtained from stable cracking experiments will vary along the X–u cracking locus (i.e. values of R given by Gurney segmental work areas vary as the crack propagates, cf. Figs. 2.7, 2.8, 2.10, 2.11 and so on). But at any instant $R = J(X,A)$. To investigate stability under $\mathrm{d}X/X > 0$, the variation of R with an incremental input X and output A is

$$\frac{\mathrm{d}R}{\mathrm{d}A} = \frac{\mathrm{d}J}{\mathrm{d}A} = \left(\frac{\partial J}{\partial A}\right)_X + \left(\frac{\partial J}{\partial X}\right)_A \frac{\mathrm{d}X}{\mathrm{d}A} \tag{8.7}$$

From Equation (2.9a), we have

$$(\partial J/\partial X)_A = (\partial u/\partial A)_X$$

so

$$\frac{\mathrm{d}R}{\mathrm{d}A} = \left(\frac{\partial J}{\partial A}\right)_X + \left(\frac{\partial u}{\partial A}\right)_X \frac{\mathrm{d}X}{\mathrm{d}A} \tag{8.8}$$

Now from lines of constant crack area on Xu plots we observe that $(\partial u/\partial A)_X$ is positive. Hence, under soft testing constraints with $\mathrm{d}X/X$ positive also, it follows that stability must correspond with

$$\frac{\mathrm{d}R}{\mathrm{d}A} \geqslant \left(\frac{\partial J}{\partial A}\right)_X = \left(\frac{\partial^2 \Omega}{\partial A^2}\right)_X \tag{8.9}$$

from relation (2.11a). Equation (8.9) is a generalised version of (8.1) to include non-linear elastic structures.

Similarly, considering incremental input u and output A, we obtain for $\mathrm{d}u/u > 0$

$$\frac{\mathrm{d}R}{\mathrm{d}A} \geqslant \left(\frac{\partial J}{\partial A}\right)_u = -\left(\frac{\partial^2 \Lambda}{\partial A^2}\right)_u \tag{8.10}$$

which is the general version of (8.2). Writing $\Omega = \Lambda = \frac{1}{2} Xu$ in (8.9) and (8.10) of course gives (8.1) and (8.2) respectively. It should be noted again that the l.h.s. of relations (8.9) and (8.10) concern possible variations in the *material* toughness as the crack propagates whereas the r.h.s. are determined entirely by the *geometry* of the cracked structure. Note also (although we do not pursue the topic in this book), that $\partial^2\Lambda/\partial A^2$ and $\partial^2\Omega/\partial A^2$ will be orientation-dependent in anisotropic solids, so energetic stability may depend upon path just as directional stability does (cf. section 3.5.2, Example 3.10 etc.).

Example 8.2 In the cracking experiments of Gurney and Ngan (1971) illustrated in Fig. 2.10, non-linear stiffness lines are given by $X = 350\ (u/X) - 1980\ (u^2/A^{3/2}) + 5250(u^3/A^2)$. ($X$ in N, u mm, A mm^2). Determine stability conditions in hard testing machines for those experiments.

Ans. $\Lambda = \int X\mathrm{d}u = 175\, u^2/A - 630\, u^3/A^{3/2} + 1312\, u^4/A^2$ (cf. Example 2.7).

$$-(\partial^2\Lambda/\partial A^2)_u = -\ 350u^2/A^3 + 1362.5\, u^3/A^{7/2} - 7872\, u^4/A^4$$

Stability is achieved when $\mathrm{d}R/\mathrm{d}A \geqslant -(\partial^2\Lambda/\partial A^2)_u$. For $u = 4$ mm, $A = 44$ mm^2 (cf. Example 2.7 and Fig. 2.10) we have $\mathrm{d}R/\mathrm{d}A \geqslant -\ 449 \times 10^9$ J/m^4. This is easily achieved unless $\mathrm{d}R/\mathrm{d}A$ is excessively negative, and we note that the constant J-loci in Fig. 2.10 are almost horizontal, which is sensibly in agreement.

8.3 TESTING MACHINE STIFFNESS

With a testing machine of compliance C, the total deflexion of the system is $u^* = u + CX$ where u is the displacement of the testpiece. Hence both $\mathrm{d}/\mathrm{d}A\ (u^*/X)$ and $\mathrm{d}^2/\mathrm{d}A^2\ (u^*/X)$ are the same as $\mathrm{d}/\mathrm{d}A\ (u/X)$ and $\mathrm{d}^2/\mathrm{d}A^2\ (u/X)$. It follows therefore from (8.1) that stability under $\mathrm{d}X/X > 0$ (soft conditions) is *unaffected* by the flexibility of the testing machine. This is perhaps not unexpected. The situation is different under $\mathrm{d}u/u > 0$ (hard conditions), however, as $X/u \neq X/u^*$.

We have

$$X/u^* = \frac{(X/u)}{1 + C(X/u)}$$

so that

$$\mathrm{d}/\mathrm{d}A\ (X/u^*) = \frac{\mathrm{d}/\mathrm{d}A\ (X/u)}{[1 + C(X/u)]^2} \tag{8.11}$$

and

$$\mathrm{d}^2/\mathrm{d}A^2\ (X/u^*) = \frac{[1 + C(X/u)]\,\mathrm{d}^2/\mathrm{d}A^2\ (X/u) - 2C\,[\mathrm{d}/\mathrm{d}A(X/u)]^2}{[1 + C(X/u)]^3} \tag{8.12}$$

which give for stability in (8.2)

$$\frac{1}{R}\frac{dR}{dA} > \frac{d^2/dA^2\,(X/u)}{d/dA\,(X/u)} - \frac{2C\,d/dA\,(X/u)}{1 + C(X/u)} \tag{8.13}$$

Since $d/dA\ (X/u)$ is negative (stiffness decreasing with increasing crack length) the r.h.s. is more positive than the r.h.s. of equation (8.2). Stability is therefore decreased by the flexibility of the testing machine. This means, for example, that longer starting cracks are required for stable cracking when C is finite. In Fig. 8.4 point B represents, for some constant G-locus in the X–u plane, the transition between crack lengths for unstable conditions under $du/u \geqslant 0$. Point D* represents the new transition point, on the same constant G-locus (but now referred to the X–u^* axes). D* lies on a longer transition crack length than does B. Note that at point D (the location of the longer crack length on the R-locus referred to X–u axes), $dX/du \neq \infty$ as at point B, because of the influence of the testing machine compliance. The strain energy stored in the testing machine is available to help feed the crack, so that the part of the G-locus between B and D (capable of being followed in an infinitely stiff machine) is no longer available. In fracture toughness testing, the choice of the machine

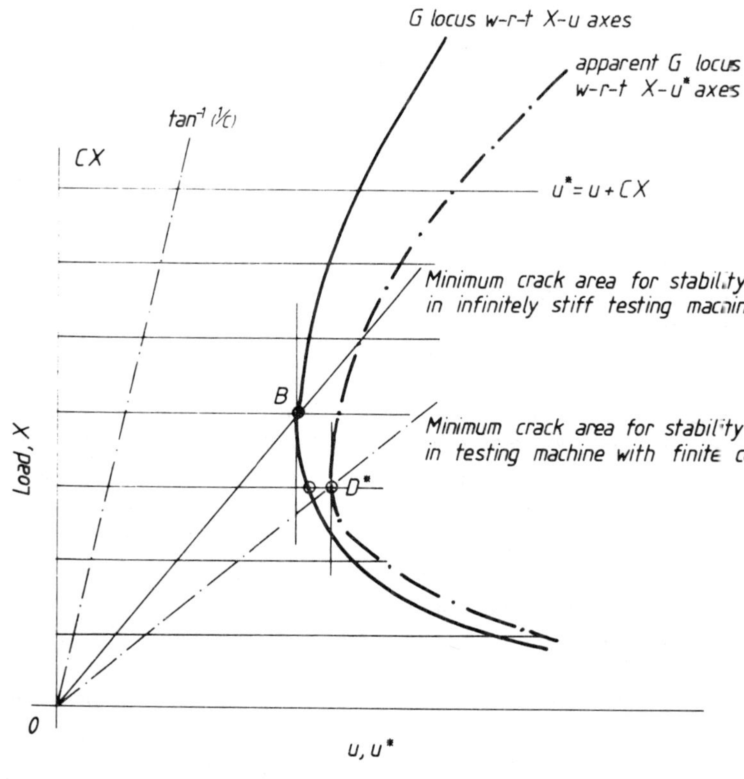

Fig. 8.4 – Effect of machine stiffness on load–deflexion diagram and on R-locus.

flexibility is very important. While it is impossible to have $C = 0$, it is best to have C very small. Several other investigators, viz. Clausing (1969) and Mirabile and Venzi (1973), have also investigated the effects of testing machine stiffness on stability of cracking, and found similar conclusions. For further details readers should consult these references. Gurney (and Clausing)-type analyses for elastic fracture appear in Appendix II of a paper of Paris *et al.* (1979) the main purpose of which was to address stability in plastic fracture (section 8.10 following).

Example 8.3 Discuss crack stability in terms of the design of nutcrackers.

Ans. A common design of nutcracker consists of two arms hinged together at their ends. The nut is inserted between these levers and fractured by squeezing the arms together in the hand. The system is load controlled, and soft, and it is well-known that it is difficult to control cracking; the kernel often gets crushed. When, sometimes, it is possible to control cracking with this design, it comes about because of non-linear behaviour of the partially-cracked shell. An alternative design consists of a screw working into a hole into which the nut is placed, which operates like a vice. This clearly is a much harder, displacement-controlled, system and there is less damage to the kernel. A related design (rather like a micrometer frame) is used as a 'nut-splitter' to break (metal) nuts rusted on to bolts.

8.4 STABILITY IN TERMS OF CRACK VELOCITY

In the unstable part of the X, u, A system, *negative* values of $\dot{A}/\dot{u}$ will be predicted, since in following the J-locus, du is negative. This is an alternative way of tackling stability. In equation (3.2) (for $\dot{A}/\dot{u}$), dX/dA is so big negatively, that the denominator becomes negative.

The cracked ring specimen shown in Fig. 8.5 is related to the 'Brazilian ring test' used by civil engineers and geologists for the strength of brittle materials, section 3.2. Gurney and Mai reported in 1972 that this testpiece displays the most general R-loci, i.e. 'completely' unstable regions at small crack lengths, regions stable for d$u/u \geqslant 0$, and then regions stable for d$X/X \geqslant 0$. We may derive $u.\mathrm{d}X/\mathrm{d}A$ from the X, u curves in Fig. 8.5(a), to give the velocity ratios shown in Fig. 8.5(b). At $a = 35.4$ mm ($A = 99.4$ mm^2) which corresponds with the vertical tangent stability point for d$u/u > 0$, it may be shown that $2R + u.\mathrm{d}X/\mathrm{d}A = 0$ and therefore $(\dot{a}/\dot{u}) \to \infty$, from relation (3.2). The negative velocities predicted for cracks shorter than this length will never be seen on conventional testing machines, and tests go 'bang'. Fig. 8.5(b) shows that in the stable regions, velocities diminish as a gets longer, and that if the starter crack length is only just greater than the critical 35.4 mm, marked reductions in velocity occur in the first stages of cracking. This is true generally for all types of specimen if the

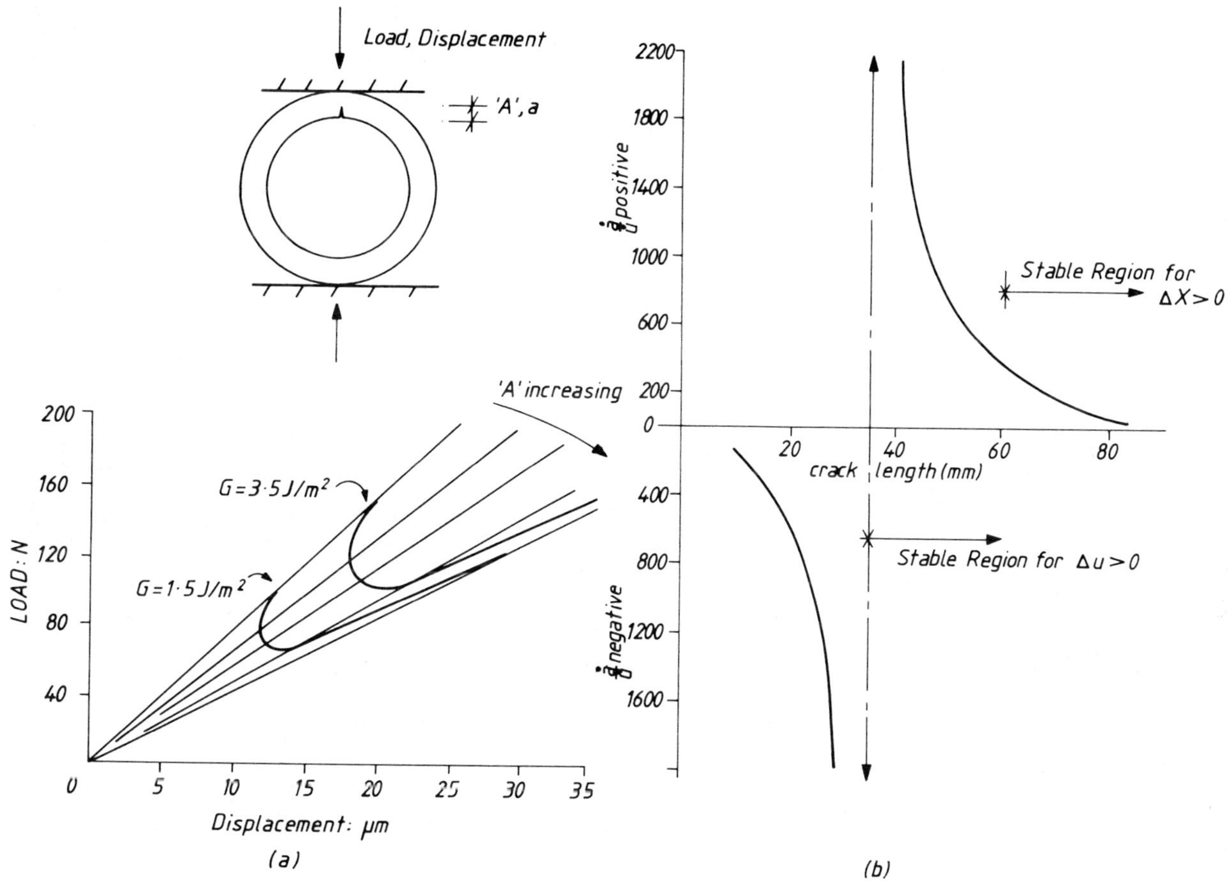

Fig. 8.5 – (a) Load–deflexion diagram for cracking of a ring under diametrical compression. (b) Plots of $(\dot{a}/\dot{u})$ vs. a for the cracked ring in (a).

loading line intersects a J-locus close to the vertical tangency point and hence should be avoided in testpiece design – especially if the material being tested is rate-dependent (Chapter 6).

8.5 *R*-CURVES AND STABILITY

In section 2.3 we showed that a material with a constant fracture toughness should display a step function R-curve with a sudden jump in the J–a diagram to a plateau given by J_C as the crack starts off (Fig. 2.15). While in practice there must always be some curvature in the transition to the J_C plateau level, for some materials it is not marked and R may be taken to be constant in the stability relations (8.1) etc. and (8.9) etc. In many cases however it is found that there is a noticeable transition region in the J–a diagram which, under equilibrium conditions, means that R changes with crack growth. Such increases in R with a during early crack growth are studied in terms of so-called R-curves which are characteristic of the material, cf. section. 4.2.4. Clearly $\mathrm{d}R/\mathrm{d}a$ is now not zero and stability via equations (8.1), (8.2) and so on must be affected.

If stable crack propagation is possible along the whole R-curve through to the plateau level (and, as we shall see, this depends on specimen geometry and is not always achieved) there may also be *subsequent* variations in R (i.e. J_C) during propagation caused by a variety of effects such as velocity-, or environmental-dependency. These variations will, in turn, affect stability.

In this section, we concentrate on stability during the *rising* part of R-curves. In thin sheet metals the increase in R during early propagation is caused mainly by the formation of slanted shear lips (although that is not the entire story, and increasing plastic flow at the crack tip plays a role, cf. Example 4.4). Rising R-curves are displayed also by viscoelastic solids owing to rate effects in R (Chapter 6) and by fibre-reinforced composite materials owing to fibre pull-out where filaments bridging the crack faces produce a stabilising effect. Since crack growth occurs *before* the J_C plateau level, and since in many (but not all) testpieces propagation at the J_C level is 'fast fracture', the early propagation is often called 'slow crack growth'.

A typical R-curve is shown in Fig. 8.6. Krafft *et al.* in 1961, and others subsequently, have postulated that build-up in the R-curve to the plateau value is unique for a given material, being independent of specimen geometry and initial crack length. Of course, R-curves may be dependent on thickness, temperature and environment. In Fig. 8.6 we indicate a threshold R_0 value below which there is no crack growth and our R-curves correspond with *actual* crack extensions. Note that some authors show R-curves curving over directly from the starting crack length (a_1) position on the abscissa, i.e. the vertical jump to R_0 represented by AB is absent. This comes about because they associate growth of the *plastic zone* at the crack tip with Δa even when the crack is *stationary* (i.e. they employ the Irwin embedded crack tip concept given in Example 2.14; this approach appears also in 'J-integral' testing methods, Chapter 4).

The rate at which strain energy may be released depends very much on the geometry of the cracked structure and the conditions of loading. As an example,

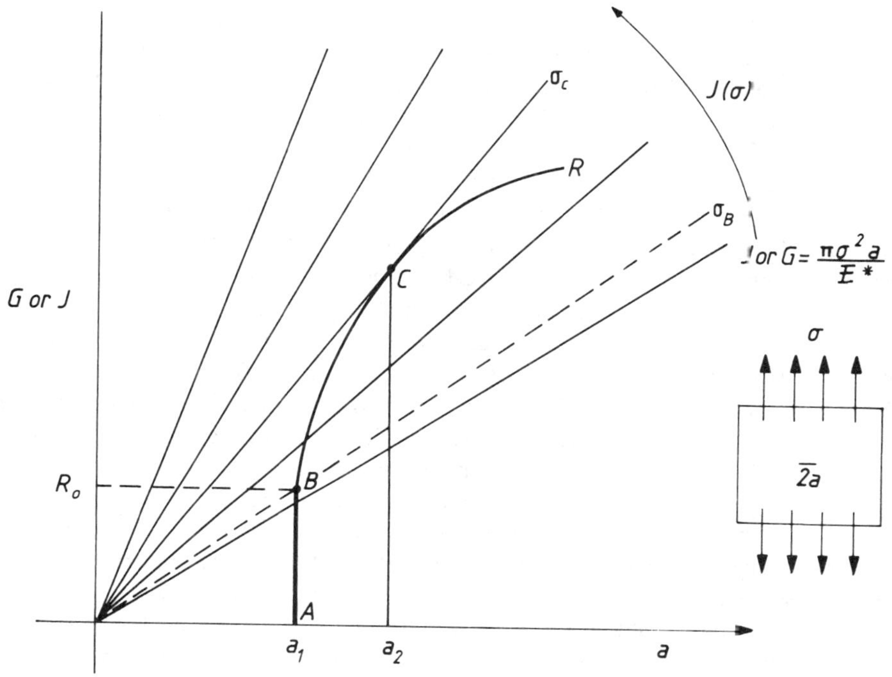

Fig. 8.6 – G vs. a curves for constant applied stress σ; R vs. a curve is also superimposed.

in the Griffith geometry of a small crack in a large sheet $G = \pi\sigma^2 a/E^*$ which may be plotted in G versus a space for constant σ as shown in Fig. 8.6 as a series of radial lines of slope $\tan^{-1}(\pi\sigma^2/E^*)$. The magnitudes of these slopes give $(\partial G/\partial a)_\sigma$. There is a complementary set of radial lines at constant strain ϵ from $G = \pi\epsilon^2 aE^*$, with $(\partial G/\partial a)_\epsilon = (\pi\epsilon^2 E^*)$. We write G because the Griffith problem is linear elastic, but in what follows J may be read for G in the general non-linear reversible case. To investigate crack stability, let us locate the crack growth resistance curve for the material along the a-axis so that R rises at the location of the end of the starter crack (i.e. at $a = a_1$ where $2a_1$ is the total length of the Griffith crack). For some other Griffith starter crack length, the R-curve would be shifted along accordingly, but the radial energy release rate lines would be *invariant*. As the stress is increased from zero under load-control $\mathrm{d}X/X > 0$, the available G increases up A to B. Those parts of the G–a-curves for $a < a_1$ play no part in the present problem. At B, which represents the minimum fracture toughness (R_0) of the material before any subsequent increase in R along the R-curve, G_B (J_B in general) $= R_0$. Cracking can thus commence, and because $\mathrm{d}R/\mathrm{d}a > (\partial G/\partial a)_{\sigma=\sigma_\mathrm{B}}$ cracking is *stable* (equation (8.9)). In this way the crack propagates quasi-statically under increasing load (or stress) until point C is reached, when the crack has lengthened to a_2. Between B and C, G is always just enough to balance R ($G = R$) and the R–a slope is always steeper than the G–a slope. Were the stress held constant somewhere between σ_B and σ_C cracking would cease. However, at C the slopes are equal, and beyond C along the R-curve, $dR/da < (\partial G/\partial a)_\sigma$. Thus, as argued by Krafft *et al.*, C represents the point of instability under increasing σ conditions and $\sigma = \sigma_C$ is the critical Griffith stress. The part of the material characteristic R–a-curve beyond C is *never* observed with this geometry.

Cracking is also unstable under increasing displacement conditions because the G–a-curves are also radial lines, with slope $\tan^{-1}(\pi E\epsilon^2) = (\partial G/\partial a)_\epsilon$, which eventually is steeper than dR/dA at some ϵ, i.e. at some u.

The amount of stable crack growth $(a_2 - a_1)$ at each end of the central crack, before instability – also called the sub-critical or slow crack growth – depends upon the shape of the R-curve and the G-curves. Clearly $(a_2 - a_1)$ is large for dR/da-positive materials. For the constant R case, with a step relationship between R and a, there is *no* sub-critical growth and instability is simultaneous with first cracking in the Griffith geometry.

Sometimes, the information is plotted in terms of stress intensity factors. Then instead of the specimen G–a curve, we have $K(=\sqrt{E^*G})$ versus a curves (which are parabolic at constant σ in the Griffith case) and instead of the R-curve, we have the so-called K_R versus a curve for the materials, given by $K_R = \sqrt{E^*R}$.

The stability of a cracked ring under compression in terms of R-curves provides an excellent illustration of all the principles involved in this section. The Xu plots, with lines of constant A and G already shown in Fig. 8.5 may be redrawn as GA plots with lines of constant X and u. In Fig. 8.7(a) the data are given as Ga plots as the ring has uniform thickness. Unlike the Griffith case, where the lines of constant σ and ϵ were straight, here the constant X and u lines have maxima (to understand this consider the changes in G-loci values cut by a vertical and a horizontal line of the Xu plots in Fig. 8.5). The maxima occurs at a_u and a_X which are the minimum crack lengths for stability in hard and soft systems given by vertical and horizontal tangent points in the Xu diagram. Now $(\partial G/\partial a)_u$ and $(\partial G/\partial a)_X$ – which are the (varying) slopes along the constant u or X lines in the J (or G) – a plane – are important in establishing stability from equations (8.9) and (8.10). Clearly to the left of the maxima the slopes are positive and to the right of their maxima at longer a they are negative.

Superimpose the R-curve for the material on the Ga diagram. Its location along the a axis depends upon the starter crack length. Consider two locations, on either side of the maxima in the G-curve, shown as the two thick lines in Fig. 8.7(a). Consider the one to the right which is located at a starter crack length of some 63 mm. As the load is increased at contant a prior to crack spreading, $G < R$ until the R-curve bends over and a increases somewhat (the crack initiating at about $X = 25$ kN, $u = 1.7$ mm). Then, since $dR/dA = (1/B)$ $dR/da > (\partial G/\partial a)_{u \text{ or } X}$ cracking is *stable,* and since $(\partial G/\partial a)_{u \text{ or } X}$ remains negative at very long a, further stable cracking is assured even though dR/da is diminishing. Even if R levels out to a constant value and $dR/da = 0$, $(\partial G/\partial a)_{u \text{ or } X}$ are still algebraically smaller. On the other hand, a starter crack some 24 mm long (to the left of the maxima) initiates at about $X = 28$ kN, $u = 1.3$ mm, and at first is stable because $dR/da > \partial G/\partial a)_{u \text{ or } X}$ – the R curve being steeper at

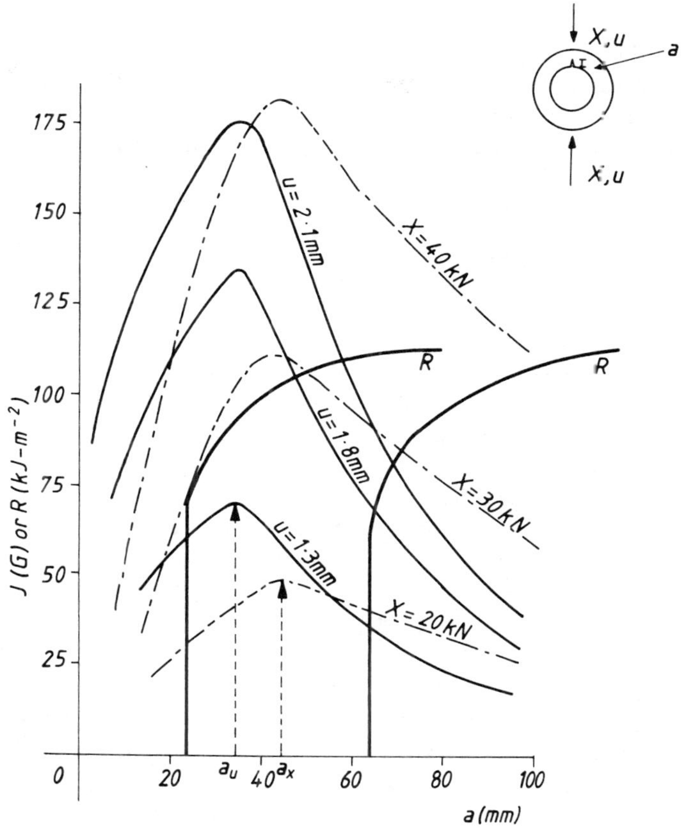

Fig. 8.7(a) – G vs. a curves for constant load and constant displacement for a steel ring fracturing under diametrial compression. R vs. a curve is also shown.

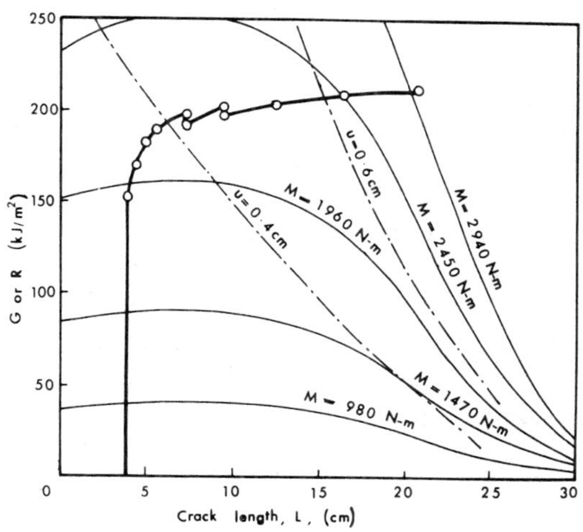

Fig. 8.7(b) – Crack growth resistance (R) and strain energy release rate (G) curves for a low carbon steel (after Mai *et al.* (1976) with permission).

that crack length than either the G lines of constant u or X. However, at only slightly longer a, the R-curve and the G lines of u or X lines become tangential, because dR/da is diminishing as a increases. In Fig. 8.7(a), the R-curve is shown tangential to the line of $X = 30$ kN at $u = 1.4$ mm and $a = 26$ mm. The R-curve is tangential to a constant u line at longer a (say at $a = 32$ mm when $u = 1.5$ mm and $X = 29$ kN). Immediately beyond these two separate points of tangency, $dR/da < (\partial G/\partial a)_{u \text{ or } X}$ so that cracking is unstable. Only a short period of stable growth, $(26 - 24) = 2$ mm in a soft machine or $(32 - 24) = 8$ mm in a hard machine is possible before the testpiece breaks catastrophically.

Clearly, to obtain in an experiment a complete R-locus to include the eventual continuous propagation part where R levels off, minimum crack lengths corresponding with the maxima in the constant u or X loci are required. In this example $a_u = 34$ and $a_X = 44$ mm. A crack with $34 < a < 44$ mm would be stable in a hard machine but unstable in a soft testing machine. The crack lengths a_u and a_X derived here correspond of course with the tangent points in the Xu diagrams (Fig. 8.5).

Were the R-locus to be constant, with no slow crack growth, the sharp corners of the locus would locate the X, u coordinates at cracking. To the left of the maxima $\partial G/\partial a > dR/da = 0$ so unstable propagation ensues; to the right, $\partial G/\partial a < dR/da = 0$ so stability is achieved.

The gsf of the specimen is given by $(\partial J/\partial A)_{u \text{ or } X}$ (i.e. $(\partial G/\partial A)_{u \text{ or } X}$ in this linear case) and we see in the case of the cracked ring that at long a the gsf are good, having appreciable negative values. Thus stability could be achieved in some materials with dR/dA negative (such as polyester and epoxy resins), so long as $dR/da > \partial G/\partial a$. For example, according to Fig. 8.7(a) (with a 2.8 mm thick plate), $(\partial G/\partial A)_u$ is at least -710 MJ/m^4 at crack lengths greater than 34 mm, and $(\partial G/\partial A)_X$ is at least -180 MJ/m^4 at lengths greater than 44 mm (taking the most shallow slopes beyond these lengths). Thus solids with dR/dA algebraically greater than these values could be satisfactorily investigated in this test geometry, even under soft testing conditions. For example, in a stiff machine, 10 mm stable crack growth beyond an initial 34 mm long starter crack in the 2.8 mm thick testpiece is possible even if R falls by $(10 \times 2.8 \times 10^{-6}) \times (710 \times 10^6) = 20$ kJ/m^2 from its starting value. Similarly, 10 mm stable growth from a 44 mm long starter crack under soft conditions is possible when R falls by 5 kJ/m^2. As percentage changes in the material R, these figures depend of course on the absolute R at crack initiation.

The trend of $\partial J/\partial A$ shows that stiff testing conditions permit more unstable materials to be tested. In practice the regions of large gsf are difficult to use in the cracked geometry owing to the very small rate of change of compliance with crack area: it is difficult sometimes to identify the load at which propagation commences on the 'upswing' of the R-locus as the loading line and the R-locus the cracked ring geometry owing to the very small rate of change of compliance with crack area: it is difficult sometimes to identify the load at which propagation have nearly equal slopes (Fig. 8.5). Similar difficulties arise with cracking under ball indentations sometimes which have Xu plots like Fig. 8.1(d).

Because of the requirements for increasing cracking loads at long a, the phenomenon of crack arrest from short starter cracks is sometimes displayed in this type of testpiece. As discussed later in section 8.9, arrest occurs when the excess energy at initially unstable fractures is fully absorbed in later propagation before the testpiece has broken right through.

Example 8.4 Demonstrate why testpieces with stable characteristics should be employed to generate R-curves, and show that unstable testpieces give instability R values that are not absolute, but vary instead with specimen geometry.

Ans. It is clear from the previous example that when the R-curve becomes tangential to the J or G loci, cracking becomes unstable. Consequently, to obtain a complete R-curve, it is required that $\mathrm{d}R/\mathrm{d}A > \partial J/\partial A$ *always* for whatever testing constraints applying (i.e. what type of machine is available). If this condition is not met, only the part of the R-curve before tangency is obtained, and for many starter crack lengths in many testpieces this is uselessly small. The plateau is not reached and the values of J at which instability occurs *vary* with testpiece geometry.

Clearly by shifting the R-curve along the a-axis in Figs. 8.6 or 8.7(a), it is readily seen that the 'cut-off' R at instability takes a variety of values. For example the constant u or X loci are steeper the shorter the starter a for some fixed G or R in Fig. 8.7(a). Thus unstable propagation from shorter crack lengths will occur sooner than from longer starter crack lengths, and the R (or K_R) at instability also varies. That is not to deny the possibility of the R-curve being unique (as has been postulated); it simply says that only small segments of the curve will be picked up.

Many early investigations of R-curves used the inherently unstable centre-notched panel type of testpiece (number 6 in Appendix 1), the behaviour of which is rather like the Griffith geometry Fig. 8.6. It is not suprising therefore that the so-called critical K_R values varied with starter crack size, and also were less than the plateau K for the fully developed crack tip zone. Also in this type of testpiece, it is possible that some data were influenced by generalised yielding in the path of the crack (net section or ligament), especially for very tough solids.

Example 8.5 The crack growth resistance curve of a material is given by $\Delta a = CK_R^m$ where Δa is the effective crack length (i.e. including the plastic zone size) and m and C are material constants. Show that the instability value is dependent on both crack length, size and specimen geometry. Consider load controlled experiments.

Ans. $\Delta a = a - a_i$ (a_i is initial crack length). Therefore $a = a_i + \Delta a$, i.e.

$$\frac{dK_R}{da} = \left(\frac{K_R^{1-m}}{Cm}\right)_{K_R=K_C} = \frac{K_C^{1-m}}{Cm}$$

K is normally expressed as $K = Y\sigma\sqrt{\pi a}$ so that

$$\left(\frac{\partial K}{\partial a}\right)_{X=X_c} = K_C\left(\frac{dY/da}{Y} + \frac{1}{2a}\right) = K_C\ [\Phi]^{-1}$$

The instability condition equivalent of equation (8.9) gives:

$$\frac{dK_R}{da} = \left(\frac{\partial K}{\partial a}\right)_X$$

i.e. $K_C^{1-m}/(Cm) = K_C/\Phi$

$$K_C = \left(\frac{\Phi}{Cm}\right)^{1/m}$$

Now since Φ is dependent on crack length, size and specimen geometry it follows that the instability K_C is dependent on these variables. (See values of Φ/W, W being specimen size, for a range of specimens and crack lengths (a/W) shown in Table 8.1). Knowing K_C it is possible to determine the amount of slow crack growth Δa in each specimen geometry/crack length combination.

Table 8.1

Variation of geometric parameter Φ with a/W

	Φ/W		
a/W	Compact tension	Single edge notch	Three-point notch bend
0.1	17.43	0.163	0.219
0.2	2.75	0.237	0.353
0.3	0.53	0.264	0.389
0.4	0.40	0.256	0.364
0.5	0.34	0.234	0.311
0.6	0.25	0.216	0.265
0.7	0.19	0.209	0.236

Note that if the crack growth resistance curve in Example 8.5 is expressed as a function of the *actual* crack extension, i.e.

$$\Delta a = C(K_R - K_0)^m$$

Table 8.2
Values of material properties for steels and aluminium alloys

Material	σ_y MPa	F (kJ/m^2)	p	B (mm)	K_m/σ_y (mm$^{1/2}$)	R_0 (kJ/m^2)	K_0 MPa$\sqrt{m}$	K_{IC} MPa$\sqrt{m}$	K_m MPa$\sqrt{m}$	K_m/K_0
4130	1190	25	0.56	1.60	4.70	60	111	NA	177	1.59
4340	1490	58	0.83	1.27	3.30				122	2.20
		28	0.435	6.86	1.96	15	56	65	92	1.64
		21	0.27	25.4	1.75				82	1.46
Ti6A14V	1060	8	0.44	1.66	3.30	10	45	49	110	2.44
PH-15-7Mo	1540	46	0.60	1.27	6.60	50	101	70	322	3.19
PH-14-8Mo	1450	60	0.687	1.27	6.80	10	45	NA	311	7.00
Low carbon steel	280	50	0.25	0.76	23.20	90	138	NA	205	1.49
7475-T761	451	28	0.85	1.53	12.00	62	67	NA	171	2.56
7475-T61	482	24	0.85	1.53	8.90	41.3	53	NA	136	2.56
7079-T6	490	10	1.00	1.53	5.75	28.6	45	34	88	2.00
7075-T3	196	20	0.225	0.76	14.20	60	66	NA	88	1.34
7075-T6	545	10.5	0.660	1.53	3.88	27.8	45	37	67	1.50
2024-T3	301	46.0	0.625	1.53	17.60	24.5	42	NA	166	3.95
1100-0	63	20	0.40	0.76	55.30	63	68	NA	110	1.62

NA: K_{IC} values not available.

then by a similar procedure it is easy to prove that at the instability point, the critical stress intensity factor is given by:

$$K_C (K_C - K_0)^{m-1} = (\Phi/Cm)$$

K_0 is of course the threshold value below which no physical crack extension occurs corresponding with R_0.

The good stability characteristics of the Gurney–Mai–Owen rig described in section 3.9.2 make the apparatus suitable for the determination of complete R-curves in ductile metals. Fig. 8.7(b) shows the crack growth resistance curve for 0.76 mm thick low carbon steel (σ_y = 280 MPa, E = 210 GPa) obtained from the Gurney rig. First cracking occurs at 90 kJ/m^2 and rises to a plateau of some 210 kJ/m^2 after some 200 mm crack extension. (The excellent gsf even improve at longer crack lengths as seen by $\mathrm{d}R/\mathrm{d}a > (\partial G/\partial a)_u$ (chain lines) or $(\partial G/\partial a)_M$ (thin lines – at constant bending moment rather than constant load in this example)).

Table 8.2 sumarises some R-curve data, where R_0 is the toughness at which cracks start to propagate ($R_0 \approx K_0^2/E$) and K_m is the plateau toughness level (given here as a stress intensity rather than R_m). Analysis shows that the geometry of R-curves may be represented by $R = R_0 + F(\Delta a)^p$ where F and p are given in Table 8.2. This is similar to a polynomial relation given by Wang and McCabe (1974); Broek (e.g. 1973) uses $R = F(\Delta a)^p$ where, as already explained in Example 8.5, Δa includes a crack tip plasticity correction even when the crack is stationary.

We should expect K_m to coincide with K_{IC} when B is thick enough to satisfy the plane strain empirical thickness requirement, i.e. $B > 2.5\,(K_{IC}/\sigma_y)^2$. It is of interest that Table 8.2 suggests that R_0 (and $K_0 = \sqrt{ER_0}$) may correspond with plane strain values ($R_0 = G_{IC}$). While it is odd at first sight that plane stress fractures should initiate at plane strain values (given the different stress state and plastic zone size), we remember that the fully developed plane stress zone does not develop until K_m is reached, so perhaps there is enough constraint in the middle of even thin sheets to initiate cracks at K_{IC}. It is significant that R_0 has a common value for the same material when only thickness is changed (cf. 4340 steel in Table 8.2) and differences in R-curves for the same material concern the degree of curvature up to the different K_m plateaux.

8.6 PRACTICAL EXAMPLES OF ENERGETICALLY STABLE AND UNSTABLE ELASTIC CRACKING

For materials having constant toughness, $\mathrm{d}R/\mathrm{d}A = 0$ and stable cracking should occur in testpieces possessing negative n_u/A or n_X/A. Unstable cracking will occur in specimens with positive gsf. In broad terms the more negative the gsf the better. Of the commonly used testpieces, the following give stable propagation with stiff testing machines: plain and contoured DCB testpieces, peeling and trousers testpieces, compact tension testpieces, three-point specimens with sufficiently long starter cracks, Outwater double torsion testpieces, and ball indentation cone cracks. The following commonly used testpieces give unstable cracking in materials of constant R: centre-cracked and edge-cracked tensile panels, part-through cracked sheets, and three-point bend specimens with short starter cracks.

For materials whose toughness increases with crack velocity (i e. dR/dA or $dR/d\dot{A} > 0$), stability may be achieved even in testpieces with positive gsf. On the other hand, for materials whose toughness decreases with A or $\dot{A}$, cracking may be unstable even when the gsf is appreciably negative. In these latter cases, the techniques described in section 8.7 are often successful in promoting stability.

8.6.1 Choice of specimen

A very unstable testpiece geometry is the centre-notched tension panel in which $2a \ll 2W$ (number 6 or 11 in Appendix 1). The stability conditions for cracking under $dX/X > 0$ and $du/u > 0$ are the same and are given by $1/R(dR/dA) > 1/A$ so that $n_X = n_u = 1.0$. Therefore, unless dR/dA is sufficiently positive, stable cracking is difficult to achieve under any loading condition. This explains why centre-notched tension panels are neither useful in general for quasi-static R measurements nor useful for picking up R-curves, as little, if any, of the slow growth can be determined before instability. Only K_C values (at some unknown

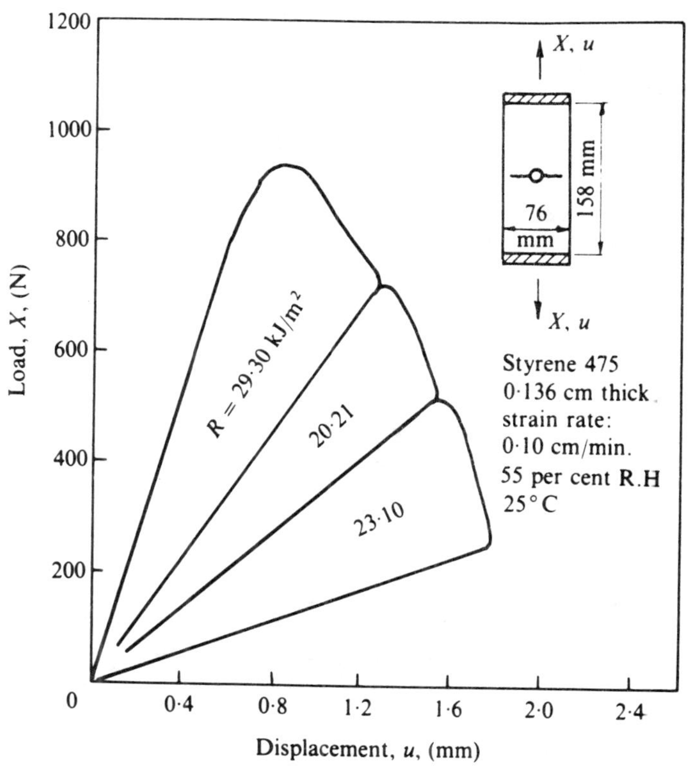

Fig. 8.8 – Stable cracking of Styrene 475 in centre-cracked specimen. (From Mai and Atkins (1980). Reprinted by permission of the Council of the Institution of Mechanical Engineers from *J. Strain Analysis.*)

fast velocity) can be calculated taking the load or displacement at which the crack goes bang, using linear elastic K solutions.

Some rubber-modified polymers, such as ABS, have $\mathrm{d}R/\mathrm{d}A$ quite strongly positive, and in such circumstances stable cracking can occur in the normally 'unstable' centre notched panel testpiece, as shown in Fig. 8.8 where the segmental work areas are used to calculate R according to Gurney's method. The crazes which develop ahead of the main crack front dissipate a considerable amount of work, and the viscoelastic nature of polymers promotes positive $\mathrm{d}R/\mathrm{d}\dot{A}$. Thus, although the specimen has considerable elastic energy available for cracking, the material's increasing need for more toughness work as the crack tries to speed up, compensates in such a way as to give stability.

Filament-reinforced composites can also give stable cracking in testpieces with unfavourable gsf, especially those composites with long pull-out lengths. There are often partially pulled-out fibres bridging the crack faces after passage of the main crack front. These filaments retard crack propagation as work has to be done in pulling them out when the main crack front has gone past, and this is the principal contribution to rising R-curves in fibrous composites. The longer the pull-out lengths, the greater the effect. Other components of composite toughness may not, of themselves, give stable cracking in testpieces with bad

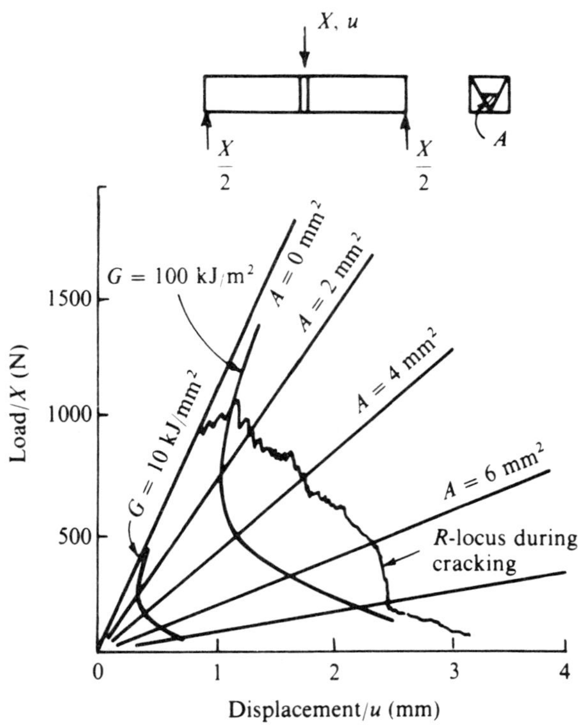

Fig. 8.9 – Stable cracking in inherently unstable Tattersall and Tappin testpiece. (From Mai and Atkins (1980). Reprinted by permisssion of the Council of the Institution of Mechanical Engineers from *J. Strain Analysis.*)

gsf – indeed epoxy matrices have $dR/d\dot{A}$ negative – but having to 'drag the bridging filaments along behind' can give stability. Thus, crack propagation is stable in Tattersall and Tappin three-point bend testpieces made from extra-tough boron-epoxy composites, even though the theoretical R-loci 'point backwards' in the early stages of crack propagation (Fig. 8.9). Notice that even though the crack propagation is overall stable, there are local precipitous load drops and arrests; this is perhaps caused by some fibres pulling right out at that instant so that the system loses their stabilising influence, only to regain stability at a slightly longer crack length when more filaments are pulled out and brought into play.

In thin sheets of ductile metals, shear lips lag the main crack front, which promotes dR/dA positive (see Chapter 2 and 4).

8.6.2 Instabilities caused by $dR/dA < 0$ or $dR/d\dot{A} < 0$

It is very easy to produce unstable fractures in epoxy resin, TPX plastics and low carbon steels where the experiments are catastrophic with precipitous load drops, like that in Fig. 8.1(a). Even specimens with quite good gsf are incapable of coping with the dR/dA-negative characteristics of these materials. Sometimes in testpieces with reasonably good gsf a sequence of stable and unstable propagation occurs in a jerky fashion (akin to Fig. 8.1(e)) where the specimen copes, then fails to cope, and so on. The work area triangles bounded by the precipitous drops correspond with mean toughness at unknown crack speeds in between every crack initiation and arrest.

A different illustration of a global instability is illustrated in Fig. 8.10. Here PMMA is being cracked using a double torsion specimen the n_u and n_X for which are −2 and 0 respectively (see Appendix 1). In air, PMMA has dR/dA-positive at room temperature, so continuous stable cracking is easily achieved at a constant load of some 400 N, since crack propagation occurs at constant velocity and constant load in this type of specimen; the R-locus of 360 J/m^2 is parallel to the u axis. When CCl_4 was squirted into the crack tip (at B in Fig. 8.10), the crack stopped moving because the environment-affected R for PMMA in CCl_4 is about 800 J/m^2 and thus requires a greater load (see Chapter 6). Loading proceeds at constant crack area along the stiffness line OBC until the appropriate cracking load of 600 N is reached and propagation starts again into the CCl_4-affected crack tip zone. However, when the crack emerges from that zone into unaffected PMMA, instability must occur as there is too much energy available at applied loads of magnitude 600 N in the region of C compared with the required load level of 400 N at B. When emerging from the CCl_4-affected zone into the virgin material, dR/dA effectively becomes negative so that stable cracking is impossible and the crack shoots ahead to break the testpiece.

A discussion of environmental effects on the stability of cracking in brittle polymers is given by Mai and Leete (1979).

Crack instabilities may also be produced in materials of low thermal conductivity, if R diminishes with temperature. Local crack tip heating caused by

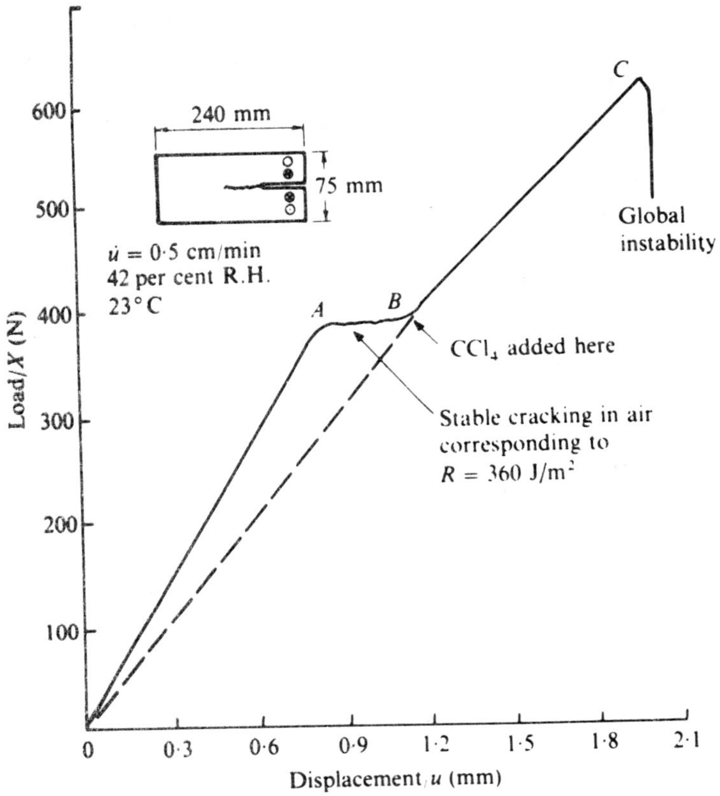

Fig. 8.10 – Cracking of a PMMA double torsion specimen: stable in air, unstable in CCl_4. (From Mai and Atkins (1980). Reprinted by Permission of the Council of the Institution of Mechanical Engineers from *J. Strain Analysis.*)

the dissipation of toughness work produces a local dR/dA-negative effect and hence cause unstable propagation in an experiment which otherwise would be stable. This is discussed in the next section.

8.6.3 Instabilities caused by local crack tip heating: limiting quasi-static crack velocities in dR/dT-negative materials

The localised heating arising at crack tips in materials of poor thermal conductivity has been discussed in section 6.2.3. According to isothermal–adiabatic transition theory the temperature rise at the crack tip process zone is not large. Even so, this local softening can cause a drop in E and hence in K_C or G_C in some solids, and thereby an effectively negative dR/dA behaviour. This clearly must affect the energetic stability of crack propagation, and suggests a limiting crack velocity for quasi-static propagation at every ambient temperature.

If we know $R(\dot{a},T)$ or $K_c(\dot{a},T)$, we can investigate the conditions under which instabilities will occur by re-writing equations (8.9) or (8.10) as

$$\frac{dR}{dA} \propto \frac{dR}{d\dot{a}}\,\frac{\ddot{a}}{\dot{a}} \leqslant (\partial J/\partial A)_{X \text{ or } u} \tag{8.14}$$

This involves a knowledge of crack acceleration $\ddot{a}$ which can only be determined experimentally. However, for some testpieces $(\partial J/\partial A) = 0$, e.g. the Outwater double torsion and tapered DCB constant rate of change of compliance geometries. Under these conditions equation (8.14) reduces simply to

$$\frac{dR}{d\dot{a}} \text{ or } \frac{dK_C}{d\dot{a}} \leqslant 0 \tag{8.15}$$

which was proposed by Williams and Marshall (1974). Note, however, that dJ/dA or $dK/dA \neq 0$ in general so that, for instance, the limiting crack velocity at instability $\dot{a}_c$ is geometry-dependent. The instability $\dot{a}_c$ so derived can be viewed as lower bound solution since the non-vanishing terms of $(\dot{a}/\ddot{a})(\partial K/\partial a)$ would help in achieving higher limiting crack speeds (see for example, Atkins *et al.*, 1975).

To obtain the $K_c - \dot{a}$ relationship, consider the fracture process zone in glassy polymers to be described by the Dugdale yield strip model (section 4.2.3) with a line zone at each crack tip of thickness (b) and length (c), Fig. 4.4. Assume now the heat generated in this process zone is due to the work dissipated in fracture, so the adiabatic temperature rise ΔT_{ad} is given by:

$$\Delta T_{ad} = \frac{R}{\rho s b} = \frac{(\delta_C \epsilon_y)^{1/2} K_C}{\rho s b} = \frac{\epsilon_y \delta_C E}{\rho s b} \tag{8.16}$$

where ρ is density and s the specific heat. The crack tip temperature rise for a given time t in presence of thermal conductivity k, using the yield strip model, is represented by (Carslaw and Jaeger, 1959):

$$\Delta T = \Delta T_{ad} \; [1 - 4i^2 \; erfc \left(\frac{b^2 \rho s}{16kt}\right) \;] \tag{8.17}$$

where i^2 *erfc* refers to the second integral of $(1 - erfx)$ with *erfx* as the error function. As b for many glassy polymers is very small, being in the range (300–1000 nm), we may sensibly investigate the lower bound solution to the limiting crack speed using $b \to 0$ which gives

$$\Delta T = \frac{\epsilon_y \delta_C E}{\sqrt{\pi \rho s k t}}$$

(Parvin (1979) has solved equation (8.17) for $b \neq 0$). The time-scale appropriate to the craze zone may be obtained by considering the crack tip strain rate in this region (by substituting $\dot{\epsilon}_m = \epsilon_y/t$ in equation (6.4) so

$$t = \frac{1}{\pi \epsilon_y^2} \left(\frac{K_C}{E}\right)^2 \frac{1}{\dot{a}} \tag{8.18}$$

and the temperature rise becomes

$$\Delta T = \frac{(\delta_C \epsilon_y)^{1/2} \; \epsilon_y E \dot{a}^{1/2}}{(\rho s k)^{1/2}} = \frac{\epsilon_y K_C}{\sqrt{\rho s k}} \; \dot{a}^{1/2} \tag{8.19}$$

The rate- and temperature-dependence of E at the crack tip region can be expressed in terms of an Arrhenius equation given in section 6.3

$$E = E_0\, t^{-n} \exp\left(\frac{nQ}{\bar{k}}\left[\frac{1}{T} - \frac{1}{T_0}\right]\right)$$

Also, K_C has already been given by equation (6.21), as

$$K_C = (\delta_C \epsilon_y)^{1/2}\, E = (\delta_C \epsilon_y)^{1/2}\, E_0 t^{-n} \exp\left(\frac{nQ}{\bar{k}}\left[\frac{1}{T} - \frac{1}{T_0}\right]\right)$$

Substituting t in terms of $\dot{a}$ from equation (8.18) gives

$$K_C = (\delta_C \epsilon_y)^{1/2}\left(\frac{\pi \epsilon_y}{\delta_C}\right)^n E_0 \dot{a}^n \exp\left(\frac{nQ}{\bar{k}}\left[\frac{1}{T} - \frac{1}{T_0}\right]\right) \qquad (8.20)$$

This is almost identical to equation (6.23) obtained earlier in section 6.3 for the rate and temperature effects on K_C, except for a slight difference in the constant of the term (ϵ_y/δ_C) in the second bracket, i.e. π as opposed to $(8/\pi)$. This discrepancy arises from equations (6.22) and (8.18) for t.

It is now possible to use equations (8.19) and (8.20) to solve the instability temperature (T_c) and limiting crack velocity ($\dot{a}_c$) for a given ambient T_0 using $\mathrm{d}K_C/\mathrm{d}\dot{a} \leqslant 0$ for those special geometries to which it applies.

Thus, differentiating K_C of equation (8.20) with respect to $\dot{a}$ and setting this to zero gives:

$$\frac{\mathrm{d}T}{\mathrm{d}\dot{a}} = \frac{\bar{k}T^2}{Q\dot{a}} = \frac{(\delta_C \epsilon_y)^{1/2} \epsilon_y E}{\sqrt{\rho s k}} \cdot \frac{1}{2\sqrt{\dot{a}}} \qquad (8.21)$$

the last expression follows from $\mathrm{d}\Delta T/\mathrm{d}\dot{a}$ in equation (8.19). At the instability condition ($\dot{a} = \dot{a}_C$ and $T = T_c$) and rearranging (8.21) we have:

$$\frac{2\bar{k}T_c^2}{Q} = \frac{(\delta_C \epsilon_y)^{1/2}\, \epsilon_y E}{\sqrt{\rho s k}} (\dot{a}_c)^{1/2} = \frac{\epsilon_y K_C (\dot{a}_c)^{1/2}}{\sqrt{\rho s k}} \qquad (8.22a)$$

and

$$\frac{2\bar{k}T_c^2}{Q} = T_c - T_0 = \Delta T \qquad (8.22b)$$

The above two equations allow T_c and $\dot{a}_c$ to be determined for any given T_0 at fracture instability. Considering the same example for PMMA given in section 6.2.3, we obtain T_c = 39.5°C for T_0 = 20°C so ΔT = 19.5°C. At this temperature the instability K_c = 1.75 MPa$\sqrt{\text{m}}$ so that the predicted limiting crack speed $\dot{a}_C$ = 12 mm/s. This agrees fairly well with an experimental measurement of 50 mm/s. ΔT and $\dot{a}_c$ for other ambient temperatures T_0 can be similarly obtained and are shown in Table 8.3.† There is close agreement between predicted

† Here $\dot{a}_C$ are predicted based on an instability K_C = 0.90 MPa$\sqrt{\text{m}}$ for all ambient temperatures. In fact the instability K_C values are not constant and depend on the ambient temperature (Chapter 6); when this is taken into account, the predicted $\dot{a}_C$ do not give as good agreement with experimental measurements as are shown in Table 8.3.

Table 8.3
Temperature rise and crack speeds at instability for PMMA
(Williams and Marshall, 1974)
[$K_C = 0.90$ MPa$\sqrt{m}$, $s = 5\ T_0$ (J/°C kg), T_0 in K]

T_0 (°C)	$\Delta T = T - T_0$ (°C)	$\dot{a}_c$ (mm/s)	Experimental $\dot{a}_c$ (mm/s)
80	29.2	126.4	240
60	25.7	92.4	140
40	22.5	66.5	85
20	19.5	46.8	50
0	16.8	32.4	40
−20	14.3	21.7	24
−40	12.0	14.0	17
−60	9.9	8.8	1.4

values of $\dot{a}_c$ and experimental results except at the lowest temperature, $T_0 = 60$°C. This seems to support the isothermal–adiabatic transition theory for fracture instability for PMMA within the temperature range −40°C to +80°C. Indeed further support for the theory has been obtained for two other glassy polymers, polycarbonate (Parvin, 1979) and polystyrene (Mai and Atkins, 1976).

Earlier it was acknowledged that $\dot{a}_c$ may be geometry-dependent owing to $(\partial K/\partial a) \neq 0$. Fig. 8.11 compares the instability $\dot{a}_c$ data for PMMA using grooved

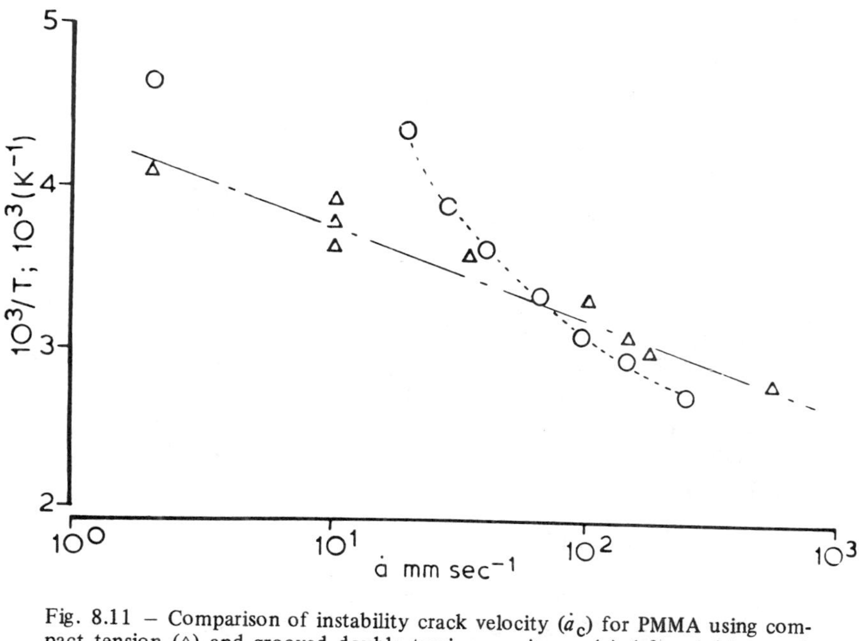

Fig. 8.11 – Comparison of instability crack velocity ($\dot{a}_c$) for PMMA using compact tension (△) and grooved double torsion specimens (○) (after Atkins *et al.* (1975), with permission).

double torsion (DT) and compact tension (CT) specimens under displacement-controlled conditions. The generally higher $\dot{a}_C$ values for DT specimens seems to agree with the theory that $(\partial K/\partial a)_u$ for a *grooved* DT geometry can be more negative than a CT geometry if the groove ratio is correct (Atkins *et al.*, 1975), cf. the extreme left-hand point of Williams and Marshall's data in Fig. 8.11. (Grooving as a means of promoting crack stability is discussed in section 8.7 following.)

***Example** 8.5* Using the isothermal–adiabatic transition theory estimate the instability crack speeds for a general-purpose-grade polystyene over a range of temperatures.

Ans. From the *Polymer Handbook* (Burrel and Immergut, 1966) we obtain the following physical constants for polystyrene: $\rho = 1.05$, $k = 0.105$ W/m°C, $s \cong 4.2\ T_0$ (J/°C kg). Experiments show that $\epsilon_y = 0.015$, $Q/\bar{k} = 2.5 \times 10^4$ and K_C is a function of temperature as shown by Mai and Atkins (1976). Using equations (8.22a) and (8.22b) the predicted instability $\dot{a}_C$ and ΔT values are:

T_0 (°C)	$\Delta T = T_C - T_0$ (°C)	$\dot{a}_C$ (mm/s)	Experimental $\dot{a}_C$ (mm/s)
25	7.40	(8.4) 1.34	3
50	8.80	(12.8) 3.26	5
70	9.00	(14.2) 6.86	10
90	12.00	(26.7) 30.00	13

Except at the highest test temperature of 90°C there is very good agreement between predicted and measured $\dot{a}_C$ values if the lowest K_C (= 2MPa$\sqrt{\text{m}}$) is used to predict $\dot{a}_C$ for all T_0, i.e. disregarding the temperature dependence of instability K_C. When $K_C(T)$ is taken into account the agreement is not as good. These results are shown in parentheses in the above table.

In Lee's 1974 study of toughness in PMMA, simultaneous pairs of K_C and R for quasi-static cracks were determined at various crack velocities and temperatures. The corresponding E's $(= K_C^2/R)$ were compared with independent uniaxial tensile Young's moduli at various strain rates and temperatures, in order to establish crack tip effective strain rates. In the range $10^{-4} < \dot{a} < 10^{-2}$ m/s it was found that

$$\dot{\epsilon} \approx 0.13\,\dot{a}$$

which was independent of temperature within the limits of experimental accuracy. For $\dot{a} > 10^{-2}$ m/s, $\dot{\epsilon}$ rose less steeply and flattened off at about $\epsilon \approx 10^{-2}$ s^{-1} (Fig. 8.12). Using equation (6.4) for $\dot{\epsilon}$ and the toughness-biased Ree–Eyring

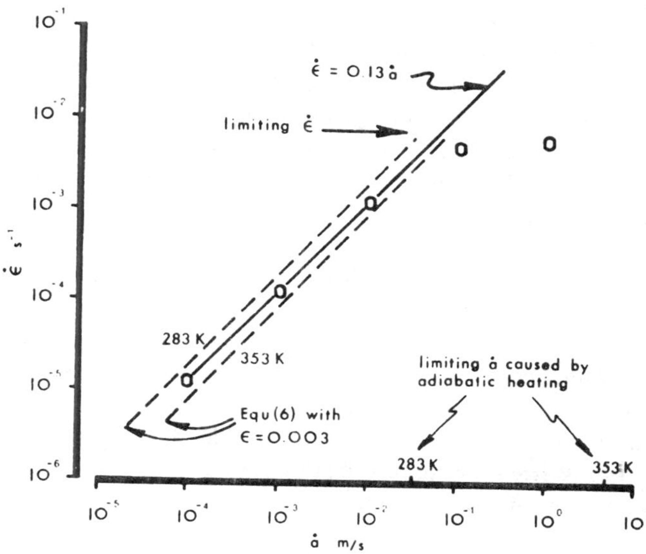

Fig. 8.12 – Relationship between crack tip strain rate ($\dot{\epsilon}$) and crack velocity ($\dot{a}$) for PMMA (after Lee *et al.* (1975), with permission).

relation for $R(\dot{a},T)$ in PMMA, equation (6.26), a theoretical relation was obtained between $\dot{\epsilon}$ and $\dot{a}$ (Problem 8.23). This was dependent on temperature and the predictions for T = 283 K and 353 K are superimposed on Fig. 8.12, using $\epsilon_y \approx 0.003$ (which is rather low as discussed by Lee *et al.* in 1975). The theoretical $\dot{\epsilon} \sim \dot{a}$ relation is continuously increasing and does not reflect the limiting instability velocity for each experimental temperature caused by crack tip adiabatic heating. However, these limiting velocities coincide with the region of the experimental $\dot{\epsilon}$ vs. $\dot{a}$ data where $\dot{\epsilon}$ level off at some 10^{-2} s^{-1}. That this strain rate level falls below the theoretical (rising) $\dot{\epsilon}$ vs. $\dot{a}$ relation makes sense in that the temperature-dependence of the relation gives lower $\dot{\epsilon}$ at higher T.

8.6.4 Instabilities caused by blunt crack tips

K_C values determined from crack initiation loads diminish as the crack tip radius diminishes, until a minimum value is reached when further sharpening of the crack does not seem to have any effect (Fig. 2.33). This type of behaviour has led to requirements for starter cracks to be fatigued (or otherwise sharpened) in standard fracture toughness tests. The reason for the larger K from blunt cracks is simply that for a given applied load the stress intensity at a blunt crack tip is smaller than for a sharp crack tip. It is thus difficult at a blunt notch to initiate a crack to its natural running tip radius, but once initiated there is more energy in the cracked structure than is required for further cracking, so an instability results. This is obviously true when the specimen has a bad gsf which would encourage unstable cracking anyway, but it is also true in testpieces with good gsf. The blunter the crack, the worse the instability effect.

In the CCl_4-affected PMMA case shown in Fig. 8.10 it is probable that the liquid affects the crazes ahead of the crack tip, softening the material immediately beyond the tip, thereby blunting the tip and relieving the local stresses. Consequently the load required to cause crack propagation is elevated. Kinloch and Williams (1980) argue that unstable cracking in certain epoxy resins is largely due to extensive plastic deformation at the crack tip region which causes a blunting effect.

Blunting of crack tips can sometimes be used to advantage, to produce effects that sharp cracks fail to display. For example, polycarbonate sometimes cracks with low propagation resistance, but at other times cracks with large crack tip zones, depending upon sheet thickness, temperature, loading rate etc. The formation of large zones can be encouraged by slight crack tip blunting. Propagation, once started, occurs at a natural crack sharpness formed by the craze-initiated crack running from the blunted starter crack, but the work of fracture is large, owing to the large crack tip zone, which is recreated ahead of the moving (sharp) crack tip.

8.7 METHODS OF PROMOTING STABLE CRACKING

It is clear from equations (8.9) and (8.10) that crack stability in a given situation is promoted by (a) maximising dR/dA and/or (b) minimising $\partial J/\partial A$ in so far as this is possible. In the subsections following, various means of achieving these ends are described. Minimisation of $\partial J/\partial A$ into the crack would mean that excess energy is diverted to places other than the crack and absorbed at those places by special means. Maximisation of dR/dA would mean that more work is made necessary to propagate the crack than normally required, by use of additional energy sinks associated with cracking. In so far as the means by which released strain energy is diverted may be the same as those by which additonal toughness work is demanded, the two approaches are similar. In this way some of the devices described below could fit into either category. Consideration of grooving testpieces as an aid to stability is dealt with subsequently in section 8.7.2.2.

8.7.1 Maximising dR/dA

Materials prossessing moderately negative dR/dA or $dR/d\dot{A}$ characteristics may be coupled with a sufficiently dR/dA positive material of known toughness to give a sandwich which displays a composite dR/dA positive and which thus should crack stably when made up into a testpiece possessing a good gsf. This means physically that as the composite crack grows, the increasing incremental work input required in the dR/dA-positive layer may be made to be greater than the excess energy being released as the dR/dA-negative layer cracks.

The fracture toughness of the unstable material (R_2 say) can be determined from the difference between the specific crack work area for the sandwich (R^*) and that of the known dR/dA-positive component (R_1), using a rule of mixtures on R, i.e.

$$R^* = R_1 B_1/(B_1 + B_2) + R_2 B_2/(B_1 + B_2)$$

or

$$R_2 = (B_1/B_2)(R^* - R_1) + R^* \qquad (8.23)$$

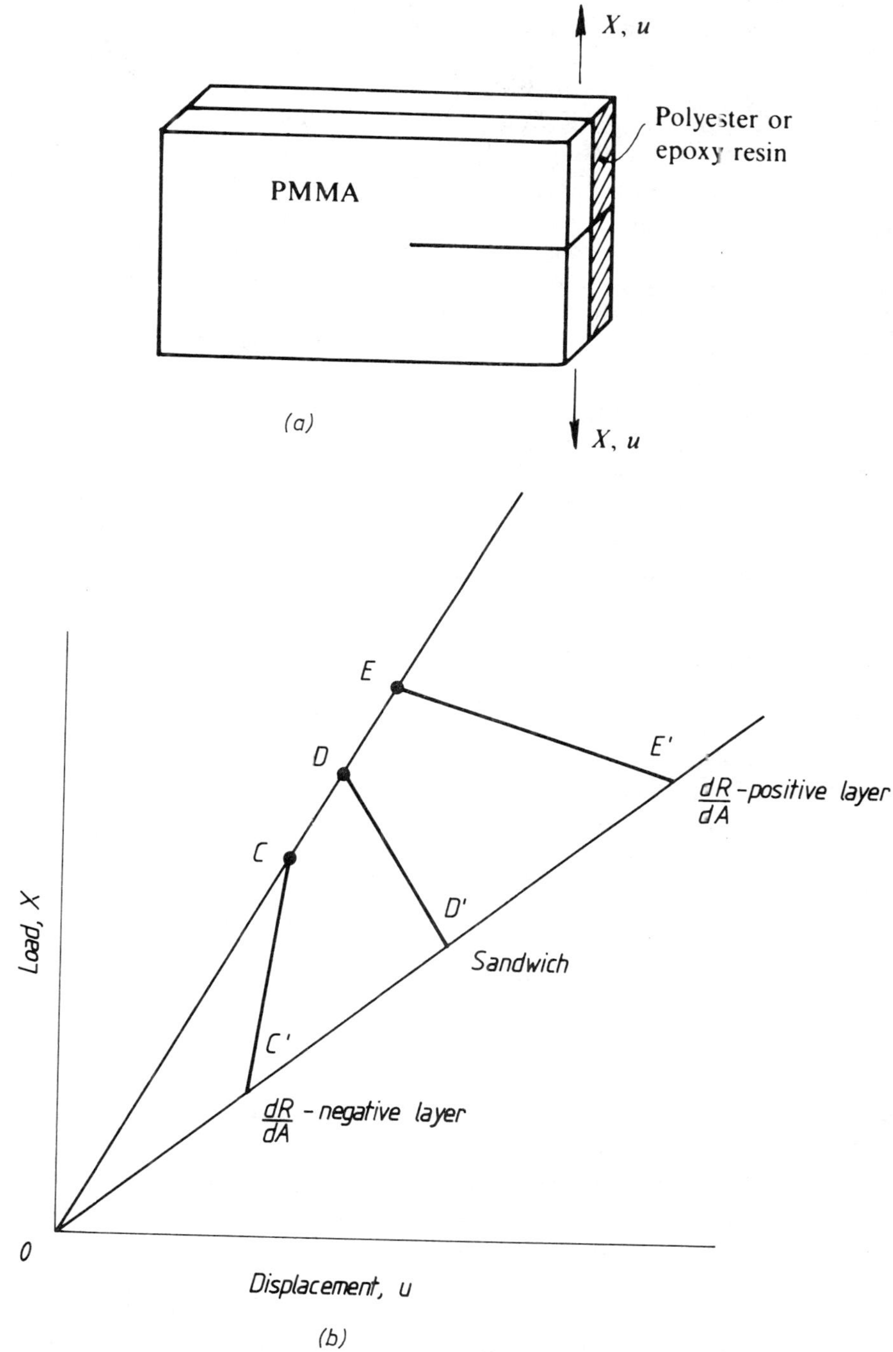

Fig. 8.13 – (a) Sandwiching unstable polyester or epoxy resins with PMMA. (From Mai and Atkins (1980). Reprinted by permission of the Council of the Institution of Mechanical Engineers from *J. Strain Analysis.*) (b) The mechanism of stable cracking in terms of the theory of fracture in layered structures (Chapter 3).

where B_1 and B_2 are layer thicknesses (see equations (3.36) and (3.40)). For example, epoxy and polyester resins which have negative $dR/d\dot{A}$ characteristics can be made to crack stably by gluing a sheet of PMMA in parallel as shown in Fig. 8.13(a). The means by which stability of cracking is attained may be explained in terms of the theory of cracking of layered structures given in section 3.6: in Fig. 8.13(b), CC' would be the energy release rate locus for the dR/dA-negative layer and EE' for the dR/dA-positive layer. The apparent behaviour for each layer in a composite is given by DD' which, by suitable choice, can be made to obey $du/u > 0$. (For simplicity layers of equal stiffness are shown in Fig. 8.13(b), but similar arguments apply to the likes of Fig. 3.29(b) where one component has a 'backwards pointing' J-locus.)

8.7.2 Minimising $(\partial J/\partial A)_u$ or X

There are two principal methods which result in $\partial J/\partial A < 0$ and hence promote stable cracking in otherwise unstable materials. The first is modification of specimen geometries in order to limit the release of strain energy. The second is the introduction of appropriate external stabilisers which absorb excess energy released from unstable testpieces.

8.7.2.1 *Stability by suitable specimen design*

As a simple example of demonstrating how the gsf of a specimen can be changed, let us consider a double-ended cantilever beam specimen with a central meridian crack of length a, (number 3 in Appendix 1). In a stiff testing machine, stable cracks should be readily produced in materials with R constant or dR/da positive, but stability in soft machines is only possible with materials possessing strongly dR/da positive.

The stability conditions under $dX/X > 0$ may be improved by drilling a hole at the end of the crack (number 22 in Appendix 1). In place of the beam theory compliance $u/X = a^3/96EI$ used formerly, we now have

$$u/X = (2a^3 b^3)/(3EI(a+b)^3)$$

and it may be shown that n_X is improved considerably (although n_u is not quite as good as formerly).

8.7.2.2 *Stability improvement using external stabilisers*

As detailed by Gurney and Mai in 1972, external stabilisers may take the form of metal springs, rate-sensitive adhesive tapes, or inertia devices placed across the most probable crack path of the fracture toughness specimen. The choice of which type of stabiliser depends on the degree of instability required to be overcome.

Consider a cracked body with springs attached around and in-line with the crack (Fig. 8.14). During propagation the springs are stretched as the crack spreads. Note that this occurs even though the machine crosshead displacement may be 'instantaneously fixed'. Work has to be performed in stretching these springs, and the concept of crack stabilisation in this instance is that the excess energy in the system at crack initiation is absorbed by the work of stretching.

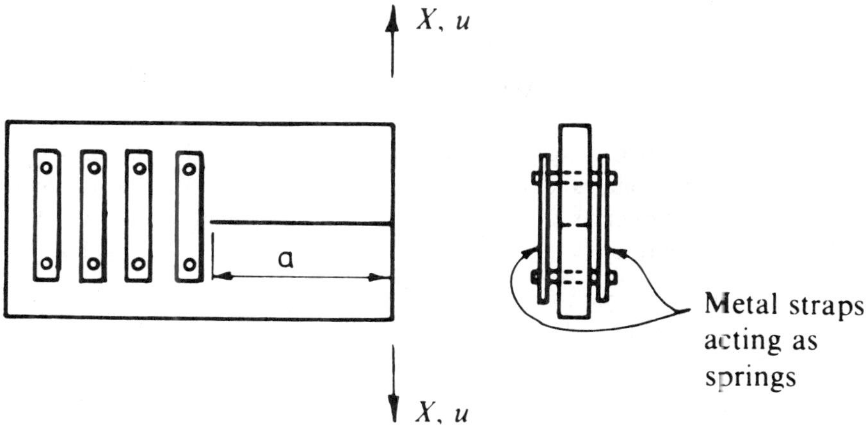

Fig. 8.14 – Stabilising unstable cracking by means of metal springs placed across crack path. (From Mai and Atkins (1980). Reprinted by permission of the Council of the Institution of Mechanical Engineers from *J. Strain. Analysis.*)

The springs must have appropriate stiffness, which absorb adequate work at the relevant displacements, and must be spaced accordingly. Clearly, the closer the spacing the better, as temporary instabilities could occur as the crack front moves along between spring locations.

When this method slows down a crack such that controlled unloading is possible, the strain energy stored in the stretched springs is completely reversible, and goes back out of the system to the testing machine. Gurney's irreversible work area method for R measurement can then be used. It is important to appreciate in this case that the stabilising springs themselves should not yield or break. Were spring fractures to occur during main crack propagation, the energy released could cause unstable cracking in the main body.

Example 8.6 A material with R = 30 kJ/m^2 displays dR/dA = – 54 MJ/m^4 over the range of interst of crack length changes. The uncreacked ligament in a 10 mm thick beam-like testpiece of this material is 50 mm long, and the starter crack is 250 mm long. The beam half height is 25 mm and the Young's modulus is 70 GPa. Show that cracking is likely to be unstable, and investigate the stabilisation of crack spreading by means of external springs attached to the testpiece. Assume simple beam theory applies.

Ans. The gsf of a DCB-type specimen loaded by end moments in a stiff machine is $(-4/aB)m^{-2}$ = – 1600 m^{-2} for a 250 mm starter crack. $(1/R)$ dR/dA = – 54 × 10^6/30 × 10^3 = – 1800 m^{-2}. Therefore as – 1800 $\not>$ – 1600, cracking is unstable. Having traversed the 50 mm crack path in the 10 mm thick testpiece, the material toughness drops to

$$30 \times 10^3 - 54 \times 10^6 \times 50 \times 10^{-3} \times 10 \times 10^{-3} \quad = 3\ \text{kJ/m}^2$$

Assuming a linear change in R, the average toughness for crack spreading is $(30 + 3)/2 = 16.5\ \text{kJ/m}^2$. Thus the total work required is

$$16.5 \times 10^3 \times 50 \times 10^{-3} \times 10 \times 10^{-3} = 8.25\ \text{J}$$

The strain energy available at crack initiation is given by $\frac{1}{2}Xu = \frac{1}{2}X^2(u/X)$, where the compliance (u/X) is given by $(2a^3/3EI)$ from the beam theory and X^2 at cracking comes from $GEIB/a^2$ for DCB specimens. Therefore the available strain energy is

$$\frac{1}{2}\left(\frac{GEIB}{a^2}\right)\left(\frac{2a^3}{3EI}\right) = \frac{GaB}{3} = \frac{30 \times 10^3 \times 250 \times 10^{-3} \times 10 \times 10^{-3}}{3} = 25\ \text{J}$$

Clearly there are $(25 - 8.25) = 16.75$ J excess to the requirements for stable cracking.

At crack initiation the displacement of the loading pins is given by

$$u = 2Xa^3/3EI = 4a^2\sqrt{R/3Eh^3}$$

$$= 4(250)^2\,(10^{-6})\sqrt{30 \times 10^3/3 \times 70 \times 10^9 \times (25)^3\,(10^{-9})} = 23\ \text{mm}$$

Assume that the crack propagates unstably to the end of the specimen at this fixed u. Then, assuming also for simplicity a linear shape for the opened faces, the displacement δ of the faces at some distance x measured from the back end of the specimen is

$$\delta = \left(\frac{x}{(250 + 50)\,10^{-3}}\right) 23\ \text{mm}$$

The work done on a spring at such a location is

$$W = \tfrac{1}{2}\,k\,\delta^2$$

where k is the spring stiffness.

If there are n springs of the same stiffness located along the crack path at $x, 2x, 3x, \ldots, nx$, measured from the back end, the total work done is

$$\text{Total } W = \tfrac{1}{2}\,k\,\delta^2\,(1 + 2^2 + 3^2 + \ldots + n^2)$$

Stability should be achieved if this absorbed work is greater than the 16.75J excess.

This allows an approximate determination of the spacing and stiffness of the required springs. For example, let there be five springs at 10 mm centres from the backface (i.e. along merely the uncracked ligament). We have

$$16.75 < \tfrac{1}{2}\,k\left(\frac{10}{300}\right)^2 (23 \times 10^{-3})^2\,(1^2 + 2^2 + 3^2 + 4^2 + 5^2)$$

$$< 16.16\,k\,10^{-6}$$

therefore

$$k > 1.04 \text{ MN/m}$$

If the springs are in the form of steel straps, say 40 mm tall between attachment pins by 8 mm width, we have

$$k = X/u = Ewt/L = 210 \times 10^9 \times 8 \times 10^{-3} \times t/40 \times 10^{-3} = 42 \times 10^9\, t$$

where t is the total thickness of the steel strap. If $k \approx 1.04$ MN/m, $t \approx (1/42)$ mm, which is quite thin.

In the fracture testing of inherently unstable materials such as certain species of wood and polymers, which have much smaller toughnesses than metals, much lower spring constant k values are required for stable cracking and thus materials other than metal springs may be used as stiffening agents. In the case of an unstable epoxy resin Gurney and Mai (1972) used 'continuous' springs formed by attaching sticky adhesive tapes all along the crack and on each side of the testpiece to obtain stable cracking. Of course, the 'solid sandwiching' of section 8.7.1. may be thought of as a thick continuous tape.

Fig. 8.15(a) shows unstable stick-slip cracking in a double-ended DCB specimen of epoxy resin. The R values correspond to some mean speed of cracking which is unknown. Fig. 8.15(b) shows the stability achieved by using adhesive tapes as springs. The deflexions of the tape are small, so that the tape acts more-or-less reversibly with little flow. In other situations where the specimen deflexions are large (because R is relatively large), a not-insignificant amount of work is done in performing viscous slip of the tapes, as the stabilised crack propagates. Gurney's work area technique is nevertheless applicable as unloading lines *do* go back to the origin, the slipped tapes forming slack loops alongside the crack (Example 4.1), the viscous work being deducted in the calculations.

Example 8.7 Fig. 8.16 shows an Xu plot at room temperature (45 per cent relative humidity) for ICI TPX which is stabilised by an adhesive tape. On unloading, loops of tape were left alongside the specimen. These were removed, and replaced by similar pieces of tape again laid smoothly across the crack. The crosshead of the testing machine was run out at the same rate to the same displacement as had occurred during the stabilised cracking experiment, and then unloaded to give slack loops as before. From the area of the hysteresis loop obtained, the irreversible work performed in the tape in the original experiment as the crack advanced, was estimated to be about 1 kJ/m^2 increase in area of crack. Determine R for TPX.

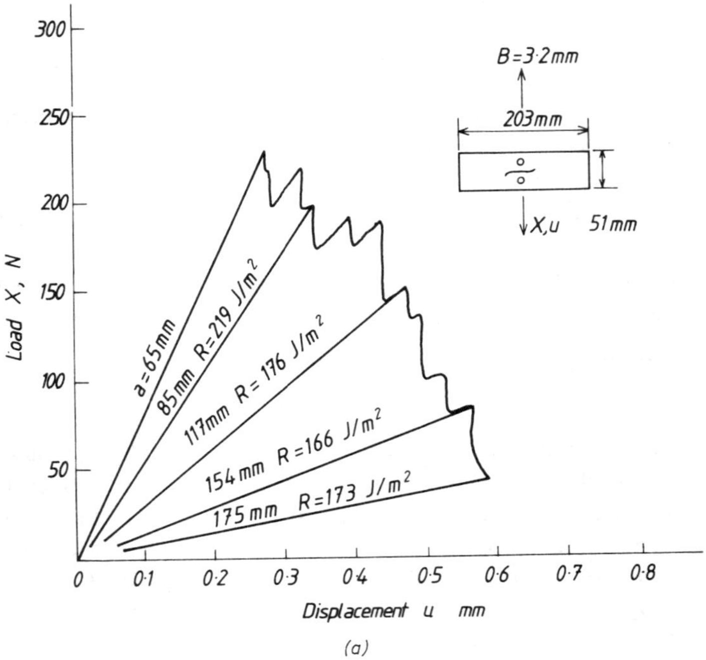

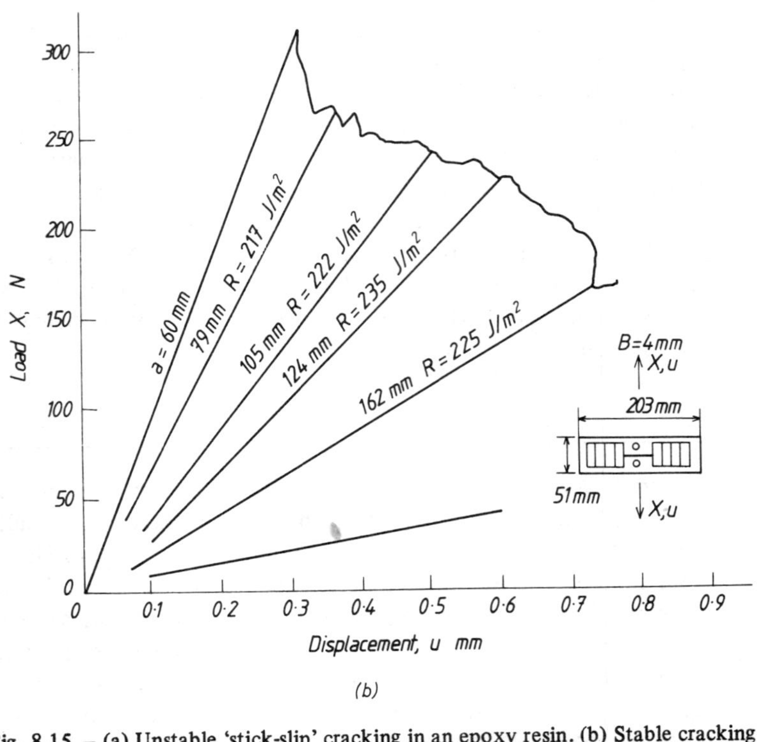

Fig. 8.15 – (a) Unstable 'stick-slip' cracking in an epoxy resin. (b) Stable cracking of an epoxy resin by attaching two layers of adhesive tape on each side of specimen (both after Gurney and Mai (1972), with permission).

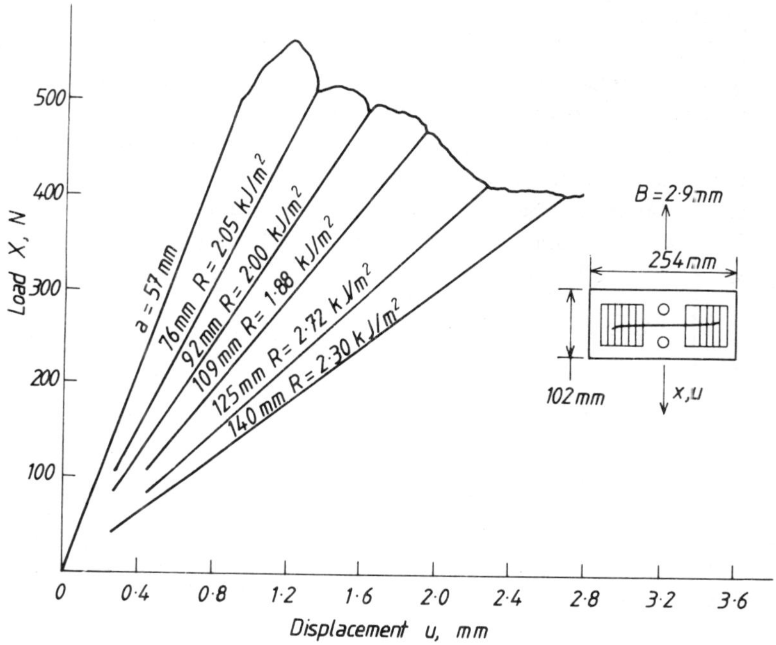

Fig. 8.16 – Stable cracking of T.P.X. when taped across the crack path (after Gurney and Mai (1972), with permission).

Ans. Measurement of work areas from the *Xu* diagram, for the associated changes in crack length, give an average total work of about 2 kJ/m². If the tape work is about 1 kJ/m², then the TPX fracture toughness is estimated to be $(2 - 1) = 1$ kJ/m².

Viscous tapes are rate-sensitive devices. Other rate-sensitive or inertia devices make use of processes that are approximately reversible. For example, Gurney and Mai in 1972 used a series of mercury-filled hypodermic syringes which straddle a testpiece on stirrups. As the crack propagates, the dense fluid is squirted out, and the associated forces on the plungers of the syringers serve to slow down and stabilise crack spreading in unstable cases. Bernoulli's theorem for the forces allows design of the location and size of the syringes. An alternative to such syringes are metal diaphragm oilcans filled with mercury, which would provide mixed spring/inertia stabilisation devices. The interested reader is referred to that original paper.

An alternative interpretation for the success of springs, viscous tapes and other devices in promoting stability is given by beam-on-elastic foundation models of crack propagation, used (originally for crack arrest purposes) by Kanninen and co-workers (e.g. Kanninen, 1974a) and Gurney and co-workers (1974). Indeed, such models explain why grooved testpieces display better gsf than ungrooved specimens (for the same R). A thin layer of material in the path of the crack is considered to be the 'foundation' on which the 'beams' (rest of testpiece above and below the crack direction) deflect, Fig. 8.17. In the presence of grooves the foundation has an obviously different bending stiffness from the rest of the cracked solid. Both the tensile stretching of the groove immediately ahead of the crack tip, and the compressive squashing of the groove far ahead of the crack, react on the 'beams' thereby affecting the overall X/u stiffness characteristics of the cracked body, the associated $d/dA\ (X/u)$ and so on, and hence $(\partial J/\partial A)_{X \text{ or } u}$.

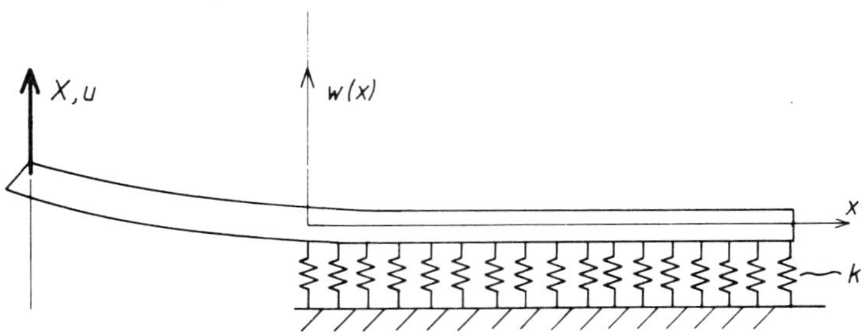

Fig. 8.17 – The beam-on-elastic-foundation fracture model.

Stabilisation produced by sticky tapes may be explained if the continuous tapes constitute an elastic foundation. The compliance of the system is approximately

$$\frac{u}{X} = \frac{\lambda}{2k}\left(\frac{\cosh \lambda a + \cos \lambda a - 2}{\sinh \lambda a + \sin \lambda a}\right) \tag{8.24}$$

(Hetenyi, 1947) where k is the stiffness per unit beam length of the tape and $\lambda = [k/4EI]^{1/4}$. n_u gsf values may be determined by two differentiations of this expression for u/X to use in equation (8.3). Gurney and Mai obtained

λa	0	1	2	3	4	5	6	6.5
n_u	−4	−4	−4.2	−4.8	−6.8	−12	−48	−212

Note that when $\lambda a = 0$, $n_u = -4$ which is the result obtained for a simple built-in beam which does not have stabilising tapes and which neglects effects ahead of the crack (section 8.2). When λa is sufficiently large (> 3 say), i.e. the springs are sufficiently stiff, stability is greatly improved, as shown by Fig. 8.15(b).

8.8 TURNER'S η FACTOR AND UPPER BOUND TOUGHNESSES

In those cases where specimens break catastrophically with a precipitous load drop (Fig. 8.1(a)), the triangular work area under the Xu plot must be an overestimate of the work of fracture, or otherwise the crack would stop within the confines of the specimen. The 'real' R is given by $-(\partial\Lambda/\partial A)_u$ and concerns rates of change of Λ. It is of interest to compare this 'real' R with an 'apparent' R obtained on the assumption that the total elastic strain energy at first fracture is just dissipated in breaking the body, i.e. in the case of a remaining ligament of length $(W-a)$ in the path of a crack, all Λ is taken up in feeding the crack which then has an apparent toughness is $\Lambda/B(W-a)$. Possible work inputs to the system during propagation are ignored.

In the case of linear bodies $\Lambda = \frac{1}{2}Xu = \frac{1}{2}X^2(u/X)$. The compliance (u/X) may be known for certain simple geometries, but may always be determined from K-formulae as already explained in section 8.2. We have

$$\mathrm{d}/\mathrm{d}A\,(u/X) = 2K^2/E^*X^2 = 2\sigma^2\pi a Y^2/E^*X^2 \qquad (3.4\text{ bis})$$

where shape factor Y is a function of the geometry of the cracked body. X and σ are related by an area term, for example $\sigma_{\text{remote}} = X/WB$ or $\sigma_{\text{net}} = X/(W-a)B$ (the Y factor will be different for different difinitions of σ). Hence

$$\mathrm{d}/\mathrm{d}A\,(u/X) = \frac{2\pi a Y^2}{E^*(W-a)^2B^2} \qquad (8.25)$$

Since $\mathrm{d}A = B\mathrm{d}a$

$$(u/X) = \int \frac{2\pi Y^2 a\,\mathrm{d}a}{E^*(W-a)^2B} + \text{constant} \qquad (8.26)$$

where the constant is the compliance of the crack-free body. Thus

$$\Lambda = \tfrac{1}{2}X^2\left\{\int \frac{2\pi Y^2 a\,\mathrm{d}a}{E^*(W-a)^2B} + \text{constant}\right\} \qquad (8.27)$$

If the compliance of the crack-free body is extremely small, the *apparent* toughness is

$$\frac{\Lambda}{B(W-a)} = \frac{\pi X^2 \int Y^2 a\mathrm{d}a}{B^2(W-a)^3E^*} \qquad (8.28)$$

But the *real* toughness is

$$R = K^2/E^* = \sigma^2\pi a\,Y^2/E^* = X^2\pi a\,Y^2/E^*(W-a)^2B^2 \qquad (8.29)$$

Consequently

$$R = \eta\Lambda/B(W-a) \qquad (8.30)$$

where

$$\eta = (W-a)\,Y^2a/\int Y^2a\mathrm{d}a \qquad (8.31)$$

This general relation (which is of course valid for both *stable* and *unstable* cases) was first derived by Turner in 1973. When Y is known as a polynomial in (a/W), η may be found via direct integration of $\int Y^2 a da$ – noting that there can be problems connected with convergence of the series. Problem 8.18 shows that, in this way, $\eta \to 2$ for deeply notched three-point bend testpieces. This means that, for this specimen, the total area under the complete *Xu*-curve for complete fracture (OCE in Fig. 8.18) is twice the triangular area OCD which represents Λ at first cracking (cf. Example 2.4). When $Y\,(a/W)$ is given as a calibration curve, $\int Y^2 a da$ may be evalulated graphically.

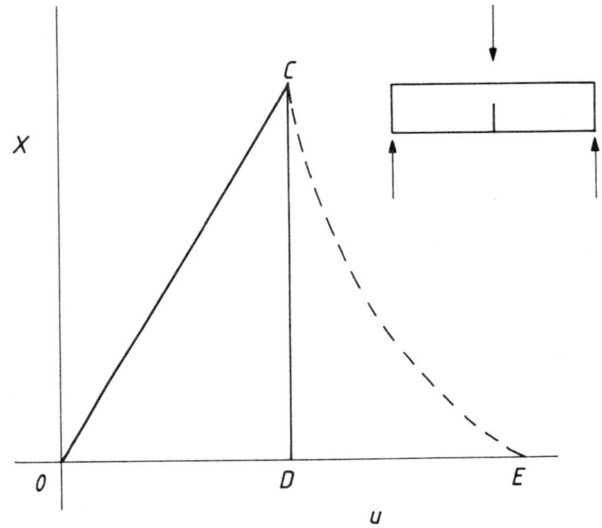

Fig. 8.18 – The work of fracture for a three-point bend specimen. OCD is for unstable cracking and OCE for stable fracture.

In unstable cases we expect $\eta < 1$ as, by definition, there is more energy available at first cracking than required for complete fracture. This is illustrated in the following Example 8.8.

Example 8.8 Determine η for a shallow-notched DEN testpiece, using

$$Y\,(a/W) = \sqrt{\frac{2W}{\pi a} \tan \frac{\pi a}{2W}}$$

where crack length is $2a$ in a strip of width $2W$.

Ans.

$$Y^2 a = (2W/\pi) \tan (\pi a/2W)$$

and

$$\int Y^2 a da = -\log_e \cos (\pi a/2W)$$

Hence

$(u/X) = [2\pi/E(W-a)^2B]\ [-\log_e \cos(\pi a/2W)] + L/2BWE$

where the compliance of the uncracked testpiece of 'gauge length, L is $L/2BWE$.

$\Lambda = \frac{1}{2} X^2\ (u/X) = \frac{1}{2} X^2\ [L/2BWE - 2\pi \log_e \cos(\pi a/2W)/E(W-a)^2 B]$

Thus

$\Lambda/B(W-a) = [X^2/2(W-a)^2B^2]\ [(W-a)L/2WE - 2\pi \log_e \cos(\pi a/2W)/E(W-a)]$

Now

$$R = \frac{\sigma^2 \pi a}{E} \left(\frac{2W}{\pi a} \tan \frac{\pi a}{2W}\right) = [X^2 \pi a/B^2\ (W-a)^2] \left(\frac{2W}{\pi a} \tan \frac{\pi a}{2W}\right)$$

For small a/W, $\Lambda/B(W-a) \to X^2L/4B^2W^2E$ and R is $X^2\pi a/B^2W^2E$ so $\eta = 4\pi a/L = 4\pi(a/W)\ (W/L)$. For $a/W = 0.1$ (say) and $L/W = 8$ (say) $\eta = 0.16$. Hence toughness derived from the triangular areas for Λ at the instant of (unstable) cracking in DEN samples are upper bounds on R, only 16 per cent of the area being required for toughness work. (Problem 8.21 shows that $\eta \ll 1$ for the well-known unstable Griffith case.)

An alternative definition of η is obtained by noting that $\eta = R \div (\Lambda/A)$ where A is crack area (cf. equation (8.30)), so

$$\eta = -\frac{A}{\Lambda}\left(\frac{\partial \Lambda}{\partial A}\right)_u = -\frac{A}{X}\left(\frac{\partial X}{\partial A}\right)_u = -\frac{(W-a)}{W}\left(\frac{1}{X}\right)\left(\frac{\partial X}{\partial(a/W)}\right)_u \tag{8.32}$$

for remaining crack area $A = B(W-a)$. This particular formulation is useful in connection with elastoplastic fracture and the J-integral as described in section 4.5.2.

We conclude this section by noting that the 'backwards-pointing' R-locus may be constructed when η and (u/X) are known as functions of crack area. Stiffness lines are drawn on an Xu diagram, and the work area $R\Delta A$ contained between the R-locus and two adjacent stiffness lines is given by $(\eta\Lambda\Delta A)/(B(W-a))$. The R locus may be plotted in this way as a series of linear segments.

When η is not known, it is sometimes sensible to plot the 'unstable' part of the R-locus from a series of (catastrophic) tests on specimens with different starter crack lengths. This precedure is particularly relevant when only small quantities of material are available for testing, and/or are available only in peculiar-shaped pieces for which K-formulae (and hence η) are not known. Thus, in a study of thermal shock fracture in tool and turbine ceramics (where toughnesses were measured in the temperature range up to 1200°C) equilateral triangular tool inserts (~ 14 mm on side by 6.4 mm thick) were used directly as testpieces by Mai and Atkins (1975) (see Fig. 8.19). Gurney segmental areas give $R = 1.65\ \text{kJ/m}^2$ over the crack length increment 1.19 mm (i.e. starter crack lengths 0.46 and 1.65 mm). The whole triangular work area for the 0.46 mm starter crack is some $6.8 \times 10^3 \times 41 \times 10^{-6}/2 \approx 140$ mJ. The ligament is some $(12.1 - 0.46) \times 6.4 \approx 77\ \text{mm}^2$. An upper bound for R is thus $140 \times 10^{-3}/77 \times 10^{-6} = 1.88\ \text{kJ/m}^2$, and η seems to be $1.65/1.88 = 0.88$.

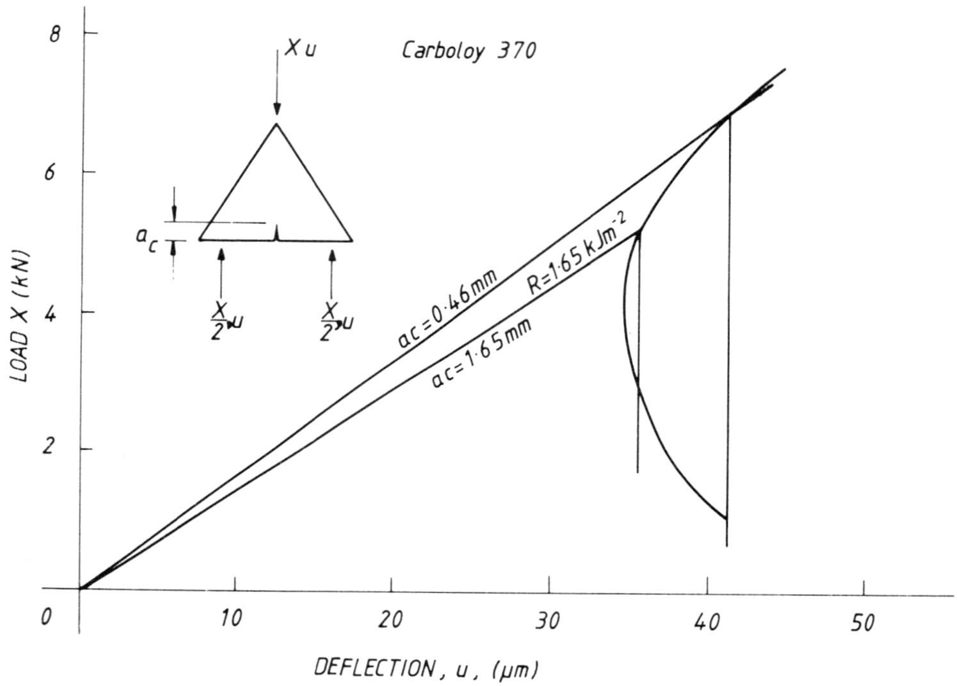

Fig. 8.19 – Determination of R using Gurney's work area method using inherently unstable testpieces with different crack lengths (after Mai and Atkins (1975), with permission).

8.9 CRACK ARREST

Sometimes, at the intant a crack starts to propagate, and for some time during later propagation, there is more than enough energy available in the system to provide the fracture toughness work required. As we have seen, instabilities result and the crack runs away in an uncontrolled manner – very often to fracture the body completely. Sometimes, however, the running crack 'runs out of steam', slows down, and stops after some propagation distance. Crack arrest can occur both in 'stable' geometries when the starter crack is blunt and in 'unstable' geometries when the starter crack is sharp. Why does this happen, and can use be made of the phenomenon in designing crack stoppers in large structures? For example, many fracture design problems concern flaws associated with welds. Application of simple fracture mechanics principles is complicated for a number of reasons, such as residual stresses (really residual strain energy) and material toughness values in the heat-affected zone (HAZ) both of which are of unknown magnitudes. A crack arrest philosophy in design is attractive in these circumstances whereby, even if crack propagation occurs, it will eventually be arrested in the regions of the body outside the HAZ.

Crack arrest is also important in the ability of materials to resist thermal shock, particularly in the case of porous ceramic materials. Hasselman (1969) has shown how cracks, which are initiated and propagated by thermal stresses in a single shock, arrest and are then 'sub-critical' for further shocks of the same magnitude and so display no further degradation of after-shock retained strength. There are interesting differences between those material parameters which resist the initiation of cracks in thermal shock and those which resist the propagation of cracks. Departures from Hasselman's model soon occur when temperature and rate-dependent properties are taken into account, as explained in Chapter 6.

Let us consider crack instability and arrest in a testpiece which, under normal circumstances in the presence of sharp starter crack, would display stable crack propagation, e.g. an Outwater double-torsion specimen made of a material having constant R which would crack at constant load. Instabilities under these conditions will occur when the starter crack is blunt, causing the crack initiation load to rise higher than it would have done had the crack been sharp. An experimental Xu trace under these conditions is shown in Fig. 8.20. The load rises to point S before suddenly dropping to A (at essentially fixed u). Simultaneously the crack jumps from the (blunt) starting length a_0 (of stiffness OS) to a longer length a_f (of stiffness OA), at which (sharp) length it arrests and further loading occurs at this constant arrested length up AB, until stable propagation commences at B. The constant load at B is the load at which the specimen would have cracked had the starter crack been adequately sharp in the first place.

For the Outwater specimen $u = kaX$, where k is constant, so $G = kX^2/2B$ from equation (2.6). At the quasi-static stable load level given by X_E. X_D or X_B, $G = \bar{R}$ where $\bar{R}$ is the 'true' toughness of the material, so $X_E = X_D = X_B = \sqrt{2B\bar{R}/k}$. At S, $G_S > \bar{R}$; at A, $G_A < \bar{R}$. The argument for crack arrest (first proposed by Hoagland and Rosenfield in 1974) hinges on preservation of energy, viz. that the excess energy at initiation which becomes kinetic energy eventually reverts to work of fracture at the crack tip. While this approach is an over-simplification because dynamic effects exist in crack arrest, it is, in fact, adequate for many problems. The point is discussed later in this section. The work required quasi-statically to change the crack length from a_0 to a_f at (constant) $\bar{R}$ is simply work area OEB $= \bar{R}\,(a_f - a_0)$B, which would occur at constant X_E over displacement u_E to u_B. In the presence of a blunt starter crack, cracking is delayed until u_A (where $u_E < u_A < u_B$), and the energy consumed for the same propagation a_0 to a_f occurs at essentially constant u_A, and is given by area OSA. While the change in crack length is the same in the two cases, the crack velocity history will be different as the one event is stable, and the other is an arrested

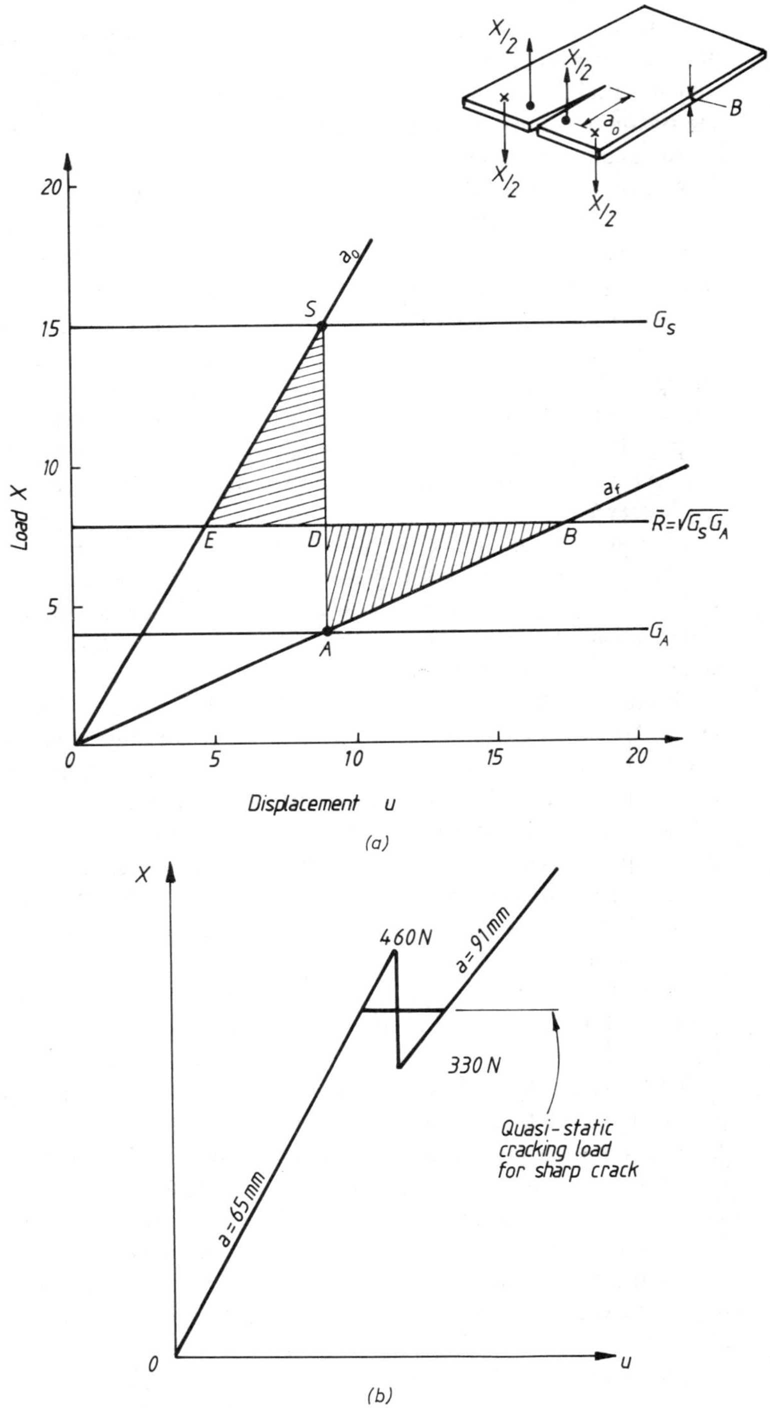

Fig. 8.20 – (a) Unstable cracking in an Outwater double-torsion testpiece under constant displacement condition. (b) Load–deflexion diagram for a PMMA double-torstion specimen with a blunt crack tip.

unstable event. Even so, arguing that the works consumed must be the same in both cases, we have OEB = OSA, or ESD = ADB (that is, the shaded areas balance). Noting that $u_E = ka_0\sqrt{2B\bar{R}/k}$, $X_S = \sqrt{2BG_S/k}$ and so on, geometry gives

$$a_0/a_f = (\bar{R} - \sqrt{\bar{R}G_A} + G_A)/(G_S - \sqrt{\bar{R}G_S} + \bar{R}) = \sqrt{G_A/G_S}$$

which becomes

$$\bar{R} = \sqrt{G_A G_S} \tag{8.33}$$

which relates in a simple way the natural toughness to the geometric mean of the initiating and arrest strain energy release rates *in this particular specimen.*

The data in Fig. 8.20(b) apply to PMMA. Experimental X_S and X_A values are about 460 N and 330 N, corresponding with $G_S = 504$ J/m^2 and $G_A = 260$ J/m^2 from which $\bar{R} = 504 \times 260 = 360$ J/m^2. Quasi-static experiments with a sharp starter crack gave for propagation from $a_0 = 65$ mm to the same arrested length $a_f = 91$ mm, the work area OEB of 55 mJ. As $B \approx 6$ mm thick, $\bar{R} = 55 \times 10^{-3}/26 \times 6 \times 10^{-6} = 353$ J/m^2 which corresponds well with R.

Example 8.9 Crack propagation in a double torsion testpiece starts at X_E in the presence of a sharp crack of length a_0. With a blunt crack of the same length, the load has to be increased to X_S before crack propagation commences. If $u = kaX$ in the specimen, determine how long in the path of the crack the specimen should be so that the arrest point lies within the specimen if $a_0 = 50$ mm and $X_S = 1.5X_E$.

Ans. Referring to Fig. 8.20 $u_E = ka_0X_E$; $u_D = u_S = ka_0X_S$; and work area EDS = ½ $ka_0(X_S - X_E)^2$. Similarly work area DBA = ½$ka_f(X_E - X_A)^2$. These areas are equal, hence $a_f = a_0(X_S - X_E)^2/(X_E - X_A)^2$. But $X_Aa_f = X_Sa_0 = u_{D,S}$, so

$$a_0/a_f = [1 - (a_0/a_f)(X_S/X_E)]^2/[X_S/X_E - 1]^2$$

giving a_f. Unless the specimen is longer in the crack path than a_f, the testpiece will simply break apart without arrest. For $a_0 = 50$ mm and $X_S = 1.5\ X_E$, $a_f = (9/4)\,a_0 \approx 113$ mm.

Similar arguments apply to crack arrest in other testpieces which, in the absence of blunt starter cracks, display stable cracking. For example, Fig. 8.21 shows the work areas that have to balance in testpieces such as DCB, CT which (with sharp starter cracks) crack under decreasing load. A general empirical expression relating X, u and A for all types of testpiece (including non-linear cracked bodies) is

$$X = (pA + q)^m\, u^n$$

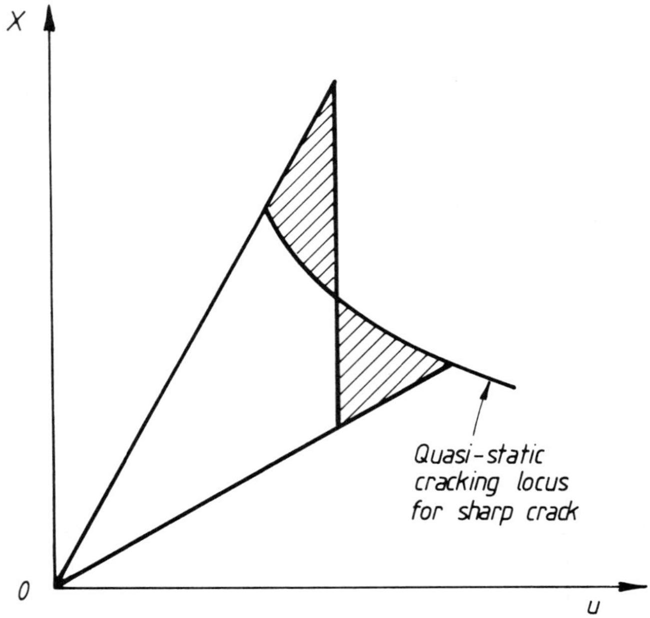

Fig. 8.21 – Unstable cracking and arrest in DCB or CT test pieces with blunt crack tip.

where p, q, n and m are numerical constants. (The Outwater geometry corresponds with $q = 0, n = +1, m = -1$. A DCB specimen for which $u/X = 2L^3/3EI$ corresponds with $q = 0$, $m = -3$, $n = 1$ and so on.) Using the above X, A, u relation in $\Lambda = \int X \mathrm{d}u$, we have

$$\Lambda = (pA + q)^m u^{n+1}/(n+1) \tag{8.34}$$

If $R(A_\mathrm{f} - A_0) = (\Lambda_{u,\, a = a_0} - \Lambda_{u,\, a = a_\mathrm{f}})$ generally, we obtain

$$\bar{R} = \frac{u^{n+1}}{n+1} \left(\frac{(pA_\mathrm{f} + q)^m - (pA_0 + q)^m}{(A_\mathrm{f} - A_0)} \right) \tag{8.35}$$

In terms of G_S and G_A, this becomes (Problem 8.15).

$$\bar{R} = \sqrt{G_\mathrm{A} G_\mathrm{S}} \left\{ \frac{1}{m} \sqrt{\frac{G_\mathrm{S}}{G_\mathrm{A}}} \left[\frac{(G_\mathrm{A}/G_\mathrm{S})^{m/(m-1)} - 1}{(G_\mathrm{A}/G_\mathrm{S})^{1/(m-1)} - 1} \right] \right\} \tag{8.36}$$

$$= Q\sqrt{G_\mathrm{A} G_\mathrm{S}}$$

and we see that the simple relation obtained from horizontal constant R-loci of the Outwater testpiece is modified by the quantity Q in the curly brackets. Note that Q does not depend on the order of non-linearity n. Values for Q for a range $m < 0$ are given in Table 8.4. In the Outwater case $m = -1$, so $Q = 1$ and we revert

Table 8.4
Ratio of fracture energy ($\bar{R}$) of average value (Q) to the geometric mean of strain energy release rates $\sqrt{(G_A G_S)}$.

$\frac{G_A}{G_S}$	$m = -1$	-2	-3	-4	-5	-6
0.05	1.00	1.127	1.156	1.238	1.275	1.288
0.1	1.00	1.075	1.114	1.137	1.153	1.165
0.2	1.00	1.036	1.055	1.066	1.073	1.079
0.4	1.00	1.012	1.017	1.021	1.023	1.025
0.6	1.00	1.004	1.005	1.007	1.007	1.008
0.8	1.00	1.001	1.001	1.001	1.001	1.001
1.0	1.00	1.000	1.000	1.000	1.000	1.000

to the simple result. As another example consider the non-linear case of a slender beam subject to large deflexions. The $X-u$ relationship at constant crack area is given by equation (8.34) with $n > 0$, p constant, $q = 0$ and $m = -3$. If, in Fig. 8.22 unstable cracking starts at S with $a_0 = 2.5$, $G_S = 1.3$ when $u = 8$, (arbitrary, but scaled, units) and if crack arrest occurs at A with $G = 0.1$ and $a_f = 4$, the measured specific energy dissipation represented by the curvilinear area OSA ≈ 0.4. Also, from equation (8.36) or by interpolation from Table 8.4, with $m = -3$ and $G_A/G_S = 0.08$, $Q \approx 1.14$. Thus $\bar{R} = \sqrt{G_A G_S}.\, Q = \sqrt{0.13} \times 1.14 = 0.42$ which agrees well with the measured value.

If the toughness is rate-dependent, but arrest still occurs, R will depend on the history of velocity in the unstable phase. This demands a full dynamic analysis of the testpiece between S and A. Hahn and co-workers at Battelle's Columbus Laboratories have studied this problem using wedge-loaded, 'duplex' DCB specimens which have a brittle steel starter section welded to the test section to provide an initially fast running crack (e.g. their 1973 paper). K_S can be made arbitrarily greater than K_{IC} so that the crack speed and the crack jump length can be systematically varied. Note that with wedge loading the crack propagates into a diminishing stress field and hence the arrest of a fast-moving crack within a DCB specimen is possible.

With cracked bodies that are inherently unstable even for a sharp starter crack, there is already an excess of energy to cause instability. Crack arrest is still possible, however, if the constant G-loci swing around 'soon enough', as shown schematically in Fig. 8.23(a). It is argued that work areas SED and ADB then balance, (for constant $\bar{R}$), the crack arresting at a_f and repropagating after loading along OAB, along the natural R locus beyond B. If R is rate-sensitive, and if arrest occurs, the arrest point A and arrested length a_f are given by the requirement that

$$\int_{a_0}^{a_f} R(\dot{A})\,dA = \bar{R}(A_f - A_0) = \text{OSA}$$

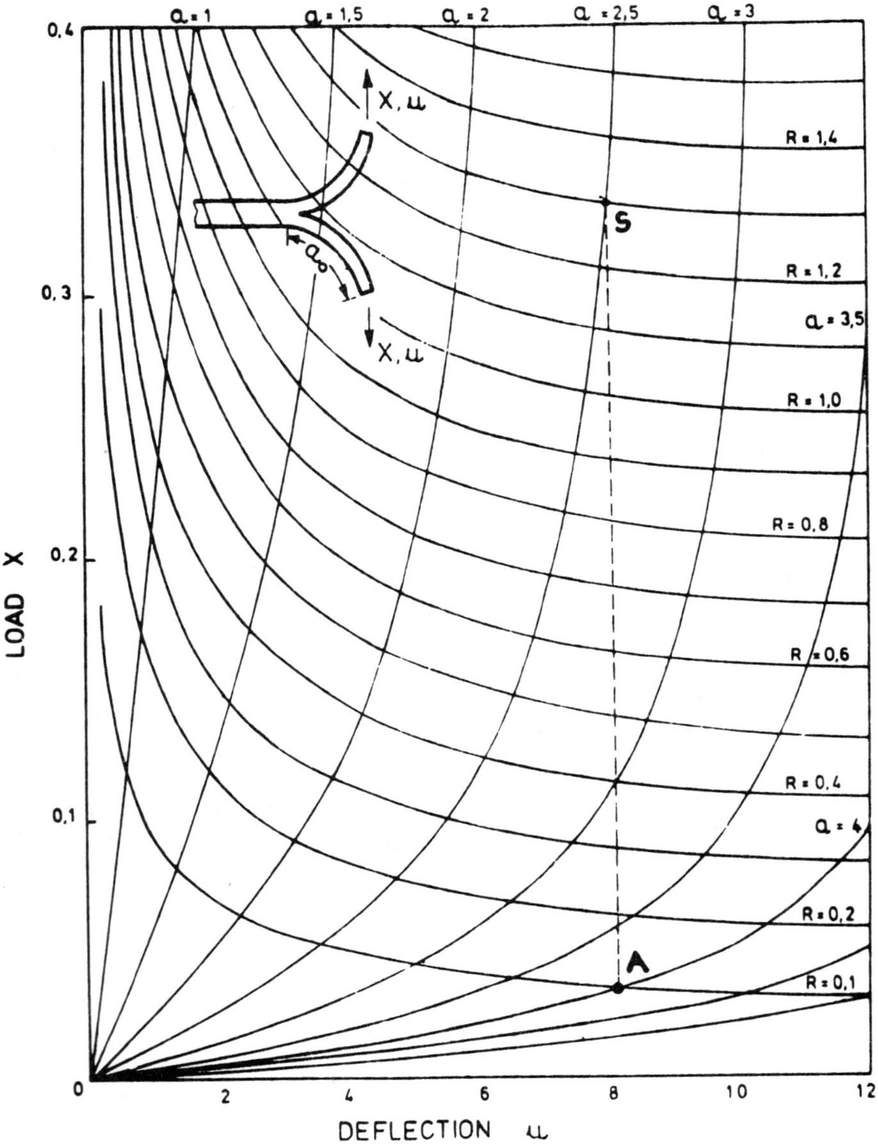

Fig. 8.22 – X–u relationship as a function of crack length (a) for splitting of slender double-cantilever beams at large deflexions when deformation becomes non-linear elastic (after Gurney and Ngan (1971), with permission).

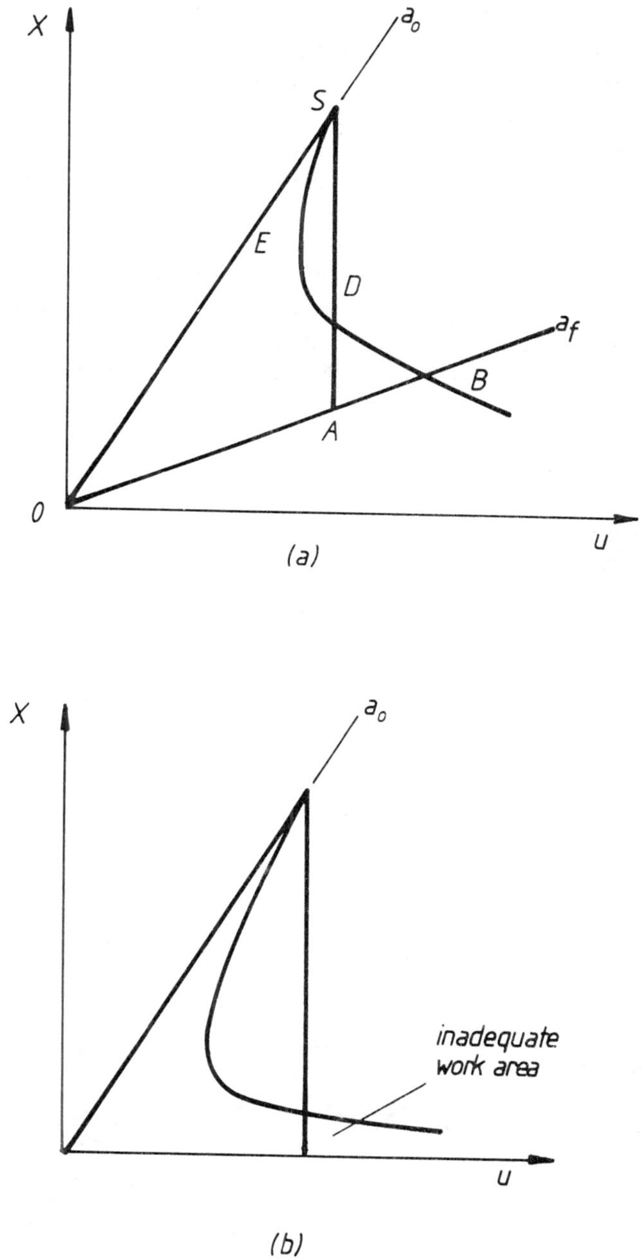

Fig. 8.23 – Cracking in inherently unstable testpiece: (a) with crack arrest at A, and (b) with no crack arrest.

Area OSA will be less than the quasi-static work area bounded by OSEDB when $\mathrm{d}R/\mathrm{d}\dot{A}$ is negative and greater when $\mathrm{d}R/\mathrm{d}\dot{A}$ is positive. If the shapes of the constant G-loci are as in Fig. 8.23(b) we see that arrest is difficult as the excess energy cannot be absorbed. In these cases the testpiece just breaks with a bang. We have already seen in Example 8.9 that even testpieces with inherently stable characteristics require an adequate length of run for the crack to arrest, or otherwise here too the testpiece simply breaks uncontrolledly.

In place of $\bar{R}$, G_A, G_S relations such as equation (8.36), relations are available between $\bar{R}$, G_S and the crack jump distance $(a_f - a_0)$ – e.g. Crosley and Ripling (1980) – cf. derivation of equation (8.33).

While the balanced work-area approach to crack arrest is easily understood, it is *not correct* fundamentally and a full dynamic treatment is required, which will incorporate inertia forces in the equation of motion of the cracked body

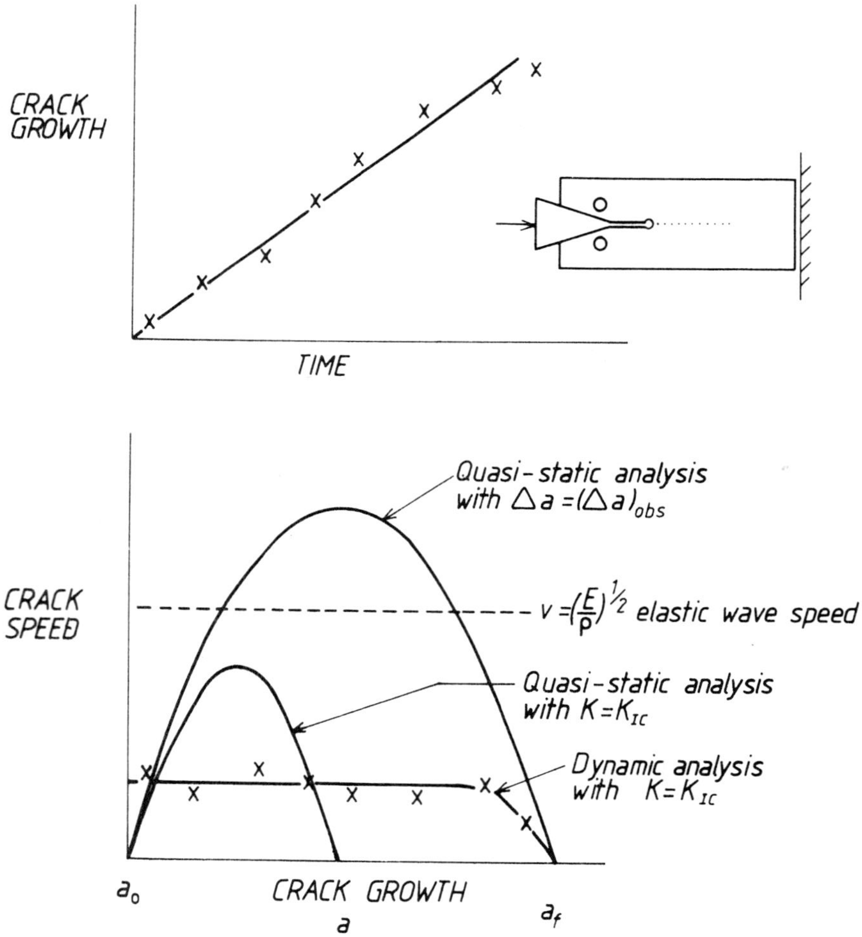

Fig. 8.24 – Quasi-static versus dynamic analyses of rapid crack propagation and arrest (after Hahn *et al.* (1973), with permission).

and stress wave interactions with the boundaries of the body during propagation. One shortcoming of the simple analysis so far presented is that, even after arrest, the arms of fracture test specimens may still be vibrating for example. Kanninen reviewed the situation in 1980. Fig. 8.24 reported by Hahn *et al.* in 1973, clearly reveals the importance of dynamic effects. Crack propagation in a 'duplex' DCB specimen, from an initially blunted crack tip under slowly inserted wedge loading, proceeds at an ostensibly constant velocity. This fact, albeit unexpected, makes possible a comparison of various possible analytical approaches. The simplest of these would suppose that the crack propagates under quasi-static conditions with a dynamic critical stress intensity factor (so-called K_{Id}) that is always equal to the initiation K_{IC}. As shown in the lower part of Fig. 8.24, for quasi-static conditions with $K_{Id} = K_{IC}$, a higher crack speed is predicted than is found experimentally. Also, the crack jump length $(a_f - a_0)$ is considerably underestimated. Hence, this approach is inadequate. If, instead of $K_{Id} = K_{IC}$, a value of K_{Id} is selected to match the observed crack arrest point, much higher crack speeds are obtained. It usually happens that, as indicated in the figure, the predicted crack speeds can exceed the elastic wave speeds for the material. As Kanninen explains, therefore, the resolution of this difficulty does not reside in the choice of a value for K_{Id}. (K_{Id}, of course, corresponds with R at some unknown mean velocity in the crack run/crack arrest event).

Results of the kind shown in Fig. 8.24 provided strong evidence that extended amounts of unstable crack propagation could not be characterised with a quasi-static computational approach. This led Kanninen and co-workers to develop in 1974 and in 1979 dynamic models of crack propagation in the DCB specimen based upon beam-on-elastic formation theory (cf. section 8.7.2). A typical result using this approach is included in Fig. 8.24 where, to a quite good approximation, the experimental results were reproduced both qualitatively (i.e. a linear crack length–time record virtually from the onset of crack growth to just prior to arrest) and quantitatively.

In finite size testpieces stress and strain waves will be reflected from the boundaries of the testpiece back to the crack tip regions, and reflected kinetic energy plays a crucial role in crack propagation. In the Battelle testpiece, it was found that the infinite medium solution (where there are no reflexions) departs from the solution for the finite size testpiece just at the time of arrival at the crack of a wave reflected from the boundary. In fact the infinite medium solution predicts arrest whereas in reality the crack does not arrest until later. Fig. 8.25 shows how the initial strain energy in the testpiece is partitioned during the crack run/crack arrest event. The kinetic energy rises to a maximum at about the statically predicted arrest point (i.e. $a - a_0 = 35$ mm). The subsequent decrease indicates that the kinetic energy reflected from the specimen boundaries is being utilised to continue crack propagation.

The magnitude of dynamic effects is likely to be different in different cracked geometries. For example, Kalthoff (1980) concluded that the SEN test specimen displays fewer dynamic effects than the DCB testpiece, since surfaces parallel to the crack plane are further away in the SEN geometry, thus diminishing the influence of reflected stress waves. Again, Ruiz and co-workers (see Corran

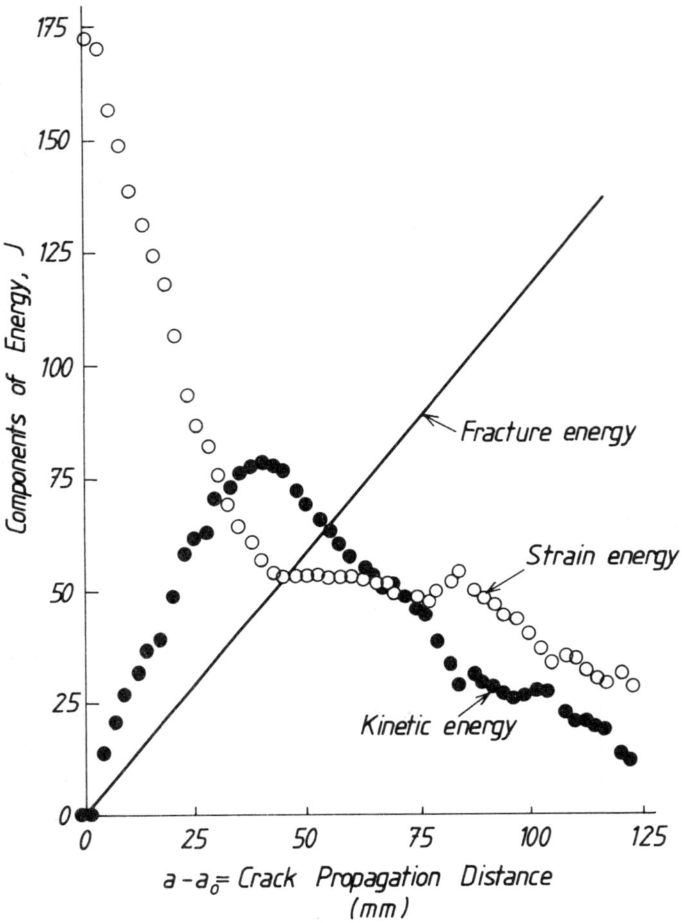

Fig. 8.25 – Distribution of energy during crack propagation in a DCB test specimen (after Hahn *et al.* (1973), with permission).

et al., 1985) have studied uniformity of crack tip stress and strain distributions both before and after propagation in three-point bend bars (as in instrumented Charpy testing) and SEN testpieces and concluded that the dynamically-loaded SEN geometries are better than testpieces in bending.

When the crack jump length is not too great, dynamic and static approaches give similar predictions for arrest. Furthermore in many practical designs, the return of kinetic energy to the crack tip is likely to be small, again leading to similar arrest predictions. However, when answers differ the *dynamic* approach should be followed, particularly since it will usually predict that cracks propagate faster and over a longer distance than corresponding static analysis. Depending on whether experiments are analysed in terms of static or dynamic models of crack propagation/crack arrest, different answers will be obtained for the 'arrest toughness' (i.e. G at A, or its equivalent K at A, in Figs. 8.20 *et seq.*). In the USA, Crosley and Ripling (1977) have developed test procedures that employ the static analysis approach to evaluate crack arrest toughness. Values obtained in this way are designated K_{Ia} by these workers. Similar procedures devised by Hoagland *et al.* (1977) use the dynamic analysis approach. In this case, the crack arrest toughness is designated K_{Im} (see Cheverton *et al.*, 1980). K_{Im} is also used for the minimum (if it exists) in a K_{IC} ($\dot{a}$) plot. There was a school of thought that argued that 'arrest toughness' was a material property. Such an approach assumes that the kinetic energy of the structure is never returned to the crack tip, and that the static stress intensity functions apply for dynamic cracking. From the foregoing discussion and Figs. 8.20 *et seq.*, however, it seems clear that G_A can take any value depending on G_S (whether analysed statically or dynamically) and that therefore a static post-arrest characterisation cannot be correct. Kendall's study (section 3.3.2) of the acceleration or retardation of cracks at interfaces (which is based on energy arguments and which includes the possibility of complete arrest) is not compatible with arrest toughness being a material property.

ASTM STP 711 discusses crack arrest and has a number of papers dealing with applications to important practical engineering problems. Among these are

(i) crack propagation in pneumatically-pressurised pipelines (e.g. Freund and Parks, 1980) which is an extremely complicated process, involving interactions between the fracturing pipe (which in ductile solids concerns important out-of-plane deformations), the escaping gas, the covering backfill in trenches, and mechanical arresters (if present), and

(ii) 'loss of coolant acident' (LOCA) in pressurised water nuclear reactor vessels (e.g. Marston *et al.*, 1980), where arrest of an interior crack can occur at a K_I level higher than the K_{IC} associated with initiation, as the crack extends in regions of increasing temperature towards the outside wall surface with consequent change in toughness properties, cracks having been initiated by cold water injection into the hot vessel.

Numerous references to earlier work on arrest will be found in this STP. Current analytical studies on arrest use rate-dependent viscoplastic constitutive laws, the crack tip stress singularities associated with which have (perhaps surprisingly) the same form as lefm (section 2.5). According to Kanninen (1985) this explains the success of lefm in arrest problems and also explains why derived K_{Ia} values at sometimes 500 MPa$\sqrt{m}$ are so much higher than might be expected, in that the crack tip zone is highly constrained and has no time to build up during rapid fracture and arrest.

From this section and Chapter 6, we observe that 'dynamic fracture mechanics' can mean two different things – either initiation of a stationary crack under rapidly applied loading or rapid propagation and possible arrest under slowly varying load.

8.10 CRACK STABILITY IN THE PRESENCE OF REMOTE PLASTIC FLOW

It is an experimental fact that crack propagation accompanied by extensive remote plastic flow is usually more stable than elastic cracking in the same geometry testpiece. This is clear from the behaviour of DEN sheets manufactured from ductile metals described in section 4.3.2: the geometric stability factor of even deeply notched DEN specimens is positive (Appendix 1, specimen number 7; $(1/R)/(\mathrm{d}R/\mathrm{d}A) > 4.5/b$ for $a/b = 0.75$) and when the deformation is globally elastic, stable fractures are difficult to achieve, yet cracking is stable when remote flow occurs. Perhaps this is not unexpected, since $\mathrm{d}\Gamma$ appears in the basic work rate equation (1.2), and $\mathrm{d}\Gamma/\mathrm{d}A$ during propagation will usually indicate a positive sink for excess energy. Remote flow thus can stabilise crack propagation, particularly when the remote flow is extensive. We note, however, that if remote plastic flow is non-uniform, residual elastic strain energy may be released on crack propagation if parts of the structure unload, and this too should really be taken account of in a full analysis of stability in the presence of remote flow.

The Gurney analysis for elastic crack stability (section 8.2) may be extented to stability in plastic fracture, i.e. the operations on equation (1.1) developed in earlier sections of this chapter may be applied to equation (1.2), (or better, equation (4.69)). Historically, however, this was not done and instead different (but complementary) lines of attack have been given by Paris *et al.* (1979) based on a J_R formulation of crack propagation and by Turner (1980) for stability of so-called 'ductile tearing'. (We note that the implication of a specific mechanism of crack propagation in the use of the words 'ductile tearing', viz. hole growth, is too restrictive as simultaneous flow and fracture could occur even if propagation in metals were by cleavage (see Chapters 3 and 10) and, of course, by other mechanisms again in polymers or biological materials.)

Of course, if the (really plastic) materials to which J-based theories are applied were really non-linear elastic, there would be no need for 'special' stability criteria to cover combined flow and fracture because Gurney and Mai analysed non-linear elastic fracture stability in 1972 (section 8.2 here). In so far as a J_R approach (within its limitations) coincides with non-linear elasticity, stability criteria presented for J-based theories must relate to the Gurney non-linear elastic stability criteria. We explore this point in a later section, but difficulties will inevitably occur as soon as total area under the Xu curve (presumed to be fully reversible elastic strain energy Λ in the theory) is partitioned into the actual reversible and irreversible components – a partitioning procedure complicated in unloading situations by the role played by residual elastic strain energy (Chapter 4).

When plastic flow is extensive and/or when unloading occurs, however, it is better conceptually to use equations (1.2) or (4.69) to describe elastoplastic fracture for which new stability criteria will emerge. Since the contour integral definition J loses its meaning as soon as unloading occurs, it may be expected that difficulties will arise in comparing criteria based on equations (1.2) or (4.69) and those based on J_R. To anticipate one such difficulty, we note that, under constant displacement conditions, inequality (4.70) gives *ranges* for

$G \leqslant [R + (\mathrm{d}\Gamma/\mathrm{d}A) + (\mathrm{d}\Lambda_r/\mathrm{d}A)] \leqslant J$. Stability criteria under constant displacement will concern the *rates of change* of these quantities.

We present a Gurney-type analysis based on equation (4.69) first as it follows logically from what has gone before in this chapter, then the other approaches based on J and Turner's I.

8.10.1 Gurney-type analysis

This analysis is new. We have

$$X\mathrm{d}u = \mathrm{d}\Lambda + R\mathrm{d}A + \mathrm{d}\Gamma + \mathrm{d}\Lambda_r$$

or

$$X\frac{\mathrm{d}u}{\mathrm{d}A} = \frac{\mathrm{d}\Lambda}{\mathrm{d}A} + R + \frac{\mathrm{d}\Gamma}{\mathrm{d}A} + \frac{\mathrm{d}\Lambda_r}{\mathrm{d}A} \tag{4.69 bis}$$

where u has both elastic and plastic components.

Under hard testing conditions with $\mathrm{d}u \geqslant 0$ we have during equilibrium quasi-static fracture

$$R + (\partial\Gamma/\partial A)_u = -\left(\frac{\partial(\Lambda + \Lambda_r)}{\partial A}\right)_u \tag{8.37}$$

and, following the same line of argument as for elastic fractures, section 8.2, stability demands that

$$\frac{\mathrm{d}}{\mathrm{d}A}[R + (\partial\Gamma/\partial A)_u] > -\left(\frac{\partial^2(\Lambda + \Lambda_r)}{\partial A^2}\right)_u \tag{8.38}$$

Similarly, under load control with $\mathrm{d}X \geqslant 0$, equilibrium quasi-static fracture corresponds with

$$R + (\partial\Gamma/\partial A)_X = \left(\frac{\partial\Omega}{\partial A}\right)_X + X\left(\frac{\partial u_p}{\partial A}\right)_X - \left(\frac{\partial\Lambda_r}{\partial A}\right)_X \tag{8.39}$$

where the complementary elastic strain energy $\Omega = Xu_{el} - \Lambda$ and where the total displacement u has elastic components u_{el} and plastic components u_p. (Notice that term-by-term comparison of the left- and right-hand sides of equations (8.37) and (8.39) suggests the sort of partitioning of loaded work areas employed in Chapter 4). Stability with $\mathrm{d}X \geqslant 0$ corresponds with

$$\frac{\mathrm{d}}{\mathrm{d}A}\left[R + (\partial\Gamma/\partial A)_X\right] > \left(\frac{\partial^2\Omega}{\partial A^2}\right)_X + X\left(\frac{\partial^2 u_p}{\partial A^2}\right)_X - \left(\frac{\partial^2\Lambda_r}{\partial A^2}\right)_X \tag{8.40}$$

Notice that, in both equations (8.38) and (8.40), there are separate material- and geometry-dependent terms exactly as in the relations for stability during elastic fracture. The geometry-dependent terms involve Λ_r.

Other things being equal, we see that if $\mathrm{d}\Gamma/\mathrm{d}A$ is positive, stability is improved under displacement control, but that the situation is not clear under load control where, possibly, stability may not be improved. Since

$$\frac{\mathrm{d}\Gamma}{\mathrm{d}A} = W^{\mathrm{P}}\frac{\mathrm{d}V^{\mathrm{P}}}{\mathrm{d}A} + V^{\mathrm{P}}\frac{\mathrm{d}W^{\mathrm{P}}}{\mathrm{d}A}$$

we note that $\mathrm{d}\Gamma/\mathrm{d}A$ could become negative if the volume of material being plastically deformed dropped rapidly with crack area increase (e.g. along diminishing lengths of ligament in the path of a crack in some situations); $\mathrm{d}W^{\mathrm{P}}/\mathrm{d}A$ would be expected to increase probably as A increased, owing to increasing plastic strains.

The influence of testing machine stiffness on stability of elastoplastic fracture may be tackled as in section 8.3 for elastic fracture (cf. Fig. 8.4).

8.10.2 J_{R} formulation and the tearing modulus

The J_{R} formulation for crack propagation in extensive irreversible flow fields was explained in section 4.5.4. J during propagation is calculated from contour integrals, or from $-\mathrm{d}\Pi/\mathrm{d}A = -(\partial U_{\mathrm{T}}/\partial A)_u$ with U_{T} the total work area under the *Xu* plot which (in practice) includes irreversible as well as reversible components, or from other methods in section 4.5.3. On the basis of the HRR model of crack tip characterisation, it is argued that a J-based analysis can adequately describe elastoplastic crack propagation – at least for comparatively small amounts of crack advance – providing certain criteria are met which relate to the size of the remaining ligament and to $\mathrm{d}J/\mathrm{d}A$ (see section 4.5.3).

We recall the use of R-curves in section 8.5 to determine stability in elastic fracture, in particular the *RA* diagrams in Figs. 8.6 and 8.7(a). There, in the *linear* elastic case comparison was between (i) the elastic strain energy release rate G from the cracked body available to drive the crack, and the specific work of fracture R required by the material for propagation and especially (ii) the rates of change of R and G with crack advance subject to chosen constraints. The same principles apply to non-linear elastic fracrure, leading to the general equations (8.9) and (8.10) which, were the areas beneath *Xu* plots in elastoplastic fracture really to represent only elastic and reversible strain energy, could be applied directly to elastoplastic fracture stability. Two difficulties arise, (i) the areas, of course, include irreversible plastic work, 'unavailable' to drive cracks, and (ii) the magnitude of $\mathrm{d}J/\mathrm{d}A$ is in doubt since the validity of J_{R} during propagation beyond J_{C} is questionable (in most cases, J_{R} includes irreversible plastic flow which is quite unconnected with specific work of fracture leading to J_{R} being geometry-dependent, unlike R-curves determined in truly elastic cases which are taken to be material characteristics). Even so, propagation and stability in elastoplastic fracture has been tackled in the same way (within the limitations of a J-based formulation) by (i) viewing the calculated J during propagation as a crack driving 'force' available from the cracked body and comparing it with J_{R} which describes the requirements of the material and especially (ii) comparing the rates of change of the material J_{R} and the calculated J from the cracked body, subject to chosen constraints, i.e. investigate $\mathrm{d}J_R/\mathrm{d}a \gtrless (\partial J/\partial a)_{u \text{ or } X}$ *exactly* like equations (8.9) and (8.10) for *non-linear elastic* fracture, but where now $J_R = J_e + J_p$ (cf. Equations (4.42–47)).

Fig. 8.26(a) shows a JA diagram for the elastoplastic fracture of a 25 per cent side-grooved 100-mm thick compact tension specimen made from A533B steel at 93°C, taken from the GE/EPRI studies (e.g. EPRI Report NP-1931, 1981). Lines of constant X and u, relating to the J crack driving 'force', are shown on the diagram and we see a direct correspondence with Fig. 8.6 and 8.7(a). Superimposed on Fig. 8.26(a) is a J_R curve relating to the material (often questionably assumed independent of geometry). Points of tangency with the constant u, or constant X, J-loci locate instabilities under displacement- or load-control respectively. In this particular case, instability will occur under load control after a small amount of stable crack growth under rising load. The testpiece will always be stable under displacement control, for the particular shaped J_R-curve shown.

The predicted Xu diagram, corresponding with the JA diagram in Fig. 8.26(a), is shown in Fig. 8.26(b), it is obtained by simple cross-plotting. Two predicted Xu curves are given in Fig. 8.26(b), based on full finite element calculation of the problem and on the 'estimation scheme' for interpolation between elastic and collapse solutions of the problem (see section 4.5.4). The experimental results shown in the figure were obtained under displacement control.

Prior to the availability of many finite element solutions for J-based elasto-plastic fracture, the problem of stability and instability of crack propagation in extensive remote flow fields was tackled by Paris and co-workers (1979) on the basis of changes in the length of specimens during propagation. In *some* specimens, there is competition between a plastic extension associated with crack propagation and an elastic contraction caused by unloading as the crack propagates. In such specimens, it is argued that, under displacement control, instability results when the magnitude of the elastic contraction exceeds the corresponding plastic extension. That is, in $X\mathrm{d}u = X(\mathrm{d}u_{el} + \mathrm{d}u_{pl})$, $\mathrm{d}u < 0$ when $|\mathrm{d}u_{el}| > |\mathrm{d}u_{pl}|$. The relative values of $\mathrm{d}u_{el}$ and $\mathrm{d}u_{pl}$ were determined from simple flow fields in basic testpieces as illustrated in Example 8.10 following.

As with all criteria for crack stability, separate material- and geometry-dependent terms emerge. In the Paris approach, the material-dependent term appears not merely as $\mathrm{d}J_R/\mathrm{d}A$ but as the so-called 'tearing modulus' T given by

$$T = (E/\sigma_y^2)\,(\mathrm{d}J_R/\mathrm{d}a) \tag{8.41}$$

with E the Young's modulus and σ_y the flow stress. T is dimensionless. Since $\mathrm{d}J_R/\mathrm{d}a$ varies with a (e.g. Fig. 8.26 (a)) as propagation proceeds, T is a variable quantity. However, early results suggested that T was constant (i.e. $\mathrm{d}J_R/\mathrm{d}a$ was independent of crack growth). That may be a useful approximation for small amounts of crack advance, but is unlikely to be true for large amounts of propagation in extensive flow fields (remember J_R for rigid-plastic solids corresponds with $R + \mathrm{d}\Gamma/\mathrm{d}A$). Even so, on that basis, the experimental range for $\mathrm{d}J_R/\mathrm{d}a$ for many steels is $10 \sim 300$ MJ/m³ which gives $T \approx 20$–200. For materials which

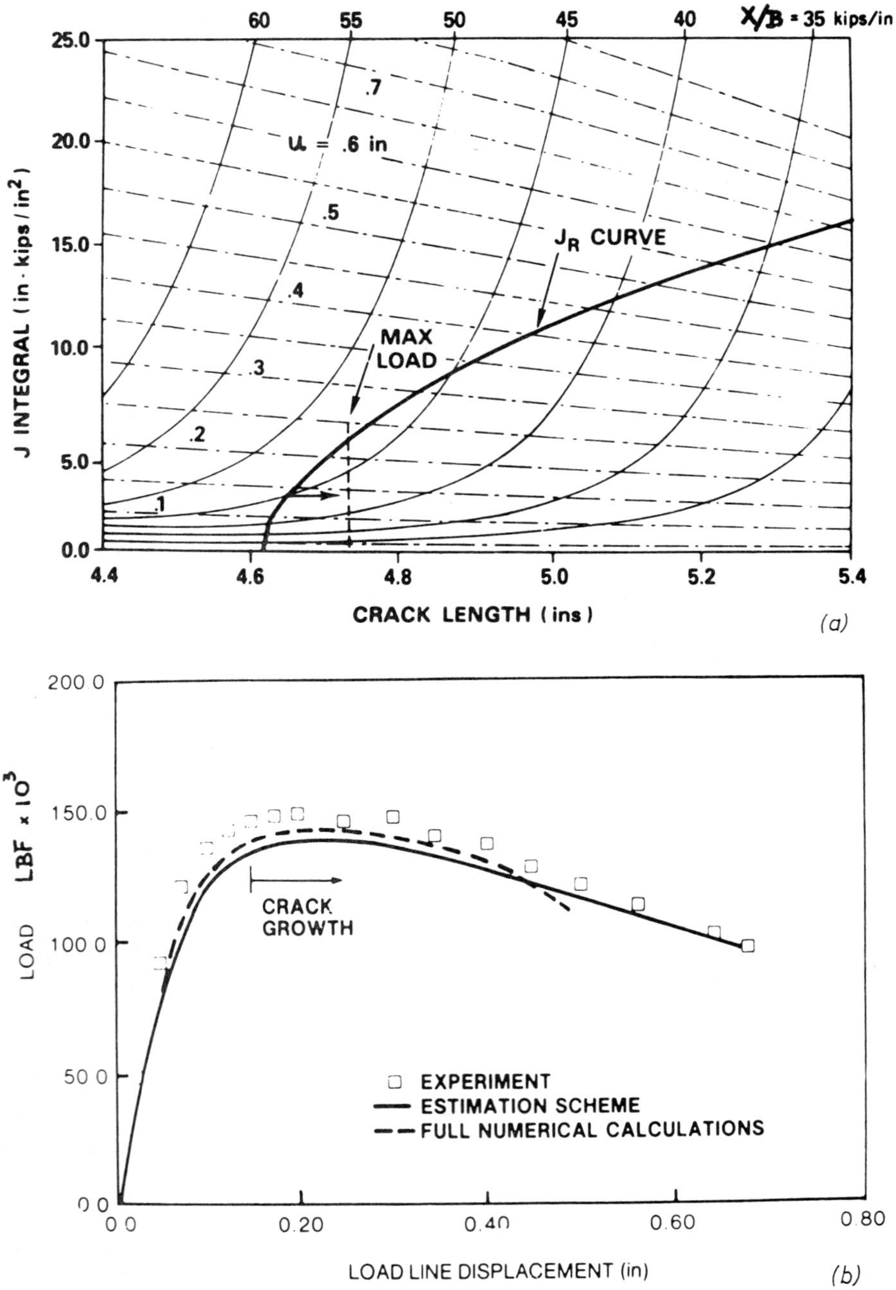

Fig. 8.26 – (a) A J-integral crack driving force diagram for a 4T, T-52, plane strain compact specimen of A533B steel. The constant load/thickness curves are shown by solid lines and the constant displacement curves by chain lines. The material J_R-curve, indicated by the heavy solid lines, is superimposed at the initial crack length of 4.615 inches. (b) Comparison of predicted and experimentally measured load–displacement relationship for 4T, T-52, A533B steel, 25 per cent side-grooves plane strain compact specimen with initial crack length of 4.615 inches. (Both after Kumar *et al.* (1981), with permission.)

work harden appreciably during a test, σ_y is often arbitrarily replaced by $(\sigma_y + \sigma_{UTS})/2$ in the calculation of T. Appendix III of Paris *et al.* (1979) gives T (at first crack growth) for a number of steels and aluminiums. On that basis also, we recall from section 4.3 that dJ_R/da was four times the slope of the Cotterell–Mai normalised work vs. ligament length plots in DEN testpieces. From Fig. 4.13, the slope for the annealed low carbon steel is some 22 MJ/m^3, so $dJ_R/da \approx 90$ MJ/m^3. For materials with prior cold work, the slope diminishes and eventually, in highly workhardended samples, stability would be difficult to achieve. This is not unexpected since highly pre-strained samples have little capacity for plastic flow and tend towards an elastic response, and in globally *elastic* fracture the DEN testpiece is unstable.

There once was also a curious belief that J_R-curves (and by implication, dJ_R/da) would be geometry-independent (section 4.6). It is now clear that different J_R-curves apply to different testpieces (even under conditions of strict J-control); information on this point may be found in, for example, Garwood (1979), Kanninen (1980) and ASTM STP 803 (1983). That J_R is likely to vary with cracked geometry arises from the fact that J_R (as usually expressed) contains the work of remote flow $d\Gamma/dA$ and that depends on the magnitude and extent of the irreversible flow field remote from the crack tip (consider, for example, simple tearing of a rigid-plastic ductile metal sheet where $d\Gamma$ changes with the width being torn, but RdA remains the same, for corresponding amounts of crack advance).

Example 8.10 Determine the stability criterion for DEN samples using the approach of Paris *et al.* (1979).

Ans. The fully constrained flow stress on the central ligament of a DEN sample (e.g. Fig. 4.12) is some $3\sigma_y$, so for a ligament of length L, $X = 3\sigma_y LB$. During plastic flow, at the site of the original starter cracks, plastic stretch $\Delta\delta$ occurs which will be associated with an irreversible increase in the length of specimen Δh_p. The relation between $\Delta\delta$ and Δh depends on the slip line field; for a DEN testpiece with Prandtl–Hill field across the ligament, we have $\Delta h_p = (\frac{1}{2})\,\Delta\delta$. Since $J = m\,\sigma_y\delta_C$, the increase in $\Delta\delta$ at the *original* position of the crack tip implies a corresponding increase in J given by $\Delta J = m\,\sigma_y\Delta\delta$. Hence $\Delta h_P = (\frac{1}{2})\,\Delta J/m\,\sigma_y$. Now the existence of ΔJ implies a crack extension Δa (cf. procedure (ii) for J_C in section 4.5.2) which shortens the ligament and reduces the load. Since $\Delta L = -2\Delta a$, $\Delta X = 6\,\sigma_y B\Delta a$, and a reduction in X further implies an elastic shortening of the specimen given by $\Delta h_e = \Delta Xh/BWE$ where W is the total width of the specimen (change in remote stress $= \Delta X/BW$, change in remote strain $= \Delta h/h$, E = Young's modulus). Thus $\Delta h_e = -6\sigma_y h\Delta a/WE$. So for stability under $du > 0$,

$$(\Delta h_p + \Delta h_e) > 0 \text{ or } (\tfrac{1}{2}\frac{\Delta J}{m\sigma_y} - \frac{6\sigma_y\,h\,\Delta a}{WE}) > 0,$$

i.e.

$$T = (E/\sigma_y^2)\left(\frac{dJ_R}{da}\right) > \frac{12hm}{W} \approx 12h/W \text{ taking } m \approx 1.$$

It is of interest to relate the expression derived in Example 8.10 with the DEN ductile fracture experiments on brass sheet shown in Fig. 4.15. For a tesetpiece with 200 mm long gauge length by 75 mm wide, the geometrical stability term is 12 (200/75) = 32. The slope of the curve for annealed brass in Fig. 4.15 is 22.86 MJ/m^3 so that $dJ_R/da = 4 \times 22.86 \approx 91$ MJ/m^3. With $E \approx 105$ GPa and $\sigma_y \approx 200$ MPa, $T \approx 238$ and stability is assured. For the samples cold worked by 35 per cent reduction of area, the slope is about (0.13) of the annealed value and $\sigma_y \approx 300$ MPa, so T becomes about 14 and, according to this criterion, instability should result; cracking was, however, stable in the experiments.

The tearing modulus approach, therefore, looks to plastic work to stabilise cracked geometries which would be unstable in elastic fracture, i.e. stabilise those geometries which have u negative along a quasi-static R-locus as in Fig. 8.3, for example. The argument that $d\Gamma$ serves to make the net increment displacement $(du_{pl} + du_{el})$ positive, 'pushing forward' the $(R + \Gamma)$ locus and hence to make the system stable, is rather like the use of dR/dA-positive materials sandwiched with materials whose dR/dA is highly negative in order to achieve stability during elastic cracking (e.g. Fig. 8.13). We note, however, that many geometries *already* display stable elastic cracking with u_{el} positive under displacement control (and in stiff systems), e.g. DCB testpieces, Outwater testpieces and many others listed in Appendix 1, and that in these cases there is *no* 'elastic shortening' (or its equivalent). The presence of positive u_{pl} during accompanying plastic flow in such geometries can only increase stability further under displacement control (e.g. the compact tension testpiece in Fig. 8.26(a); the elastoplastic glued cantilever beam problem of Chapter 4). Under load control, the situation is not clear since the requirement of $dX > 0$ has to be achieved and this will depend on work-hardening and the extent of the plastic flow field.

Since the (J, J_R) line of attack for stability illustrated in Fig. 8.26(a) involves dJ_R/da, the criteria for instability under load-, displacement-, or chosen machine-stiffness-control, may be couched in terms of T. The material- and driving 'force' J's are distinguished by two alternative pairs of subscripts by different authors, viz. 'mat' and 'app', or 'J' or 'JR', and the instability condition is given by

$$T_{app} \geqslant T_{mat}$$

or

$$T_J \geqslant T_{JR}$$

where $T_{mat} = T_{JR} = (E/\sigma_y^2)(dJ_R/da)$ and $T_{app} = T_J = (E/\sigma_y^2)(\partial J/\partial a)_{X \text{ or } u}$.

In place of the JA 'crack driving force' diagram to assess elastoplastic stability, an alternative plot has been suggested by Paris (1983) in which the J–A information is cross-plotted into a J–T plane with two separate curves representing J_R vs. T_{mat} for the material and J vs. T_{app}. The point of intersection of the lines defines the transition to instability. Varying dJ_R/da behaviour seems to have been the inspiration of this so-called 'failure assessment' JT diagram.

We conclude this section by noting that everything such as T that has been written in terms of J may be interpreted in terms of COD. We note also that the associated use of $\mathrm{d}\delta/\mathrm{d}a$ term (corresponding to $\mathrm{d}J/\mathrm{d}a$) leads to the idea of crack tip opening *angles* during propagation, either at the original or moving crack tip locations.

8.10.3 Comparison between J_{R}-based stability criteria and 'separable-R'-based criteria

The 'separable-R'-based criteria for elstoplastic fracture in section 8.10.1, and true linear or non-linear elastic fracture stability criteria in sections 8.1 and 8.2 involve different *energy* interchanges. In contrast, the original tearing modulus line of attack illustrated in Example 8.10 concerned merely incremental *displacements.* It is of interest to compare the two formulations particularly, since as highlighted above, $\mathrm{d}u_{\mathrm{el}}$ remains positive when fracture occurs under decreasing load for some cracked geometries and there is *no* elastic shortening. If we multiply equation (8.38) throughout by E/σ_{y}^2 we have

$$(E/\sigma_{\mathrm{y}}^2)\ \frac{\mathrm{d}}{\mathrm{d}A}\left[(R+(\partial\Gamma/\partial A)_u\right] > -(E/\sigma_{\mathrm{y}}^2)\ \left[\frac{\partial^2(\Lambda+\Lambda_{\mathrm{r}})}{\partial A^2}\right]_u \qquad (8.38\mathrm{a})$$

or, alternatively,

$$(E/\sigma_{\mathrm{y}}^2)\ \frac{\mathrm{d}}{\mathrm{d}A}\left[(R+(\partial\Gamma/\partial A)_u+(\partial\Lambda_{\mathrm{r}}/\partial A)_u\right] > -(E/\sigma_{\mathrm{y}}^2)\,(\partial^2\Lambda/\partial A^2)_u \qquad (8.38\mathrm{b})$$

for stability where the right-hand sides depend on the geometry and loading arrangement of the crack body.

In the J-based approach we have

$$(E/B\sigma_{\mathrm{y}}^2)\ (\mathrm{d}J_{\mathrm{R}}/\mathrm{d}a) > T_{\mathrm{app}} \text{ or } T_{\mathrm{J}} \qquad (8.42)$$

for stability where, again, the right-hand side depends on geometry and loading, but is $-(E/\sigma_{\mathrm{y}}^2)\,(\partial^2 U_{\mathrm{T}}/\partial A^2)_u$ with U_{T} the *total* work area under load beneath the partially irreversible $X\dot{u}$ plot rather than the similar terms involving only *reversible* elastic strain energy Λ and residual elastic strain energy Λ_{r} in equations (8.38a) and (8.38b).

These differences on the right-hand sides of equations (8.38a), (8.38b) and (8.42) highlight the differences in problem statement depending on whether a specific work of fracture term $R\mathrm{d}A$ is used for initiation and during propagation (which is separable from work of remote flow $\mathrm{d}\Gamma$), or whether a J_{R} approach is used in which fracture work, remote flow and changes in strain energy during propagation are lumped together.

The left-hand sides of the criteria relate to the material behaviour, but are not quite identical (again owing to the difference in formulation). Inequality (4.70) for constant displacement conditions tells us that

$$G \leqslant R + \frac{\mathrm{d}\Gamma}{\mathrm{d}A} + \frac{\mathrm{d}\Lambda_{\mathrm{r}}}{\mathrm{d}A} \leqslant J_{\mathrm{R}} \qquad (4.70 \text{ bis})$$

but, for comparison between criteria, we need *rates of change* of these types of quantity, about which there is no information in the general case.

We note that Ernst (1983) has proposed a modified J-integral, called J_M, which has applied to elastoplastic instability problems (using equivalent T_Ms) in which the amount of crack growth extended well beyond the usual 'J-controlled' growth regime. It is difficult, again, to make a comparison with R-separable-based criteria, but cursory inspection seems to indicate that the Ernst proposal brings equations (8.38) and (8.42) closer together. (It is of interest to compare the definition of J_M with that for the generalised G for elastoplastic fracture given in equation (4.64).)

It must be recalled that the use of J_R, and its differentiation w.r.t. A in the tearing modulus approach, implies that $R\mathrm{d}A$ and $\mathrm{d}\Gamma$ are coupled via A, this cannot always be the case and J_R is likely to be geometry-dependent itself. Also, the lack of partitioning between $R\mathrm{d}A$ and $\mathrm{d}\Gamma$ prevents the tearing modulus approach being applied to situations such as the elastoplastic glued cantilever beam problem in Chapter 4, where $\mathrm{d}R/\mathrm{d}A$ could be negative for the glue line which could cancel out the favourable contribution to stability of possible positive $\mathrm{d}\Gamma/\mathrm{d}A$ in the plastically deforming adherand.

8.10.4 Turner's instability criterion

As explained in section 4.6.5, in Turner's studies of elastoplastic fracture, he identifies a parameter I which is the rate of change with respect to crack area of unloaded work areas such as EBCD in Fig. 4.34. Such unloaded work areas represent work of fracture, work of remote flow and residual elastic strain energy. I coincides with G if all unloading lines are linear and return to the origin of load and displacement; I coincides with J if the lines are curved and return to the origin. In general, $I = (R + \mathrm{d}\Gamma/\mathrm{d}A + \mathrm{d}\Lambda_r/\mathrm{d}A)$.

Turner (1979) has defined instability during ductile crack growth as the point where I exceeds the work dissipation rate. If the latter includes elastic energy, the work dissipation rate is $\mathrm{d}U_T/\mathrm{d}A$; if elastic energy is omitted, it is $\mathrm{d}\Gamma/\mathrm{d}A$ (cf. equations (4.38) and (4.41)). Turner has calculated I for a number of geometries under displacement loading at plastic collapse.

We note that Gurney-type and J-based criteria for elastoplastic fracture concern second derivatives of U_T and what amounts to first derivatives of I, rather than first derivatives of U_T and I itself. The Gurney analysis is subject to chosen constraints ($\mathrm{d}u \geqslant 0$; $\mathrm{d}X \geqslant 0$) during which the basic energy rate balance (equations (1.2),(4.69) etc.) is *obeyed*. Turner's approach is a statement of stability in terms of the basic energy rate balance *not* being obeyed; in another he defines 'final instability' in terms similar to those in equation (8.38) (see Turner 1979).

Chell and Milne (1983) give a comparison between a number of these different formulations for crack stability in the presence of remote plastic flow.

8.11 STABILITY IN FATIGUE

Chapter 7 noted that fatigue crack propagation is stable in soft testing machines in which the same specimen would fail unstably in monotonic loading. This came about because of the screening action of the plastic zone of 'damage' surrounding the crack tip during fatigue cracking.

There is not a lot to say about crack stability in fatigue, except perhaps that in designing fatigue testpieces, dK/da-negative geometries should be avoided whenever possible. We also note the possibility of crack arrest of propagating cracks in special circumstances during fatigue (Example 7.8).

PROBLEMS

8.1 Show that, for a given size testing machine, the strain energy stored in the testing machine at fracture is a relatively larger proportion of the strain energy stored in the testpiece, the smaller the size of the testpiece. Hence show that stable cracking is more difficult to achieve in small-size testpieces than large.

8.2 Show that with a given specimen, stability is easier to achieve with a hard testing machine than with a soft.

8.3 Using the Gross–Srawley expressions given in section 8.2 for the CT specimen, show that the stability condition under hard conditions is $(1/R)(dR/da) > -4.3/a$. Compare n_u with the simple beam value.

8.4 Derive the stability conditions for cracking in a DCB testpiece under dynamic considerations, i.e. kinetic energy has to be taken into account. Relate your answer to stability under quasi-static conditions, where $\dot{a} = 0$. Show that, perhaps surprisingly, stability is apparently improved at high $\dot{a}$. [Hint. Use $X du = d\Lambda + R dA + dH$; use $H = \dfrac{3}{70}\rho h B \dfrac{\dot{a}^2 u^2}{a}$ for $du > 0$ and $H = \dfrac{\rho h B}{3E^2 I^2} X^2 a^5 \dot{a}$ for $dX > 0$, where ρ is the density of the beam material, h the height of one arm, B thickness, a crack length.]

8.5 Show that in theory, for stability to be achieved by a Griffith crack, the minimum starter crack length is always shorter in displacement-controlled machines than in load-controlled machines. [Hint. Assume that $X = ku/(l + ma^2)$ – an empirical relation first given by Berry – from which $a_u = \sqrt{l/3m}$ and a_X is non-existent.]

8.6 Show that stability of cone cracking beneath punch indentation is governed by

$$\frac{1}{R} \cdot \frac{dR}{dr} \geqslant \frac{-3}{r} \quad \text{(for load control, } dX/X > 0\text{)}$$

and

$$\frac{1}{R} \cdot \frac{dR}{dr} \geqslant \frac{-1}{r} \left[\frac{3(r/r_0) - 1}{(r/r_0) - 1}\right] \text{(for displacement control, } du/u > 0\text{)}$$

[Hint. Use $R \propto X^2/r^3$.]

Plot out the r.h.s. against r/r_0 and show that the gsf is always better for the displacement controlled case.

8.7 Is 'fast fracture' necessarily the same as 'unstable fracture'?

8.8 Write down stability conditions for fracture when residual (elastic) displacements occur, i.e. when crack propagation is controlled by equations (2.21) or (2.23).

8.9 Following 8.7 above discuss the limiting speeds for quasi-static cracking in double-cantilever beam specimens for typical metals and polymers. [Hint. Refer to Gurney and Ngan, 1971, *Proc. Roy. Soc.*, **A325**, 207.]

8.10 Show that $u\mathrm{d}X/\mathrm{d}A = -4.38R$ in a three-point bend testpiece having $a/W = 0.5$. Determine therefrom $\dot{a}/\dot{u}$ and comment on the sign of your answer. (Use $Y(a/W) = 1.09 - 1.74\,(a/W) + 8.20\,(a/W)^2\ 14.18\,(a/W)^3 + 14.57\,(a/W)^4$ which applies when the span-to-beam depth ratio $S/W = 4$; $\sigma = 3XS/2BW^2$.)

8.11 Crack propagation in grooved testpieces is usually more stable both directionally and energetically than in ungrooved specimens. Discuss.

8.12 In Example 8.6 it was suggested that five external springs be employed alongside the starting ligament in order to achieve stability. Determine the stiffness required of five springs spaced at 25 mm intervals (in place of 10 mm intervals) which thus encompass the starter crack region flanks as well as the ligament.

8.13 Explain why quasi-static analyses overestimate the likelihood of crack arrest.

8.14 Discuss crack arrest in the cracked annulus testpiece loaded in compression (Fig. 8.5).

8.15 Derive the expression $\bar{R} = Q\sqrt{G_S G_A}$ for the mean toughness in a crack arrest for the general expression $X = (pA + q)^m\ u^n$.

8.16 Show that an alternative expression for crack arrest, i.e. $\bar{R} = Q\sqrt{G_S G_A}$ is $Q(X_S/X_{\bar{R}})^2\ (A_f/A_0)^m\ \sqrt{\mathrm{d}/\mathrm{d}A\ (u/X)_f/\mathrm{d}/\mathrm{d}A\ (u/X)_0} = 1$ where $X_{\bar{R}}$ is the load corresponding with crack propagation at crack area A_0 at toughness $\bar{R}$; other symbols are defined in section 8.9.

8.17 Show in crack arrest problems that the statically computed post-arrest toughness value varies systematically with crack jump length. [Hint. Use the algebra of Example 8.9.]

8.18 Show how the value of η sets limits on the location of the R-locus during crack propagation.

8.19 Determine the value of η for a cracked three-point bend testpiece where, for a span/height of beam (S/W) ratio of 4, $Y\,(a/W) = 1.09 - 1.74\,(a/W) + 8.20\,(a/W)^2 - 14.18\,(a/W)^3 + 14.57\,(a/W)^4$. Show that $\eta \to 2$ for deep notches.

8.20 Show that the ratio of cracked and uncracked compliances for a body is given by $2\ Y^2 a\ (W - a)/LW\eta$ where L is either gauge length (for a tension sample) or $S/9$ (S = span) for a three-point bend testpiece.

8.21 Show that $\eta = 2\pi a/h$ for the Griffith geometry and that therefore cracking is unstable (crack length = $2a$; height of large sheet = h; $u/X = \dfrac{h(1 + 2\pi a^2/hW)}{EBW}$; W = width of large sheet; B = thickness).

8.22 Show that crack instability in the presence of plastic collapse in the ligaments of a centre-cracked strip under tension is given by $T = (\mathrm{d}J/\mathrm{d}a)\ \times$

$(E/\sigma_y^2) < 2L/W$ where L is the overall length and W the total width of the strip.

8.23 Using equation (6.4) for $\dot{\epsilon}$ and equation (6.26) for $R(\dot{a},T)$, where T is temperature, together with $E = 12.12\dot{\epsilon}^{0.0087} - 0.0268T$ (GPa), obtain the following relation between $\dot{\epsilon}$ and $\dot{a}$

$$\dot{a} \approx (U/\lambda)\ \dot{\epsilon}/\pi\epsilon^3\ 10^9\ (12.12\dot{\epsilon}^{0.0087} - 0.00268T)$$

8.24 Discuss the notion of 'crispness' in foodstuffs in terms of crack stability.

8.25 Discuss the reasons why fatigue crack propagation is stable in testpieces which have unfavourable geometric stability factors for monotonic loading.

REFERENCES

Atkins, A. G., Lee, C. S. and Caddell, R. M. (1975) *J. Mater. Sci.* **10**, 1981.

Bazant, Z. P. and Ohtsubo, H. (1977) *Mech. Res. Comm.* **4**, 353.

Broek, D. (1973) *Engr. Fract. Mech.* **5**, 35.

Burrel, H. and Immergut, E. H. (1966) *Polymer Handbook*. New York: Interscience.

Carslaw, H. S. and Jaeger, J. C. (1959). *Conduction of Heat in Solids* Oxford: University Press.

Chell, G. G. and Milne, I. (1983) ASTM STP No. 803, II-179. Philadelphia: ASTM.

Cheverton, R. D., Gehlen, P. C., Hahn, G. T. and Iskander, S. K. (1980) ASTM STP No. 711. Philadelphia: ASTM.

Chow, C. L. and Lam, P. M. (1974) *J. Engr. Mater. & Tech.* **96**, 41.

Clausing, D. P. (1969) *Int. J. Fract.* **5**, 211.

Corran, R. S. J., Mines, R. A. W. and Ruiz, C. (1985) *Int. J. Fract.* (in press).

Crosley, P. B. and Ripling, E. J. (1977) ASTM STP No. 627 Philadelphia: ASTM.

Crosley, P. B. and Ripling, E. J. (1980) ASTM STP No. 711. Philadelphia: ASTM.

Ernst, H. (1983) ASTM STP No. 803. Philadelphia: ASTM.

Freund, L. B. and Parks, D. M. (1980) ASTM STP No. 711. Philadelphia: ASTM.

Garwood, S. J. (1979) ASTM STP 677. Philadelphia: ASTM.

Gehlen, P. C., Popelar, C. H. and Kanninen, M. F. (1979) *Int. J. Fract.* **15**, 281.

Gurney, C. and Hunt, J. (1967) *Proc. Roy. Soc.* **A299**, 508.

Gurney, C. and Mai, Y. W. (1972) *Engr. Fract. Mech.* **4**, 853.

Gurney, C., Mai, Y. W. and Owen, R. C. (1974) *Proc. Roy. Soc.* **A340**, 213.

Gurney, C. and Ngan, K. M. (1971) *Proc. Roy. Soc.* **A325**, 207.

Hahn, G. T., Hoagland, R. G., Kanninen, M. F. and Rosenfield, A. R. (1973) in *Dynamic Crack Propagation*, G. C. Sih (ed.). Leiden: Noordhoff, p. 649.

Hasselman, D. P. H. (1969) *J. Amer. Ceram. Soc.* **52**, 600.

Hetenyi, M. (1974) *Beams on Elastic Foundations.* Ann Arbor, Michigan: University of Michigan Press.

Hoagland, R. G. and Rosenfield, A. R. (1974) *Int. J. Fract.* **10**, 299.

Hoagland, R. G. *et al.* (1977) ASTM Committee E 24.03.04 Paper, October.

Kalthoff, J. F., Beinert, J., Winkler, S. and Klemm, W. (1980) ASTM STP No. 711. Philadelphia: ASTM.

Kanninen, M. F. (1974a) *Int. J. Fract.* **10**, 415.

Kanninen, M. F. (1974b) in *Prospects of Fracture Mechanics*, G. C. Sih *et al.* (eds.). Leiden: Noordhoff, p. 251.

Kanninen, M. F. (1980) *Proc. 2nd Int. Conf. Numerical Methods in Fracture Mechanics*, D. R. J. Owen and A. R. Luxmore (eds.). Swansea: Pineridge Press, p. 433.

Kanninen, M. F. (1985) private communication.

Kinloch, A. J. and Williams, J. G. (1980) *J. Mater. Sci.* **15**, 987.

Krafft, J. M., Sullivan, A. M. and Boyle, R. W. (1961) *Proc. Crack Propagation Symposium*, Cranfield, Vol. 1, Paper 8.

Kumar, V., German, M. D. and Shih, C. F. (1981) EPRI Report NP-1913.

Lee, C. S. (1974) Ph.D. dissertation, University of Michigan, Ann Arbor (see also (1975) *Int. J. Fract.* **11**, 1052).

Mai, Y. W. (1976) *Int. J. Fract.* **12**, 107.

Mai, Y. W. and Atkins, A. G. (1975) *J. Mater. Sci.* **10**, 1904.

Mai, Y. W. and Atkins, A. G. (1976) *J. Mater. Sci.* **11**, 677.

Mai, Y. W. and Atkins, A. G. (1980) *J. Strain Analysis* **15**, 63.

Mai, Y. W., Atkins, A. G. and Caddell, R. M. (1976) *Int. J. Fract.* **12**, 391.

Mai, Y. W. and Leete, N. B. (1979) *J. Mater. Sci.* **14**, 2264.

Mamaros, T. C. and Jones, A. T. (1974) *Closed-Loop*, Fall/Winter edition.

Marston, T. U., Smith, E. and Stahlkopf, K. E. (1980) ASTM STP No. 711. Philadelphia: ASTM.

Mirabile, M. and Venzi, S. (1973) in *Dynamic Crack Propagation*, G. C. Sih (ed.). Leiden: Noordnoff, p. 693.

Paris, P. C., Tada, H., Zahoor, A. and Ernst, H. (1979) ASTM STP No. 668, 5. Philadelphia: ASTM.

Paris, P. C. (1983). ASTM STP No. 803, II 50,136. Philadelphia: ASTM.

Parvin, M. (1979) *Int. J. Fract.* **15**, 397.

Tattersall, H. G. and Tappins, G. (1966) J. Mater. Sci. 1, 296.

Turner, C. E. (1973) *Mater. Sci. & Engr.* **11**, 275.

Turner, C. E. (1979) ASTM STP No. 677. Philadelphia: ASTM, pp. 614–628.

Turner, C. E. (1980) ASTM STP No. 700. Philadelphia: ASTM, p. 314.

Turner, C. E. (1981) *Proc. ICF-5* (Cannes), D. Francois (ed.). Oxford: Pergamon, p. 1167.

Wang, D. Y. and McCabe, D. E. (1974) *Proc. 8th Nat. Symp.* Providence, Phode Island: Brown University.

Williams, J. G. and Marshall, G. P. (1974) *Polymer* **15**, 251.

UNIVERSITY OF BRADFORD

MEMORANDUM

To

From

Date

Subject

9

Scaling in static fracture and fatigue fracture

'Curiouser and curiouser!' cried Alice. . . . 'Now I'm opening out like the largest telescope that ever was!'

9.1 INTRODUCTION

When load/deflexion tests are performed on different size bodies of nominally crack-free material it is possible to normalise the different load/extension curves into one stress/strain curve, from which elastic moduli and plastic flow properties are traditionally determined. Variations sometimes do occur depending on the size of the body (e.g. segregation in castings, strengthening in fine wires), but it is customary to consider that elastic moduli and yield stresses are independent of the size of the material, and that is translated into traditional 'strength of materials' design practice.

Load–extension curves given by globally elastic *cracked* bodies are used directly to determine J_C, G_C or K_C, consistent values of which will be obtained from the appropriate formula (to correspond with the specific work R absorbed in the process zone), irrespective of the overall size of the crack body (noting of course that the toughness may vary with rate, temperature and environment as discussed in Chapter 6). However, unlike nominally crack-free bodies, the loads at fracture in larger bodies do *not* change in proportion to the changed cross-sectional area compared with smaller cracked bodies. In fact, as shown later, cracking loads usually increase less rapidly than areas. Thus, at constant fracture toughness *the stresses to cause fracture* can get *smaller* as the size of the cracked body increases. This is dangerous, as the cracking stress may be brought below the safety-factored yield stress in sufficiently large structures, and the structure will unexpectedly break in a brittle fashion even though small samples of the same material will be ductile in the laboratory (cf. Fig. 1.22). It is an example of cube/square law scaling: the energy *available* to feed the crack increases as the cube of some characteristic length, whereas the toughness work *required* depends on the area of the crack which increases as the square of the length. Brunel successfully got the *Great Western* paddle wheeler from Bristol to New York in 1843 with coal to spare as he built her large enough (bunker capacity proportional to volume of hull, resistance through the water proportional to the water-line surface area).

Fracture in the plastic range also does not follow the laws of geometric scaling. Size-effects in plastic flow itself are small, but are significant for initiation and propagation. The 'ductility' (i.e. extension before fracture) of scaled bodies undergoing elastoplastic fracture changes with absolute size. Larger bodies display reduced deformation at fracture than smaller, and very large bodies may deform merely in the elastic range before cracking, as mentioned earlier. This behaviour goes against the conventional wisdom of using a size-independent safety-factored flow stress in traditional design procedures.

In fatigue fracture, any growth rate law which incorporates the stress intensity factor K (such as the Paris equation) must exhibit a scale effect, whereby (usually) the fatigue life at a given stress is smaller the larger the body.

Irwin in 1954 recognised that explanations of size effects prior to that time neglected the influences of elastic energy in the cracked bodies: he remarked that,

> disregarding size effects due to changes in materials properties with size caused by problems of metallurgical processing, previous studies have shown that fracture size effects of significant magnitude exist in solids whether they are brittle or ductile and whether the material is a metal or a plastic. The most elaborate precautions to eliminate differences due to material quality, lack of precise similarity of dimensional ratios, specimen surface preparation, and the experimental loading devices have failed to remove fracture size effects. It is apparent that these are fundamental to the nature of the severing process.

The particular symposium, at which he spoke, was concerned with low-temperature effects but he cautioned that reduced ductility with temperature (the usual '*metallurgical*' explanation of World War II T2 tanker and Liberty ship failures), was *not* the sole reason for brittle fractures in large structures. There are *mechanics* reasons for scale-effects which are the subject of the following sections, which are concerned with predicting the relationships between critical loads, stresses, displacements and crack lengths in small and large components under monotonic elastic and plastic loading, and in fatigue.

9.2 GEOMETRICALLY SIMILAR LINEAR ELASTIC STRUCTURES UNDER MONOTONIC LOADING

Consider two cracked rectangular structures m and p as shown in Fig. 9.1. They are geometrically similar with respect to shape and crack dimensions in the linear ratio λ, i.e. sizes in p are λ times those of m. The smaller cracked piece may be thought of as a 'model' (m) and the larger as a 'prototype' (p), along

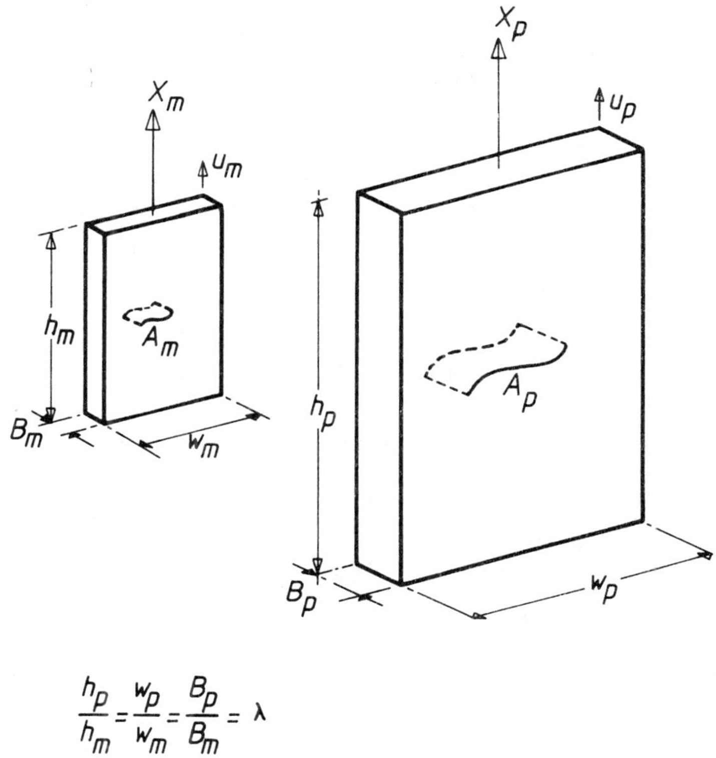

Fig. 9.1 – Two geometrically similar cracked bodies.

the lines of modelling experiments. Typical stresses and strains at corresponding points in the two cracked bodies are given by:

$$\begin{aligned} \sigma_m &= X_m/B_m w_m & \epsilon_m &= u_m/h_m & \bar{E}_m &= X_m h_m/B_m w_m u_m \\ \sigma_p &= X_p/B_p w_p & \epsilon_p &= u_p/h_p & \bar{E}_p &= X_p h_p/B_p w_p u_p \end{aligned} \tag{9.1}$$

with the assumption of linear elastic behaviour (i.e. the load/displacement Xu relationships at constant crack area follow straight lines). $\bar{E}$ are *effective* Young's moduli, taking into account the presence of cracks, and are less than the plane stress or plane strain moduli E^*. For geometrically similar bodies, $\bar{E}_p = \bar{E}_m$ (see section 9.2.1). Considering the compliance (u/X) of model and prototype, it follows that

$$\frac{u_p}{X_p} = \frac{1}{\lambda}\left[\frac{u_m}{X_m}\right] \tag{9.2}$$

and that the rates of change of compliance with respect to crack area are

$$\frac{d}{dA_p}\left[\frac{u_p}{X_p}\right] = \frac{1}{\lambda^2}\frac{d}{dA_m}\left[\frac{u_p}{X_p}\right] = \frac{1}{\lambda^3}\frac{d}{dA_m}\left[\frac{u_m}{X_m}\right] \tag{9.3}$$

From $X^2 = \dfrac{2R}{\dfrac{d}{dA}\left[\dfrac{u}{X}\right]}$ for equilibrium propagation (equation (2.6) we have

$$X_p^2 = \frac{2R_p}{\dfrac{d}{dA_p}\left[\dfrac{u_p}{X_p}\right]} = \frac{2R_p}{\dfrac{1}{\lambda^3}\dfrac{d}{dA_m}\left[\dfrac{u_m}{X_m}\right]} \tag{9.4}$$

so that

$$X_p = \lambda^{3/2}\left[\frac{R_p}{R_m}\right]^{1/2} X_m = \lambda^{3/2} X_m \left[\frac{K_{Cp}}{K_{Cm}}\right] \tag{9.5}$$

writing $K_C^2 = E^*R$ with E^* the same in model and prototype.
The areas, over which these forces act, differ by λ^2. It follows that

$$\sigma_p = \left[\frac{1}{\sqrt{\lambda}}\right] \sigma_m \left[\frac{K_{Cp}}{K_{Cm}}\right] \tag{9.6}$$

Again, from (9.2)

$$u_p = \sqrt{\lambda}\left[\frac{K_{Cp}}{K_{Cm}}\right] u_m \tag{9.7}$$

Equations (9.5) to (9.7) are scaling laws in the mechanics of fracture for geometrically similar structures (Gurney and Hunt, 1967). For many materials it is reasonable to assume that $R_m = R_p$ (so that $K_{Cm} = K_{Cp}$) whence

$$\frac{X_p}{X_m} = \lambda^{3/2}; \quad \frac{\sigma_p}{\sigma_m} = \frac{1}{\sqrt{\lambda}}; \quad \frac{u_p}{u_m} = \sqrt{\lambda} \tag{9.8}$$

The second relation says that the stresses to propagate cracks in large structures are *smaller* than those stresses required in geometrically similar structures. Thus a material which appears ductile in a small-size laboratory testpiece may behave in a brittle manner in a sufficiently large piece. In other words cracking can occur before any yielding takes place in large structures, whereas the same material can display general yielding without cracking in sufficiently small pieces. The second of (9.8) gives the equation for the falling curve of fracture stress vs. size in Fig. 1.22 (see also Chapter 10).

Example 9.1 Discuss methods of non-dimensionalising load/displacement data obtained from geometrically similar testpieces in (a) conventional tensile tests of uncracked material, (b) propagation tests on cracked bodies.

Ans. (a) Stress (σ) and strain (ϵ) are the quantities into which load and extension are converted for uncracked pieces in order to determine elastic moduli, yield strengths, etc. In geometrically similar pieces, the load increases as λ^2 and the extension by λ, which when reduced to σ and ϵ, lead to a *single* curve. The area under this curve is the work done/unit volume; in a large piece the done is

$$\lambda^2 \text{ (small load)} \times \lambda \text{ (small extension)} = \lambda^3 \text{ (work on small piece)}.$$

Since the volume of the larger piece is λ^3 times the volume of the smaller, the work done per unit volume is constant in both elastic and plastic ranges *so long as* fracture does not occur.

(b) Consider the analogous treatment of linear elastic cracking data. Are there 'master curves', (constant G_C-loci) for a given shape testpiece, for which segmental areas represent the work done in quasi-static cracking? Evidently for pieces of the same material, if loads are divided by $\lambda^{3/2}$ and the extensions by $\lambda^{1/2}$ (see equation (9.8)) we get 'universal' toughness loci. The radial lines from the origin representing constant crack areas are normalised at the same time, since the normalised compliance is

$$\frac{\left[\dfrac{u_p}{\lambda^{1/2}}\right]}{\left[\dfrac{X_p}{\lambda^{3/2}}\right]} = \lambda \left[\frac{u_p}{X_p}\right] = \lambda \left[\frac{1}{\lambda}\frac{u_m}{X_m}\right] = \frac{u_m}{X_m}$$

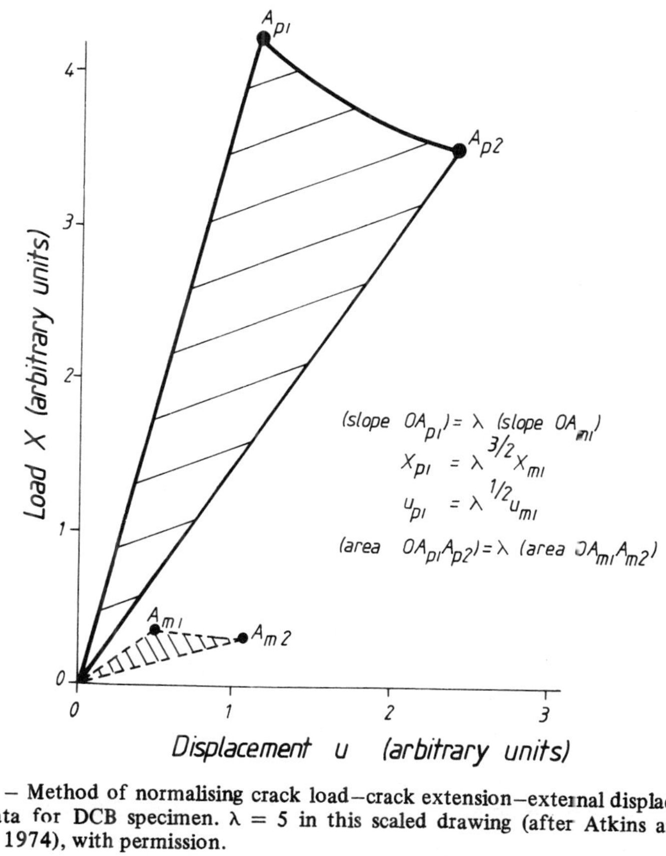

Fig. 9.2 – Method of normalising crack load–crack extension–external displacement data for DCB specimen. $\lambda = 5$ in this scaled drawing (after Atkins and Caddell, 1974), with permission.

using equation (9.2). Regarding normalised work done,

$$\left[\frac{X_p}{\lambda^{3/2}}\right]\left[\frac{u_p}{\lambda^{1/2}}\right] = \left[\frac{1}{\lambda^2}\right] X_p\, u_p = X_m\, u_m$$

using equation (9.8). $X_p\, u_p$ 'feeds' on the area λ^2 bigger than does $X_m\, u_m$ so the fracture toughness work is equivalent for both cases. These effects are shown schematically in Fig. 9.2.

9.2.1 Application to ice-tank towing studies

It is well known that in conventional towing tank experiments, with models of prototype ships or structures, it is generally impossible to arrange complete dynamic similarity, i.e. the independent force ratios concerning inertia forces, viscous forces and gravity forces in model and prototype do not have the same scaling factors.

The reasons are as follows. Inertia forces are given by $X_I = M\ddot{s}$, viscous forces by $X_v = \mu A(\partial s/\partial y)$, gravity forces by $X_G = Mg$ and, for completeness, elasticity forces by $X_e = \sigma A = E\,\epsilon\,A$. Thus it may be shown that (Szabo, 1961)

$$\left.\begin{aligned} \frac{(X_I)_p}{(X_I)_m} &= \frac{\rho_p\lambda^4}{\rho_m\tau^2}\,; \qquad \frac{(X_v)_p}{(X_v)_m} = \frac{\nu_p}{\nu_m}\,\frac{\rho_p}{\rho_m}\,\frac{\lambda^2}{\tau}\,; \\ \frac{(X_G)_p}{(X_G)_m} &= \frac{\rho_p}{\rho_m}\,\lambda^3; \qquad \frac{(X_e)_p}{(X_e)_m} = \frac{E_p}{E_m}\,\lambda^2 \end{aligned}\right\} \tag{9.9}$$

Consequently, to satisfy the same scaling factor for inertia forces and viscous (friction) forces,

$$\frac{\rho_p\lambda^4}{\rho_m\tau^2} = \frac{\nu_p}{\nu_m}\,\frac{\rho_p}{\rho_m}\,\frac{\lambda^2}{\tau^2} \tag{9.10}$$

so that the time scale between model and prototype is

$$\tau = \frac{\nu_m}{\nu_p}\,\lambda^2 \tag{9.11}$$

The velocity ratio between model and prototype is

$$\frac{v_p}{v_m} = \frac{\lambda}{\tau} = \frac{\nu_p}{\nu_m}\,\frac{1}{\lambda} = \frac{\nu_p}{\nu_m}\,\frac{L_m}{L_p} \tag{9.12}$$

Therefore, we have

$$\frac{v_p L_p}{\nu_p} = \frac{v_m L_m}{\nu_m} = R_n \quad \text{(Reynolds number)} \tag{9.13}$$

Likewise we can show that for equal scaling factors between inertia and gravity (wave-making) forces

$$\frac{v_p}{\sqrt{L_p}} = \frac{v_m}{\sqrt{L_m}} \quad \text{(speed: length ratio)} \tag{9.14}$$

or

$$\frac{v_p}{\sqrt{(gL_p)}} = \frac{v_m}{\sqrt{(gL_m)}} = F_n \quad \text{(Froude number)} \tag{9.15}$$

Again, for simultaneous satisfaction of inertia and elastic forces

$$v_p \left[\frac{\rho_p}{E_p}\right]^{1/2} = v_m \left[\frac{\rho_m}{E_m}\right]^{1/2} = C_n \text{ (Cauchy number)} \tag{9.16}$$

or since $(E/\rho)^{1/2} = c_0$ (the sonic velocity), we have

$$\frac{v_p}{c_{0p}} = \frac{v_p}{c_{0m}} = M_n \quad \text{(Mach number).} \tag{9.17}$$

Since tank fluids are incompressible for all practical purposes, C_n is neglected in open water towing studies, and M_n appears more in wind tunnel work. Regarding F_n and R_n, complete dynamic similitude is impossible unless a tank fluid can be found for which $\nu_p/\nu_m = \lambda^{3/2}$. It is customary to scale according to Foude's law and correct the test results by independent experimental 'plate' data based upon Reynold's law.

When dealing with towing basins covered in ice to model the ice-breaking capabilities of prototype ships or ice-affected structures, for example, the situation is complicated beyond the above considerations since the ice-breaking force ratios must satisfy some scaling law. Some workers argue that the Cauchy number for the ice sheet is the relevant scaling parameter, but since we have shown that cracking forces should not be identified continuum elastic forces, C_n is not appropriate. Rather a new number has to be satisfied, based upon the force realtionship $X_p = \lambda^{3/2} (K_p/K_m)X_m$ (equation (9.5)), as follows.

In order to satisfy the same scaling law for inertia forces and cracking forces, we have from arguments analogous to those presented earlier,
i.e.

$$\frac{\rho_p}{\rho_m}\frac{\lambda^4}{\tau^2} = \lambda^{3/2}\left[\frac{K_p}{K_m}\right]^{1/2} \tag{9.18}$$

$$\tau = \left[\frac{\rho_p}{\rho_m}\,\lambda^{5/2}\,\frac{K_m}{K_p}\right]^{1/2} \tag{9.19}$$

and

$$\frac{v_p}{v_m} = \frac{\lambda}{\tau} = \left(\frac{\rho_m K_p}{\rho_p K_m}\right)^{1/2} \frac{1}{\lambda^{1/4}} \tag{9.20}$$

so that

$$v_p \left[\frac{\rho_p L_p^{1/2}}{K_p}\right]^{1/2} = v_m \left[\frac{\rho_m L_m^{1/2}}{K_m}\right]^{1/2} \tag{9.21}$$

or

$$\frac{v_p^2 \rho_p L_p^{1/2}}{K_{Cp}} = \frac{v_m^2 \rho_m L_m^{1/2}}{K_{Cm}} = I_n \text{ (new 'ice' number).} \tag{9.22}$$

This result is due to Atkins (1973). Notice that $I_n = C_n^2(EL/R)^{1/2}$, so that I_n appears to be the Cauchy number generalised to include the effects of a 'defective' (cracked) continuum.

Scaling laws for ice tank work were ill-understood and the techniques used to translate from model to prototype were quasi-empirical. The foregoing analysis gave a new scaling law for the ice fracture part of the resistance of a ship to passage through ice or, conversely, to the force on dams, etc. caused by ice jams. It is customary to weaken the strength of tank ice relative to sea or lake ice. This is accomplished by copiously salting the tank water to reduce the ice strength. Brine pockets in the frozen ice are the cause of weakening. It is known that the mechanical properties of such ice sheets markedly differ from 'prototype' ice. Not only is the strength diminished, but the ice moduli are also altered, the tank ice being almost rubber-like in flexibility. The 'rules' by which model ice 'strength' should be diminished were ill-understood. (The fracture 'strength' of sea or lake ice is determined by *in situ* cantilever beam tests but, of course, since failure loads really depend on R or K_C and the size of the brine-pockets which act as starter cracks, the use of ice bend 'strength' should be discouraged for the same reasons that the use of the 'compressive strength' of rocks was discouraged in Chapter 3).

In general terms, the density of salt ice is greater than prototype, but the toughness and Young's modulus of the tank ice are less than prototype, so that $K_{Cm} < K_{Cp}$.

Hence, in equation (9.20), $\rho_p < \rho_m$ and $K_{Cm} < K_{Cp}$. Depending on the behaviour of the ρK terms and the magnitude of λ, v_m could be less than v_p. Indeed we may argue that if the ice strength (strictly toughness) be scaled according to some, as yet unknown, factor, both Foude's law and the cracking law can be simultaneously satisfied.

The time scale for Foude's law is $\tau = \sqrt{\lambda}$. For cracking, it is given by equation (9.19). If these scales are equated, we get

$$\frac{\rho_p L_p^{3/2}}{K_{Cp}} = \frac{\rho_m L_m^{3/2}}{K_{Cm}}\text{; that is, } \frac{K_{Cp}}{K_{Cm}} = \frac{\rho_p}{\rho_m}\lambda^{3/2} \tag{9.23}$$

When the ice toughness is scaled according to this condition, model testing at Froude speeds is possible. It follows (using equation (9.6)) that

$$\sigma_m = \frac{1}{\lambda}\cdot\frac{\rho_m}{\rho_p}\cdot\sigma_p \tag{9.24}$$

is the rule by which model ice strength should be scaled in order to satisfy both Froude's law and the correct cracking law. If the differences in density between model and prototype ice are ignored, it is seen that the tank ice strength should be reduced directly as the linear scaling factor.

A theoretical basis was thus provided for an empirical rule used by some workers in which the model ice strength was scaled linearly.

9.2.2 Non-proportionately scaled linear bodies

Most large prototype engineering structures are not actually proportionally scaled-up versions of model testpieces. For example, ships' plates can be tested for toughness in full-thickness in the laboratory, but the height and width dimensions of a real hull are impossible to duplicate; again in pipe steel, annular cross-sections can readily be tested, but the considerable length of pipelines cannot. Clearly geometric similarity between laboratory model and prototype does not hold, yet brittle fracture in such structures made of ductile materials is well known. This section concerns such problems where there are different scaling factors for height, width, thickness and crack length.

A generalised approach for scaling in non-proportionately scaled elastic structures does not exist (Mai and Atkins, 1978), so for purposes of illustration consider centre-cracked panels (CCP) as in Fig. 9.1, but where λ is not the same for all dimensions. We have

$$\frac{h_p}{h_m} = \lambda_h; \quad \frac{B_p}{B_m} = \lambda_B; \quad \frac{w_p}{w_m} = \lambda_w \; ; \text{ and } \quad \frac{a_p}{a_m} = \lambda_a$$

Following a similar line of attack as before, it may be shown that

$$\frac{d}{dA_p}\left[\frac{u_p}{X_p}\right] = \frac{\lambda_h}{\lambda_B^2\,\lambda_w\,\lambda_a}\,\frac{\bar{E}_m}{\bar{E}_p}\,\frac{d}{dA}\left[\frac{u_m}{X_m}\right] \tag{9.25}$$

Hence,

$$\frac{\sigma_p}{\sigma_m} = \frac{1}{\lambda_B\,\lambda_w}\left[\frac{X_p}{X_m}\right] = \left[\frac{\lambda_a}{\lambda_h\,\lambda_w}\right]^{1/2}\sqrt{\frac{\bar{E}_p\,R_p}{\bar{E}_m\,R_m}} \tag{9.26}$$

Unlike geometrically similar centre-cracked panels, $\bar{E}_p \neq \bar{E}_m$ (as in section 9.2) owing to the different scaling factors in different directions. However, we may readily determine their relationship by invoking the boundary conditions for fracture in CCP (i.e. $\sigma\sqrt{\pi a} = K_C$), from which $\sigma_p = \sigma_m/\sqrt{\lambda_a}$. Equation (9.26) must reduce to this expression when $R_p = R_m$, so we conclude that $\bar{E}_m/\bar{E}_p = \lambda_a^2/\lambda_h\,\lambda_w$ (cf. when $\lambda_a = \lambda_h = \lambda_w$, $\bar{E}_p = \bar{E}_m$ as in section 9.2). The scaling laws are thus given by,

$$\left.\begin{aligned} \frac{X_p}{X_m} &= \left[\frac{\lambda_B\,\lambda_w}{\sqrt{\lambda_a}}\right]\sqrt{\frac{R_p}{R_m}} \\ \frac{\sigma_p}{\sigma_m} &= \left[\frac{1}{\sqrt{\lambda_a}}\right]\sqrt{\frac{R_p}{R_m}} \\ \frac{u_p}{u_m} &= \left[\frac{\lambda_a^{3/2}}{\lambda_w}\right]\sqrt{\frac{R_p}{R_m}} \end{aligned}\right\} \tag{9.27}$$

which for invariant R (or K_C) become

$$\frac{X_p}{X_m} = \frac{\lambda_B \lambda_w}{\sqrt{\lambda_a}}, \quad \frac{u_p}{u_m} = \frac{\lambda_a^{3/2}}{\lambda_w}, \quad \frac{\sigma_p}{\sigma_m} = \frac{1}{\sqrt{\lambda_a}} \tag{9.28}$$

(When all λs are equal, we obtain the expressions in section 9.2.) Notice that only λ_a comes into σ_p/σ_m, which says that all small cracks in large plates behave like Griffith cracks.

Burst tests on mild steel pipes provide examples of the appliction of these scaling laws. The hoop stress is responsible for causing fracture in pipelines containing central cracks orientated in the longitudinal axis. It is reasonable to employ the CCP scaling laws for pipelines, but a relation must be found between the remotely applied stress and the hoop stress. Folias (1965) gives for relatively large crack lengths,

$$\sigma^2 \approx \left[\frac{(4\nu + 1)}{2}\right] \left[\frac{5\pi a^2 \sqrt{12(1-\nu^2)}}{32\rho B}\right] \sigma_{\text{hoop}}^2 \tag{9.29}$$

where ρ is the radius of curvature and B the pipe thickness. The bracketed terms represent a correction factor for the effect of curvature in the pipe for use in the flat panel expressions. (Note that for small crack lengths equation (9.29) is rather conservative; see Hahn, 1969). Writing out this expression for both model and prototype pipelines made of the same material, considering the 'height' of the equivalent CCP as the circumference (equivalently radius) direction of the pipe, we obtain by division

$$\frac{\sigma_p^2}{\sigma_m^2} = \left[\frac{\sigma_{\text{hoop p}}^2}{\sigma_{\text{hoop m}}^2}\right] \frac{\lambda_a^2}{\lambda_h \lambda_B} \tag{9.30}$$

(Notice that λ_B has reappeared in the two-dimensional analysis as a result of the correction for plate curvature.) The scaling relation is obtained by combining equations (9.28) and (9.30), viz.

$$\frac{\sigma_{\text{hoop p}}}{\sigma_{\text{hoop m}}} = \frac{\sqrt{\lambda_B \lambda_h}}{\lambda_a^{3/2}} \tag{9.31}$$

Experimental results were given by Duffy *et al.* in 1969 from burst tests on mild steel pipes of difference diameters (168–194 mm), wall thicknesses (6–22 mm) and crack lengths (127–254 mm); all pipes had the same length. For those combinations of sizes for which $\lambda_B = \lambda_h = 1$ (i.e. the product ρB is constant) we expect from relation (9.31) that $(\sigma_{\text{hoop p}}/\sigma_{\text{hoop m}}) = \lambda_a^{-3/2}$. Let us choose as the reference 'model' their experimental result with the smallest starter crack $a_m = 127$ mm, for which $\sigma_{\text{hoop m}} = 172$ MPa (converted from the burst pressure). A set of data, all for $\rho B = 536$ mm^2 is given in Table 9.1 and shows that there is reasonable agreement between the hoop stress ratios and $\lambda_a^{-3/2}$.

Table 9.1
Scaling laws of fracture in pipes

σ_{hoop} (MPa)	172	138	117	103	83	69
a_p (mm)	127	152	178	203	229	254
$\sigma_{\text{hoop p}}/\sigma_{\text{hoop m}}$	1	0.83	0.68	0.60	0.48	0.40
$(a_p/a_m)^{-3/2}$	1	0.76	0.61	0.50	0.42	0.36

***Example** 9.2* Show that hoop stresses at fracture in full-scale pipelines are *larger* than those in laboratory CCP coupons containing the same crack length.

Ans. For $\lambda_a = 1$, $\sigma_{\text{hoop p}}/\sigma_{\text{hoop m}} = \sqrt{\lambda_B \lambda_h}$ and if B or ρ (or both) are larger in the practical pipeline, the bursting stress is greater.

The conclusion of larger stresses in Example 9.2 is supported by data from Duffy *et al.* (1969) regarding tests with fixed $a_m = a_p = 254$ mm but varying ρB. Table 9.2 gives some results. There is again reasonable agreement between experiment and theory. Notice that the theoretical results are systematically higher in Table 9.2 and vice versa in Table 9.1. It may be shown that slight non-linearity in the *Xu* relation (section 9.3) gives closer agreement.

Table 9.2
Scaling laws for fracture in pipes

σ_{hoop} (MPa)	69	124	179	193	207
a_p (mm)	536	2362	3626	4710	10 000
$\sigma_{\text{hoop p}}/\sigma_{\text{hoop m}}$	1.0	1.8	2.6	2.8	3.0
$\sqrt{\rho_p B_p/\rho_m B_m}$	1.0	2.1	2.6	3.0	4.3

This pipeline example (showing that much larger pressures can be sustained in larger pipes) demonstrates that cracking stresses in larger structures are *not necessarily* always lower than those in smaller structures as predicted by geometrically similar scaling which gives $\sigma_p = \sigma_m/\sqrt{\lambda}$ (equation (9.8)). When λ varies in different directions σ_p may even be greater than σ_m. Hence some large structures containing flaws may not fail in a brittle fashion and design based on a criterion of plastic yielding and collapse may be adequate. Even so, it is important to realise that a size-effect does exist in fracture, unlike yielding, and that unexpected brittle failures can occur.

For a factor of safety f, the strength-of-materials design working stress is σ_Y/f. Brittle fracture will occur if $\sigma_p < \sigma_Y/f$ where σ_p is the stress to cause cracking without yielding. For plate structures containing some sort of central crack, $\sigma_p = \sigma_m/\sqrt{\lambda_a}$ from relation (9.28). Also $\sigma_m = K_{Cm}/\sqrt{\pi a_m}$. Thus for brittle fracture,

$$\lambda_a > \left[\frac{K_{Cm}}{\sigma_y}\right]^2 \left[\frac{f^2}{\pi a_m}\right] \tag{9.32}$$

For structural steels with $K_C/\sigma_y \approx 0.5\sqrt{\text{m}}$ at room temperature, we conclude that brittle fracture occurs when a_p ($= \lambda_a a_m$) $\geqslant 0.75$ m for $f = 3$. Ship's hatches larger than this have been known to initiate brittle fracture. For K_C/σ_y reduced to 0.05 (say, at lower temperatures) $a_p \geqslant 8$ mm only for brittle fracture.

9.3 NON-LINEAR ELASTIC STRUCTURES

Cracking is governed by equations (2.11). If we assume that $X = k(A)u^n$, it follows that for quasi-static fracture,

$$R = \frac{-k' k^{-(n+1)/n} X^{(n+1)/n}}{(n+1)} \tag{9.33}$$

where $k'(A) = \mathrm{d}k(A)/\mathrm{d}A$ (Mai and Atkins, 1975).

Let us first consider geometrically similar bodies with the common scaling factor λ. Using subscripts m and p for model and prototype, and employing routes for solution similar to those in section 9.2, we obtain the scaling laws for geometrically similar non-linear bodies having invariant R as:

$$\frac{X_p}{X_m} = \lambda^{(n+2)/(n+1)}; \quad \frac{\sigma_p}{\sigma_m} = \lambda^{-n/(n+1)};$$
$$\frac{u_p}{u_m} = \lambda^{n/(n+1)} \tag{9.34}$$

(Problem 9.3). For $n = 1$, these reduce to equations (9.8).

We see that the stresses to cause cracking in large non-linear structures are even smaller than those in large linear structures. When the structure is highly non-linear (n is large) $\sigma_p = \sigma_m/\lambda$ rather than $\sigma_p = \sigma_m/\sqrt{\lambda}$.

Similar procedures may be applied to particular non-proportionally scaled, non-linear, bodies as in section 9.2.1. In the case of centre-cracked panels of bodies with invariant R, it may be shown (Problem 9.4) that:

$$\left.\begin{aligned}
\frac{X_p}{X_m} &= \left[\frac{\lambda_B^{n+1}\lambda_w^2}{\lambda_h^{n-1}\lambda_a^{2-n}}\right]^{1/(n+1)} \\
\frac{\sigma_p}{\sigma_m} &= \left[\frac{\lambda_a^{n-2}}{(\lambda_h\lambda_w)^{n-1}}\right]^{1/(n+1)} \\
\frac{u_p}{u_m} &= \left[\frac{\lambda_h^{n+1}\lambda_a^3}{\lambda_w^2}\right]^{1/(n+1)}
\end{aligned}\right\} \tag{9.35}$$

which are the most general expressions (Mai and Atkins, 1978).

9.4 SCALE EFFECTS ON ELASTIC CRACK VELOCITY

Equation (3.2) gives the ratio of the crack velocity to the driving velocity $(\dot{a}/\dot{u})$ for quasi-static cracking of a linear elastic body. This may be recast as

$$\frac{\dot{a}}{\dot{u}} = \frac{2k'}{uk''B} \tag{9.36}$$

using $X = k(A)u$, from which $R = -\,k'X^2/2k^2$ (equation (2.7)). The prime notation denotes the derivatives with respect to area A. The ratio of $\dot{a}_p/\dot{a}_m$ (for the same $\dot{u}_m = \dot{u}_p$) is thus given by:

$$\frac{\dot{a}_p}{\dot{a}_m} = \left[\frac{k'_p}{k'_m}\right]\left[\frac{u_m}{u_p}\right]\left[\frac{B_m}{B_p}\right]\left[\frac{k''_m}{k''_p}\right] \tag{9.37}$$

For geometrically similar linear bodies, equation (9.8) gives $u_m/u_p = \lambda^{-1/2}$, $B_m/B_p = \lambda^{-1}$, $k'_p/k'_m = \lambda^{-1}$ and $k''_p/k''_m = \lambda^{-3}$, therefore

$$\frac{\dot{a}_p}{\dot{a}_m} = \lambda^{1/2} \tag{9.38}$$

Equation (9.38) predicts for linear elastic structures that the crack velocity in the prototype is always faster than that in the model by a factor of $\sqrt{\lambda}$. Hence not only do large structures sometimes unexpectedly fail in a brittle fashion, but they do so at fast crack speeds. This result has been arrived at for E, R or K_C being constant. When these properties are rate dependent, the model and prototype velocities may be closer or wider apart, depending upon the sign of $K_C(\dot{a})$. If K decreases with increase of crack velocity, the readier cracking of prototype structures over model structures is made even more easy.

In the case of non-linear elastic structures following $X = k(A)u^n$, Mai and Atkins (1975) obtained by similar procedures (Problem 9.5).

$$\frac{\dot{a}}{\dot{u}} = \frac{nX}{\left[RB(n+1) + \dfrac{u\,dX}{da}\right]} = -\frac{k'(n+1)}{k''uB} \tag{9.39}$$

so that

$$\frac{\dot{a}_p}{\dot{a}_m} = \lambda^{1/(n+1)} \tag{9.40}$$

Notice that when n is large, the crack velocity of the large component is closer to that of the small component, in contrast to the linear case where $n = 1$.

In the case of non-proportionately scaled, non-linear CCP it may be shown (Problem 9.2) that

$$\frac{\dot{a}_p}{\dot{a}_m} = \left[\lambda_a^{(n-2)}\,\lambda_w^2\,\lambda_h^{(1-n)}\right]^{1/(n+1)} \tag{9.41}$$

(Mai and Atkins, 1978). For all λ identical, this reduces to (9.40) and for $n = 1$, to (9.38).

For a non-proportionately scaled, but linear CCP, we have $\dot{a}_p/\dot{a}_m = \lambda_w/\sqrt{\lambda_a}$. For identical flaw sizes in model and prototype, $\dot{a}_p$ is $\lambda_w \dot{a}_m$ so that 'fast fracture' in a ship's hull under bending is readily explained. (Note that the crack velocity is increased, even though $\sigma_p = \sigma_m$ if $\lambda_a = 1$ according to equation (9.28).) Even if the prototype has a larger flaw than the model, $\sqrt{\lambda_a}$ will not rise as quickly as λ_w, so fast fracture still results.

Example 9.3 Determine the fracture velocity between full-scale pipelines and laboratory small coupon specimens of the same thickness, diameter and initial flaw size (assume linear behaviour). Show that $\dot{a}_p$ is much larger than that given by geometrically similar scaling expressions.

Ans. From equation (9.41), $\dot{a}_p/\dot{a}_m = \lambda_w$ when $\lambda_a = 1$. As λ_w is the scale factor for the length of the pipe, it is clear that enormous prototype velocities can be attained. Cracks in pipelines therefore run long distances – hundreds of metres and more – before arresting (Fig. 9.3). A full dynamic analysis of pipeline

Fig. 9.3 – A long crack in a pipeline (courtesy British Gas).

fracture is very complicated (section 8.9). The increase in velocity is much larger than for geometrically similar scaling, where $\dot{a}_p = \sqrt{\lambda}\dot{a}_m$ (equation (9.38)). It must be pointed out there, however, that according to Wilkowski (1979) actual measurements of brittle fracture speeds have shown a limiting maximum speed, $\dot{a}_{p_{max}}$ (100 m/s), in steel pipelines. This is believed to be caused by elastic flexural stress waves in the pipe causing a sinusoidal fracture path. Fracture speeds in laboratory testing of flat specimens (e.g. flat plates, DCB etc.) can be higher due to the absence of flexural stress waves ($\dot{a}_{p_{max}} \approx 2800$ m/s).

9.5 FRACTURE PREDICTIONS USING EQUIVALENT ENERGIES FROM SCALED TESTPIECES

Valid fracture toughness determination requires that no generalised yielding be occurring in regions of testpieces remote from the crack tip. For materials possessing relatively high fracture toughness and low yield strength, as explained in section 3.9, this necessarily requires huge testpieces which are not practicable in laboratories (because of limited loading capacities of test rigs if nothing else). One method of determining the fracture toughness of materials possessing large (ER/σ_y^2) ratios relates to scaling laws. It is the 'equivalent energy method' proposed by Witt and Berggren in 1971.

For purposes of illustration, consider two geometrically similar three-point bend cracked testpieces as shown in Fig. 9.4(a) and let crack extensions da and λda occur in the testpieces under constant displacements u. Fig. 9.4(b) shows graphically the corresponding load/displacement curves for the cracking events. Denoting $d\Lambda_m$ and $d\Lambda_p$ as the total (elastic) strain energy change in the m, p testpieces, we have

$$\left.\begin{aligned} d\Lambda_m &= \frac{1}{2}\, u_m\, dX_m \\ d\Lambda_p &= \frac{1}{2}\, u_p\, dX_p \end{aligned}\right\} \qquad (9.42)$$

Thus

$$\frac{d\Lambda_m}{d\Lambda_p} = \left[\frac{dX_m}{dX_p}\right]\left[\frac{u_m}{u_p}\right] = \left[\frac{X_m}{X_p}\right]^2 \lambda$$

from equation (9.2), and using $dX_m/dX_p = X_m/X_p$. From equation (9.5), we have

$$\left[\frac{X_m}{X_p}\right]^2 = \left[\frac{1}{\lambda^3}\right]\left[\frac{R_m}{R_p}\right] \qquad (9.43)$$

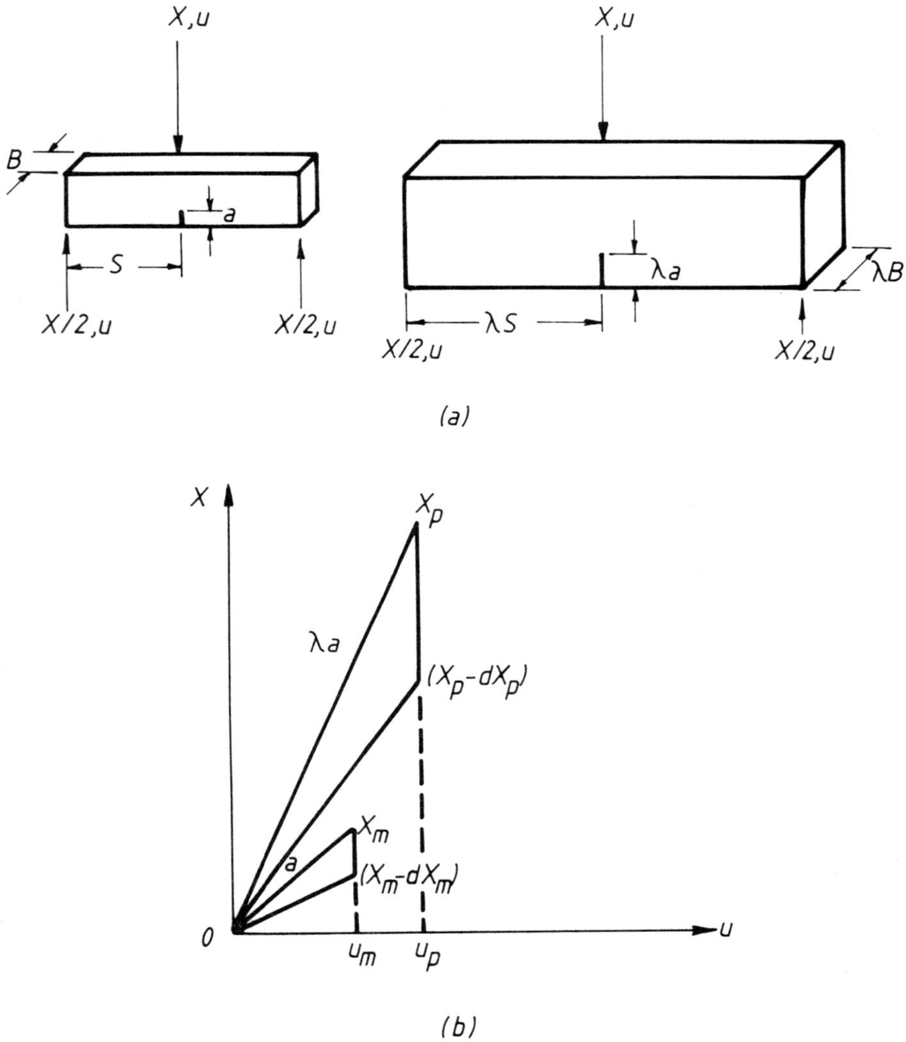

Fig. 9.4 – (a) Two geometrically similar three-point notched bend specimens. (b) Load–displacement diagrams for fracture in these specimens.

Thus, a combination of equations (9.42) and (9.43) yields

$$\left[\frac{R_m}{R_p}\right]^2 \doteq \left[\frac{d\Lambda_m}{d\Lambda_p}\right] \lambda^2$$

and since $E^*R = K_C^2$, we obtain

$$\frac{K_{Cm}}{K_{Cp}} = \lambda \sqrt{\frac{\Lambda_m}{\Lambda_p}} \tag{9.44}$$

for the crack extending quasi-statically across the whole cross-section of the testpiece at constant external displacement.

If $K_{Cm} = K_{Cp}$ (i.e. the critical stress intensity factor at fracture does not change with size), a series of experiments with geometrically similar testpieces of varying scale factor should produce a slope of (−2) in a log-log plot of Λ_m/Λ_p versus λ.

Example 9.4 Show that so long as cracking always precedes yielding in two geometrically similar structures of the same material, the fracture toughness obtained from the work areas should be equivalent.

Ans. Consider, for illustrative purposes, model and prototype DCB specimens as shown in the inset of Fig. 9.2. OA_{p1} represents the compliance of the prototype with crack area $\lambda^2 A_1$ and OA_{m1} that for the model of crack area A_1. From equation (9.2) we have

$$\frac{u_p}{X_p} = \left[\frac{1}{\lambda}\right]\left[\frac{u_m}{X_m}\right]$$

When the crack areas in the model and prototype have spread to A_2 and $\lambda^2 A_2$ the ratio of their slopes is also given by $(1/\lambda)$. Regarding work, we also have from equations (9.8)

$$X_p \, du_p = \lambda^2 X_m \, du_m$$

If we apply the energy equation to the prototype and model respectively, we have for the prototype,

$$R_p \, dA_p = X_p \, du_p - d\Lambda_p$$

$$= \lambda^2 X_m \, du_m - d\left[\frac{1}{2} X_{p2} u_{p2} - \frac{1}{2} X_{p1} u_{p1}\right]$$

$$= \lambda^2 \left(X_m \, du_m - \frac{1}{2} d\left[X_{m2} u_{m2} - X_{m1} u_{m1}\right]\right)$$

and for the model.

$$R_m \, dA_m = X_m \, du_m - \frac{1}{2} d\left[X_{m2} u_{m2} - X_{m1} u_{m1}\right]$$

Thus

$$R_m \, dA_m = \frac{R_p \, dA_p}{\lambda^2}$$

since

$$\lambda^2 \, dA_m = dA_p, \; R_m = R_p$$

It is important to appreciate that this result is true *only* if there have been no irreversibilities at regions remote from the crack faces, i.e. that the cracked structures remain globally reversible. In the laboratory, it is often very difficult to prevent specimens from yielding as well as cracking; under such experimental circumstances it is erroneous to infer large-scale, yield-free, cracking behaviour from such tests and proper elastoplastic scaling laws should be employed (section 9.6).

It has been customary in the literature to rewrite equation (9.44) in terms of 'strain energy density' (S) rather than strain energy alone, where $S = \mathrm{d}\Lambda/\mathrm{d}(\text{volume})$, which represents the physical variation of stored energy in a cracked body through the volume of that body. Geometrically similar bodies have volumes that differ by a factor λ^3, thus $S_\mathrm{m}/S_\mathrm{p} = \lambda^3(\Lambda_\mathrm{m}/\Lambda_\mathrm{p})$, and equation (9.44) can be written

$$\frac{K_{\mathrm{Cp}}}{K_{\mathrm{Cm}}} = \sqrt{\lambda \frac{S_\mathrm{p}}{S_\mathrm{m}}} \tag{9.45}$$

Written in this form, if the critical stress intensity at fracture is the same in both bodies, experiments should produce log-log plots of $S_\mathrm{p}/S_\mathrm{m}$ versus λ with a slope of (-1). Determinations of S_m in the laboratory allow K_{Cp} to be estimated. Equation (9.45) is usually referred to as 'the equivalent energy method for toughness determination' of large structures using small-size geometrically similar specimens.

For non-linear elastic bodies obeying, say, $X = k(A)u^n$, the ratio of 'equivalent' $K_\mathrm{p}/K_\mathrm{m}$ is identical with equation (9.45), and does not contain n (Problem 9.1). This interesting result suggests that the equivalent energy method applies to both linear and non-linear *elastic* bodies containing cracks (cf. J_C formulation for elastoplastic cracking, Chapter 4).

The fact that the equivalent energy method applies to non-linear bodies has misled people to apply the method to situations where curvature in the Xu diagram is caused by irreversible plasticity rather than non-linear reversible elasticity (indeed relations identical with (9.44) and (9.45) have been derived assuming a relation between stress and strain of the form $\sigma = E\epsilon^{n^*}$ where n^* is said to be a workhardening exponent; cf. again a J-based theory of elastoplastic fracture). An example is given in Fig. 9.5 which shows the variation of $S_\mathrm{p}/S_\mathrm{m}$ with λ for A533, grade B, class 1 steel at the upper shelf temperature. The results are obtained from standard dynamic tear (DT) specimens with the reference model taken as a 25 mm thick testpiece; (a DT specimen is essentially a three-point bend specimen with an embrittled starter crack loaded dynamically by dropped weights). The data in Fig. 9.5 show that $S_\mathrm{p}/S_\mathrm{m} \geqslant \lambda^{-1}$ and that $K_{\mathrm{Cp}} \geqslant K_{\mathrm{Cm}}$. The equivalent energy method is not being obeyed. All K_{Cp} inferred from these experiments are lower bounds on the true value of crack resistance of the large prototype body; only when the models get sufficiently large would $K_{\mathrm{Cm}} = K_{\mathrm{Cp}}$. How 'large' is 'large' depends on $(K_{\mathrm{IC}}/\sigma_\mathrm{y})$ as explained in Chapter 10.

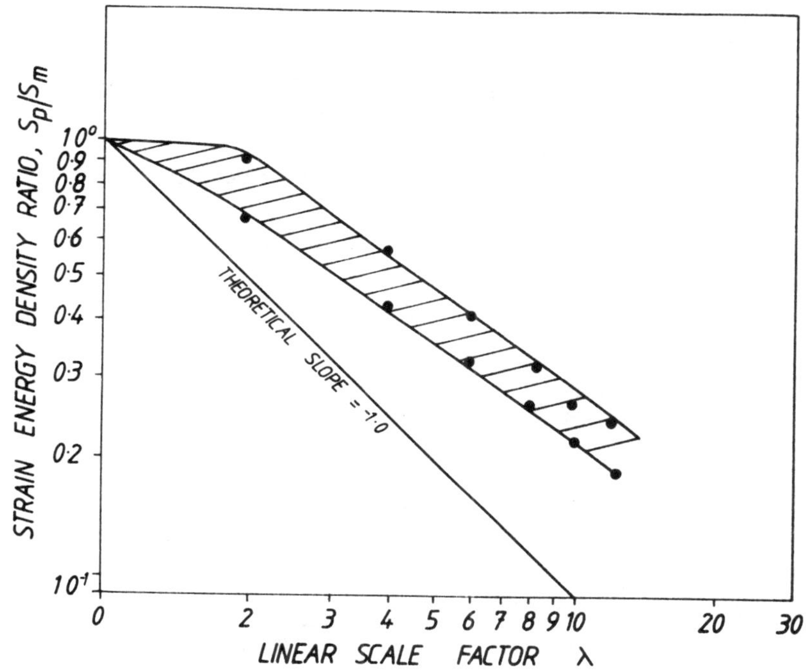

Fig. 9.5 – Experimental data for A533B steel with log (S_p/S_m) plotted against log (λ). (Unit thickness = 25 mm.) (After Mai and Atkins (1975), with permission.)

When bodies do behave in a globally elastic fashion, equation (9.45) *is* obeyed, as shown by Mai *et al.* (1980) for fracture in asbestos cement.

It seems clear that the small testpieces in the series of experiments on steel are yielding before cracking (Fig. 9.6(a)) and that the loads X_y being used to calculate K_{Cm} are yielding loads rather than loads at first cracking. As the size of the specimens increases, the relative importance of prior yielding diminishes and eventually $K_{Cm} = K_{Cp}$; this is what the size-effect in fracture is all about.

Clearly $K_{Cp} = K_{Cm}\sqrt{\lambda S_p/S_m}$ is *only* valid for displacement-reversible cracking (Figs 9.6(b) and (c)), and should not be expected to apply when generalised yielding occurs. As already discussed in Chapter 1 and in detail in Chapters 4 and 5, there is a distinct difference on the basis of energy arguments between Figs. 9.6(a) and (c). In Fig. 9.6(c) the work area ODE represents toughness work and nothing else; in Fig. 9.6(a) the work area represents a combination of toughness work and yielding work – in fact OYPQ might represent only yielding if cracking had not commenced by P. Appeal is often made to so-called 'small-scale yielding' to gloss over this difficulty, which is tantamount to saying that the permanent yielding displacement OQ is not large. Fig. 9.5 demonstrates that in many cases of practical importance this assumption is just not true. The saving grace of the equivalent energy method when applied to yielding situations is that estimates for K_p are conservative (but R_p would be

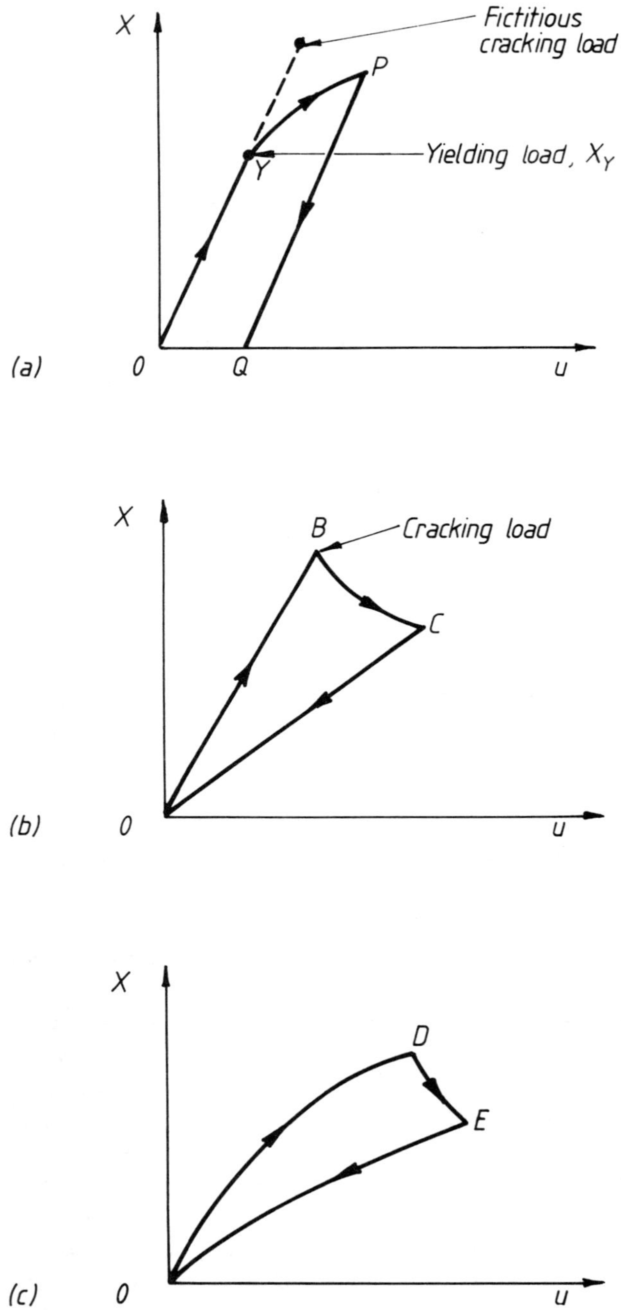

Fig. 9.6 – Load–displacement relationship for cracking with (a) yielding before crack initiation, (b) linear elasticity, and (c) non-linear elasticity.

an over-estimate). For those situations where equivalent energy may be correctly applied a valid answer would just as well be obtained from a single specimen.

9.6 SCALING IN ELASTOPLASTIC FRACTURE

There is a number of early reports of size-effects in the fracture of metal testpieces undergoing plastic flow before rupture. Stanton and Batson (1921) and Docherty (1932, 1935) used geometrically-similar notched specimens of various steels, Monel metal (Cu/Ni), and phsophor-bronze, and observed two principal effects, that

(i) the extent of plastic flow before cracking is smaller in larger testpieces;
(ii) the relative amount of total work done after peak load is smaller in larger testpieces.

Fig. 9.7(a) (taken from Shearin *et al.* (1948)) shows similar results in notched, geometrically-similar three-point bend bars of low carbon steel. Shearin and co-workers noted the difficulty of obtaining the same material properties at different scales in this sort of study but paid particular attention to, e.g., heat treatments of thin and thick specimens. The results of Fig. 9.7(a) show that smaller specimens undergo much more extensive plastic displacement before final fracture; the larger specimens show much less plasticity before final fracture. In Fig. 9.7(a), the load ordinate is normalised by specimen cross-sectional area to give a stress, and the displacement abscissa by a characteristic specimen size to give a measure of strain. We see that prior to final fracture all testpieces essentially follow the same curve, but the load drop associated with the final break occurs at different normalised displacements. Similar results are shown in Fig. 9.7(b), taken from Hagiawara *et al.* (1983) on tests on a variety of notched and unnotched scaled I-beams. In this figure, the load ordinate is normalised by λ^2 (where λ is the scaling factor) to give stress which, in turn, is non-dimensionalised by the different flow stresses σ_Y of the various steels from which the beams were made; again, the displacement is normalised to (engineering) strain by dividing by λ (cf. Example 9.1).

There are other results which support these general trends, including data from dynamic tests. For example, Booth *et al.* (1983) reported drop tests on one-quarter to full-scale fabricated mild steel and stainless steel plate structures. It was observed that weld fracture and tearing were considerably more pronounced in full-scale 'eggbox' and plate girder specimens than in the smaller.

In both Figs. 9.7(a) and (b), larger testpieces fracture earlier than smaller, which accords with general ideas on size-effects described in Chapter 1, namely that bodies which are 'large enough' may fracture in a brittle (globally elastic) fashion, even though small bodies of the same material may undergo extensive plastic flow before fracture. In Figs. 9.7(a) and (b), no bodies fail in a truly elastic fashion (i.e. in the early linear reversible portion of the curves), but we make the important observation that there is a continuing scale-effect within the elastoplastic fracture range, and that larger bodies undergo less elastoplastic flow before final fracture than scaled smaller bodies.

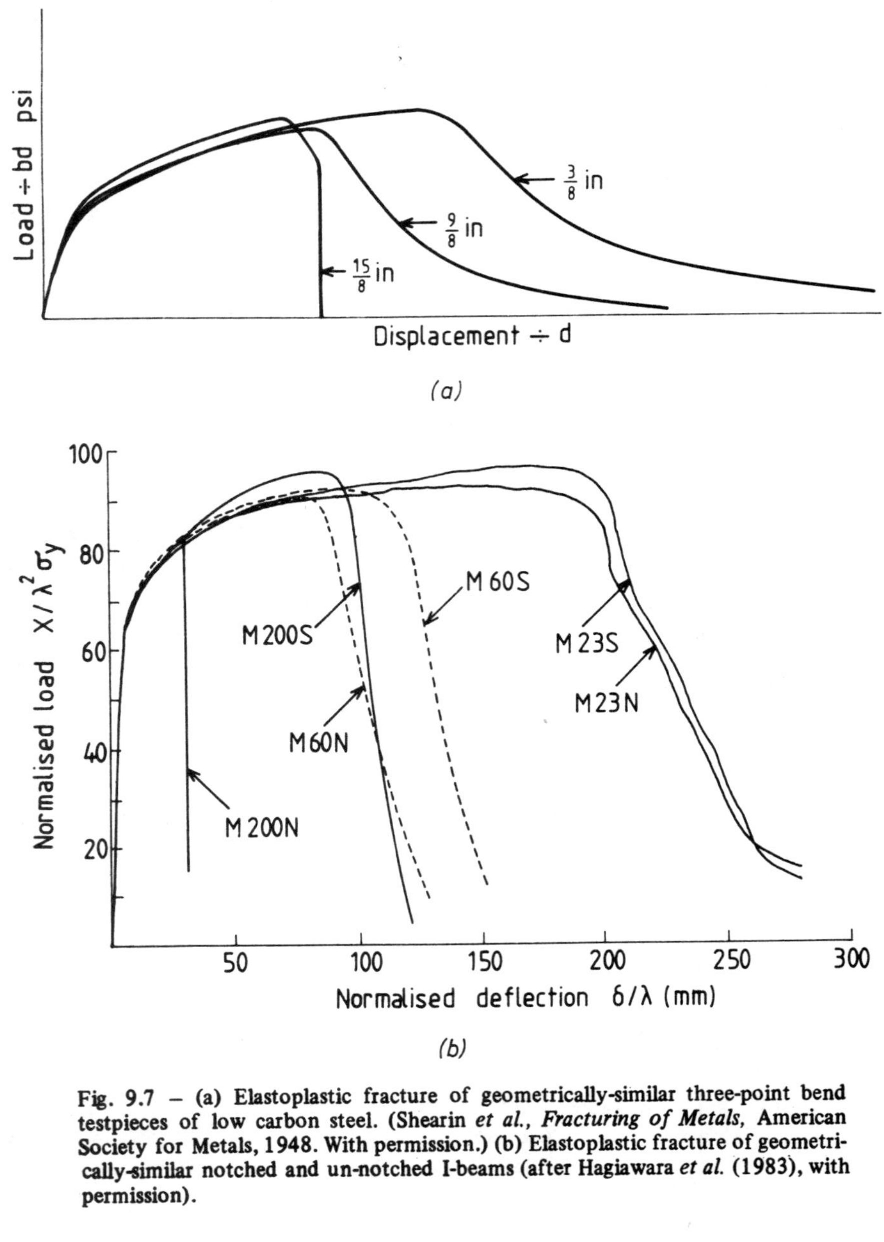

Fig. 9.7 – (a) Elastoplastic fracture of geometrically-similar three-point bend testpieces of low carbon steel. (Shearin *et al.*, *Fracturing of Metals*, American Society for Metals, 1948. With permission.) (b) Elastoplastic fracture of geometrically-similar notched and un-notched I-beams (after Hagiawara *et al.* (1983), with permission).

It is clear from Figs. 9.7(a) and (b) that until the final load drop, data from all sizes of cracked body follow the same master curve. The master curve is the effective stress-effective strain $\bar{\sigma}$–$\bar{\epsilon}$ curve for elastoplastic flow for the material (since $\sigma = X/\text{area}$ and area changes as λ^2; and since $e = u/L_0$ or $\epsilon = \ln(1 + u/L_0)$). That all the results follow the same master curve (within the limits of experimental measurement and material reproducibility) is another way of saying that there is no scale-effect in simple plastic flow.

It is explained in Shearin *et al.* (1948) that cracks begin to grow *before* the final drop in load. Consequently, the different normalised displacements at final fracture in specimens of different sizes are connected with crack *instabilities* in elastoplastic fracture in those different testpieces, rather than crack *initiation.* Indeed, application of a J_C or critical plastic work/volume criterion of crack initiation in a rigid-plastic body would predict approximately the *same* normalised u at fracture, independent of size, since the area under a $\bar{\sigma}$–$\bar{\epsilon}$ curve is the plastic work/volume, cf. Chapter 4 and void growth theories in Chapter 5. (We note that the scaling laws in section 9.3 for non-linear elastic fracture must apply for crack initiation in elastoplastic fracture if a J_C criterion is used.) Thus for crack initiation in Figs. 9.7(a) and (b) there is a size-effect on the amount of stable slow crack growth before elastoplastic crack instability.

Elastoplastic crack instabilities are investigated in section 8.10. One approach to the determination of the amount of stable crack growth is to use the J_R or tearing modulus line of attack, as in Fig. 8.26. The *Xu* plot corresponding to first elastoplastic flow/crack initiation/crack propagation/instability may be cross-plotted from a J–A diagram, or determined directly from, for example, the GE/EPRI estimation procedures.

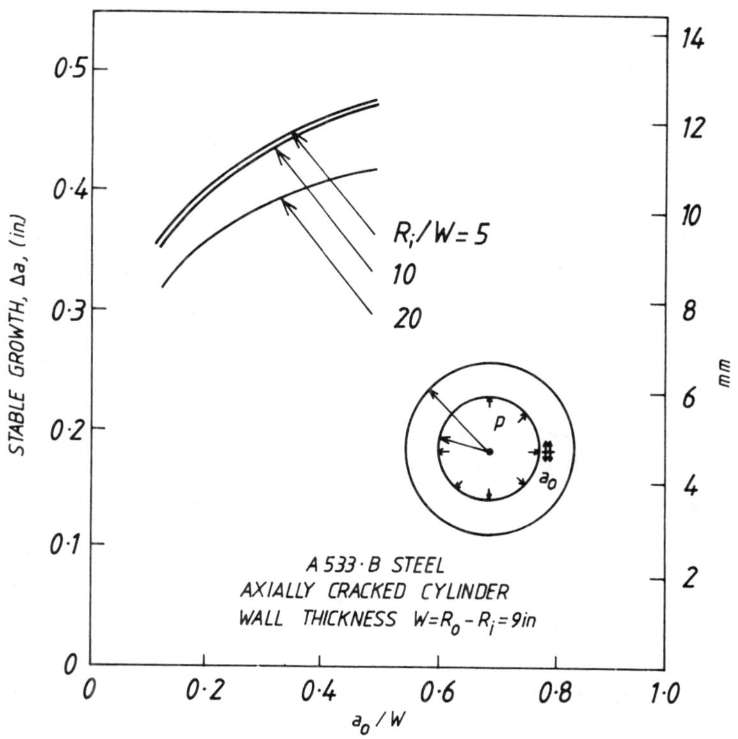

Fig. 9.8 – Magnitude of slow crack growth between initiation and instability for axially cracked cyclinders of A533B steel showing a scale effect. Calculations by Kumar *et al.* (1982).

Results from the J_R-based finite element and estimation procedures of Shih *et al.* for elastoplastic fracture problems provide evidence of size-effects. Solutions are available for a variety of cracked geometries including compact tension testpieces, centre-cracked plates, singe- and double-edge cracked plates, three-point bend bars, internally- and externally-cracked cylinders (see Chapter 4 and various EPRI documents and ASTM STPs such as STP No. 803, 1983). By way of illustration, Fig. 9.8 shows the amount of slow crack growth (absolute) from initiation to instability for axially-cracked cylinders made from A533B steel. The absolute Δa appears to be the same for geometrically similar bodies (cf. the effect or sliding the J_R-curve along the abscissa in Fig. 8.26(a) where it appears that the stable growth is not much changed for different starting a). Hence when normalised by the scale factor, the stable growth will be smaller in larger bodies. Similar remarks apply to the normalised displacement at which instability occurs (again consider sliding the J_R-curve along the abscissa in Fig. 8.26(a) and the relative change in u at the same a/W). It is noteworthy that Kumar *et al.* (1982) remark that, in the case of internally-pressurised, circumferentially-cracked cylinders of A533B steel, cracking occurs earlier at larger diameters.

It should be possible to predict the behaviour similar to that shown in Fig. 9.7(a) from the GE/EPRI J_R three-point bend analysis.

Example 9.5 Show how elastoplastic scaling laws may be couched in terms of crack opening displacements.

Ans. The COD design curve described in section 4.2.3 may be written

$$\delta_C/2\pi\epsilon_y\, a_m = (\epsilon/\epsilon_y) - 0.25 \text{ for } \epsilon/\epsilon_y \geqslant 0.5$$

where $a_m = a/3$ is the equivalent through-thickness crack size corresponding to a 50% probability of fracture (with a the critical through-thickness crack size). For constant δ_C and ϵ_y between model and prototype, it follows that as the characteristic size of the cracked body (given by a, say) increases, the applied strain ϵ decreases. That is (loosely speaking), the displacement at fracture divided by the size of the body is smaller in larger bodies than smaller, which corresponds with illustrations in Fig. 9.7(a) and (b). (We note that Lubahn and Yukawa (1958) found in notched bend tests that the notch-root strain at fracture decreased with increasing size of geometrically-similar specimens.)

Pettersen and Valsgard (1983) employed the COD design curve in numerical studies of ship collisions, and Sumpter (1983) has also used δ in scale model testing of ductile fracture.

We note, in passing, that Shearin *et al.* (1948) mention empirical relations for the effect of specimen size on the total work required to effect fracture. One of the simplest expressions for rupture work in geometrically similar specimens was

$$W_T = C_1 d^3 + C_2 d^2 \tag{9.46}$$

where C_1 and C_2 are constants and d is some characteristic dimension. We see that this corresponds with the Cotterell–Mai-type relation for specific essential work of fracture in plastic flow fields given in section 4.4.1. With this more recent interpretation of the plots, values for R in the range 600–900 kJ/m^2 for various alloy steels in various heat treatments and 60 kJ/m^2 for 24-ST aluminium are inferred. The plots pick up noticeable anisotropy in both fracture and plastic work in certain metal plates.

The work analysis in Chapter 4 concerned varying a/W produced by different crack lengths with W fixed (or, if altered, only over a small range). In contrast, equation (9.46) comes from fixed (a/W) with varying absolute W. Owing to the change from fully-plastic, to elastoplastic, to elastic fracture as the absolute size of the body increases, it is surprising that the plastic work/volume (represented by C_1 in the equation) remains constant with size. It seems that equation (9.46) must only apply over a range of elastoplastic fracture where the remote plastic strains do not alter much.

Scaling of dynamic elastoplastic fracture results were mentioned earlier in this section. Rather like the ice-breaker modelling problem in section 9.2.1, additional complicating factors have to be considered (e.g. the impact velocity in the small and large bodies). Also, in this sort of investigation, it is known that material strain rate effects do not scale and hence a model is stronger than a prototype made from the same solid (e.g. Jones (1983), Calladine (1983) and other authors in the *Structural Crashworthiness* book referenced at the end of the chapter). Such effects can confuse interpretation of tests.

The field of elastoplastic fracture scaling (particularly in dynamic tests) is relevant to the impact behaviour of air, land and sea vehicles. Prototype testing is expensive and difficult and rarely, if ever, undertaken for ships, offshore platforms and other large structures. There is therefore great need for a better understanding of the scaling laws of elastoplastic fracture as given in this section.

9.7 SCALE EFFECTS IN FATIGUE

Size-effects in fatigue are well-documented (see, for example, Heywood, 1962). Larger specimens may display lower fatigue strengths than geometrically smaller specimens (both plain and notched specimens), and also display smaller fatigue lives for the same applied stress level. Until recently, these effects have not had an adequate explanation (statistical arguments regarding distribution of flaw sizes in common ductile engineering metals are not convincing) and it is of interest to explore the role of crack propagation rates to total fatigue life and/or strength. In what follows we consider the consequences of the Paris law applying to small- and large-scaled bodies. Of course, fatigue lives of plain samples relate to *initiation* as well as propagation (and, as described in Chapter 7, the initiation phase can take up the major portion of fatigue life in high cycle fatigue). Consequently, as Paris-type laws relate to Stage II fatigue propagation only, the conclusions can therefore tell only part of the story for *plain* fatigue samples. In the case of *welds*, however, the analyses are likely to be directly applicable since the inherent tiny intrusion defects associated with welds act as initiation regions for Stage II propagation to occur directly (see Chapter 7).

Consider two geometrically similar cracked bodies undergoing Stage II fatigue. If the Paris equation (7.6) is obeyed by both, we have

$$[da/dN]_{small} = C(\Delta K)^n_{small} = C\left[\Delta\sigma Y\left(\frac{a}{W}\right)\sqrt{\pi a}\right]^n_{small}$$
$$[da/dN]_{large} = C(\Delta K)^n_{large} = C\left[\Delta\sigma Y\left(\frac{a}{W}\right)\sqrt{\pi a}\right]^n_{large} \quad (9.47)$$

So, if C is independent of size, and if the two bodies are being fatigued at the same $\Delta\sigma$ levels,

$$[da/dN]_{large} = (\sqrt{\lambda})^n \; [da/dN]_{small} \quad (9.48)$$

at corresponding (a/W), where λ is the scaling factor (given here by $\lambda = a_{large}/a_{small}$). Crack growth rates vary according to a power law which depends on the material (i.e. value of n) and are *faster* in the larger body at the same *relative* position (same a/W) in the path of fracture. What effect that has on the number of cycles to failure as between large and small bodies depends on the values of a_i/W and a_f/W in large and small bodies in the integration of equation (9.47). That is, the crack growth per cycle may be faster in large bodies but the distance that the crack has to run in stage II fatigue depends on the (a_i/W) at the start and (a_f/W) at final fracture.

Various assumptions may be made: if a_i is to do with microstructure or weld defects, say, then it may be reasonable to say that a_i is independent of the size of the body undergoing fatigue. The value of a_f is given by $a_f = [K_C^2 \div Y^2(a_f/W)\,\sigma_{max}^2\pi]$ which is an implicit relation for a_f. In the *special* case of $Y(a/W) = 1$ for all (a/W), i.e. Inglis-type geometries of small cracks in extensive bodies, a_f has the same absolute value given by $a_f = K_C^2/\sigma_{max}^2\pi$, irrespective of specimen size. In all other cases, $(a_f)_{large} > (a_f)_{small}$ for the same K_C and σ_{max}. For example, consider geometrically-similar SEN-type specimens of a solid with $K_{IC} = 65$ MPa$\sqrt{m}$ subjected to $\sigma_{max} = 100$ MPa. Consider two extensive edge-cracked panels of width 1000 mm (the smaller) and 10,000 mm (the larger). We have for both $(a_f)_{small}$ and $(a_f)_{large}$

$$a_f = \frac{1}{\pi}\left[\frac{65}{100Y}\right]^2 = 0.134/Y^2$$

but in the one case a_f/W, on which Y depends, is $a_f/1000$ and in the other is $a_f/10{,}000$ (a_f in mm). Graphical or iteration solutions give $(a_f)_{small} = 90$ mm and $(a_f)_{large} = 150$ mm. Note however that $(a_f/W)_{small} = 0.09$ but $(a_f/W)_{large} = 0.0105$ which is *less*.

Other cases may be evaluated. For larger σ_{max}, the difference between (a_f) for small and large specimens is less. Conversely, for very small σ_{max}, the difference between a_fs will be greater. For example, if, in the case above $\sigma_{max} = 20$ MPa and everything else is as before,

$$(a_f)_{small} = 480 \text{ mm and } (a_f/W)_{small} = 0.48$$

but

$$(a_f)_{large} = 1800 \text{ mm and } (a_f/W)_{large} = 0.18$$

These calculations are irrelevant in a sense, however, as we recall from section 7.2.1 that when $a_f \gg a_i$, a_f drops out of the equation and we have

$$N_f = 1/[a_i^{(n/2)-1}\, C(\Delta\sigma)^n \pi^{n/2}\, (n/2 - 1)\, Y^n] \tag{7.12 bis}$$

If a_i is taken as constant, differences in N_f between small and large samples then depend entirely on the differences in $Y(a_i/W)$ in small and large bodies. For many testpieces $Y(a/W)$ decreases as (a/W) decreases as W increases. Then, according to equation (7.12)

$$(N_f)_{large} = \left[\frac{Y(a_i/W)_{small}}{Y(a_i/W)_{large}}\right]^n (N_f)_{small} \tag{9.49}$$

and, equivalently, the $\Delta\sigma$ required for the same N_f is

$$(\Delta\sigma)_{large} = \left[\frac{Y(a_i/W)_{small}}{Y(a_i/W)_{large}}\right] (\Delta\sigma)_{small} \tag{9.50}$$

These relations suggest that $(N_f)_{large} > (N_f)_{small}$ and $(\Delta\sigma)_{large} > (\Delta\sigma)_{small}$.

If $a_i \ll W$ in both small and large bodies, $(N_f)_{large} = (N_f)_{small}$ and $(\Delta\sigma)_{large} = (\Delta\sigma)_{small}$ and there is no scale effect in Paris-law-controlled fatigue propagation.

An alternative proposition is that a_i at the commencement of Stage II propagation is proportional to size, so that (a_i/W) and Y are constant. This is merely an assumption for algebraic convenience; it may not be realistic in practice. Then, using equation (7.12) again,

$$(N_f) \propto 1/a_i^{(n/2)-1} \propto W^{1-(n/2)}$$

or

$$(N_f)_{large} = \lambda^{1-(n/2)}\, (N_f)_{small} \tag{9.51}$$

where λ is the scaling factor W_{large}/W_{small}.

Equivalently, for the same N_f,

$$(\Delta\sigma)_{large} = \lambda^{(2-n)/2n}\, (\Delta\sigma)_{small} \tag{9.52}$$

For the common range of values for n, i.e. $2 \leqslant n \leqslant 4$, $(-0.25) \leqslant (2-n)/2n \leqslant 0$ and $-1 \leqslant 1-(n/2) \leqslant 0$ and, according to this assumption (of a_i increasing with W), therefore fatigue life and fatigue strength where controlled by Stage II growth should exhibit a size dependence of a different sense from the case when a_i is constant, i.e. $(N_f)_{large} < (N_f)_{small}$ and $\Delta\sigma_{large} < (\Delta\sigma)_{small}$ now.

There appear to be no reliable data concerning crack growth rates (da/dN) at corresponding crack positions for different size specimens which would enable relation (9.48), for example, to be assessed. As already mentioned, however, it has long been known that the applied stress (however defined) required to fracture a specimen depends on its size, so that the absolute level of stress at a given lifetime, or the life at constant stress, are functions of some linear dimension. This statement applies particularly to inhomogeneous stress fields, as in bending of plain testpieces or any loading of notched bars; whether there is a size-effect in axial loading of plain specimens is a matter of debate. The most extensive literature relates to fatigue *limits* of steels, usually of rotating cantilever beam Wöhler testpieces, but it is clear from many studies of S–N curves that the fatigue limit correlates closely with fatigue strengths at lifetimes greater than 10^3 cycles (e.g. Heywood's 1962 text).

In order to compare the predictions of a simple Stage II-dominated model of fatigue life, Puttick and Atkins (1983) correlated the fatigue limits $\Delta\sigma_{fL}$ and the diameter D of plain Wöhler testpieces for a wide range of 27 steels in the literature. It was found that

$$\Delta\sigma_{fL} \propto D^{-0.051 \pm 0.009}$$

that is, fatigue limits are smaller in larger testpieces. We note that these data are at least consistent with one of the sets of predictions for $\Delta\sigma$ in small and large testpieces, since the result corresponds with $n \approx 2.23$ in equation (9.52). The conclusion is perhaps surprising if a_i at the commencement of Stage II-controlled fatigue crack growth is believed to be determined by microstructural features and not proportional to size (i.e. $a_i \propto W$) on which equation (9.52) is founded.

In non-proportionately scaled structures, whether Paris-controlled crack growth rates (da/dN) depend on size depends on possible differences between $(\Delta K)_{large}$ and $(\Delta K)_{small}$, cf. examples for K formulae with different λ in monotonic fracture earlier in this chapter (section 9.2.2). Burdekin's finite element calculations for fatigue of fillet-welded joints reported in 1981 show (i) that for a given attachment thickness and weld size, increasing the main plate thickness gives an increasing tolerance for initial defects but (ii) that when the attachment thickness and weld size are scaled up in proportion to the main plate thickness, increase in the size of the joint leads to a dramatic drop in tolerance for initial defects at larger thickness (cf. virtually all the fatigue life of welded structures is occupied in Stage II as the inherent tiny intrusion defects associated with welds act as initiation regions). (There are no simple algebraic relations for these geometries.)

Note from Example 7.17 and Fig. 3.38 that the Smith–Miller (1978) analysis for notches predicts a size-effect for geometrically similar specimens of the same K_T, i.e. K_f increases as the size of specimen increases. Data in Chapter 6 of Heywood (1962) lend support to a scale effect varying as $1/\sqrt{\alpha}$, where α is the size of the notch.

PROBLEMS

9.1 For a non-linear elastic structure whose load–deflexion relationship is given by $X = k(A)u^n$ show that the scaling law for fracture

$$\frac{K_{\mathrm{Cp}}}{K_{\mathrm{Cm}}} = \sqrt{\lambda \frac{S_{\mathrm{p}}}{S_{\mathrm{m}}}}$$

is independent of n and is thus applicable for toughness determination for both linear and non-linear elastic structures using equivalent energies.

9.2 Consider two non-linear elastic centre-crack panels with varying scale ratios for areal and crack dimensions. Show that if they are driven with the same displacement rate ($\dot{u}$)

$$\frac{\dot{a}_{\mathrm{p}}}{\dot{a}_{\mathrm{m}}} = \left[\lambda_{\mathrm{a}}^{(n-2)} \lambda_{\mathrm{w}}^{2} \lambda_{\mathrm{h}}^{(1-n)}\right]^{1/(n+1)}$$

where n is the degree of non-linearity for the panels.

9.3 Show that for a non-linear elastic structure whose X–u relationship is given by $X = k(A)u^n$ equation (9.34) gives the required scaling laws of fracture.

9.4 Prove that equation (9.35) gives the scaling laws of fracture for non-linear, non-proportionately scaled, centre-cracked elastic panel structures.

9.5 For the conditions given in Problem 9.4, show that the crack velocity ratio for prototype (p) and model (m) centre cracked panels is given by

$$\dot{a}_{\mathrm{p}}/\dot{a}_{\mathrm{m}} = \lambda^{1/(n+1)}$$

if $\dot{u}_{\mathrm{p}} = \dot{u}_{\mathrm{m}}$.

9.6 Show that for non-proportionately scaled DCB specimens, which are also non-linear (obeying $X = k(A)u^n$), the scaling laws are $\sigma_{\mathrm{p}}/\sigma_{\mathrm{m}} = (1/\lambda_{\mathrm{h}})^{n/(n+1)}$; $X_{\mathrm{p}}/X_{\mathrm{m}} = (\lambda_{\mathrm{B}}/\lambda_{\mathrm{a}})\,\lambda_{\mathrm{h}}^{(n+2)/(n+1)}$ $u_{\mathrm{p}}/u_{\mathrm{m}} = \lambda_{\mathrm{a}}^{2}/\lambda_{\mathrm{h}}^{(2+n)/(n+1)}$ where λ are the scaling factors for thickness (B), crack length (a) and height of arm (h). You may assume that the specific work of fracture R is the same in both model (m) and prototype (p) testpiece.

9.7 For the situation described in Problem 9.6, show that

$$\dot{a}_{\mathrm{p}}/\dot{a}_{\mathrm{m}} = \lambda_{\mathrm{h}}^{(n+2)/(n+1)}/\lambda_{\mathrm{a}}$$

when $\dot{u}_{\mathrm{p}} = \dot{u}_{\mathrm{m}}$.

9.8 In Hagiawara *et al.*'s experiments described in section 9.6, a transition from ductile to brittle behaviour occurs in their notched beams somewhere between a beam thickness of 6 mm and 20 mm. Assess the thickness at which this should occur using an appropriate value for $(ER/\sigma_{\mathrm{y}}^{2})$.

9.9 Discuss scaling in terms of Turner's η factors.

9.10 Discuss scaling in terms of equations (2.21) or (2.23) which include elastic unloading displacements.

9.11 Discuss scaling in terms of equation (4.68) which includes both elastic and plastic unloading displacements.

REFERENCES

Atkins, A. G. (1975) *J. Ship Res.* **19**, 40. (presented at Advanced Arctic Marine Technology Conference, Warrenton, Va. 1973).

Atkins, A. G. and Caddell, R. M. (1974) *Int. J. Mech. Sci.* **16**, 541.

Booth, E., Collier, D. and Miles, J. (1983) in *Structural Crashworthiness*, N. Jones and T. Wierzbicki (eds.). London and Boston: Butterworths, p. 136.

Burdekin, F. M. (1981) *Proc. Inst. Mech. Engrs.* **195**, 73.

Calladine, C. R. (1983) in *Structural Crashworthiness*, N. Jones and T. Wierzbicki (eds.). London and Boston: Butterworths, p. 169.

Docherty, J. G. (1932) *Engineering* **133**, 645.

Docherty, J. G. (1935) *Engineering* **139**, 211.

Duffy, A. R., McClure, G. M., Eiber, R. J. and W. A. Maxey (1969) in *Fracture*, Vol. 5, H. Liebowitz (ed.). New York: Academic Press.

Folias, E. S. (1965) *Int. J. Fract.* **1**, 104.

Gurney, C. and Hunt, J. (1967) *Proc. Roy. Soc.* **A299**, 508.

Hagiawara, K., Takanabe, H. and Kawano, H. (1983) *Int. J. Impact Engineering* **1**, 257.

Hahn, G. T. (1969) *Int. J. Fract.* **5**, 187.

Heywood, R. B. (1962) *Design against Fatigue*, London: Chapman & Hall.

Irwin, G. R. (1954) *Symp. on Effects of Temperature on Brittle Behaviour*, Philadelphia: ASTM, p. 176.

Jones, N. (1983) in *Structural Crashworthiness*, N. Jones and T. Wierzbicki (eds.). London and Boston: Butterworths, p. 308.

Kumar, V., Wilkening, W. W., Andrews, W. R., German, M. D., De Lorenzi, H. G. and Mowbray, D. F. (1982) EPRI/GE Report SRD-82-048.

Lubahn, J. D. (1955) *Weld. Res. Suppl.* 518 p.

Lubahn, J. D. and Yukawa, S. (1958) *Proc. ASTM* **58**, 661.

Mai, Y. W. and Atkins, A. G. (1975) *Int. J. Mech. Sci.* **17**, 673.

Mai, Y. W. and Atkins, A. G. (1978) *Int. J. Mech. Sci.* **20**, 437.

Mai, Y. W., Foote, R. M. L. and Cotterell, B. (1980) *Int. J. Cement Composites* **2**, 23.

Pettersen, E. and Valsgard, S. (1983) in *Structural Crashworthiness*, N. Jones and T. Wierzbicki (eds.). London and Boston: Butterworths, p. 338.

Puttick, K. E. and Atkins, A. G. (1983) *Int. J. Fract.* **23**, R51.

Shearin, P. E., Ruarak, A. E. and Trimble, R. M. (1948) in *Fracturing of Metals*. Cleveland: ASM, p. 167.

Smith, R. A. and Miller, K. J. (1978) *Int. J. Mech. Sci.* **20**, 201.

Stanton, T. E. and Batson, R. G. C. (1921) *Proc. Inst. Civil Engrs.* **211**, 67.

Sumpter, J. D. G. (1983) *Int. J. Fract.* **23**, R121.

Szabo, I. (1961) *Einfuhrung in die Technische Mechanik*. Berlin: Springer.

Wilkowski, G. M. (1979) Private communication.

Witt, F. J and Berggren, R. G. (1971) Expt. Mech. **11**, 1.

10

Deformation transitions

'How am I to get in?' asked Alice again, in a louder tone.
'Are you to get in at all?' said the Footman' 'That's the first question you know.'

10.1 FLOW OR FRACTURE?

In Chapter 1 it was remarked that a given type of solid may behave quite differently depending on different applied loading systems, and that a given type of loading system may produce quite different responses in different solids. Simple elasticity, simple plasticity, elastic fracture or plastic fracture may occur; possible responses increase in number if rate, temperature and environmental effects are important. The behaviour of laboratory-size testpieces has served historically to classify materials as brittle, ductile and so on, but we know that traditional classifications are inadequate – large structures made from ductile metals *do* suffer brittle fracture, diamond and other brittle solids *can* be plastically deformed, there is a minimum size below which powders can no further be comminuted and so on.

Is it possible to say beforehand how a given solid, of given geometry subjected to given applied loads, will deform? As far as time-independent deformation is concerned, two pairs of questions have to be answered, viz.

(i) Is the strain state elastic or plastic?
(ii) Is it favourable for cracks to initiate and propagate or not?

The first question is easily answered by the magnitude of the strains induced by the applied loads, $\epsilon_{applied}$. If they are greater than the yield strain ϵ_Y, then clearly the body has become plastic. Calculations are straightforward if the deformation is uniform (e.g. uniaxial tension of an homogeneous isotropic bar, and also multiaxial uniform loading using effective stresses and strains and the plasticity yield function). For non-uniform loading, $\epsilon_{applied}$ will be non-uniform and there may be mixed regions of elastic fields ($\epsilon_{\text{applied}} < \epsilon_y$), elastoplastic fields ($\epsilon_{\text{applied}} \approx \epsilon_y$) or fully plastic fields ($\epsilon_{\text{applied}} \gg \epsilon_y$), cf. Fig. 1.19.

The answer to the second question is more difficult to tackle and should really be subdivided depending upon whether or not there are pre-existing cracks in the body (which of course will produce non-uniform deformation fields and affect the answer to question (i)). It is helpful to consider the ratio of the specific work *required* for cracking W_r to the energy rate W_a in the loaded body which may be *released* for cracking. In elastic fracture in a body already containing a sharp crack, W_r is identified with the material fracture toughness R; under the same circumstances W_a is $-(\partial\Lambda/\partial A)_u = J$ or G the potential energy release rate, and $W_r/W_a = R/G$ in a linear system. In plastic fracture of metals W_r is the specific essential work of initiation J_i which is given by $-(\partial U_T/\partial A)_u$ or related expressions and linked to microstructure as explained in Chapter 5; in this case W_a is the history-dependent $\int(\sigma_H/\bar{\sigma})d\bar{\epsilon}$ or some version of it, or related quantities such as the crack opening displacement.

When $\epsilon_{\text{applied}} < \epsilon_y$ and $W_r > W_a$ we have simple elastic deformation.
When $\epsilon_{\text{applied}} < \epsilon_y$ and $W_r < W_a$ we have elastic brittle fracture.
When $\epsilon_{\text{applied}} > \epsilon_y$ and $W_r > W_a$ we have simple plastic flow.
When $\epsilon_{\text{applied}} > \epsilon_y$ and $W_r < W_a$ we have combined flow and fracture.

These inequalities are shown schematically in Fig. 10.1. Deformation transitions concern movement over the boundaries of the 'cross'. The well-known Charpy transition in steels concerns moving from the lower left quadrant to the upper left quadrant. The size-effect in brittle fracture concerns moving from the lower right quadrant to the lower left quadrant as the body increases in size and from the lower right quadrant to the upper right quadrant when the body becomes small. Vice versa, limiting sizes in comminution trace the reverse route, i.e. lower right to lower left when the particles are large and break down, but lower right to upper right when the limiting size has been reached. Similar considerations apply for the thickness requirement in plane-strain K_{IC} testing. Hydrostatic compression applied to brittle rocks causes the deformation to fall in the upper left quadrant (or even upper right) in place of the usual lower left quadrant. Vice versa, notch brittleness caused by hydrostatic tension at notch roots in otherwise ductile solids occurs in the lower left quadrant instead of the upper right. (Simply brittle solids remain in the lower left quandrant, and simply ductile solids in the upper right quadrant, and are unaffected by the plastic constraint of the notch.) 'Semi-brittle' solids sit on the boundaries of the cross. Transitions from normal wire-drawing of a ductile metal to the formation of a bulge ahead of the drawing die to the 'scalping' of chips from the surface of the rod, concern moving to the left from the right in the upper part of the diagram. High density polyethylene prefers to flow in air rather than crack, but in water or other liquid environments, elastic fracture occurs (upper right to lower left). Ashby deformation-mechanism and fracture-mechanism maps for the same solid may be viewed together to show transitions from flow to fracture and vice versa (see Chapter 1 and Frost and Ashby (1982)). For example, at a fixed temperature, the relative values of vertical stress intercepts will indicate likely events;

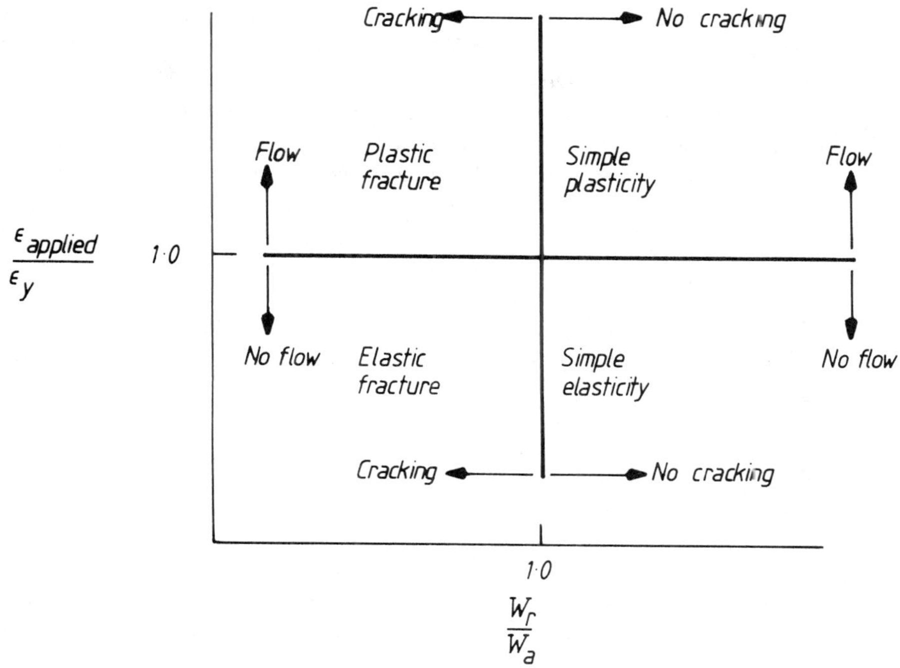

Fig. 10.1 – Deformation transition regimes.

alternatively, at fixed applied stress, horizontal intercepts will indicate transitions with temperature. Since fracture events are often controlled by energies, Ashby maps could usefully have a third axis of strain to give flow and failure surfaces, rather like (but in a different manner) Fig. 7.54.

Xu diagrams are displacement reversible in the lower half of Fig. 10.1 but are displacement irreversible in the upper half. In the lower left quadrant, broken pieces may be refitted to regain the original object; in the upper left refitting fails to reproduce the original object owing to extensive distortion and shape change.

It would be extremely useful to be able to predict quantitatively into which quadrant a given (material, geometry, applied loading) combination will finally fall (after the initial elastic response which all bodies display). With that information it would be possible either (a) to avoid moving into an undesirable region or (b) to ensure that deformation takes place in a favoured region. The material properties which are likely to influence the behaviour are the yield strength σ_y, the Young's modulus E, and the resistance to cracking (given by a variety of parameters, depending on the circumstances, such as the fracture toughness R, the critical stress intensity factor K_C (which includes E of course), δ_C the COD, the specific essential work of initiation J_i etc.). Geometrical effects, and differences in applied loading, are represented by two inter-linked parameters, biz. (i) some characteristic length, which may be the length of a pre-existing crack or the size of the body (in elastic fracture mechanics formulae these are connected by a/W ratios), together with (ii) a factor related to the level of strain in the resulting deformation field.

For illustrative purposes, consider a DCB testpiece loaded by end forces (Fig. 2.3). In the elastic quadrant of Fig. 10.1, the maximum applied stress is located at the root of the cantilever and is given by '$My/I = Xay/I$' from bending theory. As the ratio $\epsilon_{\text{applied}}/\epsilon_{\text{y}} = \sigma_{\text{applied}}/\sigma_{\text{y}}$ so long as the deformation is elastic, we have

$$\epsilon_{\text{applied}}/\epsilon_{\text{y}} = Xay/I\sigma_{\text{y}} \tag{10.1}$$

This defines the horizontal arm of the cross in Fig. 10.1. The *lower* part of the vertical boundary is obtained as follows: $G = -(\partial\Lambda/\partial A)_u = X^2a^2/EIB$ so

$$W_{\text{r}}/W_{\text{a}} = R/G = REIB/X^2a^2 \tag{10.2}$$

The *upper* part of the vertical boundary (for ductile fracture by void coalescence and growth, for example) is defined by

$$W_{\text{r}}/W_{\text{a}} \doteq D/(\int(\sigma_{\text{H}}/\bar{\sigma})\mathrm{d}\bar{\epsilon}) \tag{10.3}$$

where $D = \ln(l/2r)$ the McClintock damage function with l the spacing between voids of radius r, or some other criterion for a different micromechanism of plastic fracture.

Thus, armed with material properties (E, σ_{y} R, D and so on) and the value of X applied, we may decide *ab initio* the response of this particular type of specimen with geometrical details (a, B, h).

Example 10.1 A high yield stress variety of the Ti-6A1-4V alloy has the following properties. $E = 108$ GPa, $\sigma_{\text{y}} = 1.1$ GPa, $K_{\text{IC}} = 38$ MPa$\sqrt{\text{m}}$. If the starter crack in a 4 mm thick DCB specimen of the material is 100 mm long, and each arm is 12 mm high, will it yield or fracture first as the applied load is increased?

Ans. From equation (10.1), the load to cause first yielding when $\epsilon_{\text{applied}} = \epsilon_{\text{y}}$ is given by $X_{\text{y}} = I\sigma_{\text{y}}/ay = Bh^2\sigma_{\text{y}}/6a = (4 \times 10^{-3})(12 \times 10^{-3})^2\,(1.1 \times 10^9)/6 \times 100 \times 10^{-3} = 1.06$ kN. From equation (10.2) for $R = G_{\text{C}}$ the load to cause elastic fracture is $X_{\text{C}} = \sqrt{REIB/a} = KB\sqrt{h^3/12}/a = 38 \times 10^{-6} \times 4 \times 10^{-3}\sqrt{12^3 \times 10^{-9}/12}/100 \times 10^{-3} = 580$ N. Hence elastic fracture will occur first since $X_{\text{C}} < X_{\text{y}}$.

The answer to Example 10.1 could have been obtained by remembering from Chapter 3 that the transition from elastic cracking to first yielding in a DCB testpiece takes place when $h < 3ER/\sigma_{\text{y}}^2 = 3(K_{\text{IC}}/\sigma_{\text{y}})^2$. For the alloy described, $3(K_{\text{IC}}/\sigma_{\text{y}})^2 = 3\,(38/1100)^2 = 3.6$ mm. As $h = 12$ mm, elastic fracture is guaranteed. Note also that the plane strain thickness requirement (which we have tacitly assumed is satisfied by using K_{IC}) is given by $2.5\,(K_{\text{IC}}/\sigma_{\text{y}})^2 = 3.1$ mm. The specimen is 4 mm thick.

It is noteworthy that the two separate ratios $\epsilon_{\text{applied}}/\epsilon_y$ and W_r/W_a seem to be combined into one lumped parameter involving ER/σ_y^2 and h. In fact

$$ER/\sigma_y^2 h = (W_r/W_a)\,(\epsilon_{\text{applied}}/\epsilon_y)^2 \tag{10.4}$$

The proof follows from multiplying both sides of equation (10.1) and (10.2), i.e.

$$\left(\frac{REIB}{X^2a^2}\right)\left(\frac{Xay}{I\sigma_y}\right)^2 = \frac{3RE}{h\sigma_y^2} \tag{10.5}$$

using $I = Bh^3/12$ and $y = h/2$. Values of $ER/\sigma_y^2 h$ thus correspond with *areas* in Fig. 10.1.

While we have determined whether merely plasticity or elastic fracture occurs first in this problem, we have not explored all possibilities.† In particular we have not determined whether the initial plastic flow develops to only a limited extent before fracture still occurs, or whether the plastic flow becomes extensive before eventual fracture, or whether plastic collapse occurs with no fracture. The complete range of possibilities has been discussed for the glued cantilever beam problem in sections 4.3.3 and 4.6 in which plastic flow may occur in the adherand. Equation (4.57) gives the geometry/material relationship which must apply when cracking along the glue line takes place with a partially plastic beam in which the remaining elastic core of the beam is defined by ξh (Fig. 4.10). We have

$$h = \left[\frac{3RE}{\sigma_y^2}\right]\left[\frac{1}{3-2\xi}\right] \tag{4.57 bis}$$

Elastic fracture occurs for $\xi = 1$, when equation (4.57) reduces to equation (10.5); plastic collapse with no fracture occurs when $\xi = 0$, i.e. for $h \leqslant RE/\sigma_y^2$. Intermediate h correspond with partial plastic hinging in the beam.

In addition, then, to the transition between elastic fracture and elastoplastic flow, there is a *second* transition between elastoplastic fracture and plastic collapse. In terms of strains, the second transition corresponds (in the case of a perfectly plastic solid) with $\epsilon_{\text{applied}}/\epsilon_y = \infty$; for example, in the case of the glued cantilever the strain on the outside layers is $\sigma_y/E\xi$ (section 4.3.3), so $\epsilon/\epsilon_y = 1/\xi \to \infty$ for $\xi = 0$). As an alternative to strains, ξ could be employed to indicate the degree of elastoplasticity.

Puttick in a series of papers in the late 1970s termed the two transitions the *lower transition* (elastic to elastoplastic) and the *upper transition* (elastoplastic to fully-plastic). We may write generally

$$h \gtrless \alpha \cdot ER/\sigma_y^2 \tag{10.6}$$

† Even in the transition from simple elasticity to elastic fracture, there may be the possibility of different elastic fracture modes, e.g. in the embedded rod problem (Chapter 2, Problem 2.35) either (a) mode II interfacial crack propagation caused by adhesive failure or alternatively (b) simple tensile snapping of the rod.

with different values of α for the lower and upper transitions. Clearly there are critical *lengths* or sizes (h, W etc.) at which transitions occur in given geometries, which depend on the product of a *material* factor (ER/σ_y^2) and *test geometry* factors α.

Calculations of α for the lower transition are, in principle, readily performed as the critical lengths (being those of the elastic strain energy fields) are given in linear elastic fracture mechanics by the characteristic lengths of the stress fields (i.e. via the $Y(a/W)$ factor). For example, in the case of centre-cracked plates it may be shown that $\alpha = 4$ at $a = W/4$, by equating the net-section stresses at elastic cracking (i.e. $K_C/\sqrt{W \tan(\pi a/W)}$ to the yield strength σ_y over the same ligaments.

Calculations for α at the upper transition are less reliable as elastoplastic solutions for cracking problems are formidable and usually not available. This is so for elastic–perfectly plastic solids and further complications are introduced by workhardening.

In the absence of a proper solution, one approach for α at the upper transition is to consider the size of the crack tip zone and enquire when it may reach an outside boundary of the cracked body. Consider first the size of the crack tip zones at the *lower* transition in a centre-cracked plate. For $a/W = 0.25$ and with $K_C = \sigma_y\sqrt{W \tan(\pi a/W)}$, the size of the crack tip zones is $(1/\pi)(K_C/\sigma_y)^2 \approx 0.08W$ in plane stress and about $0.03W$ in plane strain (section 2.8). The crack tip zone size, as a proportion of the crack length ($2a = 0.5W$) is $0.08W/0.5W = 0.16$ (plane stress) and some $0.03W/0.05W = 0.06$ (plane strain). While the latter size is greater than that recommended for plane strain testing† (ratio ≈ 0.02, Example 2.15), the zones are still contained within the extent of the net-section ligament ($0.25W$ on each side of the crack). However, were the extent of the crack tip zone to reach an outside boundary of the cracked body, this would correspond with the upper transition. Hence for CCP with $a/W = 0.25W$ ligament between the starter crack tip and the free boundary of the specimen, d_y (presuming that it is still given by equation (2.36) when the zone is large) equals $0.25W$ when

$$(1/\pi)(ER/\sigma_y^2) = 0.25W$$

or

$$W < (4/\pi)(ER/\sigma_y^2) \tag{10.7}$$

for the upper transition, i.e. $\alpha = (4/\pi)$.

This type of calculation is crude, as it assumes that the crack tip zone is circular when in fact it may be lobe-shaped or take the form of a Dugdale-type line zone (sections 2.8 and 4.2.3), and may also depend on strain rate, the form

---

† Experiments show that measured values of K_C in centre-cracked panels are affected by impending general yield in the ligament (Kenny and Campbell, 1968). It is recommended that the net-section stress be kept below $0.8\sigma_y$ for displacement reversible cracking; this reduces the size of the crack tip zones towards recommended values. A limiting size of zone, as a proportion of crack length, is an alternative way of setting α values for the lower transition.

of the yield criterion and so on. Furthermore, and for related reasons, alternative lines of attack (e.g. in terms of loads or stresses) for the upper α do not always give the same answer, as illustrated in the following Example.

Example 10.2 Derive values for α at the upper transition in the DCB specimen (i) from considerations of the extent of the crack tip zone (ii) from the loads required to cause plastic hinging.

Ans. The crack tip zone will encompass the depth of an arm of a DCB testpiece when

$$h < (1/\pi)\,(ER/\sigma_y^2) = (1/\pi)\,(K_{IC}/\sigma_y)^2 \approx 0.32\,(K_{IC}/\sigma_y)^2$$

assuming that eqution (2.36) still gives d_y when the zone gets large. On the other hand the bending moment to give a fully-plastic hinge is $Bh^2\sigma_y/4$. So if the bending moment for fracture is still given by $\sqrt{REIB}$ (an assumption) the upper transition occurs when $h < 4ER/3\sigma_y^2$ and then $\alpha \approx 1.3$. (α for the lower transition in the DCB geometry is 3 (equation (10.5) – or equation (4.61) with $\xi = 1$).

Table 10.1 gives lower and upper α-values (calculated in various ways) for a number of different geometries. We see from the table that situations with compressive major principal stresses (e.g. Hertzian indentation) require large critical characteristic sizes at the transitions, i.e. large α. The lowest values of α are associated with tension stress fields in notched bars which experience high σ_H and hence promote fracture.

In the general workhardening case, it is difficult to assign a level for $\epsilon_{applied}/\epsilon_y$ which would give a second horizontal line for the upper transition in Fig. 10.1. Sometimes the plastic zone comprises a field of not-too-large plastic strains diffused over a wide region, but at other times high strains may be concentrated in a small volume (cf. shear cutting processes in Chapter 5); in both cases, however, the total incremental plastic work done in the deformation could be comparable with each other and with the work of fracture. Even so, we may qualitatively refine Fig. 10.1 to accommodate both the lower and upper fracture transitions, as shown in Fig. 10.2 where the plastic flow $\epsilon_{applied}/\epsilon_y$ ordinate is labelled 'non', 'limited' and 'extensive'. It may also be possible to think in terms of 'limited' and 'extensive' cracking on the abscissa too.

Example 10.3 A notch-brittle material is one that is ductile in an ordinary laboratory test, but brittle in notched tests with large plastic constraint factors, i.e. in tests with deep notches. Discuss the material properties that determine notch brittleness in terms of upper and lower transitions.

Table 10.1

Process	Lower transition elastic–elastoplastic	Upper transition elastoplastic–fully plastic
Ball indentation	$a \approx 2{,}300\ ER/\sigma_y^2$ a = radius of Hertz ring crack (friction and slip at interface reduces α to around 200 for steel ball) *also* $a \approx 100\ ER/\sigma_y^2$ for yielding directly beneath ball during later growth of cone crack. (Problem 10.8)	$a = 15\ ER/\sigma_y^2$ a = impression radius *or* equivalently $c = 25\ ER/\sigma_y^2$ c = plastic zone radius around indentation
Embedded rod	$d = 8ER/\sigma_y^2$	$d = 8ER/(\mathrm{UTS})^2$ (necking)
Unloading indentation cracks (residual stress fields) (scratching)	$a \approx 6\ ER/\sigma_y^2$	
DCB compression (comminution)	$d \approx (32/3)\ ER/\sigma_y^2$	
Centre-cracked plate	$W \approx 4ER/\sigma_y^2$ $(a = W/4)$	
DCB	$h \approx 3ER/\sigma_y^2$	$h \approx (4/3)\ ER/\sigma_y^2$ or $h \approx (1/\pi)\ ER/\sigma_y^2$
Glued elastoplastic beam	$h = 3RE/\sigma_y^2$	$h = RE/\sigma_y^2$
Three-point bending Charpy	$W \approx ER/\sigma_y^2$	$W \approx 0.15\ ER/\sigma_y^2$
Deeply edge-notched plate	$W \approx 0.8\ ER/\sigma_y^2$	$W \approx 0.1\ ER/\sigma_y^2$

Table 10.1 – *continued*

Machining (fragmentation-to-curly chip formation)	$h \approx 0.6\, ER/\sigma_y^2$ h = depth of cut	
Guillotining (ductile metals)		$\delta_{cr}^{n+1} = Rw^n/\tau_0 \left[\frac{1}{n+1} - f\right]$ δ_{cr} critical depth of indentation; w width of shear zone n workhardening index $\tau = \tau_0\, \gamma^n$ f friction factor

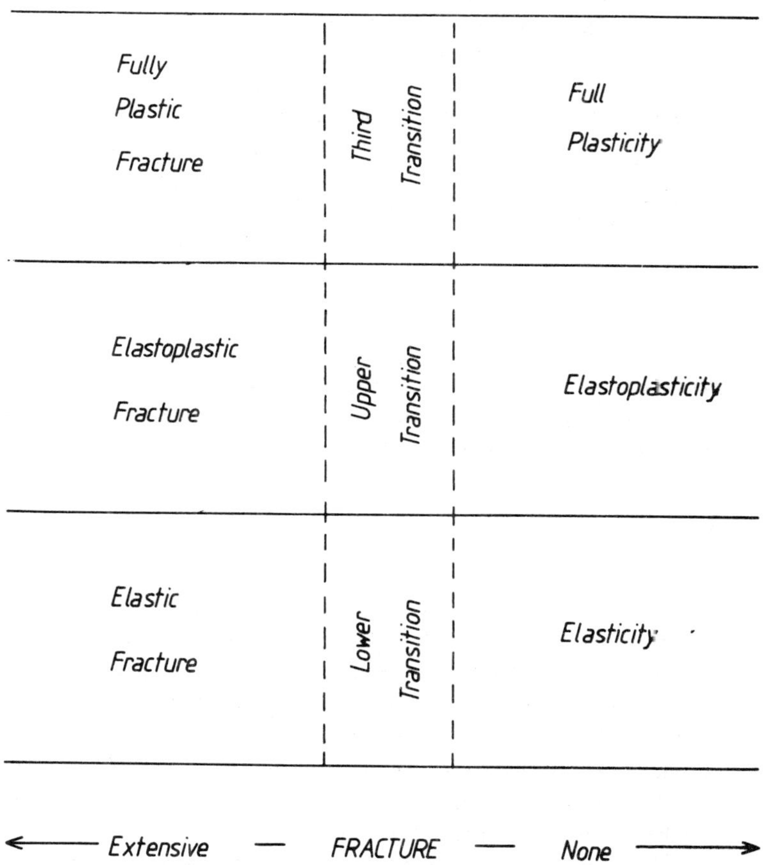

Fig. 10.2 – Upper, lower and third transitions for fracture and plastic flow.

Ans. A material is simply brittle if $\sigma_{crack} < \sigma_y$; notch-brittle if $\sigma_y < \sigma_{crack} < 3\sigma_y$; simply ductile if $\sigma_{crack} > 3\sigma_y$ (Chapter 1). Since the cracking stress will be given by a formula such as $\sigma^2_{crack} = (ER)/(Y^2 \pi a)$ where Y is a constant for the specimen geometry, the performance of a material in this test depends on the relative values of σ_y, $(1/Y)\sqrt{ER/\pi a}$, and $3\sigma_y$. That is, a simply brittle solid has $ER/\sigma_y^2 < Y^2 \pi a$, a simply ductile material has $ER/\sigma_y^2 > 9Y^2 \pi a$, and a notch-brittle solid has ER/σ_y^2 in between. $(1/Y^2 \pi)$ and $(1/9Y^2 \pi)$ correspond with α for the lower and upper transitions respectively. Knowing the appropriate $Y^2 a$ for a chosen type of testpiece permits the value of ER/σ_y^2 for the solid to be bracketed and different types of notched specimen can, in principle, refine the answer. This was a useful concept (like UTS/σ_y ratio) years ago to quantify 'toughness' but these days K_C or R is measured independently from standard toughness tests and rarely from notch brittleness tests.

A number of different physical interpretations may be put on the ER/σ_y^2 or $ER/\sigma_y^2 h$ parameters. We have already shown for the DCB specimen in section 3.9 that $ER/\sigma_y^2 h = (K_{IC}/\sigma_y)^2/h$ represents the ratio of cracking force to the force required to produce plastic flow. The relevant lower and upper α values determine whether the flow is limited or extensive, i.e. α for the lower transition is larger than α for the upper, so as $ER/\sigma_y^2 h = 1/\alpha$, this is the same as saying that cracking will occur if the force to crack is lower than the force to yield. This approach has been useful in consideration of transitions in machining (section 10.5). The following example gives another interpretation.

Example 10.4 Show that $ER/\sigma_y^2 = \delta_C/\epsilon_y$.

Ans. Writing $R \approx \sigma_y \delta_C$ it follows that $K_C^2/E\sigma_y = \delta_C$ or $(K_C^2/\sigma_y^2)(\sigma_y/E) = \delta_C$ i.e $(K_C/\sigma_y)^2 \epsilon_y = \delta_C$ (cf. COD design curve in Chapter 4 which, for $\epsilon/\epsilon_y \leqslant 0.5$ is $\delta_C/2\pi\epsilon_y a = (\epsilon/\epsilon_y)^2$ which transposes to $\delta_C = 2(K_C/\sigma_y)^2 \epsilon_y$; there is a deliberately built-in factor of two.

As $ER/\sigma_y^2 = \delta_C/\epsilon_y$, we have at the transitions $h \gtreqless \alpha\, \delta_C/\epsilon_y$, and this may sometimes be a more convenient way of establishing α, particularly at the upper transition. Hence even materials with large δ_C can still behave in a globally elastic reversible fashion if h is big enough, but the large δ_C inevitably has to be accommodated in an extensive plastic field if h is small (Chapter 1). Clearly the situation is changed in different cracked geometries (different α): when α is small (at notch roots in tensile specimens, Table 10.1), elastic fracture occurs in smaller specimens, other things being equal.

There is a third α-value which determines when, in a given geometry, plastic fracture supervenes plastic flow. As illustrated schematically below, the three

α-values determine the behaviour when successive choices have to be made between

(a) elastic to either elastic fracture or elastoplastic flow,
(b) elastoplastic flow to either elastoplastic fracture or plastic flow,
(c) plastic flow to either plastic fracture or unlimited plastic flow.

elastic deformation
- elastic fracture
- elastoplastic deformation
  - elastoplastic fracture
  - plastic flow
    - plastic fracture
    - unlimited plastic flow

α = large	α = intermediate	α = small
lower transition	upper transition	third transition

The trend in Fig. 10.2 is that the transitions are set by critical values of $ER/\sigma_y^2 h = 1/\alpha$, and that $1/\alpha$ is smallest for the lower transition, larger for the upper transition and (if it can be used) presumably largest for the third transition. Of course, it may be argued that the third transition is meaningless as all materials will *eventually* fracture as the level of deformation is increased unless they have peculiar (time-dependent) flow properties. Information that the third transition does supply, though, is the level of extensive large-strain plastic deformation which precedes fracture.

Fig. 10.3 presents the transitions in terms of the diagram used to explain the size-effect in fracture in Chapters 1 and 3. The single line depicting yield in Fig. 1.22 is now replaced in Fig. 10.3 by two stress levels, one EFG for initial yield and a second BCD for extensive yielding (which corresponds with unlimited plastic flow or collapse in a rigid-perfectly plastic solid). As in Fig. 10.2, it is not obvious how to set the level of BCD in a workhardening solid. One point of view is to say that BCD corresponds with the least stress level at which the plastic zones reach an external boundary of the cracked body. The fracture stress curve M . . . N will be given by the appropriate algebra for the fracture stress at *all* sizes, i.e. which expresses elastic fracture, elastoplastic fracture and plastic fracture. For example, in the case of the glued beam discussed in section 4.3.3, elastic fracture is governed by $M^2 = 2REIw$ (see Example 4.5) and elastoplastic fracture is governed by equation (4.20b). Let us define the 'fracture stress' for this problem as (X/hw) where w is the width of the beam and $2h$ its

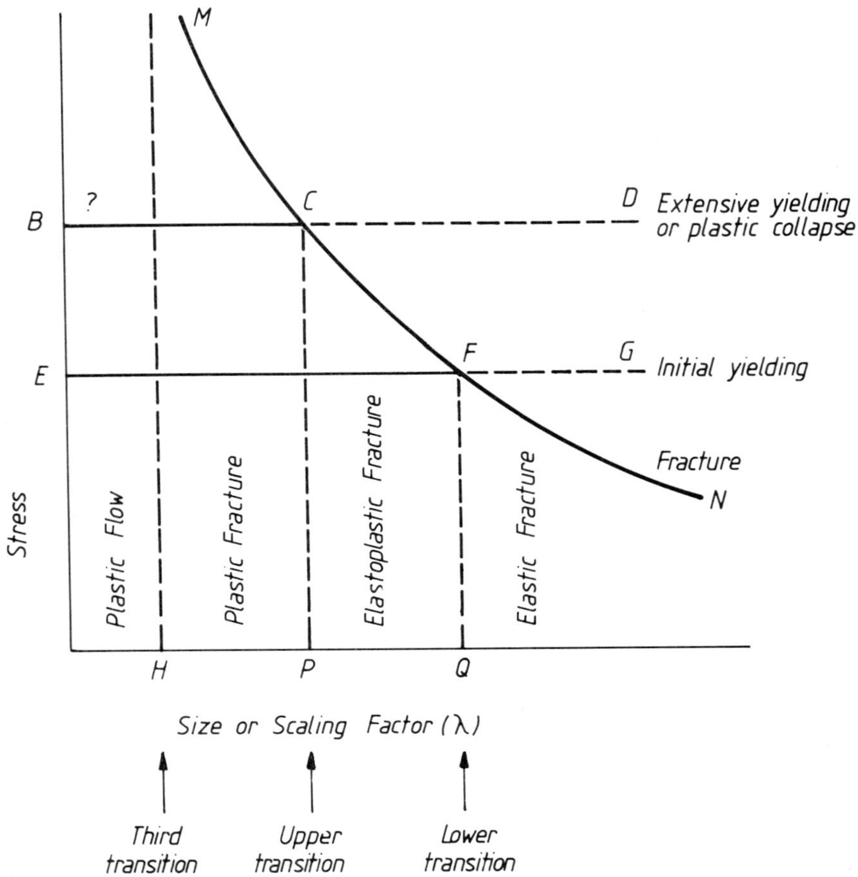

Fig. 10.3 – Size-effect in fracture incorporating lower, upper and third transitions.

depth, with X the equivalent end-load which, with a cantilever length a, gives the cracking moment $M = Xa$. Then, for elastic fracture

$$X/wh = \sqrt{4RE(h/a)/3}\,(1/\sqrt{a}) \tag{10.8}$$

and for elastoplastic fracture

$$X/wh = (h/a)\,\sigma_y\,[1-(3/4)\ \{1-RE/\sigma_y^2(h/a)\,(a)\}^2] \tag{10.9}$$

In both cases, for geometrically similar specimens where h/a is constant (assumed for simplicity) the fracture stress diminishes as the 'size' λ (represented by a here) increases and the fracture stress curve falls from left to right in Fig. 10.3.

It is clear the intersection points F and C, between the fracture stress and flow stress curves, correspond with Puttick's lower and upper transitions. The fracture stress curve is continuous through point F, with the portion NF being given by equation (10.8) and portion FC by equation (10.9). At point F, $(X/wh) = 2RE/\sigma_y a$ by both relations. For rigid–perfectly plastic solids, the portion CM does not exist in theory as an unlimited rotation occurs to the left of point C. For practical work-hardening solids CM does exist, if BC is considered (in this beam problem) simply as the stress level at which the whole beam depth is encompassed by plastic flow, i.e. $\xi = 0$ in the nomenclature of section 4.3.3, and where increased rotation requires increasing loading. The albebra for portion FC of the fracture stress curve may be viewed as the algebra for the *elastic* curve, but with a progressively changing 'enhanced' or 'apparent' toughness R^* given for the beam problem by $R^* = R\,(3 - \xi^2)^2/4(3 - 2\xi)$ (cf. equation (4.54)). It is likely that many elastoplastic cracking problems may, in general, be viewed in this way where the connexion between R^* and R depends upon process zone size, accompanying irreversibilities inherent to propagation in a given geometry and so on (see Chapter 4).

Hence, in broadest terms, the fracture stress will usually have the shape given in Fig. 10.3 and the following conclusions may then be drawn:

(i) As the stress is increased in a body where $\lambda > \lambda_Q$, the portion FN of the fracture curve is intersected first and hence elastic fracture ensues.

(ii) For a body in which $\lambda_P < \lambda < \lambda_Q$, the initial yielding boundary is intersected first, followed by the fracture curve between C and F. For bodies with $\lambda_H < \lambda < \lambda_P$, initial yielding is followed by extensive yielding and eventual fracture on the MC boundary.

(iii) If λ_H exists, and $h < \lambda_H$ it seems that unlimited flow can take place with no fracture at all. The position of H would be such that the fracture curve NFCM would be asymptotic to a vertical line through H.

The fracture stress required for propagation, and the stresses associated with yiedling, may change as propagation ensues which means the boundaries may change and the events may change. For example, plastic flow may occur after some elastic fracture propagation in ligaments in the path of the crack, which obviously decrease in size as cracking progresses. Questions of non-proportional scaling, crack stability, rate, temperature and environment all come into play in the full interpretation of events in Fig. 10.3.

10.2 EXPERIMENTAL EVIDENCE

There is plenty of experimental evidence to demonstrate deformation transitions; we concluded section 10.1 by discussing cracking or plastic collapse in structures. Alteration of any one variable in the lumped parameter $\alpha ER/\sigma_y^2 h$ may put the deformation into different regions of Figs. 10.1, 10.2 or 10.3. Remember in this context that E and particularly R and σ_y are likely to be affected by rate, temperature and environment.

Transitions caused simply by changes in *size* (h) have been discussed in Chapters 1 and 3 in connexion with brittle fracture of large ships, pipelines and so on and in Chapter 9 on scaling. In Chapter 1 also a transition in the behaviour of zinc owing to microstructural grain size was mentioned. When the size of the deformation zone is small enough, even the most brittle solids will flow. Kendall (1978) demonstrated the impossibility of comminuting

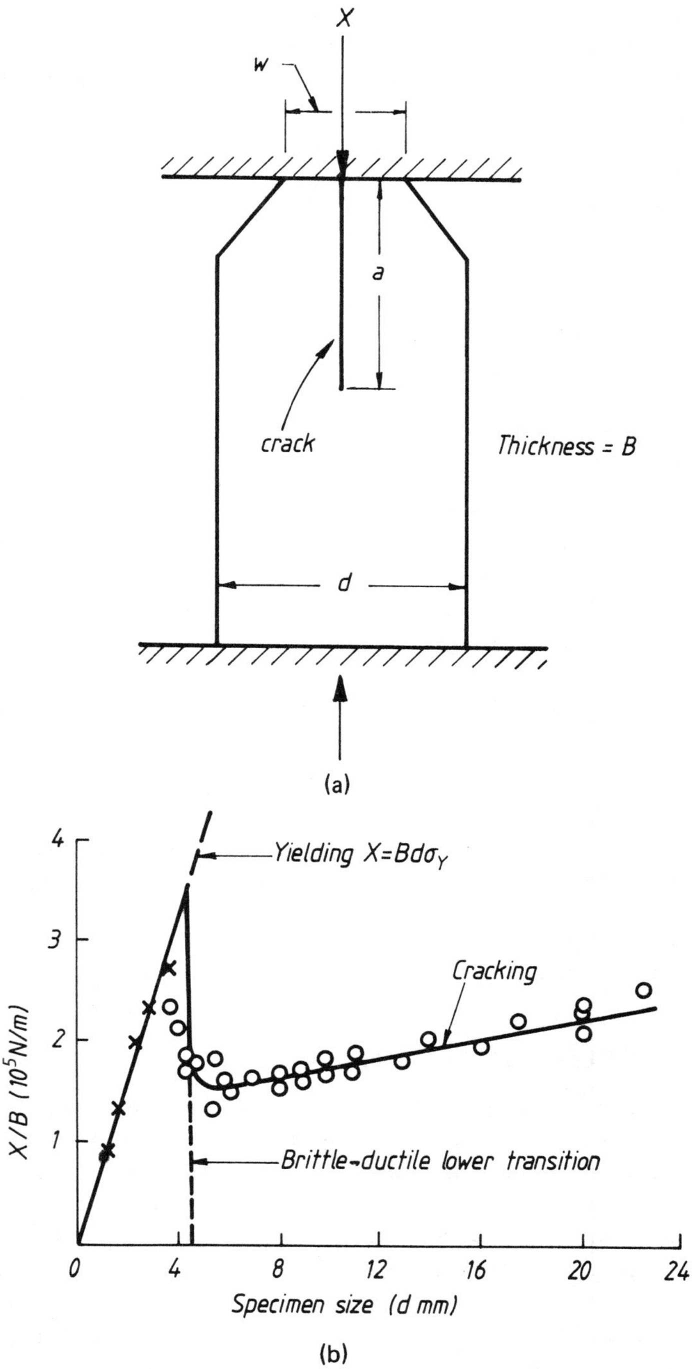

Fig. 10.4 – (a) Axial splitting a DCB specimen by compression. (b) Compression results for a range of specimen sizes showing yielding of small samples and cracking of large ones. (Kendall, 1978. Reprinted by permission from *Nature*, Macmillan Journals Ltd.).

small particles by compression, using the 'buckling' model of the DCB specimen described in Chapter 3 in which the arms are loaded axially in compression along their lengths. The axial cracking force X is related to the geometry of the beam (Fig. 10.4(a)) and mechanical properties of the solid by

$$X_{\mathrm{crack}} = [B/(1 - w/d)] \sqrt{2ERd/3} \tag{3.13 bis}$$

The axial load to cause yielding is given by

$$Y_{\mathrm{yield}} = \sigma_y Bw \tag{10.10}$$

where w is the current width of an originally-sharp edge.

The lower transition is given when $X_{\mathrm{crack}} = X_{\mathrm{yield}}$, i.e.

$$(1 - w/d)\, w = \sqrt{2ERd/3\sigma_y^2} \tag{10.11}$$

For different (w, d) size combinations, equation (10.11) gives the conditions for a lower transition.

The value of w depends on the applied X and the plastic flow properties given by σ_y. If w is eliminated from equation (3.13) using equation (10.10), X_{crack} is seen to be governed by the following quadratic:

$$(X/B)^2\ (1/\sigma_y d) - (X/B) + \sqrt{2ERd/3} = 0 \tag{10.12}$$

The solution appears in two parts in Fig. 10.4. For large d, the first term in equation (10.12) may be neglected and X_{crack} depends on $d^{1/2}$ as in equation (3.13), but as the particle size is reduced towards a critical value given by

$$d = 32ER/3\sigma_y^2 \tag{10.13}$$

(when in the quadratic solution '$\sqrt{b^2 - 4ac} = 0$' i.e. $\sqrt{1 - (4/\sigma_y d)\sqrt{2ERd/3}} = 0$), the cracking force rises very rapidly. Eventually it exceeds the yield stress (however high this may be) and below the critical size, cracking is impossible and flow occurs instead. We see that $\alpha = 32/3$ for equation (10.13). Fig. 10.4(b) includes Kendall's model results on polystyrene for which $E = 2.8$ GPa, $\sigma_y = 80$ MPa, $R \approx 970$ J/m^2. The critical d at transition is about 4.5 mm from equation (10.13); the experimental transition is some 3.6 mm. Given the simplifications in the analysis the agreement is very good.

A delightful example of deformation transitions concerns the quality control test for spot welds. A cruciform testpiece formed by spot welding two strips is pulled apart normal to the sheet thickness and depending on the relative size of weld diameter and sheet thickness, failure occurs either by cracking across the weld junction at lower loads or by pulling out a plug of (ductile) metal in shear at higher loads. Interfacial fracture occurs at small weld diameter-to-sheet thickness (d/B) ratios, and vice versa for plug formation (Fig. 10.5(a)).

Smith analysed the transition in 1980. For small (d/B) and low loads, deformations are small, the initial geometry is maintained, and elastic fracture may be modelled as a sharp circular neck surrounding the embrittled weld nugget and heat-affected zone. The stress intensity factor formula employed was

$$K = 0.8\ XB/d^{5/2} \tag{10.14}$$

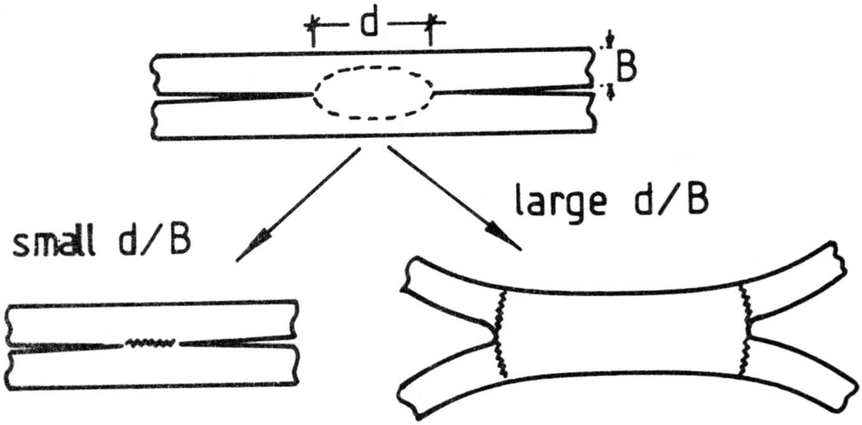

Fig. 10.5(a) – Fracture transitions in spot weld cruciform tests. At small d/B elastic fracture occurs between the sheets of metal; at large d/B plastic yielding leads to fracture by plug formation through the thickness.

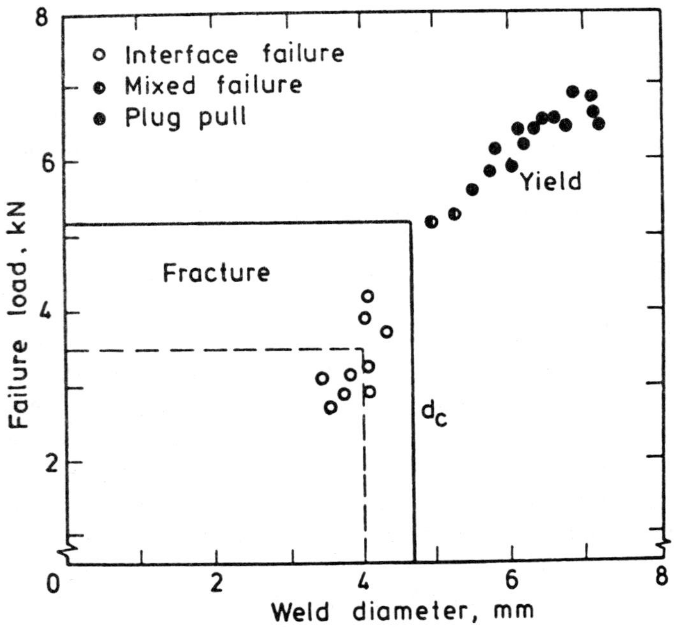

Fig. 10.5(b) – Comparison between Rivett's experimental data for cold-rolled mild steel and Smith's predictions. (Both after Smith, 1980, reproduced with permission.)

which is an appropriate simplification of a more complicated expression. Hence

$$X_{\text{crack}} = 1.25 K_C d^{5/2}/B \tag{10.15}$$

For 'failure' by shear around the *two* circumferential areas (above and below the weld) we have

$$X_{\text{yield}} = 2(\pi d)B\tau_y \tag{10.16}$$

where τ_y was taken as the flow stress. (Of course yielding is only a precursor of fracture, the actual fracture load in shear being different – see comments on guillotining etc. in Chapter 4 and 5.)

The critical weld size at the transition is given when $X_{\text{crack}} = X_{\text{yield}}$, i.e. from

$$\frac{B^4}{d^3} = 0.16\left[\frac{K_C}{\sigma_y}\right]^2$$

or (10.17)

$$B = 0.16\left[\frac{d}{B}\right]^3\left[\frac{ER}{\sigma_y^2}\right]$$

(taking $2\tau_y = \sigma_y$). Notice that the length dimension for Puttick's transition expression is a combination of B and d together, not one simple size parameter.

Fig. 10.5(b) gives Smith's comparison between the analysis and data of Rivett (1979) for spot welds of 1.18 mm thick cold-rolled mild steel. The transition weld diameter was predicted to be 4.65 mm which, together with the predicted transition load of 5.1 kN, clearly separates the elastic and elastoplastic fracture modes. Alexander and Brewer (1963) describe a Belgian cruciform testpiece which is loaded biaxially: when tested at temperatures slightly below the brittle/ductile transition temperature, cracks propagate across the diagonal of the centre part of the cross; at temperatures above transition ductile failure tends to sever off one arm of the cross.

As is well-known from the Charpy test, change in *temperature* is a powerful means of producing a transtion, as it alters R and σ_y in steels (and would alter E as well in many polymers around room temperature). Indeed it is true to say that this particular type of transition, which is closely related to changes in the micromode of fracture of structural steels, was the first to be studied in detail. Notch bend tests on silicon iron by Griffiths and Oates 1969 provide evidence of the correlation between fracture transitions and the ER/σ_y^2 parameter. The fracture load, general yield load and the length of the plastic zone at the notch root (as a function of applied load/yield load) were measured at a variety of temperatures. The results in Fig. 10.6 show that the data follow a straight line relation with α (through the origin) of about 0.2. The line of best fit has a slope of ~0.15.

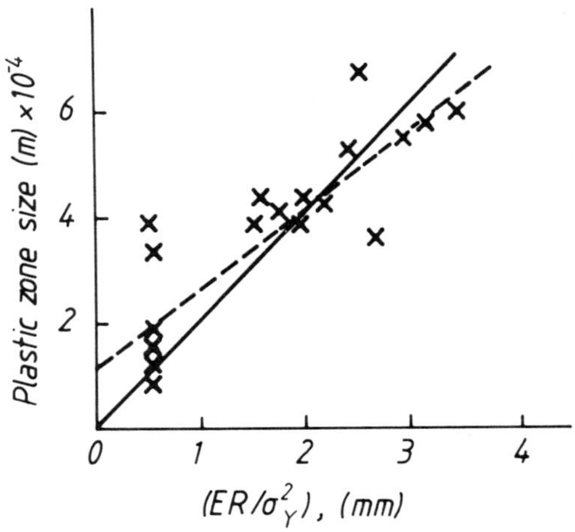

Fig. 10.6 – Charpy notched bar fracture of silicon iron: notch root plastic zone length at fracture for values of ratio fracture load/general yield load < 0.64. – – – – Line of best fit (slope 0.15); ——— line through origin of slope 0.2. (Data from Griffiths and Oates (1969); after Puttick (1978), with permission.)

Example 10.5 Estimate the vlue of α for the brittle–ductile transition at general yield which occurred at 40°C in the experiments of Griffiths and Oates (1969). At that temperature for silicon iron $E = 200$ GPa, $R = 5$ kJ/m^2 and $\sigma_Y = 470$ MPa, and the plastic zone was about 1 mm in extent.

Ans. $ER/\sigma_Y^2 = 200 \times 10^9 \times 5 \times 10^3/(470 \times 10^6)^2 = 4.5 \times 10^{-3}$. Hence with a plastic zone size of 1 mm, $\alpha = 10^{-3}/4.5 \times 10^{-3} = 0.2$ which agrees with the slope in Fig. 10.6.

Fig. 10.7(a) shows (upper) transitions between elastoplastic flow and elastoplastic fracture caused by change of temperature in a different test, viz. when balls of various sizes are indented to a depth of one radius in blocks of PMMA at different temperatures. The critical radius of the indentation plastic zone at which at least one radial fracture was observed correlates well with ER/σ_Y^2; the experimental slope of about 30 in Fig. 10.7(b) should be contrasted with the theoretical α of 25 in Table 10.1 (independent values for E, R and σ_Y as a function of temperature were used in the calculations). The different size balls produce different degrees of deformation and hence change the transition temperature: remember that Charpy transitions, for example, are also altered by the geometry of the notch (e.g. keyhole notches and so on).

Fig. 10.8 shows transitions in similar ball indentation tests caused by *environmental* effects. The critical radius changes with solvent solubility parameter δ_S when the surface being indented is flooded with solvent. The bulk flow stress σ_Y is comparatively unaffected by the solvents, so Fig. 10.8 represents changing

K_C with δ_S; independent measurements of K_C in the presence of solvents show identical behaviour (Chapter 6). A different environmentally induced transition is the steaming of wood in order to bend it without splitting; ER/σ_Y^2 is changed by moisture. Water has a marked influence on the properties of biological materials: a soggy biscuit is an example of a deformation transition from the crisp, dry state; loss of turgor pressure in plants causes wilting (cf flaccid carrot or celery).

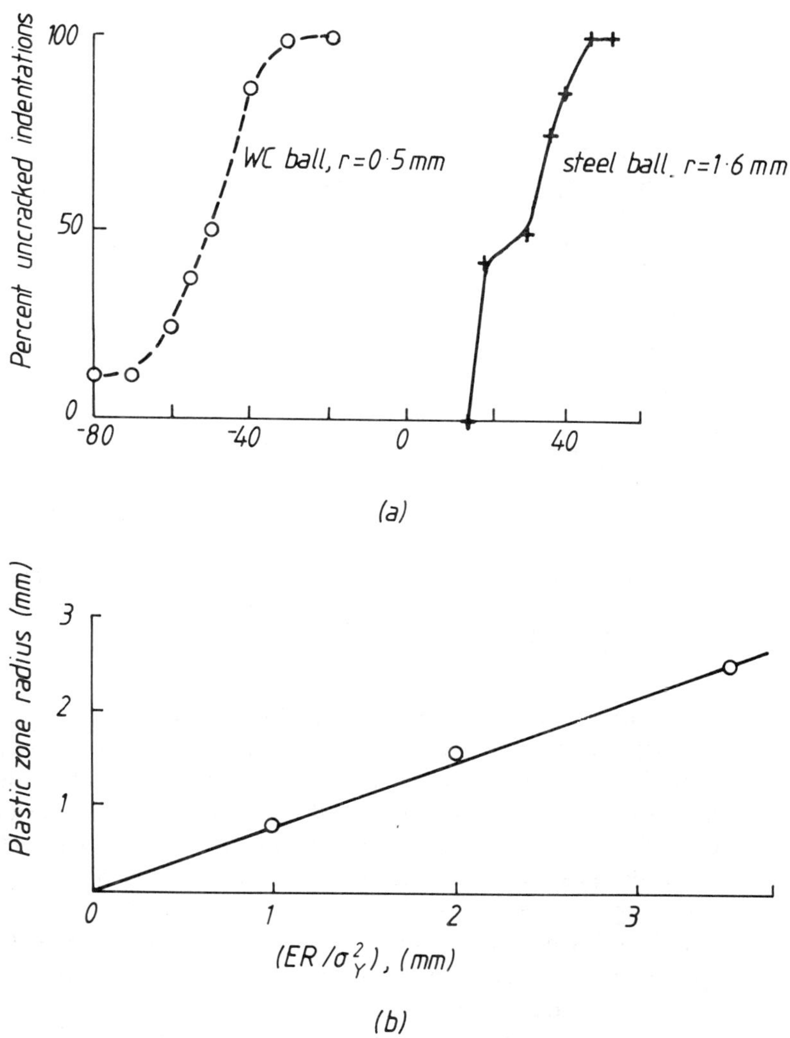

Fig. 10.7 – (a) Upper transitions between elastoplastic flow and elastoplastic fracture caused by change of temperature in indentation tests on PMMA. • WC balls of radius 0.5 mm and ○ steel ball of radius 1.6 mm (b) Critical radius of the indentation plastic zone as a function of (ER/σ_Y^2). Line of best fit gives slope of 30 (after Puttick *et al.* (1979), with permission).

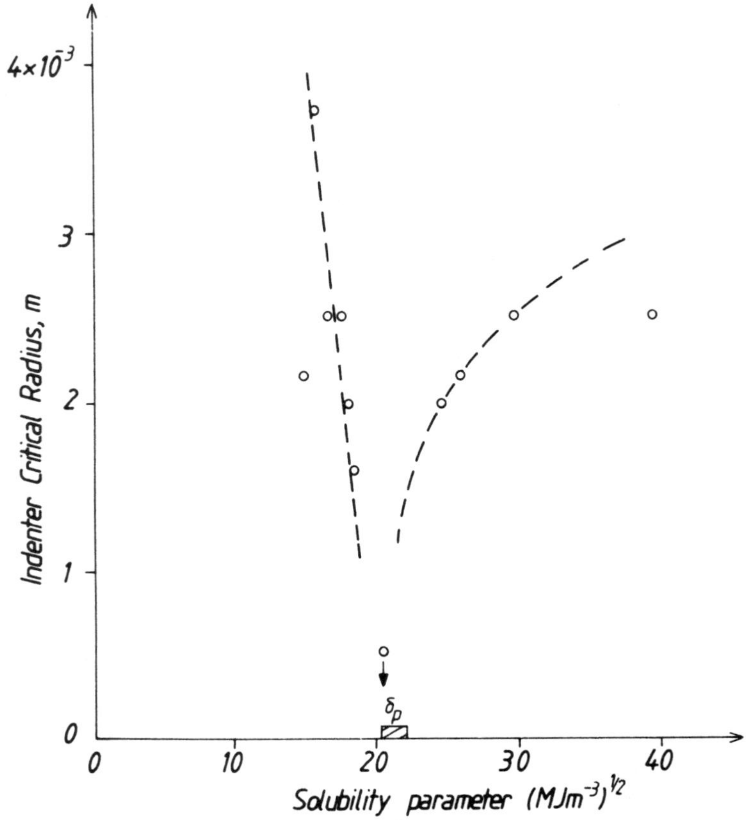

Fig. 10.8 – Critical radius of the indentation plastic zone separating flow and fracture beneath a ball in PMMA as a function of solubility parameter of liquid environment (courtesy M. K. Omar).

Hydrostatic pressure can also cause brittle–ductile transitions; this is well-known for rocks (von Karman) and zinc (Boker) – see, for example, Nadai 1950). Polyethylene usually behaves in a ductile fashion, but Ward and co-workers (e.g. see Truss *et al.*, 1981) have shown that the pressure dependence of the fracture stress of 'Rigidex 50' (a high density homopolymer manufactured by BP Chemicals Ltd) is significantly lower than the pressure dependence of its yield stress. Thus with the application of sufficiently high hydrostatic pressure, the yield stress may be made higher than the fracture stress and brittle behaviour results (Fig. 10.9(a)). Again in this material, transitions can be produced by changing the *strain-rate*, since the strain-rate dependence of the fracture stress is less than that of the yield stress (particularly at high strain rates, high pressures or low tempratures (Fig. 10.9(b)). Similar behaviour is known in PMMA (Rabinowitz *et al.*, 1970) and PC (Christiansen *et al.*, 1971), and parallels those studies in which strain-to-fracture has been noted to decrease with increasing pressure (e.g. Mears *et al.*, 1969; Trent *et al.*, 1978). It should be noted that in some of Ward *et al.*'s

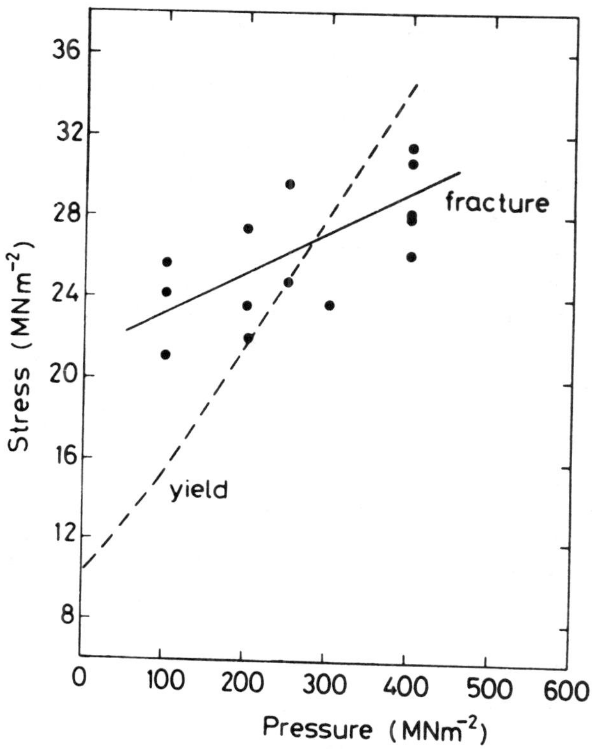

(a)

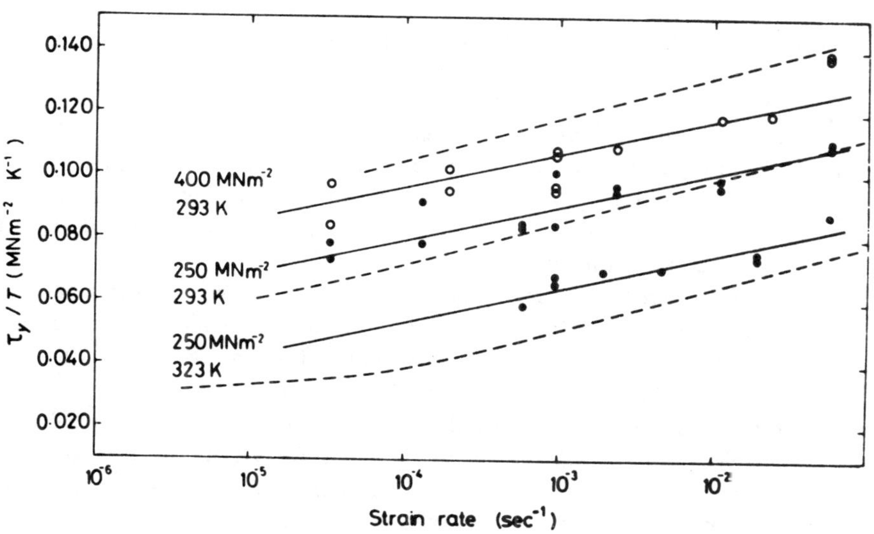

(b)

Fig. 10.9 – (a) Fracture stress as a function of pressure for notched samples of Rigidex 50 tested in Igepal solution at 293 K and strain rate of 9.4×10^{-4} s^{-1}. (b) Fracture and yield stresses as a function of strain rate, temperature and pressure for Rigidex 50 tested in Igepal solution. (Both reproduced from Truss *et al.*, 1981. Chapman & Hall, with permission).

experiments, brittle failure of polyethylene could only be induced by notching the samples and subjecting them to an aggressive environment under pressure (the Igepal solution mentioned in the captions to Figs. 10.9(a) and (b) is the environment recommended by ASTM for standard tests on the susceptibility of polyethylenes to environmental stress cracking). It would appear, therefore, that one effect of hydrostatic pressure is to reduce the time scale within which environmental stress cracking is observed.

We also note the copolymer versions of polyethylene (e.g. BP Rigidex 002-40) have lower yield stresses than the homopolymer grades mentioned above, and therefore display ductile behaviour over wider ranges of conditions. Such greater resistance to brittle fracture explains why it is the copolymer material that is used for gas pipes.

Changing ER/σ_y^2, and changing α with contact geometry, explain transitions in *hardness* tests (see Chapter 3).

The interactions between material and geometry of body produce what might be thought of as *fracture stability transitions,* Chapter 8, in that the same material may display slow or fast cracks depending upon the shape and size of the test-pieces; and the same shape and size testpiece, under the same loading conditions, may involve slow or fast cracking depending upon the material from which it is made.

10.3 ABRASIVE MACHINING

Deformation transitions are important for the efficiency of abrasive machining processes, which range from grinding at the coarse end of the spectrum, through the action of SiC-type abrasive papers to, at the fine end, mechanical polishing using abrasive powder slurries. Samuels reviewed in 1978 the extensive studies by himself and co-workers on the action of single pyramidal tools which simulate the behaviour of real abrasive machining (abrasive particles when magnified look like the Matterhorn). The experiments are hardness tests with superimposed tractions across the indented surface.

It is found that the cross-sectional area of the groove formed as an abrasive particle slides along on the surface of a ductile metal shows a transition with orientation and geometry of the scratching tool. For given frictional conditions and load, there is a critical 'attack angle' for the tool below which a chip or ribbon fails to be cut; instead a prow of material is formed ahead of the tool which is moved into side ridges alongside the groove as the tool 'rubs along' (Figs 10.10 and 10.11). The attack angle β is measured from the horizontal and corresponds with $\beta = (90 + \bar{\alpha})$ where $\bar{\alpha}$ is the tool rake angle as conventionally defined from the vertical. The critical attack angles for a range of metals being scratched with pyramidal tools are roughly as follows: copper, 45°; α-brass 55°; lead, 60°; nickel, 70°; aluminium, 85°. At larger attack angles when ribbons of material are machined away, the size of groove increases rapidly and eventually seizure occurs as the tool 'digs-in' (Fig. 10.10).

The critical attack angle corresponds with an upper transition and it should be possible to correlate β_{crit} with (ER/σ_y^2) but lack of all the information required prevents this. There is also a lower transition between cutting a ribbon and 'knocking lumps out', i.e. fragmentation of the workpiece in the track of the tool. Lower transitions are more readily seen in less ductile solids than those studied by Samuels and corworkers. For example, if PMMA is planed with the sorts of tools commonly found on shaping machines, extensive cracking supersedes continuous (serrated) chip formation when the depth of the cut exceeds about 150 μm: scallops are removed quite violently and the deformation and crack patterns go below the cut surface. Wood, when planed with too deep a cut will split, which is the same phenomenon but, going to smaller depths of cut, we compare the action of a joiner's scraper.

Effects of size of cut are well-known on a smaller scale in the scratching of brittle solids. Fig. 10.12 shows a scratch in glass. The occurrence of these cracks is closely bound up with relief of residual stresses and the critical size of track for fracture to occur has been derived by Puttick *et al.* (1979) as some $6(ER/\sigma_y^2)$, using a model of the fields of residual stress around indentations proposed by Swain and Hagan in 1976. Experiments on single crystals of silicon confirm this expression.

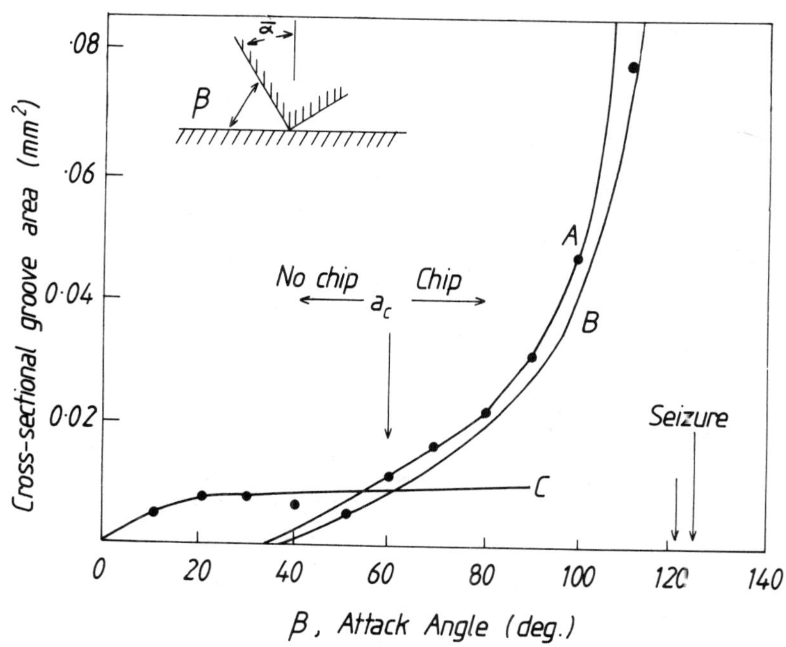

Fig. 10.10 – Transition between chipping and no chipping in abrasive machining (after Sedricks and Mulhearn (1964), with permission).

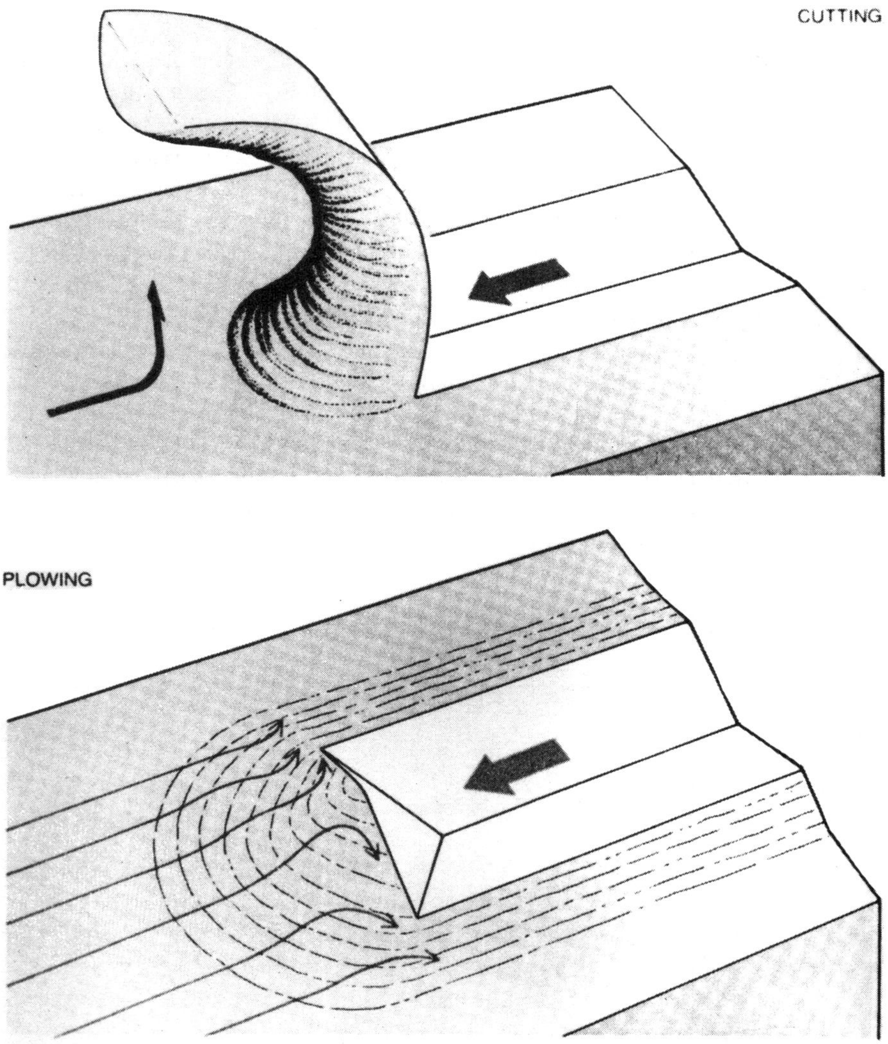

CUTTING AND PLOWING are portrayed for a *V*-point tool. In cutting (*top*) a ribbon of material is separated from the surface of the workpiece. The material moves upward past the rake face of the tool continuously. In plowing (*bottom*) the material first moves upward ahead of the rake face and then moves around it and into side ridges. Cutting is an efficient method of removing material; if any material is removed in plowing, the process is highly inefficient.

Fig. 10.11 – Cutting and ploughing observed in abrasive machining (reproduced from Samuels (1978), with permission).

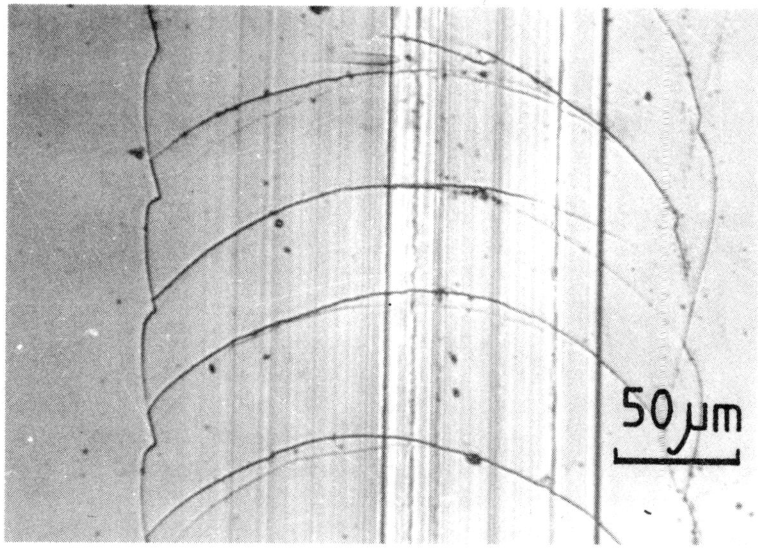

Fig. 10. 12 – A scratch in glass formed by sliding a 1 mm radius hard steel ball under a load of 7.5 kg across the surface at a speed of 200 μm/s. Many partial ring cracks are observed trailing the direction of motion. The lines along the length of the track are plastically-formed grooves produced either by asperities on the ball or by trapped debris (from Howes (1968), with permission).

Example 10.6 In cutting PMMA a lower transition occurs at a depth of cut of about 150 μm. At what values would corresponding transitions take place in (a) glasses and ceramics and (b) low carbon steel.

Ans. Assume that the transition occurs at a critical value of $(ER/\sigma_y^2 h)$ where h is the depth of cut. For PMMA $(K_C/\sigma_y) \approx 0.02\sqrt{m}$, for glasses and ceramics say $0.01\sqrt{m}$, and for low carbon steel say $0.5\sqrt{m}$. Hence for ceramics and glasses the lower transition occurs at $h = (0.01/0.02)^2\ 150 \approx 35\ \mu$m. For low carbon steel it is $h = (0.05/0.02)^2\ 150 = 95$ mm. For h greater than these values, extensive brittle cracking would occur. Thus machining of ceramics is done by abrasive grinding to avoid fragmentation, and brittle cracking in the machining of low carbon steels is not observed as such large depths of cut are never taken. (It was once suggested to one of the authors that he should perform an experiment with a piece of steel plate attached to a railway platform, and a cutting tool attached to the buffer beam of a locomotive, to check the latter prediction!)

All solids can in principle be machined, abraded or polished, but some present difficulties – for example brittle solids are more susceptible to fragmentation and uncontrolled cracking which causes problems for surface finish and surface integrity (e.g. residual microcracks that may be sites for later fatigue fracture). It is clear that of the four time-independent modes of deformation of a surface produced by a tool of some sort (i.e. (a) elasticity alone (b) elastic fracture (c) plastic fracture, (d) flow alone), material is removed only in (b) and (c) in theory. Furthermore, of the two processes, more material is removed in (b) than (c) as the deformation and elastic cracking extends over a much greater region ahead and below the tool tip than when plastic ribbons are formed local to the tool. However, there is less dimensional control in process (b) than (c), cf. drilling 'semi-brittle' solids with too high a feed rate and/or blunt drills.

Samuels (1978) reports that the debris produced in abrasive machining of *ductile* solids corresponds with the ribbons produced by single-point tools. Fig. 10.13(a), (b), (c) illustrates that the particles removed when steel is ground or abraded have the same characteristics as the curved ribbon-like chips produced by turning steel on a lathe. Ductile solids may also be polished and Fig. 10.13(d) shows that the very small particles removed when copper is polished are ribbons and have the periodic structure typical of a machining chip. Polished surfaces produced by the action of abrasives have closely spaced shallow grooves differing only in degree from the grooves produced by grinding and abrasion. Therefore it seems clear that grinding, abrasion and polishing are all similar material-removal processes and differ only in degree, i.e. size of cut. Furthermore Puttick *et al.*, (1979) suggests that the initiation of cracks, by indentations or scratches, marks the transition from polishing to grinding of *brittle* solids as a function of abrasion particle size. Similar thoughts apply on a larger scale to the sculpting of stone. Materials with hard and soft components (e.g. iron carbide in ferrite in steels) will display a range of behaviour along the crack traversed by the tool as the microstructure changes. Bowden and Tabor (1964) contrast Samuel's viewpoint with that of Beilby who argued that the mechanism of polishing related mainly to surface flow. The importance of local surface temperatures in polishing is highlighted.

The effeciency of material removal may be defined in terms of the volume swept out by the tool and the amount of material actually removed. When fragmentation occurs, *more* material is often removed owing to sub-surface cracking and 'scalloping'. When ribbons are produced, the metal removed and the swept volume match pretty well. When ribbons are absent and prows are formed ahead of tools having $\beta < \beta_{crit}$, no material is removed in theory.

The efficiency of grinding and abrasion produced by wheels, papers and slurry powders is not particularly high – often lower than 20%. This comes about because only a proportion of all the abrasive points in contact with the workpiece have rake angles suitable for cutting a chip (i.e. attack angles $> \beta_{crit}$). Abrasive papers have been photographed in silhouette in the SEM to determine the distribution curve of attack angles (Samuels, 1978). For a paper coated with 220-mesh SiC abrasive, and with steel as the workpiece (where $\beta_{crit} \approx 80°$) only about 25 per cent of the points are favourably orientated. In a grinding wheel perhaps only 10 per cent grits would actually be cutting, and in polishing where there is little constraint on particles which can roll about in the slurry it is probable that few of the many particles in a polishing mixture do anything useful. Actually, a small amount of material in the form of small chunky fragments is removed by

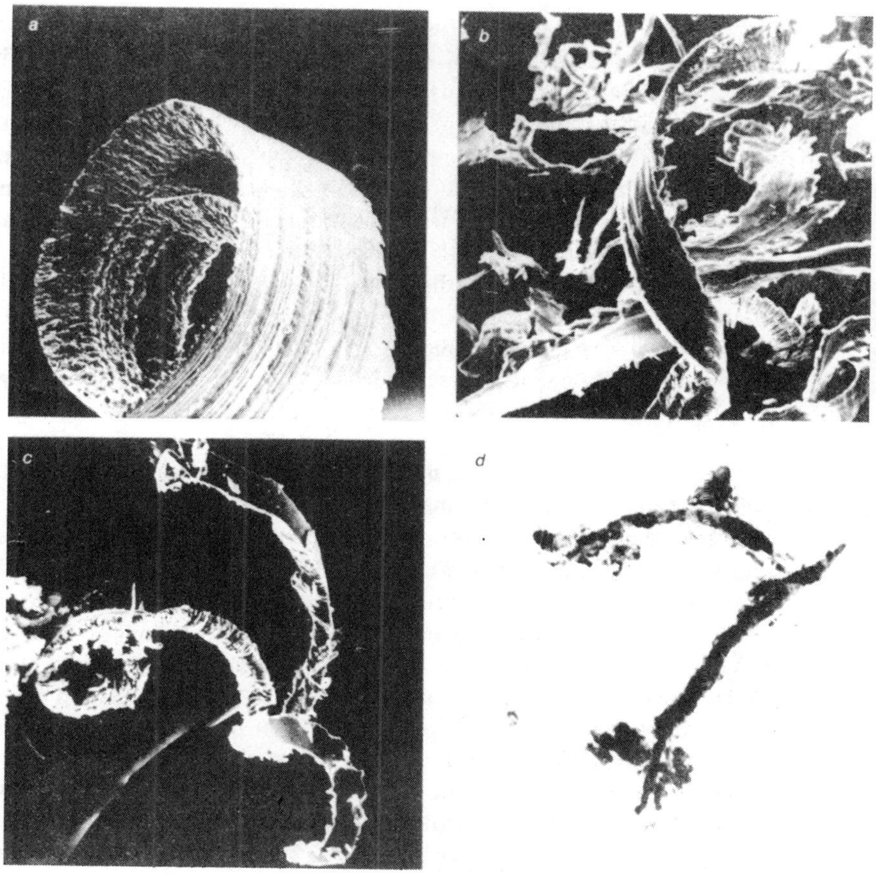

Fig. 10.13 – Debris produced in abrasive machining of ductile metals. (a) A chip produced by turning steel on a lathe; (b) particles produced by grinding steel; (c) particles produced by abrasion; (d) particles removed when copper is polished (reproduced from Samuels (1978), with permission).

particles with attack angles less than β_{crit} (in steel and brass down to $\beta \approx 20°$ where $\beta_{crit} \approx 80°$ and 45° respectively). They are prows that break off when the abrasive point leaves the workpiece at a free edge. However, the removal of material by this mechanism is, of course, highly inefficient compared with removal in ribbons or brittle fragmentation.

The parameter ER/σ_y^2 may be used in grinding to define transitions from continuous chip formation to knocking lumps out which corresponds to reducing the parameter from the upper left quadrant of Fig. 10.1 down to the lower left quadrant. The influence of changes in imposed strain, in otherwise similar situations, is shown by the fact that in abrasive wear experiments with single-point pyramidal tools, chips are produced when scratching with indenters in a facet-first orientation (upper left quadrant, Fig. 10.1) but a piled-up groove is produced by a cube-edge-first orientation (upper right quadrant). The transition from negative rake micromachining to rubbing in this type of situation is linked to transitions between bulging and scalping in metal forming. It is also worth noting that a ratio of imposed strain to yield strain, i.e. ϵ_i/ϵ_y, is implicit in the works of Blok (1952) and Halliday (1955) on the shape of surface asperities that may be pressed into the general surface level of contacting bodies without incurring plastic deformation in those asperities. Similarly, plasticity indices for Greenwood–Williamson and Whitehouse–Archard models of surface roughness also relate to the same ratio of strains (see, for example, Archard, 1974). Johnson (1970–71) has pointed out that the low efficiency of material removal in abrasion and grinding is affected by accomodation of the elastic hinterland to blunt grits which corresponds to ϵ_i/ϵ_y being small.

10.4 WEAR

Abrasive machining is but one branch of the wide field of cutting, the object of which is definitely to *remove* material from a surface, as it is in abrasive cleaning processes such as pot-scouring (which is not really a frivolous example). In contrast, when material is *undesirably* removed from a surface, we have *wear.* Since we do our best to avoid wear of machine parts by suitable choice of contacting materials and by lubrication, wear rates are low and processes of wear occur on a fine scale. Even so, the micromechanisms of material removal will be the same as just described – certainly in abrasive wear (where one surface is much harder than the other) but possibly also in adhesive wear (which takes place when the contacting surfaces have comparable yield strengths and material is transferred by adhesion from one surface to another), in which crack growth by fatigue plays an important role in particle removal.

The volume wear rate per unit sliding distance (V/L) is usually related to the hardness (H) of the material being worn and the load (X) by the Archard equation (1952)

$$(V/L) = k(X/H) \tag{10.18}$$

where k is the 'wear coefficient' which takes into account the contact geometry, the fact that only a proportion of the grit seems to remove material, and that only a proportion of the scratch volume is removed. Various workers have shown that the wear rate of many materials is inversely proportional to hardness; but there are anomalies. In particular, Kruschov and Babichev in 1954 and Buttery and Archard in 1970–71 showed that the rate of wear of various heat-treated steels of increasing hardness decreased less rapidly than expected. Again, for metals strengthened by workhardening, the wear rate was almost independent of hardness, according to Kruschov and Babichev; they cite similar findings reported by Tonn for Armco iron and copper cold-rolled by various amounts. In addition to hardness as a controlling parameter, it has long been known that the area under the stress–strain curve to fracture (conveniently measured by the product of $\sigma_{UTS} \times \epsilon_f$) may sometimes be correlated with wear rates, and this hints at a role played by fracture toughness.

Fig. 10.14 illustrates the wear resistance (the reciprocal of wear rate) versus bulk hardness of the solid being worn away. Materials both following the simple relation with hardness and deviating from it are evident. Careful measurements by Mutton and Watson in 1978 showed that the wear resistance of quenched and tempered steels seems to fit sigmoid-shaped curves which depart from the pure metals line at a point corresponding to the behaviour of pure iron and are displaced to higher wear resistance the higher the carbon content (Kruschov and Babichev drew sets of second straight lines through similar data). Murray *et al.* (1979) considered mechanisms of abrasive wear (brittle chipping, micromachining and prow formation) and suggested relations that followed three straight lines emanating from the origin (Fig. 10.15). The sigmoidal curves in Fig. 10.15

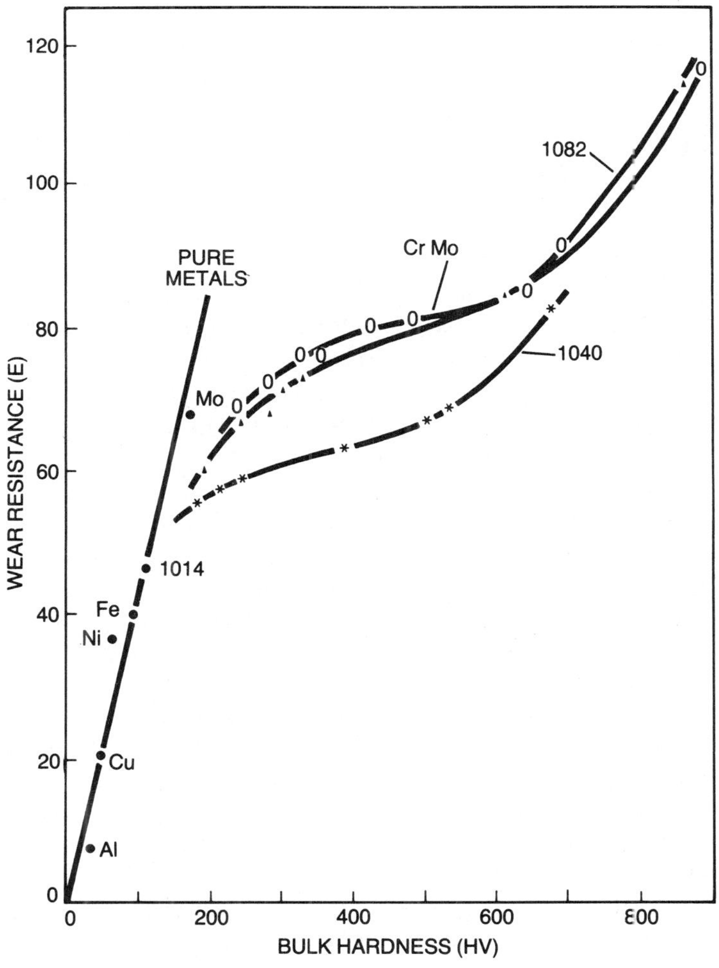

Fig. 10.14 – Wear resistance as a function of hardness for three steels as indicated together with points on the pure metals line (after Murray *et al.* (1979), with permission).

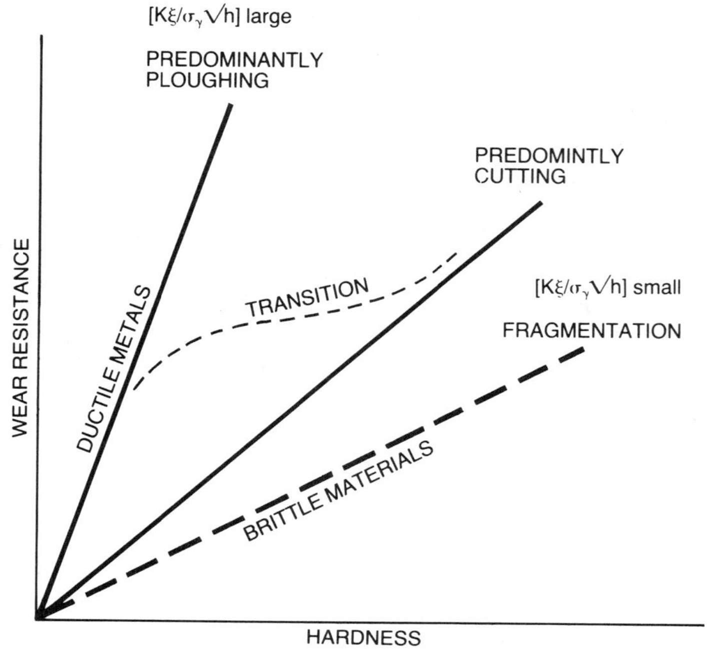

Fig. 10.15 – Three possible linear relations between wear resistance and hardness. The broken curve shows the transitional behaviour found for steels (as in Fig. 10.15) (after Murray *et al.* (1979); taken from Atkins (1980), with permission).

represent transitions in the major means of material removal as the microstructures change from ductile to brittle. Clearly this discussion exactly parallels that by Samuels given in section 10.3 for abrasive machining.

Since the type of material removal is determined by upper and lower transitions, Atkins (1979) suggested that it is likely that abrasive machining and wear rates should correlate with $ER/\sigma_y^2 h$. Following earlier arguments, the rate of material removal should vary inversely with the parameter, i.e. maximum wear when brittle fragmentation occurs, followed by smaller wear rates at the same h when ribbons are produced. No material is removed when the deformation is simply elastic, nor when simply plastic, but as already noted for the latter case, mounds and ridges alongside tool tracks do sometimes become detached in practice. Support for an inverse $ER/\sigma_y^2 h$ relation is found in the work of Selwood (1961) where experiments showed that the abrasive wear of brittle solids was greater than metals of the same hardness, but the wear of tough elastic solids such as rubber and some polymers was less at the same value of σ_y. Recently Omar (1984) has shown that the wear rate of polymers varies inversely with K_C^2, i.e. with $1/ER$. It is possible that transitions from mild to severe wear are linked to changes in the magnitude of $ER/\sigma_y^2 h$. It is known, for example, that the wear of leaded 60/40 brass is much worse than the corresponding unleading metal and, as with free-machining brass, it is likely that liquid metal embrittlement in the hot contact zone reduces the fracture toughness. While the variation of R with temperature is not available, it is known that Charpy values of leaded brasses show an embrittlement trough round about 327°C (the melting point of lead) and that the best free-machining properties are obtained with higher volume fractions of lead particles which give lower toughness (e.g. Lunn and Christensen, 1970). Note that the presence of h in $ER/\sigma_y^2 h$ could be interpreted as a possible dependence on surface roughness.

It would seem that the separate lines in Fig. 10.15 could be condensed into a single line were the data plotted against $ER/\sigma_y^2 h$ instead of against bulk hardness (i.e. σ_y) alone. Unfortunately there are no independent values of the component terms (particularly R) to test this hypothesis, and a full treatment may be quite involved owing to changes in mechanical properties with microstructure along tool tracks, and with number of repeated traversals where material may be fatigued away.

Evidence for the positive role played by toughness in wear, however, is shown in Fig. 10.16. Evans and Lancaster in 1976 showed that the wear rates of polymers depend markedly on solvents and are maximum when the solvent solubility matches that of the solid. Independent measurements of crack resistance demonstrate that R is least when the solvent solubility parameters match (Chapter 6) and Fig. 10.8 shows that the K_C/σ_y ratio is also least since σ_y is comparatively unaffected by solvents. An inverse $K_C^2/\sigma_y^2 h$ relation would therefore predict maximum wear rate when $\delta = \delta_s$, exactly as shown in Fig. 10.16. However, this is not the whole story, since Evans and Lancaster's results relate more to *adhesive/fatigue* wear than to abrasive (micromachining) wear for which the $ER/\sigma_y^2 h$ parameter applies. In adhesive/fatigue wear, debris is produced ultimately by fatigue crack growth after many repeated loadings of the wearing surfaces (gear teeth for example). It is reasonable to envisage that the wear rate should vary inversely as the number of fatigue cycles N_f required to propagate inherent (microstructural defect) cracks to the critical crack lengths at which the peak cycling load will detach a wear particle. Hence wear rate should depend on the product $C(n/2 - 1)$ according to equation (7.12), where C and n are the parameters of the Paris equation for high cycle fatigue crack growth rate (equation (7.6)). Fig. 10.17(a) shows how C and n vary with solvent solubility for PES (polyethersulphone): C is a maximum, and n is a minimum, when $\delta_s = \delta_p$. Owing to the large change in C, the product $C(n/2 - 1)$ goes through a maximum at $\delta_s = \delta_p$ exactly like the wear rate curve in Fig. 10.16. Fig. 10.17(b) shows a histogram comparison of experimental wear rates in PES and theoretical predictions; the predicted rates are fitted to the acetone result because neither the value of a_0 nor the constant of proportionality are known (Atkins *et al.*, 1984). It is noteworthy that Fleming and Suh (1977) found that the wear rate of 7075-T6 aluminium alloy was greater than that of 2024-T3 aluminium alloy, even though the former was some 25 per cent harder than the latter; the important difference between the alloys was that the fatigue crack propagation rates were faster in the 7056-T6 alloy. While, of course, the mode of crack propagation in wear is unlikely to be simply mode I (for which most C,n data are generated), environments are likely to alter rates of crack advance in other modes in a similar manner to those in mode I.

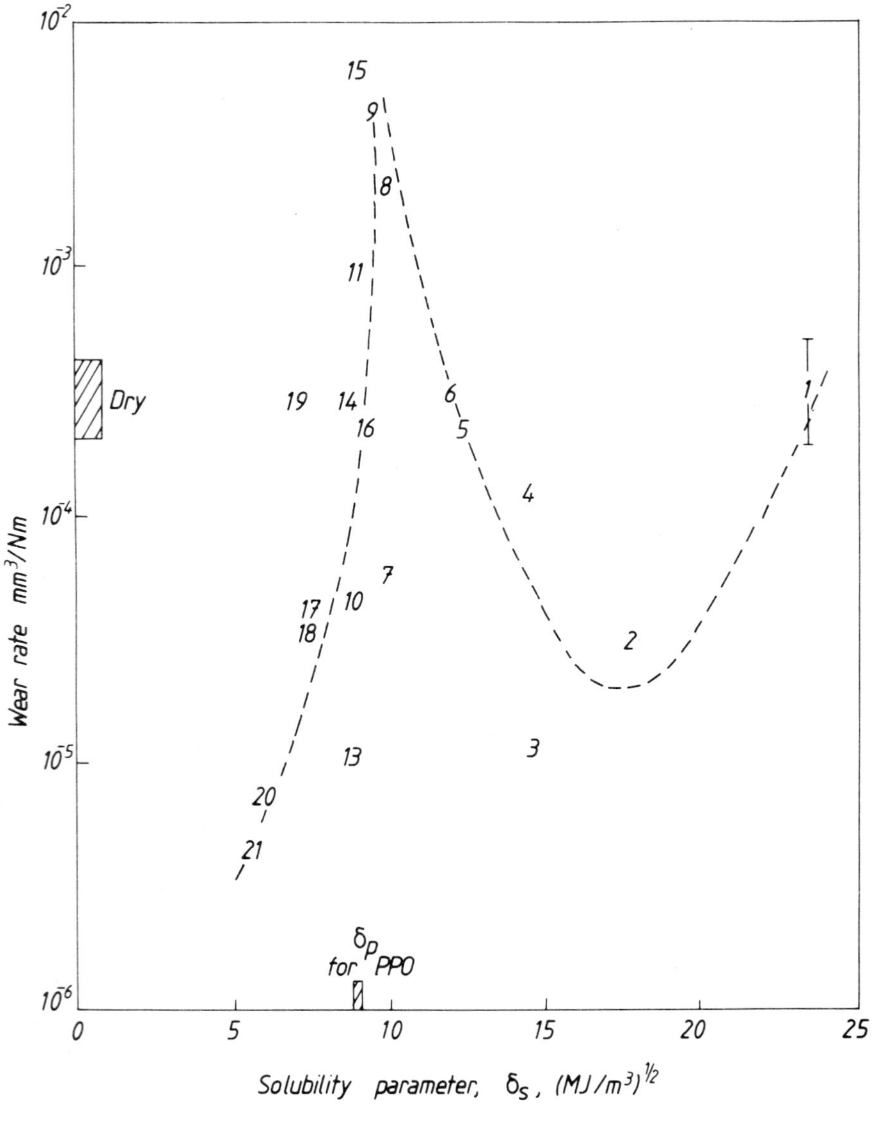

Fig. 10.16 – Wear rate as a function of solubility parameter for PPO. (Solubility parameters lised in Table 6.3.) Note ordinate is mm^3/Nm, the (usually defined) wear rate being divided by the load. (After Evans and Lancaster (1976), with permission).

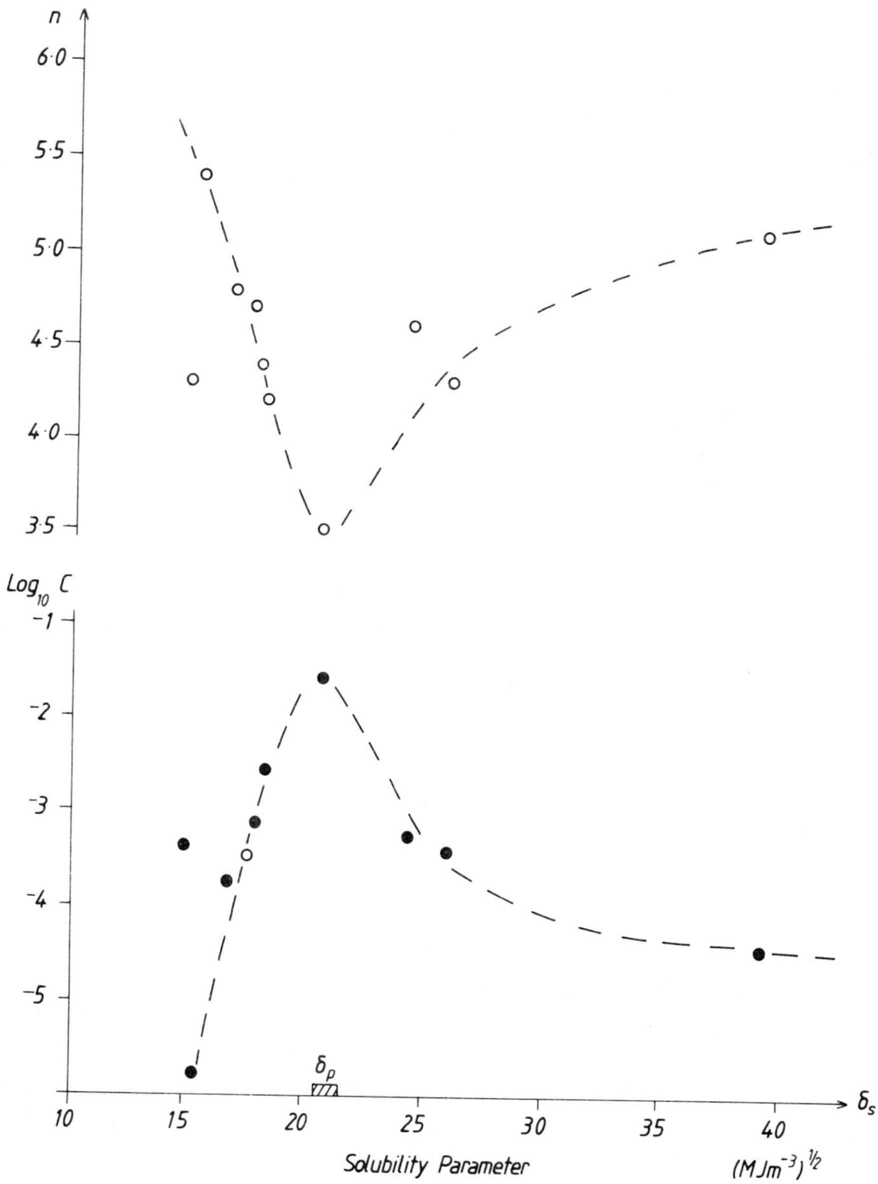

Fig. 10.17 (a) – Variation of the Paris fatigue equation constants C (upper graph) and n (lower graph) as a function of solubility parameter for PES.

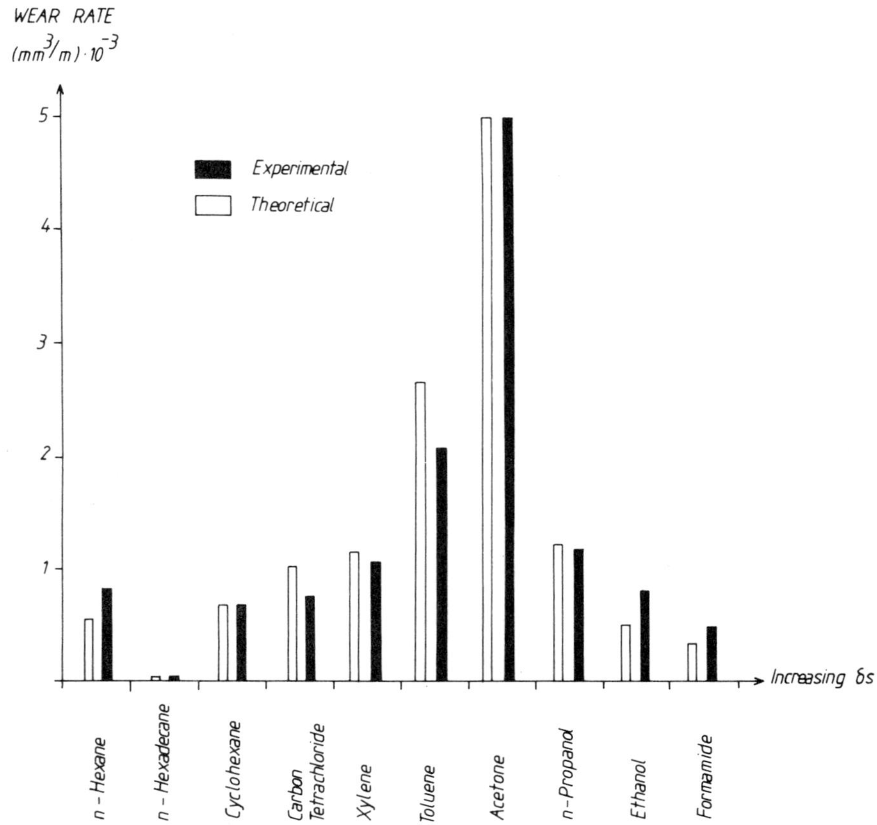

Fig. 10.17(b) – Histogram comparing experimental data the theoretical predictions using equation (7.12) (courtesy M. K. Omar).

In the field of erosive wear, which is related to abrasive wear, Hutchings *et al.* (1976) classified the oblique impacts of irregular particles on surfaces into ploughing and cutting categories (cf. Fig. 10.11) and Springer (1975) successfully correlated optical transimission losses in materials (caused by repeated impacts of water droplets) with an $ER/\sigma_y^2 h$-type parameter. Arguing that sub-surface cracks grow in fatigue under the action of impinging raindrops and that optical transmittance T varies inversely with crack length, it may be shown (Problem 10.15) that the relative transmittance

$$T/T_0 \approx \frac{1}{1 + \Theta\ [p^2\ a_0/ER]^2 N} \tag{10.19}$$

where Θ is a constant, p is the impingement pressure, a_0 is the starter crack size and N the number of impacts. Fig. 10.18 shows that the relationship is obeyed for a number of materials with widely different a_0/ER values, namely 10^{-10} Pa^{-2} (calcium aluminate, magnesium oxide) to 10^{-18} Pa^{-2} (polycarbonate,

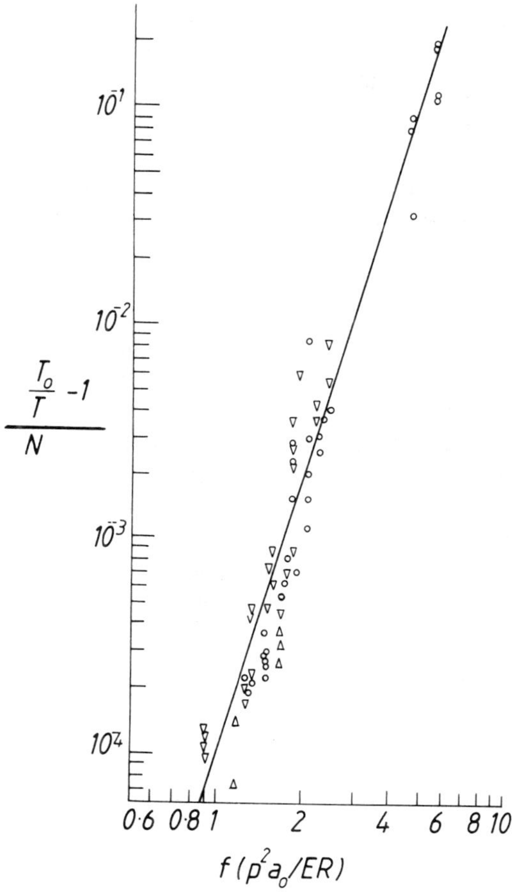

Fig. 10.18 – Correlation of optical transmission loss with $(ER/\sigma_y^2 h)$ (after Springer (1975), with permission).

polysulphone, quartz). Mass loss rate in erosion also obeys a relation similar to equation (10.19). Furthermore as the pressure at the liquid–solid interface and the number of impacts are both proportional to impact velocity, it follows that mass loss rate and relative transmittance loss rate both are proportional to $(\text{velocity})^5$.

10.5 CUTTING

Even though quantitative application of the $ER/\sigma_y^2 h$ parameter is not yet possible for abrasive wear rates in metals, the parameter does seem to be successful quantitatively for models of the forces required to cut metal on lathes etc. Traditional approaches to the mechanics of machining consider only plastic work and friction work, and although apparently reasonably successful when applied to continuous chip formation in metals, such models fail to explain discontinuous chips in brittle metals and the cutting behaviours of many polymers, ceramics and other materials. The rate of performing work in cutting really has three parts, (i) plastic work, (ii) friction work and (iii) fracture work. It has been customary to measure experimentally the components of cutting force and relate them to models incorporating the first two items, where the plastic flow occurs on a shear band AB, and friction occurs on the tool rake face (Fig. 10.19). No *independent* measurements of plastic and friction work are generally made, and factors such as material flow stress and coefficient of friction are derived to make analysis and experiment agree. Such models are not wholly satisfactory. For example, the coefficient of friction is often greater than 1/2, (the value predicted by a Tresca yield criterion for sticking friction along all the rake face). Also, the material flow stresses are very much greater than quasi-static test values; this is attributed to large strain rates in the shear zone, even for materials usually considered rate-insensitive. Again, values for the shear plane angle ϕ in Fig. 10.19 given by minimisation of energy, or maximisation of shear stress, in plasticity and friction analyses do not agree with experiment (see, for example, Alexander and Brewer, 1963). In many cutting operations load fluctuations are present which are sometimes quite violent. Although such chatter, with its associated surface finish effects, is normally explained in terms of tool–workpiece stiffness or recording system response, it does seem inherent in the cutting process for some materials at low speeds in stiff systems. When experimental cutting forces are inserted into traditional analyses, the values are averaged from the oscillatory loads measured in practice.

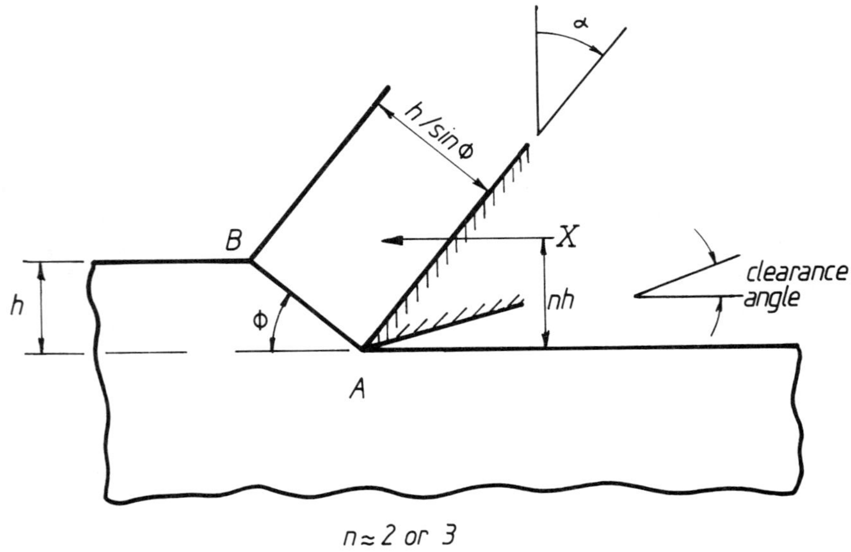

Fig. 10.19 – Geometry of tool and workpiece in machining.

From the discussion of guillotining and similar processes given in Chapter 5, it follows that metal cutting is an operation that adjusts itself to the simultaneous actions of cracking and plastic flow, with friction playing a superimposed role. Cracking in metal cutting was contemplated by Mallock in 1881 and Rosenhain and Sturney in 1925, but the concept has been out of fashion since then, principally because free-standing cracks ahead of the tool are rarely seen when turning ductile metals on the lathe, or when milling or shaping etc., except right at the end of a cut and in semi-brittle metals such as molybdenum. But following the arguments for cropping in Chapter 5, free-standing cracks are *not* necessary for toughness to be important as the *whole* machined surface is one side of a crack and the swarf the other. Whether fracture of the chip occurs into discontinuous parts is a subsidiary issue. According to this Atkins viewpoint, fracture is inherent in *continuous* chip formation and all material-removal (separation) process.

A dimensional analysis for machining which incorporates fracture toughness leads to the ubiquitous $(ER/\sigma_y^2 h)$ group, as shown by Atkins in 1973 and given in the next example.

Example 10.7 Perform a dimensional analysis for the force required to orthogonally cut a chip of depth h and width w (as in Fig. 10.19).

Ans. Let the force be a function of h, w, σ_y, E, R and friction. Let friction be represented by some constant shear stress at the tool rake face/chip interface, which is some fraction of the shear yield stress (first used by Rowe and Spick in 1967). Friction is then incorporated in the σ_y term. Then

$$\text{force}, X \propto \text{function}\,(h, w, \sigma_y, R, E)$$

$$= (\text{constant})\, h^a\, w^b\, \sigma_y^c\, R^d\, E^e$$

i.e.

$$\frac{\text{kg m}}{\text{s}^2} = (\text{constant})\ (m^a)\,(m^b)\left[\frac{\text{kg}}{\text{m s}^2}\right]^c \left[\frac{\text{kg}}{\text{s}^2}\right]^d \left[\frac{\text{kg}}{\text{m s}^2}\right]^e$$

which, upon grouping parameters, becomes

$$\text{force}, X = (\text{constant}, P)\, h^2\, (w/h)^b\, \sigma_y\, [R/\sigma_y h]^d\, (E/\sigma_y)^e \qquad (10.20\text{a})$$

(Studies on *machinability* (defined in terms of tool life) have incorporated other parameters such as thermal conductivity in dimensional analyses of cutting (Datsko and Henkin, 1964), but these are omitted here since tool life is not of concern and we attempting to find the influential parameters in the cutting process *per se*. Clearly, temperature build-up, coolants etc. affect machining, but only in so far as they alter σ_y, R and E as functions of temperature and rate).

The work required to cut unit volume in unit time follows from equation (10.20a) as

$$(\text{work/volume–time}) = P(h/w)^{1-b}\sigma_y\, [R/\sigma_y h]^d\, (E/\sigma_y)^e \qquad (10.21\text{a})$$

(Problem 10.16), since the tangential component of cutting force is the major contributor to total work of cutting. The work/volume–time is the 'unit horsepower' of traditional metal cutting analyses using imperial units. An extensive body of data exists in the literature that correlates unit horsepower with hardness of the cut material (or equivalently with σ_y) illustrated in Figs. 10.20(a), (b). While a general agreement is displayed, there is considerable scatter and some anomalies. At constant (h, w) equation (10.20a) predicts a dependence on ${\sigma_y}^{1-d-e} R^d E^e$ rather than the assumed σ_y alone. If fracture toughness is a significant parameter in cutting, it is clear that log-plotting of merely unit horsepower and hardness does not take the whole picture into account.

If E and R can be combined into K_C, it follows that $d = e = f/2$ say, and then

$$X = P\,h^2\,(w/h)^b\,\sigma_y\,(K_C/\sigma_y\sqrt{h})^f \qquad (10.20b)$$

and

$$(\text{work/volume–time}) = P(h/w)^{(1-b)}\,\sigma_y\,(K_C/\sigma_y\sqrt{h})^f \qquad (10.21b)$$

As with wear and abrasive machining, were appropriate R or K_C values known, the analysis could be tested. Progress may still be made by appealing to a possible relation between K_C and σ_y. Krafft in 1961 reported that, approximately, K_C varied inversely with σ_y according to

$$K_C = Q/\sigma_y^q \qquad (10.22)$$

where $2 \approx q \approx 3$ (for example for SAE 4340 steel, $q \approx 2.1$) and Q is constant for a given solid. Then at constant h, w

$$(\text{work/volume–time}) \propto \sigma_y^{1-f(q+1)} \qquad (10.23)$$

The slope of the log-plots in Figs. 10.20(a), (b) should be of magntidue $(1-f\times(q+1))$. For 4340 steel data alone the slope is 0.35, and if $q = 2.1$, it follows that $f = 0.21$.

The form of equation (10.20) should be contrasted with old-established empirical relations for metal cutting forces such as

$$X = (\text{constant})\,h^x w^y \qquad (10.24)$$

where $0.75 < x < 1.0$ and $0.96 < y < 1.0$ and where the constant varies with material and tool signature (Boston, 1951). For simplicity let us take $y = 1$, in which case $b = 1$ in equation (10.20b) and then $x = (1 - \tfrac{1}{2}f)$. For 4340 steel therefore, for which $f = 0.21$, $x = 0.9$ which is a sensible value which agrees with experience. The constant P in equation (10.20b) may be determined, and compared with the empirical constant of equation (10.24) from unit horsepower data, i.e. equation (10.21b) becomes (with $b = 1$ and using relation (10.22))

$$(\text{work/volume–time}) = PQ^f\,h^{-1/2\,f}\,{\sigma_y}^{1-f(q+1)}$$

f is known from the slope, Q, q are given, leaving P to be solved (in machining tests using imperial units a cut of $h = 0.0125$ in was often standardised).

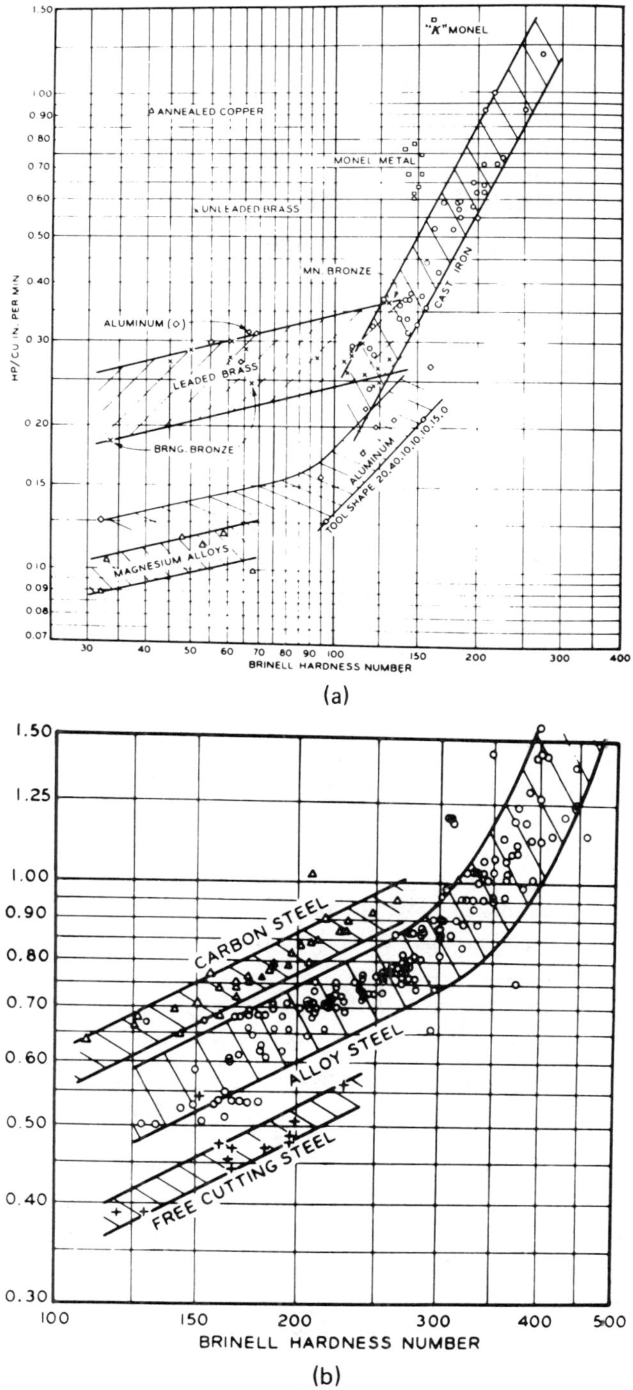

Fig. 10.20 – Unit horsepower as a function of Brinell hardness number when turning (a) various metals and (b) steels (reproduced from ASME (1952), with permission).

Problem 10.20 shows that it is possible to predict from equation (10.20) the cutting force for a low carbon steel to be $X \approx 120{,}000\, h^{0.86}\, w$ lbf, which should be contrasted with Boston's result for a hot-rolled SAE 1020 steel of $X = 133{,}000\, h^{0.85}\, w^{0.98}$ lbf. The agreement is quite good.

10.6 DEFORMATION TRANSITIONS AND DESIGN

We have seen that a body, when loaded, may behave merely elastically, may crack elastically, may behave elastoplastically, may crack in the elastoplastic regime, may experience widespread flow, or may fracture after extensive remote flow. Which (if any) of these various possibilities constitutes 'failure' to an engineering designer? The answer clearly depends upon the purpose of the component or structure being designed, its size and shape, the loading, and the material from which it is made.

Most traditional design relates to the elastic range of deformation with, usually, high structural loading coefficients (i.e. comparatively high loadings over the extent of the body) using stiff materials; in contrast tents, pneumatic structures and the like, and many natural structures (plants and animals) have low structural loading coefficients and low stiffness materials. Traditional design is based on a safety-factored yield stress, as explained in section 1.5, employing 'ductile' materials. It is an acceptable design philosophy, particularly in small, well-manufactured components and structures, and relates to keeping the working stresses below line EF in the HP size range of Fig. 10.3 for monotonic loadings. If the article is defective (cracks, bad welds and so on) unexpected failures can occur. But deformation transitions, produced by increased size of components and structures, led to brittle fractures (FN line in Fig. 10.3) in even well-manufactured bodies, and hence to the birth of modern elastic fracture mechanics in order to explain these unexpected failures and to design against them. Furthermore, the ability to perform 'strength-of-materials' elastic calculations for *cracked* bodies (for that is what LEFM is all about as far as a designer is concerned) has permitted the use, in some applications, of high-strength materials whose K_{IC}/σ_y ratios are relatively much lower than the traditional ductile solids of engineering design; use of these materials implies appropriate crack inspection procedures. Of course, an application of fracture mechanics principles is central to design with brittle materials such as glasses and ceramics whose K_{IC}/σ_y ratios are very low indeed.

Design rules for material/geometry/size combinations in the region from the origin to P, and in the region beyond Q, in Fig. 10.3, seem to have been covered therefore, viz. yielding or plastic collapse in the first case and reversible elastic fracture in the other. What of the region PQ in between, in which elastoplastic fracture is possible, i.e. cracking after only limited irreversible flow? This is difficult: if plasticity is not very extensive (just to the left of point Q, say) the J-integral approach may be used (section 4.5.1); alternatively, the COD design curve may be used (section 4.2.3). However, if design concerns the region of the diagram just to the right of P, the solution is not clear, and this is the subject of current research.

One approach to the problem is to *bound* the answer by LEFM and limit analysis. Crack initiation may precede overall yielding or vice versa. Hence LEFM and limit analysis will provide either upper or lower bounds depending on the circumstances, i.e. in Fig. 10.3 line MCFN or line BCD may be reached first, depending on where on the abscissa the particular material/size/cracked geometry combination lies. Ruiz (1982) draws the analogy of the design of a strut with the safe design of a structure containing a flaw: with a strut, the possibilities of either Euler crippling or plastic squashing have to be considered; design against the failure of a structure made of material with intermediate-to-high K_{IC}/σ_y and possessing flaws needs consideration of fracture in the elastic field on the one hand or general yielding on the other. That the possibility of plastic flow may be as important (or more important) a consideration as elastic fracture in these circumstances is borne out by the observations of Soete in 1977 who noted that the safety of structures built with ordinary commercial steels depends more on their plastic preperties than on their fracture toughness (i.e. extensive yielding nearly always precedes fracture). Furthermore, in 1978 Ruiz showed that ductile crack growth in cracked cylindrical shells can in fact be predicted by limit analysis alone, i.e. the predicted pressure to cause widespread yielding in the presence of longitudinal notches or through-thickness slits agrees with reported experimental 'failure' pressures in pipelines and boilers made of medium-to-high K_{IC}/σ_y materials. LEFM and related analyses were not appropriate.

It is clear that deformation transitions are central to the use of extremal elastic and fully plastic solutions for the design of structures made from materials possessing high K_{IC}/σ_y ratios, and also for the assessment of defects in such structures, i.e. what cracks may be tolerated *vis a vis* realistic levels of inspection. Fig. 10.21 is taken from Corran *et al.* (1982) and illustrates the competition between fracture and flow for a centre-cracked panel. Elastic fracture occurs when $\sigma\sqrt{\pi a} = K_{IC}$, where $\sigma \approx X/B(W - 2a)$. Assuming the material to be rigid–perfectly plastic with flow stress σ_y, plastic collapse in plane stress occurs when $X = \sigma_y B(W - 2a)$. These two relations for X (normalised by $\sigma_y BW$) are shown in Fig. 10.21(b), the former for two values of $(K_{IC}/\sigma_y\sqrt{(\pi W)})$. The intersection between the 'ductile failure' line and 'brittle fracture' curve corresponds to the transition between the two modes of failure, governed by ductile (limit analysis) behaviour to the left of the intersection and by brittle (LEFM) behaviour to the right. Note the similarity of the diagram to Fig. 10.3. Thus, in Fig. 10.21(c), a plate with a small defect, at A, will first yield as the load reaches X_A. As the crack grows, reducing the net cross-sectional area, the applied load drops until at C, the crack grows in a brittle (i.e. elastic) manner, a transition that may be repeated if the notch toughness of the material is sufficiently high as in Fig. 10.21(d). In practice K_{IC} corresponds to crack initiation and, depending on the loading condition, strain rate and plastic constraint, failure within the brittle region $C_1 C_2$ of Fig. 10.21(d) may take place as illustrated in Fig. 10.21(e): a crack of initial length (a/W_0) starts growing at D, along the path DE in the transitional region until at E fully plastic behaviour sets in. The various intersection points correspond with Puttick's α equalling $(1/\pi)$ in value – that is, when $\sigma = K_{IC}/\sqrt{\pi a} = \sigma_y, a = (1/\pi)(K_{IC}/\sigma_y)^2$.

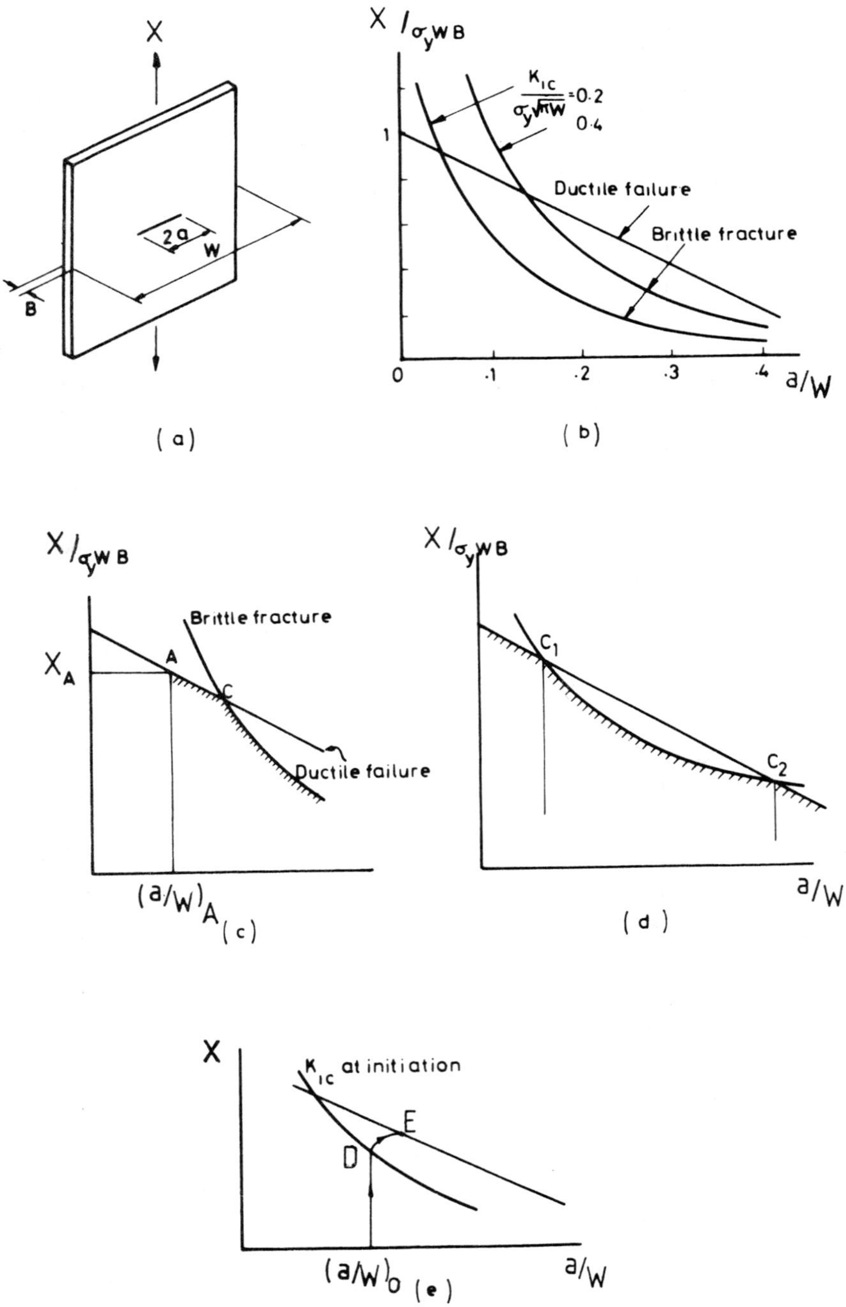

Fig. 10.21 – Failure of a plate with a central crack, (a) general configuration; (b) failure analysis diagram; (c) ductile/brittle transition; (d) ductile/brittle/ductile transition; (e) growth of crack into transitional region (after Corran *et al.* (1982), with permission).

The investigations of Ruiz and co-workers into the elastoplastic behaviour of compact tension specimens of maraging steel ($K/\sigma_y \approx 0.1\sqrt{m}$) and polycarbonate ($K/\sigma_y \approx 0.07\sqrt{m}$) show that although crack initiation is predicted by LEFM, the maximum load is predicted by limit analysis; and that in specimens with relatively large cracks, in which substantial plastic deformation precedes crack initiation, the two approaches coincide. (To account for practical strain hardening in such analyses it is customary to use $(\sigma_{UTS} + \sigma_y)/2$ or some such in place of σ_y). Corran *et al.* (1982) applied the results to model flawed cylindrical pressure vessels made from polycarbonate (PC), using the Folias expression for crack tip stress intensity (equation (9.13)) and the Ruiz 1978 limit analysis. They obtained excellent agreement between theory and experiment – pressure versus crack growth diagrams following the shape of Fig. 10.21(e) when failure was by widespread yielding which occurred with small vessels; of note, a set of large model vessels failed in a brittle manner, which accords with the size effect shown in Fig. 10.3.

Example 10.8 Merkle (1979) investigated the elastic-plastic behaviour of nozzle reinforcements with corner cracks (Fig. 10.22(a)). The nozzle reinforced openings are defined by the following dimensions: R_i = 13.5 in (343 mm); T = 6 in (152 mm); R = 16.5 in (419 mm); r_i = 4.5 in (114 mm); t = 6 in (152 mm); r = 7.5 in (190 mm). Merkle's analysis was based on LEFM, including experimental and numerical data to which correction factors were added to account for steep strain gradients in the nozzle corner regions, localised yielding at the crack tip and plastic constraint intermediate between plane stress and plane strain. Re-analyse the results using LEFM and limit analysis (C. Ruiz).

Ans. Reinforcements around holes in shells are customarily dimensioned according to 'area replacement' rules, whereby the resistance to widespread yielding is made approximately equal to that of the plain, unperforated, shell. In general, however, limit analysis will give an estimate of the weakening factor in the form of a dimensionless failure pressure, p^*, where

$$p^* = \frac{\text{pressure for overall yielding in shell with reinforced opening}}{\text{pressure for overall yielding in plain unperforated shell}}$$

which approaches unity as the cross-sectional area of reinforcement increases and/or as the diameter of the hole decreases. p^* has been calculated by a number of authors employing limit analyses (see Ruiz and Corran (1982) for references) and is expressed as a function of the two parameters t/T and $\rho = r/\sqrt{RT}$ as shown in Fig. 10.22(b) for the particular case of flush-nozzle-reinforced openings. From Fig. 10.22(a), $\rho = 190/\sqrt{419 \times 152} = 0.75$ and $p^* \approx 0.8$ from Fig. 10.22(b).

The presence of a crack of cross-sectional area (ΔA) in a reinforcement of

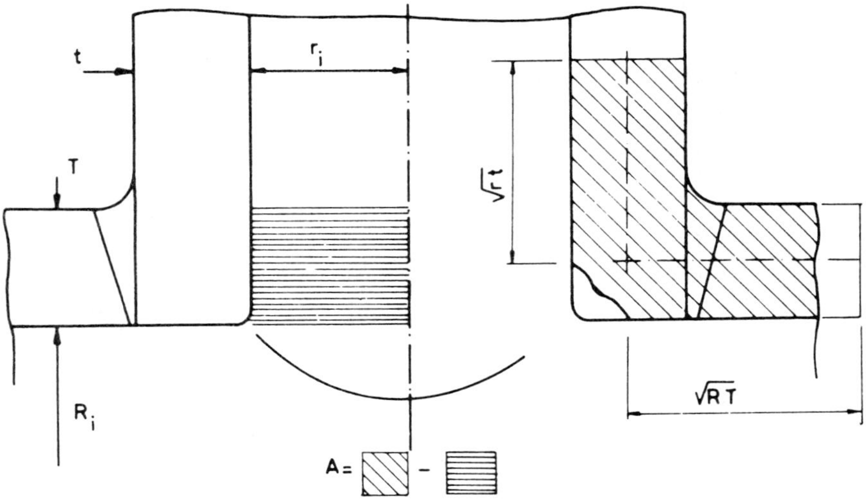

(a)

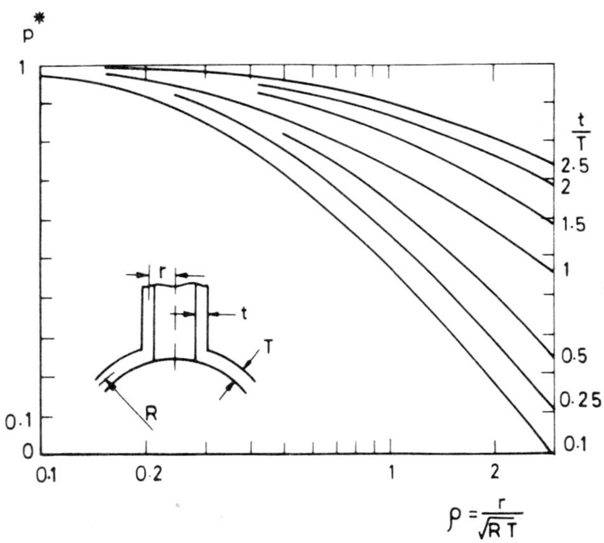

(b)

Fig. 10.22 – (a) Nozzle reinforced opening tested by Merkle (1979). (b) Limit pressure (weakening factor) for nozzle-reinforced openings. (Both reproduced from Ruiz and Corran (1982), with permission.)

area A would cause a further reduction in pressure-withstanding capacity. This may be conveniently expressed as

$$(p^*)_{\text{crack}} = p^* \left(1 - \frac{\Delta A}{A}\right)$$

In Fig. 10.22(a) the cross-sectional area of the crack $\Delta A \cong 1{,}500\ \text{mm}^2$, while $A \cong 45{,}000\ \text{mm}^2$, hence:

$$(p^*)_{\text{crack}} \cong 0.80 \times \left(1 - \frac{1{,}500}{45{,}000}\right) \cong 0.77$$

The pressure for yielding of an unperforated vessel is given by thin cylinder theory (and the Tresca yield criterion) as

$$p_{\text{y}} = \sigma_{\text{y}}\, T/R$$

Taking a flow stress intermediate between σ_{y} and σ_{u} to account for work-hardening, and of value $\sigma_{\text{F}} = 425$ to 533 MPa (Merkle's Vessel V5), the failure pressure calculated from limit analysis is

$$p_{\text{failure}} = 0.77 \times \frac{\sigma_{\text{F}} T}{R_{\text{i}}} = 145 \text{ to } 189 \text{ MPa.}$$

An LEFM analysis requires Y (in $K_{\text{IC}} = \sigma\sqrt{\pi a} \cdot Y$) for cracks in reinforced nozzles. Values for Y have been determined by numerical analysis and by photoelasticity (see Ruiz and Corran (1982) for references). In this particular problem $Y \approx 2.5$. From Merkle's data, the fracture toughness of the material may be estimated to be between 159 and 298 MPa$\sqrt{\text{m}}$ (it is not known with more certainty). Therefore, for brittle cracking caused by the hoop stress σ_θ,

$$\sigma_\theta = \frac{pR}{T} = \frac{K_{\text{IC}}}{2.5\sqrt{\pi a}}$$

which, using $a = 30$ mm, gives

$$\begin{aligned} p &= 0.473\, K_{\text{IC}} \\ &= 75 \text{ to } 141 \text{ MPa} \end{aligned}$$

That is, brittle fracture is predicted rather than plastic collapse. Note, however, that in this sort of problem (which is of great practical importance) there is a deal of uncertainty about the value of K_{IC} and $Y(a/W)$. Note also that Kitching and Zarrabi (1981, 1982) have provided lower bound limit pressures for cylindrical sheels with part-though slots, which suggests collapse rather than fracture in some cases.

Of course, extremal solutions provide only independent estimates for the two limits to the initiation of 'failure', i.e. brittle fracture or wide-spread yielding. (Fracture subsequent to widespread yielding, i.e. the third transition of section 10.1, is not usually considered in this type of structural calculation.) Bounding solutions give no indication of the transitional behaviour, and (when formulated in terms of rigid–perfectly-plastic collapse) do not account for workhardening deformation. Dowling and Townley in 1975 proposed to combine the two extremal solutions, establishing an interaction diagram, which leads to the CEGB 'two-criteria' approach for design (sometimes called the R-6 failure assessment procedure). Plasticity is that associated with a Dugdale zone extending ahead of the crack tip, and (although not actually presented this way originally) degrees of elastoplastic fracture may be visualised to arise from what proportion of a remaining ligament ahead of the crack is taken up by the Dugdale zone. The diagram has axes of

$$\left.\begin{aligned} K_r &= \frac{K_I\,(a,\,X,\,Y(a/W))}{K_{IC}} \\ &\text{and} \\ S_r &= \frac{X}{X_{\text{collapse}}} \end{aligned}\right\} \qquad (10.25)$$

for the ordinate and abscissa respectively. For a given crack geometry, and a given loading condition, the diagram identifies how close the situation is to either elastic fracture, elastoplastic fracture or plastic collapse. The diagram is a useful means, therefore, of assessing the safety margin of a flawed structure. Before describing developments of the diagram and how K_r, S_r may be calculated, or approximated for, in practical structures, the following example gives an *exact* algebraic solution for an elastoplastic fracture R-6 diagram.

Example 10.9 Construct a 'failure assessment' or 'interaction' diagram for the glued cantilever beam problem of section 4.3.3.

Ans. For elastoplastic deformation of the beam coupled with fracture along the glue line (i.e. fracture in the transition region between elastic fracture and plastic collapse), we have from equation (4.20a) $M^2 = (REI\,w/2)\,(3-\xi^2)^2/(3-2\xi)$. The moment when the beam just remains completely elastic, i.e. $\xi = 1$ is $M_{el}^2 = 2REI\,w$. For the purpose of the diagram, the ordinate is M/M_{el} and we see that $(M/M_{el})^2 = (3-\xi^2)^2/4(3-2\xi)$. This expression must be transposed so that the function of ξ becomes a function of M/M_{collapse}. Now $M_{\text{collapse}} = wh^2\sigma_y$, and from equation (4.16) an alternative expression for M_{elpl} is $wh^2\sigma_y\,(3-\xi^2)/3$. Thus $M/M_{\text{collapse}} = (3-\xi^2)/3$ and $\xi^2 = 3(1-M/M_{\text{collapse}})$. Substituting back for ξ, and calling $M/M_{el} = y$ and $M/M_{\text{collapse}} = x$, we have $y^2 = 9x^2/4\,(3-2\sqrt{3(1-x)})$ as the equation of the interaction diagram. Fig. 10.23 gives the plot. Note that $M/M_{el} = 1$ until $M/M_{\text{collapse}} = 2/3$ (for rectangular beams). The interaction curve corresponds with ξ changing as shown. The bounding radial lines for $\xi = 0$ and $\xi = 1$ give Puttick's lower and upper transitions in this problem.

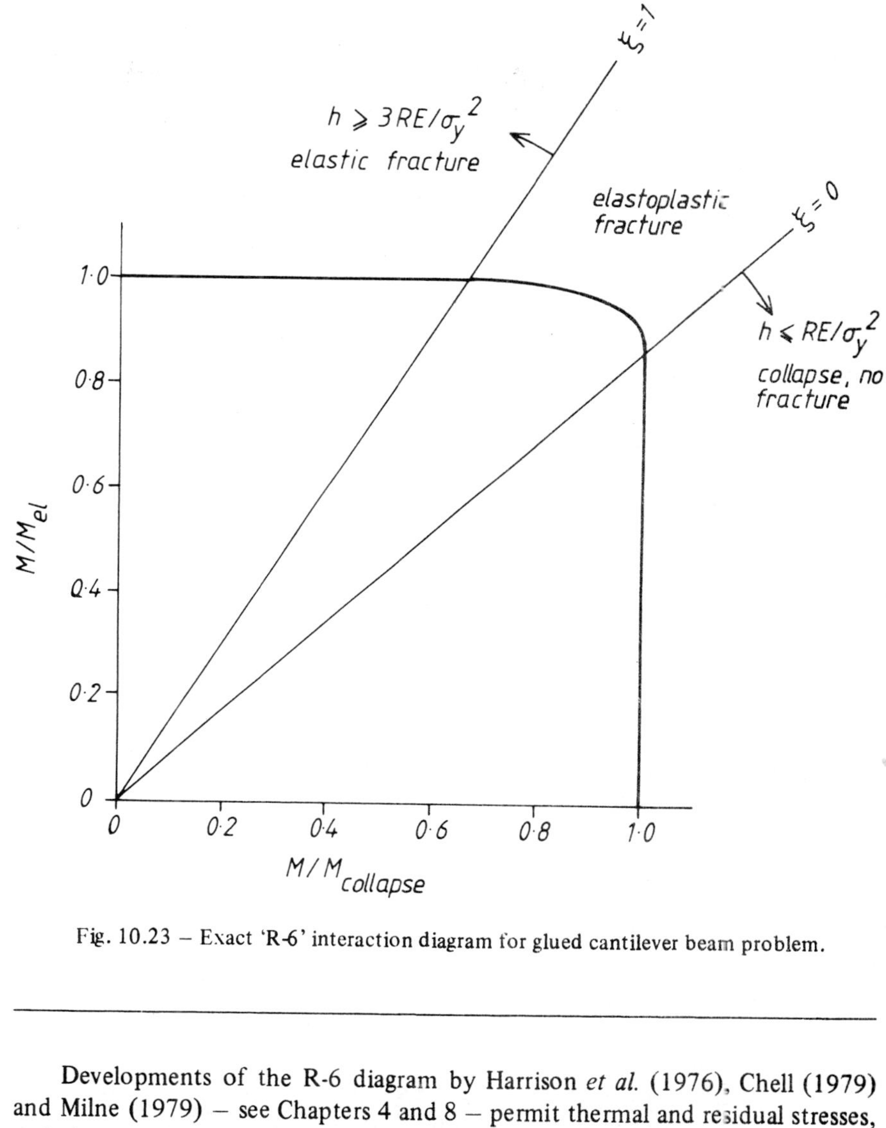

Fig. 10.23 – Exact 'R-6' interaction diagram for glued cantilever beam problem.

Developments of the R-6 diagram by Harrison *et al.* (1976), Chell (1979) and Milne (1979) – see Chapters 4 and 8 – permit thermal and residual stresses, and slow crack growth to be incorporated. The original derivation of the R-6 diagram was based on a plane-stress Dugdale solution for an infinite sheet subjected to uniform tension (since as in Problem 10.25, an algebraic expression could be obtained for the relationship between K_r and S_r, viz.

$$\frac{1}{K_r^2} = \frac{1}{S_r^2}\left[\frac{8}{\pi^2}\ \ln\left(\sec\frac{\pi}{2}\cdot S_r\right)\right]$$

which gives an 'elliptical quadrant' curve shaped like that in Fig. 10.23). It is clear that the effects of material properties and cracked geometry do not appear in this expression, i.e. $Y(a/W) = 1$ always in the definition of K_r in equation

(10.25), and σ_y is constant with no allowance for workhardening. Also, there are questions to be asked about the validity of the algebraic manipulations leading to the K_r–S_r relation given above (see Corran *et al.*, 1982; Problem 10.25).

The EPRI/GE J_R-based finite-element and estimation procedures for elastoplastic fracture now permit R-6 diagrams to be constructed for work-hardening solids in various cracked geometries. The procedure is as follows.

We have from section 4.5.4,

$$J = J^e(a) + J^p(a, N) \tag{4.47 bis}$$

and in quasi-static propagation therefore (Kumar *et al.*, 1981)

$$J_R(\Delta a) = \sigma_y \epsilon_y a \left[\frac{X}{X_y}\right]^2 J^e\left(\frac{a}{W}\right) + \alpha\sigma_y\epsilon_y a \left[\frac{X}{X_y}\right]^{N+1} J^p\left(\frac{a, N}{W}\right) \tag{10.26}$$

where X_y is the initial load for constrained yielding (with $\sigma_y = E\epsilon_y$). Owing to the incorporation of workhardening, and the use of a J_R-based assumption for elastoplastic fracture, the K_r ordinate is defined here as

$$K_r = \frac{K_I}{K_R} = \left[\frac{J^e}{J_R}\right]^{1/2} \tag{10.27}$$

where J^e is the elastic 'driving force' given by the first of equation (4.47). It follows, by manipulation of the expression for J^e/J^R using the right-hand side of equation (10.26) as J_R, that

$$K_r^2 = \frac{S_r^2}{S_r^2 + \alpha H_N S_r^{N+1}} = \frac{1}{1 + \alpha H_N S_r^{N-1}} \tag{10.28}$$

where $H_N = [J^p(a/W, N)]/[J^e(a/W)]$. (If J^e is based on the Irwin-modified crack length rather than the actual starter length, a slightly different expression arises see, for example Kumar *et al.*, 1981).

A variety of R-6 diagrams has been produced using the EPRI method. Fig. 10.24(a) shows the plots for a large centre-cracked plate in plane stress for various values of work hardening index N. We note that for low workhardening rates (N large in the Ramberg–Osgood $\sigma\epsilon$ constitutive relation) the shape of the curve approaches the perfectly-plastic R-6 curve for the same geometry, but that there is always a 'tail' to the diagram where the assessment line approaches the S_r abscissa asymptotically. (This is an artefact of the Ramberg–Osgood equation; the associated implication that the cracked body can support an infinite load arises because the 'third transition' relating to cracking subsequent to widespread yielding is not considered.) Fig. 10.24(b) shows the effect of different (a/W) where this comes into play, in this case for compact tension specimens in plane strain. The material is A533D steel for which $N = 10$ (i.e.

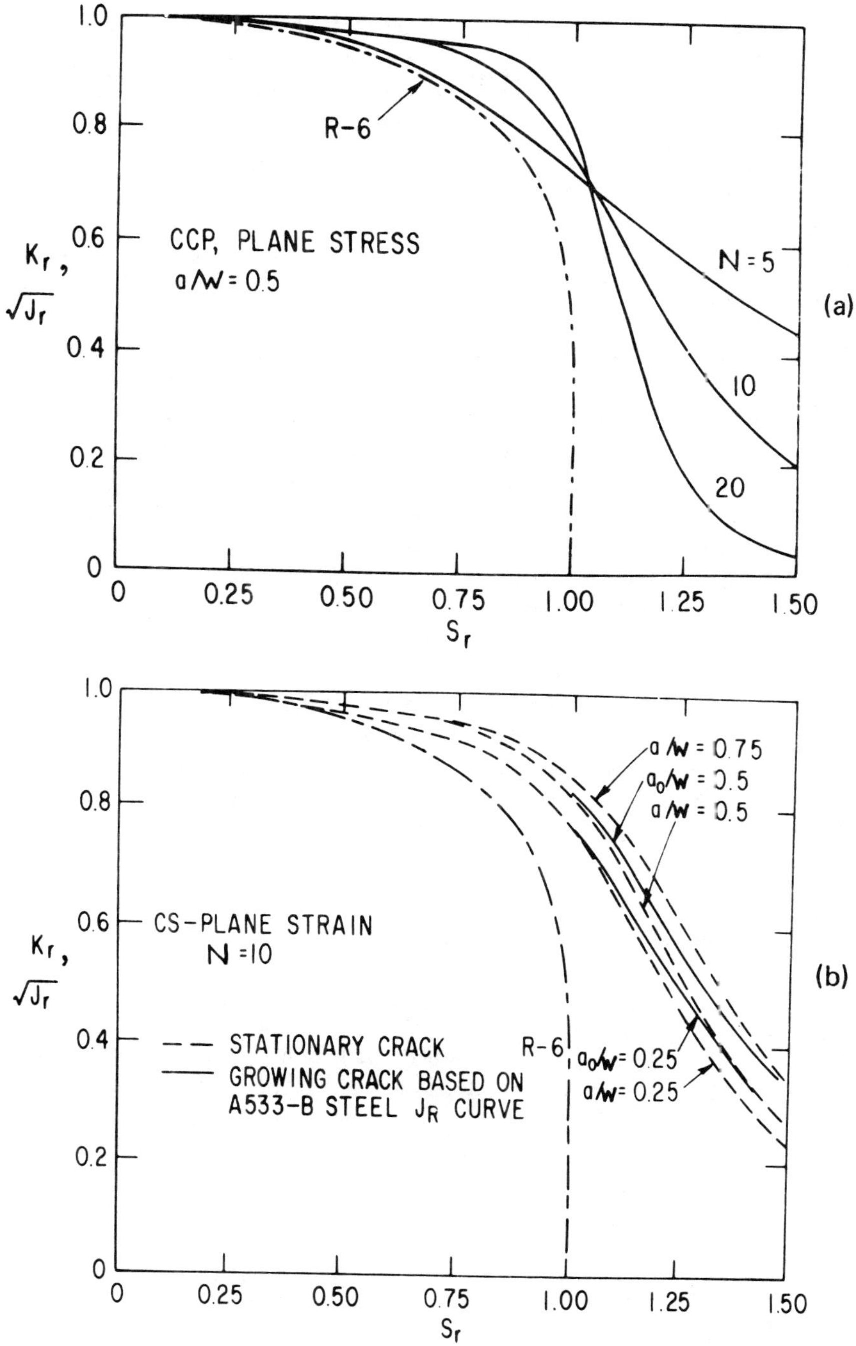

Fig. 10.24 – R-6 diagram for centre-cracked plate, for several values of work-hardening index in Ramberg–Osgood relation, based on EPRI-GE J_R computations. (b) R-6 diagram for plane strain crack growth in compact specimens of A533B steel with initial crack length-to-width (a/W) ratios of 0.25 and 0.50, based on EPRI-GE J_R computations. (Both after Kumar *et al.* (1981), with permission.)

$n = 0.1$ in $\sigma = \sigma_0 \epsilon^n$). Although, again, in this illustration the basic R-6 curve is a lower bound, examples have been produced where the GE/EPRI J_R-based failure assessment curve lies *inside* the R-6 locus (e.g. for circumferentially and axially cracked cyclinders of 304 stainless steel which workhardens considerably).

The failure assessment diagrams are usually constructed for fixed (a/W) rather than for increasing (a/W) as stable crack growth occurs. In Fig. 10.24(b) the dashed lines relate to fixed starting (a/W) values, but the solid lines relate to the grown crack size. Such modified lines are obtained by adjusting H_N in equation (10.28) according to the revised crack length determined from that portion of the complete J_R-curve which is encompassed during slow crack growth prior to instability. (In Fig. 10.23 of course, since R is constant and the cracking moment is constant as the problem is steady-state, the K_r, S_r coordinate is fixed at one point on the assessment diagram for all fracture.) We note that the GE/EPRI calculations are J_R-based (and hence limited to comparatively small amounts of crack growth) and we recall that J_R is a geometry-dependent curve for different materials. J_R should, of course, be properly determined from independent material tests, but there is a slight chicken-and-egg situation for J and J_R in the application of the results. We are not aware of any calculations along these lines done on the basis of the (RdA, dΓ) partition of work input (using, say, the sort of modified J-integral, proposed by Ernst, described in place of the lumped J_R formulation.

Further considerations of work-hardening R-6 diagrams, and experimental validation of the different theoretical bases on which they are constructed, may be found in, for example, Bloom (1983), Chell and Milne (1983).

We note that a different form of two-parameter 'failure' criterion, involving brittle fracture or collapse, was proposed by Newman (1975).

PROBLEMS

10.1 What is a 'semi-brittle' material?

10.2 Discuss how (a) temperature, (b) loading rate and (c) notch sharpness affect the ductile-to-brittle transition in metals and polymers.

10.3 Show that the factor α in equation (10.6) for upper and lower fracture transitions is mainly determined by the proportion of strain energy (as a fraction of the total) associated with tensile stresses.

10.4 Using simple beam theory, determine α for both the lower and upper transitions in notched three-point bend testpieces.

10.5 For the embedded rod problem in Chapter 2 (Problem 2.37) show that the transition from adhesive interfacial fracture in mode II to tensile yielding of the rod is given by $d_{crit} = 8ER/\sigma_Y$.

10.6 Show that the lower transition α for a centre-cracked plate (Table 10.1) is 5.14 if both a correction is made for the plastic zone size on the crack length and the net section stress is limited to 0.8 σ_Y.

10.7 Explain why Auerbach found in 1891 that the critical stress for initiation of a cone crack in glass beneath a ball indenter decreased with increasing diameter of the ball. What physical quantity *is* constant at fracture in such tests?

10.8 Show that after the development of a cone crack, the contact region below a ball indenter will yield when $a = 100\,(ER/\sigma_y^2)$.

10.9 Show that amorphous polymers such as PMMA will be notch-brittle below 40°C, but indentation-ductile for ball indenters of diameters smaller than 14 mm at 40°C (ER/σ_y^2 for PMMA at 40°C $\approx 10^{-3}$ m; at −40°C it is $\approx 10^{-4}$ m).

10.10 Show that a size-dependent ductile-to-brittle transition should be expected between +40°C and −40°C for ball indentations in PMMA.

10.11 American Bureau of Shipping hull plate has $ER/\sigma_y^2 \approx 0.3$ m at 0°C and 0.02 m at −40°C. From these data show that standard Charpy testing should give a ductile-to-brittle transition between 0°C and −40°C, but that the same steel is indentation-ductile throughout the same temperature range.

10.12 SAE 4340 steel exhibits $ER/\sigma_y^2 \approx 7 \times 10^{-3}$ m at room temperature; water-quenched high carbon EN44B steel shows $ER/\sigma_y^2 \approx 5 \times 10^{-5}$ m. Explain why 4340 steel is notch-brittle at room temperature, yet indentation ductile, whereas the quenched high carbon steel is both notch-brittle *and* indentation-brittle for indentation greater than 0.5 mm radius.

10.13 The energy required for comminution seems to be utilised more effectively when large particles are being broken down rather than small. Why is this?

10.14 Derive equation (10.12) for a DCB specimen loaded axially.

10.15 Derive equation (10.19) for the loss in relative optical transmittance T under the action of repeated impacts of water droplets. (Hint: $T \propto 1/a$; $dL/dN = C_1(\sigma\sqrt{a})^n$; take an average stress due to impacts and single effective crack of length a.)

10.16 Derive equation (10.21) using (10.20).

10.17 Why are ceramics and glasses often machined by grinding?

10.18 Why are cracks rarely seen at the tips of tools when cutting ductile metals, even though machining is a process of controlled crack propagation?

10.19 The lower cutting forces of 'free machining' metals are usually explained in terms of improved lubrication. What other factor is probably playing as big a role in the improvement of machinability?

10.20 The unit horsepower for 1015 steel is 0.62 for $w = 0.1$ in and $h = 0.0125$ in; log-plotting of HP_u vs. BHN has a slope of about 0.17. Determine the cutting force as a function of h and w for this steel, assuming $K = 100$ ksi$\sqrt{\text{in}}$ and $\sigma_y = 30$ ksi.

10.21 Discontinuous chips occur at low speed, in the unlubricated cutting of ductile metals, when the shear plane angle ϕ is very small. In leaded brass or magnesium on the other hand, discontinuous chips occur when $\phi \approx 45°$. Explain why this might be so.
(For leaded brass $K_{IC}/\sigma_y \approx 0.13\sqrt{\text{m}}$; for ductile steel $K_{IC}/\sigma_y \approx 0.5\sqrt{\text{m}}$.)

10.22 Show that the ratio of cracking force to shearing force in cutting depends upon $K_C \sin 2\phi/\sigma_y\sqrt{h}$ where ϕ is the shear plane angle. Using this parameter, discuss transitions between continuous chip formation, discontinuous chip formation and brittle chipping in the machining of different solids.

10.23 Prolonged milling of calcium carbonate produces an average particle size of about 1 μm. Plastic yielding, as opposed to cracking, has been observed in compressed calcium carbonate particles about 4 μm in size. Again, calcuim carbonate may be indented by a sharp tool without cracking when the size of indentation is smaller than some 3 μm. Link these observations together via a theory of comminution.

10.24 Explain the observation that ductile swarf, as found in the cutting of metals, is found when glass is machined at depths of cut smaller than about 0.5 μm.

10.25 Using the Burdekin–Stone expression for COD given by equation (4.10), and expanding the ln(sec) term, derive the relation between K_r ($= \sigma\sqrt{\pi a}/K_{IC}$) and S_r ($= \sigma/\sigma_y$) given in section 10.6 for the Dowling–Townley interaction diagram. For what range of σ/σ_y is the expansion valid? For what range of σ/σ_y is the expression used?

10.26 In design with solids having large K_{IC}/σ_y ratios, where plastic collapse may occur before fracture, which asset is of more value (a) high toughness or (b) high workhardening index?

10.27 Construct a fracture stress – size of body diagram for a given cracked geometry by normalising appropriate formulae for fracture stresses in the elastic, elastoplastic and plastic ranges by the theoretical elastic fracture stress $\sigma_{cr} = \sqrt{2E\gamma/a_0}$ (where a_0 is the interatomic spacing, Example 3.2) and plotting against $a/a_0 = (a/W)\,(W/a_0)$ with a the crack length and W the size of the body.

10.28 Desirable criteria for forgeability amd machinability are listed as (T. Childs/SERC)

forgeability	machinability
low specific loads	low cutting forces
resistance to cracking	chip flow and ease of disposal
ease of die filling	surface integrity
surface integrity	tool life
die life	consistency
consistency	

Discuss whether these criteria are mutually exclusive and whether it is possible to have easily machinable forging solids or easily forgeable free-cutting solids.

REFERENCES

Alexander, J. M. and Brewer, R. C. (1963) *Manufacturing Properties of Materials.* London: van Nostrand.

Archard, J. F. (1952) *Research* **5**, 395; (1953) *J. Appl. Phys.* **24**, 981.

Archard, J. F. (1974) *Tribol. Int.* **7**, 213.

ASME (1952) *Manual on Cutting Mets.* 2nd edn. New York: ASME.

Atkins, A. G. (1974) *Int. J. Prod. Res.*, **12**, 263.

Atkins, A. G. (1974) *Proc. NAMRC-II, Madison Wisconsin.* Dearborn: SME, p. 398.

Atkins, A. G. (1979) *Proc. 4th Tewksbury Symp. on Fracture, Melbourne,* Australia, D. S. Mansell and G. H. Varsey (eds.) Melbourne: University Press, paper 17.

Atkins, A. G. (1980) *Wear,* **61**, 183.

Atkins, A. G. (1982) *Metal Science* **16**, 127.

Atkins, A. G. and Omar, M. K. (1985) *Wear of Materials* p. 405 New York: ASME.

Atkins, A. G., Omar, M. K. and Lancaster, J. K. (1984) *J. Mater. Sci. Letters* **3**, 779.

Auerbach, A. J. (1891) *Pogg. Ann. Phys.* **43**, 60.

Blok, H. (1952) *Proc. Roy. Soc.* **A212**, 480.

Bloom, J. M (1983) STP No. 803,11-206. Philadelphia: ASTM.

Boston, O. W. (1951) *Metal Processing,* 2nd edn. New York: Wiley.

Bowden, F. P. and Tabor, D. (1964) *The Friction and Lubrication of Solids,* Part II. Oxford: Clarendon Press.

British Standards Institution, 1980, Document PD 6493.

Buttery, T. C. and Archard, J. F. (1970–71) *Proc. I. Mech. E.* **185**, 537.

Chell, G. G. and Milne, I. (1983) STP No. 803,11-179. Philadelphia: ASTM.

Christiansen, A. W., Baer, E. and Radcliffe, S. V. (1971) *Phil Mag.* **24**, 451.

Chukwejekwu, S. E. and Ruiz, C. (1969) *Trans. ASME J. Engr. Industry* **91B**, 710.

Corran, R. S. J., Davies, P. H. and Ruiz, C. (1982) *Engr. Fract. Mech.* **16** (4), 585.

Datsko, J. and Henkin, A. (1964) *Trans. SAE,* 486.

Dowling, A. R. and Townley, C. H. A. (1975) *Int. J. Pres. Ves. & Piping,* **3**, 77.

Evans, D. C. and Lancaster, J. K. (1976) Royal Aircraft Establishment Tech. Rept. 76099 (see *Proc. 3rd Leeds–Lyons Tribology Symp. 1978.* London: I. Mech. E., p. 47).

Fleming, J. R. and Suh, N. P. (1977) *Wear* **44**, 57.

Frost, H. J. and Ashby, M. F. (1982) *Deformation – Mechanism Maps.* Oxford: Pergamon.

Goodall, I. W. and Griffiths, J. E. (1982) *I.J.M.S.* **24** (10), 635.

Griffiths, J. R. and Oates, G. (1969) *ICF-2, Brighton,* Paper 19.

Halliday, J. S. (1955) *Proc. Inst. Mech. Engrs.* **169**, 777.

Howes, M. G. J. W. (1968) Ph. D. Dissertation, University of Reading.

Hutchings, I. M., Winter, R. E. and Field, J. E. (1976) *Proc. Roy. Soc.* **A348**, 379.

Johnson, K. L. (1970–71) *Proc. Inst. Mech. Engrs.* **185**, 43/71.
Kelly, A. (1966) *Strong Solids.* Oxford: Clarendon Press.
Kendall, K. (1978) *Nature,* **272**, 710.
Kenny, P. and Campbell, J. D. (1968) *Progress in Materials Science* **13**, 135.
Kitching, R. and Zarrabi, K. (1981) *Int. J. Mech. Sci.* **23**, 31.
Kitching, R. and Zarrabi, K. (1982) *Int. J. Pres. Ves. & Piping* **10**, 235.
Krafft, J. M. (1961) *Mater. Res and Standards* **1**, 11.
Kruschov, M. M. and Babichev, M. A. (1954) *Vestn. Mashinostr.* **34** (9), 3.
Kumar, V., German, M. D. and Shih, C. F. (1981) *EPRI/GE Plastic Fracture Handbook,* Palo Alto: EPRI.
Lunn, B. and Christensen, J. V. (1970) *J. Inst. Metals* **98**, 164.
Mallock, A. (1981) *Proc. Roy. Soc.* **33**, 127.
Mears, D. R., Pae, K. D. and Sauer, J. A. (1969) *J. Appl. Phys.* **40**, 4229.
Merkle, J. G. (1979) STP 668,675. Philadelphia: ASTM.
Mulhearn, T. O. and Samuels, L. E. (1962) *Wear* **5**, 478.
Murray, M. J., Watson, J. D. and Mutton, P. J. (1979) *'Wear of Materials' Conference.* New York: ASME, p. 257.
Mutton, P. J. and Watson, J. D. (1978) *Wear* **48**, 385.
Nadai, A. (1950) *Theory of Flow and Fracture of Solids,* 2nd edn. New York: McGraw-Hill.
Newman, J. C. (1975) NASA Rept. No. TMX 72709.
Omar, M. K. (1984) Ph. D. dissertation, University of Reading.
Puttick, K. E. (1978) *J. Phys. D. (Appl. Phys.)* **11**, 595; *ibid.* L69.
Puttick, K. E. (1979) *J. Phys. D. Letters* **12**, L19.
Puttick, K. E., Shahid, M. A. and Hosseini, M. M. (1979) *J. Phys. D.* **12**, 195.
Puttick, K. E., Smith, L. S. A. and Miller, L. E. (1977) *J. Phys. D.* **10**, 617.
Puttick, K. E. (1980) *J. Phys. D.* **13**, 2249.
Rabinowitz, S., Ward, I. M. and Parry, J. S. C. (1970) *J. Mater. Sci.* **5**, 29.
Rivett, R. M. (1979) *Welding Inst. Res. Bull* **20**, 235.
Rosenhain, W. and Sturney, A. C. (1925) *Proc. Inst. Mech. Engrs.* **1**, 141.
Rowe, G. W. and Spick, P. (1967) *Trans. ASME* **89**, 530.
Rowe, G. W. and Wetton, A. G. (1965) *Wear* **8**, 448.
Royal Society (1984) Report of Working Group on Subcritical Crack Growth (Royal Society British National Committee on Theoretical and Applied Mechanics).
Ruiz, C. (1978) *Int. J. Mech. Sci.* **20**, 277.
Ruiz, C. (1984) Private communication.
Ruiz, C. and Corran, R. S. J. (1982) *Int. J. Pres. Ves & Piping* **10**, 361.
Samuels, L. E. (1978) *Scientific American,* 132.
Sedricks, A. J. and Mulhearn, T. O. (1964) *Wear* **7**, 451.
Selwood, A. (1961) *Wear* **4**, 311.
Smith, R. A. (1980) in *Fracture and Fatigue* J. C. Radon (ed.) 3rd European Colloquium on Fracture (London), p. 49.
Soete, W. 1977 Proceedings ICF – 4 Waterloo Canada, Pergamon.
Springer, G. S. (1975) *Erosion by Liquid Impact.* New York: Wiley.
Swain, M. V. and Hagan, J. T. (1976) *J. Phys. D.* **9**, 2201.

Trent, J. S., Moet, A. Y., Miles, M. J. and Baer, E. (1978) *Polymer Engr. Sci.* **18**, 1235.

Truss, R. W., Duckett, R. A. and Ward, I. M. (1981) *J. Mater. Sci.* **16**, 1689.

Williams, J. A. (1975) Ph. D. dissertation, University of Cambridge.

Appendices

Appendix 1
Toughness equations and stability conditions for various specimen geometries

No.	Crack geometry	Toughness equation	Displacement-controlled: Stability condition	Displacement-controlled: Experimental result	Load-controlled: Stability condition	Load-controlled: Experimental result
1		$M^2 = REIt$	$\frac{1}{R}\frac{dR}{da} \geqslant \frac{-2}{a}$	Stable	$\frac{1}{R}\frac{dR}{da} \geqslant 0$	Stable
2	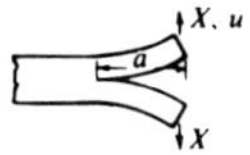	$M^2 = REIt$ $M = Xa$	$\frac{1}{R}\frac{dR}{da} \geqslant \frac{-4}{a}$	Stable	$\frac{1}{R}\frac{dR}{da} \geqslant \frac{2}{a}$	Unstable
3	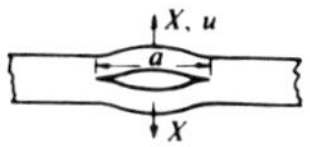	$M^2 = REIt$ $M = Xa/8$	$\frac{1}{R}\frac{dR}{da} \geqslant \frac{-4}{a}$	Stable	$\frac{1}{R}\frac{dR}{da} \geqslant \frac{2}{a}$	Unstable
4	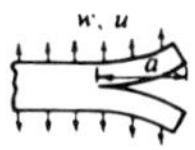	$M^2 = REIt$ $M = \frac{1}{2}wa^2$	$\frac{1}{R}\frac{dR}{da} \geqslant \frac{-6}{a}$	Stable	$\frac{1}{R}\frac{dR}{da} \geqslant \frac{4}{a}$	Unstable
5	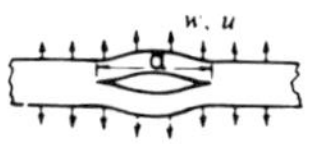	$M^2 = REIt$ $M = \frac{1}{12}wa^2$	$\frac{1}{R}\frac{dR}{da} \geqslant \frac{-6}{a}$	Stable	$\frac{1}{R}\frac{dR}{da} \geqslant \frac{4}{a}$	Unstable
6	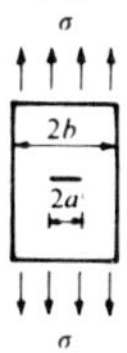	$E^*R\,\sigma^2 2b \tan\left(\frac{\pi a}{2b}\right)$	$\frac{1}{R}\frac{dR}{da} \geqslant \frac{\pi}{b} csc\left(\frac{\pi a}{b}\right)$	Unstable	$\frac{1}{R}\frac{dR}{da} \geqslant \frac{\pi}{b} csc\left(\frac{\pi a}{b}\right)$	Unstable
7	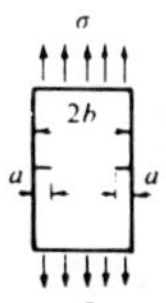	$E^*R = \sigma^2 2b\left[\tan\left(\frac{\pi a}{2b}\right) + 0.1 \sin\left(\frac{\pi a}{b}\right)\right]$	$\frac{1}{R}\frac{dR}{da} \geqslant \frac{\pi}{b} \times \left[\frac{1/2 \sec^2(\pi a/2b) + 0.1 \cos(\pi a/b)}{\tan(\pi a/2b) + 0.1 \sin(\pi a/b)}\right]$	Unstable	Same as column 4	Unstable

8	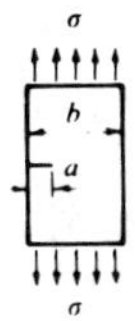	$E^*R = \sigma^2 b[7.59\,(a/b) - 32(a/b)^2 + 117(a/b)^3]$	$\frac{1}{R}\frac{dR}{da} \geqslant \frac{1}{b} \times \left[\frac{7.59 - 64(a/b) + 351(a/b)^2}{7.59(a/b) - 32(a/b)^2 + 117(a/b)^3}\right]$	Unstable	Same as column 4	Unstable
9	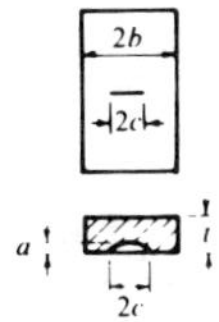	$E^*R = \frac{1.2\pi\sigma^2 a}{\Phi^2 - 0.212(\sigma/\sigma_{ys})^2}$ $\Phi = \int_0^{\pi/2} \sqrt{\left(1 - \frac{c^2 - a^2}{c^2}\sin^2\theta\right)}\, d\theta$ $a \leqslant t/2$ and $c \leqslant b/3$	$\frac{1}{R}\frac{dR}{da} \geqslant \frac{1}{a}$	Unstable	$\frac{1}{R}\frac{dR}{da} \geqslant \frac{1}{a}$	Unstable
10	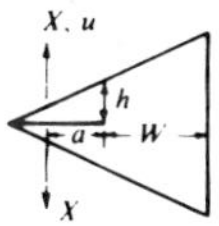	$E^*R = \frac{4X^2}{t} m$ $m = \frac{1}{H} + \frac{3a^2}{H^2} = \text{constant}$	$\frac{1}{R}\frac{dR}{da} \geqslant \frac{-2}{a}$	Stable	$\frac{1}{R}\frac{dR}{da} \geqslant 0$	Stable
11	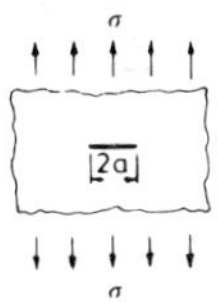	$E^*R = \sigma^2 \pi a$	$\frac{1}{R}\frac{dR}{da} \geqslant \frac{1}{a}$	Unstable	$\frac{1}{R}\frac{dR}{da} \geqslant \frac{1}{a}$	Unstable
12	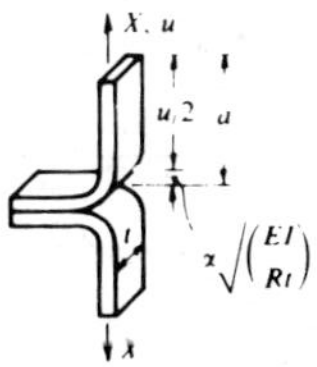	$R = 2X/t$	$\frac{1}{R}\frac{dR}{da} \geqslant \frac{-2}{(a - u/2)}$ $(a - u/2) \leqslant a$	Stable	$\frac{1}{R}\frac{dR}{da} \geqslant 0$	Stable
13	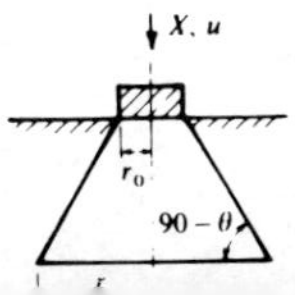	$ER = \frac{K(\nu)X^2}{r^3}$ $K(\nu)$ = function of Poisson's ratio	$\frac{1}{R}\frac{dR}{da} \geqslant \frac{-1}{r}\left[\frac{3(r/r_0) - 1}{(r/r_0) - 1}\right]$	Stable	$\frac{1}{R}\frac{dR}{da} \geqslant \frac{-3}{r}$	Unstable

Appendix 1 *continued*

Toughness equations and stability conditions for various specimen geometries

No.	Crack geometry	Toughness equation	Displacement-controlled: Stability condition	Displacement-controlled: Experimental result	Load-controlled: Stability condition	Load-controlled: Experimental result
14	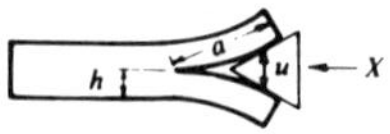	$R = \frac{3}{4}\frac{Eu^2H^3}{a^4[1+0.64H/a]^4}$	$\frac{1}{R}\frac{dR}{da} \geqslant \frac{-4}{a\{1+0.64(H/a)\}}$	Stable	$\frac{1}{R}\frac{dR}{da} \geqslant \frac{2}{a\{1+0.64(H/a)\}}$	Unstable
15	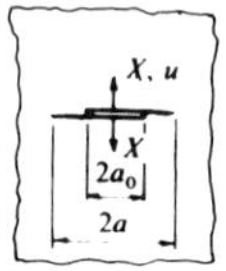	$E^*R = \frac{P^2}{\pi a}$	$\frac{1}{R}\frac{dR}{da} \geqslant \frac{-1}{a}\left[1+\frac{2}{\ln(a/a_0)}\right]$	Stable	$\frac{1}{R}\frac{dR}{da} \geqslant \frac{-1}{a}$	Stable
16	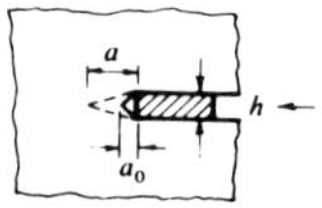	$R = \frac{E^*h^2}{2\pi a}$	$\frac{1}{R}\frac{dR}{da} \geqslant \frac{-1}{a}\left[1+\frac{2}{\ln(a/a_0)}\right]$	Stable	$\frac{1}{R}\frac{dR}{da} \not\geqslant \frac{-1}{a}$	Stable
17	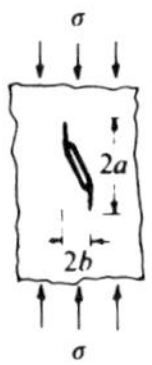	$E^*R = \frac{\sigma^2\pi b}{8}$	$\frac{1}{R}\frac{dR}{da} \geqslant \frac{-2}{a}$	Stable	$\frac{1}{R}\frac{dR}{da} \geqslant 0$	Stable
18	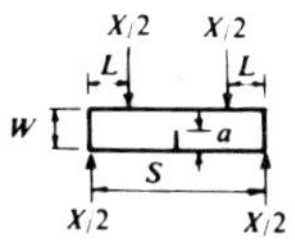	$E^*R = \frac{36M^2a}{t^2W^4}Y^2$ $M = XL/2$ $Y = 1.992 - 2.468(a/W)$ $+ 12.97(a/W)^2 - 23.17(a/W)^3$ $+ 24.80(a/W)^4$	$\frac{1}{R}\frac{dR}{da} \geqslant \frac{n_u}{a}$ n_u: see footnote	Stable for large a/W and unstable for small a/W	$\frac{1}{R}\frac{dR}{da} \geqslant \frac{n_X}{a}$ n_X is positive	Unstable

19	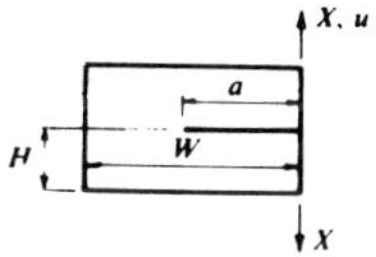 	$E^*R = \frac{X^2 a}{t^2 W^2} Y^2$ Y: see footnote	$\frac{1}{R}\frac{dR}{da} \geqslant \frac{n_u}{a}$ n_u: see footnote	Stable	$\frac{1}{R}\frac{dR}{da} \geqslant \frac{n_X}{a}$ n_X is positive	Unstable
20	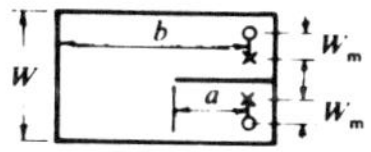 	$E^*R = X^2 W_m^2 \left[\frac{3(1+\nu)}{Wt^4}\right]$	$\frac{1}{R}\frac{dR}{da} \geqslant \frac{-2}{a}$	Stable	$\frac{1}{R}\frac{dR}{da} \geqslant 0$	Stable
21	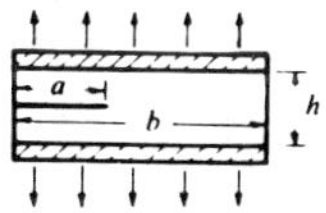 	$E^*R = \frac{\sigma^2 h}{2}$	$\frac{1}{R}\frac{dR}{da} \geqslant \frac{-2}{a}$	Stable	$\frac{1}{R}\frac{dR}{da} \geqslant 0$	Stable
22	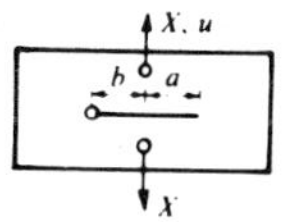 	$E^*R = \frac{X^2 b^4 a^2}{2It(b+a)^4}$	$\frac{1}{R}\frac{dR}{da} \geqslant \frac{-4}{a} + \frac{2}{(b+a)}$	Stable	$\frac{1}{R}\frac{dR}{da} \geqslant \frac{-2}{a^2}\left(\frac{a-b}{a+b}\right)$	Stable if $a \geqslant b$
23	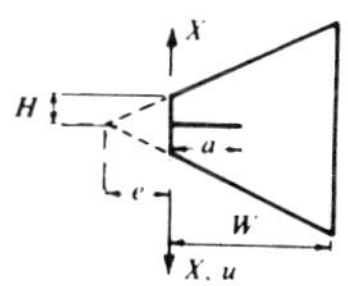 	E^*R expressions: see footnote	n_u: see footnote $\frac{1}{R}\frac{dR}{da} \geqslant \frac{n_u}{a}$	Stable	n_X: see footnote $\frac{1}{R}\frac{dR}{da} \geqslant \frac{n_u}{a}$	Unstable
24	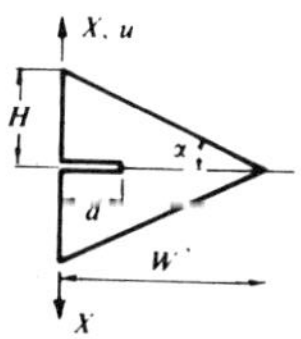 	E^*R expression: see footnote	$\frac{1}{R}\frac{dR}{da} \geqslant \frac{n_u}{a}$ n_u: see footnote	Stable	$\frac{1}{R}\frac{dR}{da} \geqslant \frac{n_X}{a}$ n_X is positive	Unstable

Footnote:
For further general information see Mai, Y. W. and Atkins, A. G. (1980) *J. Strain Analysis* **15**, 63, from which this appendix is adapted; for particular information on testpiece number 18 see Clausing, D. P. (1969) *Int. J. Fract.* **5**, 211; for no. 19 see Clausing loc. cit. and Srawley, J. E. and Brown, W. F. (1966) ASTM STP No. 410, 1; for nos. 23 and 24 see Mai, Y. W. Atkins A. G. and Caddell, R. M. (1975) *Int. J. Fract.* **11**, 939.

Appendix 2
Representative Toughness Values for Various Materials

(Caution: values can depend markedly on rate, temperature and environment).

Material	R (kJ/m^2)	E (GPa)	K_{IC} ($MPa\sqrt{m}$)	σ_y (MPa)	K_{IC}/σ_y ($\sqrt{m}$)	R/σ_y (μm)
Steels						
Medium carbon	14	210	54	260	0.21	54
Pressure vessel (ASTM 533B, Q + T)	206	210	208	470	0.44	438
High strength alloy	45	210	98	1,460	0.07	31
Maraging steel	28	210	76	1,800	0.04	16
AFC 77 stainless	33	210	83	1,530	0.05	22
AISI M^2 tool steel (hardened and tempered)	2	210	22	2,600	0.008	0.8
Aluminium alloys						
2024-T8	10	72	27	420	0.06	24
7075-T6	13	72	30	540	0.05	24
7178-T6	7	72	23	560	0.04	13
Titanium alloys						
Ti-6Al-4V	49	110	73	1060	0.07	46
Other metals						
Copper	100	119	110	60	1.83	1,670
α-Brass	60	105	80	200	0.40	300
Cast iron	0.2–3	130–180	6–20	220–1030	0.03	1–3
Lead	30	14	20	13	1.5	2,290
Ceramics						
Magnesia	0.07	315	5	100[a]	0.005	0.7
Boron carbide	0.01	500	6	5,000	0.001	0.002
Zirconia (PSZ)	0.02	170	7	1,000	0.007	0.02
Silicon nitride	0.10	320	4–5	690	0.01	0.14
Silicon carbide	0.05	450	3	660	0.005	0.08
Electrical porcelains	0.01	70	1	70[a]	0.02	0.14
Alumina	0.10	371	4	560[a]	0.01	0.18
Bulk soda glass	0.01	70	0.8	70[a]	0.005	0.14
Tungsten carbide	0.30	560	13	1,680[a]	0.01	0.18
Titanium carbide	0.11	446	7	500[a]	0.01	0.22
Westerly granite	2.00	40	18	140	0.13	0.01
Cement paste	0.01	20	0.5	5[a]	0.10	2
Mortar	0.06	30	1.3	4[a]	0.33	15

Appendix 2 – *continued*
Representative Toughness Values for Various Materials

(Caution: values can depend markedly on rate, temperature and environment.)

Material	R (kJ/m^2)	E (GPa)	K_{IC} (MPa$\sqrt{m}$)	σ_y (MPa)	K_{IC}/σ_y ($\sqrt{m}$)	R/σ_y (μm)
Polymers						
Polycarbonate (PC)	0.4–1	2.6	1–2.6	60	0.02–0.04	67–17
Polyvinylchloride (PVC)	1.4	2.5	1.8	60	0.33	23
Nylon	2.4	2~4	3.0	60	0.35	33~66
Polypropylene (PP)	8	1	3.0	20	0.15	400
High density polyethylene (HPE)	6–7	0.7	2.0	20	0 10	300–500
Polymethyl-methacrylate (PMMA)	0.64	3.5	1.50	60	0.03	10.7
Polystyrene (PS)	0.33	3	1.00	40	0.03	8.3
High impact polystyrene (HIPS)	1.00	2.5	1.60	20	0.08	10
Teflon (PTFE)	8.0	1	3.0	10	0.3	0.8
Acrylonitrile-butadiene-styrene (ABS)	3.0	2	2.5	40	0.06	75
Polyuretane Foam (80 kg/m^3 density)	0.06	0.01	0.03	0.5	0.05	0.12
Composites[b] (volume fractions ca. 40%)						
GFRP (cracks ⊥ fibres)	≮5	40	–	1100	–	4.5
CFRF (cracks ⊥ fibres)	<80	150	–	1200	–	67
CFRP (cracks // fibres)	0.2	15	–	300	–	0.7
Boron–epoxy (cracks ⊥ fibres)	40	160	–	1400	–	29
Boron–epoxy (cracks ⊥ fibres) (intermittent bond, v_f = 25%)	200	95		600		333
Concrete	0.08	30	<1.5	4 (tension)	0.4	20
WC–Co	1.7	100	13	300	0.04	567
Biological materials						
Woods						
Douglas fir (cracks // grain)	0.03	10–16	–	50	–	60
(cracks ⊥ grain)	0.2	1	–		–	
Teak (cracks ⊥ grain)	12	10–16	–	100	–	120
Papers						
Newsprint	6	4.8	–	26.3	–	224
Writing paper	14	5.8	–	32.6	–	442
Tracing paper	26	7.7	–	47.5	–	547

Appendix 2 – *continued*
Representative Toughness Values for Various Materials

(Caution: values can depend markedly on rate, temperature and environment.)

Material	R (kJ/m^2)	E (GPa)	K_{IC} (MPa$\sqrt{m}$)	σ_y (MPa)	K_{IC}/σ_y ($\sqrt{m}$)	R/σ_y (μm)
Kraft paper (unbleached)	21	5.4	–	37.3	–	563
Bone	2	13.5				
Antler	7	7.4				
Teeth						
dentine ⊥	0.3	12				
dentine //	0.5	50				
enamel ⊥	0.2	50				
enamel //	0.01	50				
Grass	0.4	–	–	15	–	3
Carrot	0.20					
Pig thoracic aorta	1.0	0.001				
Rabbit skin	20	100	–	5–15	–	2,000
Apple	6	2				
Rat skin	14	0.006				
Anemone (Mesogloea)	1.4	170×10^{-6}				
Potato	0.2					
Triatomata cuticle	1.2	0.05				
Meat (along perimysial tracts)	0.5~1.0					
Seaweed (*Laminaria digitata*)						
(across width)	0.45					
(along length)	0.001					

Representative fatigue crack growth data may be found in Tables 7.1 and 7.3 in Chapter 7. Additional fracture toughness data, including both fatigue and slow crack growth equations, for other materials are available in: (1) 'A Computerised Fracture Mechanics Data Base for Oxide Glasses', S. W. Freiman, T. L. Baker and J. B. Wachtman, Jr; (2) 'Linear Elastic Fracture Mechanics Parameters of Concrete', R. N. Swamy; and (3) 'A Compendium of Sources of Fracture Toughness and Fatigue Crack Growth Data for Metallic Alloys – Pts I & II', C. M. Hudson and S. K. Seward.

(1) NBS Technical Note 1212, US Dept. of Commerce, June 1975.
(2) In 'Fracture Mechanics of Concrete', (F. Wittmann, ed.), Oxford: Elsevier, pp. 411–462, 1983.
(3) Int. J. Fract., *14* (1978) R151–R184; *20* (1982) R59–R117.

[a] Fracture bend strengths only. Otherwise in this section from indentation hardness.

[b] In the case of anisotropic solids, where E varies with orientation and shear modulus can vary independently, care has to be taken in conversion between R and K_{IC}. K-formulae for continua displaying certain types of anisotropy are available (see comments in section 3.5.2) but their use for fibre-reinforced solids, say, implies certain levels of interfacial bonding which may not occur in practice. Furthermore, in those highly extensible biological materials displaying a J-shaped $\sigma\epsilon$ curve, small-strain linear elastic theory (upon which $K^2 = E^*R$ for isotropic continua is based) no longer applies.

Appendix 3

Relations between Stresses and Strains in Plasticity Incremental and Deformation Theories

Textbooks, such as those referenced at the end of this appendix, show by work arguments that the 'effective stress' $\bar{\sigma}$ and 'effective strain' $\bar{\epsilon}^{\mathrm{P}}$ for a Maxwell–von Mises solid, experiencing general plastic flow $(\sigma_{ij}, \epsilon_{ij})$, are defined as

$$\bar{\sigma} = (1/\sqrt{2})\ [(\sigma_x - \sigma_y)^2 + (\sigma_y - \sigma_z)^2 + (\sigma_z - \sigma_x)^2 + 6(\tau_{xy}^2 + \tau_{yz}^2 + \tau_{zx}^2)]^{1/2} \tag{A3.1}$$

and

$$d\bar{\epsilon}^{\mathrm{P}} = (\sqrt{2})/3\ [(d\epsilon_x^{\mathrm{P}} - d\epsilon_y^{\mathrm{P}})^2 + (d\epsilon_y^{\mathrm{P}} - d\epsilon_z^{\mathrm{P}})^2 + (d\epsilon_z^{\mathrm{P}} - d\epsilon_x^{\mathrm{P}})^2 + 6(d\epsilon_{xy}^{\mathrm{P}2} + d\epsilon_{zx}^{\mathrm{P}2} + d\epsilon_{zx}^{\mathrm{P}2})]^{1/2} \tag{A3.2}$$

with $d\epsilon_{ij}^{\mathrm{P}}$ the plastic strain *increment*.

The relations between strain increments and stresses are given by the 'flow rules'. Textbooks explain that all plasticity problems are really incremental in nature, and plasticity flow rules relate strain *increments* to the current full state of stress (not to stress increments) i.e. plastic flow is path-dependent. In elasticity, in contrast, there is unique path-independent one-to-one relation between total strain and total stress. The Prandtl–Reuss set of equations for incremental elastoplastic flow include

$$d\epsilon_x = \frac{1}{E}\ [d\sigma_x - \nu(d\sigma_y + d\sigma_z)] + \frac{d\bar{\epsilon}^{\mathrm{P}}}{\bar{\sigma}}\ [\sigma_x - \tfrac{1}{2}(\sigma_y + \sigma_z)] \tag{A3.3}$$

and

$$d\gamma_{yz} = \frac{d\tau_{yz}}{2G} + \frac{3d\bar{\epsilon}^{\mathrm{P}}}{2\bar{\sigma}}\ \tau_{yz} \tag{A3.4}$$

where ϵ_x, γ_{yz} are total strains (elastic plus plastic), σ_x etc. are normal stresses, τ_{yz} etc. shear stresses, E is Young's modulus, and ν Poisson's ratio. Similarly for $d\epsilon_y$, $d\epsilon_z$, $d\gamma_{xy}$ and $d\gamma_{zx}$. These equations should be integrated over the whole history of loading (and perhaps localised unloading) to obtain the final components of strain.

Hencky suggested that the total strains themselves (not their increments) are functions of the current stress, i.e.

$$\epsilon_x = \frac{1}{E}\ [\sigma_x - \nu(\sigma_y + \sigma_z)] + \frac{\bar{\epsilon}^{\mathrm{P}}}{\bar{\sigma}}\ [\sigma_x - \tfrac{1}{2}(\sigma_y + \sigma_z)] \tag{A3.5}$$

and so on. This set of equations uniquely relates stress and strain, or load and displacement, along a non-linear path (non-linearity coming in through the $\bar{\epsilon}^{\mathrm{P}}/\bar{\sigma}$ term which varies under strain). The Hencky relations are path-independent, and the outward response is the same as a non-linear elastic body appearing to have variable compliance. There is clearly a similarity between the flow rules and Castigliano relations derived from energy methods for cracked and uncracked bodies. It is clear that in the special case of proportional loading, (where the stress or strain components keep constant ratios), the incremental Prandtl–Reuss theory reduces to the Hencky deformation theory. In general, however, they produce different results.

Incremental theory is the 'correct' theory, as the Hencky relation implies that once the stress at a point is defined, the total strain is immediately defined – *whatever the previous history*. This is clearly untrue for plastic deformation during loading and is certainly untrue the moment any unloading occurs.

In connexion with calculations, particularly of Chapter 5, the rigid–plastic continuum plasticity route for the determination of stresses from strain increments (or total strains if linear straining paths are assumed) in plane stress is as follows:

(i) knowing the increments in principal strains $d\epsilon_1$ . . . at some point, the effective strain increment is given by equation (A3.2) with $d\epsilon_x = d\epsilon_1$ etc. $d\epsilon_{xy} = 0$ etc. Since also $d\epsilon_3 = -(d\epsilon_1 + d\epsilon_2)$ for plastic flow at constant volume, there results

$$d\bar{\epsilon} = (2/\sqrt{3})\,(d\epsilon_1^2 + d\epsilon_1 d\epsilon_2 + d\epsilon_2^2)^{1/2} \tag{A3.6}$$

For linear deformation paths defined by $L = d\epsilon_1/d\epsilon_2 = \epsilon_1/\epsilon_2$ or $M = \sigma_1/\sigma_2$ this becomes

$$\begin{aligned}\bar{\epsilon} &= (2/\sqrt{3})\,(1 + 1/L + 1/L^2)^{1/2}\,\epsilon_1\\ &= 2(M^2 - M + 1)^{1/2}\,\epsilon_1/(2M - 1)\end{aligned} \tag{A3.7}$$

(ii) The associated effective stress $\bar{\sigma}$ in a solid workhardening according to $\bar{\sigma} = K\bar{\epsilon}^n$ then follows by substituting $\bar{\epsilon}$.

(iii) $\bar{\sigma}$ is broken down into its components σ_1 and σ_2 using equation (A3.1) with $\sigma_x = \sigma_1$, $\sigma_y = \sigma_2$, $\sigma_z = 0$ (plane stress) and τ_{xy} etc. = 0, which gives,

$$\begin{aligned}\bar{\sigma} &= (1 - 1/M + 1/M^2)^{1/2}\sigma_1\\ &= \sqrt{3}(L^2 + L + 1)^{1/2}\,\sigma_1/(2L + 1)\end{aligned} \tag{A3.8}$$

(iv) The Levy–Mises flow rules relate L and M by

$$M = (2L + 1)/(L + 2)$$

or

$$L = (2M - 1)/(2 - M) \tag{A3.9}$$

which follows from

$$\frac{d\epsilon_1}{2\sigma_1 - \sigma_2} = \frac{d\epsilon_2}{2\sigma_2 - \sigma_1} = \frac{-d\epsilon_3}{(\sigma_1 + \sigma_2)} = \frac{d\bar{\epsilon}}{d\bar{\sigma}} \tag{A3.10}$$

These are plane stress relations useful for analysing sheet forming problems and also upsetting, insofar as the stress state on the surface of a barrelled specimen where $\sigma_{radial} = 0$ may be thought of as plane stress. In Fig. 5.2, ϵ_θ corresponds with ϵ_1 and ϵ_z with ϵ_2. Calculations are readily performed for this sort of process when the practical curved loading paths out of the origin are replaced by a series of radial lines having different L.

The hydrostatic (mean) stress σ_H figures prominently in the analysis of Chapter 5:

$$\sigma_H = (\sigma_1 + \sigma_2 + \sigma_3)/3 \tag{A3.11}$$

and is *positive* when *tensile* according to the convention we employ.

For $\sigma_3 = 0$ and $\sigma_1 = M\sigma_2$

$$\begin{aligned}\sigma_H &= \sigma_1 (1 + 1/M)/3 \\ &= \sigma_1 (L + 1)/(2L + 1)\end{aligned} \tag{A3.12}$$

Some McClintock-type calculations employ σ_H normalised by $\bar{\sigma}$ or by shear stress τ. Clearly

$$\begin{aligned}\sigma_H/\bar{\sigma} &= (M + 1)/3 (M^2 - M + 1)^{1/2} \\ &= (L + 1)/\sqrt{3}(L^2 + L + 1)^{1/2}\end{aligned} \tag{A3.13}$$

in plane stress. For $\tau_{13} = (\sigma_1 - \sigma_3)/2 = \sigma_1 \div 2$ in plane stress,

$$\begin{aligned}\sigma_H/\tau &= 2(1 + 1/M)/3 \\ &= 2(L + 1)/(2L + 1)\end{aligned} \tag{A3.14}$$

Similarly for τ_{12} and τ_{23} if appropriate.

REFERENCES

Ford, H. and Alexander, J. M. (1977) *Advanced Mechanics of Materials,* 2nd edn. Chichester: Ellis Horwood.

Hill, R. (1950) *The Mathematical Theory of Plasticity.* Oxford: Clarendon Press.

Johnson, W. and Mellor, P. B. (1973) *Engineering Plasticity.* London: van Nostrand.

Mendelson, A. (1968) *Plasticity: Theory and Application.* New York: Macmillan.

Author Index

Abou-Sayed, I. S., 367
Abrams, D., 186, 265
Adkins, J. E., 155, 165
Ahmad, J., 58, 73, 242, 265, 365
Albrecht, P., 335, 365
Alexander, J. M., 430, 753, 772, 789
Altman, P. L., 164
Anderson, T. L., 367
Anderson, W. E., 545, 635
Andrade, E. N. da C., 43, 73
Andrews, E. H., 67, 73(3), 309, 365, 405, 406, 428
Andrews, W. R., 365, 367, 368, 736
Aoi, K., 376, 377, 378, 430
Arad, S., 570, 572, 633
Archard, J. F., 764, 789(3)
Argon, A., 177, 267, 407, 428
Ashby, M. F. 34, 41, 73, 164, 477, 524, 537, 616, 633, 738, 789
Atkins, A. G., 49, 73(2), 104, 105, 108, 128, 165(3), 184, 224, 265(2), 266, 314, 365, 367(2), 382, 391, 397, 409, 428(3), 429(2), 431, 435, 471, 472, 510, 513, 515, 524(4), 525(4), 635, 647, 667, 668, 682, 705, 706(4), 714–5, 718–9, 734, 736(5), 766–7, 773, 789(7)
Atkinson, C., 165
Auverbach, F., 180, 265
Avila, J., 165

Babichev, M. A., 764, 765, 790
Backofen, W. A., 428
Baer, E., 789, 791
Bai, Y. L., 411, 428
Banasiak, D. H., 587, 633
Baratta, F. I., 365
Barenblatt, G. I., 242, 265, 283, 365
Barker, L. M., 140, 165, 274, 365(2)
Barquins, M., 189, 266
Barsom, J. M., 275, 365, 526, 570, 592, 594, 599, 633(4)
Barton, A. F. M., 497, 524
Batson, R. G. C., 727, 736
Bazant, Z. P., 705
Beachem, C. D., 428(2)
Beaumont, P. W. R., 524
Beck, E. J., 595, 633
Begley, J. A., 322, 367(2), 472, 524, 525
Beinert, J., 706
Benbow, J. J., 204, 266
Berg, C. A., 430
Berggren, R. G., 721, 736
Betz, E., 165
Bevan, L., 405, 406, 428
Bevis, M., 60, 73
Bhatty, J. I., 309, 365
Biggs, W. D., 70, 73
Bilby, B. A., 114, 165, 205, 266, 283, 365(4)
Billington, E. W., 73
Birch, M. W., 217, 218, 268
Birchall, J. D., 184, 186, 266(2)
Bishop, S. M., 266(2)
Bisshopp, K. E., 153, 165
Blok, H., 764, 789

Bloom, J. M., 789
Bluhm, J. I., 222, 266
Board, R. G., 524
Boddington, P., 429
Boker, R., 175, 266
Bond, G. M., 524
Booth, E., 727, 736
Boston, O. W., 789
Bowden, F. P., 762, 789
Bowden, P. B. 404, 428
Boyd, G. M., 70, 73
Boyle, R. W., 166, 280, 367, 706
Bradt, R. C., 486, 488, 525
Brandrup, P., 525
Brewer, R. C., 753, 772, 789
Bridgman, P. W., 175, 266, 392, 429
Broek, D., 365, 367, 401, 660, 705
Brown, D. K., 391, 429
Brown, L. M., 429
Brown, M. W., 601, 602, 603, 633
Brozzo, P., 382, 429
Bucci, R., 633
Buchalet, C., 524
Burdekin, F. M., 283, 284, 365, 366(2), 633, 734, 736
Burns, D. J., 571, 635(2)
Burns, S. J., 165, 241, 267
Burrel, H., 524, 668, 705
Buttery, T. C., 764, 789

Caddell, R. M., 73, 103, 111, 156, 165(2), 379, 391, 397, 420–2, 429(3), 469, 524(2), 705, 706, 736
Calladine, C. R., 731, 736
Campbell, J. D., 166, 790
Carew, T. E., 166, 205, 266
Carslaw, H. S., 665, 705
Cartwright, D. J., 166
Castino, F., 447, 524, 525
Chan, M. K. V., 326, 366, 454, 524
Chang, M. D., 296, 366
Chang, T. M., 409, 429
Charles, E. A., 183, 266
Chell, G. G., 702, 705, 783, 786, 789
Cherry, B. W., 216, 238, 266(2)
Cheverton, R. D., 693, 705
Chow, C. L., 89, 130, 165(2), 238, 266, 705
Chow, C.-Lun, 165
Christensen, J. V., 767, 790
Christensen, R. M., 587, 633
Christiansen, A. W., 756, 789
Chukwejekwu, S. E., 789
Clark, A. B. J., 472, 486, 525, 529, 634
Clark, W. G. Jr., 547, 633
Clarke, G. A., 330, 367
Clausing, D. P., 650, 705
Clift, S., 429
Cockcroft, M. G., 382, 429
Coffin, L. E., 430, 579, 616, 633, 636(2)
Coker, E. G., 165
Collier, D., 736
Constable, I., 610, 633
Cook, J., 49, 73, 165, 216, 266
Cook, N. G. W., 175, 267
Cooke, R. G., 524
Cooper, G. A., 216, 266
Corran, R. S. J., 691, 705, 779, 781, 789, 790
Corten, H. T., 367
Cottrell, A. H., 35, 37, 48, 73(3), 104, 114, 149, 266, 281, 283, 365, 366
Cotterell, B., 165(2), 167, 179, 204, 205, 266(4), 268(2), 300–11, 316, 337, 342, 366, 367(7), 368, 419–21, 425, 543, 633
Coutts, L. H., 525(2)
Cox, T. B., 401, 429
Crosley, P. B., 524, 690, 693 705(2)
Culver, L. E., 267, 525, 633(2), 634, 635
Currey, J. D., 54, 73, 104, 165
Czboly, E., 214, 268, 420, 430

Dannenburg, H., 187, 266
Datsko, J., 773, 789
Davidge, R. W., 507, 524(2)
Davidson, J. R., 627, 628, 634
Davies, P. H., 789
Dawes, M. G., 284, 366, 367
Debaise, G. R., 524
de Lorenzi, H. G., 367, 368, 736
Deluca, B., 429
de Villiers, J. W. R., 603
de Vries, K. L., 296, 366
deWit, R., 367
Dillamore, I. L., 382, 429
Dittmer, D. S., 164
Dobbs, H. S., 187, 266
Docherty, J. G., 727, 736(2)
Dodd, B., 429(2)
Dowling, A. R., 789, 782
Dowling, N. E., 633
Drucker, D. C., 152, 165, 385, 429
Duckett, R. A., 791
Duffy, A. R., 716, 717, 736
Dugdale, D. S., 282, 283, 366
Duncan, J. L., 429, 431
Duquette, D. J., 634

Eastabrook, J. N., 565, 587, 634(2)
Edington, J. W., 525
Eftis, J., 191, 192, 195, 195, 266(2), 634
Eiber, R. J., 268, 736
Elber, W., 587, 611, 627, 628, 634(3)
El-Hakeem, H. A., 626, 634
Elliott, D., 286, 367
Ellison, E. G., 473, 524
Embury, J. D., 429
Engle, R. M., 634
Erdogan, F., 266

Ernst, H., 367, 705, 706
Evans, D. C., 767, 789
Ewing, P. D., 205, 268(2), 524
Erturk, T., 377, 397, 429(2)
Eschweiter, J., 525
Evans, A. G., 183, 184, 266(2), 267, 515, 524(3), 533, 634

Fearnehough, G. D., 286, 366
Felbeck, D. K., 73(2), 510, 524
Feret, R., 186, 266
Ferguson, W. G., 280, 367
Field, J. E., 266, 789
Finnie, I., 205, 214, 242, 266(2), 268
Fleming, J. R., 767, 789
Folias, E. S., 716, 736
Foote, R. M. L., 736
Ford, F. P., 493, 502, 524(2)
Foreman, R. G., 634
Forsyth, P. J. E., 536, 634
Freiman, S. W., 267, 526
French, I. E., 399, 429
Freund, L. B., 693, 705
Frost, H. J., 789
Frost, N. E., 610, 634(2)
Fukahori, Y., 73
Fuller, E. R., 526, 533, 634
Fuller, K. N. G., 524

Garwood, S. J., 335, 366(2), 699, 705
Gehlen, P. C., 705(2)
Gent, A. N., 104, 165, 404, 405, 429(2), 499, 524
German, M. D., 367(3), 368, 706, 736, 790
Ghosh, A. K., 382, 388, 399, 429
Gifkins, R. C., 73
Gillemot, F., 214, 268, 576, 634
Gillemot, L. F., 382, 429, 576, 634
Glenny, I., 524
Goetz, C. G., 192–5, 266
Goicolea, J. M., 367
Goldman, N. L., 339, 366, 472, 524, 634
Gomez, M. P., 545, 635
Goodall, I. W., 789
Goodier, J. E., 114, 167
Goods, S. H., 429
Gordon, J. E., 49, 54, 56, 69, 73(4), 150, 165(2), 216, 266
Grandt, A. F., 633, 635
Green, A. E., 155, 165
Griffith, A. A., 72, 73(2), 94, 165, 194
Griffiths, J. E., 789
Griffiths, J. R., 753, 754, 789
Gudas, J. P., 365
Guild, F. J., 266
Guillen-Preckler, A., 399, 429
Gurney, C., 73, 84, 152, 165(2), 169, 208, 209, 250, 256, 266(3), 366(2), 433, 460, 524, 525, 634, 635, 645, 648, 650, 672, 675, 677, 678, 694, 705(4), 710, 736

Hagan, J. T., 759, 790
Hagedorn, K. E., 429
Hagiawara, K., 727, 736
Hagiawara, Y., 366
Hahn, G. T., 736, 687, 691, 705(2), 716, 736
Halliday, J. S., 764, 789
Hancock, J. W., 382, 391, 429, 431
Hang, P. T., 238, 266
Harper, M. P., 473, 524
Harrison, N. L., 216, 217, 266(2)
Harrison, R. P., 783, 789
Hartley, P., 368, 430
Hasselman, D. P. H., 429, 510, 513, 519, 524(2), 683, 705
Head, A. K., 634
Head, P. T., 569, 634
Helmuth, R. A., 186, 266
Henkin, A., 773, 789
Hennephof, J., 316, 366
Henry, L. F., 498, 524
Henshall, J. L., 525
Hertzberg, R. W., 525, 617, 634, 637
Hetenyi, M., 705
Heywood, R. B., 634, 731, 734, 736
Hill, R., 366
Hillberry, B. M., 585, 635
Hillerton, J. E., 165
Himsolt, G., 525
Hoagland, R. G., 693, 705, 706(2)
Hockey, B. J., 267, 525
Hodgkinson, J. M., 165
Hodgkinson, P. H., 366
Hoffmanner, A. L., 429
Hosford, W. F., 379, 422, 429
Hoshide, T., 634, 636
Howard, A. J., 266(2)
Howard, I. C., 489, 525
Howes, M. G. J. W., 789
Hudak, S. J. Jr., 547, 633
Hudson, C. M., 634(2)
Hull, D., 60, 73
Hunt, J., 73, 84, 165, 208, 209, 266, 645, 705, 710, 736
Hutchings, I. M., 770, 789
Hutchinson, J. W., 321, 339, 362(2), 368, 524, 634

Ibrahim, M. F. E., 576, 582, 634
Iida, S., 205, 267
Immergut, E. H., 524, 525, 668, 705
Inglestrom, N., 267
Inglis, C. E., 116, 165, 267
Irving, P. E., 544, 634
Irwin, G. R., 90, 165(5), 166, 268, 464, 636, 708, 736
Isherwood, D. P., 103, 165
Iskander, S. K., 705

Jacques, C. H. M., 498, 525
Jaeger, J. C., 175, 267, 665, 705

Jain, S. C., 429
Janicki, J. S., 166
Jebe, E. J., 635
Jeronimidis, G., 50, 51, 54, 73(2), 525
Jevons, J. D., 316, 366
Johnson, K. L., 764, 790
Johnson, M. A., 426, 430
Johnson, W., 394, 411, 428, 429
Jones, A. T., 143, 165(2), 706
Jones, D. L., 191, 192, 195, 196, 266, 634
Jones, N., 731, 736

Kallenberg, G. P., 526
Kals, J. G., 391, 429
Kalsi, G. S., 367(2)
Kalthoff, J. F., 691, 706
Kambour, R. P., 429(2)
Kanazawa, T., 339, 366(2)
Kanninen, M. F., 58, 73, 242, 265, 337, 339, 365, 367, 677, 693, 691, 699, 705(2), 706(4)
Kaplan, S. M., 430
Kapp, J. A., 143, 165
Kawano, H., 736
Kazaoglu, E., 377, 429
Kearney, V. E., 634
Keife, H., 382, 429
Kelly, A., 48, 73, 216, 266, 790
Kendall, K., 54, 73, 104 165(2), 171, 174, 176, 190, 266(2), 267(3), 693, 749, 790
Kenny, P., 166, 790
Kfouri, A. P., 218, 267, 576, 634(2), 635
Kies, J. A., 90, 165, 472, 486, 525, 529, 634
Kim, H. J., 104, 165
Kim, K., 634
Kinloch, A. J., 670, 706
Kitagawa, K., 634
Kitching, R., 781, 790(2)
Kivivuori, S., 429
Klemm, W., 706
Klepaczko, J. R., 435, 525
Klippstein, K. H., 585, 634
Knauss, W. G., 197, 212, 267
Knott, J. F., 140, 155, 166, 286, 367, 547, 566, 636(2), 634
Kobayashi, A. S., 205, 267
Kobayashi, J., 366
Kobayashi, S., 397, 400, 429(2), 430(2)
Kochendorfer, A., 429
Koehl, M. A. R., 166
Kopp, R. W., 429
Krafft, J. M., 166, 280, 367, 654, 706, 774, 790
Kramer, E. J., 525
Kreher, W., 187, 267
Kruschov, M. M., 764, 765, 790
Kuczynski, K., 382, 430
Kudo, H., 376, 377, 378, 396, 430(2)
Kuhn, H. A., 377, 378, 430(2)
Kuhn, P., 89, 166
Kumar, V., 367(3), 706, 730, 736, 790

Lai, M. O., 280, 367
Lake, G. J., 155, 158, 166(3), 572, 634
Lam, P. M., 705
Lancaster, J. K., 767, 789(2)
Landes, J. D., 322, 330, 331, 332, 367(3), 472, 524, 525, 620, 637
Lange, F. F., 515, 524
Latham, D. J., 382, 429
Lau, K. J., 130
Lawn, B. R., 149, 166, 183, 267(3), 472, 487, 488, 525
Lazzeri, L., 89, 166
Lee, C. S., 472, 524, 668, 669, 705, 706
Lee, E., 366
Lee, J. D., 634
Leech, A. B., 430
Leete, N. B., 663, 706
Leevers, P. S., 192, 267(2)
LeRoy, G. H., 429
Liebowitz, H., 594, 634
Lin, I. H., 284, 367
Lindley, P. B., 155, 166, 429, 572, 634
Lindley, T. C., 634(2)
Little, R. E., 635
Liu, H. W., 543, 635
Lloyd, G. J., 635
Loss, F. J., 365
Love, A. E. H., 166, 430
Low, J. R., 401, 429, 431
Lubahn, J. D., 730, 736(2)
Ludema, K. C., 429
Ludwik, P., 31, 71, 73
Lunn, B., 767, 790

Macey, G. E., 367
MacGregor, C. W., 195, 267
Machida, S., 366
Mai, Y. W., 104, 105, 108, 128, 130, 166(2), 167, 229, 230, 240, 241, 266, 267(2), 268(2), 301–11, 314–6, 337, 342, 365, 366(4), 367(7), 368, 419–21, 425, 460, 462, 497, 498, 513, 515, 516, 525(11), 570, 572, 587, 626, 635(4), 647, 650, 663, 667, 668, 672, 675–8, 682, 694, 705(2), 706(6), 715, 718, 719, 725, 736(3)
Maiti, S. K., 205, 214, 267(3)
Maitland, A. H., 266
Mallock, A., 773, 790
Mamaros, T. C., 706
Manson, J. A., 617, 634
Manson, S. S., 513, 525, 535
Marciniak, Z., 382, 430
Marin, J., 394, 430
Marsh, K. J., 634

Marshall, C. W., 367
Marshall, D. B., 267
Marshall, G. P., 407, 431, 460, 464, 471, 487, 525(3), 526, 665, 706
Marston, T. U., 49, 693, 706
Marti, J., 338, 367(2), 397
Matsoukas, G., 284, 367(2)
Maynard, R. A., 382
Maxey, W. A., 268, 736
Mayer, T. R., 524
Maynard, R. A., 429
McCabe, D. E., 365, 660, 706
McCartney, L. N., 544, 634
McClintock, F. A., 177, 267, 367, 382, 386–9, 397, 422 430(3), 544, 578, 635(3)
McClure, G. M., 736
McDiarmid, D. L., 601, 635
McEvily, A. J., 578, 635
Mears, D. R., 756, 790
Mecholsky, J. J., 267
Mellor, P. B., 382, 430(2), 431
Merchant, R. H., 431
Merkle, J. G., 367, 368, 779–81, 790
Michener, J. R., 241, 267
Miles, J., 736
Miles, M. J., 791
Miller, K. J., 218, 244, 267, 268, 565, 576, 577, 582, 594, 595, 601, 602, 605, 608, 633, 634(3), 635(3), 636(2), 734
Miller, L. E., 790
Milne, I., 702, 705, 783, 786, 789
Miner, M. A., 581, 635
Mines, R. A. W., 705
Minnear, W. P., 486, 488, 525
Mirabile, M., 650, 706
Moet, A. Y., 791
Montulli, L. J., 633, 635
Moran, B., 430
Morley, J. G., 50, 73
Morozon, E. M., 367
Mostovoy, S., 166, 525, 570, 635
Mowbray, D. F., 367, 368, 736
Mubeen, A., 634
Mukherjee, B., 571, 635(2)
Mukherjee, J., 547, 549, 635
Mulhearn, T. O., 790(2)
Munz, D., 525
Murray, M. J., 228, 267, 765, 790
Mutton, P. J., 765, 790

Nadai, A., 756, 790
Nagtegaal, J. C., 338, 367
Nayeb-Hashemi, H., 578, 635
Newman, J. C., 125, 166(2), 267, 786, 790
Newton, C., 368
Neuber, H., 166
Ngan, K. M., 152, 165, 169, 238, 266(2), 366, 648, 705
Nisitani, H., 244, 267
Noble, C. F., 381, 382, 430
Nordberg, H., 267
Norris, D. M., 382, 430
Novak, S. R., 526

Oates, G., 753, 754, 789
Oh, S. I., 400, 429, 430(2)
Ohtsubo, H., 705
Okinoto, K., 430
Old, C. F., 499, 525
Omar, M. K., 766, 789(2), 790
Ongchin, L., 403, 430
Orowan, E., 32, 166(2)
Owen, R. C., 266
Oxborough, R. J., 404, 428
Oxley, P. B., 381, 382, 430
Oyane, M., 382, 397, 430(2)
Pachman, P. F., 635
Pae, K. D., 790
Palaniswamy, K., 197, 212, 267
Palmgren, A. Z., 581, 635
Paris, P. C., 166, 268, 367, 368, 543, 545, 589, 635(2), 636, 650, 694, 697, 699, 706
Parkins, R. N., 492, 525(2)
Parks, D. M., 367, 693, 705
Parmar, A., 382, 430(2)
Parry, G. W., 89, 166
Parry, J. S. C., 796
Parton, V. Z., 367
Parvin, M., 665, 667, 706
Pascoe, K. J., 595, 601, 603, 635, 636(3)
Patel, D. J., 166
Pattersen, E., 730, 736
Pearson, S., 634
Pentoney, R. E., 524
Percy, M. J., 431
Peters, R. W., 89, 166
Peterson, R. E., 166
Pflegel, G. A., 165
Pillinger, I., 368, 430
Pitoniak, F. J., 587, 635
Pittman, J. F. T., 430
Plumbridge, W. J., 544, 635
Pohianka, R. C., 267
Polakowski, N. H., 175, 267
Pollett, J.-C., 165
Pompe, W., 187, 267
Pook, L. P., 585, 634, 635(3)
Porter, A. W., 524
Prasad, K. R. S. K., 205
Preece, C. M., 526
Probst, E. P., 585, 635
Purslow, P. P., 151, 154, 165, 166, 526
Puttick, K. E., 267, 635, 734, 736, 741, 759, 762, 790(5)

Quinones, D. F., 430

Rabinowitz, S., 756, 790
Radcliffe, S. V., 789
Radon, J. C., 192, 267(2), 420, 430, 526, 633, 635, 636
Raha, S., 498, 526
Raj, R., 477, 524
Reaugh, J. E., 430
Reddell, J. K., 300, 366
Rendina, R., 429
Rhodes, D., 636
Rice, J. R., 66, 74, 144, 166, 205, 266, 318, 321, 323, 335, 367(5), 368, 382, 385, 420, 426, 430(3), 543, 636
Rice, R. W., 186, 267
Richards, C. E., 634(2)
Rimrott, F. P. J., 394, 430
Ripling, E. J., 92, 166, 175, 267, 524, 525, 570, 635, 690, 693, 705(2)
Ritchie, R. O., 426, 430, 578, 634, 635(2), 636
Rivett, R. M., 790
Rivlin, R. S., 103, 155, 166(2)
Roberts, R., 368, 637
Roesler, F. C., 181, 204, 266, 267
Rogers, H. C., 35, 74, 430(2)
Rolfe, S. T., 275, 365, 526(2)
Rooke, D. P., 166
Rosenfield, A. R., 706
Rosengren, G. P., 321, 367
Rosenhain, W., 790
Rowcliffe, D. J., 525
Rowe, G. W., 338, 368, 430, 773, 790(2)
Royston, H., 524
Ruarak, A. E., 736
Ruiz, C., 691, 705, 777, 779, 781, 789(2), 790(3)
Russel, R. R., 429
Ryskewitch, T., 186, 267

Sack, R. A., 194, 267
Sadananda, K., 636
Saith, A., 205, 266
Salama, M. M., 407, 428
Samuels, L. E., 759, 762, 790(2)
Sato, T., 430
Sauer, J. A., 790
Savin, G. N., 166
Schigve, J., 636
Schilling, C. G., 585, 634
Schmidt, D. W., 365
Schmidt, R. A., 589, 636
Schonert, K., 526
Schulze, H. D., 429
Scoble, W. A., 165
Scott, V. D., 524
Selwood, A., 766
Seman, D. J., 526
Sedricks, A. J., 790
Selwood, A., 790
Server, W. L., 430
Seth, R. S., 250, 256, 267
Seward, S. K., 634(2)
Shabbits, W. O., 526
Shaffman, P. G., 407, 430
Shah, S. N., 430
Shah, S. P., 167
Shahid, M. A., 790
Shahinian, P., 636
Shank, M. E., 70, 74
Shearin, P. E., 727, 729–30, 736
Shih, C. F., 335, 339, 367(2), 368(3), 706, 730, 790
Shima, S., 430
Shoemaker, A. K., 526
Sia, T. K., 268
Sih, G. C., 166, 214, 217, 265, 268(3)
Simmons, C. J., 526
Sinay, G., 576, 634
Smelser, R. E., 165
Smith, D. J., 477, 526
Smith, E., 493, 526, 706
Smith, E. W., 595, 601, 636(3)
Smith, L. S. A., 267, 790
Smith, R. A., 214, 244, 267, 268, 563, 599, 605, 608, 636(5), 734, 751–3, 790
Sneddon, I. N., 187, 268
Soete, W., 790
Solomon, H. O., 577, 616, 636
Sowerby, R., 429, 431
Speidel, L. M., 526
Spick, P., 773, 790
Springer, G. S., 770, 790
Srawley, J. E., 329, 368
Stahlberg, U., 382, 429
Stahlkopf, K. E., 706
Stanton, T. E., 727, 736
Sternstein, S. S., 403, 430
Stone, D. E. W., 283, 366
Sturgess, C. E. N., 368, 430
Sturney, A. C., 773, 790
Suh, N. P., 767, 789
Sullivan, A. M., 166, 280, 367, 706
Sulonen, S. I., 429
Sumner, G., 636, 637
Sumpter, J. D. G., 339, 368(2), 730, 736
Sutton, S. A., 636
Swain, M. V., 759, 790
Sweldlow, J. L., 179, 191, 192, 194–6, 268(2)
Swift, H. W., 409, 429
Swinden, K. H., 114, 165, 283, 365
Szabo, I., 712, 736

Tabor, D., 368, 762, 789
Tada, H., 166, 367, 636, 706
Taira, S., 636
Takanabe, H., 736
Takahashi, S., 634
Tanaka, K., 605, 634, 636(3)
Tappin, G., 524, 663, 706

Tattersall, H. G., 663, 706
Tavernelli, J. F., 579, 636
Taylor, D., 566, 636(2)
Taylor, G. I., 407, 430
Theocaris, P. S., 166
Thomas, A. G., 103, 155, 166(2)
Thomason, P. F., 382, 431
Thompson, R. D., 382, 431
Timoshenko, S. P., 167(3), 291, 368
Tipper, C. F., 70, 74
Tomkins, B., 37, 57, 73, 74, 368, 394, 397, 431(2), 472, 477, 489, 502, 526, 537, 543–4, 575, 577, 616, 633, 636(8), 637
Towner, R. J., 526
Townley, C. H. A., 782, 789
Tracey, D. M., 382, 385, 430
Treloar, L. R. G., 499, 526
Trent, J. S., 756, 791
Trimble, R. M., 736
Truss, R. W., 756, 791
Tsien, L. C., 43, 73
Turk, D. A., 186, 266
Turner, C. E., 329–30, 335, 339, 360, 368(3), 431, 526, 694, 702, 706(4)

Uete, H. K., 111, 165, 167
Underwood, J. H., 165

Vaidyanathan, S., 242, 268
Valsgard, S., 730, 736
Van der Sluys, W. A., 365
Van Minh, H., 382, 431
Van Stone, R. H., 368, 401, 431
Venter, R. D., 429
Venzi, S., 650, 706
Vermilyea, D. A., 490, 526
Vincent, J. F. V., 56, 74, 108, 165
Vincent, P. I., 498, 526
von Euw, E. F. J., 585, 637
von Karman, T., 267

Waldren, N. E., 431
Walker, H., 330, 367
Wang, D. Y., 660, 706
Ward, I. M., 756, 790, 791
Wareing, J., 576–7, 635, 636(2), 637
Watson, J. D., 765, 790(2)
Webster, G. A., 477, 526
Wecharatana, M., 167
Wei, R. P., 479, 526, 620, 626, 637(2)
Weichart, R., 526
Weinrich, P. F., 399, 429
Weiss, H. D., 214, 266
Weitzmann, R. H., 250, 256, 268
Wells, A. A., 281, 368
Wessel, E. T., 268
West, J. M., 490–1, 526(2)
Westergaard, H. M., 268
Westwood, A. R. C., 526
Wetton, A. G., 790
Wheeler, O. E., 587, 637
Wiederhorn, S. M., 267, 478, 524(2), 525, 526
Wilhem, D. P., 143
Wilkening, W. W., 367, 736
Wilkowski, G. M., 268(2), 721, 736
Williams, D. P., 526
Williams, J. A., 791
Williams, J. G., 103, 165(2), 205, 217–9, 225, 268(4), 326, 366, 407, 431, 454, 464, 466, 487, 524(2), 525(3), 526(4), 570, 572, 626, 633, 635, 637, 665, 670, 706(2)
Williams, M. L., 120, 167, 187, 189, 192, 268(2), 297, 366
Wilshaw, T. R., 149, 166, 267
Winkler, S., 706
Winter, R. E., 789
Witt, F. J., 721, 736
Wolf, K., 268
Wood, R. D., 430
Wu, C. Cm., 267
Wullaert, R. A., 430
Wyzgoski, M. G., 498, 525

Yamada, A., 634, 636
Yap, O. F., 137, 167, 225, 268, 308, 368, 549, 637
Yeoh, O. H., 158, 166
Yoder, G. R., 428
Young, D. H., 167
Young, R. J., 524
Yukawa, S., 730

Zacharia, K. P., 585, 637
Zahoor, A., 367, 706
Zarrabi, K., 781, 790(2)
Zienkiewicz, O. C., 430

Subject Index

Accumulative damage
in fatigue, 541
see also Palmgren–Miner rule
Activation energy, 470, 472
Adhesive fracture
elastic peel test, 104
elastoplastic peel test, 289, 292
Angled crack, 177, 211
Anisotropic solids
crack path in, 214
Arrhenius equation, 469, 486

Beam on elastic foundation model
analysis of crack stability, 677
quasi-static analysis of fracture, 253
Biaxial stress
fracture, 191
fatigue, 595
Biomaterials, *see* Fracture in biomaterials
Biot modulus, 512
Blister test, 187
Blunt crack tip
crack instabilities, 669
effect on K_C, 143
Brazilian ring specimen
stability in, 650, 654–7
Brittle, definition of, 29
Brittle–ductile transition, 30, 32, 219
scaling and size effect, 71
Brittle intergranular fracture (BIF), 34

C^*–integral, 473
Can opening, plastic fracture, 291
CEGB
'two criteria' design, 782
Ceramics, *see* Fracture in ceramics and glasses
Charpy energy
fracture transition, 38
relationship to K_{IC}, 275
Coffin–Manson law for plastic fatigue, 579
Combined flow and fracture
partition of work areas, 341, 357
Compact tension specimen
stability in, 646
toughness testing, 124
Compliance
change of, 91
definition of, 77
method, 90
specimens with constant rate of
Compliance calibration equation, 90
with residual displacement, 108, 353
Cone (Hertzian) cracks, *see* Indentation cracks
Constraint factors, 32
Contact angle, 488
Cook–Gordon debonding, 49, 54, 216
Crack arrest, 682
Crack growth
definition of quasi-static, 169
in fatigue, 529 *et seq.*
in forming operations, 409 *et seq.*

Crack growth – continued
in highly deformable solids, 150
in hostile environments, 478
relaxation controlled, 469, 487
resistance curves, 98, 286, 334
stability of, 640 *et seq.*
stability analysis, 644, 647, 695
under constant displacement, 80
under constant load, 82, 92
Crack nucleation, *see* Void growth model
Crack opening displacement
connexion with fatigue, 544
design curve, 285
fracture criterion, as 145, 283
relationship to $G_C(R)$, 145
relationship to J_C, 332, 425
relationship to K_{IC}, 426
Crack path, *see* Angled crack
Crack tip
irreversible zone, 130
softening, 466
strain rate, 463
temperature rise, 466, 665
Crack velocity
at interfaces, 171
effect on fracture toughness, 471, 665
in quasi-static cracking, 169
scaling law for, 719
Cracks
angled, 177, 211
elliptical, 116
in fatigue, 541, 582
in layered structures, 219 *et seq.*
interfacial, 104, 173, 289
from notches, 244, 605
self-similar propagation, 197
surface, 126
under biaxial loading, 191 *et seq.*
under biaxial fatigue, 594 *et seq.*
Cracks in compression, 176, 179
transition to flow, 750
Cracks in metalforming processes
blanking, 380
cropping, 398, 409
deep drawing, 374
extrusion, 374
guillotining, 411
rolling, 371
upsetting, 373, 376
Craze growth, 407
Crazing
definition of, 44, 401
in environments, 493
relationship to solubility parameter, 496
theories and mechanisms, 403
Creep fatigue, 615
in stress corrosion, 617
Creep fracture
in metals, 472
in polymers, 469
Criterion for fracture initiation, J_C, 321
Criteria for void growth, 381
Critical stress intensity factor, K_C
experimental determination of, 123 *et seq.*
plane strain, 124
plane stress, 124
relationship to $R(G_C)$, 128
see also Toughness
Cumulative damage in fatigue, 581
Cutting
of extensible materials, 157
fracture transitions in, 771
Cylindrical shells
non-linear elastic fracture, 88

Damage-tolerance, *see* Fatigue design methods
Deformation transitions, 749
Delay fracture, 451, 527
see also Static fatigue
Design against fatigue, 626
Displacements around crack tip, 118
Dissolution processes
anodic, 489
cathodic, 490
Double Cantilever Beam (DCB)
determination of R, 79, 92
stability of cracks in, 645–6
with arms of unequal depths, 160
with tapered (contoured) arms, 91
Double torsion (DT) specimen (also known as Outwater specimen), 91
crack arrest in, 684
crack growth in mineral glass, 529
crack growth in Portland cement, 111
determination of G, 521
slow crack growth in PMMA, 441
Ductile, definition of, 29
Ductile fracture
criteria for, 381
definition of, 14, 269, 738
essential work of, 300
in HIPS, 309
in low strength metals, 298, 312
in UHMWPE, 307
see also Plane stress
Ductile intergranular fracture, 35
Dugdale model
crack tip zone, 282, 470, 782
Dynamic stress intensity factor, K_{Id}, 275, 691

Elastic fracture
definition of, 13, 60, 738
linear, 80, 89
non-linear, 86, 95
scaling laws of, 710, 718
Elasto-plastic fracture
definition of, 14
see also Ductile fracture

Electrochemical effects in stress corrosion cracking, 489
Elliptical cracks, 116
Energy
complementary, 95
potential, 317
residual strain, 230, 343
strain, 75, 95
Energy balance, 65, 80
Energy release rate, 60, 82, 90, 318
Environmental effects
influence of solubility parameters, 496, 767
on crack stability, 663
on fatigue crack growth, 618
on fracture toughness, 434, 441, 498
on slow crack growth, 487
on wear resistance, 767
Environmental stress cracking of polymers, 486, 492
Epoxy resin, stick–slip fracture, 225, 675
Equivalent crack length
in fatigue from notches, 605
in fracture, 244
Equivalent energy concept, 721
Essential work of ductile fracture
determination of, 298
effect of prestrain, 303, 421
in mixed mode, 309
in polymers, 307
relationship to fracture limits, 419
relationship to J, 337, 699
theoretical calculation, 301–3
Eta (η) factor
in elastoplastic fracture, 330, 335
upper bound toughness in elastic fracture, 329, 679

Fail-safe, *see* Fatigue design methods
Failure assessment diagram, *see* CEGB and R-6 interaction diagram
Faraday constant, 490, 492
Fatigue
biaxial stress, 594
constant cyclic loading, *see* Paris power law
crack closure model in, 587
crack growth from notches, 605 *et seq.*
crack growth interpretation, 542, 558
crack growth laws in, 547, 552
crack growth retardation, 587
crack growth rate prediction, 530, 620 *et seq.*
cummulative damage, 581
elastic analysis, 543
elasto-plastic analysis, ΔJ, 575 *et seq.*
environmental effects, 618
life prediction, 550
mean stress effect, 569, 589
mechanisms, 536 *et seq.*
non-propagating cracks, 610
Fatigue – continued
overload effects, 585
residual stress effect on, 611
short cracks, 565
stability of cracks in, 702
striations, 537–9, 544
temperature effects, 615
threshold, 555–6
variable amplitude loading, 592
Fatigue crack growth, 538, 601
Fatigue design methods
safe-life, 627
damage-tolerance, 627
fail-safe, 627
Fatigue life prediction
constant cyclic loading, 550
variable amplitude loading, 592
Fatigue stress concentration factor, K_f, 113, 606
Fatigue threshold
crack length dependence, 563
mean stress dependence, 555–6
Fibre-reinforced composites
Cook–Gordon debonding, 49, 216
intermittent bonding, 50, 447
stability of cracks in, 662
Fibrils in a craze, 401
Forming limit diagrams
in bulging, 379
in upsetting, 376
Fracture in adhesive joints
elastic system, 171, 456
elasto-plastic system, 239 *et seq.*
Gurney's technique, 85
Fracture in biomaterials, 51 *et seq.*
mother of pearl (nacre), 53–4
rat skin, 151, 447–50
Fracture in cement mortar, 112
in MDF cements, 185
Fracture in ceramics and glasses
aluminas, 481
silicon carbide, 515
silicon nitride, 444, 485, 515
soda-lime-silicate glasses, 455, 481, 529
tungsten carbide, 443, 682
Fracture in composites
boron-epoxy resin, 50
glass-filled PS, 498
Kevlar-epoxy resin, 448
Fracture in polymers
epoxy resin, 676
HDPE, 458, 483
HIPS, 136, 309
PC, 109–11
PMMA, 87, 438–42, 472, 664
PS, 45, 498
PVC, 222
rubber, 155
SBR vulcanisate, 445
Styrene, 475, 661
TPX, 677

Fracture in metals
aluminium alloys, 133, 255, 257, 311, 315
brasses, 305, 434
low carbon steels, 255, 300, 306, 311
AFC 77, 461
Cr-Mo-V, 473, 503
Discaloy, 451, 475
EN3B, 259
Ti-5Al-2.5Sn, 482
Fracture in papers, 105, 256
Fracture in wood, 54, 236, 446, 447, 453
Fracture mechanism
amorphous material, 42
biological material, 51
composites, 46
crystalline solids, 33
fibre debonding (Cook–Gordon), 49
fibre pull-out, 47
work of fracture values, 46
Fracture mechanism map
tungsten, 40
Nimonic 80A, 42, 43
Fracture toughness measurement
codes of practice for K_{IC}, 124, 138, 148
fibre-reinforced composites, 110, 216, 662
for ductile materials, 250 *et seq.*
for elastic materials, 82, 87
for highly extensible materials, 150
for large K_C/σ_y materials, 253
in guillotining, 105
in microtoming, 106

Geometric stability factors
in hard machines ($du > 0$), 645, 794
in soft machines ($dX > 0$), 645, 794
Glass transition, 405
Griffith
crack in biaxial stress, 191 *et seq.*
crack stability, 653
theory, 94

Hardness
indentation, 179
relationship to wear resistance, 765
Heat generated at moving crack-tip, 465, 665
Hydrostatic stress
on craze nucleation, 405
on ductile–brittle transition, 42, 756
on void growth, 392

Ice number, 714
Impact tests
Charpy and Izod, 31
correlation with K_{IC}, 275
Indentation cracks, 179
Instability, *see* Stability

Interface cracks, 171, 215
see also Cook–Gordon debonding
Irwin–Orowan correction, 94
Isothermal–adiabatic transition, 468, 667

J–integral
connexion with void growth, 422
definition of, 316
determination of J_C, 324 *et seq.*
in low cycle fatigue, 575
in terms of crack opening displacement, 332, 422
for CN, DEN specimens, 326
for CT specimen, 329
for three-point notched bend specimen, 328
relationship to G, 318, 359
relationship to K, 319
relationship to η factor, 329, 341
J-shape $\sigma-\epsilon$ curve
for biological materials, 55
J_R-controlled growth criteria, 336
J_R crack growth resistance curve, 334, 696
see also Resistance curve
Jacobian, 162

K_R-curve, *see* Resistance curve
Kinetic energy
in dynamic analysis, 691
in stability analysis, 703

Layered structures, 219
Leak-before-burst, 127
Limit analysis, 777
Limiting velocities, 169, 666
Line zone, 282
see also Dugdale model
Liquid metal embrittlement, 499
Loss factor ($\tan \delta$), relationship to fracture toughness, 469

Macro-defect-free cements (MDF), 185
Metalforming and cracks, 370, 408
Microstructure link to mechanics, 67, 397
Microtoming fracture analysis, 106
Miner rule, *see* Accumulative damage
Mixed mode fracture
crack propagation direction in, 212
essential work in ductile solids, 309
failure criteria, 197
Modes of fracture, 122
Moving crack tip
localised heating at, 465
strain rate in, 463
Multiaxial loading
elastic and plastic fatigue, 594
fracture under, 187

Nil ductility temperature, 435
Notch-brittleness, 32
Notch-sensitivity, 32
Notches
  crack propagation from, 244
  fatigue from, 605

Overload, effect on fatigue, 585

Palmgren–Miner rule, 581
Paris power law, fatigue crack growth rate, 547, 552
Path independence for J, 318
Plane strain
  relationships for $K-G$, 128
  relationship to shear lip size, 280
  toughness determination method, 138, 148
Plane stress
  essential work of ductile fracture, 298
  mixed-mode ductile fracture, 309
Plastic bending
  in peel test, 289
  in trousers test, 312
Plastic fracture
  definition of, 14
  see also Ductile fracture
Plastic zone
  correction, 137
  plane strain, 138, 146
  plane stress, 138, 146
  shape, 146–8
Plasticisation theories, 492
Plasticity-incremental vs. deformation, 801
Pop in
  in plane strain toughness determination, 111
  in sandwich structures, 220
Pores in a craze, 401
Porosity
  effect on fracture toughness, 186
  effect on strength, 186
Potential energy release rate, *see* Strain energy release rate (G)
Principle of
  Least Work, 208
  Minimum Potential Energy, 208
Process zone in fracture, 59, 298, 302
Proof testing, 503

Quasi-static crack propagation technique, 61, 80–89

R-6 interaction diagram
  effect of workhardening index, 784
  exact solution for a glued beam, 783
Ramberg–Osgood/Ludwik
  stress-strain relationship, 472
Random loading fatigue, 580
Rate effects in fracture, 460
  see also Slow crack growth
Relationship between G and K
  for anisotropic linear materials, 217
  for plane strain, 128
  for plane stress, 128
Relaxation at crack tip, 439
Relaxation controlled crack growth, 469
Residual displacements
  in elastic fracture, 100 *et seq.*
  in plastic fracture, 353 *et seq.*
Residual stresses
  effect on fracture, 232, 240, 242
  effect on fatigue, 611
Residual strain energy
  in elastic fracture of adhesive joints, 230
  in elasto-plastic fracture of adhesive joints, 343
Resistance (R) curve
  elastic fracture, 99, 653
  elasto-plastic fracture, 287
  $J_R - \Delta a$ curves, 334, 693
  prediction of crack instability, 653 *et seq.*
Rubber-modified polystyrene
  crack stability in, 661
  essential work of fracture, 309
  thickness effect on toughness, 136
Rubbers
  cavitation in, 405
  fracture mechanics of, 154
  slow crack growth in, 445

Safe-life, *see* Fatigue design methods
Saw-tooth curves, *see* Stick–slip in epoxy resin
Scaling laws
  crack velocity, 719
  elasto-plastic fracture, 727 *et seq.*
  for ice-breaking, 713
  fracture in pipes, 716, 720
  geometrically similar structures, 710
  linear elastic fracture, 708
  non-linear elastic fracture, 718
  non-proportionately scaled structures, 715
Shear decohesion model
  for fatigue crack growth, 544
Shear lip size
  relationship to K_{IC}, 281, 287
Short cracks
  fatigue, 562
  fracture, 246
Short-rod specimen
  for elastic fracture, 140
  for elasto-plastic fracture, 274

Side grooves
effect on crack stability, 677
in DCB specimen, 204
in DT specimen, 521
Slow crack growth
environmental enhanced, 478, 487
relationships for K_c–a, 485
viscoelastic analysis, 469, 472
Solid metal embrittlement, 499
Solubility parameter, 496
Specific work of fracture (R), 16, 60
see also Toughness
Stability
at maxima in K_c–a, 665
criteria for elastic cracking, 645, 647
criteria for elasto-plastic cracking, 695
definition of, 641
displacement-controlled (hard) machines, 645
effect of machine stiffness on, 649
elastic analysis, linear, 645
elastic analysis, non-linear, 647
elasto-plastic analysis, 695
environmental effects on, 663
geometric stability factors, 645, 794
in fatigue, 702
in Gurney–Mai–Owen rig, 660
load controlled (soft) machines, 645
methods for promotion of, 670
stick–slip, 639, 676
Static fatigue, 527
Stick–slip in epoxy resin, 225, 640, 675
Stiffness, 76
Strain crystallisation, 44
Strain energy release rate, 60, 90
relationship to potential energy release rate, 317
Stress concentration factor (K_t, K_T), 113, 606
Stress corrosion fatigue
linear superposition principle, 620
threshold, 620
Stress corrosion cracking
crack growth rate in, 492
determination of K_{ISCC}, 480
electrochemical effects, 489
threshold, 479, 491
Stress intensity factor, K
critical plane strain values, 124, 138
definition of, 118
for compact tension (CT), 124
for centre notched (CN), 121
for double edge notched (DEN), 121
for double cantilever beam (DCB), 92
for double torsion (DT), 521
for short rod specimen, 40
for single edge notched (SEN), 121
for part-through cracks, 126
geometry correction factor Y, 120, 124, 126
handbooks for, 120, 166

Stress intensity factor, K – continued
relationship to G, 128
superposition principle, 197, 280
Stress ratio (r) in fatigue, 569
analysis in terms of ΔG, 569
effect on crack closure, 589
Stress–strain curves
Hookean materials, 55
rubbery materials, 55
extensible biological materials, 55
Structural life prediction
in fatigue, 550, 592
in time-dependent fracture, 66 *et seq.*
Suppression of plastic flow, 348
by grooving, 250
by Gurney–Mai–Owen rig, 251
by sandwiching, 259
Surface energy, 94

Tearing modulus, 697
for DEN samples, 699
stability analysis, 700
Temperature effects
in fatigue, 616
on crack stability, *see* Isothermal–adiabatic transition
on fracture toughness, 460, 436 *et seq.*
Thermal shock
critical temperature, 513, 519
definition of 510
in alumina, 518
in titanium carbide, 514
in tungsten carbide, 517
resistance parameters, 510
Thickness effects
in HIPS, 136, 309
on fracture toughness, 133, 257
Threshold
in fatigue, 555–6
in stress corrosion cracking, 479, 491
in stress corrosion fatigue, 620
variation with fatigue stress ratio, 555–6
Time-temperature equivalence, 666
see also Arrhenius equation
Toughness
definition of, 60
see also Specific work of fracture
equations, 794
upper bound, 678
values for materials, 46, 106, 659, 798
Toughness determination test
ASTM E-399 K_{IC}, 138, 148
compact tension (CT), 124
compliance method, 90–93
deep edge notched (DEN), 298
double-contilever-beam (DCB), 129
double torsion (DT), also Outwater, 111, 529, 683
guillotining, 105

Toughness determination test – continued
Gurney's quasi-static technique, 80, 87
part-through cracks, 126
peel, 103
short rod, 140
trousers tear, 101, 155
Transitions
glass, 405
in abrasive machining, 759
in cutting, 772
in fracture, 739, 744 *et seq.*
in indentation, 754
in spot welds, 751
in wear, 766
isothermal–adiabatic, 468, 664
lower and upper, 744
relationship to design methodology, 776
see also Brittle–ductile transition
Trousers tear test
elastic fracture, 101, 155
plastic fracture, 312

Velocity
limits for quasi-static cracks, 169
scaling laws for, 719
Viscoelasticity, effects on fracture, 469
Void coalescence, *see* Void growth model
Volume fraction of a craze, 401
Void growth model
application to forming limits, 389
connexion to J, 422 *et seq.*
craze, 405
in metals, 385 *et seq.*

Wear resistance, relationship to hardness, 765
Wedge opening loaded (WOL) specimen, 256, 479
Weibull strength distribution, 505
Workhardening index
cyclic, 574
effect on plane stress ductile fracture, 301, 304, 309
effect on R-6 diagram, 784
Work theorem of statics, 61

Yield strip model, *see* Dugdale model
Yielding fracture
small-scale, 271
large-scale, 271